MW00844377

Fundamentals of Vibrations

Leonard Meirovitch

Virginia Polytechnic Institute and State University

Long Grove, Illinois

For information about this book, contact:
Waveland Press, Inc.
4180 IL Route 83, Suite 101
Long Grove, IL 60047-9580
(847) 634-0081
info@waveland.com
www.waveland.com

Reissued 2010 by Waveland Press, Inc.

10-digit ISBN 1-57766-691-7
13-digit ISBN 978-1-57766-691-2

Printed in the United States of America

7 6 5 4

To My Wife and to the

Memory of My Parents and Eldest Brother

ABOUT THE AUTHOR

Leonard Meirovitch, a well-known researcher and educator, is a University Distinguished Professor Emeritus at Virginia Polytechnic Institute and State University (VPI&SU). He is the author of a very large number of journal publications in the areas of analytical vibrations, computational structural dynamics and control of structures and of the books *Analytical Methods in Vibrations* (Macmillan, 1967), *Methods of Analytical Dynamics* (McGraw-Hill, 1970), *Elements of Vibration Analysis*, first edition (McGraw-Hill, 1975), *Computational Methods in Structural Dynamics* (Sijthoff & Noordhoff, 1980), *Introduction to Dynamics and Control* (Wiley, 1985), *Elements of Vibration Analysis*, second edition (McGraw-Hill, 1986), *Dynamics and Control of Structures* (Wiley-Interscience, 1990) and *Principles and Techniques of Vibrations* (Prentice-Hall, 1997). Dr. Meirovitch is a Fellow of the American Institute of Aeronautics and Astronautics (AIAA) and the recipient of the VPI&SU Alumni Award for Research Excellence (1981), the AIAA Structures, Structural Dynamics, and Materials Award (1983), the AIAA Pendray Aerospace Literature Award (1984), the AIAA Mechanics and Control of Flight Award (1987), the Japan Society of Mechanical Engineers Award (1989), an Alexander von Humboldt Award for Senior U.S. Scientists (Germany, 1991) and the American Society of Mechanical Engineers J. P. Den Hartog Award (1999).

PREFACE

This book presents material fundamental to a modern treatment of vibrations, placing the emphasis on analytical developments and computational solutions. It is intended as a textbook for a number of courses on vibrations ranging from the junior level to the second-year graduate level; the book can also serve as a reference for practicing engineers. Certain material from pertinent disciplines was included to render the book self-contained, and hence suitable for self-study. Consistent with this, the book begins with very elementary material and raises the level gradually. A large number of examples and homework problems, as well as computer programs written in MATLAB[1], are provided.

The following review is designed to help the reader decide how best to use the book:

Chapter 1. Concepts from Vibrations—Sections 1.1–1.6 are devoted to a review of basic concepts from Newtonian mechanics. Issues concerning the modeling of mechanical systems, from components to assembled systems, are discussed in Secs. 1.7 to 1.9, and the differential equations of motion for such systems are derived in Sec. 1.10. Sections 1.11 and 1.12 are concerned with the nature of the excitations, the system characteristics and the nature of the response; the concept of linearity and the closely related principle of superposition are discussed. Finally, in Sec. 1.13, the concepts of equilibrium points and motions about equilibrium points are introduced.

The whole chapter is suitable for a first course on vibrations at the undergraduate level, but Secs. 1.1–1.6 may be omitted from a first course at the graduate level.

Chapter 2. Response of Single-Degree-of-Freedom Systems to Initial Excitations—This chapter is concerned with the free vibration of undamped, viscously damped and Coulomb damped systems to initial displacements and velocities. It includes a MATLAB program for plotting the response of viscously damped systems.

This chapter is essential to a first course on vibrations at any level.

Chapter 3. Response of Single-Degree-of-Freedom Systems to Harmonic and Periodic Excitations—In Secs. 3.1 and 3.2, the response to harmonic excitations is represented in the frequency domain, through magnitude and phase angle frequency response plots. Sections 3.3–3.7 discuss applications such as systems with rotating eccentric masses, systems with harmonically moving support, vibration isolation and vibration measuring instruments. In Sec. 3.8, structural damping is treated by means of an analogy with viscous damping. Finally, in Sec. 3.9, the approach to the response of systems to harmonic excitations is extended to periodic excitations through the use of Fourier series. A MATLAB program generating frequency response plots is provided in Sec. 3.10.

The material in Secs. 3.1–3.6 is to be included in a first course on vibrations, but the material in Secs. 3.7–3.9 is optional.

[1]MATLAB ® is a registered trademark of The MathWorks, Inc.

Chapter 4. Response of Single-Degree-of-Freedom Systems to Nonperiodic Excitations—Sections 4.1–4.3 introduce the unit impulse, unit step function and unit ramp function and the respective response. Then, regarding arbitrary excitations as a superposition of impulses of varying magnitude, the system response is represented in Sec. 4.4 as a corresponding superposition of impulse responses, becoming the convolution integral in the limit. Section 4.5 discusses the concept of shock spectrum. Sections 4.6 and 4.7 are devoted to the system response by the Laplace transformation; the concept of transfer function is introduced. Next, in Sec. 4.8, the response is obtained by the state transition matrix. Numerical solutions for the response are carried out in discrete time by the convolution sum in Sec. 4.9 and by the discrete-time transition matrix in Sec. 4.10. A MATLAB program for the response using the convolution sum is given in Sec. 4.11 and another program using the discrete-time transition matrix is given in Sec. 4.12.

Sections 4.1–4.4 are to be included in a first course on vibrations at all levels. Section 4.5 is optional, but recommended for a design-oriented course. Sections 4.6–4.10 are optional for a junior course, recommended for a senior course and to be included in a first course at the graduate level.

Chapter 5. Two-Degree-of-Freedom Systems—Sections 5.1–5.6 present in a simple fashion such topics as the eigenvalue problem, natural modes, response to initial excitations, coupling, orthogonality of modes and modal analysis. Section 5.7 is concerned with the beat phenomenon, Sec. 5.8 derives the response to harmonic excitations and Sec. 5.9 discusses vibration absorbers. The response to nonperiodic excitations is carried out in continuous time in Sec. 5.10 and in discrete time in Sec. 5.11. Three MATLAB programs are included, the first in Sec. 5.12 for the response to initial excitations, the second in Sec. 5.13 for producing frequency response plots and the third in Sec. 5.14 for the response to a rectangular pulse by the convolution sum.

The material belongs in an undergraduate course on vibrations, but is not essential to a graduate course, unless a gradual transition to multi-degree-of-freedom systems is deemed desirable.

Chapter 6. Elements of Analytical Dynamics—Sections 6.1–6.3 provide the prerequisite material for the development in Sec. 6.4 of the extended Hamilton principle, which permits the derivation of all the equations of motion. In Sec. 6.5, the principle is used to produce a generic form of the equations of motion, namely, Lagrange's equations.

This chapter is suitable for a senior course on vibrations and is a virtual necessity for a first-year graduate course.

Chapter 7. Multi-Degree-of-Freedom System—Sections 7.1–7.4 are concerned with the formulation of the equations of motion for linear and linearized systems, as well as with some basic properties of such systems. In Secs. 7.5–7.7, some of the concepts discussed in Ch. 5, such as linear transformations, coupling, the eigenvalue problem, natural modes and orthogonality of modes, are presented in a more compact manner by means of matrix algebra. Then, in Sec. 7.8, the question of rigid-body motions is addressed. In Secs. 7.9 and 7.10, modal analysis is first developed in a rigorous manner and then used to obtain the response to initial excitations. Certain issues associated with the eigenvalue problem are discussed in Secs. 7.11 and 7.12. Section 7.13 is devoted to Rayleigh's quotient, a concept of great importance in vibrations. The response to external excitations is obtained in continuous time in Secs. 7.14 and 7.15 and in discrete time in Sec. 7.17. MATLAB programs are provided as follows: the solution of the eigenvalue problem for conservative systems and for nonconservative systems, both in Sec. 7.18, the response to initial excitations in Sec. 7.19 and the response to external excitations by the discrete-time transition matrix in Sec. 7.20.

This chapter, in full or in part, is suitable for a senior course on vibrations, and should be considered as an alternative to Ch. 5. The material rightfully belongs in a first-year graduate course.

Chapter 8. Distributed-Parameter Systems: Exact Solutions—In Sec. 8.1, the equations of motion for a set of lumped masses on a string are first derived by the Newtonian approach and then transformed in the limit into a boundary-value problem for a distributed string. The same boundary-value problem is derived in Sec. 8.2 by the extended Hamilton principle. In Sec. 8.3, the boundary-value problem for a beam in bending is derived by both the Newtonian approach and the extended Hamilton principle. Sections 8.4–8.8 are devoted to the differential eigenvalue problem and its solution. Rayleigh's quotient is used in Sec. 8.8 to develop the variational approach to the differential eigenvalue problem. The response to initial excitations and external excitations by modal analysis is considered in Secs. 8.9 and 8.10, respectively. A modal solution to the problem of a rod subjected to a boundary force is obtained in Sec. 8.11. The wave equation and its solution in terms of traveling waves and standing waves are introduced in Sec. 8.12, and in Sec. 8.13 it is shown that a traveling wave solution matches the standing waves solution obtained in Sec. 8.11.

Sections 8.1–8.5, 8.9 and 8.10 are suitable for a senior course or a first-year graduate course on vibrations. The balance of the chapter belongs in a second-year graduate course.

Chapter 9. Distributed-Parameter Systems: Approximate Methods—Sections 9.1–9.4 discuss four lumped-parameter methods, including Holzer's method and Myklestad's method. The balance of the chapter is concerned with series discretization techniques. Section 9.5 presents Rayleigh's principle, which is the basis for the variational approach to the differential eigenvalue problem identified with the Rayleigh-Ritz method, as expounded in Secs. 9.6–9.8. Sections 9.9 and 9.10 consider two weighted residuals methods, Galerkin's method and the collocation method, respectively. A MATLAB program for the solution of the eigenvalue problem for a nonuniform rod by the Rayleigh-Ritz method is provided in Sec. 9.11.

The material is suitable for a senior or a first-year graduate course on vibrations, with the exception of the second half of Sec. 9.6 and the entire Sec. 9.7, which are more suitable for a second-year graduate course.

Chapter 10. The Finite Element Method—Section 10.1 presents the formalism of the finite element method. Sections 10.2 and 10.3 consider strings, rods and shafts in terms of linear, quadratic and cubic interpolation functions. Then, Sec. 10.4 discusses beams in bending. Estimates of errors incurred in using the finite element method are provided in Sec. 10.5. In Secs. 10.6 and 10.7, trusses and frames are treated as assemblages of rods and beams, respectively. Then, system response by the finite element method is discussed in Sec. 10.8. A MATLAB program for the solution of the eigenvalue problem for a nonuniform pinned-pinned beam is provided in Sec. 10.9.

This chapter is suitable for a senior or a first-year graduate course on vibrations, with the exception of Sec. 10.3, which is optional, and Secs. 10.6 and 10.7, which are more suitable for a second-year graduate course.

Chapter 11. Nonlinear Oscillations—Sections 11.1–11.3 are concerned with qualitative aspects of nonlinear systems, such as equilibrium points, stability of motion about equilibrium, trajectories in the neighborhood of equilibrium and motions in the large. Section 11.4 discusses the van der Pol oscillator and the concept of limit cycle. Sections 11.5–11.7 introduce the perturbation approach and how to obtain periodic perturbation solutions by Lindstedt's method. Using the perturbation approach, the jump phenomenon is discussed in Sec. 11.8, subharmonic solutions in Sec. 11.9 and linear systems with time-dependent coefficients in Sec. 11.10. Section 11.11 is devoted to numerical integration of differential equations of motion by the Runge-Kutta methods. A MATLAB program for plotting trajectories for the van der Pol oscillator is provided in Sec. 11.12.

The material is suitable for a senior or a graduate course on nonlinear vibrations.

Chapter 12. Random Vibrations—Sections 12.1–12.3 introduce such concepts as random process, stationarity, ergodicity, mean value, autocorrelation function, mean square value and standard deviation. Sections 12.4 and 12.5 are concerned with probability density functions. Properties of the autocorrelation function are discussed in Sec. 12.6. Sections 12.7–12.11 are devoted to the response to random excitations using frequency domain techniques. Sections 12.12–12.15 are concerned with joint properties of two random processes. The response of multi-degree-of-freedom systems and distributed systems to random excitations is discussed in Secs. 12.16 and 12.17, respectively.

The material is suitable for a graduate course on random vibrations.

Appendix A. Fourier Series—The material is concerned with the representation of periodic functions by Fourier series. Both the real form and the complex form of Fourier series are discussed.

Appendix B. Laplace Transformation—The appendix contains an introduction to the Laplace transformation and its use to solve ordinary differential equations with constant coefficients, such as those encountered in vibrations.

Appendix C. Linear Algebra—The appendix represents an introduction to matrices, vector spaces and linear transformations. The material is indispensable to an efficient and rigorous treatment of multi-degree-of-freedom systems.

In recent years, computational algorithms of interest in vibrations have matured to the extent that they are now standard. Examples of these are the QR method for solving algebraic eigenvalue problems and the method based on the discrete-time transition matrix for computing the response of linear systems. At the same time, computers capable of handling such algorithms have become ubiquitous. Moreover, the software for the implementation of these algorithms has become easier to use. In this regard, MATLAB must be considered the software of choice. It is quite intuitive, it can be used interactively and it possesses an inventory of routines, referred to as functions, which simplify the task of programming even more. This book contains 14 MATLAB programs solving typical vibrations problems; they have been written using Version 5.3 of MATLAB. The programs can be used as they are, or they can be modified as needed, particularly the data. In addition, a number of MATLAB problems are included. Further information concerning MATLAB can be obtained from:

The MathWorks, Inc.
3 Apple Hill Drive
Natick, MA 01760

It should be stressed that the book is independent of the MATLAB material and can be used with or without it. Of course, the MATLAB material is designed to enhance the study of vibrations, and its use is highly recommended.

The author wishes to express his appreciation to William J. Atherton, Cleveland State University; Amr M. Baz, University of Maryland; Itzhak Green, Georgia Institute of Technology; Robert H. Lipp, University of New Orleans; Hayrani Ali Öz, Ohio State University; and Alan B. Palazzolo, Texas A&M University, for their extensive review of the manuscript and their many useful suggestions. He also wishes to thank Timothy J. Stemple, Virginia Polytechnic Institute and State University, for producing the computer-generated figures and for reviewing an early version of the manuscript. Special thanks are due to İlhan Tuzcu, Virginia Polytechnic Institute and State University, for his major role in developing the MATLAB programs, as well as for his thorough review of the manuscript. Last but not least, the author would like to thank Norma B. Guynn for typing the book essentially as it appears in its final form; the book places in evidence the excellent quality of her work.

Leonard Meirovitch

Contents

INTRODUCTION

Dynamics is the branch of physics concerned with the motion of bodies under the action of forces. For the problems of interest in this text, relativistic effects are extremely small, so that the motions are governed by the laws of Newtonian mechanics. *Vibrations*, or *oscillations*, can be regarded as a subset of dynamics in which a system subjected to restoring forces swings back and forth about an equilibrium position, where a system is defined as an assemblage of parts acting together as a whole. The restoring forces are due to elasticity, or due to gravity.

For the most part, engineering systems are so complex that their response to stimuli is difficult to predict. Yet, the ability to predict system behavior is essential. In such cases, it is necessary to construct a simplified model acting as a surrogate for the actual system. The process consists of identifying constituent components, determining the dynamic characteristics of the individual components, perhaps experimentally, and assembling the components into a model representative of the whole system. Models are not unique, and for a given system it is possible to construct a number of models. The choice of a model depends on its use and on the system mass and stiffness properties, referred to as parameters. For example, in preliminary design, a simple model predicting the system behavior reasonably well may suffice. On the other hand, in advanced stages of design, a very refined model capable of predicting accurately the behavior of the actual system may be necessary. Many systems can be simulated by models whose behavior is described by a single ordinary differential equation of motion, i.e., by *single-degree-of-freedom* models. This is the case when the model consists of a single mass undergoing translation in one direction, or rotation about one axis. Many other systems must be modeled by an array of masses connected elastically. The behavior of such models is described by a set of ordinary differential equations, and are known as *multi-degree-of-freedom* models. They are commonly referred to as *discrete systems*, or *lumped-parameter systems*. Then, there are systems with distributed mass and stiffness properties. They can be represented by lumped-parameter models, or by *distributed-parameter models*, where the behavior of the latter is described by partial differential equations. Occasionally, we encounter systems with both lumped and distributed properties. Modeling is an important part of engineering vibrations.

The response of a system to given excitations depends on the system characteristics, as reflected in the differential equations of motion. If the response increases proportionally to the excitation, then the system is said to be *linear;* otherwise it is *nonlinear*. Linearity is of paramount importance to a system, as it dictates the approach to the solution of the equations of motion. Indeed, in the case of linear systems the *principle of superposition* applies, which can simplify the solution greatly. The superposition principle does not apply to nonlinear systems.

Different types of excitations call for different methods of solution, particularly the external excitations. By virtue of the superposition principle, the response of linear systems to initial excitations and to external excitations can be obtained separately and then combined linearly. Because for all practical purposes the response to initial excitations decays with time, it is referred to as *transient*. In the case of sinusoidal excitations, it is more advantageous to treat the response in the frequency domain, through frequency response plots, rather than in the time domain. Periodic excitations can be represented as a combination of sinusoidal functions by means of Fourier series, and the response can be obtained as a corresponding combination of sinusoidal responses. Because in both cases time plays no particular role, the response to sinusoidal excitations and the response to periodic excitations are said to be *steady state*. Arbitrary excitations can be regarded as superpositions of impulses of varying magnitude, so that the response can be obtained as corresponding superpositions of impulse responses. The Laplace transformation method yields the same results, perhaps in a less intuitive manner. In linear system theory, the most common approach to the response is to cast the equations of motion in state form and then solve them by a technique based on the state transition matrix. For the most part, the response to arbitrary excitations must be obtained numerically on a computer, which implies discrete-time processing. Random excitations require entirely different approaches, and the response can be obtained in terms of statistical quantities.

Although the preceding discussion applies to all types of models, multi-degree-of-freedom systems and distributed-parameter systems require further elaboration. The equations of motion for multi-degree-of-freedom systems are more efficiently derived by means of Lagrange's equations than by direct application of Newton's second law. Linear, or linearized equations of motion are best expressed in matrix form. Because these are simultaneous equations, the coefficient matrices, albeit symmetric, are fully populated. Their solution can only be carried out by rendering the equations independent by means of modal analysis. This involves the solution of an algebraic eigenvalue problem and an orthogonal transformation using the modal matrix, all made possible by developments in linear algebra. The independent modal equations resemble those for a single-degree-of-freedom and can be solved accordingly. Although different in appearance, partial differential equations describing distributed-parameter systems can be solved in an analogous manner, the primary difference being that they require the solution of a differential eigenvalue problem instead of an algebraic one.

For the most part, differential eigenvalue problems do not admit analytical solutions, so that they must be solved approximately, which amounts to reducing them to algebraic eigenvalue problems. This implies the construction of a discrete model approximating the distributed-parameter system, which can be done through parameter lumping or series discretization. Among series discretization methods, we include the Rayleigh-Ritz method, the Galerkin method and the finite element method, the latter being perhaps the most important development in structural dynamics in the last half a century.

The fact that the superposition principle does not hold for nonlinear systems causes difficulties in producing solutions. If the interest lies only in qualitative stability characteristics, rather than in the system response, then such information can be obtained by linearizing the equations of motion about a given equilibrium point, solving the corresponding eigenvalue problem and reaching stability conclusions from the nature of the eigenvalues. For systems with small nonlinearities, more quantitative results can be obtained by means of perturbation techniques, which permit solutions using once again methods of linear analysis. For nonlinearities of arbitrary magnitude, solutions can only be obtained numerically on a computer. To this end, the Runge-Kutta methods are quite effective.

Difficulties of a different kind arise in the case of random excitations. The response to random excitations is also random and can only be defined in terms of statistical quantities. This situation is much better for Gaussian random processes, for which the probability that the response will remain below a certain value can be defined by means of two statistics alone: the mean value and the standard deviation. The latter is the more important one and can be computed working in the frequency domain using Fourier transforms, rather than in the time domain.

Finally, it should be noted that the numerical work involved in this vibrations study can be programmed for computer evaluation using MATLAB software. In fact, this book contains MATLAB programs for a variety of vibrations problems, which can be regarded as the foundation for a vibrations toolbox.

CHAPTER 1

CONCEPTS FROM VIBRATIONS

This text is concerned with systems executing oscillatory motion, where a system is defined as an aggregation of components acting together as a whole. For mechanical systems the oscillatory motion is generally referred to as vibration. The basic question in vibrations is how systems respond to various stimuli, or excitations. As a preliminary to our vibrations study, in this chapter we consider such topics as fundamental concepts from Newtonian mechanics, component modeling, system modeling, derivation of system differential equations of motion, general excitation and response characteristics and motion stability.

The derivation of the equations of motion can be carried out by means of methods of Newtonian mechanics or by methods of analytical dynamics, also known as Lagrangian mechanics. Newtonian mechanics uses such concepts as force, momentum, velocity and acceleration, all of which are vector quantities. For this reason, Newtonian mechanics is referred to as *vectorial mechanics*. The basic tool in deriving the equations of motion is the free-body diagram, namely, a diagram for each mass in the system showing all the forces acting upon the mass. Newtonian mechanics is physical in nature and considers constraints explicitly. By contrast, analytical dynamics is more abstract in nature and eliminates constraints automatically. We discuss Newtonian mechanics in this chapter and analytical dynamics in Ch. 6.

A model consists of a collection of either individual components, or groups of components, or both. Before modeling can be carried out, it is necessary to identify and characterize the various types of system components, which implies establishing the excitation-response relation, or input-output relation, for individual components or for groups of components, either from experience or through testing.

The task of system modeling amounts to devising a simplified model capable of simulating the behavior of an actual physical system. For vibrating systems, the behavior is governed by equations of motion. Derivation of explicit equations of motion consists of applying the laws of physics to generate a mathematical formulation relating the response to the excitation, or the output to the input. The formulation is commonly in the form of differential equations obtained by methods presented in the beginning of this chapter.

To derive the system response, it is necessary to solve the equations of motion. This is by far the largest part of the study of vibrations, which can be traced to the fact that there is a large variety of excitations, and each type of excitations tends to require a different approach to the solution. We begin this study in Ch. 2.

In this chapter, we begin with a brief review of Newtonian mechanics and then discuss the excitation-response characteristics of various system components with a view to the derivation of the differential equations governing the behavior of vibrating systems. Next, we examine the nature of the excitations and response in a general way, and finally introduce such concepts as equilibrium positions and stability of motion about equilibrium.

1.1 NEWTON'S LAWS

Newton's laws were formulated for single particles and can be extended to systems of particles and rigid bodies. Actually, the laws can be extended to elastic bodies as well, as shown later in this text. Newton's laws can be stated as follows:

First Law. *If there are no forces acting upon a particle, then the particle will move in a straight line with constant velocity.*
The first law states mathematically that, if $\mathbf{F} = \mathbf{0}$, then $\mathbf{v} =$ constant, where $\mathbf{F}$ is the *force vector* and $\mathbf{v}$ the *velocity vector* measured relative to a set of *inertial axes* xyz (Fig. 1.1), defined as a reference frame either at rest or moving with uniform velocity relative to an average position of the "distant stars."

Second Law. *A particle acted upon by a force moves so that the force vector is equal to the time rate of change of the linear momentum vector.*
The mathematical expression of the second law is

$$\mathbf{F} = \frac{d\mathbf{p}}{dt} \tag{1.1}$$

where

$$\mathbf{p} = m\mathbf{v} = m\dot{\mathbf{r}} \tag{1.2}$$

is the *linear momentum vector*, in which m is the *mass* of the particle, a positive quantity whose value does not depend on time, and $\mathbf{r}$ is the *position vector* of m relative to the inertial space xyz. If we insert Eq. (1.2) into Eq. (1.1), we obtain Newton's second law in its most familiar form

$$\mathbf{F} = m\dot{\mathbf{v}} = m\ddot{\mathbf{r}} = m\mathbf{a} \tag{1.3}$$

in which $\mathbf{a}$ is the *acceleration vector* of the particle relative to the inertial space. Note that all kinematical quantities measured relative to an inertial space are referred to as

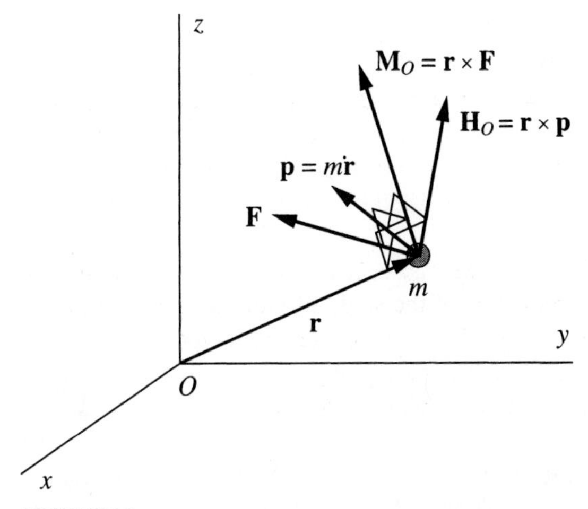

FIGURE 1.1
Motion of a particle relative to an inertial reference frame

absolute. Equation (1.1), or Eq. (1.3), represents the *equations of motion* for particle m. In SI units, the unit of mass is the kilogram (kg) and the unit of force is the newton (N). The kilogram is a basic unit and the newton is a derived unit, $1\ \mathrm{N} = 1\ \mathrm{kg} \cdot \mathrm{m/s^2}$.

Third Law. *When two particles exert forces upon one another, the forces lie along the line joining the particles and the corresponding force vectors are the negative of each other.*
This law is also known as the *law of action and reaction*. Denoting by $\mathbf{f}_{ij}$ the force exerted by particle j on particle i, the law can be stated mathematically as

$$\mathbf{f}_{ij} = -\mathbf{f}_{ji}, \; i \neq j \tag{1.4}$$

where the vectors $\mathbf{f}_{ij}$ and $\mathbf{f}_{ji}$ are clearly collinear. Electromagnetic forces are exceptions to this law, but they are of no concern in this text.

Note that the first law, known as *Galileo's inertial law*, is a special case of the second law in which the force $\mathbf{F}$ is zero. In this case, we conclude from Eq. (1.1) that the linear momentum $\mathbf{p}$, and hence the velocity $\mathbf{v}$, is constant. Such a constant quantity is the result of the integration of Eq. (1.1), for which reason it is referred to as an *integral of motion*. The statement $\mathbf{p} =$ constant is commonly known as the *conservation of linear momentum*.

It should be pointed out that, in using Newton's second law to derive the equations of motion, it is necessary to draw a *free-body diagram*, which is a diagram of the isolated particle m showing all forces acting upon m. If in the process of isolating the particle it becomes necessary to cut through internal forces, then these forces acquire the role of externally applied forces. In this regard, it must be made clear that the symbol $\mathbf{F}$ in Eq. (1.1), or Eq. (1.3), stands for the resultant of all forces acting on m.

Example 1.1. A simple pendulum consists of a bob of mass m suspended on a string of length L (Fig. 1.2a). Derive the differential equation for the angular displacement $\theta(t)$ of the pendulum, as well as an expression for the tension T in the string.

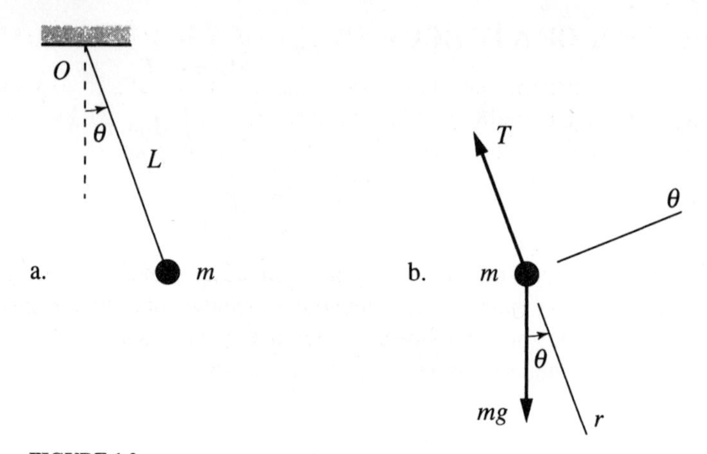

FIGURE 1.2
a. Simple pendulum, b. Free-body diagram

The equation of motion and the tension T can be obtained conveniently by means of Newton's second law in terms of radial and transverse components. Figure 1.2b shows the necessary free-body diagram, in which the tension T, an internal force, plays the role of an externally applied force. The only other force is the weight mg, which in the context of this problem can be regarded as an applied force. From Fig. 1.2b, we can write the equations of motion in terms of the radial and transverse components r and θ as follows:

$$\Sigma F_r = mg\cos\theta - T = ma_r$$
$$\Sigma F_\theta = -mg\sin\theta = ma_\theta \tag{a}$$

Recognizing that $r = L =$ constant, so that $\dot{r} = \ddot{r} = 0$, and recalling from kinematics the expressions of the radial and transverse components of the acceleration (Ref. 11, Sec. 2.3), we can write

$$a_r = \ddot{r} - r\dot{\theta}^2 = -L\dot{\theta}^2$$
$$a_\theta = r\ddot{\theta} + 2\dot{r}\dot{\theta} = L\ddot{\theta} \tag{b}$$

Inserting the second of Eqs. (b) into the second of Eqs. (a), we obtain the differential equation of motion

$$\ddot{\theta} + \frac{g}{L}\sin\theta = 0 \tag{c}$$

from which we conclude that the motion of the pendulum does not depend on m, a fact known to the ancient Greeks. On the other hand, the first of Eqs. (a) and (b) give the tension in the string

$$T = mg\cos\theta + mL\dot{\theta}^2 \tag{d}$$

which does depend on the value of m, in addition to the angular displacement and angular velocity of the pendulum.

1.2 MOMENT OF A FORCE AND ANGULAR MOMENTUM

We consider a particle of mass m moving under the action of a force $\mathbf{F}$ and denote its position relative to the origin O of the reference frame xyz by $\mathbf{r}$ and its absolute velocity by $\mathbf{v} = \dot{\mathbf{r}}$. By definition, the *moment of the force* $\mathbf{F}$ about point O is a vector given by the cross product (vector product)

$$\mathbf{M}_O = \mathbf{r} \times \mathbf{F} \tag{1.5}$$

and it represents a vector normal to the plane defined by $\mathbf{r}$ and $\mathbf{F}$ (Fig. 1.1). In a similar fashion, the *moment of momentum*, or *angular momentum* of m with respect to point O is defined as the moment of the linear momentum about O and is a vector represented mathematically by the cross product of the radius vector $\mathbf{r}$ and the linear momentum $\mathbf{p} = m\dot{\mathbf{r}}$, or

$$\mathbf{H}_O = \mathbf{r} \times \mathbf{p} = \mathbf{r} \times m\dot{\mathbf{r}} \tag{1.6}$$

and we note that $\mathbf{H}_O$ is a vector normal to the plane defined by $\mathbf{r}$ and $\mathbf{p}$ (Fig. 1.1).

Next, we consider the time rate of change of $\mathbf{H}_O$, recall that m is constant and write

$$\dot{\mathbf{H}}_O = \dot{\mathbf{r}} \times m\dot{\mathbf{r}} + \mathbf{r} \times m\ddot{\mathbf{r}} = \mathbf{r} \times m\ddot{\mathbf{r}} \tag{1.7}$$

where, by the definition of the cross product, $\dot{\mathbf{r}} \times m\dot{\mathbf{r}} = m(\dot{\mathbf{r}} \times \dot{\mathbf{r}}) = \mathbf{0}$. But, by Newton's second law, Eq. (1.3),

$$m\ddot{\mathbf{r}} = \mathbf{F} \tag{1.8}$$

Hence, inserting Eq. (1.8) into Eq. (1.7) and considering Eq. (1.5), we conclude that

$$\mathbf{M}_O = \dot{\mathbf{H}}_O \tag{1.9}$$

or, *the moment of a force about a fixed point O is equal to the time rate of change of the angular momentum about O.*

When the moment about O is zero, $\mathbf{M}_O = \mathbf{0}$, it follows from Eq. (1.9) that

$$\mathbf{H}_O = \text{constant} \tag{1.10}$$

which represents the *principle of conservation of angular momentum*, stating that, *in the absence of moments about O, the angular momentum about O is constant.* Note that it is not necessary that the force resultant be zero for the angular momentum to be conserved, but only that the moment about O be zero, which is the case when the force resultant passes through O.

It should be pointed out that the developments of this section were carried out for the general three-dimensional case. In the special case of planar motions, the vectors $\mathbf{M}_O$, $\mathbf{H}_O$ and $\dot{\mathbf{H}}_O$ are all normal to the plane of motion.

Example 1.2. Derive the equation of motion for the simple pendulum of Fig. 1.3 using the moment, angular momentum relation, Eq. (1.9).

From Fig. 1.3, the position, force and linear momentum vectors can be written in terms of radial and transverse components as follows:

$$\mathbf{r} = L\mathbf{u}_r, \ \mathbf{F} = mg(\cos\theta\mathbf{u}_r - \sin\theta\mathbf{u}_\theta), \ \mathbf{p} = mL\dot{\theta}\mathbf{u}_\theta \tag{a}$$

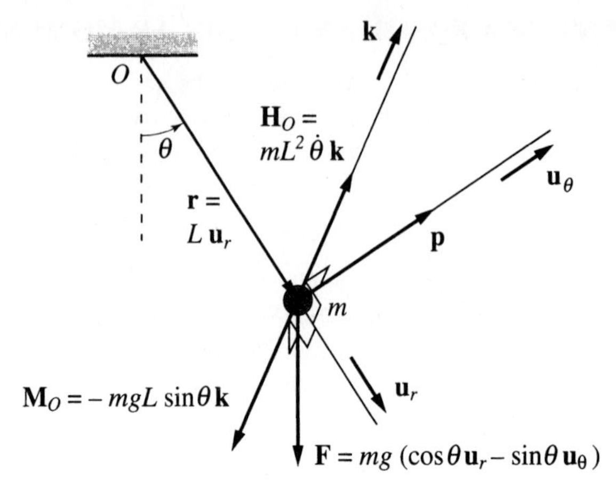

FIGURE 1.3
Simple pendulum

so that, using Eqs. (1.5) and (1.6), the moment and angular momentum about O are

$$\begin{aligned} \mathbf{M}_O &= \mathbf{r}\times\mathbf{F} = L\mathbf{u}_r \times mg(\cos\theta\mathbf{u}_r - \sin\theta\mathbf{u}_\theta) = -mgL\sin\theta\mathbf{k} \\ \mathbf{H}_O &= \mathbf{r}\times\mathbf{p} = L\mathbf{u}_r \times mL\dot{\theta}\mathbf{u}_\theta = mL^2\dot{\theta}\mathbf{k} \end{aligned} \tag{b}$$

respectively, where $\mathbf{k}$ is a unit vector normal to $\mathbf{u}_r$ and $\mathbf{u}_\theta$. Inserting Eqs. (b) into Eq. (1.9), omitting the unit vector $\mathbf{k}$ and dividing through by mL^2, we obtain the desired equation of motion in the form

$$\ddot{\theta} + \frac{g}{L}\sin\theta = 0 \tag{c}$$

which coincides with Eq. (c) of Example 1.1.

1.3 WORK AND ENERGY

We consider a particle of mass m moving along curve S under the action of a given force $\mathbf{F}$ (Fig. 1.4). By definition, the *increment of work* performed by the force $\mathbf{F}$ in moving the particle from position $\mathbf{r}$ to position $\mathbf{r}+d\mathbf{r}$ is given by the dot product (scalar product)

$$\overline{dW} = \mathbf{F}\cdot d\mathbf{r} \tag{1.11}$$

where the overbar indicates that $\overline{dW}$ is an incremental expression rather than the differential of a function W. Clearly, $\overline{dW}$ is a scalar quantity. But, from kinematics $d\mathbf{r} = \dot{\mathbf{r}}dt$, so that using Newton's second law, Eq. (1.3), we can write

$$\overline{dW} = m\ddot{\mathbf{r}}\cdot\dot{\mathbf{r}}dt = m\frac{d\dot{\mathbf{r}}}{dt}\cdot\dot{\mathbf{r}}dt = m\dot{\mathbf{r}}\cdot d\dot{\mathbf{r}} = d\left(\frac{1}{2}m\dot{\mathbf{r}}\cdot\dot{\mathbf{r}}\right) \tag{1.12}$$

in which we recognized that the order of the terms in the multiplication is immaterial in a dot product. At this point, we define the *kinetic energy* of mass m as

$$T = \frac{1}{2}m\dot{\mathbf{r}}\cdot\dot{\mathbf{r}} \tag{1.13}$$

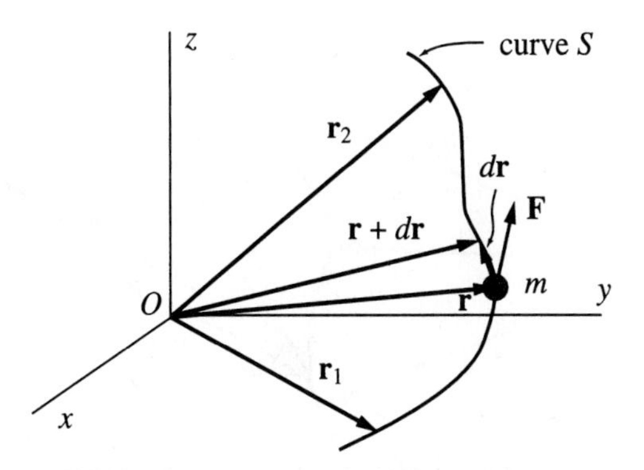

FIGURE 1.4
Particle moving under the action of a force

so that Eq. (1.12) can be rewritten in the form

$$\overline{dW} = dT \tag{1.14}$$

and we note that, unlike $\overline{dW}$, dT does represent the differential of a function, namely, the kinetic energy function T.

Next, we consider the work performed by $\mathbf{F}$ in moving the particle from position $\mathbf{r}_1$ to position $\mathbf{r}_2$, as shown in Fig. 1.4. Integrating Eq. (1.14) and using Eq. (1.11), we obtain

$$\int_{\mathbf{r}_1}^{\mathbf{r}_2} \mathbf{F} \cdot d\mathbf{r} = \int_{T_1}^{T_2} dT = T_2 - T_1 \tag{1.15}$$

in which T_i is the kinetic energy in the position $\mathbf{r}_i$ $(i = 1, 2)$. Hence, *the work performed by the force* $\mathbf{F}$ *in moving the particle m from position* $\mathbf{r}_1$ *to position* $\mathbf{r}_2$ *is responsible for a change in the kinetic energy from* T_1 *to* T_2.

A very important class of forces is the class of *conservative forces* for which *the work depends only on the initial position* $\mathbf{r}_1$ *and the final position* $\mathbf{r}_2$, and not on the path taken from $\mathbf{r}_1$ to $\mathbf{r}_2$. Denoting two distinct paths from $\mathbf{r}_1$ to $\mathbf{r}_2$ by I and II (Fig. 1.5), we can express the preceding statement in the mathematical form

$$\underset{\text{path I}}{\int_{\mathbf{r}_1}^{\mathbf{r}_2} \mathbf{F} \cdot d\mathbf{r}} = \underset{\text{path II}}{\int_{\mathbf{r}_1}^{\mathbf{r}_2} \mathbf{F} \cdot dr} \tag{1.16}$$

Equation (1.16) can be given a different interpretation by writing

$$\underset{\text{path I}}{\int_{\mathbf{r}_1}^{\mathbf{r}_2} \mathbf{F} \cdot d\mathbf{r}} - \underset{\text{path II}}{\int_{\mathbf{r}_1}^{\mathbf{r}_2} \mathbf{F} \cdot d\mathbf{r}} = \underset{\text{path I}}{\int_{\mathbf{r}_1}^{\mathbf{r}_2} \mathbf{F} \cdot d\mathbf{r}} + \underset{\text{path II}}{\int_{\mathbf{r}_2}^{\mathbf{r}_1} \mathbf{F} \cdot d\mathbf{r}} = \oint \mathbf{F} \cdot d\mathbf{r} = 0 \tag{1.17}$$

in which $\oint$ denotes an integral over a closed path. In view of this, we can state that *the work performed by conservative forces over a closed path is zero.* In the following discussions, we identify conservative forces by the subscript c.

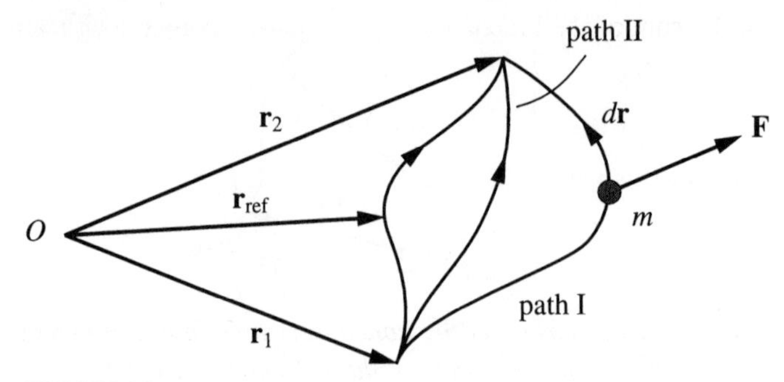

FIGURE 1.5
Motion on a path passing through a reference position

At this point, we consider a conservative force $\mathbf{F}_c$, choose a path from $\mathbf{r}_1$ to $\mathbf{r}_2$ passing through the reference position $\mathbf{r}_{\text{ref}}$, as shown in Fig. 1.5, and define the *potential energy* as the work performed by conservative forces in moving a particle from position $\mathbf{r}$ to the reference position $\mathbf{r}_{\text{ref}}$, or

$$V(\mathbf{r}) = \int_{\mathbf{r}}^{\mathbf{r}_{\text{ref}}} \mathbf{F}_c \cdot d\mathbf{r} \tag{1.18}$$

where V is a scalar function depending on $\mathbf{r}$ alone, as $\mathbf{r}_{\text{ref}}$ is arbitrary and hence immaterial. Indeed, because the interest lies in changes in the potential energy as the particle changes positions, rather than in the potential energy at a given point alone, when the difference in the potential energy between two points is considered, any contribution to the potential energy from $\mathbf{r}_{\text{ref}}$ cancels out. In view of definition (1.18), we can express the work performed by conservative forces in moving a particle from position $\mathbf{r}_1$ to position $\mathbf{r}_2$ in the form

$$\int_{\mathbf{r}_1}^{\mathbf{r}_2} \mathbf{F}_c \cdot d\mathbf{r} = \int_{\mathbf{r}_1}^{\mathbf{r}_{\text{ref}}} \mathbf{F}_c \cdot d\mathbf{r} + \int_{\mathbf{r}_{\text{ref}}}^{\mathbf{r}_2} \mathbf{F}_c \cdot d\mathbf{r} = \int_{\mathbf{r}_1}^{\mathbf{r}_{\text{ref}}} \mathbf{F}_c \cdot d\mathbf{r} - \int_{\mathbf{r}_2}^{\mathbf{r}_{\text{ref}}} \mathbf{F}_c \cdot d\mathbf{r}$$
$$= V(\mathbf{r}_1) - V(\mathbf{r}_2) = -(V_2 - V_1) \tag{1.19}$$

in which $V_i = V(\mathbf{r}_i)$ $(i = 1, 2)$. Equation (1.19) states that *the work performed by conservative forces in moving a particle from* $\mathbf{r}_1$ *to* $\mathbf{r}_2$ *is equal to the negative of the change in the potential energy from* V_1 *to* V_2.

In general, forces can be divided into two classes, conservative and nonconservative, where the latter are denoted by $\mathbf{F}_{nc}$. Consistent with this, the work can be expressed as the sum of work performed by conservative forces and nonconservative forces, or

$$\int_{\mathbf{r}_1}^{\mathbf{r}_2} \mathbf{F} \cdot d\mathbf{r} = \int_{\mathbf{r}_1}^{\mathbf{r}_2} \mathbf{F}_c \cdot d\mathbf{r} + \int_{\mathbf{r}_1}^{\mathbf{r}_2} \mathbf{F}_{nc} \cdot d\mathbf{r} \tag{1.20}$$

Inserting Eqs. (1.15) and (1.19) into Eq. (1.20), we obtain

$$T_2 - T_1 = -(V_2 - V_1) + \int_{\mathbf{r}_1}^{\mathbf{r}_2} \mathbf{F}_{nc} \cdot d\mathbf{r} \tag{1.21}$$

Then, defining the sum of the kinetic energy and potential energy as the *total energy* function

$$E = T + V \tag{1.22}$$

Eq. (1.21) can be rewritten as

$$\int_{\mathbf{r}_1}^{\mathbf{r}_2} \mathbf{F}_{nc} \cdot d\mathbf{r} = E_2 - E_1 \tag{1.23}$$

which states that *the work performed by nonconservative forces in moving a particle from* $\mathbf{r}_1$ *to* $\mathbf{r}_2$ *is responsible for a change in the total energy from* E_1 *to* E_2. Equation (1.23) can be expressed in the incremental form

$$\mathbf{F}_{nc} \cdot d\mathbf{r} = dE \tag{1.24}$$

so that, dividing through by dt, we obtain

$$\mathbf{F}_{nc} \cdot \dot{\mathbf{r}} = \dot{E} \tag{1.25}$$

But, in general the scalar product $\mathbf{F} \cdot \dot{\mathbf{r}}$ represents the rate of work and is known as the *power*. Hence, Eq. (1.25) states that *the power associated with nonconservative forces is equal to the time rate of change of the total energy*. From Eq. (1.25), we conclude that nonconservative forces can add or dissipate energy, depending on whether the product $\mathbf{F}_{nc} \cdot \dot{\mathbf{r}}$ is positive or negative, respectively. Physically, this depends on whether the projection of the nonconservative force vector $\mathbf{F}_{nc}$ on the velocity vector $\dot{\mathbf{r}}$ is in the same direction as $\dot{\mathbf{r}}$ or in the opposite direction, respectively.

When there are only conservative forces present, $\mathbf{F}_{nc} = \mathbf{0}$, Eq. (1.25) reduces to

$$\dot{E} = 0 \tag{1.26}$$

which yields

$$E = \text{constant} \tag{1.27}$$

Hence, *in the absence of nonconservative forces the total energy is conserved*, a statement known as the *conservation of energy principle*. This statement provides the justification for the term "conservative forces" introduced earlier in this section, before a satisfactory explanation of the term could be given. The total energy is a different type of integral of motion than the linear momentum integral or angular momentum integral; it represents a relation between displacements and velocities, and it can provide a great deal of information concerning the motion of the particle. The value of the integral depends on the initial displacements and velocities.

To ascertain the existence of an energy integral, or the absence of one, it is necessary to identify the type of forces acting on the particle. To this end, we observe that the class of conservative forces contains constant forces and forces depending on the position $\mathbf{r}$ alone. On the other hand, the class of nonconservative forces includes forces depending explicitly on time, or on the velocity $\dot{\mathbf{r}}$, or on both.

For conservative systems defined by a single coordinate, Eq. (1.26) can be used to derive the equation of motion.

Example 1.3. Consider the simple pendulum of Examples 1.1 and 1.2, ascertain the existence of a motion integral and determine its expression for the initial conditions $\theta(0) = 0,\ \dot{\theta}(0) = \omega_0$. Then, use Eq. (1.26) to derive the equation of motion.

The only external force acting upon the bob, the weight mg, destroys the conservation of both the linear momentum and the angular momentum about the point of support O. On the other hand, the force mg is constant, and hence conservative. As a result, there is a motion integral in the form of the total energy.

From Eq. (1.13), we can write the kinetic energy

$$T = \frac{1}{2}mv^2 = \frac{1}{2}mL^2\dot{\theta}^2 \tag{a}$$

To obtain the potential energy, we insert the first two of Eqs. (a) of Example 1.2 into Eq. (1.18) with $\mathbf{r}_{\text{ref}} = \mathbf{0}$, recognize that the only change in the vector $\mathbf{r}$ is due to a change $d\theta$ in direction and that the change $Ld\theta$ is normal to $\mathbf{r}$ (Ref. 11, Sec. 2.3), so that

$$d\mathbf{r} = d(L\mathbf{u}_r) = Ld\theta\mathbf{u}_\theta \tag{b}$$

and write

$$V(\theta) = \int_\theta^0 mg(\cos\theta\mathbf{u}_r - \sin\theta\mathbf{u}_\theta)\cdot Ld\theta\mathbf{u}_\theta = -mgL\int_\theta^0 \sin\theta d\theta = mgL(1-\cos\theta) \tag{c}$$

Equation (c) can be obtained in a simpler manner by using a scalar form of the integral (1.18) in terms of the vertical component of force and displacement differential. An even more direct approach consists of writing, on physical grounds,

$$V(\theta) = mg\Delta h = mgL(1-\cos\theta) \tag{d}$$

where Δh is the rise of the bob above the reference position $\theta = 0$, in which the bob is at its lowest level. Hence, considering the initial conditions, the energy integral is

$$E = T + V = \frac{1}{2}mL^2\dot{\theta}^2 + mgL(1-\cos\theta) = \frac{1}{2}mL^2\omega_0^2 = \text{constant} \tag{e}$$

which clearly represents a relation between the angular displacement θ and angular velocity $\dot{\theta}$. As a matter of interest, we obtain the maximum angle reached by the pendulum by letting $\dot{\theta} = 0$ and writing

$$\theta_{\max} = \cos^{-1}(1 - L\omega_0^2/2g),\ \omega_0 < 2\sqrt{g/L} \tag{f}$$

If $\omega_0 > 2\sqrt{g/L}$, then $\dot{\theta} > 0$, which implies that the pendulum rotates continuously, never reaching an equilibrium position.

Next, we take the time derivative of Eq. (e) and write according to Eq. (1.26)

$$\dot{E} = mL^2\dot{\theta}\ddot{\theta} + mgL\sin\theta\dot{\theta} = (mL^2\ddot{\theta} + mgL\sin\theta)\dot{\theta} = 0 \tag{g}$$

so that, for $\dot{\theta} \neq 0$, we must have

$$mL^2\ddot{\theta} + mgL\sin\theta = 0 \tag{h}$$

which represents the equation of motion.

1.4 DYNAMICS OF SYSTEMS OF PARTICLES

Newton's laws were formulated for single particles, but in this text the interest lies in the vibration of flexible bodies and to a smaller extent in the oscillation of rigid bodies.

Newton's second law can be used to derive equations of motion for rigid bodies and deformable bodies, which can be done conveniently by first deriving the equations for systems of particles.

We consider a system of N particles of mass m_i $(i = 1, 2, \ldots, N)$, as shown in Fig. 1.6, in which $\mathbf{F}_i$ denotes the force acting on m_i and $\mathbf{f}_{ij}$ denotes the force exerted by m_j on m_i $(i, j = 1, 2, \ldots, N;\ j \neq i)$. According to Newton's second law, Eq. (1.3), the equation of motion for particle m_i is

$$\mathbf{F}_i + \sum_{j=1}^{N} \mathbf{f}_{ij} = m_i \ddot{\mathbf{r}}_i = m_i \mathbf{a}_i, \quad i = 1, 2, \ldots, N \tag{1.28}$$

where $\ddot{\mathbf{r}}_i = \mathbf{a}_i$ is the acceleration of particle m_i relative to an inertial space. To derive an equation for the motion of the entire system of particles, we sum up Eqs. (1.28) and write

$$\sum_{i=1}^{N} \mathbf{F}_i + \sum_{i=1}^{N} \sum_{j=1}^{M} \mathbf{f}_{ij} = \sum_{i=1}^{N} m_i \ddot{\mathbf{r}}_i = \sum_{i=1}^{N} m_i \mathbf{a}_i \tag{1.29}$$

Then, recognizing that, by virtue of Eqs. (1.4), the internal forces $\mathbf{f}_{ij}$ and $\mathbf{f}_{ji}$ cancel out in pairs and letting

$$\sum_{i=1}^{N} \mathbf{F}_i = \mathbf{F} \tag{1.30}$$

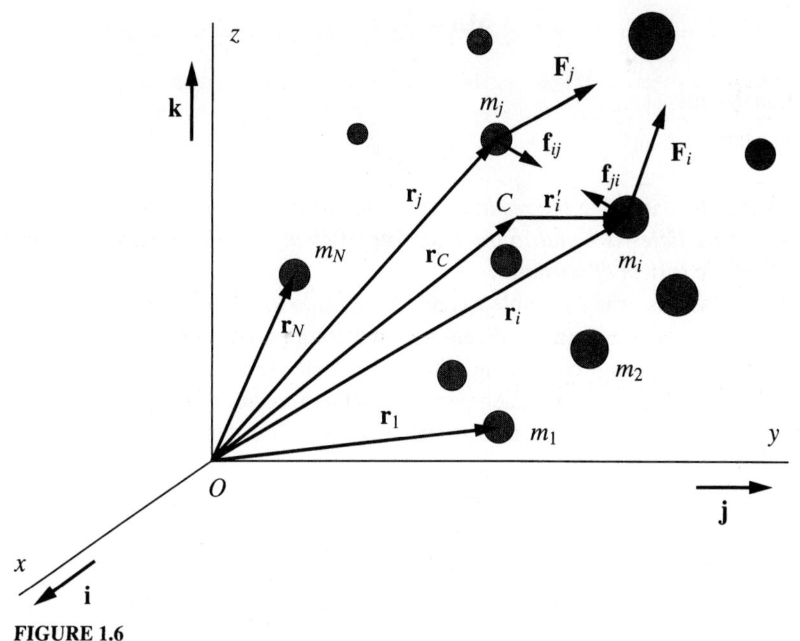

FIGURE 1.6
System of particles

be the *resultant* of all forces, the *equation of motion for the system of particles* is

$$\mathbf{F} = \sum_{i=1}^{N} m_i \ddot{\mathbf{r}}_i = \sum_{i=1}^{N} m_i \mathbf{a}_i \tag{1.31}$$

Next, we define the moment of momentum of the system of particles about the fixed point O as

$$\mathbf{H}_O = \sum_{i=1}^{N} \mathbf{r}_i \times m_i \dot{\mathbf{r}}_i \tag{1.32}$$

Taking the time derivative of Eq. (1.32) and recognizing that $\dot{\mathbf{r}}_i \times \dot{\mathbf{r}}_i = \mathbf{0}$, we obtain

$$\dot{\mathbf{H}}_O = \sum_{i=1}^{N} (\dot{\mathbf{r}}_i \times m_i \dot{\mathbf{r}}_i + \mathbf{r}_i \times m_i \ddot{\mathbf{r}}_i) = \sum_{i=1}^{N} \mathbf{r}_i \times m_i \ddot{\mathbf{r}}_i \tag{1.33}$$

But, inserting Eqs. (1.28) into Eq. (1.33) and recognizing that the moments due to the internal forces add up to zero (Ref. 11, Sec. 9.5), we can write

$$\dot{\mathbf{H}}_O = \sum_{i=1}^{N} \mathbf{r}_i \times \mathbf{F}_i \tag{1.34}$$

Then, denoting the moment of all forces about O by

$$\mathbf{M}_O = \sum_{i=1}^{N} \mathbf{r}_i \times \mathbf{F}_i \tag{1.35}$$

Eq. (1.34) yields

$$\mathbf{M}_O = \dot{\mathbf{H}}_O \tag{1.36}$$

Equation (1.36) states that *the resultant of the moments about a fixed point O acting on a system of particles is equal to the time rate of change of the moment of momentum about O of the system of particles.*

On occasions, it is advisable to refer the motion to a moving point, rather than to a fixed point. A point playing a special role in dynamics is the *mass center*, denoted by C and defined as a point coinciding with *a weighted average position of all particles*, where the weighting factor for each particle i is the mass m_i of the particle. The radius vector from the origin O to the mass center C (Fig. 1.6) is defined as

$$\mathbf{r}_C = \frac{\sum_{i=1}^{N} m_i \mathbf{r}_i}{\sum_{i=1}^{N} m_i} = \frac{1}{m} \sum_{i=1}^{N} m_i \mathbf{r}_i \tag{1.37}$$

in which $m = \sum_{i=1}^{N} m_i$ is the *total mass* of the system of particles. Then, we can express the absolute position, velocity and acceleration of particle m_i as

$$\begin{aligned}
\mathbf{r}_i &= \mathbf{r}_C + \mathbf{r}'_i, \ i = 1, 2, \ldots, N \\
\dot{\mathbf{r}}_i &= \mathbf{v}_i = \dot{\mathbf{r}}_C + \dot{\mathbf{r}}'_i = \mathbf{v}_C + \mathbf{v}'_i, \ i = 1, 2, \ldots, N \\
\ddot{\mathbf{r}}_i &= \mathbf{a}_i = \ddot{\mathbf{r}}_C + \ddot{\mathbf{r}}'_i = \mathbf{a}_C + \mathbf{a}'_i, \ i = 1, 2, \ldots, N
\end{aligned} \tag{1.38}$$

where $\mathbf{r}_C$, $\dot{\mathbf{r}}_C = \mathbf{v}_C$ and $\ddot{\mathbf{r}}_C = \mathbf{a}_C$ are the absolute position, velocity and acceleration of point C, respectively, and $\mathbf{r}'_i$, $\dot{\mathbf{r}}'_i = \mathbf{v}'_i$ and $\ddot{\mathbf{r}}'_i = \mathbf{a}'_i$ are the corresponding position, velocity and acceleration vectors of particle m_i relative to C. Using the first of Eqs. (1.38), we can write

$$\mathbf{r}_C = \frac{1}{m} \sum_{i=1}^{N} m_i (\mathbf{r}_C + \mathbf{r}'_i) = \mathbf{r}_C + \frac{1}{m} \sum_{i=1}^{N} m_i \mathbf{r}'_i \tag{1.39}$$

from which we conclude that

$$\sum_{i=1}^{N} m_i \mathbf{r}'_i = \mathbf{0} \tag{1.40}$$

Hence, *the mass center C can also be defined as a point such that the weighted average position relative to C is zero.*

Inserting Eqs. (1.38) into Eq. (1.31) and observing that $\ddot{\mathbf{r}}_C = \mathbf{a}_C$ is independent of i, we obtain the force equation in the form

$$\begin{aligned}
\mathbf{F} &= \sum_{i=1}^{N} m_i \ddot{\mathbf{r}}_i = \sum_{i=1}^{N} m_i (\ddot{\mathbf{r}}_C + \ddot{\mathbf{r}}'_i) = m\ddot{\mathbf{r}}_C + \sum_{i=1}^{N} m_i \ddot{\mathbf{r}}'_i \\
&= \sum_{i=1}^{N} m_i \mathbf{a}_i = \sum_{i=1}^{N} m_i (\mathbf{a}_C + \mathbf{a}'_i) = m\mathbf{a}_C + \sum_{i=1}^{N} m_i \mathbf{a}'_i
\end{aligned} \tag{1.41}$$

Introducing Eq. (1.40) in Eq. (1.41), we obtain the simple force equation

$$\mathbf{F} = m\mathbf{a}_C \tag{1.42}$$

which can be interpreted as stating that *the motion of the system of particles is equivalent to the motion of a single body of mass equal to the total mass m of the system of particles and whose acceleration under the resultant force $\mathbf{F}$ is equal to the acceleration $\mathbf{a}_C$ of the mass center.*

Now, we define the moment of momentum about C as

$$\mathbf{H}_C = \sum_{j=1}^{N} \mathbf{r}'_i \times m_i \dot{\mathbf{r}}_i \tag{1.43}$$

consider Eq. (1.28) and the second of Eqs. (1.38), use the same argument as that leading to Eq. (1.34), recall Eq. (1.40) and write the time derivative of $\mathbf{H}_C$ in the form

$$\dot{\mathbf{H}}_C = \sum_{i=1}^{N} (\dot{\mathbf{r}}_i' \times m_i \dot{\mathbf{r}}_i + \mathbf{r}_i' \times m_i \ddot{\mathbf{r}}_i) = \left(\sum_{i=1}^{N} m_i \dot{\mathbf{r}}_i' \right) \times \dot{\mathbf{r}}_C + \sum_{i=1}^{N} \mathbf{r}_i' \times \mathbf{F}_i = \sum_{i=1}^{N} \mathbf{r}_i' \times \mathbf{F}_i \tag{1.44}$$

Then, observing that

$$\sum_{i=1}^{N} \mathbf{r}_i' \times \mathbf{F}_i = \mathbf{M}_C \tag{1.45}$$

is the moment of all forces about C, the moment equation about C reduces to the same simple form

$$\mathbf{M}_C = \dot{\mathbf{H}}_C \tag{1.46}$$

as the moment equation about a fixed point O, Eq. (1.34). Moreover, introducing the second of Eqs. (1.38) in Eq. (1.43) and using Eq. (1.40), we obtain

$$\mathbf{H}_C = \sum_{i=1}^{N} \mathbf{r}_i' \times m_i \dot{\mathbf{r}}_i' \tag{1.47}$$

so that the moment of momentum about the mass center C also has the same simple form as the moment of momentum about a fixed point O.

Note that if the motion is referred to an arbitrary moving point, rather than to the mass center, both the force and moment equations are more involved. Hence, if the motion is to be referred to a moving point, then *it is a good policy to choose the moving point as the mass center C.*

1.5 DYNAMICS OF RIGID BODIES

Rigid bodies can be regarded as systems of particles, so that the developments of Sec. 1.4 apply equally well to rigid bodies. Still, there are some basic differences between rigid bodies and arbitrary systems of particles, which require certain extensions of the developments of Sec. 1.4. In the first place, rigid bodies are characterized by continuous mass, i.e., the mass is distributed over the entire body, instead of being concentrated at discrete points. As a result, the mass properties are described by means of a *mass density function* $\rho(x, y, z)$, representing mass per unit volume at a given point in the body identified by the spatial variables x, y and z, which are the coordinates of the given point relative to a set of axes x, y, z fixed in the body and known as *body axes*. This is in contrast with the discrete masses m_i in the case of collections of particles, which are identified by the index i. Hence, to apply the developments of Sec. 1.4 to continuous bodies, we must replace the discrete mass m_i by the differential element of mass $dm(x, y, z)$ and the summation over the collection of particles by integration over the body. Another difference lies in the fact that the distance between any two points in a rigid body is constant. As a result, any motion of one point in a rigid body relative to another is due entirely to rotation, which reduces drastically the number of variables required for a description of the motion of rigid bodies.

In deriving the equations of motion for rigid bodies, we consider the following cases:

1.5.1 Pure Translation Relative to the Inertial Space

In the case of pure translation, the equations of motion are all force equations. In view of the preceding discussion, the force equation can be obtained from Eq. (1.31) by writing in general

$$\mathbf{F}(t) = \int_{\text{body}} \mathbf{a}(x, y, z, t)dm(x, y, z) \tag{1.48}$$

where $\mathbf{F}$ is the resultant force vector, $\mathbf{a}$ is the acceleration vector of a point in the rigid body and $dm(x, y, z) = \rho(x, y, z)dV$ is the differential element of mass, in which m is the total mass and dV is the differential element of volume. But, in pure translation the acceleration is the same for every point of the body, $\mathbf{a}(x, y, z, t) = \mathbf{a}(t)$, so that Eq. (1.48) yields simply

$$\mathbf{F} = m\mathbf{a} \tag{1.49}$$

In the case of planar motions, the force equation, Eq. (1.49), has only two scalar components, or

$$F_x = ma_x, \; F_y = ma_y \tag{1.50}$$

where F_x and F_y are the resultant forces in the x- and y-direction, respectively.

Note that on occasions the force equations are not sufficient to solve the problem, so that a moment equation must be invoked, even though the angular acceleration is zero. On such occasions, it is generally advisable to take moments about the mass center.

Example 1.4. Derive the equations of motion for the body in horizontal translation shown in Fig. 1.7a. The horizontal reactions at the points of contact are proportional to the vertical reactions at these points, where the proportionality constant is the friction coefficient μ. Then, use the parameter values $\beta = 30°$, $\mu = 0.5$, $H = 0.6\text{L}$ and $D = 0.1\text{L}$ and determine the magnitude F of the force $\mathbf{F}$ as a fraction of the weight mg when the body is on the verge of tipping over, as well as the acceleration of the body as a fraction of g.

Figure 1.7b shows the corresponding free-body diagram. Just before tipping over, the reaction N_A reduces to zero. Hence, using Eqs. (1.50), we obtain simply

$$\begin{aligned} F_x &= F\cos\beta - \mu N_B = ma_x \\ F_y &= F\sin\beta + N_B - mg = 0 \end{aligned} \tag{a}$$

We observe that there are three unknowns, F, N_B and a_x, and only two equations, so that we must have another equation. We can write another equation by considering the fact that the body undergoes no rotations, so that the moment about the mass center is zero. Hence, taking moments about the mass center C, we have

$$M_C = -F\cos\beta\frac{H}{2} + (N_B - F\sin\beta)\frac{L}{2} - \mu N_B\left(D + \frac{H}{2}\right) = 0 \tag{b}$$

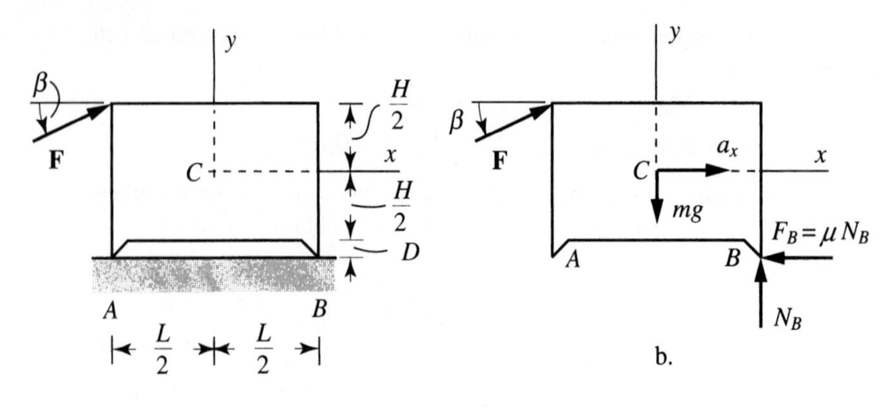

FIGURE 1.7
a. Rigid body in horizontal translation, b. Free-body diagram

Solving the second of Eqs. (a) and Eq. (b) for F/mg and using the given parameter values, we can write

$$\frac{F}{mg} = \frac{L - 2\mu D - \mu H}{H\cos\beta + (2L - 2\mu D - \mu H)\sin\beta}$$

$$= \frac{1 - 2\times 0.5\times 0.1 - 0.5\times 0.6}{0.6(\sqrt{3}/2) + (2 - 2\times 0.5\times 0.1 - 0.5\times 0.6)/2} = 0.4547 \qquad \text{(c)}$$

Then, inserting N_B from the second of Eqs. (a) into the first and using Eq. (c), we obtain the nondimensional acceleration

$$\frac{a_x}{g} = \frac{F\cos\beta - \mu(mg - F\sin\beta)}{mg} = 0.4547\frac{\sqrt{3}}{2} - 0.5\left(1 - \frac{1}{2}0.4547\right)$$

$$= 0.75\times 10^{-2} \qquad \text{(d)}$$

1.5.2 Pure Rotation About a Fixed Point

In the case of pure rotation about a fixed point O, the motion is described by the moment equation given by Eq. (1.36), in which, from Eq. (1.35),

$$\mathbf{M}_O = \int_{\text{body}} \mathbf{r}\times d\mathbf{F} \qquad (1.51)$$

is the resultant torque about point O (Fig. 1.8) and, from Eq. (1.32),

$$\mathbf{H}_O = \int_{\text{body}} \mathbf{r}\times \mathbf{v}dm \qquad (1.52)$$

is the moment of momentum, or *angular momentum* of the body about O. From kinematics (Ref. 11, Sec. 2.5), we can write the velocity vector of a point on a rigid body in pure planar rotation in the form

$$\mathbf{v} = r\omega\mathbf{u}_\theta \qquad (1.53)$$

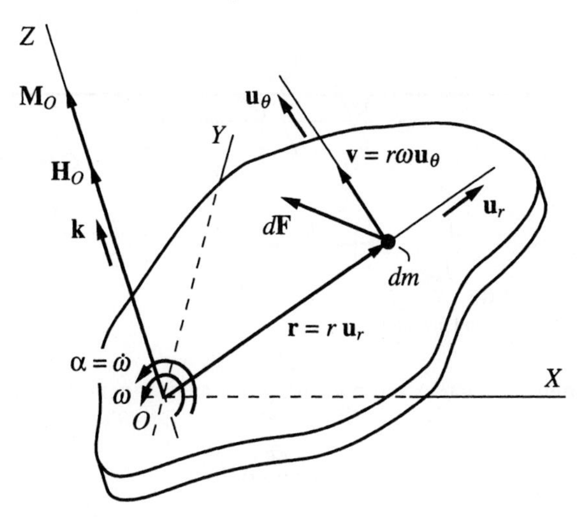

FIGURE 1.8
Pure rotation about a fixed point

Then, letting $\mathbf{r} = r\mathbf{u}_r$ and inserting Eq. (1.53) into Eq. (1.52), the angular momentum for planar motion becomes

$$\mathbf{H}_O = \int_{\text{body}} r\mathbf{u}_r \times r\omega\mathbf{u}_\theta dm = \left(\int_{\text{body}} r^2 dm\right)\omega\mathbf{k} = I_O\omega\mathbf{k} \tag{1.54}$$

where $\mathbf{k}$ is a unit vector normal to the plane of motion and

$$I_O = \int_{\text{body}} r^2 dm \tag{1.55}$$

is the moment of inertia of the body about O. Hence, for planar motions, the angular momentum has only one component. Consistent with this, the torque vector has only one component also, namely

$$\mathbf{M}_O = M_O\mathbf{k} \tag{1.56}$$

But, from Eq. (1.54),

$$\dot{\mathbf{H}}_O = I_O\alpha\mathbf{k} \tag{1.57}$$

where $\alpha = \dot{\omega}$ is the angular acceleration magnitude. Hence, inserting Eqs. (1.56) and (1.57) into Eq. (1.36), we obtain the single scalar moment equation

$$M_O = I_O\alpha \tag{1.58}$$

Example 1.5. A rigid body suspended from a point other than the mass center and free to oscillate is known as a "compound pendulum." Figure 1.9 depicts a compound pendulum consisting of a uniform bar of total mass m hinged at point O at a distance $L/6$ from the mass center C. If the pendulum is released from rest in the horizontal position, determine the angular acceleration of the pendulum immediately after release.

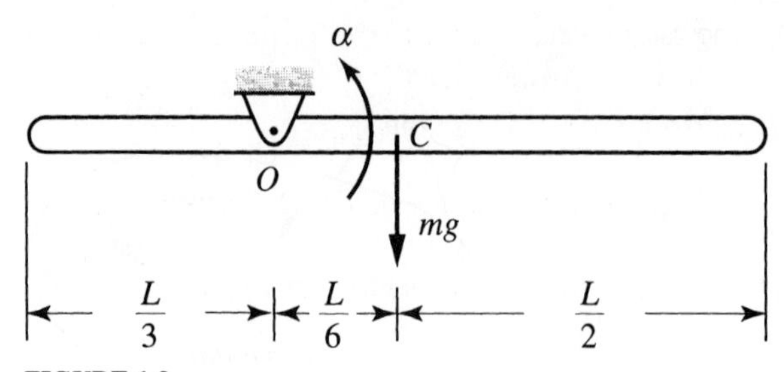

FIGURE 1.9
Compound pendulum

Regarding counterclockwise moments and angular motions as positive, Eq. (1.58) yields

$$M_O = -mg\frac{L}{6} = I_O\alpha \tag{a}$$

where the mass moment of inertia of the pendulum about O can be obtained through integration as follows:

$$I_O = \int_m x^2 dm = \frac{m}{L}\int_{-L/3}^{2L/3} x^2 dx = \frac{m}{L}\frac{x^3}{3}\bigg|_{-L/3}^{2L/3} = \frac{mL^2}{9} \tag{b}$$

Hence, inserting Eq. (b) into Eq. (a), we conclude that the angular acceleration immediately after release is

$$\alpha = -\frac{3g}{2L} \tag{c}$$

where the minus sign indicates that the acceleration is in the opposite sense to that indicated in Fig. 1.9, i.e., it is in the clockwise sense.

1.5.3 General Planar Motion Referred to the Mass Center

In the general case, the body is capable of both translation and rotation relative to the inertial space, so that it is necessary to refer the motion to a moving point. In this case, it is advantageous for the most part to refer the motion to the moving mass center C. The force and moment equations for rigid bodies retain the same general form as for collections of particles. Hence, from Eq. (1.42), we write the force equation

$$\mathbf{F} = m\mathbf{a}_C \tag{1.59}$$

which states that the force equation of motion of a rigid body is the same as that of a fictitious body with the entire mass concentrated at the mass center C. Moreover, from Eq. (1.46), the moment equation about the mass center C is

$$\mathbf{M}_C = \dot{\mathbf{H}}_C \tag{1.60}$$

where, by analogy with Eq. (1.45), the moment about C of a rigid body in planar motion is defined as

$$\mathbf{M}_C = \int_{\text{body}} \mathbf{r}' \times d\mathbf{F} = M_C\mathbf{k} \tag{1.61}$$

in which $\mathbf{r}'$ is the radius vector from C to dm, and, by analogy with Eq. (1.54), the angular momentum about C is given by

$$\mathbf{H}_C = I_C \omega \mathbf{k} \tag{1.62}$$

where I_C is the moment of inertia of the rigid body about C and ω is the angular velocity of the body relative to the inertial space.

For planar motions, the force equation, Eq. (1.59), has the two scalar components

$$F_x = ma_{Cx}, \quad F_y = ma_{Cy} \tag{1.63}$$

and, by analogy with Eq. (1.58) the moment equation about the mass center C has the single component

$$M_C = I_C \alpha \tag{1.64}$$

where α is the angular acceleration of the body relative to the inertial space. It should be pointed out that, to obtain the motion of the rigid body, Eqs. (1.63) and (1.64) must in general be solved simultaneously.

Example 1.6. A uniform rigid bar of total mass m and length L_2, suspended at point O by a string of length L_1, is acted upon by the horizontal force F, as shown in Fig. 1.10a. Use the angular displacements θ_1 and θ_2 to define the position, velocity and acceleration of the mass center C in terms of body axes and then derive the equations of motion for the translation of C and the rotation about C.

Referring to Fig. 1.10b, we can write the position, velocity and acceleration of the mass center C in the form

$$\mathbf{r}_C = \mathbf{r}_A + \mathbf{r}_{AC} = L_1 \mathbf{u}_{r1} + \frac{L_2}{2}\mathbf{u}_{r2} \tag{a}$$

$$\mathbf{v}_C = \mathbf{v}_A + \mathbf{v}_{AC} = L_1 \dot{\theta}_1 \mathbf{u}_{\theta 1} + \frac{L_2}{2}\dot{\theta}_2 \mathbf{u}_{\theta 2} \tag{b}$$

and

$$\mathbf{a}_C = \mathbf{a}_A + \mathbf{a}_{AC} = -L_1 \dot{\theta}_1^2 \mathbf{u}_{r1} + L_1 \ddot{\theta}_1 \mathbf{u}_{\theta 1} - \frac{L_2}{2}\dot{\theta}_2^2 \mathbf{u}_{r2} + \frac{L_2}{2}\ddot{\theta}_2 \mathbf{u}_{\theta 2} \tag{c}$$

respectively. Equations (a) - (c) are in terms of two sets of unit vectors. To obtain expressions in terms of the body axes r_2, θ_2, we observe from Fig. 1.10b that the two sets of unit vectors are related by

$$\begin{aligned} \mathbf{u}_{r1} &= \cos(\theta_2 - \theta_1)\mathbf{u}_{r2} - \sin(\theta_2 - \theta_1)\mathbf{u}_{\theta 2} \\ \mathbf{u}_{\theta 1} &= \sin(\theta_2 - \theta_1)\mathbf{u}_{r2} + \cos(\theta_2 - \theta_1)\mathbf{u}_{\theta 2} \end{aligned} \tag{d}$$

Inserting Eqs. (d) into Eqs. (a) - (c), we obtain the position, velocity and acceleration of the mass center C in terms of components along the body axes, as follows:

$$\mathbf{r}_C = \left[L_1 \cos(\theta_2 - \theta_1) + \frac{L_2}{2}\right]\mathbf{u}_{r2} - L_1 \sin(\theta_2 - \theta_1)\mathbf{u}_{\theta 2} \tag{e}$$

$$\mathbf{v}_C = L_1 \dot{\theta}_1 \sin(\theta_2 - \theta_1)\mathbf{u}_{r2} + \left[L_1 \dot{\theta}_1 \cos(\theta_2 - \theta_1) + \frac{L_2}{2}\dot{\theta}_2\right]\mathbf{u}_{\theta 2} \tag{f}$$

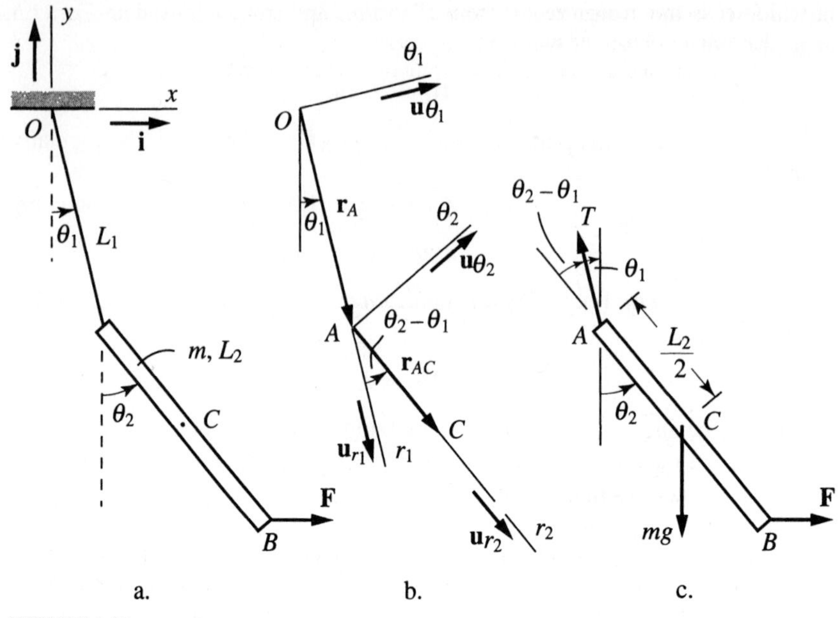

FIGURE 1.10
a. Rigid bar on a string, b. Position of the mass center, c. Free-body diagram

and

$$\mathbf{a}_C = \left[L_1\ddot{\theta}_1 \sin(\theta_2-\theta_1) - L_1\dot{\theta}_1^2 \cos(\theta_2-\theta_1) - \frac{L_2}{2}\dot{\theta}_2^2\right]\mathbf{u}_{r2}$$

$$+\left[L_1\ddot{\theta}_1 \cos(\theta_2-\theta_1) + L_1\dot{\theta}_1^2 \sin(\theta_2-\theta_1) + \frac{L_2}{2}\ddot{\theta}_2\right]\mathbf{u}_{\theta 2} \qquad \text{(g)}$$

respectively.

To derive the equations of motion, we refer to the free-body diagram of Fig. 1.10c, in which T is the tension in the string and mg the weight of the bar. By analogy with Eqs. (1.63), the force equations in terms of body axes components are

$$F_{r2} = ma_{Cr2}, \; F_{\theta_2} = ma_{C\theta 2} \qquad \text{(h)}$$

and from Eq. (1.64) the moment equation is

$$M_C = I_C\alpha = I_C\ddot{\theta}_2 \qquad \text{(i)}$$

But, from Fig. 1.10c, the force resultants and moment about C are

$$\begin{aligned} F_{r2} &= F\sin\theta_2 + mg\cos\theta_2 - T\cos(\theta_2-\theta_1) \\ F_{\theta 2} &= F\cos\theta_2 - mg\sin\theta_2 + T\sin(\theta_2-\theta_1) \\ M_C &= F\frac{L_2}{2}\cos\theta_2 - T\frac{L_2}{2}\sin(\theta_2-\theta_1) \end{aligned} \qquad \text{(j)}$$

Moreover, we recognize that the acceleration components a_{Cr2} and $a_{C\theta 2}$ are simply the coefficients of $\mathbf{u}_{r2}$ and $\mathbf{u}_{\theta 2}$ in Eq. (g), respectively, and that for a thin bar $I_C = mL_2^2/12$. In view of this, the desired equations of motion take the explicit form

$$F\sin\theta_2 + mg\cos\theta_2 - T\cos(\theta_2-\theta_1) = m\left[L_1\ddot{\theta}_1\sin(\theta_2-\theta_1) - L_1\dot{\theta}_1^2\cos(\theta_2-\theta_1) - \frac{L_2}{2}\dot{\theta}_2^2\right]$$

$$F\cos\theta_2 - mg\sin\theta_2 + T\sin(\theta_2-\theta_1) = m\left[L_1\ddot{\theta}_1\cos(\theta_2-\theta_1) + L_1\dot{\theta}_1^2\sin(\theta_2-\theta_1) + \frac{L_2}{2}\ddot{\theta}_2\right]$$

$$F\frac{L_2}{2}\cos\theta_2 - T\frac{L_2}{2}\sin(\theta_2-\theta_1) = \frac{mL_2^2}{12}\ddot{\theta}_2 \tag{k}$$

We observe that the motion of the bar is fully defined by the angular displacements θ_1 and θ_2, and there are three equations of motion. Hence, there must be a third unknown, besides θ_1 and θ_2, which is the tension T. If the value of T presents no interest, then it can be eliminated from Eqs. (k) and, in the process, reduce the number of equations to two. For example, the first two of Eqs. (k) can be used to solve for T. Then, introducing this expression for T into any two of the three equations, we obtain two equations of motion in terms of θ_1 and θ_2 and their time derivatives alone. This elimination process is left as an exercise to the reader.

1.6 KINETIC ENERGY OF RIGID BODIES IN PLANAR MOTION

In Sec. 1.3, we defined the kinetic energy for a single particle, Eq. (1.13). The definition can be extended to collections of particles and rigid bodies. Because in this text we have no particular interest in the kinetic energy of arbitrary collections of particles, we consider the kinetic energy of rigid bodies directly. To this end, we use the pattern of Sec. 1.5, as follows:

1.6.1 Pure Translation Relative to the Inertial Space

By analogy with Eq. (1.13), the kinetic energy for a continuous body in planar motion can be written as

$$T(t) = \frac{1}{2}\int_{\text{body}} \mathbf{v}(x,y,t)\cdot\mathbf{v}(x,y,t)dm(x,y) \tag{1.65}$$

But, for a rigid body in pure translation, the velocity is the same for every point in the body, $\mathbf{v}(x,y,t) = \mathbf{v}(t)$, so that the kinetic energy reduces to

$$T = \frac{1}{2}m\mathbf{v}\cdot\mathbf{v} \tag{1.66}$$

which has the same form as the kinetic energy of a single particle, except that here m represents the total mass of the body.

1.6.2 Pure Rotation About a Fixed Point

Inserting Eq. (1.53) into Eq. (1.65), the kinetic energy in pure planar rotation about the fixed point O can be written as

$$T = \frac{1}{2}\int_{\text{body}} r\omega\mathbf{u}_\theta \cdot r\omega\mathbf{u}_\theta dm = \frac{1}{2}\omega^2\int_{\text{body}} r^2 dm = \frac{1}{2}I_O\omega^2 \tag{1.67}$$

in which I_O is recognized as the mass moment of inertia about O, Eq. (1.55), and ω is the magnitude of the angular velocity vector $\boldsymbol{\omega}$.

1.6.3 General Planar Motion Referred to the Mass Center

As in Sec. 1.5, it is advantageous to work with a set of body axes with the origin at the mass center C of the rigid body, so that the velocity of any point in the rigid body is the sum of the velocity of translation of the mass center and the velocity of rotation about the mass center. Hence, substituting r' for r in Eq. (1.53), the velocity in planar motion is

$$\mathbf{v} = \mathbf{v}_C + r'\omega\mathbf{u}_\theta \tag{1.68}$$

where r' is the distance from C to dm. Inserting Eq. (1.68) into Eq. (1.65) and recognizing that, by analogy with Eq. (1.40),

$$\int_{\text{body}} r' dm = 0 \tag{1.69}$$

we can write the kinetic energy of a rigid body in general planar motion in the form

$$\begin{aligned} T &= \frac{1}{2}\int_{\text{body}} (\mathbf{v}_C + r'\omega\mathbf{u}_\theta)\cdot(\mathbf{v}_C + r'\omega\mathbf{u}_\theta) dm \\ &= \frac{1}{2}m\mathbf{v}_C\cdot\mathbf{v}_C + \mathbf{v}_C\cdot\omega\mathbf{u}_\theta\int_{\text{body}} r' dm + \frac{1}{2}\int_{\text{body}} r'\omega\mathbf{u}_\theta\cdot r'\omega\mathbf{u}_\theta dm \\ &= T_{\text{tr}} + T_{\text{rot}} \end{aligned} \tag{1.70}$$

in which

$$T_{\text{tr}} = \frac{1}{2}m\mathbf{v}_C\cdot\mathbf{v}_C \tag{1.71}$$

is the *kinetic energy of translation* as if the entire body were concentrated at the mass center C and

$$T_{\text{rot}} = \frac{1}{2}\omega^2\int_{\text{body}} (r')^2 dm = \frac{1}{2}I_C\omega^2 \tag{1.72}$$

is the *kinetic energy of rotation about the mass center*, where I_C is the mass moment of inertia of the body about the mass center. Clearly, the advantage of choosing the mass center C as the origin of the body axes is that the kinetic energy separates into two parts, one due to translation of point C and one due to rotation about C, and there is no coupling between the translation and rotation.

Example 1.7. Derive the kinetic energy of the bar on a string of Example 1.6.

The bar translates and rotates, with the velocity of the mass center C being given by Eq. (f) of Example 1.6 and the velocity of rotation being $\dot{\theta}_2$. Hence, the kinetic energy consists of two parts, one due to translation of C and one due to rotation about C. Inserting Eq. (f) of Example 1.6 into Eq. (1.71), we obtain the kinetic energy of translation

$$T_{\text{tr}} = \frac{1}{2} m \mathbf{v}_C \cdot \mathbf{v}_C = \frac{1}{2} m \left\{ L_1 \dot{\theta}_1 \sin(\theta_2 - \theta_1) \mathbf{u}_{r2} + \left[L_1 \dot{\theta}_1 \cos(\theta_2 - \theta_1) + \frac{L_2}{2} \dot{\theta}_2 \right] \mathbf{u}_{\theta 2} \right\}$$

$$\cdot \left\{ L_1 \dot{\theta}_1 \sin(\theta_2 - \theta_1) \mathbf{u}_{r2} + \left[L_1 \dot{\theta}_1 \cos(\theta_2 - \theta_1) + \frac{L_2}{2} \dot{\theta}_2 \right] \mathbf{u}_{\theta 2} \right\}$$

$$= \frac{1}{2} m \left[L_1^2 \dot{\theta}_1^2 + L_1 L_2 \dot{\theta}_1 \dot{\theta}_2 \cos(\theta_2 - \theta_1) + \frac{L_2^2}{4} \dot{\theta}_2^2 \right] \tag{a}$$

On the other hand, using Eq. (1.72) and recalling that the mass moment of inertia of a thin uniform bar about C is $I_C = mL^2/12$, the kinetic energy of rotation about C is simply

$$T_{\text{rot}} = \frac{1}{2} I_C \omega^2 = \frac{1}{2} \frac{m L_2^2}{12} \dot{\theta}_2^2 \tag{b}$$

1.7 CHARACTERISTICS OF DISCRETE SYSTEM COMPONENTS

Vibrating systems represent assemblages of individual components acting together as a whole. Before we can produce the equations of motion for a given system, it is necessary to establish the excitation-response characteristics of the constituent components. The components can be broadly divided into three classes according to whether the component forces are proportional to displacements, proportional to velocities, or proportional to accelerations. Correspondingly, they can be divided into components that store and release potential energy, dissipate energy and store and release kinetic energy. This section is devoted to such component characterization.

In the first class, the components possess the characteristic that, when displaced from equilibrium, they generate forces seeking to restore the system to equilibrium. For the most part, but not exclusively, this property is due to elasticity. All *elastic components store potential energy* as displacements increase, and *release potential energy* as displacements decrease. A typical component in this group is the *helical spring* depicted in Fig. 1.11 and shown schematically in Fig. 1.12a. Although this is only approximately true, springs are generally assumed to be *massless*, so that a force F_s at one end must be balanced by a force F_s acting at the other end. A tensile force F_s, such as that shown in Fig. 1.12a, causes the spring to undergo an elongation δ equal to the difference $x_2 - x_1$ between the displacements x_2 and x_1 of the two end points. A typical plot of the force F_s

FIGURE 1.11
Helical Spring

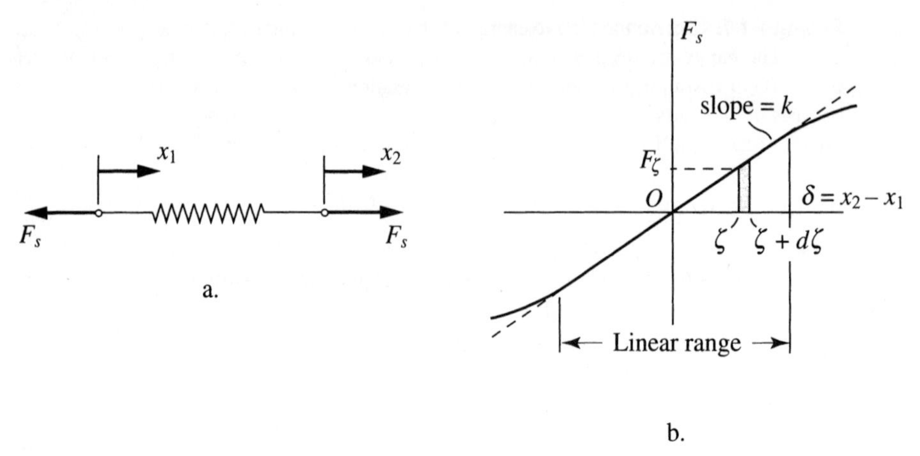

FIGURE 1.12
a. Spring under a tensile force, b. Force versus elongation

as a function of the elongation δ is as depicted in Fig. 1.12b. For a given range, known as the *linear range*, δ is proportional to F_s, where the constant of proportionality k is equal to the slope of the curve F_s versus δ. Hence, in the linear range, the relation between the force and elongation is simply

$$F_s = k\delta = k(x_2 - x_1) \tag{1.73}$$

A spring operating in the linear range is said to be *linear*, in which case the constant k is referred to as the *spring constant*, or *spring stiffness*. It is customary to identify a linear spring by its stiffness k. We note that the units of k are newtons per meter (N/m). It should be pointed out that Figs. 1.12a and 1.12b show the force F_s external to the spring. At every point inside the spring there is an *elastic force* $-F_s$ which tends to return the spring to the undeformed configuration, and hence it represents a *restoring force*. In many cases the undeformed configuration corresponds to the static equilibrium position (Sec. 1.10). Clearly, because they depend on the elongation alone, *spring forces are conservative* (see Sec. 1.3).

Next, we derive the potential energy expression. Recognizing that in the linear range there is a spring restoring force equal to $-k\zeta$ corresponding to an elongation ζ and taking the undeformed configuration as the reference position, we can use Eq. (1.18) and write the potential energy

$$V(\delta) = \int_\delta^0 F_\zeta d\zeta = \int_\delta^0 (-k\zeta)d\zeta = \frac{1}{2}k\delta^2 \tag{1.74}$$

The potential energy can be interpreted geometrically as the integral of the shaded area in Fig. 1.12b.

Beyond the linear range elongations are no longer proportional to the force, in which case the spring is said to be *nonlinear*. If the force F_s increases at a slower rate than the elongation δ, then the spring is said to be a "softening spring." This is the case shown in Fig. 1.12b. On the other hand, if the force F_s increases at a faster rate than the elongation δ, then the spring is referred to as a "stiffening spring."

Spring forces are conservative, regardless of whether a spring is linear or nonlinear, as they depend on the elongation alone. Hence using Eq. (1.18), we can write the potential energy for springs operating beyond the linear range in the form

$$V(\delta) = \int_{\delta}^{0} F_{\zeta} d\zeta \tag{1.75}$$

However, before the integral can be evaluated, it is necessary to specify how F_{ζ} varies with ζ in the nonlinear range. If Fig. 1.12b was obtained experimentally, and no analytical expression for F_{ζ} is available, then the potential energy can be evaluated by determining the area under the curve numerically.

The second type of component relates forces to velocities. This is the *viscous damper*, or the *dashpot*, and it consists of a piston fitting loosely in a cylinder filled with oil or water so that the viscous fluid can flow around the piston inside the cylinder, as depicted in Fig. 1.13. Alternatively, the piston has holes permitting fluid to flow through them. As with the spring, the viscous damper is assumed to be massless, so that a force F_d at one end must be balanced by a corresponding force F_d at the other end, as shown schematically in Fig. 1.14a. It is also assumed that the forces F_d cause smooth shear in the viscous fluid, so that the plot F_d versus $\dot{\delta}$ is linear, as depicted in Fig. 1.14b, where $\dot{\delta} = \dot{x}_2 - \dot{x}_1$ is the velocity of separation of the end points. The relation between the force

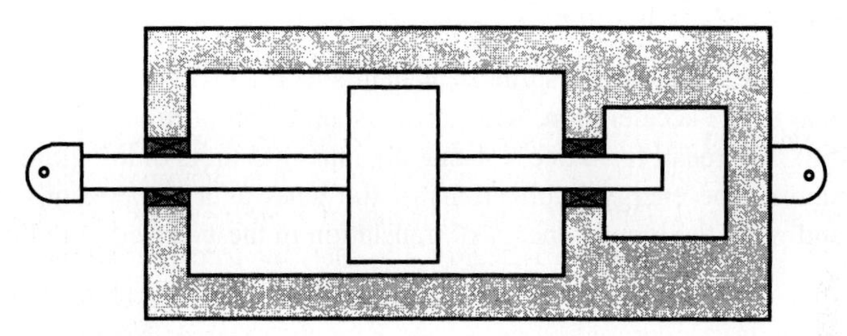

FIGURE 1.13
Viscous damper

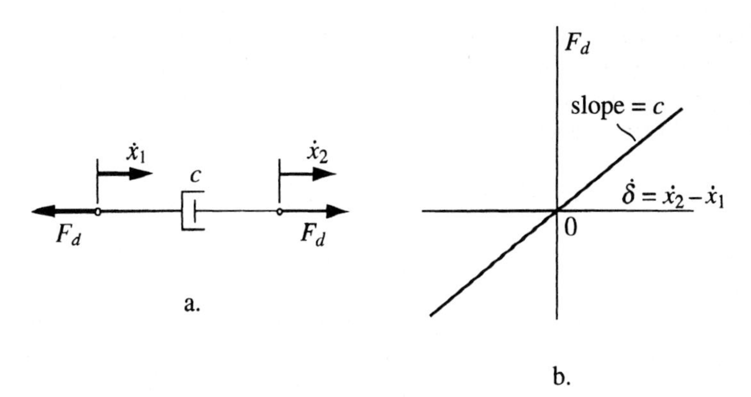

FIGURE 1.14
a. Damper under a tensile force, b. Force versus separation velocity

F_d and the velocity of separation $\dot{\delta}$ is simply

$$F_d = c\dot{\delta} = c(\dot{x}_2 - \dot{x}_1) \tag{1.76}$$

where the proportionality constant c, which is merely the slope of the curve F_d versus $\dot{\delta}$, is known as the *coefficient of viscous damping*. We identify a viscous damper by the coefficient c. The units of c are newton · second per meter (N · s/m).

At this point, we consider the energy implications of viscous dampers. The force in the damper is opposed to the external force. By virtue of the assumption that the damper is massless, the force has the same magnitude as the external force, so that it is equal to $-c\dot{\delta}$. Clearly, *the damper force is nonconservative*, as it depends on the velocity and not on the position. Regarding the damper as part of a system and using Eq. (1.25), we can write

$$\dot{E} = (-c\dot{\delta})\dot{\delta} = -c\dot{\delta}^2 \tag{1.77}$$

where E is the total energy of the system. But, the right side of Eq. (1.77) is negative as long as $\dot{\delta} \neq 0$, and is equal to zero only when $\dot{\delta} = 0$. Hence, we must conclude that the system loses energy steadily, so that *viscous dampers dissipate energy*.

The third and final type of component is the rigid mass in translation. For motion in the x-direction only, as shown in Fig. 1.15a, Newton's second law, the first of Eqs. (1.50), yields

$$F_m = ma_x = m\ddot{x} \tag{1.78}$$

Consistent with the discussion of springs and dampers, Eq. (1.78) states that the force F_m is proportional to the acceleration, with the constant of proportionality being the mass m (Fig. 1.15b). We recall from Sec. 1.1 that the mass has units of kilograms (kg).

To examine the energy implications of the mass as a component, we consider Eq. (1.66) and write the kinetic energy of translation in the x-direction in the form

$$T = \frac{1}{2}m\dot{x}^2 \tag{1.79}$$

from which we conclude that *masses store kinetic energy* as velocities increase, and *release kinetic energy* as velocities decrease.

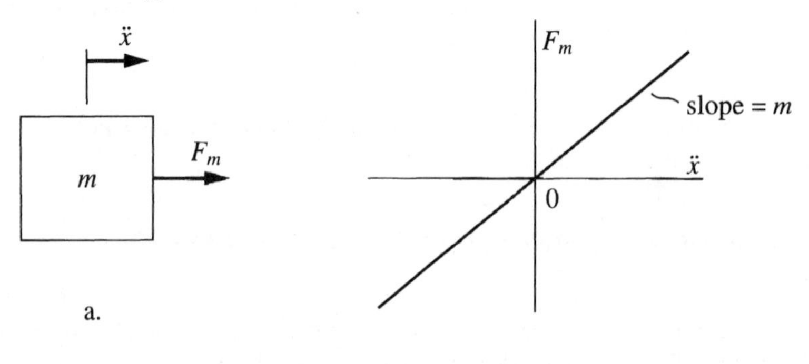

FIGURE 1.15
a. Mass acted upon by a force, b. Force versus acceleration

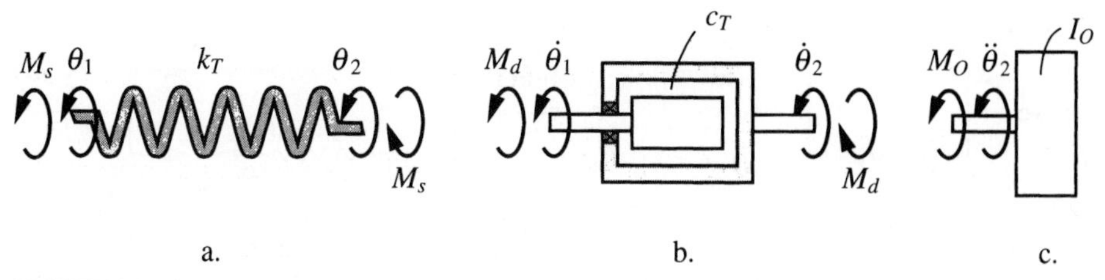

FIGURE 1.16
a. Torsional spring, b. Torsional damper, c. Rotating mass

The spring constant k, the coefficient of viscous damping c and the mass m represent *parameters* of a system. Because they do not require spatial variables to describe their location and can be regarded as being located at discrete points, they are referred to as *lumped*, or *discrete* parameters. We reiterate that, unless otherwise stated, springs and dampers possess no mass and masses are rigid. Later in this text, we relax these assumptions.

Similar types of components relate torques to rotational motions. For the torsional spring of Fig. 1.16a, the relation between the torque M_s and the angle of twist $\theta_2 - \theta_1$ is

$$M_s = k_T(\theta_2 - \theta_1) \tag{1.80}$$

where k_T is the torsional spring constant and θ_1 and θ_2 are the angular displacements of the end points. The units of k_T are newtons · meter per radian (N · m/rad). In a like fashion, for the torsional viscous damper of Fig. 1.16b, the relation is

$$M_d = c_T(\dot{\theta}_2 - \dot{\theta}_1) \tag{1.81}$$

in which M_d is the torque acting on the damper and c_T is the torsional coefficient of viscous damping, where the units of c_T are newtons · meter · second per radian (N · m · s/rad). Finally, from Eq. (1.58), the relation between the torque M_O about the fixed point O and the acceleration $\ddot{\theta}$ of a rigid body about O, as shown in Fig. 1.16c, is

$$M_O = I_O\ddot{\theta} \tag{1.82}$$

where I_O is the mass moment of inertia of the body about O, and we note that I_O has units kilogram · meter^2 (kg · m^2).

1.8 EQUIVALENT SPRINGS, DAMPERS AND MASSES

On occasion springs and dampers occur in certain combinations. In such cases, the analysis can be simplified appreciably by using *equivalent springs* and *dampers* to simulate the action of the combinations in question. To illustrate the idea, we consider springs connected in parallel and springs connected in series. We confine ourselves to linear springs, as the concept does not apply to nonlinear ones. Figure 1.17a shows a system of two *springs in parallel* under the action of the tensile force F_s and Fig. 1.17b

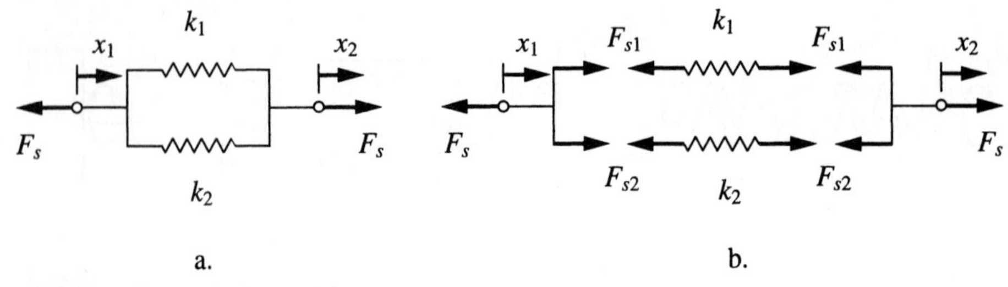

FIGURE 1.17
a. Springs in parallel, b. Force diagram

shows a diagram with the individual springs and the corresponding internal forces. From Fig. 1.17b, by analogy with Eq. (1.73), we have the relations

$$F_{s1} = k_1(x_2 - x_1), \quad F_{s2} = k_2(x_2 - x_1) \tag{1.83}$$

where F_{s1} and F_{s2} are the forces acting on the springs k_1 and k_2, respectively. Also from Fig. 1.17b, we conclude that the spring forces F_{s1} and F_{s2} must add up to the total force F_s, or

$$F_s = F_{s1} + F_{s2} \tag{1.84}$$

Inserting Eqs. (1.83) into Eq. (1.84), we obtain

$$F_s = k_{\text{eq}}(x_2 - x_1) \tag{1.85}$$

in which

$$k_{\text{eq}} = k_1 + k_2 \tag{1.86}$$

denotes the stiffness of an *equivalent spring* representing the combined effect of k_1 and k_2 arranged in parallel. If a number n of springs k_i $(i = 1, 2, \ldots, n)$ are arranged in parallel, then it is not difficult to show that the equivalent spring is

$$k_{\text{eq}} = \sum_{i=1}^{n} k_i \tag{1.87}$$

For two *springs in series*, as depicted in Fig. 1.18a, we first recognize that the same force F_s acts throughout both springs. Then, from Figs. 1.18b, we have the relations

$$F_s = k_1(x_0 - x_1), \quad F_s = k_2(x_2 - x_0) \tag{1.88}$$

so that, eliminating x_0 from Eqs. (1.88), we can write

$$F_s = k_{\text{eq}}(x_2 - x_1) \tag{1.89}$$

in which

$$k_{\text{eq}} = \left(\frac{1}{k_1} + \frac{1}{k_2}\right)^{-1} \tag{1.90}$$

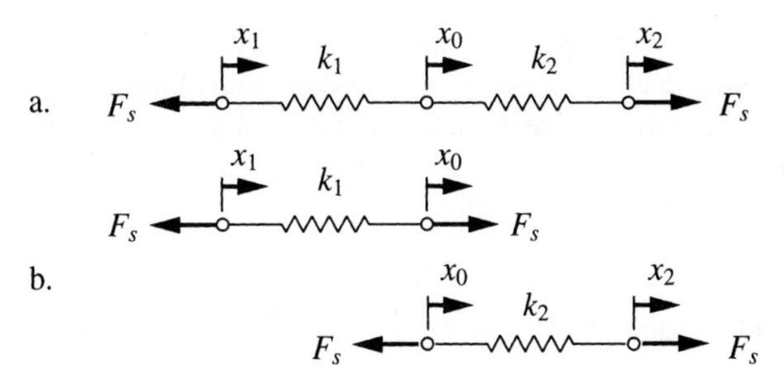

FIGURE 1.18
a. Springs in series, b. Force diagram

is the equivalent spring constant for two springs connected in series. If there are n springs connected in series, then the equivalent stiffness is

$$k_{\text{eq}} = \left(\sum_{i=1}^{n} \frac{1}{k_i} \right)^{-1} \tag{1.91}$$

Expressions for equivalent spring constants for *torsional springs in parallel* and *in series* can be shown to resemble Eqs. (1.87) and (1.91), respectively. Moreover, equivalent coefficients of viscous damping for *dashpots in parallel* and *in series* can be derived in an analogous manner. They have the same structure as Eqs. (1.87) and (1.91), respectively, except that the symbol k is replaced by the symbol c.

Under certain circumstances, distributed elastic components can be treated as equivalent discrete springs. As an illustration, we consider a component in the form of a thin rod fixed at $x = 0$ and with the tensile force F at $x = L$ (Fig. 1.19). If the mass of the rod is negligibly small compared to any other masses in the system, then by analogy with Eq. (1.73), the rod can be regarded as a spring with the equivalent spring constant

$$k_{\text{eq}} = \frac{F}{\delta} \tag{1.92}$$

where $\delta = u(L)$ is the axial displacement of the tip of the rod. But, from mechanics of materials (Ref. 1, Sec. 2.8), the static axial displacement $u(x)$ of a point at a distance x from the left end must satisfy the differential equation

$$EA(x)\frac{du(x)}{dx} = F(x),\ 0 < x < L \tag{1.93}$$

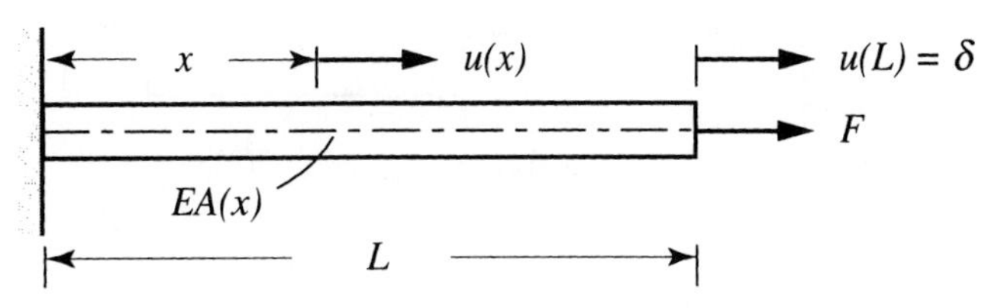

FIGURE 1.19
Rod under a tensile force

in which E is Young's modulus, or modulus of elasticity, and $A(x)$ is the cross-sectional area of the rod. Note that the product $EA(x)$ is commonly known as the *axial stiffness*. Moreover, $F(x)$ is the axial force, which in this case is the same for any point x and is equal to the force F at $x = L$, $F(x) = F$. Because the rod is fixed at $x = 0$, the solution $u(x)$ must satisfy the boundary condition

$$u(0) = 0 \tag{1.94}$$

Integrating Eq. (1.93) and considering Eq. (1.94), we obtain simply

$$u(x) = F \int_0^x \frac{d\xi}{EA(\xi)} \tag{1.95}$$

But, at $x = L$ the displacement must be equal to δ, so that

$$u(L) = \delta = F \int_0^L \frac{d\xi}{EA(\xi)} \tag{1.96}$$

Hence, inserting Eq. (1.96) into Eq. (1.92), we can write the *equivalent axial spring constant* for the rod of Fig. 1.19 in the form

$$k_{\text{eq}} = \frac{F}{\delta} = \left[\int_0^L \frac{d\xi}{EA(\xi)} \right]^{-1} \tag{1.97}$$

and we note that the units of k_{eq} are newtons per meter (N/m). In the case of a *uniform rod*, $EA(x) = EA =$ constant, Eq. (1.97) reduces to

$$k_{\text{eq}} = \frac{EA}{L} \tag{1.98}$$

Another case of interest is that in which the axial stiffness of the rod is sectionally uniform, as shown in Fig. 1.20. Such a rod can be treated as two axial springs in series with the stiffnesses

$$k_i = \frac{EA_i}{L_i}, \quad i = 1, 2 \tag{1.99}$$

Then, inserting Eqs. (1.99) into Eq. (1.90), we obtain the equivalent spring constant for the sectionally uniform rod in the form

$$k_{\text{eq}} = \left(\frac{1}{k_1} + \frac{1}{k_2} \right)^{-1} = \left(\frac{L_1}{EA_1} + \frac{L_2}{EA_2} \right)^{-1} \tag{1.100}$$

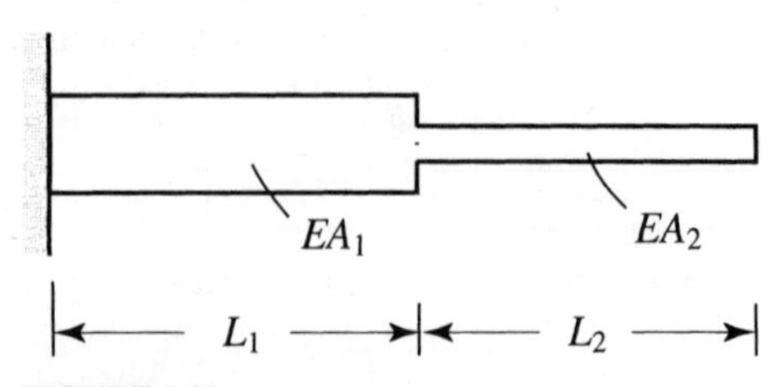

FIGURE 1.20
Rod with sectionally constant stiffness

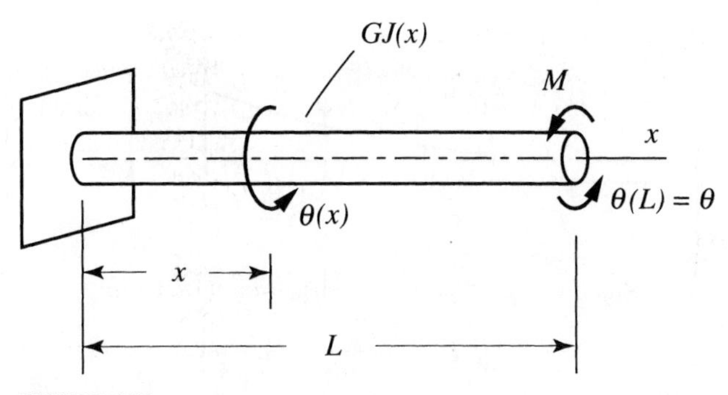

FIGURE 1.21
Shaft in torsion subjected to a torque

Shafts in torsion are entirely analogous to rods deforming axially. Figure 1.21 shows a shaft clamped at $x = 0$ and subjected to a torque M at $x = L$. Indeed, all the developments covered by Eqs. (1.93)–(1.98) remain the same except that the torsional displacement $\theta(x)$ replaces the axial displacement $u(x)$, the moment M replaces the force F and the torsional stiffness $GJ(x)$ replaces the axial stiffness $EA(x)$, where G is the shear modulus and $J(x)$ is the polar moment of inertia of the cross-sectional area. Hence, from Eq. (1.98) the equivalent torsional spring constant for a uniform shaft is

$$k_{\text{eq}} = \frac{M}{\theta} = \frac{GJ}{L} \tag{1.101}$$

in which $\theta = \theta(L)$ is the tip angular displacement. We observe that, unlike axial springs, the units of torsional springs are newton · meter per radian (N · m/rad). Clearly, the analogy extends to sectionally uniform shafts in torsion. Indeed, to obtain the equivalent torsional spring constant for two shafts in series clamped at $x = 0$, we simply replace EA_i by GJ_i $(i = 1, 2)$ in Eq. (1.100).

Next, we derive the equivalent spring constant for a beam in bending clamped at $x = 0$ and acted upon by a transverse force F at $x = L$, as shown in Fig. 1.22. From mechanics of materials (Ref. 1, Sec. 8.3), according to the elementary beam theory, the transverse displacement $w(x)$ satisfies the differential equation

$$EI(x)\frac{d^2w(x)}{dx^2} = M(x),\ 0 < x < L \tag{1.102}$$

where $EI(x)$ is the *flexural rigidity*, in which $I(x)$ is the cross-sectional area moment of inertia, and $M(x)$ is the bending moment. In the case at hand, the bending moment is equal to $F(L - x)$. At the clamped end, $x = 0$, the displacement and the slope of the displacement curve must be zero, so that the boundary conditions are

$$w = 0,\ \frac{dw}{dx} = 0 \text{ at } x = 0 \tag{1.103}$$

Integrating Eq. (1.102) twice and considering Eqs. (1.103), we obtain the solution

$$w(x) = F\int_0^x \left[\int_0^\zeta \frac{L-\xi}{EI(\xi)}d\xi\right]d\zeta \tag{1.104}$$

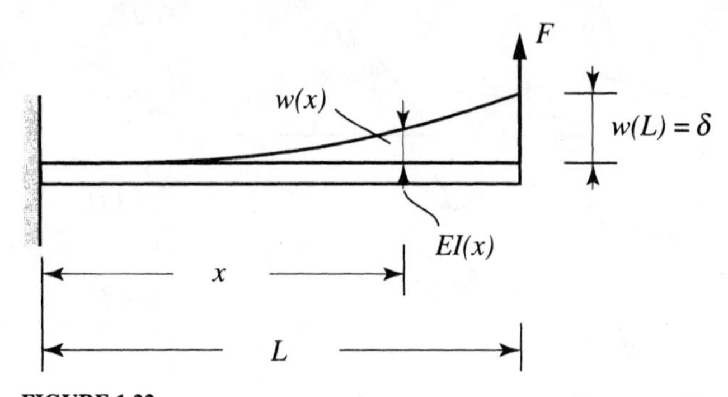

FIGURE 1.22
Beam in bending under a transverse force

so that, letting $w(L) = \delta$ and using the same approach as for the rod, the *equivalent spring constant for a cantilever beam* is

$$k_{\text{eq}} = \frac{F}{\delta} = \frac{F}{w(L)} = \left\{ \int_0^L \left[\int_0^\zeta \frac{L-\xi}{EI(\xi)} d\xi \right] d\zeta \right\}^{-1} \tag{1.105}$$

In the case of a *uniform cantilever beam*, $EI(x) = EI = \text{constant}$, Eq. (1.105) reduces to

$$k_{\text{eq}} = \left\{ EI \int_0^L \left[\int_0^\zeta (L-\xi) d\xi \right] d\zeta \right\}^{-1} = \left[EI \int_0^L \left(L\zeta - \frac{1}{2}\zeta^2 \right) d\zeta \right]^{-1} = \frac{3EI}{L^3} \tag{1.106}$$

It should be pointed out that this is not the only equivalent spring constant possible, although it is the most common one. Indeed, another equivalent spring constant can be defined as the ratio of the bending moment M to the slope dw/dx at $x = L$ (Problem 1.16).

Determination of the equivalent spring constant for a uniform cantilever beam is relatively easy. When the beam is not uniform it may be possible to determine the equivalent spring constant by performing the indicated double integration in Eq. (1.105). Except for some simple cases, however, the integrations must be carried out numerically. Moreover, in more complex cases even this task can prove very difficult. In such cases, other approaches are advisable. One such approach is known as the *moment-area method* (Ref. 1, Ch. 9), which is based on two theorems. To demonstrate these theorems, we consider the displacement curve for a beam in bending shown in Fig. 1.23a. Then, we denote the slope at any point x by

$$\theta(x) = \frac{dw(x)}{dx} \tag{1.107}$$

and rewrite Eq. (1.102) in the form

$$d\theta(x) = \frac{M(x)}{EI(x)} dx \tag{1.108}$$

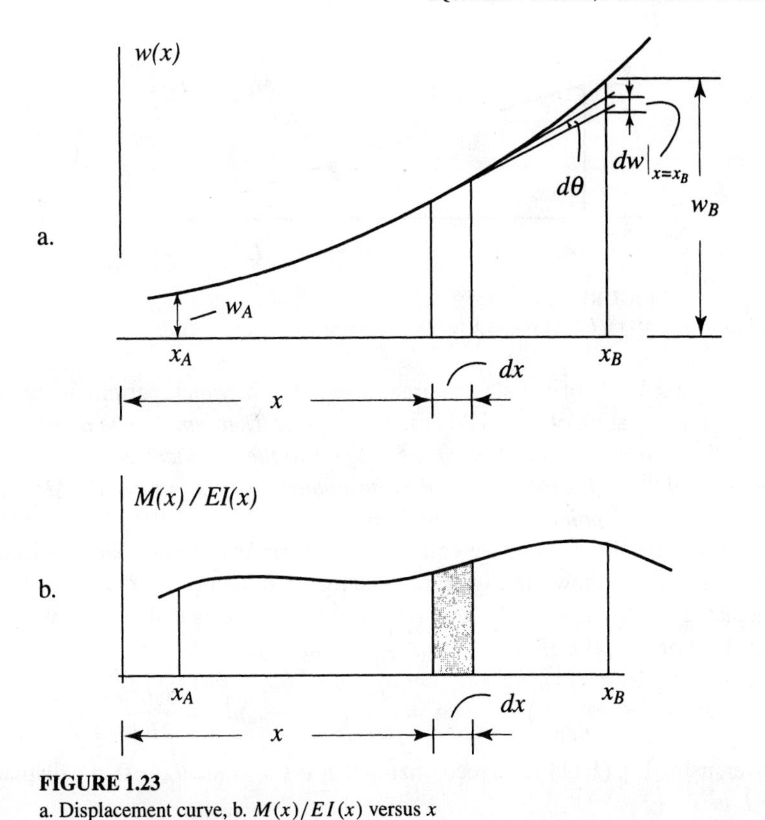

FIGURE 1.23
a. Displacement curve, b. $M(x)/EI(x)$ versus x

Equation (1.108) can be interpreted geometrically as the shaded area in the diagram $M(x)/EI(x)$ versus x of Fig. 1.23b. Integrating Eq. (1.108) between the points $x = x_A$ and $x = x_B$, we obtain simply

$$\theta_B - \theta_A = \int_{x_A}^{x_B} \frac{M(x)}{EI(x)} dx \tag{1.109}$$

Equation (1.109) can be stated in words in the form of *Theorem 1: The difference in slopes between the points x_A and x_B is equal to the area of the $M(x)/EI(x)$ diagram between these two points.*

Next, we refer to Fig. 1.23a, consider Eq. (1.108) and express the differential element of displacement at $x = x_B$ in the form

$$dw\big|_{x=x_B} = (x_B - x)d\theta = (x_B - x)\frac{M(x)}{EI(x)}dx \tag{1.110}$$

so that, integrating Eq. (1.110) between $x = x_A$ and $x = x_B$, we obtain

$$w_B - w_A - \theta_A(x_B - x_A) = \int_{x_A}^{x_B} (x_B - x)\frac{M(x)}{EI(x)}dx \tag{1.111}$$

Equation (1.111) lends itself not only to an interesting but also a useful geometric interpretation. Indeed, we observe that the quantity inside the integral sign represents the

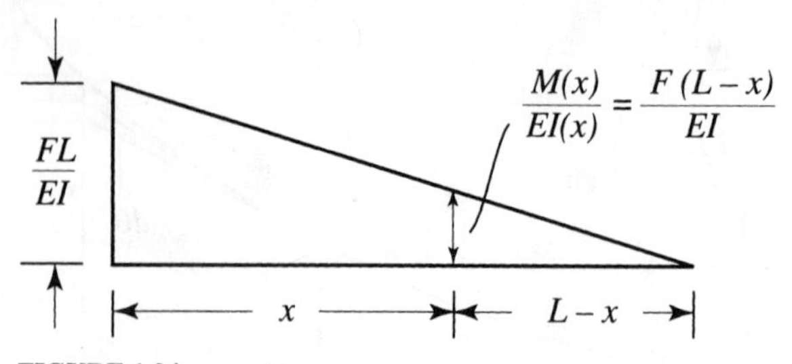

FIGURE 1.24
$M(x)/EI(x)$ versus x for a cantilever beam

moment about $x = x_B$ of the shaded area in Fig. 1.23b, which permits us to state the geometric interpretation of Eq. (1.111) in the form of *Theorem 2: The displacement of point B relative to the intersection of the tangent to the displacement curve at $x = x_A$ and the vertical through $x = x_B$ is equal to the moment about $x = x_B$ of the $M(x)/EI(x)$ diagram between the points $x = x_A$ and $x = x_B$.*

As a simple illustration, we consider a uniform cantilever beam subjected to a force F at $x = L$, as shown in Fig. 1.22. The diagram $M(x)/EI(x)$ has the triangular form depicted in Fig. 1.24. Using Eq. (1.109) and recognizing that $\theta_A = 0$, we conclude that the slope of the deflection curve at $x = L$ is simply

$$\theta(L) = \theta_B = \int_0^L \frac{M(x)}{EI(x)}dx = \frac{F}{EI}\int_0^L (L-x)dx = \frac{FL^2}{2EI} \tag{1.112}$$

Moreover, using Eq. (1.111) and recognizing that $w_A = 0$ and $\theta_A = 0$, the displacement at $x = L$ is

$$w(L) = w_B = \int_0^L (L-x)\frac{M(x)}{EI(x)}dx = \frac{F}{EI}\int_0^L (L-x)^2 dx = \frac{FL^3}{3EI} \tag{1.113}$$

This permits us to determine the equivalent spring constant by writing

$$k_{\text{eq}} = \frac{F}{\delta} = \frac{F}{w(L)} = \frac{3EI}{L^3} \tag{1.114}$$

which is the same as that given by Eq. (1.106). Note that in this particular example it is perhaps simpler to obtain the results by applying the two theorems presented above than carrying out integrations. Indeed, from Theorem 1, we conclude that the slope at $x = L$ is simply the area of the diagram of Fig. 1.24. Moreover, from Theorem 2, the displacement at $x = L$ is equal to the moment of the diagram of Fig. 1.24 with respect to the end $x = L$, which is simply equal to the area of the diagram multiplied by the distance $2L/3$ between the geometric center of the diagram and the vertical through $x = L$.

The preceding process for the calculation of the slope and displacement can be better explained perhaps by conceiving of a fictitious beam free at $x = 0$ and clamped at $x = L$ and loaded with a distributed load in the form of the diagram $M(x)/EI(x)$, as shown in Fig. 1.25. Then, recalling from mechanics of materials that the integral of a distributed load f is equal to the shearing force Q, we conclude that *the slope $\theta(x) = dw(x)/dx$ of the actual beam at point x is equal to the shearing force $\overline{Q}$ of the fictious beam at x. Moreover, the displacement $w(x)$ of the actual beam at x is equal to*

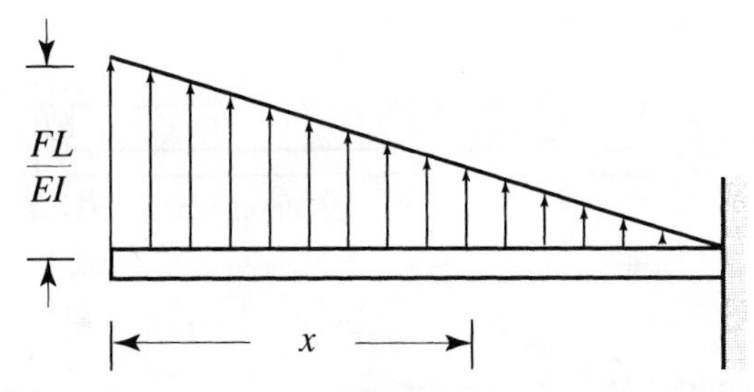

FIGURE 1.25
Cantilever beam loaded by the $M(x)/EI(x)$ diagram

Table 1.1 - End Types for Conjugate Beams

Actual Beam			Conjugate Beam		
End Type	Boundary Conditions		Boundary Conditions		End Type
Clamped	$w=0$	$dw/dx=0$	$\overline{M}=0$	$\overline{Q}=0$	Free
Free	$w\neq 0$	$dw/dx\neq 0$	$\overline{M}\neq 0$	$\overline{Q}\neq 0$	Clamped
Pinned	$w=0$	$dw/dx\neq 0$	$\overline{M}=0$	$\overline{Q}\neq 0$	Pinned

the moment $\overline{M}$ *of the fictitious beam at* x. The fictitious free-clamped beam loaded with the diagram $M(x)/EI(x)$ is known as the *conjugate beam* corresponding to the actual clamped-free beam, and the procedure for determining the slope and the displacement just outlined is called the *conjugate beam method.* The approach can be extended to other types of beams by replacing an end type of the actual beam by an end type of a conjugate beam as shown in Table 1.1. Of course, the first two entries in Table 1.1 merely represent the case just discussed. On the other hand, the third entry states that the conjugate beam for a pinned-pinned beam is a pinned-pinned beam, which has very useful implications.

As an example of the use of the conjugate beam method, we propose to calculate the equivalent spring constant for a uniform pinned-pinned beam acted upon by a transverse force F at $x=a$, as shown in Fig. 1.26a. To this end, we consider the displacement curve of Fig. 1.26b, in which we identify θ_A as the angle between the tangent to the displacement curve at $x=0$ and the horizontal and δ as the displacement at $x=a$. The conjugate beam is also pinned-pinned and loaded by the diagram $M(x)/EI(x)$, as shown in Fig. 1.26c. The angle θ_A, which can be identified as the shearing force of the conjugate beam at $x=0$, is

$$\theta_A=\overline{Q}_A=\frac{Fab}{EIL}\frac{a}{2}\left(\frac{a}{3}+b\right)\frac{1}{L}+\frac{Fab}{EIL}\frac{b}{2}\frac{2b}{3L}=\frac{Fab}{6EIL^2}[a(a+3b)+2b^2] \tag{1.115}$$

Then, the displacement of the actual beam at $x=a$ is simply the moment of the conjugate beam at $x=a$ (Fig. 1.26d), or

$$\delta=w(a)=\overline{M}(a)=\frac{Fab}{6EIL^2}[a(a+3b)+2b^2]a-\frac{Fab}{EIL}\frac{a}{2}\frac{a}{3}=\frac{Fa^2b^2}{3EIL} \tag{1.116}$$

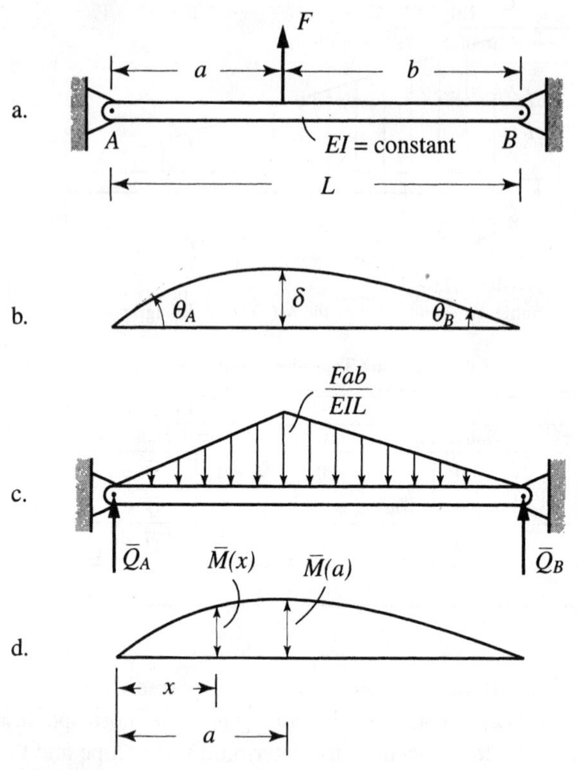

FIGURE 1.26
a. Pinned-pinned beam under transverse force, b. Displacement curve, c. Conjugate beam, d. Moment diagram for the conjugate beam

Finally, the equivalent spring constant is simply

$$k_{\text{eq}} = \frac{F}{\delta} = \frac{F}{w(a)} = \frac{3EIL}{a^2b^2} \tag{1.117}$$

The possibilities for defining equivalent spring constants are endless. Some of the more common ones are given in Table 1.2.

Using the same approach, it is possible to define equivalent coefficients of viscous damping for distributed components. However, the concept is not as useful as that of equivalent spring constants.

Next, we reconsider the assumption that springs are massless. To this end, we refer to a uniformly distributed spring fixed at $x = 0$ and free at $x = L$, as shown in Fig. 1.27,

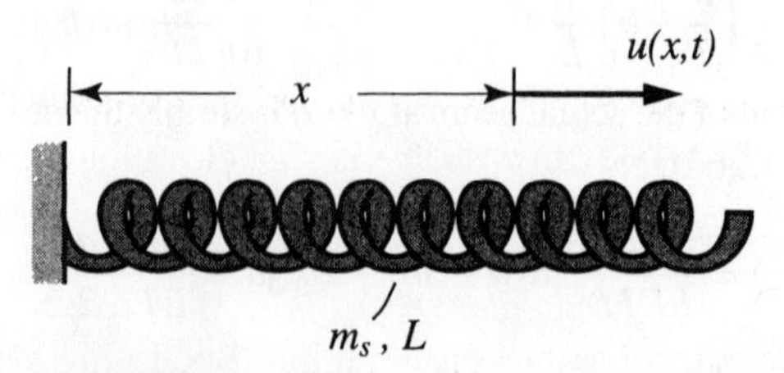

FIGURE 1.27
Spring with distributed mass

Table 1.2 - Equivalent Spring Constants

Component Sketch and Description		k_{eq}
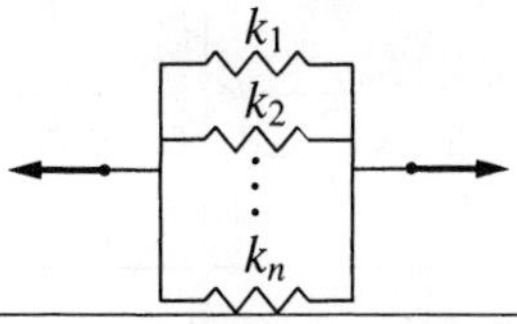 k_1, k_2, ⋮, k_n	Springs in parallel	$\sum_{i=1}^{n} k_i$
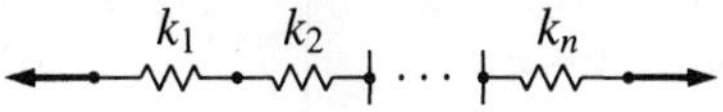 k_1 k_2 … k_n	Springs in series	$\frac{1}{\sum_{i=1}^{n}(1/k_i)}$
	Torsional spring	$\frac{EI}{L}$
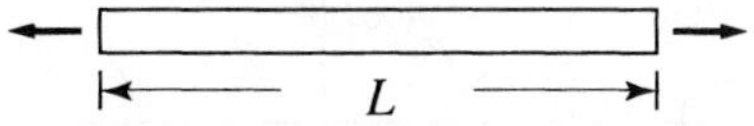 L	Rod in axial deformation	$\frac{EA}{L}$
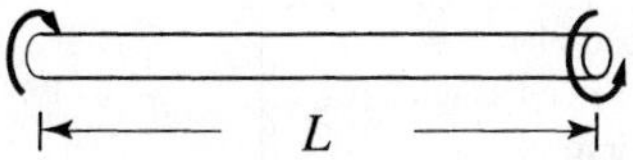 L	Shaft in torsion	$\frac{GJ}{L}$
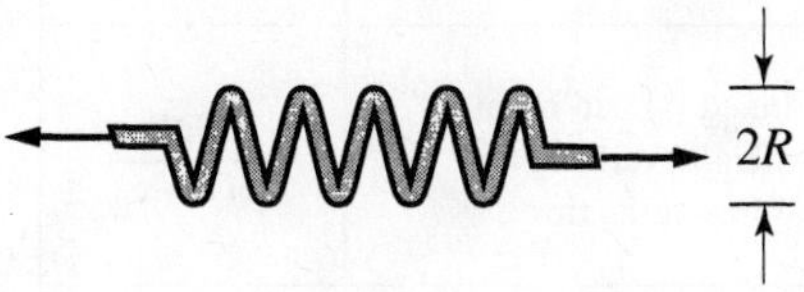 $2R$	Helical spring d = diameter of coil cross section n = number of coils	$\frac{Gd^4}{64nR^3}$
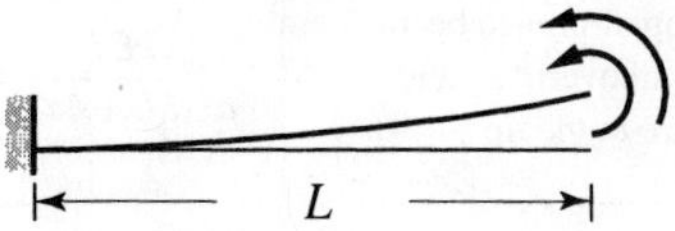 L	Cantilever beam with a moment at the tip	$\frac{EI}{L}$
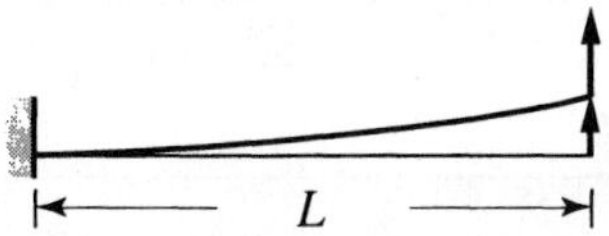 L	Cantilever beam with a force at the tip	$\frac{3EI}{L^3}$

and propose to derive an equivalent mass for the spring. Denoting the mass of the spring by m_s and using Eq. (1.65), we can write the kinetic energy in the form

$$T(t) = \frac{1}{2}\int_{m_s} \dot{u}^2(x,t)dm_s \tag{1.118}$$

where $\dot{u}(x,t)$ is the velocity of point x on the spring. But, from Eq. (1.95) with $EA(\xi) = EA =$ constant, it is easy to see that the static displacement of a massless spring is a

Table 1.2 - Equivalent Spring Constants (continued)

Component Sketch and Description	k_{eq}
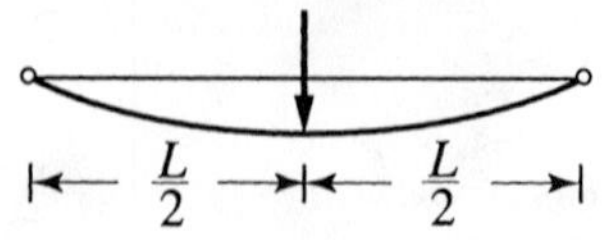 Pinned-pinned beam with a force at midspan ($\frac{L}{2}$, $\frac{L}{2}$)	$\frac{48EI}{L^3}$
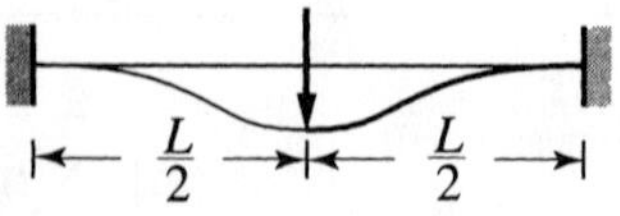 Clamped-clamped beam with a force at midspan ($\frac{L}{2}$, $\frac{L}{2}$)	$\frac{192EI}{L^3}$
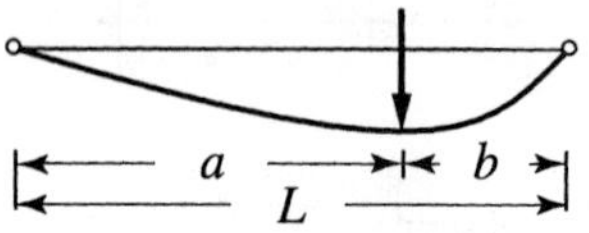 Pinned-pinned beam with an off-center force (a, b, L)	$\frac{3EIL}{a^2b^2}$
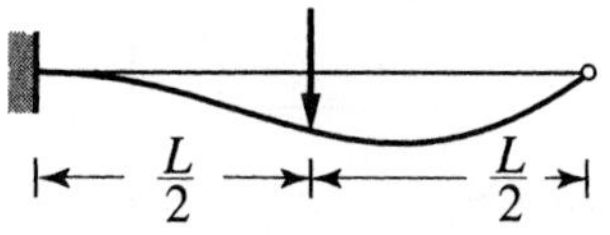 Clamped-pinned beam with a force at midspan ($\frac{L}{2}$, $\frac{L}{2}$)	$\frac{768EI}{7L^3}$
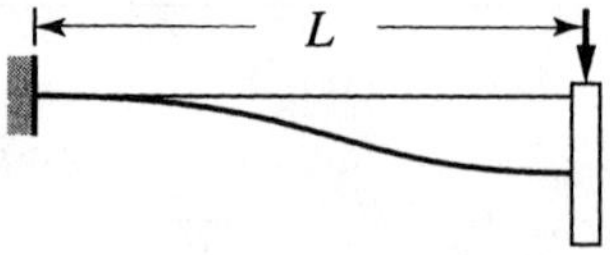 Clamped-clamped beam with one end sagging under a force (L)	$\frac{12EI}{L^3}$
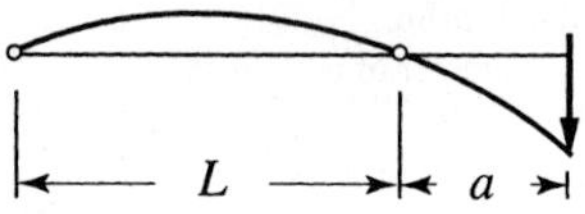 Pinned-pinned beam with an overhang and a force at the tip (L, a)	$\frac{3EI}{a^2(L+a)}$
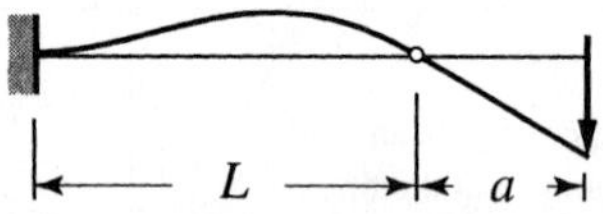 Clamped-pinned beam with an overhang and a force at the tip (L, a)	$\frac{12EI}{a^2(3L+4a)}$

E = modulus of elasticity
I = cross-sectional area moment of inertia
A = cross-sectional area
G = shear modulus
J = cross-sectional area polar moment of inertia

---

linear function of x. Hence, assuming that the mass of the spring does not affect this displacement characteristic in a meaningful way, we approximate the displacement at any point on the spring as follows:

$$u(x,t) = \frac{x}{L}u(L,t) \tag{1.119}$$

Inserting Eq. (1.119) into Eq. (1.118) and recognizing that for a uniform spring $dm_s = (m_s/L)dx$, we obtain

$$T(t) = \frac{1}{2}\frac{m_s}{L^3}\dot{u}^2(L,t)\int_s^L x^2 dx = \frac{1}{2}\frac{m_s}{3}\dot{u}^2(L,t) = \frac{1}{2}m_{\text{eq}}\dot{u}^2(L,t) \tag{1.120}$$

in which

$$m_{\text{eq}} = \frac{1}{3}m_s \tag{1.121}$$

represents the *equivalent mass* for the spring.

1.9 MODELING OF MECHANICAL SYSTEMS

In many ways, modeling is more of an art than an exact science. Indeed, more often than not a physical system is so complex that an exact description is not feasible. Fortunately, in many cases an exact description is not really necessary. This is certainly the case in preliminary design, in which the objective is primarily to verify whether a certain system is capable of meeting given performance criteria. In other cases, the interest lies in checking only certain properties of the system, so that the same physical system can be modeled in different ways. A model represents only an approximation of the actual physical system, and a good model must retain all the essential dynamic characteristics of the system. The implication is that the behavior predicted by the model must match the observed behavior of the actual system reasonably well.

Broadly speaking, models of vibrating mechanical systems fall into two classes, lumped-parameter, frequently referred to as discrete, and distributed-parameter systems. On occasion, models contain both lumped and distributed parameters. The classification is often a subjective matter, and the same physical system can at one time be modeled as discrete and at another time as distributed. In this section, we consider a number of physical systems and corresponding models.

Figure 1.28a represents a washing machine mounted on rubber supports and with the drum rotating in the vertical plane with the constant angular velocity ω relative to the body of the machine. In the first place, we assume that the body of the machine and the drum undergo no elastic deformations. Moreover, we assume that the clothes are spread uniformly around the drum. Because the mass of the drum and of the clothes is symmetric with respect to the axis of rotation, the inertial properties do not change with time. It follows that the combined mass of the body of the machine, the drum and the clothes, denoted by M, is constant and behaves as if it were rigid. Hence, the motion of the system is fully defined by the vertical displacement $x(t)$ of mass M. The rubber supports can be assumed to behave viscoelastically, which implies that they act as springs and dashpots in parallel. For simplicity of notation, we denote the spring constants and coefficients of viscous damping of the left and right supports by $k/2$ and $c/2$ each, so that the corresponding model is as shown in Fig. 1.28b. The situation is different when the clothes are spread nonuniformly around the drum. In this case, it is convenient to represent the nonuniformity by an excess mass m concentrated at a distance e from the

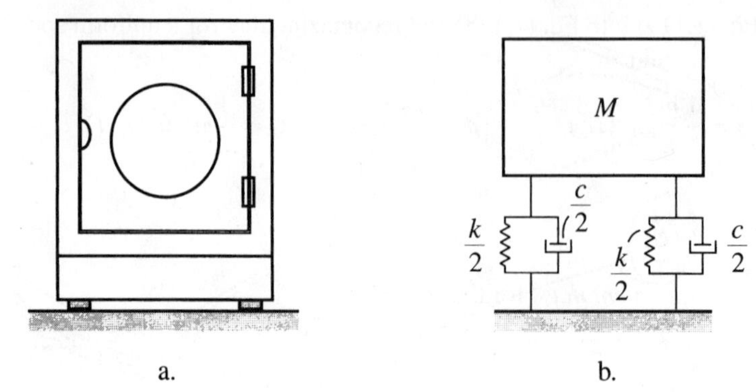

FIGURE 1.28
a. Washing machine, b. Model of washing machine

axis of rotation, where e is known as the eccentricity. Because the motion of the eccentric mass m relative to the body of the machine is prescribed, the motion of the system is fully defined by the vertical displacement $x(t)$ in this case as well. As we shall see later in this chapter, the effect of the rotating eccentric mass is to exert an inertial force upon the system.

Another system of interest is the automobile shown in Fig. 1.29a. Although the body/chassis structure is capable of elastic deformations, as a first approximation, it is reasonable to assume that the body/chassis structure can be treated as a rigid slab. The mass of the slab is supported by primary suspension systems at each of the four wheels, where each system consists of a relatively soft spring and a hydraulic shock absorber representing a viscous damper. The suspension systems transmit the load to the axles and tires, where the latter transmit the load to the ground. The axle and wheels, including the tires, possess some mass and the tires can be regarded as stiff springs. Although the tires provide some viscoelastic damping, the magnitude is relatively small and can be ignored. These considerations lead to the model depicted in Fig. 1.29b, from which we identify the motions of the system as the vertical translation $x_b(t)$ of the body, the rotations $\theta_y(t)$ and $\theta_z(t)$ of the body about axes y and z, respectively, and the vertical translations $x_{wi}(t)$ $(i = 1, 2, 3, 4)$ of the wheels. The system parameters can be identified as the mass of the body m_b, the mass moments of inertia I_y and I_z of the body about axes y and z, respectively, the coefficients of viscous damping c_{si} and spring constants k_{si} of the primary suspension systems, the wheel masses m_{wi} and the tire spring constants k_{ti} $(i = 1, 2, 3, 4)$. Of course, the two front suspensions parameters are likely to have equal values, and the same can be said about the rear suspensions. Moreover, the wheel masses and tire spring constants are the same for all wheels and tires, respectively. The distance between the axles is L and between the left and right wheels is B.

Another example is the missile in free flight depicted in Fig. 1.30a. The missile tends to be a slender elastic body capable of bending about two transverse axes. It is common, however, to assume that the vibration takes place in one plane only, namely, in the plane of the missile trajectory. A discrete model of the missile can be conceived

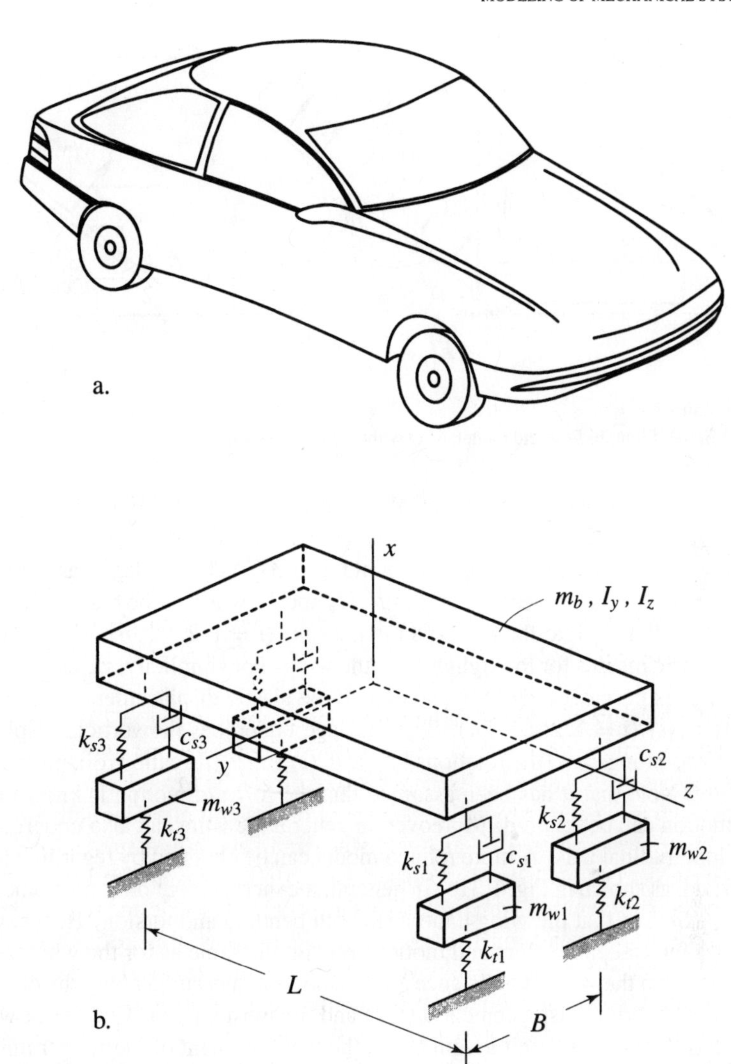

FIGURE 1.29
a. Automobile, b. Model of automobile

by dividing the mass into n lumps of mass m_i $(i = 1, 2, \ldots, n)$ connected by massless segments of length Δx_i and bending stiffness EI_i $(i = 1, 2, \ldots, n-1)$, as shown in Fig. 1.30b. Then, the vibration of the missile can be defined in terms of the transverse displacements $w_i(t)$ of the masses m_i $(i = 1, 2, \ldots, n)$. The missile can also be modeled as a distributed-parameter system in the form of a beam of length L, free at both ends and undergoing bending vibration $w(x, t)$ in the transverse direction (Fig. 1.30c). The system parameters are the mass per unit length $m(x)$ and the bending stiffness $EI(x)$.

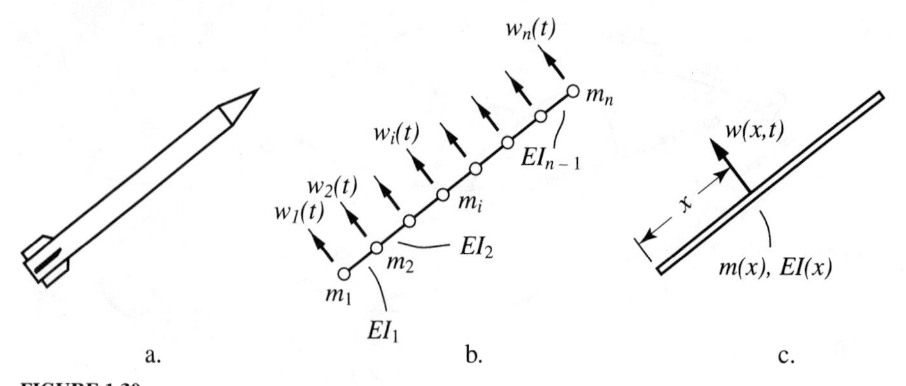

FIGURE 1.30
a. Missile in free flight, b. Discrete model, c. Distributed-parameter model

Finally, we consider the modeling of the aircraft displayed in Fig 1.31a. We assume that the fuselage is rigid and that the wing is flexible and construct first a discrete model by lumping the mass of the wing as shown in Fig. 1.31b. The fuselage has the mass m_f and the principal mass moments of inertia I_x, I_y and I_z, whereas the mass of the left half of the wing is divided into the n lumped masses m_i $(i = 1, 2, \ldots, n)$. Of course, there are n symmetric masses for the right half of the wing. For simplicity, we assume that the wing undergoes pure bending only and denote the elastic displacements of the lumped masses by $w_i(t)$ $(i = 1, 2, \ldots, 2n)$. The fuselage undergoes the vertical displacement $z(t)$, known as plunge, and the rotations $\theta_x(t)$, $\theta_y(t)$ and $\theta_z(t)$, called roll, pitch and yaw, respectively. Note that it has been assumed that the forward motion is known and that the side motion can be ignored. Moreover, in general the wing can also undergo torsion about its longitudinal axis. A more refined model can be obtained by regarding the wing as distributed, as shown in Fig. 1.31c. In general, the inertia axis,[1] does not coincide with the elastic axis,[2] so that the wing undergoes both bending and torsion. Hence, whereas the fuselage inertial parameters and motions remain the same as for the wholly discrete model, a point on the wing at a distance ξ from the root, measured along the elastic axis, undergoes the bending displacement $w(\xi, t)$ and the twist $\psi(\xi, t)$. Consistent with this, the wing has the mass per unit length $m(\xi)$, the mass moment of inertia per unit length $I_\psi(\xi)$, the bending stiffness $EI(\xi)$, where $I(\xi)$ is the cross-sectional area moment of inertia, and the torsional stiffness $GJ(\xi)$, where G is the shear modulus and $J(\xi)$ is the cross-sectional area polar moment of inertia. Clearly, the model of Fig. 1.31c is part discrete and part distributed.

Next, we return to the automobile model of Fig. 1.29b and observe that when the automobile travels in straight forward motion, it is reasonable to assume that the front wheels on the one hand and the rear wheels on the other hand undergo the same

[1]The inertia axis is the locus of the mass centers of the cross sections.

[2]The elastic axis is defined as the locus of the shear centers of the cross sections, where a shear center is a point such that a shearing force acting through it produces pure bending (with no torsion) and a moment about it produces pure torsion (with no bending).

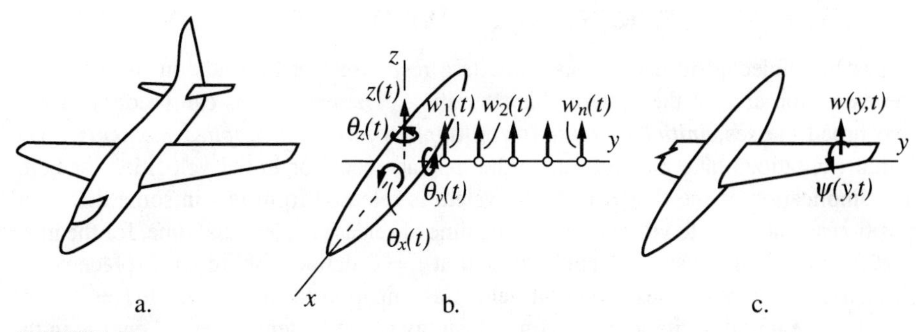

FIGURE 1.31
a. Aircraft in flight, b. Discrete model, c. Distributed-parameter model

displacement, which implies that the rotation θ_z is zero. Under these circumstances, the model can be simplified as shown in Fig. 1.32a. Finally, assuming that the tire stiffness is infinitely large, the model can be further simplified by ignoring the mass of the wheels and the tire springs, as shown in Fig. 1.32b.

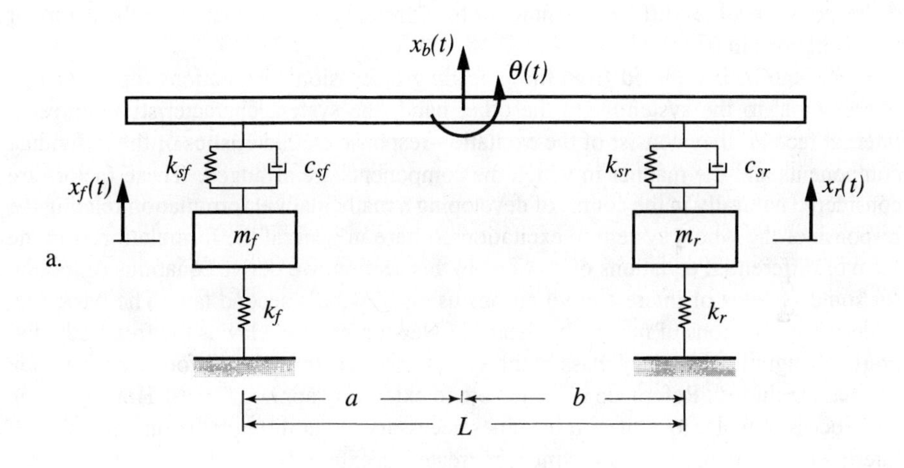

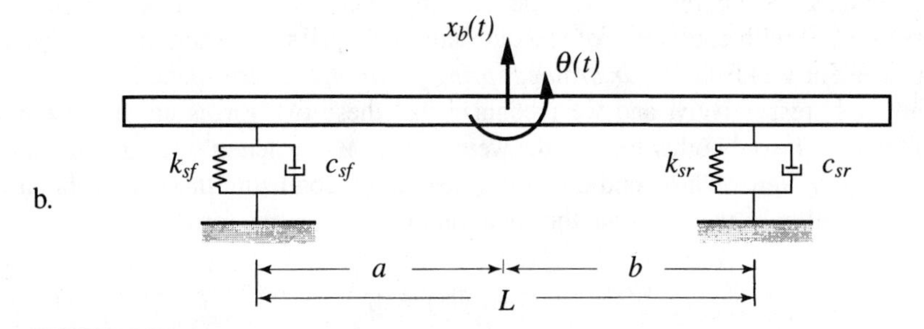

FIGURE 1.32
a. Model of automobile in planar motion, b. Simplified automobile model

1.10 SYSTEM DIFFERENTIAL EQUATIONS OF MOTION

A system subjected to excitations exhibits a response that depends on the nature of the excitation and on the system characteristics. The excitations can be divided into two broad classes, *initial excitations* and *applied forces*, or *applied moments*. The initial excitations take the form of initial displacements or initial velocities, or both. The implication of the first is that the system is released from rest in some displaced position and allowed to vibrate freely. The time of release is the initial time, for the most part $t = 0$, and the displaced configuration at $t = 0$ defines the initial displacements. Similarly, initial velocities represent velocities imparted to the masses at $t = 0$. The effect of the initial excitations is to impart energy to the system, potential energy in the case of initial displacements and kinetic energy in the case of initial velocities. After the energy has been imparted to the system, there are no longer any external factors affecting the system, for which reason the subsequent motion is referred to as *free vibration*, or *free response*. On the other hand, the response to applied forces and/or applied moments is called *forced vibration*, or *forced response*. Note that applied forces (moments) are also known as *external* or *impressed* forces (moments). Whereas the initial excitations require little further discussion, the applied forces (moments) require a great deal of elaboration. Indeed, there are many types of external forces (moments), and determining the response involves different techniques for different types. We examine the nature of the excitations in Sec. 1.11.

As can be concluded from the preceding discussions, excitations represent factors external to the system. On the other hand, the system characteristics represent internal factors; they consist of the excitation-response characteristics of the individual components and the manner in which the components are arranged. These factors are considered naturally in the course of developing a mathematical formulation relating the response of the whole system to excitations, where in general the formulation is in the form of differential equations of motion. In this section, we derive equations of motion for some systems of interest in vibrations using Newton's second law. The basic tool in deriving equations of motion by means of Newton's second law is the free-body diagram, a diagram with every mass in the system isolated and with all forces acting upon the mass included. Reference is made here to externally applied forces. However, if in the process of isolating a mass it becomes necessary to cut through the line of action of internal forces, then these forces must be treated as external.

As a simple illustration, we consider the model of the washing machine of Fig. 1.28b. The corresponding free-body diagram is depicted in Fig. 1.33, in which the two springs have been combined into one with spring constant equal to k and the two dampers into one with coefficient of viscous damping equal to c. Then, if we measure the displacement $y(t)$ from the *unstrained spring position*, the corresponding forces are $-ky$ and $-c\dot{y}$, respectively, and we recognize that these two forces are not genuine applied forces. The only other force is the weight $W = Mg$, where g is the gravitational constant. Using Newton's second law, Eq. (1.3), and recognizing that the problem is one-dimensional only, we can write the equation of motion in the form

$$-Mg - ky(t) - c\dot{y}(t) = M\ddot{y}(t) \tag{1.122}$$

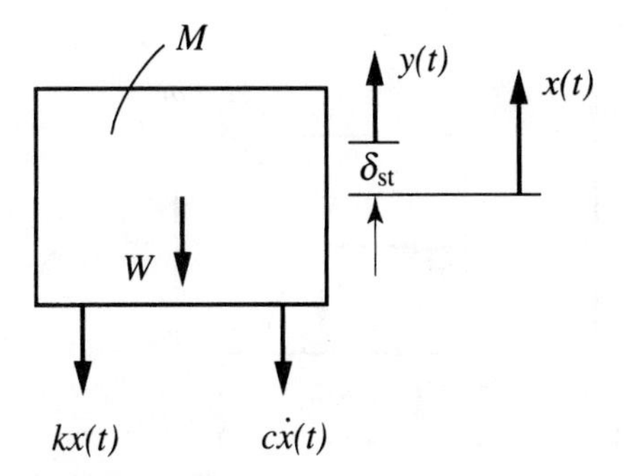

FIGURE 1.33
Free-body diagram for a washing machine model

which can be rewritten as

$$M\ddot{y}(t) + c\dot{y}(t) + ky(t) = -Mg \tag{1.123}$$

and we note that the term $-Mg$ on the right side renders Eq. (1.123) nonhomogeneous. As it turns out, a simple transformation can render the equation homogeneous. To this end, we denote by $x(t)$ the displacement of M from the *static equilibrium position*, which differs from the unstrained spring position $y(t)$ by the static equilibrium displacement defined as

$$\delta_{st} = \frac{W}{k} = \frac{Mg}{k} \tag{1.124}$$

But, from Fig. 1.33, the various displacements are related by

$$y(t) = x(t) - \delta_{st} \tag{1.125}$$

Inserting Eq. (1.125) into Eq. (1.123), recognizing that δ_{st} is constant, so that $\ddot{y}(t) = \ddot{x}(t)$, and canceling the term $-Mg$ on both sides of the equation, we obtain the equation of motion relative to the equilibrium position in the form

$$M\ddot{x}(t) + c\dot{x}(t) + kx(t) = 0 \tag{1.126}$$

which is homogeneous. The question may be asked as to how the weight disappeared. The fact is that the weight did not really disappear. Indeed, the weight Mg is balanced at all times by a constant force $k\delta_{st}$ in the spring. The conclusion is that it is possible to simplify the equation of motion by measuring the displacement from equilibrium, a conclusion which is true in general. Equation (1.126) represents the *free vibration equation*; it will be studied in Ch. 2.

Next, we turn our attention to the case in which there is some imbalance in the system, as shown in Fig. 1.34a. To derive the equation of motion, it is convenient to consider two free-body diagrams, one for $M - m$ and one for m. The two free-body diagrams are given by Figs. 1.34b and 1.34c, respectively, in which F_H and F_V represent hinge reactions. Measuring $x(t)$ from the static equilibrium position, which permits us to ignore the weight Mg, and using Newton's second law, the equations of motion in the

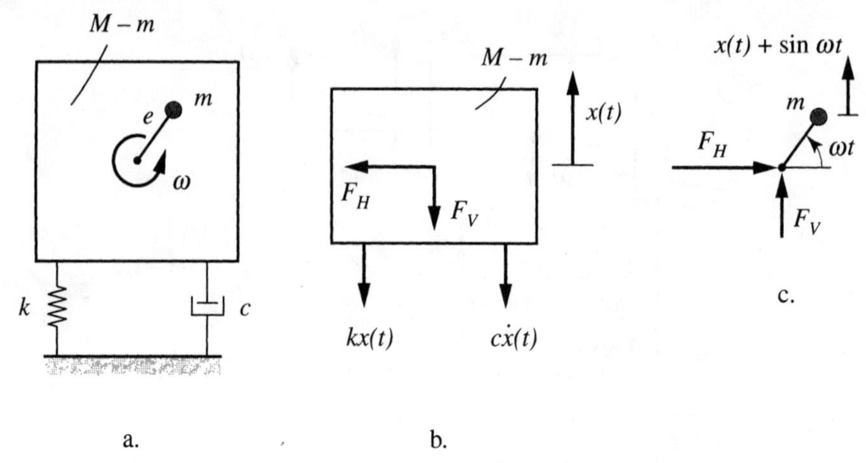

FIGURE 1.34
a. System with rotating eccentric mass, b. Free-body diagram for main mass, c. Free-body diagram for eccentric mass

vertical direction are

$$\begin{aligned} -F_V - kx - c\dot{x} &= (M-m)\ddot{x} \\ F_V &= m\frac{d^2}{dt^2}(x + e\sin\omega t) = m(\ddot{x} - e\omega^2 \sin\omega t) \end{aligned} \tag{1.127}$$

Eliminating the vertical reaction F_V and rearranging, we obtain the single equation of motion

$$M\ddot{x} + c\dot{x} + kx = F = me\omega^2 \sin\omega t \tag{1.128}$$

where $F = F(t)$ represents a force acting upon the system. Equation (1.128) confirms the statements made in Sec. 1.9 that a single displacement defines the motion of the system fully and that the rotating eccentric mass m exerts an inertial force on the system. We will study the behavior of the system described by Eq. (1.128) in Ch. 3.

Finally, we propose to derive the equations of motion for the automobile model of Fig. 1.32a with a vertical force added. This requires three free-body diagrams, one for each mass, as shown in Fig. 1.35, and we note that gravitational forces were ignored on the assumption that displacements are measured from equilibrium. To define the displacements of mass m_b, we must choose a reference point. To this end, we recall from Sec. 1.5 that the most indicated choice is the mass center, as the equations of motion in terms of the mass center have a relatively simple form. Hence, we define x_b as the vertical displacement of the mass center C. Moreover, it is reasonable to assume that the angular displacement θ is relatively small, so that $\sin\theta \simeq \theta$. It follows that the front and rear suspension systems undergo the elongations $x_b - a\theta - x_f$ and $x_b + b\theta - x_r$, respectively. Then, referring to the free-body diagrams of Fig. 1.35 and using the first

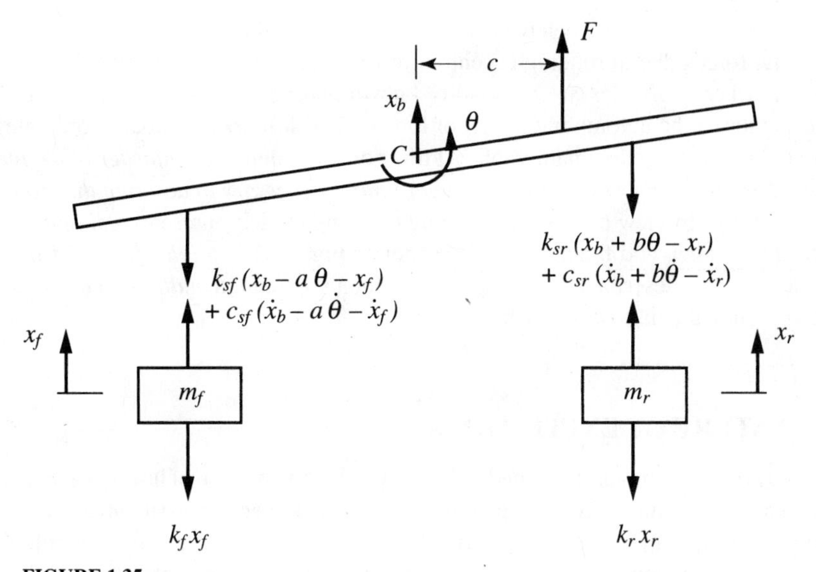

FIGURE 1.35
Free-body diagram for the system of Figure 1.32a

of Eqs. (1.63) and Eq. (1.64), we obtain the system equations of motion

$$\begin{aligned}
&F - k_{sf}(x_b - a\theta - x_f) - c_{sf}(\dot{x}_b - a\dot{\theta} - \dot{x}_f) - k_{sr}(x_b + b\theta - x_r) \\
&\quad - c_{sr}(\dot{x}_b + b\dot{\theta} - \dot{x}_r) = m_b\ddot{x}_b \\
&Fc + [k_{sf}(x_b - a\theta - x_f) + c_{sf}(\dot{x}_b - a\dot{\theta} - \dot{x}_f)]a - [k_{sr}(x_b + b\theta - x_r) \\
&\quad + c_{sr}(\dot{x}_b + b\dot{\theta} - \dot{x}_r)]b = I_C\ddot{\theta} \\
&k_{sf}(x_b - a\theta - x_f) + c_{sf}(\dot{x}_b - a\dot{\theta} - \dot{x}_f) - k_f x_f = m_f\ddot{x}_f \\
&k_{sr}(x_b + b\theta - x_r) + c_{sr}(\dot{x}_b + b\dot{\theta} - \dot{x}_r) - k_r x_r = m_r\ddot{x}_r
\end{aligned} \tag{1.129}$$

which can be rewritten as follows:

$$\begin{aligned}
&m_b\ddot{x}_b + (c_{sf} + c_{sr})\dot{x}_b - (c_{sf}a - c_{sr}b)\dot{\theta} - c_{sf}\dot{x}_f - c_{sr}\dot{x}_r \\
&\quad + (k_{sf} + k_{sr})x_b - (k_{sf}a - k_{sr}b)\theta - k_{sf}x_f - k_{sr}x_r = F \\
&I_C\ddot{\theta} - (c_{sf}a - c_{sr}b)\dot{x}_b + (c_{sf}a^2 + c_{sr}b^2)\dot{\theta} + c_{sf}a\dot{x}_f - c_{sr}b\dot{x}_r \\
&\quad - (k_{sf}a - k_{sr}b)x_b + (k_{sf}a^2 + k_{sr}b^2)\theta + k_{sf}ax_f - k_{sr}bx_r = Fc \\
&m_f\ddot{x}_f - c_{sf}\dot{x}_b + c_{sf}a\dot{\theta} + c_{sf}\dot{x}_f - k_{sf}x_b + k_{sf}a\theta + k_{sf}x_f = 0 \\
&m_r\ddot{x}_r - c_{sr}\dot{x}_b - c_{sr}b\dot{\theta} + c_{sr}\dot{x}_r - k_{sr}x_b - k_{sr}b\theta + k_{sr}x_r = 0
\end{aligned} \tag{1.130}$$

Equations (1.130) represent a set of four simultaneous second-order ordinary differential equations in terms of the four unknowns x_b, θ, x_f and x_r. Clearly, the automobile model is considerably more involved than the washing machine model.

We observe that the behavior of the washing machine model, whether subjected to inertial forces due to rotating eccentric masses or not, is fully described by a single variable, where variables are commonly known as coordinates. On the other hand, the behavior of the automobile model of Fig. 1.35 is described by four coordinates, as indicated in the preceding paragraph. At this point, we define the *number of degrees of freedom as the number of independent coordinates necessary to describe the motion of a system fully*. In view of this, the washing machine model represents a *single-degree-of-freedom system* and the automobile model of Fig. 1.35 is a *four-degree-of-freedom system*. Systems described by two or more variables are called *multi-degree-of-freedom systems*. They are discussed in Ch. 7.

1.11 NATURE OF EXCITATIONS

The study of vibrations is concerned essentially with the question of how systems behave in response to stimuli. To answer this question, it is necessary to solve the system equations of motion, such as those derived in Sec. 1.10. The choice of methodology for obtaining the solution and the solution itself depend on the type of excitation and on the system characteristics. We examine the nature of the excitations in this section and of the system characteristics in Sec. 1.12.

As indicated in Sec. 1.10, we distinguish between initial excitations and applied forces, or moments. This distinction is not as airtight as it may seem, because initial velocities are really caused by a special type of forces, namely, impulsive forces, as we shall verify later in this text. As far as solving the differential equations of motion is concerned, however, the distinction is important, as in the case of initial excitations the equations are homogeneous and in the case of applied forces the equations are nonhomogeneous.

Initial excitations consist of initial displacements and initial velocities and they are generated by imparting potential energy and kinetic energy to a system, respectively. The initial excitations set the system in a motion known as *free vibration*. If the system is conservative, this motion persists ad infinitum, at least in theory. Whereas the total energy remains constant, the balance between the potential energy and kinetic energy fluctuates. On the other hand, if there is damping in the system, then energy is dissipated, causing the total energy to go down continuously until it reaches zero, at which point the motion stops. Of course, in practice all systems dissipate energy, even those assumed to be conservative. The main difference is that conservative systems dissipate energy very slowly. Still, all motions caused by initial excitations come to rest eventually. For this reason, initial excitations are referred to at times as *transient excitations*.

In contrast with initial excitations, there is a large variety of applied forces. A very important class of forces consists of *harmonic excitations*, which are simply forces proportional to the trigonometric function $\sin \omega t$, or to $\cos \omega t$, or to a combination of the two. Examining Eq. (1.128), we conclude that the rotating eccentric mass exerts a harmonic force upon the washing machine. This is but one of many examples of harmonic forces occurring in real life, as we shall have the opportunity to see throughout this text. To examine the nature of harmonic functions, we consider a combination of

$\sin \omega t$ and $\cos \omega t$ of the form

$$F(t) = A_1 \sin \omega t + A_2 \cos \omega t = A \cos(\omega t - \psi) \tag{1.131}$$

where ω is the *frequency* of the harmonic function; it has units of radians per second (rad/s). Moreover,

$$A = \sqrt{A_1^2 + A_2^2} \tag{1.132}$$

is known as the *amplitude* and

$$\psi = \tan^{-1} \frac{A_1}{A_2} \tag{1.133}$$

as the *phase angle*. The function $F(t)$ can be interpreted geometrically as the vertical projection of a vector A rotating with the angular velocity ω, as shown in Fig. 1.36a. The angle $\omega t - \psi$ between A and the vertical axis increases linearly with time, so that the vertical projection varies harmonically with time. The plot $F(t)$ versus t is displayed in Fig. 1.36b, and we observe that the function repeats itself every time interval T, where T has units of seconds (s) and is known as the *period*. Because $\cos(\omega t - \psi) = \cos[\omega(t + T) - \psi]$, we conclude that the period is related to the frequency by

$$T = \frac{2\pi}{\omega} \tag{1.134}$$

Also from Fig. 1.36b, we conclude that there is a time interval ψ/ω between $F(0)$ and the first peak, so that the phase angle is a measure of this interval. Whereas the amplitude and frequency of the harmonic excitation function $F(t)$ are important factors, the phase angle of the excitation is largely irrelevant. Indeed, as we shall demonstrate in Ch. 3, depending on the system damping, there is in general a phase angle between the excitation and response, which is a characteristic of the response, and is not affected by the phase angle ψ of the excitation. This implies that the location of the origin $t = 0$ of the time axis has no particular meaning. In fact, because the general shape of harmonic functions, such as the cosine function of Eq. (1.131), is well defined, the only two pieces of information necessary for the characterization of a given harmonic function are the

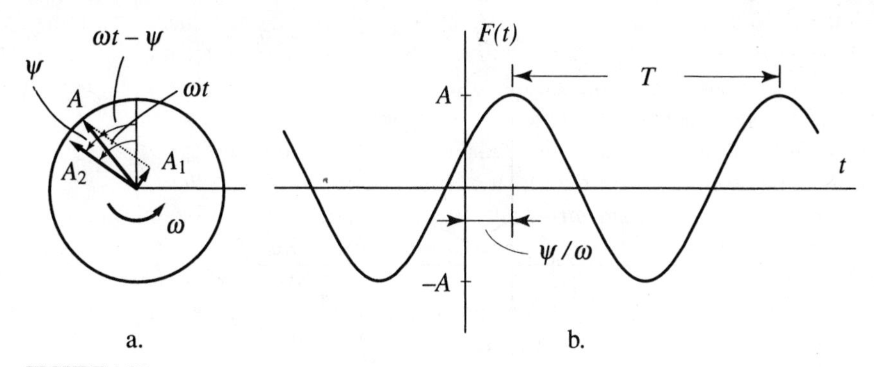

FIGURE 1.36

a. Rotating vector, b. Harmonic force as a function of time

amplitude and frequency, with time playing a secondary role only. Indeed, harmonic excitations have the same characteristics for all times, $-\infty < t < \infty$, for which reason they are known as *steady-state excitations*. They are distinctly different from transient excitations, such as initial displacements and velocities, for which the origin $t = 0$ of the time axis defines the time when the response starts.

In deriving the response to harmonic excitations, it is convenient to work with a different form than the trigonometric form given by Eq. (1.131), namely, the exponential form. To this end, we consider the series

$$\begin{aligned} e^{i\omega t} &= 1 + i\omega t + \frac{1}{2!}(i\omega t)^2 + \frac{1}{3!}(i\omega t)^3 + \frac{1}{4!}(i\omega t)^4 + \frac{1}{5!}(i\omega t)^5 + \ldots \\ &= 1 - \frac{1}{2!}(\omega t)^2 + \frac{1}{4!}(\omega t)^4 \ldots + i[\omega t - \frac{1}{3!}(\omega t)^3 + \frac{1}{5!}(\omega t)^5 - \ldots] \\ &= \cos\omega t + i \sin\omega t, \; i = \sqrt{-1} \end{aligned} \tag{1.135}$$

Equation (1.135) can be given a geometric interpretation similar to that of Fig. 1.36a. Indeed, as shown in Fig. 1.37, the exponential function $e^{i\omega t}$ can be represented in the complex plane as *a vector of unit magnitude and making an angle* ωt *with respect to the real axis*. Clearly, the projection of the vector on the real axis is $\cos\omega t$ and that on the imaginary axes is $i \sin\omega t$. As time increases, the vector rotates in the complex plane with the angular velocity ω, causing the two projections to vary harmonically with time. From Eq. (1.135), we can write

$$\text{Re}\, e^{i\omega t} = \cos\omega t, \; \text{Im}\, e^{i\omega t} = \sin\omega t \tag{1.136}$$

where Re and Im denote the real part and imaginary part, respectively. Equations (1.136) can be used to express either $\cos\omega t$ or $\sin\omega t$ in exponential form, as the case may be. Hence, ignoring the excitation phase angle ψ, we can replace the trigonometric form of the harmonic force, Eq. (1.131), by the exponential form

$$F(t) = \text{Re}\, Ae^{i\omega t} \tag{1.137}$$

The advantage of expressing a harmonic force in exponential form is that the response is significantly simpler to obtain. Of course, the response will also have an exponential

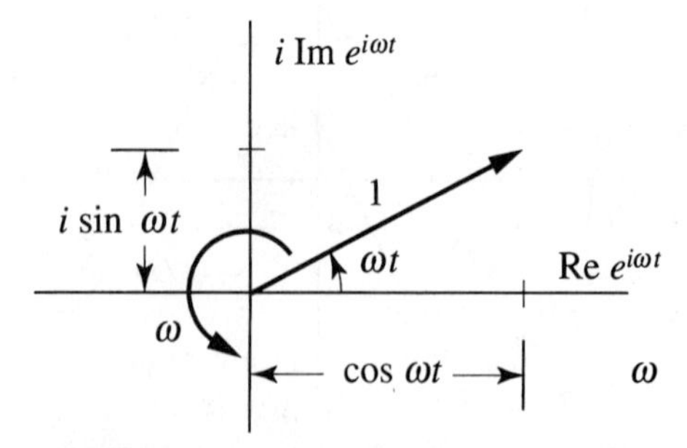

FIGURE 1.37
Unit vector rotating in the complex plane

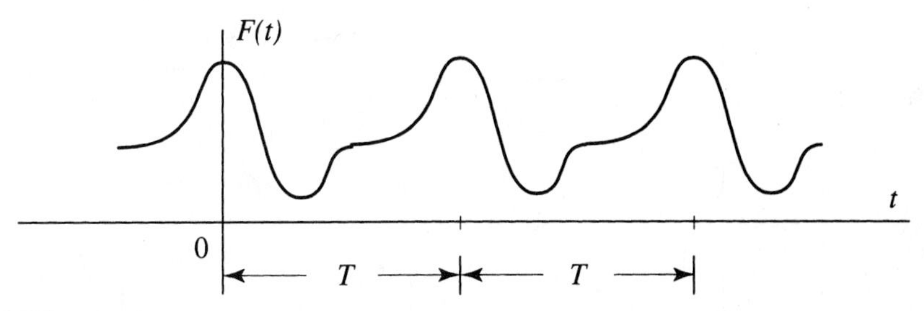

FIGURE 1.38
Periodic excitation

form, and hence it will be complex. Then, if the excitation is proportional to $\cos \omega t$, we retain the real part of the response, and if the excitation is proportional to $\sin \omega t$, we retain the imaginary part of the response. The process may appear as an unnecessary complication at this point, but its efficiency will be amply demonstrated in Ch. 3.

Harmonic excitations belong to a larger class of functions characterized by the fact that the functions repeat themselves every time interval T. This is the class of *periodic functions*, such as the function $F(t)$ of Fig. 1.38, in which T is the *period*. We observe that, as with harmonic functions, in the case of periodic functions time plays only a secondary role. Hence, periodic excitations represent *a more general class of steady-state excitations*. Whereas harmonic functions are periodic, periodic functions are not necessarily harmonic. However, periodic functions can be expressed as linear combinations of harmonic functions known as Fourier series (Appendix A). As with harmonic functions, periodic functions can be expressed in terms of trigonometric functions or exponential functions, referred to at times as the real form or the complex form of Fourier series, respectively. The frequency of each harmonic function in a Fourier series is an integer multiple of the lowest frequency, which is known as the *fundamental frequency*. A great deal of information concerning the nature of a periodic function $F(t)$ is revealed by a plot of the amplitude of each of the constituent harmonic functions in the Fourier series as a function of the frequency, as it displays in one diagram the degree of participation of each of these harmonic functions in $F(t)$, a diagram known as a *frequency spectrum*. The frequency spectrum represents a *frequency domain description* of a periodic function. Hence, although Fig. 1.38 permits easy visualization of the periodic function $F(t)$ as a function of time, a frequency domain description of $F(t)$ is more useful. Because the plot of the frequency spectrum of a periodic function consists of mere points at the individual frequencies, rather than being a continuous plot, this is a *discrete frequency spectrum*. We study the response to periodic forces in Ch. 3.

The remaining types of excitations clearly belong in the class of *nonperiodic excitations*, which includes a large variety of forces. Many of these forces represent known functions of time, two of the most important ones being the impulse function and the step function, and many other forces can be expressed as combinations of known functions. In general, nonperiodic forces represent *arbitrary excitations*, such as the force $F(t)$ depicted in Fig. 1.39. Interestingly, even such completely arbitrary forces can be represented as combinations of known functions, and in particular as combinations of impulse forces of different amplitude and applied at different times. This is in contrast

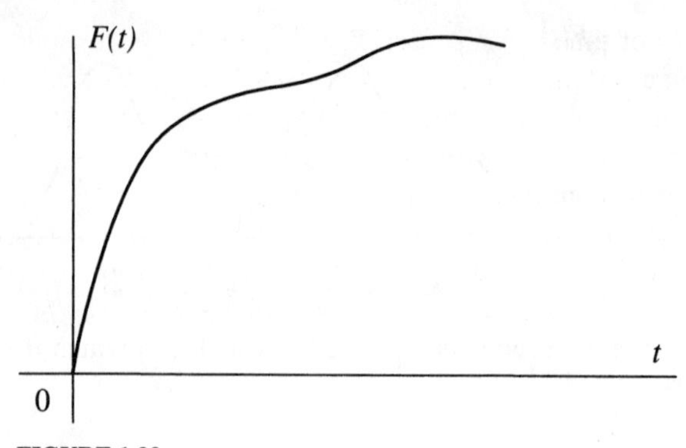

FIGURE 1.39
Nonperiodic excitation

with periodic forces, which can be represented by means of combinations of harmonic functions. The representation in terms of impulse forces is very convenient in deriving the response of systems to arbitrary excitations. This subject is discussed in great detail in Ch. 4.

The three types of forces discussed above, namely, harmonic, periodic and nonperiodic, have one thing in common, namely, their value is given in advance for any time t. Such excitation forces are said to be *deterministic*. There are many excitation forces, however, that do not lend themselves to such explicit time description. Examples of such excitations are the forces exerted by an earthquake on a building, by a rough runway on a taxiing aircraft, by a rocket engine on a structure, etc. The implication is that the value of the force at some future time cannot be predicted. The reason for this is that there are too many factors affecting the force. Of course, it may be possible to measure these forces as functions of time, but records from different dates may differ from one another. Forces of this type are classified as *nondeterministic*, and are commonly referred to as *random*. A typical random excitation is displayed in Fig. 1.40. Clearly, a description of random

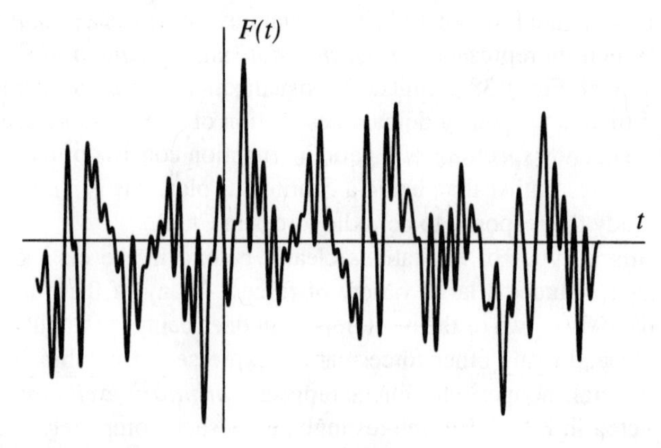

FIGURE 1.40
Random excitation

forces as functions of time is not particularly meaningful. Many random phenomena, but not all, exhibit a certain pattern known as *statistical regularity*, which permits their description in terms of certain averages, such as the *mean value* and the *mean square value*. In evaluating the response to random excitations, it turns out that a frequency domain description is more useful than a time domain description. This amounts to decomposing the random function $F(t)$ of Fig. 1.40 into harmonic components. Because $F(t)$ is nonperiodic, a plot showing the contribution from each harmonic component will have an entry at every frequency, resulting in a *continuous frequency spectrum*. The subject of random vibrations is presented in Ch. 12, in which these concepts are discussed in detail.

1.12 SYSTEM AND RESPONSE CHARACTERISTICS. THE SUPERPOSITION PRINCIPLE

As indicated in Sec. 1.10, the manner in which a system responds to excitations depends on the nature of the excitations and on the system characteristics. In Sec. 1.11, we examined various types of excitations, and in this section we wish to investigate how the system characteristics affect the response of the system. To this end, we consider the symbolic block diagram of Fig. 1.41, in which the system is represented by a "black box" containing the system characteristics. The meaning of the block diagram is that a system subjected to an excitation $F(t)$ exhibits a certain response $x(t)$.

A system is broadly defined as an aggregation of components working together as a single unit. The system characteristics are determined not only by the excitation-response relations of the individual components but also by the manner in which these components are connected to one another within the framework of the system. The characteristics of a whole system are determined naturally in the process of deriving the system equations of motion, as can be concluded from the developments in Sec. 1.10.

One of the most fundamental questions in vibrations is whether a system is linear or nonlinear, as the answer has profound implications as far as the solution of the equations of motion is concerned. To answer this question, we assume that a given system, when acted upon by two distinct forces $F_1(t)$ and $F_2(t)$, exhibits the responses $x_1(t)$ and $x_2(t)$, respectively. Then, if we subject the system to a force of the form

$$F(t) = c_1 F_1(t) + c_2 F_2(t) \tag{1.138}$$

where c_1 and c_2 are constants, and the response to $F(t)$ is

$$x(t) = c_1 x_1(t) + c_2 x_2(t) \tag{1.139}$$

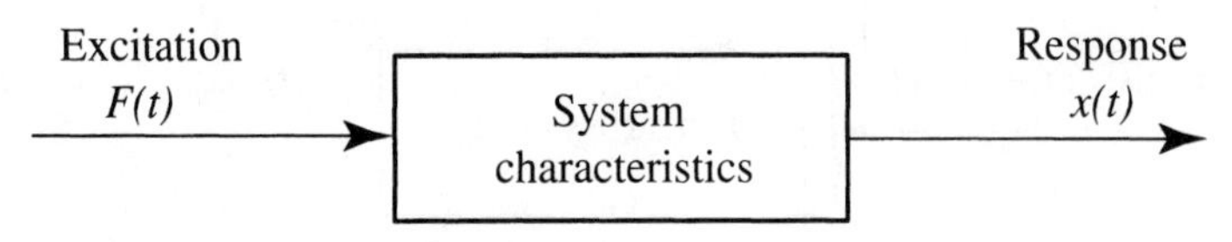

FIGURE 1.41
Symbolic block diagram relating the excitation and the response

the system is linear. This situation is depicted in Fig. 1.42. On the other hand, if

$$x(t) \neq c_1 x_1(t) + c_2 x_2(t) \tag{1.140}$$

the system is nonlinear. Equations (1.138) and (1.139) can be extended to the case in which $F(t)$ and $x(t)$ are the sum of any number of excitations and responses, respectively. The equations represent the *principle of superposition* and can be stated as follows: *if a linear system is acted upon by a linear combination of individual excitations, the individual responses can be first obtained separately and then combined linearly to obtain the total response.*

As an illustration, we consider a system described by the differential equation

$$m\frac{d^2x}{dt^2} + c\frac{dx}{dt} + kx = F \tag{1.141}$$

and denote the response to F_1 by x_1 and the response to F_2 by x_2, so that

$$\begin{aligned} m\frac{d^2x_1}{dt^2} + c\frac{dx_1}{dt} + kx_1 &= F_1 \\ m\frac{d^2x_2}{dt^2} + c\frac{dx_2}{dt} + kx_2 &= F_2 \end{aligned} \tag{1.142}$$

Then, we assume an excitation in the form of Eq. (1.138), multiply the first of Eqs. (1.142) by c_1 and the second by c_2, add up the results and write

$$\begin{aligned} & c_1\left(m\frac{d^2x_1}{dt^2} + c\frac{dx_1}{dt} + kx_1\right) + c_2\left(m\frac{d^2x_2}{dt^2} + c\frac{dx_2}{dt} + kx_2\right) \\ & \quad = m\frac{d^2}{dt^2}(c_1x_1 + c_2x_2) + c\frac{d}{dt}(c_1x_1 + c_2x_2) + k(c_1x_1 + c_2x_2) \\ & \quad = c_1F_1 + c_2F_2 = F \end{aligned} \tag{1.143}$$

Comparing Eqs. (1.141) and (1.143), we conclude that Eq. (1.139) holds, so that the system is linear.

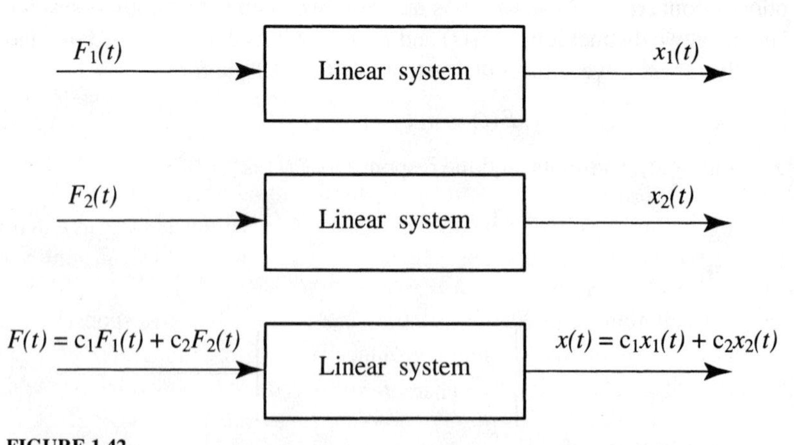

FIGURE 1.42
Excitation-response relation for linear systems

Next, we consider the system described by

$$m\frac{d^2x}{dt^2} + c\frac{dx}{dt} + k(x + \epsilon x^3) = F \tag{1.144}$$

Following the same process, we write

$$\begin{aligned} m\frac{d^2x_1}{dt^2} + c\frac{dx_1}{dt} + k(x_1 + \epsilon x_1^3) &= F_1 \\ m\frac{d^2x_2}{dt^2} + c\frac{dx_2}{dt} + k(x_2 + \epsilon x_2^3) &= F_2 \end{aligned} \tag{1.145}$$

multiply the first of Eqs. (1.145) by c_1 and the second by c_2 and obtain

$$\begin{aligned} &c_1\left[m\frac{d^2x_1}{dt^2} + c\frac{dx_1}{dt} + k(x_1 + \epsilon x_1^3)\right] + c_2\left[m\frac{d^2x_2}{dt^2} + c\frac{dx_2}{dt} + k(x_2 + \epsilon x_2^3)\right] \\ &= m\frac{d^2}{dt^2}(c_1x_1 + c_2x_2) + c\frac{d}{dt}(c_1x_1 + c_2x_2) + k(c_1x_1 + c_2x_2) + k\epsilon(c_1x_1^3 + c_2x_2^3) \\ &= c_1F_1 + c_2F_2 = F \end{aligned} \tag{1.146}$$

But, because

$$c_1x_1^3 + c_2x_2^3 \neq (c_1x_1 + c_2x_2)^3 \tag{1.147}$$

we conclude that Eq. (1.139) does not hold, so that the system described by Eq. (1.144) is nonlinear. This can be explained by the fact that Eq. (1.144) represents the equation of motion of a single-degree-of-freedom system with a nonlinear spring. For $\epsilon > 0$ it is a stiffening spring, and for $\epsilon < 0$ it is a softening spring.

Comparing Eqs. (1.141) and (1.144), we see that the only difference between the two lies in the cubic term in x in Eq. (1.144). Hence, we can draw the conclusion that *a system is linear if the dependent variable $x(t)$ and all its time derivatives appear in the equation of motion to the first power or zero power only*, where zero power implies that the corresponding term is constant. Based on this statement, it is possible for the most part to ascertain whether a system is linear or nonlinear by merely inspecting the differential equation, and tests such as the preceding ones are not really necessary. Although we reached this conclusion on the basis of a single-degree-of-freedom system, a similar conclusion can be reached for multi-degree-of-freedom systems and for distributed-parameter systems. Indeed, it is sufficient for a single dependent variable or one of its derivatives to be nonlinear for the whole system to be nonlinear.

The distinction between linear and nonlinear systems is not as sharp as it may seem, and the same system can be regarded as linear over a certain range and as nonlinear over another. To illustrate the idea, we consider Eq. (1.144) and assume that ϵ is a small quantity. Then, in the range in which $\epsilon x^3 << x$ the system can be regarded as linear. On the other hand, if x reaches amplitudes such that ϵx^3 is of the same order of magnitude as x, the system must be treated as nonlinear. Clearly, there is a point beyond which the system becomes nonlinear. This point bounds the linear range, as shown in Fig. 1.12b, and quite often the point is not well defined; it depends to a large extent on the desired accuracy. Nonlinear systems are discussed in Ch. 11.

Before we discuss response characteristics, it is necessary to introduce an additional concept. To this end, we refer to the block diagram of Fig. 1.43 and consider such excitations $F(t)$ that, if $F(t)$ is delayed by an amount of time τ, the response $x(t)$ is delayed by the same amount τ. This condition is automatically satisfied by linear systems for which the coefficients multiplying the dependent variable $x(t)$ and its time derivatives do not depend explicitly on time. Such systems are known as *linear time-invariant systems*, or more commonly as *linear systems with constant coefficients*. An example of a time-invariant system is that given by Eq. (1.141), in which the constant coefficients are the system parameters m, c and k. On the other hand, in the case in which the system is described by the differential equation

$$m\ddot{x}(t) + k(1 + a\cos\omega t)x(t) = F(t) \tag{1.148}$$

the excitation-response relation is not as depicted by the block diagram of Fig. 1.43. The reason for this is that Eq. (1.148) represents a *time-varying system*, or a *system with time-dependent coefficients*. The treatment of time-varying systems is significantly more difficult than that of time-invariant systems. The subject is discussed in Ch. 11. Unless otherwise stated, we can assume that we are dealing with linear systems with constant coefficients.

The response to initial excitations is the simplest problem in vibrations. It amounts to letting $F(t) = 0$ in the differential equation of motion and assuming that the solution $x(t)$ of the resulting homogeneous equation has exponential form, which leads to a *characteristic equation* for the exponents. Then, the coefficients of the exponential terms are determined by letting $x(t)$ and $\dot{x}(t)$ evaluated at $t = 0$ match the initial displacement and velocity, respectively. We discuss this subject in Ch. 2.

The response to harmonic excitations is also harmonic and has the same frequency as the excitation frequency, but it differs in magnitude and possesses a phase angle relative to the excitation, both magnitude and phase angle depending on the driving frequency ω. The response to harmonic excitations is a *steady-state response* and, as in the case of the excitation, it is best treated in the frequency domain. Plots of the magnitude and phase angle versus ω are known as *frequency response* plots, and provide a great deal of information concerning the nature of the system response, much more than time domain plots. The response to harmonic forces is presented in Ch. 3.

The power of the superposition principle becomes evident in the response to periodic excitations. From Sec. 1.11, we recall that periodic excitations can be represented by Fourier series, i.e., series of harmonic functions. The response to each of these harmonic excitations is also harmonic, as indicated in the preceding paragraph. Then, invoking the superposition principle, the response to periodic excitations can be expressed in the

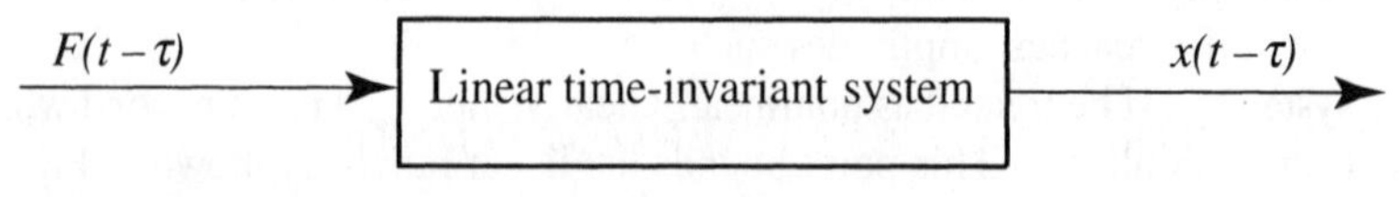

FIGURE 1.43
Excitation-response relation for linear time-invariant systems

form of a series of these harmonic responses. Hence, the response to periodic excitations is a steady-state response as well. The response to periodic excitations is also discussed in Ch. 3.

As pointed out in Sec. 1.11, an arbitrary excitation can be regarded as a superposition of impulse forces of different magnitude and applied at different times. But, the response to a unit impulse applied at $t = 0$ defines the *impulse response* and it represents a characteristic of the system. Indeed, it is a function of time reflecting the system inertia, damping and stiffness properties. Assuming that the impulse response is known, the response of a linear system with constant coefficients can be expressed as a superposition of impulse responses of different magnitudes and applied at different times. This superposition is called the *convolution integral*, or the *superposition integral*. A more detailed discussion of the convolution integral and of practical ways of evaluating it on a digital computer is presented in Ch. 4.

The principle of superposition lies at the basis of linear analysis and is largely responsible for the theory of vibrations of linear systems being so well developed. Indeed, the consequences of the principle are so pervasive that many of them are taken for granted. A prime example is the fact that *the solution of the equations of motion to initial excitations*, or the homogeneous solution, *and the solution to applied forces*, or the nonhomogeneous solution, *can be obtained separately and then combined linearly to obtain the complete solution*. This fact applies to linear systems alone. At this point, a word of caution is in order. Whereas the superposition of solutions is valid for linear systems without restrictions, there are cases in which the rationale for superposing solutions must be questioned. In this regard, we recall that initial displacements and velocities are transient excitations, with the response to such excitations best described in the time domain beginning at $t = 0$. On the other hand, constant, harmonic and periodic forces are steady-state excitations, the latter two more meaningfully described in the frequency domain. But, responses to steady-state harmonic and periodic excitations are also steady state, so that they too are better described in the frequency domain than in the time domain. Hence, although the principle of superposition permits it, from a physical point of view it is difficult to justify the addition of the response to initial excitations to a steady-state response.

There remains the question of the response to random excitations. To answer this question, it is necessary to introduce a whole variety of new concepts concerning the nature of random functions, such as the *mean value*, *mean square value*, *autocorrelation function*, *power spectral density function*, etc. Clearly, if the excitation is a random function, so is the response. To obtain the various quantities just mentioned for the response, it is convenient to use *Fourier transforms*, which implies working in the frequency domain rather than the time domain. Still, the results are defined neither in the frequency domain nor in the time domain but in terms of *probability distributions*. The entire Ch. 12 is devoted to the response of linear systems to random excitations.

1.13 VIBRATION ABOUT EQUILIBRIUM POINTS

In Sec. 1.10, we introduced the concepts of equilibrium and displacements from equilibrium and used them to simplify the equations of motion. These concepts have signif-

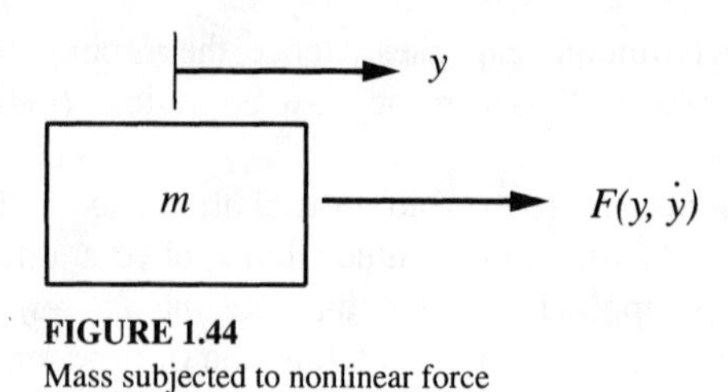

FIGURE 1.44
Mass subjected to nonlinear force

icantly wider implications than it may appear. In this section, we propose to examine these implications, before we proceed with the actual solution of the equations of motion in the following chapters.

We consider the single-degree-of-freedom system shown in Fig. 1.44 and described by the generic differential equation of motion

$$m\ddot{y} = F(y, \dot{y}) \tag{1.149}$$

where m is the mass and $F(y, \dot{y})$ is in general a nonlinear function of the displacement y and velocity $\dot{y}$. We assume that general solutions of Eq. (1.149) are not possible, and our interest lies in *special solutions* capable of shedding some light on the system behavior. To this end, we assume that Eq. (1.149) admits the special constant solutions

$$y = y_e = \text{constant}, \;\; \dot{y} = \ddot{y} = 0 \tag{1.150}$$

Because the velocity and acceleration are zero, the constant solutions defined by Eqs. (1.150) represent *equilibrium points*, not unlike the static equilibrium of Sec. 1.10; they can be obtained by inserting Eqs. (1.150) into Eq. (1.149) and solving the *equilibrium equation*

$$F(y_e, 0) = 0 \tag{1.151}$$

Equation (1.151) represents an algebraic equation in y_e. If $F(y_e, 0)$ is a polynomial, there are as many solutions as the degree of the polynomial, and if $F(y_e, 0)$ is linear, then there is just one solution. On the other hand, if $F(y_e, 0)$ is a transcendental function, then mathematically there could be an infinite number of solutions. Physically, however, there is only a finite number of equilibrium points, as many of these solutions represent the same point, as shown in Example 1.9. If $y_e = 0$ is a solution of Eq. (1.151), then the corresponding equilibrium point is said to be *trivial.*

A problem of considerable importance in vibrations is how the system behaves when disturbed from equilibrium, and in particular whether the subsequent motion remains confined to the neighborhood of the equilibrium point or not. This is the same as asking whether the equilibrium point is stable or not. The subject is discussed in a rigorous manner in Ch. 11. At this point, we are content with some simple definitions, as follows:

1. If a system disturbed from an equlibrium point returns to the same equilibrium point, then the motion (or the equilibrium point) is said to be *asymptotically stable*.

2. If a system disturbed from an equlibrium point oscillates about the same equilibrium point without exhibiting any secular trend, i.e., the system neither returns to the equilibrium point nor moves away from it with time, then the motion (or the equilibrium point) is merely *stable*.
3. If a system disturbed from an equlibrium point moves away from it with time, then the motion (or the equilibrium point) is *unstable*.

To lend the discussion some quantitative substance, we let the solution of Eq. (1.149) have the form

$$y(t) = y_e + x(t) \tag{1.152}$$

where $x(t)$ is a relatively small displacement from equilibrium. In view of Eqs. (1.150), it follows that

$$\dot{y}(t) = \dot{x}(t),\ \ddot{y}(t) = \ddot{x}(t) \tag{1.153}$$

Next, we expand $F(y, \dot{y})$ in a Taylor series about an equilibrium point y_e, consider Eq. (1.152) and the first of Eqs. (1.153) and write

$$F(y, \dot{y}) = F(y_e, 0) + \left.\frac{\partial F(y, \dot{y})}{\partial y}\right|_{\substack{y = y_e \\ \dot{y} = 0}} x + \left.\frac{\partial F(y, \dot{y})}{\partial \dot{y}}\right|_{\substack{y = y_e \\ \dot{y} = 0}} \dot{x} + O(x^2) \tag{1.154}$$

in which $O(x^2)$ denotes terms of second order and higher in x and $\dot{x}$, i.e., nonlinear terms. Then, inserting the second of Eqs. (1.153) and Eq. (1.154) into Eq. (1.149), considering Eq. (1.151), introducing the notation

$$\frac{1}{m}\left.\frac{\partial F(y, \dot{y})}{\partial y}\right|_{\substack{y = y_e \\ \dot{y} = 0}} = -b, \qquad \frac{1}{m}\left.\frac{\partial F(y, \dot{y})}{\partial \dot{y}}\right|_{\substack{y = y_e \\ \dot{y} = 0}} = -a \tag{1.155}$$

and assuming that displacements from equilibrium are sufficiently small that the nonlinear terms can be ignored, we obtain

$$\ddot{x} + a\dot{x} + bx = 0 \tag{1.156}$$

Equation (1.156) represents the *linearized equation of motion about equlibrium*, and the assumption permitting linearization of Eq. (1.149) is called the *small motions assumption*. We propose to use Eq. (1.156) to investigate the motion characteristics in the neighborhood of equilibrium. Of course, these characteristics depend on the parameters a and b, which differ from one equilibrium point to another.

Equation (1.156) is linear with constant coefficients, so that its solution has the exponential form

$$x(t) = Ae^{st} \tag{1.157}$$

where A is an inconsequential amplitude and s is a constant exponent. Clearly, the behavior of the system in the neighborhood of equilibrium is dictated by the values of s. Note that, because Eq. (1.156) is of second order, there are two such values. To obtain these values, we introduce Eq. (1.157) into Eq. (1.156), divide through by Ae^{st} and conclude that the exponent s must satisfy the algebraic equation

$$s^2 + as + b = 0 \tag{1.158}$$

which is known as the *characteristic equation.* The roots of Eq. (1.158) are

$$\begin{matrix} s_1 \\ s_2 \end{matrix} = -\frac{a}{2} \pm \sqrt{\left(\frac{a}{2}\right)^2 - b} \tag{1.159}$$

so that the solution of Eq. (1.156) is

$$x(t) = A_1 e^{s_1 t} + A_2 e^{s_2 t} \tag{1.160}$$

and we note that s_1 and s_2 are in general complex. The nature of the motion in the neighborhood of an equlibrium point can be investigated by considering the s-plane of Fig. 1.45, a complex plane containing s_1 and s_2. To this end, we observe that the roots s_1 and s_2 of the characteristic equation can be real, pure imaginary, or complex. Because $x(t)$ must be real, if s_1 and s_2 are either pure imaginary or complex, then they are the complex conjugates of one another, and so are A_1 and A_2. From Eq. (1.160), we see that when s_1 and s_2 are both real and negative the solution approaches zero asymptotically. If the roots s_1 and s_2 are complex, the magnitude of the solution is controlled by the real part of the roots. Indeed, an exponential function with complex exponent can be expressed as the product of two factors, one corresponding to the real part of the exponent and the other corresponding to the imaginary part. The factor corresponding to the real part plays the role of a time-dependent amplitude and the factor corresponding to the imaginary part varies harmonically with time, as can be concluded from Eq. (1.135). Hence, if s_1 and s_2 are complex conjugates with negative real part, the solution approaches zero in an oscillatory fashion as $t \to \infty$. It follows that *in all cases in which* s_1 *and* s_2 *are both real and negative or complex conjugates with negative real part the motion*

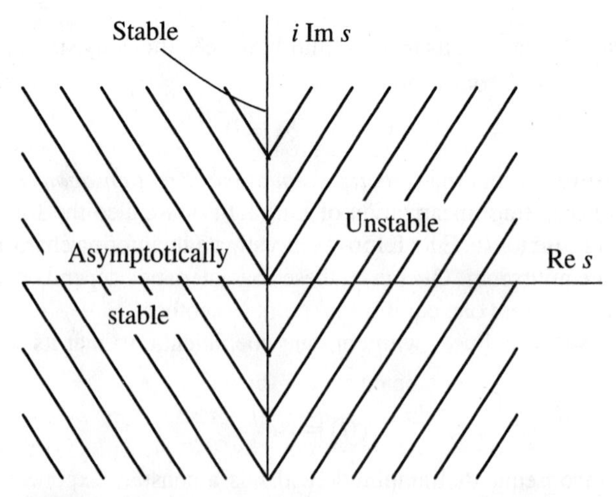

FIGURE 1.45
Stability statements in the complex s-plane

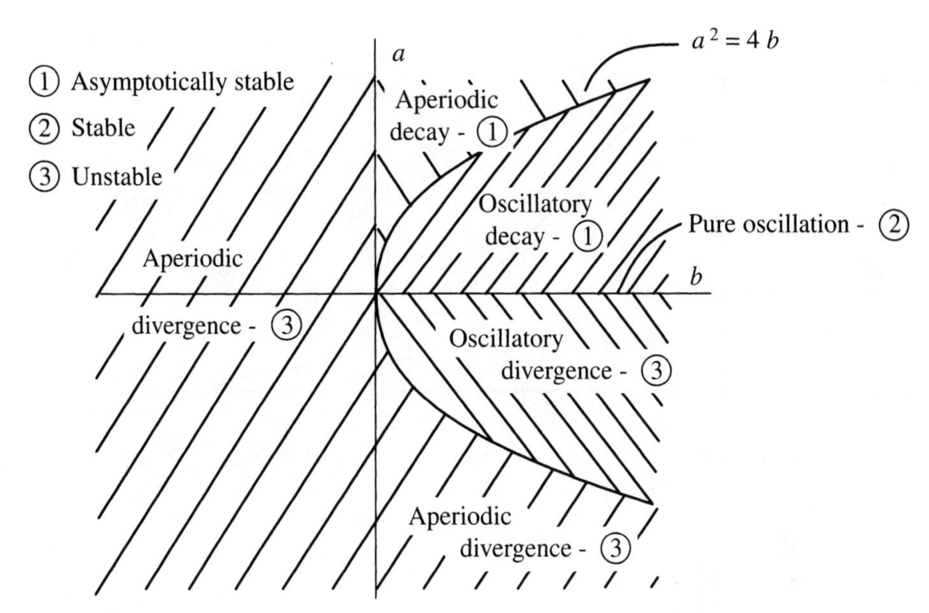

FIGURE 1.46
Stability statements in the parameter plane

in the neighborhood of an equilibrium point is asymptotically stable. This situation corresponds to the cases in which both s_1 and s_2 lie in the left half of the s-plane of Fig. 1.45. When s_1 and s_2 are pure imaginary complex conjugates, the solution is harmonic, so that the system neither tends to the equilibrium point nor does it move away from it as $t \to \infty$. Hence, *in all cases in which s_1 and s_2 are pure imaginary the motion is merely stable.* These cases are represented by the imaginary axis in Fig. 1.45. Finally, *if either s_1 or s_2 is real and positive, or both s_1 and s_2 are real and positive, or s_1 and s_2 are complex conjugates with positive real part*, the solution diverges, so that *the motion is unstable.* This situation corresponds to the cases in which at least one of the roots of the characteristic equation lies in the positive half of the s-plane, as shown in Fig. 1.45.

The above stability statements can be rendered more explicit by using Eq. (1.159) to tie them directly to the system parameters a and b. The whole range of stability possibilities and of the corresponding types of motion are displayed in the parameter plane a versus b of Fig. 1.46. The following review of the information contained in Fig. 1.46 should prove rewarding:

1. **Asymptotically stable motion**. This region covers the first quadrant of the parameter plane, $a > 0,\ b > 0$. The parabola $a^2 = 4b$ separates the region into two subregions. In the subregion above the parabola, $a^2 > 4b$ and the roots s_1 and s_2 are real and negative, so that the motion decays aperiodically. A typical plot is shown in Fig. 1.47. In the subregion below the parabola, $a^2 < 4b$ and the roots s_1 and s_2 are complex conjugates with negative real part. As explained above, the motion in this case is a decaying oscillation, as depicted in Fig. 1.48. The parabola $a^2 = 4b$ represents a

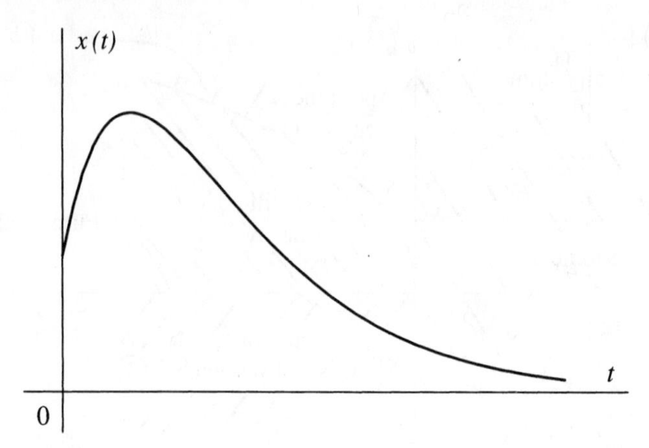

FIGURE 1.47
Aperiodically decaying motion

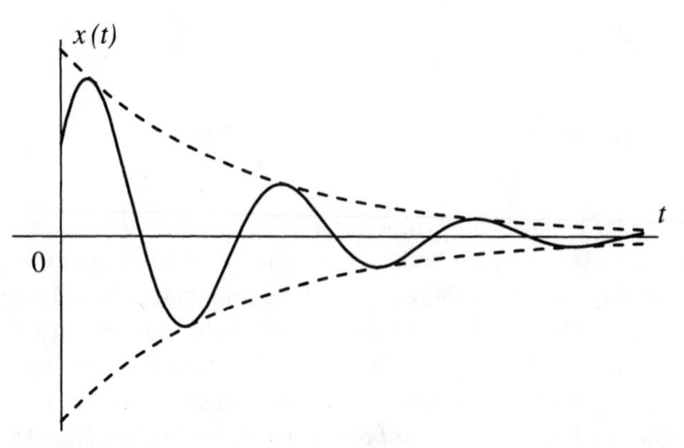

FIGURE 1.48
Decaying oscillation

borderline case corresponding to the repeated root $s_1 = s_2 = a/2$; the motion in this case also decays aperiodically.

2. **Stable motion**. This region is simply the line $a = 0$, $b > 0$. In this case the roots s_1 and s_2 are pure imaginary complex conjugates, so that the motion is pure harmonic oscillation, as shown in Fig. 1.49. Note that this harmonic motion is different from the steady-state harmonic response of the type encountered in Sec. 1.12.
3. **Unstable motion**. This region covers the remaining three quadrants, $b < 0$ and $a < 0$, $b > 0$. In the region between the parabola $a^2 = 4b$ and the positive b-axis, the roots s_1 and s_2 are complex conjugates with positive real part, so that the motion represents divergent oscillation. A typical plot is shown in Fig. 1.50. In the region below the parabola $a^2 = 4b$ in the fourth quadrant and in the second and third quadrants, both roots are real with at least one root being positive, so that the motion diverges aperiodically, as depicted in Fig. 1.51. The statements concerning the motions displayed

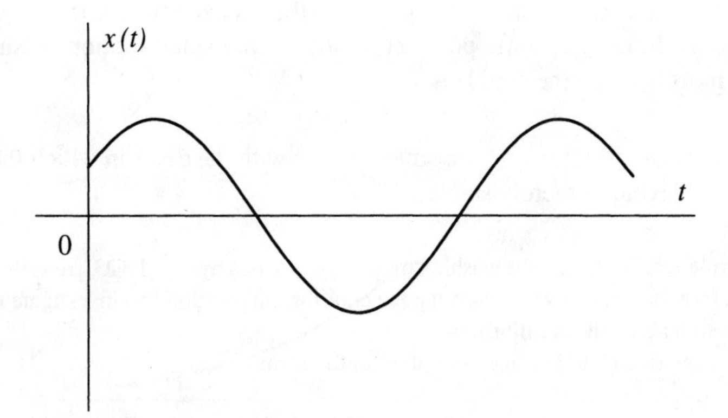

FIGURE 1.49
Pure harmonic oscillation

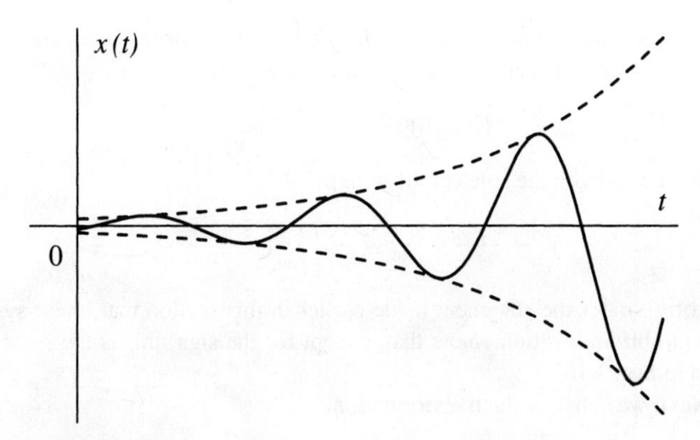

FIGURE 1.50
Diverging oscillation

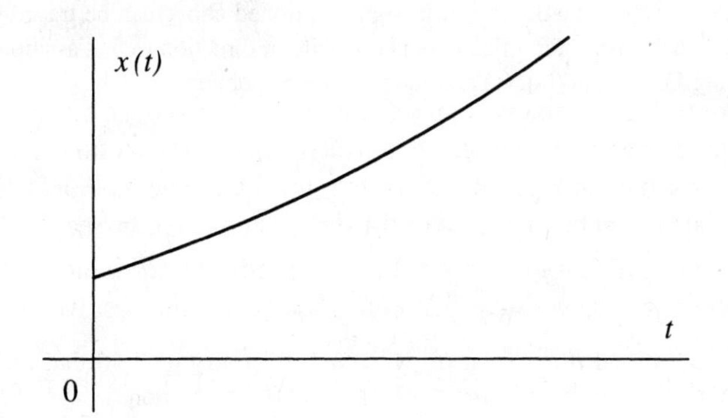

FIGURE 1.51
Aperiodically diverging motion

in Figs. 1.50 and 1.51 must be tempered by the realization that the plots represent only trends. Indeed, at some point $x(t)$ violates the small motions assumption, so that the plots become meaningless.

The study of vibrations is concerned mainly with the cases in which the motion is asymptotically stable or merely stable.

Example 1.8. Consider the washing machine described by Eq. (1.123), rewrite the equation in the form of Eq. (1.149), determine the equilibrium position and investigate the nature of the motion about the equilibrium.

Equation (1.123) can be rewritten in the form

$$M\ddot{y} = -c\dot{y} - ky - Mg \tag{a}$$

so that, comparing Eq. (a) to Eq. (1.149), we conclude that $m = M$ and

$$F(y, \dot{y}) = -c\dot{y} - ky - Mg \tag{b}$$

Hence, in this particular case $F(y, \dot{y})$ is linear and no Taylor series expansion is necessary. Consistent with this, the equilibrium equation is

$$F(y_e, 0) = -ky_e - Mg = 0 \tag{c}$$

from which we obtain the sole equilibrium point

$$y_e = -\frac{Mg}{k} \tag{d}$$

which corroborates the statement made earlier in this section that linear systems admit a single equilibrium position. Note that, except for the sign, this is the same result as that obtained in Sec. 1.10.

Next, we consider the transformation

$$y(t) = y_e + x(t) = -\frac{Mg}{k} + x(t) \tag{e}$$

and observe that the right side of Eq. (e) is identical to Eq. (1.125) with δ_{st} as given by Eq. (1.124). Hence, the difference in sign mentioned above can be traced to the fact that the equilibrium position implied here is opposite in direction to that assumed in Sec. 1.10. Introducing Eq. (e) into Eq. (a) and rearranging, we obtain

$$\ddot{x} + \frac{c}{M}\dot{x} + \frac{k}{M}x = 0 \tag{f}$$

which is the same as Eq. (1.156), provided that

$$a = \frac{c}{M}, \quad b = \frac{k}{M} \tag{g}$$

Because both a and b are positive, we conclude from Fig. 1.46 that the motion about equilibrium is *asymptotically stable*. If $c^2 > 4kM$, the motion decays aperiodically, and if $c^2 < 4kM$ the motion represents oscillatory decay. In the borderline case $c^2 = 4kM$ the motion also decays aperiodically. As we shall see in Ch. 2, the three cases represent overdamping, underdamping and critical damping, respectively.

Example 1.9. From Example 1.1, the equation of motion of a simple pendulum is

$$\ddot{\theta}+\frac{g}{L}\sin\theta=0 \tag{a}$$

Determine the equilibrium points and investigate the nature of the motion in the neighborhood of equilibrium.

Comparing Eq. (a) to Eq. (1.149), we conclude that $m=1,\ y=\theta$ and

$$F(y,\dot{y})=F(\theta,\dot{\theta})=F(\theta,0)=-\frac{g}{L}\sin\theta \tag{b}$$

so that the function does not depend on $\dot{\theta}$. The equilibrium equation is simply

$$F(\theta_e,0)=-\frac{g}{L}\sin\theta_e=0 \tag{c}$$

which has the solutions

$$\theta_e=0,\pm\pi,\pm2\pi,\ldots \tag{d}$$

Hence, mathematically there is an infinite number of solutions. Physically, however, many of these solutions represent the same equilibrium points, and in fact there are only two equilibrium positions

$$\theta_{e1}=0,\ \theta_{e2}=\pi \tag{e}$$

Of course, the first one is recognized as the trivial one, in which the pendulum is at rest hanging down. In the second equilibrium point, the pendulum is at rest in the upright position.

Introducing the transformation

$$\theta(t)=\theta_e+\phi(t) \tag{f}$$

in Eq. (a), we can write the linearized equation, Eq. (1.149), in the form

$$\ddot{\phi}+b\phi=0 \tag{g}$$

so that $a=0$. Moreover, from the first of Eqs. (1.155),

$$b=-\left.\frac{\partial F(\theta,0)}{\partial\theta}\right|_{\theta=\theta_e}=\frac{g}{L}\cos\theta_e \tag{h}$$

Hence, in the case of the trivial equilibrium, $\theta_{e1}=0$, we have

$$b=\frac{g}{L}>0 \tag{i}$$

From Fig. 1.46, we conclude that the parameters lie on the positive b-axis, so that *motion in the neighborhood of the trivial equilibrium is stable*. On the other hand, for $\theta_{e2}=\pi$ we obtain

$$b=-\frac{g}{L}<0 \tag{j}$$

so that, from Fig. 1.46, we conclude that the parameters lie on the negative b-axis, so that *motion in the neighborhood of the upright equilibrium position is unstable*.

The above results conform to expectations. Any small disturbance from the equilibrium position in which the pendulum hangs down results in oscillation about the equilibrium. On the other hand, any small disturbance of the pendulum from the upright equilibrium position causes the pendulum to move away from equilibrium, soon violating the small motions assumption. The case in which the pendulum oscillates about $\theta_{e1}=0$ is by far the most important one, which explains why the equilibrium position $\theta_{e2}=\pi$ is seldom mentioned in vibrations.

1.14 SUMMARY

The study of vibrations is concerned with the motion of a variety of systems, ranging from the oscillation of a simple pendulum to the vibration of a complex structure. These systems have one thing in common, namely, they all involve restoring forces. In the case of a simple pendulum the restoring force is due to gravity, and in the case of a structure the restoring forces are due to elasticity.

The motion of vibrating systems is governed by laws of mechanics, and in particular by Newton's second law in one form or another. Although such material is generally taught in a sophomore course on dynamics, the equations of motion are of such importance in vibrations that the ability to derive them cannot be taken for granted. Hence, the inclusion of material on the derivation of the equations of motion for vibrating systems is highly desirable, and has the added advantage of making this text self-contained.

For the most part, particularly in applications from aerospace, civil and mechanical engineering, vibration is undesirable and is to be avoided, or at least reduced. This can be done through proper design, or by means of controls. To this end, it is necessary to be able to predict how the system responds to various stimuli. When the system is complex, this response must be predicted on the basis of a simplified model acting as a surrogate for the actual system. Such a model must be sufficiently accurate to retain the essential dynamic characteristics of the actual system and yet sufficiently simple to lend itself to reasonable mathematical description. The main factors affecting the behavior of vibrating systems are the mass and stiffness properties, as well as the damping properties. Implicit is the manner in which these quantities are distributed throughout the system. In general, a model can be regarded as an aggregation of individual components. Modeling amounts to identifying the individual components and their inertia, stiffness and damping properties, as well as how the individual components are connected to one another so as to act together as a whole system. It should be said that modeling is more of an art than an exact science, as there are no particular guidelines to rely on. In fact, a model is not unique for a system, and the same system can be modeled in various ways so as to reflect different objectives.

The equations describing the vibration of lumped models are in general ordinary differential equations. In deriving the equations by some suitable form of Newton's second law, it is necessary to draw one free-body diagram for each mass in the system, i.e., a diagram of a given mass isolated from all other masses in the system and showing all forces acting upon the mass, which includes both externally applied forces and internal forces. Note that, in cutting through internal forces in the process of isolating the mass, these internal forces are to be treated as external. The number of ordinary differential equations for a system generally coincides with the number of degrees of freedom of the system, where the latter represents the minimum number of coordinates required to describe the motion of the system fully. On occasion, when internal forces are carried as unknowns, the number of equations exceeds the number of degrees of freedom by the number of unknown forces.

To derive the system response, it is necessary to solve the differential equations of motion. The nature of the response depends on the excitations and on the system characteristics. The excitations represent external factors and consist of initial displacements and velocities and applied forces and/or moments. For linear systems, it is possible to invoke the principle of superposition and determine the response to initial excitations and the response to applied forces separately and combine them linearly to obtain the system total response. In the absence of applied forces, the response to initial excitations can be expressed in exponential form, where the exponents are in general complex, with the real part defining the amplitude of the response and the imaginary part defining the frequency. The type of applied forces determines not only the nature of the response but also the choice of method by which the response is to be obtained. The simplest type consists of harmonic forces, in which the response is also harmonic and having the same frequency as the excitation. They are referred to as steady-state excitation and steady-state response, respectively. Periodic forces can be expanded in Fourier series, which are series of harmonic functions with frequencies that are integer multiples of the lowest frequency, where the latter is known as the fundamental frequency. Then, by virtue of the superposition principle, the individual responses to these harmonic components can be combined linearly to obtain the response to the periodic force. In a somewhat similar approach, a nonperiodic force can be regarded as a linear combination of impulses and, invoking the superposition principle, the response can be obtained in the form of a linear combination of impulse responses, where the combination becomes in the limit the convolution integral. Clearly, linearity is a system property of crucial importance, because the superposition principle applies only to linear systems, which makes solutions for nonlinear systems much more difficult to obtain than solutions for linear systems.

On occasion, particularly in preliminary design, explicit expressions for the system response are not really necessary, and a statement concerning system stability suffices. This is particularly true if the system is nonlinear. To produce such qualitative statements, it is necessary to identify special solutions of the equations of motion; they are constant solutions known as equilibrium points. Then, assuming small motions in the neighborhood of a given equilibrium point, the equations of motion can be linearized about that equilibrium point, and a stability statement can be based for the most part on the eigenvalues of the linearized system. Cases in which the analysis based on linearized equations is not valid are discussed in Ch. 11.

PROBLEMS

1.1. Two masses sliding on smooth inclined planes are each connected to a massless pulley (Fig. 1.52). The two pulleys are rigidly attached to one another and the diameter of one is twice the diameter of the other. Use Newton's second law to derive an expression for the acceleration of mass m_2.

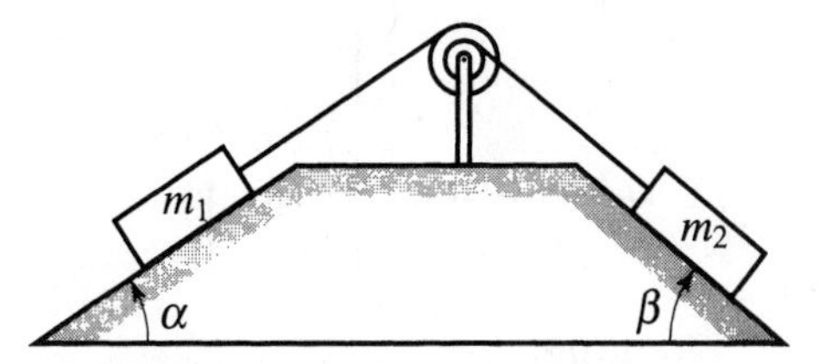

FIGURE 1.52
Masses sliding on inclined planes

1.2. A bead of mass m is free to slide along a smooth circular hoop rotating about a vertical axis with the constant angular velocity Ω (Fig. 1.53). Use Newton's second law to derive the equation of motion for the angle θ working with the transverse component of force and acceleration. Hint: the rotation of the hoop gives rise to a centripetal acceleration perpendicular to the vertical axis, in addition to the transverse component of acceleration $R\ddot{\theta}$. Note that the radial component $-R\dot{\theta}^2$ does not enter the picture.

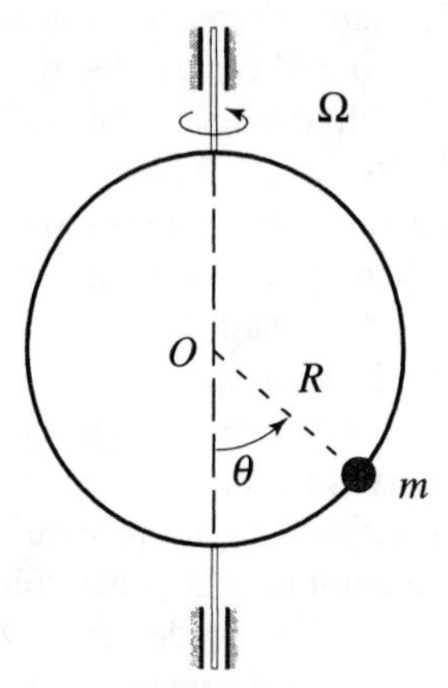

FIGURE 1.53
Bead sliding along a rotating hoop

1.3. A simple pendulum of mass $m = 5$ kg and length $L = 2$ m is released from rest in a position defined by the angle $\theta_0 = 60°$ with respect to the vertical. Assuming that the string is inextensible, determine the tension in the string in the positions $\theta_1 = 30°$ and $\theta_2 = 0$.

1.4. A compound pendulum in the form of a uniform bar hinged at point O is hanging at rest when struck at a point a distance h from O by a horizontal force F, as shown in Fig. 1.54. Determine h so that the horizontal reaction at O is zero. Note that such a point is called the center of percussion.

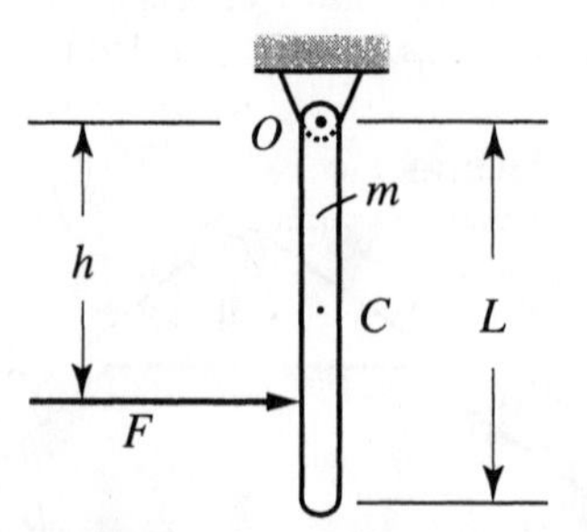

FIGURE 1.54
Compound pendulum struck by a force

1.5. A uniform rectangular door hangs at an angle α with respect to the vertical (Fig. 1.55). Assume that the door is displaced initially with respect to the vertical plane and then allowed to oscillate. Derive the equation for the oscillatory motion $\theta(t)$ of the door by Newton's second law; the angle θ can be arbitrarily large.

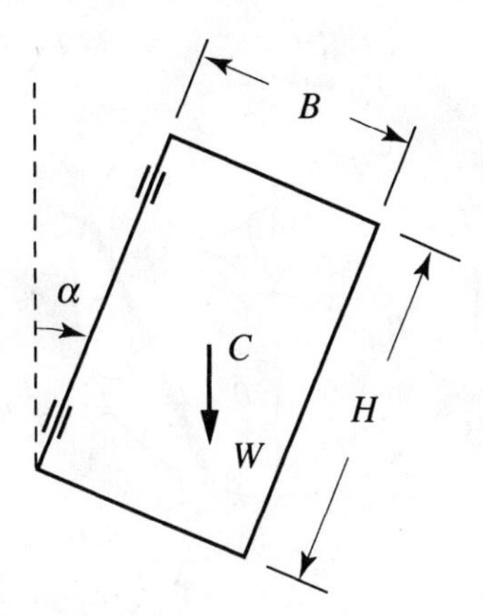

FIGURE 1.55
Oscillating door

1.6. Use Newton's second law to derive the equation of motion for a compound pendulum consisting of a uniform rod of total length L and mass per unit length m and a disk of radius R and total mass M, as shown in Fig. 1.56.

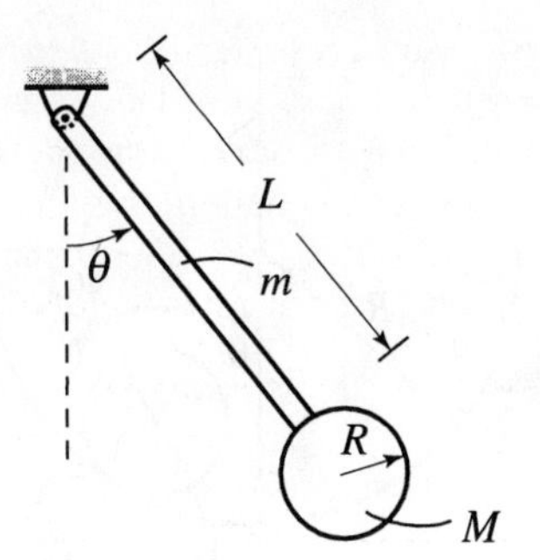

FIGURE 1.56
Compound pendulum

1.7. A uniform bar of mass m and length $\sqrt{2}R$ slides inside a smooth circular surface of radius R (Fig. 1.57). Use Newton's second law to derive the equation of motion for arbitrarily large angles θ and determine the forces exerted by the surface on the bar.

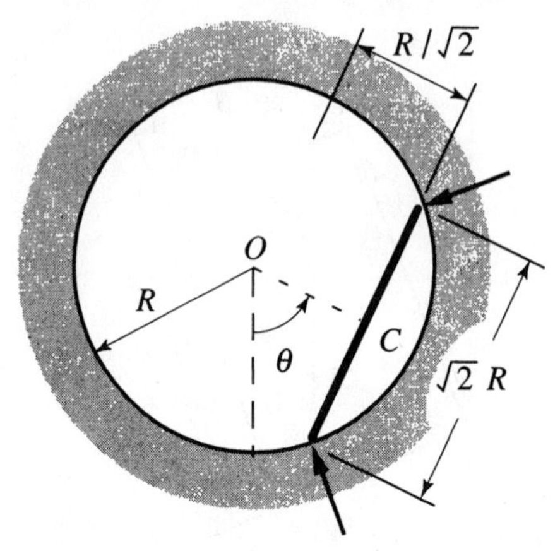

FIGURE 1.57
Bar sliding inside a smooth circular surface

1.8. A disk of mass m and radius r rolls without slip inside a rough circular surface of radius R, as shown in Fig. 1.58. Derive the differential equation for the angular motion θ by writing two equations of motion, one for the translation of C and one for the rotation of the disk about C, and then eliminating the force at the point of contact A. The angle θ can be arbitrarily large.

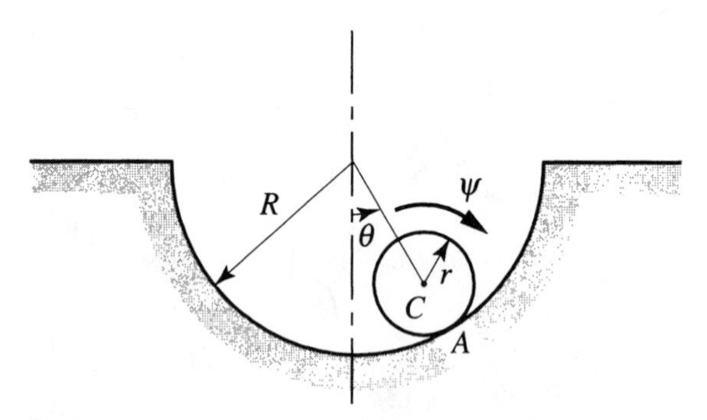

FIGURE 1.58
Disk rolling without slip

1.9. A double pendulum consists of two bobs of mass m_1 and m_2 suspended by inextensible, massless strings of length L_1 and L_2 (Fig. 1.59). Use Newton's second law to derive four equations of motion for the rectangular coordinates x_1, y_1, x_2 and y_2. Then, express x_1, y_1, x_2 and y_2 in terms of the angles θ_1 and θ_2, eliminate the tensions T_1 and T_2 in the strings and obtain two equations of motion for θ_1 and θ_2.

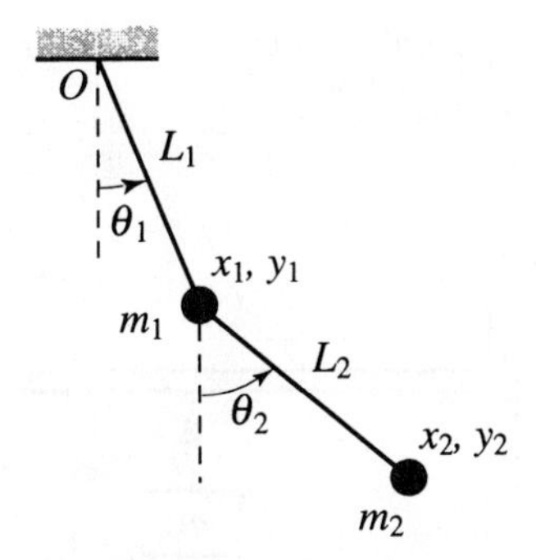

FIGURE 1.59
Double pendulum

1.10. Solve Problem 1.9 by defining the motion directly in terms of the angular displacements θ_1 and θ_2.

1.11. The system depicted in Fig. 1.60 consists of two uniform rigid links of mass m_1 and m_2 and length L_1 and L_2. Use Eq. (1.58) for link 1 and Eqs. (1.63) and (1.64) for link 2 to obtain four equations of motion. Then, eliminate the constraint forces between the links and derive two equations of motion in terms of θ_1 and θ_2.

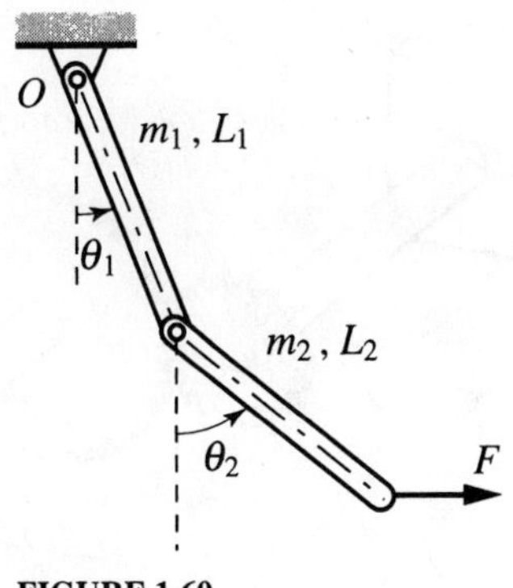

FIGURE 1.60
Two-link system

1.12. Determine the equivalent spring constant for the system of Fig. 1.61.

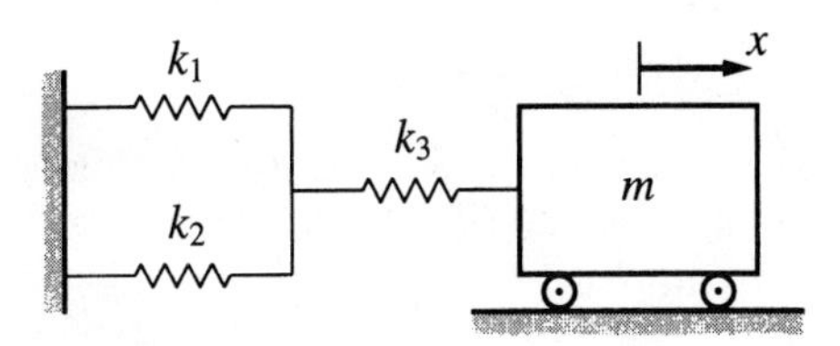

FIGURE 1.61
System with springs in parallel and in series

1.13. Derive the equivalent spring constant for the system of Fig. 1.62.

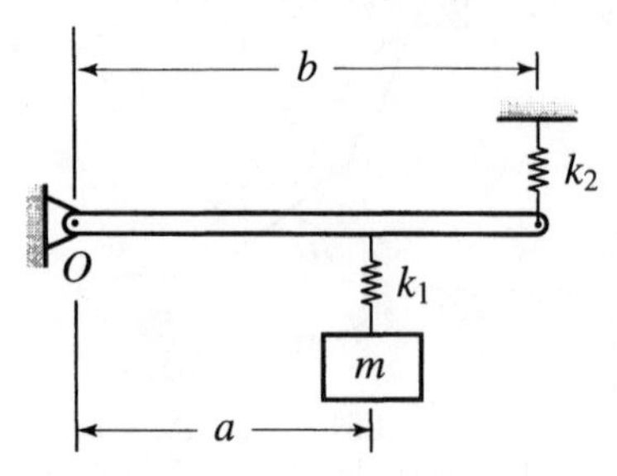

FIGURE 1.62
System supported by springs through a rigid link

1.14. The system shown in Fig. 1.63 consists of two gears A and B mounted on uniform circular shafts of equal stiffness GJ/L; the gears are capable of rolling on each other without slip. Derive an expression for the equivalent spring constant of the system for the radii ratio $R_A/R_B = n$.

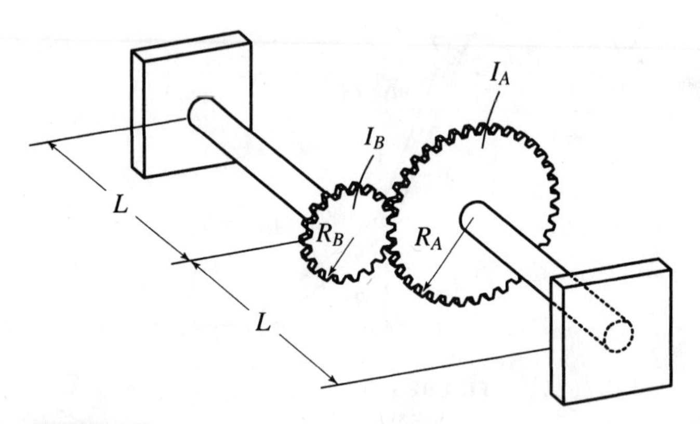

FIGURE 1.63
Two gears rolling on one another

1.15. The circular shaft of Fig. 1.64 has the torsional stiffness $GJ(x) = GJ[1 - \frac{1}{2}(x/L)^2]$. Obtain the equivalent spring constant corresponding to a torque at $x = L$.

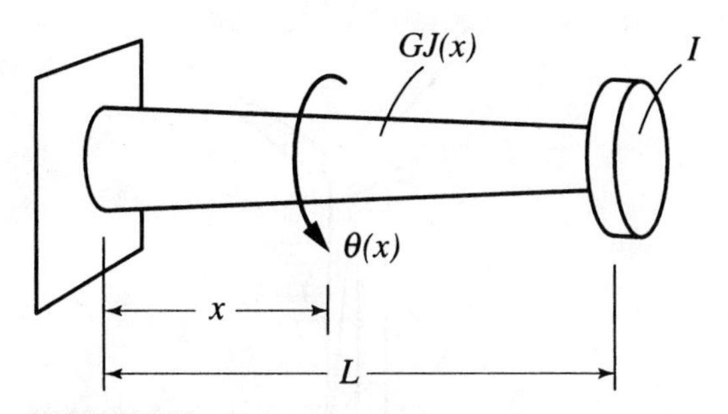

FIGURE 1.64
Nonuniform shaft acting as a torsional spring

1.16. Consider a uniform cantilever beam of bending stiffness EI and obtain the equivalent spring constant corresponding to a bending moment applied at the free end $x = L$.

1.17. A cantilever beam in bending is made of two uniform sections, as shown in Fig. 1.65. Obtain the equivalent spring constant corresponding to a vertical force applied at the free end $x = L$.

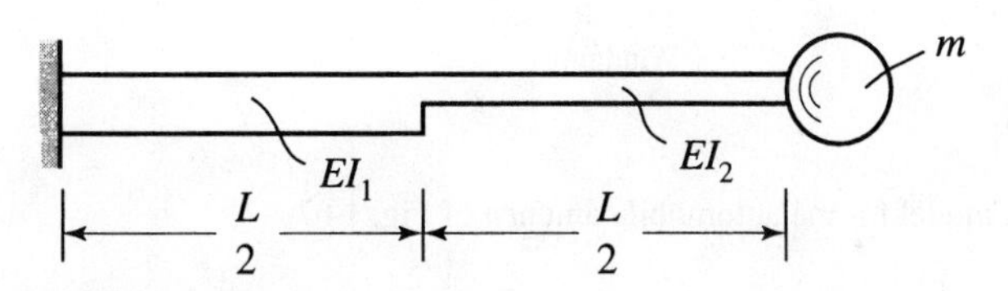

FIGURE 1.65
Nonuniform beam acting as a spring

1.18. Verify the expression in Table 1.2 for the equivalent spring constant for a uniform pinned-pinned beam with an overhang and a force at the tip.

1.19. Verify the expression in Table 1.2 for the equivalent spring constant for a uniform clamped-pinned beam with an overhang and a force at the tip. Hint: Regard the problem as a combination of two problems, one of a cantilever beam with the load at the tip $x = L + a$ and the other of a cantilever beam loaded with the pin reaction at $x = L$. Then, determine the pin reaction and subsequently the spring constant by setting the displacement at $x = L$ equal to zero.

1.20. The two gears of the system of Fig. 1.63 have mass polar moments of inertia I_A and I_B. Derive an expression for the equivalent mass polar moment of inertia for the radii ratio $R_A/R_B = n$. Hints: (1) The reaction forces on the gears at the point of contact are equal in magnitude and opposite in direction, (2) the angular acceleration of gear B is n times the angular acceleration of gear A and (3) write one torque equation for each of the gears separately with the shafts absent.

1.21. Devise a model for the windmill of Fig. 1.66.

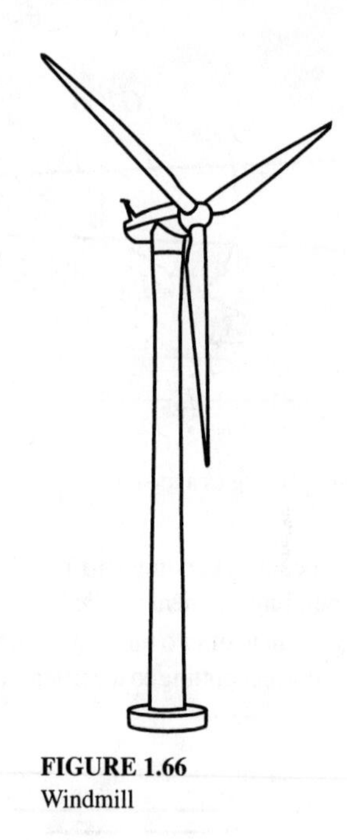

FIGURE 1.66
Windmill

1.22. Devise a model for the automobile antenna of Fig. 1.67.

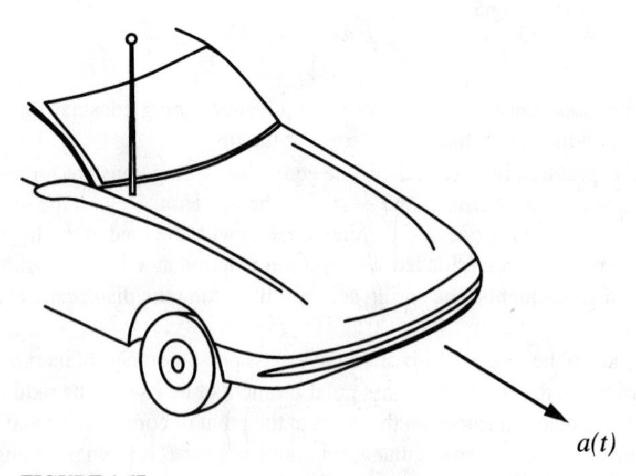

FIGURE 1.67
Automobile antenna

1.23. Devise two different models for the boat shaft and propeller of Fig. 1.68.

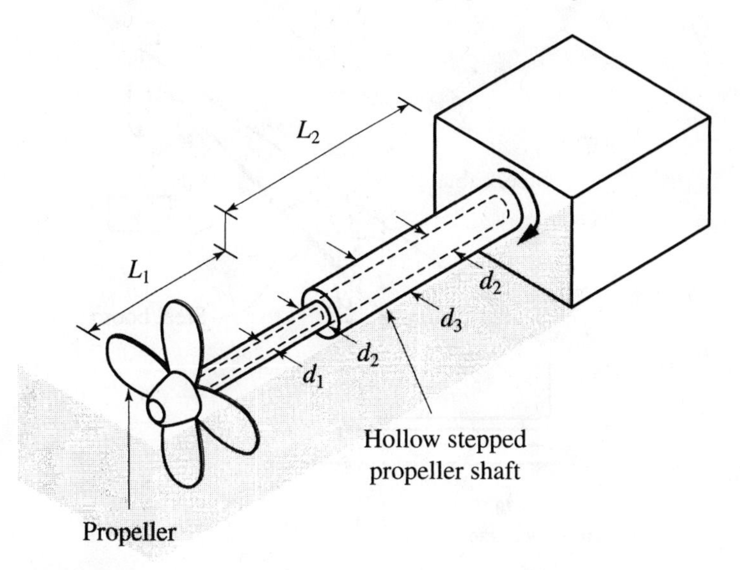

FIGURE 1.68
Boat shaft and propeller

1.24. Devise a model for the radar antenna tower of Fig. 1.69.

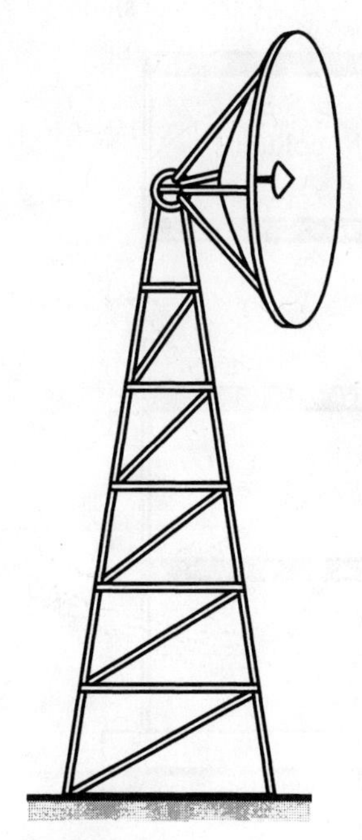

FIGURE 1.69
Radar antenna tower

1.25. Devise a model for the construction crane of Fig. 1.70.

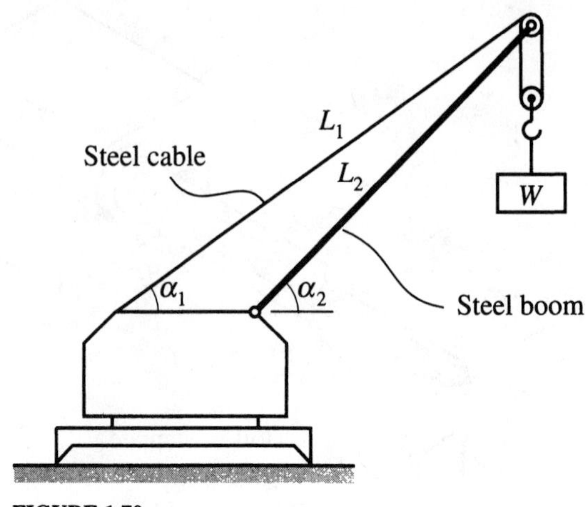

FIGURE 1.70
Construction crane

1.26. Devise a lumped model for the n-story building subjected to a horizontal earthquake excitation (Fig. 1.71).

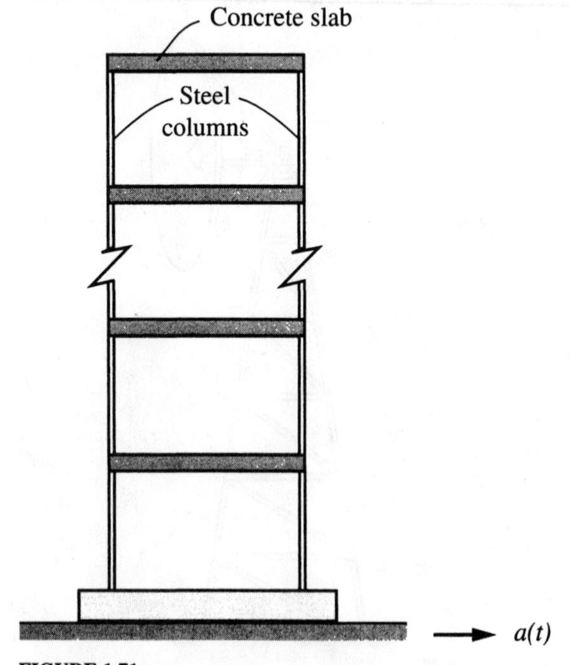

FIGURE 1.71
n-story building subjected to earthquake excitation

1.27. Devise a model for a motorcycle and rider (Fig. 1.72).

FIGURE 1.72
Motorcycle and rider

1.28. Derive the differential equation of motion for the system of Fig. 1.62 and verify the expression for the equivalent spring constant derived in Problem 1.13. Hint: Write an equation for the rotation θ about O and an equation for the translation x of mass m and then eliminate θ to obtain a single equation.

1.29. A mass m is suspended on a massless beam of uniform bending stiffness EI, as shown in Fig. 1.73. Derive the differential equation of motion for the system.

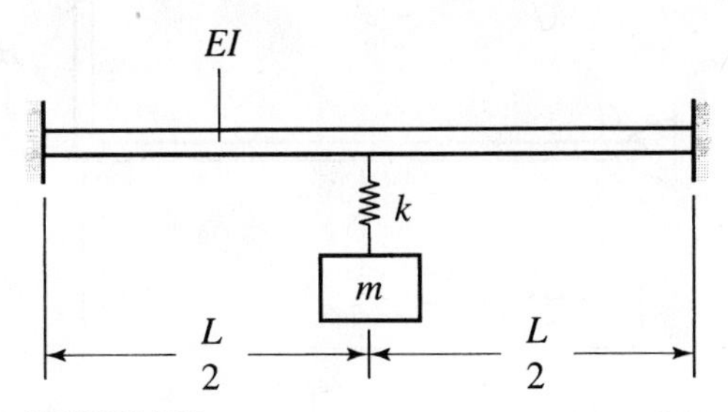

FIGURE 1.73
Mass suspended on a massless beam through a spring

1.30. A mass m is connected to a spring of stiffness k through a string wrapping around a rigid pulley of radius R and mass moment of inertia I (Fig. 1.74). Derive the equation of motion for the system.

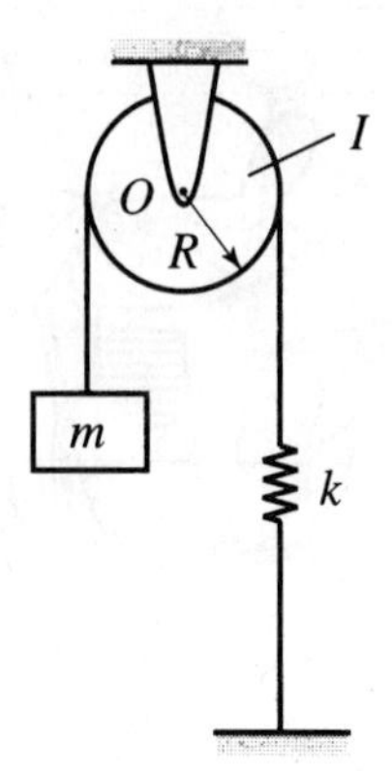

FIGURE 1.74
Mass connected to a spring through a pulley

1.31. An L-shaped massless rigid member with a mass m at the tip and supported by a spring of stiffness k is hinged at point O, as shown in Fig. 1.75. It is required to:

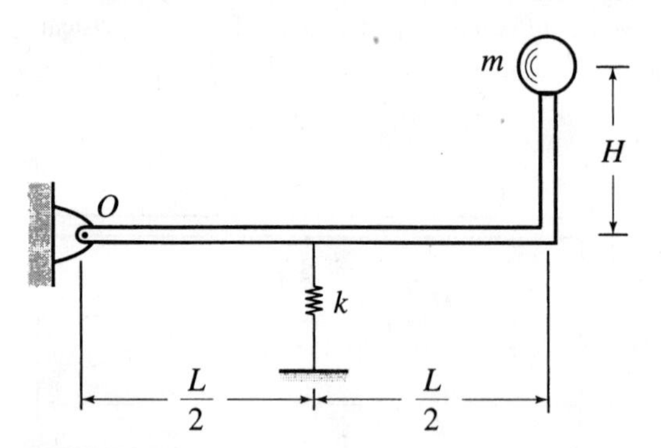

FIGURE 1.75
Mass supported by a spring through an L-shaped rigid member

(a) Derive the equation for the angular motion $\theta(t)$ about O.
(b) Determine the equilibrium angle θ_e.
(c) Derive the differential equation for small angular motions $\theta_1(t)$ about θ_e.
(d) Determine the height H for which the system becomes unstable.

1.32. An inverted pendulum is supported by a linear spring, as shown in Fig. 1.76. It is required to:

(a) Derive the equation for the angular motion $\theta(t)$ about O.
(b) Determine the equilibrium positions.

(c) Derive the differential equation for small angular motions $\theta_1(t)$ about θ_e for each equilibrium position.
(d) Determine the stability nature of each equilibrium position.

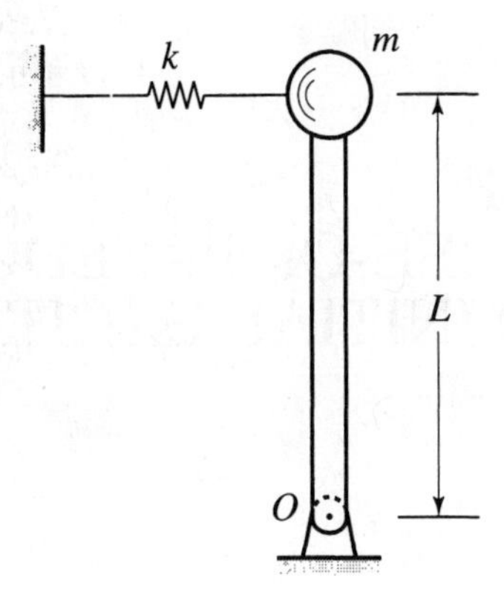

FIGURE 1.76
Inverted pendulum supported by a spring

1.33. A uniform rigid bar of total mass m and length L is hinged at point O to a shaft rotating with the constant angular velocity Ω about a vertical axis, as shown in Fig. 1.77. It is required to:
(a) Derive the equation for the angular motion $\theta(t)$ about O.
(b) Determine the equilibrium positions, assuming that the angle θ can range from 0 to π.
(c) Derive the differential equation for small angular motions $\theta_1(t)$ about θ_e for each equilibrium position.
(d) Determine the stability nature of each equilibrium position.

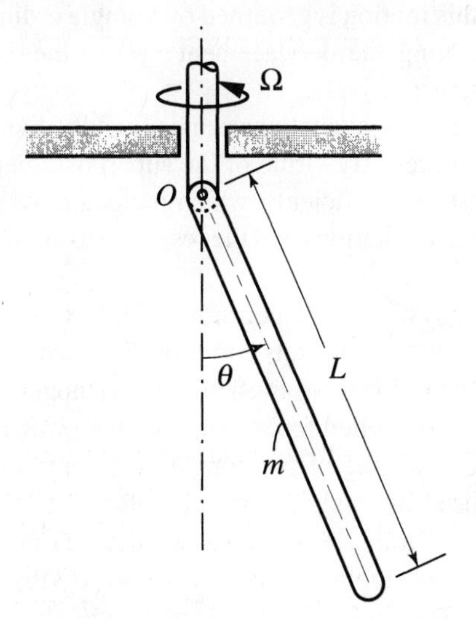

FIGURE 1.77
Rigid bar hinged to a rotating shaft

CHAPTER 2

RESPONSE OF SINGLE-DEGREE-OF-FREEDOM SYSTEMS TO INITIAL EXCITATIONS

The most basic mechanical system is the *single-degree-of-freedom system*, which is characterized by the fact that its motion is described by a single variable, or coordinate. As shown in Sec. 1.10, this motion is governed by a single ordinary differential equation, such as Eq. (1.128), relating the displacement $x(t)$ to the force $F(t)$, referred to as response and excitation, respectively.

As indicated in Sec. 1.11, excitations can be broadly divided into two types, initial excitations and applied forces. By virtue of the superposition principle (Sec. 1.12), for linear systems with constant coefficients, which include most systems discussed in this text, the response to initial excitations and the response to applied forces can be obtained separately and combined linearly.

The vibration of a system in response to initial excitations, consisting of initial displacements and/or initial velocities, is commonly known as free vibration. To obtain the response to initial excitations, we must solve a homogeneous ordinary differential equation, i.e., one with zero applied forces, such as that given by Eq. (1.126). We study the free vibration of single-degree-of-freedom systems in this chapter.

The vibration caused by applied forces is referred to as forced vibration, and it represents a problem considerably wider in scope than the free vibration problem, which is due to the large variety of applied forces. The forced vibration of single-degree-of-freedom systems is discussed in Chs. 3 and 4.

2.1 UNDAMPED SINGLE-DEGREE-OF-FREEDOM SYSTEMS. HARMONIC OSCILLATOR

We consider the free vibration of an undamped single-degree-of-freedom system of the type shown in Fig. 2.1. The system represents a special case of the model of the washing machine of Fig. 1.28b; its equation of motion was derived in Sec. 1.10. Hence, letting $M = m$ and $c = 0$ in Eq. (1.126), the equation of motion for the free vibration of an undamped single-degree-of-freedom system is

$$m\ddot{x}(t) + kx(t) = 0 \tag{2.1}$$

where $x(t)$ is the displacement from the static equilibrium position, m the mass and k the spring constant. Dividing through by m, Eq. (2.1) can be written in the form

$$\ddot{x}(t) + \omega_n^2 x(t) = 0 \tag{2.2}$$

in which

$$\omega_n = \sqrt{k/m} \tag{2.3}$$

is a real constant. The solution of Eq. (2.2) is subject to the initial conditions

$$x(0) = x_0, \quad \dot{x}(0) = v_0 \tag{2.4}$$

where x_0 and v_0 are the initial displacement and initial velocity, respectively.

Equation (2.1), or Eq. (2.2), represents a system with constant coefficients of the type studied in Sec. 1.13. Its solution has the exponential form

$$x(t) = Ae^{st} \tag{2.5}$$

Inserting Eq. (2.5) into Eq. (2.2) and dividing through by Ae^{st}, we obtain the *characteristic equation*

$$s^2 + \omega_n^2 = 0 \tag{2.6}$$

which has the two pure imaginary complex conjugate roots

$$\begin{matrix} s_1 \\ s_2 \end{matrix} = \pm i\omega_n \tag{2.7}$$

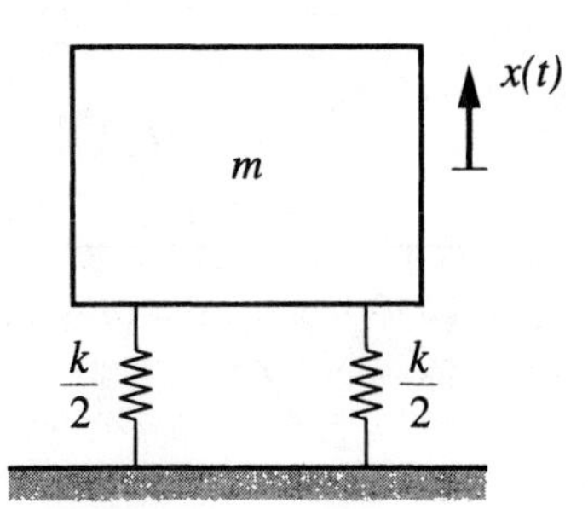

FIGURE 2.1
Force-free undamped system

where $i = \sqrt{-1}$. Introducing Eq. (2.7) in Eq. (2.5), the general solution of Eq. (2.2) can be written as

$$x(t) = A_1 e^{i\omega_n t} + A_2 e^{-i\omega_n t} \tag{2.8}$$

in which A_1 and A_2 are constants of integration, both complex quantities. Because $x(t)$ must be real and $e^{-i\omega_n t}$ is the complex conjugate of $e^{i\omega_n t}$, it follows that A_2 is the complex conjugate $\overline{A}_1$ of A_1. But, any complex number can be expressed as the product of its magnitude multiplied by an exponential with pure imaginary exponent. For convenience, we use the notation

$$A_1 = \frac{C}{2} e^{-i\phi}, \quad A_2 = \overline{A}_1 = \frac{C}{2} e^{i\phi} \tag{2.9}$$

where C and ϕ are real constants. Inserting Eqs. (2.9) into Eq. (2.8) and recalling that $e^{i\alpha} + e^{-i\alpha} = 2\cos\alpha$, we obtain the response

$$x(t) = \frac{C}{2}\left[e^{i(\omega_n t - \phi)} + e^{-i(\omega_n t - \phi)}\right] = C\cos(\omega_n t - \phi) \tag{2.10}$$

so that now the constants of integration are C and ϕ.

Equation (2.10) represents harmonic oscillation, for which reason a system described by Eq. (2.2) is called a *harmonic oscillator*. In Sec. 1.10, we discussed the nature of harmonic functions as excitations. Whereas many of the concepts and definitions remain the same, a discussion of harmonic functions as response to initial excitations is in order. To this end, we plot the response given by Eq. (2.10), as shown in Fig. 2.2. There are three quantities defining the response, the *amplitude* C, the *phase angle* ϕ and the *frequency* ω_n, the first two depending on external factors, namely, the initial excitations, and the third depending on internal factors, namely, the system parameters. It follows that the amplitude and phase angle of the response differ from case to case, according to the initial conditions. On the other hand, for a given system, the frequency of the response is a characteristic of the system that stays always the same, independently of the initial excitations, as can be concluded from Eq. (2.3). For this reason, ω_n is called the *natural frequency* of the harmonic oscillator.

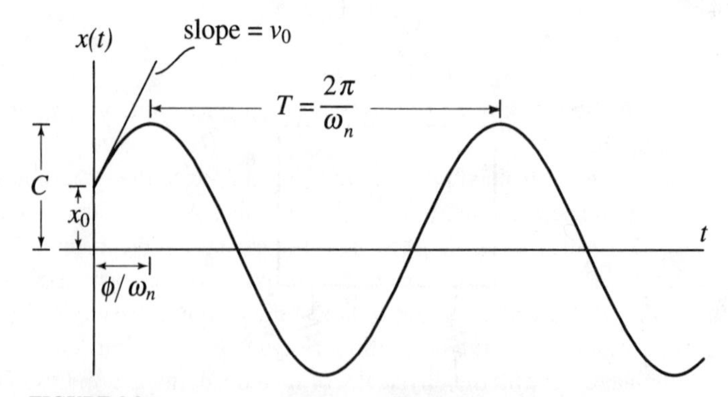

FIGURE 2.2
Response of a harmonic oscillator to initial excitations

To determine the constants of integration C and ϕ, we must insist that solution (2.10) match the initial conditions, Eqs. (2.4). Hence, we write

$$x(0) = x_0 = C\cos\phi,\ \dot{x}(0) = v_0 = \omega_n C\sin\phi \tag{2.11}$$

Equations (2.11) can be solved for the amplitude and phase angle, with the result

$$C = \sqrt{x_0^2 + \left(\frac{v_0}{\omega_n}\right)^2},\quad \phi = \tan^{-1}\frac{v_0}{x_0\omega_n} \tag{2.12}$$

Equation (2.10) defines the harmonic oscillation fully for given initial conditions x_0 and v_0 and natural frequency ω_n, where the dependence on x_0 and v_0 is only implicit, through Eqs. (2.12). An explicit expression can be obtained by recalling from trigonometry that $\cos(\alpha-\beta) = \cos\alpha\cos\beta + \sin\alpha\sin\beta$, considering Eqs. (2.11) and writing

$$\begin{aligned} x(t) &= C\cos(\omega_n t - \phi) = C(\cos\omega_n t\cos\phi + \sin\omega_n t\sin\phi) \\ &= x_0\cos\omega_n t + \frac{v_0}{\omega_n}\sin\omega_n t \end{aligned} \tag{2.13}$$

Although Eq. (2.13) has a simpler appearance than Eq. (2.10), it represents the sum of two trigonometric functions and is not so easy to plot as Eq. (2.10).

The amplitude is identified in Fig. 2.2 as the maximum displacement, and the phase angle as a measure of the amount of time necessary for the displacement to reach its peak. Moreover, we recognize that the slope of the curve at $t = 0$ represents the initial velocity v_0. Also identified in Fig. 2.2 is the *period* T, defined as the time necessary for the system to complete one vibration cycle, or as the time between two consecutive peaks. The period of oscillation is also a characteristic of the system, in the sense that it is determined by the system parameters and not by external factors. It is related to the natural frequency by

$$T = \frac{2\pi}{\omega_n} \tag{2.14}$$

where T has units of seconds (s) and ω_n has units of radians per second (rad/s). Note that the natural frequency can also be defined as the reciprocal of the period, or

$$f_n = \frac{1}{T} \tag{2.15}$$

in which case it has units of cycles per second (cps), where one cycle per second is known as one hertz (Hz).

As can be concluded from Eq. (2.10) and Fig. 2.2, once the system has been set in motion by the initial excitations, the oscillation will continue indefinitely with the same amplitude. This is because the system neither dissipates nor gains energy, so that a harmonic oscillator is a *conservative system*. It corresponds to the positive imaginary axis in the s-plane of Fig. 1.45 and the positive b-axis of the parameter plane of Fig. 1.46, and it represents stable motion and pure oscillation, respectively. The harmonic oscillator is an idealized system at odds with the physical world. Indeed, in reality physical systems, even those assumed to be conservative, tend to dissipate energy to some degree, so that free vibration comes eventually to rest. Still, the concepts of harmonic oscillator and natural frequency are very useful and can be justified when the rate of energy dissipation is so small that it takes many oscillation cycles before a reduction in the amplitude can be discerned.

A large variety of dynamical systems behave like harmonic oscillators, quite often when restricted to small motions. A typical example is a simple pendulum oscillating about the trivial equilibrium $\theta = 0$, such as that of Example 1.9. For small motions, the differential equation is

$$\ddot{\theta} + \omega_n^2 \theta = 0 \tag{2.16}$$

which describes a harmonic oscillator with the natural frequency

$$\omega_n = \sqrt{\frac{g}{L}} \tag{2.17}$$

We note that Eq. (2.16) represents the linearized version of the nonlinear equation

$$\ddot{\theta} + \frac{g}{L} \sin\theta = 0 \tag{2.18}$$

where the linearization is valid as long as $\sin\theta \simeq \theta$, which is approximately true for surprisingly large values of θ. For example, for $\theta = 30° = 0.5236$, $\sin\theta = \sin 30° = 0.5000$, there is less than 5% error in using θ instead of $\sin\theta$, and for $\theta = 20°$ the error drops to about 2%.

In the following, we illustrate the variety of systems that can be modeled as harmonic oscillators by means of several examples.

Example 2.1. Show that the water tank of Fig. 2.3a can be modeled as a harmonic oscillator and determine the natural frequency on the assumption that the tank and water act as a rigid body and that the support column is a massless uniform cantilever beam of bending stiffness EI.

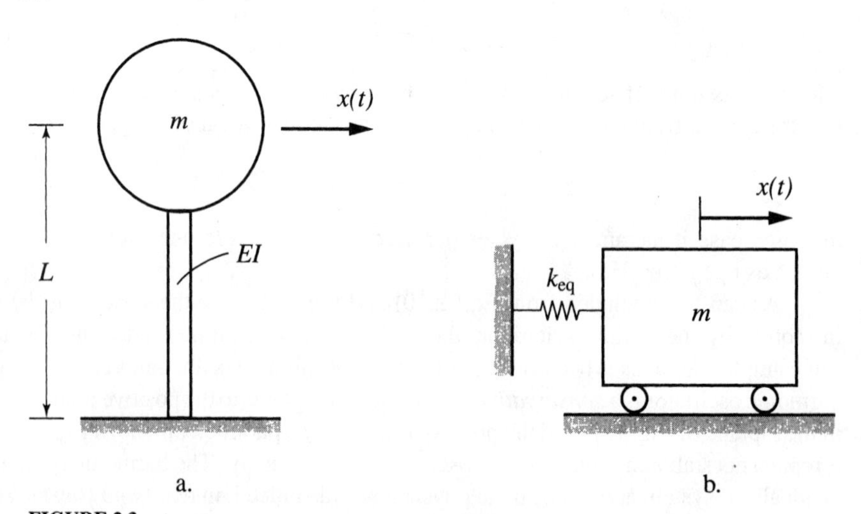

FIGURE 2.3
a. Model of a water tank, b. Equivalent single-degree-of-freedom system

The stated assumptions permit us to treat the tank as a single-degree-of-freedom system, as shown in Fig. 2.3b. From Sec. 1.8, the support column can be regarded as an equivalent spring with the spring constant given by Eq. (1.114), or

$$k_{\text{eq}} = \frac{3EI}{L^3} \tag{a}$$

Hence using the analogy with the system of Fig. 2.1 and Eq. (2.1), the equation of motion is

$$m\ddot{x}(t) + k_{\text{eq}}x(t) = 0 \tag{b}$$

Example 2.2. A manometer is a device for measuring gas or liquid pressure. Derive the differential equation of motion for the U-tube manometer of Fig. 2.4 and obtain the period of oscillation. Denote the density of the liquid by ρ, the cross-sectional area of the manometer by A and the total length of the column of liquid by L.

Assuming that the viscosity of the liquid is negligibly small, the system is conservative, so that we can use Eq. (1.26) to derive the equation of motion. To this end, we observe that the potential energy is due to the weight dislocated from equilibrium. Hence, regarding the weight above the reference line as positive and the one below the reference line as negative, we can write the potential energy

$$V = \rho g A x \frac{x}{2} + (-\rho g A x)\left(-\frac{x}{2}\right) = \rho g A x^2 \tag{a}$$

where g is the gravitational constant and A the cross-sectional area of the tube. Moreover, the kinetic energy is simply

$$T = \frac{1}{2}m\dot{x}^2 = \frac{1}{2}\rho A L \dot{x}^2 \tag{b}$$

so that the total energy is

$$E = T + V = \frac{1}{2}\rho A L \dot{x}^2 + \rho g A x^2 = \rho A L\left(\frac{1}{2}\dot{x}^2 + \frac{g}{L}x^2\right) \tag{c}$$

Hence, using Eq. (1.26), we obtain

$$\dot{E} = \rho A L(\dot{x}\ddot{x} + \frac{2g}{L}x\dot{x}) = \rho A L \dot{x}\left(\ddot{x} + \frac{2g}{L}x\right) = 0 \tag{d}$$

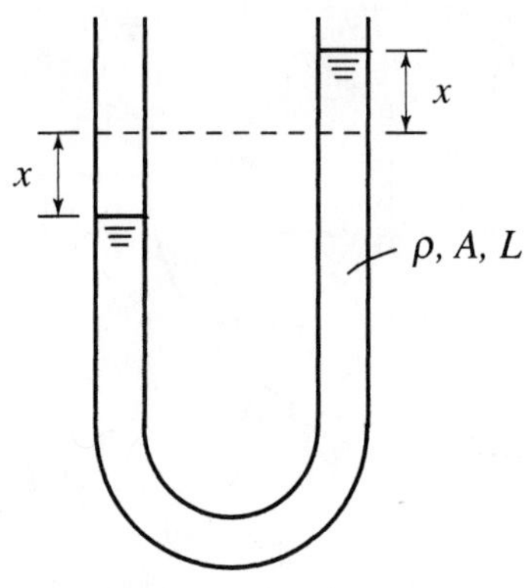

FIGURE 2.4
U-tube manometer

from which we obtain the differential equation of motion

$$\ddot{x} + \frac{2g}{L}x = 0 \tag{e}$$

It follows that the manometer behaves like a harmonic oscillator with the natural frequency

$$\omega_n = \sqrt{\frac{2g}{L}} \tag{f}$$

and the period

$$T = \frac{2\pi}{\omega_n} = 2\pi\sqrt{\frac{L}{2g}} \tag{g}$$

No confusion should arise from the fact that we used the same notation for the kinetic energy and the period.

Example 2.3. A uniform rigid disk of radius r rolls without slipping inside a circular track of radius R, as shown in Fig. 2.5a. Derive the equation of motion for arbitrarily large angles θ. Then, show that in a small neighborhood of the trivial equilibrium, $\theta = 0$, the system behaves like a harmonic oscillator, and determine the natural frequency of oscillation.

To derive the equation of motion, we refer to the free-body diagram of Fig. 2.5b, as well as to Eqs. (1.63) and (1.64) and write the force equation in the transverse direction

$$F_\theta = -F - mg\sin\theta = ma_{C\theta} = m(R-r)\ddot{\theta} \tag{a}$$

and the moment equation about the mass center

$$M_C = Fr = I_C\alpha = \frac{mr^2}{2}\ddot{\phi} \tag{b}$$

respectively. At this point, we are faced with the problem of having two equations and three unknowns, F, θ and ϕ, and there is only one degree of freedom. But, the two equations can be combined into one by eliminating the reaction force F. Another unknown can be eliminated by observing that the angular velocities $\dot{\theta}$ and $\dot{\phi}$ are related, so that one of them is redundant. To this end, we calculate the velocity v_C of the mass center of the disk in two different ways, first by regarding point C as moving on a circular path about point O, and then by recognizing that point A on the disk is instantaneously at rest. Hence, we write

$$v_C = (R-r)\dot{\theta} = r\dot{\phi} \tag{c}$$

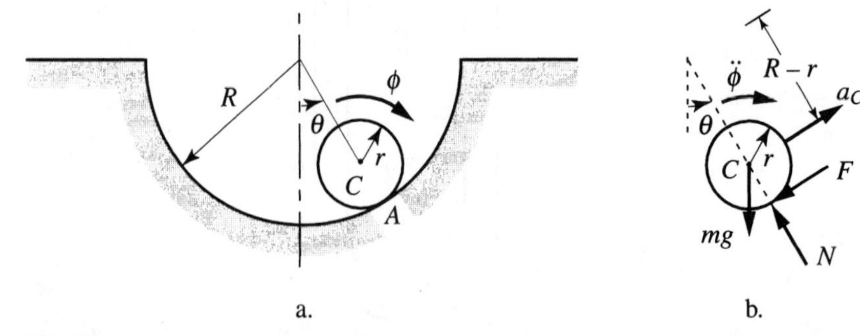

FIGURE 2.5
a. Disk rolling inside a circular track, b. Free-body diagram

Then, combining Eqs. (a), (b) and (c), we obtain the desired equation of motion

$$\begin{aligned} m(R-r)\ddot{\theta}+F+mg\sin\theta &= m(R-r)\ddot{\theta}+\frac{mr}{2}\ddot{\phi}+mg\sin\theta \\ &= \frac{3}{2}m(R-r)\ddot{\theta}+mg\sin\theta=0 \end{aligned} \tag{d}$$

which can be rewritten in the form

$$\ddot{\theta}+\frac{2g}{3(R-r)}\sin\theta=0 \tag{e}$$

In a small neighborhood of $\theta=0$, $\sin\theta$ can be approximated by θ, so that Eq. (e) reduces to

$$\ddot{\theta}+\frac{2g}{3(R-r)}\theta=0 \tag{f}$$

which represents the equation of a harmonic oscillator with the natural frequency

$$\omega_n=\sqrt{\frac{2g}{3(R-r)}} \tag{g}$$

2.2 VISCOUSLY DAMPED SINGLE-DEGREE-OF-FREEDOM SYSTEMS

A typical model of a viscously damped single-degree-of-freedom system in free vibration about the static equilibrium is as shown in Fig. 2.6. We encountered this system in Sec. 1.9 in connection with the model of the washing machine depicted in Fig. 1.28b. Then, in Sec. 1.10 we derived the equation of motion, Eq. (1.126), which we rewrite here in the form

$$m\ddot{x}(t)+c\dot{x}(t)+kx(t)=0 \tag{2.19}$$

where m is the mass, c the coefficient of viscous damping and k the spring constant. It is convenient to divide Eq. (2.19) by m and write

$$\ddot{x}(t)+2\zeta\omega_n\dot{x}(t)+\omega_n^2x(t)=0 \tag{2.20}$$

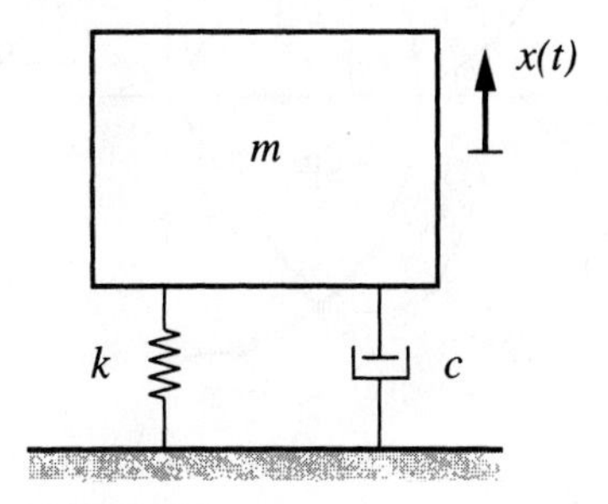

FIGURE 2.6
Force-free, viscously damped system

in which ω_n is the natural frequency of undamped oscillation having the form given by Eq. (2.3) and

$$\zeta = \frac{c}{2m\omega_n} \tag{2.21}$$

is a nondimensional quantity known as the *viscous damping factor*. The solution of Eq. (2.20) must satisfy the initial conditions

$$x(0) = x_0, \; \dot{x}(0) = v_0 \tag{2.22}$$

As in the case of undamped systems, Eq. (2.20) has the exponential solution

$$x(t) = Ae^{st} \tag{2.23}$$

so that, following the same steps as in Sec. 2.1, it is not difficult to verify that the exponent s must satisfy the *characteristic equation*

$$s^2 + 2\zeta\omega_n s + \omega_n^2 = 0 \tag{2.24}$$

Equation (2.24) is quadratic in s and has the roots

$$\begin{matrix} s_1 \\ s_2 \end{matrix} = -\zeta\omega_n \pm \sqrt{\zeta^2 - 1}\,\omega_n \tag{2.25}$$

As discussed in Sec. 1.13, the nature of the motion about equilibrium depends on the roots s_1 and s_2, which in turn depend on the value of the parameter ζ. As in Sec. 1.13, we can use the s-plane to display this dependence, except that here we can be more explicit. From Eq. (2.21), we recognize that $\zeta \geq 0$, so that we can plot the locus of the roots s_1 and s_2 as functions of the parameter ζ and for a given value of ω_n, as shown in Fig. 2.7. This s-plane diagram provides a complete picture of the manner in which the roots s_1 and s_2 change with ζ. Even more insight into the system behavior can be gained by connecting

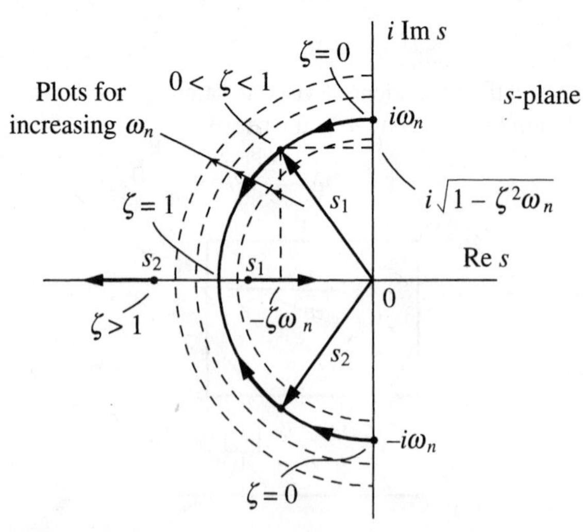

FIGURE 2.7
Root-locus diagram for a viscously damped system

ζ and ω_n to the parameter plane of Fig. 1.46. Indeed, contrasting Eqs. (1.158) and (2.24), we conclude that the parameter plane in terms of ζ and ω_n is $2\zeta\omega_n$ versus ω_n^2. In our particular case, the interest lies only in the first quadrant, as shown in Fig. 2.8, so that the system is guaranteed to be stable and the only question to be addressed is how the system responds.

For $\zeta = 0$, the roots s_1 and s_2 correspond to the points $i\omega_n$ and $-i\omega_n$ on the imaginary axis of the s-plane, Fig. 2.7, and to all the horizontal axis of the parameter plane of Fig. 2.8. Clearly, in this case the motion represents *harmonic oscillation* with the natural frequency ω_n. This case was discussed in detail in Sec. 2.1.

For $0 < \zeta < 1$, the roots s_1 and s_2 are complex conjugates and they correspond to pairs of points in the s-plane symmetrically located with respect to the real axis. As ζ changes, the pairs of points move on a semicircle of radius ω_n, as can be seen from Fig. 2.7. The same points correspond to a region in the parameter plane of Fig. 2.8 between the horizontal axis and the parabola $\zeta = 1$. From Fig. 2.8, we conclude that in this case the motion represents *oscillatory decay*. The case in which $0 < \zeta < 1$ is commonly referred to as *underdamping*.

For $\zeta = 1$, the roots s_1 and s_2 coalesce at the point $-\omega_n$ on the real axis of the s-plane, as shown in Fig. 2.7, and they correspond to the parabola $\zeta = 1$ in the parameter plane of Fig. 2.8. This case is known as *critical damping* and the motion represents *aperiodic decay*.

Finally, for $\zeta > 1$, both roots, s_1 and s_2, are located on the negative real axis of the s-plane, with s_1 between the point $-\omega_n$ and the origin and with s_2 to the left of $-\omega_n$, as can be seen from Fig. 2.8. As ζ increases, s_1 tends to 0 and s_2 to $-\infty$. This case corresponds to the region between the parabola $\zeta = 1$ and the vertical axis in the parameter plane of Fig. 2.8, and is characterized by *aperiodic decay*. The case is referred to as *overdamping*. In retrospect, we must conclude that the term "critical damping" is a misnomer, as there is nothing critical about it. Indeed, it merely corresponds to the lowest value of ζ for which the motion decays aperiodically.

In the preceding analysis, the emphasis has been on the effect of the viscous damping factor ζ on the nature of motion. This is fully justified by the fact that the natural frequency does not affect the type of response. In this regard, we should mention that the parameter plane of Fig. 2.8 contains more information than the s-plane of

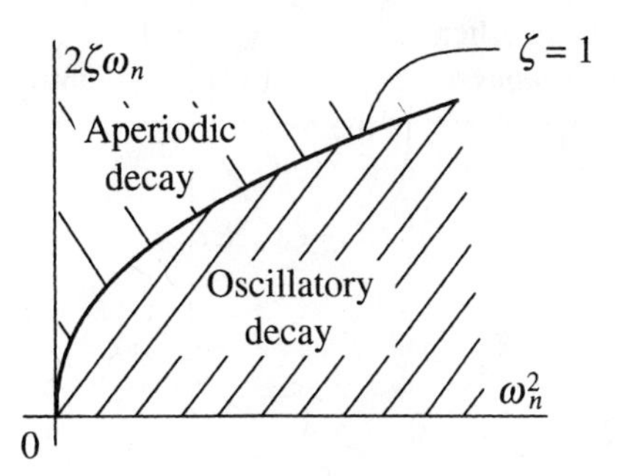

FIGURE 2.8
Parameter plane for a viscously damped system

Fig. 2.7, as the parameter plane is in terms of both ζ and ω_n and the root locus in the s-plane is a plot of the roots s_1 and s_2 as functions of ζ alone, with ω_n being regarded as a given constant. To include the effect of ω_n in the s-plane, we must plot the root locus for different values of ω_n; some of these plots are shown in dashed line in Fig. 2.7. Hence, the complete root locus picture consists of the imaginary axis for $\zeta = 0$, concentric semicircles of increasing radius with increasing ω_n for $0 < \zeta < 1$ and the negative real axis for $\zeta \geq 1$.

The foregoing discussion was qualitative in nature, in the sense that we could only establish how the parameters, and in particular the viscous damping factor, dictate the type of motion. To obtain a more quantitative picture, it is necessary to solve the equation of motion, Eq. (2.20), for the response. Here too, the parameter ζ is the dominant factor, as the nature of the solution and the type of motion change with ζ. Before considering the different types of motion, however, we carry out a number of steps toward the solution common to all types. To this end, we recognize that the general solution, Eq. (2.23), becomes

$$x(t) = A_1 e^{s_1 t} + A_2 e^{s_2 t} \tag{2.26}$$

where A_1 and A_2 are constants of integration depending on the initial conditions. To determine these constants, we insert Eq. (2.26) into Eqs. (2.22) and write

$$\begin{aligned} x(0) &= A_1 + A_2 = x_0 \\ \dot{x}(0) &= s_1 A_1 + s_2 A_2 = v_0 \end{aligned} \tag{2.27}$$

which can be solved for A_1 and A_2, with the result

$$A_1 = \frac{-s_2 x_0 + v_0}{s_1 - s_2}, \quad A_2 = \frac{s_1 x_0 - v_0}{s_1 - s_2} \tag{2.28}$$

Hence, substituting these values in Eq. (2.26), we obtain the solution

$$x(t) = \frac{-s_2 x_0 + v_0}{s_1 - s_2} e^{s_1 t} + \frac{s_1 x_0 - v_0}{s_1 - s_2} e^{s_2 t} \tag{2.29}$$

which is valid for all cases. In the following, we specialize this solution to the various types of damping identified earlier.

For *underdamped systems*, $0 < \zeta < 1$, it is convenient to express the roots of the characteristic equation, Eqs. (2.25), in the form

$$\begin{matrix} s_1 \\ s_2 \end{matrix} = -\zeta\omega_n \pm i\omega_d \tag{2.30}$$

where

$$\omega_d = \sqrt{1 - \zeta^2}\, \omega_n \tag{2.31}$$

is known as the *frequency of damped vibration*, for reasons that will become obvious shortly. Inserting Eqs. (2.30) into Eq. (2.29) and recalling from complex analysis that

$e^{i\alpha} - e^{-i\alpha} = 2i \sin\alpha$ and $e^{i\alpha} + e^{-i\alpha} = 2\cos\alpha$, we obtain

$$\begin{aligned}
x(t) =& \frac{e^{-\zeta\omega_n t}}{2i\omega_d}\left\{[-(-\zeta\omega_n - i\omega_d)x_0 + v_0]e^{i\omega_d t} + [(-\zeta\omega_n + i\omega_d)x_0 - v_0]e^{-i\omega_d t}\right\} \\
=& \frac{e^{-\zeta\omega_n t}}{2i\omega_d}\left[(\zeta\omega_n x_0 + v_0)(e^{i\omega_d t} - e^{-i\omega_d t}) + i\omega_d x_0(e^{i\omega_d t} + e^{-i\omega_d t})\right] \\
=& e^{-\zeta\omega_n t}\left(\frac{\zeta\omega_n x_0 + v_0}{\omega_d}\sin\omega_d t + x_0\cos\omega_d t\right) \\
=& Ce^{-\zeta\omega_n t}\cos(\omega_d t - \phi)
\end{aligned} \tag{2.32}$$

where C and ϕ represent the amplitude and phase angle of the response, respectively, having the values

$$C = \sqrt{x_0^2 + \left(\frac{\zeta\omega_n x_0 + v_0}{\omega_d}\right)^2}, \quad \phi = \tan^{-1}\frac{\zeta\omega_n x_0 + v_0}{\omega_d x_0} \tag{2.33}$$

To plot the response $x(t)$ as a function of time, it is convenient to regard Eq. (2.32) as the product of two factors, $Ce^{-\zeta\omega_n t}$ and $\cos(\omega_d t - \phi)$. The factor $Ce^{-\zeta\omega_n t}$ represents an exponentially decaying function and $\cos(\omega_d t - \phi)$ is a harmonic function similar to that depicted in Fig. 2.2, except that the amplitude is 1 and the frequency is ω_d instead of ω_n. Hence, the product of the two can be interpreted as an oscillation with a decaying amplitude. This interpretation is consistent with our earlier qualitative analysis of the motion, and can be used for plotting purposes. As shown in Fig. 2.9, the time-varying amplitude provides the exponentially narrowing envelope $\pm Ce^{-\zeta\omega_n t}$ modulating the harmonic function $\cos(\omega_d t - \phi)$. This also explains the term "frequency of damped vibration" for ω_d.

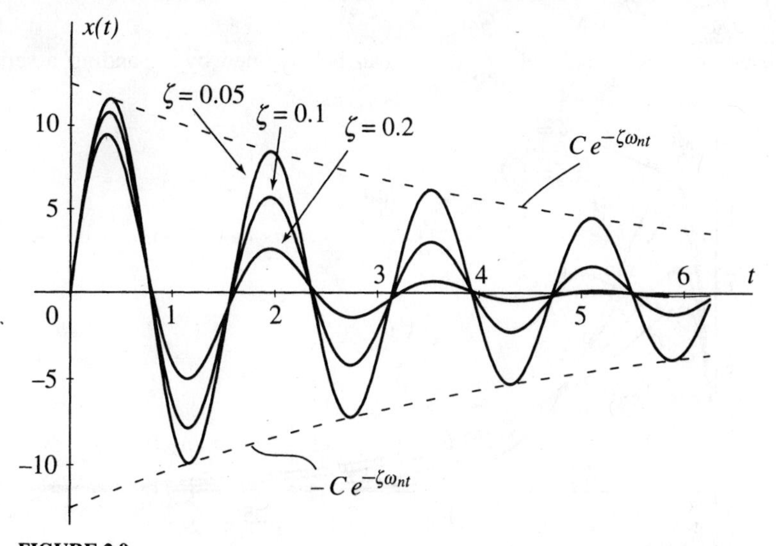

FIGURE 2.9
Response of an underdamped ($0 < \zeta < 1$) system

Next, we consider the case of *overdamping*, $\zeta > 1$, in which case the roots s_1 and s_2 are as given by Eqs. (2.25), and they are both real and negative. Hence, inserting Eqs. (2.25) into Eq. (2.29) and recognizing that $e^{\alpha} - e^{-\alpha} = 2\sinh\alpha$ and $e^{\alpha} + e^{-\alpha} = 2\cosh\alpha$, we obtain the response

$$
\begin{aligned}
x(t) =& \frac{e^{-\zeta\omega_n t}}{2\sqrt{\zeta^2-1}\,\omega_n}\left\{\left[-\left(-\zeta\omega_n - \sqrt{\zeta^2-1}\,\omega_n\right)x_0 + v_0\right]e^{\sqrt{\zeta^2-1}\omega_n t}\right.\\
&\left.+\left[\left(-\zeta\omega_n + \sqrt{\zeta^2-1}\,\omega_n\right)x_0 - v_0\right]e^{-\sqrt{\zeta^2-1}\omega_n t}\right\}\\
=& \frac{e^{-\zeta\omega_n t}}{2\sqrt{\zeta^2-1}\,\omega_n}\left[(\zeta\omega_n x_0 + v_0)\left(e^{\sqrt{\zeta^2-1}\omega_n t} - e^{-\sqrt{\zeta^2-1}\omega_n t}\right)\right.\\
&\left.+\sqrt{\zeta^2-1}\,\omega_n x_0\left(e^{\sqrt{\zeta^2-1}\omega_n t} + e^{-\sqrt{\zeta^2-1}\omega_n t}\right)\right]\\
=& e^{-\zeta\omega_n t}\left(\frac{\zeta\omega_n x_0 + v_0}{\sqrt{\zeta^2-1}\,\omega_n}\sinh\sqrt{\zeta^2-1}\,\omega_n t + x_0\cosh\sqrt{\zeta^2-1}\,\omega_n t\right)
\end{aligned}
\tag{2.34}
$$

which represents aperiodic decay. Figure 2.10 shows the response for several values of ζ for given ω_n, x_0 and v_0, from which we conclude that the peaks decrease in magnitude and the decay slows down as ζ increases.

Finally, we turn our attention to *critically damped systems*, $\zeta = 1$, characterized by the double root $s_1 = s_2 = -\omega_n$. In this case, the response can be obtained conveniently by the Laplace method (Appendix B). It is simpler, however, to obtain the response as a limiting case of Eq. (2.34) obtained by letting ζ approach 1. In particular, we observe that

$$
\lim_{\zeta\to 1}\frac{\sinh\sqrt{\zeta^2-1}\,\omega_n t}{\sqrt{\zeta^2-1}\,\omega_n} = t,\quad \lim_{\zeta\to 1}\cosh\sqrt{\zeta^2-1}\,\omega_n t = 1 \tag{2.35}
$$

The result on the left side of Eq. (2.35) can be obtained by expanding a series for

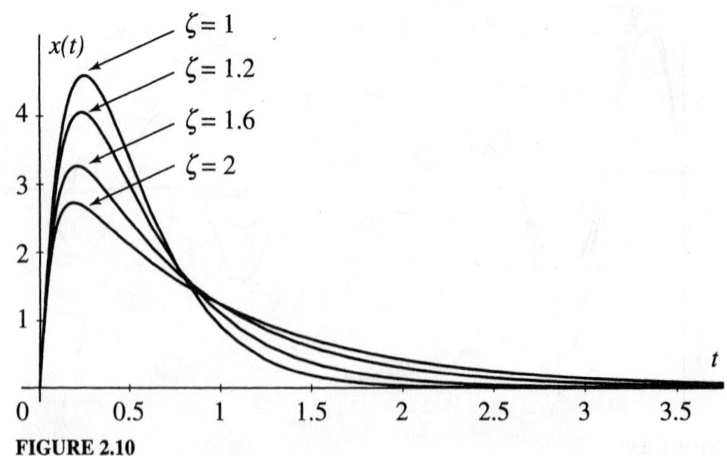

FIGURE 2.10
Response of a critically damped ($\zeta = 1$) and overdamped ($\zeta > 1$) system

$\sinh\sqrt{\zeta^2-1}\,\omega_n t$ or by using L'Hospital's rule.[1] In view of this, the response is simply

$$x(t) = [x_0 + (\omega_n x_0 + v_0)t]e^{-\omega_n t} \tag{2.36}$$

which represents *aperiodic decay*, as in the case of overdamping. The response is also plotted in Fig. 2.10, from which we observe that it reaches a higher peak and approaches zero faster than any overdamped system, indeed the highest peak and fastest decay of all systems for which $\zeta \geq 1$. Hence, critical damping is merely a borderline case, separating aperioding decay for overdamping from oscillatory decay for underdamping.

Example 2.4. The damped single-degree-of-freedom system of Fig. 2.6 is subjected to the initial conditions $x(0) = 0,\ \dot{x}(0) = v_0 = 50$ cm/s. Plot the response of the system for the following cases: i) $\omega_n = 4$ rad/s, $\zeta = 0.05,\ 0.1,\ 0.2$, ii) $\omega_n = 4$ rad/s, $\zeta = 1.2,\ 1.6,\ 2.0$ and iii) $\omega_n = 4$ rad/s, $\zeta = 1$.

In view of the fact that the initial displacement is zero and considering Eqs. (2.32) and (2.33), the response in case i, which represents underdamping, has the form

$$x(t) = \frac{v_0}{\omega_d} e^{-\zeta\omega_n t} \sin\omega_d t \tag{a}$$

where

$$\omega_d = \sqrt{1-\zeta^2}\,\omega_n \tag{b}$$

Using the given parameter values, the response is simply

$$x(t) = \frac{12.5}{\sqrt{1-\zeta^2}} e^{-4\zeta t} \sin 4\sqrt{1-\zeta^2}\,t \tag{c}$$

The responses for $\zeta = 0.05,\ 0.1$ and 0.2, together with the envelope for $\zeta = 0.05$, are plotted in Fig. 2.9. It is clear that the responses represent damped oscillation, with amplitudes decreasing as damping increases.

Using Eq. (2.34) and the given parameter values, the response in case ii, which represents overdamping is

$$x(t) = \frac{v_0}{\sqrt{\zeta^2-1}\,\omega_n} e^{-\zeta\omega_n t} \sinh\sqrt{\zeta^2-1}\,\omega_n t = \frac{12.5}{\sqrt{\zeta^2-1}} e^{-4\zeta t} \sinh 4\sqrt{\zeta^2-1}\,t \tag{d}$$

The responses for $\zeta = 1.2,\ 1.6$ and 2.0 are shown in Fig. 2.10; they represent aperiodic decay with peak amplitudes decreasing as damping increases. Note, however, that the rate of decay is slower for higher damping.

Finally, in case iii, the case of critical damping, we obtain from Eq. (2.36)

$$x(t) = v_0 t e^{-\omega_n t} = 50te^{-4t} \tag{e}$$

The response is also plotted in Fig. 2.10 and, as expected, it experiences the highest peak and the largest decay rate of all cases for which $\zeta \geq 1$.

---

[1] Pipes, L. A., *Applied Mathematics for Engineers and Physicists*, 2nd ed., McGraw-Hill, New York, 1958, p. 31.

2.3 MEASUREMENT OF DAMPING

The single-degree-of-freedom system described by Eq. (2.19) is defined by three parameters, the mass m, the coefficient of viscous damping c and the spring stiffness k. At times, the parameters are not known and it is necessary to measure them. The value of the mass can be obtained by simply weighing it and writing $m = W/g$, where W is the weight. On the other hand, the value of the spring stiffness can be obtained by pulling the spring with an increasing force, measuring the elongation and taking the slope of the force-elongation diagram. The same approach can be used to measure the coefficient of viscous damping, except that in this case it is necessary to measure the time rate of change of the elongation of a dashpot rather than the elongation of a spring. In general, however, damping is a much more complex phenomenon than generally assumed. This may be due to a variety of factors difficult to account for, such as the nature of the connection between individual components, air resistance, etc. The process commonly used to measure damping differs from the simple process mentioned above in two respects: it measures the damping factor ζ instead of the damping coefficient c and it uses the whole system instead of the damper alone.

A convenient measure of the amount of damping in a single-degree-of-freedom system is the drop in amplitude at the completion of one cycle of vibration. To illustrate the idea, we consider an experimental record of the displacement given by Fig. 2.11 and assume that the curve is representative of Eq. (2.32). Then, we let t_1 and t_2 be the times corresponding to the first and second peak, denote the associated peak displacements by x_1 and x_2, respectively, and form the ratio

$$\frac{x_1}{x_2} = \frac{x(t_1)}{x(t_2)} = \frac{Ce^{-\zeta\omega_n t_1}\cos(\omega_d t_1 - \phi)}{Ce^{-\zeta\omega_n t_2}\cos(\omega_d t_2 - \phi)} = \frac{e^{-\zeta\omega_n t_1}\cos(\omega_d t_1 - \phi)}{e^{-\zeta\omega_n t_2}\cos(\omega_d t_2 - \phi)} \tag{2.37}$$

But, because $t_2 = t_1 + T$, where $T = 2\pi/\omega_d$ is the period of damped oscillation, we can use Eq. (2.31) and obtain

$$\frac{x_1}{x_2} = \frac{e^{-\zeta\omega_n t_1}\cos(\omega_d t_1 - \phi)}{e^{-\zeta\omega_n (t_1+T)}\cos[\omega_d(t_1+T) - \phi]} = \frac{e^{\zeta\omega_n T}\cos(\omega_d t_1 - \phi)}{\cos(\omega_d t_1 - \phi + 2\pi)}$$

$$= e^{\zeta\omega_n T} = e^{2\pi\zeta\omega_n/\omega_d} = e^{2\pi\zeta/\sqrt{1-\zeta^2}} \tag{2.38}$$

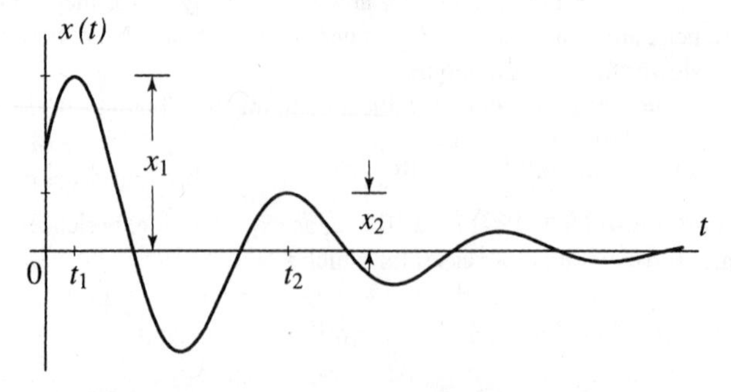

FIGURE 2.11
Experimental record of an underdamped system response

In view of the exponential form of the right side of Eq. (2.38), we take the natural logarithm on both sides and write

$$\delta = \ln\frac{x_1}{x_2} = \frac{2\pi\zeta}{\sqrt{1-\zeta^2}} \tag{2.39}$$

where δ is known as the *logarithmic decrement*. Equation (2.39) can be solved for the damping factor ζ, with the result

$$\zeta = \frac{\delta}{\sqrt{(2\pi)^2 + \delta^2}} \tag{2.40}$$

For small damping, such that $\zeta << 1$, Eq. (2.39) yields directly

$$\zeta \cong \frac{\delta}{2\pi} \tag{2.41}$$

The damping factor ζ can be determined, perhaps more accurately, by measuring the displacements at two different times separated by a given number of periods. Letting x_1 and x_{j+1} be the peak displacements corresponding to the times t_1 and $t_{j+1} = t_1 + jT$, where j is an integer, and recognizing that the value given by the extreme right of Eq. (2.38) is the same for the ratio of any two consecutive peak displacements, not only for x_1/x_2, we conclude that

$$\frac{x_1}{x_{j+1}} = \frac{x_1}{x_2}\frac{x_2}{x_3}\cdots\frac{x_j}{x_{j+1}} = \left(e^{2\pi\zeta/\sqrt{1-\zeta^2}}\right)^j = e^{j2\pi\zeta/\sqrt{1-\zeta^2}} \tag{2.42}$$

from which we obtain the logarithmic decrement

$$\delta = \frac{2\pi\zeta}{\sqrt{1-\zeta^2}} = \frac{1}{j}\ln\frac{x_1}{x_{j+1}} \tag{2.43}$$

which can be inserted in Eq. (2.40) or Eq. (2.41), as the case may be, and obtain the viscous damping factor ζ.

Equation (2.43) bases the calculation of the damping factor on two measurements alone, x_1 and x_{j+1}. Whereas this may be better than using measurements of two consecutive peaks, it can still lead to errors, particularly for small damping, when differences between peak amplitudes are difficult to measure accurately. In such cases, accuracy may be improved by using Eq. (2.43) and writing

$$\ln x_j = \ln x_1 - \delta(j-1), \quad j = 1, 2, \ldots, \ell \tag{2.44}$$

On semilog paper, a plot $\ln x_j$ versus j based on Eq. (2.44) has the form of a straight line with the slope $-\delta$. Because measurements are never exact, some of the intermediate points, $\ln x_2$, $\ln x_3, \ldots, \ln x_{\ell-1}$, may not fall exactly on this line, as shown in Fig. 2.12. Hence, rather than let the line pass through $\ln x_1$ and $\ln x_\ell$, a more accurate value of δ may be obtained by choosing the line so as to minimize the error, perhaps through a least squares fit. Such an error minimizing line is shown as a dashed line in Fig. 2.12.

The foregoing discussion was based on the assumption that damping is viscous, and hence proportional to the velocity. As indicated earlier in this section, some factors may cause this assumption to be only approximately true. It is in this case that the error-minimizing approach described in the preceding paragraph is most useful, as the net

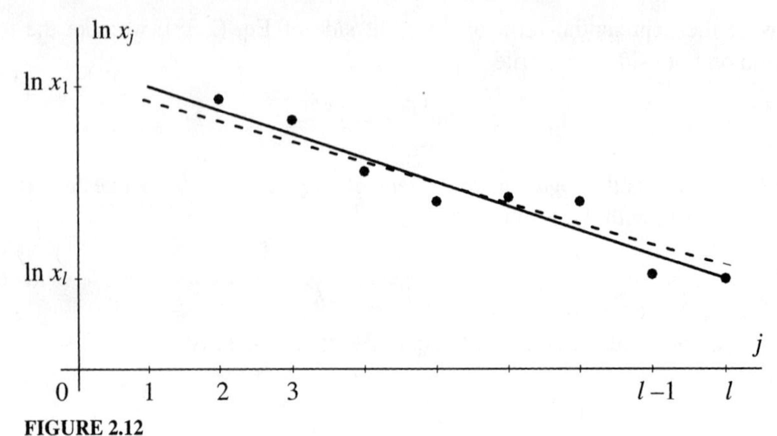

FIGURE 2.12
Natural logarithm of peak amplitude measurement versus measurement number

result is to obtain a "best" viscous damping model of some damping that is not exactly viscous.

Example 2.5. Measurement of the peak amplitudes of a vibrating lightly damped single-degree-of-freedom system has yielded the values $x_1, x_2, \ldots, x_6$ listed in the second column of Table 2.1. Develop a least squares approach based on Eq. (2.44) to calculate a corresponding viscous damping factor.

We base the approach on Fig. 2.12, which amounts to finding a straight line to fit the set of six points representing the natural logarithm of the measurements $x_1, x_2, \ldots, x_6$. We express the equation of the straight line in the discrete form

$$z_j = ay_j + b \tag{a}$$

and we note that z_j corresponds to $\ln x_j$, a to $-\delta$, y_j to $j-1$ and b to $\ln x_1$. Hence, the approach amounts to determining the constants a and b by minimizing the sum of the squares of the differences between the natural logarithm of the measured displacements and the straight line. To this end, we define the sum of the squares of the differences as the error

$$\epsilon = \sum_{j=1}^{6} (\ln x_j - z_j)^2 = \sum_{j=1}^{6} (\ln x_j - ay_j - b)^2 \tag{b}$$

Table 2.1

j	x_j	$\ln x_j$	y_j	$(\ln x_j)y_j$	z_j
1	11.5837	2.4496	0	0	2.4211
2	8.0406	2.0845	1	2.0845	2.1080
3	5.9489	1.7832	2	3.5664	1.7949
4	4.4039	1.4825	3	4.4475	1.4818
5	3.1677	1.1530	4	4.6120	1.1687
6	2.4032	0.8768	5	4.3840	0.8556

To minimize the error, we write

$$\frac{\partial \epsilon}{\partial a} = 2\sum_{j=1}^{6}(\ln x_j - ay_j - b)(-y_j) = 0$$
$$\frac{\partial \epsilon}{\partial b} = 2\sum_{j=1}^{6}(\ln x_j - ay_j - b)(-1) = 0 \tag{c}$$

Equations (c) represent two algebraic equations in the unknowns a and b, which can be rewritten in the more explicit form

$$\left(\sum_{j=1}^{6} y_j^2\right) a + \left(\sum_{j=1}^{6} y_j\right) b = \sum_{j=1}^{6}(\ln x_j) y_j$$
$$\left(\sum_{j=1}^{6} y_j\right) a + 6b = \sum_{j=1}^{6} \ln x_j \tag{d}$$

Inserting the values from Table 2.1 into Eqs. (d), we obtain

$$55a + 15b = 19.0944$$
$$15a + 6b = 9.8296 \tag{e}$$

which have the solution

$$a = -0.3131, \; b = 2.4211 \tag{f}$$

Hence, inserting these values into Eq. (a), we obtain the straight line

$$z_j = 2.4211 - 0.3131 y_j, \; j = 1, 2, \ldots, 6 \tag{g}$$

The values of z_j are listed in the last column of Table 2.1. The plots of $\ln x_j$ versus y_j and z_j versus y_j are given in Fig. 2.13 using values from Table 2.1.

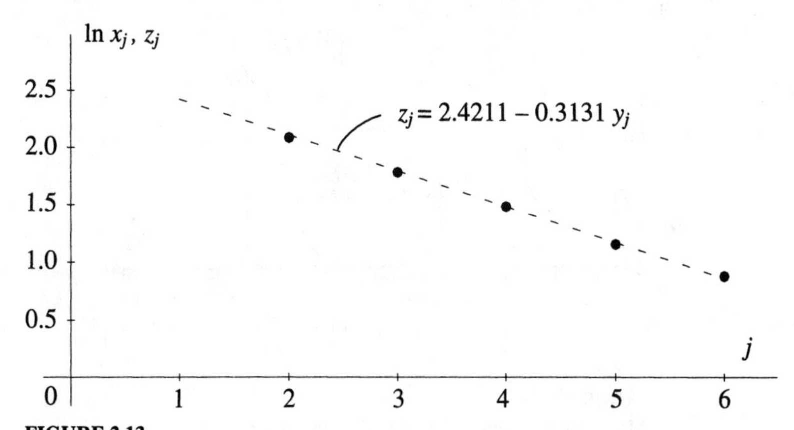

FIGURE 2.13
Determination of viscous damping factor by the least squares method

The conclusion is that damping that is only approximately viscous can be treated as if it were viscous with the logarithmic decrement

$$\delta = -a = 0.3131 \tag{h}$$

and with the viscous damping factor

$$\zeta = \frac{\delta}{\sqrt{(2\pi)^2 + \delta^2}} = \frac{0.3131}{\sqrt{(2\pi)^2 + 0.3131^2}} = 0.0498 \tag{i}$$

2.4 COULOMB DAMPING. DRY FRICTION

Coulomb damping arises when bodies slide on dry surfaces. For motion to begin, there must be a force acting upon the body that overcomes the resistance to motion caused by friction. The dry friction force is parallel to the surface and proportional to the force normal to the surface; in the case of the mass-spring system of Fig. 2.14, the normal force is equal to the weight W. The constant of proportionality is the static friction coefficient μ_s, a number varying between 0 and 1 depending on the surface materials. Once motion is initiated, the force drops to $\mu_k W$, where μ_k is the kinetic friction coefficient, whose value is generally smaller than that of μ_s. The friction force is opposite in direction to the velocity, and remains constant in magnitude as long as the forces acting on the mass m, namely, the inertia force and the restoring force due to the spring, are sufficient to overcome the dry friction. When these forces are no longer sufficient, the motion simply stops.

Denoting by F_d the magnitude of the damping force, where $F_d = \mu_k W$, the equation of motion can be written in the form

$$m\ddot{x} + F_d \operatorname{sgn}(\dot{x}) + kx = 0 \tag{2.45}$$

where the symbol "sgn" denotes the *signum function*, or the *sign of*, and represents a function having the value $+1$ if the argument $\dot{x}$ is positive and the value -1 if the

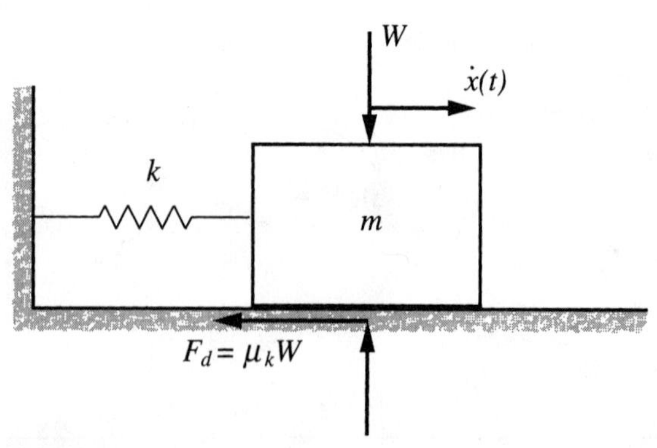

FIGURE 2.14
Mass-spring system subjected to Coulumb damping

argument is negative. Mathematically, the function can be expressed as

$$\mathrm{sgn}(\dot{x}) = \frac{\dot{x}}{|\dot{x}|} \tag{2.46}$$

Equation (2.45) is nonlinear, but it can be separated into two linear equations, one for positive and another one for negative $\dot{x}$, as follows:

$$\begin{aligned} m\ddot{x} + kx &= -F_d \qquad \text{for } \dot{x} > 0 \\ m\ddot{x} + kx &= F_d \qquad \text{for } \dot{x} < 0 \end{aligned} \tag{2.47}$$

Although Eqs. (2.47) are nonhomogeneous, so that they can be regarded as representing forced vibration, the damping forces are passive in nature, so that discussion of these equations in this chapter is in order.

The solution of Eqs. (2.47) can be obtained for one time interval at a time, depending on the sign of $\dot{x}$. Without loss of generality, we assume that the motion starts from rest with the mass m in the displaced position $x(0) = x_0$, where the initial displacement x_0 is sufficiently large that the restoring force in the spring exceeds the static friction force. Because the ensuing motion is from right to left, we conclude that the velocity is negative, so that we must solve the second of Eqs. (2.47) first, where the equation can be written in the form

$$\ddot{x} + \omega_n^2 x = \omega_n^2 f_d \qquad \omega_n^2 = \frac{k}{m} \tag{2.48}$$

in which $f_d = F_d/k$ represents an equivalent displacement. Equation (2.48) is subject to the initial conditions $x(0) = x_0$, $\dot{x}(0) = 0$, so that its solution is simply

$$x(t) = (x_0 - f_d)\cos\omega_n t + f_d \tag{2.49}$$

which represents harmonic oscillation superposed on the average response f_d. Equation (2.49) is valid for $0 \le t \le t_1$, where t_1 is the time at which the velocity reduces to zero and the motion is about to reverse direction. Differentiating Eq. (2.49) with respect to time, we obtain

$$\dot{x}(t) = -\omega_n(x_0 - f_d)\sin\omega_n t \tag{2.50}$$

so that the lowest nontrivial value satisfying the condition $\dot{x}(t_1) = 0$ is $t_1 = \pi/\omega_n$, at which time the displacement is $x(t_1) = -(x_0 - 2f_d)$. If $x(t_1)$ is sufficiently large in magnitude to overcome the static friction, then the mass starts moving from left to right, so that the velocity becomes positive and the motion must satisfy the first of Eqs. (2.47), namely,

$$\ddot{x} + \omega_n^2 x = -\omega_n^2 f_d \tag{2.51}$$

where $x(t)$ is subject to the initial conditions $x(t_1) = -(x_0 - 2f_d)$, $\dot{x}(t_1) = 0$. The solution of Eq. (2.51) is

$$x(t) = (x_0 - 3f_d)\cos\omega_n t - f_d \tag{2.52}$$

Compared to solution (2.49), the amplitude of the harmonic component in solution (2.52) is smaller by $2f_d$ and the average response is $-f_d$. Solution (2.52) is valid in the time

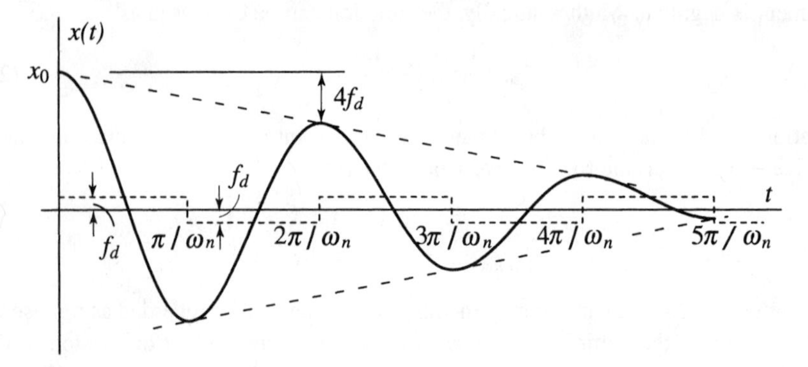

FIGURE 2.15
Response of system subjected to Coulomb damping

interval $t_1 \leq t \leq t_2$, where t_2 is the next value of time at which the velocity reduces to zero. This value is $t_2 = 2\pi/\omega_n$, at which time the velocity is ready to reverse direction once again, this time from right to left. The displacement at $t = t_2$ is $x(t_2) = x_0 - 4f_d$.

The above procedure can be repeated for $t > t_2$, every time switching back and forth between the first and second of Eqs. (2.47). However, a pattern seems to emerge, rendering this task unnecessary. Over each half-cycle the motion consists of a constant component equal to the average value of the solution and a harmonic component with frequency equal to the natural frequency ω_n of the simple mass-spring system, where the duration of every half-cycle is equal to π/ω_n. The average value of the solution alternates between f_d and $-f_d$, and at the end of each half-cycle the displacement magnitude is reduced by $2f_d = 2F_d/k$. It follows that for Coulomb damping the decay is linear with time, as opposed to the exponential decay for viscous damping. The motion stops abruptly when the displacement at the end of a given half-cycle is not sufficiently large for the restoring force in the spring to overcome the static friction. This occurs at the end of the half-cycle for which the amplitude of the harmonic component is smaller than $2f_d$. Denoting by n the half-cycle just prior to the cessation of motion, we conclude that n is the smallest integer satisfying the inequality

$$x_0 - (2n-1)f_d < \left(1 + \frac{\mu_s}{\mu_k}\right) f_d \tag{2.53}$$

The plot $x(t)$ versus t can be obtained by combining solutions (2.49), (2.52), etc. Such a plot is shown in Fig. 2.15.

Example 2.6. Let the parameters of the system of Fig. 2.14 have the values $m = 400$ kg, $k = 14 \times 10^4$ N/m, $\mu_s = 0.11$ and $\mu_k = 0.1$ and calculate the decay per cycle and the number of half-cycles until oscillation stops for the initial conditions $x(0) = x_0 = 3$ cm, $\dot{x}(0) = 0$.

The magnitude of the average value of the solution is

$$f_d = \frac{F_d}{k} = \frac{\mu_k mg}{k} = \frac{0.1 \times 400 \times 9.81}{14 \times 10^4} = 0.2803 \text{ cm} \tag{a}$$

so that the decay per cycle is

$$4f_d = 4 \times 0.2803 \times 10^{-2} \text{ m} = 1.1212 \text{ cm} \tag{b}$$

Moreover, n must be the smallest integer satisfying the inequality

$$x_0 - (2n-1)f_d = 3 - (2n-1) \times 0.2803 < \left(1 + \frac{\mu_s}{\mu_k}\right) f_d = 2.1 \times 0.2803 \text{ cm} \tag{c}$$

from which we conclude that the oscillation stops after the half-cycle $n = 5$ with m in the position $x(t_5) = -(x_0 - 10f_d) = -(3 - 10 \times 0.2803) = -0.197$ cm.

2.5 PLOTTING THE RESPONSE TO INITIAL EXCITATIONS BY MATLAB

MATLAB can be used to produce numerical solutions to vibration problems not permitting analytical solutions, as well as to evaluate analytical solutions numerically. As an example, Eq. (2.32) represents an analytical expression for the response of an underdamped single-degree-of-freedom systems to given initial displacement and velocity. If numerical values for the initial displacement x_0 and initial velocity v_0, as well as for the natural frequency of undamped oscillation ω_n and damping factor ζ, are given, then MATLAB can be used in conjunction with Eq. (2.32) to compute the response $x(t)$ at discrete values of time. A listing of these values may not be very meaningful. Indeed, much more insight can be gained from plots of $x(t)$ versus t for various values of ζ, as can be concluded from Fig. 2.9. Such plots can be generated by means of a computer program written in MATLAB. Following is such a program designed to duplicate the plots of Fig. 2.9 using the data for case i of Example 2.4:

```
% The program 'rspin1.m' plots the response of an underdamped single-degree-of-
% freedom system to initial excitations for given values of the damping factor zeta
clear
clf

wn=4; % natural frequency of undamped oscillation
zeta = [0.05; 0.1; 0.2]; % damping factors arranged as a three-dimensional vector
x0=0; % initial displacement
v0=50; % initial velocity
t0=0; % initial time
deltat = 0.01; % time increment
tf=6; % final time

t=[t0: deltat: tf];
for i=1:length(zeta),
  wd = sqrt(1-zeta(i)^2)*wn; % frequency of damped oscillation
  x=exp(-zeta(i)*wn*t).*(((zeta(i)*wn*x0+v0)/wd)*sin(wd*t)+x0* cos(wd*t));

  plot (t,x)
  hold on

end
```

```
title('Response to Initial Excitations')
xlabel('t(s)')
ylabel('x(t)')
grid
```

The program can be used to carry out a parametric study by plotting the response corresponding to various values of the frequency ω_n and the damping factor ζ. Note that the program was based on the third line of Eq. (2.32). The reader is encouraged to modify the program so as to base it on the bottom line of Eq. (2.32), as suggested in Problem 2.28

2.6 SUMMARY

Excitations can be broadly divided into two classes, initial excitations, which consist of initial displacements and velocities, and applied forces. Linear systems possess an extremely important property, namely, they satisfy the superposition principle. As a result, if an excitation can be represented as a linear combination of excitations, the response can be obtained by first determining the responses to the individual excitations in the linear combination separately and then combining these responses linearly. In view of this, it is convenient to treat the response to initial excitations separately from the response to applied forces.

The response of an undamped single-degree-of-freedom system to initial excitations is harmonic with amplitude and phase angle depending on the initial displacement and/or initial velocity, which vary from one case to another, and with a frequency depending on the system parameters, which are constant. As a result, the frequency is always the same for a given system, for which reason it is referred to as the natural frequency. Undamped single-degree-of-freedom systems are commonly known as harmonic oscillators. Although harmonic oscillators represent a mathematical idealization, as all physical systems possess some measure of damping, the concept is quite useful when damping is negligibly small and the time interval of interest is long relative to the natural period. A surprisingly large number of physical systems can be regarded in the first approximation as harmonic oscillators.

Damping is a very complex phenomenon. In the first place, there are various types of damping, and in many cases damping cannot be readily identified and must be inferred from measurements; in some cases it is even nonlinear. The most common type is viscous damping, which manifests itself in the form of a force proportional to the velocity and opposite in direction to the velocity. For underdamping the motion decays in an oscillatory fashion, and for overdamping the motion decays aperiodically. In both cases the decay is exponential. Measurement of viscous damping is commonly based on the concept of logarithmic decrement.

Another common type of damping is due to dry friction and occurs when bodies slide relative to one another; it is generally known as Coulomb damping. Although Coulomb damping is nonlinear, the equation of motion can be separated into two linear equations, one valid when the velocity is positive and the other valid when the velocity is negative. As a result, the equation of motion can be solved in closed form. Coulomb damping causes the motion to decay linearly.

The response problems in this chapter admit analytical solutions, even in the case of Coulomb damping, which is nonlinear. For parametric studies, plotting the response seems advised. This can be done efficiently by means of MATLAB.

PROBLEMS

2.1. A cylindrical buoy of cross-sectional area A and total mass m is first depressed from equilibrium and then allowed to oscillate (Fig. 2.16). Denote the mass density of the liquid by γ and determine the natural frequency of oscillation.

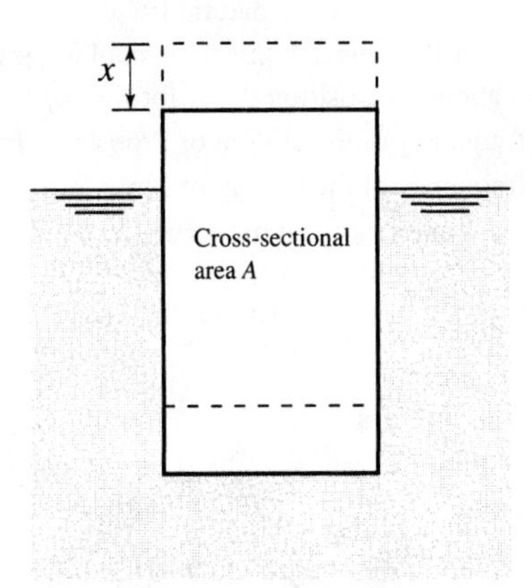

FIGURE 2.16
Oscillating buoy

2.2. Determine the natural frequency of the system of Fig. 1.61 for $k_1 = k_2 = 0.8 \times 10^5$ N/m, $k_3 = 2.4 \times 10^5$ N/m and $m = 240$ kg.

2.3. A given system of unknown mass m and spring constant k was observed to oscillate harmonically in free vibration with the natural period $T = 2\pi \times 10^{-2}$ s. When a mass $M = 0.9$ kg was added to the system (Fig. 2.17) the new period rose to $T^* = 2.5\pi \times 10^{-2}$ s. Determine the system parameters m and k.

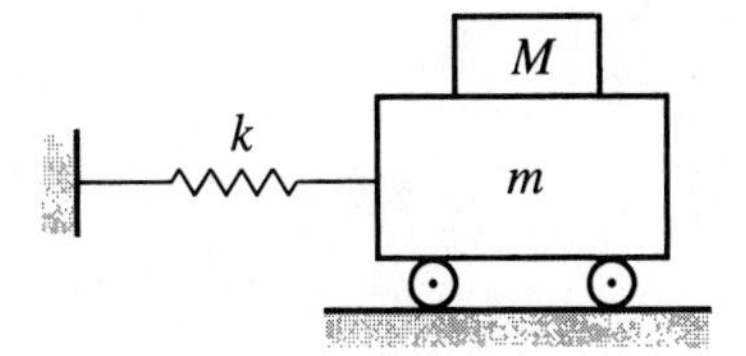

FIGURE 2.17
System with unknown mass m and spring k

2.4. Consider the system of Fig. 1.62 in conjunction with the parameters $a = 2\text{m}$, $b = 2.5\text{m}$, $k_1 = 5 \times 10^5\,\text{N/m}$, $k_2 = 1.8 \times 10^5\,\text{N/m}$ and $m = 200\,\text{kg}$, assume small motions, solve for the equilibrium position and determine the natural frequency of oscillation about the equilibrium position.

2.5. Determine the natural frequency of the system of Fig. 1.63 for the gear ratio $R_A/R_B = n = 2$. The gears are made of the same material and have the same thickness.

2.6. The door of Problem 1.5 has the width $B = 0.8\text{m}$ and hangs at an angle $\alpha = 5°$ with respect to the vertical. Assume small angles θ and determine the natural frequency of oscillation.

2.7. The rod in the compound pendulum of Problem 1.6 has mass per unit length $m = 1\,\text{kg/m}$ and total length $L = 2\text{m}$ and the disk has the total mass $M = 5\,\text{kg}$ and radius $R = 0.25\text{m}$. Assume small angles θ and determine the natural frequency of oscillation.

2.8. Consider the rolling disk of Problem 1.8, assume small angles θ and determine the natural frequency of oscillation about the position $\theta = 0$ for $r = R/4$.

2.9. Determine the natural frequency of the system of Problem 1.15.

2.10. Determine the natural frequency of the system of Problem 1.17.

2.11. To determine the centroidal mass moment of inertia I_C of a tire mounted on a hub, the wheel was suspended on a knife edge, as shown in Fig. 2.18, and the natural period T was measured. Derive a formula for I_C in terms of the mass m, the natural period T and the radius r from the center C to the support.

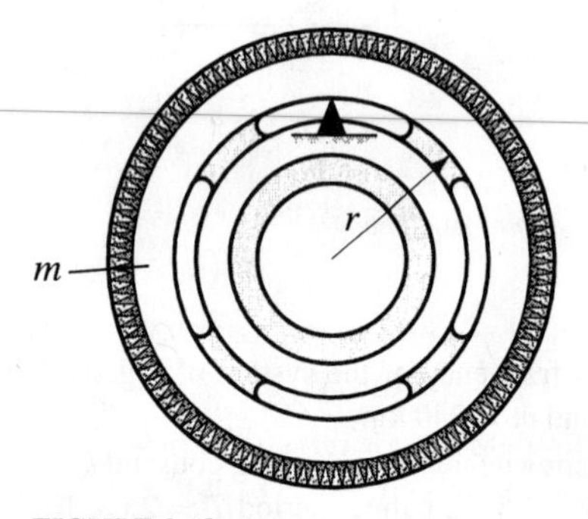

FIGURE 2.18
Wheel suspended on a knife edge

2.12. A connecting rod of mass $m = 3 \times 10^{-3}$ kg and centroidal mass moment of inertia $I_C = 0.432 \times 10^{-4}\,\mathrm{kg\,m^2}$ is suspended on a knife edge about the upper inner surface of a wrist-pin bearing, as shown in Fig. 2.19. When disturbed slightly, the rod was observed to oscillate harmonically with the natural frequency $\omega_n = 6\,\mathrm{rad/s}$. Determine the distance h between the support and the mass center C. Hint: the distance h must be smaller than the length of a corresponding simple pendulum of mass m, because the mass of the connecting rod is distributed and that of the pendulum is lumped.

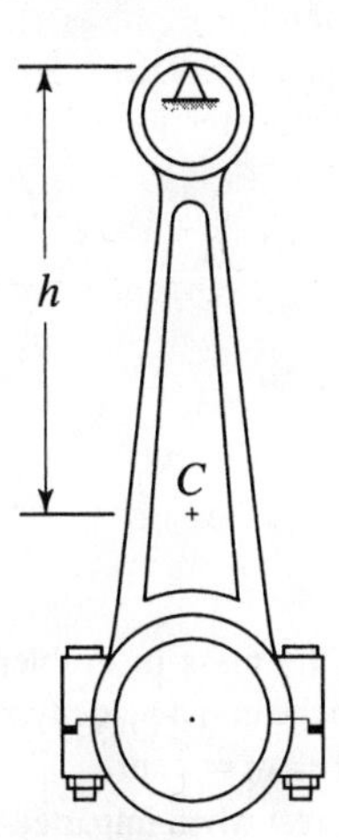

FIGURE 2.19
Connecting rod suspended on a knife edge

2.13. A bead of mass m is suspended on a massless string, as shown in Fig. 2.20. Assume that the string is subjected to the tension T and that the tension remains constant throughout the vertical motion of the bead and derive the equation for small motions $y(t)$ from equilibrium, as well as the natural frequency of oscillation.

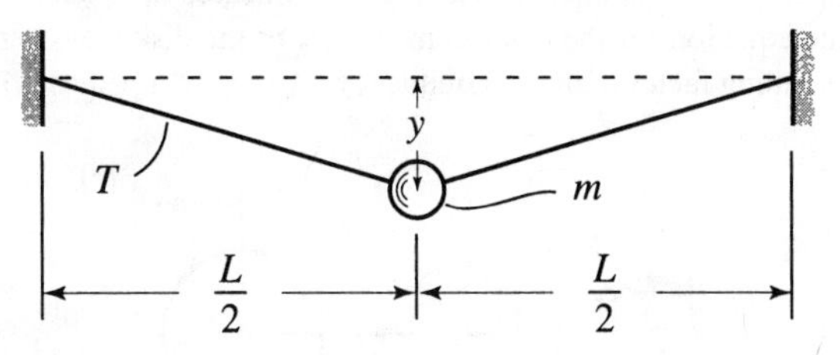

FIGURE 2.20
Mass suspended on a string

2.14. The one-story structure shown in Fig. 2.21 can be modeled in the first approximation as a single-degree-of-freedom system by regarding the columns as massless beams clamped at both ends and the roof as a rigid slab. Derive the equation for the horizontal translation of the slab and determine the natural frequency. The mass of the slab is denoted by M and the flexural rigidity of the columns by EI.

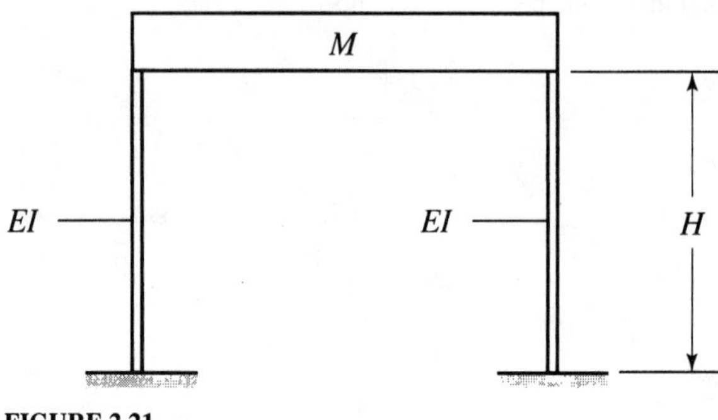

FIGURE 2.21
One-story structure

2.15. The maximum acceleration of the mass in Problem 2.2 was observed to have the value $a_{max} = 1,000 \text{ cm/s}^2$. Determine the initial velocity $\dot{x}(0) = v_0$ for the case in which the initial displacement has the value $x(0) = x_0 = 2\,\text{cm}$.

2.16. The system of Problem 2.2 is at rest when imparted the initial velocity v_0. Plot on the same graph the response for the values $v_0 = 10, 20, 30$ and 40 cm/s over the time interval $0 \leq t \leq 1$s and draw conclusions concerning the nature of the response.

2.17. The upper right corner of the door of Problem 2.6 is released from rest after being held at a distance $x_0 = 2$ cm from the vertical plane containing the hinge axis. Plot the response over 2.5 periods.

2.18. A simple pendulum is immersed in viscous fluid so that there is a resisting force of magnitude $cL\dot{\theta}$ acting on the bob, where c is the coefficient of viscous damping, L the length of the pendulum and θ the angular displacement. In the case in which the mass of the bob has the value $m = 1$ kg and the length is $L = 1$m the period of small-amplitude oscillations was observed to be $T = 2.02$ s. Determine the damping coefficient c.

2.19. A disk of mass m and radius R rolls without slip while restrained by a dashpot with coefficient of viscous damping c in parallel with a spring of stiffness k, as shown in Fig. 2.22. Derive the differential equation for the displacement $x(t)$ of the disk mass center C and determine the viscous damping factor ζ and the frequency ω_n of undamped oscillation.

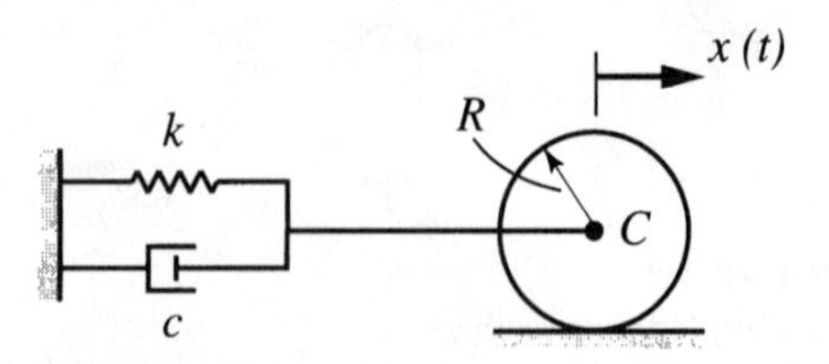

FIGURE 2.22
Rolling disk restrained by a spring and a dashpot

2.20. Calculate the frequency of damped oscillation of the system shown in Fig. 2.23 for the values $m = 1,750$ kg, $c = 3,500$ N · s/m, $k = 7 \times 10^5$ N/m, $a = 1.25$ m and $b = 2.5$ m. Determine the value of the critical damping.

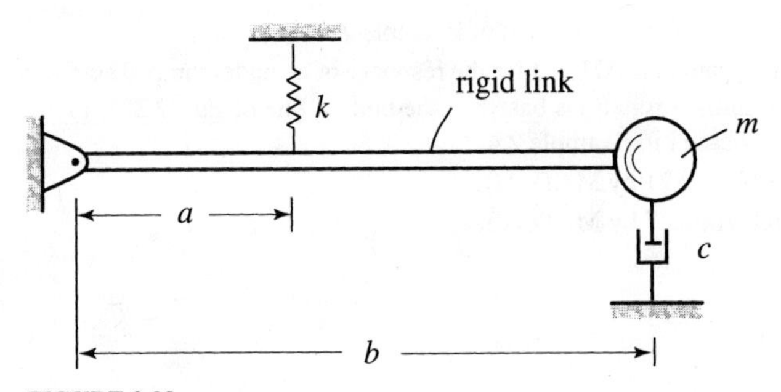

FIGURE 2.23
Mass supported by a spring and dashpot through a rigid bar

2.21. Consider the system of Fig. 2.6 and plot the response to the initial conditions $x(0) = 2$ cm, $\dot{x}(0) = 0$ for the values of the damping factor $\zeta = 0.1$, 1 and 2. Let the frequency of undamped oscillation have the value $\omega_n = 5$ rad/s and plot the response over the interval $0 \leq t \leq 5$ s.

2.22. A projectile of mass $m = 10$ kg traveling with the velocity $v = 50$ m/s strikes and becomes embedded in a massless board supported by a spring of stiffness $k = 6.4 \times 10^4$ N/m in parallel with a dashpot with the coefficient of viscous damping $c = 400$ N · s/m (Fig. 2.24). Determine the time required for the board to reach the maximum displacement and the value of the maximum displacement.

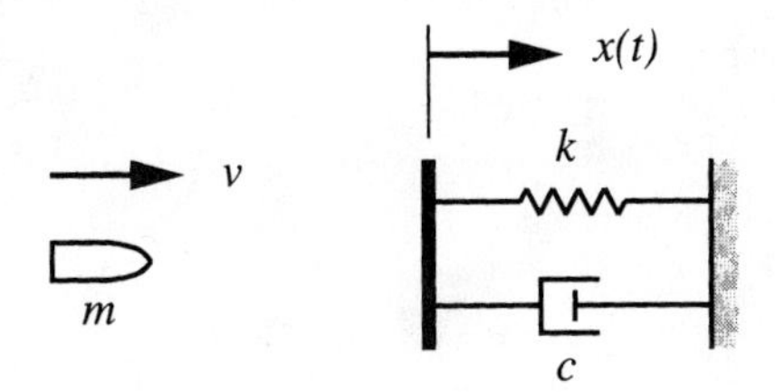

FIGURE 2.24
Projectile striking a board restrained by a spring and dashpot

2.23. Devise a vector construction representing Eq. (2.32).

2.24. From the observation of the response of an underdamped single-degree-of-freedom system, it was determined that the maximum displacement amplitude during the second cycle is 75% of the first. Calculate the damping factor ζ and determine the maximum displacement amplitude during cycle 4 1/2 as a fraction of the first.

2.25. Measurements of the response peak amplitudes of a vibrating single-degree-of-freedom system are as follows: $x_1 = 24.86, x_2 = 22.98, x_3 = 21.49, x_4 = 20.55, x_5 = 19.14, x_6 = 17.80$, $x_7 = 17.03$ and $x_8 = 15.97$. Use the least squares method described in Sec. 2.3 to determine the "best" viscous damping factor ζ.

2.26. Prove inequality (2.53).

2.27. Plot $x(t)$ versus t for the system of Example 2.6.

2.28. Write a program in MATLAB for the response of an underdamped single-degree-of-freedom system to initial excitations based on the bottom line of Eq. (2.32). Plot the response using the data for case i in Example 2.4.

2.29. Solve Problem 2.21 by MATLAB.

2.30. Solve Problem 2.27 by MATLAB.

CHAPTER 3

RESPONSE OF SINGLE-DEGREE-OF-FREEDOM SYSTEMS TO HARMONIC AND PERIODIC EXCITATIONS

As indicated in Sec. 1.11, harmonic and periodic forces belong to a very important class of excitations, namely, the class of steady-state excitations. The importance of steady-state excitations, and in particular harmonic excitations, is due to the fact that they occur frequently in various areas of engineering. Hence, it is only fitting that an entire chapter be devoted to the response to harmonic and periodic excitations.

Steady-state harmonic excitations differ from all other types of excitations in that more information concerning the behavior of the response to such excitations can be extracted by means of frequency domain rather than time domain techniques. Particularly useful are frequency-response plots, which are two companion diagrams showing how the amplitude and phase angle of the response vary as the excitation frequency changes. They provide a broader picture of the response characteristics than that provided by a time response, which is limited to a given excitation frequency.

In many systems, rotating components give rise to harmonic excitations. We encountered such an example in the model of a washing machine depicted in Fig. 1.28b. Turbines, automobile tires and vibration measuring instruments can be counted among other typical examples.

Periodic excitations are also steady state. But, whereas harmonic functions are periodic by definition, periodic functions are not necessarily harmonic. They can be represented, however, by infinite series of harmonic functions. As with the harmonic response, frequency domain techniques are more suitable for treating the response to periodic excitations than time domain techniques.

3.1 RESPONSE OF SINGLE-DEGREE-OF-FREEDOM SYSTEMS TO HARMONIC EXCITATIONS

According to Eq. (1.141), the equation of motion of a damped single-degree-of-freedom system of the type shown in Fig. 3.1 is

$$m\ddot{x}(t) + c\dot{x}(t) + kx(t) = F(t) \tag{3.1}$$

where m is the mass, c the coefficient of viscous damping, k the spring constant, $x(t)$ the displacement and $F(t)$ the force. The solution of Eq. (3.1) is subject to the initial displacement and velocity, $x(0) = x_0$ and $\dot{x}(0) = v_0$, respectively. But, according to the principle of superposition, the response to the applied force $F(t)$ and to the initial excitations x_0 and v_0 can be obtained separately and combined linearly. Chapter 2 was devoted entirely to the response to initial excitations. This chapter and Ch. 4 are devoted to the response to applied forces.

In this section, we are concerned with the case in which $F(t)$ represents a harmonic force, which can be expressed for convenience in the form

$$F(t) = kf(t) = kA\cos\omega t \tag{3.2}$$

where

$$f(t) = A\cos\omega t \tag{3.3}$$

in which ω is the *excitation frequency*, or *driving frequency*, and note that $f(t)$ and A have units of displacement. Introducing Eq. (3.2) in Eq. (3.1) and dividing through by m, we obtain

$$\ddot{x}(t) + 2\zeta\omega_n\dot{x}(t) + \omega_n^2 x(t) = \omega_n^2 A\cos\omega t \tag{3.4}$$

where $\zeta = c/2\omega_n m$ is the viscous damping factor and $\omega_n = \sqrt{k/m}$ is the natural frequency of undamped oscillation.

As indicated in Sec. 1.11, harmonic forces are *steady-state excitations*, for which time plays only a secondary role. Consistent with this, the response is also steady state.

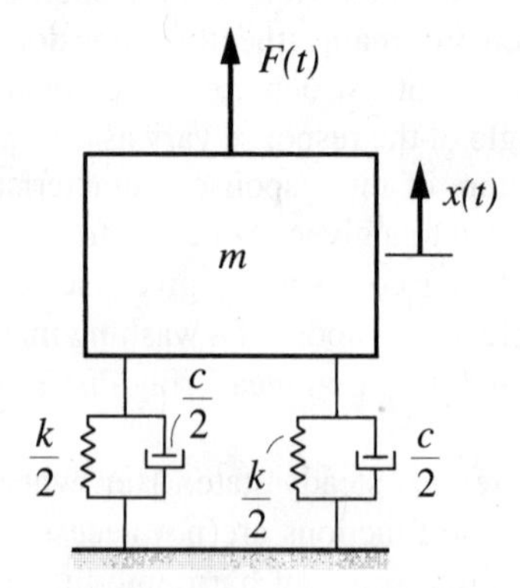

FIGURE 3.1
Damped single-degree-of-freedom system

In particular, the response is harmonic and has the same frequency as the excitation frequency. Hence, we assume a solution of Eq. (3.4) in the form

$$x(t) = C_1 \sin\omega t + C_2 \cos\omega t \tag{3.5}$$

in which C_1 and C_2 are constants yet to be determined. Inserting Eq. (3.5) into Eq. (3.4), we obtain

$$\begin{aligned} &-\omega^2(C_1 \sin\omega t + C_2 \cos\omega t) + \omega 2\zeta\omega_n(C_1 \cos\omega t - C_2 \sin\omega t) \\ &\qquad + \omega_n^2(C_1 \sin\omega t + C_2 \cos\omega t) \\ &= \left[(\omega_n^2 - \omega^2)C_1 - 2\zeta\omega\omega_n C_2\right] \sin\omega t + \left[2\zeta\omega\omega_n C_1 + (\omega_n^2 - \omega^2)C_2\right] \cos\omega t \\ &= \omega_n^2 A \cos\omega t \end{aligned} \tag{3.6}$$

Equation (3.6) can be satisfied provided the coefficients of $\sin\omega t$ on both sides of the equation are equal, and the same can be said about the coefficients of $\cos\omega t$, which yields two algebraic equations in C_1 and C_2, as follows:

$$\begin{aligned} (\omega_n^2 - \omega^2)C_1 - 2\zeta\omega\omega_n C_2 &= 0 \\ 2\zeta\omega\omega_n C_1 + (\omega_n^2 - \omega^2)C_2 &= \omega_n^2 A \end{aligned} \tag{3.7}$$

Using Cramer's rule (Ref. 18, p. 233), the solution of Eqs. (3.7) is

$$\begin{aligned} C_1 &= \frac{\begin{vmatrix} 0 & -2\zeta\omega\omega_n \\ \omega_n^2 A & \omega_n^2 - \omega^2 \end{vmatrix}}{\begin{vmatrix} \omega_n^2 - \omega^2 & -2\zeta\omega\omega_n \\ 2\zeta\omega\omega_n & \omega_n^2 - \omega^2 \end{vmatrix}} = \frac{2\zeta\omega\omega_n^3 A}{(\omega_n^2 - \omega^2)^2 + (2\zeta\omega\omega_n)^2} \\ &= \frac{2\zeta\omega/\omega_n}{[1 - (\omega/\omega_n)^2]^2 + (2\zeta\omega/\omega_n)^2} A \\ C_2 &= \frac{\begin{vmatrix} \omega_n^2 - \omega^2 & 0 \\ 2\zeta\omega\omega_n & \omega_n^2 A \end{vmatrix}}{\begin{vmatrix} \omega_n^2 - \omega^2 & -2\zeta\omega\omega_n \\ 2\zeta\omega\omega_n & \omega_n^2 - \omega^2 \end{vmatrix}} = \frac{(\omega_n^2 - \omega^2)\omega_n^2 A}{(\omega_n^2 - \omega^2)^2 + (2\zeta\omega\omega_n)^2} \\ &= \frac{1 - (\omega/\omega_n)^2}{[1 - (\omega/\omega_n)^2]^2 + (2\zeta\omega/\omega_n)^2} A \end{aligned} \tag{3.8}$$

Introducing Eqs. (3.8) in Eq. (3.5), we obtain the steady-state solution

$$x(t) = \frac{A}{[1 - (\omega/\omega_n)^2]^2 + (2\zeta\omega/\omega_n)^2} \left\{ \frac{2\zeta\omega}{\omega_n} \sin\omega t + \left[1 - \left(\frac{\omega}{\omega_n}\right)^2\right] \cos\omega t \right\} \tag{3.9}$$

Solution (3.9) can be cast in a form more suitable for physical interpretation. To this end, we introduce the notation

$$\frac{2\zeta\omega/\omega_n}{\{[1-(\omega/\omega_n)^2]^2+(2\zeta\omega/\omega_n)^2\}^{1/2}} = \sin\phi$$
$$\frac{1-(\omega/\omega_n)^2}{\{[1-(\omega/\omega_n)^2]^2+(2\zeta\omega/\omega_n)^2\}^{1/2}} = \cos\phi \tag{3.10}$$

so that the harmonic response can be written in the compact form

$$x(t) = X\cos(\omega t - \phi) \tag{3.11}$$

where

$$X = X(\omega) = \frac{A}{\left\{[1-(\omega/\omega_n)^2]^2+(2\zeta\omega/\omega_n)^2\right\}^{1/2}} \tag{3.12}$$

is the magnitude and

$$\phi = \phi(\omega) = \tan^{-1}\frac{2\zeta\omega/\omega_n}{1-(\omega/\omega_n)^2} \tag{3.13}$$

is the phase angle of the steady-state response. As pointed out in Sec. 1.12, a great deal more information can be extracted from plots $X(\omega)$ versus ω and $\phi(\omega)$ versus ω than from plots $x(t)$ versus t, where the first two are *frequency response* plots and the third is a *time domain* plot. In this section, we propose to study the two frequency response plots. But, before that, it will prove useful for future studies to examine a more expeditious way of deriving the steady-state response.

We recall from Sec. 1.11 that trigonometric functions can be expressed in exponential form. Hence, using Eqs. (1.136) and recognizing that A is real, we can rewrite Eq. (3.3) as

$$f(t) = Ae^{i\omega t} \tag{3.14}$$

and Eq. (3.4) in the form

$$\ddot{x}(t) + 2\zeta\omega_n\dot{x}(t) + \omega_n^2 x(t) = \omega_n^2 Ae^{i\omega t} \tag{3.15}$$

with the understanding that, if the excitation is $f(t) = A\cos\omega t$, the response is Re $x(t)$ and if $f(t) = A\sin\omega t$, the response is Im $x(t)$. The advantage of the complex notation is that the solution of Eq. (3.15) is much easier to obtain than a solution using real notation. Indeed, the solution is simply

$$x(t) = X(i\omega)e^{i\omega t} \tag{3.16}$$

Inserting Eq. (3.16) into Eq. (3.15), we have

$$Z(i\omega)X(i\omega)e^{i\omega t} = \omega_n^2 Ae^{i\omega t} \tag{3.17}$$

where

$$Z(i\omega) = \omega_n^2 - \omega^2 + i2\zeta\omega\omega_n \tag{3.18}$$

is the *impedance function*. Dividing Eq. (3.17) through by $Z(i\omega)e^{i\omega t}$ and considering Eq. (3.18), we have

$$X(i\omega) = \frac{\omega_n^2 A}{Z(i\omega)} = \frac{\omega_n^2 A}{\omega_n^2 - \omega^2 + i2\zeta\omega\omega_n} = \frac{A}{1-(\omega/\omega_n)^2 + i2\zeta\omega/\omega_n} \tag{3.19}$$

It will prove convenient to introduce the nondimensional ratio

$$G(i\omega) = \frac{X(i\omega)}{A} = \frac{\omega_n^2}{Z(i\omega)} = \frac{1}{1-(\omega/\omega_n)^2 + i2\zeta\omega/\omega_n} \tag{3.20}$$

where $G(i\omega)$ is known as the *frequency response*, a very important concept in vibrations. Inserting Eq. (3.20) into Eq. (3.16), we can express the harmonic response in the general form

$$x(t) = AG(i\omega)e^{i\omega t} \tag{3.21}$$

Clearly, $G(i\omega)$ is a measure of the system response to a harmonic excitation of frequency ω, which explains why the function is called frequency response.

In general, the frequency response is a complex function, which makes it difficult to study the nature of the response using Eq. (3.21) in the indicated form. To obtain a more suitable form of the harmonic response, we write

$$G(i\omega) = \text{Re}\ G(i\omega) + i\ \text{Im}\ G(i\omega) \tag{3.22}$$

and recognize that, as any complex quantity, the frequency response can be expressed as

$$G(i\omega) = |G(i\omega)|e^{-i\phi(\omega)} \tag{3.23}$$

where

$$|G(i\omega)| = \left[G(i\omega)\overline{G}(i\omega)\right]^{1/2} = \{[\text{Re}\ G(i\omega)]^2 + \left[\text{Im}\ G(i\omega)\right]^2\}^{1/2} \tag{3.24}$$

is the *magnitude* of $G(i\omega)$, in which $\overline{G}(i\omega)$ is the complex conjugate of $G(i\omega)$, and

$$\phi(\omega) = \tan^{-1}\left[\frac{-\text{Im}\ G(i\omega)}{\text{Re}\ G(i\omega)}\right] \tag{3.25}$$

is the *phase angle* of $G(i\omega)$. Hence, inserting Eq. (3.23) into Eq. (3.21), we obtain

$$x(t) = A|G(i\omega)|e^{i(\omega t - \phi)} \tag{3.26}$$

Equation (3.26) is much more convenient for deriving the actual harmonic response than Eq. (3.21). Indeed, according to our earlier understanding, if the excitation is $f(t) = A\ \cos\omega t$, then the response is

$$x(t) = \text{Re}\ A|G(i\omega)|e^{i(\omega t-\phi)} = A|G(i\omega)|\cos(\omega t - \phi) \tag{3.27}$$

and if the excitation is $f(t) = A\sin\omega t$, the response is

$$x(t) = \text{Im}\ A|G(i\omega)|e^{i(\omega t-\phi)} = A|G(i\omega)|\sin(\omega t - \phi) \tag{3.28}$$

The representation of harmonic excitations and the response to harmonic excitations by complex vectors can be given an interesting geometric interpretation by means

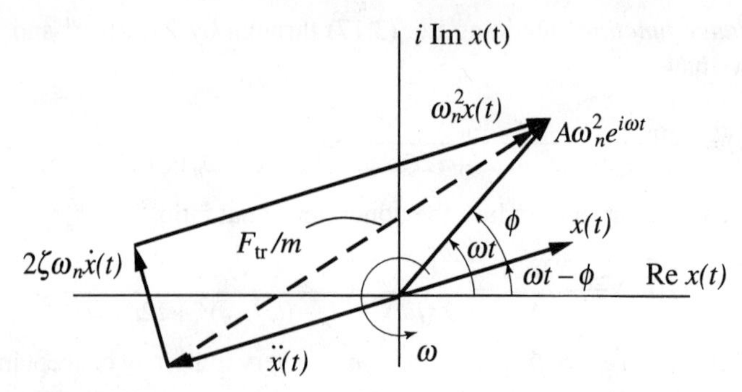

FIGURE 3.2
Representation of harmonic excitation and response in the complex plane

of a diagram in the complex plane. To this end, we differentiate Eq. (3.26) with respect to time and obtain

$$\begin{aligned}\dot{x}(t) &= i\omega A|G(i\omega)|e^{i(\omega t-\phi)} = i\omega x(t)\\ \ddot{x}(t) &= (i\omega)^2 A|G(i\omega)|e^{i(\omega t-\phi)} = -\omega^2 x(t)\end{aligned} \tag{3.29}$$

Because i can be written as $i = \cos\pi/2 + i\sin\pi/2 = e^{i\pi/2}$, we conclude that the velocity leads the displacement by the phase angle $\pi/2$ and that its magnitude is equal to the magnitude of the displacement multiplied by the factor ω. Moreover, because -1 can be expressed as $-1 = \cos\pi + i\sin\pi = e^{i\pi}$, it follows that the acceleration leads the displacement by the phase angle π and that its magnitude is equal to the magnitude of the displacement multiplied by the factor ω^2.

In view of the above, we can represent Eq. (3.15) in the complex plane by the diagram shown in Fig. 3.2. The diagram can be interpreted as stating that the sum of the complex vectors $\ddot{x}(t)$, $2\zeta\omega_n\dot{x}(t)$ and $\omega_n^2 x(t)$ balances $\omega_n^2 Ae^{i\omega t}$, which is precisely what Eq. (3.15) states. Consistent with the nature of the complex vector $e^{i\omega t}$ (Fig. 1.37), we conclude that, as t increases, the entire diagram of Fig. 3.2 rotates in the complex plane with the angular velocity ω. It is clear that considering only the real part of the response is equivalent to projecting the diagram onto the real axis. We can just as easily retain the projections on the imaginary axis, or any other axis, without affecting the nature of the response. In this regard, we observe that projecting the excitation and response on an axis making an angle ψ with respect to the real axis is equivalent to multiplying both sides of Eq. (3.15) by the constant factor $e^{-i\psi}$. Clearly, this has no effect on the magnitude $|G(i\omega)|$ and phase angle $\phi(\omega)$ of the harmonic response.

3.2 FREQUENCY RESPONSE PLOTS

Equation (3.27), or Eq. (3.28), defines the harmonic response in the time domain for any excitation frequency ω. A broader picture of the harmonic response can be obtained by examining how the magnitude $|G(i\omega)|$ and phase angle $\phi(\omega)$ of the frequency response $G(i\omega)$ vary with ω. Insertion of Eq. (3.20) into Eq. (3.24) yields the magnitude of the frequency response

$$
\begin{aligned}
|G(i\omega)| &= [G(i\omega)\overline{G}(i\omega)]^{1/2} \\
&= \left[\frac{1}{1-(\omega/\omega_n)^2+i2\zeta\omega/\omega_n}\,\frac{1}{1-(\omega/\omega_n)^2-i2\zeta\omega/\omega_n}\right]^{1/2} \\
&= \frac{1}{\left\{[1-(\omega/\omega_n)^2]^2+(2\zeta\omega/\omega_n)^2\right\}^{1/2}}
\end{aligned}
\tag{3.30}
$$

Before we attempt to derive the phase angle $\phi(\omega)$, it is advisable to cast $G(i\omega)$ in a form exhibiting the real and imaginary parts of the frequency response explicitly. To this end, we multiply the top and bottom of Eq. (3.20) by the complex conjugate $\overline{G}(i\omega)$, use Eq. (3.30) and write the desired form

$$
G(i\omega) = G(i\omega)\frac{\overline{G}(i\omega)}{\overline{G}(i\omega)} = \frac{|G(i\omega)|^2}{\overline{G}(i\omega)} = \frac{1-(\omega/\omega_n)^2-i2\zeta\omega/\omega_n}{[1-(\omega/\omega_n)^2]^2+(2\zeta\omega/\omega_n)^2}
\tag{3.31}
$$

so that

$$
\begin{aligned}
\operatorname{Re} G(i\omega) &= \frac{1-(\omega/\omega_n)^2}{\left[1-(\omega/\omega_n)^2\right]^2+(2\zeta\omega/\omega_n)^2} \\
\operatorname{Im} G(i\omega) &= -\frac{2\zeta\omega/\omega_n}{\left[1-(\omega/\omega_n)^2\right]^2+(2\zeta\omega/\omega_n)^2}
\end{aligned}
\tag{3.32}
$$

Hence, introducing Eqs. (3.32) in Eq. (3.25), we obtain the phase angle of the frequency response

$$
\phi(\omega) = \tan^{-1}\left[\frac{-\operatorname{Im} G(i\omega)}{\operatorname{Re} G(i\omega)}\right] = \tan^{-1}\frac{2\zeta\omega/\omega_n}{1-(\omega/\omega_n)^2}
\tag{3.33}
$$

Considerable insight into the system behavior can be gained by examining how the magnitude and phase angle of the frequency response vary with the excitation frequency. The corresponding nondimensional plots, $|G(i\omega)|$ versus ω/ω_n and $\phi(\omega)$ versus ω/ω_n, are known as *frequency response plots*. The plots are made even more informative by using the viscous damping factor ζ as a parameter. The plot $|G(i\omega)|$ versus ω/ω_n is shown in Fig. 3.3, from which we observe that damping tends to reduce amplitudes and to shift the peaks to the left of the vertical through $\omega/\omega_n = 1$. To find the location and value of the peaks, we use the standard technique of calculus for finding the maximum of a function, namely, setting the derivative equal to zero. Hence, we write

$$
\frac{d|G(i\omega)|}{d(\omega/\omega_n)} = -\frac{1}{2}\frac{2\left[1-(\omega/\omega_n)^2\right](-2\omega/\omega_n)+2(2\zeta\omega/\omega_n)2\zeta}{\{[1-(\omega/\omega_n)^2]^2+(2\zeta\omega/\omega_n)\}^{3/2}} = 0
\tag{3.34}
$$

from which we conclude that peaks occur at

$$
\frac{\omega}{\omega_n} = \sqrt{1-2\zeta^2}
\tag{3.35}
$$

thus corroborating the statement that they occur for $\omega/\omega_n < 1$. Clearly, for $\zeta > 1/\sqrt{2}$ the response has no peaks. We observe that in the undamped case, $\zeta = 0$, the response

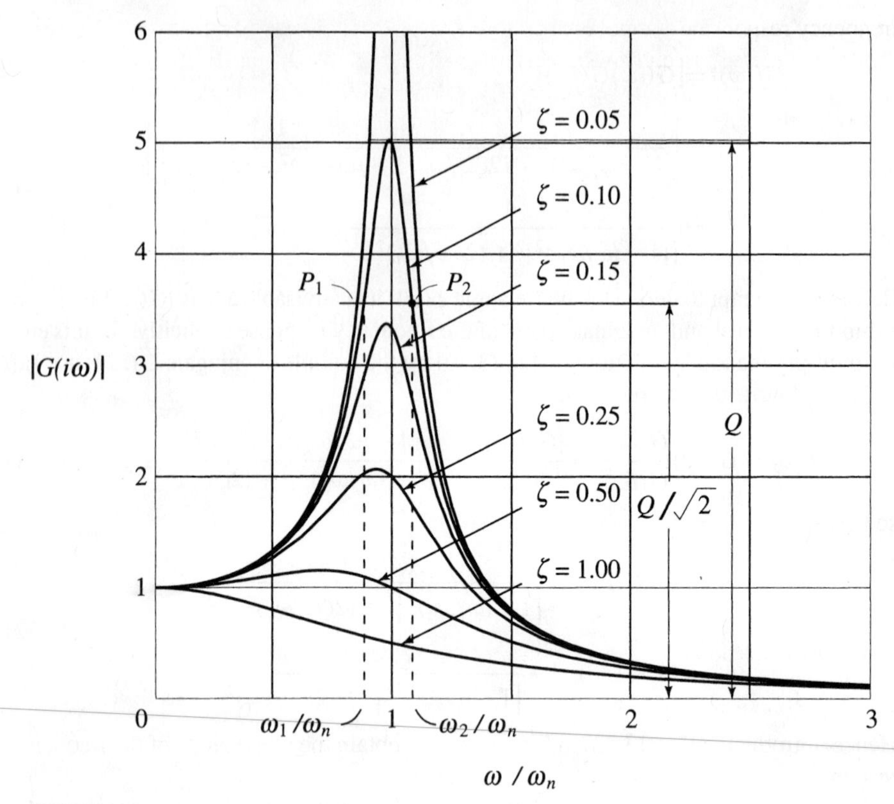

FIGURE 3.3
Magnitude of the frequency response

increases indefinitely as the driving frequency ω approaches the natural frequency ω_n. In this case, the system experiences a *resonance condition* characterized by violent vibration. Clearly solution (3.27), or (3.28), is no longer valid at resonance, and a new solution of Eq. (3.4) corresponding to $\zeta = 0$ and $\omega = \omega_n$ must be generated; we discuss such a solution later in this section. Inserting Eq. (3.35) back into Eq. (3.30), we obtain the value of the peak amplitudes

$$|G(i\omega)|_{\max} = \frac{1}{2\zeta\sqrt{1-\zeta^2}} \tag{3.36}$$

There is considerable interest in the case of light damping, such as when $\zeta < 0.05$, in which case peaks occur in the immediate neighborhood of $\omega/\omega_n = 1$. Moreover, for small values of ζ, Eq. (3.36) yields the approximation

$$|G(i\omega)|_{\max} = Q \cong \frac{1}{2\zeta} \tag{3.37}$$

where Q is known as the *quality factor*, because in many electrical engineering applications, such as the tuning circuit in a radio, the interest lies in amplitudes at resonance as large as possible. The symbol is often referred to as the *Q factor* of the circuit. Equation (3.37) can be used as a quick way of estimating the viscous damping factor of a system by producing the plot $|G(i\omega)|$ versus ω/ω_n experimentally, measuring the peak amplitude Q and writing

$$\zeta \cong \frac{1}{2Q} \tag{3.38}$$

The points P_1 and P_2, where the amplitude of $|G(i\omega)|$ falls to $Q/\sqrt{2}$, are called *half-power points*, because the power absorbed by the resistor in an electric circuit or by the damper in a mechanical system subjected to a harmonic force is proportional to the square of the amplitude (see Sec. 3.8). To obtain the driving frequencies corresponding to P_1 and P_2, we use Eqs. (3.30) and (3.37) and write

$$|G(i\omega)| = \frac{1}{\{[1-(\omega/\omega_n)^2]^2 + (2\zeta\omega/\omega_n)^2\}^{1/2}} = \frac{1}{\sqrt{2}}Q = \frac{1}{2\sqrt{2}\,\zeta} \tag{3.39}$$

which yields the quadratic equation in $(\omega/\omega_n)^2$

$$\left(\frac{\omega}{\omega_n}\right)^4 - 2(1-2\zeta^2)\left(\frac{\omega}{\omega_n}\right)^2 + 1 - 8\zeta^2 = 0 \tag{3.40}$$

Ignoring the term in ζ^3, Eq. (3.40) has the solutions

$$\begin{matrix}(\omega_1/\omega_n)^2 \\ (\omega_2/\omega_n)^2\end{matrix} = (1-2\zeta^2) \mp \sqrt{(1-2\zeta^2)^2 - (1-8\zeta^2)} \cong 1 - 2\zeta^2 \mp 2\zeta \tag{3.41}$$

where ω_1 and ω_2 are the excitation frequencies corresponding to P_1 and P_2, respectively, so that

$$\left(\frac{\omega_2}{\omega_n}\right)^2 - \left(\frac{\omega_1}{\omega_n}\right)^2 = 4\zeta \tag{3.42}$$

Then recognizing that for light damping $\omega_1 + \omega_2 \cong 2\omega_n$, Eq. (3.42) yields

$$\Delta\omega = \omega_2 - \omega_1 \cong 2\zeta\omega_n \tag{3.43}$$

The increment of frequency $\Delta\omega$ associated with the half-power points P_1 and P_2 is referred to as the *bandwidth* of the system. Inserting Eq. (3.43) into Eq. (3.37), we obtain

$$Q \cong \frac{1}{2\zeta} \cong \frac{\omega_n}{\Delta\omega} = \frac{\omega_n}{\omega_2 - \omega_1} \tag{3.44}$$

so that the requirement of high quality factor is equivalent to small bandwidth.

At this point, we turn our attention to the frequency response plot for the phase angle. To this end, we use Eq. (3.33) and plot $\phi(\omega)$ versus ω/ω_n with ζ acting as a parameter, as shown in Fig. 3.4. We observe that all curves pass through the point $\phi = \pi/2$, $\omega/\omega_n = 1$. Moreover, for $\omega/\omega_n < 1$, the phase angle tends to zero as $\zeta \to 0$, and for $\omega/\omega_n > 1$ the phase angle tends to π as $\zeta \to 0$. For $\zeta = 0$, the phase angle is zero for $\omega/\omega_n < 1$, it experiences a discontinuity at $\omega/\omega_n = 1$, jumping from 0 to $\pi/2$

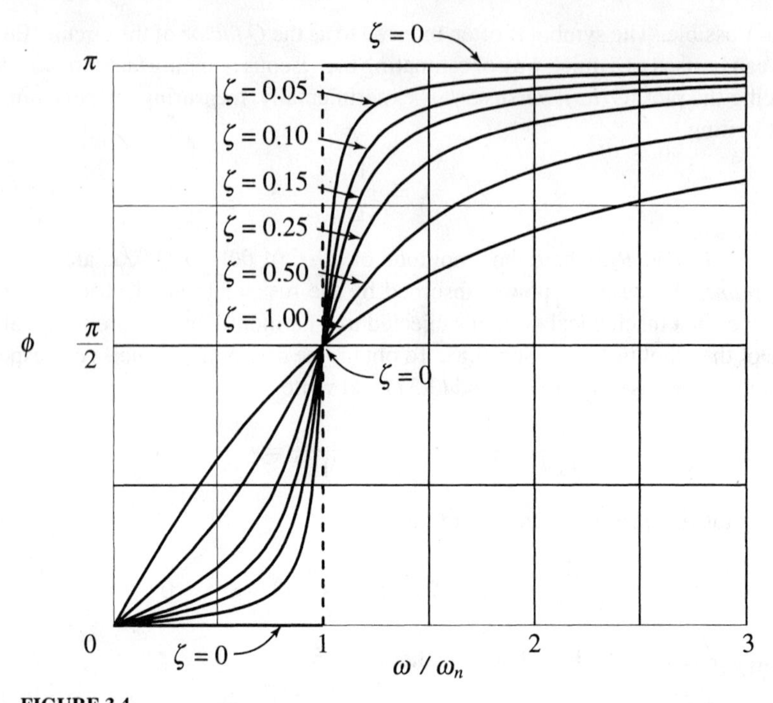

FIGURE 3.4
Phase angle of the frequency response

and then to π, and it continues with the value π for $\omega/\omega_n > 1$. Note that, letting $\zeta \to 0$, we can conclude from Fig. 3.4 that the phase angle at resonance is $\pi/2$.

The case in which the system is undamped, $\zeta = 0$, can be perhaps better explained by considering it separately. In fact, because the differential equation does not contain odd-order derivatives, there is no reason to use complex notation. Hence, letting $\zeta = 0$ in Eq. (3.4), we can write

$$\ddot{x}(t) + \omega_n^2 x(t) = \omega_n^2 A \cos \omega t \tag{3.45}$$

Then, assuming a solution in the form (3.5) with $C_1 = 0$, it is easy to verify that the response is

$$x(t) = \frac{1}{1 - (\omega/\omega_n)^2} A \cos \omega t = A G(\omega) \cos \omega t \tag{3.46}$$

where now the frequency response is simply

$$G(\omega) = \frac{1}{1 - (\omega/\omega_n)^2} \tag{3.47}$$

Hence, in contrast with Eq. (3.20), in the undamped case the frequency response, Eq. (3.47), is a real function. Consistent with this, it is not really necessary to have two frequency response plots, magnitude and phase angle, and a single plot suffices. Indeed, the plot $G(\omega)$ versus ω/ω_n displayed in Fig. 3.5 contains all the needed information.

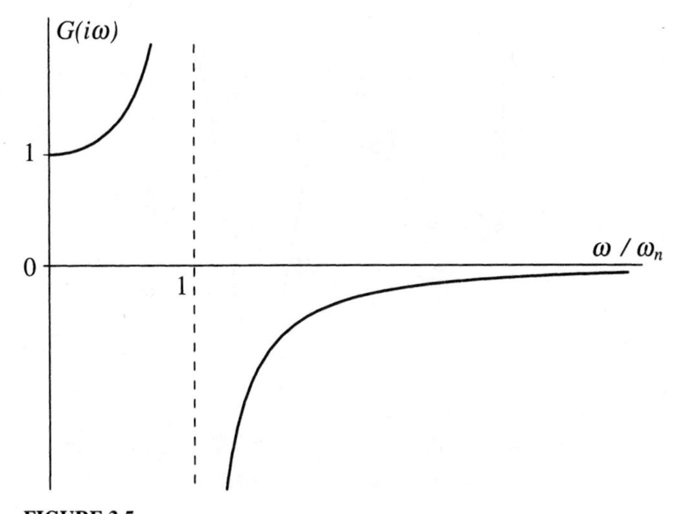

FIGURE 3.5
Frequency response for an undamped system

It shows that $G(\omega)$ is positive for $\omega/\omega_n < 1$, so that the response is in phase with the excitation when the driving frequency is below resonance, and is negative for $\omega/\omega_n > 1$, so that the response is 180° out of phase with the excitation when the driving frequency is above resonance. Moreover, for $\omega/\omega_n = 1$, the response jumps from $+\infty$ to $-\infty$. Of course, this is the case in which the system experiences *resonance*, and solution (3.46) is invalid, as it violates the small motions assumption implied by the linear restoring force in the spring. Hence, when the driving frequency approaches resonance, $\omega \to \omega_n$, the magnitude frequency response plot $G(\omega)$ versus ω/ω_n permits us only to make the qualitative statement that the system experiences violent vibration, rather than to obtain an exact vibration amplitude.

The question remains as to how to obtain a more quantitative statement concerning the system behavior at resonance. To answer this question, we let $\omega = \omega_n$ in Eq. (3.45) and write

$$\ddot{x}(t) + \omega_n^2 x(t) = \omega_n^2 A \cos \omega_n t \tag{3.48}$$

Clearly, a steady-state solution in the form of Eq. (3.5) with $\omega = \omega_n$ will not work here, because it leads to an infinite response, which at some point violates the assumption that the spring behaves linearly. Indeed, here the interest lies in a solution showing how the response builds up before it exceeds the linear range, which implies a transient solution. Techniques for solving Eq. (3.48) are studied in Ch. 4 in conjunction with transient solutions. At this point, we must be content with the statement that the particular solution of Eq. (3.48) has the form

$$x(t) = \frac{A}{2} \omega_n t \sin \omega_n t \tag{3.49}$$

which can be verified by substitution. Equation (3.49) represents oscillatory response with an amplitude increasing linearly with time. The response $x(t)$ is displayed as a

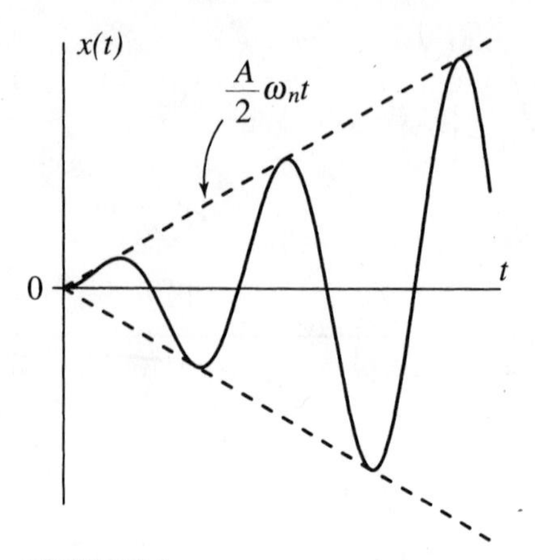

FIGURE 3.6
Response of an undamped system at resonance

function of time in Fig. 3.6, in which the straight lines $\pm(A/2)\omega_n t$ provide a linearly widening envelope. This implies that the response undergoes increasingly wild fluctuations as time increases, as expected. Here too, the validity of the solution is limited by the small motions assumption implicit in a linear system. Still, the solution provides a more quantitative picture of the response not available before. In this regard, we observe that the excitation is proportional to $\cos\omega_n t$ and the response is proportional to $\sin\omega_n t$. But, $\sin\omega_n t = \cos(\omega_n t - \pi/2)$, so that we conclude that the response experiences a 90° phase angle with respect to the excitation. This is entirely consistent with the conclusion reached from the phase angle plot of Fig. 3.4 for the case $\zeta = 0$, $\omega/\omega_n = 1$.

3.3 SYSTEMS WITH ROTATING UNBALANCED MASSES

There is a variety of engineering systems subjected to harmonic excitations, many of them involving rotating unbalanced masses. Of course, our first encounter with such systems was in Sec. 1.10, in which we used the model of a washing machine with nonuniformly distributed clothes to demonstrate that the effect of a rotating unbalanced mass is to exert a harmonic force upon the body of the machine. Moreover, the equation of motion for the model of Fig. 1.34a, Eq. (1.128), was shown to be

$$M\ddot{x}(t) + c\dot{x}(t) + kx(t) = me\omega^2 \sin\omega t \tag{3.50}$$

where $x(t)$ is the displacement, M the total mass of the system, c the coefficient of viscous damping of the support, k the spring constant of the support, m the rotating mass and e the eccentricity of m. Dividing Eq. (3.50) through by M and introducing the notation $c/M = 2\zeta\omega_n$, $k/M = \omega_n^2$, where ζ is the viscous damping factor and ω_n the natural frequency of undamped oscillation, we can rewrite Eq. (3.50) in the form

$$\ddot{x}(t) + 2\zeta\omega_n\dot{x}(t) + \omega_n^2 x(t) = \frac{m}{M} e\omega^2 \sin\omega t \tag{3.51}$$

which is similar to Eq. (3.4), except that the harmonic force is proportional to $\sin\omega t$ instead of $\cos\omega t$ and the proportionality factor is $(m/M)e\omega^2$ instead of $\omega_n^2 A$. Hence, with these modifications, we can use the solution obtained in Sec. 3.1.

The general solution of Eq. (3.4) was shown in Sec. 3.1 to be given by Eq. (3.27), where $|G(i\omega)|$ is the magnitude and $\phi(\omega)$ the phase angle of the frequency response, Eqs. (3.30) and (3.33), respectively. But, because the harmonic force is proportional to $\sin\omega t$ instead of $\cos\omega t$, we must use the response given by Eq. (3.28) instead of that given by Eq. (3.27). Moreover, comparing $(m/M)e\omega^2$ with $\omega_n^2 A$, we conclude that the constant A in Eq. (3.28) must be replaced by $(m/M)e\omega^2/\omega_n^2$. Hence, the solution of Eq. (3.51) is simply

$$x(t) = \frac{m}{M}e\left(\frac{\omega}{\omega_n}\right)^2 |G(i\omega)|\sin(\omega t - \phi) \tag{3.52}$$

To determine a nondimensional ratio capable of describing the magnitude of the response as a function of the driving frequency ω, we invoke once again the analogy with the approach of Sec. 3.1 and write the response in the form

$$x(t) = |X|\sin(\omega t - \phi) \tag{3.53}$$

so that, contrasting Eqs. (3.52) and (3.53), the indicated nondimensional ratio for the problem at hand is

$$\frac{M|X|}{me} = \left(\frac{\omega}{\omega_n}\right)^2 |G(i\omega)| \tag{3.54}$$

instead of $|G(i\omega)|$ alone. Hence, Fig. 3.3 is not applicable. Indeed, in the case at hand, the magnitude of the response is described by plots $(\omega/\omega_n)^2|G(i\omega)|$ versus ω/ω_n with ζ as a parameter; such plots are shown in Fig. 3.7. On the other hand, the phase angle plots $\phi(\omega)$ versus ω/ω_n for different values of ζ remain as in Fig. 3.4.

Comparing Fig. 3.7 to Fig. 3.3, we observe that the response of the system with the rotating eccentric mass, Fig. 3.7, differs from the response of the system subjected to a mere harmonic force, Fig. 3.3, in three respects: i) all curves begin at 0, as opposed to beginning at 1, ii) the peaks occur for $\omega/\omega_n > 1$, rather than for $\omega/\omega_n < 1$, and iii) as ω/ω_n becomes very large, the magnitude of the response tends to 1, instead of tending to 0. The latter has some interesting implications. To explain this statement, we propose to determine the position of the system mass center for large ω/ω_n. Referring to Figs. 1.34b and 1.34c, we note that the main mass $M - m$ undergoes the displacement x and the eccentric mass m the vertical displacement $x + e\sin\omega t$, so that using Eq. (1.39) we conclude that the position of the mass center relative to the equilibrium position is given by

$$x_C = \frac{1}{M}[(M-m)x + m(x + e\sin\omega t)] = \frac{1}{M}(Mx + me\sin\omega t) \tag{3.55}$$

But, from Figs. 3.7 and 3.4, as ω/ω_n becomes very large, we can write

$$\lim_{\omega/\omega_n\to\infty}\left(\frac{\omega}{\omega_n}\right)^2 |G(i\omega)| = 1, \quad \lim_{\omega/\omega_n\to\infty}\phi(\omega) = \pi \tag{3.56}$$

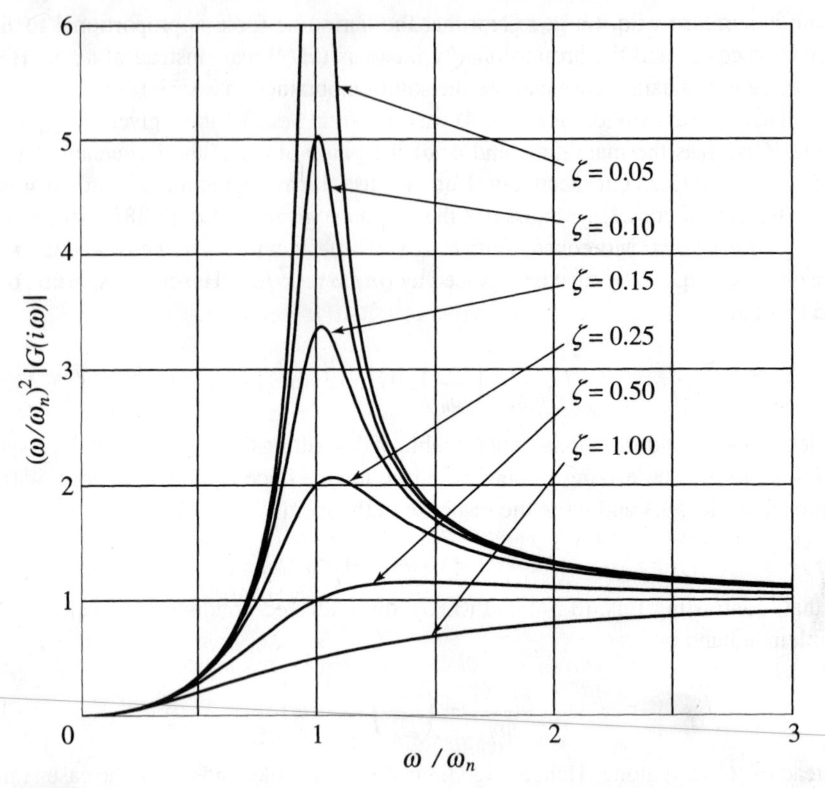

FIGURE 3.7
Magnitude of the frequency response for a system with rotating unbalanced masses

so that, using Eq. (3.52),

$$\lim_{\omega/\omega_n\to\infty} x = \frac{m}{M} e \sin(\omega t - \pi) = -\frac{m}{M} e \sin \omega t \tag{3.57}$$

Hence, inserting Eq. (3.57) into Eq. (3.55), we conclude that

$$\lim_{\omega/\omega_n\to\infty} x_C = \frac{1}{M}\left(-M\frac{m}{M} e \sin \omega t + me \sin \omega t\right) = 0 \tag{3.58}$$

or, for large driving frequencies ω, the masses $M - m$ and m move in such a way that the system mass center remains at rest. This statement is true regardless of the amount of damping.

3.4 WHIRLING OF ROTATING SHAFTS

Many mechanical systems involve a heavy rotating disk, known as a rotor, attached to a flexible shaft mounted on bearings. Typical examples are electric motors, turbines, compressors, etc. If the rotor has some eccentricity, i.e., if the mass center of the disk does not coincide with the geometric center, then the rotation produces a centrifugal force causing the shaft to bend. The rotation of the plane containing the bent shaft about the bearing axis is known as *whirling*. For certain rotational velocities, the system experiences violent vibrations, a phenomenon we propose to investigate.

Figure 3.8a shows a shaft rotating with the constant angular velocity ω relative to the inertial axes x, y. The shaft carries a disk of total mass m at midspan and is assumed to be massless. Hence, the motion of the system can be described by the displacements x and y of the geometric center S of the disk. Although this implies a two-degree-of-freedom system, the x and y motions are independent, so that the solution can be carried out as if there were two systems with one degree of freedom each.

As a preliminary to the derivation of the equations of motion, we wish to calculate the acceleration of the mass center. To this end, we denote the origin of the inertial system x, y by O and the center of mass of the disk by C. Due to some imperfection of the rotor, the mass center C does not coincide with the geometric center S. We denote the distance between S and C by e, as shown in Fig. 3.8b, where e represents the eccentricity. To calculate the acceleration $\mathbf{a}_C$ of the mass center C, we first write the radius vector $\mathbf{r}_C$ from O to C in terms of rectangular components as follows:

$$\mathbf{r}_C = (x + e\cos\omega t)\mathbf{i} + (y + e\sin\omega t)\mathbf{j} \tag{3.59}$$

where $\mathbf{i}$ and $\mathbf{j}$ are constant unit vectors along axes x and y, respectively. Then, differentiating Eq. (3.59) twice with respect to time, we obtain the acceleration of C in the form

$$\mathbf{a}_C = (\ddot{x} - e\omega^2\cos\omega t)\mathbf{i} + (\ddot{y} - e\omega^2\sin\omega t)\mathbf{j} \tag{3.60}$$

To derive the equations of motion, we assume that the only forces acting on the disk are restoring forces due to the elasticity of the shaft and resisting forces due to viscous damping, such as caused by air friction. The elastic effects are represented by equivalent spring constants k_x and k_y associated with the deformation of the shaft in the

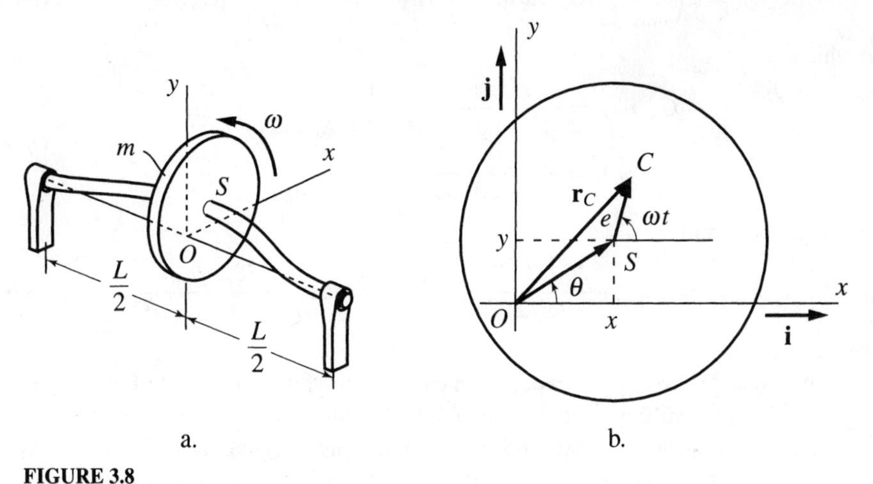

FIGURE 3.8
a. Rotor on a rotating shaft, b. Diagram showing mass center C and geometric center S

x and y directions, respectively. Moreover, we assume that the coefficient of viscous damping is the same in both directions and equal to c. The elastically restoring forces and the viscous damping forces are acting at point S. Considering Eq. (3.60), the x and y components of Newton's second law, Eqs. (1.63), are

$$\begin{aligned} -k_x x - c\dot{x} &= m(\ddot{x} - e\omega^2 \cos\omega t) \\ -k_y y - c\dot{y} &= m(\ddot{y} - e\omega^2 \sin\omega t) \end{aligned} \tag{3.61}$$

which can be rearranged in the form

$$\begin{aligned} \ddot{x} + 2\zeta_x\omega_{nx}\dot{x} + \omega_{nx}^2 x &= e\omega^2 \cos\omega t \\ \ddot{y} + 2\zeta_y\omega_{ny}\dot{y} + \omega_{ny}^2 y &= e\omega^2 \sin\omega t \end{aligned} \tag{3.62}$$

where

$$\begin{aligned} \zeta_x &= \frac{c}{2m\omega_{nx}}, \qquad \omega_{nx} = \sqrt{\frac{k_x}{m}} \\ \zeta_y &= \frac{c}{2m\omega_{ny}}, \qquad \omega_{ny} = \sqrt{\frac{k_y}{m}} \end{aligned} \tag{3.63}$$

are viscous damping factors and natural frequencies.

Equations (3.62) are of the same type as Eq. (3.51) of Sec. 3.3. This should come as no surprise, as a rotating flexible shaft carrying an unbalanced rotor is equivalent to a system with a rotating unbalanced mass. Hence, the steady-state solution of Eqs. (3.62) can be obtained by the pattern established in Sec. 3.3. Indeed, following that pattern, we can write the solutions

$$x(t) = |X(\omega)|\cos(\omega t - \phi_x), \qquad y(t) = |Y(\omega)|\sin(\omega t - \phi_y) \tag{3.64}$$

where the individual amplitudes are

$$|X(\omega)| = e\left(\frac{\omega}{\omega_{nx}}\right)^2 |G_x(i\omega)|, \qquad |Y(\omega)| = e\left(\frac{\omega}{\omega_{ny}}\right)^2 |G_y(i\omega)| \tag{3.65}$$

in which

$$\begin{aligned} |G_x(i\omega)| &= \frac{1}{\{[1-(\omega/\omega_{nx})^2]^2 + (2\zeta_x\omega/\omega_{nx})^2\}^{1/2}} \\ |G_y(i\omega)| &= \frac{1}{\{[1-(\omega/\omega_{ny})^2]^2 + (2\zeta_y\omega/\omega_{ny})^2\}^{1/2}} \end{aligned} \tag{3.66}$$

are magnitudes and

$$\phi_x = \tan^{-1}\frac{2\zeta_x\omega/\omega_{nx}}{1-(\omega/\omega_{nx})^2}, \qquad \phi_y = \tan^{-1}\frac{2\zeta_y\omega/\omega_{ny}}{1-(\omega/\omega_{ny})^2} \tag{3.67}$$

are phase angles.

We consider first the most common case, namely, that of a shaft of circular cross section, so that the stiffness is the same in both directions, $k_x = k_y = k$. In this case, the two natural frequencies coincide and so do the viscous damping factors, or

$$\omega_{nx} = \omega_{ny} = \omega = \sqrt{\frac{k}{m}}, \qquad \zeta_x = \zeta_y = \zeta = \frac{c}{2m\omega_n} \tag{3.68}$$

Moreover, in view of Eqs. (3.68), we conclude from Eqs. (3.66) and (3.67) that the magnitudes on the one hand and the phase angles on the other hand are the same, or

$$|G_x(i\omega)| = |G_y(i\omega)| = |G(i\omega)| = \frac{1}{\{[1-(\omega/\omega_n)^2]^2 + (2\zeta\omega/\omega_n)^2\}^{1/2}}$$
$$\phi_x = \phi_y = \phi = \tan^{-1}\frac{2\zeta\omega/\omega_n}{1-(\omega/\omega_n)^2} \tag{3.69}$$

It follows immediately, from Eqs. (3.64), that the amplitudes of the motions x and y are equal to one another, or

$$|X(\omega)| = |Y(\omega)| = e\left(\frac{\omega}{\omega_n}\right)^2 |G(i\omega)| \tag{3.70}$$

But, from Fig. 3.9 and Eqs. (3.64), we can write

$$\tan\theta = \frac{y}{x} = \tan(\omega t - \phi) \tag{3.71}$$

from which we conclude that

$$\theta = \omega t - \phi \tag{3.72}$$

and that

$$\dot{\theta} = \omega \tag{3.73}$$

Hence, in this case the shaft whirls with the same angular velocity as the rotation of the disk, so that the shaft and the disk rotate together as a rigid body. This case is known as *synchronous whirl.* It is easy to verify that in synchronous whirl the radial distance from O to S for a given ω is constant, or

$$r_{OS} = \sqrt{x^2+y^2} = e\left(\frac{\omega}{\omega_n}\right)^2 |G(i\omega)| = \text{constant} \tag{3.74}$$

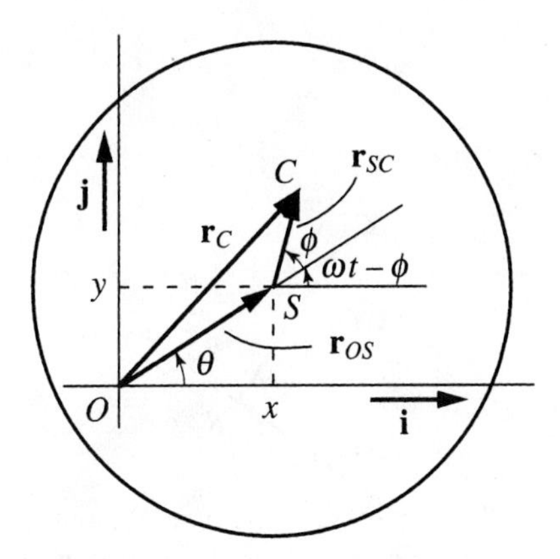

FIGURE 3.9
Diagram showing the relation between the whirling angle θ of the shaft and the rotation angle $\omega t - \phi$ of the rotor

so that point S describes a circle about point O. To determine the position of C relative to the whirling plane, we consider Eq. (3.72). The relation between the angles θ, ωt and ϕ is depicted in Fig. 3.9. Indeed, from Fig. 3.9, we can interpret the phase angle ϕ as the angle between the radius vectors $\mathbf{r}_{OS}$ and $\mathbf{r}_{SC}$. Hence, recalling the second of Eqs. (3.69), we conclude that $\phi < \pi/2$ for $\omega < \omega_n$, $\phi = \pi/2$ for $\omega = \omega_n$ and $\phi > \pi/2$ for $\omega > \omega_n$. The three configurations are shown in Fig. 3.10.

As a final remark concerning synchronous whirl, we note from Eqs. (3.69) that the magnitude and the phase angle have the same expressions as in the case of the rotating unbalanced mass discussed in Sec. 3.3, which corroborates our earlier statements that the two systems are analogous.

Next, we return to the case in which the two stiffnesses are different and consider the undamped case, $c = 0$. In this case, solutions (3.64) can be written as

$$x(t) = X(\omega)\cos\omega t, \qquad y(t) = Y(\omega)\sin\omega t \tag{3.75}$$

where

$$X(\omega) = \frac{e(\omega/\omega_{nx})^2}{1-(\omega/\omega_{nx})^2}, \qquad Y(\omega) = \frac{e(\omega/\omega_{ny})^2}{1-(\omega/\omega_{ny})^2} \tag{3.76}$$

Dividing the first of Eqs. (3.75) by $X(\omega)$ and the second by $Y(\omega)$, squaring and adding the results, we obtain

$$\frac{x^2}{X^2} + \frac{y^2}{Y^2} = 1 \tag{3.77}$$

which represents the equation of an ellipse. Hence, as the shaft whirls, point S describes an ellipse with point O as its geometric center. To gain more insight into the motion, we consider Eqs. (3.75) and write

$$\tan\theta = \frac{y}{x} = \frac{Y}{X}\tan\omega t \tag{3.78}$$

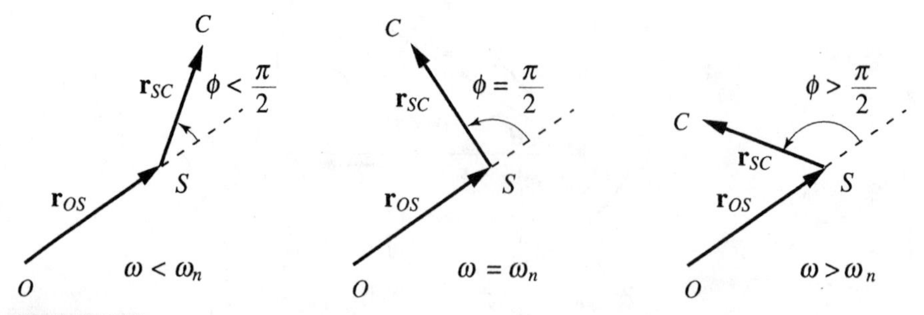

FIGURE 3.10
Phase angle ϕ for synchronous whirl

Differentiating both sides of Eq. (3.78) with respect to time and considering Eqs. (3.75), we obtain

$$\dot{\theta} = \frac{XY}{X^2 \cos^2 \omega t + Y^2 \sin^2 \omega t}\,\omega \tag{3.79}$$

But, the denominator on the right side of Eq. (3.79) is always positive, so that the sign of $\dot{\theta}$ depends on the sign of XY. By convention, the sign of ω is assumed as positive, i.e., the disk rotates in the counter-clockwise sense. We can distinguish the following cases:

1. $\omega < \omega_{nx}$ and $\omega < \omega_{ny}$. In this case, we conclude from Eqs. (3.76) that $XY > 0$, so that point S moves on the ellipse in the same sense as the rotation ω.
2. $\omega_{nx} < \omega < \omega_{ny}$ or $\omega_{ny} < \omega < \omega_{nx}$. In either of these two cases $XY < 0$, so that S moves in the opposite sense from ω.
3. $\omega > \omega_{nx}$ and $\omega > \omega_{ny}$. In this case $XY > 0$, so that S moves in the same sense as ω.

The three cases are displayed in Fig. 3.11.

Examining solutions (3.75) and (3.76) for the undamped case, we conclude that the possibility of resonance exists. In fact, there are two frequencies for which resonance is possible, namely, $\omega = \omega_{nx}$ and $\omega = \omega_{ny}$. Of course, in the case of resonance, solutions (3.75) and (3.76) are no longer valid. Following the approach of Sec. 3.1, it is easy to verify by substitution that the particular solutions in the two cases of resonance are

$$x(t) = \frac{1}{2} e\omega_{nx} t \sin \omega_{nx} t, \qquad y(t) = -\frac{1}{2} e\omega_{ny} t \cos \omega_{ny} t \tag{3.80}$$

The plot $x(t)$ versus t resembles that of Fig. 3.6. In fact, it is the same for $A = e$. The plot $y(t)$ versus t also resembles that of Fig. 3.6 except that ω_{nx} and $\sin \omega_{nx} t$ must be replaced by ω_{ny} and $\sin(\omega_{ny} t - \pi/2)$, respectively. This is easily explained by the fact that $\sin(\omega_{ny} t - \pi/2) = -\cos \omega_{ny} t$. The two frequencies $\omega = \omega_{nx}$ and $\omega = \omega_{ny}$ represent *critical frequencies*, more commonly known as *critical speeds*.

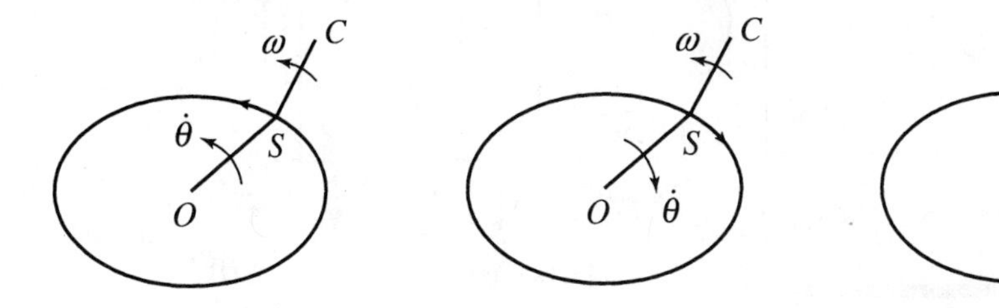

$\omega < \omega_{nx}$ and $\omega < \omega_{ny}$ $\qquad$ $\omega_{nx} < \omega < \omega_{ny}$ or $\omega_{ny} < \omega < \omega_{ny}$ $\qquad$ $\omega > \omega_{nx}$ and $\omega > \omega_{ny}$

FIGURE 3.11
Diagrams showing the shaft angular velocity $\dot{\theta}$ and the rotor angular velocity ω

3.5 HARMONIC MOTION OF THE BASE

On occasions, sensitive equipment must be placed on a foundation undergoing undesirable vibration. To protect the equipment, it is necessary to isolate it from the damaging effects of the vibrating foundation. This can be achieved through rubber mounts acting both as springs and dampers in the same manner as the supports of washing machines discussed in Sec. 1.10. A similar example is a vehicle traveling on a wavy road, in which case the suspension must isolate the body from the motion induced by the wavy road. Yet another example is an engine mounted on a vibrating aircraft wing.

We assume that each of the various systems mentioned above can be modeled as a single-degree-of-freedom system mounted on a base undergoing the displacement $y(t)$, as depicted in Fig. 3.12a. The corresponding free-body diagram is shown in Fig. 3.12b. Using Newton's second law, we obtain the equation of motion

$$-c(\dot{x} - \dot{y}) - k(x - y) = m\ddot{x} \tag{3.81}$$

Dividing through by m and rearranging, we can write

$$\ddot{x} + 2\zeta\omega_n\dot{x} + \omega_n^2 x = 2\zeta\omega_n\dot{y} + \omega_n^2 y \tag{3.82}$$

We are interested in the case in which the motion of the base is harmonic. Hence, following the procedure of Sec. 3.1, we express the displacement of the base in the exponential form

$$y(t) = \text{Re } Ae^{i\omega t} \tag{3.83}$$

and the response in the corresponding form

$$x(t) = X(i\omega)e^{i\omega t} \tag{3.84}$$

with the understanding that, if the excitation is $A\cos\omega t$, the response is Re $x(t)$ and if the excitation is $A\sin\omega t$, the response is Im $x(t)$. Inserting Eqs. (3.83) and (3.84) into

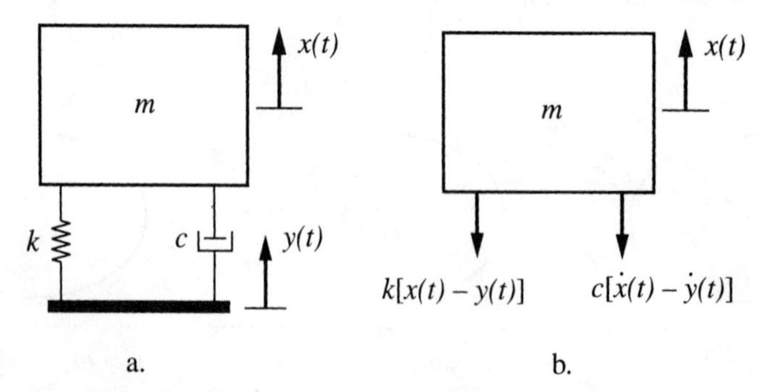

FIGURE 3.12
a. Mass-damper-spring system with moving base, b. Free-body diagram

Eq. (3.82), dividing through by $e^{i\omega t}$ and solving for $X(i\omega)$, we obtain

$$X(i\omega) = \frac{1 + i2\zeta\omega/\omega_n}{1 - (\omega/\omega_n)^2 + i2\zeta\omega/\omega_n} A = (1 + i2\zeta\omega/\omega_n)G(i\omega)A \tag{3.85}$$

in which $G(i\omega)$ is the frequency response, Eq. (3.20).

As in Sec. 3.1 we will find it convenient to express $X(i\omega)$ in the form

$$X(i\omega) = |X(i\omega)|e^{-i\phi(\omega)} \tag{3.86}$$

so that the response can be written as

$$x(t) = |X(i\omega)|e^{i(\omega t - \phi)} \tag{3.87}$$

Using the analogy with Eq. (3.24), the magnitude of $X(i\omega)$ can be determined by writing

$$|X(i\omega)| = \sqrt{X(i\omega)\overline{X}(i\omega)} = \sqrt{(1 + i2\zeta\omega/\omega_n)G(i\omega)(1 - i2\zeta\omega/\omega_n)\overline{G}(i\omega)}\ A$$

$$= \left[1 + \left(\frac{2\zeta\omega}{\omega_n}\right)^2\right]^{1/2} |G(i\omega)|A \tag{3.88}$$

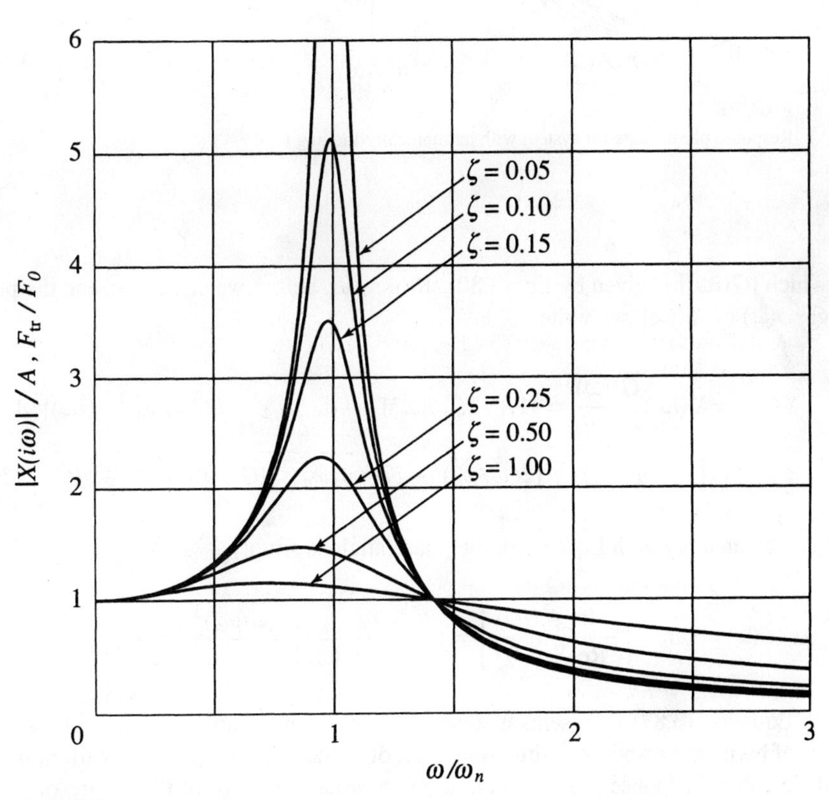

FIGURE 3.13
Nondimensional response magnitude for system with harmonically-moving base

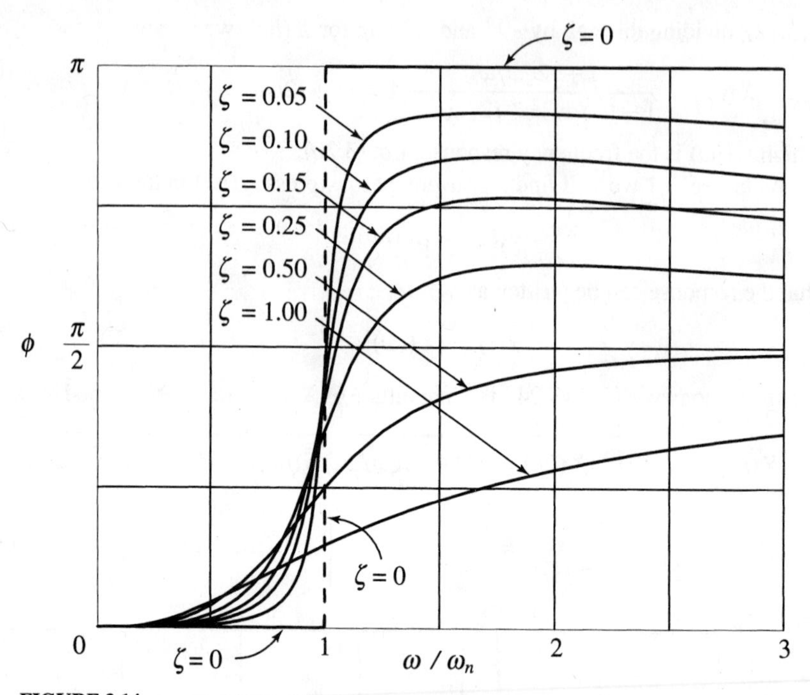

FIGURE 3.14
Response phase angle for system with harmonically-moving base

in which $|G(i\omega)|$ is given by Eq. (3.30). Moreover, before we can determine the phase angle $\phi(\omega)$ of $X(i\omega)$, we write

$$\begin{aligned} X(i\omega) =& X(i\omega)\frac{\overline{G}(i\omega)}{\overline{G}(i\omega)} = (1 + i2\zeta\omega/\omega_n)[1 - (\omega/\omega_n)^2 - i2\zeta\omega/\omega_n]|G(i\omega)|^2 A \\ =& \left[1 - (\omega/\omega_n)^2 + (2\zeta\omega/\omega_n)^2 - i2\zeta(\omega/\omega_n)^3\right]|G(i\omega)|^2 A \end{aligned} \tag{3.89}$$

Hence, by analogy with Eq. (3.33), the phase angle is simply

$$\phi(\omega) = \tan^{-1}\left[\frac{-\text{Im } X(i\omega)}{\text{Re } X(i\omega)}\right] = \tan^{-1}\frac{2\zeta(\omega/\omega_n)^3}{1 - (\omega/\omega_n)^2 + (2\zeta\omega/\omega_n)^2} \tag{3.90}$$

Equation (3.87) represents a steady-state harmonic response. As in all previous cases of harmonic response, a time-domain plot of the response is not very illuminating, and frequency-response plots provide a much broader picture of the nature of the response. From Eq. (3.88), the indicated nondimensional ratio descriptive of the response magnitude for the case at hand is

$$\frac{|X(i\omega)|}{A} = \left[1 + \left(\frac{2\zeta\omega}{\omega_n}\right)^2\right]^{1/2} |G(i\omega)| = \left\{\frac{1 + (2\zeta\omega/\omega_n)^2}{\left[1 - (\omega/\omega_n)^2\right]^2 + (2\zeta\omega/\omega_n)^2}\right\}^{1/2} \tag{3.91}$$

where the ratio $|Xi\omega)|/A$ is known as *transmissibility* for reasons to be explained in Sec. 3.6. Curves $|X(i\omega)|/A$ versus ω/ω_n with ζ as a parameter are plotted in Fig. 3.13. We observe by substitution into Eq. (3.91) that for $\omega/\omega_n = \sqrt{2}$ the response has the same magnitude as the excitation, and it is magnified relative to the excitation for $\omega/\omega_n < \sqrt{2}$ and reduced for $\omega/\omega_n > \sqrt{2}$, the amount of magnification and reduction depending on the viscous damping factor ζ. These are facts to be considered in designing the isolation parameters ζ and ω_n. Moreover, curves $\phi(\omega)$ versus ω/ω_n with ζ acting as a parameter are plotted in Fig. 3.14. As in previous cases, such as in Fig. 3.4, for $\zeta = 0$ the response is in phase with the excitation for $\omega/\omega_n < 1$ and 180° out of phase with the excitation for $\omega/\omega_n > 1$.

3.6 VIBRATION ISOLATION

In many systems of the type shown in Fig. 3.1, we are interested in transmitting as little vibration as possible to the base. This problem can become critical when the excitation is harmonic. Clearly, the force is transmitted to the base through springs and dampers. From Fig. 3.2, we conclude that the amplitude of that force is

$$F_{\text{tr}} = m\left[(2\zeta\omega_n\dot{x})^2 + (\omega_n^2 x)^2\right]^{1/2} \tag{3.92}$$

where the amplitude of the velocity is simply ωx. Hence, because $m\omega_n^2 = k$, we have

$$F_{\text{tr}} = kx\left[1 + \left(\frac{2\zeta\omega}{\omega_n}\right)^2\right]^{1/2} \tag{3.93}$$

But from Eq. (3.26), if we recall that the phase angle is of no consequence as far as the force amplitude is concerned, we conclude that

$$F_{\text{tr}} = Ak\left[1 + \left(\frac{2\zeta\omega}{\omega_n}\right)^2\right]^{1/2} |G(i\omega)| \tag{3.94}$$

In view of the fact that Ak is the amplitude of the actual excitation force, which we denote here by F_0, the nondimensional ratio F_{tr}/F_0 is a measure of the force transmitted to the base. The ratio can be written as

$$\frac{F_{\text{tr}}}{F_0} = \left[1 + \left(\frac{2\zeta\omega}{\omega_n}\right)^2\right]^{1/2} |G(i\omega)| \tag{3.95}$$

and is recognized as the *transmissibility* given by Eq. (3.91). Hence, the plots F_{tr}/F_0 versus ω/ω_n are the same as the plots $|X|/A$ versus ω/ω_n shown in Fig. 3.13. From Sec. 3.5, when $\omega/\omega_n = \sqrt{2}$ the full force is transmitted to the base, $F_{\text{tr}}/F_0 = 1$. For $\omega/\omega_n > \sqrt{2}$ the force transmitted tends to decrease with increasing driving frequency ω, regardless of ζ. Interestingly, damping does not alleviate the situation and in fact, for $\omega/\omega_n > \sqrt{2}$, the transmitted force increases as damping increases. However, because in increasing the driving frequency to values $\omega > \sqrt{2}\omega_n$ we must go through resonance, we conclude that a certain amount of damping is necessary to prevent displacements from becoming unduly large.

3.7 VIBRATION MEASURING INSTRUMENTS

The most common vibration measuring instruments are used to measure displacements and accelerations. Many instruments consist of a case containing a mass-damper-spring system of the type depicted in Fig. 3.15 and a transducer measuring the displacement of the mass relative to the case. We note that a transducer is a device converting one form of energy into another, generally mechanical into electrical in the case of instruments measuring vibration. The mass, referred to as a *seismic mass*, or a *proof mass*, is constrained to move along a given axis and damping may be provided by a viscous fluid inside the case. It is easy to see that the system of Fig. 3.15, except for the transducer, is of the type discussed earlier in this chapter.

The displacement of the case, the displacement of the mass relative to the case and the absolute displacement of the mass are denoted by $y(t)$, $z(t)$ and $x(t)$, respectively, so that

$$x(t) = y(t) + z(t) \tag{3.96}$$

The system of Fig. 3.15 resembles the system of Fig. 3.12a entirely, so that from Eq. (3.81), we can write the equation of motion

$$m\ddot{x}(t) + c[\dot{x}(t) - \dot{y}(t)] + k(x(t) - y(t)] = 0 \tag{3.97}$$

The objective is to determine the motion $y(t)$ of the case from measurements of the relative displacement $z(t)$. To this end, we use Eq. (3.96) to eliminate $x(t)$ and rewrite

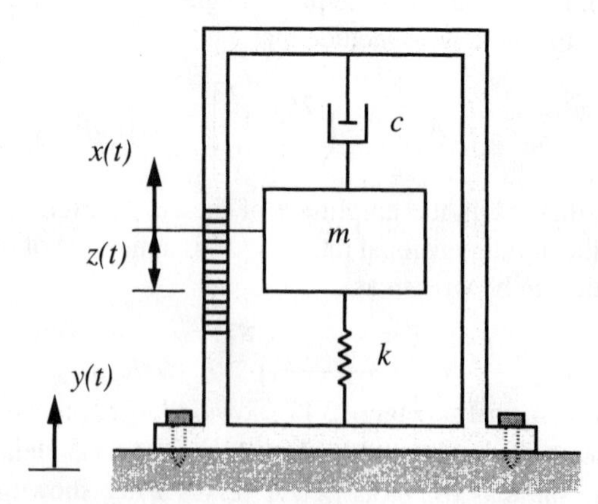

FIGURE 3.15
Vibration measuring instrument

Eq. (3.97) as

$$m\ddot{z}(t) + c\dot{z}(t) + kz(t) = -m\ddot{y}(t) \tag{3.98}$$

Next, we assume that the vibration to be measured is harmonic and of the form

$$y(t) = Y_0 \cos \omega t \tag{3.99}$$

so that Eq. (3.98) becomes

$$m\ddot{z} + c\dot{z} + kz = Y_0 m\omega^2 \cos \omega t \tag{3.100}$$

which is of the same type as Eq. (3.50). By analogy with Eq. (3.52), the response is

$$z(t) = Y_0 \left(\frac{\omega}{\omega_n}\right)^2 |G(i\omega)| \cos(\omega t - \phi) \tag{3.101}$$

where the phase angle ϕ is given by Eq. (3.33). Introducing the notation

$$z(t) = Z_0 \cos(\omega t - \phi) \tag{3.102}$$

in which Z_0 is the *measurement amplitude*, we conclude that

$$\frac{Z_0}{Y_0} = \left(\frac{\omega}{\omega_n}\right)^2 |G(i\omega)| \tag{3.103}$$

so that the plot Z_0/Y_0 versus ω/ω_n is identical to that given in Fig. 3.7. The plot is shown again in Fig. 3.16 on a scale more suitable for our current application. The phase angle plot ϕ versus ω/ω_n is as given by Fig. 3.4.

3.7.1 Accelerometers—high frequency instruments

For small values of the ratio ω/ω_n the magnitude $|G(i\omega)|$ is nearly unity, so that the measurement amplitude Z_0 can be approximated by

$$Z_0 \cong Y_0 \left(\frac{\omega}{\omega_n}\right)^2 \tag{3.104}$$

Because $Y_0\omega^2$ represents the acceleration of the case, it follows that the measurement amplitude Z_0 is proportional to the acceleration of the case, where the proportionality constant is $1/\omega_n^2$. Hence, if the natural frequency ω_n of the measuring instrument is sufficiently high relative to the frequency ω of the harmonic motion to be measured that the amplitude ratio Z_0/Y_0 can be approximated by the parabola $(\omega/\omega_n)^2$ (see dashed line in Fig. 3.16), the instrument is known as an *accelerometer*. Because the range of ω/ω_n in which the amplitude ratio can be approximated by $(\omega/\omega_n)^2$ is the same as the range in which $|G(i\omega)|$ is approximately equal to unity, it is advantageous to refer to the plot $|G(i\omega)|$ versus ω/ω_n instead of the plot Z_0/Y_0 versus ω/ω_n to determine the range of utility of the instrument. Figure 3.17 shows enlarged plots $|G(i\omega)|$ versus ω/ω_n corresponding to the magnitude range $0.95 \leq |G(i\omega)| \leq 1.05$ and the frequency ratio range $0 \leq \omega/\omega_n \leq 1$ with ζ acting as a parameter. From Fig. 3.17, we conclude that the range in which $|G(i\omega)|$ is approximately unity is very small for light damping, which implies that the natural frequency of lightly damped accelerometers must be appreciably

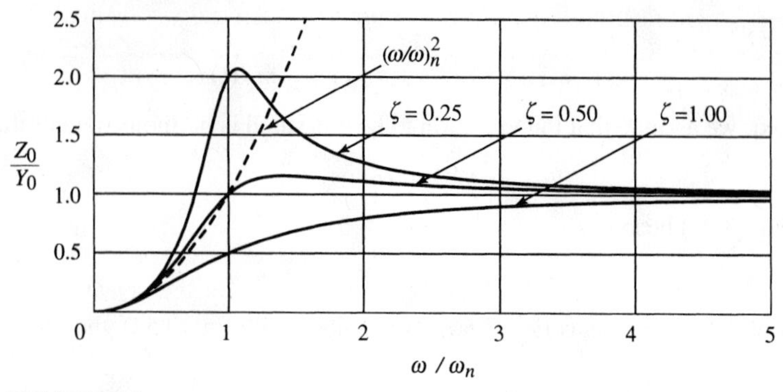

FIGURE 3.16
Detailed frequency response magnitude plots for accelerometers

larger than the frequency of the harmonic motion to be measured. To increase the range of utility of the instrument, larger damping is necessary. It is clear from Fig. 3.17 that the approximation is valid for a larger range of ω/ω_n if $0.65 < \zeta < 0.70$. Indeed, for $\zeta = 0.7$ the accelerometer can be used in the range $0 \leq \omega/\omega_n \leq 0.4$ with less than 1 percent error, and the range can be extended to $\omega/\omega_n \leq 0.7$ if proper corrections, based on the instrument calibration, are made.

In the preceding discussion, it was tacitly assumed that the accelerometer axis is parallel to the local vertical (defined as the direction of a radius vector from the center of the earth to the location of the instrument) and hence parallel to the force due to gravity, so that the weight mg causes the seismic mass to undergo the static displacement $\delta_{st} = mg/k$, as discussed in Sec. 1.10. In general, the instrument will interpret the reading erroneously as an acceleration of the case equal to $\ddot{y} = g$. This problem can be disposed of easily by calibrating the instrument so as to measure $x(t)$ from the static equilibrium. The situation is not so simple if the accelerometer is used to

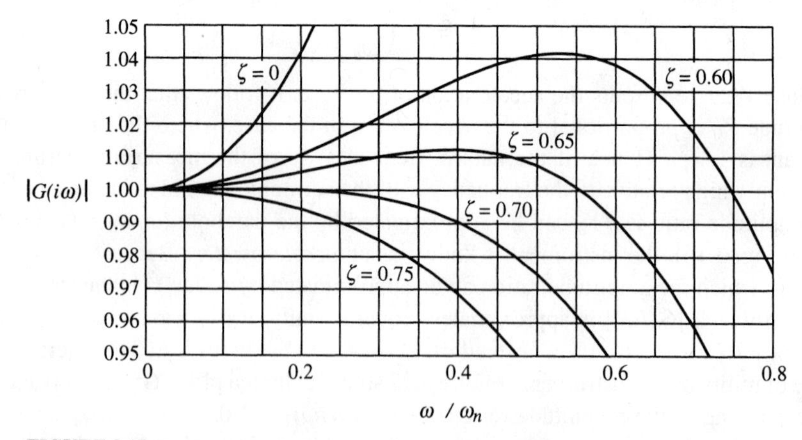

FIGURE 3.17
Diagram showing the accuracy of accelerometer measurements for various viscous damping factors

measure vibration in aircraft and missile structures, as in such cases the accelerometer axis is not guaranteed to coincide with the direction of the gravitational force. If the accelerometer axis tilts by an angle θ with respect to the local vertical, then the deflection is only $mg \cos\theta/k$, causing a false reading of the instrument. This problem can be taken care of by using a precision tilt table, thus permitting a continuous calibration of the instrument.

The most commonly used accelerometers are the compression-type piezoelectric accelerometers. They consist of a mass resting on a piezoelectric ceramic crystal, such as quartz, barium titanate, or lead zirconium titanate, with the crystal acting both as spring and sensor. The accelerometers have a preload providing a compressive stress exceeding the highest dynamic stress expected. Any acceleration increases or decreases the compressive stress in the piezoelectric element, thus generating an electric charge appearing at the accelerometer terminals. Piezoelectric accelerometers have negligible damping and they typically have a frequency range from 0 to 10,000 Hz (and beyond) and a natural frequency range from 30,000 to 50,000 Hz. They tend to be very light, weighing less than 20 g, and are relatively small, measuring less than 2 cm in diameter.

3.7.2 Seismometers—low frequency instruments

Also from Fig. 3.16, we notice that for very large values of ω/ω_n, the ratio $Z_0/Y_0 = (\omega/\omega_n)^2|G(i\omega)|$ approaches unity, regardless of the amount of damping. Hence, if the object is to measure displacements, then we should make the natural frequency of the instrument very low relative to the frequency to be measured, in which case the instrument is called a *seismometer*. For a seismometer, which is an instrument designed to measure ground displacements, such as those caused by earthquakes, or underground nuclear explosions, the requirement for a low natural frequency dictates that the spring be very soft and the mass relatively heavy, so that, in essence, the mass remains nearly stationary in inertial space while the case, being attached to the ground, moves relative to the mass. Displacement-measuring instruments typically have a frequency range from 10 to 500 Hz and a natural frequency between 1 and 5 Hz.

To examine the utility range of displacement-measuring instruments, we consider enlarged plots of $(\omega/\omega_n)^2|G(i\omega)|$ versus ω/ω_n, Fig. 3.16, for several values of ζ. The enlarged plots correspond to the amplitude range $0.95 \le (\omega/\omega_n)^2|G(i\omega)| \le 1.05$ and the frequency ratio range $1 \le \omega/\omega_n \le 10$, as shown in Fig. 3.18. We see that when $\zeta = 0$ the error remains below 5% for frequency ratios $\omega/\omega_n > 5$. The error falls to 1 % for $\zeta = 0.6$ and $\omega/\omega_n > 5$, $\zeta = 0.65$ and $\omega/\omega_n > 3.25$ and $\zeta = 0.7$ and $\omega/\omega_n > 2.6$. If errors of 5% are acceptable, then, the range for $\zeta = 0.6$ extends down to $\omega/\omega_n > 1.2$, for $\zeta = 0.65$ to $\omega/\omega_n > 1.4$ and for $\zeta = 0.7$ to $\omega/\omega_n > 1.75$. Significantly, the instrument is virtually free of errors for $\zeta = 0.7$ and $\omega/\omega_n > 5$.

Depending on the transducer, the instrument can measure displacements or velocities. Indeed, we see from Eqs. (3.99), (3.102) and (3.103) that

$$\frac{|\dot{z}(t)|}{|\dot{y}(t)|} = \frac{|\omega z(t)|}{|\omega y(t)|} = \frac{Z_0}{Y_0} = \left(\frac{\omega}{\omega_n}\right)^2 |G(i\omega)| \tag{3.105}$$

so that the ratio of velocity amplitudes is the same as the ratio of displacement ampli-

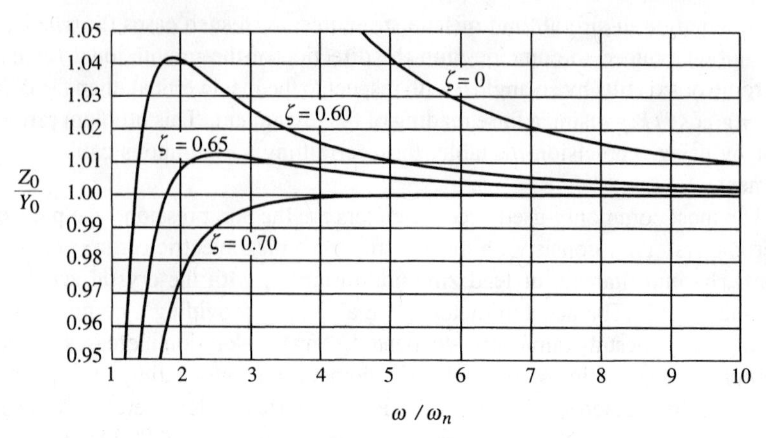

FIGURE 3.18
Detailed frequency response magnitude plots for seismometers

tudes. If the transducer measures velocities, the instrument is known as a *vibrometer*, or *velometer*. In such cases, the instrument output can be integrated once to obtain displacements, or differentiated once to obtain accelerations.

Because seismometers require a much larger mass than accelerometers and the relative motion of the mass in a seismometer is nearly equal in magnitude to the motion to be measured, seismometers are considerably larger in size than accelerometers. In view of this, if the interest lies in displacements, it may prove more desirable to use an accelerometer to measure the acceleration of the case, and then integrate twice with respect to time to obtain the displacement.

The above discussion has focused on the measurement of harmonic motion. In measuring more complicated motions, not only the amplitude but also the phase angle comes into play. As an example, if the motion consists of two harmonics, or

$$y(t) = Y_1 \cos \omega_1 t + Y_2 \cos \omega_2 t \tag{3.106}$$

and the accelerometer output is

$$y_a(t) = Y_1 \cos(\omega_1 t - \phi_1) + Y_2 \cos(\omega_2 t - \phi_2) \tag{3.107}$$

where ϕ_1 and ϕ_2 are two distinct phase angles, then the accelerometer fails to reproduce the motion $y(t)$, because the two harmonic components of the motion are shifted relative to one another. There are two cases in which the accelerometer output is able to reproduce the motion $y(t)$ without distortion. The first is the case of an undamped accelerometer, $\zeta = 0$, in which case the phase angle is zero. The second is the case in which the phase angle is proportional to the frequency, or

$$\phi_1 = c\omega_1, \qquad \phi_2 = c\omega_2 \tag{3.108}$$

Indeed, introducing Eqs. (3.108) into Eq. (3.107), we obtain

$$y_a(t) = Y_1 \cos \omega_1 (t - c) + Y_2 \cos \omega_2 (t - c) \tag{3.109}$$

so that both harmonics are shifted to the right on the time scale by the same time interval, thus retaining the nature of the motion $y(t)$. To explore the possibility of eliminating the phase distortion, we consider the case of small ω/ω_n, in which case the phase angle ϕ is small, as can be concluded from Eq. (3.33). Then, assuming that the phase angle increases linearly with the frequency, we can write

$$\sin\phi \cong \phi = c\omega, \qquad \cos\phi \cong 1 - \frac{1}{2}\phi^2 = 1 - \frac{1}{2}(c\omega)^2 \tag{3.110}$$

Inserting Eqs. (3.110) into Eq. (3.33), we obtain

$$\tan\phi = \frac{2\zeta\omega/\omega_n}{1-(\omega/\omega_n)^2} \cong \frac{c\omega}{1-(c\omega)^2/2} \tag{3.111}$$

which is satisfied provided

$$c = \sqrt{2}/\omega_n, \qquad \zeta = \sqrt{2}/2 = 0.707 \tag{3.112}$$

In general, any arbitrary motion can be regarded as a superposition of harmonic components. Hence, an accelerometer can be used for measuring arbitrary motions if the damping factor ζ is either equal to zero or equal to 0.707.

3.8 ENERGY DISSIPATION. STRUCTURAL DAMPING

Damped systems dissipate energy, so that such systems are nonconservative. In Ch. 2, we considered two types of damping, viscous damping and Coulomb damping. Of the two, viscous damping is more important for a variety of reasons. In the first place, viscous damping is widely used for vibration reduction devices, such as shock absorbers and base isolation. Another reason is that viscous damping is linear in velocity and permits easy analytical treatment, which makes it very desirable as a mathematical model for damping.

Experience shows that energy is dissipated in all real systems, including those regarded as conservative. For example, energy is dissipated in real springs as a result of internal friction. In contrast to viscous damping, damping due to internal friction does not depend on velocity. In general, this type of damping does not lend itself to easy modeling. Under certain circumstances, however, it is possible to model damping due to internal friction as viscous damping, thus making an analytical treatment of systems with internal damping possible.

The analogy between viscous damping and damping caused by internal friction is based on energy dissipation. To show this, we begin by determining the energy loss in viscous damping. In Sec. 3.1, we have shown that the response of a mass-damper-spring system subjected to a harmonic excitation equal to the real part of

$$F(t) = Ake^{i\omega t} \tag{3.113}$$

in which A is a real constant and k is the spring constant, is given by the real part of

$$x(t) = A|G(i\omega)|e^{i(\omega t-\phi)} = Xe^{i(\omega t-\phi)} \tag{3.114}$$

where

$$X = A|G(i\omega)| \tag{3.115}$$

is the response amplitude, in which $|G(i\omega)|$ is the magnitude of the frequency response, Eq. (3.30). Clearly, because of damping, the system is not conservative, and indeed energy is dissipated. Because this energy dissipation must be equal to the work done by the external force, we can refer to Eq. (1.24) and write the expression for the energy dissipated per cycle of vibration in the form

$$\Delta E_{\text{cyc}} = \int_{\text{cyc}} F\,dx = \int_0^{2\pi/\omega} F\dot{x}\,dt \tag{3.116}$$

where we recognize that we must retain the real part of F and $\dot{x}$ only. Inserting the first of Eqs. (3.29) and Eq. (3.113) into Eq. (3.116), we obtain

$$\begin{aligned}\Delta E_{\text{cyc}} &= -kA^2|G(i\omega)|\omega \int_0^{2\pi/\omega} \cos\omega t \sin(\omega t - \phi)dt \\ &= m\omega_n^2 A^2|G(i\omega)|\pi \sin\phi \end{aligned} \tag{3.117}$$

From Eqs. (3.30) and (3.33), it is not difficult to show that

$$\sin\phi = 2\zeta\frac{\omega}{\omega_n}|G(i\omega)| = \frac{c\omega}{m\omega_n^2}|G(i\omega)| \tag{3.118}$$

where it is recalled that $\zeta = c/2m\omega_n$. Inserting Eqs. (3.115) and (3.118) into Eq. (3.117), we obtain the simple expression

$$\Delta E_{\text{cyc}} = c\pi\omega X^2 \tag{3.119}$$

from which it follows that the energy dissipated per cycle is directly proportional to the viscous damping coefficient c, the driving frequency ω and the square of the response amplitude X.

Experiments performed on a large variety of materials show that energy loss per cycle due to internal friction is roughly proportional to the square of the displacement amplitude,[1]

$$\Delta E_{\text{cyc}} = \alpha X^2 \tag{3.120}$$

where α is a constant independent of the frequency of the harmonic oscillation. This type of damping, called *structural damping*, is attributed to the *hysteresis phenomenon* associated with cyclic stress in elastic materials. The energy loss per cycle of stress is equal to the area inside the hysteresis loop shown in Fig. 3.19. Hence, comparing

[1]See L. Meirovitch, *Principles and Techniques of Vibrations*, Prentice Hall, New York, 1997.

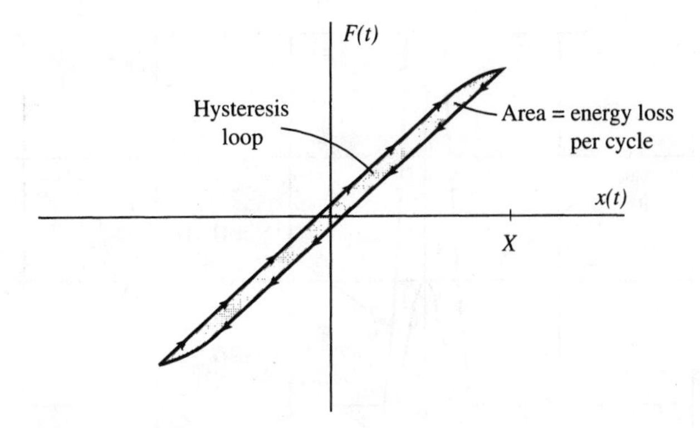

FIGURE 3.19
Hysteresis loop

Eqs. (3.119) and (3.120), we conclude that systems possessing structural damping and subjected to harmonic excitation can be treated as if they possessed viscous damping with the equivalent coefficient

$$c_{eq} = \frac{\alpha}{\pi\omega} \tag{3.121}$$

Under these circumstances, we can write Eq. (3.1) in the form

$$m\ddot{x}(t) + \frac{\alpha}{\pi\omega}\dot{x}(t) + kx(t) = F(t) = Ake^{i\omega t} \tag{3.122}$$

where consideration has been given to Eqs. (3.113) and (3.121). Because $\dot{x} = i\omega x$, we can rewrite Eq. (3.122) as

$$m\ddot{x}(t) + k(1 + i\gamma)x(t) = Ake^{i\omega t} \tag{3.123}$$

where

$$\gamma = \frac{\alpha}{\pi k} \tag{3.124}$$

is called the *structural damping factor*. The quantity $k(1+i\gamma)$ is called *complex stiffness*, or *complex damping*. Dividing Eq. (3.123) through by m, we obtain

$$\ddot{x}(t) + \omega_n^2(1 + i\gamma)x(t) = \omega_n^2 Ae^{i\omega t} \tag{3.125}$$

Following the procedure of Sec. 3.1 and recalling that the excitation is Re $Ake^{i\omega t}$, the steady-state solution of Eq. (3.125) can be shown to be

$$x(t) = \text{Re}[AG^*(\omega)e^{i\omega t}] = \text{Re}[A|G^*(\omega)|e^{i(\omega t - \phi^*)}] = A|G^*(\omega)|\cos(\omega t - \phi^*) \tag{3.126}$$

where

$$G^*(\omega) = \frac{1}{1 - (\omega/\omega_n)^2 + i\gamma} \tag{3.127}$$

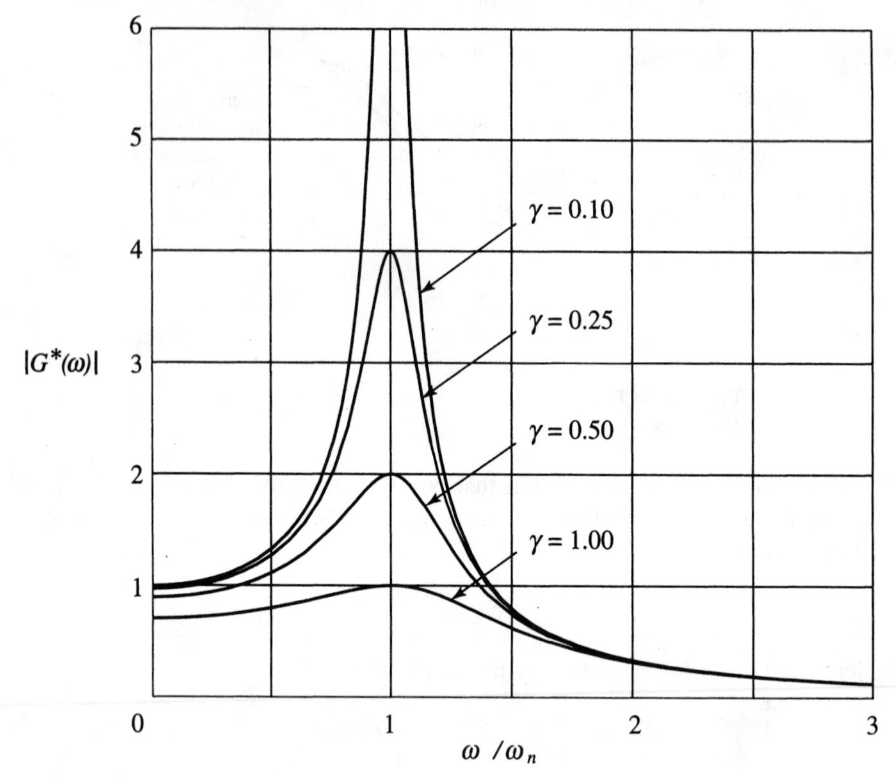

FIGURE 3.20
Magnitude of frequency response for structural damping

is the equivalent frequency response for structural damping,

$$|G^*(\omega)| = \frac{1}{\left\{[1-(\omega/\omega_n)^2]^2+\gamma^2\right\}^{1/2}} \tag{3.128}$$

is its magnitude and

$$\phi^*(\omega) = \tan^{-1}\frac{\gamma}{1-(\omega/\omega_n)^2} \tag{3.129}$$

is its phase angle. The plots $|G^*(\omega)|$ versus ω/ω_n and $\phi^*(\omega)$ versus ω/ω_n are shown in Figs. 3.20 and 3.21, respectively. We observe that, in contrast with the case of viscous damping, for structural damping the maximum response amplitude is obtained exactly for $\omega = \omega_n$.

One word of caution is in order. *The analogy between structural and viscous damping is valid only for harmonic excitation*, because the response of a system to harmonic excitation with the driving frequency ω is implied throughout the foregoing developments.

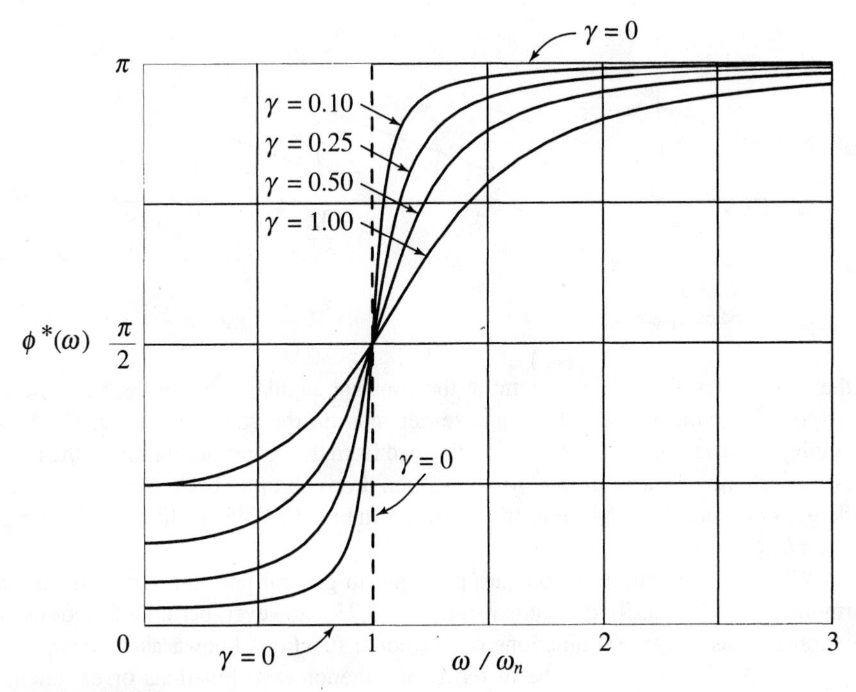

FIGURE 3.21
Phase angle of frequency response for structural damping

3.9 RESPONSE TO PERIODIC EXCITATIONS. FOURIER SERIES

Certain excitations repeat themselves every time interval T. Such excitations are said to be *periodic*, and the time interval T is known as the *period*. A classical example of a mechanism generating a periodic excitation is the cam and follower depicted in Fig. 3.22. It is not difficult to verify that the equation of motion for the system is

$$m\ddot{x} + c\dot{x} + (k_1 + k_2)x = k_2 y \tag{3.130}$$

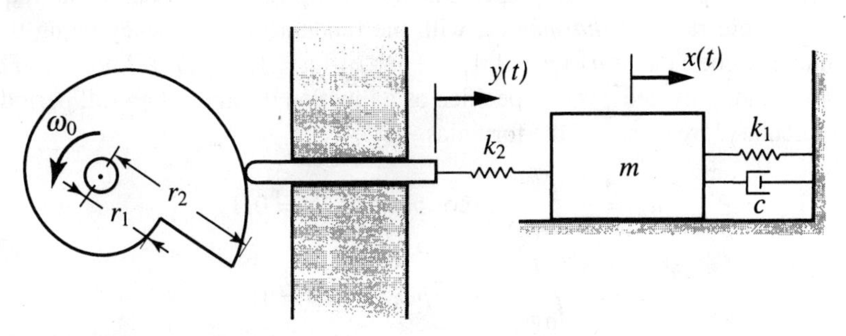

FIGURE 3.22
Cam and follower exerting a periodic force

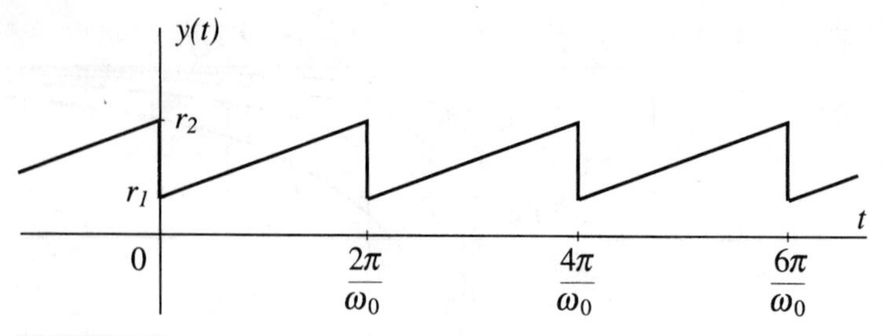

FIGURE 3.23
Periodic displacement

If the cam rotates about a fixed point at the constant angular velocity ω_0, the follower undergoes periodic motion whose nature depends on the shape of the cam. As an example, in the case in which the radial distance from the center of rotation to the rim of the cam increases linearly from r_1 to r_2 and then drops abruptly back to r_1, the follower undergoes the periodic displacement $y(t)$ shown in Fig. 3.23, where the period is simply $T = 2\pi/\omega_0$.

Whereas harmonic functions are periodic, in general periodic functions are not harmonic, as can be easily concluded from Fig. 3.23. However, periodic functions can be expressed as linear combinations of harmonic functions known as *Fourier series* (Appendix A). The series can be in terms of trigonometric functions or exponential functions, referred to at times as the real form or the complex form of Fourier series, respectively. We observe that, as with pure harmonic functions, in the case of periodic functions time plays only a secondary role. Hence, *periodic excitations represent a more general class of steady-state excitations.*

From Appendix A, we conclude that a given periodic function $f(t)$ with period T, such as that displayed in Fig. 3.23, can be expanded in the *trigonometric form of the Fourier series*

$$f(t) = \frac{1}{2}a_0 + \sum_{p=1}^{\infty} (a_p \cos p\omega_0 t + b_p \sin p\omega_0 t), \quad \omega_0 = 2\pi/T \tag{3.131}$$

where ω_0 is called the *fundamental frequency* and $p = 1, 2, \ldots$ are integers. Because the frequencies $p\omega_0$ $(p = 1, 2, \ldots)$ represent integral multiples of the fundamental frequency ω_0, they are referred to as *harmonics*, with the fundamental frequency being the first harmonic. The coefficients a_p $(p = 0, 1, \ldots)$ and b_p $(p = 1, 2, \ldots)$ are known as *Fourier coefficients* and, provided $f(t)$ is specified as a function of time over a full period, they can be calculated by means of the formulas

$$\begin{aligned} a_p &= \frac{2}{T}\int_0^T f(t)\cos p\omega_0 t\, dt, \quad p = 0, 1, \ldots \\ b_p &= \frac{2}{T}\int_0^T f(t)\sin p\omega_0 t\, dt, \quad p = 1, 2, \ldots \end{aligned} \tag{3.132}$$

As shown in Appendix A, derivation of the formulas for the Fourier coefficients, Eqs. (3.132), is based on the orthogonality of the trigonometric functions. The Fourier coefficients represent a measure of the participation of the harmonic components $\cos p\omega_0 t$ and $\sin p\omega_0 t$ $(p = 1, 2, \ldots)$ in the function $f(t)$, respectively, and $a_0/2$ is the average value of $f(t)$. We note that the limits of integration can be replaced by more convenient ones, as long as the interval of integration is exactly one period. The Fourier series representation is possible provided the integrals in Eqs. (3.132) exist, which can be safely assumed to be the case for the physical problems of interest in this text.

Under certain circumstances, the Fourier series, Eq. (3.131), can be simplified. To this end, we plot the trigonometric functions $\cos p\omega_0 t$ and $\sin p\omega_0 t$, as shown in Fig. 3.24, and observe that $\cos p\omega_0 t$ $(p = 0, 1, \ldots)$ are *even functions*, defined mathematically as

$$f(t) = f(-t) \tag{3.133}$$

and $\sin p\omega_0 t$ $(p = 1, 2 \ldots)$ are *odd functions*, defined as

$$f(t) = -f(-t) \tag{3.134}$$

Then, it is intuitively obvious, and can be demonstrated mathematically that, if the periodic function is an even function of time, the Fourier series cannot contain any $\sin p\omega_0 t$ terms, so that $b_p = 0$ $(p = 1, 2, \ldots)$. It follows that even periodic functions can be expanded in the *Fourier cosine series*

$$f(t) = \frac{1}{2}a_0 + \sum_{p=1}^{\infty} a_p \cos p\omega_0 t \tag{3.135}$$

where the coefficients a_p $(p = 0, 1, \ldots)$ are given by the first of Eqs. (3.132). Using a similar argument, we conclude that odd periodic functions can be expanded in the

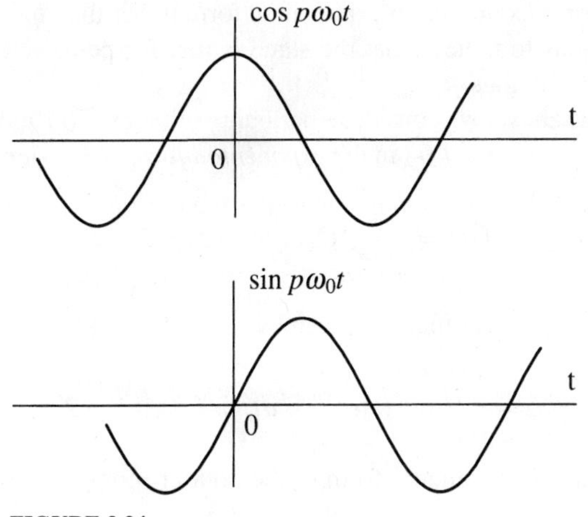

FIGURE 3.24
Even function and odd function

Fourier sine series

$$f(t) = \sum_{p=1}^{\infty} b_p \sin p\omega_0 t \tag{3.136}$$

in which the coefficients b_p $(p = 1, 2, \ldots)$ are given by the second of Eqs. (3.132).

A question of interest is how to convey the information concerning the participation of the individual harmonic components in the periodic function $f(t)$ in the most effective manner. Of course, the coefficients a_p and b_p provide this information, so that the question reduces to how to express it. Because $\sin p\omega_0 t$ and $\cos p\omega_0 t$ represent the same harmonic, we recognize that the amplitude of the pth harmonic is simply $c_p = \sqrt{a_p^2 + b_p^2}$. Clearly, the plot c_p versus p displays in one diagram the degree of participation of each harmonic in $f(t)$, a diagram known as a *frequency spectrum*. Because the diagram has entries only at the discrete frequencies $\omega = \omega_0,\ 2\omega_0,\ 3\omega_0, \ldots$, it represents a *discrete frequency spectrum*. In using frequency spectra to characterize a periodic function, time is relegated to a secondary role, with the primary role being played by the amplitude and frequency of each harmonic, thus justifying a statement to this effect made earlier in this section. We show frequency spectra in Example 3.1 at the end of this section.

Although the expansion of a periodic function in a Fourier series yields useful information concerning the frequency content of the function, this information does not represent an end in itself. Indeed, our interest in expanding periodic functions in series of harmonic functions is motivated by a desire to obtain the response to periodic excitations with relative ease. In this regard, we recall from Sec. 3.1 that the response to a harmonic excitation of frequency ω is a harmonic function having the same frequency ω and an amplitude proportional to the excitation amplitude, where the proportionality factor is the frequency response function. Hence, invoking the superposition principle, we can express the response to a periodic function as a superposition of harmonic responses. We also recall from Sec. 3.1 that, in deriving the system response, it is more advantageous to express a harmonic excitation in exponential form rather than in trigonometric form, and there are reasons to believe that the same is true for periodic excitations and the response to such excitations.

In view of the above, we consider a periodic excitation $f(t)$ and use developments from Appendix A to express $f(t)$ in the *exponential form of Fourier series*

$$f(t) = \sum_{p=-\infty}^{\infty} C_p e^{ip\omega_0 t}, \quad \omega_0 = 2\pi/T \tag{3.137}$$

in which C_p are complex coefficients given by

$$C_p = \frac{1}{T}\int_0^T f(t)e^{-ip\omega_0 t}\,dt, \quad p = 0, \pm 1, \pm 2, \ldots \tag{3.138}$$

As in the case of the trigonometric form of the Fourier series, the coefficient C_0 is real and can be recognized as the average value of $f(t)$.

There are two objections to working with Eqs. (3.137) and (3.138), namely, they contain negative frequencies and are not in a form that can be used readily to obtain the system response. To remove these objections, we use the analogy with Eq. (3.14) and rewrite the exponential form of the Fourier series as follows:

$$f(t) = \frac{1}{2}A_0 + \text{Re}\left(\sum_{p=1}^{\infty} A_p e^{ip\omega_0 t}\right) \tag{3.139}$$

where

$$A_p = \frac{2}{T}\int_0^T f(t)e^{-ip\omega_0 t}dt, \ p = 0, 1, 2, \ldots \tag{3.140}$$

in which A_0 is a real coefficient, equal to twice the average value of $f(t)$, and A_p ($p = 1, 2, \ldots$) are in general complex coefficients. Here too, the integration limits can be changed, provided the integral covers one full period. For this exponential form of the Fourier series, the frequency spectrum is given by $|A_p|$ versus p, where $|A_p| = \sqrt{(\text{Re } A_p)^2 + (\text{Im } A_p)^2}$ is the magnitude of A_p. It is not difficult to verify that the three forms of Fourier series, Eqs. (3.131), (3.137) and (3.139), are equivalent.

At this point, we turn our attention to the response to periodic excitations. In Sec. 3.1, we have shown that if a system described by the differential equation

$$m\ddot{x}(t) + c\dot{x}(t) + kx(t) = F(t) = kf(t) \tag{3.141}$$

is acted upon by the harmonic excitation

$$f(t) = \text{Re}(Ae^{i\omega t}) \tag{3.142}$$

then the response is

$$x(t) = \text{Re}[AG(i\omega)e^{i\omega t}] = \text{Re}[A|G(i\omega)|e^{i(\omega t - \phi)}] \tag{3.143}$$

where $G(i\omega)$ is the frequency response, Eq. (3.20), and $|G(i\omega)|$ and ϕ are the magnitude and phase angle of $G(i\omega)$, Eqs. (3.30) and (3.33), respectively. Moreover, from Eq. (3.141), it is easy to see that the constant excitation $A_0/2$ produces the constant response $A_0/2$. Hence, invoking the superposition principle (Sec. 1.12), we conclude that the response to a periodic excitation $f(t)$ in the form of Eq. (3.139) is simply

$$x(t) = \frac{1}{2}A_0 + \text{Re}\left[\sum_{p=1}^{\infty} A_p G_p e^{ip\omega_0 t}\right] = \frac{1}{2}A_0 + \text{Re}\left[\sum_{p=1}^{\infty} A_p |G_p| e^{i(p\omega_0 t - \phi_p)}\right] \tag{3.144}$$

in which, by analogy with Eq. (3.20),

$$G_p = \frac{1}{1 - (p\omega_0/\omega_n)^2 + i2\zeta p\omega_0/\omega_n} \tag{3.145}$$

and, by analogy with Eqs. (3.30) and (3.33),

$$|G_p| = \frac{1}{\{[1 - (p\omega_0/\omega_n)^2]^2 + (2\zeta p\omega_0/\omega_n)^2\}^{1/2}} \tag{3.146}$$

and

$$\phi_p = \tan^{-1} \frac{2\zeta p\omega_0/\omega_n}{1-(p\omega_0/\omega_n)^2} \tag{3.147}$$

We observe that, even though each harmonic component in $x(t)$, Eq. (3.144), is shifted by the phase angle ϕ_p, the response remains periodic and with the same period $T = 2\pi/\omega_0$ as the excitation. We also observe that, as p increases, $|G_p|$ tends to a value inversely proportional to p^2, so that the participation of the higher harmonics in the response decreases rapidly.

As pointed out earlier in this section, frequency domain descriptions of periodic functions reveal a great deal more information than time domain descriptions. Of course, for periodic excitations and for the response to periodic excitations, these descriptions are in the form of discrete frequency spectra. From Eq. (3.139), the excitation frequency spectrum consists of the plot $|A_p|$ versus p and, from Eq. (3.144), the response frequency spectrum is given by the response magnitude plot $|A_p||G_p|$ versus p. Some information can be gained from the response phase angle plot ϕ_p versus p, but this information is not as significant as that obtained from the magnitude plot.

We recall from Eq. (3.141) that $f(t)$ is only a normalized excitation having units of displacement, and the actual force excitation is $F(t) = kf(t)$. In this connection, it should be pointed out that the equation describing the cam and follower of Fig. 3.22, Eq. (3.130), differs somewhat from Eq. (3.141), so that if the follower displacement $y(t)$ is given by Eq. (3.139), the response $x(t)$ is given by Eq. (3.144) multiplied by $k_2/(k_1+k_2)$. Of course, in this case the parameters ζ and ω_n must also be modified so as to read

$$\zeta = c/2\sqrt{m(k_1+k_2)} \tag{3.148}$$

and

$$\omega_n = \sqrt{(k_1+k_2)/m} \tag{3.149}$$

Example 3.1. Calculate the response of a damped single-degree-of-freedom system to the periodic excitation $f(t)$ depicted in Fig. 3.25 by means of the exponential form of the Fourier series. Plot the excitation and response frequency spectra, the latter for the case $\zeta = 0.1$ and $\omega_n = 4\omega_0$. Plot $x(t)$ versus t, thus verifying that the response is periodic.

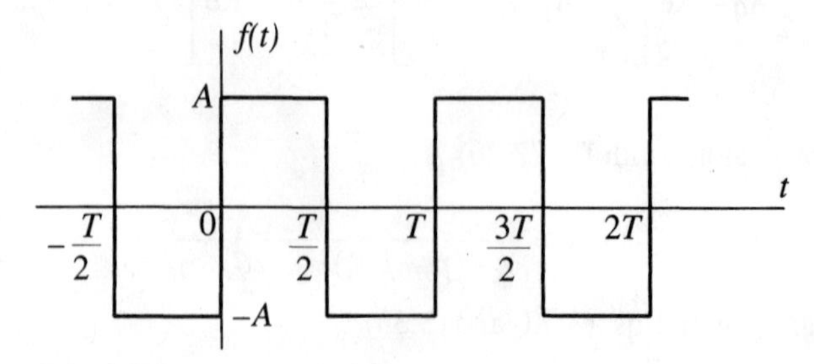

FIGURE 3.25
Periodic excitation

We observe from Fig. 3.25 that the average value of the excitation is zero, so that $A_0 = 0$. Hence the exponential form of the Fourier series is given by Eq. (3.139) with $A_0 = 0$ and the coefficients A_p are given by Eq. (3.140). Over the period $0 < t < T$, the excitation is simply

$$f(t) = \begin{cases} A \text{ for } 0 < t < T/2 \\ -A \text{ for } T/2 < t < T \end{cases} \tag{a}$$

so that, inserting Eq. (a) into Eq. (3.140), we can calculate the Fourier coefficients

$$\begin{aligned} A_p &= \frac{2}{T}\int_0^T f(t)e^{-ip\omega_0 t}dt = \frac{2A}{T}\left(\int_0^{T/2} e^{-ip\omega_0 t}dt - \int_{T/2}^T e^{-ip\omega_0 t}dt\right) \\ &= \frac{2A}{T}\left(\left.\frac{e^{-ip\omega_0 t}}{-ip\omega_0}\right|_0^{T/2} - \left.\frac{e^{-ip\omega_0 t}}{-ip\omega_0}\right|_{T/2}^T\right) = \frac{2iA}{p\omega_0 T}\left(2e^{-ip\omega_0 T/2} - 1 - e^{-ip\omega_0 T}\right) \\ &= \frac{iA}{p\pi}\left(2e^{-ip\pi} - 1 - e^{-i2p\pi}\right) = -\frac{2iA}{p\pi}\left[1-(-1)^p\right] = \begin{cases} -4iA/p\pi \text{ for } p = 1, 3, \ldots \\ 0 \text{ for } p = 2, 4, \ldots \end{cases} \end{aligned} \tag{b}$$

Introducing Eqs. (b) in Eq. (3.139) with $A_0 = 0$, we obtain the Fourier series for the excitation

$$f(t) = \text{Re}\left(\sum_{p=1,3,\ldots}^{\infty} A_p e^{ip\omega_0 t}\right) = \frac{4A}{\pi}\sum_{p=1,3,\ldots}^{\infty}\frac{1}{p}\sin p\omega_0 t \tag{c}$$

The response is given by Eq. (3.144) with $A_0 = 0$ and with the magnitude and phase angle of the frequency response to be determined by Eqs. (3.146) and (3.147), respectively, Hence, letting $\zeta = 0.1$ and $\omega_n = 4\omega_0$ in Eqs. (3.146) and (3.147), we can write simply

$$|G_p| = \frac{1}{\{[1-(p\omega_0/\omega_n)^2]^2 + (2\zeta p\omega_0/\omega_n)^2\}^{1/2}} = \frac{1}{\{[1-(p/4)^2]^2 + (p/20)^2\}^{1/2}} \tag{d}$$

and

$$\phi_p = \tan^{-1}\frac{2\zeta p\omega_0/\omega_n}{1-(p\omega_0/\omega_n)^2} = \tan^{-1}\frac{p/20}{1-(p/4)^2} \tag{e}$$

respectively. Inserting Eqs. (b) and (d) into Eq. (3.144) with $A_0 = 0$, we obtain the response Fourier series

$$\begin{aligned} x(t) &= \text{Re}\left[\sum_{p=1,3,\ldots}^{\infty} A_p|G_p|e^{i(p\omega_0 t-\phi_p)}\right] = \sum_{p=1,3,\ldots}^{\infty}|A_p||G_p|\sin(p\omega_0 t-\phi_p) \\ &= \frac{4A}{\pi}\sum_{p=1,3,\ldots}^{\infty}\frac{1}{p\{[1-(p/4)^2]^2+(p/20)^2\}^{1/2}}\sin(p\omega_0 t-\phi_p) \end{aligned} \tag{f}$$

The excitation frequency spectrum $|A_p|$ versus p is depicted in Fig. 3.26a and the response frequency spectra $|A_p||G_p|$ versus p and ϕ_p versus p are displayed in Figs. 3.26b and 3.26c, respectively. The response $x(t)$ is plotted as a function of time in Fig. 3.27 and is indeed periodic.

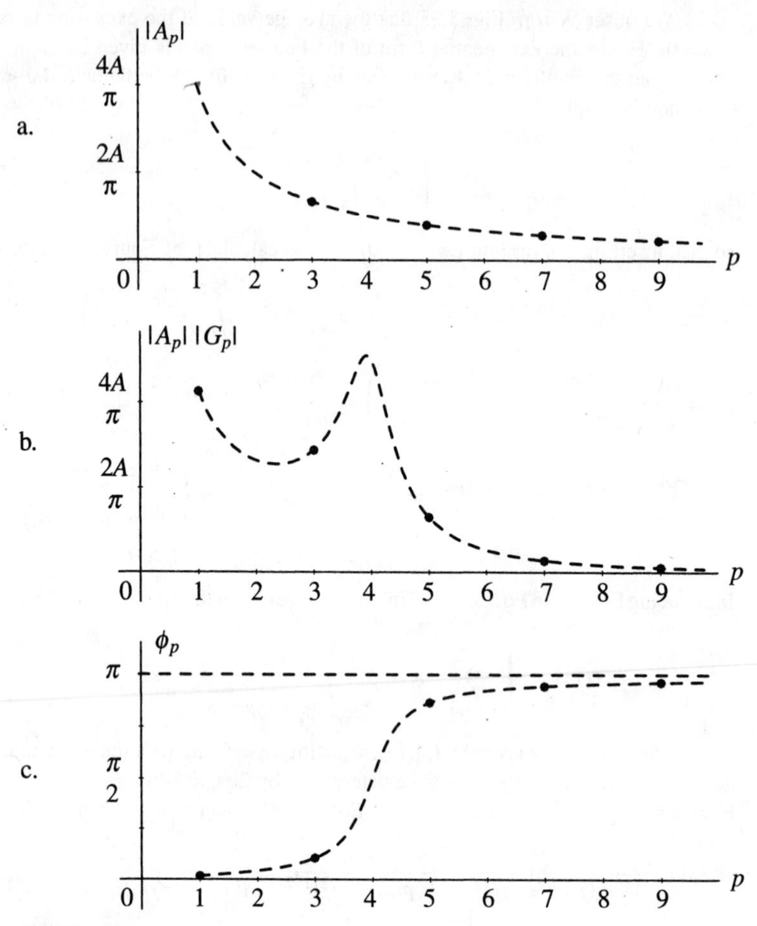

FIGURE 3.26
a. Excitation frequency spectrum, b. Response magnitude frequency spectrum, c. Response phase angle frequency spectrum

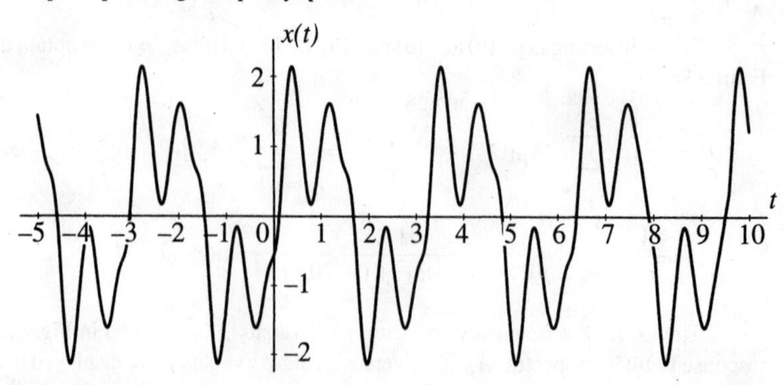

FIGURE 3.27
Periodic response

3.10 FREQUENCY RESPONSE PLOTS BY MATLAB

In Sec. 3.1, we derived an analytical expression for the frequency response $G(i\omega)$ in the form of Eq. (3.20). The frequency response is a very important concept in vibrations, as well as in other areas of engineering, most notably in controls. It relates the harmonic response to harmonic excitations. Of particular interest are the magnitude $|G(i\omega)|$ and phase angle $\phi(\omega)$ of the frequency response given by Eqs. (3.30) and (3.33) and plotted in Figs. 3.3 and 3.4, respectively. The plots reveal in two diagrams how the magnitude and phase angle of the harmonic response vary with the driving frequency for various values of the damping factor ζ. The plots can be conveniently obtained by MATLAB, and can be programed in several ways. One way is as follows:

```
% The program 'frqrsp1.m' plots the magnitude and phase angle of the frequency
% response for a single-degree-of-freedom system

clear
figure (1); clf
figure (2); clf
zeta=[0.05; 0.1; 0.15; 0.25; 0.5; 1.00; 1.25]; % damping factors arranged as a seven-
% dimensional vector
r=[0: 0.01: 3]; % ratio of the excitation frequency to the natural frequency of un-
% damped oscillation
for i=1:length (zeta),
  G=1./sqrt((1-r.^2).^2+(2*zeta(i)*r).^2); % magnitude of the frequency response
  phi=atan2(2*zeta(i)*r,1-r.^2); % phase angle of the frequency response

  figure (1)
  plot (r,G)
  hold on
  figure (2)
  plot (r,phi)
  hold on
end
figure (1)
title ('Frequency Response Magnitude')
xlabel ('\omega/\omega_n')
ylabel ('|G(i\omega)|')
grid

figure (2)
title ('Frequency Response Phase Angle')
xlabel ('\omega/\omega_n')
ylabel ('\phi(\omega)')
ha=gca;

set(ha, 'ytick', [0: pi/2: pi])
```

```
grid
set(ha, 'FontName' , ' Symbol ' , ' ylim', [0 pi])
set(ha, 'yticklabel' , {' 0 '; 'p/2'; 'p'})
```

Another way of plotting the frequency response magnitude and phase angle by MATLAB is by using complex notation. In this regard, we recognize that the desired complex form for the frequency response is given by Eq. (3.31).

3.11 SUMMARY

Harmonic excitations are quite common in engineering applications, and in particular in systems with rotating unbalanced masses. Because harmonic excitations keep repeating themselves continuously, so that they can be regarded as having no beginning and no end, they are said to be steady-state excitations. This, of course, is a physical impossibility, but if the excitations repeat themselves over a time interval considerably larger than their period, then the steady-state concept can be justified. In view of the fact that the shape of sine functions, or cosine functions, is well known, harmonic excitations can be regarded as being fully defined by the amplitude and frequency. The implication is that time is not really necessary to describe harmonic excitations.

The response to a harmonic excitation is also harmonic, and hence steady state. The response frequency is the same as the excitation frequency, but the response amplitude differs from the excitation amplitude. Moreover, the response tends to lag the excitation by a phase angle. Both the amplitude and phase angle of the harmonic response can be obtained by expressing the frequency response function in exponential form, where the frequency response can be broadly defined as the ratio of the harmonic response to the harmonic excitation. The frequency response function depends on the excitation frequency, as well as on the system parameters. A great deal of information can be extracted by examining how the amplitude and phase angle of the harmonic response vary as the excitation frequency changes. This can be done conveniently through two plots, namely, the magnitude and phase angle of the frequency response versus the driving frequency. These two plots provide such a broad picture of the harmonic response that any time description appears very limited in scope and very inefficient in conveying information. This picture can be enhanced by using the viscous damping factor as a parameter to generate families of plots of the frequency response magnitude and phase angle. The magnitude plot reveals significantly more information than the phase plot. In particular, it highlights the violent vibration likely to ensue when the driving frequency is in the proximity of the natural frequency. Note that the frequency response magnitude and phase angle can be plotted conveniently using MATLAB.

In many systems with rotating parts, such as the washing machine of Fig. 1.34a, the rotation is produced by an electric motor. When the system is turned on, the driving frequency starts from zero and increases until it reaches a constant operational frequency. Hence, if possible, the system should be designed so that the operational frequency remains below the natural frequency, thus avoiding violent vibration. If this is not possible, then the magnitude plot can be used to decide on the amount of damping necessary to mitigate the undesirable effects of the vibration.

Portions of the frequency response magnitude plot can be used in conjunction with instruments measuring vibration. In particular, for accelerometer measurements the interest lies in the portion corresponding to low ratios of the driving frequency to the natural frequency of the instrument. Because low ratios can be achieved by making the natural frequency large, the accelerometer is a high frequency instrument. By contrast, for seismometer measurements, the portion corresponding to the high frequency ratio is used, which makes the seismometer a low frequency instrument.

A different type of damping is structural damping. For simplicity, it is common to treat structural damping as if it were viscous. However, the analogy is valid only in harmonic vibration.

Periodic excitation is a more general type of steady-state excitation than harmonic excitation. But, periodic excitations can be expanded in Fourier series of harmonic functions. Hence, using the superposition principle, the response to periodic excitations can be expressed in the form of infinite series of harmonic responses.

Finally, we raise the question of the role of the response to initial excitations in the context of steady-state response. We recall from Chs. 1 and 2 that the response to initial excitations is a transient response, which begins at $t = 0$ and tends to fade away with time, even for conservative systems. On the other hand, the steady-state response persists forever, entirely oblivious to time. It follows that, in the case of steady-state excitation, the use of the principle of superposition to combine the response to initial excitations with the response to applied forces is meaningless.

PROBLEMS

3.1. A control tab of an airplane elevator is hinged about an axis in the elevator, shown as point O in Fig. 3.28, and activated by a control linkage behaving like a torsional spring of stiffness k_T. The mass moment of inertia of the control tab is I_O, so that the natural frequency of the system is $\omega_n = \sqrt{k_T/I_O}$. Because k_T cannot be calculated exactly, it is necessary to

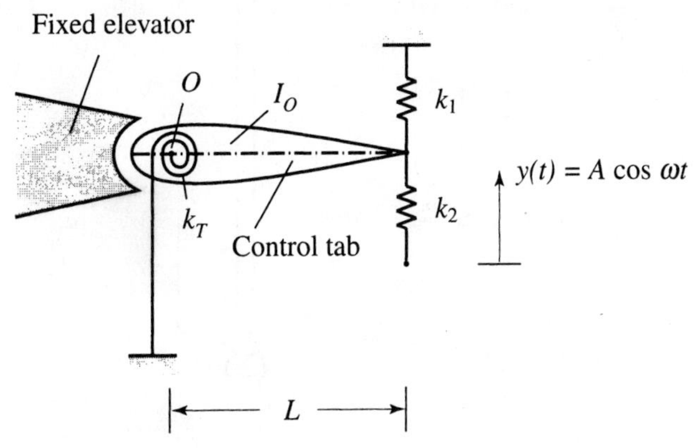

FIGURE 3.28
Control tab of airplane elevator

obtain the natural frequency ω_n experimentally. To this end, the elevator is held fixed and the tab is excited harmonically by means of the spring k_2 while restrained by the spring k_1, as shown in Fig. 3.28, and the excitation frequency ω is varied until the resonance frequency ω_r is reached. Calculate the natural frequency ω_n of the control tab in terms of ω_r and the parameters of the experimental setup.

3.2. A machine of mass M rests on a massless elastic floor, as shown in Fig. 3.29. If a unit load is applied at midspan, the floor undergoes a deflection x_{st}. A shaker having total mass m_s and carrying two rotating unbalanced masses (similar to the rotating mass shown in Fig. 1.34a) produces a vertical harmonic force $ml\omega^2 \sin \omega t$, where the frequency of rotation can vary. Show how the shaker can be used to derive a formula for the natural frequency of flexural vibration of the structure.

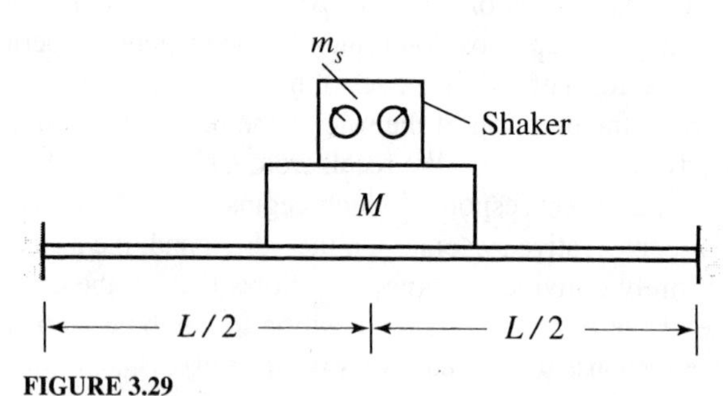

FIGURE 3.29
Shaker for measuring the natural frequency of structures

3.3. Derive the differential equation of motion for the inverted pendulum of Fig. 3.30, where $A \cos \omega t$ represents a displacement excitation. Then assume small amplitudes and solve for the angle θ as a function of time.

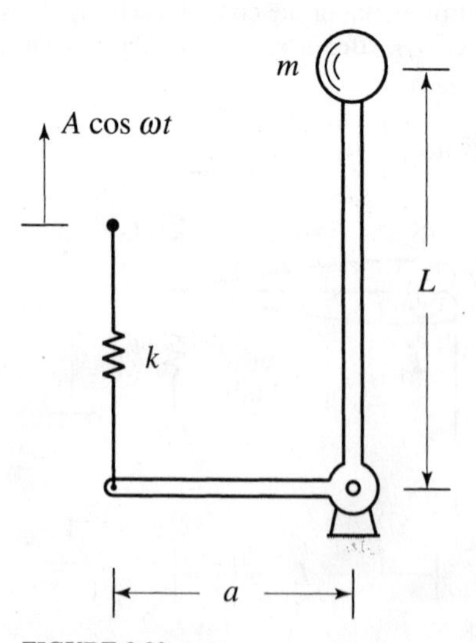

FIGURE 3.30
Inverted pendulum excited harmonically

3.4. One side of the manometer tube of Example 2.2 is subjected to the pressure $p(t) = p_0 \cos \omega t$, where p_0 has units of newtons per square meter (N/m^2). Derive the differential equation of motion, and obtain the resonance frequency.

3.5. The left end of the cantilever beam shown in Fig. 3.31 undergoes the harmonic motion $x(t) = A \cos \omega t$. Derive the differential equation for the motion of the mass M and determine the resonance frequency. Assume that the beam is massless and that its bending stiffness EI is constant.

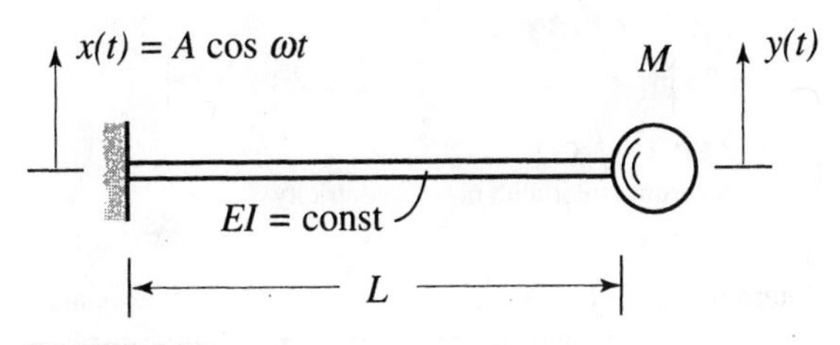

FIGURE 3.31
Cantilever beam with harmonically-excited end

3.6. The foundation of the building in Problem 2.14 undergoes the horizontal motion $y(t) = y_0 \sin \omega t$. Derive the system response.

3.7. Gear A in Problem 1.14 is subjected to the torque $M_A = M_0 \cos \omega t$. Derive an expression for the angular motion of gear B.

3.8. Solve the differential equation

$$m\ddot{x}(t) + c\dot{x}(t) + kx(t) = kA \sin \omega t$$

describing the motion of a damped single-degree-of-freedom system subjected to a harmonic force. Assume a solution in the form $x(t) = X(\omega)\sin(\omega t - \phi)$ and derive expressions for X and ϕ by equating coefficients of $\sin \omega t$ and $\cos \omega t$ on both sides of the equation.

3.9. Assume a solution of Eq. (3.15) in the form $x(t) = X(\omega)e^{i(\omega t - \phi)}$ and show that this form contains the solutions to both $f(t) = A \cos \omega t$ and $f(t) = A \sin \omega t$.

3.10. A mass-damper-spring system of the type shown in Fig. 3.1 has been observed to achieve a peak magnification factor $Q = 5$ at the driving frequency $\omega = 10$ rad/s. It is required to determine: (1) the damping factor, (2) the driving frequencies corresponding to the half-power points and (3) the bandwidth of the system.

3.11. A piece of machinery can be regarded as a rigid mass with two reciprocating rotating unbalanced masses such as that in Fig. 1.34a. The total mass of the system is 12 kg and each of the unbalanced masses is equal to 0.5 kg. During normal operation, the rotation of the masses varies from 0 to 600 rpm. Design a support system so that the maximum vibration amplitude will not exceed 10 percent of the rotating masses' eccentricity.

3.12. The rotor of a turbine having the form of a disk is mounted at the midspan of a uniform steel shaft, as shown in Fig. 3.32. The mass of the disk is 15 kg and its diameter is 0.3 m. The disk has a circular hole of diameter 0.03 m at a distance of 0.12 m from the geometric center. The bending stiffness of the shaft is $EI = 1{,}600\,\mathrm{N \cdot m^2}$. Determine the amplitude of vibration if the turbine rotor rotates with the angular velocity of 6,000 rpm. Assume that the shaft bearings are rigid.

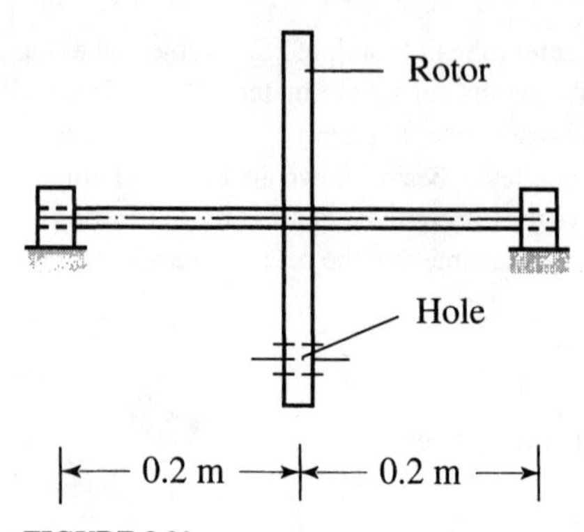

FIGURE 3.32
Turbine rotor with mass eccentricity

3.13. Consider the system of Fig. 3.15. When the support is fixed, $y = 0$, and the mass is allowed to vibrate freely, the ratio between two consecutive maximum displacement amplitudes is $x_2/x_1 = 0.8$. On the other hand, when the mass is in equilibrium, the spring is compressed by an amount $x_{st} = 2.5$ mm. The weight of the mass is $mg = 100$ N. Let $y(t) = A\cos\omega t$, $x(t) = X\cos(\omega t - \phi)$ and plot X/A versus ω/ω_n and ϕ versus ω/ω_n for $0 < \omega/\omega_n < 2$.

3.14. The system shown in Fig. 3.33 simulates a vehicle traveling on a rough road. Let the vehicle velocity be uniform, $v = \text{const}$, and calculate the response $z(t)$, as well as the force transmitted to the vehicle.

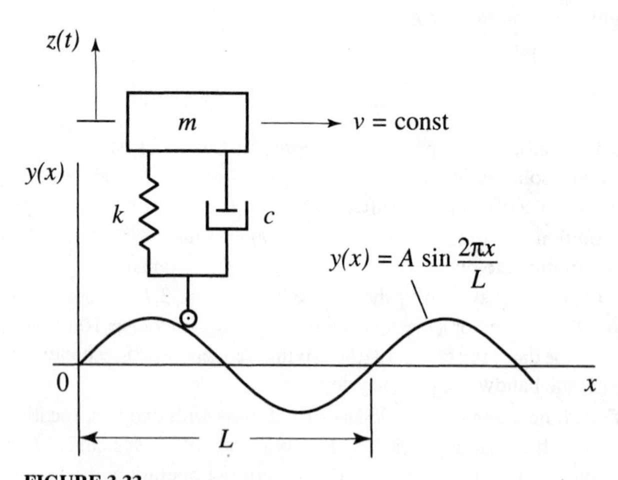

FIGURE 3.33
Vehicle traveling on a rough road

3.15. The support of the viscously damped pendulum shown in Fig. 3.34 undergoes harmonic oscillation. Derive the differential equation of motion of the system, then assume small amplitudes and solve for $\theta(t)$.

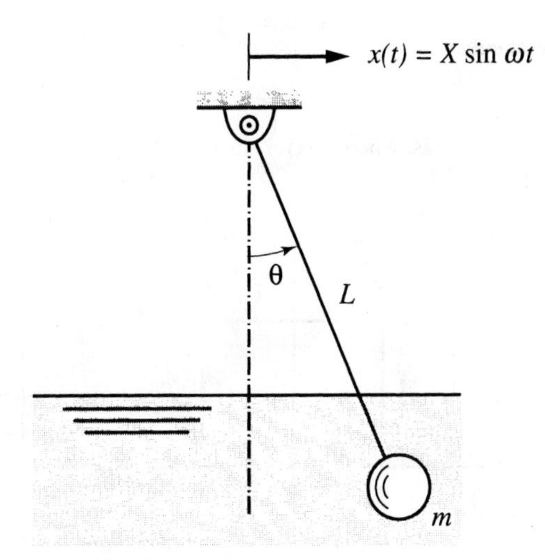

FIGURE 3.34
Pendulum with harmonically-moving support

3.16. The system of Fig. 1.34a has the following parameters: $M = 80$ kg, $m = 5$ kg, $k = 8,000$ N/m, $e = 0.1$ m. Design a viscous damper so that at the rotating speed $\omega = 4\omega_n$ the force transmitted to the support does not exceed 250 N.

3.17. It is observed that during one cycle of vibration a structurally damped single-degree-of-freedom system dissipates energy in the amount of 1.2 percent of the maximum potential energy. Calculate the structural damping factor γ.

3.18. The cam and follower of Fig. 3.35a impart a displacement $y(t)$ in the form of a periodic sawtooth function to the lower end of the system, where $y(t)$ is shown in Fig. 3.35b. Derive an expression for the response $x(t)$ by means of a Fourier analysis.

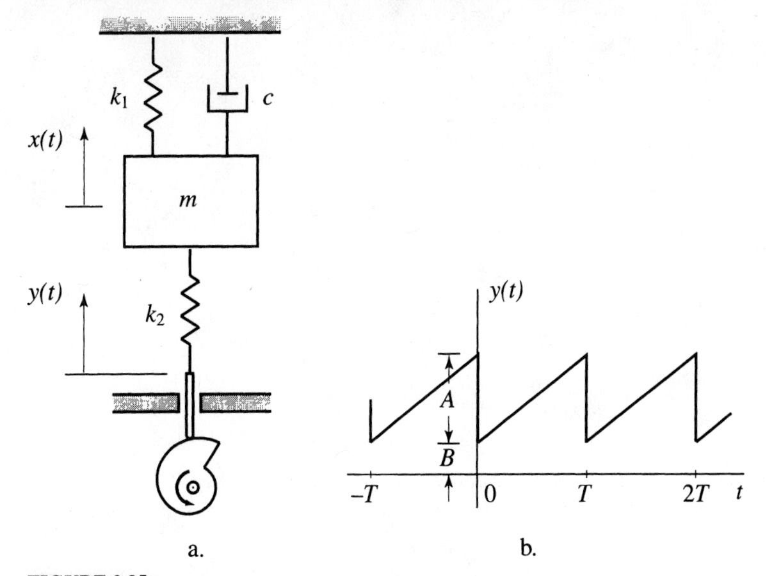

FIGURE 3.35
a. System excited by cam and follower, b. Periodic displacement generated by the follower

3.19. Solve the differential equation

$$m\ddot{x}(t) + c\dot{x}(t) + kx(t) = kf(t)$$

by means of a Fourier analysis, where $f(t)$ is the periodic function shown in Fig. 3.36.

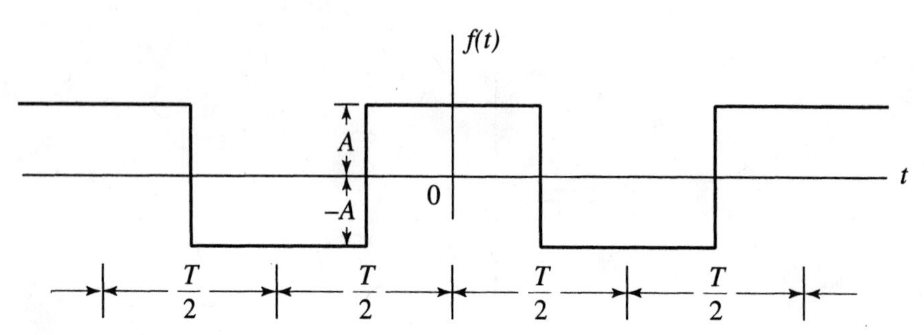

FIGURE 3.36
Periodic excitation

3.20. Write a MATLAB program for plotting the frequency response magnitude and phase angle based on Eq. (3.31) and using complex notation.

3.21. Use MATLAB to plot Fig. 3.7.

3.22. Use MATLAB to plot Figs. 3.13 and 3.14.

CHAPTER 4

RESPONSE OF SINGLE-DEGREE-OF-FREEDOM SYSTEMS TO NONPERIODIC EXCITATIONS

Chapter 3 was devoted entirely to the response of single-degree-of-freedom systems to periodic excitations, of which the harmonic excitations are the most important ones. Harmonic and mere periodic forces represent steady-state excitations; they are conveniently described in the frequency domain, and the same can be said about the response to steady-state excitations. Although frequency domain techniques can be extended to the treatment of nonperiodic excitations, these techniques are different from those encountered in Ch. 3; they will be used in Ch. 12 in conjunction with random vibrations. This chapter is devoted to time domain techniques, which are more suitable for systems subjected to deterministic nonperiodic excitations than frequency domain techniques.

Nonperiodic excitations are often referred to as transient, although some of them can last a long time. The term *transient* is to be interpreted in the sense that nonperiodic excitations are not steady state. In fact, nonperiodic excitations are assumed for the most part to start at $t = 0$. This point can be best illustrated by the example of a force that is zero for $t < 0$ and has the form of a sine function for $t \geq 0$ (Fig. 4.1). The one-sided sine function is regarded as transient even though it is harmonic for $t \geq 0$ and lasts indefinitely.

By virtue of the superposition principle (Sec. 1.12), the response of linear systems to nonperiodic excitations can be combined with the response to initial excitations to obtain the total response, which is consistent with the fact that both types of excitations are regarded as transient. Of course, we discussed the response to initial excitations in Ch. 2, so that in this chapter we concentrate on the response to applied forces.

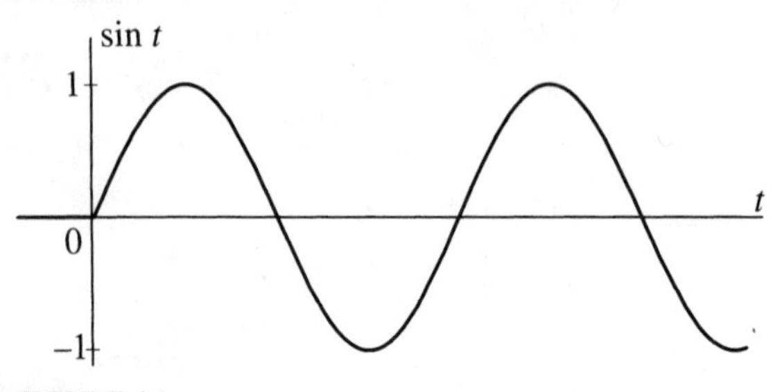

FIGURE 4.1
One-sided sine function

We begin this chapter with a discussion of three important and useful functions, the unit impulse, the unit step function and the unit ramp function. The response to these functions, known as impulse response, step response and ramp response, respectively, can be used to synthesize the response to a variety of complicated forces, including entirely arbitrary forces. Indeed, the response of linear systems with constant coefficients to arbitrary excitations can be expressed as a superposition of impulse responses by means of the well-known convolution integral.

In studying the behavior of linear systems with constant coefficients, as well as in obtaining the response of such systems, the Laplace transformation method and the concept of state transition matrix prove very useful. However, except for some simple cases, analytical evaluation of the response is not possible, so that numerical evaluation on a digital computer is an absolute necessity. This requires that time be regarded as a discrete rather than a continuous variable; the corresponding formalism for computing the response is referred to as discrete-time systems. All these subjects are discussed in this chapter. Moreover, programs written in MATLAB for discrete-time computation of the response of single-degree-of-freedom systems to arbitrary excitations are included.

4.1 THE UNIT IMPULSE. IMPULSE RESPONSE

Among nonperiodic functions, there is a family of functions of special interest in vibrations, namely, the unit impulse, the unit step function and the unit ramp function. They are not only useful in their own right, but they can also be used to synthesize more complicated functions. We discuss the unit impulse in this section and the other two in the following two sections.

The *unit impulse*, also known as the *Dirac delta function*, is defined mathematically as

$$\delta(t-a) = 0 \text{ for } t \neq a$$
$$\int_{-\infty}^{\infty} \delta(t-a)dt = 1 \tag{4.1}$$

The unit impulse is depicted in Fig. 4.2; it has units of seconds^{-1}, which can be easily explained by the fact that the value of the integral in (4.1) is nondimensional. The implication of Eqs. (4.1) is that the function is zero everywhere, except in the neighborhood

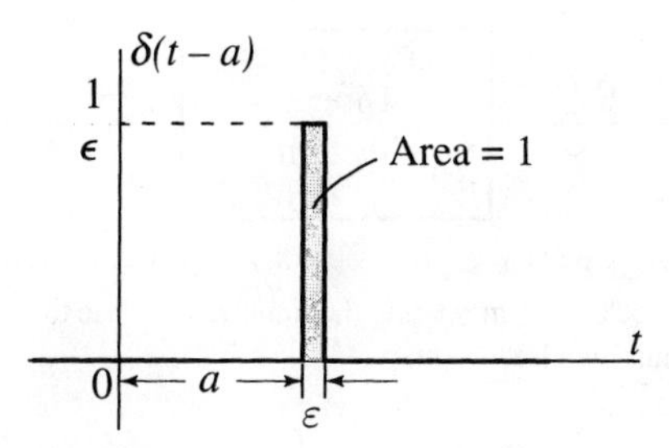

FIGURE 4.2
Unit impulse, or Dirac delta function

of $t = a$, where it reaches very high amplitudes over a very short time interval ε in such a way that the area under the graph $\delta(t - a)$ versus t is equal to 1.

The Dirac delta function is particularly suited for describing impulsive forces. Indeed, if a very large force acts over a very short time interval at $t = a$, then such a force can be expressed in the form

$$F(t) = \hat{F}\delta(t - a) \tag{4.2}$$

where $\hat{F}$ is the magnitude of the impulse and has units newtons per second (N/s).

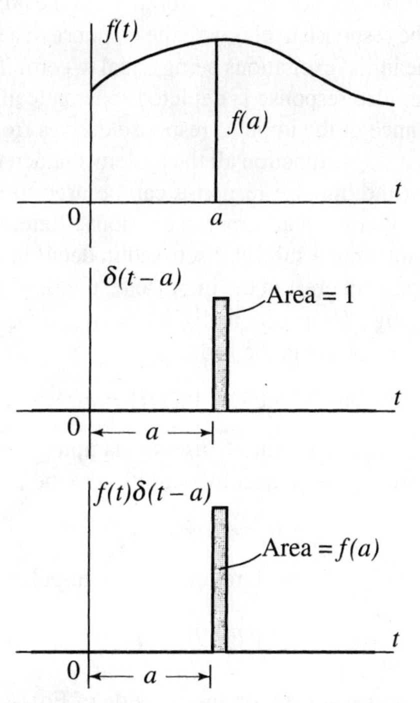

FIGURE 4.3
Sampling property of Dirac delta function

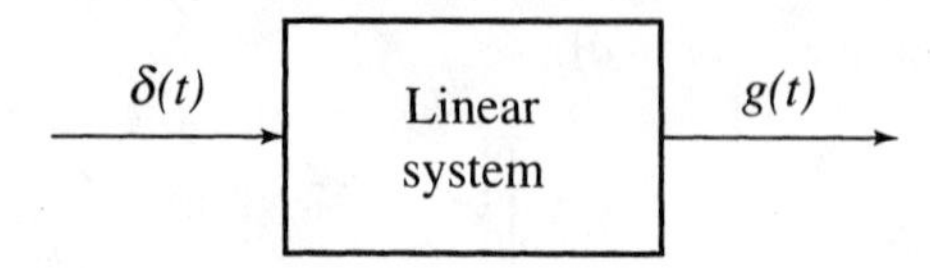

FIGURE 4.4
Block diagram relating the impulse response to the unit impulse

The unit impulse has a very interesting and useful property. Multiplying a continuous but otherwise arbitrary function $f(t)$ by $\delta(t-a)$, integrating with respect to time and recognizing that for all practical purposes $f(t)$ does not change over the duration of the impulse, we can write

$$\int_{-\infty}^{\infty} f(t)\delta(t-a)dt \cong f(a)\int_{-\infty}^{\infty} \delta(t-a)dt = f(a) \tag{4.3}$$

so that the effect of integrating a function $f(t)$ weighted by the Dirac delta function applied at $t=a$ is simply to evaluate $f(t)$ at $t=a$. The process is illustrated in Fig. 4.3. This property of the delta function is sometimes referred to as the *sampling property* and it represents a simple way of evaluating integrals involving delta functions. Clearly, the above result is valid also for finite limits of integration as long as these limits bracket the time $t=a$.

One of the most important concepts in vibrations is the *impulse response*, denoted by $g(t)$ and defined as the response to $\delta(t)$, i.e., the response of a system to a unit impulse applied at $t=0$, with the initial excitations being equal to zero. The relation between the unit impulse and the impulse response is depicted schematically by the block diagram of Fig. 4.4. The importance of the impulse response derives from two facts. In the first place, $g(t)$ embodies in a single function all the system characteristics. More important, however, is the fact that the impulse response can be used to synthesize the response of linear time-invariant systems to arbitrary excitations. Reference is made here to the celebrated convolution integral, a subject discussed in detail in Sec. 4.4.

We have considerable interest in the impulse response of the mass-damper-spring system of Fig. 3.1. Letting $x(t)=g(t)$ and $F(t)=\delta(t)$ in Eq. (3.1), we can write the equation for the impulse response in the form

$$m\ddot{g}(t)+c\dot{g}(t)+kg(t)=\delta(t) \tag{4.4}$$

where m is the mass, c the coefficient of viscous damping and k the spring constant. But, the impulse response $g(t)$ is subject by definition to the initial conditions

$$g(0)=0,\ \dot{g}(0)=0 \tag{4.5}$$

so that, integrating Eq. (4.4) over the duration ε of the impulse, we have

$$\int_{0}^{\varepsilon}(m\ddot{g}+c\dot{g}+kg)dt=\int_{0}^{\varepsilon}\delta(t)dt=1 \tag{4.6}$$

Next, we take the limit of the integral on the left side of Eq. (4.6) as ε approaches zero and evaluate the integral. To this end, we assume that $g(t)$ is continuous and $\dot{g}(t)$ is not, consider Eqs. (4.5) and write

$$\begin{aligned}
\lim_{\varepsilon\to 0}\int_0^{\varepsilon} m\ddot{g}(t)dt &= \lim_{\varepsilon\to 0} m\dot{g}(t)\Big|_0^{\varepsilon} = \lim_{\varepsilon\to 0} m[\dot{g}(\varepsilon)-\dot{g}(0)] = m\dot{g}(0+) \\
\lim_{\varepsilon\to 0}\int_0^{\varepsilon} c\dot{g}(t)dt &= \lim_{\varepsilon\to 0} cg(t)\Big|_0^{\varepsilon} = \lim_{\varepsilon\to 0} c[g(\varepsilon)-g(0)] = 0 \\
\lim_{\varepsilon\to 0}\int_0^{\varepsilon} kg(t)dt &= \lim_{\varepsilon\to 0} kg(0)t\Big|_0^{\varepsilon} = \lim_{\varepsilon\to 0} kg(0)\varepsilon = 0
\end{aligned} \tag{4.7}$$

where $\dot{g}(0+)$ denotes the slope of the impulse response curve at the termination of the impulse, as opposed to $\dot{g}(0)=0$ at the initiation of the impulse, according to the second of Eqs. (4.5). Hence, inserting Eqs. (4.7) into Eq. (4.6) and taking the limit as $\varepsilon\to 0$, we obtain

$$\lim_{\varepsilon\to 0}\int_0^{\varepsilon}(m\ddot{g}+c\dot{g}+kg)dt = m\dot{g}(0+) = 1 \tag{4.8}$$

from which we conclude that the effect of a unit impulse at $t=0$ is to produce an equivalent initial velocity

$$\dot{g}(0+) = \frac{1}{m} \tag{4.9}$$

which explains a statement made in Sec. 1.11 that impulsive forces produce initial velocities.

At this point, the derivation of the impulse response is relatively simple. Indeed, instead of considering a nonhomogeneous system subjected to a unit impulse, we consider a homogeneous system subjected to an equivalent initial velocity as given by Eq. (4.9). We recall that we derived the response of a mass-damper-spring system to initial excitations in Sec. 2.2. Hence, inserting Eq. (4.9) into Eqs. (2.33), we have

$$C = \frac{1}{m\omega_d},\quad \phi = \tan^{-1}\infty = \pi/2 \tag{4.10}$$

so that, using Eq. (2.32) with $x(t)$ replaced by $g(t)$, we obtain the impulse response of a mass-damper-spring system in the form

$$g(t) = \begin{cases} \dfrac{1}{m\omega_d}e^{-\zeta\omega_n t}\sin\omega_d t \ \text{ for } t>0 \\ 0 \ \text{ for } t<0 \end{cases} \tag{4.11}$$

where we recognized that the response must be zero before the impulse has occurred. Note that the symbols appearing in Eq. (4.11) were defined in Sec. 2.2 as the viscous damping factor $\zeta = c/2m\omega_n$, the natural frequency $\omega_n = \sqrt{k/m}$ and the frequency of damped oscillation $\omega_d = \omega_n\sqrt{1-\zeta^2}$. Equation (4.11) implies an underdamped system, $\zeta<1$. A typical plot of the impulse response based on Eq. (4.11) is shown in Fig. 4.5.

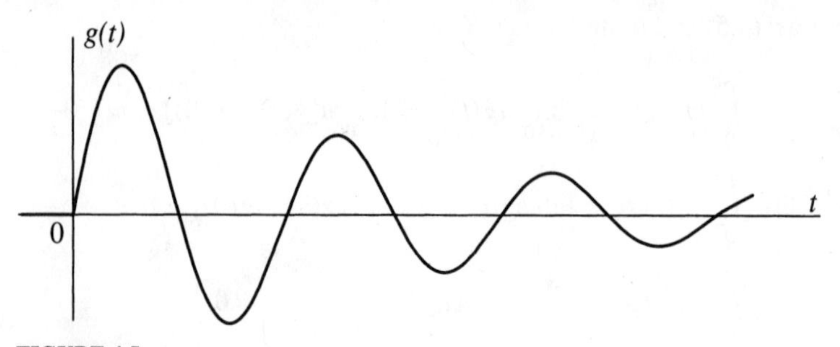

FIGURE 4.5
Impulse response of a mass-damper-spring system

4.2 THE UNIT STEP FUNCTION. STEP RESPONSE

Another function of great importance in vibrations is the *unit step function*, depicted in Fig. 4.6 and defined mathematically as

$$\mathscr{u}(t-a) = \begin{cases} 0 & \text{for } t < a \\ 1 & \text{for } t > a \end{cases} \tag{4.12}$$

The function is clearly discontinuous at $t = a$, at which points its value jumps from 0 to 1. If the discontinuity occurs at $t = 0$, then the unit step function is denoted simply by $\mathscr{u}(t)$. The unit step function is dimensionless.

An interesting feature of the unit step function is that multiplication of an arbitrary function $f(t)$ by $\mathscr{u}(t-a)$ annihilates the portion of $f(t)$ corresponding to $t < a$ and leaves unaffected the portion for $t > a$. We can put this feature to use immediately by rewriting the impulse response, Eq. (4.11), in the compact form

$$g(t) = \frac{1}{m\omega_d} e^{-\zeta\omega_n t} \sin\omega_d t \, \mathscr{u}(t) \tag{4.13}$$

There is a close relationship between the unit step function $\mathscr{u}(t-a)$ and the unit impulse $\delta(t-a)$. In particular, the unit step function is the integral of the unit impulse, or

$$\mathscr{u}(t-a) = \int_{-\infty}^{t} \delta(\tau - a) d\tau \tag{4.14}$$

where τ is a mere dummy variable of integration. Conversely, the unit impulse is the

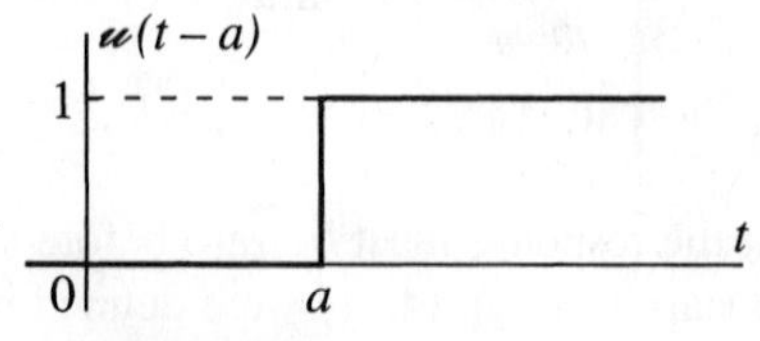

FIGURE 4.6
Unit step function

time derivative of the unit step function, or

$$\delta(t-a) = \frac{d\mathscr{u}(t-a)}{dt} \tag{4.15}$$

The *step response*, denoted by $\mathscr{s}(t)$, is defined as the response of a system to $\mathscr{u}(t)$, i.e., to a unit step function applied at $t = 0$, with the initial conditions being equal to zero. The step response is related to the impulse response in the same manner as the unit step function is related to the unit impulse. To demonstrate this relation, we rewrite Eq. (4.4) in the symbolic form

$$m\frac{d^2g(t)}{dt^2} + c\frac{dg(t)}{dt} + kg(t) = \left(m\frac{d^2}{dt^2} + c\frac{d}{dt} + k\right)g(t) = \delta(t) \tag{4.16}$$

Similarly, the relation between the step response and the unit step function can be expressed as follows:

$$\left(m\frac{d^2}{dt^2} + c\frac{d}{dt} + k\right)\mathscr{s}(t) = \mathscr{u}(t) \tag{4.17}$$

Integrating Eq. (4.16) with respect to time, assuming that the order of the integration and differentiation processes is interchangeable and using Eq. (4.14), we obtain

$$\int_{-\infty}^{t}\left(m\frac{d^2}{d\tau^2} + c\frac{d}{d\tau} + k\right)g(\tau)d\tau = \left(m\frac{d^2}{dt^2} + c\frac{d}{dt} + k\right)\int_{-\infty}^{t} g(\tau)d\tau$$

$$= \int_{-\infty}^{t} \delta(\tau)d\tau = \mathscr{u}(t) \tag{4.18}$$

Hence, comparing Eqs. (4.17) and (4.18), we conclude that

$$\mathscr{s}(t) = \int_{-\infty}^{t} g(\tau)d\tau \tag{4.19}$$

or, the step response is the integral of the impulse response. Although we demonstrated relation (4.19) by means of a mass-damper-spring system, the relation is valid for any linear time-invariant system.

Equation (4.19) can be used as a convenient way of determining the step response of a system. As an illustration, we propose to use Eq. (4.19) to derive the step response of a mass-damper-spring system. To this end, we introduce Eq. (4.13) into Eq. (4.19) and write

$$\mathscr{s}(t) = \frac{1}{m\omega_d}\int_{-\infty}^{t} e^{-\zeta\omega_n\tau}\sin\omega_d\tau\,\mathscr{u}(\tau)d\tau = \frac{1}{m\omega_d}\int_{0}^{t} e^{-\zeta\omega_n\tau}\sin\omega_d\tau\,d\tau \tag{4.20}$$

The integration can be carried out with relative ease by recognizing that

$$\sin\omega_d\tau = \frac{e^{i\omega_d\tau} - e^{-i\omega_d\tau}}{2i} \tag{4.21}$$

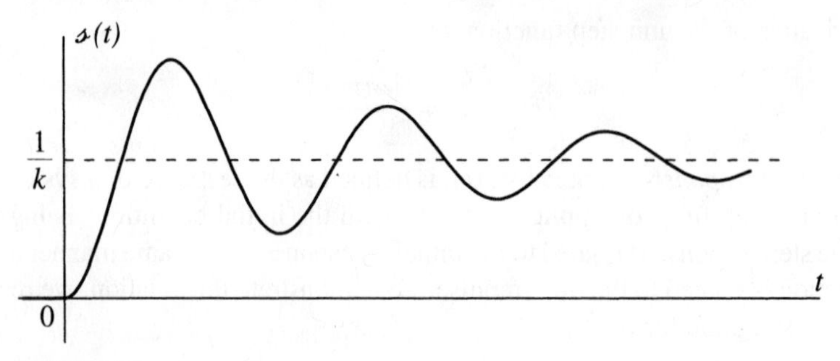

FIGURE 4.7
Step response of a mass-damper-spring system

Hence, inserting Eq. (4.21) into Eq. (4.20), we can write

$$\begin{aligned}
\mathscr{s}(t) =& \frac{1}{2im\omega_d}\int_0^t e^{-\zeta\omega_n\tau}\left(e^{i\omega_d\tau} - e^{-i\omega_d\tau}\right)d\tau \\
=& \frac{1}{2im\omega_d}\int_0^t \left[e^{-(\zeta\omega_n - i\omega_d)\tau} - e^{-(\zeta\omega_n + i\omega_d)\tau}\right]d\tau \\
=& \frac{1}{2im\omega_d}\left[\frac{e^{-(\zeta\omega_n - i\omega_d)\tau}}{-(\zeta\omega_n - i\omega_d)} - \frac{e^{-(\zeta\omega_n + i\omega_d)\tau}}{-(\zeta\omega_n + i\omega_d)}\right]\Bigg|_0^t \\
=& \frac{1}{2im\omega_d\omega_n^2}\left\{(\zeta\omega_n + i\omega_d)\left[1 - e^{-(\zeta\omega_n - i\omega_d)t}\right]\right. \\
& \left. -(\zeta\omega_n - i\omega_d)\left[1 - e^{-(\zeta\omega_n + i\omega_d)t}\right]\right\} \\
=& \frac{1}{k}\left[1 - e^{-\zeta\omega_n t}\left(\cos\omega_d t + \frac{\zeta\omega_n}{\omega_d}\sin\omega_d t\right)\right]\mathscr{u}(t)
\end{aligned} \tag{4.22}$$

in which we used the relations $m\omega_n^2 = k$, $\omega_d^2 = (1-\zeta^2)\omega_n^2$, as well as the formula

$$\cos\omega_d t = \frac{e^{i\omega_d t} + e^{-i\omega_d t}}{2} \tag{4.23}$$

Moreover, we multiplied the result by the unit step function $\mathscr{u}(t)$ to account for the fact that $\mathscr{s}(t) = 0$ for $t < 0$. A typical plot $\mathscr{s}(t)$ versus t is shown in Fig. 4.7.

Certain excitation functions, such as a rectangular pulse, can be represented as linear combinations of step functions. Then, using the superposition principle, the response to a force in the form of a rectangular pulse can be expressed as like combinations of step responses, as shown in the following example.

Example 4.1. Use the concept of step response to calculate the response $x(t)$ of an undamped single-degree-of-freedom system to the rectangular pulse shown in Fig. 4.8. Plot $x(t)$ versus t.

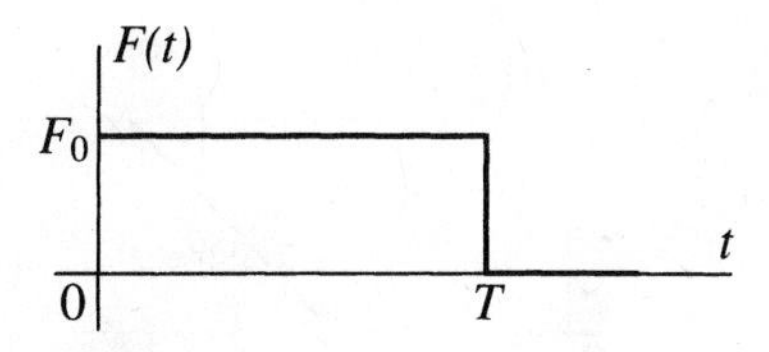

FIGURE 4.8
Rectangular pulse

It is easy to verify that the rectangular pulse depicted in Fig. 4.8 can be expressed as a combination of step functions of the form

$$F(t) = F_0\,[\mathscr{u}(t) - \mathscr{u}(t-T)] \tag{a}$$

so that the response can be expressed as a like combination of step responses, as follows:

$$x(t) = F_0[\mathscr{s}(t) - \mathscr{s}(t-T)] \tag{b}$$

But, from Eq. (4.22) with $\zeta = 0$ and $\omega_d = \omega_n$, the response of an undamped single-degree-of-freedom system to a unit step function initiated at $t = a$ is

$$\mathscr{s}(t-a) = \frac{1}{k}[1 - \cos\omega_n(t-a)]\mathscr{u}(t-a) \tag{c}$$

Hence, inserting Eq. (c) into Eq. (b), we conclude that the response of an undamped single-degree-of-freedom system to the rectangular pulse of Fig. 4.8 is simply

$$x(t) = \frac{F_0}{k}\{(1 - \cos\omega_n t)\mathscr{u}(t) - [1 - \cos\omega_n(t-T)]\mathscr{u}(t-T)\} \tag{d}$$

The plot $x(t)$ versus t is displayed in Fig. 4.9.

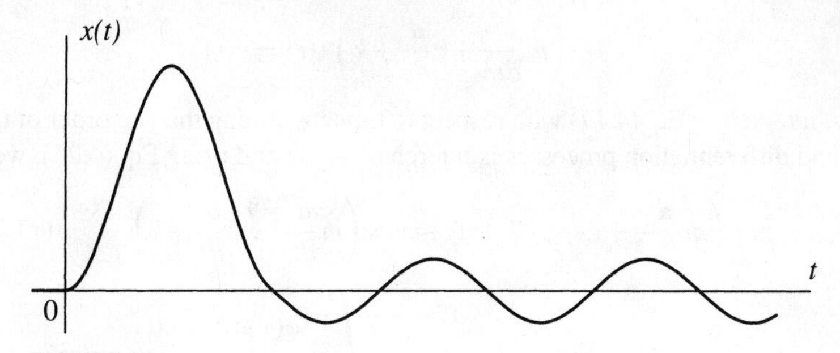

FIGURE 4.9
Response of a mass-spring system to a rectangular pulse

4.3 THE UNIT RAMP FUNCTION. RAMP RESPONSE

Yet another function of interest in vibrations is the *unit ramp function*, defined as

$$r(t-a) = (t-a)\mathscr{u}(t-a) \tag{4.24}$$

The function is shown in Fig. 4.10. Clearly, the unit ramp function has units of second(s).

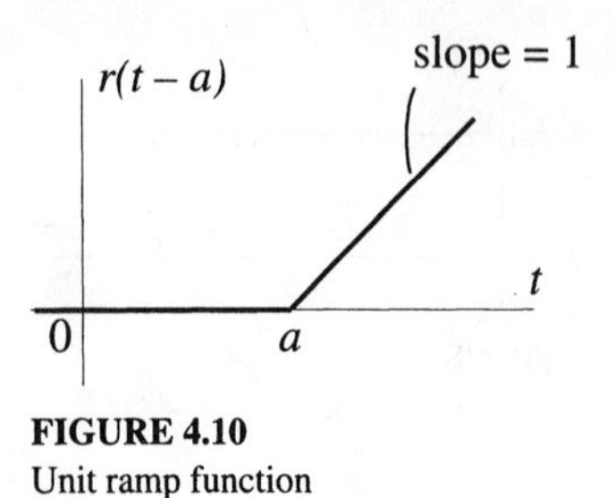

FIGURE 4.10
Unit ramp function

The unit ramp function is closely related to the unit step function. Indeed, it can be expressed as the integral of the unit step function, or

$$r(t-a) = \int_{-\infty}^{t} \mathscr{u}(\tau - a)d\tau \tag{4.25}$$

Consistent with this, the unit step function can be expressed as the time derivative of the unit ramp function, or

$$\mathscr{u}(t-a) = \frac{dr(t-a)}{dt} \tag{4.26}$$

The *ramp response*, denoted by $\mathscr{r}(t)$, is defined as the response of a system to $r(t)$, i.e., to a unit ramp function beginning at $t = 0$, and with zero initial conditions. Using the same approach as in Sec. 4.2, we can demonstrate that the ramp response is closely related to the step response. To this end, we use the mass-damper-spring system and write the relation between the ramp response and the unit ramp function in the symbolic form

$$\left(m\frac{d^2}{dt^2} + c\frac{d}{dt} + k\right)\mathscr{r}(t) = r(t) \tag{4.27}$$

Then, integrating Eq. (4.17) with respect to time, assuming that the order of the integration and differentiation processes is interchangeable and using Eq. (4.25), we obtain

$$\begin{aligned}\int_{-\infty}^{t}\left(m\frac{d^2}{d\tau^2} + c\frac{d}{d\tau} + k\right)\mathscr{s}(\tau)d\tau &= \left(m\frac{d^2}{dt^2} + c\frac{d}{dt} + k\right)\int_{-\infty}^{t}\mathscr{s}(\tau)d\tau \\ &= \int_{-\infty}^{t}\mathscr{u}(\tau)d\tau = r(t)\end{aligned} \tag{4.28}$$

Contrasting Eqs. (4.27) and (4.28), we arrive at the conclusion that

$$\mathscr{r}(t) = \int_{-\infty}^{t} \mathscr{s}(\tau)d\tau \tag{4.29}$$

or, the ramp response is equal to the integral of the step response. Relation (4.29) holds true for any linear time-invariant system, and is not restricted to mass-damper-spring systems. Equation (4.29) can be used as a convenient way of deriving the ramp response of a system, as demonstrated in Example 4.2.

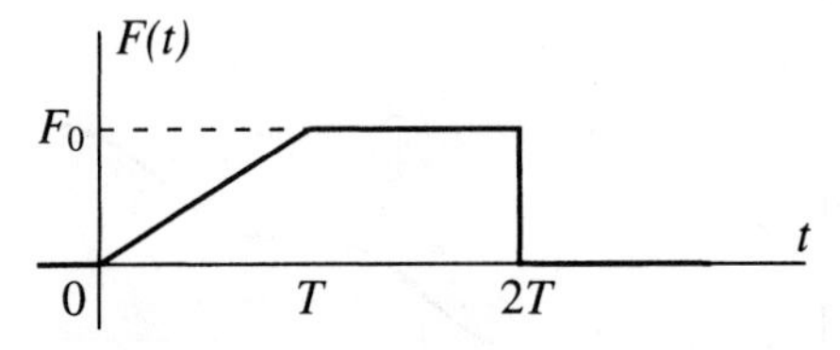

FIGURE 4.11
Trapezoidal pulse

The unit ramp function, sometimes in conjunction with the unit step function, can be used to express certain excitation functions in a simple form, thus permitting the derivation of the response in the same simple form. As an illustration, the trapezoidal pulse of Fig. 4.11 can be described as follows:

$$F(t) = \begin{cases} 0 \text{ for } t < 0 \\ F_0 t/T \text{ for } 0 < t < T \\ F_0 \text{ for } T < t < 2T \\ 0 \text{ for } t > 2T \end{cases} \tag{4.30}$$

It can be expressed more conveniently, and more compactly, in the form

$$F(t) = F_0 \left[\frac{1}{T} r(t) - \frac{1}{T} r(t-T) - u(t-2T) \right] \tag{4.31}$$

Then, the response of a system to $F(t)$ is simply

$$x(t) = F_0 \left[\frac{1}{T} \imath(t) - \frac{1}{T} \imath(t-T) - \mathscr{s}(t-2T) \right] \tag{4.32}$$

where $\imath(t)$ is the ramp response and $\mathscr{s}(t)$ the step response of the system.

Example 4.2. Use Eq. (4.29) to derive the ramp response of the mass-spring system of Example 4.1.

From Eq. (c) of Example 4.1, the step response of a mass-spring system is

$$\mathscr{s}(t) = \frac{1}{k}(1 - \cos \omega_n t) u(t) \tag{a}$$

Hence, inserting Eq. (a) into Eq. (4.29) and carrying out the integration, we obtain

$$\begin{aligned} \imath(t) &= \int_{-\infty}^{t} \mathscr{s}(\tau) d\tau = \frac{1}{k} \int_{-\infty}^{t} (1 - \cos \omega_n \tau) u(\tau) d\tau = \frac{1}{k} \int_{0}^{t} (1 - \cos \omega_n \tau) d\tau \\ &= \frac{1}{k} \left(\tau - \frac{\sin \omega_n \tau}{\omega_n} \right) \Bigg|_0^t = \frac{1}{k\omega_n} (\omega_n t - \sin \omega_n t) u(t) \end{aligned} \tag{b}$$

where we multiplied the result by $u(t)$ in consideration of the fact that the response is zero for $t < 0$. The ramp response is displayed in Fig. 4.12.

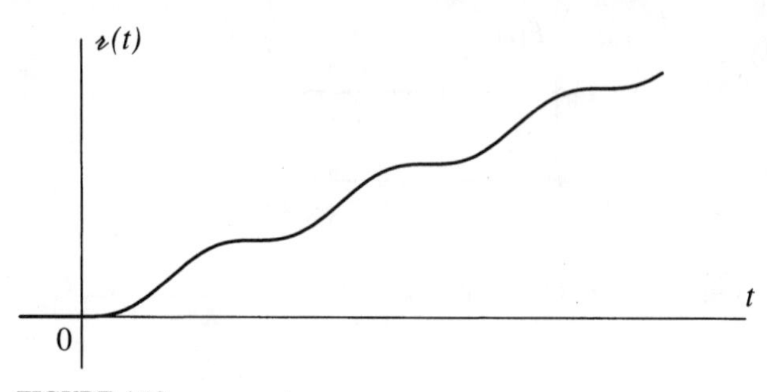

FIGURE 4.12
Ramp response of a mass-spring system

4.4 RESPONSE TO ARBITRARY EXCITATIONS. THE CONVOLUTION INTEGRAL

In Ch. 3, we discussed the response of linear time-invariant systems to harmonic and periodic excitations. Then, earlier in this chapter, we considered the response to a unit impulse, a unit step function, a unit ramp function and linear combinations of the latter two. In one form or another, all these excitations have one thing in common, namely, they can all be described as explicit functions of time. The question remains as to how to obtain the response to arbitrary excitations.

For complicated excitations, the general approach is to express them as linear combinations of simpler excitations, sufficiently simple that the response is readily available, or can be produced without much difficulty. In this regard, we should point out that the harmonic response, impulse response, step response and ramp response fall in this category. We used this approach in Sec. 3.9, in which we expressed periodic excitations as Fourier series of harmonic components and then the response to periodic excitations as linear combinations of harmonic responses. Then, in Sec. 4.2 we represented a trapezoidal pulse as a linear combination of step and ramp functions and the response to the trapezoidal pulse as a corresponding linear combination of step and ramp responses. It turns out that the same approach can also be used in the case of arbitrary excitations.

There are two ways of deriving the response to arbitrary excitations, depending on the manner in which the excitation function is described. One way is to regard the arbitrary excitation as periodic and represent it by a Fourier series. Then, using a limiting process whereby the period is allowed to approach infinity, so that in essence the function ceases to be periodic and becomes arbitrary, the Fourier series becomes a Fourier integral. This is the frequency-domain representation of functions, which is more suitable for random excitations than for deterministic excitations. This approach is discussed in detail in Ch. 12. The second approach consists of regarding the arbitrary excitation as a superposition of impulses of varying magnitude and applied at different times. This is the time-domain representation of functions, and is the one used in this section.

We consider an arbitrary excitation $F(t)$, such as that depicted in Fig. 4.13, and focus our attention on the contribution to the response of an impulse corresponding to the time interval $\tau < t < \tau + \Delta\tau$. Assuming that the time increment $\Delta\tau$ is sufficiently small

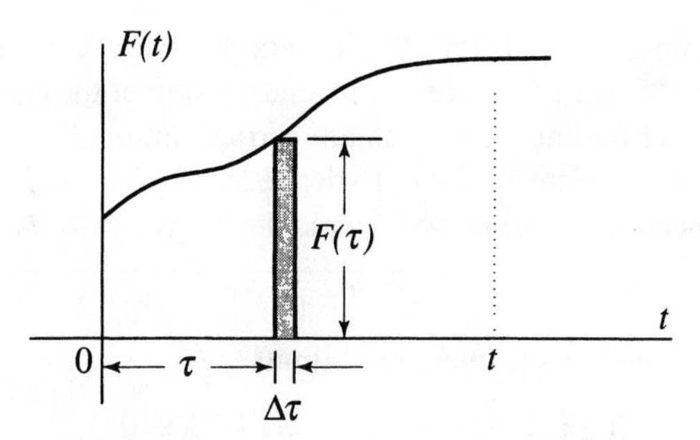

FIGURE 4.13
Arbitrary excitation

that $F(t)$ does not change very much over this time increment, the shaded area in Fig. 4.13 can be regarded as an impulse acting over $\tau < t < \tau + \Delta\tau$ and having the magnitude $F(\tau)\Delta\tau$. Hence, recalling Eq. (4.2), the excitation corresponding to the shaded area can be treated as an impulsive force having the form

$$\hat{F}(\tau)\delta(t-\tau) = F(\tau)\Delta\tau\delta(t-\tau) \tag{4.33}$$

But, as shown in Fig. 4.14, the response of a linear time-invariant system to the impulsive force given by Eq. (4.33) is simply

$$\Delta x(t,\tau) = F(\tau)\Delta\tau g(t-\tau) \tag{4.34}$$

where $g(t-\tau)$ is the impulse response delayed by the time interval τ. Then, regarding the excitation $F(t)$ as a superposition of impulsive forces, we can approximate the response by writing

$$x(t) = \sum_{\tau} F(\tau)\Delta\tau g(t-\tau) \tag{4.35}$$

In the limit, as $\Delta\tau \to 0$, we can replace the summation by integration and obtain the exact response

$$x(t) = \int_0^t F(\tau)g(t-\tau)d\tau \tag{4.36}$$

Equation (4.36) is known as the *convolution integral*, and expresses the response as a superposition of impulse responses. For this reason, Eq. (4.36) is also referred to at times as the *superposition integral.*

We observe that the impulse response in the convolution integral is a function of $t-\tau$, rather than of τ, where τ is the variable of integration. As demonstrated later in

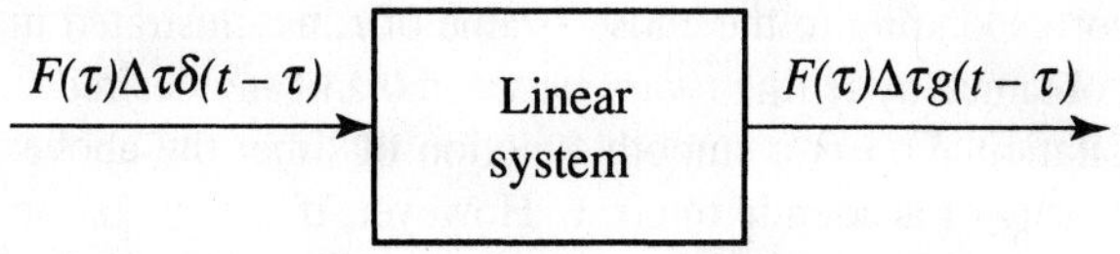

$F(\tau)\Delta\tau\delta(t-\tau)$ → Linear system → $F(\tau)\Delta\tau g(t-\tau)$

FIGURE 4.14
Block diagram relating the response to an excitation in the form of an impulse of magnitude $F(\tau)\Delta\tau$

this section, to obtain $g(t-\tau)$ from $g(\tau)$, it is necessary to carry out two operations, namely, shifting and "folding." There is a second version of the convolution integral in which the shifting and folding operations are carried out on $F(\tau)$ instead of on $g(\tau)$, which may be more convenient at times. To derive the second version of the convolution integral, we introduce a transformation of variables from τ to λ, as follows:

$$t-\tau=\lambda,\ \tau=t-\lambda,\ d\tau=-d\lambda \tag{4.37}$$

which requires the change in the integration limits

$$\tau=0\rightarrow\lambda=t,\ \tau=t\rightarrow\lambda=0 \tag{4.38}$$

Introducing Eqs. (4.37) and (4.38) in Eq. (4.36), we obtain

$$x(t)=\int_t^0 F(t-\lambda)g(\lambda)(-d\lambda)=\int_0^t F(t-\lambda)g(\lambda)d\lambda \tag{4.39}$$

which is the second form of the convolution integral. Recognizing that τ in Eq. (4.36) and λ in Eq. (4.39) are mere dummy variables of integration, we can combine Eqs. (4.36) and (4.39) into

$$x(t)=\int_0^t F(\tau)g(t-\tau)d\tau=\int_0^t F(t-\tau)g(\tau)d\tau \tag{4.40}$$

from which we conclude that the convolution integral is symmetric in the excitation $F(t)$ and the impulse response $g(t)$, in the sense that the result is the same regardless of which of the two functions is shifted and folded. The question can be raised as to which form of the convolution integral to use. The choice depends on the nature of the functions $F(t)$ and $g(t)$, and must be the one making the integration task simpler.

The convolution integral lends itself to a geometric interpretation that is not only interesting but at times also quite useful. This interpretation involves the various steps implicit in the evaluation of the integral. To review these steps, we consider the first version of the convolution integral, Eq. (4.36). Figure 4.15a shows an arbitrary excitation $F(\tau)$ and Fig. 4.15b a typical impulse response $g(\tau)$ corresponding to an underdamped mass-damper-spring system, both with t replaced by the variable of integration τ. The first step is to shift the impulse response backward by the time interval t, which yields $g(\tau+t)$, as shown in Fig. 4.15c. The second step is the "folding," which results in $g(t-\tau)$, as depicted in Fig. 4.15d. The step consists of taking the mirror image of $g(\tau+t)=g(t+\tau)$ with respect to the vertical axis, which amounts to replacing τ by $-\tau$. The third step is to multiply $F(\tau)$ by $g(t-\tau)$, yielding the curve shown in Fig. 4.15e. The final step is the integration of the curve $F(\tau)g(t-\tau)$, which is the same as determining the area under the curve in Fig. 4.15e. The result is one point on the response $x(t)$ corresponding to the chosen value of t, as illustrated in Fig. 4.15f. The full response is obtained by letting t vary between 0 and any desired value.

If the excitation $F(t)$ is a smooth function of time, the above geometric interpretation is primarily of academic interest. However, if the excitation function is only sectionally smooth, such as the rectangular pulse of Fig. 4.8, then the limits of integration in the convolution integral must be chosen judiciously. In this regard, the preceding geometric interpretation is vital to a successful determination of the response, as the

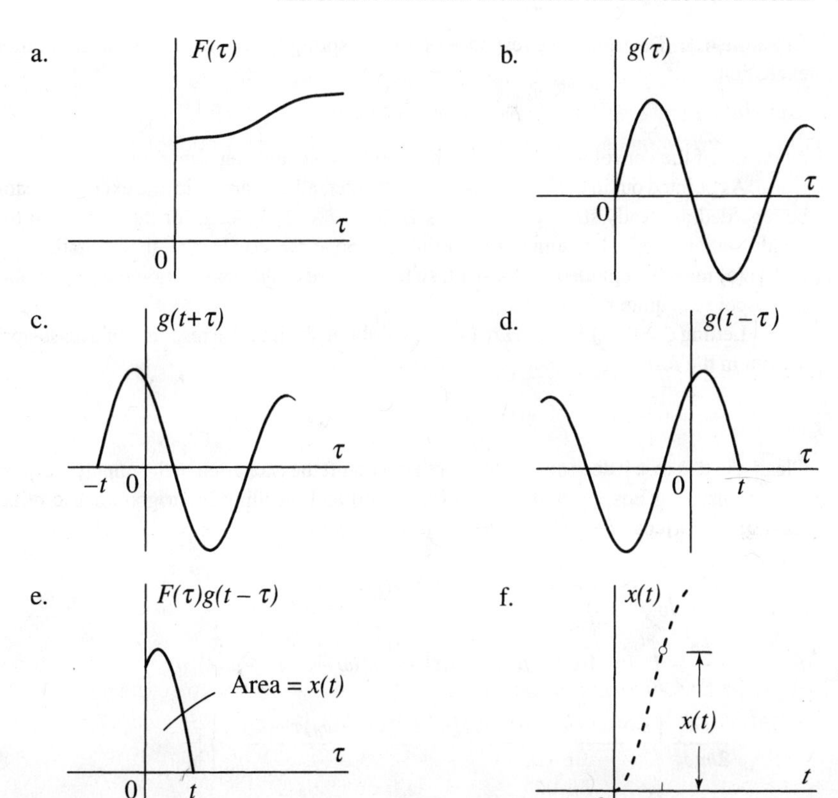

FIGURE 4.15
a. Arbitrary excitation, b. Impulse response, c. Impulse response shifted backward, d. Impulse response shifted backward and "folded," e. Product of $F(\tau)$ and $g(t-\tau)$, f. Point on the response curve

choice of limits is not immediately obvious. These points are demonstrated in Examples 4.3 and 4.4.

The convolution integral given by Eq. (4.40) represents a special case in the sense that one of the functions involved is the excitation $F(t)$, the other is the impulse response $g(t)$ and the result is the response $x(t)$. Indeed, a general definition of the convolution integral involves two arbitrary functions $f_1(t)$ and $f_2(t)$, not necessarily the excitation $F(t)$ and the impulse response $g(t)$, and the result does not necessarily represent a response. The general form of the convolution integral can be demonstrated by means of the Laplace transformation method, and is presented in Appendix B.

Quite often, the excitation is such a complicated function of time that closed-form evaluation of the convolution integral is likely to cause considerable difficulties. In other cases, the excitation cannot even be expressed in terms of known functions and is given either in the form of a graph or a list of values at discrete times. In all these cases, evaluation of the convolution integral must be carried out numerically. Later in this chapter, we present a formal algorithm for the numerical processing of the convolution integral on a digital computer.

Example 4.3. Determine the response of a mass-spring system to the one-sided harmonic excitation

$$F(t) = F_0 \sin \omega t\, u(t) \tag{a}$$

by means of the convolution integral, where $u(t)$ is the unit step function.

As pointed out in the beginning of this chapter, albeit harmonic, the excitation cannot be regarded as steady state, because it is zero for $t < 0$. Indeed, for the excitation to be steady state it should be defined for all times, $-\infty < t < \infty$. Hence, the excitation given by Eq. (a) must be regarded as transient, so that the use of the convolution integral to obtain the response is quite appropriate.

Letting $\zeta = 0$, $\omega_d = \omega_n$ in Eq. (4.13), we obtain the impulse response of a mass-spring system in the form

$$g(t) = \frac{1}{m\omega_n} \sin \omega_n t\, u(t) \tag{b}$$

Clearly, in this case it does not matter whether we shift the excitation or the impulse response. Hence, inserting Eqs. (a) and (b) into Eq. (4.36) and recalling the trigonometric relation $\sin\alpha \sin\beta = \frac{1}{2}[\cos(\alpha-\beta) - \cos(\alpha+\beta)]$, we can write

$$\begin{aligned}
x(t) &= \frac{F_0}{m\omega_n}\int_0^t \sin\omega\tau\, u(\tau) \sin\omega_n(t-\tau)\, u(t-\tau)d\tau \\
&= \frac{F_0}{2m\omega_n}\int_0^t \{\cos[(\omega+\omega_n)\tau - \omega_n t] - \cos[(\omega-\omega_n)\tau + \omega_n t]\}\, d\tau \\
&= \frac{F_0}{2m\omega_n}\left\{\frac{\sin[(\omega+\omega_n)\tau - \omega_n t]}{\omega+\omega_n} - \frac{\sin[(\omega-\omega_n)\tau + \omega_n t]}{\omega-\omega_n}\right\}\Bigg|_0^t \\
&= \frac{F_0}{2m\omega_n(\omega^2-\omega_n^2)}[(\omega-\omega_n)(\sin\omega t + \sin\omega_n t) - (\omega+\omega_n)(\sin\omega t - \sin\omega_n t)] \\
&= \frac{F_0}{k[1-(\omega/\omega_n)^2]}(\sin\omega t - \frac{\omega}{\omega_n}\sin\omega_n t) u(t)
\end{aligned} \tag{c}$$

where we multiplied the result by $u(t)$ in recognition of the fact that the response must be zero for $t < 0$.

Example 4.4. Determine the response of a mass-damper-spring system to the rectangular pulse of Fig. 4.8 by means of the convolution integral in conjunction with the geometric interpretation of Fig. 4.15, but with the excitation shifted instead of the impulse response. Show that straight application of the convolution integral formula may yield erroneous results in the case of discontinuous excitations.

The problem statement calls for the use of the second version of the convolution integral, Eq. (4.40), or

$$x(t) = \int_0^t F(t-\tau) g(\tau) d\tau \tag{a}$$

Because the rectangular pulse is discontinuous, the geometric interpretation is more involved than that of Fig. 4.15. We begin by redrawing Fig. 4.8, but with t replaced by τ, as shown in Fig. 4.16a. From Eq. (4.13), the impulse response for a mass-damper-spring system is

$$g(t) = \frac{1}{m\omega_d} e^{-\zeta\omega_n t} \sin\omega_d t\, u(t) \tag{b}$$

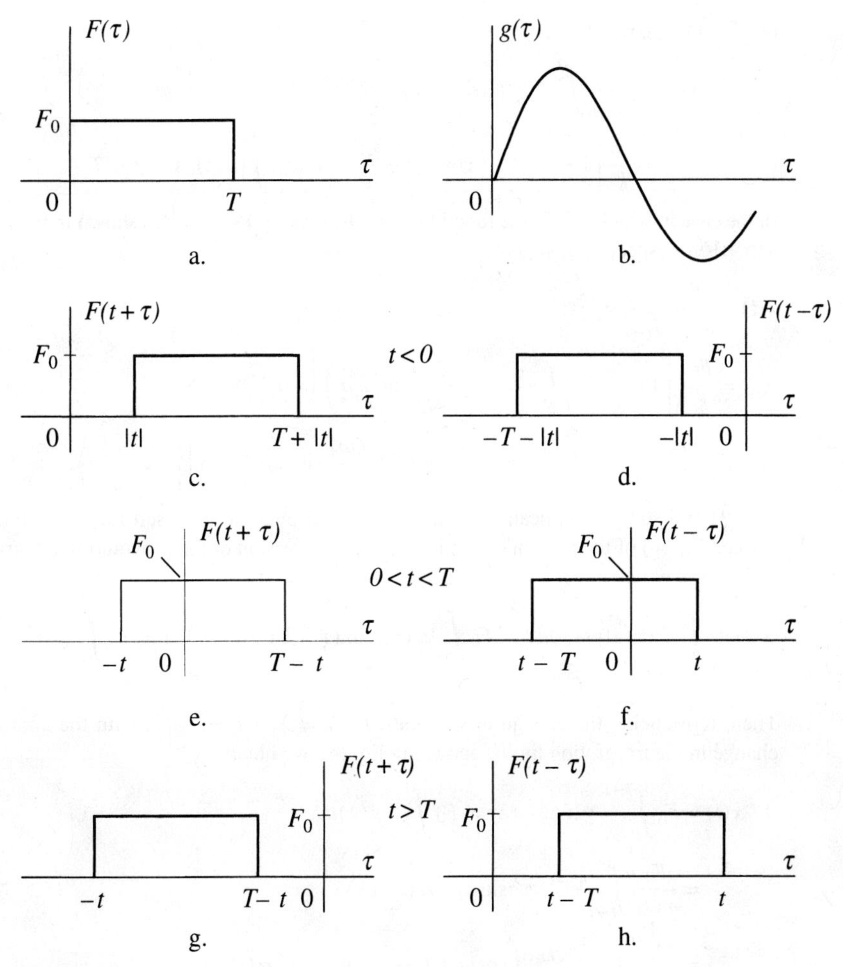

FIGURE 4.16
a. Rectangular pulse, b. Impulse response, c. Shifted pulse for $t < 0$, d. Shifted and foleded pulse for $t < 0$, e. Shifted pulse for $0 < t < T$, f. Shifted and folded pulse for $0 < t < T$, g. Shifted pulse for $t > T$, h. Shifted and folded pulse for $t > T$

It is displayed in Fig. 4.16b with t replaced by τ. In view of the discontinuous nature of $F(\tau)$, the shifting and folding require careful consideration. We first consider the case $t < 0$, in which case $F(\tau+t)$ and $F(t-\tau)$ are as shown in Figs. 4.16c and 4.16d, respectively. It is clear that for $t < 0$ there is no overlap between $F(t-\tau)$ and $g(\tau)$, so that the product of the two is zero, yielding

$$x(t) = 0,\ t < 0 \tag{c}$$

as is to be expected. For $0 < t < T$, the functions $F(\tau+t)$ and $F(t-\tau)$ are as depicted in Figs. 4.16e and Fig. 4.16f, respectively. Hence, using Eqs. (a) and (b), together with Eqs.

(4.20) - (4.22), we obtain

$$x(t) = \frac{F_0}{m\omega_d}\int_0^t e^{-\zeta\omega_n\tau}\sin\omega_d\tau\,\mathscr{u}(\tau)d\tau$$

$$= \frac{F_0}{k}\left[1 - e^{-\zeta\omega_n t}\left(\cos\omega_d t + \frac{\zeta\omega_n}{\omega_d}\sin\omega_d t\right)\right]\mathscr{u}(t),\ 0 < t < T \tag{d}$$

In the case in which $t > T$, the functions $F(\tau + t)$ and $F(t - \tau)$ are as shown in Figs. 4.16g and 4.16h, respectively, so that

$$x(t) = \frac{F_0}{m\omega_d}\int_{t-T}^t e^{-\zeta\omega_n\tau}\sin\omega_d\tau\,\mathscr{u}(\tau)d\tau$$

$$= \frac{F_0}{k}\left\langle\left[1 - e^{-\zeta\omega_n t}\left(\cos\omega_d t + \frac{\zeta\omega_n}{\omega_d}\sin\omega_d t\right)\right]\mathscr{u}(t)\right.$$

$$\left. - \left\{1 - e^{-\zeta\omega_n(t-T)}\left[\cos\omega_d(t-T) + \frac{\zeta\omega_n}{\omega_d}\sin\omega_d(t-T)\right]\right\}\mathscr{u}(t-T)\right\rangle,\ t > T \tag{e}$$

For a straight application of the convolution integral, we insert Eq. (a) of Example 4.1 and Eq. (b) of the present example into the first version of the convolution integral, Eq. (4.36), and write

$$x(t) = \int_0^t F(\tau)g(t-\tau)d\tau = F_0\int_0^t [\mathscr{u}(\tau) - \mathscr{u}(\tau - T)]g(t-\tau)d\tau = F_0\int_0^T g(t-\tau)d\tau \tag{f}$$

Then, introducing the change in variables, $t - \tau = \lambda,\ d\tau = -\,d\lambda$, with the appropriate change in the integration limits, and using Eq. (e), we obtain

$$x(t) = F_0\int_t^{t-T} g(\lambda)(-d\lambda) = F_0\int_{t-T}^t g(\lambda)d\lambda$$

$$= \frac{F_0}{m\omega_d}\int_{t-T}^t e^{-\zeta\omega_n\lambda}\sin\omega_d\lambda\,\mathscr{u}(\lambda)d\lambda$$

$$= \frac{F_0}{k}\left\langle\left[1 - e^{-\zeta\omega_n t}\left(\cos\omega_d t + \frac{\zeta\omega_n}{\omega_d}\sin\omega_d t\right)\right]\mathscr{u}(t)\right.$$

$$\left. - \left\{1 - e^{-\zeta\omega_n(t-T)}\left[\cos\omega_d(t-T) + \frac{\zeta\omega_n}{\omega_d}\sin\omega_d(t-T)\right]\right\}\mathscr{u}(t-T)\right\rangle \tag{g}$$

Comparing Eqs. (e) and (g), we conclude that straight application of the convolution integral yielded only the response following the termination of the pulse, which demonstrates the fact that, for functions $F(t)$ not defined by a single expression for all times $t > 0$, guidance must be sought from the geometric interpretation of the convolution integral in determining the proper integration limits.

4.5 SHOCK SPECTRUM

Many structures are subjected on occasions to relatively large forces applied suddenly and over periods of time that are short relative to the natural period of the structure. Such forces can produce local damage, or they can excite undesirable vibration of the structure. Indeed, at times the vibration results in large cyclic stress damaging the structure or impairing its performance. A force of this type has come to be known as a *shock*. The response of structures to shocks is of vital importance in design. The severity of the shock is generally measured in terms of the maximum value of the response. For comparison purposes, it is customary to use the response of an undamped single-degree-of-freedom system. The plot of the peak response of a mass-spring system to a given shock as a function of the natural frequency of the system is known as *shock spectrum*, or *response spectrum*.

A shock $F(t)$ is generally characterized by its maximum value F_0, its duration T and its shape, or alternatively the impulse $\int_0^T F(t)dt$. These characteristics depend on the force-producing mechanism and on the properties of the interface material. A reasonable approximation for the force is the half-sine pulse shown in Fig. 4.17; we propose to derive the associated shock spectrum.

The half-sine pulse of Fig. 4.17 can be regarded as the superposition of two one-sided sine functions, one initiated at $t = 0$ and the second initiated at $t = T = \pi/\omega$. Hence, using the procedure introduced in Sec. 4.2, we can describe the half-sine pulse for all times in the form

$$F(t) = F_0[\sin\omega t\,\omega(t) + \sin\omega(t-T)\omega(t-T)] \tag{4.41}$$

But, the response of an undamped single-degree-of-freedom system to the one-sided sine function initiated at $t = 0$ was obtained earlier in the form of Eq. (c) of Example 4.3. Using this result, it follows that the response to the half-sine pulse of Fig. 4.17 is simply

$$x(t) = \frac{F_0}{k}\frac{1}{1-(\omega/\omega_n)^2}\left\{\left(\sin\omega t - \frac{\omega}{\omega_n}\sin\omega_n t\right)\omega(t) + \left[\sin\omega(t-T) - \frac{\omega}{\omega_n}\sin\omega_n(t-T)\right]\omega(t-T)\right\} \tag{4.42}$$

To generate the shock spectrum, we must find the maximum response. To this end, we recognize that, although Eq. (4.42) describes the response for all times, the response during the pulse differs from the response after the termination of the pulse, so that we must consider the response for each time interval separately. From Eq. (4.42), the response *during the pulse* is

$$x(t) = \frac{F_0}{k[1-(\omega/\omega_n)^2]}\left(\sin\omega t - \frac{\omega}{\omega_n}\sin\omega_n t\right), \quad 0 < t < \frac{\pi}{\omega} \tag{4.43}$$

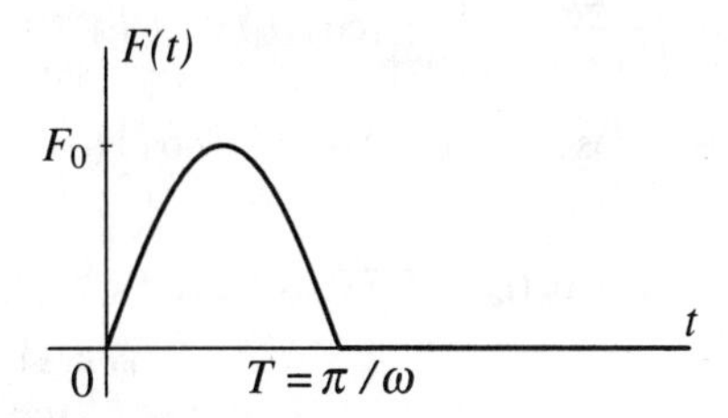

FIGURE 4.17
Half-sine pulse

To obtain the maximum response, we must solve for the time t_m for which $\dot{x} = 0$ and then substitute this value of t_m in Eq. (4.43). Differentiating Eq. (4.43) with respect to time, we obtain

$$\dot{x}(t) = \frac{F_0 \omega}{k[1-(\omega/\omega_n)^2]}(\cos\omega t - \cos\omega_n t) \tag{4.44}$$

so that, using the relation $\cos\alpha - \cos\beta = -2\sin\frac{1}{2}(\alpha+\beta)\sin\frac{1}{2}(\alpha-\beta)$, we conclude that t_m must satisfy the equation

$$\sin\tfrac{1}{2}(\omega_n+\omega)t_m \sin\tfrac{1}{2}(\omega_n-\omega)t_m = 0 \tag{4.45}$$

which has two families of solutions

$$\left.\begin{matrix} t'_m \\ t''_m \end{matrix}\right\} = \frac{2i\pi}{\omega_n \pm \omega}, \qquad i = 1, 2, \ldots \tag{4.46}$$

Substituting the above values in Eq. (4.43), we obtain

$$\begin{aligned} x(t'_m) &= \frac{F_0}{k(1-\omega/\omega_n)} \sin\frac{2i\pi\omega/\omega_n}{1+\omega/\omega_n} \\ x(t''_m) &= \frac{F_0}{k(1+\omega/\omega_n)} \sin\frac{2i\pi\omega/\omega_n}{1-\omega/\omega_n} \end{aligned} \tag{4.47}$$

It is obvious from Eqs. (4.47) that the response corresponding to $t = t'_m$ achieves higher values than the response corresponding to $t = t''_m$. The question remains as to how to determine the value of the integer i. To answer this question, we recall that t'_m must occur during the pulse, so that from Eqs. (4.46) we must have $[2i\pi/(\omega_n+\omega)] < \pi/\omega$. Hence, we conclude that for $0 < t < \pi/\omega$ we have the maximum response

$$x_{\max} = \frac{F_0\omega_n/\omega}{k[(\omega_n/\omega)-1]} \sin\frac{2i\pi}{1+\omega_n/\omega}, \quad i < \frac{1}{2}\left(1+\frac{\omega_n}{\omega}\right) \tag{4.48}$$

Also from Eq. (4.42), the response for *any time after the termination of the pulse* can be verified to be

$$x(t) = \frac{F_0\omega_n/\omega}{k[1-(\omega_n/\omega)^2]}[\sin\omega_n t + \sin\omega_n(t-T)], \; t > \pi/\omega \tag{4.49}$$

As before, to obtain the maximum response, we must first determine $t = t_m$, at which time $\dot{x}(t) = 0$. To this end, we write first

$$\dot{x}(t) = \frac{F_0\omega_n^2/\omega}{k[1-(\omega_n/\omega)^2]}[\cos\omega_n t + \cos\omega_n(t-T)] \tag{4.50}$$

Then, recalling that $\cos\alpha + \cos\beta = 2\cos\frac{1}{2}(\alpha+\beta)\cos\frac{1}{2}(\alpha-\beta)$, we conclude that t_m must satisfy the equation

$$\cos\omega_n(t_m - \tfrac{1}{2}T)\cos\tfrac{1}{2}\omega_n T = 0 \tag{4.51}$$

which yields the solutions

$$t_m = (2i-1)\frac{\pi}{2\omega_n} + \frac{T}{2}, \; i = 1, 2, \ldots \tag{4.52}$$

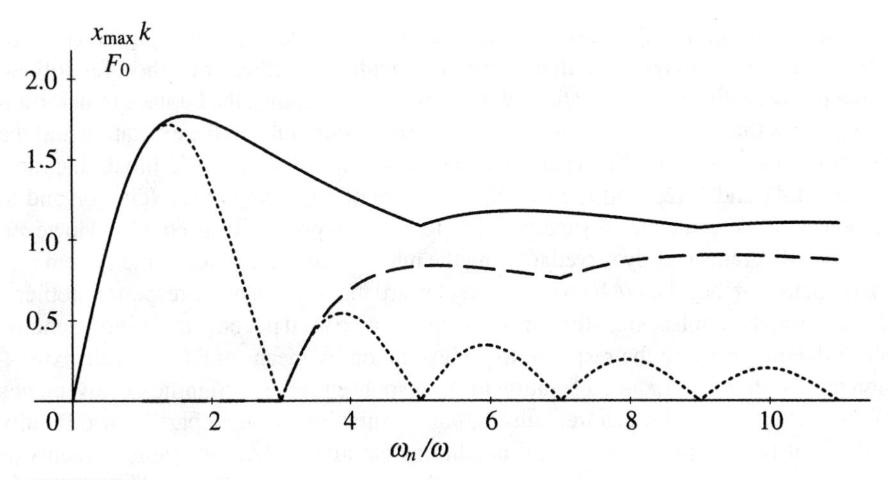

FIGURE 4.18
Shock spectrum

Introducing $t = t_m$ in Eq. (4.49), we obtain the maximum response for $t > \pi/\omega$ in the form

$$x_{max} = \frac{2F_0\omega_n/\omega}{k[1-(\omega_n/\omega)^2]} \cos\frac{\pi}{2}\frac{\omega_n}{\omega} \tag{4.53}$$

The response spectrum is simply the plot x_{max} versus ω_n/ω, in which both Eqs. (4.48) and (4.53) must be considered. Of course, only the larger of the two values must be used. We note that solution (4.48) is not valid for $\omega_n < \omega$, but both solutions are valid for $\omega_n > \omega$. It turns out that the maximum response is given by Eq. (4.53) for $\omega_n < \omega$ and by Eq. (4.48) for $\omega_n > \omega$. The response spectrum is shown in Fig. 4.18 in the form of the nondimensional plot $x_{max}k/F_0$ versus ω_n/ω.

For different pulse shapes, different shock spectra can be anticipated. For a rectangular pulse, or a triangular pulse, the ratio ω_n/ω has no meaning, because there is no ω in the definition of these pulses. However, these pulses can be defined in terms of their duration T, so that in these cases the shock spectrum is given by $x_{max}k/F_0$ versus T/T_n, or $x_{max}k/F_0$ versus $2T/T_n$, where $T_n = 2\pi/\omega_n$ is the natural period of the mass-spring system.

4.6 SYSTEM RESPONSE BY THE LAPLACE TRANSFORMATION METHOD. TRANSFER FUNCTION

The solution of many problems in vibrations by direct means can cause serious difficulties, and may not even be possible, unless some type of transformation is used. There is a large variety of transformations, but the general idea behind all of them is the same, namely, transform a difficult problem into a simple one, solve the simple problem and inverse transform the solution of the simple problem to obtain the solution of the original difficult problem. In this section, we consider the most widely used transformation in vibrations, namely, the Laplace transformation.

The Laplace transformation method has gained wide acceptance in the study of linear time-invariant systems. In addition to providing an efficient method for solving linear ordinary differential equations with constant coefficients, the Laplace transformation permits the writing of a simple algebraic expression relating the excitation and the response of systems. In this regard, we are reminded of the symbolic block diagrams of Figs. 1.41 and 1.42, used to relate the response of a system to the excitation and to demonstrate the principle of superposition for linear systems, respectively. However, the block diagrams merely served to bring the inherent characteristics of the system into sharp focus, but they did not help in any way toward the solution of the response problem. By contrast, the Laplace transformation method can be used not only to define a genuine block diagram relating the response to the excitation by means of an algebraic expression but also to help produce a solution to the response problem. Significant advantages of the method are that it can treat discontinuous functions without particular difficulty and that it takes into account initial conditions automatically. Sufficient elements of the Laplace transformation to provide us with a working knowledge of the method are presented in Appendix B. Here we concentrate mainly on using the method to study response problems.

The *(one-sided) Laplace transformation* of $x(t)$, written symbolically as $X(s) = \mathscr{L}x(t)$ is defined by the definite integral

$$X(s) = \mathscr{L}x(t) = \int_0^\infty e^{-st} x(t) dt \tag{4.54}$$

where s is in general a complex quantity referred to as a *subsidiary variable*. The function e^{-st} is known as the kernel of the transformation. Because this is a definite integral, with t as the variable of integration, the transformation yields a function of s. To solve Eq. (3.1) by the Laplace transformation method, it is necessary to evaluate the transforms of the derivatives dx/dt and d^2x/dt^2. A simple integration by parts leads to

$$\begin{aligned}\mathscr{L}\frac{dx(t)}{dt} &= \int_0^\infty e^{-st}\frac{dx(t)}{dt}dt = e^{-st}x(t)\Big|_0^\infty + s\int_0^\infty e^{-st}x(t)dt \\ &= sX(s) - x(0)\end{aligned} \tag{4.55}$$

where $x(0)$ is the value of the function $x(t)$ at $t = 0$. Physically, it represents the initial displacement of the mass m. Similarly, it is not difficult to show that

$$\begin{aligned}\mathscr{L}\frac{d^2x(t)}{dt^2} &= \int_0^\infty e^{-st}\frac{d^2x(t)}{dt^2}dt = e^{-st}\frac{dx(t)}{dt}\Big|_0^\infty + s\int_0^\infty e^{-st}\frac{dx(t)}{dt}dt \\ &= -\dot{x}(0) + s\mathscr{L}\frac{dx(t)}{dt} = s^2X(s) - sx(0) - \dot{x}(0)\end{aligned} \tag{4.56}$$

where $\dot{x}(0)$ is the initial velocity of m. The Laplace transformation of the excitation function is simply

$$F(s) = \mathscr{L}F(t) = \int_0^\infty e^{-st} F(t) dt \tag{4.57}$$

Transforming both sides of Eq. (3.1) and rearranging, we obtain

$$(ms^2 + cs + k)X(s) = F(s) + m\dot{x}(0) + (ms + c)x(0) \tag{4.58}$$

In the following discussion, we concentrate on the effect of the forcing function $F(t)$, although we could have just as easily regarded the right side of Eq. (4.58) as a general transformed excitation. Hence, ignoring the homogeneous solution, which is equivalent to letting $x(0) = \dot{x}(0) = 0$, we can write the ratio of the transformed excitation to the transformed response in the form

$$Z(s) = \frac{F(s)}{X(s)} = ms^2 + cs + k \tag{4.59}$$

where the function $Z(s)$ is known as the *generalized impedance* of the system. The concept of impedance was introduced first in Sec. 3.2 in connection with the steady-state response of systems.

In the study of systems, we encounter a more general concept relating the transformed response to the transformed excitation. This general concept is known as the *system function*, or *transfer function*. For the special case of the second-order system described by Eq. (3.1), the transfer function has the form

$$G(s) = \frac{X(s)}{F(s)} = \frac{1}{ms^2 + cs + k} = \frac{1}{m(s^2 + 2\zeta\omega_n s + \omega_n^2)} \tag{4.60}$$

and it represents an *algebraic expression* in the s-plane, namely, a complex plane sometimes referred to as the *Laplace plane*, or *Laplace domain*. Note that, by letting $s = i\omega$ in $G(s)$ and multiplying by m, we obtain the frequency response $G(i\omega)$, Eq. (3.20).

Equation (4.60) can be rewritten as

$$X(s) = G(s)F(s) \tag{4.61}$$

Equation (4.61) represents the solution to the simple problem mentioned in the introduction to this section and can be expressed schematically by the block diagram of Fig. 4.19. This is a genuine block diagram, stating that the transformed response $X(s)$ can be obtained by merely multiplying the transformed excitation $F(s)$ by the transfer function $G(s)$. The final step is to carry out an inverse transformation to obtain the solution to the original problem, namely, the response $x(t)$. To this end we simply evaluate the *inverse Laplace transformation* of $X(s)$, defined symbolically by

$$x(t) = \mathscr{L}^{-1}X(s) = \mathscr{L}^{-1}G(s)F(s) \tag{4.62}$$

The operation $\mathscr{L}^{-1}$ involves in general a line integral in the complex s-domain. For all practical purposes, however, it is not necessary to go so deeply into the theory of the Laplace transformation method. Indeed, in virtually all cases solutions can be carried out by means of tables of Laplace transforms. The tables consist of pairs of entries $f(t)$

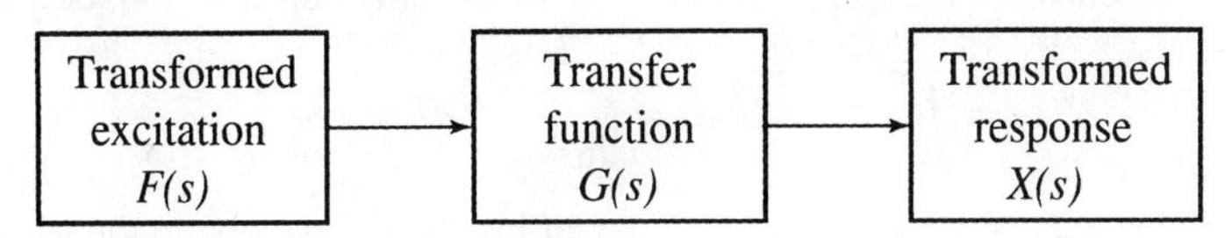

FIGURE 4.19
Block diagram relating the transformed response to the transformed excitation by means of the transfer function

and $F(s)$ arranged in two columns, so that to an entry $f(t)$ in the left column corresponds the Laplace transform $F(s) = \mathscr{L}f(t)$ in the right column, where $f(t)$ and $F(s)$ represent a Laplace transform pair. Conversely, to an entry $F(s)$ in the right column corresponds the inverse Laplace transform $f(t) = \mathscr{L}^{-1}F(s)$ in the left column. The process is a two-way street whereby one can begin with a given function $f(t)$ and find its Laplace transform $F(s)$ on the right, or begin with a given Laplace transform $F(s)$ and find its inverse Laplace transform $f(t)$ on the left. Of course, for the most part, the interest lies in using the tables to determine inverse Laplace tranforms, thus rendering integrations in the complex plane unnecessary. A modest table, yet covering many of the transform pairs encountered frequently in vibrations, can be found at the end of Appendix B.

Quite often, the transformed response $X(s)$ is too involved to be found readily in tables. In such cases, it is generally possible to decompose $X(s)$ into a sum of functions sufficiently simple that their inverse Laplace transforms can be found in tables. The procedure for carrying out such decompositions is known as the *method of partial fractions*, and is presented in Appendix B. Also in Appendix B we discuss a theorem for the inversion of a function $X(s)$ having the form of a product of two functions of s, such as Eq. (4.62). This is the convolution theorem, which is applicable to the product of any two functions $F_1(s)$ and $F_2(s)$, not necessarily the transfer function $G(s)$ and the Laplace transform $F(s)$ of an excitation force.

Equation (4.62) can be used to derive the response of any linear system with constant coefficients subjected to arbitrary excitations. There are three excitations of particular interest in vibrations, namely, the unit impulse, the unit step function and the unit ramp function. These functions and the response to these functions were already discussed in Secs. 4.1 - 4.3, but in this section we wish to present the derivation of the response by means of the Laplace transformation method.

The Laplace transform of the unit impulse is

$$\Delta(s) = \int_0^\infty e^{-st}\delta(t)dt = e^{-st}\Big|_{t=0}\int_0^\infty \delta(t)dt = 1 \tag{4.63}$$

where use has been made of the sampling property of delta functions, Eq. (4.3). Inserting $x(t) = g(t)$ and $F(s) = \Delta(s) = 1$ in Eq. (4.62), we obtain the impulse response

$$g(t) = \mathscr{L}^{-1}G(s)\Delta(s) = \mathscr{L}^{-1}G(s) \tag{4.64}$$

Hence, the *impulse response is equal to the inverse Laplace transform of the transfer function*, so that the unit impulse and the transfer function represent a Laplace transform pair. Clearly, they both contain all the information concerning the dynamic characteristics of a system, the first in integrated form and the second in algebraic form.

Next, we consider the Laplace transform of the unit step function, defined as

$$\mathscr{U}(s) = \int_0^\infty e^{-st}u(t)dt = \int_0^\infty e^{-st}dt = \frac{e^{-st}}{-s}\Big|_0^\infty = \frac{1}{s} \tag{4.65}$$

Inserting $x(t) = \mathscr{s}(t)$ and $F(s) = \mathscr{U}(s) = 1/s$ in Eq. (4.62), we obtain the step response

$$\mathscr{s}(t) = \mathscr{L}^{-1}G(s)\mathscr{U}(s) = \mathscr{L}^{-1}\frac{G(s)}{s} \tag{4.66}$$

or, *the step response is equal to the inverse Laplace transform of the transfer function divided by s*.

Finally, the Laplace transform of the unit ramp function is

$$\mathscr{R}(s) = \int_0^\infty e^{-st} r(t)dt = \int_0^\infty e^{-st} t\, u(t)dt = \int_0^\infty e^{-st} t\, dt$$

$$= \frac{e^{-st}t}{-s}\bigg|_0^\infty + \frac{1}{s}\int_0^\infty e^{-st} dt = \frac{1}{s^2} \tag{4.67}$$

so that the ramp response is simply

$$r(t) = \mathscr{L}^{-1} G(s)\mathscr{R}(s) = \mathscr{L}^{-1}\frac{G(s)}{s^2} \tag{4.68}$$

or, the ramp response is equal to the inverse Laplace transform of the transfer function divided by s^2.

Example 4.5. Derive the impulse response of a damped single-degree-of-freedom system by the Laplace transformation method.

The transfer function of a damped single-degree-of-freedom system is given by Eq. (4.60), which can be rewritten in terms of partial fractions as follows:

$$G(s) = \frac{1}{m(s^2 + 2\zeta\omega_n s + \omega_n^2)} = \frac{C_1}{s - s_1} + \frac{C_2}{s - s_2} \tag{a}$$

where s_1 and s_2 are simple poles of $G(s)$, i.e., they are distinct roots of the equation

$$s^2 + 2\zeta\omega_n s + \omega_n^2 = 0 \tag{b}$$

or,

$$\begin{matrix} s_1 \\ s_2 \end{matrix} = -\zeta\omega_n \pm i\omega_d, \; \omega_d = (1 - \zeta^2)^{1/2}\omega_n \tag{c}$$

Inserting Eq. (c) into Eq. (a), we can write

$$C_1(s - s_2) + C_2(s - s_1) = (C_1 + C_2)s - C_1(-\zeta\omega_n - i\omega_d) - C_2(-\zeta\omega_n + i\omega_d) = \frac{1}{m} \tag{d}$$

which yields the coefficients

$$C_1 = -\,C_2 = \frac{1}{2im\omega_d} \tag{e}$$

so that the transfer function can be expressed in terms of partial fractions as follows:

$$G(s) = \frac{1}{2im\omega_d}\left(\frac{1}{s - s_1} - \frac{1}{s - s_2}\right) \tag{f}$$

The impulse response is obtained by introducing Eq. (f) in conjunction with Eq. (c) in Eq. (4.64) and using the table of Laplace transforms in Appendix B, with the result

$$g(t) = \mathscr{L}^{-1}\frac{1}{2im\omega_d}\left(\frac{1}{s - s_1} - \frac{1}{s - s_2}\right) = \frac{1}{2im\omega_d}\left(e^{s_1 t} - e^{s_2 t}\right)$$

$$= \frac{1}{2im\omega_d}e^{-\zeta\omega_n t}\left(e^{i\omega_d t} - e^{-i\omega_d t}\right) = \frac{1}{m\omega_d}e^{-\zeta\omega_n t}\sin\omega_d t\, u(t) \tag{g}$$

where we multiplied the result by $u(t)$ because the impulse response is by definition equal to zero for $t < 0$. As is to be expected, Eq. (g) represents the same expression for the impulse response as that obtained by classical means, Eq. (4.13).

Example 4.6. Determine the step response of a damped single-degree-of-freedom system by the Laplace transformation method.

Introducing Eq. (4.60) in Eq. (4.66), the desired step response can be written in the form

$$\mathscr{s}(t) = \mathscr{L}^{-1}\frac{G(s)}{s} = \mathscr{L}^{-1}\frac{1}{ms(s^2+2\zeta\omega_n s+\omega_n^2)} = \mathscr{L}^{-1}\left(\frac{C_1}{s-s_1}+\frac{C_2}{s-s_2}+\frac{C_3}{s-s_3}\right) \tag{a}$$

where, using results from Example 4.5, the simple poles have the values

$$s_1 = 0,\ s_2 = -\zeta\omega_n + i\omega_d,\ s_3 = -\zeta\omega_n - i\omega_d \tag{b}$$

At this point, we could follow the procedure used in Example 4.5 to obtain the coefficients $C_k (k = 1, 2, 3)$. Instead, we evaluate the coefficients by means of formula (B.19) of Appendix B, which reads

$$C_k = \left.\frac{A(s)}{B'(s)}\right|_{s=s_k},\quad k = 1, 2, 3 \tag{c}$$

in which $A(s)$ is the numerator and $B(s)$ is the denominator of the middle expression in Eq. (a) and the prime denotes the derivative with respect to s. In the case at hand,

$$\frac{A(s)}{B'(s)} = \frac{1}{m[s^2+2\zeta\omega_n s+\omega_n^2+2s(s+\zeta\omega_n)]} \tag{d}$$

Hence, using Eqs. (b) and (c), we can write

$$\begin{aligned}
C_1 &= \left.\frac{A(s)}{B'(s)}\right|_{s=s_1=0} = \frac{1}{m\omega_n^2}\\
C_2 &= \left.\frac{A(s)}{B'(s)}\right|_{s=s_2=-\zeta\omega_n+i\omega_d} = \frac{1}{2im\omega_d(-\zeta\omega_n+i\omega_d)}\\
C_3 &= \left.\frac{A(s)}{B'(s)}\right|_{s=s_3=-\zeta\omega_n-i\omega_d} = \frac{1}{-2im\omega_d(-\zeta\omega_n-i\omega_d)}
\end{aligned} \tag{e}$$

Inserting Eqs. (e) in conjunction with Eqs. (b) into Eq. (a) and using Laplace transform tables, we obtain the step response

$$\begin{aligned}
\mathscr{s}(t) &= \mathscr{L}^{-1}\left\{\frac{1}{m\omega_n^2}\frac{1}{s}+\frac{1}{2im\omega_d}\left[\frac{1}{-\zeta\omega_n+i\omega_d}\,\frac{1}{s-(-\zeta\omega_n+i\omega_d)}\right.\right.\\
&\qquad\left.\left.-\frac{1}{-\zeta\omega_n-i\omega_d}\,\frac{1}{s-(-\zeta\omega_n-i\omega_d)}\right]\right\}\\
&= \frac{1}{m\omega_n^2}+\frac{e^{-\zeta\omega_n t}}{2im\omega_d}\left(\frac{e^{i\omega_d t}}{-\zeta\omega_n+i\omega_d}-\frac{e^{-i\omega_d t}}{-\zeta\omega_n-i\omega_d}\right)
\end{aligned}$$

$$= \frac{1}{m\omega_n^2}\left\{1 + \frac{e^{-\zeta\omega_n t}}{2i\omega_d}\left[(-\zeta\omega_n - i\omega_d)e^{i\omega_d t} - (-\zeta\omega_n + i\omega_d)e^{-i\omega_d t}\right]\right\}$$

$$= \frac{1}{k}\left[1 - e^{-\zeta\omega_n t}\left(\frac{\zeta\omega_n}{\omega_d}\sin\omega_d t + \cos\omega_d t\right)\right]\mathscr{u}(t) \tag{f}$$

which is the same result as that given by Eq. (4.22). Note that, as the final step, we multiplied the result by $\mathscr{u}(t)$ to account for the fact that $\mathscr{s}(t) = 0$ for $t < 0$.

Example 4.7. Derive the response of an undamped single-degree-of-freedom system to the sawtooth pulse shown in Fig. 4.20 by the Laplace transformation method. Plot the response for the pulse duration $T_0 = 0.4$ s and the natural frequency $\omega_n = 4$ rad/s. For convenience, let $F_0/m = 40$.

Letting $c = 0$ in Eq. (3.1), the differential equation for the response is

$$m\ddot{x}(t) + kx(t) = F(t) \tag{a}$$

where, from Fig. 4.20, the force has the expression

$$F(t) = \frac{F_0}{T_0}t,\ 0 < t < T_0 \tag{b}$$

From Eq. (4.62), the response can be written in the form of the inverse Laplace transformation

$$x(t) = \mathscr{L}^{-1}G(s)F(s) \tag{c}$$

in which, for the system of Eq. (a), the transfer function is

$$G(s) = \frac{1}{ms^2 + k} = \frac{1}{m(s^2 + \omega_n^2)} \tag{d}$$

Moreover, using Eq. (4.57) and integrating by parts, the transformed excitation can be written as

$$F(s) = \int_0^\infty e^{-st}F(t)dt = \frac{F_0}{T_0}\int_0^{T_0} e^{-st}t\,dt = \frac{F_0}{T_0}\left(\frac{e^{-st}t}{-s}\bigg|_0^{T_0} + \frac{1}{s}\int_0^{T_0} e^{-st}dt\right)$$

$$= \frac{F_0}{T_0}\left(-\frac{e^{-sT_0}T_0}{s} + \frac{e^{-st}}{-s^2}\bigg|_0^{T_0}\right) = \frac{F_0}{T_0}\left[-\frac{T_0e^{-sT_0}}{s} + \frac{1}{s^2}(1 - e^{-sT_0})\right] \tag{e}$$

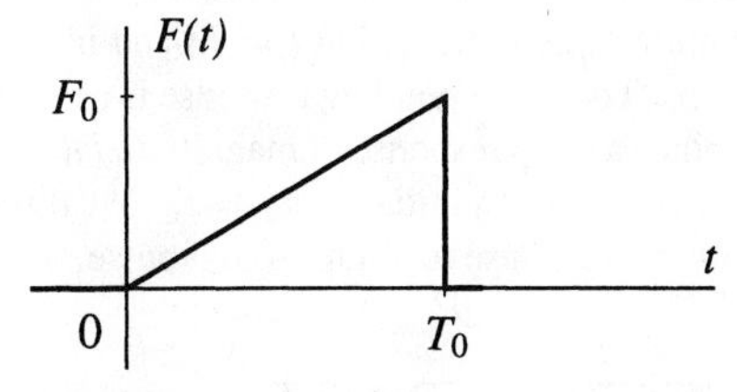

FIGURE 4.20
Sawtooth pulse

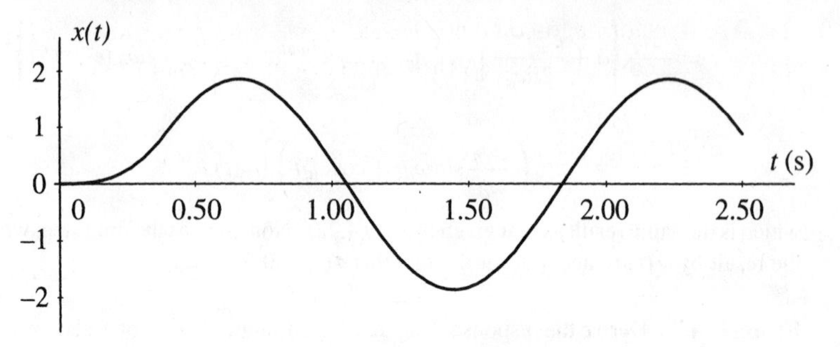

FIGURE 4.21
Response of a mass-spring system to a sawtooth pulse

In carrying out the inverse transformation, we recall that the term e^{-sT_0} can be handled in the context of the shifting theorem in the real domain (see Sec. B.5). Then, from Laplace transform tables, we have

$$\begin{aligned} \mathscr{L}^{-1}\frac{1}{s(s^2+\omega_n^2)} &= \frac{1}{\omega_n^2}(1-\cos\omega_n t) \\ \mathscr{L}^{-1}\frac{1}{s^2(s^2+\omega_n^2)} &= \frac{1}{\omega_n^3}(\omega_n t-\sin\omega_n t) \end{aligned} \tag{g}$$

Hence, using the shifting theorem just mentioned, we obtain the response to the sawtooth pulse as follows:

$$\begin{aligned} x(t) &= \frac{F_0}{mT_0}\left\{-\frac{T_0}{\omega_n^2}[1-\cos\omega_n(t-T_0)]\boldsymbol{\mathscr{u}}(t-T_0)+\frac{1}{\omega_n^3}(\omega_n t-\sin\omega_n t)\boldsymbol{\mathscr{u}}(t)\right. \\ &\qquad \left. -\frac{1}{\omega_n^3}[\omega_n(t-T_0)-\sin\omega_n(t-T_0)]\boldsymbol{\mathscr{u}}(t-T_0)\right\} \\ &= \frac{F_0}{m\omega_n^2}\left\{\frac{1}{\omega_n T_0}(\omega_n t-\sin\omega_n t)\boldsymbol{\mathscr{u}}(t)-\frac{1}{\omega_n T_0}[\omega_n(t-T_0)-\sin\omega_n(t-T_0)]\boldsymbol{\mathscr{u}}(t-T_0)\right. \\ &\qquad \left. -[1-\cos\omega_n(t-T_0)]\boldsymbol{\mathscr{u}}(t-T_0)\right\} \end{aligned} \tag{h}$$

The response is plotted in Fig. 4.21.

As a matter of interest, we observe that the solution given by Eq. (h) can be obtained by treating the force as a ramp function of magnitude F_0/T_0 initiated at $t=0$, minus a ramp function of magnitude F_0/T_0 initiated at $t=T_0$ and minus a step function of magnitude F_0 initiated at $t=T_0$. The corresponding response is a ramp response of magnitude F_0/T_0 starting at $t=0$, minus a ramp response of magnitude F_0/T_0 beginning at $t=T_0$ and minus a step response of magnitude F_0 initiated at $t=T_0$. We discussed this approach in Sec. 4.3 in conjunction with the response to a trapezoidal pulse.

4.7 GENERAL SYSTEM RESPONSE

The interest lies in deriving the general response of a damped single-degree-of-freedom system to the arbitrary excitation $F(t)$ and to the initial conditions $x(0)=x_0$, $\dot{x}(0)=v_0$

by the Laplace transformation method. Hence, transforming both sides of Eq. (3.1) and recalling Eq. (4.58), we obtain the transformed response in the form

$$X(s) = \frac{F(s)}{m(s^2+2\zeta\omega_n s+\omega_n^2)} + \frac{s+2\zeta\omega_n}{s^2+2\zeta\omega_n s+\omega_n^2}x_0 + \frac{1}{s^2+2\zeta\omega_n s+\omega_n^2}v_0 \tag{4.69}$$

The inverse transformation of $X(s)$ will be carried out by considering each term on the right side of Eq. (4.69) separately. To obtain the inverse transformation of the first term, we use the convolution theorem (see Sec. B.7). To this end, we let

$$F_1(s) = F(s),\ F_2(s) = \frac{1}{m(s^2+2\zeta\omega_n s+\omega_n^2)} \tag{4.70}$$

Clearly, $f_1(t) = F(t)$. Moreover, from the tables of Laplace transforms in Appendix B, we conclude that

$$f_2(t) = \frac{1}{m\omega_d}e^{-\zeta\omega_n t}\sin\omega_d t,\ \omega_d = (1-\zeta^2)^{1/2}\omega_n \tag{4.71}$$

and we note that $f_2(t)$ is equal to the impulse response $g(t)$, as can be seen from Eq. (g) of Example 4.5. (The reader is urged to explain why.) Hence, considering Eq. (B.7), the inverse transformation of the first term on the right side of Eq. (4.69) is

$$\begin{aligned}\mathscr{L}^{-1}F_1(s)F_2(s) &= \int_0^t f_1(\tau)f_2(t-\tau)d\tau \\ &= \frac{1}{m\omega_d}\int_0^t F(\tau)e^{-\zeta\omega_n(t-\tau)}\sin\omega_d(t-\tau)d\tau \end{aligned} \tag{4.72}$$

Also from the tables of Laplace transforms, we obtain the inverse transform of the coefficient of x_0 in Eq. (4.69) in the form

$$\mathscr{L}^{-1}\frac{s+2\zeta\omega_n}{s^2+2\zeta\omega_n s+\omega_n^2} = \frac{\omega_n}{\omega_d}e^{-\zeta\omega_n t}\cos(\omega_d t-\psi),\ \psi = \tan^{-1}\frac{\zeta\omega_n}{\omega_d} \tag{4.73}$$

Moreover, the inverse transformation of the coefficient of v_0 can be obtained by multiplying $f_2(t)$, as given by Eq. (4.71), by m. Hence, considering Eqs. (4.71) to (4.73), we obtain the general response

$$\begin{aligned}x(t) &= \frac{1}{m\omega_d}\int_0^t F(\tau)e^{-\zeta\omega_n(t-\tau)}\sin\omega_d(t-\tau)d\tau \\ &\quad + \frac{x_0\omega_n}{\omega_d}e^{-\zeta\omega_n t}\cos(\omega_d t-\psi) + \frac{v_0}{\omega_d}e^{-\zeta\omega_n t}\sin\omega_d t \end{aligned} \tag{4.74}$$

and note that the Laplace transformation method permits us to produce both the response to the initial conditions and the response to the external excitation simultaneously. We observe that Eq. (4.72) can also be written as

$$\begin{aligned}\mathscr{L}^{-1}F_1(s)F_2(s) &= \int_0^t f_1(t-\tau)f_2(\tau)d\tau \\ &= \frac{1}{m\omega_d}\int_0^t F(t-\tau)e^{-\zeta\omega_n\tau}\sin\omega_d\tau\, d\tau \end{aligned} \tag{4.75}$$

so that the general response has the alternative form

$$x(t) = \frac{1}{m\omega_d}\int_0^t F(t-\tau)e^{-\zeta\omega_n\tau}\sin\omega_d\tau\, d\tau + \frac{x_0\omega_n}{\omega_d}e^{-\zeta\omega_n t}\cos(\omega_d t - \psi) + \frac{v_0}{\omega_d}e^{-\zeta\omega_n t}\sin\omega_d t \tag{4.76}$$

We shall make repeated use of Eq. (4.74), or Eq. (4.76), throughout this text.

4.8 RESPONSE BY THE STATE TRANSITION MATRIX

Equation (4.74) gives the response of a damped single-degree-of-freedom to any arbitrary excitation $F(t)$ in terms of a convolution integral. Except for some simple excitations, however, evaluation of the integral in closed form is likely to cause serious difficulties. In such cases, it is natural to seek a numerical solution for the response, which in one form or another amounts to numerical integration of the equation of motion. The equations of motion are generally of second order, and some of the most efficient numerical integration algorithms are based on first-order differential equations. However, second-order equations can be transformed into first-order equations, as we are about to show. Although the solution of the first-order differential equations to be presented in this section is still in terms of a convolution integral, implying a closed-form solution, the integral can be used to derive an efficient numerical algorithm, as shown in Sec. 4.10.

For convenience, we rewrite the equation of motion of a damped single-degree-of-freedom system, Eq. (3.1), in the form

$$\ddot{x}(t) + 2\zeta\omega_n\dot{x}(t) + \omega_n^2 x(t) = m^{-1}F(t) \tag{4.77}$$

Then, introducing the notation $x(t) = x_1(t)$, $\dot{x}(t) = x_2(t)$, as well as the identity $\dot{x}(t) \equiv \dot{x}(t)$, we can replace the single second-order differential equation, Eq. (4.77), by the set of two first-order differential equations

$$\begin{aligned} \dot{x}_1(t) &= x_2(t) \\ \dot{x}_2(t) &= -\omega_n^2 x_1(t) - 2\zeta\omega_n x_2(t) + m^{-1}F(t) \end{aligned} \tag{4.78}$$

The pair of variables $x_1(t)$, $x_2(t)$, or $x(t)$ and $\dot{x}(t)$, are known as *state variables* and Eqs. (4.78) are called *state equations*. For any given set of initial conditions $x(0) = x_1(0)$, $\dot{x}(0) = x_2(0)$, the state equations define uniquely the state of the system for any future time.

The solution of Eqs. (4.78) can be best presented in terms of matrix notation. To this end, we introduce the *state vector*

$$\mathbf{x}(t) = \begin{bmatrix} x_1(t) \\ x_2(t) \end{bmatrix} \tag{4.79}$$

so that Eqs. (4.78) can be rewritten in the matrix form

$$\dot{\mathbf{x}}(t) = A\mathbf{x}(t) + \mathbf{b}F(t) \tag{4.80}$$

where

$$A = \begin{bmatrix} 0 & 1 \\ -\omega_n^2 & -2\zeta\omega_n \end{bmatrix}, \quad \mathbf{b} = \begin{bmatrix} 0 \\ m^{-1} \end{bmatrix} \tag{4.81}$$

are a matrix and a vector of coefficients, respectively. The derivation of the solution of Eq. (4.80) can be carried out by premultiplying both sides of the equation by a 2×2 matrix $K(t)$, obtain first a solution for $K(t)$ and then one for $K(t)\mathbf{x}(t)$ and finally premultiply the latter by $K^{-1}(t)$ to obtain $\mathbf{x}(t)$. For details of the derivation, the reader is urged to consult Ref. 13. Here, we merely give the result

$$\mathbf{x}(t) = \Phi(t)\mathbf{x}(0) + \int_0^t \Phi(t-\tau)\mathbf{b}F(\tau)d\tau \tag{4.82}$$

in which $\mathbf{x}(0) = [x_1(0)\ x_2(0)]^T$ is the initial state vector and

$$\Phi(t-\tau) = e^{A(t-\tau)} = I + (t-\tau)A + \frac{(t-\tau)^2}{2!}A^2 + \frac{(t-\tau)^3}{3!}A^3 + \ldots \tag{4.83}$$

is a matrix known as the *state transition matrix*, in the case at hand a 2×2 matrix. We observe that Eq. (4.82) contains both the homogeneous and the particular solution, where the latter has the form of a convolution integral. Hence, the problem of obtaining the system response has been reduced to that of determining the state transition matrix and carrying out the indicated matrix operations. The convolution integral in Eq. (4.82) can be expressed in an alternative form. Indeed, introducing the transformation of variables defined by Eqs. (4.37) and (4.38) in Eq. (4.82) and recognizing that λ is a mere dummy variable, we obtain

$$\begin{aligned} \mathbf{x}(t) &= \Phi(t)\mathbf{x}(0) + \int_t^0 \Phi(\lambda)\mathbf{b}F(t-\lambda)(-d\lambda) \\ &= \Phi(t)\mathbf{x}(0) + \int_0^t \Phi(\tau)\mathbf{b}F(t-\tau)d\tau \end{aligned} \tag{4.84}$$

where the second term on the right side can be identified as the alternative form of the convolution integral.

In general, the transition matrix must be computed numerically, which amounts to evaluating a matrix series. In view of the factorial $n!$ at the denominator, which increases faster with n than $(t-\tau)^n A^n$, convergence of the series is assured. However, the number of terms required for convergence is still an open question. We consider this subject in Ch. 7, in conjunction with the response of multi-degree-of-freedom systems. In the case at hand, in which the transition matrix is 2×2, it is possible to derive the matrix in closed form. To this end, we consider the homogeneous part of Eq. (4.80), or

$$\dot{\mathbf{x}}(t) = A\mathbf{x}(t) \tag{4.85}$$

which has the solution

$$\mathbf{x}(t) = e^{At}\mathbf{x}(0) = \Phi(t)\mathbf{x}(0) \tag{4.86}$$

The solution can be verified by substitution into Eq. (4.85), or it can be identified merely as the first term on the right side of Eq. (4.82), or Eq. (4.84). The Laplace transform of

Eq. (4.85) is simply

$$s\mathbf{X}(s) - \mathbf{x}(0) = A\mathbf{X}(s) \tag{4.87}$$

where $\mathbf{X}(s) = \mathscr{L}\mathbf{x}(t)$. Solving Eq. (4.87) for the transformed state vector $\mathbf{X}(s)$, we obtain

$$\mathbf{X}(s) = (sI - A)^{-1}\mathbf{x}(0) \tag{4.88}$$

Hence, inverse transforming both sides of Eq. (4.88), we can write the homogeneous part of the state vector in the form

$$\mathbf{x}(t) = \mathscr{L}^{-1}(sI - A)^{-1}\mathbf{x}(0) \tag{4.89}$$

Comparing Eqs. (4.86) and (4.89), we conclude that the transition matrix can be expressed as the inverse Laplace transform

$$\Phi(t) = \mathscr{L}^{-1}(sI - A)^{-1} \tag{4.90}$$

and we observe that the inverse Laplace transformation can be carried out entry by entry following the matrix inversion. To obtain the transition matrix for the single-degree-of-freedom system described by the state equations (4.78), we insert the first of Eqs. (4.81) into Eq. (4.90), use the Laplace transform tables in Appendix B and write

$$\Phi(t) = \mathscr{L}^{-1}\begin{bmatrix} s & -1 \\ \omega_n^2 & s+2\zeta\omega_n \end{bmatrix}^{-1} = \mathscr{L}^{-1}\frac{1}{s^2+2\zeta\omega_n s+\omega_n^2}\begin{bmatrix} s+2\zeta\omega_n & 1 \\ -\omega_n^2 & s \end{bmatrix}$$

$$= \frac{1}{\omega_d}e^{-\zeta\omega_n t}\begin{bmatrix} \omega_d\cos\omega_d t + \zeta\omega_n\sin\omega_d t & \sin\omega_d t \\ -\omega_n^2\sin\omega_d t & \omega_d\cos\omega_d t - \zeta\omega_n\sin\omega_d t \end{bmatrix} \tag{4.91}$$

Finally, introducing the second of Eqs. (4.81) and Eq. (4.91) in Eq. (4.84), we obtain the total response in the general form

$$\mathbf{x}(t) = \frac{1}{\omega_d}e^{-\zeta\omega_n t}\begin{bmatrix} \omega_d\cos\omega_d t + \zeta\omega_n\sin\omega_d t & \sin\omega_d t \\ -\omega_n^2\sin\omega_d t & \omega_d\cos\omega_d t - \zeta\omega_n\sin\omega_d t \end{bmatrix}\mathbf{x}(0)$$

$$+ \frac{1}{m\omega_d}\int_0^t e^{-\zeta\omega_n \tau}\begin{bmatrix} \sin\omega_d \tau \\ \omega_d\cos\omega_d \tau - \zeta\omega_n\sin\omega_d \tau \end{bmatrix}F(t-\tau)d\tau \tag{4.92}$$

The response given by Eq. (4.92) is complete, in the sense that it contains the response to both the initial excitations and the applied force. In fact, the top component in Eq. (4.92), representing the displacement, is identical in every way to the general response given by Eq. (4.76). In addition, going beyond Eq. (4.76), the bottom component of Eq. (4.92) represents the velocity.

We observe that Eq. (4.92) gives the part of the state vector $\mathbf{x}(t)$ attributable to external forces in the form of a convolution integral. Hence, when the external forces involve discontinuities, care must be exercised in choosing the limits of integration, as discussed in Sec. 4.4. This problem is obviated in Sec. 4.10, in which we present a numerical algorithm for solving state equations in discrete time.

4.9 DISCRETE-TIME SYSTEMS. THE CONVOLUTION SUM

In Secs. 4.1–4.8, we examined various techniques for determining the response of linear, time-invariant systems to nonperiodic excitations. Although not stated explicitly, the emphasis was placed on analytical solutions. In one form or another, the response to arbitrary forces necessitates the evaluation of convolution integrals. But, except for some relatively simple forcing functions, analytical evaluation of convolution integrals can be very difficult, if not impossible. In fact, in many cases the excitation forces cannot even be described in terms of known functions of time. Hence, for the most part, the only viable option is to obtain the response numerically on a digital computer. In this section, we present an approach suitable for the numerical evaluation of scalar convolution integrals, such as that given by Eq. (4.72), or Eq. (4.75), and in Sec. 4.10 we discuss a technique for the evaluation of vector convolution integrals, such as that in Eq. (4.82), or Eq. (4.84).

The evaluation of solutions to vibration problems on a computer involves three operations, getting the input data into the computer, carrying out the computations by means of an algorithm stored in the computer and getting the output data out of the computer. To relate these operations to our vibration problems, we recognize that the input data corresponds to the excitation, the computational algorithm may involve the numerical evaluation of a convolution integral and the output data represents the response. But, whereas vibration phenomena evolve continuously in time, digital computers cannot process information involving time as a continuous independent variable, which makes it necessary that time be converted into a discrete variable. Moreover, the input and computational algorithm must be converted to discrete time from continuous time. The output is generated by the computer in discrete time automatically. We refer to this whole process as *discretization in time*, and to the associated discrete-in-time model as a *discrete-time system*.

The process of computing the response of vibrating systems on a digital computer begins with the conversion of the continuous time t into the discrete time t_n ($n = 0, 1, 2, \ldots$), where t_n represent *sampling times*. Although in general the sampling times $t_0, t_1, t_2, \ldots$ can be spaced unevenly, for the most part they are chosen to be evenly spaced, so that $t_n = nT$ ($n = 0, 1, 2, \ldots$), in which T is known as the *sampling period*. Hence, the question of choosing the sampling times reduces to the question of choosing the sampling period. This is an important question, and we discuss it later in this section. Because the sampling period is constant, it is convenient to identify the sampling times $t_n = nT$ simply by n. The operations described in the preceding paragraph are depicted schematically in Fig. 4.22, in which every operation is represented by a block. The first block represents the conversion of the continuous-time excitation $F(t)$ into the

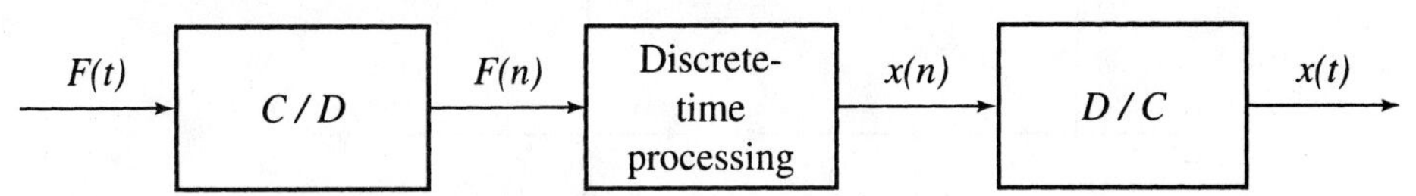

FIGURE 4.22
Computation of the response in discrete time

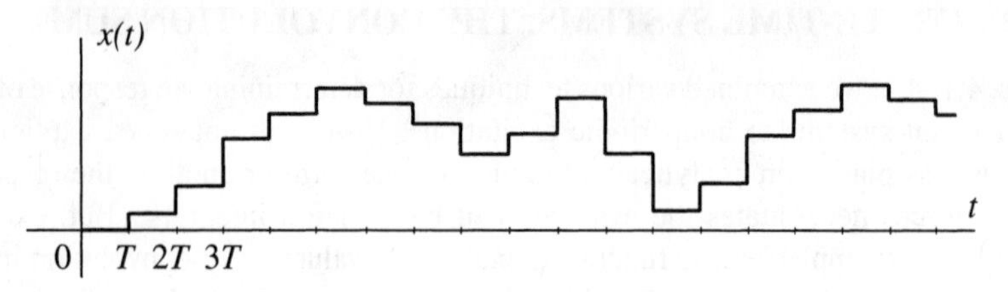

FIGURE 4.23
Response computed in discrete time and converted to continuous time

discrete-time input data $F(t_n) = F(n)$, which has the form of a sequence of numbers $F(0),\ F(1),\ F(2),\ldots$ obtained simply by sampling $F(t)$ at $t = nT$ $(n = 0, 1, 2, \ldots)$. This operation is denoted by C/D, meaning conversion from continuous to discrete. The second block represents discrete-time processing of the solution to the vibration problem on a computer. The result is the output data in the form of a sequence of numbers $x(1),\ x(2),\ldots$, representing the discrete-time response. Finally, the third block stands for the conversion of the discrete-time response into the continuous-time response, which can be done by writing simply

$$x(t) = x(n),\ nT \leq t \leq nT + T;\ n = 1, 2, \ldots \tag{4.93}$$

This operation, denoted by D/C, generates a plot in the form of a staircase, as depicted in Fig. 4.23. In this regard, it should be noted that if the output data $x(n)$ $(n = 1, 2, \ldots)$ is plotted, instead of merely listing the numbers, and if the resolution of the graph is relatively high, then the plot will appear as a continuous curve, rather than as a staircase. Hence, the net effect of choosing a small sampling time T is to obviate the operation D/C.

At this point, we turn our attention to the discrete-time algorithm for the computation of the response. One approach is to discretize the convolution integral, Eq. (4.40), in the manner discussed in Ref. 13. Perhaps a simpler approach is to regard the system as a discrete-time system from the beginning. To describe mathematically discrete-time functions, such as the input sequence, it is convenient to introduce the *discrete-time unit impulse*, or *unit sample*, as the discrete-time Kronecker delta (Ref. 13)

$$\delta(n-k) = \begin{cases} 1 & \text{for} \quad n = k \\ 0 & \text{for} \quad n \neq k \end{cases} \tag{4.94}$$

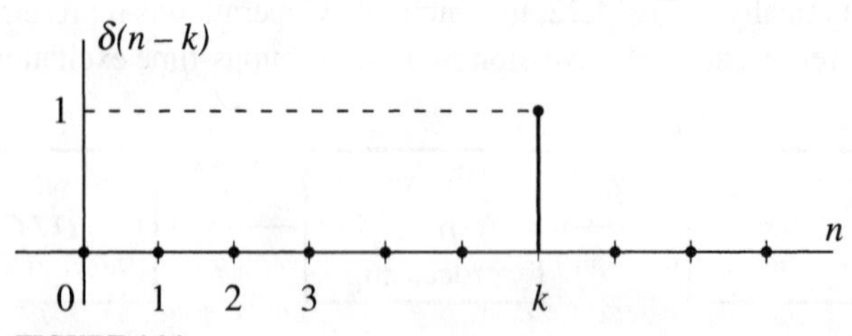

FIGURE 4.24
Discrete-time unit impulse, or unit sample

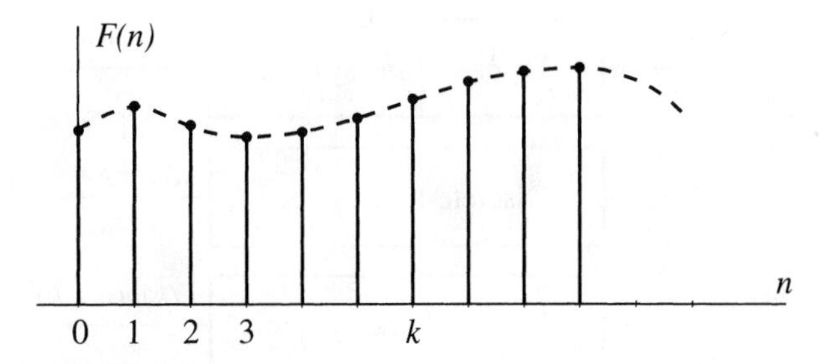

FIGURE 4.25
Discrete-time excitation function

The unit sample is shown in Fig. 4.24. Then, the discrete-time excitation can be expressed mathematically as

$$F(n) = \sum_{k=0}^{\infty} F(k)\delta(n-k) \tag{4.95}$$

where $F(0),\ F(1),\ F(2),\ldots$ is the input sequence. A typical discrete-time excitation function is shown in Fig. 4.25.

Next, by analogy with continuous-time systems, we define the *discrete-time impulse response $g(n)$ as the response of a linear discrete-time system to a discrete-time unit impulse $\delta(n)$, applied at $k = 0$, with all the initial conditions being equal to zero.* The definition implies that $g(n) = 0$ for $n < 0$, because there cannot be any response before the system is excited. The relation between $g(n)$ and $\delta(n)$ is shown schematically in the form of the first block diagram of Fig. 4.26. At this point, we must recognize that discretization in time cannot change the inherent properties of a system. Hence, a linear time-invariant system in continuous time remains so in discrete time. It follows that, for linear time-invariant systems, if the excitation is delayed by the time interval kT, then the response is delayed by the same amount of time, so that the response to $\delta(n-k)$ is $g(n-k)$, as shown in the second block diagram of Fig. 4.26. Moreover, if the excitation is multiplied by $F(k)$, then the response is multiplied by $F(k)$, as depicted in the third block diagram of Fig. 4.26. Finally, because the principle of superposition holds for discrete-time systems in the same manner as for the associated continuous-time systems, we conclude that the response to the discrete-time excitation given by Eq. (4.95) is simply

$$x(n) = \sum_{k=0}^{\infty} F(k)g(n-k) = \sum_{k=0}^{n} F(k)g(n-k) \tag{4.96}$$

where we replaced the upper limit of the series by n in recognition of the fact that $g(n-k) = 0$ for $n - k < 0$, which is the same as $k > n$. Equation (4.96) expresses the response of a linear time-invariant system in the form of a *convolution sum*, and it represents the discrete-time counterpart of the convolution integral given by Eq. (4.40). The discrete-time response given by Eq. (4.96) consists of a sequence of numbers $x(1),\ x(2),\ldots$, which can be used to plot $x(n)$ versus n.

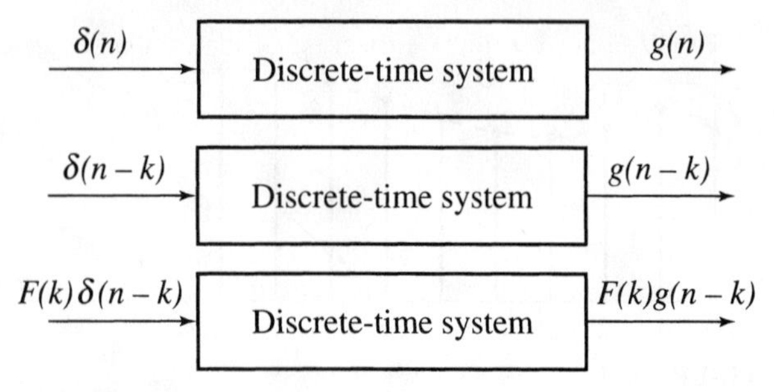

FIGURE 4.26
Block diagrams relating the discrete-time impulse response to the discrete-time unit impulse, relating the same quantities shifted by k sampling periods and relating the same quantities shifted and multiplied by $F(kT)$.

There remains the question as to how to obtain the discrete-time impulse response $g(n)$. To answer this question, we must first examine the nature of the discrete-time unit impulse $\delta(n)$. From Eq. (4.94), we observe that $\delta(n)$ is dimensionless, as opposed to $\delta(t)$, which has units of s^{-1}. Implicit in this is that $\delta(n)$ corresponds not to $\delta(t)$ but to the area under the curve $\delta(t)$ versus t. Hence, in discretizing excitations by means of Eq. (4.95), in which $F(0),\ F(1),\ F(2),\ldots$ represent the magnitude of the force $F(t)$ at $t = 0,\ T,\ 2T,\ldots$, the time interval T between the samples is implicitly allocated to $\delta(n-k)$ rather than to $F(k)$. It follows that the discrete-time impulse response $g(n)$ can be obtained from the continuous-time impulse response $g(t_n)$ by writing

$$g(n) = Tg(t_n) \tag{4.97}$$

Finally, we must address the issue of choosing a sampling period T, as the accuracy of the discrete-time solution depends on this choice. It is reasonable to expect that, for a relatively small sampling period, the discrete-time response will approximate the continuous-time response quite well, so that the question reduces to what is small. This question can only be answered in the context of the excitation and of the system characteristics. Indeed, the net effect of replacing a convolution integral by a convolution sum is to approximate curves in continuous time by staircases in discrete time, because $F(n)$ and $g(n)$ can be regarded as constant over the sampling period T, at the end of which their values jump to $F(n+1)$ and $g(n+1)$, respectively. Hence, a small sampling period for one problem may not be small for another problem, and vice versa. To elaborate on this statement, we focus our attention on the system characteristics, rather than on the nature of the excitation. To this end, we refer to a specific system, such as an underdamped single-degree-of-freedom system. From Eq. (4.11), we recall that the continuous-time impulse response is given by

$$g(t) = \frac{1}{m\omega_d} e^{-\zeta\omega_n t} \sin \omega_d t \tag{4.98}$$

and, from Eq. (4.97), we conclude that the discrete-time impulse response is

$$g(n) = \frac{T}{m\omega_d} e^{-n\zeta\omega_n T} \sin n\omega_d T \tag{4.99}$$

Hence, it is clear that the sampling period T must be sufficiently small that $g(n)$ approximate $g(t)$ with the desired degree of accuracy, bearing in mind that the multiplying constant T plays the role of a scaling factor related to the definition of the discrete-time unit impulse. Except for the scaling factor T, $g(n)$ matches $g(t)$ exactly at $t = nT$, but this does not tell the whole story. Indeed, for all practical purposes, $g(n)$ acts as the staircase approximation of $g(t)$, as pointed out above. How well $g(n)$ matches $g(t)$ depends largely on how well $\sin n\omega_d T$ matches $\sin \omega_d t$, which in turn depends on the width T of the steps in the staircase. In view of this, we conclude that an accurate approximation demands that the sampling period T be only a small fraction of the period of damped oscillation

$$T_d = \frac{2\pi}{\omega_d} \tag{4.100}$$

perhaps of the order of 10^{-2}.

The computational algorithm for the response by the convolution sum has many advantages. In the first place, the algorithm is quite simple. Moreover, it can handle excitations that in continuous time were discontinuous with the same ease as excitations that were continuous originally. Indeed, because of the discrete-in-time nature of the algorithm, the concept of continuity does not even apply. Finally, it has no difficulties in handling excitations of finite duration, such as pulses. In this regard, it is important to remember that, for excitations of finite duration, the discrete variable k, and hence the number of terms in the convolution sum, cannot exceed the number of sampling times n_0 defining the duration of the discrete-time excitation.

On the other side of the ledger, the algorithm using the convolution sum has several drawbacks. The first that comes to mind is that this is a scalar approach relating a single response to a single excitation. Then, this is not a recursive process, which implies that the computation of every number $x(n)$ in the response sequence is carried out independently of the previously computed numbers $x(1),\ x(2), \ldots,\ x(n-1)$. This implies further that all the values of $F(k)$ and $g(k)$ up to the sampling time n must be saved $(k = 0, 1, 2, \ldots, n-1)$. Indeed, in a recursive process the computation of $x(n)$ requires only the values of $x(n-1)$ and $F(n-1)$, and all the previous values can be discarded. Finally, the computation of $x(n)$ becomes progressively longer with n, as the convolution sum involves $n+1$ products of $F(k)$ and $g(n-k)$ $(k = 0, 1, \ldots, n-1)$. An exception to this is the case in which the excitation sequence consists of a finite number n_0 of samplings only, in which case the number of products $F(k)g(n-k)$ is limited to a maximum of n_0. This can be explained by the fact that there is a maximum of n_0 sampling times for which the sequences $F(k)$ and $g(n-k)$ overlap.

In writing a program for the evaluation of the convolution sum on a computer, the advantage of simplicity remains. On the other, the facts that the process is not recursive and that the computation at each step becomes increasingly longer are primarily of academic interest, as the computational burden is assumed by the computer. In Sec. 4.11,

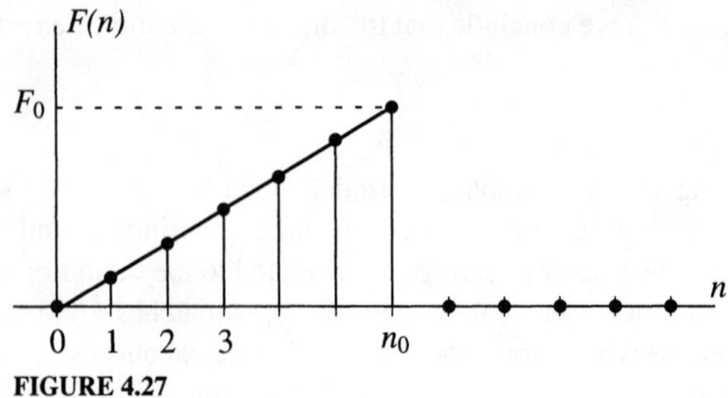

FIGURE 4.27
Sawtooth pulse discretized in time

we present a computer program for the response by the convolution sum written in MATLAB.

Example 4.8. Obtain the solution to the problem of Example 4.7 by means of the convolution sum for the two cases: 1) $T = 0.01$ s and 2) $T = 0.005$ s. Plot the discrete-time response for the two cases, as well as the continuous-time response obtained in Example 4.7, and discuss the accuracy of the results.

The discrete-time counterpart of the sawtooth pulse of Fig. 4.20 is shown in Fig. 4.27, in which $n_0 = T_0/T = 0.4/T$ is the number of sampling times. It is easy to see that the discretized sawtooth pulse can be expressed as

$$F(n) = \begin{cases} \dfrac{F_0}{n_0} n & \text{for} \quad 0 \le n \le n_0 \\ 0 & \text{for} \quad n > n_0 \end{cases} \tag{a}$$

Moreover, from Eq. (4.99) with $\zeta = 0$, the discrete-time impulse response is

$$g(n) = \frac{T}{m\omega_n} \sin n\omega_n T \tag{b}$$

It is shown in Fig. 4.28a. The shifted graph and folded graph are displayed in Figs. 4.28b and 4.28c, respectively. Hence, using Eq. (4.96), the discrete-time response is

$$x(n) = \sum_{k=0}^{n} F(k) g(n-k) = \begin{cases} \dfrac{F_0 T}{n_0 m\omega_n} \displaystyle\sum_{k=0}^{n} k \sin(n-k)\omega_n T & \text{for} \quad n \le n_0 \\ \dfrac{F_0 T}{n_0 m\omega_n} \displaystyle\sum_{k=0}^{n_0} k \sin(n-k)\omega_n T & \text{for} \quad n > n_0 \end{cases} \tag{c}$$

Case 1

In this case, $T = 0.01$ s, $n_0 = 40$, so that the response is given by the formula

$$x(n) = \begin{cases} 0.0025 \displaystyle\sum_{k=0}^{n} k \sin 0.04(n-k) & \text{for} \quad n \le 40 \\ 0.025 \displaystyle\sum_{k=0}^{40} k \sin 0.04(n-k) & \text{for} \quad n > 40 \end{cases} \tag{d}$$

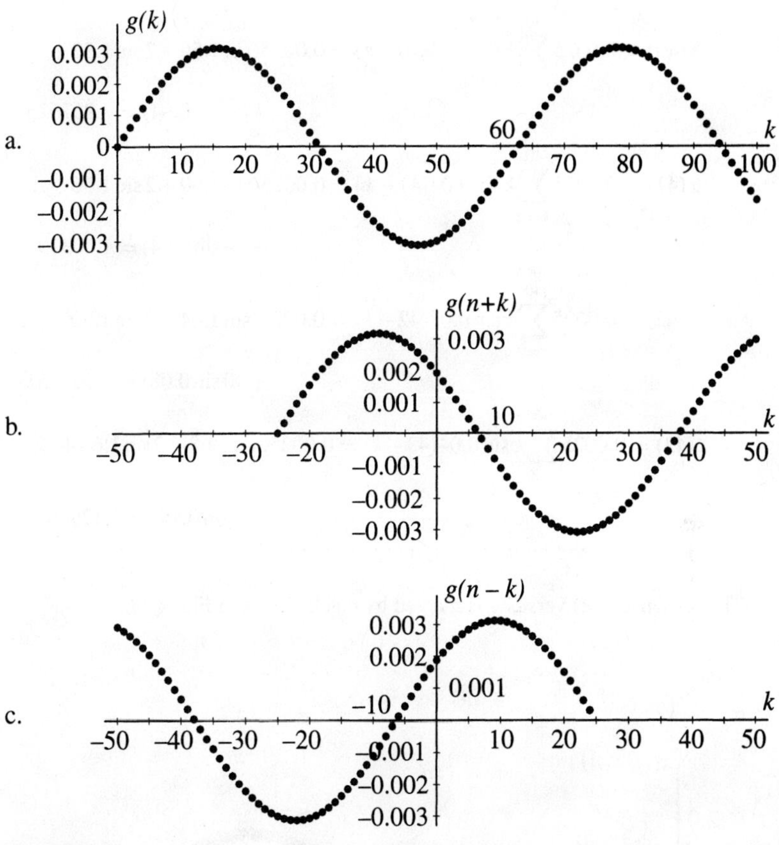

FIGURE 4.28
a. Discrete-time impulse response for a mass-spring system, b. The impulse response shifted backward by k periods, c. The impulse response shifted backward and folded.

Hence, the response sequence is as follows:

$$x(1) = 0.0025\sum_{k=0}^{1} k\sin 0.04(1-k) = 0$$

$$x(2) = 0.0025\sum_{k=0}^{2} k\sin 0.04(2-k) = 0.0025\sin 0.04 = 0.000100$$

$$x(3) = 0.0025\sum_{k=0}^{3} k\sin 0.04(3-k) = 0.0025(\sin 0.08 + 2\sin 0.04) = 0.000400$$

$$x(4) = 0.0025\sum_{k=0}^{4} k\sin 0.04(4-k) = 0.0025(\sin 0.12 + 2\sin 0.08 + 3\sin 0.04) = 0.000999$$

. . . .

$$x(40) = 0.0025 \sum_{k=0}^{40} k \sin 0.04(40-k) = 0.0025(\sin 1.56 + 2\sin 1.52 + \ldots + 39\sin 0.04) = 0.937625$$

$$x(41) = 0.0025 \sum_{k=0}^{40} k \sin 0.04(41-k) = 0.0025(\sin 1.60 + 2\sin 1.56 + \ldots + 40\sin 0.04) = 1.003191$$

$$x(42) = 0.0025 \sum_{k=0}^{40} k \sin 0.04(42-k) = 0.0025(\sin 1.64 + 2\sin 1.60 + \ldots + 40\sin 0.08) = 1.067151$$

$$x(43) = 0.0025 \sum_{k=0}^{40} k \sin 0.04(43-k) = 0.0025(\sin 1.68 + 2\sin 1.64 + \ldots + 40\sin 0.12) = 1.129405$$

$$\ldots\ldots\ldots\ldots\ldots\ldots\ldots\ldots\ldots \qquad \text{(e)}$$

The response $x(n)$ versus n is marked by black circles in Fig. 4.29.

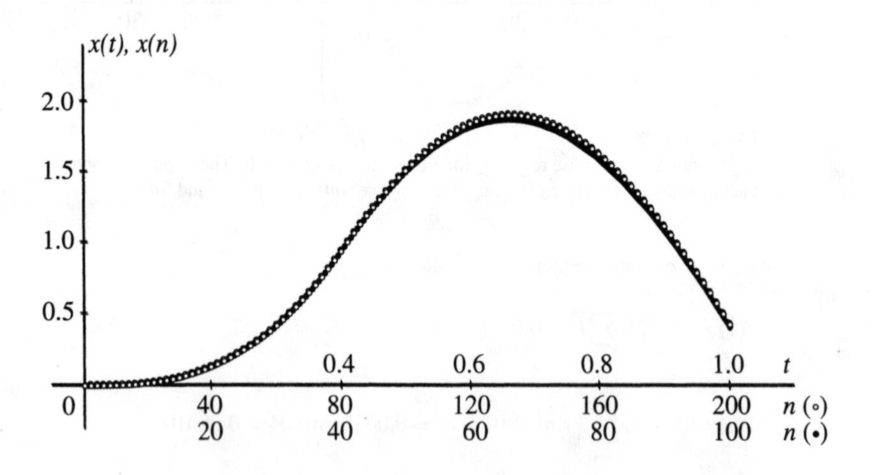

FIGURE 4.29
Discrete-time response of a mass-spring system to a sawtooth pulse by the convolution sum

Case 2

In this case, $T = 0.005$ s, $n_0 = 80$, so that the response has the explicit form

$$x(n) = \begin{cases} 6.25 \times 10^{-4} \sum_{k=0}^{n} k \sin 0.02(n-k) & \text{for} \quad n \leq 80 \\ 6.25 \times 10^{-4} \sum_{k=0}^{80} k \sin 0.02(n-k) & \text{for} \quad n > 80 \end{cases} \tag{f}$$

which yields the response sequence

$$x(1) = 6.25 \times 10^{-4} \sum_{k=0}^{1} k \sin 0.02(1-k) = 0$$

$$x(2) = 6.25 \times 10^{-4} \sum_{k=0}^{2} k \sin 0.02(2-k) = 6.25 \times 10^{-4} \sin 0.02 = 0.000012$$

$$x(3) = 6.25 \times 10^{-4} \sum_{k=0}^{3} k \sin 0.02(3-k) = 6.25 \times 10^{-4}(\sin 0.04 + 2 \sin 0.02) = 0.000050$$

$$x(4) = 6.25 \times 10^{-4} \sum_{k=0}^{4} k \sin 0.02(4-k) = 6.25 \times 10^{-4}(\sin 0.06 + 2 \sin 0.04 + 3 \sin 0.02) = 0.000125$$

. . . .

$$x(80) = 6.25 \times 10^{-4} \sum_{k=0}^{80} k \sin 0.02(80-k) = 6.25 \times 10^{-4}(\sin 1.58 + 2 \sin 1.56 + \ldots + 79 \sin 0.02) = 0.938031$$

$$x(81) = 6.25 \times 10^{-4} \sum_{k=0}^{80} k \sin 0.02(81-k) = 6.25 \times 10^{-4}(\sin 1.60 + 2 \sin 1.58 + \ldots + 80 \sin 0.02) = 0.969505$$

$$x(82) = 6.25 \times 10^{-4} \sum_{k=0}^{80} k \sin 0.02(82-k) = 6.25 \times 10^{-4}(\sin 1.62 + 2 \sin 1.60 + \ldots + 80 \sin 0.04) = 1.000591$$

$$x(83) = 6.25 \times 10^{-4} \sum_{k=0}^{80} k \sin 0.02(83-k) = 6.25 \times 10^{-4}(\sin 1.64 + 2 \sin 1.62 + \ldots + 80 \sin 0.06) = 1.031277$$

. . . . (g)

The response $x(n)$ versus n is indicated by white circles in Fig. 4.29.

For comparison purposes, the continuous-time response plot $x(t)$ versus t obtained in Example 4.7 is also shown in Fig. 4.29. It is clear that in case 1 the sampling period is too large to permit accurate results. On the other hand, in case 2 the sampling period is sufficiently small to yield results close in value to the exact solution. The discrete-time solution can be improved by further reducing the sampling period. It should be stated that the errors experienced in this example are worse than one would expect, because both the excitation and impulse response exhibit strong variations with time, thus requiring a smaller sampling period for good approximation.

It should be pointed out here that the amount of detail presented in this example is strictly for pedagogical reasons. In practical situations, the discrete-time displacements $x(1), x(2), x(3), \ldots$ are generated by a computer in the form of a sequence of numbers or in the form of a plot $x(n)$ versus n. Such a MATLAB computer program entitled 'convsum.m' is presented in Sec. 4.11.

4.10 DISCRETE-TIME RESPONSE USING THE TRANSITION MATRIX

In Sec 4.9, we developed a technique for calculating the discrete-time response of single-degree-of-freedom systems based on the convolution sum. One of the drawbacks of the technique is that it is not recursive. As a result, to compute the response at the sampling time $n+1$, it is necessary to use the value of the excitation and impulse response at all preceding sampling times. In this section, we present a technique not suffering from this drawback.

Equation (4.82) gives the continuous-time response of a single-degree-of-freedom system in state form, a two-dimensional vector with the top component equal to the displacement and the bottom component equal to the velocity. The response to external excitations is in the form of a convolution integral involving the state transition matrix $\Phi(t-\tau)$, which has the form of a matrix series, Eq. (4.83). Equations (4.82) and (4.83) can be used to derive a recursive algorithm for the response. To this end, we consider a particular sampling time $t = nT$, where T is the sampling period, and write the response at nT in the form

$$\mathbf{x}(n) = \mathbf{x}(nT) = e^{AnT}\mathbf{x}(0) + \int_0^{nT} e^{A(nT-\tau)}\mathbf{b}F(\tau)d\tau \tag{4.101}$$

Moreover, the response at $t = (n+1)T$ is

$$\mathbf{x}(n+1) = \mathbf{x}(nT+T) = e^{A(nT+T)}\mathbf{x}(0) + \int_0^{nT+T} e^{A(nT+T-\tau)}\mathbf{b}F(\tau)d\tau \tag{4.102}$$

Then, assuming that the sampling period T is sufficiently small that the excitation can be approximated by a constant over the period T, or

$$F(\tau) \cong F(nT) = F(n) = \text{constant}, \; nT \le \tau < nT + T \tag{4.103}$$

it is shown in Ref. 13 that $\mathbf{x}(n+1)$ can be expressed in terms of $\mathbf{x}(n)$ and $F(n)$ as follows:

$$\mathbf{x}(n+1) = \Phi\mathbf{x}(n) + \gamma F(n), \; n = 0, 1, 2, \ldots \tag{4.104}$$

where

$$\Phi = e^{AT} \tag{4.105}$$

is a 2×2 matrix known as the *discrete-time transition matrix* and

$$\gamma = A^{-1}(e^{AT} - I)\mathbf{b} \tag{4.106}$$

is a two-dimensional vector.

Equation (4.104) represents the desired recursive relation. Indeed, the relation permits the computation of the state vector at $(n+1)T$ based on the state vector and the excitation at nT. Equation (4.104) can be used to compute the discrete-time sequence of state vectors $\mathbf{x}(1), \mathbf{x}(2), \ldots$. Of course, before the computation of the discrete-time states can begin, it is necessary to generate the discrete-time transition matrix Φ and the vector γ. Note that Eq. (4.104) gives the response to both initial excitations and applied forces.

The algorithm is very easy to program on a digital computer. We present such a program, written in MATLAB, in Sec. 4.12.

Example 4.9. Obtain the solution to the problem of Example 4.7 by means of the approach based on the discrete-time transition matrix using the sampling period $T = 0.005$ s. The initial state vector $\mathbf{x}(0)$ is zero. Plot the top component of $\mathbf{x}(n)$ thus obtained, as well as the continuous-time response obtained in Example 4.7 and the discrete-time response obtained in Example 4.8 by means of the convolution sum for $T = 0.005$ s, and discuss the accuracy of the results.

We carry out the computations by means of the recursive process given by Eq. (4.104). To this end, we let $\zeta = 0$, $\omega_d = \omega_n = 4$ rad/s and $t = T = 0.005$ s in Eq. (4.91) and obtain the discrete-time transition matrix for an undamped single-degree-of-freedom system in the form

$$\Phi = e^{AT} = \begin{bmatrix} \cos\omega_n T & \omega_n^{-1}\sin\omega_n T \\ -\omega_n \sin\omega_n T & \cos\omega_n T \end{bmatrix} = \begin{bmatrix} 0.999800 & 0.005000 \\ -0.079995 & 0.999800 \end{bmatrix} \tag{a}$$

Moreover, from Eqs. (4.81), we have

$$A = \begin{bmatrix} 0 & 1 \\ -\omega_n^2 & 0 \end{bmatrix}, \quad \mathbf{b} = \begin{bmatrix} 0 \\ m^{-1} \end{bmatrix} \tag{b}$$

so that, using Eq. (4.106), we can write

$$\gamma = A^{-1}(e^{AT} - I)\mathbf{b} = \begin{bmatrix} 0 & 1 \\ -\omega_n^2 & 0 \end{bmatrix}^{-1} \begin{bmatrix} \cos\omega_n T & \omega^{-1}\sin\omega_n T \\ -\omega_n \sin\omega_n T & \cos\omega_n T \end{bmatrix} \begin{bmatrix} 0 \\ m^{-1} \end{bmatrix}$$

$$= \frac{1}{m\omega_n^2}\begin{bmatrix} 1-\cos\omega_n T \\ \omega_n \sin\omega_n T \end{bmatrix} = \frac{1}{16m}\begin{bmatrix} 0.000200 \\ 0.079995 \end{bmatrix} \tag{c}$$

Hence, recalling from Example 4.7 that $F_0/m = 40$ and from Example 4.8 that $n_0 = 80$, we can write

$$F(n) = \frac{F_0}{80} f(n) \tag{d}$$

where

$$f(n) = \begin{cases} n & \text{for} \quad 0 \leq n \leq 80 \\ 0 & \text{for} \quad n > 80 \end{cases} \tag{e}$$

so that, using Eq. (4.104), the computational algorithm has the form

$$\mathbf{x}(n+1) = \Phi \mathbf{x}(n) + \gamma F(n) = \begin{bmatrix} 0.999800 & 0.005000 \\ -0.079995 & 0.999800 \end{bmatrix} \mathbf{x}(n) + \frac{f(n)}{32} \begin{bmatrix} 0.000200 \\ 0.079995 \end{bmatrix} \tag{f}$$

Recalling that $\mathbf{x}(0) = \mathbf{0}$ and using Eqs. (e) and (f), we compute the first four state vectors in the response sequence. The results are

$$\mathbf{x}(1) = \begin{bmatrix} 0.999800 & 0.005000 \\ -0.079995 & 0.999800 \end{bmatrix} \mathbf{0} + \frac{0}{32} \begin{bmatrix} 0.000200 \\ 0.079995 \end{bmatrix} = \mathbf{0}$$

$$\mathbf{x}(2) = \begin{bmatrix} 0.999800 & 0.005000 \\ -0.079995 & 0.999800 \end{bmatrix} \mathbf{0} + \frac{1}{32} \begin{bmatrix} 0.000200 \\ 0.079995 \end{bmatrix} = \begin{bmatrix} 0.000006 \\ 0.002500 \end{bmatrix}$$

$$\mathbf{x}(3) = \begin{bmatrix} 0.999800 & 0.005000 \\ -0.079995 & 0.999800 \end{bmatrix} \begin{bmatrix} 0.000006 \\ 0.002500 \end{bmatrix} + \frac{2}{32} \begin{bmatrix} 0.000200 \\ 0.079995 \end{bmatrix} = \begin{bmatrix} 0.000031 \\ 0.007499 \end{bmatrix}$$

$$\mathbf{x}(4) = \begin{bmatrix} 0.999800 & 0.005000 \\ -0.079995 & 0.999800 \end{bmatrix} \begin{bmatrix} 0.000031 \\ 0.007499 \end{bmatrix} + \frac{3}{32} \begin{bmatrix} 0.000200 \\ 0.079995 \end{bmatrix} = \begin{bmatrix} 0.000087 \\ 0.014994 \end{bmatrix} \tag{g}$$

Moreover, for comparison purposes, we list another four state vectors as follows:

$$\mathbf{x}(80) = \begin{bmatrix} 0.922137 \\ 6.369809 \end{bmatrix}, \quad \mathbf{x}(81) = \begin{bmatrix} 0.954300 \\ 6.494756 \end{bmatrix}$$
$$\mathbf{x}(82) = \begin{bmatrix} 0.986580 \\ 6.417118 \end{bmatrix}, \quad \mathbf{x}(83) = \begin{bmatrix} 1.018466 \\ 6.336913 \end{bmatrix} \tag{h}$$

The top component in Eqs. (g) and (h) represents the displacement and the bottom component the velocity at the sampling times. The top component is used for the plot $x(n)$ versus n marked in Fig. 4.30 by black circles. Also in Fig. 4.30, the continuous-time response $x(t)$ versus t obtained in Example 4.7 is plotted in a solid line and the discrete-time response obtained in Example 4.8 by the convolution sum is indicated by white circles. We conclude that the approximate solution obtained by the convolution sum is more accurate than that obtained by the discrete-time transition matrix. This can be attributed to the fact that the discrete-time impulse response $g(n)$ tends to be more accurate than the discrete-time transition matrix Φ and the vector γ. These inaccuracies disappear as the sampling period T is reduced.

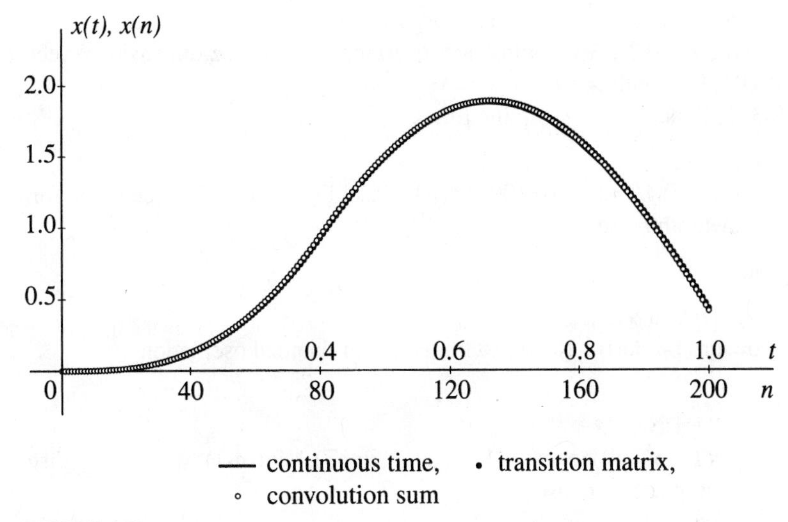

FIGURE 4.30
Response of a mass-spring system to a sawtooth pulse by the discrete-time transition matrix

The above results can be obtained by the MATLAB computer program 'transm.m' given in Sec. 4.12. In fact, the program considers a somewhat more general problem than that in the present example, as it considers the response of a damped single-degree-of-freedom system to a sawtooth pulse, in which $\zeta = 0$ represents one of several values of the damping factor.

4.11 RESPONSE BY THE CONVOLUTION SUM USING MATLAB

From Example 4.8, we conclude that the determination of the response by means of the convolution sum becomes increasingly laborious as the number of sampling times increases, so that a computer solution is an absolute necessity. In fact, the results of Example 4.8, given both as a sequence of computed displacements and in the form of a response plot, were obtained by means of a computer. The sequence of computed displacement was included merely to illustrate the process.

In this section, we present a MATLAB program for the determination of the response by the convolution sum, Eq. (4.95), adapting the MATLAB function 'conv' to the problem at hand. To this end, we consider the response of the single-degree-of-freedom to a sawtooth pulse of Example 4.8, except that here we consider a damped system with the damping factor ζ acting as a parameter and with $\zeta = 0$ being merely one of the values. The program is as follows:

```
% The program 'convsum.m' plots the response of a single-degree-of-freedom
% system to a sawtooth pulse

clear
clf

m=1; % mass
```

```
wn=4; % frequency of undamped oscillation
zeta=[0; 0.1; 0.2]; % damping factors arranged as a three-dimensional vector
T=0.01; % sampling period
N=300; % number of sampling times

  for j=1:N,
      if j<=0.4/T+1; F(j)=100*T*(j-1); else; F(j)=0; end % force in the form of a
  % sawtooth pulse

  end

for i=1:length(zeta),
  wd=sqrt(1-zeta(i)^2)*wn; % frequency of damped oscillation


      n=[1:1:N];
      g=(T/(m*wd))*exp(-(n-1)*zeta(i)*wn*T).*sin((n-1)*wd*T); % discrete-time
      % impulse response

      c0=conv(F,g); % convolution sum by MATLAB function
      c=c0(1:N); % limit the plot to 300 samples

    plot (c, '.')
    hold on

      end

title ('Response by the Convolution Sum')
xlabel ('n')
ylabel ('x(n)')
axis ([0 300 -2 2])
```

Note that the response plots obtained by means of this program consist of dots representing the displacement at discrete times. In fact, the plot for $T = 0.01$ s in Fig. 4.29 represents the first third of the plot corresponding to $\zeta = 0$ generated by this program.

4.12 RESPONSE BY THE DISCRETE-TIME TRANSITION MATRIX USING MATLAB

In Sec. 4.11, we presented a computer program for the response of a damped single-degree-of-freedom system by the convolution sum, which is a scalar approach based on a single equation of motion, albeit of second order. For the most part, the preferred approach in numerical integration of differential equations is based on first-order equations, which implies a state vector formulation, such as that given by Eq. (4.80). Of course, Eq. (4.80) is in continuous time and in numerical integration the interest lies in a discrete-time formulation, such as that given by Eq. (4.104). A computer program solving the response problem of Example 4.9 reads as follows:

```
% The program 'tramat.m' plots the response of a single-degree-of-freedom
% system to a sawtooth pulse by the discrete-time transition matrix
```

```
clear
clf
m=1; % mass
wn=4; % frequency of undamped oscillation
zeta=[0; 0.1; 0.2]; % damping factors arranged in a three-dimensional vector
b=[0 1/m]'; % coefficient vector, second of Eqs. (4.81)
T=0.005; % sampling period
N=600; % number of sampling times
for i=1:length(zeta),
  A=[0 1; -wn^2 -2*zeta(i)*wn]; % system matrix, first of Eqs. (4.81)

  Phi=eye(2)+T*A+T^2*A^2/2+T^3*A^3/6+T^4*A^4/24; % discrete-time
  % transition matrix
  gamma=inv(A)*(Phi-eye(2))*b; % two-dimensional vector of coefficients, Eq. (4.106)
  x(1, 1)=0; % initial displacement
  x(2, 1)=0; % initial velocity
  for n=1:N,
    if n<=0.4/T+1; F(n) = 100*T*(n-1); else; F(n)=0;
    end
    x(:,n+1)=Phi*x(:,n)+gamma*F(n); % discrete-time state vector
  end
  n=[0: 1: N];
  plot (n,x(1,:),'.')
  hold on
end
title ('Response by the Transition Matrix')
xlabel ('n')
ylabel ('x(n)')
grid
```

Note that the program plots $x(n)$ versus n, where $x(n)$ is the top component of the discrete-time state vector, i.e., the discrete-time displacement.

4.13 SUMMARY

This chapter is concerned with the response of linear time-invariant systems, perhaps better known as linear systems with constant coefficients, to arbitrary, or transient excitations. It marks a new direction in our study in the sense that it requires different techniques than those encountered until now. By virtue of the superposition principle, the response to arbitrary excitations can be obtained separately from the response to initial excitations (Ch. 2), and then the two can be combined linearly to obtain the total response. By contrast, the response to harmonic excitations and periodic excitations (Ch. 3) is steady state, in which time plays only a secondary role, so that a combination with the response to initial excitations, which is transient, is meaningless.

A problem of special interest is how to describe arbitrary forces, as well as certain discontinuous forces. In this regard, the unit impulse, the unit step function and the omit ramp function can be very helpful. In particular, arbitrary forces can be represented as a superposition of impulses of varying magnitude. Then, the response can be obtained in the form of a superposition of corresponding impulse responses, giving rise to the superposition integral, or convolution integral. A method particularly suited for the determination of the response of linear time-invariant systems to both initial and arbitrary excitations is the Laplace transformation method, which is capable of yielding both responses at the same time. The Laplace transformation method can be used to relate the response to the excitation through an algebraic expression known as the transfer function, a very useful concept. It can also be used to derive the convolution integral in a more general manner than by means of the concept of impulse response. A most widely used approach to the integration of the differential equation of motion for linear time-invariant systems consists of transforming the second-order differential equation describing the motion of a single-degree-of-freedom system into a set of two first-order differential equations referred to as state equations. The solution of the state equations can be obtained by an approach based on the state transition matrix, which reduces to the evaluation of a vector convolution integral.

For the most part, it is not possible to evaluate analytically convolution integrals, neither in scalar nor in vector form, so that the response must be obtained numerically by means of a digital computer. This implies evaluation of the response in discrete time, as a digital computer does not work with continuous time. Even when the response is plotted in continuous time, the computations must be carried out in discrete time. In this chapter, we develop discrete time algorithms for the determination of the response by both the scalar form and the vector form of the convolution integral. The first algorithm is the convolution sum and the second is based on the discrete-time transition matrix. Correspondingly, two MATLAB computer programs have been written, the first entitled 'convsum.m' and the second 'tramat.m'.

PROBLEMS

4.1. Use Eq. (4.11) to determine the time necessary for the impulse response of a viscously damped single-degree-of-freedom system to reach its peak value and then determine the peak value.

4.2. Use Eq. (4.22) to derive the impulse response of a viscously damped single-degree-of-freedom system.

4.3. Derive the response of a viscously damped single-degree-of-freedom system to the rectangular pulse shown in Fig. 4.8. Plot the response for the system parameters $m = 12$ kg, $c = 24\,\mathrm{N\cdot s/m}$, $k = 4{,}800$ N/m and for the pulse parameters $F_0 = 200$ N, $T = 0.5$ s.

4.4. Derive the ramp response of a viscously damped single-degree-of-freedom system.

4.5. Derive the response of a viscously damped single-degree-of-freedom system to the force shown in Fig. 4.31 by regarding the force as a superposition of ramp functions. Plot the response for the system of Problem 4.3 over the time interval $0 \leq t \leq 1$ s.

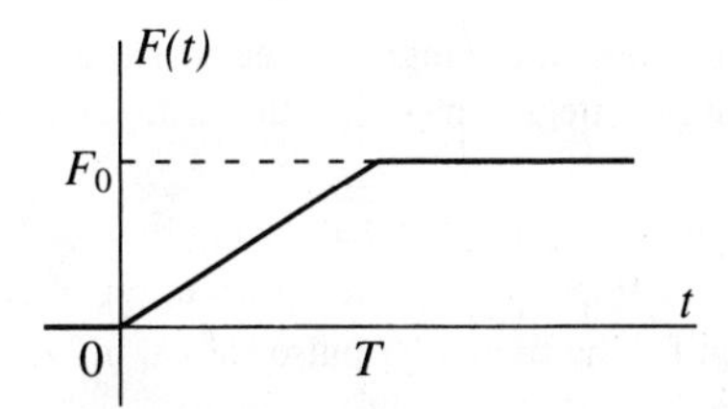

FIGURE 4.31
Force in the form of a difference of two ramp functions

4.6. Use the superposition principle to derive the response of a viscously damped single-degree-of-freedom system to the triangular pulse shown in Fig. 4.32.

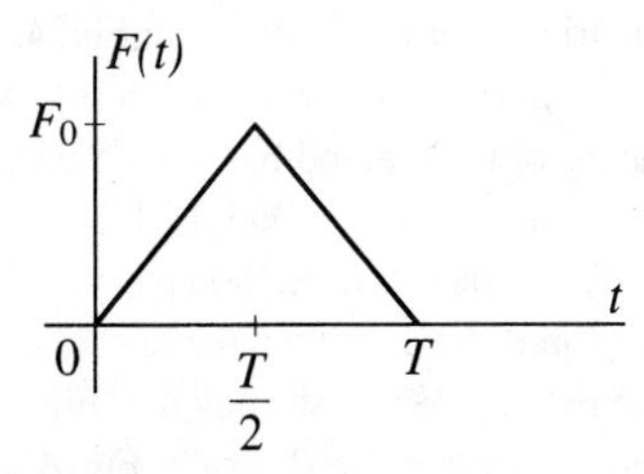

FIGURE 4.32
Force in the form of a triangular pulse

4.7. Repeat Problem 4.6 for the trapezoidal pulse shown in Fig. 4.33.

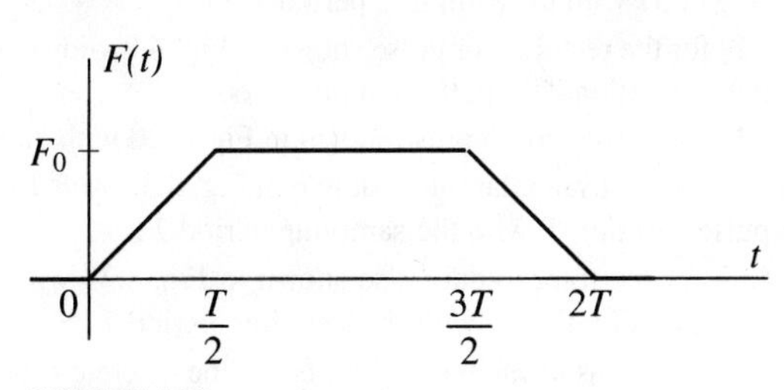

FIGURE 4.33
Force in the form of a trapezoidal pulse

4.8. Repeat Problem 4.6 for the sawtooth pulse shown in Fig. 4.20.

4.9. Derive the response of a viscously damped single-degree-of-freedom system to the force $F(t) = F_0 e^{-\alpha t} \mathscr{u}(t)$ by means of the convolution integral. Plot the response for the system parameters $m = 12$ kg, $c = 24 \text{N} \cdot \text{s/m}$, $k = 4,800$ N/m and the force parameters $F_0 = 200$ N, $\alpha = 1$.

4.10. Derive and plot the response of the system of Problem 4.9 to the force $F(t) = F_0(1 - e^{-\alpha t})\mathscr{u}(t)$ by means of the convolution integral. Then, use Eq. (4.22) to plot the response to a step function of magnitude $F_0 = 200$ N, compare results and draw conclusions.

4.11. Derive the response of an undamped single-degree-of-freedom system to the force shown in Fig. 4.31 by the convolution integral in conjunction with the geometric interpretation of Sec. 4.4.

4.12. Solve Problem 4.11 for the triangular pulse shown in Fig. 4.32.

4.13. Solve Problem 4.11 for the trapezoidal pulse shown in Fig. 4.33.

4.14. Plot the shock spectrum for the triangular pulse shown in Fig. 4.32. Compare results with those obtained in Sec. 4.5 and draw conclusions.

4.15. Plot the shock spectrum for the trapezoidal pulse of Fig. 4.33. Compare results with those obtained in Sec. 4.5 and draw conclusions.

4.16. Derive the response of a viscously damped single-degree-of-freedom system to the force $F(t) = F_0 e^{-\alpha t} \mathcal{u}(t)$ by means of the Laplace transformation method.

4.17. Solve Problem 4.16 for the rectangular pulse shown in Fig. 4.8.

4.18. Solve Problem 4.16 for the sawtooth pulse shown in Fig. 4.20.

4.19. Solve Problem 4.16 for the force shown in Fig. 4.31.

4.20. Solve Problem 4.16 for the triangular pulse shown in Fig. 4.32.

4.21. Solve Problem 4.16 for the trapezoidal pulse shown in Fig. 4.33.

4.22. Solve Problem 4.16 by means of the method based on the state transition matrix.

4.23. Derive the response of a viscously damped single-degree-of-freedom system to the force $F(t) = F_0(1 - e^{-\alpha t})\mathcal{u}(t)$ by means of the method based on the state transition matrix.

4.24. Solve Problem 4.23 for the force shown in Fig. 4.31.

4.25. Solve Problem 4.23 for the rectangular pulse shown in Fig. 4.8.

4.26. Solve Problem 4.23 for the sawtooth pulse shown in Fig. 4.20.

4.27. Solve Problem 4.23 for the triangular pulse shown in Fig. 4.32.

4.28. Solve Problem 4.23 for the trapezoidal pulse shown in Fig. 4.33.

4.29. Solve Problem 4.9 by means of the convolution sum. Plot the response using the sampling period $T = 0.003$ s and the number of sampling times $n = 200$.

4.30. Solve Problem 4.29 for the force shown in Fig. 4.31 with $T = 0.1$ s. Caution: Do not confuse the symbol T in Fig. 4.31 with the sampling period T.

4.31. Solve Problem 4.29 for the rectangular pulse shown in Fig. 4.8 with $T = 0.1$ s. Caution: Do not confuse the pulse duration T with the sampling period T.

4.32. Solve Problem 4.29 for the sawtooth pulse shown in Fig. 4.20 with $T_0 = 0.1$ s.

4.33. Solve Problem 4.29 for the triangular pulse shown in Fig. 4.32 with $T = 0.1$ s. Caution: Do not confuse the pulse duration T with the sampling period T.

4.34. Solve Problem 4.29 for the trapezoidal pulse shown in Fig. 4.33 with $T = 0.1$ s. Caution: Do not confuse the pulse duration T with the sampling period T.

4.35. Solve Problem 4.29 by means of the method based on the discrete-time transition matrix.

4.36. Solve Problem 4.30 by means of the method based on the discrete-time transition matrix.

4.37. Solve Problem 4.31 by means of the method based on the discrete-time transition matrix.

4.38. Solve Problem 4.32 by means of the method based on the discrete-time transition matrix.

4.39. Solve Problem 4.33 by means of the method based on the discrete-time transition matrix.

4.40. Solve Problem 4.34 by means of the method based on the discrete-time transition matrix.

4.41. Solve Problem 4.29 by first modifying and then using the MATLAB program of Sec. 4.11.

4.42. Solve Problem 4.30 by first modifying and then using the MATLAB program of Sec. 4.11.

4.43. Solve Problem 4.31 by first modifying and then using the MATLAB program of Sec. 4.11.

4.44. Solve Problem 4.33 by first modifying and then using the MATLAB program of Sec. 4.11.

4.45. Solve Problem 4.34 by first modifying and then using the MATLAB program of Sec. 4.11.

4.46. Solve Problem 4.35 by first modifying and then using the MATLAB program of Sec. 4.12.

4.47. Solve Problem 4.36 by first modifying and then using the MATLAB program of Sec. 4.12.

4.48. Solve Problem 4.37 by first modifying and then using the MATLAB program of Sec. 4.12.

4.49. Solve Problem 4.39 by first modifying and then using the MATLAB program of Sec. 4.12.

4.50. Solve Problem 4.40 by first modifying and then using the MATLAB program of Sec. 4.12.

CHAPTER 5

TWO-DEGREE-OF-FREEDOM SYSTEMS

Until now, our study has been concerned with the vibration of single-degree-of-freedom systems, defined as systems whose behavior can be described by a single coordinate. Whereas many systems fit this description, and many more can be approximated by a single-degree-of-freedom system, most systems require a more refined model. We encountered such an example in Sec. 1.10 in the form of the automobile model depicted in Fig. 1.32a. Indeed, a description of the motion of this model necessitated four independent coordinates, the translation x_b of the body/chassis, treated as a rigid slab, the rotation θ of the body/chassis, the translation x_f of the front tire and the translation x_r of the rear tire. We refer to such a system as a four-degree-of-freedom system, where the number of degrees of freedom of a system is defined as the number of independent coordinates required to describe the motion fully. Systems requiring two or more coordinates are called *multi-degree-of-freedom systems*.

An undamped single-degree-of-freedom system set in motion by initial excitations executes natural vibration, in the sense that the system vibrates at the system natural frequency. What sets apart natural vibration of a multi-degree-of-freedom system from that of a single-degree-of-freedom system is that for multi-degree-of-freedom systems natural vibration implies not only a certain natural frequency but also a certain natural displacement configuration assumed by the system masses during motion. Moreover, a multi-degree-of-freedom system possesses not only one natural frequency and associated natural configuration but a finite number of them. In fact, the system possesses as many natural frequencies and natural configurations, known as *natural modes of vibrations*, as the number of degrees of freedom of the system. Depending on the initial excitation, the system can be made to vibrate in any of these modes independently, which is due to an important property called *orthogonality*.

The mathematical formulation for an n-degree-of-freedom system consists of n simultaneous ordinary differential equations of motion. Hence, the motion of one mass depends on the motion of the other $n-1$ masses. For a proper choice of coordinates, known as *principal coordinates*, or *natural coordinates*, the n differential equations become independent of one another. The natural coordinates represent abstract quantities rather than actual displacements of the individual masses. In fact, they represent linear combinations of the actual displacements and, conversely, the motion of the individual masses can be represented by a superposition of the natural coordinates. The interesting and useful feature of the natural coordinates is that the differential equation for each of the independent natural coordinates resembles the equation of motion of a single-degree-of-freedom system, thus suggesting a method for the determination of the response of multi-degree-of-freedom systems. Two questions remaining are how to obtain the equations for the natural coordinates and how to combine the natural coordinates to determine the actual motion of the system.

We begin this chapter with a detailed discussion of the concept of system configuration and how it can be used to visualize the motion of multi-degree-of-freedom systems. Then, we use a two-degree-of-freedom system to introduce the concepts and techniques necessary for a study of the dynamic characteristics and for the derivation of the response of multi-degree-of-freedom systems by the approach outlined in the preceding paragraph. Certain applications typical of two-degree-of-freedom systems are also presented.

5.1 SYSTEM CONFIGURATION

The behavior of multi-degree-of-freedom systems is appreciably more complex than the behavior of single-degree-of-freedom systems. To study this behavior, it is necessary to introduce new concepts. One question relates to the manner in which the information concerning the system response is best exhibited. In the case of a single-degree-of-freedom system, the response is commonly displayed in the form of a time plot $x(t)$ versus t for transient excitations, where $x(t)$ is the displacement of the mass, and in the form of frequency response plots $|G(i\omega)|$ versus ω/ω_n and $\phi(\omega)$ versus ω/ω_n for harmonic excitations, where $|G(i\omega)|$ and $\phi(\omega)$ are the magnitude and phase angle of the frequency response, respectively. The response of multi-degree-of-freedom systems of very low order, such as two-degree-of-freedom systems, can be presented in a similar manner, but the approach becomes impractical as the number of degrees of freedom increases. Indeed, for n-degree-of-freedom systems, it is convenient to express the response by means of an n-dimensional vector $\mathbf{x}(t) = [x_1(t)\ x_2(t)\ \dots\ x_n(t)]^T$ known as the *configuration vector*, where $x_i(t)$ is the displacement of mass $m_i\,(i = 1, 2, \dots, n)$. Then, the motion of the system can be represented geometrically as a path traced by the tip of the vector $\mathbf{x}(t)$ in an n-dimensional space, as shown in Fig. 5.1, where the space is called the *configuration space*; in this representation the time t plays the role of a parameter. The problem with this approach is apparent immediately. Indeed, whereas Fig. 5.1 provides an interesting picture, the information conveyed is qualitative in nature and of little practical value. Hence, a different approach recommends itself.

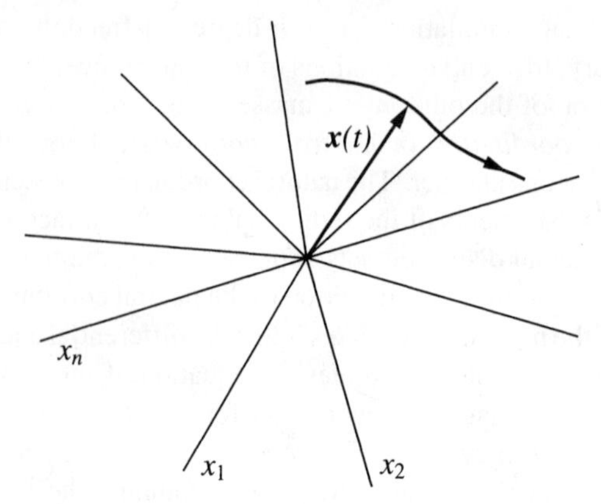

FIGURE 5.1
Geometric representation of the motion in the configuration space

In the case of lumped-parameter models, such as the automobile model of Fig. 1.32a, a logical way to display the motion is by plotting the displacement of each of the masses in the model. To introduce the idea, we consider a model whose motion is easier to visualize than that of Fig. 1.32a, namely, a set of masses on a string vibrating in a vertical plane. Figure 5.2 shows the masses m_i and the associated displacements $x_i(t)$ $(i = 1, 2, \ldots, n)$. Clearly, Fig. 5.2 represents a more practical way of displaying the system configuration vector $\mathbf{x}(t) = [x_1(t)\ x_2(t)\ \ldots\ x_n(t)]^T$ than Fig. 5.1. But, the figure illustrates in a clear fashion why it is so much more difficult to describe the motion of a multi-degree-of-freedom system than that of a single-degree-of-freedom system. Indeed, the displacements of the n masses form a geometric pattern, a concept devoid of meaning for single-degree-of-freedom systems. This geometric pattern represents the configuration of the system at a particular time t, and it plays the same role as a point on the curve $x(t)$ versus t for a single-degree-of-freedom system. Of course, in principle it is possible to conceive of a time axis normal to the plane of vibration and to plot the functions $x_i(t)$ versus t $(i = 1, 2, \ldots, n)$, thus generating the three-dimensional surface shown in Fig. 5.3, with the pattern displayed in Fig. 5.2 representing just one cross section of this surface corresponding to a given time t. This possibility was mentioned mainly to demonstrate the complexity of exhibiting the motion of a multi-degree-of-freedom system, rather than suggesting Fig. 5.3 as a practical way of displaying it. Still,

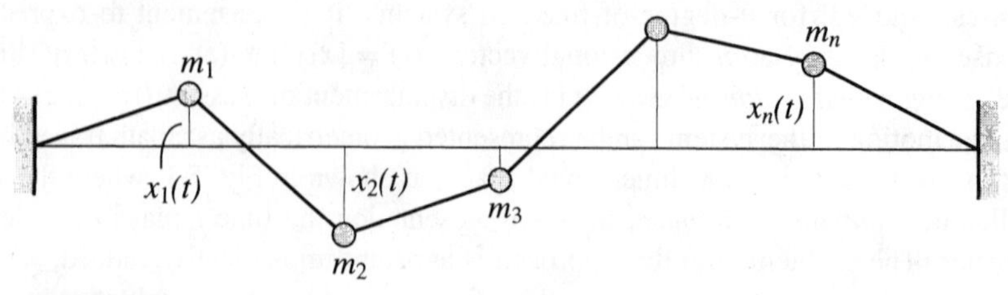

FIGURE 5.2
Displacement pattern of an n-degree-of-freedom system at time t

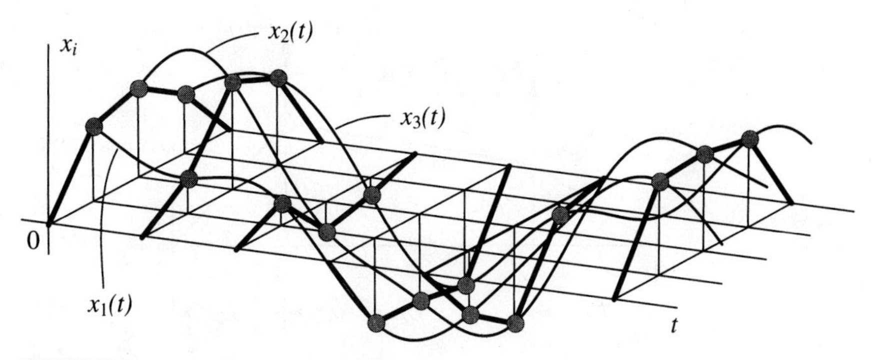

FIGURE 5.3
Evolution of the displacement pattern with time

the idea of the system configuration representing a displacement pattern is a useful one, as it forms the basis not only for displaying the motion but also for solving the system equations of motion. Reference is made here to modal analysis, whereby the motion of a multi-degree-of-freedom system can be expressed as a linear combination of certain displacement patterns referred to as modes of vibration, as stated earlier in this chapter. We discuss this subject in Sec. 5.3 and in Ch. 7.

5.2 THE EQUATIONS OF MOTION OF TWO-DEGREE-OF-FREEDOM SYSTEMS

Before we begin our study of techniques for solving vibration problems, we must have the differential equations of motion of the system under consideration; we derived such equations in Sec. 1.10. In this section, we obtain the equations of motion for several two-degree-of-freedom systems as a way of introducing the formalism for solving vibration problems.

As indicated in Sec. 5.1, a system of masses on a string has the advantage that its motion is easy to visualize. In view of this, we begin by deriving the differential equations of motion for the two-degree-of-freedom system shown in Fig. 5.4a. The system consists of two masses m_1 and m_2 suspended on a string of tension T and subjected to the external forces $F_1(t)$ and $F_2(t)$, respectively. We measure the displacements $x_1(t)$ and $x_2(t)$ of m_1 and m_2, respectively, from equilibrium and assume that they are small, so that the string tension does not change during the motion. We propose to derive the equations of motion by means of Newton's second law, Eqs. (1.28), recognizing that in the case at hand there is only one component of motion for each of the two masses. To apply Newton's second law, it is necessary to draw one free-body diagram per mass; this is done in Fig. 5.4b. Hence, summing up forces in the vertical direction, we have

$$\begin{aligned} F_1(t) - m_1 g - T\sin\theta_1(t) + T\sin\theta_2(t) &= m_1\frac{d^2x_1(t)}{dt^2} \\ F_2(t) - m_2 g - T\sin\theta_2(t) - T\sin\theta_3(t) &= m_2\frac{d^2x_2(t)}{dt^2} \end{aligned} \tag{5.1}$$

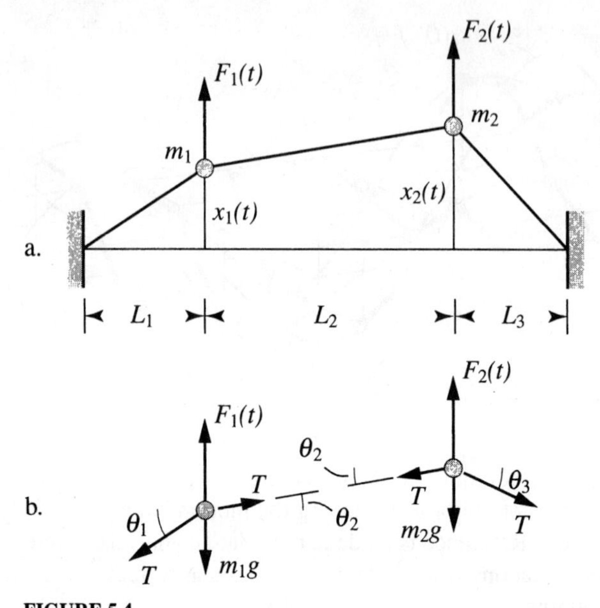

FIGURE 5.4
a. Two masses on a string, b. Free-body diagrams

But, under the assumption that the displacements are small, we can write

$$\sin\theta_1(t) \cong \frac{x_1(t)}{L_1}, \quad \sin\theta_2(t) \cong \frac{x_2(t)-x_1(t)}{L_2}, \quad \sin\theta_3(t) \cong \frac{x_2(t)}{L_3} \tag{5.2}$$

so that Eqs. (5.1) reduce to

$$\begin{aligned} m_1\frac{d^2x_1}{dt^2} + \left(\frac{T}{L_1}+\frac{T}{L_2}\right)x_1 - \frac{T}{L_2}x_2 + m_1g &= F_1 \\ m_2\frac{d^2x_2}{dt^2} - \frac{T}{L_2}x_1 + \left(\frac{T}{L_2}+\frac{T}{L_3}\right)x_2 + m_2g &= F_2 \end{aligned} \tag{5.3}$$

Equations (5.3) contain the constant terms m_1g and m_2g. These terms contribute to the displacements $x_1(t)$ and $x_2(t)$, but not to the vibration. This statement may appear paradoxical, but it simply means that each of the displacements can be expressed as the sum of a constant term representing the static equilibrium position and a time-dependent term representing the vibration, as follows:

$$x_1(t) = x_{e1} + \tilde{x}_1(t), \quad x_2(t) = x_{e2} + \tilde{x}_2(t) \tag{5.4}$$

where x_{e1} and x_{e2} are the constant equilibrium positions and $\tilde{x}_1(t)$ and $\tilde{x}_2(t)$ the vibration from equilibrium. Inserting Eqs. (5.4) into Eqs. (5.3) and separating the constant terms and the time-varying terms, we obtain the *equilibrium equations*

$$\begin{aligned}\left(\frac{T}{L_1}+\frac{T}{L_2}\right)x_{e1}-\frac{T}{L_2}x_{e2}+m_1 g=0\\-\frac{T}{L_1}x_{e1}+\left(\frac{T}{L_2}+\frac{T}{L_3}\right)x_{e2}+m_2 g=0\end{aligned}\tag{5.5}$$

which can be solved for x_{e1} and x_{e2}, and the *equations for vibration about equilibrium*

$$\begin{aligned}m_1\frac{d^2\tilde{x}_1}{dt^2}+\left(\frac{T}{L_1}+\frac{T}{L_2}\right)\tilde{x}_1-\frac{T}{L_2}\tilde{x}_2=F_1\\m_2\frac{d^2\tilde{x}_2}{dt^2}-\frac{T}{L_2}\tilde{x}_1+\left(\frac{T}{L_2}+\frac{T}{L_3}\right)\tilde{x}_2=F_2\end{aligned}\tag{5.6}$$

Another two-degree-of-freedom system of interest consists of a slab supported on two springs, as shown in Fig. 5.5. It can be regarded as a simplified model of the automobile of Fig. 1.29a. In fact, a four-degree-of-freedom model of that automobile is shown in Fig. 1.32a, and the corresponding equations of motion were derived in Sec. 1.10 in the form of Eqs. (1.130). Hence, the equations of motion for the system of Fig. 5.5 can be obtained from Eqs. (1.130) by letting $x_b=x,\ x_f=x_r=0,\ m_b=m,\ c_{sf}=c_{sr}=0,\ k_{sf}=k_1$ and $k_{sr}=k_2$. The resulting equations of motion are

$$\begin{aligned}m\ddot{x}+(k_1+k_2)x-(k_1a-k_2b)\theta=F\\I_C\ddot{\theta}-(k_1a-k_2b)x+(k_1a^2+k_2b^2)=Fc\end{aligned}\tag{5.7}$$

where overdots denote derivatives with respect to time.

Next, we consider a system consisting of two masses supported by springs and dashpots, as shown in Fig. 5.6a. The corresponding free-body diagrams are displayed in Fig. 5.6b. Using Newton's second law for each of the two masses, we obtain the equations of motion

$$\begin{aligned}F_1+k_2(x_2-x_1)+c_2(\dot{x}_2-\dot{x}_1)-k_1x_1-c_1\dot{x}_1=m_1\ddot{x}_1\\F_2-k_2(x_2-x_1)-c_2(\dot{x}_2-\dot{x}_1)-k_3x_2-c_3\dot{x}_2=m_2\ddot{x}_2\end{aligned}\tag{5.8}$$

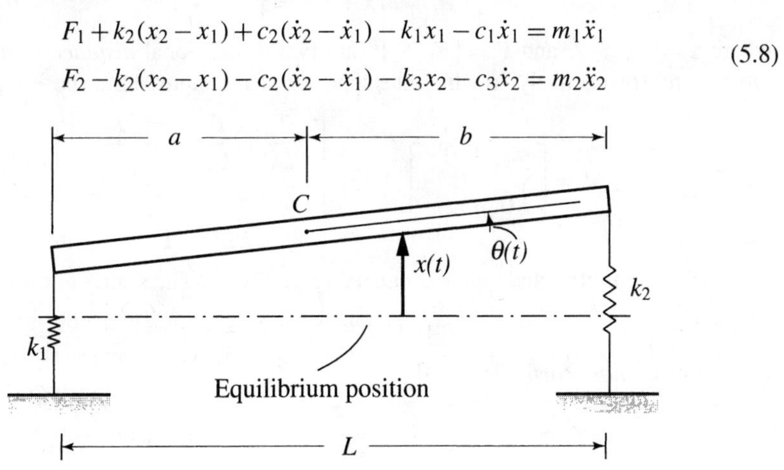

FIGURE 5.5
Simplified model of an automobile

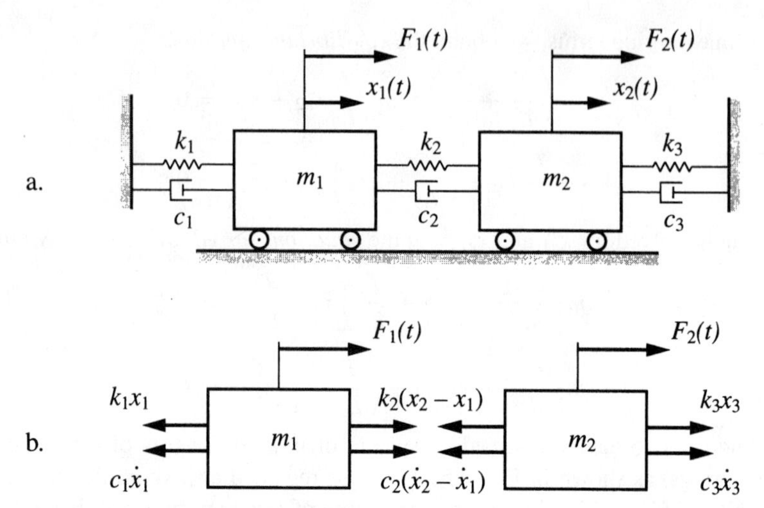

FIGURE 5.6
a. Damped two-degree-of-freedom system in horizontal vibration, b. Free-body diagrams

which can be rearranged as follows:

$$\begin{aligned} m_1\ddot{x}_1 + (c_1 + c_2)\dot{x}_1 - c_2\dot{x}_2 + (k_1 + k_2)x_1 - k_2x_2 &= F_1 \\ m_2\ddot{x}_2 - c_2\dot{x}_1 + (c_2 + c_3)\dot{x}_2 - k_2x_1 + (k_2 + k_3)x_2 &= F_2 \end{aligned} \tag{5.9}$$

In solving vibration problems, matrix methods are indispensable, so that we wish to cast the equations of motion in matrix form. To this end, we consider first Eqs. (5.6), omit the overtilde from $\tilde{x}_1$ and $\tilde{x}_2$, use the standard practice of denoting time derivatives by means of overdots and rewrite the equations in the compact matrix form

$$M\ddot{\mathbf{x}} + K\mathbf{x} = \mathbf{F} \tag{5.10}$$

where $\mathbf{x} = [x_1\ x_2]^T$ and $\mathbf{F} = [F_1\ F_2]^T$ are two-dimensional *displacement vector* and *force vector*, respectively, in which the superscript T denotes a transposed quantity, and

$$M = \begin{bmatrix} m_1 & 0 \\ 0 & m_2 \end{bmatrix}, \quad K = \begin{bmatrix} \dfrac{T}{L_1} + \dfrac{T}{L_2} & -\dfrac{T}{L_2} \\ -\dfrac{T}{L_2} & \dfrac{T}{L_2} + \dfrac{T}{L_3} \end{bmatrix} \tag{5.11}$$

are 2×2 *mass matrix* and *stiffness matrix*, respectively. The matrix entries

$$m_{11} = m_1, \ m_{22} = m_2, \ m_{12} = m_{21} = 0 \tag{5.12}$$

are known as *mass coefficients* and

$$k_{11} = \frac{T}{L_1} + \frac{T}{L_2}, \ k_{22} = \frac{T}{L_2} + \frac{T}{L_3}, \ k_{12} = k_{21} = -\frac{T}{L_2} \tag{5.13}$$

as *stiffness coefficients*. We observe that the mass matrix is diagonal, and hence symmetric by definition, and the stiffness matrix is symmetric. The matrix symmetry is reflected in the fact that the off-diagonal terms are equal, or

$$m_{12} = m_{21},\ k_{12} = k_{21} \tag{5.14}$$

and can be expressed in matrix form as

$$M = M^T,\ K = K^T \tag{5.15}$$

Equations (5.7) can be expressed in the same matrix form as that given by Eq. (5.10), except that now the displacement and force vectors are $\mathbf{x} = [x\ \theta]^T$ and $\mathbf{F} = [F\ Fc]^T$, where we note that the second component is an angle and a moment, respectively, and the mass and stiffness matrices are

$$M = \begin{bmatrix} m & 0 \\ 0 & I_C \end{bmatrix},\ K = \begin{bmatrix} k_1 + k_2 & -(k_1 a - k_2 b) \\ -(k_1 a - k_2 b) & k_1 a^2 + k_2 b^2 \end{bmatrix} \tag{5.16}$$

Similarly, Eqs. (5.9) have the matrix form

$$M\ddot{\mathbf{x}} + C\dot{\mathbf{x}} + K\mathbf{x} = \mathbf{F} \tag{5.17}$$

in which

$$C = \begin{bmatrix} c_1 + c_2 & -c_2 \\ -c_2 & c_2 + c_3 \end{bmatrix},\ K = \begin{bmatrix} k_1 + k_2 & -k_2 \\ -k_2 & k_2 + k_3 \end{bmatrix} \tag{5.18}$$

are 2×2 *damping matrix* and stiffness matrix, respectively. The remaining quantities are as defined earlier. The matrix entries

$$c_{11} = c_1 + c_2,\ c_{22} = c_2 + c_3,\ c_{12} = c_{21} = -c_2 \tag{5.19}$$

are called *damping coefficients*. Clearly, the damping matrix is symmetric, or

$$C = C^T \tag{5.20}$$

The symmetry of the mass, damping and stiffness matrices is an inherent property of vibrating systems. On the other hand, the fact that the mass matrix is diagonal and the damping and stiffness matrices are not is a reflection of the choice of coordinates used to describe the motion, rather than a characteristic of the system. Indeed, a different choice of coordinates can result in a nondiagonal mass matrix and diagonal damping and stiffness matrices. The issues of matrix symmetry and of the choice of coordinates rendering matrices diagonal are very important in vibrations, and will receive a great deal of attention in this and in the next chapter.

5.3 FREE VIBRATION OF UNDAMPED SYSTEMS. NATURAL MODES

In the study of vibrations, there is considerable interest in *conservative systems*, i.e., in systems that neither dissipate energy nor gain energy. The implication is that conservative systems are subjected to neither damping forces nor to externally applied forces. Under these circumstances, the equations of motion of a two-degree-of-freedom system reduce to

$$M\ddot{\mathbf{x}}(t) + K\mathbf{x}(t) = \mathbf{0} \tag{5.21}$$

where $\mathbf{x}(t) = [x_1(t)\ x_2(t)]^T$ is the displacement vector and

$$M = \begin{bmatrix} m_1 & 0 \\ 0 & m_2 \end{bmatrix}, \quad K = \begin{bmatrix} k_{11} & k_{12} \\ k_{12} & k_{22} \end{bmatrix} \tag{5.22}$$

are the diagonal mass matrix and the symmetric stiffness matrix, respectively. As can be concluded from Sec. 5.2, the systems of Figs. 5.4a, 5.5 and 5.6a can be described by Eq. (5.21) in the absence of damping forces and external forces. Equation (5.21) represents a set of two homogeneous ordinary differential equations of second order and their solution is subject to four initial conditions, two initial displacements $x_1(0),\ x_2(0)$ and two initial velocities $\dot{x}_1(0),\ \dot{x}_2(0)$.

Fundamental to the study of vibrations are certain special solutions in which the whole system executes the same motion in time. We refer to such motions as *synchronous*. The implication of synchronous motion in the case of the two-degree-of-freedom under consideration is that the coordinates $x_1(t)$ and $x_2(t)$ increase and decrease in the same proportion with time, so that the ratio $x_2(t)/x_1(t)$ remains constant for all times. Another interpretation is that in synchronous motion the two masses assume a certain displacement configuration, or displacement pattern, and the shape of this configuration does not change throughout the motion; only the amplitude of the displacement configuration does. This type of motion can be expressed in the form

$$\mathbf{x}(t) = f(t)\mathbf{u} \tag{5.23}$$

where $f(t)$ is the time-dependent amplitude and $\mathbf{u} = [u_1\ u_2]^T$ is a constant vector representing the displacement pattern, or displacement profile. Inserting Eq. (5.23) into Eq. (5.21), we can write simply

$$\ddot{f}(t)M\mathbf{u} + f(t)K\mathbf{u} = \mathbf{0} \tag{5.24}$$

Next, we premultiply Eq. (5.24) by $\mathbf{u}^T$ and obtain the scalar equation

$$\ddot{f}(t)\mathbf{u}^T M\mathbf{u} + f(t)\mathbf{u}^T K\mathbf{u} = 0 \tag{5.25}$$

in which we recognized that $\mathbf{u}^T M\mathbf{u}$ and $\mathbf{u}^T K\mathbf{u}$ are scalars. Then, introducing the notation

$$\frac{\mathbf{u}^T K\mathbf{u}}{\mathbf{u}^T M\mathbf{u}} = \lambda \tag{5.26}$$

Eq. (5.25) reduces to

$$\ddot{f}(t) + \lambda f(t) = 0 \tag{5.27}$$

Moreover, inserting Eq. (5.27) into Eq. (5.24), dividing through by $f(t)$ and rearranging, we obtain

$$K\mathbf{u} = \lambda M\mathbf{u} \tag{5.28}$$

Hence, the time-dependent amplitude $f(t)$ of the synchronous motion must satisfy Eq. (5.27) and the displacement configuration must satisfy Eq. (5.28).

There are several questions yet to be addressed concerning the nature and significance of synchronous motions. In the first place, there is the question as to the type of motions the system executes in time. The second question relates to the number and nature of the displacement patterns. Finally, we must address the question of the relation between the special synchronous solutions and the general solution to the free vibration problem. The answer to the first question lies in the solution of Eq. (5.27), which is a linear homogeneous differential equation. The form of this solution depends on λ. According to Eq. (5.26), λ is the ratio of two quadratic forms involving real quantities alone, so that λ is real. It follows that the nature of the solution depends on the sign of λ. To ascertain this sign, we observe that the quadratic form at the numerator in Eq. (5.26) is proportional to the potential energy, which in the case of the systems of Figs. 5.4a, 5.5 and 5.6a is a positive quantity. Moreover the quadratic form at the denominator is proportional to the kinetic energy, which is always positive. We conclude that in the cases under consideration λ is positive. In view of the fact that λ is real and positive, it is convenient to introduce the notation

$$\lambda = \omega^2 \tag{5.29}$$

so that Eq. (5.27) can be rewritten as

$$\ddot{f}(t) + \omega^2 f(t) = 0 \tag{5.30}$$

and we note that ω is a real number, which can be assumed to be positive. But, Eq. (5.30) resembles Eq. (2.2), where the latter represents the equation of a harmonic oscillator, studied extensively in Sec. 2.1. Hence, using results from Sec. 2.1, we conclude that the solution of Eq. (5.30) is harmonic and of the form

$$f(t) = C\cos(\omega t - \phi) \tag{5.31}$$

where C is an arbitrary constant amplitude, ω the frequency of the harmonic motion and ϕ a phase angle, all three quantities being the same for both displacements $x_1(t)$ and $x_2(t)$.

The constants C and ϕ differ in nature from the constant ω. Indeed, whereas C and ϕ depend on external factors, and in particular on the initial excitations, ω depends on internal factors, namely, the value of the system parameters. A question arising naturally is whether synchronous motion can take place in one frequency ω or in several frequencies. This question is closely related to the second one raised above, namely, the question concerning the number of displacement configurations capable of synchronous motion. The answer to these questions lies in Eq. (5.28), which is known as the *algebraic eigenvalue problem*, also known as the *characteristic-value problem*. The problem consists of determining the values of the parameter $\lambda = \omega^2$ for which the set of homogeneous algebraic equations, Eq. (5.28), admits nontrivial solutions. It should be pointed out here that, although the derivation of the eigenvalue problem in this section was motivated by two-degree-of-freedom systems, Eq. (5.28) applies to multi-degree-of-freedom systems as well.

In general, the algebraic eigenvalue problem can only be solved numerically, requiring methods of matrix algebra. The sole exception is that in which the system possesses just two degrees of freedom, which is the case of interest in this section. In this case, matrices M and K are 2×2, and the eigenvalue problem admits closed-form solutions. Moreover, the solution can be obtained by using elementary algebra. We use this approach here and consider the matrix approach in Ch. 7.

Letting $\lambda = \omega^2$ in Eq. (5.28) and using Eqs. (5.22), the eigenvalue problem can be expressed in the form

$$\begin{aligned} (k_{11}-\omega^2 m_1)u_1 + k_{12}u_2 &= 0 \\ k_{12}u_1 + (k_{12}-\omega^2 m_2)u_2 &= 0 \end{aligned} \tag{5.32}$$

Equations (5.32) represent two homogeneous algebraic equations in the unknowns u_1 and u_2, with ω^2 playing the role of a parameter. From linear algebra, Eqs. (5.32) possess a nontrivial solution only if the determinant of the coefficients of u_1 and u_2 is zero, or

$$\Delta(\omega^2) = \det \begin{bmatrix} k_{11}-\omega^2 m_1 & k_{12} \\ k_{12} & k_{22}-\omega^2 m_2 \end{bmatrix} = 0 \tag{5.33}$$

where $\Delta(\omega^2)$ is known as the *characteristic determinant*, or the *characteristic polynomial*, a polynomial of second degree in ω^2. Indeed, an expansion of the determinant yields

$$\Delta\omega^2 = m_1 m_2 \left[\omega^4 - \left(\frac{k_{11}}{m_1} + \frac{k_{22}}{m_2} \right) \omega^2 + \frac{k_{11}k_{22}-k_{12}^2}{m_1 m_2} \right] = 0 \tag{5.34}$$

which represents a quadratic equation in ω^2 called the *characteristic equation*, or *frequency equation*. The equation has the roots

$$\left.\begin{matrix} \omega_1^2 \\ \omega_2^2 \end{matrix}\right\} = \frac{1}{2}\left(\frac{k_{11}}{m_1} + \frac{k_{22}}{m_2} \right) \mp \frac{1}{2}\sqrt{\left(\frac{k_{11}}{m_1} + \frac{k_{22}}{m_2} \right)^2 - 4\frac{k_{11}k_{22}-k_{12}^2}{m_1 m_2}} \tag{5.35}$$

which are known as *eigenvalues*, or *characteristic values*. The square roots ω_1 and ω_2 of the eigenvalues represent the *natural frequencies* of the system. Hence synchronous harmonic motion can take place in only two ways, one with the frequency ω_1 and the other with the frequency ω_2. The natural frequencies ω_1 and ω_2 play a role for two-degree-of-freedom systems similar to that played by the natural frequency ω_n for single-degree-of-freedom systems. Inserting ω_1 and ω_2 into Eq. (5.31), in sequence, we conclude that the time-dependent functions associated with the synchronous motions have the form

$$f_1(t) = C_1 \cos(\omega_1 t - \phi_1), \quad f_2(t) = C_2 \cos(\omega_2 t - \phi_2) \tag{5.36}$$

Having established that there are two synchronous harmonic motions, one with the frequency ω_1 and the other with the frequency ω_2, we must answer the second half of the second question, which is concerned with the shape of the displacement configuration for each case. To this end, we let $\omega^2 = \omega_i^2$, $u_1 = u_{1i}$, $u_2 = u_{2i}$ $(i = 1, 2)$ in Eqs. (5.32), so that

$$\left.\begin{aligned} (k_{11}-\omega_i^2 m_1)u_{1i} + k_{12}u_{2i} &= 0 \\ k_{12}u_{1i} + (k_{22}-\omega_i^2 m_2)u_{2i} &= 0 \end{aligned}\right\} i = 1, 2 \tag{5.37}$$

and we observe that the first subscript in the two displacements u_{1i} and u_{2i} identifies the mass and the second indicates whether the displacement configuration oscillates with the frequency ω_1 or with the frequency ω_2. Equations (5.37) represent two sets, each consisting of two homogeneous algebraic equations, one for $i=1$ and the other for $i=2$. Because the equations are homogeneous, it is not possible to solve for both u_{1i} and u_{2i} uniquely, but only for the ratios u_{2i}/u_{1i} $(i=1,2)$. To solve for these ratios, it is possible to use either the first or the second of Eqs. (5.37), as they both yield the same result. Indeed, inserting the values of ω_1^2 and ω_2^2 obtained from Eq. (5.35) into Eqs. (5.37), in sequence, we can write

$$\begin{aligned}\frac{u_{21}}{u_{11}} &= -\frac{k_{11}-\omega_1^2 m_1}{k_{12}} = -\frac{k_{12}}{k_{22}-\omega_1^2 m_2}\\ \frac{u_{22}}{u_{12}} &= -\frac{k_{11}-\omega_2^2 m_1}{k_{12}} = -\frac{k_{12}}{k_{22}-\omega_2^2 m_2}\end{aligned} \tag{5.38}$$

The ratios u_{21}/u_{11} and u_{22}/u_{12} determine uniquely the shape of the displacement profile assumed by the system while it oscillates with the frequency ω_1 and ω_2, respectively. If one element in each ratio is assigned a certain arbitrary value, then the value of the other element in the ratio is determined automatically. The resulting pairs of numbers, u_{11} and u_{21} on the one hand and u_{12} and u_{22} on the other hand, can be exhibited in the form of the vectors

$$\mathbf{u}_1 = \begin{bmatrix} u_{11} \\ u_{21} \end{bmatrix}, \quad \mathbf{u}_2 = \begin{bmatrix} u_{12} \\ u_{22} \end{bmatrix} \tag{5.39}$$

where $\mathbf{u}_1$ and $\mathbf{u}_2$ are referred to as *modal vectors*, or *eigenvectors*, and less frequently as *characteristic vectors*. The natural frequency ω_1 and modal vector $\mathbf{u}_1$ constitute what is broadly known as the *first mode of vibration*, and ω_2 and $\mathbf{u}_2$ constitute the *second mode of vibration*. It is no coincidence that a two-degree-of-freedom system possesses two modes of vibration. Indeed, it is shown in Ch. 7 that the number of modes for multi-degree-of-freedom systems coincides with the number of degrees of freedom. The *natural modes of vibrations*, i.e., the natural frequencies and modal vectors, represent a characteristic property of the system, and they are unique for a given system except for the magnitude of the modal vectors, implying that the shape of the modal vectors is unique but the length is not. Indeed, because the problem is homogeneous, a modal vector multiplied by a constant represents the same modal vector. It is often convenient to render a modal vector unique by means of a process known as *normalization*. One common normalization scheme is to assign a given value to one of the components of the modal vector, typically to assign the value one to the component largest in magnitude, which amounts to dividing all components of the vector by the value largest in magnitude. Another very convenient and frequently used normalization scheme is to assign the value one to the magnitude of the vector, which implies division of all the vector components by the magnitude of the vector. Vectors of unit magnitude are called *unit vectors*. Following normalization, the natural modes are referred to as *normal modes*. Clearly, normalization is arbitrary and does not affect the mode shape, as all components of the normalized vector are changed in the same proportion. The main reason for normalization is convenience, such as in plotting when the vector components are too large or too small, or for certain developments calling for unit modal vectors.

The complete synchronous motions are obtained by introducing Eqs. (5.36) and $\mathbf{u}=\mathbf{u}_1$ and $\mathbf{u}=\mathbf{u}_2$ in Eq. (5.23) and writing

$$\begin{aligned} \mathbf{x}_1(t) &= f_1(t)\mathbf{u}_1 = C_1\mathbf{u}_1\cos(\omega_1 t-\phi_1) \\ \mathbf{x}_2(t) &= f_2(t)\mathbf{u}_2 = C_2\mathbf{u}_2\cos(\omega_2 t-\phi_2) \end{aligned} \tag{5.40}$$

We refer to $\mathbf{x}_1(t)$ and $\mathbf{x}_2(t)$ as *natural motions*, because they represent harmonic oscillations at the natural frequencies with the system configuration in the shape of the modal vectors, i.e., they represent vibration in the natural modes. Each of these natural motions can be excited independently of the other. In general, however, the free vibration of a conservative system is a superposition of the natural motions, or

$$\mathbf{x}(t)=\mathbf{x}_1(t)+\mathbf{x}_2(t)=C_1\cos(\omega_1 t-\phi_1)\mathbf{u}_1+C_2\cos(\omega_2 t-\phi_2)\mathbf{u}_2 \tag{5.41}$$

where the amplitudes C_1 and C_2 and the phase angles ϕ_1 and ϕ_2 are determined by the initial displacements $x_1(0),\ x_2(0)$ and the initial velocities $\dot{x}_1(0),\ \dot{x}_2(0)$. We discuss this subject in Sec. 5.4.

Example 5.1. The two-degree-of-freedom system of Fig. 5.4a consists of two masses on a string vibrating in the vertical plane. Let $m_1=m,\ m_2=2m,\ L_1=L_2=L,\ L_3=0.5L$ and determine the natural modes of vibration.

The stiffness coefficients for the system under consideration are given by Eqs. (5.13). Hence, inserting the given data into Eqs. (5.13), we have

$$\begin{aligned} k_{11} &= \frac{T}{L_1}+\frac{T}{L_2}=\frac{T}{L}+\frac{T}{L}=\frac{2T}{L} \\ k_{22} &= \frac{T}{L_2}+\frac{T}{L_3}=\frac{T}{L}+\frac{T}{0.5L}=\frac{3T}{L} \\ k_{12} &= -\frac{T}{L_2}=-\frac{T}{L} \end{aligned} \tag{a}$$

Introducing the mass coefficients $m_1=m,\ m_2=2m$ and the stiffness coefficients given by Eqs. (a) into Eq. (5.35), we obtain the roots of the characteristic equation

$$\begin{aligned} \left.\begin{matrix}\omega_1^2\\ \omega_2^2\end{matrix}\right\} &= \frac{1}{2}\left(\frac{k_{11}}{m_1}+\frac{k_{22}}{m_2}\right)\mp\frac{1}{2}\sqrt{\left(\frac{k_{11}}{m_1}+\frac{k_{22}}{m_2}\right)^2-4\frac{k_{11}k_{22}-k_{12}^2}{m_1m_2}} \\ &= \frac{1}{2}\left(\frac{2T}{ml}+\frac{3T}{2ml}\right)\mp\frac{1}{2}\sqrt{\left(\frac{2T}{mL}+\frac{3T}{2mL}\right)^2-4\left(\frac{2T}{L}\frac{3T}{L}-\frac{T^2}{L^2}\right)\frac{1}{2m^2}} \\ &= \left[\frac{7}{4}\mp\sqrt{\left(\frac{7}{4}\right)^2-\frac{5}{2}}\right]\frac{T}{mL}=\begin{cases}\dfrac{T}{mL}\\[2ex] \dfrac{5T}{2mL}\end{cases} \end{aligned} \tag{b}$$

so that the natural frequencies are simply

$$\omega_1=\sqrt{\frac{T}{mL}},\ \omega_2=\sqrt{\frac{5T}{2mL}}=1.581139\sqrt{\frac{T}{mL}}\ \text{(rad/s)} \tag{c}$$

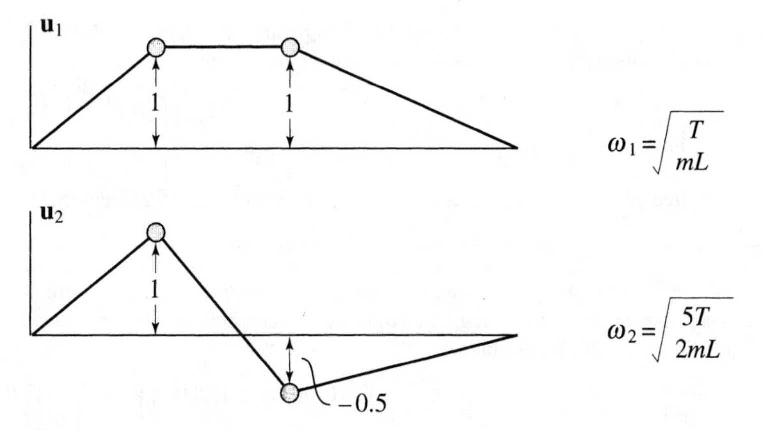

FIGURE 5.7
Modal vectors for the two-degree-of-freedom system of Fig. 5.4a

As pointed out earlier, only the shape of the modal vectors can be determined uniquely, but not the magnitude. The shape of each of the two modes is defined by the ratios u_{21}/u_{11} and u_{22}/u_{12}, as given by Eqs. (5.38). Hence, using Eqs. (5.38), we obtain

$$\frac{u_{21}}{u_{11}} = -\frac{k_{11} - \omega_1^2 m_1}{k_{12}} = -\frac{\dfrac{2T}{L} - \dfrac{T}{mL}m}{-\dfrac{T}{L}} = 1$$

$$\frac{u_{22}}{u_{12}} = -\frac{k_{11} - \omega_2^2 m_1}{k_{12}} = -\frac{\dfrac{2T}{L} - \dfrac{5T}{2mL}m}{-\dfrac{T}{L}} = -0.5 \tag{d}$$

Then, normalizing the modal vectors by letting arbitrarily $u_{11} = 1$, $u_{12} = 1$, which is convenient for plotting them, the natural modes can be written as

$$\mathbf{u}_1 = \begin{bmatrix} u_{11} \\ u_{21} \end{bmatrix} = \begin{bmatrix} 1 \\ 1 \end{bmatrix}, \quad \mathbf{u}_2 = \begin{bmatrix} u_{12} \\ u_{22} \end{bmatrix} = \begin{bmatrix} 1 \\ -0.5 \end{bmatrix} \tag{e}$$

The modal vectors are plotted in Fig. 5.7, in which the corresponding natural frequencies are also given.

We observe from Fig. 5.7 that the components of the first modal vector have the same sign. On the other hand, the components of the second modal vector have opposite signs. This implies that when the system vibrates in the second mode a point on the string between m_1 and m_2 remains at rest. This point is known as a *node*, and is typical of one-dimensional systems of the type under consideration.

Example 5.2. Consider the simplified model of an automobile shown in Fig. 5.5, let the parameters have the values $m = 1,500$ kg, $I_C = 2,000$ kg m^2, $k_1 = 36,000$ kg/m, $k_2 = 40,000$ kg/m, $a = 1.3$ m and $b = 1.7$ m, calculate the natural modes of the system and write an expression for the response.

To calculate the natural modes, we must solve the eigenvalue problem for the system, which is based on the free vibration equations, obtained by letting the force F be equal to zero in Eqs. (5.7). Hence, inserting the given parameter values into Eqs. (5.7) with $F=0$, we can write the free vibration equations in the matrix form

$$\begin{bmatrix} 1,500 & 0 \\ 0 & 2,000 \end{bmatrix} \begin{bmatrix} \ddot{x} \\ \ddot{\theta} \end{bmatrix} + \begin{bmatrix} 76,000 & 21,200 \\ 21,200 & 176,440 \end{bmatrix} \begin{bmatrix} x \\ \theta \end{bmatrix} = \begin{bmatrix} 0 \\ 0 \end{bmatrix} \tag{a}$$

But, free vibration is harmonic, so that by analogy with Eqs. (5.23) and (5.31) we can write

$$x(t) = X\cos(\omega t - \phi), \ \theta(t) = \Theta\cos(\omega t - \phi) \tag{b}$$

where X and Θ are constant amplitudes, ω is the frequency of the harmonic motion and ϕ a phase angle. Substituting Eqs. (b) in Eq. (a) and dividing through by $\cos(\omega t - \phi)$, we obtain the eigenvalue problem

$$-\omega^2 \begin{bmatrix} 1,500 & 0 \\ 0 & 2,000 \end{bmatrix} \begin{bmatrix} X \\ \Theta \end{bmatrix} + \begin{bmatrix} 76,000 & 21,200 \\ 21,200 & 176,440 \end{bmatrix} \begin{bmatrix} X \\ \Theta \end{bmatrix} = \begin{bmatrix} 0 \\ 0 \end{bmatrix} \tag{c}$$

Following the procedure described in this section, we first calculate the natural frequencies by solving the characteristic equation. Hence, according to Eq. (5.33), the characteristic equation for the problem at hand is

$$\Delta(\omega^2) = \det \begin{bmatrix} 76,000 - 1,500\omega^2 & 21,200 \\ 21,200 & 176,440 - 2,000\omega^2 \end{bmatrix}$$
$$= 3 \times 10^6(\omega^4 - 138.886667\omega^2 + 4{,}320) = 0 \tag{d}$$

which has the solutions

$$\left.\begin{matrix} \omega_1^2 \\ \omega_2^2 \end{matrix}\right\} = 69.443334 \mp \sqrt{69.443334^2 - 4{,}320}$$

$$= 69.443334 \mp 22.413758 = \begin{cases} 47.029576 \ (\text{rad/s})^2 \\ 91.857092 \ (\text{rad/s})^2 \end{cases} \tag{e}$$

so that the natural frequencies are

$$\omega_1 = 6.857811 \text{ rad/s}, \ \omega_2 = 9.584211 \text{ rad/s} \tag{f}$$

The natural modes can be obtained by replacing ω^2 by ω_1^2 and ω_2^2 in Eq. (c), in sequence, and solving for the pairs X_1, Θ_1 and X_2, Θ_2, respectively. However, as pointed out earlier in this section, because Eq. (c) is homogeneous, it is not possible to solve for X_i, Θ_i uniquely, but only for the ratios X_i/Θ_i $(i = 1, 2)$, or Θ_i/X_i $(i = 1, 2)$. To this end, we substitute $\omega = \omega_1^2$ in the top row of Eq. (c) and write

$$(76,000 - 1,500 \times 47.029576)X_1 + 21,200\Theta_1 = 0 \tag{g}$$

which yields

$$\frac{\Theta_1}{X_1} = -\frac{5{,}455.636}{21,200} = -0.257341 \text{ rad/m} \tag{h}$$

Similarly, introducing $\omega^2 = \omega_2^2$ in the top row of Eq. (c), we have

$$(76,000 - 1,500 \times 91.857092)X_2 + 21,200\Theta_2 = 0 \tag{i}$$

from which we obtain

$$\frac{\Theta_2}{X_2} = \frac{61,785.638}{21,200} = 2.914417 \text{ rad/m} \tag{j}$$

Hence, letting arbitrarily $X_1 = 1$, $X_2 = 1$, the natural modes become

$$\mathbf{u}_1 = \begin{bmatrix} X_1 \\ \Theta_1 \end{bmatrix} = \begin{bmatrix} 1 \\ -0.257341 \end{bmatrix}, \quad \mathbf{u}_2 = \begin{bmatrix} X_2 \\ \Theta_2 \end{bmatrix} = \begin{bmatrix} 1 \\ 2.914417 \end{bmatrix} \tag{k}$$

Note that the same results would have been obtained had we used the second row of Eq. (c) instead of the first. The modes are plotted in Fig. 5.8. Comparing Fig. 5.8 with Fig. 5.7, we conclude that the modes of the slab on two springs are not as easy to visualize as the modes of the two masses on a string, which can be attributed to the fact that one component of the modal vectors represents a translational displacement and the second an angular displacement.

Inserting the computed natural frequencies, Eqs. (f), and the modal vectors, Eqs. (k), into Eq. (5.41), we can express the response in the general form

$$\mathbf{x}(t) = C_1 \cos(6.857811t - \phi_1) \begin{bmatrix} 1 \\ -0.257341 \end{bmatrix}$$

$$+ C_2 \cos(9.584211t - \phi_2) \begin{bmatrix} 1 \\ 2.914417 \end{bmatrix} \tag{l}$$

where the amplitudes C_1, C_2 and the phase angles ϕ_1, ϕ_2 must be determined from the initial displacements $x(0)$, $\theta(0)$ and the initial velocities $\dot{x}(0)$, $\dot{\theta}(0)$, as will be shown in Sec. 5.4.

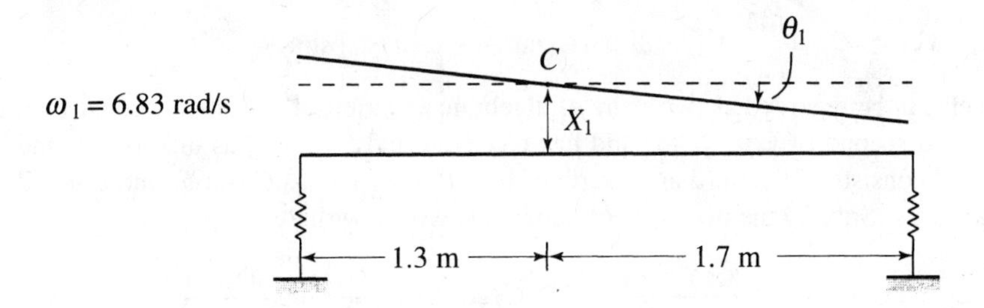

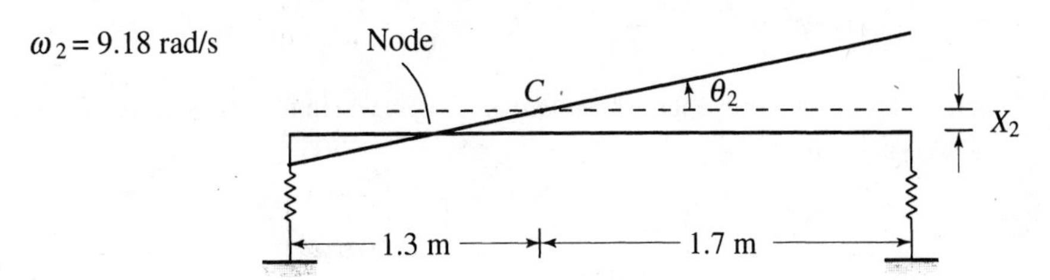

FIGURE 5.8
Natural modes for the automobile model of Fig. 5.5

5.4 RESPONSE TO INITIAL EXCITATIONS

Equation (5.41) gives the free response of conservative two-degree-of-freedom systems in the form of a linear combination of natural motions, defined as products of the modal vectors $\mathbf{u}_1$ and $\mathbf{u}_2$ and the harmonic functions $f_1(t)$ and $f_2(t)$, respectively, where the latter have frequencies equal to the natural frequencies. The modal vectors, and in particular the mode shapes, and the natural frequencies represent a characteristic of the system, in the sense that they are unique for a given system, and they are a reflection of the system parameters, as can be concluded from Eqs. (5.35) and (5.38). The harmonic functions $f_1(t)$ and $f_2(t)$ contain the constants C_1, ϕ_1 and C_2, ϕ_2, respectively, where C_1 and C_2 represent amplitudes and ϕ_1 and ϕ_2 represent phase angles. Unlike the natural frequencies and natural modes, which can be regarded as depending on internal factors, the constants C_1, C_2, ϕ_1 and ϕ_2 depend on external factors, namely, the initial excitations.

To obtain the response to the initial excitations, it is necessary to determine the value of the constants C_1, C_2, ϕ_1 and ϕ_2. To this end, we introduce the notation

$$\mathbf{x}(0) = \begin{bmatrix} x_{10} \\ x_{20} \end{bmatrix}, \quad \dot{\mathbf{x}}(0) = \mathbf{v}(0) = \begin{bmatrix} v_{10} \\ v_{20} \end{bmatrix} \tag{5.42}$$

Then, letting $t = 0$ in Eq. (5.41) and its time derivative, recalling Eqs. (5.39) and inserting the results into Eqs. (5.42), we obtain

$$\begin{aligned} x_{10} &= u_{11}C_1 \cos\phi_1 + u_{12}C_2 \cos\phi_2 \\ x_{20} &= u_{21}C_1 \cos\phi_1 + u_{22}C_2 \cos\phi_2 \\ v_{10} &= \omega_1 u_{11}C_1 \sin\phi_1 + \omega_2 u_{12}C_2 \sin\phi_2 \\ v_{20} &= \omega_1 u_{21}C_1 \sin\phi_1 + \omega_2 u_{22}C_2 \sin\phi_2 \end{aligned} \tag{5.43}$$

which can be regarded as two pairs of algebraic equations, the first pair consists of the first and second of Eqs. (5.43) and has $C_1 \cos\phi_1$ and $C_2 \cos\phi_2$ as unknowns, and the second consists of the third and fourth of Eqs. (5.43) and has $C_1 \sin\phi_1$ and $C_2 \sin\phi_2$ as unknowns. Solving the two pairs of equations, we can write

$$\begin{aligned} C_1 \cos\phi_1 &= \frac{u_{22}x_{10} - u_{12}x_{20}}{|U|}, \quad C_2 \cos\phi_2 = \frac{u_{11}x_{20} - u_{21}x_{10}}{|U|} \\ C_1 \sin\phi_1 &= \frac{u_{22}v_{10} - u_{12}v_{20}}{\omega_1 |U|}, \quad C_2 \sin\phi_2 = \frac{u_{11}v_{20} - u_{21}v_{10}}{\omega_2 |U|} \end{aligned} \tag{5.44}$$

where $|U|$ is the determinant of the *modal matrix* U, which for a two-degree-of-freedom system is defined as

$$U = \begin{bmatrix} \mathbf{u}_1 & \mathbf{u}_2 \end{bmatrix} = \begin{bmatrix} u_{11} & u_{12} \\ u_{21} & u_{22} \end{bmatrix} \tag{5.45}$$

Equations (5.44) can be solved for C_1, C_2, ϕ_1 and ϕ_2 explicitly. It turns out that this is not really necessary. Indeed, expanding $\cos(\omega_i t - \phi_i)$ $(i = 1, 2)$ in Eq. (5.41) and using

Eqs. (5.44), we obtain the response to the initial excitations directly as follows:

$$\mathbf{x}(t) = C_1(\cos\omega_1 t\cos\phi_1 + \sin\omega_1 t\sin\phi_1)\mathbf{u}_1 + C_2(\cos\omega_2 t\cos\phi_2 + \sin\omega_2\sin\phi_2)\mathbf{u}_2$$

$$= \frac{1}{|U|}\left\{\left[(u_{22}x_{10} - u_{12}x_{20})\cos\omega_1 t + \frac{u_{22}v_{10} - u_{12}v_{20}}{\omega_1}\sin\omega_1 t\right]\mathbf{u}_1 + \left[(u_{11}x_{20} - u_{21}x_{10})\cos\omega_2 t + \frac{u_{11}v_{20} - u_{21}v_{10}}{\omega_2}\sin\omega_2 t\right]\mathbf{u}_2\right\} \quad (5.46)$$

A MATLAB program for plotting the response of two-degree-of-freedom systems to initial excitations based on Eq. (5.46) is presented in Sec. 5.12.

The approach of this section becomes impractical for systems with more than two degrees of freedom. A more general approach to the response to initial excitations, one based on modal analysis and capable of accommodating systems with an arbitrary number of degrees of freedom, is discussed in Sec. 7.10.

Example 5.3. Obtain the response of the two-degree-of-freedom system of Example 5.1 to the initial displacement $x_{10} = 1.2$ cm. The other initial conditions are zero.

From Example 5.1, we obtain the natural frequencies

$$\omega_1 = \sqrt{\frac{T}{mL}}, \quad \omega_2 = 1.581139\sqrt{\frac{T}{mL}} \text{ (rad/s)} \quad \text{(a)}$$

and the modal matrix

$$U = \begin{bmatrix} u_{11} & u_{12} \\ u_{21} & u_{22} \end{bmatrix} = \begin{bmatrix} 1 & 1 \\ 1 & -0.5 \end{bmatrix} \quad \text{(b)}$$

The determinant of the modal matrix is simply

$$|U| = u_{11}u_{22} - u_{12}u_{21} = -1.5 \quad \text{(c)}$$

Hence, inserting Eqs. (a)–(c) into Eq. (5.46), we can write the response in the form

$$\mathbf{x}(t) = \begin{bmatrix} x_1(t) \\ x_2(t) \end{bmatrix} = \frac{1}{-1.5}\left\{(-0.5)\times 1.2\cos\sqrt{\frac{T}{mL}}t\begin{bmatrix} 1 \\ 1 \end{bmatrix} - 1\times 1.2\cos 1.581139\sqrt{\frac{T}{mL}}t\begin{bmatrix} 1 \\ -0.5 \end{bmatrix}\right\}$$

$$= \begin{bmatrix} 0.4\cos\sqrt{\frac{T}{mL}}t + 0.8\cos 1.581139\sqrt{\frac{T}{mL}}t \\ 0.4\cos\sqrt{\frac{T}{mL}}t - 0.4\cos 1.581139\sqrt{\frac{T}{mL}}t \end{bmatrix} \text{ (cm)} \quad \text{(d)}$$

The responses $x_1(t)$ versus t and $x_2(t)$ versus t for the system of this example can be plotted using the MATLAB program entitled 'tdotin.m' given in Sec. 5.12.

5.5 COORDINATE TRANSFORMATIONS. COUPLING

The equations of motion of a conservative two-degree-of-freedom system given by Eqs. (5.21) and (5.22) are characterized by the fact that the mass matrix is diagonal. This is typical of many dynamical systems when the coordinates used to describe the motion represent the displacements of the masses. On the other hand, the stiffness matrix is not diagonal. Were the stiffness matrix also diagonal, then the two equations of motion would have been independent, an ideal state of affairs, as in this case the two equations could be solved as if they were representing two separate single-degree-of-freedom systems. Because the stiffness matrix is not diagonal, the two equations are simultaneous, i.e., they are *coupled.*

To explore the concept of coupling a little further, we consider the two-degree-of-freedom system of Fig. 5.9. It is not difficult to show that the equations of motion in this case can be written in the form

$$M\ddot{\mathbf{x}}(t) + K\mathbf{x}(t) = \mathbf{0} \tag{5.47}$$

where $\mathbf{x}(t) = [x_1(t)\ x_2(t)]^T$ and

$$M = \begin{bmatrix} m_1 & 0 \\ 0 & m_2 \end{bmatrix}, \quad K = \begin{bmatrix} k_1 + k_2 & -k_2 \\ -k_2 & k_2 \end{bmatrix} \tag{5.48}$$

so that once again the mass matrix is diagonal and the stiffness matrix is not. Next, we wish to describe the motion of the system by means of a different set of coordinates, namely, the elongations of the springs k_1 and k_2, denoted by $z_1(t)$ and $z_2(t)$, respectively. Because they describe the motion of the same system, the two sets of coordinates $x_1(t),\ x_2(t)$ and $z_1(t),\ z_2(t)$ are related. Indeed, the displacement of m_1 is equal to the elongation of k_1 and the displacement of m_2 is equal to the sum of elongations of k_1 and k_2, or

$$x_1(t) = z_1(t), \quad x_2(t) = z_1(t) + z_2(t) \tag{5.49}$$

Equations (5.49) describe a *coordinate transformation*, which can be expressed in the matrix form

$$\mathbf{x}(t) = T\mathbf{z}(t) \tag{5.50}$$

where

$$T = \begin{bmatrix} 1 & 0 \\ 1 & 1 \end{bmatrix} \tag{5.51}$$

represents the *transformation matrix*. Next, we propose to derive the equations of motion in terms of the elongations $z_1(t)$ and $z_2(t)$. To this end, we introduce Eqs. (5.50) and

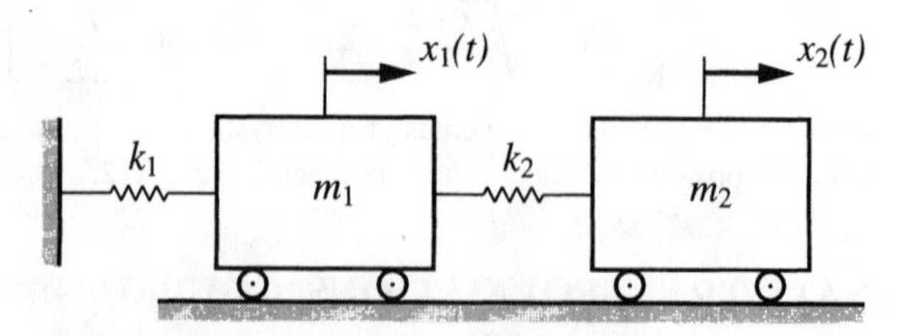

FIGURE 5.9
Undamped two-degree-of-freedom system

(5.51) in Eq. (5.48). Then, to retain the symmetry of the formulation, we premultiply the result by T^T, which yields

$$M'\ddot{\mathbf{z}}(t) + K'\mathbf{z}(t) = \mathbf{0} \tag{5.52}$$

in which the new mass and stiffness matrices are given by

$$\begin{aligned} M' = T^T M T &= \begin{bmatrix} 1 & 1 \\ 0 & 1 \end{bmatrix} \begin{bmatrix} m_1 & 0 \\ 0 & m_2 \end{bmatrix} \begin{bmatrix} 1 & 0 \\ 1 & 1 \end{bmatrix} = \begin{bmatrix} m_1 + m_2 & m_2 \\ m_2 & m_2 \end{bmatrix} \\ K' = T^T K T &= \begin{bmatrix} 1 & 1 \\ 0 & 1 \end{bmatrix} \begin{bmatrix} k_1 + k_2 & -k_2 \\ -k_2 & k_2 \end{bmatrix} \begin{bmatrix} 1 & 0 \\ 1 & 1 \end{bmatrix} = \begin{bmatrix} k_1 & 0 \\ 0 & k_2 \end{bmatrix} \end{aligned} \tag{5.53}$$

Hence, when the motion is expressed in terms of the elongations of the springs the stiffness matrix is diagonal and the mass matrix is not.

It will prove of interest to examine the matter of coupling in the context of the two-degree-of-freedom system consisting of the slab supported by two springs shown in Fig. 5.5. We recall that the equations of motion for this system were derived in Sec. 5.2 in the form of Eqs. (5.7), the corresponding mass and stiffness matrices being given by Eqs. (5.16). Hence, with the coordinate $x(t)$ representing the vertical translation of the mass center C and $\theta(t)$ denoting the rotation of the slab, the mass matrix is diagonal and the stiffness matrix is not. Next, we define the motion in terms of the vertical translation $x_1(t)$ of point O on the slab and the rotation $\theta(t)$, where O lies at distances a_1 and b_1 from the springs k_1 and k_2, respectively. Point O is not arbitrary but chosen so that a vertical force acting at O causes the slab to undergo pure translation, as depicted in Fig. 5.10a. For this to happen, the moment about O must be zero, which implies that a_1 and b_1 must satisfy the condition

$$k_1 x_1 a_1 = k_2 x_1 b_1 \tag{5.54}$$

To obtain the equations of motion in terms of $x_1(t)$ and $\theta(t)$, we transform Eqs. (5.7) by means of the procedure just used. To this end, we refer to Fig. 5.10b and observe that the relation between $x(t)$ and $x_1(t)$ is

$$x(t) = x_1(t) + e\theta(t) \tag{5.55}$$

where e is the distance between C and O. Then, adjoining the identity $\theta(t) = \theta(t)$, we can write the transformation between the coordinates $x(t)$, $\theta(t)$ and $x_1(t)$, $\theta(t)$ in the matrix form

$$\mathbf{x}(t) = T\mathbf{x}_1(t) \tag{5.56}$$

in which the coordinate vectors are $\mathbf{x} = [x\ \theta]^T$, so that $\mathbf{x}_1 = [x_1\ \theta]^T$, so that the transformation matrix is

$$T = \begin{bmatrix} 1 & e \\ 0 & 1 \end{bmatrix} \tag{5.57}$$

Hence, by analogy with Eqs. (5.52) and (5.53), if we use Eqs. (5.16) and (5.57), the free vibration equations can be written in the compact form

$$M_1\ddot{\mathbf{x}}_1(t) + K_1\mathbf{x}_1(t) = \mathbf{0} \tag{5.58}$$

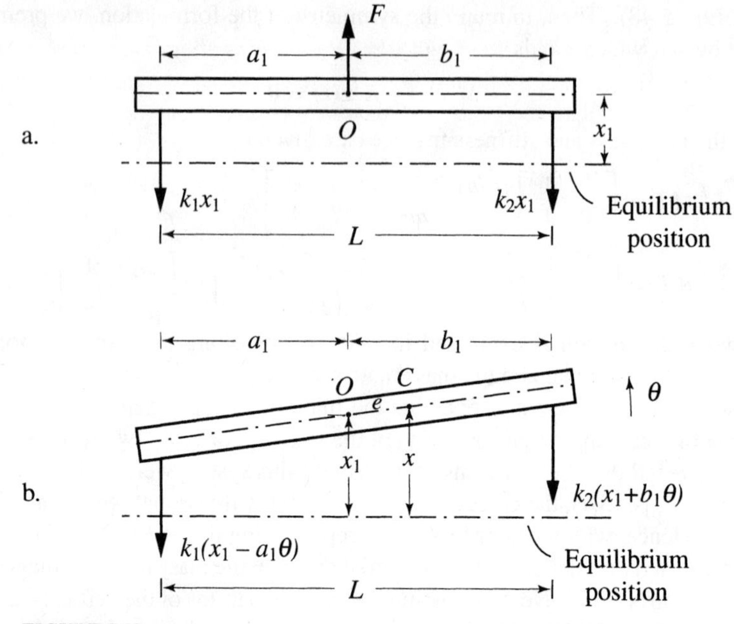

FIGURE 5.10
a. The automobile model of Fig. 5.5 in pure translation, b. The same model in translation and rotation

where, recognizing that $a-e=a_1$, $b+e=b_1$ and using Eq. (5.54),

$$M_1 = T^T MT = \begin{bmatrix} 1 & 0 \\ e & 1 \end{bmatrix} \begin{bmatrix} m & 0 \\ 0 & I_C \end{bmatrix} \begin{bmatrix} 1 & e \\ 0 & 1 \end{bmatrix} = \begin{bmatrix} m & em \\ em & I_O \end{bmatrix}$$

$$K_1 = T^T KT = \begin{bmatrix} 1 & 0 \\ e & 1 \end{bmatrix} \begin{bmatrix} k_1+k_2 & -(k_1a-k_2b) \\ -(k_1a-k_2b) & k_1a^2+k_2b^2 \end{bmatrix} \begin{bmatrix} 1 & e \\ 0 & 1 \end{bmatrix} \tag{5.59}$$

$$= \begin{bmatrix} k_1+k_2 & 0 \\ 0 & k_1a_1^2+k_2b_1^2 \end{bmatrix}$$

in which $I_O = I_C + me^2$ is the mass moment of inertia of the slab about point O. It follows that in this case also the new stiffness matrix is diagonal and the new mass matrix is not.

We refer to the case in which the mass matrix is diagonal and the stiffness matrix is not as *elastically coupled.* On the other hand, when the stiffness matrix is diagonal and the mass matrix is not the system is said to be *inertially coupled.* In this sense, the effect of using the coordinate transformation given by Eqs. (5.50) and (5.51), or that given by Eqs. (5.56) and (5.57), is to transform coupling from elastic to inertial. As far as solving the equations of motion, we are not better off with a system of equations defined by matrices M' and K', or by matrices M_1 and K_1, than with a system in terms of matrices M and K, as in all cases the system is coupled. However, this exercise helped us recognize one fact, namely, *coupling is not an inherent characteristic property of the system, but of the coordinates used to describe the motion of the system.* For a coordinate transformation to justify the effort it must facilitate the solution of the equations of motion, which implies that it must remove both the dynamic and the elastic coupling from the system at the same time. In mathematical terms, this is the same as saying that the coordinate transformation must diagonalize the mass and stiffness matrices simultaneously. We state here, and demonstrate in Sec. 5.6, that such a coordinate transformation does indeed exist and that the transformation matrix is the modal matrix, Eq. (5.45). Moreover, the coordinates corresponding to the independent equations of motion are known as *natural coordinates*, or *principal coordinates*. Unlike any other coordinates, *the natural coordinates are unique* for a given system. In the special case of free vibration, the natural coordinates coincide with the harmonic functions $f_1(t)$ and $f_2(t)$ given by Eqs. (5.36).

5.6 ORTHOGONALITY OF MODES. NATURAL COORDINATES

The importance of the modal vectors, first introduced in Sec. 5.3, goes well beyond the fact that they represent configuration vectors experienced by conservative systems vibrating freely in synchronous motion. Indeed, they are indispensable to the solution of the forced vibration problem for conservative systems. This statement extends to the free vibration solution for systems with more than two degrees of freedom. The modal vectors owe this pivotal position to a remarkable property known as *orthogonality*. There are many sets of orthogonal vectors, but the modal vectors are unique in that they are the only ones orthogonal with respect to both the mass matrix and the stiffness matrix. As a result, a coordinate transformation based on the modal vectors is capable of diagonalizing the mass and stiffness matrices simultaneously, thus decoupling the differential equations of motion. This permits a solution of the equations of motion for multi-degree-of-freedom systems with the same ease as the solution of the differential equation of a single-degree-of-freedom system.

From Eqs. (5.38) and (5.39), the modal vectors for the two-degree-of-freedom system defined by Eqs. (5.21) and (5.22) can be expressed as

$$\mathbf{u}_1 = u_{11}\begin{bmatrix} 1 \\ -(k_{11}-\omega_1^2 m_1)/k_{12} \end{bmatrix}, \quad \mathbf{u}_2 = u_{12}\begin{bmatrix} 1 \\ -(k_{11}-\omega_2^2 m_1)/k_{12} \end{bmatrix} \tag{5.60}$$

where ω_1^2 and ω_2^2 are given by Eq. (5.35). Next, we use the first of Eqs. (5.22) and form the matrix product

$$\begin{aligned} \mathbf{u}_2^T M \mathbf{u}_1 &= u_{11}u_{12}\begin{bmatrix} 1 \\ -(k_{11}-\omega_2^2 m_1)/k_{12} \end{bmatrix}^T \begin{bmatrix} m_1 & 0 \\ 0 & m_2 \end{bmatrix} \begin{bmatrix} 1 \\ -(k_{11}-\omega_1^2 m_1)/k_{12} \end{bmatrix} \\ &= \frac{u_{11}u_{12}}{k_{12}^2}\left[m_1 k_{12}^2 + m_2(k_{11}-\omega_1^2 m_1)(k_{11}-\omega_2^2 m_1)\right] \\ &= \frac{u_{11}u_{12}}{k_{12}^2}\left[m_1^2 m_2 \omega_1^2 \omega_2^2 - m_1 m_2 k_{11}(\omega_1^2+\omega_2^2) + m_1 k_{12}^2 + m_2 k_{11}^2\right] \end{aligned} \tag{5.61}$$

which is a scalar. But, from Eqs. (5.35), we can write

$$\omega_1^2\omega_2^2 = \frac{k_{11}k_{22} - k_{12}^2}{m_1 m_2}, \ \omega_1^2 + \omega_2^2 = \frac{m_1 k_{22} + m_2 k_{11}}{m_1 m_2} \tag{5.62}$$

Inserting Eqs. (5.62) into Eq. (5.61), we conclude that

$$\mathbf{u}_2^T M \mathbf{u}_1 = \mathbf{u}_1^T M \mathbf{u}_2 = 0 \tag{5.63}$$

where the second equality is true because M is diagonal, and hence symmetric by definition. Equation (5.63) states that *the modal vectors* $\mathbf{u}_1$ *and* $\mathbf{u}_2$ *are orthogonal with respect to the mass matrix* M, where M plays the role of a weighting matrix. We point out that orthogonality with respect to a weighting matrix is different from ordinary orthogonality, which involves no weighting matrix. Similarly, we use the second of Eqs. (5.22) and write

$$\mathbf{u}_2^T K \mathbf{u}_1 = u_{11}u_{12} \begin{bmatrix} 1 \\ -(k_{11} - \omega_2^2 m_1)/k_{12} \end{bmatrix}^T \begin{bmatrix} k_{11} & k_{12} \\ k_{12} & k_{22} \end{bmatrix} \begin{bmatrix} 1 \\ -(k_{11} - \omega_1^2 m_1)/k_{12} \end{bmatrix}$$

$$= \frac{1}{k_{12}^2}\left[m_1^2 k_{22}\omega_1^2\omega_2^2 - m_1(k_{11}k_{22} - k_{12}^2)(\omega_1^2 + \omega_2^2) + k_{11}(k_{11}k_{22} - k_{12}^2)\right] \tag{5.64}$$

Then, using Eqs. (5.62), it can be verified that

$$\mathbf{u}_2^T K \mathbf{u}_1 = \mathbf{u}_1^T K \mathbf{u}_2 = 0 \tag{5.65}$$

so that *the modal vectors* $\mathbf{u}_1$ *and* $\mathbf{u}_2$ *are orthogonal with respect to the stiffness matrix* K *as well.* We observe that the second equality follows from the symmetry of K.

Next, we insert Eq. (5.29) into Eq. (5.28) and write the solutions of the algebraic eigenvalue problem as follows:

$$\begin{aligned} K\mathbf{u}_1 &= \omega_1^2 M \mathbf{u}_1 \\ K\mathbf{u}_2 &= \omega_2^2 M \mathbf{u}_2 \end{aligned} \tag{5.66}$$

Then, we premultiply the first of Eqs. (5.66) by $\mathbf{u}_1^T$ and the second by $\mathbf{u}_2^T$ and write the scalar relations

$$\begin{aligned} \mathbf{u}_1^T K \mathbf{u}_1 &= \omega_1^2 \mathbf{u}_1^T M \mathbf{u}_1 \\ \mathbf{u}_2^T K \mathbf{u}_2 &= \omega_2^2 \mathbf{u}_2^T M \mathbf{u}_2 \end{aligned} \tag{5.67}$$

It will prove convenient to introduce the notation

$$\begin{aligned} \mathbf{u}_1^T M \mathbf{u}_1 = m'_{11}, \ \mathbf{u}_2^T M \mathbf{u}_2 = m'_{22} \\ \mathbf{u}_1^T K \mathbf{u}_1 = k'_{11}, \ \mathbf{u}_2^T K \mathbf{u}_2 = k'_{22} \end{aligned} \tag{5.68}$$

which permits us to express the natural frequencies squared in the the form of the ratios

$$\omega_1^2 = \frac{k'_{11}}{m'_{11}}, \ \omega_2^2 = \frac{k'_{22}}{m'_{22}} \tag{5.69}$$

At this point, we are ready to demonstrate some of the statements made in the beginning of this section. To this end, it is convenient to rewrite Eq. (5.21) as

$$M\ddot{\mathbf{x}}(t) + K\mathbf{x}(t) = \mathbf{0} \tag{5.70}$$

where M and K are given by Eqs. (5.22), and express the solution as the linear combination

$$\mathbf{x}(t) = q_1(t)\mathbf{u}_1 + q_2(t)\mathbf{u}_2 \tag{5.71}$$

where $q_1(t)$ and $q_2(t)$ are two functions of time yet to be determined. Inserting Eq. (5.71) into Eq. (5.70), we have

$$M[\ddot{q}_1(t)\mathbf{u}_1 + \ddot{q}_2(t)\mathbf{u}_2] + K[q_1(t)\mathbf{u}_1 + q_2(t)\mathbf{u}_2] = \mathbf{0} \tag{5.72}$$

Next, we premultiply Eq. (5.72) by $\mathbf{u}_1^T$ and $\mathbf{u}_2^T$, in sequence, consider Eqs. (5.63), (5.65), (5.68) and (5.69), divide through the results by m'_{11} and m'_{22}, respectively, and obtain the two independent equations

$$\begin{aligned} \ddot{q}_1(t) + \omega_1^2 q_1(t) &= 0 \\ \ddot{q}_2(t) + \omega_2^2 q_2(t) &= 0 \end{aligned} \tag{5.73}$$

which are known as *modal equations*. Hence, a linear transformation in terms of the modal vectors, Eq. (5.71), in conjunction with the orthogonality of the modal vectors, enables us to transform a set of simultaneous equations of motion, Eq. (5.70), into a set of independent modal equations, Eqs. (5.73), each one resembling the equation of a harmonic oscillator, Eq. (2.2). In essence, the process is tantamount to simultaneous diagonalization of the mass and stiffness matrices, i.e., to both inertial and elastic decoupling of the equations of motion. The variables $q_1(t)$ and $q_2(t)$ defining the decoupled equations, Eqs. (5.73), are called *natural coordinates*, or *principal coordinates*.

In view of the fact that Eqs. (5.73) resemble Eq. (2.2), their solution can be obtained by invoking the analogy with Eq. (2.10), representing the free response of a harmonic oscillator, and writing

$$q_1(t) = C_1 \cos(\omega_1 t - \phi_1), \; q_2(t) = C_2 \cos(\omega_2 t - \phi_2) \tag{5.74}$$

where C_1, C_2 are amplitudes and ϕ_1, ϕ_2 are phase angles. Comparing Eqs. (5.74) with Eqs. (5.36), we conclude that the harmonic functions $f_1(t)$ and $f_2(t)$ multiplying the configuration vectors in synchronous motions are precisely the natural coordinates $q_1(t)$ and $q_2(t)$. Moreover, inserting Eqs. (5.73) into Eq. (5.71), we obtain the actual displacement vector in the form

$$\mathbf{x}(t) = C_1 \cos(\omega_1 t - \phi_1)\mathbf{u}_1 + C_2 \cos(\omega_2 t - \phi_2)\mathbf{u}_2 \tag{5.75}$$

which coincides with the solution given by Eq. (5.41) obtained earlier. Equation (5.75) expresses the fact that *the free response of conservative systems is a superposition of the natural modes multiplied by the natural coordinates*.

The same modal approach can be used to solve for the response of undamped systems to applied forces. In this more general case the coordinates $q_1(t)$ and $q_2(t)$ are known as *modal coordinates*. This process is known as *modal analysis* and is discussed in Sec. 5.10.

It was indicated in the end of Sec. 5.5 that a coordinate transformation capable of decoupling the equations of motion both inertially and elastically exists. This is the same as stating that the transformation can diagonalize the mass matrix M and the stiffness matrix K simultaneously. In view of the foregoing discussion, it is now clear that the transformation matrix postulated in Sec. 5.5 is $T = U = [\mathbf{u}_1 \;\; \mathbf{u}_2]$, where U is the *modal matrix*. The coordinate transformation using the modal matrix forms the basis for modal analysis. The real power of modal analysis becomes evident in the case of multi-degree-of-freedom systems. This is demonstrated in Ch. 7, in which modal analysis is presented in a more formal manner than here.

Example 5.4. Obtain the response to initial excitations of the system of Examples 5.1 and 5.3 by means of modal analysis.

Using data from Example 5.1, the equations of motion can be written in the matrix form

$$M\ddot{\mathbf{x}}(t) + K\mathbf{x}(t) = \mathbf{0} \tag{a}$$

where the mass and stiffness matrices are given by

$$M = \begin{bmatrix} m_1 & 0 \\ 0 & m_2 \end{bmatrix} = m\begin{bmatrix} 1 & 0 \\ 0 & 2 \end{bmatrix}, \quad K = \begin{bmatrix} k_{11} & k_{12} \\ k_{12} & k_{22} \end{bmatrix} = \frac{T}{L}\begin{bmatrix} 2 & -1 \\ -1 & 3 \end{bmatrix} \tag{b}$$

respectively. Equation (a) is subject to the initial conditions

$$\mathbf{x}(0) = \begin{bmatrix} 1.2 \\ 0 \end{bmatrix}, \; \dot{\mathbf{x}}(0) = \mathbf{0} \tag{c}$$

Moreover, the natural frequencies and modal vectors are

$$\omega_1 = \sqrt{\frac{T}{mL}}, \; \omega_2 = \sqrt{\frac{5T}{2mL}} = 1.581139\sqrt{\frac{T}{mL}} \tag{d}$$

and

$$\mathbf{u}_1 = \begin{bmatrix} 1 \\ 1 \end{bmatrix}, \; \mathbf{u}_2 = \begin{bmatrix} 1 \\ -0.5 \end{bmatrix} \tag{e}$$

respectively.

Introducing the linear transformation

$$\mathbf{x}(t) = q_1(t)\mathbf{u}_1 + q_2(t)\mathbf{u}_2 = q_1(t)\begin{bmatrix} 1 \\ 1 \end{bmatrix} + q_2(t)\begin{bmatrix} 1 \\ -0.5 \end{bmatrix} \tag{f}$$

in Eq. (a) and premultiplying the result by $\mathbf{u}_1^T$ and $\mathbf{u}_2^T$, in sequence, we obtain the modal equations

$$\begin{aligned} \ddot{q}_1(t) + \sqrt{\frac{T}{mL}}q_1(t) &= 0 \\ \ddot{q}_2(t) + \sqrt{\frac{5T}{2mL}}q_2(t) &= 0 \end{aligned} \tag{g}$$

which are subject to the initial modal displacements $q_i(0)$ $(i = 1, 2)$; the initial modal velocities are zero by virtue of the fact that $\dot{\mathbf{x}}(0) = \mathbf{0}$.

To determine the initial modal displacements, we let $t = 0$ in Eq. (f), consider the first of Eqs. (c) and write

$$\mathbf{x}(0) = \begin{bmatrix} 1.2 \\ 0 \end{bmatrix} = q_1(0)\mathbf{u}_1 + q_2(0)\mathbf{u}_2 = q_1(0)\begin{bmatrix} 1 \\ 1 \end{bmatrix} + q_2(0)\begin{bmatrix} 1 \\ -0.5 \end{bmatrix} \tag{h}$$

Equation (h) represents two algebraic equations in the unknowns $q_1(0)$ and $q_2(0)$. Their solution is simply

$$q_1(0) = 0.4, \; q_2(0) = 0.8 \tag{i}$$

In the case of a two-degree-of-freedom system, which is the case at hand, it is relatively easy to obtain the initial modal conditions from the actual initial conditions. In the case of multi-degree-of-freedom systems, the initial modal conditions are obtained by a more formal approach, namely, one that uses the orthogonality of the modal vectors with respect to the mass matrix. This approach is presented in Ch. 7.

Considering Eqs. (d) and (i), recalling that $\dot{q}_1(0) = \dot{q}_2(0) = 0$ and using the analogy with Eq. (2.13), the solution of Eqs. (g) is

$$\begin{aligned} q_1(t) &= q_1(0)\cos\omega_1 t = 0.4\cos\sqrt{\frac{T}{mL}}t \\ q_2(t) &= q_2(0)\cos\omega_2 t = 0.8\cos 1.581139\sqrt{\frac{T}{mL}}t \end{aligned} \tag{j}$$

The response to the specified initial excitation is obtained by inserting Eqs. (j) into Eq. (f), which yields

$$\mathbf{x}(t) = 0.4\cos\sqrt{\frac{T}{mL}}t\begin{bmatrix} 1 \\ 1 \end{bmatrix} + 0.8\cos 1.581139\sqrt{\frac{T}{mL}}t\begin{bmatrix} 1 \\ -0.5 \end{bmatrix} \tag{k}$$

It can be easily verified that this is the same result as that obtained in Example 5.3.

5.7 BEAT PHENOMENON

When two identical single-degree-of-freedom systems are connected by means of a weak spring the resulting two-degree-of-freedom system is characterized by natural frequencies very close in value. The response of such a two-degree-of-freedom system to a certain initial excitation exhibits a phenomenon known as the *beat phenomenon*, whereby the displacement of one mass decreases from some maximum value to zero, while the displacement of the other mass increases from zero to the same maximum value, and then the roles are reversed, with the pattern repeating itself continuously.

To illustrate the beat phenomenon, we consider a system consisting of two identical pendulums connected by a spring, as shown in Fig. 5.11a. For the time being we make no assumption concerning the value of the spring constant. The corresponding free-body diagrams for the two masses are shown in Fig. 5.11b, in which the assumption of small angles θ_1 and θ_2 is implied. The moment equations of pendulums 1 and 2 about the points O and O', respectively, yield the differential equations of motion

$$\begin{aligned} mL^2\ddot{\theta}_1 + mgL\theta_1 + ka^2(\theta_1 - \theta_2) &= 0 \\ mL^2\ddot{\theta}_2 + mgL\theta_2 - ka^2(\theta_1 - \theta_2) &= 0 \end{aligned} \tag{5.76}$$

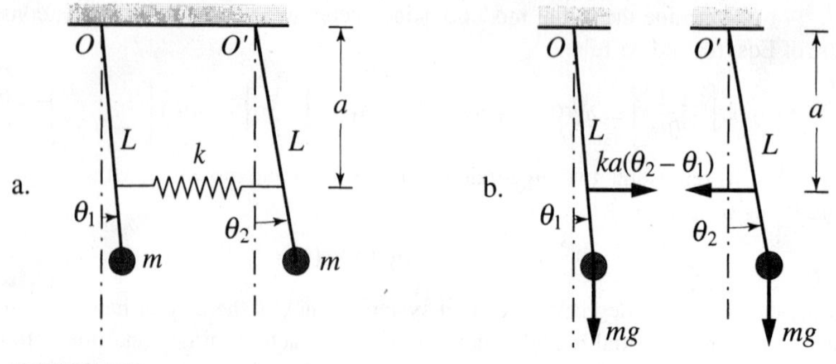

FIGURE 5.11
a. Two identical pendulums connected by a spring, b. Free-body diagrams

which can be arranged in the matrix form

$$\begin{bmatrix} mL^2 & 0 \\ 0 & mL^2 \end{bmatrix} \begin{bmatrix} \ddot{\theta}_1 \\ \ddot{\theta}_2 \end{bmatrix} + \begin{bmatrix} mgL+ka^2 & -ka^2 \\ -ka^2 & mgL+ka^2 \end{bmatrix} \begin{bmatrix} \theta_1 \\ \theta_2 \end{bmatrix} = \begin{bmatrix} 0 \\ 0 \end{bmatrix} \tag{5.77}$$

indicating that the system is coupled elastically. As expected, when the spring stiffness k reduces to zero the coupling disappears and the system reduces to two independent simple pendulums with identical natural frequencies equal to $\sqrt{g/L}$.

Next, we recall that free vibration is harmonic and assume a solution of the equations of motion in the form

$$\theta_1(t) = \Theta_1 \cos(\omega t - \phi), \; \theta_2(t) = \Theta_2 \cos(\omega t - \phi) \tag{5.78}$$

where Θ_1 and Θ_2 are amplitudes, ω is the frequency of harmonic oscillation and ϕ is a phase angle. Inserting Eqs. (5.78) into Eq. (5.77) and dividing through by $\cos(\omega t - \phi)$, we obtain the eigenvalue problem

$$-\omega^2 \begin{bmatrix} mL^2 & 0 \\ 0 & mL^2 \end{bmatrix} \begin{bmatrix} \Theta_1 \\ \Theta_2 \end{bmatrix} + \begin{bmatrix} mgL+ka^2 & -ka^2 \\ -ka^2 & mgL+ka^2 \end{bmatrix} \begin{bmatrix} \Theta_1 \\ \Theta_2 \end{bmatrix} = \begin{bmatrix} 0 \\ 0 \end{bmatrix} \tag{5.79}$$

leading to the characteristic equation

$$\det \begin{bmatrix} mgL+ka^2-\omega^2 mL^2 & -ka^2 \\ -ka^2 & mgL+ka^2-\omega^2 mL^2 \end{bmatrix}$$

$$= (mgL+ka^2-\omega^2 mL^2)^2 - (ka^2)^2 = 0 \tag{5.80}$$

which is equivalent to

$$mgL+ka^2-\omega^2 mL^2 = \pm ka^2 \tag{5.81}$$

Hence, the two natural frequencies are

$$\omega_1 = \sqrt{\frac{g}{L}}, \; \omega_2 = \sqrt{\frac{g}{L} + 2\frac{k}{m}\frac{a^2}{L^2}} \tag{5.82}$$

The natural modes are obtained by letting $\omega^2 = \omega_i^2$ in Eq. (5.79) and writing

$$-\omega_i^2 \begin{bmatrix} mL^2 & 0 \\ 0 & mL^2 \end{bmatrix} \begin{bmatrix} \Theta_1 \\ \Theta_2 \end{bmatrix}_i + \begin{bmatrix} mgL + ka^2 & -ka^2 \\ -ka & mgL + ka^2 \end{bmatrix} \begin{bmatrix} \Theta_1 \\ \Theta_2 \end{bmatrix}_i = \begin{bmatrix} 0 \\ 0 \end{bmatrix},$$

$$i = 1, 2 \qquad (5.83)$$

Inserting $\omega_1^2 = g/L$ and $\omega_2^2 = g/L + 2(k/m)(a^2/L^2)$ into Eqs. (5.83), in sequence, and solving for the ratios Θ_{21}/Θ_{11} and Θ_{22}/Θ_{12}, we obtain

$$\frac{\Theta_{21}}{\Theta_{11}} = 1, \quad \frac{\Theta_{22}}{\Theta_{12}} = -1 \qquad (5.84)$$

so that in the first natural mode the two pendulums move as if they were a single pendulum with the spring k unstretched, which is consistent with the fact that the first natural frequency of the system is that of the simple pendulum, $\omega_1 = \sqrt{g/L}$. On the other hand, in the second natural mode the two pendulums are 180° out of phase. The two modes are shown in Fig. 5.12.

As pointed out in Sec. 5.6, the free vibration of a conservative two-degree-of-freedom system can be expressed as a superposition of the two natural modes multiplied by associated harmonic natural coordinates, or

$$\begin{bmatrix} \theta_1(t) \\ \theta_2(t) \end{bmatrix} = C_1 \cos(\omega_1 t - \theta_1)\mathbf{u}_1 + C_2 \cos(\omega_2 t - \phi_2)\mathbf{u}_2$$

$$= C_1 \cos(\omega_1 t - \phi_1) \begin{bmatrix} \Theta_{11} \\ \Theta_{21} \end{bmatrix} + C_2 \cos(\omega_2 t - \phi_2) \begin{bmatrix} \Theta_{12} \\ \Theta_{22} \end{bmatrix} \qquad (5.85)$$

Choosing arbitrarily $\Theta_{11} = \Theta_{12} = 1$ and using Eqs. (5.84), Eq. (5.85) can be rewritten in the scalar form

$$\begin{aligned} \theta_1(t) &= C_1 \cos(\omega_1 t - \phi_1) + C_2 \cos(\omega_2 t - \phi_2) \\ \theta_2(t) &= C_1 \cos(\omega_1 t - \phi_1) - C_2 \cos(\omega_2 t - \phi_2) \end{aligned} \qquad (5.86)$$

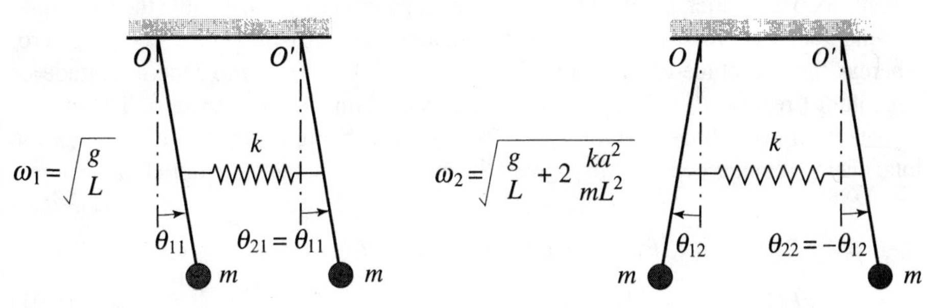

FIGURE 5.12
Natural modes for the system of Fig. 5.11a

Then, considering the initial conditions

$$\theta_1(0) = \theta_0,\ \theta_2(0) = 0,\ \dot{\theta}_1(0) = \dot{\theta}_2(0) = 0 \tag{5.87}$$

and using the trigonometric relations $\cos(\alpha \pm \beta) = \cos\alpha\cos\beta \mp \sin\alpha\sin\beta$, in which $\alpha = (\omega_2 - \omega_1)t/2$, $\beta = (\omega_2 + \omega_1)t/2$, Eqs. (5.86) become

$$\begin{aligned} \theta_1(t) &= \tfrac{1}{2}\theta_0 \cos\omega_1 t + \tfrac{1}{2}\theta_0 \cos\omega_2 t = \theta_0 \cos\frac{\omega_2 - \omega_1}{2} t \cos\frac{\omega_2 + \omega_1}{2} t \\ \theta_2(t) &= \tfrac{1}{2}\theta_0 \cos\omega_1 t - \tfrac{1}{2}\theta_0 \cos\omega_2 t = \theta_0 \sin\frac{\omega_2 - \omega_1}{2} t \sin\frac{\omega_2 + \omega_1}{2} t \end{aligned} \tag{5.88}$$

At this point, we consider the case in which the spring constant satisfies the inequality $k \ll mgL/a^2$, so that the coupling provided by the spring is very weak. Then, using Eqs. (5.82) and introducing the approximations

$$\frac{\omega_B}{2} = \frac{\omega_2 - \omega_1}{2} \cong \frac{1}{2}\frac{k}{m}\frac{a^2}{\sqrt{gL^3}},\quad \omega_{\text{ave}} = \frac{\omega_2 + \omega_1}{2} \cong \sqrt{\frac{g}{L}} + \frac{1}{2}\frac{k}{m}\frac{a^2}{\sqrt{gL^3}} \tag{5.89}$$

Eqs. (5.88) can be rewritten in the form

$$\theta_1(t) \cong \theta_0 \cos\tfrac{1}{2}\omega_B t \cos\omega_{\text{ave}} t,\ \theta_2(t) \cong \theta_0 \sin\tfrac{1}{2}\omega_B t \sin\omega_{\text{ave}} t \tag{5.90}$$

Hence, $\theta_1(t)$ and $\theta_2(t)$ can be regarded as being harmonic functions with frequency ω_{ave} and with amplitudes varying slowly according to $\theta_0 \cos\frac{1}{2}\omega_B t$ and $\theta_0 \sin\frac{1}{2}\omega_B t$, respectively. The plots $\theta_1(t)$ versus t and $\theta_2(t)$ versus t are shown in Figs. 5.13, with the slowly varying amplitudes indicated by the dashed-line envelopes. Geometrically, Fig. 5.13a (or Fig. 5.13b) implies that if two harmonic functions possessing equal amplitudes and nearly equal frequencies are added, then the resulting function is an *amplitude-modulated* harmonic function with a frequency equal to the average of the two frequencies. At first, when the two harmonic waves reinforce each other, the amplitude doubles, and later, as the two waves cancel each other, the amplitude reduces to zero. The phenomenon is known as the *beat phenomenon*, and the frequency of modulation ω_B, which in this particular case is equal to $ka^2/m\sqrt{gL^3}$, is called the *beat frequency*. From Fig. 5.13a, we conclude that the time between two peaks is $T/2 = 2\pi/\omega_B$, whereas the period of the amplitude-modulated envelope is $T = 4\pi/\omega_B$.

We observe from Figs. 5.13 that there is a 90° phase angle between $\theta_1(t)$ and $\theta_2(t)$. At $t = 0$, pendulum 1 begins to swing with the amplitude θ_0 while pendulum 2 is at rest. Soon thereafter, pendulum 2 is entrained, gaining amplitude while the amplitude of pendulum 1 decreases. At $t_1 = \pi/\omega_B$, the amplitude of pendulum 1 becomes zero, whereas the amplitude of pendulum 2 reaches θ_0. At $t_2 = 2\pi/\omega_B$, the amplitude of pendulum 1 reaches θ_0 once again and that of pendulum 2 reduces to zero. The motion repeats itself every time interval $T/2 = 2\pi/\omega_B$. This being a conservative system, the total energy remains constant throughout the motion and equal to the initial total energy, or

$$\begin{aligned} E &= \tfrac{1}{2}mL^2\dot{\theta}_1^2 + \tfrac{1}{2}mL^2\dot{\theta}_2^2 + mgL(1 - \cos\theta_1) + mgL(1 - \cos\theta_2) + \tfrac{1}{2}ka^2(\theta_2 - \theta_1)^2 \\ &= mgL(1 - \cos\theta_0) + \tfrac{1}{2}ka^2\theta_0^2 \end{aligned} \tag{5.91}$$

At $t = 0$, the energy of pendulum 1 is $mgL(1 - \cos\theta_0)$ and the energy of pendulum 2 is zero, with the energy in the spring being equal to $\frac{1}{2}ka^2\theta_0^2$. At $t = t_1$, the energy of

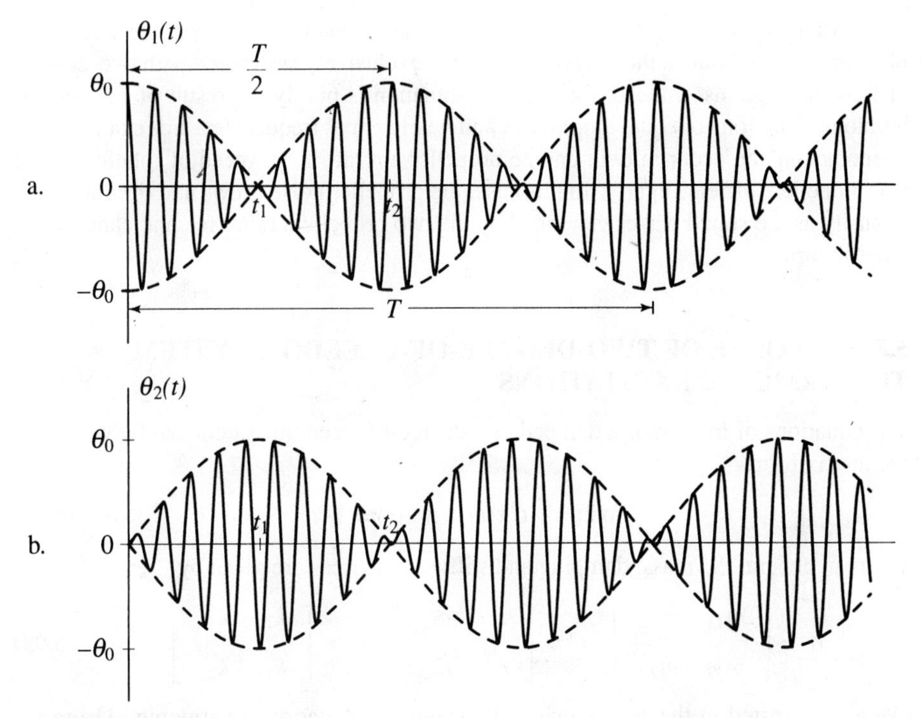

FIGURE 5.13
Response of the pendulums of Fig. 5.11a demonstrating the beat phenomenon

pendulum 1 reduces to zero and the energy of pendulum 2 reaches $mgL(1-\cos\theta_0)$, while the energy in the spring is the same as at $t=0$. At $t=t_2$, the situation reverts to that at $t=0$. As the motion keeps repeating itself, there is a complete transfer of energy from one pendulum to another every time interval $T/4=\pi/\omega_B$.

Another example of a system exhibiting the beat phenomenon is the "Wilberforce spring," which consists of a mass of finite dimensions suspended by a helical spring such that the frequencies of extensional and torsional vibrations are very close in value. In this case, the kinetic energy changes from pure translational in the vertical direction to pure rotational about the vertical axis, as shown in Fig. 5.14.

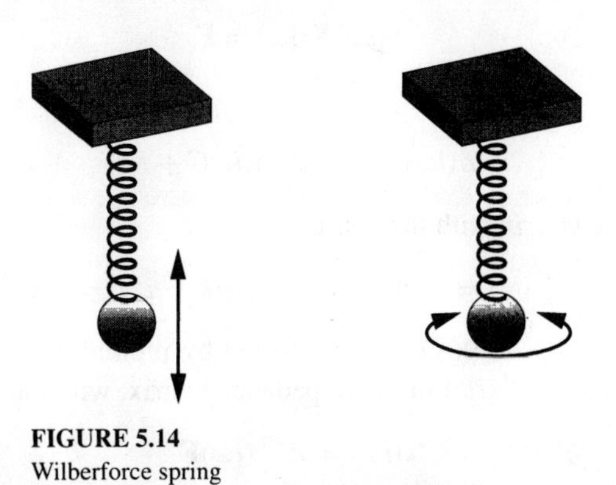

FIGURE 5.14
Wilberforce spring

Although in our particular case the beat phenomenon results from the weak coupling of two pendulums, the phenomenon is not exclusively associated with two-degree-of-freedom systems. Indeed, the beat phenomenon is purely the result of adding two harmonic functions of equal amplitudes and nearly equal frequencies. For example, the phenomenon can occur in twin-engine propeller aircraft if the speed of rotation of the two propellers differs slightly. In this case, the propeller noise grows and diminishes in intensity as the sound waves generated by the two propellers reinforce and cancel each other in turn.

5.8 RESPONSE OF TWO-DEGREE-OF-FREEDOM SYSTEMS TO HARMONIC EXCITATIONS

The equations of motion of a damped two-degree-of-freedom system can be written in the matrix form

$$M\ddot{\mathbf{x}}(t)+C\dot{\mathbf{x}}(t)+K\mathbf{x}(t)=\mathbf{F}(t) \tag{5.92}$$

where in general the mass, damping and stiffness matrices are given by

$$M=\begin{bmatrix} m_{11} & m_{12} \\ m_{12} & m_{22} \end{bmatrix},\ C=\begin{bmatrix} c_{11} & c_{12} \\ c_{12} & c_{22} \end{bmatrix},\ K=\begin{bmatrix} k_{11} & k_{12} \\ k_{12} & k_{22} \end{bmatrix} \tag{5.93}$$

We are interested in the case in which the external excitation is harmonic. Using the analogy with Eq. (3.14), it is convenient to express the force vector in the exponential form

$$\mathbf{F}(t)=\mathbf{F}e^{i\omega t} \tag{5.94}$$

where $\mathbf{F}$ is a constant amplitude vector. Then, by analogy with Eq. (3.16), the steady-state response can be expressed as

$$\mathbf{x}(t)=\mathbf{X}(i\omega)e^{i\omega t} \tag{5.95}$$

in which $\mathbf{X}(i\omega)$ is generally a complex vector depending on the driving frequency ω and the system parameters. Inserting Eqs. (5.94) and (5.95) into Eq. (5.92) and dividing through by $e^{i\omega t}$ we obtain

$$Z(i\omega)\mathbf{X}(i\omega)=\mathbf{F} \tag{5.96}$$

where

$$Z(i\omega)=-\omega^2 M+i\omega C+K \tag{5.97}$$

is a 2×2 *impedance matrix* with the entries

$$z_{ij}(i\omega)=-\omega^2 m_{ij}+i\omega c_{ij}+k_{ij},\ i,j=1,2 \tag{5.98}$$

The solution of Eq. (5.96) can be obtained by premultiplying both sides of the equation by the inverse $Z^{-1}(i\omega)$ of the impedance matrix, with the result

$$\mathbf{X}(i\omega)=Z^{-1}(i\omega)\mathbf{F} \tag{5.99}$$

where the inverse has the explicit form

$$Z^{-1}(i\omega) = \frac{1}{|Z(i\omega)|}\begin{bmatrix} z_{22}(i\omega) & -z_{12}(i\omega) \\ -z_{12}(i\omega) & z_{11}(i\omega) \end{bmatrix}$$

$$= \frac{1}{z_{11}(i\omega)z_{22}(i\omega) - z_{12}^2(i\omega)}\begin{bmatrix} z_{22}(i\omega) & -z_{12}(i\omega) \\ -z_{12}(i\omega) & z_{11}(i\omega) \end{bmatrix} \tag{5.100}$$

Introducing Eq. (5.100) in Eq. (5.99) and carrying out the matrix multiplication, we obtain

$$X_1(i\omega) = \frac{z_{22}(i\omega)F_1 - z_{12}(i\omega)F_2}{z_{11}(i\omega)z_{22}(i\omega) - z_{12}^2(i\omega)}, \quad X_2(i\omega) = \frac{-z_{12}(i\omega)F_1 + z_{11}(i\omega)F_2}{z_{11}(i\omega)z_{22}(i\omega) - z_{12}^2(i\omega)} \tag{5.101}$$

where F_1 and F_2 are the components of the constant vector $\mathbf{F}$, and we note that the functions $X_1(i\omega)$ and $X_2(i\omega)$ are analogous to the frequency response functions first encountered in Sec. 3.1.

In the case of undamped systems, such as that described by Eqs. (5.10) and (5.11), the impedance functions are real, or

$$z_{11}(\omega) = k_{11} - \omega^2 m_1, \; z_{22}(\omega) = k_{22} - \omega^2 m_2, \; z_{12}(\omega) = k_{12} \tag{5.102}$$

Introducing Eqs. (5.102) into Eqs. (5.101), we conclude that the frequency response functions are also real, or

$$X_1(\omega) = \frac{(k_{22} - \omega^2 m_2)F_1 - k_{12}F_2}{(k_{11} - \omega^2 m_1)(k_{22} - \omega^2 m_2) - k_{12}^2}$$
$$X_2(\omega) = \frac{-k_{12}F_1 + (k_{11} - \omega^2 m_1)F_2}{(k_{11} - \omega^2 m_1)(k_{22} - \omega^2 m_2) - k_{12}^2} \tag{5.103}$$

For a given set of system parameters, Eqs. (5.103) can be used to plot $X_1(\omega)$ versus ω and $X_2(\omega)$ versus ω, thus obtaining the complete frequency response. Indeed, both the amplitude and phase angle are included in $X_1(\omega)$ and $X_2(\omega)$, where for undamped systems the phase angle is either 0 or 180°. Consistent with this, the amplitude is positive and negative, respectively.

Example 5.5. Consider the system of Example 5.1 and plot the frequency-response curves for the case in which $F_2 = 0$.

Using the parameter values of Example 5.1, Eqs. (5.103) become

$$X_1(\omega) = \frac{(3k - 2m\omega^2)F_1}{2m^2\omega^4 - 7mk\omega^2 + 5k^2}, \quad X_2(\omega) = \frac{kF_1}{2m^2\omega^4 - 7mk\omega^2 + 5k^2} \tag{a}$$

where we used the notation $T/L = k$. But the denominator of X_1 and X_2 is recognized as the characteristic determinant, which can be written as

$$\Delta(\omega^2) = 2m^2\omega^4 - 7mk\omega^2 + 5k^2 = 2m^2(\omega^2 - \omega_1^2)(\omega^2 - \omega_2^2) \tag{b}$$

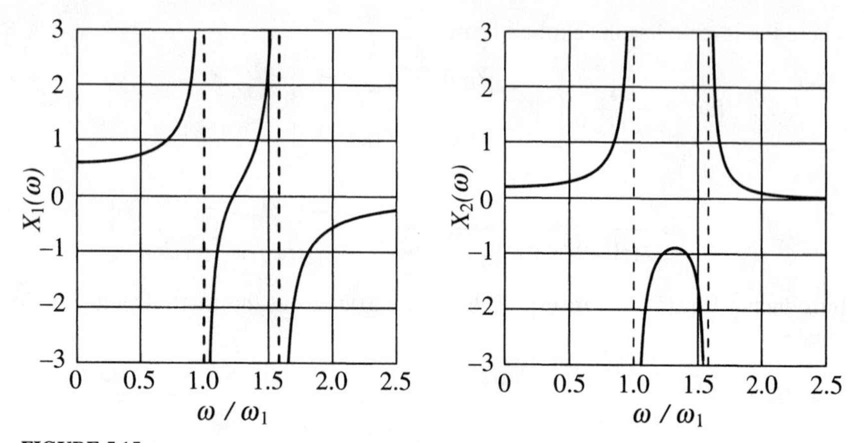

FIGURE 5.15
Frequency response curves for an undamped two-degree-of-freedom system

where

$$\omega_1^2 = \frac{k}{m}, \quad \omega_2^2 = \frac{5}{2}\frac{k}{m} \tag{c}$$

are the squares of the natural frequencies. Hence, Eqs. (a) can be written in the form

$$X_1(\omega) = \frac{F_1}{5k} \frac{3 - 2(\omega/\omega_1)^2}{[1 - (\omega/\omega_1)^2][1 - (\omega/\omega_2)^2]}$$
$$X_2(\omega) = \frac{F_1}{5k} \frac{1}{[1 - (\omega/\omega_1)^2][1 - (\omega/\omega_2)^2]} \tag{d}$$

The frequency response curves $X_1(\omega)$ versus ω/ω_1 and $X_2(\omega)$ versus ω/ω_1 are plotted in Fig. 5.15.

5.9 UNDAMPED VIBRATION ABSORBERS

When rotating machinery operates at a frequency close to resonance, violent vibration is induced. Assuming that the system can be represented by a single-degree-of-freedom system subjected to harmonic excitation, the situation can be alleviated by changing either the mass or the spring constant. At times, however, this may not be possible. In such a case, a second mass and spring can be added to the system, where the added mass and spring are so designed as to produce a two-degree-of-freedom system whose frequency response is zero at the excitation frequency. We note from Fig. 5.15a that a point at which the frequency response is zero does indeed exist. The new two-degree-of-freedom system has two resonant frequencies, but these frequencies generally present no problem because they are reasonably far removed from the operating frequency.

We consider the system of Fig. 5.16, where the original single-degree-of-freedom system, referred to as the *main system*, consists of the mass m_1 and the spring k_1, and the added system, referred to as the *absorber*, consists of the mass m_2 and the spring k_2.

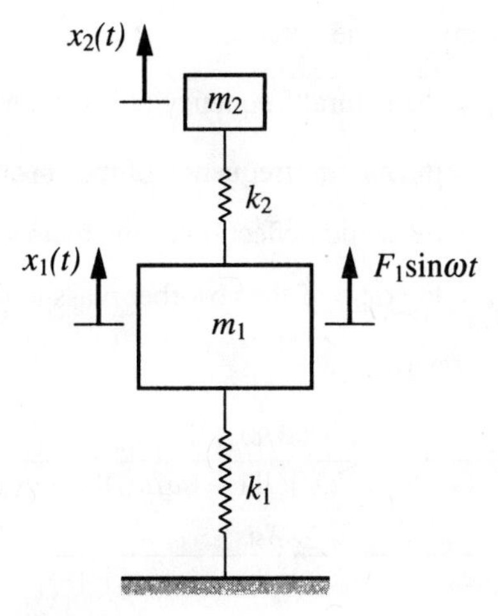

FIGURE 5.16
Vibration absorber

The equations of motion of the combined system can be shown to be

$$\begin{aligned} m_1\ddot{x}_1 + (k_1+k_2)x_1 - k_2x_2 &= F_1 \sin\omega t \\ m_2\ddot{x}_2 - k_2x_1 + k_2x_2 &= 0 \end{aligned} \tag{5.104}$$

and we note that, because the system is undamped, the complex notation is not necessary. Letting the solution of Eqs. (5.104) be

$$x_1(t) = X_1 \sin\omega t, \quad x_2(t) = X_2 \sin\omega t \tag{5.105}$$

and following the steps outlined in Sec. 5.8, we obtain two algebraic equations in X_1 and X_2 having the matrix form

$$\begin{bmatrix} k_1+k_2-\omega^2 m_1 & -k_2 \\ -k_2 & k_2-\omega^2 m_2 \end{bmatrix} \begin{bmatrix} X_1 \\ X_2 \end{bmatrix} = \begin{bmatrix} F_1 \\ 0 \end{bmatrix} \tag{5.106}$$

Hence, using Eqs. (5.103), the solution of Eq. (5.106) is

$$\begin{aligned} X_1 &= \frac{(k_2-\omega^2 m_2)F_1}{(k_1+k_2-\omega^2 m_1)(k_2-\omega^2 m_2)-k_2^2} \\ X_2 &= \frac{k_2 F_1}{(k_1+k_2-\omega^2 m_1)(k_2-\omega^2 m_2)-k_2^2} \end{aligned} \tag{5.107}$$

It is customary to introduce the notation:

$$\omega_n = \sqrt{k_1/m_1} = \text{the natural frequency of the main system alone}$$

$$\omega_a = \sqrt{k_2/m_2} = \text{the natural frequency of the absorber alone}$$

$$x_{\text{st}} = F_1/k_1 = \text{the static deflection of the main system}$$

$$\mu = m_2/m_1 = \text{the ratio of the absorber mass to the main mass}$$

so that Eqs. (5.107) can be rewritten as

$$X_1 = \frac{[1-(\omega/\omega_a)^2]x_{\text{st}}}{[1+\mu(\omega_a/\omega_n)^2-(\omega/\omega_n)^2][1-(\omega/\omega_a)^2]-\mu(\omega_a/\omega_n)^2}$$
$$X_2 = \frac{x_{\text{st}}}{[1+\mu(\omega_a/\omega_n)^2-(\omega/\omega_n)^2][1-(\omega/\omega_a)^2]-\mu(\omega_a/\omega_n)^2} \tag{5.108}$$

From the first of Eqs. (5.108), we conclude that for $\omega_a = \omega$ the amplitude X_1 of the main mass reduces to zero. Hence, the absorber can indeed perform the task for which it was designed, namely, to eliminate the vibration of the main mass, provided the natural frequency of the absorber is the same as the operating frequency of the machinery. Moreover, for $\omega_a = \omega$, the second of Eqs. (5.108) reduces to

$$X_2 = -\left(\frac{\omega_n}{\omega_a}\right)^2 \frac{x_{\text{st}}}{\mu} = -\frac{F_1}{k_2} \tag{5.109}$$

so that, inserting Eq. (5.109) into the second of Eqs. (5.105), we obtain

$$x_2(t) = -\frac{F_1}{k_2}\sin\omega t \tag{5.110}$$

from which we conclude that the force in the absorber spring at any time is

$$k_2 x_2(t) = -F_1 \sin\omega t \tag{5.111}$$

Hence, the absorber exerts on the main mass a force $-F_1 \sin\omega t$ which balances exactly the applied force $F_1 \sin\omega t$. Because the same effect is obtained by any absorber provided its natural frequency is equal to the operating frequency, there is a wide choice of absorber parameters. The actual choice is generally dictated by space limitations, which restricts the amplitude X_2 of the absorber motion.

Although a vibration absorber is designed for a given operating frequency ω, the absorber can perform satisfactorily for operating frequencies close in value to ω. In this case, the motion of m_1 is not zero, but its amplitude is very small. This statement can be verified by using the first of Eqs. (5.108) and plotting $X_1(\omega)/x_{\text{st}}$ versus ω/ω_a. Figure 5.17 shows such a plot for $\mu = 0.2$ and $\omega_n = \omega_a$. The shaded area indicates the region in which the performance of the absorber can be regarded as satisfactory. As pointed out earlier, one disadvantage of the vibration absorber is that two new resonant frequencies are created, as can be seen from Fig. 5.17. To reduce the amplitude at the resonant frequencies, damping can be added, but this results in an increase in amplitude in the

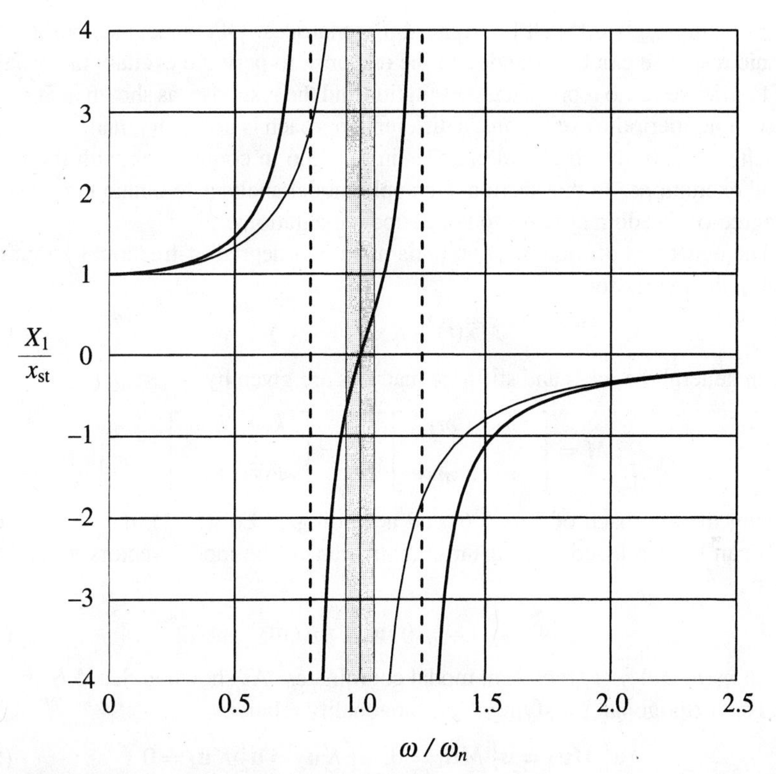

FIGURE 5.17
Frequency response curve for the main mass

neighborhood of the operating frequency $\omega = \omega_a$. It should be recalled that, for any rotating machinery, the frequency increases from zero to a steady operating frequency, so that the system is likely to go through the first resonant frequency. As a matter of interest, the plot X_1/x_{st} versus ω/ω_n corresponding to the main system alone is also shown in Fig. 5.17.

5.10 RESPONSE OF TWO-DEGREE-OF-FREEDOM SYSTEMS TO NONPERIODIC EXCITATIONS

Until now our discussion of two-degree-of-freedom systems has been concentrated on the response to initial and to harmonic excitations. In both cases, it is possible to obtain the response by elementary means. Whereas the elementary approach is suitable for two-degree-of-freedom systems, it does not extend very well to multi-degree-of-freedom systems. In the case of free vibration, the approach was based on the solution of the characteristic equation, which is an easy task if the equation is quadratic, but solving characteristic equations is not advisable when the degree of the characteristic polynomial exceeds two. Then, in the case of harmonic excitations the approach required an analytical matrix inversion, which can be carried out with ease for 2×2 matrices, but becomes increasingly difficult for higher-order matrices. Of course, the approach to the harmonic response can be extended to the response to periodic excitations through the use of Fourier series to represent the excitation and the response, as shown in Sec. 3.9. In the case of nonperiodic excitations, a different approach is necessary, namely, the modal approach. We introduced modal analysis in Sec. 5.6 in conjunction with the response to initial excitations. In this section, we apply modal analysis to obtain the response of two-degree-of-freedom systems to nonperiodic excitations.

The equations of motion of an undamped two-degree-of-freedom system can be written in the matrix form

$$M\ddot{\mathbf{x}}(t) + K\mathbf{x}(t) = \mathbf{F}(t) \tag{5.112}$$

where in general the mass and stiffness matrices are given by

$$M = \begin{bmatrix} m_{11} & m_{12} \\ m_{12} & m_{22} \end{bmatrix}, \quad K = \begin{bmatrix} k_{11} & k_{12} \\ k_{12} & k_{22} \end{bmatrix} \tag{5.113}$$

Following the approach of Sec. 5.6, and according to Eq. (5.71), the solution of Eq. (5.112) can be expressed as a linear combination of the modal vectors $\mathbf{u}_1$ and $\mathbf{u}_2$, as follows:

$$\mathbf{x}(t) = \eta_1(t)\mathbf{u}_1 + \eta_2(t)\mathbf{u}_2 \tag{5.114}$$

in which $\eta_1(t)$ and $\eta_2(t)$ represent modal coordinates. As shown in Sec. 5.6, the modal vectors are orthogonal, satisfying the orthogonality relations

$$\mathbf{u}_1^T M\mathbf{u}_2 = \mathbf{u}_2^T M\mathbf{u}_1 = 0, \quad \mathbf{u}_1^T K\mathbf{u}_2 = \mathbf{u}_2^T K\mathbf{u}_1 = 0 \tag{5.115}$$

Moreover, recalling Eqs. (5.69), we can rewrite Eqs. (5.68) as

$$\begin{aligned} &\mathbf{u}_1^T M\mathbf{u}_1 = m'_{11}, \quad \mathbf{u}_2^T M\mathbf{u}_2 = m'_{22} \\ &\mathbf{u}_1^T K\mathbf{u}_1 = k'_{11} = \omega_1^2 m'_{11}, \quad \mathbf{u}_2^T K\mathbf{u}_2 = k'_{22} = \omega_2^2 m'_{22} \end{aligned} \tag{5.116}$$

where ω_1 and ω_2 are the natural frequencies.

Equation (5.114) in conjunction with the orthogonality relations, Eqs. (5.115), can be used to reduce the set of two simultaneous equations, Eq. (5.112), to a set of two independent modal equations. Indeed, inserting Eq. (5.114) into Eq. (5.112), premultiplying the result by $\mathbf{u}_1^T$ and $\mathbf{u}_2^T$, in sequence, and using Eqs. (5.115) and (5.116), we obtain the modal equations

$$\begin{aligned} m'_{11}\ddot{\eta}_1(t) + m'_{11}\omega_1^2\eta_1(t) &= N_1(t) \\ m'_{22}\ddot{\eta}_2(t) + m'_{22}\omega_2^2\eta_2(t) &= N_2(t) \end{aligned} \tag{5.117}$$

in which

$$N_1(t) = \mathbf{u}_1^T \mathbf{F}(t), \quad N_2(t) = \mathbf{u}_2^T \mathbf{F}(t) \tag{5.118}$$

are known as *modal forces*, some abstract forces representing linear combinations of the actual forces $F_1(t)$ and $F_2(t)$. The nature of the modal forces $N_1(t)$ and $N_2(t)$ depends on the nature of the modal coordinates. For example, if $\eta_1(t)$, or $\eta_2(t)$, represents an angle, then $N_1(t)$, or $N_2(t)$, represents a moment.

The modal equations, Eqs. (5.117), represent two independent equations for the modal coordinates $\eta_1(t)$ and $\eta_2(t)$ and they resemble the equation of motion of an undamped single-degree-of-freedom system, Eq. (3.1) with $c = 0$. Using the analogy with the single-degree-of-freedom system, the solution of Eqs. (5.117) to applied forces alone can be obtained by means of the convolution integral, Eq. (4.40), in the form

$$\eta_1(t) = \int_0^t N_1(t-\tau)g_1(\tau)d\tau$$
$$\eta_2(t) = \int_0^t N_2(t-\tau)g_2(\tau)d\tau \tag{5.119}$$

where $g_1(t)$ and $g_2(t)$ represent the impulse response corresponding to $\eta_1(t)$ and $\eta_2(t)$, respectively. But, from Eq. (b) of Example 4.3, the impulse response can be written as

$$g_i(t) = \frac{1}{m'_{ii}\omega_i}\sin\omega_i t\,u(t), \ i = 1, 2 \tag{5.120}$$

in which $u(t)$ is the unit step function. Hence, inserting Eqs. (5.120) into Eqs. (5.119), the modal coordinates can be expressed in the general form

$$\eta_1(t) = \frac{1}{m'_{11}\omega_1}\int_0^t N_1(t-\tau)\sin\omega_1\tau\,u(\tau)d\tau = \frac{1}{m'_{11}\omega_1}\int_0^t N_1(t-\tau)\sin\omega_1\tau d\tau$$
$$\eta_2(t) = \frac{1}{m'_{22}\omega_2}\int_0^t N_2(t-\tau)\sin\omega_2\tau\,u(\tau)d\tau = \frac{1}{m'_{22}\omega_2}\int_0^t N_2(t-\tau)\sin\omega_2\tau d\tau \tag{5.121}$$

The solution process begins by solving the eigenvalue problem, Eq. (5.28), and computing the natural frequencies ω_i and modal vectors $\mathbf{u}_i$ $(i = 1, 2)$. Then, using the first two of Eqs. (5.116), it is possible to determine the constants m'_{11} and m'_{22}. Next, for any given force vector $\mathbf{F}(t)$, Eqs. (5.118) yield the modal forces $N_1(t)$ and $N_2(t)$. The process continues by inserting $N_1(t)$ and $N_2(t)$ into Eqs. (5.121) and evaluating the convolution integrals for the modal coordinates $\eta_1(t)$ and $\eta_2(t)$. The formal solution is completed by inserting $\eta_1(t)$ and $\eta_2(t)$, together with the modal vectors $\mathbf{u}_1$ and $\mathbf{u}_2$, into Eq. (5.114).

It should be pointed out that, although the modal analysis solution presented in this section was developed with arbitrary forces in mind, the same modal analysis can be used to decouple the equations of motion in the case of harmonic forces as well. Indeed, assuming that the force vector is given by $\mathbf{F}(t) = \mathbf{F}\cos\omega t$, where $\mathbf{F}$ is a constant vector and ω is the driving frequency, then the modal forces $N_1(t)$ and $N_2(t)$ given by Eqs. (5.118) are proportional to $\cos\omega t$, so that the modal equations, Eqs. (5.117), can be solved directly by the methods of Ch. 3.

In the event the force vector $\mathbf{F}(t)$ is such that analytical evaluation of the convolution integrals in Eqs. (5.121) is not possible, one must be content with a numerical solution in discrete time. To this end, there are several alternatives. One of them is to use the approach of Sec. 4.9 to replace the convolution integrals for $\eta_1(t)$ and $\eta_2(t)$ by convolution sums, where the latter can be evaluated with ease. This approach is discussed in Sec. 5.11. Another approach, which bypasses the eigenvalue problem altogether, is to determine the response by means of the recursive relations using the discrete-time transition matrix, as discussed in Sec. 4.10. Note that, to use the latter approach, it is necessary to extend the formulation so as to accommodate multi-degree-of-freedom systems, which amounts to replacing the vectors γ and $\mathbf{b}$ in Eqs. (4.104) and (4.106) by matrices Γ and B, respectively. This extension is discussed in Ch. 7.

Example 5.6. The two-degree-of-freedom system of Example 5.1 is acted upon by the rectangular pulse

$$F_2(t) = F_0[\mathscr{u}(t) - \mathscr{u}(t-a)] \tag{a}$$

applied to mass m_2. Determine the response by modal analysis.

From Example 5.1, we obtain the mass and stiffness matrices

$$M = m\begin{bmatrix} 1 & 0 \\ 0 & 2 \end{bmatrix}, \quad K = \frac{T}{L}\begin{bmatrix} 2 & -1 \\ -1 & 3 \end{bmatrix} \tag{b}$$

as well as the natural frequencies

$$\omega_1 = \sqrt{\frac{T}{mL}}, \quad \omega_2 = \sqrt{\frac{5T}{2mL}} \tag{c}$$

and the modal vectors

$$\mathbf{u}_1 = \begin{bmatrix} 1 \\ 1 \end{bmatrix}, \quad \mathbf{u}_2 = \begin{bmatrix} 1 \\ -0.5 \end{bmatrix} \tag{d}$$

Hence, from the first two of Eqs. (5.116), we have

$$\begin{aligned} m'_{11} &= \mathbf{u}_1^T M \mathbf{u}_1 = \begin{bmatrix} 1 \\ 1 \end{bmatrix}^T m \begin{bmatrix} 1 & 0 \\ 0 & 2 \end{bmatrix} \begin{bmatrix} 1 \\ 1 \end{bmatrix} = 3m \\ m'_{22} &= \mathbf{u}_2^T M \mathbf{u}_2 = \begin{bmatrix} 1 \\ -0.5 \end{bmatrix}^T m \begin{bmatrix} 1 & 0 \\ 0 & 2 \end{bmatrix} \begin{bmatrix} 1 \\ -0.5 \end{bmatrix} = 1.5m \end{aligned} \tag{e}$$

Moreover, recognizing that $\mathbf{F}(t) = [0 \quad F_2(t)]^T$ and using Eqs. (5.118), we can write the modal forces

$$\begin{aligned} N_1(t) &= \mathbf{u}_1^T \mathbf{F}(t) = \begin{bmatrix} 1 \\ 1 \end{bmatrix}^T \begin{bmatrix} 0 \\ F_2(t) \end{bmatrix} = F_2(t) \\ N_2(t) &= \mathbf{u}_2^T \mathbf{F}(t) = \begin{bmatrix} 1 \\ -0.5 \end{bmatrix}^T \begin{bmatrix} 0 \\ F_2(t) \end{bmatrix} = -0.5F_2(t) \end{aligned} \tag{f}$$

Then, inserting Eqs. (a), (c), (e) and (f) into Eqs. (5.121), we determine the modal coordinates as follows:

$$\begin{aligned} \eta_1(t) &= \frac{F_0}{m'_{11}\omega_1}\int_0^t [\mathscr{u}(t-\tau) - \mathscr{u}(t-a-\tau)]\sin\omega_1\tau \, d\tau \\ &= \frac{F_0}{m'_{11}\omega_1^2}\{(1-\cos\omega_1 t)\mathscr{u}(t) - [1-\cos\omega_1(t-a)]\mathscr{u}(t-a)\} \end{aligned}$$

$$= \frac{F_0 L}{3T}\left\{\left(1-\cos\sqrt{\frac{T}{mL}}t\right)\mathscr{u}(t)-\left[1-\cos\sqrt{\frac{T}{mL}}(t-a)\right]\mathscr{u}(t-a)\right\}$$

(g)

$$\eta_2(t) = \frac{-0.5F_0}{m'_{22}\omega_2}\int_0^t [\mathscr{u}(t-\tau)-\mathscr{u}(t-a-\tau)]\sin\omega_2\tau\, d\tau$$

$$= \frac{-0.5F_0}{m'_{22}\omega_2^2}\{(1-\cos\omega_2 t)\mathscr{u}(t)-[1-\cos\omega_2(t-a)]\mathscr{u}(t-a)\}$$

$$= -\frac{F_0 L}{7.5T}\left\{\left(1-\cos\sqrt{\frac{5T}{2mL}}t\right)\mathscr{u}(t)-\left[1-\cos\sqrt{\frac{5T}{2mL}}(t-a)\right]\mathscr{u}(t-a)\right\}$$

Finally, introducing Eqs. (d) and (g) in Eq. (5.114), we obtain the response by components

$$x_1(t) = \eta_1(t)u_{11}+\eta_2(t)u_{12}$$

$$= \frac{F_0 L}{3T}\left\langle\left\{\left(1-\cos\sqrt{\frac{T}{mL}}t\right)\mathscr{u}(t)-\left[1-\cos\sqrt{\frac{T}{mL}}(t-a)\right]\mathscr{u}(t-a)\right\}\right.$$
$$\left.-\frac{1}{2.5}\left\{\left(1-\cos\sqrt{\frac{5T}{2mL}}t\right)\mathscr{u}(t)-\left[1-\cos\sqrt{\frac{5T}{2mL}}(t-a)\right]\mathscr{u}(t-a)\right\}\right\rangle$$

$$x_2(t) = \eta_1(t)u_{21}+\eta_2(t)u_{22} \tag{h}$$

$$= \frac{F_0 L}{3T}\left\langle\left\{\left(1-\cos\sqrt{\frac{T}{mL}}t\right)\mathscr{u}(t)-\left[1-\cos\sqrt{\frac{T}{mL}}(t-a)\right]\mathscr{u}(t-a)\right\}\right.$$
$$\left.+\frac{1}{5}\left\{\left(1-\cos\sqrt{\frac{5T}{2mL}}t\right)\mathscr{u}(t)-\left[1-\cos\sqrt{\frac{5T}{2mL}}(t-a)\right]\mathscr{u}(t-a)\right\}\right\rangle$$

The response is plotted in Fig. 5.18, in which the values $T/L = 1$ N/m, $m = 1$ kg, $a = 2$ s were used.

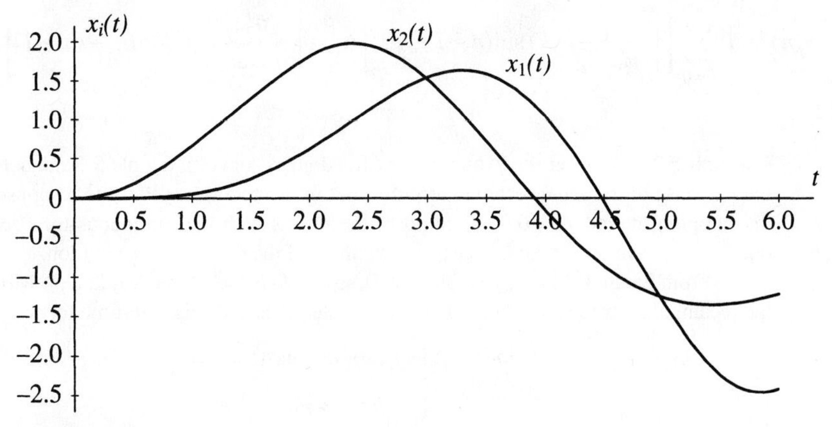

FIGURE 5.18
Response of the system of Fig. 5.4a to a rectangular pulse computed in continuous time

5.11 RESPONSE TO NONPERIODIC EXCITATIONS BY THE CONVOLUTION SUM

As discussed in Sec. 4.9, when the nature of the force vector $\mathbf{F}(t)$ precludes analytical evaluation of convolution integrals for the response, a reasonable alternative is to obtain the response numerically by means of convolution sums. In the case of two-degree-of-freedom systems, the discrete-time solution parallels the continuous-time modal solution presented in Sec. 5.10.

By analogy with Eq. (5.114), the discrete-time response vector can be written in the form of the sequence

$$\mathbf{x}(n) = \eta_1(n)\mathbf{u}_1 + \eta_2(n)\mathbf{u}_2,\ n = 1, 2, \ldots \tag{5.122}$$

where $\eta_1(n)$ and $\eta_2(n)$ are the modal coordinates at the discrete times $t = nT$ $(n = 1, 2, \ldots)$, in which T is the sampling period. Considering Eqs. (5.117) and (5.118), we conclude that the discrete-time modal coordinates can be expressed in the form of the convolution sums

$$\begin{aligned}\eta_1(n) &= \sum_{k=0}^{n} N_1(k)g_1(n-k) = \mathbf{u}_1^T \sum_{k=0}^{n} \mathbf{F}(k)g_1(n-k)\\ \eta_2(n) &= \sum_{k=0}^{n} N_2(k)g_2(n-k) = \mathbf{u}_2^T \sum_{k=0}^{n} \mathbf{F}(k)g_2(n-k)\end{aligned} \tag{5.123}$$

in which $g_1(n)$ and $g_2(n)$ are the discrete-time impulse responses associated with the modal equations, Eqs. (5.117). Hence, using Eqs. (5.120) and recalling Eq. (4.97), we can write

$$g_i(n) = \frac{T}{m'_{ii}\omega_i} \sin n\omega_i T,\ i = 1, 2 \tag{5.124}$$

Inserting Eqs. (5.124) into Eqs. (5.123) and the result into Eq. (5.122), the discrete-time response vector for a two-degree-of-freedom conservative system is simply

$$\mathbf{x}(n) = T\sum_{k=0}^{n}\left\{\left[\frac{\mathbf{u}_1^T}{m'_{11}\omega_1}\mathbf{F}(k)\sin(n-k)\omega_1 T\right]\mathbf{u}_1 + \left[\frac{\mathbf{u}_2^T}{m'_{22}\omega_2}\mathbf{F}(k)\sin(n-k)\omega_2 T\right]\mathbf{u}_2\right\} \tag{5.125}$$

Example 5.7. Consider the two-degree-of-freedom system of Example 5.6 and determine the response to the same rectangular pulse, but in discrete time. Plot the response using the sampling period $T = 0.01$ s, compare the results with the continuous-time response obtained in Example 5.6 and discuss the accuracy of the discrete-time solution.

From Example 5.6, we have $F_1(t) = 0$, so that $F_1(n) = 0$ $(n = 0, 1, 2, \ldots)$. Moreover, the rectangular pulse given by Eq. (a) of Example 5.6 has the discrete-time form

$$F_2(n) = \begin{cases} F_0 \text{ for } n \le n_0 \\ 0 \text{ for } n > n_0 \end{cases} \tag{a}$$

where $n_0 = a/T$, in which a is the pulse duration. Then, considering the analogy with the procedure in Example 4.8 for treating pulses in discrete time, inserting Eqs. (c)–(e) of

Example 5.6 into Eq. (5.125) and letting $T/L = 1$ N/m, $m = 1$ kg, $a = 2$ s, we can write the response in the form

$$\mathbf{x}(n) = \begin{cases} T\displaystyle\sum_{k=0}^{n}\left\{\left[\frac{u_{21}F_2(k)}{m'_{11}\omega_1}\sin(n-k)\omega_1 T\right]\mathbf{u}_1 + \left[\frac{u_{22}F_2(k)}{m'_{22}\omega_2}\sin(n-k)\omega_2 T\right]\mathbf{u}_2\right\} & \text{for } n \le n_0 \\ T\displaystyle\sum_{k=0}^{n_0}\left\{\left[\frac{u_{21}F_2(k)}{m'_{11}\omega_1}\sin(n-k)\omega_1 T\right]\mathbf{u}_1 + \left[\frac{u_{22}F_2(k)}{m'_{22}\omega_2}\sin(n-k)\omega_2 T\right]\mathbf{u}_2\right\} & \text{for } n > n_0 \end{cases}$$

$$= \begin{cases} \dfrac{0.01F_0}{3}\displaystyle\sum_{k=0}^{n}\left\{\sin 0.01(n-k)\begin{bmatrix}1\\1\end{bmatrix} - \sqrt{\frac{2}{5}}\sin 0.01\sqrt{\frac{5}{2}}(n-k)\begin{bmatrix}1\\-0.5\end{bmatrix}\right\} & \text{for } n \le 200 \\ \dfrac{0.01F_0}{3}\displaystyle\sum_{k=0}^{200}\left\{\sin 0.01(n-k)\begin{bmatrix}1\\1\end{bmatrix} - \sqrt{\frac{2}{5}}\sin 0.01\sqrt{\frac{5}{2}}(n-k)\begin{bmatrix}1\\-0.5\end{bmatrix}\right\} & \text{for } n > 200 \end{cases} \quad \text{(b)}$$

which yields the response sequence

$$\mathbf{x}(1) = \frac{0.01F_0}{3}\left\{\sin 0.01\begin{bmatrix}1\\1\end{bmatrix} - \sqrt{\frac{2}{5}}\sin 0.01\sqrt{\frac{5}{2}}\begin{bmatrix}1\\-0.5\end{bmatrix}\right\} = \begin{bmatrix}0.000000\\0.000150\end{bmatrix}$$

$$\mathbf{x}(2) = \frac{0.01F_0}{3}\left\{(\sin 0.02 + \sin 0.01)\begin{bmatrix}1\\1\end{bmatrix} - \sqrt{\frac{2}{5}}\left(\sin 0.02\sqrt{\frac{5}{2}} + \sin 0.01\sqrt{\frac{5}{2}}\right)\begin{bmatrix}1\\-0.5\end{bmatrix}\right\} = \begin{bmatrix}0.000000\\0.000450\end{bmatrix}$$

$$\mathbf{x}(3) = \frac{0.01F_0}{3}\left\{(\sin 0.03 + \sin 0.02 + \sin 0.01)\begin{bmatrix}1\\1\end{bmatrix} - \sqrt{\frac{2}{5}}\left(\sin 0.03\sqrt{\frac{5}{2}} + \sin 0.02\sqrt{\frac{5}{2}} + \sin 0.01\sqrt{\frac{5}{2}}\right)\begin{bmatrix}1\\-0.5\end{bmatrix}\right\} = \begin{bmatrix}0.000001\\0.000900\end{bmatrix}$$

$$\mathbf{x}(200) = \frac{0.01F_0}{3}\left\{(\sin 2.00 + \sin 1.99 + \ldots + \sin 0.01)\begin{bmatrix}1\\1\end{bmatrix} - \sqrt{\frac{2}{5}}\left(\sin 2.00\sqrt{\frac{5}{2}} + \sin 1.99\sqrt{\frac{5}{2}} + \ldots + \sin 0.01\sqrt{\frac{5}{2}}\right)\begin{bmatrix}1\\-0.5\end{bmatrix}\right\} = \begin{bmatrix}0.620849\\1.820598\end{bmatrix}$$

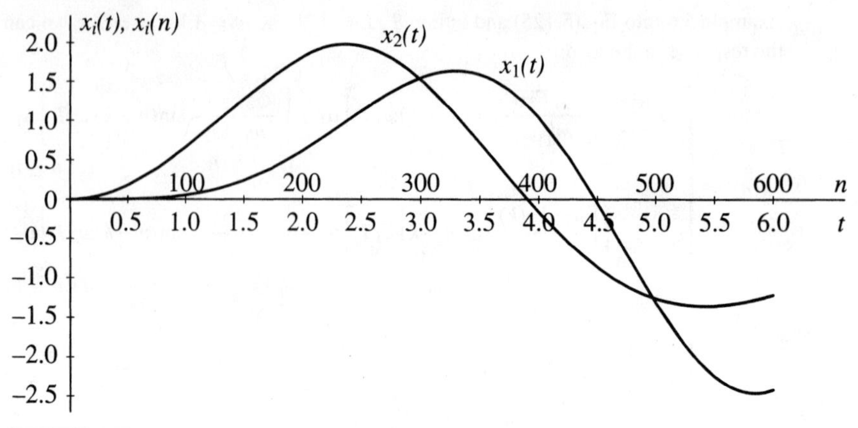

FIGURE 5.19
Response of the system of Fig. 5.4a to a rectangular pulse computed in discrete time

$$\mathbf{x}(201) = \frac{0.01F_0}{3}\left\{(\sin 2.01 + \sin 2.00 + \ldots + \sin 0.01)\begin{bmatrix} 1 \\ 1 \end{bmatrix} - \sqrt{\frac{2}{5}}\left(\sin 2.01\sqrt{\frac{5}{2}} + \sin 2.00\sqrt{\frac{5}{2}} + \ldots + \sin 0.01\sqrt{\frac{5}{2}}\right)\begin{bmatrix} 1 \\ -0.5 \end{bmatrix}\right\} = \begin{bmatrix} 0.630131 \\ 1.829533 \end{bmatrix}$$

$$\mathbf{x}(202) = \frac{0.01F_0}{3}\left\{(\sin 2.02 + \sin 2.01 + \ldots + \sin 0.02)\begin{bmatrix} 1 \\ 1 \end{bmatrix} - \sqrt{\frac{2}{5}}\left(\sin 2.02\sqrt{\frac{5}{2}} + \sin 2.01\sqrt{\frac{5}{2}} + \ldots + \sin 0.02\sqrt{\frac{5}{2}}\right)\begin{bmatrix} 1 \\ -0.5 \end{bmatrix}\right\} = \begin{bmatrix} 0.639469 \\ 1.838226 \end{bmatrix}$$

$$\mathbf{x}(203) = \frac{0.01F_0}{3}\left\{(\sin 2.03 + \sin 2.02 + \ldots + \sin 0.03)\begin{bmatrix} 1 \\ 1 \end{bmatrix} - \sqrt{\frac{2}{5}}\left(\sin 2.03\sqrt{\frac{5}{2}} + \sin 2.02\sqrt{\frac{5}{2}} + \ldots + \sin 0.03\sqrt{\frac{5}{2}}\right)\begin{bmatrix} 1 \\ -0.05 \end{bmatrix}\right\} = \begin{bmatrix} 0.648864 \\ 1.846675 \end{bmatrix}$$

- -

(c)

The response is plotted in Fig. 5.19. For comparison purposes, the continuous-time response obtained in Example 5.6 is also plotted. As can be seen, the agreement is excellent.

5.12 RESPONSE TO INITIAL EXCITATIONS BY MATLAB

The response of a two-degree-of-freedom system to initial excitations can be obtained in closed form, as can be seen from Eq. (5.46). Our interest lies in plotting the response, which consists of two plots, $x_1(t)$ versus t and $x_2(t)$ versus t. They can be obtained conveniently by a MATLAB computer program of Eq. (5.46) in conjunction with the system of Example 5.3, as follows:

```
% The program 'rspin2.m' plots the response of a two-degree-of-freedom system
% to initial excitations

clear
clf

M=[1 0;0 2]; % mass matrix
K=[2 -1;-1 3]; % stiffness matrix
[u,W]=eig(K,M); % solution of the eigenvalue problem:u=matrix of eigenvectors,
% W=matrix of eigenvalues
u(:,1)=u(:,1)/max(u(:,1)); % normalization of the
u(:,2)=u(:,2)/max(u(:,2)); % eigenvectors
[w(1),I1]=min(max(W)); % relabeling of the eigenvalues so that the lowest is the
[w(2),I2]=max(max(W)); % first and the highest is the second
w(1)=sqrt(w(1)); % lowest natural frequency
w(2)=sqrt(w(2)); % highest natural frequency
U(:,1)=u(:,I1); % relabeling of the eigenvectors so as to
U(:,2)=u(:,I2); % correspond to the natural frequencies
x0=[1.2; 0]; % initial displacement
v0=[0; 0]; % initial velocity
t=[0: 0.1: 50]; % initial time, time increment, final time

% displacement components from Eq. (5.46)
x1=(((U(2,2)*x0(1)-U(1,2)*x0(2))*cos(w(1)*t)+(U(2,2)*v0(1)-U(1,2)*v0(2))*sin(w(1)
*t)/w(1))*U(1,1)+((U(1,1)*x0(2)-U(2,1)*x0(1))*cos(w(2)*t)+(U(1,1)*v0(2)-U(2,1)
*v0(1))*sin(w(2)*t)/w(2))*U(1,2))/det(U);

x2=(((U(2,2)*x0(1)-U(1,2)*x0(2))*cos(w(1)*t)+(U(2,2)*v0(1)-U(1,2)*v0(2))*sin(w(1)
*t)/w(1))*U(2,1)+((U(1,1)*x0(2)-U(2,1)*x0(1))*cos(w(2)*t)+(U(1,1)*v0(2)-U(2,1)
*v0(1))*sin(w(2)*t)/w(2))*U(2,2))/det(U);

plot(t, x1, t, x2)
title('Response to Initial Excitations')
ylabel('x_1(t), x_2(t)')
xlabel ('t(s)')
```

Note that the program plots both $x_1(t)$ versus t and $x_2(t)$ versus t on the same diagram. We observe that the expressions for $x_1(t)$ and $x_2(t)$ are relatively long, well exceeding one line. In this regard, it is perhaps worth mentioning that a lengthy expression must not be broken and is to be typed as one continuous line, no matter how long it is, and let the computer break it into individual lines. Indeed, if a lengthy expression is broken into individual lines, then the computer is likely to interpret these lines as separate expressions, and give erroneous results.

5.13 FREQUENCY RESPONSE PLOTS FOR TWO-DEGREE-OF-FREEDOM SYSTEMS BY MATLAB

The frequency response functions for a two-degree-of-freedom system are given by Eqs. (5.103). Using the parameter values given in Example 5.5, they reduce to Eqs. (d) of the same example. Using Eqs. (d) in conjunction with the ratio $F_1/k = 1$, the plots $X_1(\omega)$ versus ω/ω_1 and $X_2(\omega)$ versus ω/ω_1 can be obtained by the following MATLAB program:

```
% The program 'frqrsp2.m' produces frequency response plots for a two-degree-of-
% freedom system

clear
clf

a=2/5; % square of the ratio of the lowest natural frequency w1
% to the highest natural frequency w2
r=[0: 0.0015: 3]; % ratio of the driving frequency w
% to the lowest natural frequency w1

X1=(3-2*r.^2)./(5*(1-r.^2).*(1-a*r.^2)); % frequency response of mass m1
X2=1./(5*(1-r.^2).*(1-a*r.^2)); % frequency response of mass m2

axes('position', [0.1 0.15 0.35 0.45]) % positions of the corner points of X1 diagram
% as fractions of the workspace perimeter dimensions
plot(r, X1)

title('Frequency Response')
ylabel('X_1(\omega)')
xlabel('\omega/\ omega_1')

axis([0 3 -3 3])
grid

axes('position',[0.55 0.15 0.35 0.45]) % positions of the corner points of X2 diagram
% as fractions of the workspace perimeter dimensions
plot(r, X2)

title('Frequency Response')
ylabel('X_2(\ omega)')
xlabel('\omega/\ omega_1')

axis([0 3 -3 3])
grid
```

Note that the vertical lines corresponding to resonance at the frequency ratios $\omega/\omega_1 = 1$ and $\omega/\omega_1 = \sqrt{5/2}$ in the plots are to be ignored.

5.14 RESPONSE TO A RECTANGULAR PULSE BY THE CONVOLUTION SUM USING MATLAB

In Sec. 5.11, we derived the discrete-time response of a two-degree-of-freedom system to arbitrary excitations by means of the convolution sum. Then, in Example 5.7, we applied the general formulation, Eq. (5.125), to an excitation in the form of a rectangular pulse. A MATLAB computer program solving the problem of Example 5.7 reads as follows:

```
% The program 'convsum2.m' plots the response of a two-degree-of-freedom
% system to a rectangular pulse by the convolution sum

clear

M=[1 0;0 2]; % mass matrix
K=[2 -1;-1 3]; % stiffness matrix
[u,W]=eig(K,M); % solution of the eigenvalue problem: u=matrix of eigenvectors
% W=matrix of eigenvalues
u(:,1)=u(:,1)/max(u(:,1)); % normalization of
u(:,2)=u(:,2)/max(u(:,2)); % the eigenvectors
[w(1), I1]=min(max(W)); % relabeling of the eigenvalues so that the lowest
[w(2), I2]=max(max(W)); % is the first and the highest is the second
w(1)=sqrt(w(1)); % lowest natural frequency
w(2)=sqrt(w(2)); % highest natural frequency
U(:,1)=u(:, I1); % relabeling of the eigenvectors so as to
U(:,2)=u(:, I2); % correspond to the natural frequencies
m1=U(:,1)'*M*U(:,1); % mass quantities for the two
m2=U(:,2)'*M*U(:,2); % modes, top of Eqs. (5.116)
T=0.01; % sampling period
N=600; % number of sampling times
% force on the first mass is equal to zero

for n=1:N,
  sum1=0; sum2=0;
  if n<=2/T+1; F2(n)=1; else; F2(n)=0; % force on the second mass, equal to the
  % rectangular pulse
  end

  for k=1:n,
    sum1=sum1+(U(2,1)*F2(k)*sin((n-k)*w(1)*T)/(m1*w(1)))*U(1,1)+(U(2,2)
  *F2(k)*sin((n-k)*w(2)*T)/(m2*w(2)))*U(1,2);
    sum2=sum2+(U(2,1)*F2(k)*sin((n-k)*w(1)*T)/(m1*w(1)))*U(2,1)+(U(2,2)
  *F2(k)*sin((n-k)*w(2)*T)/(m2*w(2)))*U(2,2);

  end

  x1(n)=T*sum1; % displacement of the left mass
  x2(n)=T*sum2; % displacement of the right mass

end

n=[0: 1: N-1];

plot(n, x1, '.', n, x2, '.')
title('Response by the Convolution Sum')
```

```
ylabel('x_1(n), x_2(n)')
xlabel('n')
grid
```

The program can be used to plot $x_1(n)$ versus n and $x_2(n)$ versus n. Note that, because the points are very close together, the plots will appear as being continuous rather than discrete.

5.15 SUMMARY

The motion of a single-degree-of-freedom system is described by a single ordinary differential equation. When an undamped linear system is set in motion by some initial excitations, the system vibrates at a given frequency, where the frequency depends on the system parameters alone, and is independent of the initial excitations. For this reason, the frequency is known as the natural frequency. By contrast, the motion of multi-degree-of-freedom systems is described by as many ordinary differential equations as the number of degrees of freedom. These equations are simultaneous, or coupled, and it is not possible in general to solve one equation independently of the other. Undamped multi-degree-of-freedom linear system differ from single-degree-of-freedom systems in two major respects, namely, they possess as many natural frequencies as the number of degrees of freedom of the system and to each natural frequency corresponds a certain natural mode, namely, a displacement configuration unique to that mode. The natural modes possess the important property of orthogonality, which permits the reduction of the set of simultaneous equations to a set of independent equations, where the latter can be treated as if they were single-degree-of-freedom systems.

This chapter is concerned almost exclusively with two-degree-of-freedom systems. It plays the role of an introduction to multi-degree-of-freedom systems in which various topics such as the eigenvalue problem, modes of vibration, coordinate transformations and coupling, orthogonality of modes and system decoupling, are discussed with only a modest amount of mathematics. General multi-degree-of-freedom are presented in Ch. 7.

Consistent with the idea of shifting the computational burden to the computer, three MATLAB programs are included, the first concerned with the response to initial excitations, the second with the frequency response and the third with the response to nonperiodic excitations by the convolution sum.

PROBLEMS

5.1. Two disks of mass polar moments of inertia I_1 and I_2 are mounted on a circular massless shaft consisting of two segments of torsional stiffness GJ_1 and GJ_2, as shown in Fig. 5.20. Derive the differential equations for the angular displacements of the disks.

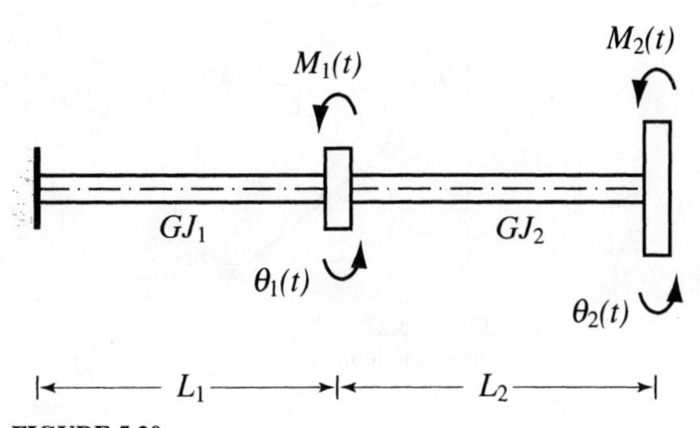

FIGURE 5.20
Two-degree-of-freedom torsional system

5.2. A rigid bar of mass per unit length m carries a point mass M at its right end. The bar is supported by two springs, as shown in Fig. 5.21. Derive the differential equations for the translation and rotation of the mass center. Assume small motions.

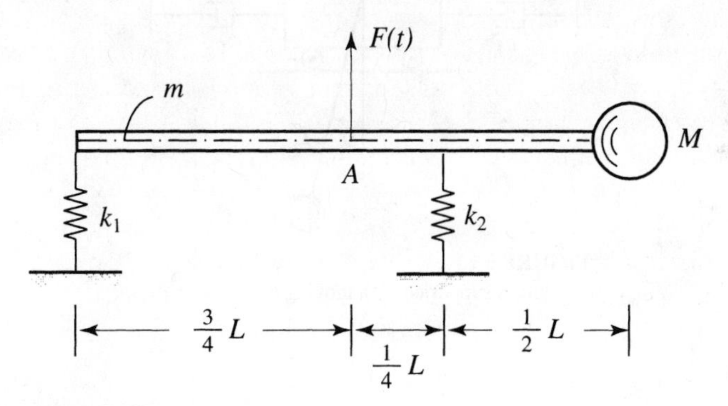

FIGURE 5.21
Mass supported by springs through a rigid bar

5.3. Derive the differential equations of motion for the double pendulum shown in Fig. 5.22. The angles θ_1 and θ_2 can be arbitrarily large.

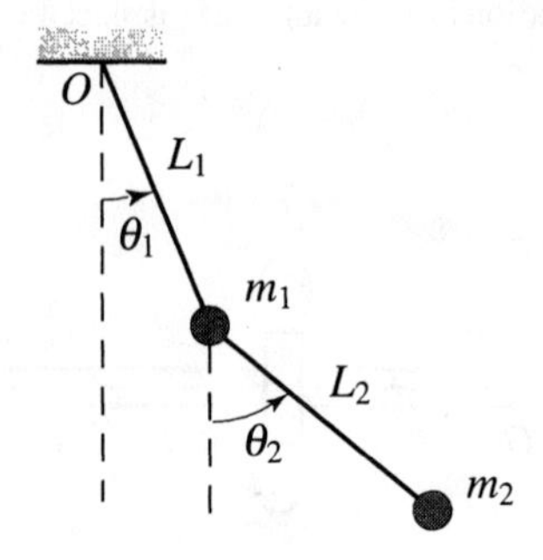

FIGURE 5.22
Double pendulum

5.4. Derive the differential equations of motion for the system shown in Fig. 5.23. Let the angle θ be small.

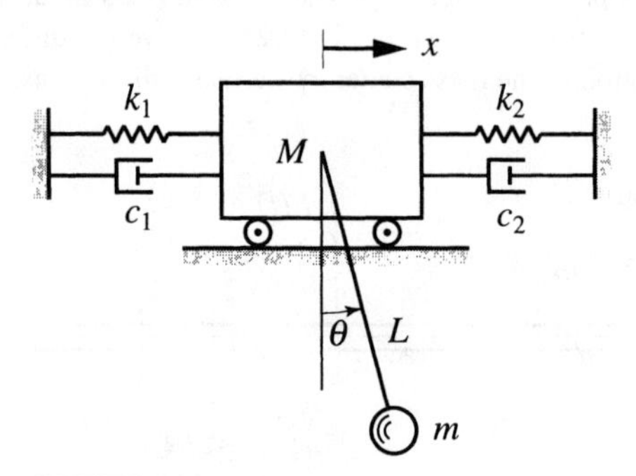

FIGURE 5.23
Pendulum supported by a moving mass

5.5. The system of Fig. 5.24 represents an airfoil section being tested in a wind tunnel. Let the airfoil have total mass m and mass moment of inertia I_C about the mass center C, and derive the differential equations of motion.

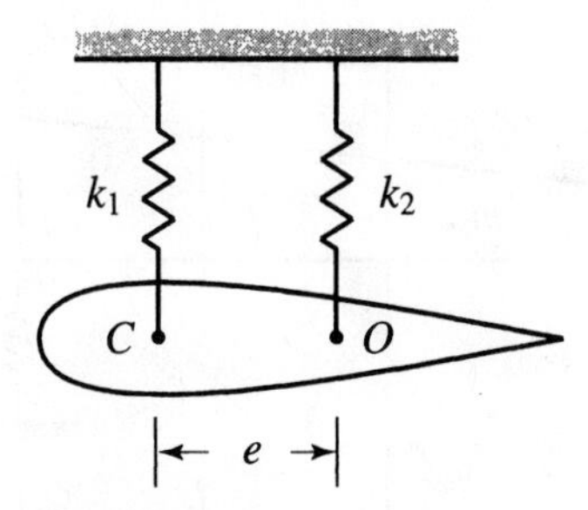

FIGURE 5.24
Airfoil section supported by springs

5.6. A uniform thin rod is suspended by a string, as shown in Fig. 5.25. Derive the differential equations of motion of the system for arbitrarily large angles.

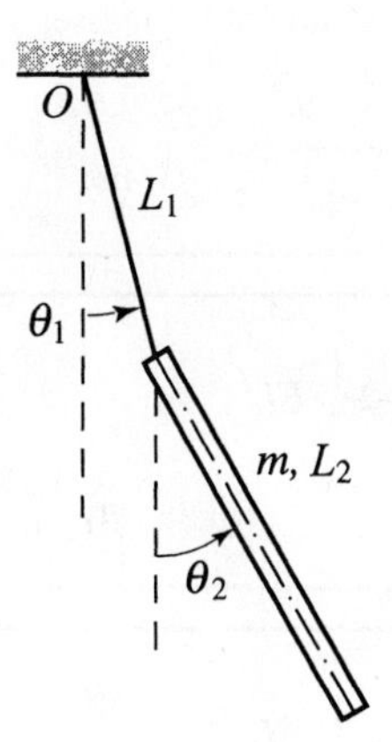

FIGURE 5.25
Rigid rod suspended by a string

5.7. A rigid bar of mass per unit length $\rho(\eta) = \rho_0(1+\eta/L)$ is supported by two springs, as shown in Fig. 5.26. Assume small motions and derive the differential equations of motion.

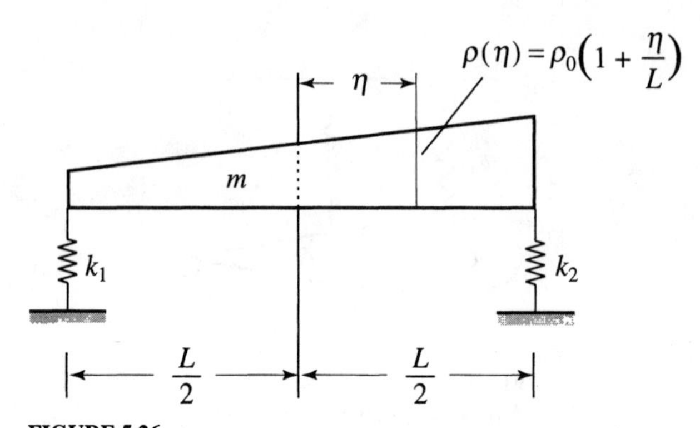

FIGURE 5.26
Nonuniform rigid bar supported by springs

5.8. Figure 5.27 depicts a two-story building. Assume that the horizontal members are rigid and that the columns are massless beams clamped at both ends and derive the differential equations for the horizontal translation of the masses.

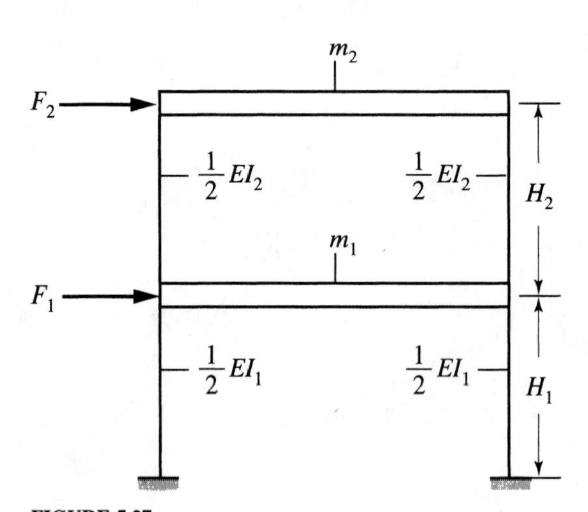

FIGURE 5.27
Two-story building

5.9. A rigid uniform bar is supported by two translational springs and one torsional spring (Fig. 5.28). Derive the differential equations of motion.

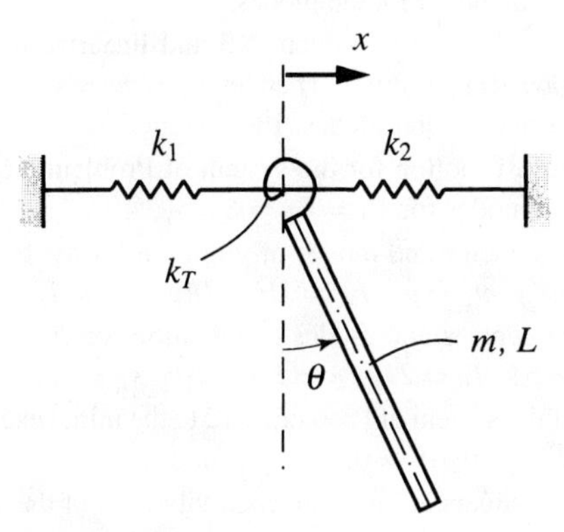

FIGURE 5.28
Rigid rod supported by springs

5.10. Figure 5.29 shows a system of gears mounted on shafts. The radii of gears A and B are related by $R_A/R_B = n$. Derive the differential equations for the torsional motion of the system.

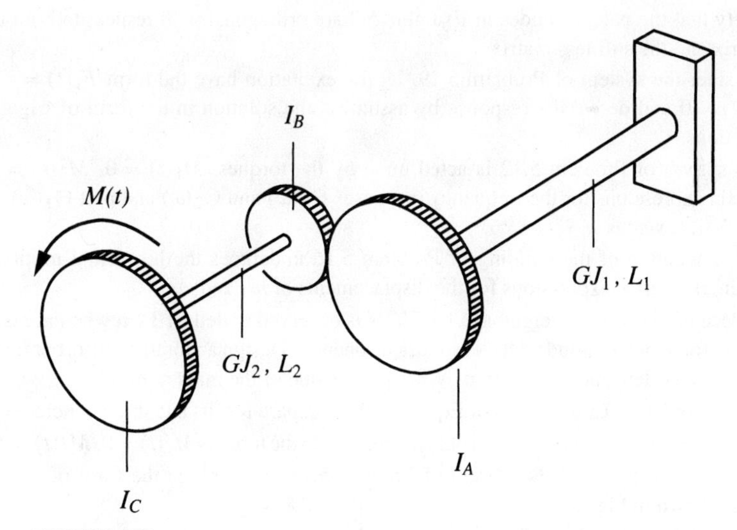

FIGURE 5.29
System of gears mounted on shafts

5.11. Use Eq. (5.35) and prove that the two ratios for u_{21}/u_{11} and the two ratios for u_{22}/u_{21}, Eqs. (5.38), are identical.

5.12. Consider the system of Problem 5.1, let $I_1 = I_2 = I$, $GJ_1 = GJ_2 = GJ$, $L_1 = L_2 = L$, and calculate the natural frequencies and natural modes. Plot the modes.

5.13. Consider the system of Problem 5.2, let $k_1 = k$, $k_2 = 2k$, $M = mL$, and calculate the natural frequencies and natural modes. Plot the modes.

5.14. Consider the double pendulum of Problem 5.3 and linearize the equations of motion by assuming that $\theta_1(t)$ and $\theta_2(t)$ are small. Then let $m_1 = m_2 = m$, $L_1 = L_2 = L$, and calculate the natural frequencies and natural modes. Plot the modes.

5.15. Linearize the equations of motion for the system of Problem 5.6 and calculate the natural frequencies and natural modes for $L_1 = L_2 = L$.

5.16. Obtain the natural frequencies and modes of vibration for the building of Problem 5.8 and plot the modes. Let $m_1 = m_2 = m$, $H_1 = H_2 = H$ and $I_1 = I_2 = I$.

5.17. Obtain the natural frequencies and modes of vibration for the system of gears of Problem 5.10. Let $n = 2$, $I_A = 5I$, $I_B = 2I$, $I_C = I$ and $k_1 = k_2 = k$.

5.18. Obtain the response of the system of Problem 5.12 to the initial excitation $\theta_1(0) = 0$, $\theta_2(0) = 1.5$, $\dot{\theta}_1(0) = 1.8\sqrt{GJ/IL}$, $\dot{\theta}_2(0) = 0$.

5.19. Determine the natural frequencies and modes of vibration of the system of Example 5.1 for the parameters $m_1 = m_2 = m$, $L_1 = L_2 = L_3 = L$. Then, obtain the response to the initial excitation $x_1(0) = 1$, $x_2(0) = -1$, $\dot{x}_1(0) = \dot{x}_2(0) = 0$. Explain your results.

5.20. Consider the system of Problem 5.2 and find a set of coordinates for which the system is elastically uncoupled. Then, let $k_1 = k$, $k_2 = 2k$, $M = mL$, calculate the natural frequencies and natural modes and plot the modes. Compare the results with those obtained in Problem 5.13 and draw conclusions.

5.21. Consider Example 5.2 and use Eqs. (e) in conjunction with the second row of Eq. (c) to derive the natural modes.

5.22. Verify that the natural modes in Example 5.2 are orthogonal with respect to both the mass matrix and the stiffness matrix.

5.23. Consider the system of Problem 5.19, let the excitation have the form $F_1(t) = F_1 \cos \omega t$, $F_2(t) = 0$ and derive the response by assuming the solution in the form of trigonometric functions.

5.24. The system of Problem 5.12 is acted upon by the torques, $M_1(t) = 0$, $M_2(t) = M_2 e^{i\omega t}$. Obtain expressions for the frequency responses $\Theta_1(\omega)$ and $\Theta_2(\omega)$ and plot $\Theta_1(\omega)$ versus ω and $\Theta_2(\omega)$ versus ω.

5.25. The foundation of the building of Problem 5.16 undergoes the horizontal motion $y(t) = Y_0 \sin \omega t$. Derive expressions for the displacements of m_1 and m_2.

5.26. A piece of machinery weighing 2.1×10^4 N is observed to deflect 3 cm when at rest. A harmonic force of magnitude 440 N induces resonance. Design a vibration absorber undergoing a maximum deflection of 2.5 mm. What is the value of the mass ratio μ?

5.27. Solve Problem 5.23 by means of Eqs. (5.117), compare results and draw conclusions.

5.28. Derive the response of the system of Problem 5.12 to the torques $M_1(t) = 0$, $M_2(t) = M_2 e^{-\alpha t}$.

5.29. Derive the response of the system of Problem 5.13 to a force in the form of the sawtooth pulse shown in Fig. 4.20.

5.30. Solve Problem 5.29 by defining the motion in terms of the vertical displacements $y_1(t)$ and $y_2(t)$ of the points of attachment of springs k_1 and k_2, respectively, compare results and draw conclusions.

5.31. Derive the response of the building of Problem 5.25 for the case in which the horizontal motion of the support resembles the triangular pulse shown in Fig. 4.32.

5.32. Solve Problem 5.28 in discrete time and plot the response.

5.33. Solve Problem 5.30 in discrete time and plot the response.

5.34. Solve Problem 5.31 in discrete time and plot the response.

5.35. Write a MATLAB program and plot the response of the system of Problem 5.18.

5.36. Write a MATLAB program and plot the response of the system of Problem 5.19.

5.37. Write a MATLAB program and plot the frequency responses from the system of Problem 5.24.

5.38. Write a MATLAB program and plot the frequency responses for the system of Problem 5.25.

5.39. Solve Problem 5.32 by MATLAB.

5.40. Solve Problem 5.33 by MATLAB.

5.41. Solve Problem 5.34 by MATLAB.

CHAPTER 6

ELEMENTS OF ANALYTICAL DYNAMICS

Newton's laws were formulated for a single particle and can be extended to systems of particles and rigid bodies, as well as to systems of rigid bodies. The equations of motion are expressed in terms of physical coordinates and forces, both quantities conveniently represented by vectors. For this reason, *Newtonian mechanics* is often referred to as *vectorial mechanics*. The main drawback of Newtonian mechanics is that it requires one free-body diagram for each of the masses in the system, thus necessitating the inclusion of reaction forces and interacting forces, the latter resulting from kinematical constraints ensuring that the individual bodies act together as a system. These reaction and constraint forces play the role of unknowns, which makes it necessary to work with a surplus of equations of motion, one additional equation for every unknown force.

A different approach to mechanics, referred to as *analytical mechanics*, or *analytical dynamics*, considers the system as a whole, rather than the individual components separately, a process that excludes the reaction and constraint forces automatically. This approach, due to Lagrange, permits the formulation of problems of dynamics in terms of two scalar functions, the kinetic energy and the potential energy, and an infinitesimal expression, the virtual work performed by the nonconservative forces. Analytical mechanics represents a broader and more abstract approach, as the equations of motion are formulated in terms of generalized coordinates and generalized forces, which are not necessarily physical coordinates and forces, although in certain cases they can be chosen as such. In this manner, the mathematical formulation is rendered independent of any special system of coordinates. The development of analytical mechanics required the introduction of the concept of virtual displacements, which in turn led to the development of the calculus of variations. For this reason, analytical mechanics is often referred to as the *variational approach to mechanics*.

We begin this chapter with a discussion of such concepts as constraints, degrees of freedom and generalized coordinates, thus paving the way from Newtonian mechanics to analytical mechanics. Then, we introduce the concept of virtual displacements, followed by the virtual work principle and d'Alembert's principle, thus providing the groundwork for the real object of this chapter, namely, the extended Hamilton's principle and Lagrange's equations, both extremely efficient methods for deriving equations of motion.

6.1 DEGREES OF FREEDOM AND GENERALIZED COORDINATES

To derive the equations of motion for a system of masses by the Newtonian approach, it is necessary to isolate the masses and draw one free-body diagram for each of the masses. We recall that a free-body diagram is a drawing of a given mass with all the forces acting upon it. These include applied forces, reaction forces and internal forces, where the latter become external to the mass when, in the process of isolating the mass, it is necessary to cut through the line of action of the internal force. This process tends to result in more equations and unknowns than necessary, which is the case when forces presenting no particular interest act as unknowns. An illustration of a force internal to the system being treated as external to the mass, and one of no interest in general, is the tensile force T in the string of Example 1.6. Another source of possible difficulties in using Newton's equations is that the motion is described in terms of physical coordinates, which may not always be independent. As an example, we consider a dumbbell consisting of two masses m_1 and m_2 connected by a massless rigid bar of length L, as shown in Fig. 6.1. Assuming that the dumbbell moves in the x, y-plane, we can define the motion by the position vectors

$$\mathbf{r}_1 = x_1\mathbf{i} + y_1\mathbf{j}, \; \mathbf{r}_2 = x_2\mathbf{i} + y_2\mathbf{j} \tag{6.1}$$

which involve four coordinates, x_1, y_1, x_2 and y_2. Clearly, they are not independent, as the length of the bar cannot change. Indeed, the four coordinates are related by the

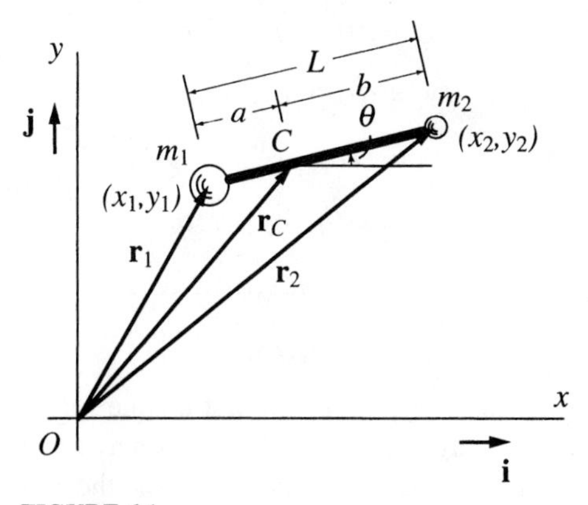

FIGURE 6.1
A dumbbell in planar motion

equation

$$(x_2 - x_1)^2 + (y_2 - y_1)^2 = L^2 \tag{6.2}$$

which represents a *constraint equation*. Because Eq. (6.2) can be solved for one of the coordinates in terms of the remaining three, it follows that only three coordinates are independent. Hence, if the problem is formulated in terms of the coordinates x_1, y_1, x_2 and y_2, then it is necessary to supplement the corresponding equations of motion by the constraint equation, Eq. (6.2).

In most vibration problems, a better choice of coordinates obviates the difficulties involved in working with surplus coordinates and constraint equations. As an illustration, in the case of the dumbbell of Fig. 6.1, it is much simpler to work with the position $\mathbf{r}_C = \mathbf{r}_C(x_C, y_C)$ of the mass center C (Sec. 1.4) and the angle θ between the rigid bar and the x-axis. Indeed, the motion can be described directly in terms of the three independent coordinates x_C, y_C and θ, and no constraint equation is needed. From Fig. 6.1, we observe that the two sets of coordinates are related by

$$\mathbf{r}_1 = \mathbf{r}_C - a(\cos\theta\mathbf{i} + \sin\theta\mathbf{j}),\ \mathbf{r}_2 = \mathbf{r}_C + b(\cos\theta\mathbf{i} + \sin\theta\mathbf{j}) \tag{6.3}$$

where, from Eq. (1.37),

$$a = \frac{m_2 L}{m_1 + m_2},\ b = \frac{m_1 L}{m_1 + m_2} \tag{6.4}$$

The problem can be generalized to a system of N mass particles with positions defined by the radius vectors $\mathbf{r}_i(x_i, y_i, z_i)$ in a three-dimensional space. If the particles are subject to a number c of constraints, then only

$$n = 3N - c \tag{6.5}$$

coordinates are independent, where n is known as the *number of degrees of freedom* of the system. It is customary to denote the independent coordinates by $q_1, q_2, \ldots, q_n$ and refer to them as *generalized coordinates*. The relation between the dependent and independent coordinates can be expressed in the form of the *coordinate transformation*

$$\begin{aligned} x_1 &= x_1(q_1, q_2, \ldots, q_n) \\ y_1 &= y_1(q_1, q_2, \ldots, q_n) \\ z_1 &= z_1(q_1, q_2, \ldots, q_n) \\ x_2 &= x_2(q_1, q_2, \ldots, q_n) \\ &\cdots\cdots\cdots\cdots \\ z_N &= z_N(q_1, q_2, \ldots, q_n) \end{aligned} \tag{6.6}$$

The generalized coordinates $q_1, q_2, \ldots, q_n$ are not unique for a given system, as there can be several such sets, although only one or two sets may represent a suitable choice. In vibrations, this choice is obvious for the most part, and the coordinate transformation (6.6) is primarily of academic interest. For example, in the case of the dumbbell of Fig. 6.1, any three of the four coordinates x_1, y_1, x_2, y_2 can serve as a set of generalized coordinates, but none of them would be a suitable choice. Clearly, the choice $q_1 = x_C$, $q_2 = y_C$, $q_3 = \theta$ is by far the most convenient one, and indeed the only suitable one, as the equations of motion in terms of these coordinates have the simplest form (see Secs. 1.4 and 1.5).

With the introduction of the concept of generalized coordinates, we begin the transition from Newtonian mechanics to Lagrangian mechanics.

6.2 THE PRINCIPLE OF VIRTUAL WORK

The principle of virtual work, due to Johann Bernoulli, is basically a statement of the static equilibrium of a mechanical system. It represents the first variational principle of mechanics. Our interest in the principle is not as a method for determining equilibrium positions but as a tool for effecting the transition from Newtonian mechanics to Lagrangian mechanics. To derive the principle, it is necessary to introduce several new concepts, such as virtual displacements and constraint forces.

We consider a system of N particles in a three-dimensional space and define the *virtual displacements* δx_1, δy_1, δz_1, $\delta x_2, \ldots,$ δz_N as *infinitesimal changes* in the coordinates x_1, y_1, z_1, $x_2, \ldots, z_N$. The virtual displacements must be *consistent with the system constraints*, but are otherwise *arbitrary*. As an example, if a particle in a real situation is confined to a surface, then the virtual displacement must be parallel to the surface, as the particle cannot penetrate the surface, nor can it leave the surface. The virtual displacements represent small variations in the coordinates resulting from imagining the system in a slightly displaced position. Implied in this process is the assumption that *the virtual displacements take place instantaneously*, i.e., they do not necessitate any time to materialize, $\delta t = 0$. The symbol δ was introduced by Lagrange to emphasize the virtual character of the instantaneous variations, as opposed to the symbol d designating actual differentials of position coordinates taking place in the time interval dt, during which time interval forces can change. *The virtual displacements*, being infinitesimal, *obey the rules of differential calculus.*

We assume that every one of the N particles in the system is acted upon by the resultant force

$$\mathbf{R}_i = \mathbf{F}_i + \mathbf{f}_i, \; i = 1, 2, \ldots, N \tag{6.7}$$

where $\mathbf{F}_i$ is an *applied force* and $\mathbf{f}_i$ is a *constraint force*. Examples of applied forces are gravitational forces, aerodynamic lift and drag, magnetic forces, etc. On the other hand, an example of a constraint force is the force that keeps a particle confined to a given surface, as mentioned earlier in this section. For a system in equilibrium every particle must be at rest, so that the resultant force on each particle must vanish.

$$\mathbf{R}_i = \mathbf{F}_i + \mathbf{f}_i = \mathbf{0}, \; i = 1, 2, \ldots, N \tag{6.8}$$

from which it follows that

$$\overline{\delta W}_i = \mathbf{R}_i \cdot \delta \mathbf{r}_i = 0, \; i = 1, 2, \ldots, N \tag{6.9}$$

is also true. The scalar product in Eq. (6.9) represents the *virtual work* performed by the resultant force vector $\mathbf{R}_i$ over the virtual displacement vector $\delta \mathbf{r}_i$ of particle i. Summing up over i, it follows that the virtual work for the entire system must vanish, or

$$\overline{\delta W} = \sum_{i=1}^{N} \mathbf{R}_i \cdot \delta \mathbf{r}_i = 0 \tag{6.10}$$

so that, inserting Eq. (6.8) into Eq. (6.10), we obtain

$$\overline{\delta W} = \sum_{i=1}^{N} \mathbf{F}_i \cdot \delta \mathbf{r}_i + \sum_{i=1}^{N} \mathbf{f}_i \cdot \delta \mathbf{r}_i = 0 \tag{6.11}$$

Next, we limit ourselves to systems for which the virtual work performed by the constraint forces is zero. An example of this is a particle confined to a smooth surface, as shown in Fig. 6.2a. In this case, the constraint force is normal to the surface and the virtual displacement is parallel to the surface, so that the virtual work is zero, because the scalar product of two vectors normal to one another is zero. On the other hand, if the particle is confined to a rough surface, in addition to the normal component of force, there is a tangential component of force due to friction, as shown in Fig. 6.2b. Hence, the virtual work performed by the constraint force is not zero. Ruling out friction forces, as well as any other forces for which the virtual work is not zero, we can write

$$\sum_{i=1}^{N} \mathbf{f}_i \cdot \delta \mathbf{r}_i = 0 \tag{6.12}$$

Inserting Eq. (6.12) into Eq. (6.11), we conclude that

$$\overline{\delta W} = \sum_{i=1}^{N} \mathbf{F}_i \cdot \delta \mathbf{r}_i = 0 \tag{6.13}$$

or *the work performed by the applied forces through infinitesimal virtual displacements compatible with the system constraints is zero*. This is the statement of the *principle of virtual work*.

When the virtual displacements are all independent, the principle of virtual work can be used to determine the conditions of static equilibrium of a system. Indeed, when all $\delta \mathbf{r}_i$ $(i = 1, 2, \ldots, N)$ are independent, we can invoke the arbitrariness of the virtual displacements and conclude that Eq. (6.13) can be satisfied for all possible values of $\delta \mathbf{r}_i$ only if

$$\mathbf{F}_i = \mathbf{0}, \; i = 1, 2, \ldots, N \tag{6.14}$$

Equations (6.14) represent the *equilibrium conditions*.

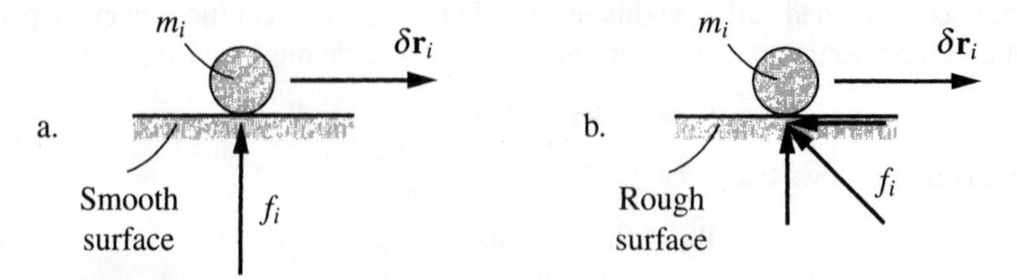

FIGURE 6.2
a. Particle on a smooth surface, b. Particle on a rough surface

The situation is entirely different when the coordinates $\mathbf{r}_i$ $(i = 1, 2, \ldots, N)$ are not independent, but related by constraint equations. As indicated in Sec. 6.1, in this case it is more convenient to switch to a set of generalized coordinates $q_1, q_2, \ldots, q_n$, which are independent by definition. To this end, we rewrite the coordinate transformation given by Eqs. (6.6) in the more compact vector form

$$\mathbf{r}_i = \mathbf{r}_i(q_1, q_2, \ldots, q_n), \; i = 1, 2, \ldots, N \tag{6.15}$$

Then, using rules of differential calculus, we obtain the virtual displacements

$$\delta\mathbf{r}_i = \frac{\partial \mathbf{r}_i}{\partial q_1}\delta q_1 + \frac{\partial \mathbf{r}_i}{\partial q_2}\delta q_2 + \ldots + \frac{\partial \mathbf{r}_i}{\partial q_n}\delta q_n = \sum_{k=1}^{n} \frac{\partial \mathbf{r}_i}{\partial q_k}\delta q_k, \; i = 1, 2, \ldots, N \tag{6.16}$$

where $\delta q_1, \delta q_2, \ldots, \delta q_n$ are *virtual generalized displacements*. Unlike $\delta\mathbf{r}_i$ $(i = 1, 2, \ldots, N)$, however, δq_k $(k = 1, 2, \ldots, n)$ are all independent. Inserting Eqs. (6.16) into Eq. (6.13) and changing the summation order, we can write

$$\overline{\delta W} = \sum_{i=1}^{N} \mathbf{F}_i \cdot \delta\mathbf{r}_i = \sum_{i=1}^{N} \mathbf{F}_i \cdot \sum_{k=1}^{n} \frac{\partial \mathbf{r}_i}{\partial q_k}\delta q_k = \sum_{k=1}^{n}\left(\sum_{i=1}^{N} \mathbf{F}_i \cdot \frac{\partial \mathbf{r}_i}{\partial q_k}\right)\delta q_k = \sum_{k=1}^{n} Q_k \delta q_k = 0 \tag{6.17}$$

in which

$$Q_k = \sum_{i=1}^{N} \mathbf{F}_i \cdot \frac{\partial \mathbf{r}_i}{\partial q_k}, \; k = 1, 2, \ldots, n \tag{6.18}$$

are known as *generalized forces*. The situation is different now, because all δq_k $(k = 1, 2, \ldots, n)$ are independent, and hence entirely arbitrary, so that they can be assigned values at will. Letting first $\delta q_1 = 1, \; \delta q_2 = \delta q_3 = \ldots = \delta q_n = 0$, we conclude that Eq. (6.17) can be satisfied only if $Q_1 = 0$. Repeating the same argument, but with $k = 2, 3, \ldots, n$, in sequence, instead of $k = 1$, we obtain the equilibrium conditions

$$Q_k = 0, \; k = 1, 2, \ldots, n \tag{6.19}$$

6.3 THE PRINCIPLE OF D'ALEMBERT

The principle of virtual work is concerned with the static equilibrium of systems. By itself, it cannot be used to formulate problems in vibrations, which are basically problems of dynamics. However, the virtual work principle can be extended to dynamics, in which form it is known as d'Alembert's principle.

Consistent with Eq. (6.7), we assume that a typical mass particle m_i in a system of particles $(i = 1, 2, \ldots, N)$ is acted upon by the applied force $\mathbf{F}_i$ and the constraint force $\mathbf{f}_i$ and, if any internal forces are negligibly small, we can rewrite Newton's second law for particle m_i, Eq. (1.28), in the form

$$\mathbf{F}_i + \mathbf{f}_i - m_i\ddot{\mathbf{r}}_i = \mathbf{0}, \; i = 1, 2, \ldots, N \tag{6.20}$$

where $-m_i\ddot{\mathbf{r}}_i$ can be regarded as an *inertia force*, which is simply the negative of the time rate of change of the momentum vector, $\mathbf{p}_i = m_i\dot{\mathbf{r}}_i$. Equation (6.20) is often referred to as d'Alembert's principle, and it permits us to regard problems of dynamics as if they were problems of statics. However, our interest in Eq. (6.20) is not for the purpose of deriving equations of motion but for extending the principle of virtual work to the dynamical case. Indeed, using Eq. (6.20) and following the same approach as in Sec. 6.2, we can write the virtual work for particle m_i as

$$(\mathbf{F}_i + \mathbf{f}_i - m_i \ddot{\mathbf{r}}_i) \cdot \delta \mathbf{r}_i = 0, \ i = 1, 2, \ldots, N \tag{6.21}$$

Then, confining ourselves to constraint forces $\mathbf{f}_i$ for which the virtual work is zero and summing up over the system of particles, we obtain

$$\sum_{i=1}^{N} (\mathbf{F}_i - m_i \ddot{\mathbf{r}}_i) \cdot \delta \mathbf{r}_i = 0 \tag{6.22}$$

Equation (6.22) embodies both the virtual work principle of statics and d'Alembert's principle, and is referred to as the *generalized principle of d'Alembert*. It is also referred to at times as the *Lagrange version of d'Alembert's principle*. The sum of the applied force $\mathbf{F}_i$ and the inertia force $-m_i \ddot{\mathbf{r}}_i$, i.e., $\mathbf{F}_i - m_i \ddot{\mathbf{r}}_i$, is sometimes called the *effective force* acting on particle m_i. Hence, we can state the generalized principle of d'Alembert as follows: *The virtual work performed by the effective forces through infinitesimal virtual displacements compatible with the system constraints is zero.*

D'Alembert's principle, Eq. (6.22), can be used to derive all the equations of motion of the system, provided the position vectors $\mathbf{r}_i$ $(i = 1, 2, \ldots, N)$ are all independent. Otherwise, it is necessary to carry out a coordinate transformation from the dependent coordinates $\mathbf{r}_i$ $(i = 1, 2, \ldots, N)$ to the independent generalized coordinates q_k $(k = 1, 2, \ldots, n)$ as given by Eqs. (6.15). Whereas this would provide d'Alembert's principle with a clear advantage over the Newtonian approach in deriving equations of motion, the process becomes increasingly inefficient as the number of degrees of freedom of the system increases, so that an approach using generalized coordinates directly, i.e., without the need for coordinate transformations, demands itself. In this regard, it should be stated that we never really intended to use d'Alembert's principle to derive equations of motion, but only as a means for deriving another variational principle, namely, the extended Hamilton's principle. The latter principle can be used to derive all the system equations of motion from three scalar quantities, the kinetic energy, the potential energy and the virtual work of the nonconservative forces. It can also be used to derive the celebrated Lagrange's equations.

6.4 THE EXTENDED HAMILTON'S PRINCIPLE

The extended Hamilton's principle is arguably the most powerful variational principle of mechanics. Its derivation from the generalized d'Alembert's principle, Eq. (6.22), is a relatively easy task. The extended Hamilton's principle is as useful as it is powerful. Indeed, it yields results where other approaches encounter difficulties, particularly in problems associated with distributed-parameter systems.

We begin with the case in which the position vectors $\mathbf{r}_i$ $(i = 1, 2, \ldots, N)$ are all independent. With reference to Eq. (6.22), we first recognize that

$$\sum_{i=1}^{N} \mathbf{F}_i \cdot \delta\mathbf{r}_i = \overline{\delta W} \tag{6.23}$$

is simply the virtual work of all the applied forces, including both conservative and nonconservative forces. On the other hand, to reduce the second term in Eq. (6.22) to a form suitable for our purposes, we consider the following:

$$\begin{aligned} \frac{d}{dt}(m_i\dot{\mathbf{r}}_i \cdot \delta\mathbf{r}) &= m_i\ddot{\mathbf{r}}_i \cdot \delta\mathbf{r}_i + m_i\dot{\mathbf{r}}_i \cdot \delta\dot{\mathbf{r}}_i = m_i\ddot{\mathbf{r}}_i \cdot \delta\mathbf{r}_i + \delta(\tfrac{1}{2}m_i\dot{\mathbf{r}}_i \cdot \dot{\mathbf{r}}_i) \\ &= m_i\ddot{\mathbf{r}}_i \cdot \delta\mathbf{r}_i + \delta T_i \end{aligned} \tag{6.24}$$

where T_i is the kinetic energy of particle m_i. Rearranging Eq. (6.24) and integrating with respect to time over the interval $t_1 \le t \le t_2$, we can write

$$\begin{aligned} -\int_{t_1}^{t_2} m_i\ddot{\mathbf{r}}_i \cdot \delta\mathbf{r}_i dt &= \int_{t_1}^{t_2} \delta T_i dt - \int_{t_1}^{t_2} \frac{d}{dt}(m_i\dot{\mathbf{r}}_i \cdot \delta\mathbf{r}_i)dt \\ &= \int_{t_1}^{t_2} \delta T_i dt - m_i\dot{\mathbf{r}}_i \cdot \delta\mathbf{r}_i\Big|_{t_1}^{t_2} \end{aligned} \tag{6.25}$$

But, the virtual displacements are arbitrary. Hence, it is convenient to choose them so as to satisfy $\delta\mathbf{r}_i = \mathbf{0}$ at $t = t_1$ and $t = t_2$, in which case Eq. (6.25) reduces to

$$-\int_{t_1}^{t_2} m_i\ddot{\mathbf{r}}_i \cdot \delta\mathbf{r}_i dt = \int_{t_1}^{t_2} \delta T_i dt, \ \delta\mathbf{r}_i = \mathbf{0}, \ t = t_1, t_2; \ i = 1, 2, \ldots, N \tag{6.26}$$

Summing up over i and integrating with respect to t over the interval $t_1 \le t \le t_2$ the second term in Eq. (6.22) integrated over t becomes

$$-\int_{t_1}^{t_2} \sum_{i=1}^{N} m_i\ddot{\mathbf{r}}_i \cdot \delta\mathbf{r}_i dt = \int_{t_1}^{t_2} \delta T dt, \ \delta\mathbf{r}_i = \mathbf{0}, \ i = 1, 2, \ldots, N; \ t = t_1, t_2 \tag{6.27}$$

in which T is the system kinetic energy. Finally, integrating Eq. (6.22) with respect to time over the interval $t_1 \le t \le t_2$ and using Eqs. (6.23) and (6.27), we obtain

$$\int_{t_1}^{t_2} (\delta T + \overline{\delta W})dt = 0, \ \delta\mathbf{r}_i = \mathbf{0}, \ i = 1, 2, \ldots, N; \ t = t_1, t_2 \tag{6.28}$$

which represents the mathematical statement of the *extended Hamilton's principle*. Equation (6.28) can be used to derive all the equations of motion. Although the equation may appear intimidating, we hasten to point out that no integrations are actually necessary. The only operations involved are some generic integrations by parts, invocations of the arbitrariness of the virtual displacements and due consideration to the auxiliary conditions on the virtual displacements listed with Eq. (6.28).

It is often convenient to divide the virtual work into two parts, one due to conservative forces and another one due to nonconservative forces. Hence, by analogy with

Eqs. (1.19) and (1.20), we can write

$$\overline{\delta W} = \overline{\delta W_c} + \overline{\delta W}_{nc} = -\delta V + \overline{\delta W}_{nc} \tag{6.29}$$

where V is the potential energy. Inserting Eq. (6.29) into Eq (6.28), we can rewrite the extended Hamilton's principle in the form

$$\int_{t_1}^{t_2} (\delta T - \delta V + \overline{\delta W}_{nc}) dt = 0, \; \delta \mathbf{r}_i = \mathbf{0}, \; i = 1, 2, \ldots, N; \; t = t_1, t_2 \tag{6.30}$$

so that all the equations of motion can be obtained from three scalar quantities, the kinetic energy, the potential energy and the virtual work due to nonconservative forces.

Next, we consider the case in which the position vectors $\mathbf{r}_i$ $(i = 1, 2, \ldots, N)$ are not independent, but related through some constraint equations of the type encountered in Sec. 6.1. In this regard, we observe that the extended Hamilton's principle involves the kinetic energy T, potential energy V and virtual work of the nonconservative forces $\overline{\delta W}_{nc}$, all three quantities being independent of the coordinates used. It follows that, although derived for a system of particles with the motion described in terms of the rectangular coordinates $\mathbf{r}_i$ $(i = 1, 2, \ldots, N)$, the extended Hamilton's principle retains its form for all sets of coordinates. In fact, the form is the same for all types of dynamical systems. In view of this, it is convenient to express T, V and $\overline{\delta W}_{nc}$ directly in terms of the independent generalized coordinates q_k $(k = 1, 2, \ldots, n)$. Moreover, recognizing from Eqs. (6.17) that $\delta \mathbf{r}_i = \mathbf{0}$ $(i = 1, 2, \ldots, N)$ implies that $\delta q_k = 0$ $(k = 1, 2, \ldots, n)$, we can replace the auxiliary conditions in Eq. (6.30) and rewrite the extended Hamilton's principle in the form

$$\int_{t_1}^{t_2} (\delta T - \delta V + \overline{\delta W}_{nc}) dt = 0, \; \delta q_k = 0, \; k = 1, 2, \ldots, n; \; t = t_1, t_2 \tag{6.31}$$

The extended Hamilton's principle, Eq. (6.31), being in terms of generalized coordinates, can be used to derive all the system equations of motion, regardless of whether the system is subjected to constraints or not. The only qualification is that the constraint forces perform no work. The process of deriving the system equations of motion by means of Eq. (6.31) hinges on the fact that the virtual generalized coordinates δq_k $(k = 1, 2, \ldots, n)$ are independent, and hence entirely arbitrary. We demonstrate this process in Example 6.1 at the end of this section. Indeed, perhaps more than for any other method, an example is absolutely essential to understanding the process.

For conservative systems $\overline{\delta W}_{nc} = 0$, so that Eq. (6.31) reduces to

$$\int_{t_1}^{t_2} \delta L \, dt = 0, \; \delta q_k = 0, \; k = 1, 2, \ldots, n; \; t = t_1, t_2 \tag{6.32}$$

where

$$L = T - V \tag{6.33}$$

is known as the *Lagrangian*. Equation (6.32) is referred to as *Hamilton's principle*.

Example 6.1. Use the extended Hamilton's principle to derive the equations of motion for the two-degree-of-freedom system consisting of a rigid bar suspended on a string, as shown in Fig. 1.10a. Note that the equations of motion for the bar were derived in Example 1.6 by means of the Newtonian approach and the kinetic energy was derived in Example 1.7. Although not stated explicitly, the angles θ_1 and θ_2 used in both examples do represent generalized coordinates, so that we propose to use $q_1 = \theta_1$, $q_2 = \theta_2$ in the present example.

From Example 1.7, we obtain the kinetic energy

$$T = T_{tr} + T_{rot} = \frac{1}{2} m \left[L_1^2 \dot{\theta}_1^2 + L_1 L_2 \dot{\theta}_1 \dot{\theta}_2 \cos(\theta_2 - \theta_1) + \frac{L_2^2}{4} \dot{\theta}_2^2 \right] + \frac{1}{2} \frac{m L_2^2}{12} \dot{\theta}_2^2$$

$$= \frac{1}{2} m \left[L_1^2 \dot{\theta}_1^2 + L_1 L_2 \dot{\theta}_1 \dot{\theta}_2 \cos(\theta_2 - \theta_1) + \frac{L_2^2}{3} \dot{\theta}_2^2 \right] \tag{a}$$

The potential energy is due to the weight mg, which is constant; it is the only conservative force acting on the system. Referring to Fig. 1.10a and choosing $\theta_1 = \theta_2 = 0$ as the reference position, Eq. (1.18) yields

$$V = \int_{\mathbf{r}_C}^{\mathbf{r}_{C\text{ref}}} (-mg\mathbf{j}) \cdot d\mathbf{r}_C = -mg\mathbf{j} \cdot \mathbf{r}_C \Big|_{\mathbf{r}_C}^{\mathbf{r}_{C\text{ref}}}$$

$$= -mg\mathbf{j} \cdot \left\{ -\left(L_1 + \frac{L_2}{2} \right) \mathbf{j} - \left[\left(L_1 \sin\theta_1 + \frac{L_2}{2} \sin\theta_2 \right) \mathbf{i} - \left(L_1 \cos\theta_1 + \frac{L_2}{2} \cos\theta_2 \right) \mathbf{j} \right] \right\}$$

$$= mg \left[L_1 (1 - \cos\theta_1) + \frac{L_2}{2} (1 - \cos\theta_2) \right] \tag{b}$$

Note that the potential energy could have been obtained with greater ease by writing

$$V = mg\Delta h \tag{c}$$

where

$$\Delta h = L_1 (1 - \cos\theta_1) + \frac{L_2}{2} (1 - \cos\theta_2) \tag{d}$$

is the height of the mass center C above its reference position.

Referring to Fig. 1.10a, using Eq. (6.23) and recognizing that F is the only nonconservative force, the associated virtual work can be written as

$$\overline{\delta W}_{nc} = \mathbf{F} \cdot \delta \mathbf{r}_B = F\mathbf{i} \cdot \delta \left[(L_1 \sin\theta_1 + L_2 \sin\theta_2) \mathbf{i} - (L_1 \cos\theta_1 + L_2 \cos\theta_2) \mathbf{j} \right]$$

$$= F (L_1 \cos\theta_1 \delta\theta_1 + L_2 \cos\theta_2 \delta\theta_2) = \Theta_1 \delta\theta_1 + \Theta_2 \delta\theta_2 = Q_1 \delta q_1 + Q_2 \delta q_2 \tag{e}$$

in which

$$Q_1 = \Theta_1 = F L_1 \cos\theta_1, \quad Q_2 = \Theta_2 = F L_2 \cos\theta_2 \tag{f}$$

represent the generalized nonconservative forces.

The extended Hamilton's principle, Eq. (6.31), calls for the variations δT and δV, rather than for T and V themselves. To obtain them, we recognize that the variation process is analogous to the differentiation process. Hence, the variation in the kinetic energy is

$$\begin{aligned}\delta T =& mL_1^2\dot{\theta}_1\delta\dot{\theta}_1 + \frac{mL_1L_2}{2}[\dot{\theta}_2\cos(\theta_2-\theta_1)\delta\dot{\theta}_1 + \dot{\theta}_1\cos(\theta_2-\theta_1)\delta\dot{\theta}_2 \\ &- \dot{\theta}_1\dot{\theta}_2\sin(\theta_2-\theta_1)\delta(\theta_2-\theta_1)] + \frac{mL_2^2}{3}\dot{\theta}_2\delta\dot{\theta}_2 \\ =& \frac{mL_1L_2}{2}\dot{\theta}_1\dot{\theta}_2\sin(\theta_2-\theta_1)\delta\theta_1 - \frac{mL_1L_2}{2}\dot{\theta}_1\dot{\theta}_2\sin(\theta_2-\theta_1)\delta\theta_2 \\ &+ mL_1\left[L_1\dot{\theta}_1 + \frac{L_2}{2}\dot{\theta}_2\cos(\theta_2-\theta_1)\right]\delta\dot{\theta}_1 + mL_2\left[\frac{L_1}{2}\dot{\theta}_1\cos(\theta_1-\theta_1) + \frac{L_2}{3}\dot{\theta}_2\right]\delta\dot{\theta}_2\end{aligned} \tag{g}$$

and the variation in the potential energy is simply

$$\delta V = mg\left(L_1\sin\theta_1\delta\theta_1 + \frac{L_2}{2}\sin\theta_2\delta\theta_2\right) \tag{h}$$

Inserting Eqs. (e)–(g) into Eq. (6.31) and collecting terms, we have

$$\begin{aligned}\int_{t_1}^{t_2}(\delta T - \delta V + \delta W_{nc})dt =& \int_{t_1}^{t_2}\left\{\left[\frac{mL_1L_2}{2}\dot{\theta}_1\dot{\theta}_2\sin(\theta_2-\theta_1) - mgL_1\sin\theta_1\right.\right. \\ &\left.+ FL_1\cos\theta_1\right]\delta\theta_1 + \left[-\frac{mL_1L_2}{2}\dot{\theta}_1\dot{\theta}_2\sin(\theta_2-\theta_1) - \frac{mgL_2}{2}\sin\theta_2 + FL_2\cos\theta_2\right]\delta\theta_2 \\ &\left.+ mL_1\left[L_1\dot{\theta}_1 + \frac{L_2}{2}\dot{\theta}_2\cos(\theta_2-\theta_1)\right]\delta\dot{\theta}_1 + mL_2\left[\frac{L_1}{2}\dot{\theta}_1\cos(\theta_2-\theta_1) + \frac{L_2}{3}\dot{\theta}_2\right]\delta\dot{\theta}_2\right\}dt = 0\end{aligned} \tag{i}$$

At this point, we observe that Eq. (i) involves both the virtual displacements $\delta\theta_1$ and $\delta\theta_2$ and the virtual velocities $\delta\dot{\theta}_1$ and $\delta\dot{\theta}_2$, and only the virtual displacements are arbitrary. Hence, before we can derive the equations of motion, we must transform the terms in $\delta\dot{\theta}_1$ and $\delta\dot{\theta}_2$ into terms in $\delta\theta_1$ and $\delta\theta_2$, respectively. To this end, we carry out the following integrations by parts:

$$\begin{aligned}\int_{t_1}^{t_2} mL_1\left[L_1\dot{\theta}_1 + \frac{L_2}{2}\dot{\theta}_2\cos(\theta_2-\theta_1)\right]\delta\dot{\theta}_1 dt =& mL_1\left[L_1\dot{\theta}_1 + \frac{L_2}{2}\dot{\theta}_2\cos(\theta_2-\theta_1)\right]\delta\theta_1\Big|_{t_1}^{t_2} \\ &- \int_{t_1}^{t_2} mL_1\frac{d}{dt}\left[L_1\dot{\theta}_1 + \frac{L_2}{2}\dot{\theta}_2\cos(\theta_2-\theta_1)\right]\delta\theta_1 dt \\ =& -\int_{t_1}^{t_2} mL_1\left[L_1\ddot{\theta}_1 + \frac{L_2}{2}\ddot{\theta}_2\cos(\theta_2-\theta_1) - \frac{L_2}{2}\dot{\theta}_2(\dot{\theta}_2-\dot{\theta}_1)\sin(\theta_2-\theta_1)\right]\delta\theta_1\, dt \\ \int_{t_1}^{t_2} mL_2\left[\frac{L_1}{2}\dot{\theta}_1\cos(\theta_2-\theta_1) + \frac{L_2}{3}\dot{\theta}_2\right]\delta\dot{\theta}_2 dt =& mL_2\left[\frac{L_1}{2}\dot{\theta}_1\cos(\theta_2-\theta_1) + \frac{L_2}{3}\dot{\theta}_2\right]\delta\theta_2\Big|_{t_1}^{t_2} \\ &- \int_{t_1}^{t_2} mL_2\frac{d}{dt}\left[\frac{L_1}{2}\dot{\theta}_1\cos(\theta_2-\theta_1) + \frac{L_2}{3}\dot{\theta}_2\right]\delta\theta_2 dt \\ =& -\int_{t_1}^{t_2} mL_2\left[\frac{L_1}{2}\ddot{\theta}_1\cos(\theta_2-\theta_1) - \frac{L_1}{2}\dot{\theta}_1(\dot{\theta}_2-\dot{\theta}_1)\sin(\theta_2-\theta_1) + \frac{L_2}{3}\ddot{\theta}_2\right]\delta\theta_2 dt\end{aligned} \tag{j}$$

where we recalled the auxiliary conditions $\delta\theta_1 = \delta\theta_2 = 0$ at $t = t_1,\ t_2$. Introducing Eqs. (j) in Eq. (i) and collecting terms, we can write

$$\int_{t_1}^{t_2} \left\{ -\left[mL_1^2\ddot{\theta}_1 + \frac{mL_1L_2}{2}\ddot{\theta}_2\cos(\theta_2-\theta_1) - \frac{mL_1L_2}{2}\dot{\theta}_2^2\sin(\theta_2-\theta_1) \right.\right.$$
$$\left. + mgL_1\sin\theta_1 - FL_1\cos\theta_1 \right]\delta\theta_1 - \left[\frac{mL_1L_2}{2}\ddot{\theta}_1\cos(\theta_2-\theta_1) + \frac{mL_2^2}{3}\ddot{\theta}_2 \right.$$
$$\left.\left. + \frac{mL_1L_2}{2}\dot{\theta}_1^2\sin(\theta_2-\theta_1) + \frac{mgL_2}{2}\sin\theta_2 - FL_2\cos\theta_2 \right]\delta\theta_2 \right\} dt = 0 \tag{k}$$

Finally, the integrand is in a form permitting the extraction of the equations of motion. To this end, we invoke the arbitrariness of $\delta\theta_1$ and $\delta\theta_2$, and assign different values to $\delta\theta_1$, while we set $\delta\theta_2 = 0$. Because the resulting equation must hold for all values of $\delta\theta_1$, we conclude that this is possible only if the coefficient of $\delta\theta_1$ is zero. A similar argument, but with the roles of $\delta\theta_1$ and $\delta\theta_2$ reversed, causes us to conclude that the coefficient of $\delta\theta_2$ must be zero as well. Hence, setting the coefficients of $\delta\theta_1$ and $\delta\theta_2$ equal to zero, we obtain the equations of motion

$$\begin{aligned} mL_1^2\ddot{\theta}_1 + \frac{mL_1L_2}{2}\left[\ddot{\theta}_2\cos(\theta_2-\theta_1) - \dot{\theta}_2^2\sin(\theta_2-\theta_1)\right] + mgL_1\sin\theta_1 &= FL_1\cos\theta_1 \\ \frac{mL_1L_2}{2}\left[\ddot{\theta}_1\cos(\theta_2-\theta_1) + \dot{\theta}_1^2\sin(\theta_2-\theta_1)\right] + \frac{mL_2^2}{3}\ddot{\theta}_2 + \frac{mgL_2}{2}\sin\theta_2 &= FL_2\cos\theta_2 \end{aligned} \tag{l}$$

We observe from Eqs. (l) that there are two equations of motion in the unknowns θ_1 and θ_2, as there should be for a two-degree-of-freedom system, and the equations are free of the string tension T. By contrast, Eqs. (k) of Example 1.6 are three in number and there are three unknowns, θ_1, θ_2 and T. Hence, the extended Hamilton's principle not only yields the correct number of equations of motion, but the equations themselves are not encumbered by quantities that may present no interest, such as internal forces and reaction forces. Of course, by eliminating the string tension T, Eqs. (k) of Example 1.6 can be reduced to Eqs. (l) of this example, but this requires extra work. Clearly, the advantages of analytical mechanics over Newtonian mechanics for more complex problems are compelling. One exception is the case in which the system includes friction forces of the type shown in Fig. 6.2b, as such systems do not fall within the confines of analytical dynamics (Sec. 6.2). Another case in which Newtonian mechanics has the edge is that in which reaction forces and forces internal to the system, such as the string tension T, are of interest. These cases are not very frequent, so that analytical mechanics is to be preferred for the vast majority of systems. In this regard, it should be mentioned that the method of choice for obtaining equations of motion is actually Lagrange's equations, which can be derived by means of the extended Hamilton's principle, as shown in Sec. 6.5.

6.5 LAGRANGE'S EQUATIONS

The extended Hamilton's principle permits the derivation of all the equations of motion of a system from three scalar expressions, the kinetic energy, the potential energy and the virtual work due to nonconservative forces. The principle is extremely versatile, as it enables one to obtain results where other methods experience difficulties, or even fail. On the other hand, for many problems the extended Hamilton's principle is not the most efficient method for deriving equations of motion, as it involves certain routine operations that must be carried out every time the principle is applied, such as the integrations by parts. In this section, we use the extended Hamilton's principle to generate a more expeditious method for deriving equations of motion, one that obviates the need for the routine operations in question. Reference is made here to the celebrated *Lagrange's equations*. Although it is possible to derive Lagrange's equations directly from d'Alembert's principle, i.e., without the use of the extended Hamilton's principle, the approach used here is arguably the simplest and most elegant way of deriving Lagrange's equations.

The kinetic energy of a generic dynamical system can be expressed in terms of generalized displacements and velocities in the functional form

$$T = T(q_1, q_2, \dots, q_n, \dot{q}_1, \dot{q}_2, \dots, \dot{q}_n) \tag{6.34}$$

so that the variation in the kinetic energy is simply

$$\delta T = \sum_{k=1}^{n} \left(\frac{\partial T}{\partial q_k} \delta q_k + \frac{\partial T}{\partial \dot{q}_k} \delta \dot{q}_k \right) \tag{6.35}$$

Similarly, the potential energy has the functional form

$$V = V(q_1, q_2, \dots, q_n) \tag{6.36}$$

so that the variation in the potential energy is

$$\delta V = \sum_{k=1}^{n} \frac{\partial V}{\partial q_k} \delta q_k \tag{6.37}$$

Moreover, from Eq. (6.17), the virtual work of the nonconservative forces has the expression

$$\overline{\delta W}_{nc} = \sum_{k=1}^{n} Q_k \delta q_k \tag{6.38}$$

where Q_k $(k = 1, 2, \dots, n)$ are the generalized nonconservative forces.

Next, we insert Eqs. (6.35), (6.37) and (6.38) into the extended Hamilton's principle, Eq. (6.31), and write

$$\int_{t_1}^{t_2} (\delta T - \delta V + \overline{\delta W}_{nc}) dt = \int_{t_1}^{t_2} \sum_{k=1}^{n} \left[\left(\frac{\partial T}{\partial q_k} - \frac{\partial V}{\partial q_k} + Q_k \right) \delta q_k + \frac{\partial T}{\partial \dot{q}_k} \delta \dot{q}_k \right] dt = 0,$$

$$\delta q_k = 0, \; k = 1, 2, \dots, n; \; t = t_1, t_2 \tag{6.39}$$

Then, following the approach of Example 6.1, we carry out the following integration by parts:

$$\int_{t_1}^{t_2} \frac{\partial T}{\partial \dot{q}_k} \delta \dot{q}_k dt = \int_{t_1}^{t_2} \frac{\partial T}{\partial \dot{q}_k} \frac{d}{dt} \delta q_k \, dt = \frac{\partial T}{\partial \dot{q}_k} \delta q_k \bigg|_{t_1}^{t_2} - \int_{t_1}^{t_2} \frac{d}{dt} \left(\frac{\partial t}{\partial \dot{q}_k} \right) \delta q_k \, dt$$

$$= - \int_{t_1}^{t_2} \frac{d}{dt} \left(\frac{\partial T}{\partial \dot{q}_k} \right) \delta q_k \, dt, \; k = 1, 2, \dots, n \tag{6.40}$$

in which we considered the auxiliary conditions $\delta q_k = 0$ $(k = 1, 2, \ldots, n)$ at $t = t_1$ and $t = t_2$, as indicated in Eq. (6.39). Introducing Eqs. (6.40) in Eq. (6.39), we have

$$\int_{t_1}^{t_2} \sum_{k=1}^{n} \left[\frac{\partial T}{\partial q_k} - \frac{\partial V}{\partial q_k} + Q_k - \frac{d}{dt}\left(\frac{\partial T}{\partial \dot{q}_k}\right) \right] \delta q_k \, dt = 0 \tag{6.41}$$

At this point, we use the standard argument concerning the arbitrariness of the virtual generalized displacements δq_k $(k = 1, 2, \ldots, n)$. By analogy with the procedure used in Example 6.1, we assign arbitrary values to δq_1 while setting $\delta q_k = 0$ $(k = 2, 3, \ldots, n)$. Under these circumstances, Eq. (6.41) can be satisfied only if the coefficient of δq_1 is zero. Using the same argument but with $\delta q_2, \delta q_3, \ldots, \delta q_n$ playing the role of δq_1, in sequence, we conclude that the coefficient of every virtual generalized displacement δq_k $(k = 1, 2, \ldots, n)$ in Eq. (6.41) must be zero, which yields Lagrange's equations

$$\frac{d}{dt}\left(\frac{\partial T}{\partial \dot{q}_k}\right) - \frac{\partial T}{\partial q_k} + \frac{\partial V}{\partial q_k} = Q_k, \; k = 1, 2, \ldots, n \tag{6.42}$$

Equations (6.42) represent the most general form of Lagrange's equations; any other form is a mere special case.

The extended Hamilton's principle, Eq. (6.31), and Lagrange's equations, Eqs. (6.42), represent entirely equivalent formulations and, for the same generalized coordinates, they yield identical equations of motion. Because they obviate certain operations and arguments used in their own derivation, Lagrange's equations, Eqs. (6.42), are more expeditious than the extended Hamilton's principle for producing equations of motion for discrete systems, and they represent the method of choice. However, the extended Hamilton's principle is more versatile and can be used for a variety of nonroutine problems lying beyond the scope of Lagrange's equations, such as problems involving distributed-parameter systems (Ch. 8).

Example 6.2. Derive Lagrange's equations of motion for the system of Example 6.1.

As in Example 6.1, we use the angles θ_1 and θ_2 (see Fig. 1.10a) as generalized coordinates, $q_1 = \theta_1$, $q_2 = \theta_2$, so that Lagrange's equations, Eqs. (6.42), take the form

$$\frac{d}{dt}\left(\frac{\partial T}{\partial \dot{\theta}_k}\right) - \frac{\partial T}{\partial \theta_k} + \frac{\partial V}{\partial \theta_k} = \Theta_k, \; k = 1, 2 \tag{a}$$

where Θ_k $(k = 1, 2)$ are the generalized nonconservative forces. From Example 6.1, we obtain the kinetic energy

$$T = \frac{1}{2} m \left[L_1^2 \dot{\theta}_1^2 + L_1 L_2 \dot{\theta}_1 \dot{\theta}_2 \cos(\theta_2 - \theta_1) + \frac{L_2^2}{3} \dot{\theta}_2^2 \right] \tag{b}$$

the potential energy

$$V = mg \left[L_1 (1 - \cos\theta_1) + \frac{L_2}{2}(1 - \cos\theta_2) \right] \tag{c}$$

and the virtual work of the nonconservative forces

$$\overline{\delta W}_{nc} = F L_1 \cos\theta_1 \delta\theta_1 + F L_2 \cos\theta_2 \delta\theta_2 \tag{d}$$

The derivatives with respect to the angular velocities are as follows:

$$\begin{aligned}\frac{\partial T}{\partial \dot{\theta}_1} &= mL_1^2\dot{\theta}_1 + \frac{mL_1L_2}{2}\dot{\theta}_2\cos(\theta_2-\theta_1)\\ \frac{\partial T}{\partial \dot{\theta}_2} &= \frac{mL_1L_2}{2}\dot{\theta}_1\cos(\theta_2-\theta_1) + \frac{mL_2^2}{3}\dot{\theta}_2\end{aligned} \tag{e}$$

so that

$$\begin{aligned}\frac{d}{dt}\left(\frac{\partial T}{\partial \dot{\theta}_1}\right) &= mL_1^2\ddot{\theta}_1 + \frac{mL_1l_2}{2}\left[\ddot{\theta}_2\cos(\theta_2-\theta_1) - \dot{\theta}_2(\dot{\theta}_2-\dot{\theta}_1)\sin(\theta_2-\theta_1)\right]\\ \frac{d}{dt}\left(\frac{\partial T}{\partial \dot{\theta}_2}\right) &= \frac{mL_1L_2}{2}\left[\ddot{\theta}_1\cos(\theta_2-\theta_1) - \dot{\theta}_1(\dot{\theta}_2-\dot{\theta}_1)\sin(\theta_2-\theta_1)\right] + \frac{mL_2^2}{3}\ddot{\theta}_2\end{aligned} \tag{f}$$

Moreover, the derivatives with respect to the angular displacements are

$$\begin{aligned}&\frac{\partial T}{\partial \theta_1} = \frac{mL_1L_2}{2}\dot{\theta}_1\dot{\theta}_2\sin(\theta_2-\theta_1), \quad \frac{\partial T}{\partial \theta_2} = -\frac{mL_1L_2}{2}\dot{\theta}_1\dot{\theta}_2\sin(\theta_2-\theta_1)\\ &\frac{\partial V}{\partial \theta_1} = mgL_1\sin\theta_1, \quad \frac{\partial V}{\partial \theta_2} = \frac{mgL_2}{2}\sin\theta_2\end{aligned} \tag{g}$$

In addition, the generalized nonconservative forces are recognized as the coefficients of $\delta\theta_1$ and $\delta\theta_2$ in the virtual work, Eq. (d), or

$$\Theta_1 = FL_1\cos\theta_1, \quad \Theta_2 = FL_2\cos\theta_2 \tag{h}$$

Inserting Eqs. (f)–(h) into Eqs. (a), we obtain the desired Lagrange's equations

$$\begin{aligned}&mL_1^2\ddot{\theta}_1 + \frac{mL_1L_2}{2}[\ddot{\theta}_2\cos(\theta_2-\theta_1) - \dot{\theta}_2^2\sin(\theta_2-\theta_1)] + mgL_1\sin\theta_1 = FL_1\cos\theta_1\\ &\frac{mL_1L_2}{2}[\ddot{\theta}_1\cos(\theta_2-\theta_1) + \dot{\theta}_1^2\sin(\theta_2-\theta_1)] + \frac{mL_2^2}{3}\ddot{\theta}_2 + \frac{mgL_2}{2}\sin\theta_2 = FL_2\cos\theta_2\end{aligned} \tag{i}$$

We observe that Eqs. (i) just derived are identical to Eqs. (l) of Example 6.1, obtained by the extended Hamilton's principle, as was to be expected. Clearly, Lagrange's equations reduce the derivation of the equations of motion to a routine series of differentiations. Still, one word of caution is in order. This example makes the task of deriving equations of motion appear simpler than it really is, because we merely availed ourselves to the kinetic energy, potential energy and virtual work derived earlier. In this regard, it must be pointed out that the major task in producing equations of motion by Lagrange's equations is the very derivation of the kinetic energy, potential energy and virtual work, and the same is true about equations of motion obtained by the extended Hamilton's principle.

6.6 SUMMARY

Newtonian mechanics formulates the equations of motion in terms of physical coordinates and forces, which are in general vector quantities. It requires one free-body diagram for each mass and it includes reaction forces and constraint forces in the equations of motion. These forces play the role of unknowns, which makes it necessary to work with more equations of motion than the number of degrees of freedom of the system. As a result, as the number of degrees of freedom increases, Newtonian mechanics rapidly loses its appeal as a way of deriving equations of motion.

Analytical mechanics, or Lagrangian mechanics, does not have the disadvantages cited above, and must be regarded as the method of choice for deriving equations of motion for multi-degree-of-freedom systems, as well as for distributed-parameter systems. It permits the derivation of all the equations of motion from three scalar quantities, namely, the kinetic energy, potential energy and virtual work of the nonconservative forces. It does not require free-body diagrams, and in fact it considers the system as a whole, rather than the individual components. As a result, reaction forces and constraint forces do not appear in the formulation, and the number of equations of motion coincides with the number of degrees of freedom. The process of deriving the equations of motion is rendered almost routine by the use of Lagrange's equations.

The transition from Newton's laws to Lagrange's equations can be carried out using a transformation from physical coordinates to more abstract generalized coordinates, and it involves simple rules of differential calculus. A more satisfying approach, and the one followed in this text, is to derive first the extended Hamilton principle and then to use the principle to derive Lagrange's equations. The advantage of this approach becomes obvious when we consider the fact that, whereas Lagrange's equations are more efficient, the extended Hamilton principle is more versatile. In fact, it can produce results in cases in which Lagrange's equations cannot, most notably in the case of distributed-parameter systems.

PROBLEMS

6.1. The system of Fig. 6.3 consists of a uniform rigid link of total mass m and length L and two linear springs of stiffnesses k_1 and k_2. When the springs are unstretched the link is horizontal. Use the principle of virtual work to calculate the angle θ corresponding to the equilibrium position.

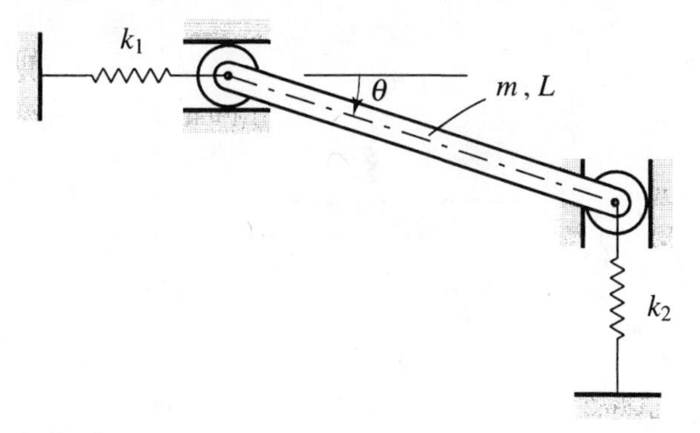

FIGURE 6.3
Rigid link supported by springs

6.2. Two masses, $m_1 = 0.5m$ and $m_2 = m$, are suspended on a massless string of constant tension T (Fig. 6.4). Assume small displacements y_1 and y_2 and use the principle of virtual work to calculate the equilibrium position of the masses sagging under their own weight.

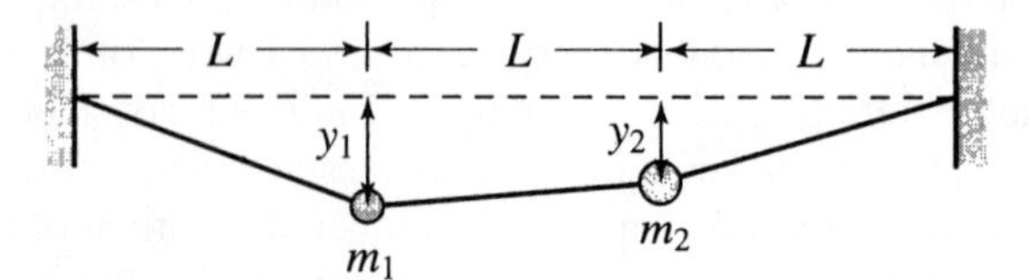

FIGURE 6.4
Two masses on a string

6.3. Use the principle of virtual work to determine the equilibrium equation for the system consisting of a mass on a rotating hoop, as described in Problem 1.2.

6.4. Derive the equation of motion for a simple pendulum by means of d'Alembert's principle.

6.5. Derive the equation of motion for the system of Problem 1.1 by means of d'Alembert's principle.

6.6. Derive the equation of motion for the system of Problem 1.2 by means of d'Alembert's principle.

6.7. Derive the equation of motion for the system of Problem 6.1 by means of d'Alembert's principle.

6.8. Derive the equation of motion for the system of Problem 1.2 by means of Hamilton's principle.

6.9. Derive the equation of motion for the system of Problem 6.1 by means of Hamilton's principle.

6.10. Derive the equations of motion for the system of Problem 6.2 by means of Hamilton's principle.

6.11. The upper end of a pendulum is attached to a linear spring of stiffness k, where the spring is constrained so as to move in the vertical direction (Fig. 6.5). Derive the equations of motion for the vertical displacement u and the angular displacement θ by means of Hamilton's principle. Assume that u is measured from the equilibrium position and that θ is arbitrarily large.

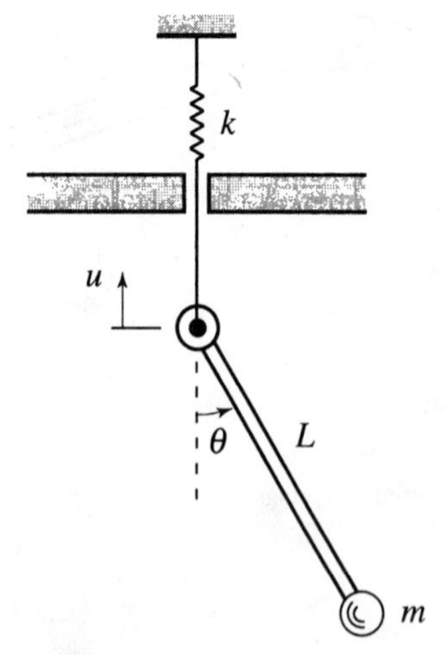

FIGURE 6.5
Pendulum attached to a spring

6.12. Derive the Lagrange equation of motion for the system of Problem 1.2.

6.13. Derive the three Newton equations of motion and the single Lagrange equation for the system of Problem 6.1 and show how Newton's equations can be reduced to Lagrange's equation.

6.14. Derive the equations of motion for the system of Problem 6.2 by means of Lagrange's equations.

6.15. Derive Newton's and Lagrange's equation of motion for the system of Problem 6.11, discuss differences and show how Newton's equations can be reduced to Lagrange's equations.

CHAPTER 7

MULTI-DEGREE-OF-FREEDOM SYSTEMS

In Ch. 1, and again in Chs. 5 and 6, we defined a *multi-degree-of-freedom system* as a system whose motion is described by more than one coordinate. In particular, to describe the motion of an n-degree-of-freedom system it is necessary to use n coordinates, where coordinates are time-dependent variables, such as the displacements of the masses. Two-degree-of-freedom systems represent special cases of multi-degree-of-freedom systems. The entire Ch. 5 was devoted to two-degree-of-freedom systems, which can be justified on pedagogical grounds. In the first place, the vibration of two-degree-of-freedom systems can be treated by elementary means, based primarily on physical considerations. Moreover, a number of systems of interest in vibrations can be modeled as two-degree-of-freedom systems. Perhaps the most persuasive argument for a separate chapter on two-degree-of-freedom systems is that it provides us with the opportunity to introduce the important concepts involved in multi-degree-of-freedom systems without being encumbered by abstract formulations.

For $n \geq 3$, the situation changes markedly from the case in which $n = 2$, as the physical world must be augmented by a more abstract mathematical world. Whereas many of the fundamental concepts are the same, the mathematical methodology is more demanding. As demonstrated in Ch. 5, to solve the differential equations of motion, it is necessary to solve first the algebraic eigenvalue problem. To this end, we recall that in Ch. 5 we solved the eigenvalue problem for a two-degree-of-freedom system by an elementary approach consisting of three steps, namely, derivation of the characteristic equation, solution of the characteristic equation for the natural frequencies and solution of corresponding algebraic equations for the natural modes. We note that the characteristic equation for two-degree-of-freedom systems is quadratic, which can be solved in closed form. As n exceeds two, however, the solution of the eigenvalue problem using the characteristic equation becomes impractical. Fortunately, the rapid increase in the capability of digital computers has led to the development of extremely efficient numerical algorithms for the solution of the eigenvalue problem, particularly for problems defined by real symmetric matrices, such as those arising in vibrations. These algorithms are based on certain developments from linear algebra and matrix theory. But, whereas the mathematical tools tend to be more advanced, the fundamental concepts involved in the derivation of the system response, such as coupling, orthogonality of modal vectors and modal analysis for decoupling the equations of motion, remain essentially the same as in Ch. 5. The basic facts are that multi-degree-of-freedom systems require a more mature treatment than two-degree-of-freedom systems and matrix algebra is ideally suited for such a treatment.

This chapter blends physical with mathematical considerations to present a rigorous approach to the vibration of multi-degree-of-freedom systems. It includes a generalization and extension of the various fundamental concepts introduced in Ch. 5 and it provides the necessary tools for efficient solutions of free and forced response problems. In studying the vibration of linear systems, certain developments from matrix theory and linear algebra permit a deeper understanding not only of the intrinsic properties of vibrating systems but also of the modern computational algorithms for the efficient solutions mentioned above. In most practical problems, analytical solutions are not possible or feasible, so that computer-generated numerical solutions are a way of life. In this regard, state-space methods in conjunction with discrete-time techniques, first introduced in Ch. 4 for single-degree-of-freedom systems, are extended to multi-degree-of-freedom systems. Several MATLAB programs dealing with eigenvalue problems and with the response to initial excitations and external forces are included.

As is to be expected, there is a certain amount of duplication between this chapter and Ch. 5. The two chapters are really intended for different readerships, however, with Ch. 5 to be included in a beginning course on vibrations, and perhaps excluded from a somewhat more advanced one based on this chapter. Of course, in the latter case, an occasional reference to material from Ch. 5 can only enhance understanding. In particular, the examples from Ch. 5 can serve this purpose well.

7.1 EQUATIONS OF MOTION FOR LINEAR SYSTEMS

We are interested in the motion of an n-degree-of-freedom system in the neighborhood of an equilibrium position, where the motion is described by the generalized coordinates $q_1(t), q_2(t), \ldots, q_n(t)$. In general the equilibrium position can be obtained by extending definition (1.151) to multivariable systems. Without loss of generality, we assume that the equilibrium position is given by the trivial solution $q_1 = q_2 = \ldots = q_n = 0$. Moreover, we assume that the displacements from the equilibrium position are sufficiently small that the force-displacement and force-velocity relations are linear, so that the generalized coordinates and their time derivatives appear in the differential equations of motion at most to the first power. This represents, in essence, the so-called *small motions assumption* (Sec. 1.13), leading to a system of linear equations.

In this section, we derive the differential equations of motion by means of Newton's second law. To this end, we consider the linear system consisting of n masses m_i $(i = 1, 2, \ldots, n)$ connected by springs and dampers, as shown in Fig. 7.1a, and draw the free-body diagram associated with the typical mass m_i (Fig. 7.1b). Because the motion takes place in one dimension, the total number of degrees of freedom of the system coincides with the number of masses. Applying Newton's second law to mass m_i $(i = 1, 2, \ldots, n)$ we can write the differential equation of motion

$$\begin{aligned} Q_i(t) + c_{i+1}[\dot{q}_{i+1}(t) - \dot{q}_i(t)] + k_{i+1}[q_{i+1}(t) - q_i(t)] \\ - c_i[\dot{q}_i(t) - \dot{q}_{i-1}(t)] - k_i[q_i(t) - q_{i-1}(t)] = m_i\ddot{q}_i(t) \end{aligned} \tag{7.1}$$

where $Q_i(t)$ represents the generalized force. Equation (7.1) can be rearranged in the form

$$\begin{aligned} m_i\ddot{q}_i(t) - c_{i+1}\dot{q}_{i+1}(t) + (c_i + c_{i+1})\dot{q}_i(t) - c_i\dot{q}_{i-1}(t) \\ - k_{i+1}q_{i+1}(t) + (k_i + k_{i+1})q_i(t) - k_iq_{i-1}(t) = Q_i(t) \end{aligned} \tag{7.2}$$

Equation (7.2) can be extended to the full system. To this end, we introduce the notation

$$\begin{aligned} &m_{ij} = \delta_{ij}m_i \\ &c_{ij} = 0, && k_{ij} = 0, && j = 1, 2, \ldots, i-2,\ i+2, \ldots, n \\ &c_{ij} = -c_i, && k_{ij} = -k_i, && j = i-1 \\ &c_{ij} = c_i + c_{i+1}, && k_{ij} = k_i + k_{i+1}, && j = i \\ &c_{ij} = -c_{i+1}, && k_{ij} = -k_{i+1}, && j = i+1 \end{aligned} \tag{7.3}$$

where m_{ij}, c_{ij} and k_{ij} are *mass, damping* and *stiffness coefficients*, respectively, and δ_{ij} is the *Kronecker delta*, defined as being equal to unity for $i = j$ and equal to zero for

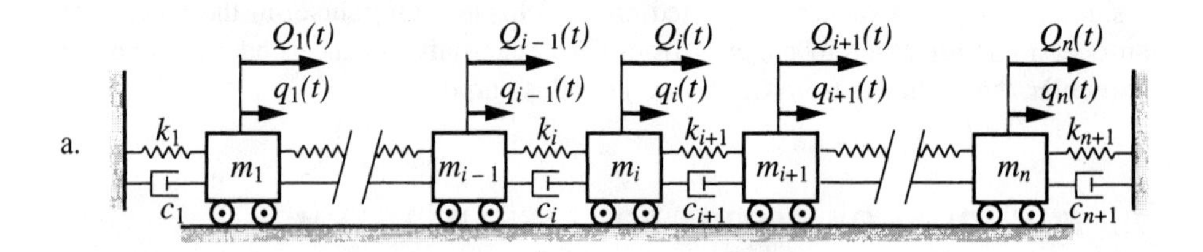

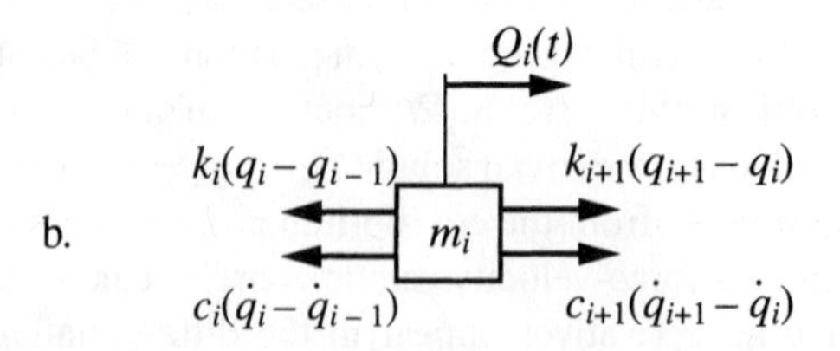

FIGURE 7.1
a. Damped n-degree-of-freedom system, b. Free-body diagram for a typical mass

$i \neq j$. Then, the equations of motion for the full system can be expressed as follows:

$$\sum_{j=1}^{n}[m_{ij}\ddot{q}_j(t)+c_{ij}\dot{q}_j(t)+k_{ij}q_j(t)] = Q_i(t),\ i=1,2,\ldots,n \tag{7.4}$$

which constitutes a set of n simultaneous second-order ordinary differential equations for the displacements $q_i(t)$ $(i=1,2,\ldots,n)$. We note that Eqs. (7.4) are quite general, in the sense that they can accommodate other end conditions as well. For example, if the right end is free instead of being fixed, then we can simply set $c_{n+1}=k_{n+1}=0$ in Eqs. (7.3). Although at this particular point the notation (7.3) appears as an unnecessary complication, its advantage lies in the fact that the use of a double index for the coefficients enables us to write Eqs. (7.4) in matrix notation. We shall have ample opportunity to work with the coefficients m_{ij}, c_{ij} and k_{ij} and to study their interesting and useful properties. In particular, it will be shown that the mass, damping and stiffness coefficients are symmetric, or

$$m_{ij}=m_{ji},\ c_{ij}=c_{ji},\ k_{ij}=k_{ji},\ i,j=1,2,\ldots,n \tag{7.5}$$

and that these coefficients control the system behavior, especially in the case of free vibration. Note that we encountered these coefficients for the first time in Sec. 5.2 in conjunction with two-degree-of-freedom systems.

In spite of the fact that Eqs. (7.4) possess constant coefficients, a general analytical solution of the equations is difficult to obtain because of the coupling introduced by the damping coefficients c_{ij}. Under special circumstances, however, a solution of Eqs. (7.4) can be obtained with the same ease as one for undamped systems. For future reference, it is convenient to write Eqs. (7.4) in matrix form. To this end, we arrange the coefficients m_{ij}, c_{ij} and k_{ij} in the following square matrices:

$$[m_{ij}]=M, \qquad [c_{ij}]=C, \qquad [k_{ij}]=K \tag{7.6}$$

where, as in Sec. 5.2, M is the *mass matrix*, C the *damping matrix* and K the *stiffness matrix*. Then, the symmetry of the coefficients translates into the symmetry of the associated matrices, or

$$M=M^T, \qquad C=C^T, \qquad K=K^T \tag{7.7}$$

in which the superscript T denotes the transpose of the matrix in question. Moreover, the displacements $q_i(t)$ and the forces $Q_i(t)$ can be regarded as the components of n-dimensional vectors $\mathbf{q}(t)$ and $\mathbf{Q}(t)$, respectively. Hence, using simple rules of matrix multiplication, Eqs. (7.4) can be written in the compact matrix form

$$M\ddot{\mathbf{q}}(t)+C\dot{\mathbf{q}}(t)+K\mathbf{q}(t)=\mathbf{Q}(t) \tag{7.8}$$

In this particular case, the mass matrix is diagonal because the coordinates used represent actual displacements of the masses. For a different set of coordinates M is not necessarily diagonal.

In Sec. 7.4, we use the Lagrangian approach to derive the equations of motion for a generic linear system, which permits us to develop the vibration theory in a general way, without reference to any particular system, such as that shown in Fig. 7.1a.

Example 7.1. Consider the three-degree-of-freedom system of Fig. 7.2a and derive the system differential equations of motion by Newton's second law. The springs exhibit linear behavior and the dampers are viscous.

As shown in Fig. 7.2a, the coordinates $q_1(t)$, $q_2(t)$ and $q_3(t)$ represent the horizontal translations of masses m_1, m_2 and m_3, respectively, and $Q_1(t)$, $Q_2(t)$ and $Q_3(t)$ are the associated externally applied forces. To derive the equations of motion by Newton's second law, we draw three free-body diagrams, associated with masses m_1, m_2 and m_3. They are all shown in Fig. 7.2b, where the forces in the springs and dampers between masses m_1 and m_2 are the same in magnitude but opposite in direction, and the same can be said about the forces between masses m_2 and m_3. Application of Newton's second law to masses m_i $(i = 1, 2, 3)$ yields the equations of motion

$$\begin{aligned} Q_1 + c_2(\dot{q}_2 - \dot{q}_1) + k_2(q_2 - q_1) - c_1\dot{q}_1 - k_1 q_1 &= m_1\ddot{q}_1 \\ Q_2 + c_3(\dot{q}_3 - \dot{q}_2) + k_3(q_3 - q_2) - c_2(\dot{q}_2 - \dot{q}_1) - k_2(q_2 - q_1) &= m_2\ddot{q}_2 \\ Q_3 - c_3(\dot{q}_3 - \dot{q}_2) - k_3(q_3 - q_2) &= m_3\ddot{q}_3 \end{aligned} \tag{a}$$

which can be rearranged in the form

$$\begin{aligned} m_1\ddot{q}_1 + (c_1 + c_2)\dot{q}_1 - c_2\dot{q}_2 + (k_1 + k_2)q_1 - k_2 q_2 &= Q_1 \\ m_2\ddot{q}_2 - c_2\dot{q}_1 + (c_2 + c_3)\dot{q}_2 - c_3\dot{q}_3 - k_2 q_1 + (k_2 + k_3)q_2 - k_3 q_3 &= Q_2 \\ m_3\ddot{q}_3 - c_3\dot{q}_2 + c_3\dot{q}_3 - k_3 q_2 + k_3 q_3 &= Q_3 \end{aligned} \tag{b}$$

It is not difficult to see that Eqs. (b) can be expressed in the matrix form (7.8), where the

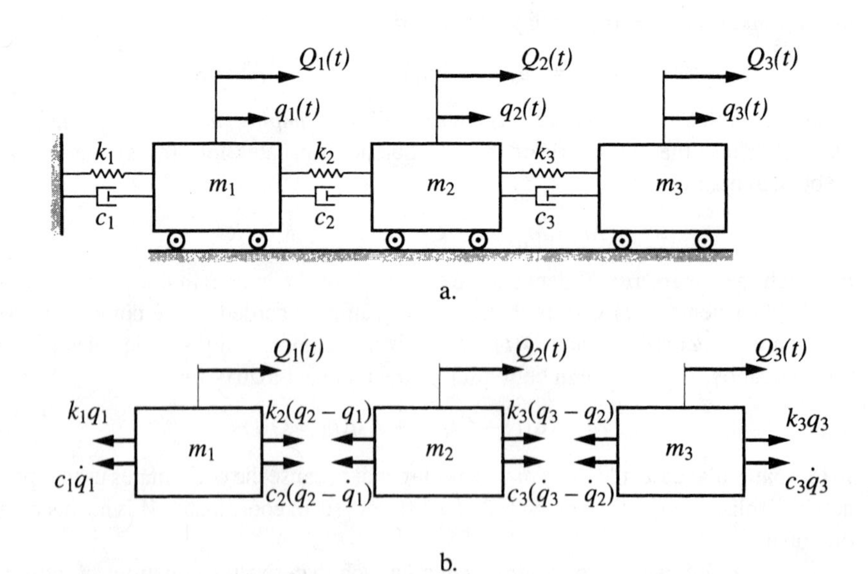

FIGURE 7.2
a. Damped three-degree-of-freedom system, b. Free-body diagrams

coefficient matrices are given by

$$M = \begin{bmatrix} m_1 & 0 & 0 \\ 0 & m_2 & 0 \\ 0 & 0 & m_3 \end{bmatrix} \tag{c}$$

$$C = \begin{bmatrix} c_1 + c_2 & -c_2 & 0 \\ -c_2 & c_2 + c3 & -c_3 \\ 0 & -c_3 & c_3 \end{bmatrix} \tag{d}$$

$$K = \begin{bmatrix} k_1 + k_2 & -k_2 & 0 \\ -k_2 & k_2 + k_3 & -k_3 \\ 0 & -k_3 & k_3 \end{bmatrix} \tag{e}$$

which are clearly symmetric. Moreover, M is diagonal.

7.2 FLEXIBILITY AND STIFFNESS INFLUENCE COEFFICIENTS

From Sec. 7.1, we conclude that the characteristics of a system are determined by its inertial, damping and stiffness properties. For linear systems, these properties enter explicitly in the differential equations through the mass coefficients m_{ij}, damping coefficients c_{ij} and stiffness coefficients k_{ij} $(i, j = 1, 2, \ldots, n)$, respectively. Of the three, the elastic properties are those causing a dynamic system to vibrate, as they are the ones inducing restoring forces.

The stiffness coefficients can be obtained by other means, not necessarily involving the equations of motion. In fact, the stiffness coefficients are more properly known as *stiffness influence coefficients*, and can be derived by using a definition to be introduced shortly. There is one more type of influence coefficients, namely, *flexibility influence coefficients*. They are intimately related to the stiffness influence coefficients, which is to be expected, because both types of coefficients can be used to describe the manner in which the system deforms under forces.

In Secs. 1.7 and 1.8, we examined springs exhibiting linear behavior. In particular, we introduced the spring constant concept for a single spring and the equivalent spring constant for a given combination of springs. In this section we introduce the concept of influence coefficients by expanding on the approach of Secs. 1.7 and 1.8.

We consider the linear discrete system shown in Fig. 7.3, which consists of n point masses m_i occupying the positions $x = x_i$ $(i = 1, 2, \ldots, n)$ when in equilibrium. In general, there are forces F_i $(i = 1, 2, \ldots, n)$ acting upon each point mass m_i, causing the masses to undergo displacements u_i, respectively. In the following, we propose to establish relations between the forces acting upon the system and the resulting displacements in terms of both flexibility and stiffness influence coefficients. To this end, we first assume that the system is acted upon by a single force F_j at $x = x_j$ and consider the displacement of any arbitrary point $x = x_i$ $(i = 1, 2, \ldots, n)$ due to the force F_j. With this in mind, we *define the flexibility influence coefficient a_{ij} as the displacement of point $x = x_i$ due to a unit force, $F_j = 1$, applied at $x = x_j$*. Because the system is linear, displacements increase proportionally with forces, so that the displacement corresponding to a force of arbitrary magnitude F_j is $a_{ij} F_j$. Moreover, for a linear system, we can

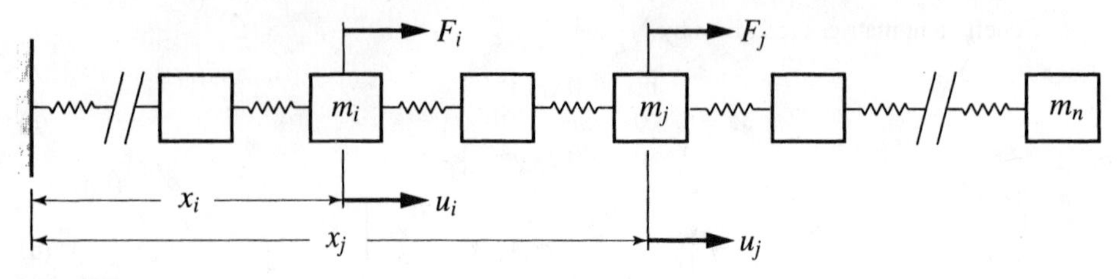

FIGURE 7.3
Linear discrete system

invoke the principle of superposition and obtain the displacement u_i at $x = x_i$ resulting from all forces F_j $(j = 1, 2, \ldots, n)$ by simply summing up the individual contributions, with the result

$$u_i = \sum_{j=1}^{n} a_{ij} F_j \tag{7.9}$$

Note that in this particular case the coefficients a_{ij} have units m/N. In cases involving torques and angular displacements, they have units rad/m · N.

By analogy, we can *define the stiffness influence coefficient k_{ij} as the force required at $x = x_i$ to produce a unit displacement, $u_j = 1$, at point $x = x_j$, and such that the displacements at all points for which $x \neq x_j$ are zero*. To obtain zero displacements at all points defined by $x \neq x_j$, the forces must simply hold these points fixed. Hence, the force at $x = x_i$ producing a displacement of arbitrary magnitude u_j at $x = x_j$ is $k_{ij}u_j$. In reality the points for which $x \neq x_j$ are not fixed, so that, invoking once again the superposition principle, the force at $x = x_i$ producing displacements u_j at $x = x_j$ $(j = 1, 2, \ldots, n)$ is simply

$$F_i = \sum_{j=1}^{n} k_{ij} u_j \tag{7.10}$$

It should be pointed out here that the stiffness coefficients as defined above, represent a special type of coefficients given in a more general form in Ch. 9. The coefficients k_{ij} defined here have units N/m. Stiffness coefficients relating angular displacements to torques have units m · N/rad.

At this juncture, it seems appropriate to point out that, although our interest in the flexibility and stiffness influence coefficients can be traced to vibration problems, the influence coefficients are really static rather than dynamic concepts. Indeed, the flexibility and stiffness influence coefficient merely relate forces to displacements alone, and the definition of the coefficients do not involve masses at all. The inclusion of masses in Fig. 7.3 was only for presenting the influence coefficients in the context of vibrating systems, and the discussion would have remained the same had we regarded points $x = x_i$ $(i = 1, 2, \ldots, n)$ as massless.

We note that for a single-degree-of-freedom system with only one spring the stiffness influence coefficient is merely the spring constant, whereas the flexibility influence coefficient is its reciprocal. An analogous conclusion can be reached in a more general context for multi-degree-of-freedom systems. In this regard, matrix operations turn out to be most useful. Hence, we arrange the flexibility and stiffness influence coefficients in the square matrices

$$[a_{ij}] = A, \ [k_{ij}] = K \tag{7.11}$$

where A is known as the *flexibility matrix* and K as the *stiffness matrix*. Then, using simple rules of matrix multiplication, Eqs. (7.9) and (7.10) can be written in the compact matrix forms

$$\mathbf{u} = A\mathbf{F} \tag{7.12}$$

and

$$\mathbf{F} = K\mathbf{u} \tag{7.13}$$

respectively, in which $\mathbf{u}$ and $\mathbf{F}$ are n-dimensional displacement and force vectors with components u_i $(i = 1, 2, \ldots, n)$ and F_j $(j = 1, 2, \ldots, n)$. Equation (7.12) represents a linear transformation, with matrix A playing the role of an operator that operates on $\mathbf{F}$ to produce the vector $\mathbf{u}$. In view of this Eq. (7.13) can be regarded as the inverse transformation leading from $\mathbf{u}$ to $\mathbf{F}$. Because Eqs. (7.12) and (7.13) relate the same vectors $\mathbf{u}$ and $\mathbf{F}$, matrices A and K must clearly be related. Indeed, introducing Eq. (7.13) in Eq. (7.12), we obtain

$$\mathbf{u} = A\mathbf{F} = AK\mathbf{u} \tag{7.14}$$

with the obvious conclusion that

$$AK = I \tag{7.15}$$

where $I = [\delta_{ij}]$ is the identity, or unit, matrix of order n, with all its elements equal to the Kronecker delta δ_{ij} $(i, j = 1, 2, \ldots, n)$. Equation (7.15) implies that

$$A = K^{-1}, \ K = A^{-1} \tag{7.16}$$

or the *flexibility and stiffness matrices are the inverse of each other.*

It should be pointed out that, although the definition of the stiffness coefficients k_{ij} may seem intimidating, the stiffness coefficients are often easier to evaluate than the flexibility coefficients a_{ij}, as can be concluded from Example 7.2 at the end of this section. Moreover, quite frequently many of the stiffness coefficients have zero values. Nevertheless, the calculation of the stiffness coefficients by the definition given above is not very efficient. Indeed, it is frequently possible to calculate the stiffness coefficients in a much simpler manner, namely, by means of the potential energy, as demonstrated in Sec. 7.3.

There is one important case in which the first of Eqs. (7.16) cannot be used to calculate the flexibility coefficients, namely, when the stiffness matrix K is singular, in which case the flexibility matrix does not exist. Physically, this implies that the system admits rigid-body motions, in which the system undergoes no elastic deformations. This can happen when the supports do not fully restrain the system from moving. Clearly, in the absence of adequate supports, the definition of the flexibility coefficients cannot be applied, so that the coefficients are not defined. We discuss this case in Sec. 7.8.

It must be emphasized at this point that the developments in this section were merely to gain some insight into the free vibration characteristics of conservative systems, and are in no way to be construed as a way of solving eigenvalue problems. In fact, the opposite is true, as the approach based on the characteristic determinant is not recommended when the number of degrees of freedom is three and higher. Indeed, by now there is a large variety of numerical algorithms capable of solving eigenvalue problems for systems with degrees of freedom reaching into the thousands. We consider the question of solving algebraic eigenvalue problems in Sec. 7.18, which also includes a MATLAB computer program.

Example 7.2. Consider the three-degree-of-freedom system shown in Fig. 7.4a and use the definitions to calculate the flexibility and stiffness matrices.

To calculate the flexibility influence coefficients a_{ij}, we apply unit forces $F_j = 1$ ($j = 1, 2, 3$), in sequence, as shown in Figs. 7.4b, c and d, respectively. In each case, the same unit force is acting at all points to the left of the point of application $x = x_j$ of the unit force. On the other hand, the force is zero to the right of $x = x_j$. It follows that the elongation of every spring is equal to the reciprocal of the spring constant to the left of x_j and to zero to the right of x_j. Hence, displacements are equal to the sum of the elongations of the springs to the left of x_j, and including x_j, and to u_j to the right of x_j, so that, from Figs. 7.4b, c and d, we conclude that

$$a_{11} = u_1 = \frac{1}{k_1}, \quad a_{21} = u_2 = u_1 = \frac{1}{k_1}, \quad a_{31} = u_3 = u_2 = u_1 = \frac{1}{k_1} \tag{a}$$

$$a_{12} = u_1 = \frac{1}{k_1}, \quad a_{22} = u_2 = \frac{1}{k_1} + \frac{1}{k_2}, \quad a_{32} = u_3 = u_2 = \frac{1}{k_1} + \frac{1}{k_2} \tag{b}$$

and

$$a_{13} = u_1 = \frac{1}{k_1}, \quad a_{23} = u_2 = \frac{1}{k_1} + \frac{1}{k_2}, \quad a_{33} = u_3 = \frac{1}{k_1} + \frac{1}{k_2} + \frac{1}{k_3} \tag{c}$$

respectively. The coefficients given by Eqs. (a)–(c) can be exhibited in the matrix form

$$A = \begin{bmatrix} \frac{1}{k_1} & \frac{1}{k_1} & \frac{1}{k_1} \\ \frac{1}{k_1} & \frac{1}{k_1} + \frac{1}{k_2} & \frac{1}{k_1} + \frac{1}{k_2} \\ \frac{1}{k_1} & \frac{1}{k_1} + \frac{1}{k_2} & \frac{1}{k_1} + \frac{1}{k_2} + \frac{1}{k_3} \end{bmatrix} \tag{d}$$

and we note that the flexibility matrix A is symmetric. This is no coincidence, as we shall learn in Sec. 7.3.

The stiffness influence coefficients k_{ij} are obtained from Figs. 7.4e, f and g, in which the coefficients are simply the shown forces, where forces opposite in direction to the unit displacements must be assigned negative signs. From Fig. 7.4e, we conclude that, corresponding to $u_1 = 1$, $u_2 = u_3 = 0$, there are the reaction forces $F_0 = -k_1$ and $F_2 = -k_2$ and because for equilibrium we must have

$$F_0 + F_1 + F_2 = 0 \tag{e}$$

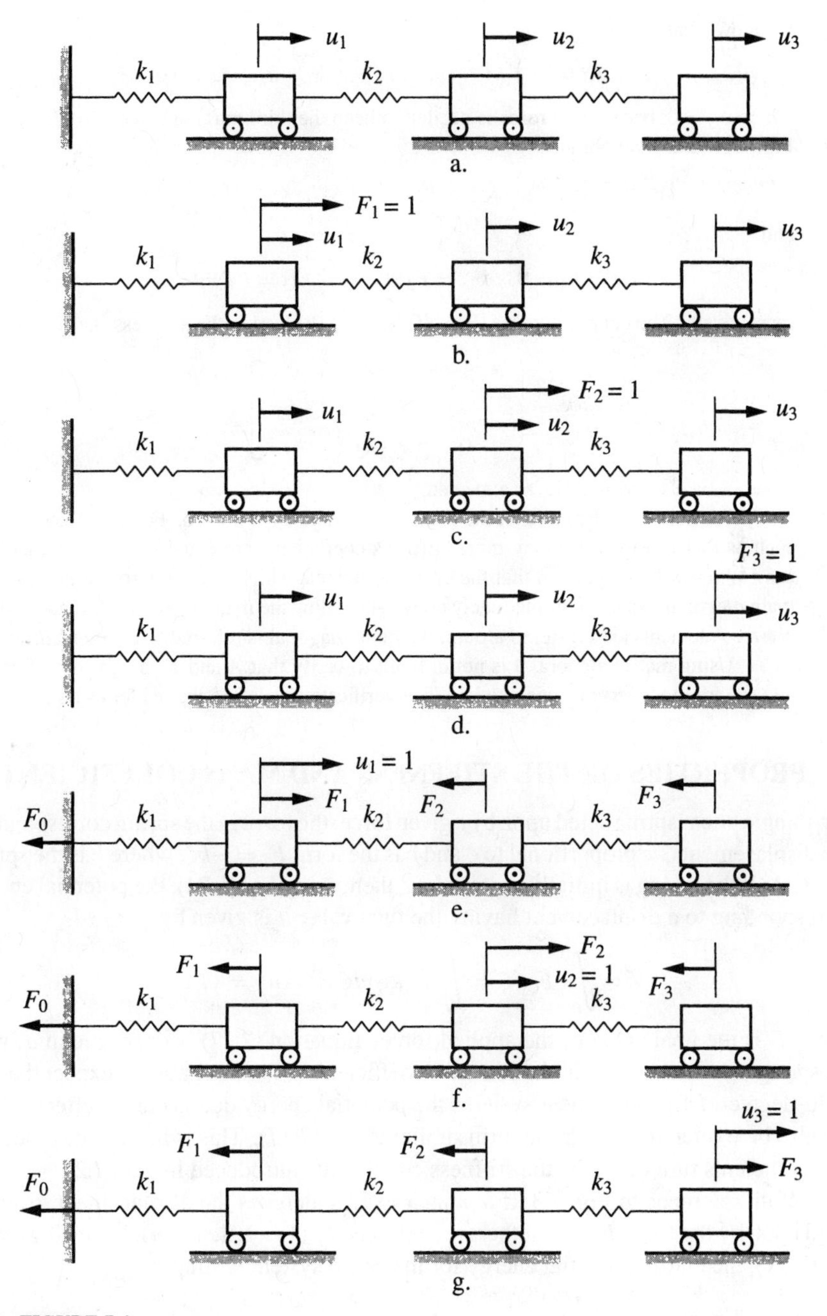

FIGURE 7.4

a. Three-degree-of-freedom system, b. Force pattern: $F_1 = 1,\ F_2 = F_3 = 0$, c. Force pattern: $F_2 = 1,\ F_1 = F_3 = 0$, d. Force pattern: $F_3 = 1,\ F_1 = F_2 = 0$, e. Displacement pattern: $u_1 = 1,\ u_2 = u_3 = 0$, f. Displacement pattern: $u_2 = 1,\ u_1 = u_3 = 0$, g. Displacement pattern: $u_3 = 1,\ u_1 = u_2 = 0$

it follows that

$$k_{11} = F_1 = k_1 + k_2, \quad k_{21} = F_2 = -k_2, \quad k_{31} = F_3 = 0 \tag{f}$$

where $F_3 = 0$, because no force is needed to keep the third mass in place. Similarly, from Figs. 7.4f and g we obtain

$$k_{12} = F_1 = -k_2, \quad k_{22} = F_2 = k_2 + k_3, \quad k_{32} = F_3 = -k_3 \tag{g}$$

and

$$k_{13} = F_1 = 0, \quad k_{23} = F_2 = -k_3, \quad k_{33} = F_3 = k_3 \tag{h}$$

respectively. The coefficients k_{ij}, Eqs. (f), (g) and (h) lead to the stiffness matrix

$$K = \begin{bmatrix} k_1 + k_2 & -k_2 & 0 \\ -k_2 & k_2 + k_3 & -k_3 \\ 0 & -k_3 & k_3 \end{bmatrix} \tag{i}$$

where K is also symmetric, as expected.

We note from Eq. (i) that the elements k_{13} and k_{31} are zero. For chainlike systems such as that of Fig. 7.3, many more stiffness coefficients are equal to zero. In fact, it is easy to verify by inspection that the only coefficients which are not zero are those on the main diagonal and those immediately above and below the main diagonal. A matrix whose nonzero elements are clustered around the main diagonal is referred to as *banded*.

Using matrix algebra, it is not difficult to verify that A and K, as given by Eqs. (d) and (i), are the inverse of one another. The verification is left to the reader as an exercise.

7.3 PROPERTIES OF THE STIFFNESS AND MASS COEFFICIENTS

For a single linear spring acted upon by a given force, the force in the spring corresponding to a displacement ζ is proportional to ζ and has the form $F_\zeta = -k\zeta$, where k is the spring constant. If the spring is initially unstretched, then, using Eq. (1.74), the potential energy corresponding to a displacement having the final value u is given by

$$V = \int_u^0 F_\zeta d\zeta = \int_u^0 (-k\zeta)d\zeta = \tfrac{1}{2}ku^2 = \tfrac{1}{2}Fu \tag{7.17}$$

where F is the final value of the applied force. Equation (7.17) is quadratic in u, with the spring constant k playing the role of a coefficient. It is reasonable to expect that for multi-degree-of-freedom linear systems the potential energy due to elastic effects alone can also be written in a quadratic form similar to Eq. (7.17). This is indeed the case, and the coefficients turn out to be the stiffness coefficients introduced in Sec. 7.2.

With reference to Fig. 7.3, if a point $x = x_i$ undergoes the displacement u_i when acted upon by the force F_i, and if there are n forces F_i $(1 = 1, 2, \ldots, n)$, by analogy with Eq. (7.17), the elastic potential energy for the entire system is simply

$$V = \sum_{i=1}^{n} V_i = \frac{1}{2}\sum_{i=1}^{n} F_i u_i \tag{7.18}$$

Note that the elastic potential energy is often referred to as *strain energy*. But the force F_i is related to the displacements u_j $(j = 1, 2, \ldots, n)$ according to Eq. (7.10). Inserting

Eq. (7.10) into Eq. (7.18), we obtain

$$V = \frac{1}{2}\sum_{i=1}^{n} u_i \left(\sum_{j=1}^{n} k_{ij} u_j\right) = \frac{1}{2}\sum_{i=1}^{n}\sum_{j=1}^{n} k_{ij} u_i u_j \tag{7.19}$$

where k_{ij} $(i, j = 1, 2, \dots, n)$ are the stiffness influence coefficients. On the other hand, Eq. (7.9) relates the displacement u_i to the forces F_j $(j = 1, 2, \dots, n)$, so that inserting Eq. (7.9) into Eq. (7.18), we have

$$V = \frac{1}{2}\sum_{i=1}^{n} F_i \left(\sum_{j=1}^{n} a_{ij} F_j\right) = \frac{1}{2}\sum_{i=1}^{n}\sum_{j=1}^{n} a_{ij} F_i F_j \tag{7.20}$$

where a_{ij} $(i, j = 1, 2, \dots n)$ are the flexibility influence coefficients.

The flexibility coefficients a_{ij} and the stiffness coefficients k_{ij} have a very important property, namely, they are symmetric. This statement is true for the flexibility and stiffness coefficients for any linear multi-degree-of-freedom mechanical system. The proof is based on the principle of superposition. Considering Fig. 7.3, we assume that only the force F_i is acting on the system and denote by $u'_i = a_{ii} F_j$ the displacement produced at $x = x_i$ and by $u'_j = a_{ji} F_i$ that produced at $x = x_j$, where primes indicate that the displacements are produced by F_i alone. It follows that the potential energy due to the force F_i is

$$\tfrac{1}{2} F_i u'_i = \tfrac{1}{2} a_{ii} F_i^2 \tag{7.21}$$

Next, we apply a force F_j at $x = x_j$, resulting in additional displacements $u''_i = a_{ij} F_j$ and $u''_j = a_{jj} F_j$ at $x = x_i$ and $x = x_j$, respectively, where double primes denote displacements due to F_j alone. Because the force F_i does not change during the application of F_j, the total potential energy has the expression

$$\tfrac{1}{2} F_i u'_i + F_i u''_i + \tfrac{1}{2} F_j u''_j = \tfrac{1}{2} a_{ii} F_i^2 + a_{ij} F_i F_j + \tfrac{1}{2} a_{jj} F_j^2 \tag{7.22}$$

Now, we apply the same forces F_i and F_j but in reverse order. Applying first a force F_j at $x = x_j$ and denoting by $u''_j = a_{jj} F_j$ the displacement produced at $x = x_j$ and by $u''_i = a_{ij} F_j$ that produced at $x = x_i$, the potential energy due to F_j alone is

$$\tfrac{1}{2} F_j u''_j = \tfrac{2}{2} a_{jj} F_j^2 \tag{7.23}$$

Then, we apply a force F_i at $x = x_i$ and denote the resulting displacement at $x = x_i$ by $u'_i = a_{ii} F_i$ and that at $x = x_j$ by $u'_j = a_{ji} F_i$. This time we recognize that it is F_j that does not change during the application of F_i, so that the potential energy is

$$\tfrac{1}{2} F_j u''_j + F_j u'_j + \tfrac{1}{2} F_i u'_i = \tfrac{1}{2} a_{jj} F_j^2 + a_{ji} F_j F_i + \tfrac{1}{2} a_{ii} F_i^2 \tag{7.24}$$

But the potential energy must be the same regardless of the order in which the forces F_i and F_j are applied. Hence, Eqs. (7.23) and (7.24) must have the same value, which yields

$$a_{ij} F_i F_j = a_{ji} F_j F_i \tag{7.25}$$

with the obvious conclusion that *the flexibility influence coefficients are symmetric*,

$$a_{ij} = a_{ji} \tag{7.26}$$

Equation (7.26) is the statement of *Maxwell's reciprocity theorem* and can be proved for more general linear systems than that of Fig. 7.3. Considering Eqs. (7.16), it is easy to verify that *the stiffness influence coefficients are also symmetric*,

$$k_{ij} = k_{ji}, \; i, j = 1, 2, \ldots, n \tag{7.27}$$

In matrix notation, the symmetry of the stiffness and flexibility influence coefficients is equivalent to the statement that *the flexibility matrix* and *the stiffness matrix are symmetric*, or

$$A = A^T, \; K = K^T \tag{7.28}$$

The potential energy can be written in the form of a triple matrix product. Indeed, in matrix notation, Eq. (7.19) has the form

$$V = \tfrac{1}{2}\mathbf{u}^T K \mathbf{u} \tag{7.29}$$

whereas Eq. (7.20) can be written as

$$V = \tfrac{1}{2}\mathbf{F}^T A \mathbf{F} \tag{7.30}$$

where $\mathbf{u}$ and $\mathbf{F}$ are the n-dimensional displacement and force vectors, respectively.

Another matrix of special interest in vibrations is the mass matrix, which is associated with the kinetic energy. Considering a multi-degree-of-freedom system and denoting by $\dot{u}_i$ the velocity of mass m_i $(i = 1, 2, \ldots, n)$, the kinetic energy is simply

$$T = \frac{1}{2}\sum_{i=1}^{n} m_i \dot{u}_i^2 \tag{7.31}$$

which can be written in the form of the triple matrix product

$$T = \tfrac{1}{2}\dot{\mathbf{u}}^T M \dot{\mathbf{u}} \tag{7.32}$$

in which M is the *mass matrix*, or *inertia matrix*. In this particular case the matrix M is diagonal. In general, M need not be diagonal (see Sec. 7.5), although it is symmetric. We assume that this is the case with M in Eq. (7.32). It is worth pointing out here that the matrices K and M in Eqs. (7.29) and (7.32), respectively, are precisely the stiffness and mass matrices appearing in the differential equations of motion for a discrete linear system, as derived in Sec. 7.1.

Equations (7.29) and (7.32) are merely quadratic forms in matrix notation, the first representing the potential energy and the second the kinetic energy. It will prove of interest to study some properties of quadratic forms from which we can infer certain motion characteristics of multi-degree-of-freedom systems. Quadratic forms represent a special type of functions, so that we first present certain definitions concerning functions in general and then apply these definitions to quadratic functions of particular interest in vibrations, namely, the potential energy and the kinetic energy.

A function of several variables is said to be *positive (negative) definite* if it is never negative (positive) and is equal to zero if and only if all the variables are zero. A function of several variables is said to be *positive (negative) semidefinite* if it is never negative (positive) and can be zero even when some or all the variables are not zero. A function of several variables is said to be *sign-variable* if it can take either positive or negative values. For the most part, the sign properties of quadratic functions encountered in vibrations can be ascertained on physical grounds.

For quadratic forms, the sign properties are governed by the corresponding constant coefficients. In the particular case of the kinetic energy T and the potential energy V, these coefficients are m_{ij} and k_{ij}, respectively. In view of the definitions concerning the sign properties of functions given in the preceding paragraph, we can define a matrix whose elements are the coefficients of a positive (negative) definite quadratic form as a *positive (negative) definite matrix.* Likewise, a matrix whose elements are the coefficients of a positive (negative) semidefinite quadratic form is said to be a *positive (negative) semidefinite matrix.* Sometimes a positive (negative) semidefinite function is referred to as merely *positive (negative).*

The kinetic energy is always positive definite, so that the mass matrix M is always positive definite. The question remains as to the sign properties of the potential energy and the associated stiffness matrix K. Two cases of particular interest in the area of vibrations are that in which K is positive definite and that in which K is only positive semidefinite. When both M and K are positive definite, *the system is said to be positive definite*, and all the eigenvalues are positive. This case is discussed in Sec. 7.6. When M is positive definite and K is only positive semidefinite, *the system is positive semidefinite*, and all the eigenvalues are nonnegative, i.e., some of them are zero and all the remaining ones are positive. This is due to the fact that semidefinite systems are capable of moving as if they were rigid, i.e., in pure translation, pure rotation, or both. Such motions have zero frequency. This is equivalent to saying that the elastic potential energy in rigid-body motions is zero, so that there is no vibration. This case was first mentioned in Sec. 7.2 and will be discussed in detail in Sec. 7.8.

Example 7.3. Derive the stiffness matrix for the system of Example 7.2 by means of the potential energy.

Considering Fig. 7.4a and recognizing that the elongations of the springs k_1, k_2 and k_3 are u_1, $u_2 - u_1$ and $u_3 - u_2$, respectively, the potential energy is simply

$$\begin{aligned} V &= \tfrac{1}{2}[k_1u_1^2 + k_2(u_2-u_1)^2 + k_3(u_3-u_2)^2] \\ &= \tfrac{1}{2}[(k_1+k_2)u_1^2 + (k_2+k_3)u_2^2 + k_3u_3^2 - 2k_2u_1u_2 - 2k_3u_2u_3] \end{aligned} \tag{a}$$

which can be rewritten in the matrix form

$$V = \tfrac{1}{2}\mathbf{u}^T K \mathbf{u} \tag{b}$$

where

$$\mathbf{u} = \begin{bmatrix} u_1 \\ u_2 \\ u_3 \end{bmatrix}, \quad K = \begin{bmatrix} k_1+k_2 & -k_2 & 0 \\ -k_2 & k_2+k_3 & -k_3 \\ 0 & -k_3 & k_3 \end{bmatrix} \tag{c}$$

are the displacement vector and stiffness matrix, respectively. Clearly, the stiffness matrix is the same as that obtained in Example 7.2. It is also clear that the derivation of the stiffness matrix via the potential energy is appreciably simpler than by using the definition. This is often the case, and not merely in this particular example.

7.4 LAGRANGE'S EQUATIONS LINEARIZED ABOUT EQUILIBRIUM

In Sec. 7.1, we derived the equations of motion for a specific linear system by means of Newton's second law. In this section, we wish to derive the equations of motion for a generic system, not necessarily linear, without reference to any particular model. To this end, we must abandon Newtonian mechanics in favor of a more versatile approach, namely, Lagrangian mechanics, and in particular *Lagrange's equations* first introduced in Sec. 6.5. Hence, using Eqs. (6.42), we write Lagrange's equations for an n-degree-of-freedom system

$$\frac{d}{dt}\left(\frac{\partial T}{\partial \dot{q}_k}\right) - \frac{\partial T}{\partial q_k} + \frac{\partial V}{\partial q_k} = Q_k, \; k = 1, 2, \ldots, n \tag{7.33}$$

where T is the kinetic energy, V is the potential energy, q_k are generalized coordinates and Q_k are generalized nonconservative forces ($k = 1, 2, \ldots, n$). The kinetic energy and potential energy have the general functional form

$$T = T(q_1, q_2, \ldots, q_n, \; \dot{q}_1, \dot{q}_2, \ldots, \dot{q}_n) \tag{7.34}$$

and

$$V = V(q_1, q_2, \ldots, q_n) \tag{7.35}$$

respectively, and the generalized nonconservative forces can be obtained from the virtual work expression

$$\overline{\delta W}_{nc} = \sum_{k=1}^{n} Q_k \delta q_k \tag{7.36}$$

in which δq_k are generalized virtual displacements.

In attempting to apply Eqs. (7.33) to systems with viscous damping, such as the system of Fig. 7.1a, we run immediately into difficulties, as Eqs. (7.33) do not account for viscous damping forces explicitly; such forces are accounted for implicitly through the generalized nonconservative forces Q_k ($k = 1, 2, \ldots, n$). However, it is shown in Ref. 13 that viscous damping forces can be accounted for explicitly in the context of Lagrange's equations by expressing them in the form

$$Q_{k\text{visc}} = -\frac{\partial \mathcal{F}}{\partial \dot{q}_k}, \; k = 1, 2, \ldots, n \tag{7.37}$$

where $\mathcal{F}$ is a function of the generalized velocities known as *Rayleigh's dissipation function.* Hence, assuming that the generalized forces Q_k include all nonconservative forces with the exception of the viscous forces, Lagrange's equations can be rewritten as

$$\frac{d}{dt}\left(\frac{\partial T}{\partial \dot{q}_k}\right) - \frac{\partial T}{\partial q_k} + \frac{\partial V}{\partial q_k} + \frac{\partial \mathcal{F}}{\partial \dot{q}_k} = Q_k, \; k = 1, 2, \ldots, n \tag{7.38}$$

We return to Rayleigh's dissipation function shortly.

As can be concluded from Example 6.2, Lagrange's equations are in general nonlinear. But, as indicated in Secs. 1.13 and 5.2, in vibrations there is considerable interest in small motions about equilibrium points, where equilibrium points are defined as constant solutions of the equations of motion. In view of this, we express the displacements in the form

$$q_k(t) = q_{ek} + \tilde{q}_k(t), \; k = 1, 2, \ldots, n \tag{7.39}$$

where the constants q_{ek} are the generalized displacements when the system is in equilibrium and $\tilde{q}_k(t)$ are small perturbations from equilibrium $(k = 1, 2, \ldots, n)$. It follows immediately from Eqs. (7.39) that the generalized velocities satisfy

$$\dot{q}_k(t) = \dot{\tilde{q}}_k(t), \; k = 1, 2, \ldots, n \tag{7.40}$$

so that they are also small quantities. Equations (7.39) and (7.40) embody the so-called *small motions assumptions* according to which the motion is confined to a small neighborhood of a given equilibrium point.

Our interest lies in identifying the equilibrium points and in deriving the equations for small motions about equilibrium in the context of Lagrange's equations. To this end, we recognize that the equilibrium points represent constant displacements satisfying Lagrange's equations, Eqs. (7.38), and that the small motions equations are linear equations obtained by linearizing Lagrange's equations. Both the equilibrium equations and the linearized equations can be produced by inserting Eqs. (7.39) and (7.40) into Lagrange's equations and ignoring terms in q_k, $\dot{q}_k$ and $\ddot{q}_k$ of degree two and higher. From the terms retained, the constant terms in q_{ek} can be separated into the equilibrium equations, and the first-order terms in $\tilde{q}_k(t)$ and their derivatives into the linearized equations of motion. Observing that Lagrange's equations contain first derivatives of the kinetic energy and potential energy with respect to $\dot{q}_k$ and q_k, we conclude that the task can be simplified by retaining only the first- and second-order terms in $\dot{\tilde{q}}_k$ and $\tilde{q}_k$ in the kinetic energy and potential energy. To this end, we expand the kinetic energy and potential energy in truncated Taylor series. Recognizing that the kinetic energy is already quadratic in velocities, we have

$$T \cong \frac{1}{2} \sum_{i=1}^{n} \sum_{j=1}^{n} \left. \frac{\partial^2 T}{\partial \dot{q}_i \partial \dot{q}_j} \right|_{\mathbf{q}=\mathbf{q}_e} \dot{\tilde{q}}_i \dot{\tilde{q}}_j = \frac{1}{2} \sum_{i=1}^{n} \sum_{j=1}^{n} m_{ij} \dot{\tilde{q}}_i \dot{\tilde{q}}_j \tag{7.41}$$

where

$$m_{ij} = m_{ji} = \left. \frac{\partial^2 T}{\partial \dot{q}_i \partial \dot{q}_j} \right|_{\mathbf{q}=\mathbf{q}_e}, \; i, j = 1, 2, \ldots, n \tag{7.42}$$

in which $\mathbf{q} = [q_1 \; q_2 \; \ldots \; q_n]^T$ and $\mathbf{q}_e = [q_{e1} \; q_{e2} \; \ldots \; q_{en}]^T$. The notation in Eqs. (7.41) and (7.42) states that the second derivatives are to be evaluated at an equilibrium point, so that the symmetric coefficients m_{ij} are constant; they represent the *mass coefficients*.

Similarly, the series for the potential energy is

$$V \cong V(\mathbf{q}_e) + \sum_{i=1}^{n} \left.\frac{\partial V}{\partial q_i}\right|_{\mathbf{q}=\mathbf{q}_e} \tilde{q}_i + \frac{1}{2}\sum_{i=1}^{n}\sum_{j=1}^{n} \left.\frac{\partial^2 V}{\partial q_i \partial q_j}\right|_{\mathbf{q}=\mathbf{q}_e} \tilde{q}_i \tilde{q}_j$$

$$= V(\mathbf{q}_e) + \sum_{i=1}^{n} \left.\frac{\partial V}{\partial q_i}\right|_{\mathbf{q}=\mathbf{q}_e} \tilde{q}_i + \frac{1}{2}\sum_{i=1}^{n}\sum_{j=1}^{n} k_{ij} \tilde{q}_i \tilde{q}_j \tag{7.43}$$

in which

$$k_{ij} = k_{ji} = \left.\frac{\partial^2 V}{\partial q_i \partial q_j}\right|_{\mathbf{q}=\mathbf{q}_e}, \; i, j = 1, 2, \ldots, n \tag{7.44}$$

are constant, symmetric *stiffness coefficients*. Moreover from Sec. 1.7, we observe that viscous damping forces are linear functions of velocities, so that Rayleigh's dissipation function contains only second-order terms in the velocities. Indeed, Rayleigh's dissipation function has the form (Ref. 13)

$$\mathcal{F} = \frac{1}{2}\sum_{i=1}^{n}\sum_{j=1}^{n} c_{ij} \dot{\tilde{q}}_i \dot{\tilde{q}}_j \tag{7.45}$$

where $c_{ij} = c_{ji}$ are constant, symmetric *damping coefficients*. Inserting Eqs. (7.41), (7.43) and (7.45) into Lagrange's equations, Eqs. (7.38), and considering Eqs. (7.39) and (7.40), we obtain

$$\sum_{j=1}^{n} (m_{kj}\ddot{\tilde{q}}_j + c_{kj}\dot{\tilde{q}}_j + k_{kj}\tilde{q}_j) + \left.\frac{\partial V}{\partial q_k}\right|_{\mathbf{q}=\mathbf{q}_e} = Q_k, \; k = 1, 2, \ldots, n \tag{7.46}$$

Then, separating the constant term from the linear terms, we obtain the *equilibrium equations*

$$\left.\frac{\partial V}{\partial q_k}\right|_{\mathbf{q}=\mathbf{q}_e} = 0, \; k = 1, 2, \ldots, n \tag{7.47}$$

The balance of the terms represent the *linearized equations about equilibrium*

$$\sum_{j=1}^{n} (m_{ij}\ddot{q}_j + c_{ij}\dot{q}_j + k_{ij}q_j) = Q_i, \; i = 1, 2, \ldots, n \tag{7.48}$$

in which we replaced the index k by i and omitted the overtilde from $q_j, \dot{q}_j$ and $\ddot{q}_j$ with the understanding that these quantities are small perturbations from equilibrium. Equations (7.48) can be written in the matrix form

$$M\ddot{\mathbf{q}}(t) + C\dot{\mathbf{q}}(t) + K\mathbf{q} = \mathbf{Q}(t) \tag{7.49}$$

where M is the *mass matrix* and K is the *stiffness matrix*, both associated with a given equilibrium point, C is the *damping matrix* and $\mathbf{Q}$ is the vector of generalized nonconservative forces.

Equations (7.48), or Eq. (7.49), were obtained by means of Lagrange's equations, which involve certain differentiations with respect to the generalized coordinates, generalized velocities and time. However, these differentiations are not really necessary, because Eq. (7.49) is fully defined as soon as the mass matrix M, damping matrix C, stiffness matrix K and the vector $\mathbf{Q}$ of generalized nonconservative forces have been determined. To this end, it is only necessary to express the kinetic energy about equilibrium, Eq. (7.41), in the matrix quadratic form

$$T = \tfrac{1}{2}\dot{\mathbf{q}}^T M \dot{\mathbf{q}} \tag{7.50}$$

Similarly, Rayleigh's dissipation function can be written as

$$\mathcal{F} = \tfrac{1}{2}\dot{\mathbf{q}}^T C \dot{\mathbf{q}} \tag{7.51}$$

and, ignoring the constant term $V(\mathbf{q}_e)$ as immaterial and considering the equilibrium equations, Eqs. (7.47), the potential energy about equilibrium has the matrix quadratic form

$$V = \tfrac{1}{2}\mathbf{q}^T K \mathbf{q} \tag{7.52}$$

As with all quadratic forms, the coefficient matrices M, C and K are symmetric. Moreover, the vector $\mathbf{Q}$ of the generalized nonconservative forces, excluding viscous damping forces, is defined by the virtual work

$$\overline{\delta W}_{nc} = \mathbf{Q}^T \delta\mathbf{q} \tag{7.53}$$

Equation (7.49) is identical in appearance to Eq. (7.8), except that the scope of Eq. (7.49) is much broader than the scope of Eq. (7.8). In fact, Eq. (7.8) is a special case of Eq. (7.49). The validity of Eqs. (7.48), or Eq. (7.49), is predicated upon the motions remaining small. If the nature of the coefficients, and in particular the nature of the stiffness coefficients, and that of the nonconservative forces are such that the small motions assumption is violated, the linearized equations cease to be valid and the original Lagrange equations, Eqs. (7.38), must be used.

7.5 LINEAR TRANSFORMATIONS. COUPLING

As demonstrated in Sec. 5.5, coupling depends on the coordinates used to describe the motion and is not a basic characteristic of the system. In this section, we discuss the ideas of coordinate transformations and coupling in broader terms than in Sec. 5.5.

The differential equations of motion for an undamped n-degree-of-freedom system can be written in the matrix form by letting $C = 0$ in Eq. (7.49), with the result

$$M\ddot{\mathbf{q}}(t) + K\mathbf{q}(t) = \mathbf{Q}(t) \tag{7.54}$$

For the purpose of this discussion, we assume that matrices M and K are arbitrary, except that they are symmetric and constant. It is clear from Eq. (7.54) that if M is not diagonal, then the equations of motion are coupled through inertial forces. On the other hand if K is not diagonal, the equations are coupled through the elastically restoring forces. Equation (7.54) represents a set of n simultaneous linear second-order ordinary differential equations with constant coefficients. The solution of such a set of equations is not a simple task, and we wish to examine efficient means of carrying it out. To this end, we express the equations of motion in a different set of generalized coordinates $\eta_j(t)$ $(j = 1, 2, \ldots, n)$ such that any displacement $q_i(t)$ $(i = 1, 2, \ldots, n)$ is a linear combination of the coordinates $\eta_j(t)$. Hence, we consider the linear transformation

$$\mathbf{q}(t) = U\boldsymbol{\eta}(t) \tag{7.55}$$

in which U is a constant nonsingular square matrix, referred to as a *transformation matrix*. The matrix U can be regarded as an operator transforming the vector $\boldsymbol{\eta}$ into the vector $\mathbf{q}$. Because U is constant, we also have

$$\dot{\mathbf{q}}(t) = U\dot{\boldsymbol{\eta}}(t), \ \ddot{\mathbf{q}}(t) = U\ddot{\boldsymbol{\eta}}(t) \tag{7.56}$$

so that the same transformation matrix U connects the velocity vectors $\dot{\boldsymbol{\eta}}$ and $\dot{\mathbf{q}}$ and the acceleration vectors $\ddot{\boldsymbol{\eta}}$ and $\ddot{\mathbf{q}}$. Inserting Eq. (7.55) and the second of Eqs. (7.56) into Eq. (7.54), we can write

$$MU\ddot{\boldsymbol{\eta}}(t) + KU\boldsymbol{\eta}(t) = \mathbf{Q}(t) \tag{7.57}$$

Then, premultiplying both sides of Eq. (7.57) by U^T, we obtain

$$M'\ddot{\boldsymbol{\eta}}(t) + K'\boldsymbol{\eta}(t) = \mathbf{N}(t) \tag{7.58}$$

where the matrices

$$M' = U^T MU = M'^T, \ K' = U^T KU = K'^T \tag{7.59}$$

are symmetric because M and K are symmetric. Moreover

$$\mathbf{N}(t) = U^T\mathbf{Q}(t) \tag{7.60}$$

is an n-dimensional vector whose elements are the generalized forces N_i associated with the generalized coordinates η_i. Note that N_i are linear combinations of Q_j $(j = 1, 2, \ldots, n)$.

The derivation of the matrices M' and K' can be carried out in a more natural manner by considering the kinetic and potential energy. Indeed, recalling Eq. (7.55) and the first of Eqs. (7.56) and recognizing that

$$\mathbf{q}^T(t) = \boldsymbol{\eta}^T(t)U^T, \ \dot{\mathbf{q}}^T(t) = \dot{\boldsymbol{\eta}}^T(t)U^T \tag{7.61}$$

the kinetic and potential energy, Eqs. (7.50) and (7.52), can be expressed in the form

$$T = \tfrac{1}{2}\dot{\boldsymbol{\eta}}^T(t)M'\dot{\boldsymbol{\eta}}(t), \ V = \tfrac{1}{2}\boldsymbol{\eta}^T(t)K'\boldsymbol{\eta}(t) \tag{7.62}$$

where M' and K' are the mass and stiffness matrices corresponding to the coordinates $\eta_j(t)$ $(j = 1, 2, \ldots, n)$ and are given by Eqs. (7.59).

At this point we return to the concept of coupling. If matrix M' is diagonal, then system (7.58) is said to be *inertially uncoupled.* On the other hand, if K' is diagonal, then the system is said to be *elastically uncoupled. The object of the transformation* (7.55) *is to produce diagonal matrices M' and K' simultaneously, because then the system consists of independent equations of motion.* Hence, if such a transformation matrix U can be found, then Eq. (7.58) represents a set of n independent equations of the type

$$M'_{jj}\ddot{\eta}_j(t) + K'_{jj}\eta_j(t) = N_j(t) \ j = 1, 2, \ldots, n \tag{7.63}$$

where M'_{jj} and K'_{jj} are the diagonal elements of M' and K', respectively. Equations (7.63) have precisely the same structure as that of an undamped single-degree-of-freedom system [see Eq. (3.1) with $c = 0$], and can be readily solved by the methods of Chs. 2–4. We state here, and prove later, that a linear transformation matrix U diagonalizing M and K simultaneously does indeed exist. This particular matrix U is the *modal matrix*, because it consists of the *modal vectors*, or *natural modes* of the system; the coordinates $\eta_j(t)$ $(j = 1, 2, \ldots, n)$ are called *natural*, or *modal coordinates*, and the independent equations are commonly referred to as *modal equations*. The procedure for solving the system of simultaneous differential equations of motion by transforming them into a set of independent equations using the modal matrix as a transformation matrix is generally referred to as *modal analysis*.

It is perhaps appropriate to pause at this point and reflect on the coordinate transformation (7.55), leading from equations of motion in terms of the generalized coordinates $q_i(t)$ $(i = 1, 2, \ldots, n)$ to equations of motion in terms of the generalized coordinates $\eta_j(t)$ $(j = 1, 2, \ldots, n)$. The new mass and stiffness matrices M' and K' are related to the original mass and stiffness matrices M and K by means of Eqs. (7.59). If U is the modal matrix, so that matrices M' and K' are diagonal, the matrix U is said to be *orthogonal* with respect to both M and K. Moreover in this case Eqs. (7.59) represent an *orthogonal transformation* (Appendix C), which *does not change the nature of the system.* Orthogonal transformations diagonalizing a real symmetric matrix are quite common in linear algebra. What is unique about the linear transformation (7.55) with the modal matrix U as the transformation matrix is that it diagonalizes both the mass matrix M and the stiffness matrix K simultaneously, which are precisely the two matrices that must be diagonalized to obtain an efficient solution of the equations of motion describing the vibration of linear conservative systems.

It remains to find a way of determining the modal matrix U for a given system. This can be accomplished by solving the algebraic eigenvalue problem associated with matrices M and K, a subject discussed in Sec. 7.6. It should be pointed out that we already used a linear transformation of the type (7.55) to uncouple equations of motion. Indeed, the vectors $\mathbf{u}_1$ and $\mathbf{u}_2$ multiplying the natural coordinates $\eta_1(t)$ and $\eta_2(t)$ in Sec. 5.10 were modal vectors, and hence the columns of the modal matrix U. But, as pointed out in Sec. 5.3, the modal vectors satisfy homogeneous algebraic equations, so that their magnitude cannot be determined uniquely; only the ratios of the components of the modal vectors can. It is often convenient to choose the magnitude of the modal vectors so as to reduce the matrix M' to the identity matrix, which automatically reduces the matrix K' to the diagonal matrix of the natural frequencies squared. This process is known as *normalization* and, under these circumstances, the modal matrix U is said to be *orthonormal* with respect to M and K. In addition, the natural or modal coordinates $\eta_j(t)$ $(j = 1, 2, \ldots, n)$ become *normal coordinates.*

Example 7.4. The mass and stiffness matrices for the two-degree-of-freedom system of Example 5.1 are given by

$$M = m\begin{bmatrix} 1 & 0 \\ 0 & 2 \end{bmatrix}, \quad K = \frac{T}{L}\begin{bmatrix} 2 & -1 \\ -1 & 3 \end{bmatrix} \tag{a}$$

and the corresponding modal matrix is

$$U = [\,\mathbf{u}_1 \quad \mathbf{u}_2\,] = \begin{bmatrix} 1 & 1 \\ 1 & -0.5 \end{bmatrix} \tag{b}$$

which was normalized by setting the top component of $\mathbf{u}_1$ and $\mathbf{u}_2$ equal to 1. (The fact that U is symmetric is a mere coincidence without any significance. Generally, U is not symmetric.) Show that the modal matrix diagonalizes the mass and stiffness matrices simultaneously. Then, normalize the modal matrix so as to satisfy $U^T MU = I$, where I is the identity matrix, and show that the associated diagonal matrix $U^T KU$ has the natural frequencies squared as the diagonal entries.

Inserting Eqs. (a) and (b) into Eqs. (7.59), we obtain

$$M' = U^T MU = m \begin{bmatrix} 1 & 1 \\ 1 & -0.5 \end{bmatrix}^T \begin{bmatrix} 1 & 0 \\ 0 & 2 \end{bmatrix} \begin{bmatrix} 1 & 1 \\ 1 & -0.5 \end{bmatrix} = m \begin{bmatrix} 3 & 0 \\ 0 & 1.5 \end{bmatrix} \tag{c}$$

and

$$K' = U^T KU = \frac{T}{L} \begin{bmatrix} 1 & 1 \\ 1 & -0.5 \end{bmatrix}^T \begin{bmatrix} 2 & -1 \\ -1 & 3 \end{bmatrix} \begin{bmatrix} 1 & 1 \\ 1 & -0.5 \end{bmatrix} = \frac{T}{L} \begin{bmatrix} 3 & 0 \\ 0 & 3.75 \end{bmatrix} \tag{d}$$

so that the modal matrix does indeed diagonalize the mass and stiffness matrices simultaneously.

The normalization of the modal matrix so as to satisfy $U^T MU = I$ can be carried out one eigenvector at a time. To this end, we write the first normalized modal vector in the form

$$\mathbf{u}_1 = \alpha_1 \begin{bmatrix} 1 \\ 1 \end{bmatrix} \tag{e}$$

where the scaling factor α_1 must be such that

$$\mathbf{u}_1^T M\mathbf{u}_1 = \alpha_1^2 m \begin{bmatrix} 1 \\ 1 \end{bmatrix}^T \begin{bmatrix} 1 & 0 \\ 0 & 2 \end{bmatrix} \begin{bmatrix} 1 \\ 1 \end{bmatrix} = 3\alpha_1^2 m = 1 \tag{f}$$

so that $\alpha_1 = 1/\sqrt{3m}$ and the first normalized modal vector is

$$\mathbf{u}_1 = \frac{1}{\sqrt{3m}} \begin{bmatrix} 1 \\ 1 \end{bmatrix} \tag{g}$$

Similarly, the second normalized modal vector can be verified to be

$$\mathbf{u}_2 = \frac{1}{\sqrt{1.5m}} \begin{bmatrix} 1 \\ -0.5 \end{bmatrix} \tag{h}$$

Hence, the normalized modal matrix is

$$U = \frac{1}{\sqrt{3m}} \begin{bmatrix} 1 & \sqrt{2} \\ 1 & -0.5\sqrt{2} \end{bmatrix} \tag{i}$$

and we note that U is no longer symmetric, as is to be expected. Clearly, in this case,

$$M' = U^T MU = \left(\frac{1}{\sqrt{3m}}\right)^2 m \begin{bmatrix} 1 & \sqrt{2} \\ 1 & -0.5\sqrt{2} \end{bmatrix}^T \begin{bmatrix} 1 & 0 \\ 0 & 2 \end{bmatrix} \begin{bmatrix} 1 & \sqrt{2} \\ 1 & -0.5\sqrt{2} \end{bmatrix}$$

$$= \begin{bmatrix} 1 & 0 \\ 0 & 1 \end{bmatrix} \tag{j}$$

so that the new mass matrix is indeed the unit matrix. Moreover, the new stiffness matrix is

$$K' = U^T K U = \left(\frac{1}{\sqrt{3m}}\right)^2 \frac{T}{L} \begin{bmatrix} 1 & \sqrt{2} \\ 1 & -0.5\sqrt{2} \end{bmatrix}^T \begin{bmatrix} 2 & -1 \\ -1 & 3 \end{bmatrix} \begin{bmatrix} 1 & \sqrt{2} \\ 1 & -0.5\sqrt{2} \end{bmatrix}$$

$$= \frac{T}{mL} \begin{bmatrix} 1 & 0 \\ 0 & 2.5 \end{bmatrix} \quad \text{(k)}$$

which, from Eq. (b) of Example 5.1, is recognized as the diagonal matrix of natural frequencies squared, or

$$K' = \begin{bmatrix} \omega_1^2 & 0 \\ 0 & \omega_2^2 \end{bmatrix} \quad \text{(l)}$$

7.6 UNDAMPED FREE VIBRATION. THE EIGENVALUE PROBLEM

In Sec. 7.5, we pointed out that, in the absence of damping, the equations of motion can be decoupled by using a transformation of coordinates, with the modal matrix acting as the transformation matrix. To determine the modal matrix, we must solve an algebraic eigenvalue problem, a problem associated with free vibration, i.e., vibration in which the external forces are zero. In this section, we show how the free vibration problem leads directly to the eigenvalue problem, the solution of the latter yielding the natural modes of vibration. Then, we show that the natural motions, defined as motions in which the system vibrates in any one of the natural modes, can be identified as special cases of free vibration. Finally, we show that in the general case of free vibration, the motion can be regarded as a linear combination of the natural motions.

In the absence of external forces, $\mathbf{Q} = \mathbf{0}$, Eq. (7.49) reduces to

$$M\ddot{\mathbf{q}}(t) + K\mathbf{q}(t) = \mathbf{0} \quad (7.64)$$

which represents a set of n simultaneous homogeneous differential equations of the type

$$\sum_{j=1}^{n} m_{ij}\ddot{q}_j(t) + \sum_{j=1}^{n} k_{ij}q_j(t) = 0, \; i = 1, 2, \ldots, n \quad (7.65)$$

We are interested in a special type of solution of Eqs. (7.65), namely, that in which all the coordinates $q_j(t)$ $(j = 1,\ 2, \ldots, n)$ execute synchronous motion. Physically, this implies a motion in which all the coordinates have the same time dependence, and the general configuration of the motion does not change, except for the amplitude, so that the ratio between any two coordinates $q_i(t)$ and $q_j(t)$, $i \neq j$, remains constant during the motion. Mathematically, this type of motion can be represented by

$$q_j(t) = u_j f(t), \; j = 1, 2 \ldots, n \quad (7.66)$$

where $u_j (j = 1,\ 2, \ldots, n)$ are constant amplitudes and $f(t)$ is a function of time, the same function for all the coordinates $q_j(t)$. The interest lies in the case in which the coordinates $q_j(t)$ represent stable oscillation, which implies that $f(t)$ must be bounded.

Inserting Eqs. (7.66) into Eq. (7.65) and recognizing that the function $f(t)$ does not depend on the index j, we obtain

$$\ddot{f}(t)\sum_{j=1}^{n} m_{ij}u_j + f(t)\sum_{j=1}^{n} k_{ij}u_j = 0,\ i = 1, 2, \ldots, n \tag{7.67}$$

Equations (7.67) can be rewritten in the form

$$-\frac{\ddot{f}(t)}{f(t)} = \frac{\sum_{j=1}^{n} k_{ij}u_j}{\sum_{j=1}^{n} m_{ij}u_j},\ i = 1, 2, \ldots, n \tag{7.68}$$

with the implication that the time dependence and the positional dependence are separable, so that the process is akin to the separation of variables for partial differential equations. Using the standard argument, we observe that the left side of Eqs. (7.68) does not depend on the index i, whereas the right side does not depend on time, so that two ratios must be equal to a constant, and in particular to the same constant. Assuming that $f(t)$ is a real function, the constant must be a real number. Denoting the constant by λ, Eqs. (7.68) yield

$$\ddot{f}(t) + \lambda f(t) = 0 \tag{7.69}$$

and

$$\sum_{j=1}^{n}(k_{ij} - \lambda m_{ij})u_j = 0,\ i = 1, 2, \ldots, n \tag{7.70}$$

We consider a solution of Eq. (7.69) in the exponential form

$$f(t) = Ae^{st} \tag{7.71}$$

Introducing solution (7.71) in Eq. (7.69) and dividing through by Ae^{st}, we conclude that s must satisfy the equation

$$s^2 + \lambda = 0 \tag{7.72}$$

which has the two roots

$$\begin{matrix} s_1 \\ s_2 \end{matrix} = \pm\sqrt{-\lambda} \tag{7.73}$$

If λ is a negative number (we have already concluded that it must be real), then s_1 and s_2 are real numbers, equal in magnitude but opposite in sign. In this case, Eq. (7.69) has two solutions, one decreasing and the other increasing exponentially with time. These solutions, however, are inconsistent with stable bounded motion, so that the possibility that λ is negative must be discarded and the one that λ is positive must be retained. Letting $\lambda = \omega^2$, where ω is real, Eq. (7.73) yields

$$\begin{matrix} s_1 \\ s_2 \end{matrix} = \pm i\omega \tag{7.74}$$

so that the solution of Eq. (7.69) becomes

$$f(t) = A_1 e^{i\omega t} + A_2 e^{-i\omega t} \tag{7.75}$$

where A_1 and A_2 are generally complex numbers constant in value. Recognizing that $e^{i\omega t}$ and $e^{-i\omega t}$ represent complex vectors of unit magnitude (Sec. 1.11), we conclude that solution (7.75) is harmonic with the frequency ω, and is the only acceptable solution of Eq. (7.69). This implies that, if synchronous motion is possible, the time dependence is harmonic. But, because $f(t)$ is a real function, A_2 must be the complex conjugate of A_1. Then, as in Sec. 4.1, solution (7.75) can be expressed in the form

$$f(t) = C\cos(\omega t - \phi) \tag{7.76}$$

where C is an arbitrary constant, ω the frequency of the harmonic motion and ϕ its phase angle, all three quantities being the same for every coordinate $q_j(t)$ $(j = 1, 2, \dots, n)$.

To complete the solution of Eqs. (7.65), we must determine the amplitudes u_j $(j = 1, 2, \dots, n)$. To this end, we turn to Eqs. (7.70), which constitute a set of n homogeneous algebraic equations in the unknowns u_j, with $\lambda = \omega^2$ playing the role of a parameter. The problem of determining the values of ω^2 for which nontrivial solutions u_j $(j = 1, 2, \dots, n)$ of Eqs. (7.70) exist is known as the *characteristic-value*, or *eigenvalue problem*. In particular, it is known as the *algebraic eigenvalue problem*.

It is convenient to write Eqs. (7.70) in the matrix form

$$K\mathbf{u} = \omega^2 M\mathbf{u} \tag{7.77}$$

Equation (7.77) represents the eigenvalue problem associated with matrices M and K and it possesses nontrivial solutions if and only if the determinant of the coefficients vanishes. This can be expressed in the form

$$\Delta(\omega^2) = \det[K - \omega^2 M] = 0 \tag{7.78}$$

where $\Delta(\omega^2)$ is called the *characteristic determinant*, or the *characteristic polynomial*, with Eq. (7.78) itself being known as the *characteristic equation*, or *frequency equation*. The characteristic polynomial is of degree n in ω^2, and possesses in general n distinct roots, referred to as *characteristic values*, or *eigenvalues*. The n roots are denoted by $\omega_1^2, \omega_2^2, \dots, \omega_n^2$ and the square roots of these quantities are the *system natural frequencies* ω_r $(r = 1, 2, \dots, n)$. The natural frequencies can be arranged in increasing order of magnitude, namely, $\omega_1 \le \omega_2 \le \dots \le \omega_n$. The lowest frequency ω_1 is referred to as the *fundamental frequency*, and for many practical problems it is the most important one. In general, all frequencies ω_r are distinct and the equality sign never holds, except in *degenerate* cases. Such cases are very rare and cannot occur in one-dimensional structures; they can occur in two-dimensional symmetric structures. It follows that there are n frequencies ω_r ($r = 1, 2, \ldots, n$) in which harmonic motion of the type (7.76) is possible.

Associated with every one of the frequencies ω_r there is a certain nontrivial vector $\mathbf{u}_r$ whose components u_{ir} are real numbers, where $\mathbf{u}_r$ is a solution of the eigenvalue problem, and hence it satisfies

$$K\mathbf{u}_r = \omega_r^2 M\mathbf{u}_r, \quad r = 1, 2, \ldots, n \tag{7.79}$$

The vectors $\mathbf{u}_r$ ($r = 1, 2, \ldots, n$) are known as *characteristic vectors*, or *eigenvectors*. The eigenvectors are also referred to as *modal vectors* and represent physically the so-called *natural modes*. These vectors are unique only in the sense that the ratio between any two components u_{ir} and u_{jr} is constant. Because Eq. (7.77) is homogeneous, if $\mathbf{u}_r$ is a solution of the equation, then $\alpha_r\mathbf{u}_r$ is also a solution, where α_r is an arbitrary constant. It follow that *the shape of the natural modes is unique but the amplitude is not.*

If the magnitude of the eigenvector $\mathbf{u}_r$ is assigned a certain value, then the eigenvector is rendered unique. The process of adjusting the magnitude of the natural modes to render them unique is called *normalization*, and the resulting eigenvectors are referred to as *normal modes.* A very convenient normalization scheme consists of setting

$$\mathbf{u}_r^T M\mathbf{u}_r = 1, \quad r = 1, 2, \ldots, n \tag{7.80}$$

which has the advantage that

$$\mathbf{u}_r^T K\mathbf{u}_r = \omega_r^2, \quad r = 1, 2, \ldots, n \tag{7.81}$$

This can be easily shown by premultiplying both sides of Eq. (7.79) by $\mathbf{u}_r^T$. Note that if this normalization scheme is adopted, and if the elements m_{ij} of the mass matrix M have units kg, then the components of $\mathbf{u}_r^T$ have units kg$^{-1/2}$. This, in turn, establishes the units of the constant C in Eq. (7.76), as can be concluded from Eqs. (7.66). Another frequently used normalization scheme consists of setting the value of the largest component of the modal vector $\mathbf{u}_r$ equal to 1, which may be convenient for plotting the modes. Clearly, *the normalization process is devoid of physical significance and should be regarded as a mere convenience.*

In view of Eqs. (7.66) and (7.76), we conclude that Eq. (7.64) has the solutions

$$\mathbf{q}_r(t) = \mathbf{u}_r f_r(t), \quad r = 1, 2, \ldots, n \tag{7.82}$$

where

$$f_r(t) = C_r \cos(\omega_r t - \phi_r), \quad r = 1, 2, \ldots, n \tag{7.83}$$

in which C_r and ϕ_r are constants of integration representing amplitudes and phase angles, respectively. Hence, the free vibration problem admits special independent solutions in which the system vibrates in any one of the natural modes. These solutions are referred to as *natural motions*. Then, invoking the superposition principle, we can write the general solution of Eq. (7.64) as a linear combination of the natural motions, or

$$\mathbf{q}(t) = \sum_{r=1}^{n} \mathbf{q}_r(t) = \sum_{r=1}^{n} \mathbf{u}_r f_r(t) = U\mathbf{f}(t) \tag{7.84}$$

where $U = [\mathbf{u}_1 \ \mathbf{u}_2 \ \dots \ \mathbf{u}_n]$ is the modal matrix and $\mathbf{f}(t) = [f_1(t) \ f_2(t) \ \dots \ f_n(t)]^T$ is a vector with components given by Eqs. (7.83). The constants C_r and ϕ_r entering into $\mathbf{f}(t)$ depend on the initial conditions $\mathbf{q}(0)$ and $\dot{\mathbf{q}}(0)$, as well as on the normalization scheme used for $\mathbf{u}_r$ $(r = 1, 2, \dots, n)$. In Sec. 7.10, we obtain solution (7.84), together with the evaluation of the constants of integration, by a more formal approach, namely, by modal analysis.

The motion characteristics described above are typical of positive definite systems, i.e., systems for which the mass and stiffness matrices are real, symmetric and positive definite. In the case in which the stiffness matrix is only positive semidefinite, there is at least one eigenvector, say $\mathbf{u}_s$, such that $K\mathbf{u}_s = \mathbf{0}$. In this case, the system is not fully restrained, and $\mathbf{u}_s$ represents a *rigid-body mode* with the corresponding natural frequency equal to zero, $\omega_s = 0$, as can be concluded from Eq. (7.79). Of course, in this case the function f_s is not harmonic as in Eq. (7.83).

It must be emphasized at this point that the developments in this section were merely to gain some insight into the free vibration characteristics of conservative systems, and are not be be construed as a way of solving eigenvalue problems. In fact, the opposite is true, as the approach based on the characteristic determinant is not recommended when the number of degrees of freedom is three and higher. Indeed, by now there is a large variety of efficient numerical algorithms capable of solving eigenvalue problems for systems with many degrees of freedom, some reaching into the thousands. We consider the question of solving algebraic eigenvalue problems in Sec. 7.18, which also includes a MATLAB computer program.

Example 7.5. Consider the three-degree-of-freedom system of Example 7.1 and solve the associated eigenvalue problem for the parameters $m_1 = m_2 = m, \ m_3 = 2m, \ c_1 = c_2 = c_3 = 0, \ k_1 = k_2 = k, \ k_3 = 2k$. Then, derive the solution to the free vibration problem for the initial excitations $\mathbf{q}(0) = q_0[1 \ 2 \ 3]^T, \ \dot{\mathbf{q}}(0) = \mathbf{0}$.

Inserting the given parameters into Eqs. (c) and (e) of Example 7.1, we can write the mass matrix

$$M = m \begin{bmatrix} 1 & 0 & 0 \\ 0 & 1 & 0 \\ 0 & 0 & 2 \end{bmatrix} \tag{a}$$

and stiffness matrix

$$K = k \begin{bmatrix} 2 & -1 & 0 \\ -1 & 3 & -2 \\ 0 & -2 & 2 \end{bmatrix} \tag{b}$$

respectively; the damping matrix C is zero. Then, introducing Eqs. (a) and (b) in Eq. (7.78) and expanding the characteristic determinant, we obtain the frequency equation

$$\begin{aligned} \Delta(\omega^2) = [K - \omega^2 M] &= \begin{bmatrix} 2k - \omega^2 m & -k & 0 \\ -k & 3k - \omega^2 m & -2k \\ 0 & -2k & 2(k - \omega^2 m) \end{bmatrix} \\ &= -2m^3 \left[\omega^6 - 6\frac{k}{m}\omega^4 + 8\left(\frac{k}{m}\right)^2 \omega^2 - \left(\frac{k}{m}\right)^3 \right] = 0 \end{aligned} \tag{c}$$

Using the Newton-Raphson method, we obtain the three roots

$$\omega_1^2 = 0.1392\frac{k}{m},\ \omega_2^2 = 1.7459\frac{k}{m},\ \omega_3^2 = 4.1149\frac{k}{m} \tag{d}$$

so that the natural frequencies are

$$\omega_1 = 0.3731\sqrt{\frac{k}{m}},\ \omega_2 = 1.3213\sqrt{\frac{k}{m}},\ \omega_3 = 2.0285\sqrt{\frac{k}{m}} \tag{e}$$

To determine the modal vectors, we insert ω_r^2 $(r = 1, 2, 3)$ into Eq. (7.79), in sequence, and solve the corresponding algebraic equations. For $r = 1$, Eq. (7.79) can be written in the scalar form

$$\begin{aligned} (k_{11} - \omega_1^2 m_1)u_1 + k_{12}u_2 + k_{13}u_3 &= 0 \\ k_{12}u_1 + (k_{22} - \omega_1^2 m_2)u_2 + k_{23}u_3 &= 0 \\ k_{13}u_1 + k_{23}u_2 + (k_{33} - \omega_1^2 m_3)u_3 &= 0 \end{aligned} \tag{f}$$

where we took into consideration that M is diagonal and K is symmetric. Equations (f) are homogeneous, so that no unique solution is possible. Indeed, a solution with all its components multiplied by the same constant is also a solution. The implication is that one of the three components can be assigned an arbitrary value, which renders the remaining two components unique. This also implies that now we have three equations and two unknowns, so that one of the equations is redundant, i.e., one is a linear combination of the other two. It follows that we need solve only two equations in two unknowns. Choosing arbitrarily $u_3 = 1$, using the indicated values from Eqs. (a), (b) and (d) and retaining the first two of Eqs. (f), we have

$$\begin{aligned} k(2 - 0.1392)u_1 - ku_2 &= 0 \\ -ku_1 + k(3 - 0.1392)u_2 &= 2k \end{aligned} \tag{g}$$

Equations (g) have the solution

$$u_1 = 0.4626,\ u_2 = 0.8608 \tag{h}$$

which can be combined with $u_3 = 1$ to yield the first modal vector

$$\mathbf{u}_1 = \begin{bmatrix} 0.4626 & 0.8608 & 1.0000 \end{bmatrix}^T \tag{i}$$

It should be pointed out that assigning an arbitrary value to any other component of the modal vector, or solving any other pair of equations for the other two components, would have yielded essentially the same modal vector. The only difference would have been that the magnitude of all three components of the modal vector would have changed proportionally.

Following the same procedure, the other two modal vectors can be shown to be

$$\mathbf{u}_2 = \begin{bmatrix} 1.0000 \\ 0.2541 \\ -0.3407 \end{bmatrix},\ \mathbf{u}_3 = \begin{bmatrix} -0.4728 \\ 1.0000 \\ -0.3210 \end{bmatrix} \tag{j}$$

and we note that all modal vectors have been normalized so that the component largest in magnitude is equal to 1. The modes are plotted in Fig. 7.5. We observe that in the first

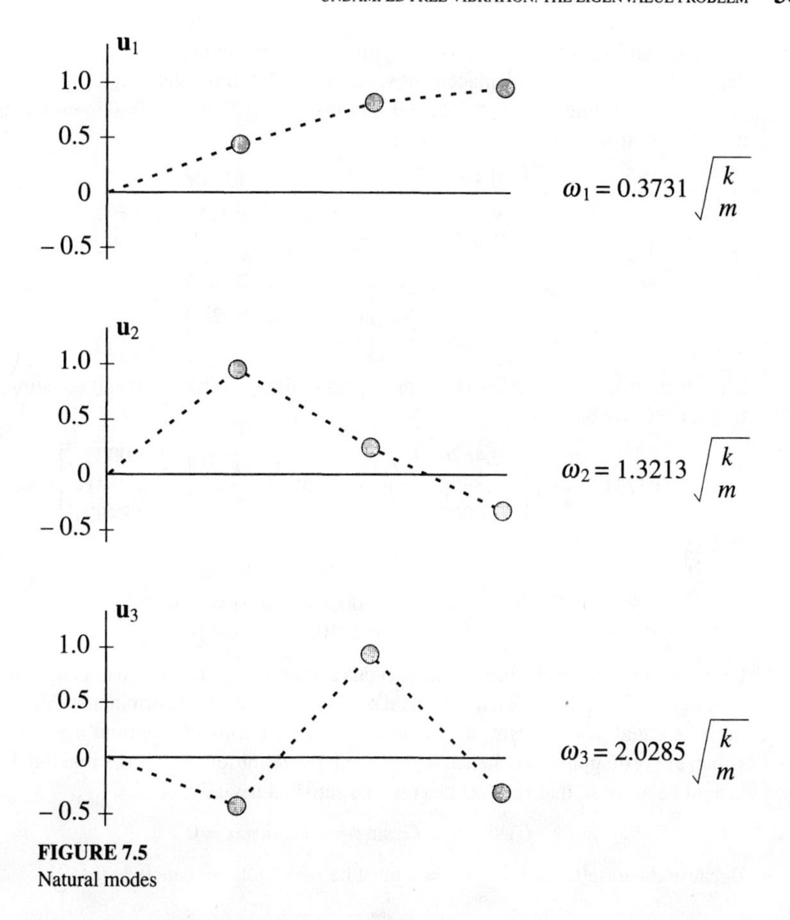

FIGURE 7.5
Natural modes

mode all displacement components have the same sign, in the second mode there is one sign change and in the third mode there are two sign changes. This is typical of modal vectors for chainlike systems of the type considered here.

According to Eq. (7.84), in conjunction with Eqs. (7.83), the solution of the free vibration problem has the general form

$$\mathbf{q}(t) = \sum_{r=1}^{3} \mathbf{u}_r f_r(t) = \sum_{r=1}^{3} C_r \mathbf{u}_r \cos(\omega_r t - \phi_r)$$

$$= C_1 \begin{bmatrix} 0.4626 \\ 0.8608 \\ 1.0000 \end{bmatrix} \cos(0.3731\sqrt{\frac{k}{m}}t - \phi_1)$$

$$+ C_2 \begin{bmatrix} 1.0000 \\ 0.2541 \\ -0.3407 \end{bmatrix} \cos\left(1.3213\sqrt{\frac{k}{m}}t - \phi_2\right)$$

$$+ C_3 \begin{bmatrix} -0.4728 \\ 1.0000 \\ -0.3210 \end{bmatrix} \cos\left(2.0285\sqrt{\frac{k}{m}}t - \phi_3\right) \tag{k}$$

where C_r and ϕ_r $(r = 1, 2, 3)$ are amplitudes and phase angles, constants of integration depending on the initial displacement vector $\mathbf{q}(0)$ and initial velocity vector $\dot{\mathbf{q}}(0)$.

To determine the constants C_r and ϕ_r $(r = 1,\ 2,\ 3)$, we let $t = 0$ in Eq. (k) and equate the result to $\mathbf{q}(0) = q_0[1\ 2\ 3]^T$, so that

$$\mathbf{q}(0) = C_1 \begin{bmatrix} 0.4626 \\ 0.8608 \\ 1.0000 \end{bmatrix} \cos\phi_1 + C_2 \begin{bmatrix} 1.0000 \\ 0.2541 \\ -0.3407 \end{bmatrix} \cos\phi_2$$

$$+ C_3 \begin{bmatrix} -0.4728 \\ 1.0000 \\ -0.3210 \end{bmatrix} \cos\phi_3 = q_0 \begin{bmatrix} 1 \\ 2 \\ 3 \end{bmatrix} \tag{l}$$

Moreover, differentiating Eq. (k) with respect to time, letting $t = 0$ and equating the result to $\dot{\mathbf{q}}(0) = \mathbf{0}$, we have

$$0.3731\sqrt{\frac{k}{m}}C_1 \begin{bmatrix} 0.4626 \\ 0.8608 \\ 1.0000 \end{bmatrix} \sin\phi_1 + 1.3213\sqrt{\frac{k}{m}}C_2 \begin{bmatrix} 1.0000 \\ 0.2541 \\ -0.3407 \end{bmatrix} \sin\phi_2$$

$$+2.0285\sqrt{\frac{k}{m}}C_3 \begin{bmatrix} -0.4728 \\ 1.0000 \\ -0.3210 \end{bmatrix} \sin\phi_3 = \begin{bmatrix} 0 \\ 0 \\ 0 \end{bmatrix} \tag{m}$$

Equation (m) represents three homogeneous algebraic equations in the unknowns $C_1 \sin\phi_1$, $C_2 \sin\phi_2$ and $C_3 \sin\phi_3$. For a nontrivial solution to exist, the determinant of the coefficients must be equal to zero. But, the columns of the determinant represent the modal vectors, which are orthogonal, and hence independent by definition. It follows that the determinant cannot be zero, so that Eq. (m) can only be satisfied trivially, or

$$C_1 \sin\phi_1 = C_2 \sin\phi_2 = C_3 \sin\phi_3 = 0 \tag{n}$$

Because the case $C_1 = C_2 = C_3 = 0$ must be ruled out, we conclude that

$$\phi_1 = \phi_2 = \phi_3 = 0 \tag{o}$$

so that all three phase angles are zero. Inserting Eqs. (o) into Eq. (l), we obtain three nonhomogeneous algebraic equations in the three unknowns $C_1,\ C_2$ and C_3, as follows:

$$\begin{aligned} 0.4626C_1 + C_2 - 0.4728C_3 &= q_0 \\ 0.8608C_1 + 0.2541C_2 + C_3 &= 2q_0 \\ C_1 - 0.3407C_2 - 0.3210C_3 &= 3q_0 \end{aligned} \tag{p}$$

Using Gaussian elimination with back substitution, we obtain the solution

$$C_1 = 2.7696q_0,\ C_2 = -0.4132q_0,\ C_3 = -0.2791q_0 \tag{q}$$

so that, introducing Eqs. (q) in Eq. (k), the solution of the free vibration problem takes the explicit form

$$\mathbf{x}(t) = q_0 \left\{ \begin{bmatrix} 1.2812 \\ 2.3841 \\ 2.7696 \end{bmatrix} \cos 0.3731\sqrt{\frac{k}{m}}t + \begin{bmatrix} -0.4132 \\ -0.1050 \\ 0.1408 \end{bmatrix} \cos 1.3213\sqrt{\frac{k}{m}}t \right.$$

$$\left. + \begin{bmatrix} 0.1320 \\ -0.2791 \\ 0.0896 \end{bmatrix} \cos 2.0285\sqrt{\frac{k}{m}}t \right\} \tag{r}$$

It should be pointed out that the computations were carried out with six decimal figures accuracy, but the results were rounded to four decimal figures to save space.

We observe that the contribution of the first mode to the response is significantly larger than the contribution of the other two modes. This can be easily explained by the fact that the initial displacement configuration resembles the first modal vector much more than the other two modal vectors.

It is perhaps appropriate at this juncture to return to the question of the arbitrariness of the modal vectors. Normalization of a vector amounts to assigning a given value to the magnitude of the vector. This in no way affects the response, because the constant representing the associated amplitude must be adjusted so that the response match the initial conditions. As an illustration, if $\mathbf{u}_1$ is normalized again by multiplying the old vector by the constant α, then the new value of C_1 would be the old value divided by α.

7.7 ORTHOGONALITY OF MODAL VECTORS

The natural modes possess a very important and useful property known as *orthogonality*. This is not an ordinary orthogonality, but an orthogonality with respect to the mass matrix M, as well as with respect to the stiffness matrix K. We introduced the concept of orthogonality briefly in Sec. 5.6 in the context of two-degree-of-freedom systems and then in Sec. 7.5. In this section, we provide a formal proof of the orthogonality of modal vectors for multi-degree-of-freedom systems, and in Sec. 7.10 we show how this orthogonality can be used to obtain the response of vibrating conservative systems.

We consider two distinct solutions of the eigenvalue problem, ω_r^2, $\mathbf{u}_r$ and ω_s^2, $\mathbf{u}_s$. These solutions satisfy

$$K\mathbf{u}_r = \omega_r^2 M\mathbf{u}_r, \quad K\mathbf{u}_s = \omega_s^2 M\mathbf{u}_s \tag{7.85}$$

Premultiplying both sides of the first of Eqs. (7.85) by $\mathbf{u}_s^T$ and of the second by $\mathbf{u}_r^T$, we have

$$\begin{aligned} \mathbf{u}_s^T K\mathbf{u}_r &= \omega_r^2 \mathbf{u}_s^T M\mathbf{u}_r \\ \mathbf{u}_r^T K\mathbf{u}_s &= \omega_s^2 \mathbf{u}_r^T M\mathbf{u}_s \end{aligned} \tag{7.86}$$

Next, we transpose the second of Eqs. (7.86), recall from Sec. 7.4 that matrices M and K are symmetric and subtract the result from the first of Eqs. (7.86) to obtain

$$(\omega_r^2 - \omega_s^2)\mathbf{u}_s^T M\mathbf{u}_r = 0 \tag{7.87}$$

Because in general the natural frequencies are distinct, $\omega_r \neq \omega_s$, Eq. (7.87) is satisfied provided

$$\mathbf{u}_s^T M\mathbf{u}_r = 0, \ r \neq s \tag{7.88}$$

which represents the *orthogonality relation* for the modal vectors $\mathbf{u}_r$ and $\mathbf{u}_s$. We note that the orthogonality is with respect to the mass matrix M, which plays the role of a *weighting matrix*. Inserting Eq. (7.88) into the first of Eqs. (7.86), it is easy to see that the modal vectors are also orthogonal with respect to the stiffness matrix K,

$$\mathbf{u}_s^T K\mathbf{u}_r = 0, \ r \neq s \tag{7.89}$$

We stress again that the orthogonality relations, Eqs. (7.88) and (7.89), are valid only if M and K are symmetric.

The above orthogonality proof is predicated upon the natural frequencies being distinct. Actually, the orthogonality holds under broader conditions. Indeed, *if the eigenvalue problem in terms of two real symmetric matrices, Eq.* (7.79), *can be reduced to an eigenvalue problem in terms of a single real symmetric matrix, then all the system eigenvectors are orthogonal*, regardless of whether the system possesses repeated eigenvalues or not (Appendix C). Such a reduction is possible when one of the two real symmetric matrices is also positive definite, which is always the case in view of the fact that the mass matrix is positive definite by definition.

If the modes are normalized, then they are called *orthonormal*, and if the normalization scheme is according to Eq. (7.80), the modes satisfy the relations

$$\mathbf{u}_r^T M \mathbf{u}_s = \delta_{rs}, \ \mathbf{u}_r^T K \mathbf{u}_s = \omega_r^2 \delta_{rs}, \ r, s = 1, 2, \ldots, n \tag{7.90}$$

where δ_{rs} is the Kronecker delta (see definition in Sec. 7.1).

The eigenvalue problem and the orthonormality relations can be cast in a single matrix form, instead of an individual matrix-vector form for each of the modal vectors. To this end, we recall the definition $U = [\mathbf{u}_1 \ \mathbf{u}_2 \ \ldots \ \mathbf{u}_n]$ of the modal matrix introduced in Sec. 7.6 in conjunction with the linear transformation (7.84) and observe that all the solutions of the eigenvalue problem, Eqs. (7.79), can be combined into the compact matrix equation

$$KU = MU\Omega \tag{7.91}$$

where $\Omega = \text{diag}\ [\omega_1^2 \ \omega_2^2 \ \ldots \ \omega_n^2]$ is a diagonal matrix of the natural frequencies squared. Moreover, the orthonormality relations, Eqs. (7.80) and (7.81), can be combined into

$$U^T M U = I, \ U^T K U = \Omega \tag{7.92}$$

in which I is the identity matrix.

The orthogonality property plays a crucial role in the vibration of multi-degree-of-freedom systems, as it forms the foundation for modal analysis whereby the response of a system can be represented as a linear combination of the natural modes.

7.8 SYSTEMS ADMITTING RIGID-BODY MOTIONS

The undamped free vibration of a multi-degree-of-freedom linear system, in which the system is capable of harmonic oscillation in any one or all of the modes of vibration, is typical of positive definite systems, i.e., systems defined by real symmetric positive definite mass and stiffness matrices. The behavior is somewhat different when the stiffness matrix is only positive semidefinite.

As indicated in Sec. 7.4, when the mass matrix M is positive definite and the stiffness matrix K is only positive semidefinite, the system is *positive semidefinite*. Physically this implies that the system is supported in such a manner that rigid-body motion is possible. When the potential energy is due to elastic effects alone, if the body undergoes pure rigid-body motion, i.e., without any elastic deformations, then the

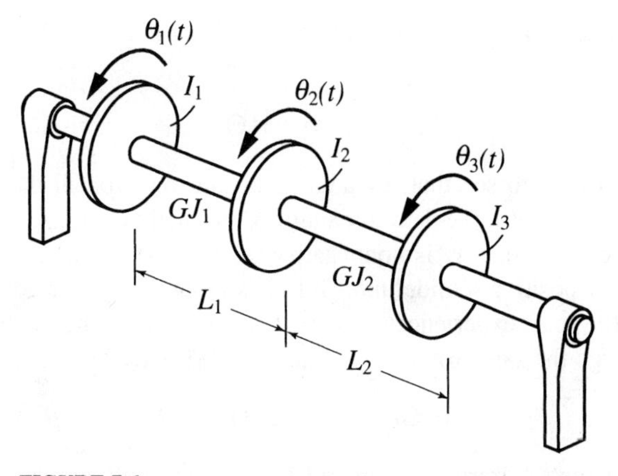

FIGURE 7.6
Three disks on a shaft in torsion unrestrained at both ends

potential energy is zero without all the coordinates being identically equal to zero. Such a semidefinite system is shown in Fig. 7.6, where the system consists of three disks of mass polar moments of inertia I_1, I_2 and I_3 connected by two segments of a massless shaft of lengths L_1 and L_2 and torsional stiffnesses GJ_1 and GJ_2, respectively. The system is supported at both ends by means of frictionless bearings in such a way that the entire system can rotate freely as a whole. Of course, torsional deformations can also be present, so that in general the motion of the system is a combination of rigid and elastic motions. Denoting by $\theta_i(t)$ $(i = 1, 2, 3)$ the angular displacements and velocities of the three disks, the kinetic energy is simply

$$T = \tfrac{1}{2}(I_1\dot{\theta}_1^2 + I_2\dot{\theta}_2^2 + I_3\dot{\theta}_3^2) = \tfrac{1}{2}\dot{\boldsymbol{\theta}}^T M \dot{\boldsymbol{\theta}} \tag{7.93}$$

where the mass matrix is diagonal,

$$M = \begin{bmatrix} I_1 & 0 & 0 \\ 0 & I_2 & 0 \\ 0 & 0 & I_3 \end{bmatrix} \tag{7.94}$$

On the other hand, the potential energy has the expression

$$V = \tfrac{1}{2}[k_1(\theta_2 - \theta_1)^2 + k_2(\theta_3 - \theta_2)^2] = \tfrac{1}{2}\boldsymbol{\theta}^T K \boldsymbol{\theta} \tag{7.95}$$

where the stiffness matrix has the form

$$K = \begin{bmatrix} k_1 & -k_1 & 0 \\ -k_1 & k_1 + k_2 & -k_2 \\ 0 & -k_2 & k_2 \end{bmatrix} \tag{7.96}$$

in which we have used the notation $k_i = GJ_i/L_i$ $(i = 1, 2)$. Using the same steps as in Sec. 7.6, we assume synchronous motion of the form

$$\theta_i(t) = \Theta_i f(t) \quad i = 1, 2, 3 \tag{7.97}$$

where Θ_i $(i = 1, 2, 3)$ are constants and $f(t)$ is harmonic, and obtain the eigenvalue problem

$$K\mathbf{\Theta} = \omega^2 M\mathbf{\Theta} \tag{7.98}$$

It is not difficult to see that, by adding all the rows (or all the columns) of the stiffness matrix, Eq. (7.96), the determinant of K is equal to zero, so that K is singular. Moreover, the potential energy is nonnegative, so that the stiffness matrix, and hence the system, is only positive semidefinite. It follows that the system admits a rigid-body mode in which the shaft experiences no elastic deformation. The implication is that all three disks undergo the same rotation, so that the rigid-body mode must have the form

$$\mathbf{\Theta} = \mathbf{\Theta}_0 = \Theta_0[1 \ \ 1 \ \ 1]^T = \Theta_0\mathbf{1} \tag{7.99}$$

in which Θ_0 is some arbitrary constant and $\mathbf{1}$ is a vector with all its components equal to unity. Inserting Eq. (7.99) into Eq. (7.98) and observing that

$$K\mathbf{\Theta}_0 = \Theta_0 \begin{bmatrix} k_1 & -k_1 & 0 \\ -k_1 & k_1 + k_2 & -k_2 \\ 0 & -k_2 & k_2 \end{bmatrix} \begin{bmatrix} 1 \\ 1 \\ 1 \end{bmatrix} = \mathbf{0} \tag{7.100}$$

independently of the values of k_1 and k_2, we conclude that the eigenvalue problem does indeed admit as a nontrivial solution the rigid-body $\Theta_0 = \mathbf{1}$ mode with the zero natural frequency, $\omega_0 = 0$. Note that the rigid-body mode $\mathbf{\Theta}_0$ is possible because both ends of the shaft are free. It is the only rigid-body mode possible for the system under consideration.

Because the rigid-body mode, defined by the constant eigenvector $\mathbf{\Theta}_0$ and zero natural frequency ω_0, is a solution of the eigenvalue problem (7.98), it follows that any other eigenvector, which can be identified as an elastic mode, must be orthogonal to it, namely, it must satisfy the condition

$$\mathbf{\Theta}_0^T M\mathbf{\Theta} = \Theta_0(I_1\Theta_1 + I_2\Theta_2 + I_3\Theta_3) = 0 \tag{7.101}$$

where Θ_i $(i = 1, 2, 3)$ are the components of $\mathbf{\Theta}$. Because Θ_0 is nonzero by definition, Eq. (7.101) implies that

$$I_1\Theta_1 + I_2\Theta_2 + I_3\Theta_3 = 0 \tag{7.102}$$

But, in view of Eqs. (7.97), Eq. (7.102) can also be written in the form

$$I_1\dot{\theta}_1(t) + I_2\dot{\theta}_2(t) + I_3\dot{\theta}_3(t) = 0 \tag{7.103}$$

which implies physically that the system angular momentum associated with the elastic motion is equal to zero, where the momentum is about the axis of the shaft. Hence, *the orthogonality of the rigid-body mode to the elastic modes is equivalent to the preservation of zero angular momentum in pure elastic motion.*

The general motion of an unrestrained system consists of a combination of elastic motions and rigid-body motions. Clearly, this type of motion is possible only for unrestrained systems, such as that shown in Fig. 7.6, because if one of the ends were to be clamped, then the reactive torque at that end would prevent rigid-body rotations from taking place. From Eq. (7.102), we conclude that, in the absence of external torques, the elastic motion must be such that the weighted average rotation of the system is zero, where the weighting factor for each rotation is the moment of inertia I_i $(i = 1, 2, 3)$. The equivalent statement for an unrestrained discrete system in translational motion is that, for no external forces, the system mass center is at rest at all times.

As pointed out earlier, det K is equal to zero, so that K is a singular matrix, with the implication that the inverse matrix K^{-1} does not exist. Recalling that $K^{-1} = A$ is the flexibility matrix, this fact can be easily explained physically by recognizing that for an unrestrained system, which is only positive semidefinite, flexibility influence coefficients cannot be defined. If the interest lies in solving a positive definite eigenvalue problem, then one can remove the singularity of K by transforming the eigenvalue problem associated with the unrestrained system into one for the elastic modes alone, as shown in the following.

Although there are three disks involved in the system of Fig. 7.6, as far as the elastic motion alone is concerned, this is not truly a three-degree-of-freedom system. Equation (7.103) can be regarded as a constraint equation and can be used to eliminate one coordinate from the problem formulation. Indeed, if we write

$$\theta_3 = -\frac{I_1}{I_3}\theta_1 - \frac{I_2}{I_3}\theta_2 \tag{7.104}$$

it becomes immediately obvious that the coordinate θ_3 is not really needed to describe the elastic motion, because it is automatically determined as soon as θ_1 and θ_2 are known. There is nothing unique about θ_3, as we could have eliminated either θ_1 or θ_2 from the problem formulation without affecting the final results. It is convenient to regard the original three-dimensional vector with components θ_1, θ_2 and θ_3 as a constrained vector $\boldsymbol{\theta} = [\theta_1 \;\; \theta_2 \;\; \theta_3]^T$ and the two-dimensional vector with components θ_1 and θ_2 as an arbitrary vector $\boldsymbol{\theta}' = [\theta_1 \;\; \theta_2]^T$. Then, considering the identities $\theta_1 \equiv \theta_1$, $\theta_2 \equiv \theta_2$ and using Eq. (7.104), we can write the relation between the two vectors in the form

$$\boldsymbol{\theta} = C\boldsymbol{\theta}' \tag{7.105}$$

where

$$C = \begin{bmatrix} 1 & 0 \\ 0 & 1 \\ \dfrac{-I_1}{I_3} & \dfrac{-I_2}{I_3} \end{bmatrix} \tag{7.106}$$

plays the role of a constraint matrix. An expression similar to (7.104) exists for the angular velocities $\dot{\theta}_i \, (i = 1, 2, 3)$, so that we can write

$$\dot{\boldsymbol{\theta}} = C\dot{\boldsymbol{\theta}}' \tag{7.107}$$

The linear transformations (7.105) and (7.107) can be used to reduce the kinetic and potential energy to expressions in terms of θ_1 and θ_2 and their time derivatives alone. Indeed, inserting Eq. (7.107) into Eq. (7.93) and recognizing that the vectors in (7.93) are constrained, we obtain the kinetic energy

$$T = \tfrac{1}{2}\dot{\theta}^T M\dot{\theta} = \tfrac{1}{2}\dot{\theta}'^T C^T MC\dot{\theta}' = \tfrac{1}{2}\dot{\theta}'^T M'\dot{\theta}' \tag{7.108}$$

where

$$M' = C^T MC = \frac{1}{I_3}\begin{bmatrix} I_1(I_1+I_3) & I_1 I_2 \\ I_1 I_2 & I_2(I_2+I_3) \end{bmatrix} \tag{7.109}$$

Moreover, introducing Eq. (7.105) in Eq. (7.95), the potential energy becomes

$$V = \tfrac{1}{2}\theta^T K\theta = \tfrac{1}{2}\theta'^T C^T KC\theta' = \tfrac{1}{2}\theta'^T K'\theta' \tag{7.110}$$

in which

$$K' = C^T KC = \frac{1}{I_3^2}\begin{bmatrix} k_1 I_3^2 + k_2 I_1^2 & -k_1 I_3^2 + k_2 I_1(I_2+I_3) \\ -k_1 I_3^2 + k_2 I_1(I_2+I_3) & k_1 I_3^2 + k_2(I_2+I_3)^2 \end{bmatrix} \tag{7.111}$$

We observe that both M' and K' are 2×2 symmetric matrices. But, unlike M and K, the transformed matrices M' and K' are both positive definite.

The eigenvalue problem associated with the transformed system is

$$K'\Theta' = \omega^2 M'\Theta' \tag{7.112}$$

which possesses all the characteristics associated with a positive definite system. Its solution consists of the natural modes Θ_1', Θ_2' and the associated natural frequencies ω_1, ω_2, respectively. The modes Θ_1' and Θ_2' give only the rotations of disks 1 and 2. The rotations of disk 3 in these elastic modes are obtained by considering Eq. (7.105) and writing

$$\Theta_1 = C\Theta_1', \quad \Theta_2 = C\Theta_2' \tag{7.113}$$

where the components of the modal vectors Θ_1 and Θ_2 are such that Eq. (7.104) is satisfied automatically.

We stress again that Eqs. (7.113) represent the elastic modes only. In addition, for this semidefinite system, we have the rigid-body mode $\Theta_0 = \Theta_0\mathbf{1}$ with the natural frequency $\omega_0 = 0$.

Example 7.6. Consider the unrestrained system of Fig. 7.6, let $k_1 = k_2 = k$ and $I_1 = I_2 = I_3 = I$ and obtain the natural modes of the system by solving a positive definite eigenvalue problem.

The natural modes are obtained by solving the eigenvalue problem (7.112), where the transformed mass and stiffness matrices M' and K' are given by Eqs. (7.109) and (7.111), respectively. Using the data given above, the two transformed matrices have the explicit form

$$M' = \frac{1}{I}\begin{bmatrix} 2I^2 & I^2 \\ I^2 & 2I^2 \end{bmatrix} = I\begin{bmatrix} 2 & 1 \\ 1 & 2 \end{bmatrix} \tag{a}$$

and

$$K' = \frac{1}{I^2}\begin{bmatrix} 2kI^2 & kI^2 \\ kI^2 & 5kI^2 \end{bmatrix} = k\begin{bmatrix} 2 & 1 \\ 1 & 5 \end{bmatrix} \tag{b}$$

It is clear that the transformed matrix K' is positive definite.

The eigenvalue problem for the system is obtained by inserting Eqs. (a) and (b) into Eq. (7.112). The solution of the eigenvalue problem is

$$\begin{aligned} \omega_1 &= \sqrt{\frac{k}{I}}, \quad \mathbf{\Theta}_1' = \begin{bmatrix} 1 \\ 0 \end{bmatrix} \\ \omega_2 &= \sqrt{\frac{3k}{I}}, \quad \mathbf{\Theta}_2' = \begin{bmatrix} 0.5 \\ -1 \end{bmatrix} \end{aligned} \tag{c}$$

Using Eq. (7.106), we can write the constraint matrix

$$C = \begin{bmatrix} 1 & 0 \\ 0 & 1 \\ -1 & -1 \end{bmatrix} \tag{d}$$

so that, from Eqs. (7.113), the constrained modal vectors corresponding to the elastic modes are

$$\begin{aligned} \mathbf{\Theta}_1 &= \begin{bmatrix} 1 & 0 \\ 0 & 1 \\ -1 & -1 \end{bmatrix}\begin{bmatrix} 1 \\ 0 \end{bmatrix} = \begin{bmatrix} 1 \\ 0 \\ -1 \end{bmatrix} \\ \mathbf{\Theta}_2 &= \begin{bmatrix} 1 & 0 \\ 0 & 1 \\ -1 & -1 \end{bmatrix}\begin{bmatrix} 0.5 \\ -1 \end{bmatrix} = \begin{bmatrix} 0.5 \\ -1 \\ 0.5 \end{bmatrix} \end{aligned} \tag{e}$$

In addition, we have the rigid-body mode

$$\omega_0 = 0, \quad \mathbf{\Theta}_0 = \begin{bmatrix} 1 \\ 1 \\ 1 \end{bmatrix} \tag{f}$$

It can be verified that the three modes are orthogonal with respect to both the mass matrix M and the stiffness matrix K. The modes are plotted in Fig. 7.7.

From Fig. 7.7, we observe that in the first elastic mode the first and third disks have displacements equal in magnitude but opposite in sense, while the center disk is at rest at all times, as it coincides with a node. This mode is what is generally called an *antisymmetric mode*. On the other hand, in the second elastic mode, the first and third disks have displacements equal in magnitude and in the same sense, while the center disk moves in opposite sense. This is a *symmetric mode*. In fact, the rigid-body mode is also a symmetric mode. Symmetric and antisymmetric modes are a common occurrence in systems with symmetrical parameter distributions, such as the system of Fig. 7.6.

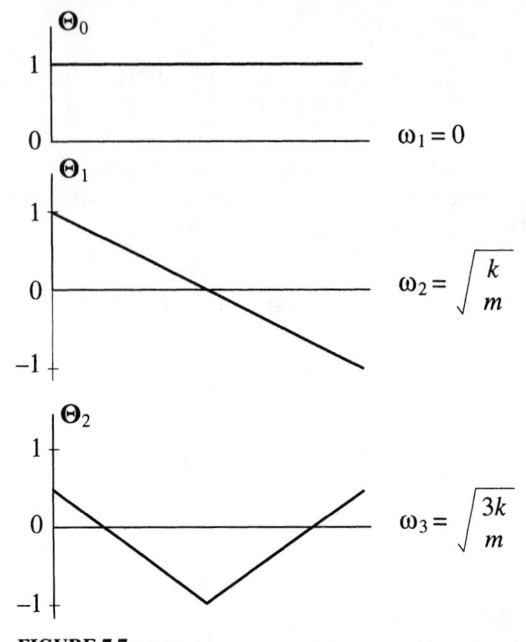

FIGURE 7.7
Natural modes for the unrestrained system of Fig. 7.6

7.9 DECOMPOSITION OF THE RESPONSE IN TERMS OF MODAL VECTORS

The orthogonality of the modal vectors is of vital importance in the vibration of multi-degree-of-freedom conservative systems, as it permits expressing the response as a superposition of the modal vectors multiplied by some time-dependent functions known as modal coordinates. To introduce the idea, we express a three-dimensional vector $\mathbf{r}$ in terms of the rectangular components $x,\ y,\ z$ as follows:

$$\mathbf{r} = x\mathbf{i} + y\mathbf{j} + z\mathbf{k} \tag{7.114}$$

where $\mathbf{i}$, $\mathbf{j}$ and $\mathbf{k}$ are unit vectors with directions coinciding with the directions of axes $x,\ y$ and z, respectively (Fig. 7.8). The components $x,\ y$ and z of vector $\mathbf{r}$ can be obtained from Eq. (7.114) by means of the dot products

$$x = \mathbf{i}\cdot\mathbf{r},\ y = \mathbf{j}\cdot\mathbf{r},\ z = \mathbf{k}\cdot\mathbf{r} \tag{7.115}$$

The above resolution of $\mathbf{r}$ in terms of rectangular components is based on the fact that axes $x,\ y$ and z are orthogonal. In fact, the unit vectors are orthonormal, as they satisfy

$$\begin{gathered} \mathbf{i}\cdot\mathbf{i} = 1,\ \mathbf{j}\cdot\mathbf{j} = 1,\ \mathbf{k}\cdot\mathbf{k} = 1 \\ \mathbf{i}\cdot\mathbf{j} = \mathbf{j}\cdot\mathbf{i} = 0,\ \mathbf{i}\cdot\mathbf{k} = \mathbf{k}\cdot\mathbf{i} = 0,\ \mathbf{j}\cdot\mathbf{k} = \mathbf{k}\cdot\mathbf{j} = 0 \end{gathered} \tag{7.116}$$

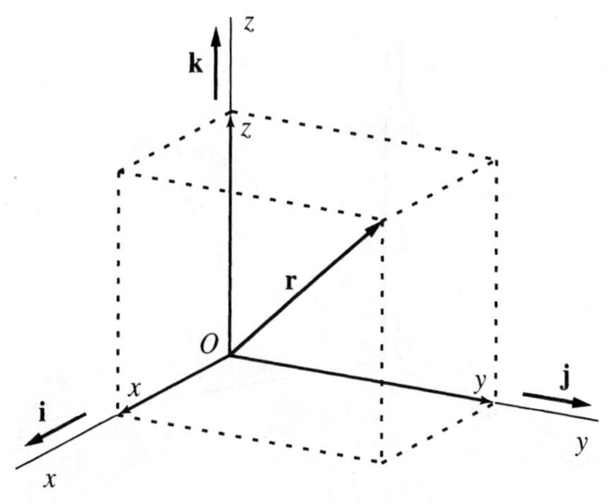

FIGURE 7.8
Resolution of a three-dimensional vector in terms of rectangular components

The same operations can be expressed in a notation more consistent with our objective. To this end, we denote the three-dimensional vector by $\mathbf{x}$ and introduce column matrices equivalent to the unit vectors $\mathbf{i}$, $\mathbf{j}$, $\mathbf{k}$, namely,

$$\mathbf{e}_1 = \begin{bmatrix} 1 \\ 0 \\ 0 \end{bmatrix}, \quad \mathbf{e}_2 = \begin{bmatrix} 0 \\ 1 \\ 0 \end{bmatrix}, \quad \mathbf{e}_3 = \begin{bmatrix} 0 \\ 0 \\ 1 \end{bmatrix} \tag{7.117}$$

so that the counterpart of Eq. (7.114) is

$$\mathbf{x} = \begin{bmatrix} x_1 \\ x_2 \\ x_3 \end{bmatrix} = x_1\mathbf{e}_1 + x_2\mathbf{e}_2 + x_3\mathbf{e}_3 = \sum_{i=1}^{3} x_i\mathbf{e}_i \tag{7.118}$$

where x_i are components along orthogonal axes coinciding with the unit vectors $\mathbf{e}_i\,(i = 1, 2, 3)$. The unit vectors $\mathbf{e}_i$ $(i = 1, 2, 3)$ are orthonormal, as they satisfy

$$\mathbf{e}_s^T\mathbf{e}_r = \delta_{rs}, \quad r, s = 1, 2, 3 \tag{7.119}$$

and we observe that Eqs. (7.119) are the counterpart of Eqs. (7.116), and they represent a much more compact form than Eqs. (7.116). Using Eqs. (7.118) and (7.119), we can express the components of $\mathbf{x}$ in the form

$$x_i = \mathbf{e}_i^T\mathbf{x}, \quad i = 1, 2, 3 \tag{7.120}$$

Figure 7.9 shows the decomposition of $\mathbf{x}$, and it represents the counterpart of Fig. 7.8.

The advantage of the new notation is in that it permits easy extension to the n-dimensional case. Hence, following the same line of thought as above, an n-dimensional

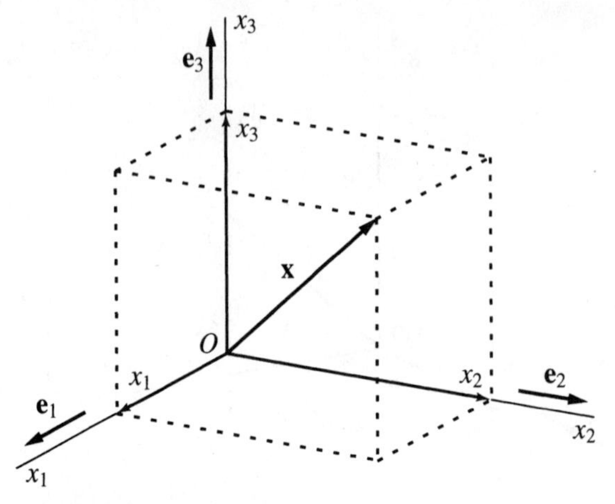

FIGURE 7.9
Decomposition of a three-dimensional vector in terms of standard orthonormal unit vectors

vector $\mathbf{x}$ can be decomposed as follows:

$$\mathbf{x} = [x_1 \ x_2 \ \dots \ x_n]^T = \sum_{i=1}^{n} x_i \mathbf{e}_i \tag{7.121}$$

where

$$\mathbf{e}_1 = \begin{bmatrix} 1 \\ 0 \\ \vdots \\ 0 \end{bmatrix}, \ \mathbf{e}_2 = \begin{bmatrix} 0 \\ 1 \\ \vdots \\ 0 \end{bmatrix}, \ \dots, \ \mathbf{e}_n = \begin{bmatrix} 0 \\ 0 \\ \vdots \\ 1 \end{bmatrix} \tag{7.122}$$

constitute a set of n orthonormal vectors known as *standard unit vectors*.

The above developments are geometric in nature, so that the question arises as to how they relate to vibrations. Of course, the decomposition of a three-dimensional vector in terms of rectangular components is widely used in mechanics, but it is mainly as a means of treating the three components of a displacement vector, or force vector, simultaneously. In the vibration of n-degree-of-freedom systems the displacement vector $\mathbf{u}$ has n components, but there would be no practical value in resolving $\mathbf{u}$ into components along the unit vectors $\mathbf{e}_1, \ \mathbf{e}_2, \dots, \mathbf{e}_n$. However, the decomposition of $\mathbf{u}$ into components along a different set of orthogonal vectors has considerable value. In this regard, we recall that the modal vectors $\mathbf{u}_r$ $(r = 1, \ 2, \dots, n)$ are orthogonal, and they can be normalized so as to constitute a set of n orthonormal vectors. Hence, we can conceive of an n-dimensional vector space with axes defined by the unit modal vectors $\mathbf{u}_1, \ \mathbf{u}_2, \dots, \mathbf{u}_n$ and resolve the displacement vector $\mathbf{u}$ into components along these axes as follows:

$$\mathbf{u} = c_1\mathbf{u}_1 + c_2\mathbf{u}_2 + \dots + c_n\mathbf{u}_n = \sum_{r=1}^{n} c_r \mathbf{u}_r \tag{7.123}$$

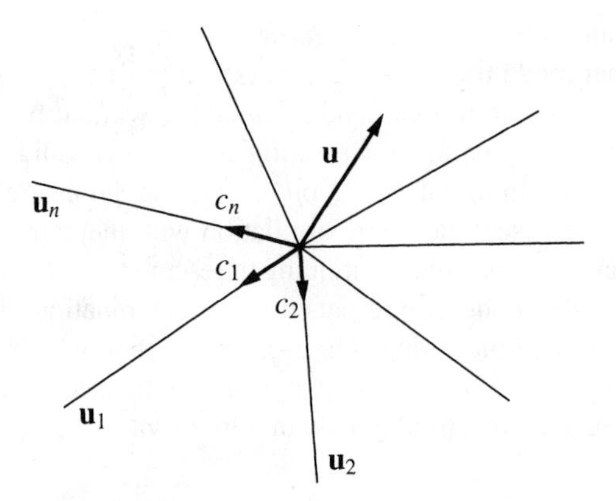

FIGURE 7.10
Decomposition of an n-dimensional displacement vector in terms of orthonormal modal vectors

where c_r $(r = 1, 2, \ldots, n)$ are the components in question. The decomposition is shown in Fig. 7.10. But, the modal vectors are not orthonormal in an ordinary sense, but orthonormal with respect to the mass matrix M, as well as with respect to the stiffness matrix K, as they satisfy Eqs. (7.90). In view of this, we can determine the coefficients c_r $(r = 1, 2, \ldots, n)$ by premultiplying Eq. (7.123) through by $\mathbf{u}_s^T M$ and using the orthonormality conditions, the first half of Eqs. (7.90), to obtain

$$c_r = \mathbf{u}_r^T M\mathbf{u}, \; r = 1, 2, \ldots, n \tag{7.124}$$

Similarly, premultiplying Eq. (7.123) through by $\mathbf{u}_s^T K$ and using the companion orthonormality conditions, the second half of Eqs. (7.90), we can write

$$\omega_r^2 c_r = \mathbf{u}_r^T K\mathbf{u}, \; r = 1, 2, \ldots, n \tag{7.125}$$

Equations (7.123)–(7.125) provide the framework for modal analysis whereby a set of simultaneous ordinary differential equations describing the response of vibrating multi-degree-of-freedom conservative systems can be transformed into a set of independent equations. We refer to Eqs. (7.123)–(7.125) as the *expansion theorem*.

The expansion theorem, Eqs. (7.123)–(7.125), can be cast in the matrix form

$$\mathbf{u} = U\mathbf{c} \tag{7.126}$$

where U is the modal matrix and $\mathbf{c}$ is an n-dimensional vector of coefficients defined by

$$\mathbf{c} = U^T M\mathbf{u}, \; \Omega\mathbf{c} = U^T K\mathbf{u} \tag{7.127}$$

in which Ω is the diagonal matrix of natural frequencies squared.

The natural frequencies and natural modes represent unique characteristics of the system, in the sense that they are fully determined by the mass matrix M and stiffness matrix K. They always appear together in pairs ω_r, $\mathbf{u}_r$ $(r = 1, 2, \ldots, n)$ and every one of these pairs can be excited independently of any other pair. For example, if the system is excited by a harmonic force with frequency ω_r, then the system will vibrate with the same frequency ω_r and with the displacement configuration resembling the natural mode $\mathbf{u}_r$, and not any other mode $\mathbf{u}_s$, $s \neq r$. Of course, in this case the system experiences resonance, in which case the motion tends to increase without bounds until the small motions assumption is violated and the solution ceases to be valid. On the other hand, if the system is imparted an initial excitation resembling the natural mode $\mathbf{u}_r$, then the ensuing motion will represent harmonic oscillation with the natural frequency ω_r and with the displacement configuration continuing to resemble $\mathbf{u}_r$. In general, however, for arbitrary excitations, the motion represents a linear combination of the natural modes with time-dependent amplitudes depending on the excitation. The determination of these amplitudes, commonly referred to as modal coordinates, is carried out by means of modal analysis, an approach used widely in vibrations.

7.10 RESPONSE TO INITIAL EXCITATIONS BY MODAL ANALYSIS

In Sec. 7.6, we have shown that the free vibration of multi-degree-of-freedom conservative systems can be expressed as a superposition of natural motions consisting of the natural modes $\mathbf{u}_r$ multiplied by harmonic functions $f_r(t)$ with frequencies equal to the natural frequencies ω_r and with amplitudes C_r and phase angles ϕ_r $(r = 1,\ 2, \dots, n)$. Whereas ω_r and $\mathbf{u}_r$ depend on internal factors, namely, the system inertia and stiffness properties, C_r and ϕ_r depend on the initial excitations, which represent external factors. To determine ω_r and $\mathbf{u}_r$ $(r = 1, 2, \dots, n)$, it is necessary to solve an algebraic eigenvalue problem defined by the mass matrix M and the stiffness matrix K. On the other hand, to determine the constants C_r and ϕ_r, it is necessary to force the general free vibration solution $\mathbf{q}(t)$ to match the initial displacement and initial velocity vectors $\mathbf{q}(0)$ and $\dot{\mathbf{q}}(0)$, respectively, as shown in Sec. 5.4 for two-degree-of-freedom systems. For multi-degree-of-freedom systems, a more efficient approach is required; modal analysis represents just such an approach.

As shown in Sec. 7.6, the free vibration of multi-degree-of-freedom systems is described by a set of simultaneous, homogeneous ordinary differential equations, which can be written in the matrix form

$$M\ddot{\mathbf{q}}(t) + K\mathbf{q}(t) = \mathbf{0} \tag{7.128}$$

where the displacement vector $\mathbf{q}(t)$ is subject to the initial conditions $\mathbf{q}(0)$ and $\dot{\mathbf{q}}(0)$, in which all vectors are of dimension n. At a given time $t = t_1$, the solution of Eq. (7.128) is $\mathbf{q}(t_1)$. But, by the expansion theorem, Eq. (7.123), the solution $\mathbf{q}(t_1)$ can be regarded as a superposition of the normal modes $\mathbf{u}_r$ $(r = 1,\ 2, \dots, n)$. Denoting the coefficients c_r for the particular configuration by $\eta_r(t_1)$ $(r = 1,\ 2, \dots, n)$, we can write

$$\mathbf{q}(t_1) = \eta_1(t_1)\mathbf{u}_1 + \eta_2(t_1)\mathbf{u}_2 + \dots + \eta_n(t_1)\mathbf{u}_n = \sum_{r=1}^{n} \eta_r(t_1)\mathbf{u}_r \tag{7.129}$$

Then, according to Eqs. (7.124) and (7.125), the coefficients in Eq. (7.129) are defined by

$$\eta_r(t_1) = \mathbf{u}_r^T M\mathbf{q}(t_1),\ \omega_r^2\eta_r(t_1) = \mathbf{u}_r^T K\mathbf{q}(t_1),\ r = 1, 2, \dots, n \tag{7.130}$$

But t_1 is arbitrary, so that its value can be changed at will. Because, Eqs. (7.129) and (7.130) must hold for all values of time, we can replace t_1 by t and write in general

$$\mathbf{q}(t) = \sum_{r=1}^{n} \eta_r(t)\mathbf{u}_r \tag{7.131}$$

and

$$\eta_r(t) = \mathbf{u}_r^T M\mathbf{q}(t),\ \omega_r^2 \eta_r(t) = \mathbf{u}_r^T K\mathbf{q}(t),\ r = 1, 2, \ldots, n \tag{7.132}$$

Equations (7.131) and (7.132) constitute the expansion theorem, written in a form directly applicable to the vibration of conservative systems. By analogy with Eqs. (7.126) and (7.127), they can be written in the compact matrix form

$$\mathbf{q}(t) = U\boldsymbol{\eta}(t) \tag{7.133}$$

where

$$\boldsymbol{\eta}(t) = U^T M\mathbf{q}(t),\ \Omega\boldsymbol{\eta}(t) = U^T K\mathbf{q}(t) \tag{7.134}$$

Next, we insert Eq. (7.133) into Eq. (7.128), premultiply through by U^T, use Eqs. (7.134) and obtain

$$\ddot{\boldsymbol{\eta}}(t) + \Omega\boldsymbol{\eta}(t) = \mathbf{0} \tag{7.135}$$

Equation (7.135) represents a set of n independent *modal equations* of the form

$$\ddot{\eta}_r(t) + \omega_r^2 \eta_r(t) = 0,\ r = 1, 2, \ldots, n \tag{7.136}$$

where $\eta_r(t)$ are known as *modal coordinates*; they are subject to the initial conditions $\eta_r(0),\ \dot{\eta}_r(0)\ (r = 1,\ 2, \ldots, n)$.

Equations (7.136) resemble entirely the equation of a harmonic oscillator, Eq. (3.2). Hence, by analogy with Eqs. (3.10) and (3.13), we can write the solution of Eqs. (7.136) as

$$\eta_r(t) = C_r \cos(\omega_r t - \phi_r) = \eta_r(0)\cos\omega_r t + \frac{\dot{\eta}_r(0)}{\omega_r}\sin\omega_r t,\ r = 1, 2, \ldots, n \tag{7.137}$$

in which, from the first of Eqs. (7.132) with $t = 0$, we obtain the initial modal coordinates and velocities

$$\eta_r(0) = \mathbf{u}_r^T M\mathbf{q}(0),\ \dot{\eta}_r(0) = \mathbf{u}_r^T M\dot{\mathbf{q}}(0),\ r = 1, 2, \ldots, n \tag{7.138}$$

If follows that the modal coordinates are

$$\eta_r(t) = \mathbf{u}_r^T M\mathbf{q}(0)\cos\omega_r t + \frac{1}{\omega_r}\mathbf{u}_r^T M\dot{\mathbf{q}}(0)\sin\omega_r t,\ r = 1, 2, \ldots, n \tag{7.139}$$

Finally introducing Eqs. (7.139) in Eq. (7.131), we obtain the response of multi-degree-of-freedom conservative systems to initial excitations in the general form

$$\mathbf{q}(t) = \sum_{r=1}^{n} [\mathbf{u}_r^T M\mathbf{q}(0)\cos\omega_r t + \frac{1}{\omega_r}\mathbf{u}_r^T M\dot{\mathbf{q}}(0)\sin\omega_r t]\mathbf{u}_r \tag{7.140}$$

At this point, we wish to demonstrate a statement made in the end of Sec. 7.9 that each of the natural modes can be excited independently of the other. To this end, we assume that the initial displacement vector resembles one of the modal vectors, say $\mathbf{u}_s$, and that the initial velocity vector is zero. Hence, inserting $\mathbf{q}(0) = \alpha\mathbf{u}_s,\ \dot{\mathbf{q}}(0) = \mathbf{0}$ into Eq. (7.140) and using the first of the orthonormality relations (7.90), we can write

$$\mathbf{q}(t) = \alpha\sum_{r=1}^{n}[\mathbf{u}_r^T M\mathbf{u}_s \cos\omega_r t]\mathbf{u}_r = \alpha\sum_{r=1}^{n}\delta_{rs}\mathbf{u}_r \cos\omega_r t = \alpha\mathbf{u}_s \cos\omega_s t \tag{7.141}$$

so that the response is indeed vibration in mode s alone.

Example 7.7. Obtain the solution of the free vibration problem for the three-degree-of-freedom system of Example 7.5 by means of modal analysis.

The free vibration response derived by means of modal analysis is given by Eq. (7.140), in which it is assumed that the modes have been normalized so as to satisfy Eqs. (7.92). The modal vectors, normalized so that the largest component is equal to 1, were computed in Example 7.5, so that they are in need of renormalization. Using the procedure of Example 7.4, the modal vectors normalized as required can be verified to be

$$\mathbf{u}_1 = m^{-1/2}\begin{bmatrix} 0.2691 \\ 0.5008 \\ 0.5817 \end{bmatrix},\ \mathbf{u}_2 = m^{-1/2}\begin{bmatrix} 0.8782 \\ 0.2231 \\ -0.2992 \end{bmatrix},\ \mathbf{u}_3 = m^{-1/2}\begin{bmatrix} -0.3954 \\ 0.8363 \\ -0.2685 \end{bmatrix} \tag{a}$$

Hence, inserting Eqs. (a), as well as the pertinent data from Example 7.5, into Eq. (7.140), we obtain the desired response

$$\begin{aligned}
\mathbf{q}(t) = q_0 \Bigg\{ & \begin{bmatrix} 0.2691 \\ 0.5008 \\ 0.5817 \end{bmatrix}^T \begin{bmatrix} 1 & 0 & 0 \\ 0 & 1 & 0 \\ 0 & 0 & 2 \end{bmatrix} \begin{bmatrix} 1 \\ 2 \\ 3 \end{bmatrix} \cos 0.3731\sqrt{\frac{k}{m}}t \begin{bmatrix} 0.2691 \\ 0.5008 \\ 0.5817 \end{bmatrix} \\
& + \begin{bmatrix} 0.8782 \\ 0.2231 \\ -0.2992 \end{bmatrix}^T \begin{bmatrix} 1 & 0 & 0 \\ 0 & 1 & 0 \\ 0 & 0 & 2 \end{bmatrix} \begin{bmatrix} 1 \\ 2 \\ 3 \end{bmatrix} \cos 1.3213\sqrt{\frac{k}{m}}t \begin{bmatrix} 0.8782 \\ 0.2231 \\ -0.2992 \end{bmatrix} \\
& + \begin{bmatrix} -0.3954 \\ 0.8963 \\ -0.2685 \end{bmatrix}^T \begin{bmatrix} 1 & 0 & 0 \\ 0 & 1 & 0 \\ 0 & 0 & 2 \end{bmatrix} \begin{bmatrix} 1 \\ 2 \\ 3 \end{bmatrix} \cos 2.0285\sqrt{\frac{k}{m}}t \begin{bmatrix} -0.3954 \\ 0.8363 \\ -0.2685 \end{bmatrix} \Bigg\} \\
= q_0 \Bigg\{ & \begin{bmatrix} 1.2812 \\ 2.3841 \\ 2.7696 \end{bmatrix} \cos 0.3731\sqrt{\frac{k}{m}}t + \begin{bmatrix} -0.4132 \\ -0.1050 \\ 0.1408 \end{bmatrix} \cos 1.3213\sqrt{\frac{k}{m}}t \\
& + \begin{bmatrix} 0.1320 \\ -0.2791 \\ 0.0896 \end{bmatrix} \cos 2.0285\sqrt{\frac{k}{m}}t \Bigg\}
\end{aligned} \tag{b}$$

which is identical to that obtained in Example 7.5.

7.11 EIGENVALUE PROBLEM IN TERMS OF A SINGLE SYMMETRIC MATRIX

In Sec. 7.6, we have shown that the algebraic eigenvalue problem associated with a conservative n-degree-of-freedom system has the form

$$K\mathbf{u} = \omega^2 M\mathbf{u} \tag{7.142}$$

where $\mathbf{u}$ is an n-dimensional displacement vector and K and M are symmetric $n \times n$ stiffness and mass matrices, respectively. Moreover, M is positive definite and K can be positive definite or positive semidefinite. For the sake of this discussion, we consider the case in which M is not diagonal.

There are many computational algorithms for solving the algebraic eigenvalue problem, but most of them are in terms of a single matrix. Eigenvalue problems in terms of a single matrix are said to be in *standard form.* Of course, it is always possible to multiply Eq. (7.142) through by M^{-1}, or by K^{-1} if K is nonsingular, which implies that K is positive definite, and obtain an eigenvalue problem in terms of a single matrix. However, by far the most efficient algorithms are for eigenvalue problems in which the single matrix is real and symmetric.

Transformations of an eigenvalue problem from one in terms of two symmetric matrices, Eq. (7.142), to one in terms of a single symmetric matrix are possible provided one of the two matrices is positive definite, which is always true in the case at hand. Indeed, the mass matrix M is real symmetric and positive definite by definition. Hence, from linear algebra, the matrix M can be decomposed as follows:

$$M = LL^T \tag{7.143}$$

where L is a nonsingular lower triangular matrix. Equation (7.143) represents a *Cholesky decomposition* and the problem amounts to determining the matrix L for a given matrix M. Details of the algorithm for carrying out the Cholesky decomposition can be found in Ref. 13. Inserting Eq. (7.143) into Eq. (7.142), we have

$$K\mathbf{u} = \omega^2 LL^T\mathbf{u} \tag{7.144}$$

Next, we premultiply Eq. (7.144) through by L^{-1} and introduce the linear transformation

$$L^T\mathbf{u} = \mathbf{v} \tag{7.145}$$

from which it follows that

$$\mathbf{u} = (L^T)^{-1}\mathbf{v} = (L^{-1})^T\mathbf{v} \tag{7.146}$$

Then, inserting Eqs. (7.145) and (7.146) into Eq. (7.142), we can write the desired eigenvalue problem in the standard form

$$A\mathbf{v} = \lambda\mathbf{v}, \; \lambda = \omega^2 \tag{7.147}$$

in which the coefficient matrix has the expression

$$A = L^{-1}K(L^{-1})^T = A^T \tag{7.148}$$

so that A is clearly symmetric.

In some computational algorithms, it is desirable that the eigenvalues be inversely proportional to the natural frequencies squared, rather than directly proportional. This is possible only if the system does not admit rigid body modes. Using the Cholesky decomposition, Eq. (7.143), premultiplying both sides of Eq. (7.142) by K^{-1} and dividing the result by ω^2, we can write

$$K^{-1}LL^T\mathbf{u} = \frac{1}{\omega^2}\mathbf{u} \tag{7.149}$$

Then, using Eq. (7.146) and premultiplying both sides of Eq. (7.149) by L^T, we obtain the eigenvalue problem

$$A\mathbf{v} = \lambda\mathbf{v}, \ \lambda = 1/\omega^2 \tag{7.150}$$

where this time the coefficient matrix has the form

$$A = L^T K^{-1} L = A^T \tag{7.151}$$

Equations (7.150) and (7.151) explain why, for this second version to be valid, the system cannot admit rigid-body modes. Indeed, in the presence of rigid-body modes the stiffness matrix is singular and the frequencies associated with the rigid-body modes are zero, so that A and λ do not exist. We note that the eigenvalue λ and the coefficient matrix A defined by Eqs. (7.150) and (7.151) are the reciprocal of those defined by Eqs. (7.147) and (7.148).

The solution of the eigenvalue problem given by Eqs. (7.147) and (7.148), or Eqs. (7.150) and (7.151), consists of the eigenvalues λ_r and the eigenvectors $\mathbf{v}_r$ $(r = 1, 2, \ldots, n)$. The eigenvectors are mutually orthogonal, as well as orthogonal with respect to the matrix A. Proof of orthogonality can be carried out by the approach used in Sec. 7.7. The eigenvectors can be conveniently normalized so as to satisfy

$$\mathbf{v}_s^T\mathbf{v}_r = \delta_{rs}, \ \mathbf{v}_s^T A\mathbf{v}_r = \lambda_r\delta_{rs}, \ r, s = 1, 2, \ldots, n \tag{7.152}$$

After the eigenvalues λ_r and eigenvectors $\mathbf{v}_r$ of A have been computed, they can be used to determine the natural frequencies ω_r and modal vectors $\mathbf{u}_r$ $(r = 1, 2, \ldots, n)$, where the latter satisfy the original eigenvalue problem, Eq. (7.142). The natural frequencies are equal to either $\sqrt{\lambda_r}$ or $1/\sqrt{\lambda_r}$, depending on whether the standard eigenvalue problem is defined by Eqs. (7.147) and (7.148), or by Eqs. (7.150) and (7.151), respectively, and from Eq. (7.146) the modal vectors are given by

$$\mathbf{u}_r = (L^T)^{-1}\mathbf{v}_r, \ r = 1, 2, \ldots, n \tag{7.153}$$

Inserting Eqs. (7.153) into Eqs. (7.152) and considering Eq. (7.142) with $\mathbf{u}$ and λ replaced by $\mathbf{u}_r$ and λ_r, respectively, we obtain

$$\mathbf{u}_s^T M\mathbf{u}_r = \delta_{rs}, \ \mathbf{u}_s^T K\mathbf{u}_r = \omega_r^2\delta_{rs}, \ r, s = 1, 2, \ldots, n \tag{7.154}$$

so that the modal vectors $\mathbf{u}_r$ $(r = 1, 2, \ldots, n)$ obtained by solving the eigenvalue problem defined by Eqs. (7.147) and (7.148), or that defined by Eqs. (7.150) and (7.151), satisfy the orthonormality conditions (7.90), as is to be expected.

The question arises as to whether one of the two versions of the eigenvalue problem, Eqs. (7.147) and (7.148), or Eqs. (7.150) and (7.151), is to be preferred over the other.

Clearly, in the case of semidefinite systems the only version possible is that given by Eqs. (7.147) and (7.148), as the version given by Eqs. (7.150) and (7.151) cannot tolerate singularities of the stiffness matrix. In view of this, the question can be asked as to why the second version is necessary at all. The answer is that, for some computational algorithms, the version given by Eqs. (7.150) and (7.151) is more desirable than that given by Eqs. (7.147) and (7.148). Indeed, some computational algorithms iterate to eigenvalues in descending order of magnitude, and accuracy is lost with each computed eigenvalue. If only a partial solution of the eigenvalue problem is of interest, then the algorithm computes the eigenvalues largest in magnitude. In vibrations, however, the interest lies in the lowest modes of vibration, i.e., those corresponding to the lowest natural frequencies, rather than the highest. These two seemingly contradictory factors are reconciled when the eigenvalues are inversely proportional to the natural frequencies, which points to the version given by Eqs. (7.150) and (7.151) as the desirable one.

The most frequently encountered case is that in which the mass matrix M is diagonal, in which case the triangular matrix L reduces to a diagonal matrix of the form

$$L = M^{1/2} = \mathrm{diag}\left[\sqrt{m_1}\ \sqrt{m_2}\ \dots\ \sqrt{m_n}\right] \tag{7.155}$$

In this case, Eq. (7.148) can be written more explicitly as

$$A = M^{-1/2} K M^{-1/2} = \left[k_{ij}/\sqrt{m_i m_j}\right] = A^T \tag{7.156}$$

where k_{ij} $(i, j = 1, 2, \dots, n)$ are the stiffness coefficients. Moreover, Eq. (7.151) can be expressed as

$$A = M^{1/2} K^{-1} M^{1/2} = \left[a_{ij}\sqrt{m_i m_j}\right] = A^T \tag{7.157}$$

where a_{ij} $(i, j = 1, 2, \dots, n)$ are the flexibility coefficients. In both cases, from Eq. (7.146), the modal vectors are related to the eigenvectors of A by

$$\mathbf{u}_r = M^{-1/2}\mathbf{v}_r,\ r = 1, 2, \dots, n \tag{7.158}$$

7.12 GEOMETRIC INTERPRETATION OF THE EIGENVALUE PROBLEM

In Sec. 7.11, we have shown that the eigenvalue problem for conservative vibrating systems can be defined in terms of a single real symmetric matrix A. This eigenvalue problem lends itself to a geometric interpretation that is not only interesting but also has practical implications.

From Sec. 7.3, we conclude that to any real symmetric matrix corresponds a given quadratic form. Hence, corresponding to the $n \times n$ matrix A describing an n-degree-of-freedom system, we can write the generic quadratic form

$$f = \mathbf{x}^T A\mathbf{x} = \sum_{i=1}^{n}\sum_{j=1}^{n} a_{ij} x_i x_j \tag{7.159}$$

To introduce the geometric interpretation mentioned above, we consider first the case $n = 2$. Then, assuming that the matrix A is not only real and symmetric but also positive definite, the expression

$$f = \mathbf{x}^T A\mathbf{x} = a_{11}x_1^2 + a_{22}x_2^2 + 2a_{12}x_1x_2 = 1 \tag{7.160}$$

represents an ellipse, as shown in Fig. 7.11. From analytic geometry, the gradient ∇f has the mathematical expression

$$\nabla f = \begin{bmatrix} \partial f/\partial x_1 \\ \partial f/\partial x_2 \end{bmatrix} = 2\begin{bmatrix} a_{11}x_1 + a_{12}x_2 \\ a_{12}x_1 + a_{22}x_2 \end{bmatrix} = 2\begin{bmatrix} a_{11} & a_{12} \\ a_{12} & a_{22} \end{bmatrix}\begin{bmatrix} x_1 \\ x_2 \end{bmatrix} = 2A\mathbf{x} \tag{7.161}$$

and it represents geometrically a vector normal to the ellipse at the tip of the vector **x**, as depicted in Fig. 7.11. In general, the direction of ∇f differs from the direction of **x**. Notable exceptions are the positions in which the vector **x** is aligned with the principal axes of the ellipse, in which positions ∇f is aligned with **x**, so that ∇f and **x** are proportional to one another. Denoting the proportionality constant by 2λ, we conclude that when the vector **x** coincides with a principal axis, the gradient can be written in the form

$$\nabla f = 2\lambda\mathbf{x} \tag{7.162}$$

Comparing Eqs. (7.161) and (7.162), we obtain

$$A\mathbf{x} = \lambda\mathbf{x} \tag{7.163}$$

which represents the statement of the eigenvalue problem, Eq. (7.147) or Eq. (7.150). It follows that *the solution of the eigenvalue problem for a real symmetric positive definite matrix A can be interpreted geometrically as the determination of the principal axes of the ellipse* $f = \mathbf{x}^T A\mathbf{x} = 1$, *and vice versa*. Although we reached this conclusion on the basis of the two-dimensional case, $n = 2$, the same conclusion is valid for $n \geq 3$ (see Ref. 13, Sec. 5.1).

Next, we solve the eigenvalue problem by finding the principle axes of the ellipse. To this end, we denote the principal axes by y_1 and y_2 and the angle from x_1 to y_1 by θ, so that finding the principal axes amounts to finding the angle θ for which the equation

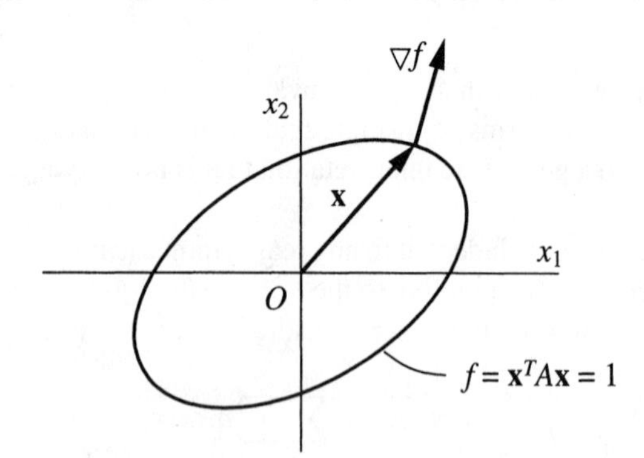

FIGURE 7.11
Geometric interpretation of the eigenvalue problem for a real symmetric matrix

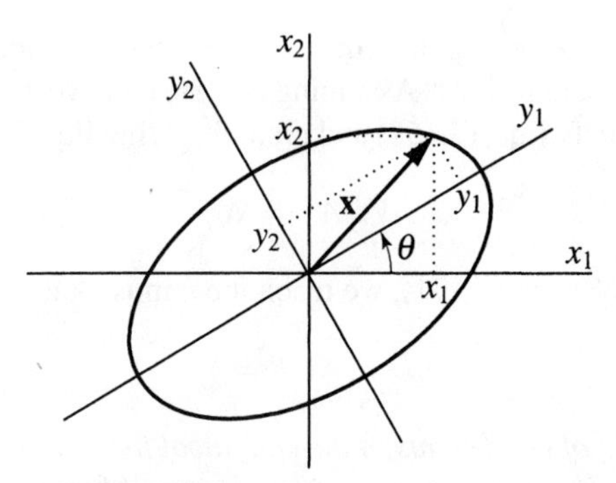

FIGURE 7.12
Coordinate transformation for finding the principal axes of the ellipse

of the ellipse reduces to its canonical form. From Fig. 7.12, this requires the coordinate transformation

$$\begin{aligned} x_1 &= y_1 \cos\theta - y_2 \sin\theta \\ x_2 &= y_1 \sin\theta + y_2 \cos\theta \end{aligned} \tag{7.164}$$

which can be expressed in the matrix form

$$\mathbf{x} = R\mathbf{y} \tag{7.165}$$

where

$$R = \begin{bmatrix} \cos\theta & -\sin\theta \\ \sin\theta & \cos\theta \end{bmatrix} \tag{7.166}$$

represents a *rotation matrix*. Inserting Eq. (7.165) into Eq. (7.160) and recognizing that reduction of f to canonical form implies elimination of cross products in the expression for f in terms of y_1 and y_2, we write

$$f = \mathbf{x}^T A\mathbf{x} = \mathbf{y}^T R^T AR\mathbf{y} = \mathbf{y}^T D\mathbf{y} = 1 \tag{7.167}$$

in which

$$D = R^T AR = \text{diag}[d_1 \ \ d_2] \tag{7.168}$$

Equation (7.168) represents an *orthogonal transformation* and the rotation matrix R is an orthogonal matrix. In fact, it is an orthonormal matrix, as it satisfies

$$R^T R = RR^T = I \tag{7.169}$$

where I is the identity matrix. Equation (7.169) is in general true for transformation matrices representing rotations of orthogonal sets of axes.

At this point, we return to the eigenvalue problem, Eq. (7.147) or Eq. (7.150), let $n = 2$ and write the two solutions in the compact matrix form

$$AV = V\Lambda \tag{7.170}$$

where $V = [\mathbf{v}_1 \ \ \mathbf{v}_2]$ is the orthogonal matrix of eigenvectors and $\Lambda = \text{diag}[\lambda_1 \ \ \lambda_2]$ is the diagonal matrix of eigenvalues. Assuming that the eigenvectors $\mathbf{v}_1$ and $\mathbf{v}_2$ have been normalized so as to satisfy Eqs. (7.152) and premultiplying Eq. (7.170) by V^T, we obtain

$$V^T A V = \Lambda \tag{7.171}$$

Contrasting Eqs. (7.168) and (7.171), we reach the unmistakable conclusion that

$$D = \Lambda, \ R = V \tag{7.172}$$

or, *the diagonal matrix of coefficients of the canonical form is the matrix of eigenvalues and the rotation matrix is the matrix of eigenvectors.* Moreover, no normalization is necessary, as the matrix of eigenvectors is already orthonormal.

There remains the question of the actual determination of Λ and V. To this end, we use Eqs. (7.166), (7.168) and the first of Eqs. (7.172) and write

$$\begin{bmatrix} \lambda_1 & 0 \\ 0 & \lambda_2 \end{bmatrix} = \begin{bmatrix} \cos\theta & -\sin\theta \\ \sin\theta & \cos\theta \end{bmatrix}^T \begin{bmatrix} a_{11} & a_{12} \\ a_{12} & a_{22} \end{bmatrix} \begin{bmatrix} \cos\theta & -\sin\theta \\ \sin\theta & \cos\theta \end{bmatrix} \tag{7.173}$$

which yields the three scalar equations

$$\begin{aligned} \lambda_1 &= a_{11}\cos^2\theta + 2a_{12}\sin\theta\cos\theta + a_{22}\sin^2\theta \\ \lambda_2 &= a_{11}\sin^2\theta - 2a_{12}\sin\theta\cos\theta + a_{22}\cos^2\theta \\ 0 &= -(a_{11} - a_{22})\sin\theta\cos\theta + a_{12}(\cos^2\theta - \sin^2\theta) \end{aligned} \tag{7.174}$$

Recalling the trigonometric relations $\sin 2\theta = 2\sin\theta\cos\theta$ and $\cos 2\theta = \cos^2\theta - \sin^2\theta$, the third of Eqs. (7.174) permits us to determine the angle θ defining the direction of the principal axes by writing

$$\tan 2\theta = \frac{2a_{12}}{a_{11} - a_{22}} \tag{7.175}$$

Our interest does not lie in the angle θ itself but in the values of $\sin\theta$ and $\cos\theta$. To determine these values directly, it is convenient to introduce the notation

$$a_{12} = b, \ \tfrac{1}{2}(a_{11} - a_{22}) = c \tag{7.176}$$

so that, introducing Eqs. (7.176) in Eq. (7.175), it is not difficult to verify that, for given values of b and c, the values of $\cos\theta$ and $\sin\theta$ can be computed, in sequence, by means of the formulas

$$\cos\theta = \left[\frac{1}{2} + \frac{c}{2(b^2 + c^2)^{1/2}}\right]^{1/2}, \ \sin\theta = \frac{b}{2(b^2 + c^2)^{1/2}\cos\theta} \tag{7.177}$$

Then, inserting the values of $\sin\theta$ and $\cos\theta$ thus computed into the first two of Eqs. (7.174), we determine the eigenvalues λ_1 and λ_2. Finally, inserting the same values of $\sin\theta$ and $\cos\theta$ into Eq. (7.166) and considering the second of Eqs. (7.172), we obtain the orthonormal eigenvectors

$$\mathbf{v}_1 = \begin{bmatrix} \cos\theta \\ \sin\theta \end{bmatrix}, \quad \mathbf{v}_2 = \begin{bmatrix} -\sin\theta \\ \cos\theta \end{bmatrix} \tag{7.178}$$

which completes the formal solution of the eigenvalue problem.

The process just described represents *diagonalization* of the matrix A through an orthogonal transformation in the form of a planar rotation annihilating the off-diagonal entry a_{12} in a single step. For $n \geq 3$, the same diagonalization process requires annihilation of all the $n(n-1)/2$ off-diagonal entries a_{ij}, $(i, j = 1, 2, \ldots, n;\ i \neq j)$, which cannot be done in a single step. Indeed, the process requires a series of planar rotations, each time in a different plane. Moreover, the number of steps in the series cannot be specified in advance. The implication is that this is a numerical process yielding a diagonal matrix iteratively. The off-diagonal terms are never reduced exactly to zero, but only approximately so, although they can be made as close to zero as desired. In fact, it is necessary to specify in advance a threshold value approximating zero, terminating the iteration process when all the off-diagonal entries fall below this value. This is the essence of the *Jacobi method* (see Ref. 13).

Example 7.8. Solve the eigenvalue problem for the two-degree-of-freedom system of Example 7.4 by finding the principal axes of the corresponding ellipse.

From Example 7.4, we obtain the mass and stiffness matrices

$$M = m\begin{bmatrix} 1 & 0 \\ 0 & 2 \end{bmatrix}, \quad K = \frac{T}{L}\begin{bmatrix} 2 & -1 \\ -1 & 3 \end{bmatrix} \tag{a}$$

Hence, including the parameters m and T/L in λ and using Eqs. (6.142), we can write the eigenvalue problem in the form given by Eq. (7.163), in which

$$A = \begin{bmatrix} 1 & 0 \\ 0 & 1/\sqrt{2} \end{bmatrix}\begin{bmatrix} 2 & -1 \\ -1 & 3 \end{bmatrix}\begin{bmatrix} 1 & 0 \\ 0 & 1/\sqrt{2} \end{bmatrix} = \begin{bmatrix} 2 & -1/\sqrt{2} \\ -1/\sqrt{2} & 3/2 \end{bmatrix} \tag{b}$$

and

$$\lambda = \omega^2 mL/T \tag{c}$$

The ellipse can be plotted by inserting the entries of A, Eq. (b), into Eq. (7.160); the plot is shown in Fig. 7.13.

Using Eqs. (7.177) in conjunction with the values

$$b = a_{12} = -\frac{1}{\sqrt{2}}, \quad c = \frac{1}{2}(a_{11} - a_{22}) = \frac{1}{4} \tag{d}$$

we obtain

$$\begin{aligned} \cos\theta &= \sqrt{\frac{1}{2} + \frac{c}{2(b^2 + c^2)^{1/2}}} = \sqrt{\frac{1}{2} + \frac{1/4}{2\left(\frac{1}{2} + \frac{1}{16}\right)^{1/2}}} = 0.816497 \\ \sin\theta &= \frac{b}{2(b^2 + c^2)^{1/2}\cos\theta} = \frac{-1/\sqrt{2}}{2\left(\frac{1}{2} + \frac{1}{16}\right)^{1/2} \times 0.816497} = -0.577350 \end{aligned} \tag{e}$$

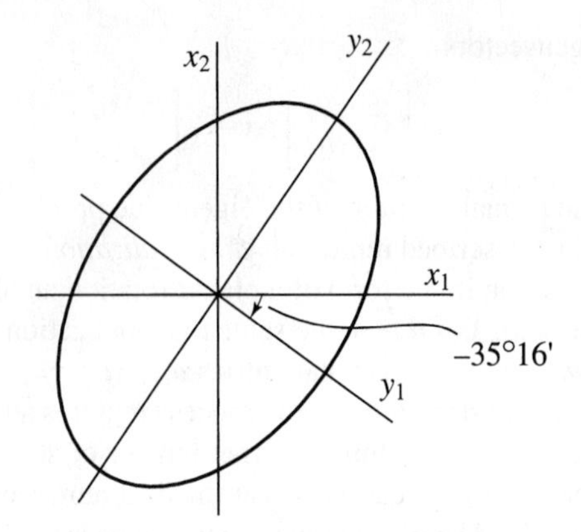

FIGURE 7.13
Plot of the ellipse using the entries of the coefficient matrix A

Hence, using the first two of Eqs. (7.174), we obtain the eigenvalues

$$\begin{aligned}\lambda_1 &= a_{11}\cos^2\theta + 2a_{12}\sin\theta\cos\theta + a_{22}\sin^2\theta \\ &= 2(-0.816497)^2 + 2\left(-\frac{1}{\sqrt{2}}\right)(-0.577350)0.816497 + \frac{3}{2}\times 0.577550^2 = 2.5 \\ \lambda_2 &= a_{11}\sin^2\theta - 2a_{12}\sin\theta\cos\theta + a_{22}\cos^2\theta \\ &= 2\times 0.577350^2 - 2\left(-\frac{1}{\sqrt{2}}\right)(-0.577350)0.816497 + \frac{3}{2}(-0.816497)^2 = 1\end{aligned} \tag{f}$$

and we note that λ_1 corresponds to the second natural frequency and vice versa. Moreover, using Eqs. (7.178), we can write the eigenvectors

$$\mathbf{v}_1 = \begin{bmatrix}\cos\theta \\ \sin\theta\end{bmatrix} = \begin{bmatrix}0.816497 \\ -0.577350\end{bmatrix},\ \mathbf{v}_2 = \begin{bmatrix}-\sin\theta \\ \cos\theta\end{bmatrix} = \begin{bmatrix}0.577350 \\ 0.816497\end{bmatrix} \tag{g}$$

so that $\mathbf{v}_1$ and $\mathbf{v}_2$ are really the second and first eigenvectors respectively. It is typical of the Jacobi method that the eigenvalues and eigenvectors may not appear in ascending order.

As a matter of interest, we compute the natural frequencies and modal vectors. Introducing λ_1 and λ_2 from Eq. (f) in Eq. (c), we obtain simply

$$\omega_1 = \sqrt{\frac{\lambda_2 T}{mL}} = \sqrt{\frac{T}{mL}},\ \omega_2 = \sqrt{\frac{\lambda_1 T}{mL}} = \sqrt{\frac{2.5T}{mL}} = 1.581139\sqrt{\frac{T}{mL}} \tag{h}$$

Moreover, inserting Eqs. (g) in conjunction with the first of Eqs. (a) into Eq. (7.158) and recalling that $\mathbf{v}_2$ and $\mathbf{v}_1$ are really the first and second eigenvector, respectively, we obtain the modal vectors

$$\begin{aligned}\mathbf{u}_1 &= M^{-1/2}\mathbf{v}_2 = m^{-1/2}\begin{bmatrix}1 & 0 \\ 0 & 1/\sqrt{2}\end{bmatrix}\begin{bmatrix}0.577350 \\ 0.816497\end{bmatrix} = m^{-1/2}\begin{bmatrix}0.577350 \\ 0.577350\end{bmatrix} \\ \mathbf{u}_2 &= M^{-1/2}\mathbf{v}_1 = m^{-1/2}\begin{bmatrix}1 & 0 \\ 0 & 1/\sqrt{2}\end{bmatrix}\begin{bmatrix}-0.816497 \\ 0.577350\end{bmatrix} = m^{-1/2}\begin{bmatrix}-0.816497 \\ 0.408248\end{bmatrix}\end{aligned} \tag{i}$$

and we observe that $\mathbf{u}_1$ and $\mathbf{u}_2$ are orthonormal with respect to the mass matrix M, $\mathbf{u}_r^T M \mathbf{u}_s = \delta_{rs}$ $(r, s = 1, 2)$.

7.13 RAYLEIGH'S QUOTIENT AND ITS PROPERTIES

Rayleigh's quotient represents a unique concept in vibrations whose importance over a broad spectrum of problems is unparalleled. Indeed, it can be used to obtain a quick estimate of the lowest natural frequency and it serves as a key component in an algorithm for computing eigensolutions for discrete systems. Moreover, it plays a central role in a theory concerned with the derivation of approximate eigensolutions for distributed-parameter systems. Equally important is the fact that the concept can be used to gain physical insights into the behavior of vibrating systems.

In Sec. 7.6, we have shown that the eigenvalue problem for a conservative system can be written in the form

$$K\mathbf{u} = \lambda M\mathbf{u}, \ \lambda = \omega^2 \tag{7.179}$$

where M and K are real symmetric mass and stiffness matrices, respectively. The mass matrix is positive definite by definition and for the most part the stiffness matrix is also positive definite, although for certain systems it is only positive semidefinite. For an n-degree-of-freedom system, Eq. (7.179) has n solutions λ_r, $\mathbf{u}_r$ $(r = 1, 2, \ldots, n)$ satisfying

$$K\mathbf{u}_r = \lambda_r M\mathbf{u}_r, \ \lambda_r = \omega_r^2, \ r = 1, 2, \ldots, n \tag{7.180}$$

Premultiplying both sides of Eq. (7.180) by $\mathbf{u}_r^T$ and dividing through by $\mathbf{u}_r^T M\mathbf{u}_r$, which represents a scalar, we can express the eigenvalues in the form of the ratios

$$\lambda_r = \omega_r^2 = \frac{\mathbf{u}_r^T K\mathbf{u}_r}{\mathbf{u}_r^T M\mathbf{u}_r}, \ r = 1, 2, \ldots, n \tag{7.181}$$

and we observe that the numerator is proportional to the potential energy and the denominator is a measure of the kinetic energy, both in the rth mode.

Equation (7.181) permits us to calculate the natural frequencies ω_r provided the modal vectors $\mathbf{u}_r$ are known $(r = 1, 2, \ldots, n)$. This is of purely academic interest, however, as the modal vectors are not known, and in fact our objective is to develop techniques for computing them. To this end, if we repeat the foregoing process, but with $\lambda = \omega^2$ and $\mathbf{u}$ replacing $\lambda_r = \omega_r^2$ and $\mathbf{u}_r$, respectively, we obtain

$$R(\mathbf{u}) = \lambda = \omega^2 = \frac{\mathbf{u}^T K\mathbf{u}}{\mathbf{u}^T M\mathbf{u}} \tag{7.182}$$

where $R(\mathbf{u})$ is a scalar whose value depends on the vector $\mathbf{u}$ for given matrices K and M. The scalar $R(\mathbf{u})$ is known as *Rayleigh's quotient* and it possesses very interesting and useful properties. To explain this statement, we regard $\mathbf{u}$ as an arbitrary vector and propose to examine the behavior of Rayleigh's quotient as the vector $\mathbf{u}$ changes. To this end, we recall from Sec. 7.8 that any n-dimensional vector $\mathbf{u}$ can be expressed as a linear combination of the system eigenvectors $\mathbf{u}_r$ $(r = 1, 2, \ldots, n)$. Hence, using Eqs. (7.123) and (7.126), we can write

$$\mathbf{u} = \sum_{r=1}^{n} c_r \mathbf{u}_r = U\mathbf{c} \tag{7.183}$$

where $U = [\mathbf{u}_1 \; \mathbf{u}_2 \; \ldots \; \mathbf{u}_n]$ is the modal matrix and $\mathbf{c} = [c_1 \; c_2 \; \ldots \; c_n]^T$ is a vector of coefficients. We assume that the modal vectors have been normalized so as to satisfy the orthonormality conditions

$$U^T M U = I, \; U^T K U = \Lambda \tag{7.184}$$

in which I is the identity matrix and $\Lambda = \text{diag}(\lambda_1 \; \lambda_2 \; \ldots \; \lambda_n)$ is the diagonal matrix of eigenvalues. Inserting Eq. (7.183) into Rayleigh's quotient, Eq. (7.182), and using Eqs. (7.184), we obtain

$$R = \frac{\mathbf{u}^T K \mathbf{u}}{\mathbf{u}^T M \mathbf{u}} = \frac{\mathbf{c}^T U^T K U \mathbf{c}}{\mathbf{c}^T U^T M U \mathbf{c}} = \frac{\mathbf{c}^T \Lambda \mathbf{c}}{\mathbf{c}^T \mathbf{c}} = \frac{\sum_{i=1}^{n} \lambda_i c_i^2}{\sum_{i=1}^{n} c_i^2} \tag{7.185}$$

Next, we let $\mathbf{u}$ wander over the entire n-dimensional space and identify c_i as the projection of $\mathbf{u}$ on the axis corresponding to the modal vector $\mathbf{u}_i$, as shown in Fig. 7.14. As long as the trial vector $\mathbf{u}$ remains reasonably far from the modal vectors $\mathbf{u}_i$, nothing significant happens to Rayleigh's quotient. Matters spring to life when $\mathbf{u}$ enters a small neighborhood of a given modal vector, say $\mathbf{u}_r$ (Fig. 7.14), as in this case all projections of $\mathbf{u}$ on axes $\mathbf{u}_i$ become small, except for the projection on $\mathbf{u}_r$. Hence, we can write

$$c_i = \epsilon_i c_r, \; i = 1, 2, \ldots, n; \; i \neq r \tag{7.186}$$

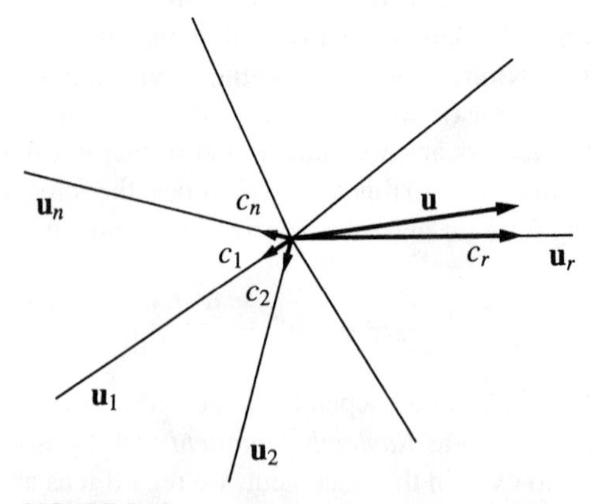

FIGURE 7.14
Trial vector $\mathbf{u}$ in the neighborhood of the modal vector $\mathbf{u}_r$

where the ratios $\epsilon_i = c_i/c_r$ represent small numbers. Introducing Eqs. (7.186) in (7.185), dividing top and bottom by c_r^2 and ignoring higher-order terms in ϵ_i^2, we obtain

$$R = \frac{\lambda_r + \sum\limits_{\substack{i=1 \\ i \neq r}}^{n} \lambda_i \epsilon_i^2}{1 + \sum\limits_{\substack{i=1 \\ i \neq r}}^{n} \epsilon_i^2} \cong \left(\lambda_r + \sum_{\substack{1=1 \\ 1 \neq r}}^{n} \lambda_i \epsilon_i^2\right)\left(1 - \sum_{\substack{i=1 \\ i \neq r}}^{n} \epsilon_i^2\right)$$

$$\cong \lambda_r + \sum_{i=1}^{n} (\lambda_i - \lambda_r)\epsilon_i^2 \tag{7.187}$$

Equations (7.186) state that the vector $\mathbf{u}$ differs from the modal vector $\mathbf{u}_r$ by a small quantity of first order in ϵ. On the other hand, Eq. (7.187) states that R differs from the eigenvalue λ_r by a small quantity of second order in ϵ. The implication is that *Rayleigh's quotient has a stationary value at an eigenvector* $\mathbf{u}_r$, *where the stationary value is the associated eigenvalue* λ_r $(r = 1, 2, \ldots, n)$.

Of special interest in vibrations is the fundamental frequency. If the eigenvalues are arranged in ascending order, $\lambda_1 \leq \lambda_2 \leq \ldots \leq \lambda_n$, then the fundamental mode $\mathbf{u}_1$ corresponds to the lowest natural frequency $\omega_1 = \sqrt{\lambda_1}$. Letting $r = 1$ in Eq. (7.187), we obtain

$$R \cong \lambda_1 + \sum_{i=2}^{n} (\lambda_i - \lambda_1)\epsilon_i^2 \tag{7.188}$$

But, because $\lambda_i \geq \lambda_1$ $(i = 2, 3, \ldots, n)$, we conclude that

$$R \geq \lambda_1 \tag{7.189}$$

where the equality sign holds only if all ϵ_i are identically zero $(i = 2, 3, \ldots, n)$. It follows that *Rayleigh's quotient has a minimum value at the fundamental modal vector, where the minimum value is the fundamental eigenvalue*. Another way of stating the same thing is that *Rayleigh's quotient is an upper bound for the lowest eigenvalue*.

The result embodied by inequality (7.189) is quite remarkable, and has far-reaching implications in vibrations. It states that, regardless of the value of the trial vector $\mathbf{u}$, Rayleigh's quotient is always larger than the lowest eigenvalue. Moreover, if it is possible to guess a trial vector $\mathbf{u}$ differing from the lowest modal vector $\mathbf{u}_1$ by a small quantity of first order in ϵ, then Rayleigh's quotient yields an estimate differing from the lowest eigenvalue λ_1 by a small quantity of second order in ϵ, where ϵ is a small number. This suggests a quick way of estimating the fundamental natural frequency, which is for the most part the most important one. Fortunately, the shape of the lowest modal vector is the easiest to guess. Indeed, quite often extremely accurate estimates can be obtained by using as a trial vector the static displacement vector resulting from the system being loaded with forces proportional to the system masses.

Inequality (7.189) can also be interpreted as stating that

$$\lambda_1 = \min_{\mathbf{u}} R(\mathbf{u}) = \min_{\mathbf{u}} \frac{\mathbf{u}^T K \mathbf{u}}{\mathbf{u}^T M \mathbf{u}} \tag{7.190}$$

where the notation implies that the minimum value of Rayleigh's quotient is obtained by allowing $\mathbf{u}$ to vary over the entire configuration space. The statement that λ_1 *is the minimum value of Rayleigh's quotient* is often referred to as *Rayleigh's principle*. The principle plays a crucial role in the derivation of approximate solutions to the eigenvalue problem for distributed-parameter systems.

Using similar arguments, it is not difficult to show that *Rayleigh's quotient has a maximum value equal to* λ_n *at* $\mathbf{u} = \mathbf{u}_n$, but this statement is not nearly as useful as that concerning the minimum value. Although at times it is important to know the range of the natural frequencies, it is not easy to come up with a good guess for $\mathbf{u}_n$.

In the foregoing discussion, we assumed that matrices M and K were given and examined the behavior of Rayleigh's quotient, Eq. (7.182), as $\mathbf{u}$ was allowed to vary over the configuration space. Equation (7.182) can also be used to examine how the natural frequencies change as the mass and stiffness properties of a system change for a given $\mathbf{u}$. Indeed, from Eq. (7.182), we see that when the entries of the stiffness matrix increase in value the natural frequencies increase, and vice versa. On the other hand, when the entries of the mass matrix increase the natural frequencies decrease, and vice versa.

Example 7.9. Consider the three-degree-of-freedom system of Examples 7.1 and 7.5 and obtain an estimate of the lowest natural frequency by means of Rayleigh's quotient.

From Example 7.5, we have the mass and stiffness matrices

$$M = m\begin{bmatrix} 1 & 0 & 0 \\ 0 & 1 & 0 \\ 0 & 0 & 2 \end{bmatrix}, \quad K = k\begin{bmatrix} 2 & -1 & 0 \\ -1 & 3 & -2 \\ 0 & -2 & 2 \end{bmatrix} \tag{a}$$

Following the suggestion made earlier in this section, we choose as a trial vector for Rayleigh's quotient the vector of static displacements obtained by loading the system with forces proportional to the masses. To this end, we first write

$$\mathbf{F} = c[m_1 \ m_2 \ m_3]^T = [1 \ 1 \ 2]^T \tag{b}$$

in which we chose $c = 1/m$ as the proportionality constant. Then, recalling that the displacement vector $\mathbf{u}$ is related to the force vector by means of the flexibility matrix A, we use Eqs. (7.12) and the first of Eqs. (7.16) and obtain the trial vector

$$\mathbf{u} = A\mathbf{F} = K^{-1}\mathbf{F} = \frac{1}{k}\begin{bmatrix} 2 & -1 & 0 \\ -1 & 3 & -2 \\ 0 & -2 & 2 \end{bmatrix}^{-1}\begin{bmatrix} 1 \\ 1 \\ 2 \end{bmatrix}$$

$$= \frac{1}{k}\begin{bmatrix} 1 & 1 & 1 \\ 1 & 2 & 2 \\ 1 & 2 & 2.5 \end{bmatrix}\begin{bmatrix} 1 \\ 1 \\ 2 \end{bmatrix} = \frac{1}{k}\begin{bmatrix} 4 \\ 7 \\ 8 \end{bmatrix} \tag{c}$$

Hence, inserting Eq. (c) into Eq. (7.182), we calculate the value of Rayleigh's quotient as follows:

$$R = \omega^2 = \frac{\mathbf{u}^T K \mathbf{u}}{\mathbf{u}^T M \mathbf{u}} = \frac{k \begin{bmatrix} 4 \\ 7 \\ 8 \end{bmatrix}^T \begin{bmatrix} 2 & -1 & 0 \\ -1 & 3 & -2 \\ 0 & -2 & 2 \end{bmatrix} \begin{bmatrix} 4 \\ 7 \\ 8 \end{bmatrix}}{m \begin{bmatrix} 4 \\ 7 \\ 8 \end{bmatrix}^T \begin{bmatrix} 1 & 0 & 0 \\ 0 & 1 & 0 \\ 0 & 0 & 2 \end{bmatrix} \begin{bmatrix} 4 \\ 7 \\ 8 \end{bmatrix}} = \frac{27k}{193m} = 0.1399\frac{k}{m} \tag{d}$$

so that the estimate of the lowest natural frequency is

$$\omega = 0.3740\sqrt{\frac{k}{m}} \tag{e}$$

As a matter of interest, we calculate the percentage error incurred in using Rayleigh's quotient. To this end, we recall from the first of Eqs. (e) in Example 7.5 that the first natural frequency is

$$\omega_1 = 0.3731\sqrt{\frac{k}{m}} \tag{f}$$

so that the percentage error is

$$\frac{\omega - \omega_1}{\omega_1} = \frac{0.3740 - 0.3731}{0.3731} = 0.002412 = 0.2412\% \tag{g}$$

Hence, the estimated first natural frequency differs from the actual one by less than one quarter of one percent, which is a remarkable result. This example demonstrates that Rayleigh's quotient is capable of yielding very accurate estimates of the lowest natural frequency. Of course, the example also demonstrates that the practice of using the static displacement vector as a trial vector is a sound one, as the static displacement vector tends to resemble closely the lowest mode of vibration. A comparison of vectors is not as simple as a comparison of scalars. To carry out such a comparison, we first normalize both vectors so that they have unit magnitude and then calculate the norm of the difference. Hence, using Eq. (c) of this example and Eq. (i) of Example 7.5, we can write the normalized vectors

$$\mathbf{u} = \begin{bmatrix} 0.3522 \\ 0.6163 \\ 0.7044 \end{bmatrix}, \quad \mathbf{u}_1 = \begin{bmatrix} 0.3309 \\ 0.6155 \\ 0.7152 \end{bmatrix} \tag{h}$$

so that we define the error in the trial vector as follows:

$$\frac{\|\mathbf{u} - \mathbf{u}_1\|}{\|\mathbf{u}_1\|} = \sqrt{(\mathbf{u} - \mathbf{u}_1)^T (\mathbf{u} - \mathbf{u}_1)} = 0.0239 = 2.39\% \tag{i}$$

As far as trial vectors are concerned, this must be regarded as a small error, so that the chosen trial vector represents a very good guess for the lowest modal vector. Comparing Eqs. (g) and (i), we conclude that the error in the estimated first natural frequency is one order of magnitude smaller than the error in the trial vector, which is consistent with the stationarity property of Rayleigh's quotient.

7.14 RESPONSE TO HARMONIC EXTERNAL EXCITATIONS

We have shown in Sec. 7.1 that the equations of motion of a viscously damped n-degree-of-freedom system can be written in the compact matrix form

$$M\ddot{\mathbf{q}}(t) + C\dot{\mathbf{q}}(t) + K\mathbf{q}(t) = \mathbf{Q}(t) \tag{7.191}$$

where M, C and K are $n \times n$ symmetric mass, damping and stiffness matrices, respectively, $\mathbf{q}(t)$ is the n-dimensional displacement vector and $\mathbf{Q}(t)$ is the n-dimensional external excitation vector.

In this section, we consider the case in which the excitation vector $\mathbf{Q}(t)$ is harmonic. Following the pattern established in Sec. 6.8, we assume that the excitation can be expressed in the form

$$\mathbf{Q}(t) = \mathbf{Q}_0 e^{i\alpha t} \tag{7.192}$$

where $\mathbf{Q}_0$ is a real vector of constant amplitudes and α is the excitation frequency. Then, the solution of Eq. (7.191), in conjunction with Eq. (7.192), can be written as

$$\mathbf{q}(t) = \mathbf{q}_0 e^{i\alpha t} \tag{7.193}$$

in which $\mathbf{q}_0$ is a complex vector. Inserting Eqs. (7.192) and (7.193) into Eq. (7.191), we have

$$(-\alpha^2 M + i\alpha C + K)\mathbf{q}_0 e^{i\alpha t} = \mathbf{Q}_0 e^{i\alpha t} \tag{7.194}$$

Dividing through by $e^{i\alpha t}$, Eq. (7.194) can be rewritten as

$$Z(i\alpha)\mathbf{q}_0 = \mathbf{Q}_0 \tag{7.195}$$

where

$$Z(i\alpha) = -\alpha^2 M + i\alpha C + K \tag{7.196}$$

is the *impedance matrix*, encountered for the first time in Sec. 5.8 in conjunction with two-degree-of-freedom systems. Equation (7.195) represents a set of nonhomogeneous algebraic equations, whose solution is simply

$$\mathbf{q}_0 = Z^{-1}(i\alpha)\mathbf{Q}_0 \tag{7.197}$$

For convenience, we introduce the notation

$$Z^{-1}(i\alpha) = G(i\alpha) \tag{7.198}$$

where $G(i\alpha)$ is a matrix of *frequency response* functions; they are related to the *transfer functions* that would result from a Laplace transformation of Eq. (7.191). Combining Eqs. (7.193), (7.197) and (7.198), we obtain

$$\mathbf{q}(t) = G(i\alpha)\mathbf{Q}_0 e^{i\alpha t} \tag{7.199}$$

Of course, as in Sec. 3.1, if the excitation is $\mathbf{Q}_0 \cos\alpha t$, we retain the real part of $\mathbf{q}(t)$, and if the excitation is $\mathbf{Q}_0 \sin\alpha t$, we retain the imaginary part.

The approach just described is feasible only for systems with a small number of degrees of freedom, such as the two-degree-of-freedom system considered in Sec. 5.8. As the number of degrees of freedom increases, it becomes necessary to adopt an approach based on the idea of decoupling the equations of motion. This approach is the modal analysis introduced in Ch. 5 and used in Sec. 7.10 to obtain the response of undamped systems to initial excitations. Of course, the modal analysis of Sec. 7.10 works only for undamped systems, $C = 0$, and systems in which the damping matrix is a linear combination of the mass and stiffness matrices, where the latter is known as proportional damping. In the case of arbitrary damping, the decoupling can be carried out by means of a complex modal analysis, as shown in Sec. 7.16

7.15 RESPONSE TO EXTERNAL EXCITATIONS BY MODAL ANALYSIS

Equation (7.191) represents a set of n simultaneous ordinary differential equations. No analytical solution of the equations in coupled form is generally possible, so that the only alternative is to decouple them. This requires a coordinate transformation in which the transformation matrix is thc modal matrix, as shown in Secs. 5.6, 5.10 and 7.10. The approach using the orthogonality properties of the modal matrix to render a set of simultaneous equations independent is known as *modal analysis*. The essence of modal analysis is to determine the response of an n-degree-of-freedom system by decomposing it in some fashion into n single-degree-of-freedom systems, determining the response of the single-degree-of-freedom systems and then combining the individual responses into the response of the original system.

There are two cases in which this classical approach is possible. The first is the case in which the system is undamped and the second is the case of proportional damping, in which the damping matrix is a linear combination of the mass matrix and the stiffness matrix.

7.15.1 Undamped systems

In this case, the damping matrix is zero, $C = 0$, so that Eq. (7.191) reduces to

$$M\ddot{\mathbf{q}}(t) + K\mathbf{q}(t) = \mathbf{Q}(t) \tag{7.200}$$

We assume that the mass matrix M is positive definite and that the stiffness matrix is either positive definite or only positive semidefinite.

As a preliminary to the use of modal analysis, we must solve the eigenvalue problem, which is given by

$$K\mathbf{u} = \omega^2 M\mathbf{u} \tag{7.201}$$

Equation (7.201) has the solutions ω_r^2, $\mathbf{u}_r$ $(r = 1, 2, \ldots, n)$, where ω_r are the natural frequencies and $\mathbf{u}_r$ are the natural modes, or modal vectors. If the modal vectors are normalized in some sense, then they are referred to as normal modes. It was shown in Sec. 7.7 that the modal vectors are orthogonal with respect to both the mass matrix and the stiffness matrix. The eigenvalues and eigenvectors can be arranged in the $n \times n$ matrices $\Omega = \text{diag}[\omega_1^2 \; \omega_2^2 \; \dots \; \omega_n^2]$, $U = [\mathbf{u}_1 \; \mathbf{u}_2 \; \dots \; \mathbf{u}_n]^T$, where U is the modal matrix. Moreover, it is convenient to normalize the modal matrix so as to satisfy the orthonormality relations

$$U^T M U = I, \; U^T K U = \Omega \tag{7.202}$$

Next, we use the analogy with the approach of Sec. 7.10 and express the solution of Eq. (7.200) as a linear combination of the modal vectors as follows:

$$\mathbf{q}(t) = \sum_{r=1}^{n} \eta_r(t)\mathbf{u}_r = U\boldsymbol{\eta}(t) \tag{7.203}$$

in which $\eta_r(t)$ represent *modal coordinates* and $\boldsymbol{\eta}(t) = [\eta_1(t) \; \eta_2(t) \; \dots \; \eta_n(t)]^T$ is the corresponding vector. Inserting Eq. (7.203) into Eq. (7.200), premultiplying the result by U^T and considering Eqs. (7.202), we obtain

$$\ddot{\boldsymbol{\eta}}(t) + \Omega\boldsymbol{\eta}(t) = \mathbf{N}(t) \tag{7.204}$$

where

$$\mathbf{N}(t) = U^T \mathbf{Q}(t) \tag{7.205}$$

is a vector of *modal forces*. Equation (7.204) represents a set of independent *modal equations*. For positive definite K, they have the scalar form

$$\ddot{\eta}_r(t) + \omega_r^2 \eta_r(t) = N_r(t), \; r = 1, 2, \dots, n \tag{7.206}$$

in which

$$N_r(t) = \mathbf{u}_r^T \mathbf{Q}(t), \; r = 1, 2, \dots, n \tag{7.207}$$

are the individual modal forces. Equations (7.206) resemble the equation of an undamped single-degree-of-freedom system entirely.

We consider first the case in which the external excitation is harmonic. Because for undamped systems there is no first derivative $\dot{\mathbf{q}}(t)$ in the equations of motion, we can dispense with the complex notation and let the excitation have the real form

$$\mathbf{Q}(t) = \mathbf{Q}_0 \cos \alpha t \tag{7.208}$$

where $\mathbf{Q}_0$ is a vector of force amplitudes and α is the excitation frequency. Inserting Eq. (7.208) into Eq. (7.207), we can write the modal forces

$$N_r(t) = \mathbf{u}_r^T \mathbf{Q}_0 \cos \alpha t, \; r = 1, 2, \dots, n \tag{7.209}$$

Then, using the analogy with the response of undamped single-degree-of-freedom systems to harmonic excitations, Eqs. (3.46) and (3.47), we conclude that the solution of the modal equations, Eqs. (7.206), in conjunction with the harmonic generalized forces given by Eqs. (7.209), is simply

$$\eta_r(t) = \frac{\mathbf{u}_r^T \mathbf{Q}_0}{\omega_r^2 - \alpha^2} \cos \alpha t, \; r = 1, 2, \dots, n \tag{7.210}$$

Note that, this being a steady-state response, it is not advisable to add to it the effect of the initial excitations, which represents a transient response. Finally, inserting Eqs. (7.210) into Eq. (7.203), we obtain the steady-state harmonic response

$$\mathbf{q}(t) = \sum_{r=1}^{n} \frac{\mathbf{u}_r^T \mathbf{Q}_0}{\omega_r^2 - \alpha^2} \mathbf{u}_r \cos \alpha t \tag{7.211}$$

Equation (7.211) permits us to conclude that a resonance condition exists if the excitation frequency α is equal to one of the natural frequencies ω_r $(r = 1, 2, \ldots, n)$. Moreover, if the excitation amplitudes vector $\mathbf{Q}_0$ is proportional to the product of the mass matrix M and one of the modal vectors, say $\mathbf{Q}_0 = M\mathbf{u}_k$, then the response is proportional to the kth mode.

Next, we turn our attention to the case in which the external excitations are arbitrary. By the superposition principle (Sec. 1.12), the response to initial excitations and the response to external excitations can be determined separately and then combined linearly to obtain the total response. In Sec. 7.10, we determined the response of multi-degree-of-freedom systems to initial excitations by modal analysis, so that in this section it is only necessary to concentrate on the response to external excitations. The solution of equations of the type (7.206) was presented in Ch. 4 in conjunction with the response of single-degree-of-freedom systems to arbitrary excitations. Hence, letting $m = 1$, $\zeta = 0$, $\omega_d = \omega_r$, $x = \eta_r$ and $F = N_r$ in Eq. (4.76) and ignoring the contributions from the initial displacement and initial velocity, we can write the particular solution of Eqs. (7.206) in the form of the convolution integral

$$\eta_r(t) = \frac{1}{\omega_r} \int_0^t N_r(t - \tau) \sin \omega_r \tau \, d\tau, \; r = 1, 2, \ldots, n \tag{7.212}$$

Then, inserting Eqs. (7.212) into Eq. (7.203) and recalling Eqs. (7.207), we can write the response of an undamped n-degree-of-freedom system to external forces as follows:

$$\mathbf{q}(t) = \sum_{r=1}^{n} \left[\frac{\mathbf{u}_r^T}{\omega_r} \int_0^t \mathbf{Q}(t - \tau) \sin \omega_r \tau \, d\tau \right] \mathbf{u}_r \tag{7.213}$$

In the case in which the stiffness matrix K is only positive semidefinite, the system is positive semidefinite and it admits rigid-body modes with zero frequency (Sec. 7.8). In this case, assuming that there are i rigid-body modes, Eqs. (7.206) with $\omega_r = 0$ reduce to

$$\ddot{\eta}_r(t) = N_r(t), \; r = 1, 2, \ldots, i \tag{7.214}$$

Equations (7.214) have the solution

$$\eta_r(t) = \int_0^t \left[\int_0^\tau N_r(\sigma) d\sigma \right] d\tau, \; r = 1, 2, \ldots, i \tag{7.215}$$

Hence, in the case in which the system is unrestrained the response to arbitrary excitations is

$$\mathbf{q}(t) = \sum_{r=1}^{i} \mathbf{u}_r^T \int_0^t \left[\int_0^\tau \mathbf{Q}(\sigma) d\sigma \right] d\tau$$
$$+ \sum_{r=i+1}^{n} \left[\frac{\mathbf{u}_r^T}{\omega_r} \int_0^t \mathbf{Q}(t-\tau) \sin\omega_r\tau \, d\tau \right] \mathbf{u}_r \tag{7.216}$$

7.15.2 Systems with proportional damping

The linear transformation expressed by Eq. (7.203) uses the modal matrix corresponding to an undamped system. As indicated by Eqs. (7.202), the modal matrix U is able to diagonalize the mass matrix M and stiffness matrix K simultaneously. In general, the modal matrix is not able to diagonalize the damping matrix C. There is a special case, however, in which the modal matrix does diagonalize the damping matrix, namely, when the damping matrix can be expressed as a linear combination of the mass and stiffness matrices of the form

$$C = \alpha M + \beta K \tag{7.217}$$

where α and β are given constant scalars. Indeed, premultiplying Eq. (7.217) by U^T, postmultiplying U and considering Eqs. (7.202), we obtain the diagonal matrix

$$U^T C U = U^T(\alpha M + \beta K)U = \alpha U^T M U + \beta U^T K U = \alpha I + \beta\Omega \tag{7.218}$$

This special case is known as *proportional damping*.

Next, we introduce Eq. (7.203) in Eq. (7.191), premultiply the result by U^T, consider Eqs. (7.202) and (7.218) and obtain

$$\ddot{\boldsymbol{\eta}}(t) + (\alpha I + \beta\Omega)\dot{\boldsymbol{\eta}}(t) + \Omega\boldsymbol{\eta}(t) = \mathbf{N}(t) \tag{7.219}$$

where $\mathbf{N}(t)$ is given by Eq. (7.205). Then, recalling that $\Omega = \text{diag}(\omega_1^2 \; \omega_2^2 \; \dots \; \omega_n^2)$ and introducing the notation

$$\alpha + \beta\omega_r^2 = 2\zeta_r\omega_r, \; r = 1, 2, \dots, n \tag{7.220}$$

in which ζ_r $(r = 1, 2, \dots, n)$ are modal viscous damping factors, Eq. (7.219) can be rewritten in the form of the independent *modal equations*

$$\ddot{\eta}_r(t) + 2\zeta_r\omega_r\dot{\eta}_r(t) + \omega_r^2\eta_r(t) = N_r(t), \; r = 1, 2, \dots, n \tag{7.221}$$

where $N_r(t)$ are modal forces having the form given by Eqs. (7.207).

Equations (7.221) resemble entirely the equation of motion of a viscously damped single-degree-of-freedom system, so that we produce their solution by adapting results obtained in Secs. 3.1 and 4.7. We begin with the case in which the system is subjected to harmonic excitations. Because Eqs. (7.221) contain the first derivative $\dot{\eta}_r$, it is convenient to work with the complex notation. Hence, assuming that the actual system is subjected to the harmonic excitation

$$\mathbf{Q}(t) = \mathbf{Q}_0 e^{i\alpha t} \tag{7.222}$$

where $\mathbf{Q}_0$ is a vector of constant force amplitudes and α is the excitation frequency, and using Eq. (7.207), we obtain the modal forces

$$N_r(t) = \mathbf{u}_r^T \mathbf{Q}_0 e^{i\alpha t}, \ r = 1, 2, \ldots, n \tag{7.223}$$

Then, using the analogy with Eqs. (3.15), (3.20) and (3.21), the steady-state solution of Eqs. (7.221), in conjunction with Eqs. (7.223), is simply

$$\eta_r(t) = \frac{\mathbf{u}_r^T \mathbf{Q}_0}{\omega_r^2 - \alpha^2 + i2\zeta_r \omega_r \alpha} e^{i\alpha t}, \ r = 1, 2, \ldots, r \tag{7.224}$$

Of course, if the excitation is $\mathbf{Q}(t) = \mathbf{Q}_0 \cos \alpha t$, we retain only the real part of $\eta_r(t)$, and if the excitations is $\mathbf{Q}(t) = \mathbf{Q}_0 \sin \alpha t$, we retain the imaginary part. Inserting Eqs. (7.224) into Eq. (7.203), we obtain the actual steady-state harmonic response

$$\mathbf{q}(t) = \sum_{r=1}^{n} \frac{\mathbf{u}_r^T \mathbf{Q}_0}{\omega_r^2 - \alpha^2 + i2\zeta_r \omega_r \alpha} \mathbf{u}_r e^{i\alpha t} \tag{7.225}$$

The notation can be simplified by introducing the *modal frequency responses*

$$G_r(i\alpha) = \frac{1}{1 - (\alpha/\omega_r)^2 + i2\zeta_r \alpha/\omega_r}, \ r = 1, 2, \ldots, n \tag{7.226}$$

so that the response can be rewritten as

$$\mathbf{q}(t) = \sum_{r=1}^{n} \frac{\mathbf{u}_r^T \mathbf{Q}_0}{\omega_r^2} |G_r(i\alpha)| \mathbf{u}_r e^{i(\alpha t - \phi_r)} \tag{7.227}$$

where

$$|G_r(i\alpha)| = \frac{1}{\{[1 - (\alpha/\omega_r)^2]^2 + (2\zeta_r \alpha/\omega_r)^2\}^{1/2}}, \ r = 1, 2, \ldots, n \tag{7.228}$$

is the magnitude of the modal frequency response $G_r(i\alpha)$ and

$$\phi_r = \tan^{-1} \frac{2\zeta_r \alpha/\omega_r}{1 - (\alpha/\omega_r)^2}, \ r = 1, 2, \ldots, n \tag{7.229}$$

is the phase angle.

To determine the solution of Eqs. (7.221) for the case in which the modal forces $N_r(t)$ arise from arbitrary excitations, we once again invoke the analogy with viscously damped single-degree-of-freedom systems. To this end, we make appropriate notation adjustments in Eq. (4.76) and ignore the response to the initial displacement and velocity. Following these adjustments, the general solution of Eqs. (7.221) can be verified to be

$$\eta_r(t) = \frac{1}{\omega_{dr}} \int_0^t N_r(t - \tau) e^{-\zeta_r \omega_r \tau} \sin \omega_{dr} \tau \, d\tau, \ r = 1, 2, \ldots, n \tag{7.230}$$

where

$$\omega_{dr} = (1-\zeta_r^2)^{1/2}\omega_r, \quad r = 1,2,\ldots,n \tag{7.231}$$

are frequencies of damped oscillation. The response of a proportionally damped system to external excitations is obtained by inserting Eqs. (7.230) into Eq. (7.203). The response to initial excitations can be obtained separately by the approach of Sec. 7.10 and added to the response to external excitations.

If the system admits rigid-body motions, then the corresponding modal coordinates satisfy equations of the type (7.214), whose particular solution is given by Eqs. (7.215).

Example 7.10. Use modal analysis to derive the response of the undamped three-degree-of-freedom system of Examples 7.1 and 7.5 to the excitation

$$Q_1(t) = Q_2(t) = 0, \; Q_3(t) = Q_0 \mathcal{u}(t) \tag{a}$$

where $\mathcal{u}(t)$ is the unit step function.

The equations of motion for the system are given in matrix form by Eq. (7.200), in which, from Example 7.5, the mass and stiffness matrices are

$$M = m\begin{bmatrix} 1 & 0 & 0 \\ 0 & 1 & 0 \\ 0 & 0 & 2 \end{bmatrix}, \quad K = k\begin{bmatrix} 2 & -1 & 0 \\ -1 & 3 & -2 \\ 0 & -2 & 2 \end{bmatrix} \tag{b}$$

Moreover, the response is given by Eq. (7.203), where the modal matrix can be shown to be

$$U = [\mathbf{u}_1 \; \mathbf{u}_2 \; \mathbf{u}_3] = \frac{1}{\sqrt{m}}\begin{bmatrix} 0.269108 & -0.878183 & 0.395443 \\ 0.500758 & -0.223145 & -0.836330 \\ 0.581731 & 0.299166 & 0.268493 \end{bmatrix} \tag{c}$$

and note that the modal matrix has been normalized so as to satisfy $U^T M U = I$, where I is the identity matrix.

Equation (7.203) calls for the modal coordinates $\eta_r(t)$ $(r = 1,2,3)$, given by Eqs. (7.212) in the form of convolution integrals in terms of the modal forces specified by Eqs. (7.207). Hence, inserting Eqs. (a) and (b) into Eqs. (7.207), we obtain the modal forces

$$\begin{aligned} N_1(t) &= \mathbf{u}_1^T \mathbf{Q}(t) = \frac{1}{\sqrt{m}}\begin{bmatrix} 0.269108 \\ 0.500758 \\ 0.581731 \end{bmatrix}^T \begin{bmatrix} 0 \\ 0 \\ Q_0\mathcal{u}(t) \end{bmatrix} = 0.581731\frac{Q_0}{\sqrt{m}}\mathcal{u}(t) \\ N_2(t) &= \mathbf{u}_2^T \mathbf{Q}(t) = \frac{1}{\sqrt{m}}\begin{bmatrix} -0.878183 \\ -0.223145 \\ 0.299166 \end{bmatrix}^T \begin{bmatrix} 0 \\ 0 \\ Q_0\mathcal{u}(t) \end{bmatrix} = 0.299166\frac{Q_0}{\sqrt{m}}\mathcal{u}(t) \\ N_3(t) &= \mathbf{u}_3^T \mathbf{Q}(t) = \frac{1}{\sqrt{m}}\begin{bmatrix} 0.395443 \\ -0.836330 \\ 0.268493 \end{bmatrix}^T \begin{bmatrix} 0 \\ 0 \\ Q_0\mathcal{u}(t) \end{bmatrix} = 0.268493\frac{Q_0}{\sqrt{m}}\mathcal{u}(t) \end{aligned} \tag{d}$$

Before we evaluate the convolution integrals in Eqs. (7.212), we list the natural frequencies as follows:

$$\omega_1 = 0.373087\sqrt{\frac{k}{m}}, \quad \omega_2 = 1.321325\sqrt{\frac{k}{m}}, \quad \omega_3 = 2.028523\sqrt{\frac{k}{m}} \tag{e}$$

Then, inserting Eqs. (d) and (e) into Eqs. (7.212) and evaluating the integrals, we can write

$$\begin{aligned}
\eta_1(t) &= \frac{0.581731 Q_0}{\sqrt{m}\omega_1}\int_0^t u(t-\tau)\sin\omega_1\tau\, d\tau = \frac{0.581731 Q_0}{\sqrt{m}\omega_1^2}(1-\cos\omega_1 t)\\
&= \frac{4.179285 Q_0\sqrt{m}}{k}\left(1-\cos 0.373087\sqrt{\frac{k}{m}}t\right)\\
\eta_2(t) &= \frac{0.299166 Q_0}{\sqrt{m}\omega_2}\int_0^t u(t-\tau)\sin\omega_2\tau\, d\tau = \frac{0.299166 Q_0}{\sqrt{m}\omega_2^2}(1-\cos\omega_2 t)\\
&= \frac{0.171353 Q_0\sqrt{m}}{k}\left(1-\cos 1.321325\sqrt{\frac{k}{m}}t\right)\\
\eta_3(t) &= \frac{0.268493 Q_0}{\sqrt{m}\omega_3}\int_0^t u(t-\tau)\sin\omega_3\tau\, d\tau = \frac{0.268493 Q_0}{\sqrt{m}\omega_3^2}(1-\cos\omega_3 t)\\
&= \frac{0.065249 Q_0\sqrt{m}}{k}\left(1-\cos 2.028523\sqrt{\frac{k}{m}}t\right)
\end{aligned} \tag{f}$$

Finally, introducing Eqs. (c) and (f) in Eq. (7.203), we obtain the response

$$\begin{aligned}
\mathbf{q}(t) = \frac{Q_0}{k}\Bigg\{ & 4.179285\left(1-\cos 0.373087\sqrt{\frac{k}{m}}t\right)\begin{bmatrix} 0.269108 \\ 0.500758 \\ 0.581731 \end{bmatrix}\\
& +0.171353\left(1-\cos 1.321325\sqrt{\frac{k}{m}}t\right)\begin{bmatrix} -0.878183 \\ -0.223145 \\ 0.299166 \end{bmatrix}\\
& +0.065249\left(1-\cos 2.028523\sqrt{\frac{k}{m}}t\right)\begin{bmatrix} 0.395443 \\ -0.836330 \\ 0.268493 \end{bmatrix}\Bigg\}
\end{aligned} \tag{g}$$

From Eq. (g), we observe that the amplitude of the second mode is only about 4% of the amplitude of the first mode and that of the third mode is less than 1.6% of the first, which can be attributed to the fact that higher modes are more difficult to excite than lower ones. This is particularly true here, because there is only one force present and this force is acting on the third mass, which, for the system at hand, tends to excite the lowest mode the most. This can be explained by observing that the first mode has no sign changes, the second mode has one sign change and the third mode has two sign changes, so that the motion of some masses in the second and third mode oppose the force.

Example 7.11. Solve Example 7.10 under the assumption that the system possesses proportional damping with the proportionality constants

$$\alpha = 0.2\sqrt{k/m},\ \beta = 0.01\sqrt{m/k} \tag{a}$$

As in the undamped case, the response of proportionally damped systems is given by Eq. (7.203), with the modal matrix in the form of Eq. (c) of Example 7.10. Moreover, the modal coordinates $q_r(t)$ $(r = 1, 2, 3)$ are given by the convolution integrals in Eqs. (7.230). Before we can evaluate the modal coordinates, we must calculate the modal damping factors

ζ_r and the frequencies of damped oscillation ω_{dr} $(r = 1, 2, 3)$. Inserting Eqs. (a) of this example and Eqs. (e) of Example 7.10 into Eqs. (7.220), we can write

$$\zeta_1 = \frac{\alpha + \beta\omega_1^2}{2\omega_1} = \frac{0.2 + 0.01 \times 0.373087^2}{2 \times 0.373087} = 0.269908$$
$$\zeta_2 = \frac{\alpha + \beta\omega_2^2}{2\omega_2} = \frac{0.2 + 0.01 \times 1.321325^2}{2 \times 1.321325} = 0.082288 \tag{b}$$
$$\zeta_3 = \frac{\alpha + \beta\omega_3^2}{2\omega_3} = \frac{0.2 + 0.01 \times 2.028523^2}{2 \times 2.028523} = 0.059440$$

so that

$$\omega_{d1} = \sqrt{1 - \zeta_1^2}\,\omega_1 = \sqrt{1 - 0.269908^2} \times 0.373087\sqrt{k/m} = 0.359240\sqrt{k/m}$$
$$\omega_{d2} = \sqrt{1 - \zeta_2^2}\,\omega_2 = \sqrt{1 - 0.082288^2} \times 1.321325\sqrt{k/m} = 1.316844\sqrt{k/m} \tag{c}$$
$$\omega_{d3} = \sqrt{1 - \zeta_3^2}\,\omega_3 = \sqrt{1 - 0.059440^2} \times 2.028523\sqrt{k/m} = 2.021356\sqrt{k/m}$$

The modal forces $N_r(t)$ $(r = 1, 2, 3)$ remain the same as in Example 7.10. Hence, introducing Eqs. (b) and (c), as well as Eqs. (d) of Example 7.10, in Eqs. (7.230) and setting the initial modal displacements and velocities to zero, we obtain the modal coordinates

$$\begin{aligned}
\eta_1(t) &= \frac{0.581731\,Q_0}{\omega_{d1}\sqrt{m}} \int_0^t \omega(t-\tau)e^{-\zeta_1\omega_1\tau} \sin\omega_{d1}\tau\,d\tau \\
&= \frac{0.581731\,Q_0}{\omega_1^2\sqrt{m}} \left[1 - e^{-\zeta_1\omega_1 t}\left(\cos\omega_{d1}t + \frac{\zeta_1\omega_1}{\omega_{d1}}\sin\omega_{d1}t\right)\right] \\
&= \frac{4.179285\,Q_0\sqrt{m}}{k} \left[1 - e^{-0.100699t}\left(\cos 0.359240\sqrt{\frac{k}{m}}t \right.\right. \\
&\qquad \left.\left. + 0.289312 \sin 0.359249\sqrt{\frac{k}{m}}t\right)\right]
\end{aligned}$$

$$\begin{aligned}
\eta_2(t) &= \frac{0.299166\,Q_0}{\omega_{d2}\sqrt{m}} \int_0^t \omega(t-\tau)e^{-\zeta_2\omega_2\tau} \sin\omega_{d2}\tau\,d\tau \\
&= \frac{0.299166\,Q_0}{\omega_2^2\sqrt{m}} \left[1 - e^{-\zeta_2\omega_2 t}\left(\cos\omega_{d2}t + \frac{\zeta_2\omega_2}{\omega_{d2}}\sin\omega_{d2}t\right)\right] \\
&= \frac{0.171353\,Q_0\sqrt{m}}{k} \left[1 - e^{-0.108729t}\left(\cos 1.316844\sqrt{\frac{k}{m}}t \right.\right. \\
&\qquad \left.\left. + 0.082568 \sin 1.316844\sqrt{\frac{k}{m}}t\right)\right]
\end{aligned} \tag{d}$$

$$\eta_3(t) = \frac{0.268493 Q_0}{\omega_{d3}\sqrt{m}} \int_0^t u(t-\tau) e^{-\zeta_3 \omega_3 \tau} \sin \omega_{d3} \tau \, d\tau$$

$$= \frac{0.268493 Q_0}{\omega_3^2 \sqrt{m}} \left[1 - e^{-\zeta_3 \omega_3 t} \left(\cos \omega_{d3} t + \frac{\zeta_3 \omega_3}{\omega_{d3}} \sin \omega_{d3} t \right) \right]$$

$$= \frac{0.065249 Q_0 \sqrt{m}}{k} \left[1 - e^{-0.120575 t} \left(\cos 2.021356 \sqrt{\frac{k}{m}} t + 0.059651 \sin 2.021356 \sqrt{\frac{k}{m}} t \right) \right]$$

Finally, inserting Eq. (c) of Example 7.10 and Eqs. (d) into Eq. (7.203), we obtain the response of the proportionally damped system

$$\mathbf{q}(t) = \frac{Q_0}{k} \left\{ 4.179285 \left[1 - e^{-0.100699 t} \left(\cos 0.359240 \sqrt{\frac{k}{m}} t + 0.289312 \sin 0.359240 \sqrt{\frac{k}{m}} t \right) \right] \begin{bmatrix} 0.269108 \\ 0.500758 \\ 0.581731 \end{bmatrix} \right.$$

$$+ 0.171353 \left[1 - e^{-0.108729 t} \left(\cos 1.316844 \sqrt{\frac{k}{m}} t + 0.082568 \sin 1.316844 \sqrt{\frac{k}{m}} t \right) \right] \begin{bmatrix} -0.878183 \\ -0.223145 \\ 0.299166 \end{bmatrix}$$

$$\left. + 0.065249 \left[1 - e^{-0.120575 t} \left(\cos 2.021356 \sqrt{\frac{k}{m}} t + 0.059651 \sin 2.021356 \sqrt{\frac{k}{m}} t \right) \right] \begin{bmatrix} 0.395443 \\ -0.836330 \\ 0.268493 \end{bmatrix} \right\} \quad \text{(e)}$$

7.16 SYSTEMS WITH ARBITRARY VISCOUS DAMPING

The equations of motion of an n-degree-of-freedom system with viscous damping are given in matrix form by Eq. (7.191). A solution of Eq. (7.191) by reducing the set of n simultaneous equations to a set of n independent equations can be obtained in the special case in which damping is of the proportional type, i.e., the damping matrix is equal to a linear combination of the mass matrix and the stiffness matrix. We discussed this case in Sec. 7.15. There are other cases in which Eq. (7.191) can be reduced to a set of independent equations, but these cases are seldom realized in practice.

In the general case of viscous damping, the modal matrix does not diagonalize the damping matrix, so that no analytical solution of the equations of motion in the configuration space is possible. However, an analytical solution is possible in the *state space*. To this end, we introduce an obvious identity, use Eq. (7.191) and write

$$\begin{aligned}\dot{\mathbf{q}}(t) &= \dot{\mathbf{q}}(t) \\ \ddot{\mathbf{q}}(t) &= -M^{-1}C\dot{\mathbf{q}}(t) - M^{-1}K\mathbf{q}(t) + M^{-1}\mathbf{Q}(t)\end{aligned} \tag{7.232}$$

Next, we define the *state vector* as the $2n$-dimensional vector $\mathbf{x}(t) = [\mathbf{q}^T(t) \;\; \dot{\mathbf{q}}^T(t)]^T$ and rewrite Eqs. (7.232) in the customary state form

$$\dot{\mathbf{x}}(t) = A\mathbf{x}(t) + B\mathbf{Q}(t) \tag{7.233}$$

where

$$A = \left[\begin{array}{c|c} 0 & I \\ \hline -M^{-1}K & -M^{-1}C \end{array}\right], \quad B = \left[\begin{array}{c} 0 \\ \hline M^{-1} \end{array}\right] \tag{7.234}$$

are $2n \times 2n$ and $2n \times n$ coefficient matrices. Equation (7.233) represents the *state equations* in matrix form.

The solution of Eq. (7.233) can be carried out by means of a modal analysis in the state space. To this end, we consider the free vibration problem, obtained by letting $\mathbf{Q}(t) = \mathbf{0}$ in Eq. (7.233), with the result

$$\dot{\mathbf{x}}(t) = A\mathbf{x}(t) \tag{7.235}$$

Because Eq. (7.235) represents a homogeneous set of ordinary differential equations with constant coefficients, its solution has the exponential form

$$\mathbf{x}(t) = e^{\lambda t}\mathbf{x} \tag{7.236}$$

where λ is a constant scalar and $\mathbf{x}$ a constant vector. Inserting Eq. (7.236) into Eq. (7.235) and dividing through by $e^{\lambda t}$, we obtain the *algebraic eigenvalue problem*

$$A\mathbf{x} = \lambda\mathbf{x} \tag{7.237}$$

which is said to be in *standard form*. But, unlike the eigenvalue problems encountered earlier in this chapter, the matrix A is nonsymmetric, $A^T \neq A$, albeit real. Hence, the desirable properties of the eigenvalues and eigenvectors associated with symmetric matrices no longer exist. In particular, the eigenvalues and eigenvectors are no longer guaranteed to be real and the eigenvectors are no longer orthogonal. The solution of Eq. (7.237) consists of the eigenvalues λ_i and the eigenvectors $\mathbf{x}_i$ $(i = 1, 2, \ldots, 2n)$. The eigenvalues can be real or they can be complex. But, because A is real, if λ_r is a complex eigenvalue, then the complex conjugate $\overline{\lambda}_r$ is also an eigenvalue. Moreover, the eigenvector $\mathbf{x}_r$ belonging to λ_r is also complex, and the eigenvector $\overline{\mathbf{x}}_r$ belonging to $\overline{\lambda}_r$ is the complex conjugate of $\mathbf{x}_r$.

Next, we consider the eigenvalue problem associated with the transposed matrix

$$A^T\mathbf{y} = \lambda\mathbf{y} \tag{7.238}$$

But, because the determinant of A is equal to the determinant of A^T (see Appendix C), the characteristic equation corresponding to Eq. (7.237) is the same as the characteristic equation associated with Eq. (7.238), or

$$\det[A - \lambda I] = \det[A^T - \lambda I] = 0 \tag{7.239}$$

It follows that *the eigenvalues of A^T are the same as the eigenvalues of A*. On the other hand, *the eigenvectors of A^T are not the same as the eigenvectors of A*. We denote the eigenvalues and eigenvectors of A^T by λ_j and $\mathbf{y}_j$ $(j = 1, 2, \dots, 2n)$, respectively. Equation (7.238) can be transposed, with the results

$$\mathbf{y}^T A = \lambda \mathbf{y}^T \tag{7.240}$$

But, because of their position relative to A, the eigenvectors $\mathbf{x}_1, \mathbf{x}_2, \dots, \mathbf{x}_{2n}$ are called *right eigenvectors* and $\mathbf{y}_1, \mathbf{y}_2, \dots, \mathbf{y}_{2n}$ are known as *left eigenvectors* of A. The eigenvalue problem for A^T is called the *adjoint eigenvalue problem*. Consistent with this, the eigenvectors $\mathbf{y}_1, \mathbf{y}_2, \dots, \mathbf{y}_{2n}$ are known as *adjoint eigenvectors* of the eigenvectors $\mathbf{x}_1, \mathbf{x}_2, \dots, \mathbf{x}_{2n}$.

The right eigenvectors $\mathbf{x}_1, \mathbf{x}_2, \dots, \mathbf{x}_{2n}$ are not mutually orthogonal, and neither are the left eigenvectors $\mathbf{y}_1, \mathbf{y}_2, \dots, \mathbf{y}_{2n}$, so that the question can be raised as to their usefulness. The two sets of eigenvectors, however, satisfy a certain type of orthogonality conditions, which makes them more useful than it may appear. To substantiate this statement, we recognize that the two sets of eigenvectors solve the eigenvalue problems

$$A\mathbf{x}_i = \lambda_i \mathbf{x}_i, \; i = 1, 2, \dots, 2n \tag{7.241}$$

and

$$\mathbf{y}_j^T A = \lambda_j \mathbf{y}_j^T, \; j = 1, 2, \dots, 2n \tag{7.242}$$

so that, premultiplying Eqs. (7.241) by $\mathbf{y}_j^T$ and postmultiplying Eqs. (7.242) by $\mathbf{x}_i$, we obtain

$$\mathbf{y}_j^T A \mathbf{x}_i = \lambda_i \mathbf{y}_j^T \mathbf{x}_i \tag{7.243}$$

and

$$\mathbf{y}_j^T A \mathbf{x}_i = \lambda_j \mathbf{y}_j^T \mathbf{x}_i \tag{7.244}$$

respectively. Next, we subtract Eq. (7.244) from Eq. (7.243) and write

$$(\lambda_i - \lambda_j)\mathbf{y}_j^T \mathbf{x}_i = 0 \tag{7.245}$$

But, if $\lambda_i \neq \lambda_j$, Eq. (7.245) can be satisfied only if

$$\mathbf{y}_j^T \mathbf{x}_i = 0, \; \lambda_i \neq \lambda_j, \; i, j = 1, 2, \dots, 2n \tag{7.246}$$

Equation (7.246) states that *the right eigenvectors of A are orthogonal to the left eigenvectors of A corresponding to distinct eigenvalues*. It should be stressed here

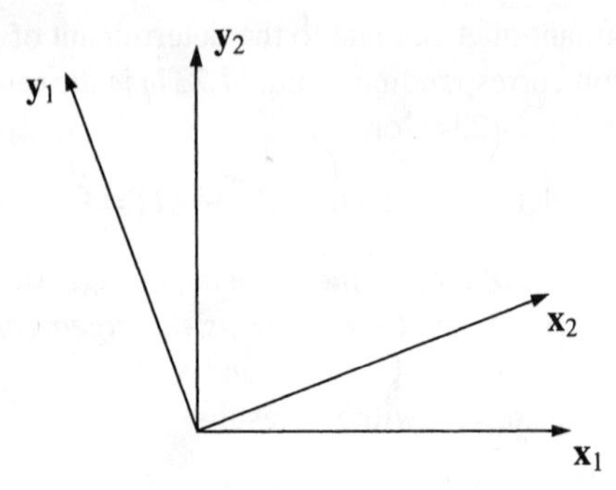

FIGURE 7.15
Geometric interpretation of biorthogonality

that this is not mutual orthogonality as for real symmetric matrices. Indeed, the type of orthogonality described by Eq. (7.246) is known as *biorthogonality*, and the right eigenvectors $\mathbf{x}_i$ $(i = 1, 2, \ldots, 2n)$ are said to be *biorthogonal* to the left eigenvectors $\mathbf{y}_j$ $(j = 1, 2, \ldots, 2n)$, and vice versa. Biorthogonality is illustrated for the two-dimensional case in Fig. 7.15. Moreover, inserting Eq. (7.246) in either Eq. (7.243) or Eq. (7.244), we conclude that

$$\mathbf{y}_j^T A\mathbf{x}_i = 0, \ \lambda_i \neq \lambda_j, \ i, j = 1, 2, \ldots, 2n \tag{7.247}$$

so that *the right eigenvectors* $\mathbf{x}_i$ *are biorthogonal with respect to the matrix A to the left eigenvectors* $\mathbf{y}_j$ *as well.* When $i = j$, the products $\mathbf{y}_i^T\mathbf{x}_i$ and $\mathbf{y}_i^T A\mathbf{x}_i$ are not zero. It is convenient to normalize the two sets of vectors by insisting that

$$\mathbf{y}_i^T\mathbf{x}_i = 1, \ i = 1, 2, \ldots, 2n \tag{7.248}$$

in which case, from Eq. (7.243) or Eq. (7.244),

$$\mathbf{y}_i^T A\mathbf{x}_i = \lambda_i, \ i = 1, 2, \ldots, 2n \tag{7.249}$$

Equations (7.246) and (7.248) on the one hand and Eqs. (7.247) and (7.249) on the other hand can be combined by writing the *biorthonormality relations*

$$\mathbf{y}_j^T\mathbf{x}_i = \delta_{ij}, \ i, j = 1, 2, \ldots, 2n \tag{7.250}$$

and

$$\mathbf{y}_j^T A\mathbf{x}_i = \lambda_i\delta_{ij}, \ i, j = 1, 2, \ldots, 2n \tag{7.251}$$

respectively, where δ_{ij} is the Kronecker delta. Because the eigenvalues λ_i and the right and left eigenvectors $\mathbf{x}_i$ and $\mathbf{y}_j$ respectively, are in general complex, the normalization process indicated by Eqs. (7.250) and (7.251) is not nearly as simple as for real symmetric matrices.

The biorthogonality property possessed by the right eigenvectors $\mathbf{x}_i$ $(i=1, 2, \ldots, 2n)$ and left eigenvectors $\mathbf{y}_j$ $(j = 1, 2, \ldots, 2n)$ of the nonsymmetric matrix A forms the basis for a modal analysis for the response of systems with arbitrary viscous damping in a manner reminiscent of the modal analysis for undamped systems. To substantiate this statement, we consider a more general expansion theorem in the state space. To this end, we assume that an arbitrary $2n$-dimensional state vector $\mathbf{v}$, generally complex can be expressed as the linear combination

$$\mathbf{v} = a_1\mathbf{x}_1 + a_2\mathbf{x}_2 + \ldots + a_{2n}\mathbf{x}_{2n} = \sum_{i=1}^{2n} a_i\mathbf{x}_i \tag{7.252}$$

Then, premultiplying Eq. (7.252) by $\mathbf{y}_j^T$ and using Eqs. (7.250), we can write

$$a_i = \mathbf{y}_i^T\mathbf{v},\ i = 1, 2, \ldots, 2n \tag{7.253}$$

Moreover, premultiplying Eq. (7.252) by $\mathbf{y}_j^T A$ and considering Eqs. (7.253), we have

$$\lambda_i a_i = \mathbf{y}_i^T A\mathbf{v},\ i = 1, 2, \ldots, 2n \tag{7.254}$$

We refer to Eqs. (7.252)–(7.254) as a *state space expansion theorem*. It is clear that, before the state space expansion theorem can be used, it is necessary to solve the eigenvalue problem for real nonsymmetric matrices twice, once for A and once for A^T.

For future reference, we recast the preceding developments in compact matrix form. To this end, we introduce the matrix of eigenvalues $\Lambda = \text{diag}[\lambda_1\ \lambda_2\ \ldots\ \lambda_{2n}]$, of right eigenvectors $X = [\mathbf{x}_1\ \mathbf{x}_2\ \ldots\ \mathbf{x}_{2n}]$ and of left eigenvectors $Y = [\mathbf{y}_1\ \mathbf{y}_2\ \ldots \mathbf{y}_{2n}]$. Then, Eqs. (7.241) can be rewritten as

$$AX = X\Lambda \tag{7.255}$$

and Eqs. (7.242) as

$$A^T Y = Y\Lambda \tag{7.256}$$

Similarly, the biorthonormality relations, Eqs. (7.250) and (7.251), take the form

$$Y^T X = I \tag{7.257}$$

and

$$Y^T AX = \Lambda \tag{7.258}$$

respectively. Postmultiplying Eq. (7.257) by X^{-1}, we have

$$Y^T = X^{-1} \tag{7.259}$$

so that, premultiplying Eq. (7.259) by X, we conclude that

$$XY^T = I \tag{7.260}$$

Then, premultiplying Eq. (7.258) by X and postmultiplying by $X^{-1} = Y^T$, we obtain

$$A = X\Lambda Y^T \tag{7.261}$$

which represents a decomposition of A in terms of a product of the matrix of right eigenvector, the matrix of eigenvalues and the matrix of left eigenvectors transposed. Finally, the state space expansion theorem, Eqs. (7.252)–(7.254) have the matrix form

$$\mathbf{v} = X\mathbf{a} \tag{7.262}$$

where the coefficient vector is given by

$$\mathbf{a} = Y^T \mathbf{v} \tag{7.263}$$

as well as by

$$\Lambda \mathbf{a} = Y^T A \mathbf{v} \tag{7.264}$$

The expansion theorem forms the basis for a state space modal analysis. To this end, we assume a solution of the state equations, Eq. (7.233), as a linear combination of the right eigenvectors multiplied by modal coordinates as follows:

$$\mathbf{x}(t) = \xi_1(t)\mathbf{x}_1 + \xi_2(t)\mathbf{x}_2 + \ldots + \xi_{2n}(t)\mathbf{x}_{2n} = \sum_{r=1}^{2n} \xi_r(t)\mathbf{x}_r = X\boldsymbol{\xi}(t) \tag{7.265}$$

Inserting Eq. (7.265) into Eq. (7.233), premultiplying through by Y^T and recognizing that X is a constant matrix, we have simply

$$Y^T X\dot{\boldsymbol{\xi}}(t) = Y^T AX\boldsymbol{\xi}(t) + Y^T B\mathbf{Q}(t) \tag{7.266}$$

Then, using the orthonormality relations, Eqs. (7.257) and (7.258), we obtain

$$\dot{\boldsymbol{\xi}}(t) = \Lambda\boldsymbol{\xi}(t) + \mathbf{n}(t) \tag{7.267}$$

where

$$\mathbf{n}(t) = Y^T B\mathbf{Q}(t) \tag{7.268}$$

Equation (7.267) is recognized as a set of independent modal equations of the form

$$\dot{\xi}_r(t) = \lambda_r \xi_r(t) + n_r(t), \ r = 1, 2, \ldots, 2n \tag{7.269}$$

in which

$$n_r(t) = \mathbf{y}_r^T B\mathbf{Q}(t), \ r = 1, 2, \ldots, 2n \tag{7.270}$$

are modal excitations.

As in Sec. 7.15, we consider first the response to harmonic excitations. To this end, we assume that the excitation is given by

$$\mathbf{Q}(t) = \mathbf{Q}_0 e^{i\alpha t} \tag{7.271}$$

where, as in Eq. (7.222), $\mathbf{Q}_0$ is a vector of force amplitudes and α is the excitation frequency. Introducing Eq. (7.271) in Eqs. (7.270), we obtain the modal forces

$$n_r(t) = \mathbf{y}_r^T B\mathbf{Q}_0 e^{i\alpha t}, \ r = 1, 2, \ldots, 2n \tag{7.272}$$

Then, we can write the solution of Eqs. (7.269) in the form

$$\xi_r(t) = \Xi_r(i\alpha)e^{i\alpha t}, \ r = 1, 2, \ldots, 2n \tag{7.273}$$

Inserting Eqs. (7.273) into Eqs. (7.269), we obtain

$$(i\alpha - \lambda_r)\Xi_r(i\alpha)e^{i\alpha t} = \mathbf{y}_r^T B\mathbf{Q}_0 e^{i\alpha t}, \ r = 1, 2, \ldots, 2n \tag{7.274}$$

from which we conclude that

$$\Xi_r(i\alpha) = \frac{\mathbf{y}_r^T B\mathbf{Q}_0}{i\alpha - \lambda_r}, \ r = 1, 2, \ldots, 2n \tag{7.275}$$

It follows that the modal coordinates are simply

$$\xi_r(t) = \frac{\mathbf{Y}_r^T B\mathbf{Q}_0}{i\alpha - \lambda_r} e^{i\alpha t}, \quad r = 1, 2, \ldots, 2n \tag{7.276}$$

Finally, inserting Eqs. (7.276) into Eq. (7.265), we obtain the state vector

$$\mathbf{x}(t) = \sum_{r=1}^{2n} \frac{\mathbf{y}_r^T B\mathbf{Q}_0}{i\alpha - \lambda_r} \mathbf{x}_r e^{i\alpha t} \tag{7.277}$$

Of course, as in Sec. 7.15, if the excitation is $\mathbf{Q}(t) = \mathbf{Q}_0 \cos \alpha t$, we retain $\mathrm{Re}\,\mathbf{x}(t)$ as the response, and if the excitation is $\mathbf{Q}(t) = \mathbf{Q}_0 \sin \alpha t$, we retain $\mathrm{Im}\,\mathbf{x}(t)$. We recall that the upper half of $\mathbf{x}(t)$ represents displacements and the lower half consists of velocities.

Next, we consider the response to arbitrary excitations. To this end, we begin with the solution of Eqs. (7.269), which can be obtained conveniently by the Laplace transformation method (Appendix B). Indeed, Laplace transforming both sides of Eqs. (7.269), we have

$$s\,\Xi_r(s) - \xi_r(0) = \lambda_r \Xi_r(s) + N_r(s), \quad r = 1, 2, \ldots, 2n \tag{7.278}$$

where $\Xi_r(s) = \mathcal{L}\xi_r(t)$ and $N_r(s) = \mathcal{L}n_r(t)$ are the Laplace transforms of $\xi_r(t)$ and $n_r(t)$, respectively, and $\xi_r(0)$ is the initial modal coordinate. The latter can be obtained by letting $t = 0$ in Eq. (7.265), premultiplying by $\mathbf{y}_r^T$ and using the orthonormality relations, Eqs. (7.250); the result is

$$\xi_r(0) = y_r^T \mathbf{x}(0), \quad r = 1, 2, \ldots, 2n \tag{7.279}$$

where $\mathbf{x}(0)$ is the initial state vector. From Eqs. (7.278), we can write the transformed modal coordinates

$$\Xi_r(s) = \frac{1}{s - \lambda_r} \xi_r(0) + \frac{N_r(s)}{s - \lambda_r}, \quad r = 1, 2, \ldots, 2n \tag{7.280}$$

Then, from Appendix B, the inverse Laplace transform is

$$\xi_r(t) = \mathcal{L}^{-1} \Xi_r(s) = e^{\lambda_r t} \xi_r(0) + \int_0^t e^{\lambda_r (t-\tau)} n_r(\tau) d\tau, \quad r = 1, 2, \ldots, 2n \tag{7.281}$$

Equations (7.281) can be combined into

$$\boldsymbol{\xi}(t) = e^{\Lambda t} \boldsymbol{\xi}(0) + \int_0^t e^{\Lambda(t-\tau)} \mathbf{n}(\tau) d\tau \tag{7.282}$$

The actual solution is obtained by inserting Eq. (7.282) into Eq. (7.265), with the result

$$\mathbf{x}(t) = X e^{\Lambda t} \boldsymbol{\xi}(0) + \int_0^t X e^{\Lambda(t-\tau)} \mathbf{n}(\tau) d\tau \tag{7.283}$$

Then, recognizing that Eqs. (7.270) and (7.279) can be written in the compact form

$$\mathbf{n}(t) = Y^T B\mathbf{Q}(t) \tag{7.284}$$

and

$$\boldsymbol{\xi}(0) = Y^T \mathbf{x}(0) \tag{7.285}$$

respectively, we obtain the state response

$$\mathbf{x}(t) = Xe^{\Lambda t}Y^T\mathbf{x}(0) + \int_0^t Xe^{\Lambda(t-\tau)}Y^T B\mathbf{Q}(\tau)d\tau \tag{7.286}$$

We refer to the process of solving the simultaneous state equations of motion, Eq. (7.233), by first solving twin state algebraic eigenvalue problems, then using the state space expansion theorem to decouple the state equations, solving the resulting independent modal equations and finally combining the modal solutions into Eq. (7.286) as the *state space modal analysis*.

Equation (7.286) can be expressed in a different form. To this end, we observe that $e^{\Lambda t}$ can be expanded in the infinite series

$$e^{\Lambda t} = I + t\Lambda + \frac{t^2}{2!}\Lambda^2 + \frac{t^3}{3!}\Lambda^3 + \ldots \tag{7.287}$$

so that, premultiplying Eq. (7.287) by X and postmultiplying by Y^T and considering Eqs. (7.257), (7.260) and (7.261), we can write

$$\begin{aligned} Xe^{\Lambda t}Y^T &= XY^T + tX\Lambda Y^T + \frac{t^2}{2!}X\Lambda Y^T X\Lambda Y^T + \frac{t^3}{3!}X\Lambda Y^T X\Lambda Y^T X\Lambda Y^T + \ldots \\ &= I + tA + \frac{t^2}{2!}A^2 + \frac{t^3}{3!}A^3 + \ldots = e^{At} \end{aligned} \tag{7.288}$$

Hence, introducing Eq. (7.288) in Eq. (7.286), we obtain the state vector in the compact form

$$\mathbf{x}(t) = \Phi(t)\mathbf{x}(0) + \int_0^t \Phi(t,\tau)B\mathbf{F}(\tau)d\tau \tag{7.289}$$

where

$$\Phi(t,\tau) = e^{A(t-\tau)} \tag{7.290}$$

is known as the *state transition matrix*; it represents a series obtained from Eq. (7.287) by replacing t by $t-\tau$.

Although we obtained solution (7.289) beginning with a modal solution, the same solution can be obtained by a more direct approach (Ref. 13), without help from modal analysis. This implies that, to obtain solution (7.289), it is not really necessary to solve eigenvalue problems.

Whereas both the modal solution, Eq. (7.286), and the transition matrix solution, Eq. (7.289), represent analytical solutions, they must in fact be evaluated numerically. Indeed, Eq. (7.286) requires the evaluation of $e^{\Lambda t}$, as well as a convolution integral involving $e^{\Lambda(t-\tau)}$ and $\mathbf{Q}(\tau)$, in addition to the solution of two algebraic eigenvalue problems for nonsymmetric matrices. Similarly, although Eq. (7.289) does not require the solution of eigenvalue problems, it does require the computation of the transition matrix e^{At} and a convolution integral involving $e^{A(t-\tau)}$ and $\mathbf{Q}(\tau)$.

The transition matrix represents an infinite matrix series, and for practical reasons its computation necessitates truncation. The number of terms to be retained in the truncated series is dictated first of all by the accuracy desired. Then, for the chosen accuracy, the number of terms depends on the eigenvalue of largest magnitude, in addition to the time t. To substantiate this statement, we observe from Eq. (7.288) that the behavior of e^{At} is the same as the behavior of $e^{\Lambda t}$, where the latter has the form

$$e^{\Lambda t} = \text{diag}[e^{\lambda_1 t} \; e^{\lambda_2 t} \; \ldots \; e^{\lambda_{2n} t}] \tag{7.291}$$

But, the eigenvalues are in general complex, so that a typical eigenvalue can be expressed as

$$\lambda_r = |\lambda_r| e^{i\phi_r}, \quad r = 1, 2, \ldots, 2n \tag{7.292}$$

where $|\lambda_r|$ is the magnitude and ϕ_r the phase angle of λ_r. Hence, the rth diagonal element of $e^{\Lambda t}$ is given by the series

$$\begin{aligned} e^{\lambda_r t} &= 1 + \lambda_r t + \frac{1}{2!}(\lambda_r t)^2 + \frac{1}{3!}(\lambda_r t)^3 + \ldots \\ &= 1 + |\lambda_r| t e^{i\phi_r} + \frac{1}{2!}(|\lambda_r| t)^2 e^{2i\phi_r} + \frac{1}{3!}(|\lambda_r| t)^3 e^{3i\phi_r} + \ldots \end{aligned} \tag{7.293}$$

Next, we consider the complex plane of Fig. 7.16 and observe that $\lambda_r = |\lambda_r| e^{i\phi_r}$, $\lambda_r^2 = |\lambda_r|^2 e^{2i\phi_r}$, $\lambda_r^3 = |\lambda_r|^3 e^{3i\phi_r}$, ... represent complex vectors of magnitude $|\lambda_r|$, $|\lambda_r|^2$, $|\lambda_r|^3$, ... and phase angle ϕ_r, $2\phi_r$, $3\phi_r$, ... relative to the real axis, respectively. Indeed, $e^{i\phi_r}$, $e^{2i\phi_r}$, $e^{3i\phi_r}$, ... are all complex unit vectors, i.e., they all have magnitude equal to 1 and they differ in direction only. Because the term $s+1$ in series (7.293) is divided by the factorial $s!$, the series is guaranteed to converge, as $s!$ will increase eventually faster than $(|\lambda_r| t)^s$, where the number of terms in series (7.293) required to achieve the desired accuracy depends on the value of $|\lambda_r| t$. It follows that, for a given time t, the number of terms required depends on the magnitude $|\lambda_{2n}|$ of the highest eigenvalue λ_{2n}.

The transition matrix possesses a very interesting property, one that has significant computational ramifications as well. This property is known as the *group property*, and

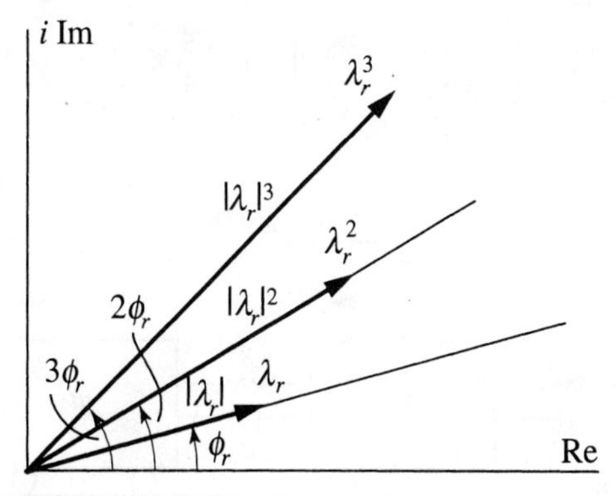

FIGURE 7.16
Geometric interpretation of the vectors, $\lambda_r, \lambda_r^2, \lambda_r^3 \ldots$ in the complex plane

can be stated as

$$\Phi(t_3, t_1) = \Phi(t_3, t_2)\Phi(t_2, t_1) \tag{7.294}$$

The validity of Eq. (7.294) can be verified by considering Eq. (7.290). The implication of Eq. (7.294) is that, if the time interval is divided into two subintervals $t_2 - t_1$ and $t_3 - t_2$, then the transition matrix corresponding to $t_3 - t_1$ is equal to the product of the transition matrices corresponding to $t_3 - t_2$ and $t_2 - t_1$. Generally, t_2 bisects the interval $t_3 - t_1$, so that $t_2 = (t_1 + t_3)/2$. Then, the computation of $\Phi(t_3, t_1)$ by computing first the matrix $\Phi(t_3, (t_1 + t_3)/2) = \Phi((t_1 + t_3)/2, t_1) = \Phi(t_3/2, t_1/2)$ and then squaring it can be shown to be more economical than by computing $\Phi(t_3, t_1)$ directly. Of course, significant computational advantage accrues by dividing the time interval $t_3 - t_1$ into a relatively large number k of very small subintervals $\Delta t = (t_3 - t_1)/k$, where Δt is sufficiently small that $e^{A\Delta t}$ can be computed with only three or four terms.

Example 7.12. Determine the response of the two-degree-of-freedom system shown in Fig. 7.17 to the excitation

$$Q_1(t) = 0,\ Q_2(t) = Q_0[t\mathcal{u}(t) - (t-4)\mathcal{u}(t-4)] \tag{a}$$

by means of the approach based on the transition matrix, Eq. (7.289), where $\mathcal{u}(t)$ is the unit step function. The initial conditions are zero and the system parameters have the values

$$m_1 = 1\ \text{kg},\ m_2 = 2\ \text{kg},\ c_1 = c_2 = 0.8\ \text{N}\cdot\text{s/m},\ k_1 = 1\ \text{N/m},\ k_2 = 4\ \text{N/m} \tag{b}$$

The equations of motion have the form given by Eq. (7.191), in which

$$\mathbf{q}(t) = [q_1(t)\ q_2(t)]^T,\ \mathbf{Q}(t) = [Q_1(t)\ Q_2(t)]^T = [0\ \ Q_0[t\mathcal{u}(t) - (t-4)\mathcal{u}(t-4)]] \tag{c}$$

are the configuration vector and force vector, respectively, and

$$M = \begin{bmatrix} m_1 & 0 \\ 0 & m_2 \end{bmatrix} = \begin{bmatrix} 1 & 0 \\ 0 & 2 \end{bmatrix}$$
$$C = \begin{bmatrix} c_1 + c_2 & -c_2 \\ -c_2 & c_2 \end{bmatrix} = \begin{bmatrix} 1.6 & -0.8 \\ -0.8 & 0.8 \end{bmatrix} \tag{d}$$
$$K = \begin{bmatrix} k_1 + k_2 & -k_2 \\ -k_2 & k_2 \end{bmatrix} = \begin{bmatrix} 5 & -4 \\ -4 & 4 \end{bmatrix}.$$

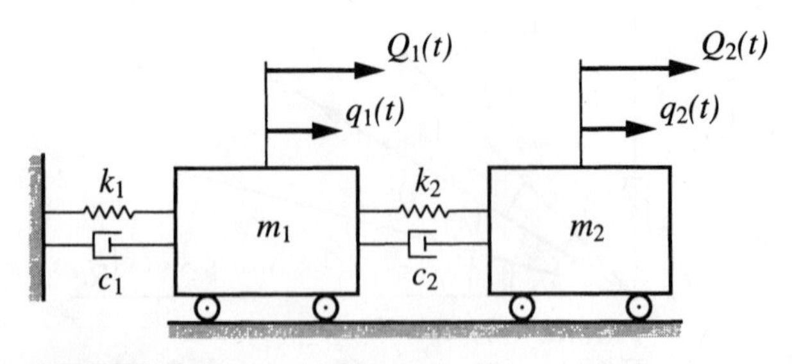

FIGURE 7.17
Damped two-degree-of-freedom system

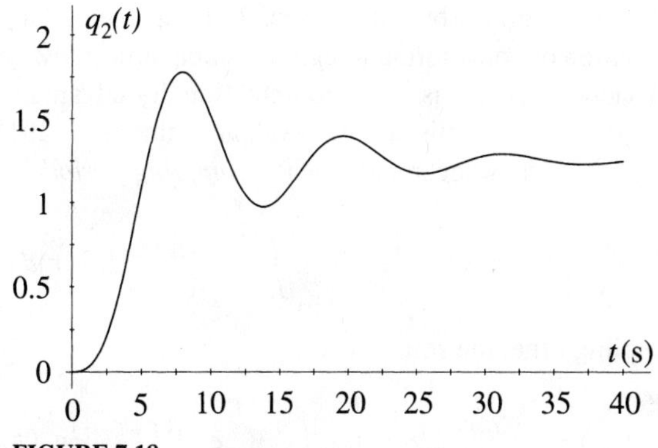

FIGURE 7.18
Response of m_2 for the system of Fig. 7.17

are the mass, damping and stiffness matrices. The equations can be rewritten in the state form given by Eq. (7.233), in which the state vector is simply

$$\mathbf{x}(t) = [q_1(t)\ q_2(t)\ \dot{q}_1(t)\ \dot{q}_2(t)]^T \tag{e}$$

Moreover, using Eqs. (7.216), the coefficient matrices are

$$A = \left[\begin{array}{c|c} 0 & I \\ \hline -M^{-1}K & -M^{-1}C \end{array}\right] = \begin{bmatrix} 0 & 0 & 1 & 0 \\ 0 & 0 & 0 & 1 \\ -5 & 4 & -1.6 & 0.8 \\ 2 & -2 & 0.4 & -0.4 \end{bmatrix}$$
$$B = \left[\begin{array}{c} 0 \\ \hline M^{-1} \end{array}\right] = \begin{bmatrix} 0 & 0 \\ 0 & 0 \\ 1 & 0 \\ 0 & 0.5 \end{bmatrix} \tag{f}$$

The response was computed using Eq. (7.289) with $\mathbf{x}(0) = \mathbf{0}$. Note that in this particular case the integral in Eq. (7.289) can be evaluated analytically. Figure 7.18 shows the plot $q_2(t)$ versus t.

7.17 DISCRETE-TIME SYSTEMS

In Sec. 7.16, we studied the response of multi-degree-of-freedom systems with arbitrary viscous damping by means of state space techniques. In particular, we obtained the response in two different ways. In the first approach, we did it by means of a state space modal analysis amounting to solving two adjoint eigenvalue problems in the state space for the right and left eigenvectors and using the two sets of eigenvector in conjunction with a state space expansion theorem to decouple the state equations. The solution of the resulting first-order equations were solved with ease by the Laplace transformation method. The second approach is based on the state transition matrix and does not require the solution of eigenvalue problems.

Although the solutions obtained in Sec. 7.16 represent analytical solutions, they involve the evaluation of convolution integrals, which almost always requires numerical integration. In view of this, it is often computationally advantageous to carry out the solutions in discrete time. To this end, we evaluate solution (7.281) at the discrete time $t = t_k = kT$, where k is an integer and T is the *sampling period* (Sec. 4.9), and write

$$\xi_r(kT) = e^{\lambda_r kT}\xi_r(0) + \int_0^{kT} e^{\lambda_r(kT-\tau)} n_r(\tau)d\tau \tag{7.295}$$

At the next sampling time, the solution is

$$\xi_r(kT+T) = e^{\lambda_r(kT+T)}\xi_r(0) + \int_0^{kT+T} e^{\lambda_r(kT+T-\tau)} n_r(\tau)d\tau \tag{7.296}$$

Then, assuming that the sampling period is sufficiently small that $n_r(\tau)$ can be regarded as being constant and equal to $n_r(kT)$ over the interval $kT < \tau < kT+T$, we can approximate $\xi_r(kT+T)$ as follows:

$$\begin{aligned}\xi_r(kT+T) &= e^{\lambda_r T}\left[e^{\lambda_r kT}\xi_r(0) + \int_0^{kT} e^{\lambda_r(kT-\tau)} n_r(\tau)d\tau\right] + \int_{kT}^{kT+T} e^{\lambda_r(kT+T-\tau)} n_r(\tau)d\tau \\ &\cong e^{\lambda_r T}\xi_r(kT) + n_r(kT)\int_{kT}^{kT+T} e^{\lambda_r(kT+T-\tau)} d\tau \end{aligned} \tag{7.297}$$

Next, we evaluate the integral on the right side of Eq. (7.297) by using the change of variables

$$\begin{aligned} &kT + T - \tau = \sigma, \; d\tau = -d\sigma \\ &\tau = kT \rightarrow \sigma = T, \; \tau = kT + T \rightarrow \sigma = 0 \end{aligned} \tag{7.298}$$

so that

$$\begin{aligned}\int_{kT}^{kT+T} e^{\lambda_r(kT+T-\tau)} d\tau &= \int_T^0 e^{\lambda_r \sigma}(-d\sigma) = \int_0^T e^{\lambda_r \sigma} d\sigma \\ &= \left.\frac{e^{\lambda_r \sigma}}{\lambda_r}\right|_0^T = \lambda_r^{-1}(e^{\lambda_r T} - 1) \end{aligned} \tag{7.299}$$

Hence, inserting Eq. (7.299) into Eq. (7.297), and omitting the sampling period T from the argument to simplify the notation, we obtain the recursive relation

$$\xi_r(k+1) = e^{\lambda_r T}\xi_r(k) + \lambda_r^{-1}(e^{\lambda_r T} - 1)n_r(k), \; k = 0, 1, 2, \ldots \tag{7.300}$$

Equation (7.300) gives the $2n$ modal coordinates ξ_r $(r = 1, 2, \ldots, 2n)$ in discrete time. Recalling the notation $\boldsymbol{\xi} = [\xi_1 \; \xi_2 \; \ldots \; \xi_{2n}]^T$ and $\Lambda = \text{diag}[\lambda_1 \; \lambda_2 \; \ldots \; \lambda_{2n}]$ from Sec. 7.16, we can rewrite the equation in the vector form

$$\boldsymbol{\xi}(k+1) = e^{\Lambda T}\boldsymbol{\xi}(k) + \Lambda^{-1}(e^{\Lambda T} - I)\mathbf{n}(k), \; k = 0, 1, 2, \ldots \tag{7.301}$$

Then, premultiplying Eq. (7.301) by the matrix X of right eigenvectors and using Eqs. (7.265), (7.268) and (7.279), we obtain

$$\mathbf{x}(k+1) = Xe^{\Lambda T}Y^T\mathbf{x}(k) + X\Lambda^{-1}(e^{\Lambda T} - I)Y^T B\mathbf{Q}(k),\ k = 0, 1, 2, \ldots \tag{7.302}$$

where Y is the matrix of left eigenvectors, B is a matrix of coefficients defined by the second of Eqs. (7.234) and $\mathbf{Q}(k)$ is the force vector evaluated at $t = kT$. Finally, recalling Eqs. (7.258)–(7.261) and (7.288), we can rewrite the discrete-time response of viscously damped multi-degree-of-freedom systems in the recursive form

$$\mathbf{x}(k+1) = \Phi\mathbf{x}(k) + \Gamma\mathbf{Q}(k),\ k = 0, 1, 2, \ldots \tag{7.303}$$

in which

$$\Phi = e^{AT} \tag{7.304}$$

is the *discrete-time transition matrix* and

$$\Gamma = A^{-1}(e^{AT} - I)B \tag{7.305}$$

where A is the coefficient matrix given by the first of Eqs. (7.234).

From the preceding discussion, we conclude that, as in continuous time, there are two options for computing the response of multi-degree-of-freedom systems in discrete time, namely, through the modal approach given by Eq. (7.302), or by the discrete-time transition matrix approach given by Eq. (7.303). Use of Eq. (7.302) requires the solution of two adjoint eigenvalue problems in the state space for X, Y and Λ, as well as the computation of $e^{\Lambda T}$. On the other hand, use of Eq. (7.303) requires the computation of the discrete-time transition matrix $\Phi = e^{AT}$. In both cases, there is the task of choosing the sampling period T. As pointed out in Sec. 7.16, for a given desired accuracy, this choice affects the number of terms necessary to compute either $e^{\Lambda T}$ or e^{AT}, as the case may be, and it depends on the magnitude $|\lambda_{2n}|$ of the highest eigenvalue. We observe that the additional task of solving eigenvalue problems in the modal approach is balanced by the significantly smaller effort required to compute $e^{\Lambda T}$ than to compute e^{AT}. Indeed, whereas $e^{\Lambda T}$ represents $2n$ scalar series, e^{AT} represents a $2n \times 2n$ matrix series. Of course, the number of terms necessary for convergence is the same for e^{AT} as for $e^{\Lambda T}$, because this number is controlled by $|\lambda_{2n}|$ in both cases.

Example 7.13. Solve the two adjoint eigenvalue problems for the viscously damped two-degree-of-freedom system of Example 7.12 and determine the sampling period T permitting the computation of $e^{\Lambda T}$, or e^{AT}, with six decimal places accuracy using only five terms in the series. Then, obtain the discrete-time response of the system and plot $q_2(k)$ versus k for $k = 0, 1, \ldots, 8/T$. Compare the discrete-time response with the continuous-time response obtained in Example 7.12 and draw conclusions.

From Example 7.12, the continuous-time coefficient matrices are as follows:

$$A = \begin{bmatrix} 0 & 0 & 1 & 0 \\ 0 & 0 & 0 & 1 \\ -5 & 4 & -1.6 & 0.8 \\ 2 & -2 & 0.4 & -0.4 \end{bmatrix}, \quad B = \begin{bmatrix} 0 & 0 \\ 0 & 0 \\ 1 & 0 \\ 0 & 0.5 \end{bmatrix} \tag{a}$$

The solution of the two adjoint eigenvalue problems associated with the matrix A consists of the eigenvalues

$$\begin{matrix}\lambda_1\\ \lambda_2\end{matrix} = -0.110404 \pm 0.538225i, \quad \begin{matrix}\lambda_3\\ \lambda_4\end{matrix} = -0.889596 \pm 2.145344i \tag{b}$$

matrix of right eigenvectors

$$X = \begin{bmatrix} 0.568530 - 0.029723i & 0.568530 + 0.029723i & -0.116178 - 0.315435i & -0.116178 + 0.315435i \\ 0.666343 + 0i & 0.666343 + 0i & 0.075734 + 0.111403i & 0.075734 - 0.111403i \\ -0.046770 + 0.309278i & -0.046770 - 0.309278i & 0.865236 + 0i & 0.865236 + 0i \\ -0.073567 + 0.358642i & -0.073567 - 0.353642i & -0.336448 + 0.083820i & -0.336448 - 0.083820i \end{bmatrix} \tag{c}$$

and matrix of left eigenvectors

$$Y = \begin{bmatrix} 0.196901 + 0.155541i & 0.196001 - 0.155541i & 0.712202 + 0i & 0.712202 + 0i \\ -0.040937 + 0.440411i & -0.040937 - 0.440411i & -0.603937 + 0.006400i & -0.603937 - 0.006400i \\ 0.337875 - 0.017664i & 0.337875 + 0.017664i & 0.111770 - 0.255805i & 0.111770 + 0.255805i \\ 0.792011 + 0i & 0.792011 + 0i & -0.037362 + 0.220595i & -0.037362 - 0.220595i \end{bmatrix} \tag{d}$$

For $T = 0.066$ s, we obtain an error in e^{AT} equal to 0.942999×10^{-6}, so that we choose $T = 0.066$ s as the sampling period. Introducing this value of T in Eqs. (7.304) and (7.305), we obtain the discrete-time coefficient matrices

$$\Phi = e^{AT} = \begin{bmatrix} 0.989581 & 0.008319 & 0.062425 & 0.001849 \\ 0.004215 & 0.995765 & 0.000925 & 0.065059 \\ -0.308425 & 0.246000 & 0.890442 & 0.057519 \\ 0.125496 & -0.126420 & 0.028760 & 0.970481 \end{bmatrix} \tag{e}$$

and

$$\Gamma = A^{-1}(e^{AT} - I)B = \begin{bmatrix} 14.2021 & 6.60002 \\ 16.2000 & 7.35108 \\ -3.93758 & -1.99908 \\ -3.99908 & -1.96747 \end{bmatrix} \tag{f}$$

respectively. Finally, corresponding to the time interval of 8 s used in Example 7.12, we conclude that $8/T \cong 121$, so that $k = 0, 1, \ldots, 121$. The discrete-time response is obtained by using Eq. (7.303) in conjunction with the values Φ and Γ given by Eqs. (e) and (f), respectively, as well as the discretized version of the continuous-time excitation vector given by the second of Eqs. (c) of Example 7.12. The discrete-time response is plotted in Fig. 7.18 and is indistinguishable from the exact continuous-time response.

7.18 SOLUTION OF THE EIGENVALUE PROBLEM. MATLAB PROGRAMS

For systems with three or more degrees of freedom, the algebraic eigenvalue problem is essentially a numerical problem. It has been the subject of intense interest on the part of linear algebraists over the last five decades. Before the development of high-speed digital computers, the emphasis has been on efficient numerical algorithms in an effort to save computer time. These efforts have been very successful, resulting in a large variety of algorithms. In this section, we first survey some of the most widely known algorithms. Then, we present two MATLAB programs, one for conservative systems and the other for nonconservative systems.

In Sec. 7.6, we introduced the eigenvalue problem for conservative systems and solved it by finding the roots of the characteristic polynomial. For a two-degree-of-freedom system, this amounts to solving a quadratic equation, a very simple task. For three-degrees-of-freedom and higher, finding the roots of a characteristic polynomial of an order equal to the number of degrees of freedom cannot compete in computational efficiency with other numerical techniques for solving the eigenvalue problem. Hence, we limit our survey to other techniques.

Virtually all algorithms for solving the algebraic eigenvalue problem are iterative in nature. We consider first conservative systems and assume that the eigenvalue problem has been reduced to one in terms of a single real symmetric matrix A, perhaps by the Cholesky decomposition (Sec. 7.11). One of the oldest methods is matrix iteration by the power method. The process iterates to the modes in descending order of magnitude of the eigenvalues and accuracy is lost with each computed mode. The method has some academic value, but is not recommended as a computational tool, except when the interest lies in a small number of modes. The Jacobi method reduces the matrix A to diagonal form through successive rotations of the type discussed in Sec. 7.12, each time annihilating an off-diagonal entry and its symmetric counterpart. At convergence, A reduces to the diagonal matrix of the eigenvalues and the product of rotation matrices reduces to the orthonormal matrix of eigenvectors. Perhaps the most widely used method for the computation of the eigenvalues of a matrix is the QR method. Before the QR method becomes competitive, it is necessary to tridiagonalize the matrix, which can be done by either Givens' or by Householder's method, and to introduce shifts in the eigenvalues. Then, the eigenvectors corresponding to the computed eigenvalues can be computed by inverse iteration, which involves solving sets of algebraic equation by means of Gaussian elimination and back substitution.

In the case of nonconservative systems, the matrix A is nonsymmetric and the eigensolutions are generally complex. One algorithm capable of solving eigenvalue problems for nonsymmetric matrices is the power method, modified to handle complex eigensolutions and to compute both right and left eigenvectors. Its main advantage is that it is able to produce partial solutions. Far more widely used is the QR method, again modified to accommodate complex eigenvalues. Note that, before use of the QR method is considered, it is virtually necessary to reduce the nonsymmetric matrix A to Hessenberg form, one in which all the entries below the first subdiagonal are zero. Then, the eigenvectors can be computed by inverse iteration suitably modified to produce right and left complex eigenvectors. Details of all these algorithms, as well as of other algorithms, can be found in Ref. 13.

At this point, with the speed of digital computers increasing at a dizzying pace, efficiency of an algorithm takes a back seat to ease of programming. In this regard, MATLAB is the likely choice, as it provides subroutines that can be readily incorporated into programs. The MATLAB program 'evpc.m' for solving the eigenvalue problem for conservative systems, characterized by real symmetric matrices A, reads as follows:

```
% The program 'evpc.m' solves the eigenvalue problem for conservative systems

clear
clf

M=[1 0 0;0 1 0;0 0 2]; % mass matrix
K=[2 -1 0; -1 3 -2;0 -2 2]; % stiffness matrix
N=3; % dimension of the eigenvalue problem
R=chol (M); % Cholesky decomposition using MATLAB function; R is upper
% triangular
L=R'; % lower triangular matrix as specified by Eq. (7.143)
A=inv(L) *K*inv(L') % coefficient matrix according to Eq. (7.148)
[x,W]=eig(A) % solution of the eigenvalue problem using MATLAB function
v=inv(L')*x % transformation to modal matrix according to Eq. (7.153)

for i=1:N,
  w1(i)=sqrt(W(i, i)); % setting the natural frequencies in an N-dimensional vector
end
[w, I]=sort(w1) % arranging the natural frequencies in ascending order

for j=1:N,
  U(:, j)=v(:, I(j)) % arranging the modal vectors in ascending order
end
```

Note that the program has been written to handle fully populated, albeit real symmetric and positive definite, mass matrices M. In the numerical example at hand, M is diagonal, so that the matrix L in the Cholesky decomposition is simply equal to $M^{1/2}$. The MATLAB subroutine $[x, W] = \text{eig}(A)$ yields unit eigenvectors $\mathbf{x}_i$ and eigenvalues W_i equal to ω_i^2 $(i = 1, 2, \ldots, N)$, but the eigenvalues are not in ascending order of magnitude. The program transforms the eigenvectors $\mathbf{x}_i$ into the modal vectors $\mathbf{u}_i$, which are orthonormal with respect to M, and computes the natural frequencies and sets them in a vector format. Then, the natural frequencies are arranged in ascending order of magnitude and the modal vectors are made to conform to that order. The program is set to solve the eigenvalue problem for the system of Example 7.5. Note that the eigenvectors obtained in Example 7.5 are yet to be rendered orthonormal with respect to the mass matrix M, which is done in Example 7.7.

For nonconservative systems, such as systems with arbitrary viscous damping, it is necessary to formulate the eigenvalue problem in the state space. Of course, the dimension of the eigenvalue problem is twice that for conservative systems and the matrix A is nonsymmetric. This presents no problem, as the same MATLAB subroutine works also for nonsymmetric matrices A, but now it is necessary to solve the eigenvalue problem for A^T as well. Moreover, the eigensolutions are generally complex. The corresponding MATLAB program, entitled 'evpnc.m', is as follows:

```
% The program 'evpnc.m' solves the eigenvalue problem for nonconservative systems

clear
clf

M=[1 0;0 2]; % mass matrix
C=[1.6 -0.8;-0.8 0.8]; % damping matrix
K=[5 -4;-4 4]; % stiffness matrix
N=4 % dimension of the eigenvalue problem
A=[zeros(size(M)) eye(size(M));-inv(M)*K -inv(M)*C] % system matrix, first
% of Eqs. (7.234)
[U, D]=eig(A) % solution of the eigenvalue problem using MATLAB function
[V, D]=eig(A') % solution of the transposed eigenvalue problem
for k=1:N,
  d(k)=(D(k, k)) % setting the eigenvalues in an N-dimensional vector
end
[d, I]=sort(d) % arranging the eigenvalues in ascending order of magnitude
for j=1:N,
  X(:,j)=U(:, I(j)) % arranging the right eigenvectors in ascending order
  Y(:,j)=V(:, I(j)) % arranging the left eigenvectors in ascending order
end

B=Y'*X % normalization matrix
X=X*inv(B) % matrix of normalized right eigenvectors
Y'*X % check of satisfaction of Eq. (7.257)
Y'*A*X % check of satisfaction of Eq. (7.258)
```

We observe that in general there is some arbitrariness in the normalization process. In the present program, only the matrix X of right eigenvectors is being adjusted, and the matrix Y of left eigenvectors remains the same. Note that, following normalization, a verification of the orthonormality relations, Eqs. (7.257) and (7.258), is provided. The program is set to solve the eigenvalue problem for the system of Example 7.13. The eigenvalues to be obtained by this program are the same as those obtained in Example 7.13. However, from a comparison of corresponding eigenvectors, it would be difficult to conclude that they are the same. Before we can conclude that the eigenvectors are the same, it is necessary to express the complex vector components in exponential form. Then, two complex eigenvectors are the same if the magnitudes of the corresponding vector components are proportional and the phase angles differ by the same amount (see, for example, Ref. 13, Sec. 6.16).

7.19 RESPONSE TO INITIAL EXCITATIONS BY MODAL ANALYSIS USING MATLAB

The MATLAB program 'rspin3.m' first solves the eigenvalue problem for a conservative system and then uses the modal information to compute and plot the response to initial excitations. The program reads as follows:

```
% the program in file 'rspin3.m' plots the response of a conservative system
% to initial excitations computed by means of modal analysis

clear
clf

M=[1 0 0;0 1 0;0 0 2]; % mass matrix
K=[2 -1 0;-1 3 -2;0 -2 2]; % stiffness matrix
N=3; % dimension of the system
R=chol(M); % Cholesky decomposition using MATLAB function; R is upper
% triangular
L=R'; % lower triangular matrix as specified by Eq. (7.143)
A=inv(L) *K*inv(L'); % coefficient matrix according to Eq. (7.148)
[x,W]=eig(A); % solution of the eigenvalue problem using MATLAB function
v=inv(L')*x; % transformation to modal matrix according to Eq. (7.153)

for i=1:N,
  w1(i)=sqrt(W(i, i)); % setting the natural frequencies in an N-dimensional vector
end
[w, I]=sort(w1); % arranging the natural frequencies in ascending order

for j=1:N,
  U(:, j)=v(:, I(j)); % arranging the modal vectors in ascending order
end

q0=[1;2;3]; qdot0=[0;0;0]; % initial displacement and velocity vectors
t=[0:0.1:25]; % initial time, time increment, final time

q=zeros (N,size(t,2));
for k=1:N,
  q=q+((U(:, k)'*M*q0*U(:, k))*cos(w(k)*t)
      +(U(:, k)'*M*qdot0*U(:, k))*sin(w(k)*t)/w(k));
  % response vector according to Eq. (7.140)
end
plot (t,q)
title ('Response to Initial Excitations')
ylabel ('q_1(t),q_2(t),q_3(t)')
xlabel ('t(s)')
```

Note that, because of the way in which MATLAB treats the time t, namely, as a vector of discrete values, the statement for $\mathbf{q}(t)$ had to be changed by having the modal vector $\mathbf{u}_r$ trade places with $\cos\omega_r t$ and $\sin\omega_r t$. In this manner, $\mathbf{q}(t)$ plays the role of a matrix of dimensions $N \times$ (dimension of t), where in this particular case dimension of $t = 250$. The implication is that the response is processed in discrete time, although it exhibits all the characteristics of a continuous-time response. This program solves the same problem as that in Example 7.5.

Finally, we observe that the equation for $\mathbf{q}$ is typed here in two lines, because it does not fit in a single line. In a MATLAB Editor/Debugger it does fit in one line, and should be typed so (see also related comment in Sec. 5.12).

7.20 RESPONSE BY THE DISCRETE-TIME TRANSITION MATRIX USING MATLAB

For systems with arbitrary viscous damping, the response must be obtained in the state space, which implies the use of the transition matrix. Of course, if the response is to be evaluated on a computer, then the state equations must be transformed to discrete time. The MATLAB program 'dtrsp3.m' plots the response of an arbitrarily damped system in discrete time, and reads as follows:

```
% The program 'dtrsp3.m' plots the response of a damped system by the discrete-time
% transition matrix

clear
clf
M=[1 0;0 2]; % mass matrix
C=[1.6 -0.8;-0.8 0.8]; % damping matrix
K=[5 -4;-4 4]; % stiffness matrix
A=[zeros(size(M))) eye(size (M)); -inv(M) *K -inv(M)*C]; % system matrix, first of
% Eqs. (7.234)
B=[zeros (size(M)); inv(M)]; % coefficient matrix, second of Eqs. (7.234)
T0=4; % rise time for the forcing function
T=0.02; % sampling period
N=2000; % number of samplings
N0=T0/T; % number of samplings during rise time

Phi=eye (size (A))+T*A+T^2*A^2/2+T^3*A^3/6+T^4*A^4/24; % discrete-time
% transition matrix, Eq. (7.304)
Gamma=inv(A)*(Phi-eye(size(A)))*B; % discrete-time coefficient matrix, Eq. (7.305)
x(:,1)=zeros((2*size(M)),1);

for k=1:N,
  if k<N0+1; Q2(k)=(k-1)/N0; else; Q2(k)=1; % force on mass m2 discretized in time
  end
  Q(:,k)=[0;Q2(k)]; % force vector
  x(:,k+1)=Phi*x(:,k)+Gamma*Q(:,k); % discrete-time state equations
end

k=[0:1:N];
plot (k,x(1,:),'.',k,x(2,:),'.')
title ('System Response')
ylabel ('q_1(k),q_2(k)')
xlabel ('k')
```

The program solves the problem of Example 7.13. Note that it plots both displacements $q_1(t)$ and $q_2(t)$, whereas in Example 7.13 only $q_2(t)$ is plotted.

7.21 SUMMARY

The treatment of general multi-degree-of-freedom systems requires more advanced techniques than the treatment of two-degree-of-freedom systems studied in Ch. 5. In this regard, a blend of methods from vibrations and from linear system theory proves very effective. Concepts such as flexibility and stiffness influence coefficients play a useful role in the treatment of linear systems. In particular, the symmetry of these coefficients renders the determination of the response of conservative systems considerably simpler. Although quite often the derivation of the equations of motion can be carried out by means of Newton's second law, the method of choice is Lagrange's equations.

The vibration of linear systems is described by sets of second-order ordinary differential equations defined by mass, damping and stiffness coefficients. In matrix form, they are defined by mass, damping and stiffness matrices, respectively. In general, the equations are simultaneous, so that it is not possible to solve one equation independently of the other. In the case of conservative systems, the basic approach to the solution is to carry out a transformation rendering the equations independent, which amounts to simultaneous diagonalization of the mass and stiffness matrices. To demonstrate the approach, it is necessary to consider first the free vibration problem and then examine the circumstances under which the system admits synchronous motions, in which the masses form a certain pattern whose shape remains the same during motion and whose amplitude varies harmonically with time. To determine these configurations and the associated frequencies, commonly known as modal vectors and natural frequencies, it is necessary to solve the eigenvalue problem. The eigenvalue problem is a classical numerical problem in linear algebra and can be solved by a variety of efficient computation algorithms. The modal vectors possess the orthogonality property, an important property that permits the simultaneous diagonalization of the mass and stiffness matrices. When the system is not completely restrained, it can undergo rigid-body motions with zero frequencies. The response of a multi-degree-of-freedom system to initial excitations and external forces can be expressed as a linear combination of the modal vectors multiplied by time-dependent modal coordinates. Then, the orthogonality of the modal vectors can be used to reduce the equations of motion to a set of independent second-order differential equations for the modal coordinates, which can be solved by the methods of Chs. 2–4. This approach is generally known as modal analysis.

For many numerical algorithms, it is necessary to transform the eigenvalue problem in terms of two symmetric matrices M and K into one in terms of a single symmetric matrix A. The latter form can be used to show that the eigenvalue problem for a symmetric matrix can be interpreted geometrically as the problem of finding the principal axes of an ellipsoid through an orthogonal transformation representing a rotation of axes. This is the essence of a computational algorithm known as the Jacobi method. One of the most important concepts in vibrations is Rayleigh's quotient, defined as the ratio of two quadratic forms, the numerator being a measure of the potential energy and the denominator a measure of the kinetic energy. Rayleigh's quotient has a stationary value in the neighborhood of an intermediate modal vector and a minimum value equal to the lowest natural frequency squared at the lowest modal vector. This property can be used to obtain a quick estimate of the lowest natural frequency by inserting into Rayleigh's quotient a trial vector resembling the lowest mode, such as the static displacement vector under forces proportional to the masses.

Systems with proportional viscous damping can be treated by the same methods as those used for conservative systems. In the case of arbitrary viscous damping, it is necessary to transform the n second-order differential equations into $2n$ first-order differential equations, which amounts to casting the problem in state form. The system matrix A in this case is a $2n \times 2n$ nonsymmetric matrix. A modal analysis in the state space is possible and it involves two solutions of the eigenvalue problem, one for A and the other for A^T. Whereas the eigenvalues are the same, the eigenvectors of A differ from the eigenvectors of A^T. The two sets of eigenvectors are biorthogonal, which forms the basis for the state space modal analysis. The response of systems with arbitrary viscous damping can also be obtained, perhaps more directly, by a method based on the transition matrix.

Except for some simple cases, virtually all multi-degree-of-freedom problems require computer solutions. In this regard, MATLAB has few peers. In this chapter, four MATLAB programs are presented, the first two for the eigenvalue problem, one for conservative and the other for nonconservative systems, the third for the response to initial excitations and the fourth for the response to external excitations by the discrete-time transition matrix.

PROBLEMS

7.1. The system shown in Fig. 7.19 consists of three lumped masses m_i connected by an inextensible string and undergoing the vertical displacements $y_i(t)$ while acted upon by the forces $F_i(t)$ $(i = 1, 2, 3)$, respectively. Assume that the displacements are small, so that the sine and the tangent of an angle can be approximated by the angle itself, and that the string tension T is constant and derive the equations of motion by means of Newton's second law.

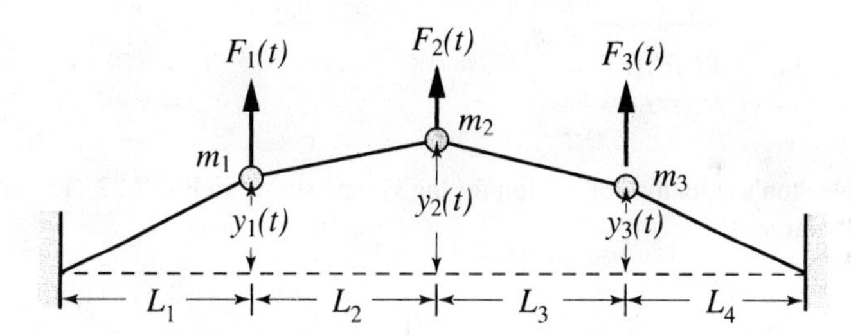

FIGURE 7.19
Three lumped masses on a string

7.2. Derive the equations of motion for the three-degree-of-freedom system shown in Fig. 7.20 by means of Newton's second law.

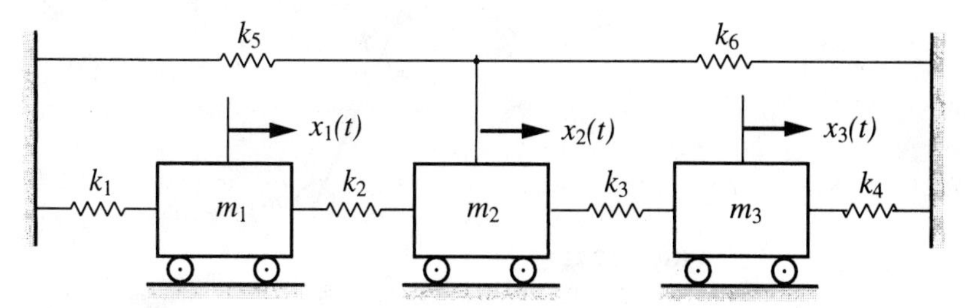

FIGURE 7.20
Three-degree-of-freedom system

7.3. The n-degree-of-freedom system shown in Fig. 7.21 represents a model of an n-story building and it consists of rigid slabs supported by columns in the form of beams clamped at both ends. Derive the equations of motion by means of Newton's second law.

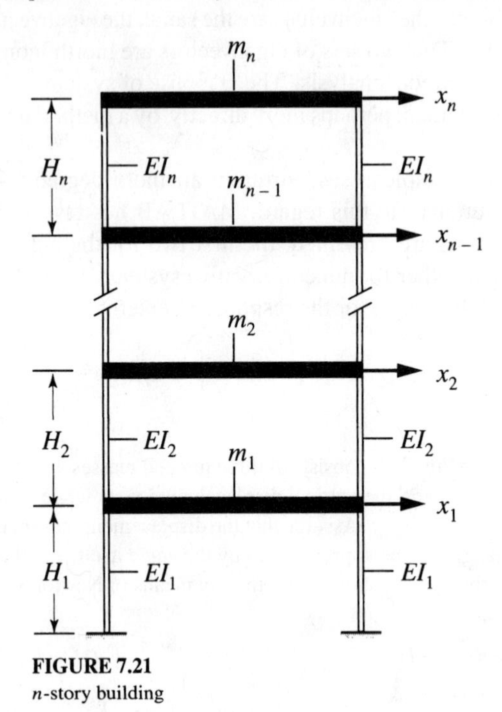

FIGURE 7.21
n-story building

7.4. Derive Newton's equations of motion for the system shown in Fig. 7.22. The angle θ can be arbitrarily large.

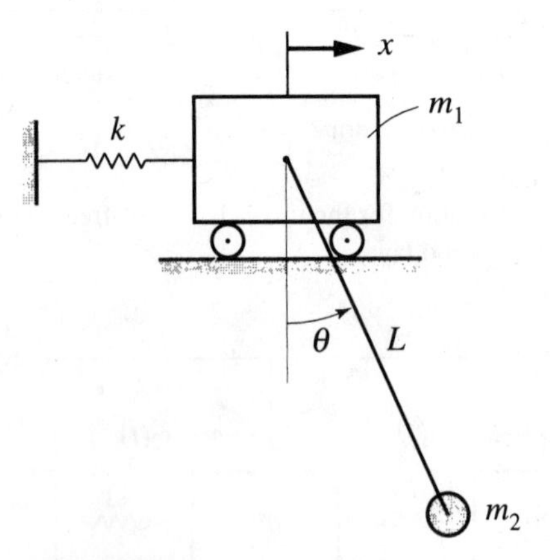

FIGURE 7.22
Pendulum supported by a moving mass

7.5. Derive Newton's equations of motion for the triple pendulum shown in Fig. 7.23. The angles θ_i $(i = 1, 2, 3)$ can be arbitrarily large.

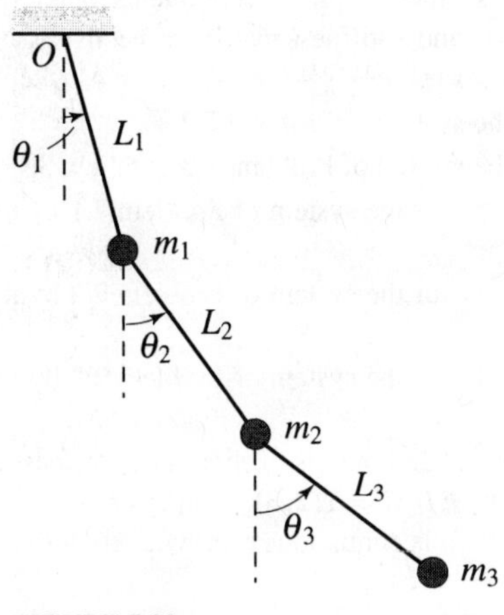

FIGURE 7.23
Triple pendulum

7.6. The four-degree-of-freedom system shown in Fig. 7.24 represents a simplified model of an automobile. Assume that mass m_0 undergoes small angular displacements and derive Newton's equations of motion in terms of the displacements $x_i(t)$ $(i = 1, 2, 3, 4)$.

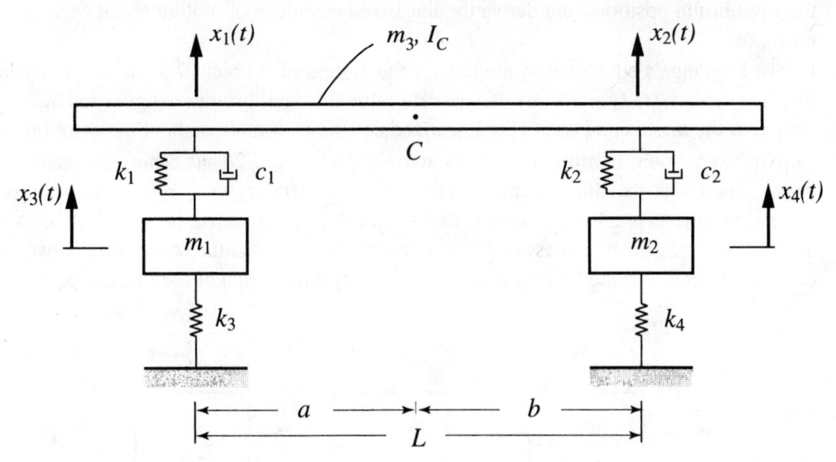

FIGURE 7.24
Simplified model of an automobile

7.7. The seven-degree-of-freedom system shown in Fig. 1.29b represents a model of an automobile. Assume that masses m_{wi} $(i = 1, 2, 3, 4)$ are symmetrically placed relative to axes y and z and that the angular displacements of mass m_b about axes y and z are small and derive Newton's equations of motion.

7.8. Derive the flexibility and stiffness influence coefficients for the system of Problem 7.1, arrange them in a flexibility matrix and a stiffness matrix, respectively, and show that the two matrices are the inverse of one another.

7.9. Solve Problem 7.8 for the system of Problem 7.2

7.10. Solve Problem 7.8 for the system of Problem 7.3 with $n = 3$.

7.11. Derive the stiffness matrix for the system of Problem 7.1 by means of the potential energy expression.

7.12. Derive the stiffness matrix for the system of Problem 7.2 by means of the potential energy expression.

7.13. Derive the stiffness matrix for the system of Problem 7.3 by means of the potential energy expression.

7.14. The system shown in Fig. 7.25 consists of three lumped masses m_i connected by massless beams of flexural rigidity EI_i $(i = 1, 2, 3)$. The system is clamped at the left end and the slope of the deflection curve is continuous everywhere. Derive the flexibility matrix.

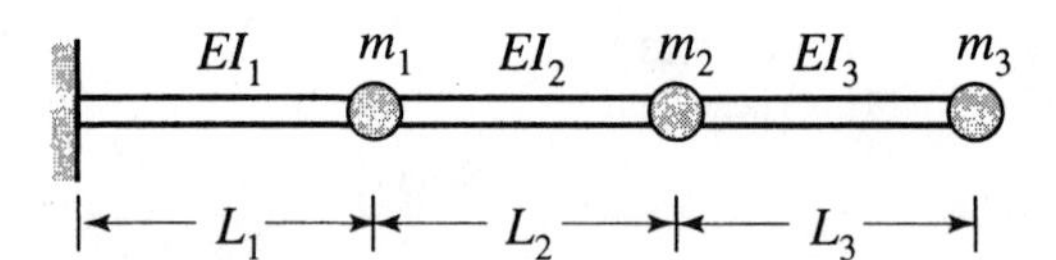

FIGURE 7.25
System consisting of three masses connected by beam segments and clamped at the left end

7.15. Derive Lagrange's equations of motion for the system of Problem 7.4. Then, determine the equilibrium positions and derive the linearized equations of motion about each of these positions.

7.16. Derive Lagrange's equations of motion for the system of Problem 7.1 for arbitrarily large displacements $y_i(t)$ $(i = 1, 2, 3)$. Then, determine the equilibrium configuration due to the weight of the masses and derive the linearized equations of motion about the equilibrium.

7.17. Derive the equations of motion for the system shown in Fig. 7.26 and arrange them in matrix form. Then, use the linear transformation $z_1(t) = x_1(t)$, $z_2(t) = x_2(t) - x_1(t)$, $z_3(t) = x_3(t) - x_2(t)$ to express the equations in terms of the coordinates $z_i(t)$ $(i = 1, 2, 3)$ as well as to symmetrize the new mass and stiffness matrices. Explain the connection between the coordinates used and the corresponding type of coupling in each of the two cases.

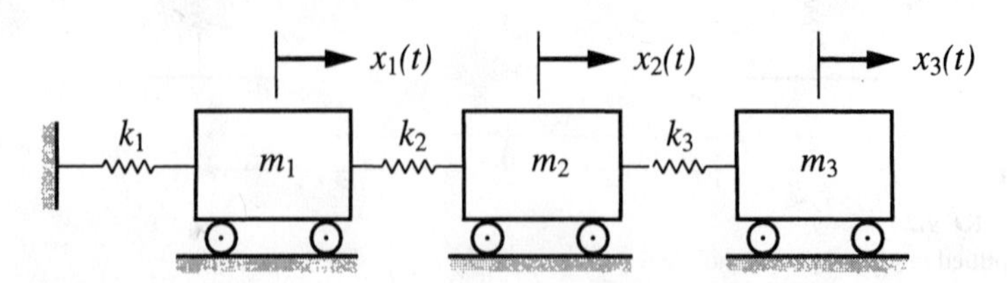

FIGURE 7.26
Three-degree-of-freedom system

7.18. The four-degree-of-freedom system shown in Fig. 1.32a and described by Eqs. (1.130) is the same as that of Problem 7.6, the only differences lying in notation and in the use of coordinates defining the motion of the rigid beam. Express Eqs. (1.130) and the equations derived in Problem 7.6 in matrix form, examine the mass, damping and stiffness matrices for each case and draw conclusions concerning the nature of coupling. Explain how one set of mass, damping and stiffness matrices can be obtained from the other.

7.19. Assume that all forces in Problem 7.1 are equal to zero, $F_i(t) = 0$, and derive and solve the eigenvalue problem for the case in which $m_i = m$ $(i = 1, 2, 3)$, $L_i = L$ $(i = 1, 2, 3, 4)$. Plot the three modes and explain the nature of the mode shapes.

7.20. Solve Problem 7.19 for the case in which $m_1 = m_2 = m$, $m_3 = 2m$, $L_i = L$ $(i = 1, 2, 3, 4)$. Compare the natural frequencies and mode shapes with those obtained in Problem 7.19 and explain the differences.

7.21. Solve the eigenvalue problem for the system of Problem 7.2 for the case in which $m_1 = m_2 = m$, $m_3 = 2m$, $k_1 = k_2 = k_3 = k$, $k_4 = k_5 = k_6 = 2k$ and plot the natural modes.

7.22. Solve the eigenvalue problem for the system of Problem 7.3 for the case in which $n = 3$ and the system has the parameters $m_1 = m_2 = m_3 = m$, $EI_1 = EI_2 = EI_3 = EI$, $H_1 = H_2 = H_3 = H$. Plot the natural modes.

7.23. Consider the triple pendulum of Problem 7.5, linearize the equations of motion by assuming small angles θ_i $(i = 1, 2, 3)$, solve the associated eigenvalue problem for the case in which $m_1 = m_2 = m_3 = m$, $L_1 = L_2 = L_3 = L$ and plot the natural modes.

7.24. Use the approach of Sec. 7.6 to determine the response of the system of Problem 7.19 to initial conditions for the two cases: 1) $\mathbf{y}(0) = [0\ 1\ 0]^T$, $\dot{\mathbf{y}}(0) = \mathbf{0}$ and 2) $\mathbf{y}(0) = [-1\ 0\ 1]^T$, $\dot{\mathbf{y}}(0) = \mathbf{0}$. Draw conclusions as to the mode participation in the response in each of the two cases.

7.25. Determine the response of the system of Problem 7.19 to the initial conditions $\mathbf{x}(0) = \mathbf{0}$, $\dot{\mathbf{x}}(0) = [0\ 1\ 0]^T$ by the approach of Sec. 7.6 and draw conclusions as to the mode participation in the response.

7.26. Determine the response of the system of Problem 7.22 to the initial conditions $\mathbf{y}(0) = [0\ 0\ 1]^T$, $\dot{\mathbf{y}}(0) = \mathbf{0}$ by the approach of Sec. 7.6.

7.27. Verify that the natural modes computed in Problem 7.19 are orthogonal. Then, normalize the modes so as the satisfy Eqs. (7.92).

7.28. Verify that the natural modes computed in Problem 7.20 are orthogonal. Then, normalize the modes so as the satisfy Eqs. (7.92).

7.29. Verify that the natural modes computed in Problem 7.21 are orthogonal. Then, normalize the modes so as the satisfy Eqs. (7.92).

7.30. Verify that the natural modes computed in Problem 7.22 are orthogonal. Then, normalize the modes so as the satisfy Eqs. (7.92).

7.31. The system shown in Fig. 7.27 consists of four masses connected by three springs. Show how the system can be reduced to a three-degree-of-freedom system for elastic displacements alone.

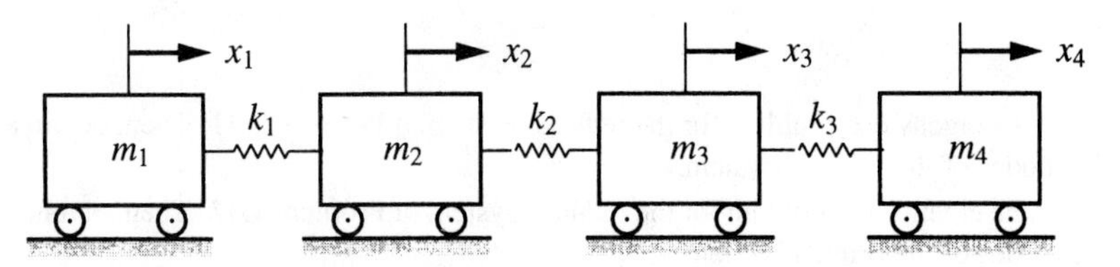

FIGURE 7.27
System consisting of four masses connected to springs and unrestrained at both ends

7.32. The system shown in Fig. 7.28 consists of three lumped masses m_i connected by massless beams of flexural rigidity EI_i and length L_i $(i = 1, 2, 3)$. The system is hinged at the left end O and the slope of the deflection curve is continuous everywhere. In this case, the system is positive semidefinite and there exists a rigid-body mode in the form of pure rotation about point O. Let $m_i = m$, $EI_i = EI$, $L_i = L$ $(i = 1, 2, 3)$ and derive the eigenvalue problem for the elastic displacements alone. **Hint:** Assume that the displacements of the masses consist of a rigid part and an elastic part, where the first is due to pure rotation about O and the second is due to flexure and is measured relative to the line of rotation. For the kinetic energy use absolute velocities, consisting of the sum of the rigid and elastic parts, and for the potential energy use only the elastic part of the displacements. Then, use the conservation of the angular momentum about O to eliminate the rigid-body rotation from the kinetic energy.

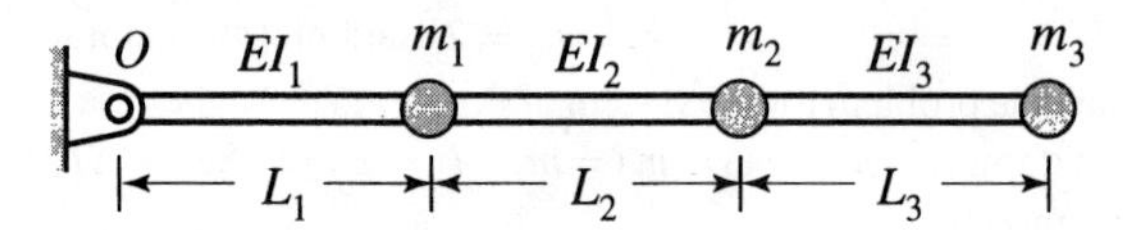

FIGURE 7.28
System consisting of three masses connected by beam segments and free to rotate at the left end

7.33. The system shown in Fig. 7.29 consists of four lumped masses m_i $(i = 1, 2, 3, 4)$ connected by massless beams of flexural rigidity EI_i and length L_i $(i = 1, 2, 3)$. The system is free at both ends and the slope of the deflection curve is continuous everywhere. In this case, the system is positive semidefinite and there exist two rigid-body modes in the form of pure transverse translation and pure rotation about the mass center C of the system. Let $m_i = m$ $(i = 1, 2, 3, 4)$ and $EI_i = EI$, $L_i = L$ $(i = 1, 2, 3)$ and derive the eigenvalue problem for the elastic displacements alone. **Hints:** Use the same approach as in Problem 7.32, except that now one must enforce the conservation of linear momentum in the transverse direction and the conservation of the angular momentum about the mass center C.

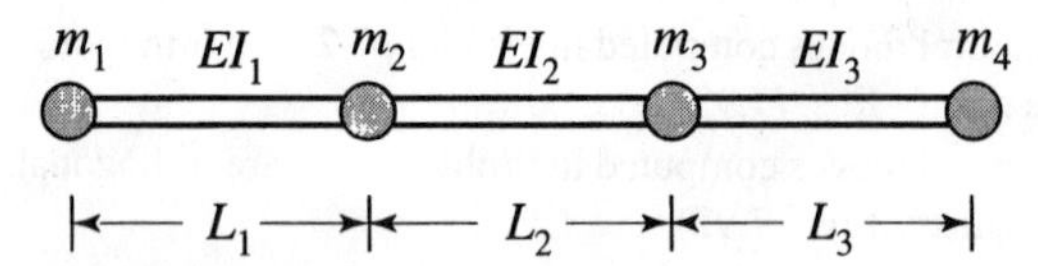

FIGURE 7.29
System consisting of four masses connected by beam segments and unrestrained at both ends

7.34. Solve the eigenvalue problem for the reduced system of Problem 7.31. Then, obtain and plot the modes of the original system.

7.35. Solve the eigenvalue problem for the reduced system of Problem 7.32. Then, obtain and plot the modes of the original system.

7.36. Solve the eigenvalue problem for the reduced system of Problem 7.33. Then, obtain and plot the modes of the original system.

7.37. Solve Problem 7.24 by modal analysis.

7.38. Solve Problem 7.25 by modal analysis.

7.39. Solve Problem 7.26 by modal analysis.

7.40. Use Rayleigh's quotient to obtain estimates of the two lowest natural frequencies of the system of Problem 7.20. Use results from Problem 7.20 to calculate the error incurred and draw conclusions concerning the suitability of the approach to estimate the second lowest natural frequency compared to the lowest.

7.41. Solve Problem 7.40 for the system of Problem 7.21.

7.42. Solve Problem 7.40 for the system of Problem 7.22.

7.43. Solve Problem 7.40 for the system of Problem 7.23.

7.44. Use the approach of Sec. 7.14 to determine the response of the system of Problems 7.1 and 7.20 to the harmonic excitation $F_2(t) = F_0 \cos 0.65t$, $F_1(t) = F_3(t) = 0$.

7.45. The system of Problem 7.44 is acted upon by the harmonic excitation $F_1(t) = F_2(t) = 0$, $F_3(t) = F_0 \cos 0.65t$. Determine the response, compare results with those obtained in Problem 7.44 and draw conclusions.

7.46. Solve Problem 7.44 for the case in which the excitation frequency is 1.2 rad/s instead of 0.65 rad/s. Compare results with those obtained in Problem 7.44 and draw conclusions.

7.47. The foundation of a three-story building obtained by letting $n = 3$ in Fig. 7.21 experiences the horizontal harmonic displacement $U(t) = u_0 \sin 4.5t$. Let $m_i = m$, $EI_i = EI$, $H_i = H$ $(i = 1, 2, 3)$ and determine the response by the approach of Sec. 7.15. Discuss the mode participation in the response.

7.48. Solve Problem 7.44 by modal analysis, compare results with those obtained in Problem 7.44 and draw conclusions. Note that the required eigenvalue problem was solved in Problem 7.20 and the natural modes were normalized in Problem 7.28.

7.49. Determine the response of the triple pendulum of Problem 7.23 to a horizontal force in the form of an impulse of magnitude $\hat{F}_0$ applied at $t = 0$ to mass m_3.

7.50. The system shown in Fig. 7.30 has the following parameters: $m_1 = m$, $m_2 = m_3 = 2m$, $k_1 = k_2 = k$, $k_3 = k_4 = 2k$. Determine the response to a force in the form of a rectangular pulse acting on m_2, $F_1(t) = 0$, $F_2(t) = F_0[\mathcal{u}(t) - \mathcal{u}(t-10)]$, $F_3(t) = 0$, where F_0 is the pulse amplitude and $\mathcal{u}(t)$ the unit step function.

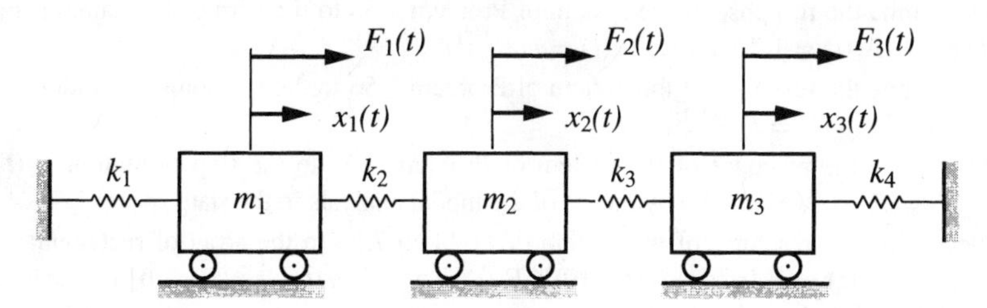

FIGURE 7.30
Undamped three-degree-of-freedom system acted upon by external forces

7.51. Determine the response of the system of Problem 7.20 to the excitation $F_1(t) = F_3(t) = (F_0/10)[r(t) - r(t-10)]$, $F_2(t) = 1.25(F_0/10)[r(t) - r(t-10)]$, where F_0 is a constant and $r(t)$ is the unit ramp function. Discuss the mode participation in the response.

7.52. Determine the response of the triple pendulum of Problem 7.49 to the same force as in Problem 7.49 under the assumption that the system is immersed in a fluid generating resisting forces proportional to the velocities of the masses, where the proportionality constants are $c_1 = c_2 = c_3 = c = 0.1$ m.

7.53. The system of Problem 7.20 is immersed in a fluid generating resisting forces proportional to the velocities of the masses, where the proportionality constants are $c_i = 0.1m_i$ $(i = 1, 2, 3)$. Determine the response to the forces $F_1(t) = 0$, $F_2(t) = F_0[\mathscr{u}(t) - \mathscr{u}(t-5)]$, $F_3(t) = 0$, where F_0 is a constant and $\mathscr{u}(t)$ is the unit step function.

7.54. The system of Fig. 7.31 has the following parameters: $m_1 = m$, $m_2 = m_3 = 2m$, $c_1 = c_2 = 0.1\sqrt{mk}$, $c_3 = c_4 = 0.2\sqrt{mk}$, $k_1 = k_2 = k$, $k_3 = k_4 = 2k$. Determine the response to the forces $F_1(t) = F_3(t) = (F_0/5)[r(t) - r(t-5)]$, $F_2(t) = 1.25(F_0/5)[r(t) - r(t-5)]$, where F_0 is a constant and $r(t)$ is the unit ramp function.

7.55. The system of Fig. 7.31 has the following parameters: $m_1 = m$, $m_2 = m_3 = 2m$, $c_1 = c_4 = 0.1\sqrt{mk}$, $c_2 = c_3 = 0.2\sqrt{mk}$, $k_1 = k_2 = k$, $k_3 = k_4 = 2k$. Set up the state equations, solve the corresponding right and left eigenvalue problems and derive the modal equations, Eqs. (7.269).

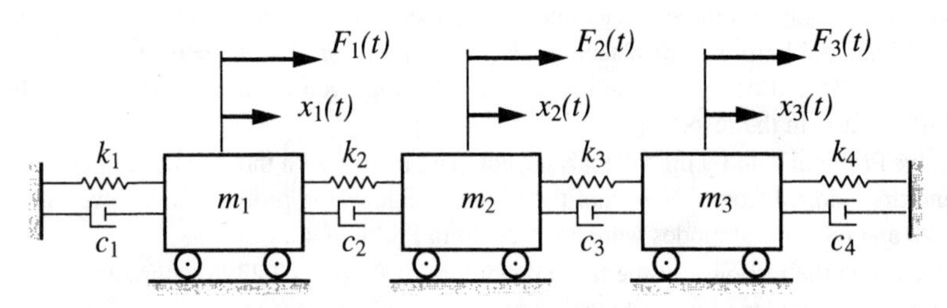

FIGURE 7.31
Damped three-degree-of-freedom system acted upon by external forces

7.56. Solve Problem 7.55 for the case in which the parameters are as follows: $m_1 = m$, $m_2 = m_3 = 2m$, $c_1 = c_2 = c_3 = c_4 = 0.1\sqrt{mk}$, $k_1 = k_2 = k$, $k_3 = k_4 = 2k$.

7.57. Determine the response of the system of Problem 7.55 to the harmonic excitation $F_1(t) = F_0 e^{i1.2t}$, $F_2(t) = 1.2F_0 e^{i1.2t}$, $F_3(t) = F_0 e^{i1.2t}$.

7.58. Determine the response of the system of Problem 7.56 to the harmonic excitation $F_1(t) = F_2(t) = 0$, $F_3(t) = F_0 e^{i1.2t}$.

7.59. Determine the response of the system of Problem 7.55 to the step excitation $F_1(t) = 0$, $F_2(t) = F_0\mathscr{u}(t)$, $F_3(t) = 0$ by means of the modal analysis in the state space.

7.60. Determine the response of the system of Problem 7.55 to the array of rectangular pulses $F_1(t) = F_2(t) = F_0[\mathscr{u}(t) - \mathscr{u}(t-10)]$, $F_3(t) = 1.25F_0[\mathscr{u}(t) - \mathscr{u}(t-10)]$ by means of the modal analysis in the state space.

7.61. Determine the response of the system of Problem 7.55 to the excitation $F_1(t) = F_2(t) = F_3(t) = (F_0/7)[r(t) - r(t-7)]$, where $r(t)$ is the unit ramp function.

7.62. Determine the response of the system of Problem 7.55 to the triangular pulse $F_1(t) = F_2(t) = 0$, $F_3(t) = (F_0/7)[r(t) - 2r(t-7) + r(t-14)]$, where $r(t)$ is the unit ramp function.

7.63. Solve Problem 7.59 for the system of Problem 7.56.

7.64. Solve Problem 7.60 for the system of Problem 7.56.

7.65. Solve Problem 7.61 for the system of Problem 7.56.

7.66. Solve Problem 7.62 for the system of Problem 7.56.

7.67. Solve Problem 7.60 using the method based on the state transition matrix.

7.68. Solve Problem 7.62 using the method based on the state transition matrix.

7.69. Solve Problem 7.64 using the method based on the state transition matrix.

7.70. Solve Problem 7.66 using the method based on the state transition matrix.

7.71. Solve Problem 7.67 in discrete time.

7.72. Solve Problem 7.68 in discrete time.

7.73. Solve Problem 7.69 in discrete time.

7.74. Solve Problem 7.70 in discrete time.

7.75. Solve Problem 7.19 by MATLAB.

7.76. Solve Problem 7.20 by MATLAB.

7.77. Solve Problem 7.21 by MATLAB.

7.78. Solve Problem 7.22 by MATLAB.

7.79. Solve Problem 7.23 by MATLAB.

7.80. Solve Problem 7.37 by MATLAB.

7.81. Solve Problem 7.38 by MATLAB.

7.82. Solve Problem 7.39 by MATLAB.

7.83. Solve Problem 7.50 by MATLAB.

7.84. Solve Problem 7.51 by MATLAB.

7.85. Write a MATLAB program to solve Problem 7.67.

7.86. Write a MATLAB program to solve Problem 7.68.

7.87. Write a MATLAB program to solve Problem 7.69.

7.88. Write a MATLAB program to solve Problem 7.70.

CHAPTER 8

DISTRIBUTED-PARAMETER SYSTEMS: EXACT SOLUTIONS

As pointed out in Sec. 1.9, models of vibrating systems can be divided into two broad classes, lumped and continuous, depending on the nature of the parameters. In the case of lumped systems, the components are discrete, with the mass assumed to be rigid and concentrated at individual points and with the stiffness in the form of massless springs connecting the rigid masses, where the masses and springs represent the system parameters. We refer to such models as *discrete systems*, or *lumped-parameter systems*. The motion of discrete systems is governed by ordinary differential equations, and there is one equation for each mass, where the number of masses generally defines the number of degrees of freedom of the system. The displacements of the masses are identified by subscripts and they depend on time alone, with the nominal position of the masses not appearing explicitly, but only implicitly through the subscript. The equations for small motions from equilibrium are conveniently displayed in matrix form and solved by techniques from linear system theory. The first seven chapters have been concerned exclusively with lumped-parameter models.

In Sec. 1.8, we have seen that rods undergoing axial deformation, shafts in torsion and beams in bending can be regarded as equivalent springs, provided their mass is negligible. The nominal position of a point on these elastic members was identified by the spatial variable x. The situation is entirely different when the mass of the elastic members is not negligible, as the members can no longer be regarded as equivalent springs, and must be treated as *continuous systems*, or *distributed-parameter systems*. Indeed, the mass and stiffness parameters are in general functions of the spatial variable x, and referred to as distributions, with the mass being given in the form of mass per unit length and representing a mass density. Moreover, the displacement depends now on two independent variables, x and t. As a result, the motion of distributed-parameter systems is governed by partial differential equations to be satisfied over the domain of the system, and is subject to boundary conditions at the end points of the domain. Such problems are known as *boundary-value problems*.

Although discrete systems and distributed systems may appear as entirely different in nature, the difference is more in form than substance. In the first place, it must be pointed out that the same physical system can be modeled as discrete or as distributed, depending on objectives. This suggests a much closer connection between discrete and distributed systems than one may be inclined to assume, which is indeed the case. Hence, it should come as no surprise that both types of systems possess natural modes, eigenvectors for discrete systems and eigenfunctions for distributed systems, and natural frequencies. The only difference is that a discrete system possesses a finite number of modes and a distributed system possesses an infinity of modes. As a result, a distributed system can be regarded as having an infinite number of degrees of freedom. Moreover, most importantly, as the modes of discrete systems, the modes of distributed systems possess the orthogonality property, so that the system response can be obtained by means of a modal analysis akin to that for discrete systems. Also of great significance is the fact that the concept of Rayleigh's quotient plays an even more important role in distributed systems than in discrete systems.

Unfortunately, for the most part, boundary-value problems do not admit exact solutions. This is particularly true when the system parameters, which appear in the form of coefficients in the partial differential equation and boundary conditions, depend explicitly on the spatial variable x. If exact solutions are possible, then almost invariably they can be obtained for systems with uniform mass and stiffness distributions. This chapter is devoted entirely to distributed-parameter systems admitting exact solutions. Although such systems are not very abundant, the theory presented here is essential to any serious study of vibrations. Indeed, the same theory applies to the cases in which only approximate solutions are possible. Moreover, exact solutions are quite useful in constructing approximate solutions. We discuss approximate solutions in Chs. 9 and 10.

In this chapter, a number of distributed systems are discussed, such as strings in transverse vibration, rods in axial vibration, shafts in torsion and beams in bending. Strings, rods and shafts are governed by second-order differential equations in x and they are analogous to one another. On the other hand, beams are governed by fourth-order differential equations. When the system parameters are uniformly distributed, second-order differential equations in the spatial variable x reduce to the so-called "wave equation." A discussion of the wave equation enables us to demonstrate the connection between traveling and standing waves.

8.1 RELATION BETWEEN DISCRETE AND DISTRIBUTED SYSTEMS. TRANSVERSE VIBRATION OF STRINGS

We pointed out in the introduction to this chapter that there is a very intimate relation between discrete and continuous systems. Indeed, they can often be regarded as two distinct mathematical models of the same physical systems. To illustrate this idea, we derive the differential equation for the transverse vibration of a string first by regarding it as a discrete system and letting it approach a distributed model in the limit. Then, we formulate the problem by regarding the system as distributed from the beginning.

We consider a system of discrete masses m_i $(i = 1, 2, \ldots, n)$ on a massless string, where the masses m_i are subjected to the external forces F_i, as shown in Fig. 8.1a. To derive the differential equation of motion for a typical mass m_i, we concentrate on the three adjacent masses m_{i-1}, m_i and m_{i+1} of Fig. 8.1b. The tensions in the string segments connecting m_i to m_{i-1} and m_{i+1} are denoted by T_{i-1} and T_i, and the horizontal projections of these segments by Δx_{i-1} and Δx_i, respectively. The displacements $y_i(t)$ $(i = 1, 2, \ldots, n)$ of the masses m_i are assumed to be small, so that the projections Δx_i remain essentially unchanged during motion. Moreover, the angles between the string segments and a horizontal reference line are sufficiently small that both the sine and tangent of the angles are approximately equal to the angles themselves. Hence, using Newton's second law, the equation of motion of mass m_i in the vertical direction has the form

$$T_i \frac{y_{i+1} - y_i}{\Delta x_i} - T_{i-1} \frac{y_i - y_{i-1}}{\Delta x_{i-1}} + F_i = m_i \frac{d^2 y_i}{dt^2} \tag{8.1}$$

Equation (8.1) is applicable to masses m_i $(i = 2, 3, \ldots, n-1)$. The equation can also be extended to masses $i = 1$ and $i = n$, but certain provisions must be made to reflect the manner in which the system is supported, as we shall see shortly. Hence, rearranging Eq. (8.1), we obtain the set of simultaneous ordinary differential equations

$$\frac{T_i}{\Delta x_i} y_{i+1} - \left(\frac{T_i}{\Delta x_i} + \frac{T_{i-1}}{\Delta x_{i-1}} \right) y_i + \frac{T_{i-1}}{\Delta x_{i-1}} y_{i-1} + F_i = m_i \frac{d^2 y_i}{dt^2}, \quad i = 1, 2, \ldots, n \tag{8.2}$$

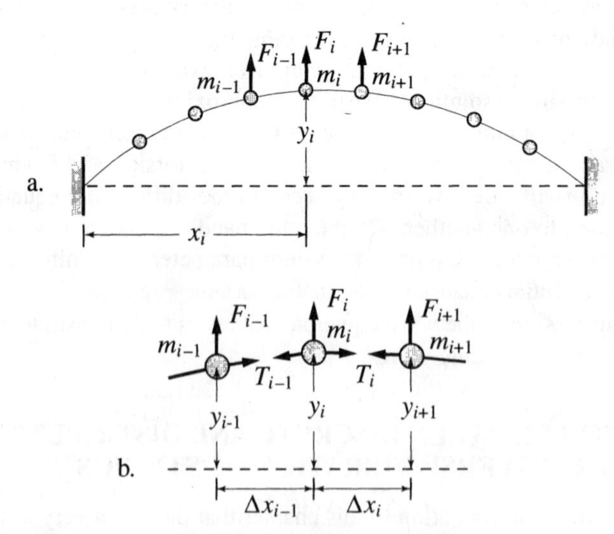

FIGURE 8.1
a. n masses on a string, b. Free-body diagram for mass m_i

in the variables y_i $(i = 1, 2, \ldots, n)$, and we observe that the equations for $i = 1$ and $i = n$ contain the displacements y_0 and y_{n+1}, respectively. If the string is fixed at both ends, as is the case with the system shown in Fig. 8.1a, then we must set

$$y_0(t) = 0, \ y_{n+1}(t) = 0 \tag{8.3}$$

in Eqs. (8.2).

Next, we introduce the notation

$$y_{i+1} - y_i = \Delta y_i, \ y_i - y_{i-1} = \Delta y_{i-1} \tag{8.4}$$

so that Eqs. (8.2) become

$$T_i \frac{\Delta y_i}{\Delta x_i} - T_{i-1} \frac{\Delta y_{i-1}}{\Delta x_{i-1}} + F_i = m_i \frac{d^2 y_i}{dt^2}, \ i = 1, 2, \ldots, n \tag{8.5}$$

But, the first two terms on the left side of Eqs. (8.5) represent the incremental difference in the vertical component of force in the string between the left side and right side of m_i. In view of this, Eqs. (8.5) can be rewritten in the form

$$\Delta \left(T_i \frac{\Delta y_i}{\Delta x_i} \right) + F_i = m_i \frac{d^2 y_i}{dt^2}, \ i = 1, 2, \ldots, n \tag{8.6}$$

Moreover, dividing both sides of Eqs. (8.6) by Δx_i, we obtain

$$\frac{\Delta}{\Delta x_i} \left(T_i \frac{\Delta y_i}{\Delta x_i} \right) + \frac{F_i}{\Delta x_i} = \frac{m_i}{\Delta x_i} \frac{d^2 y_i}{dt^2}, \ i = 1, 2, \ldots, n \tag{8.7}$$

At this time, we let the number n of masses m_i increase indefinitely, while the masses themselves and the distance between them decrease correspondingly, and replace the indexed position x_i by the independent spatial variable x. In the limit, as $\Delta x_i \to 0$, Eqs. (8.7) reduce to

$$\frac{\partial}{\partial x} \left[T(x) \frac{\partial y(x,t)}{\partial x} \right] + f(x,t) = \rho(x) \frac{\partial^2 y(x,t)}{\partial t^2} \tag{8.8}$$

which must be satisfied over the domain $0 < x < L$, where

$$f(x,t) = \lim_{\Delta x_i \to 0} \frac{F_i(t)}{\Delta x_i}, \ \rho(x) = \lim_{\Delta x_i \to 0} \frac{m_i}{\Delta x_i} \tag{8.9}$$

are the transverse force and mass at point x, respectively, both per unit length of string. We note that, by virtue of the fact that the indexed position x_i is replaced by the independent spatial variable x, there are now two independent variables. Hence, total derivatives with respect to the time t become partial derivatives with respect to t, whereas ratios of increments are replaced directly by partial derivatives with respect to x. Equation (8.8) represents the *partial differential equation of motion of the string*. Similarly, conditions (8.3) must be replaced by

$$y(0,t) = 0, \ y(L,t) = 0 \tag{8.10}$$

which are generally known as the *boundary conditions* of the problem. Equations (8.8) and (8.10) constitute the *boundary-value problem* for the string. In fact, the transverse displacement $y(x,t)$ is also subject to the initial conditions

$$y(x,0) = y_0(x), \quad \left.\frac{\partial y(x,t)}{\partial t}\right|_{t=0} = v_0(x) \tag{8.11}$$

where $y_0(x)$ is the initial displacement and $v_0(x)$ the initial velocity at every point x of the string, so that Eqs. (8.8), (8.10) and (8.11) represent a *boundary-value and initial-value problem* simultaneously. It is customary to classify a boundary-value problem according to the highest degree derivative with respect to the spatial variable x appearing in the differential equation, Eq. (8.8). Hence, in the case at hand the boundary-value-problem is of *second order*. We note that for second-order boundary-value problems there is *one boundary condition at each end.*

The problem can be formulated more directly by considering the string as a distributed system from the onset, as shown in Fig. 8.2a, where $f(x,t)$, $\rho(x)$ and $T(x)$ are, respectively, the distributed force, mass density and string tension at point x. Figure 8.2b represents the free-body diagram corresponding to an element of string of length dx. Again writing Newton's second law for the force component in the vertical direction, we obtain

$$\left[T(x) + \frac{\partial T(x)}{\partial x}dx\right]\left[\frac{\partial y(x,t)}{\partial x} + \frac{\partial^2 y(x,t)}{\partial x^2}dx\right] - T(x)\frac{\partial y(x,t)}{\partial x} + f(x,t)dx = \rho(x)dx\frac{\partial^2 y(x,t)}{\partial t^2} \tag{8.12}$$

Canceling appropriate terms and ignoring second-order terms in dx, Eq. (8.12) reduces

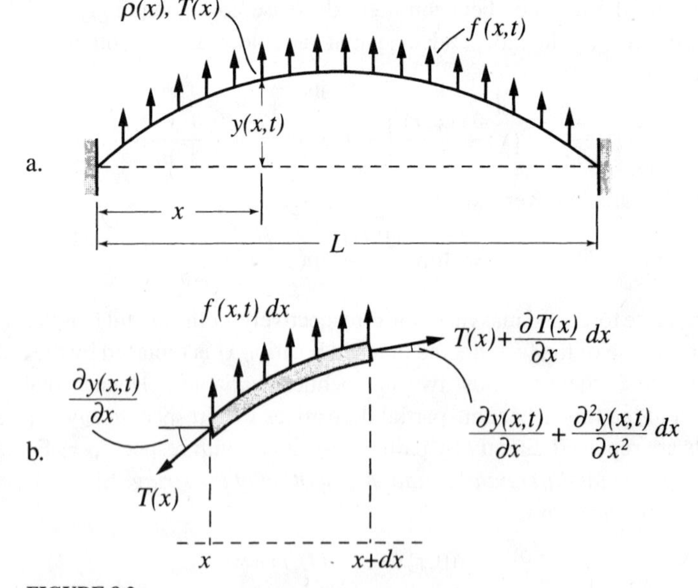

FIGURE 8.2
a. The string as a distributed-parameter system, b. Free-body diagram for an element of string

to

$$\frac{\partial T(x)}{\partial x}\frac{\partial y(x,t)}{\partial x}dx + T(x)\frac{\partial^2 y(x,t)}{\partial x^2}dx + f(x,t)dx = \rho(x)dx\frac{\partial^2 y(x,t)}{\partial t^2} \tag{8.13}$$

and, dividing through by dx and combining the first two terms, we can rewrite Eq. (8.13) in the more compact form

$$\frac{\partial}{\partial x}\left[T(x)\frac{\partial y(x,t)}{\partial x}\right] + f(x,t) = \rho(x)\frac{\partial^2 y(x,t)}{\partial t^2}, \quad 0 < x < L \tag{8.14}$$

which is identical to Eq. (8.8) in every respect. Moreover, from Fig. 8.2a, we recognize that the displacement of the string at the two ends must be zero, $y(0,t) = y(L,t) = 0$, thus duplicating boundary conditions (8.10). This completes the mathematical analogy between the discrete and continuous models for a string in transverse vibration.

The unmistakable conclusion is that Figs. 8.1a and 8.2a, although different in appearance, represent two intimately related mathematical models. In this section, we made the transition from the discrete system, Fig. 8.1a, to the distributed one, Fig. 8.2a, through a limiting process equivalent to spreading the masses over the entire string. In many practical applications, particularly if the string is nonuniform, it is more common to follow the opposite path and lump the distributed mass into discrete masses. This can be done by using the second of Eqs. (8.9) and writing $m_i = \rho(x_i)\Delta x_i$. Regardless of whether the mathematical model is discrete or distributed, it is clear that they must share similar vibrational characteristics.

The boundary-value problem described by the differential equation (8.8), or (8.14), and the boundary conditions (8.10) is a relatively simple one, because the boundary conditions are very simple. A somewhat more involved case is that in which the right end of the string is able to slide inside a vertical guide while restrained by a spring of stiffness k, as shown in Fig. 8.3a. In this case, the differential equation remains as given by Eq. (8.14) and so does the boundary condition

$$y(0,t) = 0 \tag{8.15}$$

at the left end, so that the only difference is in the boundary condition at $x = L$. To derive this boundary condition, we consider the free-body diagram shown in Fig. 8.3b,

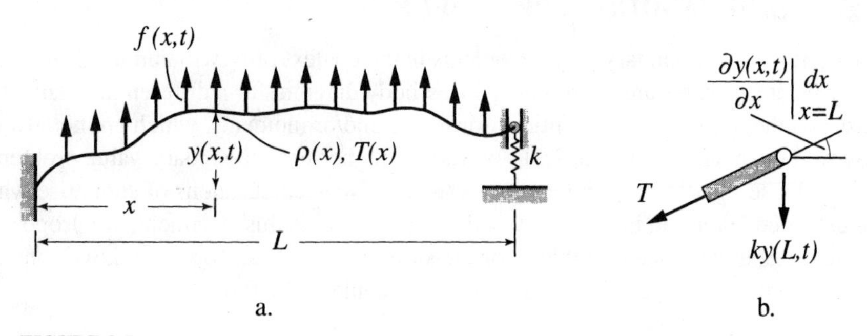

FIGURE 8.3
a. String fixed at $x = 0$ and restrained by a spring at $x = L$, b. Free-body diagram for the right end

Table 8.1 Analogous Quantities

	String	Rod	Shaft
Displacement	Transverse: $y(x,t)$	Axial: $u(x,t)$	Angular: $\theta(x,t)$
Inertia per Unit Length	Mass: $\rho(x)$	Mass: $m(x)$	Mass Polar Moment of Inertia: $I(x)$
Stiffness	Tension: $T(x)$	Axial: $EA(x)$ $E =$ modulus of elasticity $A(x) =$ cross-sectional area	Torsional: $GJ(x)$ $G =$ shear modulus $J(x) =$ polar moment of inertia of cross-sectional area
Load per Unit Length	Force: $f(x,t)$	Force: $f(x,t)$	Torque: $m(x,t)$

sum up forces in the vertical direction and obtain the boundary condition

$$-T(x)\frac{\partial y(x,t)}{\partial x} - ky(x,t) = 0, \ x = L \tag{8.16}$$

We observe that boundary condition (8.16) is significantly more involved than boundary condition (8.15). In particular, boundary condition (8.15) contains no derivatives and can be written down solely on geometric considerations. On the other hand, boundary condition (8.16) contains a spatial derivative of the first degree and it reflects the vertical force balance. For this reason, boundary condition (8.15) is said to be *geometric*, whereas boundary condition (8.16) is said to be *natural*. They are also known as *essential* and *dynamic* boundary conditions, respectively.

There are two other distributed-parameter systems described by second-order boundary-value problems, namely, thin rods in axial vibration and circular shafts in torsional vibration. In fact, the vibration of thin rods and circular shafts is governed by boundary-value problems entirely analogous to that for a string. To obtain the boundary-value problem for a thin rod and a circular shaft, it is only necessary to replace the analogous quantities indicated in Table 8.1. Some of these quantities were encountered in Sec. 1.8 in connection with equivalent spring constants for distributed elastic components.

8.2 DERIVATION OF THE STRING VIBRATION PROBLEM BY THE EXTENDED HAMILTON PRINCIPLE

The derivation of boundary-value problems in the context of Newtonian mechanics can be trying at times, because it requires a free-body diagram for a differential element of mass and the use of sign conventions for forces and/or moments, which are not always easy to remember. This is particularly true for higher-order boundary-value problems, such as those associated with beams in bending. No such questions of sign arise when the extended Hamilton principle is used. To substantiate this statement, we propose to derive the boundary-value problem for the string of Fig. 8.3a. To this end, we consider Eq. (6.31) and write the extended Hamilton principle in the form

$$\int_{t_1}^{t_2} (\delta T - \delta V + \overline{\delta W}_{nc})dt = 0, \ \delta y(x,t) = 0, \ 0 \le x \le L, \ t = t_1, t_2 \tag{8.17}$$

where

$$T(t) = \frac{1}{2}\int_0^L \rho(x)\left[\frac{\partial y(x,t)}{\partial t}\right]^2 dx \tag{8.18}$$

is the kinetic energy and

$$\overline{\delta W}_{nc}(t) = \int_0^L f(x,t)\delta y(x,t)dx \tag{8.19}$$

is the virtual work of the nonconservative distributed force; the potential energy requires some elaboration. In the first place, we observe that the potential energy arises from two sources, the first from the tendency of the tension $T(x)$ to restore the string to the equilibrium position and the second from the end spring. Hence, denoting by ds the length of a differential element dx in displaced position, referring to Fig. 8.4 and recognizing that, to restore equilibrium, the tensile force T must work through the difference in length $ds - dx$, we can write

$$V(t) = \int_0^L T(x)[ds(x,t) - dx] + \tfrac{1}{2}ky^2(L,t) \tag{8.20}$$

Then, assuming that the slope $\partial y/\partial x$ is a small quantity, we use Fig. 8.4 and write

$$ds = \left[(dx)^2 + \left(\frac{\partial y}{\partial x}dx\right)^2\right]^{1/2} = \left[1 + \left(\frac{\partial y}{\partial x}\right)^2\right]^{1/2} dx \cong \left[1 + \frac{1}{2}\left(\frac{\partial y}{\partial x}\right)^2\right] dx \tag{8.21}$$

in which we retained two terms only in the binomial expansion. Inserting Eq. (8.21) into Eq. (8.20), we obtain

$$V(t) = \frac{1}{2}\int_0^L T(x)\left[\frac{\partial y(x,t)}{\partial x}\right]^2 dx + \tfrac{1}{2}ky^2(L,t) \tag{8.22}$$

Next, we write the variation in the kinetic energy

$$\delta T = \int_0^L \rho\frac{\partial y}{\partial t}\delta\left(\frac{\partial y}{\partial t}\right)dx = \int_0^L \rho\frac{\partial y}{dt}\frac{\partial}{\partial t}\delta y\, dx \tag{8.23}$$

in which we assumed that the variation and differentiation processes are interchangeable,

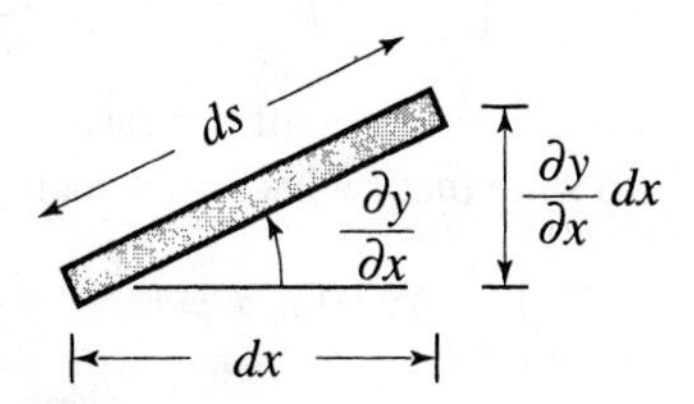

FIGURE 8.4
Differential element dx in displaced position

and carry out the following integration by parts with respect to time

$$\int_{t_1}^{t_2} \delta T dt = \int_{t_1}^{t_2} \left(\int_0^L \rho \frac{\partial y}{\partial t} \frac{\partial}{\partial t} \delta y dx \right) dt = \int_0^L \left(\int_{t_1}^{t_2} \rho \frac{\partial y}{\partial t} \frac{\partial}{\partial t} \delta y \, dt \right) dx$$

$$= \int_0^L \left(\rho \frac{\partial y}{\partial t} \delta y \Big|_{t_1}^{t_2} \right) dx - \int_0^L \left(\int_{t_1}^{t_2} \rho \frac{\partial^2 y}{\partial t^2} \delta y dt \right) dx$$

$$= - \int_{t_1}^{t_2} \left(\int_0^L \rho \frac{\partial^2 y}{\partial t^2} \delta y \, dx \right) dt \tag{8.24}$$

where we considered the fact that $\delta y = 0$ at $t = t_1$ and $t = t_2$. Similarly, we write the variation in the potential energy

$$\delta V = \int_0^L T \frac{\partial y}{\partial x} \delta \frac{\partial y}{\partial x} dx + ky(L,t)\delta y(L,t) = \int_0^L T \frac{\partial y}{\partial x} \frac{\partial}{\partial x} \delta y dx + ky(L,t)\delta y(L,t) \tag{8.25}$$

integrate by parts and obtain

$$\delta V = T \frac{\partial y}{\partial x} \delta y \Big|_0^L - \int_0^L \frac{\partial}{\partial x} \left(T \frac{\partial y}{\partial x} \right) \delta y dx + ky(L,t)\delta y(L,t)$$

$$= \left(T \frac{\partial y}{\partial x} + ky \right) \delta y \Big|_{x=L} - T \frac{\partial y}{\partial x} \delta y \Big|_{x=0} - \int_0^L \frac{\partial}{\partial x} \left(T \frac{\partial y}{\partial x} \right) \delta y dx \tag{8.26}$$

Inserting Eqs. (8.19), (8.24) and (8.26) into Eq. (8.17) and collecting terms, we have

$$\int_{t_1}^{t_2} \left\{ \int_0^L \left[-\rho \frac{\partial^2 y}{\partial t^2} + \frac{\partial}{\partial x} \left(T \frac{\partial y}{\partial x} \right) + f \right] \delta y dx - \left(T \frac{\partial y}{\partial x} + ky \right) \delta y \Big|_{x=L} + T \frac{\partial y}{\partial x} \delta y \Big|_{x=0} \right\} dt = 0 \tag{8.27}$$

At this point, we invoke the arbitrariness of the virtual displacement δy. We first set $\delta y = 0$ at $x = 0$ and $x = L$ and assign values of $\delta y(x,t)$ at will over the interval $0 < x < L$. Under these circumstances, Eq. (8.27) can be satisfied only if the coefficient of $\delta y(x,t)$ is zero over the same interval. Hence, we set

$$\frac{\partial}{\partial x} \left(T \frac{\partial y}{\partial x} \right) + f = \rho \frac{\partial^2 y}{\partial t^2}, \quad 0 < x < L \tag{8.28}$$

Then, with the integral in Eq. (8.27) disposed of, we once again invoke the arbitrariness of the virtual displacement and state that δy is chosen such that

$$T \frac{\partial y}{\partial x} \delta y = 0, \quad x = 0 \tag{8.29}$$

and

$$\left(T \frac{\partial y}{\partial x} + ky \right) \delta y = 0, \quad x = L \tag{8.30}$$

Equation (8.29) can be satisfied in two ways, namely, by setting either $T\ \partial y/\partial x$ or δy equal to zero at $x = 0$. The first is recognized as the vertical force at $x = 0$, which cannot be zero for all times at a fixed end. On the other hand, the displacement is zero at a fixed end, so that Eq. (8.29) is satisfied by setting

$$y(0, t) = 0 \tag{8.31}$$

Similarly, Eq. (8.30) is satisfied if either $T(\partial y/\partial x) + k$ or δy is zero at $x = L$. But, the displacement cannot be zero for all times at $x = L$. Hence, Eq. (8.30) can only be satisfied by setting

$$T\frac{\partial y}{\partial x} + ky = 0, \ x = L \tag{8.32}$$

This completes the derivation of the boundary-value problem, which consists of the differential equation, Eq. (8.28), and one boundary condition at each end, Eqs. (8.31) and (8.32). The results are identical to those obtained in Sec. 8.1.

We conclude that the extended Hamilton principle yields the correct boundary-value problem in almost a routine fashion. Moreover, because the kinetic energy is a positive definite quadratic form and in this particular case the potential energy is also a positive definite quadratic form, there is no room for sign errors, provided the various steps involved are carried out correctly.

8.3 BENDING VIBRATION OF BEAMS

As pointed out in Sec. 8.1, the transverse vibration of strings, axial vibration of thin rods and torsional vibration of circular shaft are all governed by entirely analogous boundary-value problems. Indeed, they all consist of a second-order partial differential equation and one boundary condition at each end. The only difference lies in the nature of the displacement, excitation and parameters, as indicated in Table 8.1.

Beams in bending represent more complex systems than strings, rods and shafts, so that it comes as no surprise that boundary-value problems for beams are also more involved. As shown shortly, they consist of a fourth-order partial differential equation and two boundary conditions at each end. Moreover, there is a larger variety of boundary conditions, and they can involve spatial derivatives up to third order. Here too, the boundary conditions can be divided between geometric and natural.

We consider the beam in bending vibration shown in Fig. 8.5a, where $y(x, t)$ denotes the transverse displacement, $f(x, t)$ the transverse force per unit length, $m(x)$ the mass per unit length and $EI(x)$ the flexural rigidity, in which E is the modulus of elasticity and $I(x)$ the cross-sectional area moment of inertia about an axis normal to x and y and passing through the center of the cross section. We propose to derive the boundary-value problem in two ways, first using the Newtonian approach and then using the extended Hamilton's principle. We begin with the Newtonian approach and, to this end, we consider the free-body diagram corresponding to a beam differential element, as shown in Fig. 8.5b, in which $M(x, t)$ is the bending moment and $Q(x, t)$ the shearing force. According to sign conventions, the bending moment and shearing force shown on both sides of the differential element are regarded as positive. We use the elementary

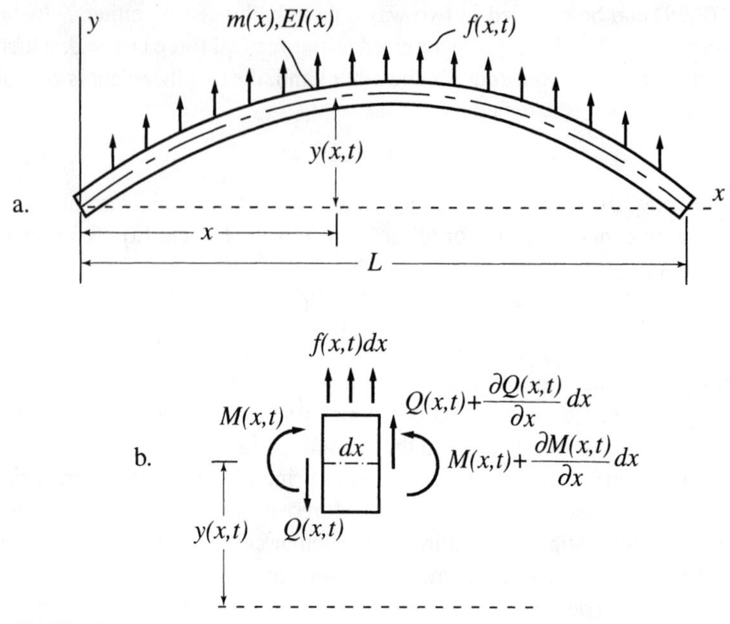

FIGURE 8.5
a. Beam in bending vibration, b. Free-body diagram for a beam element

beam theory, commonly known as the *Euler-Bernoulli beam theory*, which involves the assumptions that the rotation of the differential element is negligible compared to the translation and that the angular distortion due to shear is small in relation to the bending deformation. This theory is valid if the ratio between the length of the beam and its depth is relatively large, say more than 10, and if the beam does not become too "wrinkled" because of flexure. The Euler-Bernoulli beam theory stipulates that the rotatory inertia and shear deformation effects can be ignored.[1]

From Fig. 8.5b, the force equation of motion in the vertical direction has the form

$$\left[Q(x,t)+\frac{\partial Q(x,t)}{\partial x}dx\right]-Q(x,t)+f(x,t)dx=m(x)dx\frac{\partial^2 y(x,t)}{\partial t^2},\quad 0<x<L \tag{8.33}$$

Moreover, under the assumption that the product of the mass moment of inertia of the element and the angular acceleration is negligibly small, the moment equation of motion about an axis normal to x and y and passing through the center of the cross-sectional

[1]For a more refined theory including these effects, see L. Meirovitch, *Principles and Techniques of Vibrations*, Sec. 7.11, Prentice Hall, Englewood Cliffs, NJ, 1997.

area is

$$\left[M(x,t)+\frac{\partial M(x,t)}{\partial x}dx\right]-M(x,t)+\left[Q(x,t)+\frac{\partial Q(x,t)}{\partial x}dx\right]dx$$
$$+f(x,t)dx\frac{dx}{2}=0,\ 0<x<L \qquad (8.34)$$

Ignoring second-order terms in dx and canceling appropriate terms, the moment equation, Eq. (8.34), reduces to

$$\frac{\partial M(x,t)}{\partial x}+Q(x,t)=0,\ 0<x<L \qquad (8.35)$$

Moreover, canceling appropriate terms, dividing through by dx and using Eq. (8.35), the force equation, Eq. (8.33), becomes

$$-\frac{\partial^2 M(x,t)}{\partial x^2}+f(x,t)=m(x)\frac{\partial^2 y(x,t)}{\partial t^2},\ 0<x<L \qquad (8.36)$$

Equation (8.36) relates the bending moment $M(x,t)$ and the transverse force density $f(x,t)$ to the bending displacement $y(x,t)$. To obtain an equation in terms of $y(x,t)$ and $f(x,t)$ alone, we recall from mechanics of materials (Ref. 1) that the bending moment is related to the bending displacement by

$$M(x,t)=EI(x)\frac{\partial^2 y(x,t)}{\partial x^2} \qquad (8.37)$$

so that, using Eq. (8.35), the shearing force is related to the bending displacement by

$$Q(x,t)=-\frac{\partial}{\partial x}\left[EI(x)\frac{\partial^2 y(x,t)}{\partial x^2}\right] \qquad (8.38)$$

Inserting Eq. (8.37) into Eq. (8.36), we obtain the partial differential equation for bending vibration of a beam in the form

$$-\frac{\partial^2}{\partial x^2}\left[EI(x)\frac{\partial^2 y(x,t)}{\partial x^2}\right]+f(x,t)=m(x)\frac{\partial^2 y(x,t)}{\partial t^2},\ 0<x<L \qquad (8.39)$$

which is of order four, as anticipated.

To complete the derivation of the boundary-value problem, we must specify two boundary conditions at each end of the beam. The most common ends are *clamped*, *pinned* and *free*, shown in Figs. 8.6a, 8.6b and 8.6c, respectively. The corresponding boundary conditions are as follows:

1. *Clamped end.* At a clamped end, or fixed end, the deflection and slope of the deflection curve are zero, so that the boundary conditions are

$$y(x,t)=0,\quad \frac{\partial y(x,t)}{\partial x}=0 \qquad (8.40)$$

and we observe that both boundary conditions are geometric.

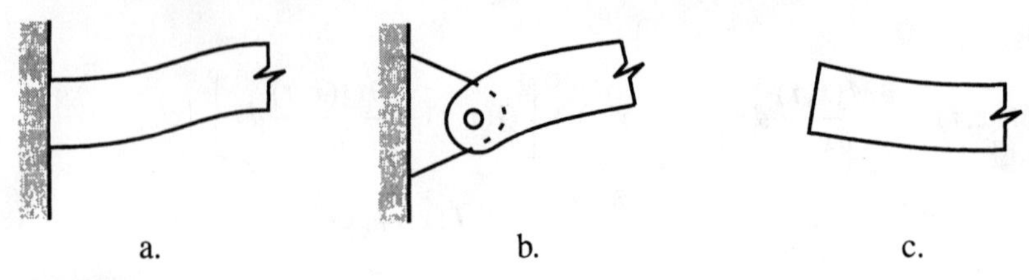

FIGURE 8.6
a. Beam clamped end, b. Beam pinned end, c. Beam free end

2. *Pinned end.* In this case, the displacement and the bending moment are zero. Hence, using Eq. (8.37), we have

$$y(x,t)=0,\ M(x,t)=EI(x)\frac{\partial^2 y(x,t)}{\partial x^2}=0 \tag{8.41}$$

so that the first boundary condition is geometric and the second is natural.

3. *Free end.* At a free end, both the bending moment and shearing force are zero. Hence, using Eqs. (8.37) and (8.38), we obtain

$$M(x,t)=EI(x)\frac{\partial^2 y(x,t)}{\partial x^2}=0,\ Q(x,t)=-\frac{\partial}{\partial x}\left[EI(x)\frac{\partial^2 y(x,t)}{\partial x^2}\right]=0 \tag{8.42}$$

so that both boundary conditions are natural.

In the three cases just discussed, it is possible to specify the boundary conditions by inspection. A case requiring elaboration is that in which the end is supported by a spring, as shown in Fig. 8.7a. To derive the corresponding boundary condition, we assume that the end $x=0$ undergoes the positive (upward) displacement $y(0,t)$, in which case there is a downward force of magnitude $ky(0,t)$. Hence, from Fig. 8.7b, and using the sign convention implied by the left side of the differential element in Fig. 8.5b, we can write

$$Q(x,t)=-\frac{\partial}{\partial x}\left[EI(x)\frac{\partial^2 y(x,t)}{\partial x^2}\right]=ky(x,t),\ x=0 \tag{8.43}$$

Note that the same sign convention implied by the right side of Fig. 8.5b, for a support

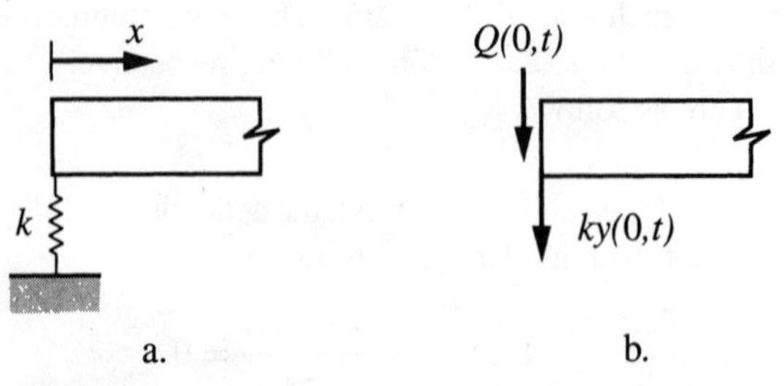

FIGURE 8.7
a. Beam end supported by a spring, b. Free-body diagram

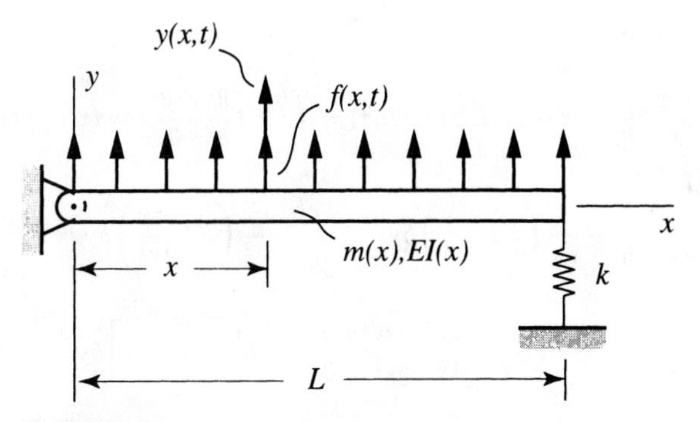

FIGURE 8.8
Beam pinned at $x = 0$ and spring supported at $x = L$

spring at $x = L$, would give the boundary condition

$$Q(x,t) = -\frac{\partial}{\partial x}\left[EI(x)\frac{\partial^2 y(x,t)}{\partial x^2}\right] = -ky(x,t),\ x = L \tag{8.44}$$

Although the derivation of boundary conditions (8.43) and (8.44) is quite straightforward, provided the sign conventions are adhered to, the results are not intuitively obvious. Of course, the other boundary condition for the end shown in Fig. 8.7a requires that the bending moment be zero.

Next, we derive the boundary-value problem by means of the extended Hamilton principle, which remains in the form of Eq. (8.17). To this end, we consider a pinned-spring supported beam, as depicted in Fig. 8.8. The kinetic energy remains in the form given by Eq. (8.18), except that the mass density now is $m(x)$ instead of $\rho(x)$. Moreover, from mechanics of materials, the potential energy is

$$V(t) = \frac{1}{2}\int_0^L EI(x)\left[\frac{\partial^2 y(x,t)}{\partial x^2}\right]^2 dx + \tfrac{1}{2}ky^2(L,t) \tag{8.45}$$

The virtual work of the nonconservative distributed force $f(x,t)$ has the same form as that for the string, Eq. (8.19).

In view of our statement concerning the kinetic energy in the preceding paragraph, if we use the analogy with Eq. (8.24), we can write

$$\int_{t_1}^{t_2} \delta T\,dt = -\int_{t_1}^{t_2}\left(\int_0^L m\frac{\partial^2 y}{\partial t^2}\delta y dx\right)dt \tag{8.46}$$

Moreover, using Eq. (8.45) and carrying two integrations by parts, we obtain the variation in the potential energy

$$\delta V = \int_0^L EI\frac{\partial^2 y}{\partial x^2}\delta\frac{\partial^2 y}{\partial x^2}dx + ky(L,t)\delta y(L,t) = \int_0^L EI\frac{\partial^2 y}{\partial x^2}\frac{\partial^2}{\partial x^2}\delta y dx + ky(L,t)\delta y(L,t)$$

$$= EI\frac{\partial^2 y}{\partial x^2}\frac{\partial}{\partial x}\delta y\bigg|_0^L - \frac{\partial}{\partial x}\left(EI\frac{\partial^2 y}{\partial x^2}\right)\delta y\bigg|_0^L + \int_0^L \frac{\partial^2}{\partial x^2}\left(EI\frac{\partial^2 y}{\partial x^2}\right)\delta y dx + ky(L,t)\delta y(L,t)$$

$$= EI\frac{\partial^2 y}{\partial x^2}\delta\frac{\partial y}{\partial x}\bigg|_{x=L} - EI\frac{\partial^2 y}{\partial x^2}\delta\frac{\partial y}{\partial x}\bigg|_{x=0} - \left[\frac{\partial}{\partial x}\left(EI\frac{\partial^2 y}{\partial x^2}\right) - ky\right]\delta y\bigg|_{x=L}$$

$$+ \frac{\partial}{\partial x}\left(EI\frac{\partial^2 y}{\partial x^2}\right)\delta y\bigg|_{x=0} + \int_0^L \frac{\partial^2}{\partial x^2}\left(EI\frac{\partial^2 y}{\partial x^2}\right)\delta y dx \tag{8.47}$$

Finally, using Eq. (8.19), the virtual work of the nonconservative distributed force is

$$\overline{\delta W}_{nc} = \int_0^L f\delta y dx \tag{8.48}$$

At this point, we have all the ingredients required for the extended Hamilton principle. Indeed, inserting Eqs. (8.46)–(8.48) into Eq. (8.17) and collecting terms, we obtain

$$\int_{t_1}^{t_2}\left\{-\int_0^L\left[m\frac{\partial^2 y}{\partial t^2} + \frac{\partial^2}{\partial x^2}\left(EI\frac{\partial^2 y}{\partial x^2}\right) - f\right]\delta y dx - EI\frac{\partial^2 y}{\partial x^2}\delta\frac{\partial y}{\partial x}\bigg|_{x=L}\right.$$

$$\left. + EI\frac{\partial^2 y}{\partial x^2}\delta\frac{\partial y}{\partial x}\bigg|_{x=0} + \left[\frac{\partial}{\partial x}\left(EI\frac{\partial^2 y}{\partial x^2}\right) - ky\right]\delta y\bigg|_{x=L} - \frac{\partial}{\partial x}\left(EI\frac{\partial^2 y}{\partial x^2}\right)\delta y\bigg|_{x=0}\right\}dt = 0 \tag{8.49}$$

Then, invoking the arbitrariness of the virtual displacement δy and using similar arguments to those used in Sec. 8.2, we conclude that Eq. (8.49) is satisfied if the differential equation

$$-\frac{\partial^2}{\partial x^2}\left(EI\frac{\partial^2 y}{\partial x^2}\right) + f = m\frac{\partial^2 y}{\partial t^2}, \quad 0 < x < L \tag{8.50}$$

is satisfied over the entire domain and the conditions

$$EI\frac{\partial^2 y}{\partial x^2}\delta\frac{\partial y}{\partial x} = 0, \quad \frac{\partial}{\partial x}\left(EI\frac{\partial^2 y}{\partial x^2}\right)\delta y = 0, \quad x = 0 \tag{8.51}$$

and

$$EI\frac{\partial^2 y}{\partial x^2}\delta\frac{\partial y}{\partial x} = 0, \quad \left[\frac{\partial}{\partial x}\left(EI\frac{\partial^2 y}{\partial x^2}\right) - ky\right]\delta y = 0, \quad x = L \tag{8.52}$$

are satisfied at the boundaries. Equations (8.51) on the one hand and Eqs. (8.52) on the other hand can be satisfied in two ways, but only one way is the correct one. Because the slope and the shearing force are not zero at a pinned end, we conclude from Eqs.

(8.51) that the boundary conditions at the left end are

$$y = 0, \; EI\frac{\partial^2 y}{\partial x^2} = 0, \; x = 0 \tag{8.53}$$

which state that the displacement and bending moment are zero at a pinned end. Similarly, because the displacement and the slope are not zero at a spring-supported end, boundary conditions at the right end must be

$$EI\frac{\partial^2 y}{\partial x^2} = 0, \; \frac{\partial}{\partial x}\left(EI\frac{\partial^2 y}{\partial x^2}\right) - ky = 0, \; x = L \tag{8.54}$$

which are consistent with the facts that the bending moment is zero and the shearing force is balanced at a spring-supported end. The boundary-value problem for a beam pinned at $x = 0$ and spring-supported at $x = L$ consists of the differential equation (8.50) and the boundary conditions (8.53) and (8.54). Hence the extended Hamilton principle yields once again the correct boundary-value problem, including boundary conditions that are correct both in form and number.

8.4 FREE VIBRATION. THE DIFFERENTIAL EIGENVALUE PROBLEM

We stressed repeatedly throughout this text, and in particular early in this chapter, that discrete and distributed systems differ more in form than in substance. Hence, in seeking solutions to vibration problems for distributed-parameter systems, we follow the same pattern as for discrete systems. Consistent with this, we consider first the free vibration problem, which leads naturally to the eigenvalue problem. For distributed-parameter systems, however, these are differential eigenvalue problems, each one depending on the system stiffness properties, as opposed to an algebraic eigenvalue problem of the same matrix form for all discrete systems.

We consider the vibrating string of Fig. 8.2a. In the case of free vibration, namely, when the distributed force $f(x,t)$ is zero, the boundary-value problem reduces to the differential equation

$$\frac{\partial}{\partial x}\left[T(x)\frac{\partial y(x,t)}{\partial x}\right] = \rho(x)\frac{\partial^2 y(x,t)}{\partial t^2}, \; 0 < x < L \tag{8.55}$$

and the boundary conditions

$$y(0,t) = 0, \; y(L,t) = 0 \tag{8.56}$$

As for discrete systems, we explore the circumstances under which the motion of the string is synchronous, namely, one in which every point of the string executes the same motion in time, passing through equilibrium at the same time and reaching the maximum excursion at the same time. The implication is that during synchronous motion the string exhibits a certain unique profile, or general shape, and the profile does not change with time, only the amplitude of the profile does. In mathematical terminology, such a solution $y(x,t)$ of the boundary-value problem, Eqs. (8.55) and (8.56), is said to be separable in the spatial variable x and time t, and can be expressed in the form

$$y(x,t) = Y(x)F(t) \tag{8.57}$$

where $Y(x)$ represents the string profile, or shape, or configuration, a function of x alone, and $F(t)$ indicates how the amplitude of the profile varies with time t.

Next, we introduce Eq. (8.57) into Eq. (8.55), divide through by $\rho(x)Y(x)F(t)$ and obtain

$$\frac{1}{\rho(x)Y(x)}\frac{d}{dx}\left[T(x)\frac{dY(x)}{dx}\right] = \frac{1}{F(t)}\frac{d^2F(t)}{dt^2}, \quad 0 < x < L \tag{8.58}$$

where, because Y depends only on x and F only on t, partial derivatives have been replaced by total derivatives. Observing that the left side of Eq. (8.58) depends on x alone and the right side on t alone, we conclude that the solution $y(x,t)$ of the boundary-value problem is indeed separable. Then, because both x and t are independent variables, we use the standard argument in the method of separation of variables (Sec. 7.6) and state that both sides of Eq. (8.58) must be equal to the same constant. In view of the fact that all quantities appearing in Eq. (8.58) are real, the constant must be real. Hence, denoting the constant by λ, Eq. (8.58) can be rewritten as

$$\frac{1}{\rho(x)Y(x)}\frac{d}{dx}\left[T(x)\frac{dY(x)}{dx}\right] = \frac{1}{F(t)}\frac{d^2F(t)}{dt^2} = \lambda, \quad 0 < x < L \tag{8.59}$$

To explore the nature of the constant λ, we consider the right side of Eq. (8.59) and write

$$\frac{d^2F(t)}{dt^2} - \lambda F(t) = 0 \tag{8.60}$$

The solution of Eq. (8.60) has the exponential form

$$F(t) = Ae^{st} \tag{8.61}$$

so that, inserting Eq. (8.61) into Eq. (8.60) and dividing through by Ae^{st}, we obtain the characteristic equation

$$s^2 - \lambda = 0 \tag{8.62}$$

which has the solutions

$$\begin{matrix} s_1 \\ s_2 \end{matrix} = \pm\sqrt{\lambda} \tag{8.63}$$

If λ is positive, then the roots are real, one positive and one negative, so that $F(t)$ is the sum of two exponential terms, one diverging and the other converging. Because diverging solutions are inconsistent with small string oscillations, the case in which λ is positive must be ruled out. Hence, we assume that λ is negative and introduce the notation $\lambda = -\omega^2$, so that Eq. (8.63) becomes

$$\begin{matrix} s_1 \\ s_2 \end{matrix} = \pm\sqrt{-\omega^2} = \pm i\omega \tag{8.64}$$

In this case, Eq. (8.61) yields

$$F(t) = A_1e^{s_1t} + A_2e^{s_2t} = A_1e^{i\omega t} + A_2e^{-i\omega t} \tag{8.65}$$

Equation (8.65) represents harmonic oscillation, which is consistent with the assumption that the string undergoes small motions. Then, following the same steps as in Sec. 3.1, solution (8.65) can be expressed in the more familiar form

$$F(t) = C\cos(\omega t - \phi) \tag{8.66}$$

where C is an amplitude, ϕ a phase angle and ω the frequency of oscillation. It follows that the string admits synchronous motion in the form of harmonic oscillation.

There are two questions remaining, the first concerns the value of the frequency ω and the second the displacement configuration $Y(x)$ assumed by the string during the harmonic oscillation. Both questions can be answered by equating the left side of Eq. (8.59) to $\lambda = -\omega^2$ and solving the differential equation

$$-\frac{d}{dx}\left[T(x)\frac{dY(x)}{dx}\right] = \omega^2 \rho(x) Y(x),\ 0 < x < L \tag{8.67}$$

where the solution must satisfy the boundary conditions, Eqs. (8.56). Inserting Eq. (8.57) into Eqs. (8.56) and dividing by $F(t)$, the boundary conditions reduce to

$$Y(0) = 0,\ Y(L) = 0 \tag{8.68}$$

The problem of determining the constant ω^2 such that Eq. (8.67) admits nontrivial solutions $Y(x)$ satisfying boundary conditions (8.68) is known as the *differential eigenvalue problem*.

Next, we turn our attention to the free vibration of beams in bending. In the absence of external excitations, $f(x,t) = 0$, the partial differential equation for the transverse displacement $y(x,t)$ of a beam in bending, Eq. (8.39) reduces to

$$-\frac{\partial^2}{\partial x^2}\left[EI(x)\frac{\partial^2 y(x,t)}{\partial x^2}\right] = m(x)\frac{\partial^2 y(x,t)}{\partial t^2},\ 0 < x < L \tag{8.69}$$

The solution $y(x,t)$ of Eq. (8.69) must satisfy two boundary conditions at each end. Some typical boundary conditions are given by Eqs. (8.40)–(8.44). For the sake of this discussion, we consider a beam supported by a spring at $x = 0$ and pinned at $x = L$ (Fig. 8.9), so that the boundary conditions are

$$EI(x)\frac{\partial^2 y(x,t)}{\partial x^2} = 0,\ -\frac{\partial}{\partial x}\left[EI(x)\frac{\partial^2 y(x,t)}{\partial x^2}\right] = ky(x,t),\ x = 0 \tag{8.70}$$

and

$$y(x,t) = 0,\ EI(x)\frac{\partial^2 y(x,t)}{\partial x^2} = 0,\ x = L \tag{8.71}$$

The solution of Eq. (8.69) has the same form as that given by Eq. (8.57), in which $F(t)$ is a harmonic function, Eq. (8.66). Inserting Eqs. (8.57) and (8.66) into Eqs. (8.69)–(8.71) and dividing through by $F(t)$, we obtain the differential eigenvalue problem consisting of the differential equation

$$\frac{d^2}{dx^2}\left[EI(x)\frac{d^2 Y(x)}{dx^2}\right] = \omega^2 m(x) Y(x),\ 0 < x < L \tag{8.72}$$

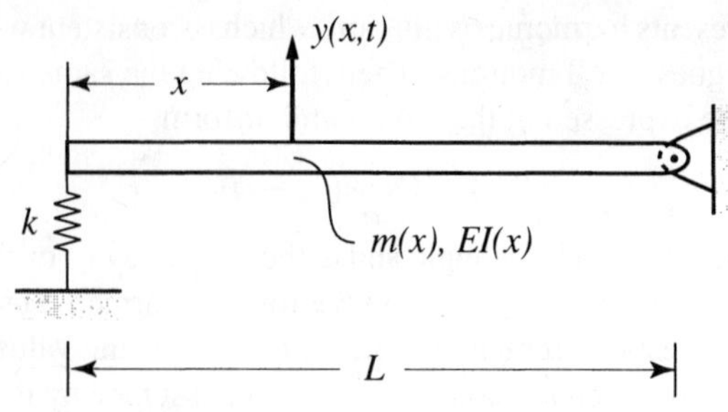

FIGURE 8.9
Beam spring supported at $x = 0$ and pinned at $x = L$

and the boundary conditions

$$EI(x)\frac{d^2Y(x)}{dx^2} = 0, \quad -\frac{d}{dx}\left[EI(x)\frac{d^2Y(x)}{dx^2}\right] = kY(x), \ x = 0 \tag{8.73}$$

and

$$Y(x) = 0, \ EI(x)\frac{d^2Y(x)}{dx^2} = 0, \ x = L \tag{8.74}$$

We observe that the sign of the stiffness differential expression on the left side of Eq. (8.72) is positive, which is consistent with a discussion of this subject in Sec. 8.3.

There is a basic difference between eigenvalue problems for discrete and distributed systems, aside from the fact that one is algebraic and the other is differential. If we confine our discussion to conservative systems, then we conclude that there is one generic algebraic eigenvalue problem describing all systems, namely, Eq. (7.77), in which M is the mass matrix and K is the stiffness matrix, a fact alluded to in the beginning of this section. The situation is quite different with distributed-parameter systems. Indeed, whereas the term describing the mass properties of the system is essentially the same for all distributed systems, the term describing the stiffness properties represents a differential expression varying from system to system, depending on the order of the system. As an illustration, the expression on the right side of Eq. (8.67) is typical of second-order systems, such as strings in transverse vibration, rods in axial vibration and shafts in torsion. Of course, the nature of the displacement and of the mass and stiffness parameters differs from case to case, as can be concluded from Table 8.1 On the other hand, beams in bending represent fourth-order systems, and their stiffness is defined by a fourth-order differential expression.

We observe that the right side of the differential equation defining the eigenvalue problem, Eq. (8.67), was assigned a positive sign, which results in a negative sign for the left side. This assignment is dictated by the physics of the problem, and in particular by the fact that the mass density $\rho(x)$ on the right side is a positive quantity by definition. On the other hand, the sign on the left side depends on the order of the stiffness differential expression under consideration; it is negative for second-order expressions and positive for fourth-order ones. This statement can be easily explained by recalling that, in deriving the string boundary-value problem by means of the extended Hamilton's principle (Sec. 8.2), the term due to the string tension in the variation in the potential energy was integrated by parts once, resulting in one sign change, as can be seen from Eq. (8.26). On the other hand, the analogous term for beams in bending (Sec. 8.3) was integrated twice, resulting in no sign change, as can be verified from Eq. (8.47).

The difference between the eigenvalue problems for discrete systems and distributed-parameter systems on the one hand and among the eigenvalue problems for various distributed systems on the other hand is more in form than in substance. The implication is that, whereas the approach to the solution of the eigenvalue problem may differ from discrete to distributed systems, the solutions and their properties exhibit a large degree of similarity, and the same can be said about the solutions for various distributed systems. In view of this, we can use the example of the string vibration, Eqs. (8.67) and (8.68), to discuss the nature of the solution of the eigenvalue problem for distributed systems in general. Because Eq. (8.67) is of second order, its solution $Y(x)$ will contain two constants of integration, in addition to the parameter ω^2, for a total of three unknowns. To evaluate the three unknowns, we must invoke the boundary conditions, Eqs. (8.68). But, because there are only two boundary conditions, it is not possible to evaluate all three unknowns uniquely. This should come as no surprise, as Eq. (8.67) is homogeneous, so that only the general shape of $Y(x)$ can be determined uniquely, but not the amplitude. Hence, the two boundary conditions can be used to derive a *characteristic equation* for ω^2 and to determine one of the constants of integration in terms of the other. The characteristic equation, generally a transcendental equation, has a denumerably infinite number of roots ω_r^2 $(r = 1, 2, \ldots)$ known as *eigenvalues*, where the term "denumerable" refers to the fact that the subscript r is identified with a given eigenvalue whose value differs in general from that of any other eigenvalue. Corresponding to each of these roots there is one function $Y_r(x)$ $(r = 1, 2, \ldots)$, where $Y_r(x)$ are known as *eigenfunctions*. Physically, the square roots ω_r of the eigenvalues ω_r^2 are recognized as the *natural frequencies* of the system and the eigenfunctions $Y_r(x)$ as the *natural modes*. Whereas the natural frequencies ω_r can be determined uniquely, the natural modes cannot. Indeed, because the problem is homogeneous, only the general shape of $Y_r(x)$ can be determined uniquely, as $A_r Y_r(x)$ represents the same natural mode, where A_r is a constant. If the constant A_r is determined uniquely through a certain normalization process, thereby determining the amplitude of the mode uniquely, then the natural modes are referred to as *normal modes*. The distinction between natural modes and normal modes is primarily of academic interest, and quite often modes are referred to as "normal" regardless whether a formal normalization process has been used or not. The natural frequencies ω_r and natural modes $Y_r(x)$ $(r = 1, 2, \ldots)$ represent a characteristic of the system, as they depend on the mass density $\rho(x)$, the tension $T(x)$ and the boundary conditions. Indeed, a change in any of these factors brings about a change in the natural frequencies and modes. We will come back to this idea later in this section. Clearly, a discrete system possesses only a finite number of natural frequencies and modes, whereas a distributed-parameter system possesses an infinite number. But, as shown in Sect. 8.1, a distributed-mass string can be obtained from a lumped-mass system through a limiting process whereby the lumped masses are spread over the entire length of the string and the indexed position x_i is replaced by the continuous independent variable x. In the process, the finite number of natural frequencies and modal vectors are replaced by an infinite number, so that the eigenfunctions $Y_r(x)$ can be regarded as eigenvectors of infinite dimension.

Finally, we return to our original exploration of the circumstances under which the motion of a distributed-parameter system in free vibration is synchronous. The results of this exploration can be summarized by stating that there is a denumerably infinite number of ways in which a distributed system can execute synchronous motion. We refer to them as *natural motions* and, in view of Eqs. (8.57) and (8.66), we can express them in the form

$$y_r(x,t) = C_r Y_r(x) \cos(\omega_r t - \phi_r), \; r = 1, 2, \ldots \tag{8.75}$$

in which $Y_r(x)$ are the natural modes and ω_r the natural frequencies, both obtained by solving the system differential eigenvalue problem. Moreover, C_r and ϕ_r are amplitudes and phase angles, respectively, constants depending on the initial displacement profile $y_0(x)$ and initial velocity profile $v_0(x)$. Clearly, $Y_r(x)$ and ω_r represent internal factors, as they reflect the system parameters and the nature of the differential equations and boundary conditions, whereas C_r and ϕ_r represent external factors. It will be shown in Sec. 8.5, that the natural modes are orthogonal, and hence by definition independent. The implication is that every one of the natural modes can be excited independently of the other. In general, however, the motion of a distributed-parameter system can be expressed as a linear combination of the natural motions of the form

$$y(x,t) = \sum_{r=1}^{\infty} y_r(x,t) = \sum_{r=1}^{\infty} C_r Y_r(x) \cos(\omega_r t - \phi_r) \tag{8.76}$$

Hence, the free vibration problem reduces to extracting the constants C_r and ϕ_r ($r = 1, 2, \ldots$) from the initial displacement profile $y_0(x)$ and initial velocity profile $v_0(x)$. We discuss this subject in Sec. 8.6.

Example 8.1. Solve the eigenvalue problem for a uniform string fixed at $x = 0$ and $x = L$ (Fig. 8.2a) and plot the first three eigenfunctions. The tension T in the string is constant.

Inserting $\rho(x) = \rho =$ constant, $T(x) = T =$ constant into Eq. (8.67), we conclude that the transverse displacement $Y(x)$ must satisfy the differential equation

$$\frac{d^2Y(x)}{dx^2} + \beta^2 Y(x) = 0, \; 0 < x < L, \; \beta^2 = \frac{\omega^2 \rho}{T} \tag{a}$$

where the solution Y of Eq. (a) is subject to boundary conditions (8.68), or

$$Y(0) = 0, \; Y(L) = 0 \tag{b}$$

Equation (a) is harmonic in x and its solution is

$$Y(x) = A \sin \beta x + B \cos \beta x \tag{c}$$

where A and B are constants of integration. Inserting the first of boundary conditions (b) into Eq. (c), we conclude that $B = 0$, so that the solution reduces to

$$Y(x) = A \sin \beta x \tag{d}$$

On the other hand, introducing the second of boundary conditions (b) in Eq. (d), we obtain

$$Y(L) = A \sin \beta L = 0 \tag{e}$$

There are two ways in which Eq. (e) can be satisfied, namely, $A = 0$ and $\sin \beta L = 0$. But $A = 0$ must be ruled out, because this would yield the trivial solution $Y(x) = 0$. It follows that we must have

$$\sin \beta L = 0 \tag{f}$$

which is recognized as the characteristic equation. Its solution consists of the infinite set of characteristic values

$$\beta_r L = r\pi, \ r = 1, 2, \ldots \tag{g}$$

to which corresponds the infinite set of eigenfunctions, or natural modes

$$Y_r(x) = A_r \sin \frac{r\pi x}{L}, \ r = 1, 2, \ldots \tag{h}$$

where A_r are undetermined amplitudes, with the implication that only the mode shapes can be determined uniquely. The first three natural modes, normalized so that $A_r = 1 (r = 1, 2, 3)$, are plotted in Fig. 8.10. We note that the first mode has no nodes, the second has one node and the third has two nodes. In general the rth mode has $r - 1$ nodes ($r = 1, 2, \ldots$).

From the second of Eqs. (a) we conclude that the system natural frequencies are

$$\omega_r = \beta_r \sqrt{\frac{T}{\rho}} = r\pi \sqrt{\frac{T}{\rho L^2}}, \ r = 1, 2, \ldots \tag{i}$$

The frequency ω_1 is called the *fundamental frequency* and the higher frequencies ω_r ($r = 2, 3, \ldots$) are referred to as *overtones*. The overtones are integer multiples of the fundamental frequency, for which reason the fundamental frequency is called the *fundamental harmonic* and the overtones are known as *higher harmonics*.

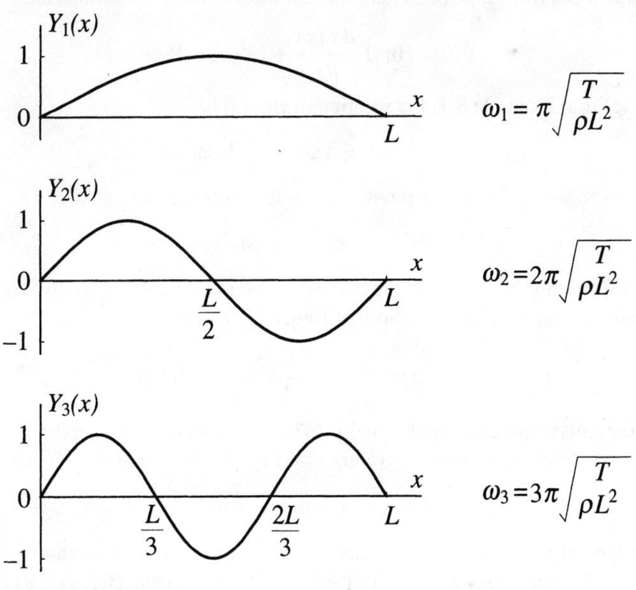

FIGURE 8.10
Natural modes of a fixed-fixed string

Vibrating systems possessing harmonic overtones are distinguished by the fact that, under certain excitations, they produce pleasant sounds. Such systems are not commonly encountered in nature but can be manufactured, particularly for use in musical instruments. It is a well-known fact that the string is the major component in a large number of musical instruments, such as the violin, the piano, the guitar and many other related instruments. For example, the violin has four strings, each possessing a different fundamental frequency. From Eq. (i), we observe that these frequencies depend on the tension T, the mass density ρ and the length L. The violinist tuning a violin merely ensures that the strings have the proper tension. This is done by comparing the pitch of a given note to that produced by a different instrument known to be tuned correctly. One must not infer from this, however, that the violin has only four fundamental frequencies and their higher harmonics. Indeed, whereas ρ and T are constant for each string, the violinist can change the pitch by adjusting the length of the strings. Hence, when fingers are run on the fingerboard, the artist merely adjusts the length L of the strings. Thus, there is a large variety of frequencies at the violinist's disposal. Generally the sounds consist of a combination of harmonics, with the lower harmonics being the predominant ones. However, a talented performer excites the proper array of higher harmonics to produce a pleasing sound.

Example 8.2. Assume that the string of Fig. 8.3a has uniform mass distribution, $\rho(x) = \rho = \text{constant}$, and constant tension, $T(x) = T = \text{constant}$, solve the eigenvalue problem for the parameter ratio $kL/T = 0.5$ and plot the first three modes. What can be said about the nature of the boundary $x = L$ as the mode number increases.

As in Example 8.1, the differential equation is

$$\frac{d^2Y(x)}{dx^2} + \beta^2 Y(x) = 0, \ 0 < x < L; \ \beta^2 = \frac{\omega^2 \rho}{T} \tag{a}$$

but this time, from Eqs. (8.31) and (8.32), the boundary conditions are

$$Y(0) = 0; \ T\frac{dY(x)}{dx} + kY(x) = 0, \ x = L \tag{b}$$

From Example 8.1, the solution of Eq. (a) is

$$Y(x) = A \sin \beta x + B \cos \beta x \tag{c}$$

Moreover, using the first of boundary conditions (b), $B = 0$, so that

$$Y(x) = A \sin \beta x \tag{d}$$

Then, inserting Eq. (d) into the second of boundary conditions (b) and rearranging, we obtain the characteristic equation, or frequency equation,

$$\tan \beta L = -\frac{T}{kL}\beta L = -2\beta L \tag{e}$$

It represents a transcendental equation to be solved numerically for the eigenvalues $\beta_r L$ ($r = 1, 2, \ldots$). Inserting these eigenvalues into Eq., (d), we obtain the eigenfunctions

$$Y_r(x) = A_r \sin \beta_r x, \ r = 1, 2, \ldots \tag{f}$$

A graphical solution of the characteristic equation, Eq. (e), is shown in Fig. 8.11. The first three eigenvalues are $\beta_1 L = 1.8366$, $\beta_2 L = 4.8158$, $\beta_3 L = 7.9171$. The first three eigenfunctions, normalized so that $A_r = 1$ ($r = 1, 2, 3$), are plotted in Fig. 8.12, and we observe that, as for the fixed-fixed string, the rth mode has $r - 1$ nodes ($r = 1, 2, \ldots$).

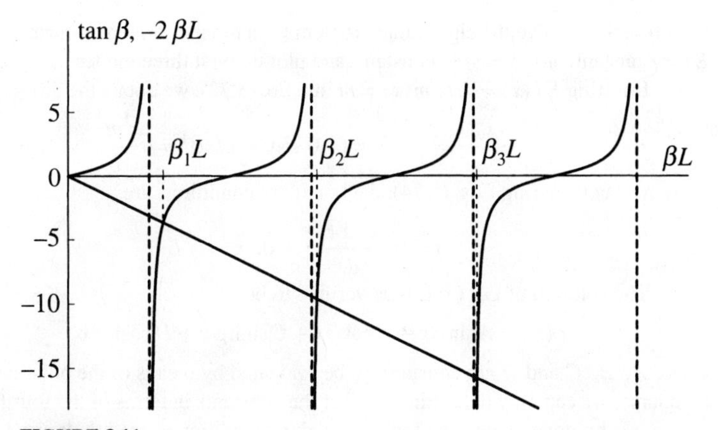

FIGURE 8.11
Graphical solution of the characteristic equation for a string fixed at $x = 0$ and spring supported at $x = L$

From Fig. 8.12, we conclude that, as $\beta_r L$ increases, the end $x = L$ behaves more and more like a free end. This observation is corroborated by Fig. 8.11, which shows that, as the mode number increases, $\beta_r L$ approaches $(2r - 1)\pi/2$, namely, the eigenvalues of a fixed-free uniform string.

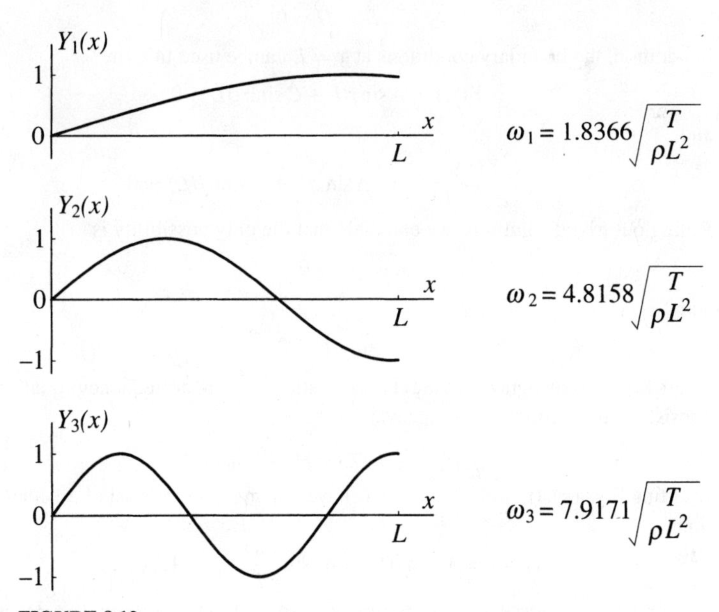

FIGURE 8.12
Natural modes for a string fixed at $x = 0$ and spring supported at $x = L$

Example 8.3. Solve the eigenvalue problem for a pinned-pinned uniform beam, $EI(x) = EI = \text{constant}$, $m(x) = m = \text{constant}$, and plot the first three modes.

Inserting $EI(x) = EI$, $m(x) = m$ into Eq. (8.72), we obtain the differential equation

$$\frac{d^4Y(x)}{dx^4} - \beta^4 Y(x) = 0, \ 0 < x < L; \ \beta^4 = \frac{\omega^2 m}{EI} \tag{a}$$

Moreover, considering Eqs. (8.74), the boundary conditions are

$$Y(x) = 0, \ \frac{d^2Y(x)}{dx^2} = 0, \ x = 0, L \tag{b}$$

The solution of Eq. (a) can be verified to be

$$Y(x) = A\sin\beta x + B\cos\beta x + C\sinh\beta x + D\cosh\beta x \tag{c}$$

where A, B, C and D are constants to be evaluated by means of the boundary conditions. Of course, we can only determine three of the constants in terms of the fourth, because the problem is homogeneous. The fourth boundary condition is used to derive a characteristic equation for β. Two of the boundary conditions involve the second derivative of Y, which is simply

$$\frac{d^2Y(x)}{dx^2} = \beta^2[-A\sin\beta x - B\cos\beta x + C\sinh\beta x + D\cosh\beta x] \tag{d}$$

The boundary conditions at $x = 0$ yield

$$Y(0) = B + D = 0 \tag{e}$$

and

$$\left.\frac{d^2Y(x)}{dx^2}\right|_{x=0} = -B + D = 0 \tag{f}$$

from which we conclude that

$$B = D = 0 \tag{g}$$

In addition, the boundary conditions at $x = L$ can be used to write

$$Y(L) = A\sin\beta L + C\sinh\beta L = 0 \tag{h}$$

and

$$\frac{d^2Y(x)}{dx^2} = \beta^2(-A\sin\beta L + C\sinh\beta L) = 0 \tag{i}$$

Ruling out trivial solutions, we conclude that the only possibility is

$$C = 0 \tag{j}$$

and

$$\sin\beta L = 0 \tag{k}$$

where Eq. (k) is recognized as the characteristic equation, or frequency equation. Its solution consists of the infinite set of eigenvalues

$$\beta_r L = r\pi, \ r = 1, 2, \ldots \tag{l}$$

Inserting Eqs. (g), (j) and (l) into Eq. (c), we obtain the infinite set of associated eigenfunctions

$$Y_r(x) = A_r \sin\beta_r x = A_r \sin\frac{r\pi x}{L}, \ r = 1, 2, \ldots \tag{m}$$

The first three normal modes, normalized so that $A_r = 1$ $(r = 1, 2, 3)$, are plotted in Fig. 8.13.

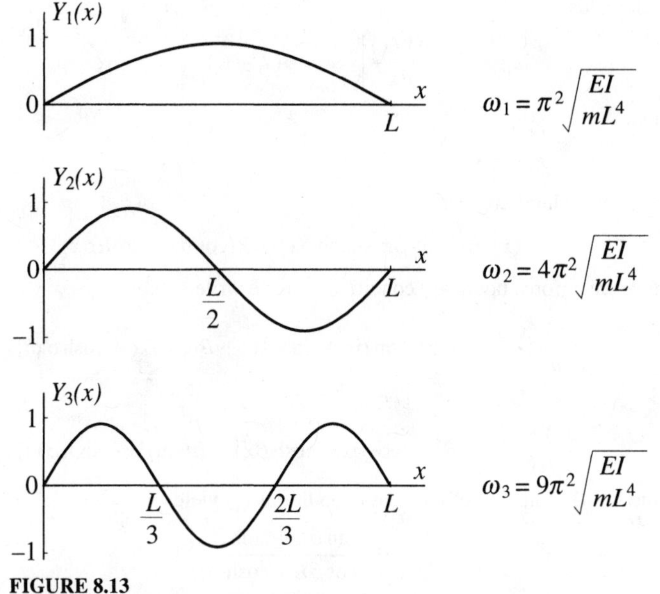

FIGURE 8.13
Natural modes for a pinned-pinned beam

Example 8.4. Solve the eigenvalue problem for a uniform cantilever beam clamped at $x = 0$ and plot the first three modes.

The differential equation is the same as in Example 8.3, namely,

$$\frac{d^4Y(x)}{dx^4} - \beta^4 Y(x) = 0,\ 0 < x < L;\ \beta^4 = \frac{\omega^2 m}{EI} \tag{a}$$

but the boundary conditions are different. From Eqs. (8.40) and (8.42), the boundary conditions are

$$Y(x) = 0,\quad \frac{dY(x)}{dx} = 0,\ x = 0 \tag{b}$$

and

$$\frac{d^2Y(x)}{dx^2} = 0,\quad \frac{d^3Y(x)}{dx^3} = 0,\ x = L \tag{c}$$

As in Example 8.3, the general solution of Eq. (a) is

$$Y(x) = A\sin\beta x + B\cos\beta x + C\sinh\beta x + D\cosh\beta x \tag{d}$$

but this time the constants are different, as should be expected. The first of boundary conditions (b) yields

$$Y(0) = B + D = 0 \tag{e}$$

so that

$$D = -B \tag{f}$$

For the second of boundary conditions (b), we write

$$\frac{dY(x)}{dx} = \beta(A\cos\beta x - B\sin\beta x + C\cosh\beta x + D\sinh\beta x) \tag{g}$$

which yields

$$\left.\frac{dY(x)}{dx}\right|_{x=0} = \beta(A+C) = 0 \tag{h}$$

so that

$$C = -A \tag{i}$$

Hence, the solution reduces to

$$Y(x) = A(\sin\beta x - \sinh\beta x) + B(\cos\beta x - \cosh\beta x) \tag{j}$$

Before we enforce boundary conditions (c), we write

$$\frac{d^2Y(x)}{dx^2} = -\beta^2[A(\sin\beta x + \sinh\beta x) + B(\cos\beta x + \cosh\beta x)] \tag{k}$$

and

$$\frac{d^3Y(x)}{dx^3} = -\beta^3[A(\cos\beta x + \cosh\beta x) - B(\sin\beta x - \sinh\beta x)] \tag{l}$$

Using Eq. (k), the first of boundary conditions (c) yields

$$B = -\frac{\sin\beta L + \sinh\beta L}{\cos\beta L + \cosh\beta L}A \tag{m}$$

Inserting Eq. (m) into Eq. (l) and using the second of boundary conditions (c), we obtain the characteristic equation

$$\cos\beta L \cosh\beta L = -1 \tag{n}$$

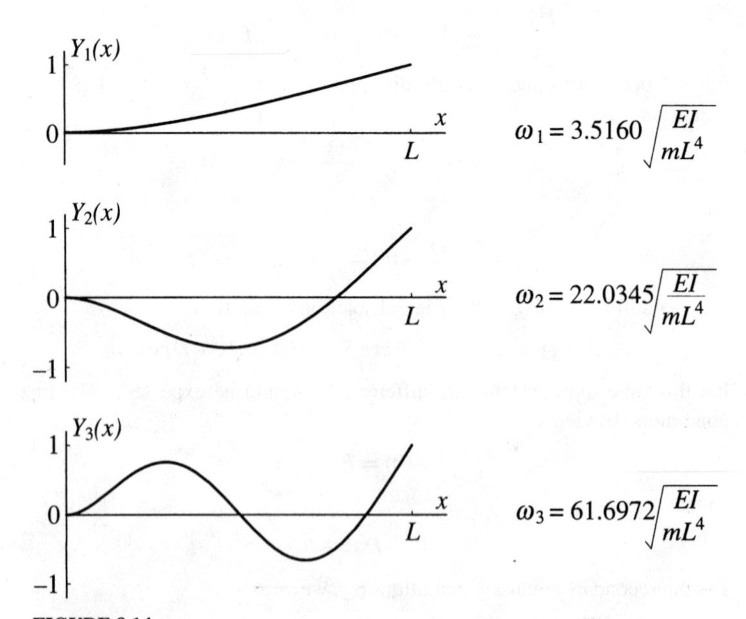

FIGURE 8.14
Natural modes for a cantilever beam

Equation (n) is transcendental, and must be solved numerically. Then, introducing Eq. (m) in Eq. (j), in conjunction with the eigenvalues $\beta_r L$ just obtained, we can write the eigenfunctions in the form

$$Y_r(x) = A_r \left[\sin \beta_r x - \sinh \beta_r x - \frac{\sin \beta_r L + \sinh \beta_r L}{\cos \beta_r L + \cosh \beta_r L} (\cos \beta_r x - \cosh \beta_r x) \right], \quad r = 1, 2, \ldots \tag{o}$$

The first three eigenvalues are $\beta_1 L = 1.8751$, $\beta_2 L = 4.6941$, $\beta_3 L = 7.8548$. The first three natural modes have been plotted in Fig. 8.14, where the modes were normalized numerically by adjusting the amplitudes so as to satisfy $Y_r(L) = 1$ $(r = 1, 2, 3)$. Inserting the eigenvalues given above into the right expression in Eq. (a), we obtain the first three natural frequencies

$$\omega_1 = 3.5160 \sqrt{\frac{EI}{mL^4}}, \quad \omega_2 = 22.0345 \sqrt{\frac{EI}{mL^4}}, \quad \omega_3 = 61.6972 \sqrt{\frac{EI}{mL^4}} \tag{p}$$

and we observe that the higher frequencies are no longer integer multiples of the fundamental frequency, as they are in the case of the pinned-pinned beam.

Example 8.5. Assume that the spring supported-pinned beam of Fig. 8.9 is uniform, $EI(x) = EI$, $m(x) = m$, and solve the eigenvalue problem for the case in which $k = 25EI/L^3$. Plot the first three modes and draw conclusions.

As in Examples 8.3 and 8.4, the differential equation is

$$\frac{d^4 Y(x)}{dx^4} - \beta^4 Y(x) = 0, \ 0 < x < L; \ \beta^4 = \frac{\omega^2 m}{EI} \tag{a}$$

and, from Eqs. (8.73) and (8.74), the boundary conditions at the spring-supported end reduce to

$$\frac{d^2 Y(x)}{dx^2} = 0, \quad -\frac{d^3 Y(x)}{dx^3} = \frac{k}{EI} Y(x), \ x = 0 \tag{b}$$

Moreover, the boundary conditions at the pinned end are

$$Y(x) = 0, \quad \frac{d^2 Y(x)}{dx^2} = 0, \ x = L \tag{c}$$

Again, as in Examples 8.3 and 8.4, the general solution of Eq. (a) is

$$Y(x) = A \sin \beta x + B \cos \beta x + C \sinh \beta x + D \cosh \beta x \tag{d}$$

where the solution must satisfy boundary conditions (b) and (c). The case at hand differs from the pinned-pinned and the clamped-free cases in that the second of boundary conditions (b) involves the parameter ratio k/EI, which is likely to complicate the determination of the constants. This complication can be avoided by using the first of Eqs. (b) and Eqs. (c) to solve for B, C and D in terms of A. Then, the second of Eqs. (b) can be used to derive the characteristic equation. Before we proceed with the evaluation of the constants, we write

$$\frac{d^2 Y}{dx^2} = \beta^2 (-A \sin \beta x - B \cos \beta x + C \sinh \beta x + D \cosh \beta x) \tag{e}$$

so that the first of boundary conditions (b) yields

$$\left.\frac{d^2Y(x)}{dx^2}\right|_{x=0} = \beta^2(-B+D) = 0 \tag{f}$$

which gives simply

$$D = B \tag{g}$$

Hence, Eq. (d) reduces to

$$Y(x) = A\sin\beta x + B(\cos\beta x + \cosh\beta x) + C\sinh\beta x \tag{h}$$

which can be used to write

$$\frac{d^2Y(x)}{dx^2} = \beta^2[-A\sin\beta x - B(\cos\beta x - \cosh\beta x) + C\sinh\beta x] \tag{i}$$

so that boundary conditions (c) yield

$$Y(L) = A\sin\beta L + B(\cos\beta L + \cosh\beta L) + C\sinh\beta L = 0 \tag{j}$$

and

$$\left.\frac{d^2Y(x)}{dx^2}\right|_{x=L} = \beta^2[-A\sin\beta L - B(\cos\beta L - \cosh\beta L) + C\sinh\beta L] = 0 \tag{k}$$

Solving Eqs. (j) and (k) for B and C in terms of A, we obtain

$$B = -\frac{\sin\beta L}{\cos\beta L}A,\ C = \frac{\sin\beta L\cosh\beta L}{\cos\beta L\sinh\beta L}A \tag{l}$$

Inserting Eqs. (l) into Eq. (h), we can write

$$Y(x) = \frac{A}{\cos\beta L\sinh\beta L}[\cos\beta L\sinh\beta L\sin\beta x$$
$$-\sin\beta L\sinh\beta L(\cos\beta x + \cosh\beta x) + \sin\beta L\cosh\beta L\sinh\beta x] \tag{m}$$

Before enforcing the second of boundary conditions (b), we write

$$\frac{d^3Y(x)}{dx^3} = \frac{A\beta^3}{\cos\beta L\sinh\beta L}[-\cos\beta L\sinh\beta L\cos\beta x$$
$$-\sin\beta L\sinh\beta L(\sin\beta x + \sinh\beta x) + \sin\beta L\cosh\beta L\cosh\beta x] \tag{n}$$

Hence, inserting Eqs. (m) and (n) into the second of Eqs. (b), we have

$$-\frac{A\beta^3}{\cos\beta L\sinh\beta L}(-\cos\beta L\sinh\beta L + \sin\beta L\cosh\beta L) = \frac{k}{EI}\left(-\frac{2A\sin\beta L\sinh\beta L}{\cos\beta L\sinh\beta L}\right) \tag{o}$$

which, for the given parameter ratio, reduces to

$$\coth\beta L - \cot\beta L = \frac{2kL^3}{EI}\frac{1}{(\beta L)^3} = \frac{50}{(\beta L)^3} \tag{p}$$

Equation (p) represents the characteristic equation, or frequency equation, and can be solved numerically for the eigenvalues $\beta_r L$ $(r = 1, 2, \dots)$. Then, omitting the inconsequential term at the denominator in Eq. (m), the eigenfunctions can be written as

$$Y_r(x) = A_r[\cos\beta_r L\sinh\beta_r L\sin\beta_r x - \sin\beta_r L\sinh\beta_r L(\cos\beta_r x + \cosh\beta_r x)$$
$$+ \sin\beta_r L\cosh\beta_r L\sinh\beta_r x],\ r = 1, 2, \dots \tag{q}$$

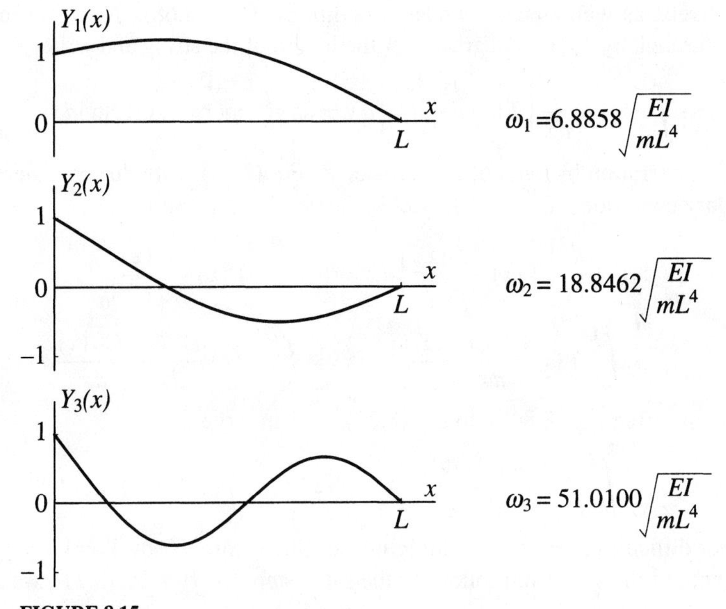

FIGURE 8.15
Natural modes for a spring supported-pinned beam

It should be pointed out that the effect of the spring k at $x = 0$ enters into the eigenfunctions implicitly through the eigenvalues $\beta_r L$. The first three eigenvalues are $\beta_1 L = 2.6241$, $\beta_2 L = 4.3412$, $\beta_3 L = 7.1421$. The first three modes, normalized so that $Y_r(0) = 1$ $(r = 1, 2, 3)$, are plotted in Fig. 8.15. We conclude that, as the mode number increases, the spring-supported end at $x = 0$ behaves more and more like a pinned end.

8.5 ORTHOGONALITY OF MODES. EXPANSION THEOREM

The similarity between discrete and distributed vibrating systems becomes striking when we consider the fact that the eigenfunctions $Y_r(x)$ $(r = 1, 2, \ldots)$ possess the *orthogonality property* in a manner analogous to the eigenvectors $\mathbf{u}_r$ $(i = 1, 2, \ldots, n)$. Moreover, an analogous *expansion theorem* exists for distributed systems as for discrete systems, and this expansion theorem forms the basis for a *modal analysis* permitting a most efficient derivation of the system response to both initial excitations and applied forces.

To derive the orthogonality relations for a string, we consider two distinct solutions of the eigenvalue problem, ω_r^2, $Y_r(x)$ and ω_s^2, $Y_s(x)$; from Eq. (8.67), they satisfy the differential equations

$$-\frac{d}{dx}\left[T(x)\frac{dY_r(x)}{dx}\right] = \omega_r^2 \rho(x) Y_r(x), \; 0 < x < L \tag{8.77}$$

and

$$-\frac{d}{dx}\left[T(x)\frac{dY_s(x)}{dx}\right] = \omega_s^2 \rho(x) Y_s(x), \; 0 < x < L \tag{8.78}$$

respectively, as well as the boundary conditions, Eqs. (8.68). Next, we multiply Eq. (8.77) through by $Y_s(x)$, integrate over the length of the string and write

$$-\int_0^L Y_s(x)\frac{d}{dx}\left[T(x)\frac{dY_r(x)}{dx}\right]dx = \omega_r^2 \int_0^L \rho(x)Y_s(x)Y_r(x)dx \tag{8.79}$$

But, an integration by parts of the left side of Eq. (8.79), with due consideration of the boundary conditions, Eqs. (8.68), yields

$$-\int_0^L Y_s(x)\frac{d}{dx}\left[T(x)\frac{dY_r(x)}{dx}\right]dx = -Y_s(x)T(x)\frac{dY_r(x)}{dx}\bigg|_0^L + \int_0^L T(x)\frac{dY_s(x)}{dx}\frac{dY_r(x)}{dx}dx = \int_0^L T(x)\frac{dY_s(x)}{dx}\frac{dY_r(x)}{dx}dx \tag{8.80}$$

Hence, inserting Eq. (8.80) into Eq. (8.79), we can write

$$\int_0^L T(x)\frac{dY_s(x)}{dx}\frac{dY_r(x)}{dx}dx = \omega_r^2 \int_0^L \rho(x)Y_s(x)Y_r(x)dx \tag{8.81}$$

It is not difficult to see that, multiplying Eq. (8.78) through by $Y_r(x)$, integrating over the length of the string and following the same steps as with Eq. (8.77), we obtain

$$\int_0^L T(x)\frac{dY_r(x)}{dx}\frac{dY_s(x)}{dx}dx = \omega_s^2 \int_0^L \rho(x)Y_r(x)Y_s(x)dx \tag{8.82}$$

But, the integrals in Eqs. (8.81) and (8.82) are unaffected by the order of the subscripts r and s, so that the integrals on the left side on the one hand and the integrals on the right side on the other hand have the same value. It follows that, subtracting Eq. (8.82) from Eq. (8.81), we have

$$(\omega_r^2 - \omega_s^2)\int_0^L \rho(x)Y_r(x)Y_s(x)dx = 0 \tag{8.83}$$

In view of the fact that the eigenvalues are distinct, we conclude that Eq. (8.83) can be satisfied only if

$$\int_0^L \rho(x)Y_r(x)Y_s(x)dx = 0,\ r,s = 1,2,\ldots;,\ \omega_r^2 \neq \omega_s^2 \tag{8.84}$$

Equations (8.84) represent the *orthogonality conditions* for the eigenfunctions of a fixed-fixed string in transverse vibration. Then, inserting Eqs. (8.84) into Eq. (8.79), we conclude that

$$-\int_0^L Y_s(x)\frac{d}{dx}\left[T(x)\frac{dY_r(x)}{dx}\right]dx = 0,\ r,s = 1,2,\ldots;\ \omega_r^2 \neq \omega_s^2 \tag{8.85}$$

constitute companion orthogonality conditions. Moreover, inserting Eqs. (8.84) into Eq. (8.81), or Eq. (8.82), we obtain an alternative set of companion orthogonality conditions in the form

$$\int_0^L T(x)\frac{dY_r(x)}{dx}\frac{dY_s(x)}{dx}dx = 0,\ r,s = 1,2,\ldots;\ \omega_r^2 \neq \omega_s^2 \tag{8.86}$$

When $s = r$, the integral in Eqs. (8.82), and hence in Eqs. (8.85) and (8.86), is not zero. In fact, the integral is a positive real number. It is convenient to normalize the natural modes by setting the value of the integral in Eqs. (8.84) equal to unity for $s = r$, or

$$\int_0^L \rho(x)Y_r^2(x)dx = 1, \; r = 1, 2, \ldots \tag{8.87}$$

Then, from Eq. (8.79) with $s = r$, it follows immediately that

$$-\int_0^L Y_r(x)\frac{d}{dx}\left[T(x)\frac{dY_r(x)}{dx}\right]dx = \omega_r^2, \; r = 1, 2, \ldots \tag{8.88}$$

and, in the alternative form, that

$$\int_0^L T(x)\left[\frac{dY_r(x)}{dx}\right]^2 dx = \omega_r^2, \; r = 1, 2, \ldots \tag{8.89}$$

Under these circumstances, we refer to the natural modes as *normal modes*. Equations (8.84) and (8.87) can be combined into the *orthonormality relations*

$$\int_0^L \rho(x)Y_r(x)Y_s(x)dx = \delta_{rs}, \; r, s = 1, 2, \ldots \tag{8.90}$$

where δ_{rs} is the Kronecker delta. Similarly, Eqs. (8.85) and (8.88) can be combined into the companion orthonormality relations

$$-\int_0^L Y_s(x)\frac{d}{dx}\left[T(x)\frac{dY_r(x)}{dx}\right]dx = \omega_r^2\delta_{rs}, \; r, s = 1, 2, \ldots \tag{8.91}$$

and Eqs. (8.86) and (8.89) yield the alternative form of orthonormality relations

$$\int_0^L T(x)\frac{dY_r(x)}{dx}\frac{dY_s(x)}{dx}dx = \omega_r^2\delta_{rs}, \; r, s = 1, 2, \ldots \tag{8.92}$$

Equations (8.91) will prove useful in the derivation of the system response and Eqs. (8.92) will be used in a variational approach to the eigenvalue problem, where the latter is convenient for obtaining approximate solutions.

The orthogonality of the natural modes is indispensable to the solution of boundary-value problems. In this regard, we recall that in Sec. 3.9 we expanded periodic functions in terms of trigonometric functions with frequencies equal to integer multiples of the fundamental frequency. This was done through the use of Fourier series, discussed in some detail in Appendix A. Figure A.1 shows such a periodic function expressed as an infinite series of the orthogonal harmonic functions $\sin rt$ $(r = 1, 2, \ldots)$. Concentrating on a single period, say $0 < t < 2\pi$, instead of the entire time domain, we conclude that a function of arbitrary shape can be expanded in an infinite series of orthogonal harmonic functions. The *expansion theorem* represents a generalization of this idea. It can be stated as follows: *Any function $Y(x)$ representing a possible displacement of the string, which implies that $Y(x)$ satisfies the boundary conditions of the problem and is such that $(d/dx)[T(x)dY(x)/dx]$ is a continuous function, can be expanded in the absolutely and* *uniformly convergent series of the eigenfunctions*

$$Y(x) = \sum_{r=1}^{\infty} c_r Y_r(x) \tag{8.93}$$

where the constant coefficients c_r are defined by

$$c_r = \int_0^L \rho(x) Y_r(x) Y(x) dx, \; r = 1, 2, \ldots \tag{8.94}$$

and

$$\omega_r^2 c_r = -\int_0^L Y_r(x) \frac{d}{dx}\left[T(x)\frac{dY(x)}{dx}\right] dx, \; r = 1, 2, \ldots \tag{8.95}$$

We note that Eqs. (8.94) and (8.95) follow directly from the orthonormality relations, Eqs. (8.90) and (8.91), respectively. It should be pointed out that the expansion theorem holds for any type of boundary conditions, and is not restricted to fixed-fixed strings, although the boundary conditions are essential to the determination of the eigenvalues and eigenfunctions of a specific system.

The derivation of the response by modal analysis is based on the expansion theorem, which means that, before the response can be obtained, it is necessary to solve the differential eigenvalue problem for the system. Clearly, the same expansion theorem is valid for thin rods in axial vibration and circular shafts in torsion.

As for strings in transverse vibration, the solution of the eigenvalue problem for beams in bending, Eqs. (8.72)–(8.74), consists of the denumerably infinite set of eigenvalues ω_r^2 and eigenfunctions, or natural modes $Y_r(x)$, where ω_r are recognized as the natural frequencies ($r = 1, 2, \ldots$). The natural modes possess the orthogonality property, which can be used to formulate an expansion theorem. However, because beams represent fourth-order systems, both the orthogonality relations and the expansion theorem differ to some extent from those for second-order systems, such as strings in transverse vibration, rods in axial vibration and shafts in torsion. To demonstrate the orthogonality relations for beams, we consider two distinct solutions of the eigenvalue problem, ω_r^2, $Y_r(x)$ and ω_s^2, $Y_s(x)$, and write

$$\frac{d^2}{dx^2}\left[EI(x)\frac{d^2 Y_r(x)}{dx^2}\right] = \omega_r^2 m(x) Y_r(x), \; 0 < x < L \tag{8.96}$$

and

$$\frac{d^2}{dx^2}\left[EI(x)\frac{d^2 Y_s(x)}{dx^2}\right] = \omega_s^2 m(x) Y_s(x), \; 0 < x < L \tag{8.97}$$

Next, we multiply Eq. (8.96) by $Y_s(x)$, integrate over the length of the beam and write

$$\int_0^L Y_s(x)\frac{d^2}{dx^2}\left[EI(x)\frac{d^2 Y_r(x)}{dx^2}\right] dx = \omega_r^2 \int_0^L m(x) Y_s(x) Y_r(x) dx \tag{8.98}$$

Integrating the left side of Eq. (8.98) by parts twice and considering boundary conditions

(8.73) and (8.74), we obtain

$$\int_0^L Y_s(x) \frac{d^2}{dx^2}\left[EI(x)\frac{d^2Y_r(x)}{dx^2}\right]dx = \left\{Y_s(x)\frac{d}{dx}\left[EI(x)\frac{d^2Y_r(x)}{dx^2}\right]\right\}\Bigg|_0^L - \left[\frac{dY_s(x)}{dx}EI(x)\frac{d^2Y_r(x)}{dx^2}\right]\Bigg|_0^L + \int_0^L EI(x)\frac{d^2Y_s(x)}{dx^2}\frac{d^2Y_r(x)}{dx^2}dx = kY_s(0)Y_r(0) + \int_0^L EI(x)\frac{d^2Y_s(x)}{dx^2}\frac{d^2Y_r(x)}{dx^2}dx \tag{8.99}$$

Inserting Eq. (8.99) into Eq. (8.98), we have simply

$$kY_s(0)Y_r(0) + \int_0^L EI(x)\frac{d^2Y_s(x)}{dx^2}\frac{d^2Y_r(x)}{dx^2}dx = \omega_r^2 \int_0^L m(x)Y_s(x)Y_r(x)dx \tag{8.100}$$

Similarly, multiplying Eq. (8.97) by $Y_r(x)$, integrating over the length of the beam and following the same steps as with Eq. (8.96), it is easy to show that

$$kY_r(0)Y_s(0) + \int_0^L EI(x)\frac{d^2Y_r(x)}{dx^2}\frac{d^2Y_s(x)}{dx^2}dx = \omega_s^2 \int_0^L m(x)Y_r(x)Y_s(x)dx \tag{8.101}$$

Then, observing that the order of the subscripts r and s is immaterial and subtracting Eq. (8.101) from Eq. (8.100), we obtain

$$(\omega_r^2 - \omega_s^2)\int_0^L m(x)Y_r(x)Y_s(x)dx = 0 \tag{8.102}$$

so that, for two distinct solutions of the eigenvalue problem, we must have

$$\int_0^L m(x)Y_r(x)Y_s(x) = 0,\ r,s = 1,2,\dots;\ \omega_r^2 \neq \omega_s^2 \tag{8.103}$$

Equations (8.103) represent the orthogonality relations for a beam in bending and, except for $m(x)$ replacing $\rho(x)$, they are essentially the same as those for a string in transverse vibration, Eqs. (8.84). Moreover, inserting Eqs. (8.103) into Eq. (8.98), we obtain the companion orthogonality relations

$$\int_0^L Y_s(x)\frac{d^2}{dx^2}\left[EI(x)\frac{d^2Y_r(x)}{dx^2}\right]dx = 0,\ r,s = 1,2,\dots;\ \omega_r^2 \neq \omega_s^2 \tag{8.104}$$

which are different from the companion orthogonality relations for a string, Eqs. (8.85). The main differences are in sign and in the order of the differential expression. Moreover, inserting Eqs. (8.103) into Eq. (8.100), or Eq. (8.101), we obtain the alternative form of the companion orthogonality relations

$$\int_0^L EI(x)\frac{d^2Y_r(x)}{dx^2}\frac{d^2Y_s(x)}{dx^2}dx + kY_r(0)Y_s(0) = 0,\ r,s = 1,2,\dots;\omega_r^2 \neq \omega_s^2 \tag{8.105}$$

We observe that not only does the integral contain second derivatives, as opposed to first derivatives for the string, but also the relations involve the explicit effect of the spring at $x = 0$. Again, for response problems, we are interested in the companion orthogonality relations, Eqs. (8.104). On the other hand, the alternative form of the companion orthogonality relations, Eqs. (8.105), is more useful in approximate solutions of the eigenvalue problem.

Using the scheme given by Eq. (8.87) with $p(x)$ replaced by $m(x)$, we can normalize the natural modes so as to satisfy $\int_0^L m(x)Y_r^2(x)dx = 1$ $(r = 1, 2, \ldots)$, in which case the modes are referred to as normal modes; they satisfy the orthonormality relations

$$\int_0^L m(x)Y_r(x)Y_s(x)dx = \delta_{rs},\ r, s = 1, 2, \ldots \tag{8.106}$$

Moreover, using Eq. (8.98), we see that they also satisfy the companion orthonormality relations

$$\int_0^L Y_s \frac{d^2}{dx^2}\left[EI(x)\frac{d^2Y_r(x)}{dx^2}\right]dx = \omega_r^2\delta_{rs},\ r, s = 1, 2, \ldots \tag{8.107}$$

and, using Eq. (8.100), they can be shown to satisfy the alternative companion orthonormality relations

$$\int_0^L EI(x)\frac{d^2Y_r(x)}{dx^2}\frac{d^2Y_s(x)}{dx^2}dx + kY_r(0)Y_s(0) = \omega_r^2\delta_{rs},\ r, s = 1, 2, \ldots \tag{8.108}$$

The above orthonormality relations, Eqs. (8.106) and (8.107), permit us to state the following *expansion theorem: Any function $Y(x)$ representing a possible displacement of the beam, which implies that $Y(x)$ satisfies the boundary conditions of the problem and is such that $(d^2/dx^2)[EI(x)d^2Y(x)/dx^2]$ is continuous, can be expanded in the absolutely and uniformly convergent series of the eigenfunctions*

$$Y(x) = \sum_{r=1}^{\infty} c_r Y_r(x) \tag{8.109}$$

where the constant coefficients c_r are defined by

$$c_r = \int_0^L m(x)Y_r(x)Y(x)dx,\ r = 1, 2, \ldots \tag{8.110}$$

and

$$\omega_r^2 c_r = \int_0^L Y_r(x)\frac{d^2}{dx^2}\left[EI(x)\frac{d^2Y(x)}{dx^2}\right]dx,\ r = 1, 2, \ldots \tag{8.111}$$

The expansion theorem forms the basis for modal analysis, which permits the derivation of the response of beams to both initial excitations and applied forces.

8.6 SYSTEMS WITH LUMPED MASSES AT THE BOUNDARIES

A situation different from any encountered until now arises when one boundary, or both, contains a lumped mass. Indeed, in such a case, the corresponding boundary condition

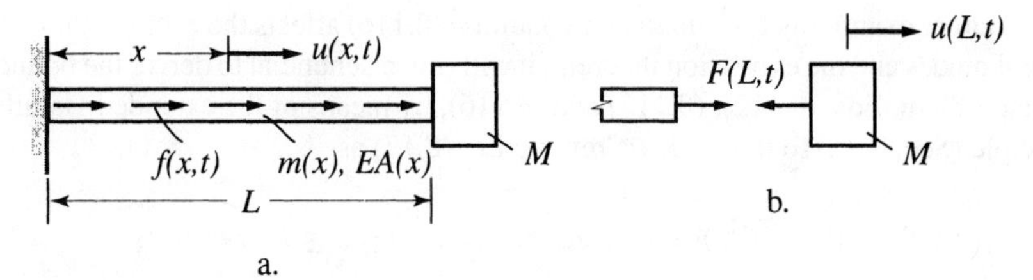

FIGURE 8.16
a. Rod fixed at $x = 0$ and with a lumped mass at $x = L$, b. Free-body diagram for the lumped mass at $x = L$

in the differential eigenvalue problem involves the eigenvalue explicitly. To substantiate this statement, we consider the axial vibration of a thin rod fixed at $x = 0$ and with a lumped mass attached at $x = L$, as shown in Fig. 8.16a. Using the analogy with the string in transverse vibration in conjunction with Eq. (8.14) and Table 8.1, the partial differential equation of motion for a thin rod in axial vibration has the form

$$\frac{\partial}{\partial x}\left[EA(x)\frac{\partial u(x,t)}{\partial x}\right] + f(x,t) = m(x)\frac{\partial^2 u(x,t)}{\partial t^2}, \quad 0 < x < L \tag{8.112}$$

where $EA(x)$ is the axial stiffness, in which E is the modulus of elasticity and $A(x)$ is the cross-sectional area of the rod, $u(x,t)$ is the axial displacement, $f(x,t)$ the axial force per unit length and $m(x)$ the mass per unit length. Moreover, the rod is fixed at $x = 0$, so that the boundary condition at the left end is

$$u(0,t) = 0 \tag{8.113}$$

To derive the boundary condition at the right end, we consider the free-body diagram shown in Fig. 8.16b in which we assume a tensile force between the end of the rod and the lumped mass M. Concentrating first on the end of the rod, we observe from Eq. (2.21) that the tensile force $F(L,t)$ is related to the deformation by

$$F(L,t) = EA(x)\left.\frac{\partial u(x,t)}{\partial x}\right|_{x=L} \tag{8.114}$$

Equation (8.114) reflects the standard convention specifying that a tensile force F produces a positive stress $E\partial u/\partial x$, as well as the assumption that this stress is distributed uniformly over the rod cross-sectional area A. Then, applying Newton's second law to mass M, we can write

$$-F(L,t) = M\left.\frac{\partial^2 u(x,t)}{\partial t^2}\right|_{x=L} \tag{8.115}$$

Hence, combining Eqs. (8.114) and (8.115), we obtain the boundary condition at the right end in the form

$$-EA(x)\frac{\partial u(x,t)}{\partial x} = M\frac{\partial^2 u(x,t)}{\partial t^2}, \quad x = L \tag{8.116}$$

Before examining how boundary condition (8.116) affects the orthogonality of the natural modes and the expansion theorem, it will prove beneficial to derive the boundary-value problem, Eqs. (8.112), (8.113) and (8.116), by means of the extended Hamilton's principle (Sec. 8.2). To this end, we rewrite Eq. (8.17) as

$$\int_{t_1}^{t_2} (\delta T - \delta V + \overline{\delta W}_{nc})dt = 0,\ \delta u(x,t) = 0,\ 0 \le x \le L,\ t = t_1, t_2 \tag{8.117}$$

where this time the kinetic energy has the expression

$$T(t) = \frac{1}{2}\int_0^L m(x)\left[\frac{\partial u(x,t)}{\partial t}\right]^2 dx + \frac{1}{2}M\left[\frac{\partial u(L,t)}{\partial t}\right]^2 \tag{8.118}$$

On the other hand, using the substitutions indicated in Table 8.1, as well as the analogy with Eq. (8.22), the potential energy becomes

$$V(t) = \frac{1}{2}\int_0^L EA(x)\left[\frac{\partial u(x,t)}{\partial x}\right]^2 dx \tag{8.119}$$

Moreover, from Eq. (8.19), the virtual work of the nonconservative distributed force has the analogous form

$$\overline{\delta W}_{nc} = \int_0^L f(x,t)\delta u(x,t)dx \tag{8.120}$$

Then, following the same steps as in Eqs. (8.23) and (8.24), we obtain

$$\begin{aligned}
\int_{t_1}^{t_2} \delta T dt &= \int_{t_1}^{t_2}\left[\int_0^L m(x)\frac{\partial u(x,t)}{\partial t}\delta\frac{\partial u(x,t)}{\partial t}dx + M\frac{\partial u(L,t)}{\partial t}\delta\frac{u(L,t)}{\delta t}\right]dt \\
&= \int_{t_1}^{t_2}\left[\int_0^L m(x)\frac{\partial u(x,t)}{\partial t}\frac{\partial}{\partial t}\delta u(x,t)dx + M\frac{\partial u(L,t)}{\partial t}\frac{\partial}{\partial t}\delta u(L,t)\right]dt \\
&= \int_0^L\left[m(x)\frac{\partial u(x,t)}{\partial t}\delta u(x,t)\Big|_{t_1}^{t_2}\right]dx - \int_0^L\left[\int_{t_1}^{t_2} m(x)\frac{\partial^2 u(x,t)}{\partial t^2}\delta u(x,t)dt\right]dx \\
&\quad + M\frac{\partial u(L,t)}{\partial t}\delta u(L,t)\Big|_{t_1}^{t_2} - \int_{t_1}^{t_2} M\frac{\partial^2 u(L,t)}{\partial t^2}\delta u(L,t)dt \\
&= -\int_{t_1}^{t_2}\left[\int_0^L m(x)\frac{\partial^2 u(x,t)}{\partial t^2}\delta u(x,t)dx + M\frac{\partial^2 u(L,t)}{\partial t^2}\delta u(L,t)\right]dt
\end{aligned} \tag{8.121}$$

Similarly, by analogy with Eqs. (8.25) and (8.26), we can write

$$\begin{aligned}
\delta V &= \int_0^L EA(x)\frac{\partial u(x,t)}{\partial x}\delta\frac{\partial u(x,t)}{\partial x}dx = \int_0^L EA(x)\frac{\partial u(x,t)}{\partial x}\frac{\partial}{\partial x}\delta u(x,t)dx \\
&= EA(x)\frac{\partial u(x,t)}{\partial x}\delta u(x,t)\Big|_0^L - \int_0^L \frac{\partial}{\partial x}\left[EA(x)\frac{\partial u(x,t)}{\partial x}\right]\delta u(x,t)dx
\end{aligned} \tag{8.122}$$

Hence, inserting Eqs. (8.120)–(8.122) into Eq. (8.117), we have

$$\int_{t_1}^{t_2}\left[-\int_0^L\left\{m(x)\frac{\partial^2 u(x,t)}{\partial t^2}-\frac{\partial}{\partial x}\left[EA(x)\frac{\partial u(x,t)}{\partial x}\right]-f(x,t)\right\}\delta u(x,t)dx\right.$$
$$-\left[EA(x)\frac{\partial u(x,t)}{\partial x}+M\frac{\partial^2 u(x,t)}{\partial t^2}\right]\delta u(x,t)\bigg|_{x=L}$$
$$\left.+EA(x)\frac{\partial u(x,t)}{\partial x}\delta u(x,t)\bigg|_{x=0}\right]dt=0 \tag{8.123}$$

Finally, invoking the arbitrariness of the virtual displacement $\delta u(x,t)$ and using the usual arguments (Sec. 8.2), we obtain the boundary-value problem for the rod under consideration in the form of the partial differential equation

$$\frac{\partial}{\partial x}\left[EA(x)\frac{\partial u(x,t)}{\partial x}\right]+f(x,t)=m(x)\frac{\partial^2 u(x,t)}{\partial t^2},\ 0<x<L \tag{8.124}$$

and the boundary conditions

$$u(0,t)=0 \tag{8.125}$$

and

$$EA(x)\frac{\partial u(x,t)}{\partial x}+M\frac{\partial^2 u(x,t)}{\partial t^2}=0,\ x=L \tag{8.126}$$

It is easy to see that the results are the same as those given earlier in this section. We once again observe that the extended Hamilton's principle yields the correct boundary-value problem in a routine fashion. In particular, whereas the derivation of boundary condition (8.116) through the use of Newton's second law requires some physical reasoning, and may cause one to ponder about the signs involved, the derivation of the same boundary condition, Eq. (8.126), by means of the extended Hamilton's principle leaves no doubts about the sign correctness. In fact, the extended Hamilton's principle is so reliable in this regard that it is a good practice to use it to verify the boundary conditions in complex cases.

Next, we consider the case of a cantilever beam in bending with a lumped mass at the free end, as shown in Fig. 8.17a. The partial differential equation and the boundary conditions at $x=0$ remain as in Eqs. (8.39) and (8.40) derived earlier, namely,

$$-\frac{\partial^2}{\partial x^2}\left[EI(x)\frac{\partial^2 y(x,t)}{\partial x^2}\right]+f(x,t)=m(x)\frac{\partial^2 y(x,t)}{\partial t^2},\ 0<x<L \tag{8.127}$$

and

$$y(x,t)=0,\quad \frac{\partial y(x,t)}{\partial x}=0,\ x=0 \tag{8.128}$$

On the other hand, one of the boundary conditions at $x=L$ was never encountered before, and we propose to derive it now. To this end, we turn our attention to Fig. 8.17b. From the left side of Fig. 8.17b and Eqs. (8.37) and (8.38), we can write the relations

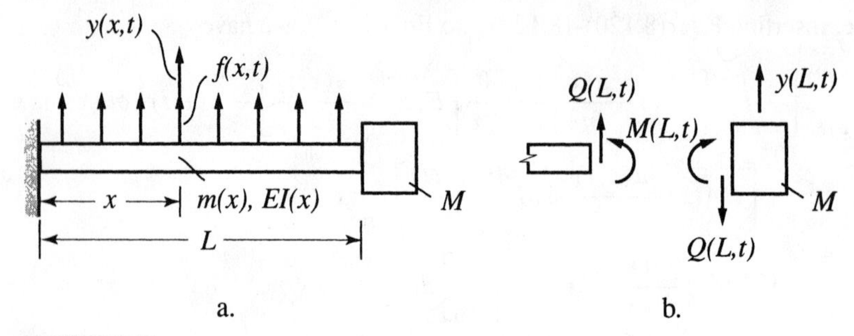

FIGURE 8.17
a. Cantilever beam with a lumped mass at the end, b. Free-body diagram for the lumped mass at the end

between the bending moment and shearing force on the one hand and the displacement on the other hand as follows:

$$M(x,t) = EI(x)\frac{\partial^2 y(x,t)}{\partial x^2}, \quad x = L \tag{8.129}$$

and

$$Q(x,t) = -\frac{\partial}{\partial x}\left[EI(x)\frac{\partial^2 y(x,t)}{\partial x^2}\right], \quad x = L \tag{8.130}$$

Then, assuming that the dimensions of the lumped mass are sufficiently small that the mass moment of inertia about an axis normal to x and y is negligible, the right side of Fig. 8.17b can be used to write

$$M(L,t) = 0 \tag{8.131}$$

In addition, the force equation in the vertical direction is

$$-Q(L,t) = M\frac{\partial^2 y(L,t)}{\partial t^2} \tag{8.132}$$

Combining Eqs. (8.129) and (8.131) on the one hand and Eqs. (8.130) and (8.132) on the other hand, we obtain the boundary conditions at the right end

$$EI(x)\frac{\partial^2 y(x,t)}{\partial x^2} = 0, \quad x = L \tag{8.133}$$

and

$$\frac{\partial}{\partial x}\left[EI(x)\frac{\partial^2 y(x,t)}{\partial x^2}\right] = M\frac{\partial^2 y(x,t)}{\partial t^2}, \quad x = L \tag{8.134}$$

respectively.

We note that, in deriving boundary condition (8.134), the sign convention for the bending moment and shearing force, established in Sec. 8.3, had to be observed scrupulously. Otherwise, a sign error can occur. No such question arises when the extended Hamilton's principle is used. To demonstrate this, we use the analogy with

Eqs. (8.18) and (8.45) and write the kinetic energy and potential energy expressions for the beam of Fig. 8.17a as follows:

$$T(t) = \frac{1}{2}\int_0^L m(x)\left[\frac{\partial y(x,t)}{\partial t}\right]^2 dx + \frac{1}{2}M\left[\frac{\partial y(L,t)}{\partial t}\right]^2 \tag{8.135}$$

and

$$V(t) = \frac{1}{2}\int_0^L EI(x)\left[\frac{\partial^2 y(x,t)}{\partial x^2}\right]^2 dx \tag{8.136}$$

respectively. The virtual work has the same expression as that for a string, Eq. (8.19).

The extended Hamilton's principle has the same form as that for a string, Eq. (8.17). Hence, following the same steps as in Sec. 8.3 and omitting details, we can write

$$\int_{t_1}^{t_2} \delta T(t)dt = -\int_{t_1}^{t_2}\left[\int_0^L m(x)\frac{\partial^2 y(x,t)}{\partial t^2}\delta y(x,t)dx + M\frac{\partial^2 y(L,t)}{\partial t^2}\delta y(L,t)\right]dt \tag{8.137}$$

and

$$\delta V(t) = EI(x)\frac{\partial^2 y(x,t)}{\partial x^2}\delta\frac{\partial y(x,t)}{\partial x}\bigg|_0^L - \frac{\partial}{\partial x}\left[EI(x)\frac{\partial^2 y(x,t)}{\partial x^2}\right]\delta y(x,t)\bigg|_0^L$$
$$+\int_0^L \frac{\partial^2}{\partial x^2}\left[EI(x)\frac{\partial^2 y(x,t)}{\partial x^2}\right]\delta y(x,t)dx \tag{8.138}$$

Inserting Eqs. (8.19), (8.137) and (8.138) into Eq. (8.17) and rearranging, we have

$$\int_{t_1}^{t_2}\left\langle -\int_0^L\left\{m(x)\frac{\partial^2 y(x,t)}{\partial t^2} - f(x,t) + \frac{\partial^2}{\partial x^2}\left[EI(x)\frac{\partial^2 y(x,t)}{\partial x^2}\right]\right\}\delta y(x,t)dx\right.$$
$$- EI(x)\frac{\partial^2 y(x,t)}{\partial x^2}\delta\frac{\partial y(x,t)}{\partial x}\bigg|_{x=L} + EI(x)\frac{\partial^2 y(x,t)}{\partial x^2}\delta\frac{\partial y(x,t)}{\partial x}\bigg|_{x=0}$$
$$+\left\{\frac{\partial}{\partial x}\left[EI(x)\frac{\partial^2 y(x,t)}{\partial x^2}\right] - M\frac{\partial^2 y(x,t)}{\partial t^2}\right\}\delta y(x,t)\bigg|_{x=L}$$
$$\left. -\frac{\partial}{\partial x}\left[EI(x)\frac{\partial^2 y(x,t)}{\partial x^2}\right]\delta y(x,t)\bigg|_{x=0}\right\rangle dt = 0 \tag{8.139}$$

Finally, invoking the arbitrariness of the virtual displacement δy and recognizing that the displacement and slope are zero at $x = 0$ and cannot be zero for all times at $x = L$, we conclude that the displacement y must satisfy the partial differential equation

$$-\frac{\partial^2}{\partial x^2}\left[EI(x)\frac{\partial^2 y(x,t)}{\partial x^2}\right] + f(x,t) = m(x)\frac{\partial^2 y(x,t)}{\partial t^2}, \quad 0 < x < L \tag{8.140}$$

and the boundary conditions

$$y(x,t) = 0, \quad \frac{\partial y(x,t)}{\partial x} = 0, \quad x = 0 \tag{8.141}$$

and

$$EI(x)\frac{\partial^2 y(x,t)}{\partial x^2} = 0, \ \frac{\partial}{\partial x}\left[EI(x)\frac{\partial^2 y(x,t)}{\partial x^2}\right] - M\frac{\partial^2 y(x,t)}{\partial t^2} = 0, \ x = L \tag{8.142}$$

which are identical to the results obtained earlier. The interesting part about boundary conditions (8.142) is that they were obtained in a natural fashion, and without any reference to the concepts of bending moment and shearing force, although the corresponding terms could be identified as such after completion of the derivation.

8.7 EIGENVALUE PROBLEM AND EXPANSION THEOREM FOR PROBLEMS WITH LUMPED MASSES AT THE BOUNDARIES

The boundary-value problem for the free axial vibration of a thin rod fixed at $x = 0$ and with a lumped mass at $x = L$ (Fig. 8.16a) can be obtained by letting $f(x,t) = 0$ in the partial differential equation of motion, Eq. (8.112), and leaving the boundary conditions, Eqs. (8.113) and (8.116), unchanged. Hence, the boundary-value problem in question is defined by the equation of motion

$$\frac{\partial}{\partial x}\left[EA(x)\frac{\partial u(x,t)}{\partial x}\right] = m(x)\frac{\partial^2 u(x,t)}{\partial t^2}, \ 0 < x < L \tag{8.143}$$

where the solution $u(x,t)$ must satisfy the boundary conditions

$$u(0,t) = 0 \tag{8.144}$$

and

$$-EA(x)\frac{\partial u(x,t)}{\partial x} = M\frac{\partial^2 u(x,t)}{\partial t^2}, \ x = L \tag{8.145}$$

As demonstrated in Sec. 8.4, free vibration is harmonic and, by analogy with Eqs. (8.57) and (8.66), can be expressed in the form

$$u(x,t) = CU(x)\cos(\omega t - \phi) \tag{8.146}$$

where C is an amplitude, $U(x)$ a displacement profile, ω the frequency of the harmonic oscillation and ϕ a phase angle. Inserting Eq. (8.146) into Eqs. (8.143)–(8.145) and dividing through everywhere by $C\cos(\omega t - \phi)$, we obtain the eigenvalue problem consisting of the differential equation

$$-\frac{d}{dx}\left[EA(x)\frac{dU(x)}{dx}\right] = \omega^2 m(x)U(x), \ 0 < x < L \tag{8.147}$$

and the boundary conditions

$$U(0) = 0 \tag{8.148}$$

and

$$EA(x)\frac{dU(x)}{dx} = \omega^2 MU(x), \ x = L \tag{8.149}$$

Examining Eq. (8.149), we conclude that the case in which there is a lumped mass at a boundary differs from all cases encountered before in that the boundary condition in question depends on the eigenvalue ω^2.

Next, we propose to investigate the orthogonality of modes for the case under consideration. To this end, we consider two distinct solutions of the eigenvalue problem, $U_r(x)$ and $U_s(x)$; they satisfy the equations

$$-\frac{d}{dx}\left[EA(x)\frac{dU_r(x)}{dx}\right] = \omega_r^2 m(x)U_r(x),\ 0 < x < L \tag{8.150}$$

and

$$-\frac{d}{dx}\left[EA(x)\frac{dU_s(x)}{dx}\right] = \omega_s^2 m(x)U_s(x),\ 0 < x < L \tag{8.151}$$

Multiplying Eq. (8.150) by $U_s(x)$ and integrating over the length of the rod, we have

$$-\int_0^L U_s(x)\frac{d}{dx}\left[EA(x)\frac{dU_r(x)}{dx}\right]dx = \omega_r^2\int_0^L m(x)U_s(x)U_r(x)dx \tag{8.152}$$

An integration by parts of the left side of Eq. (8.152), with due consideration of the boundary conditions, Eqs. (8.148) and (8.149), yields

$$\begin{aligned}
&-\int_0^L U_s(x)\frac{d}{dx}\left[EA(x)\frac{dU_r(x)}{dx}\right]dx \\
&= -U_s(x)EA(x)\frac{dU_r(x)}{dx}\bigg|_0^L + \int_0^L \frac{dU_s(x)}{dx}EA(x)\frac{dU_r(x)}{dx}dx \\
&= -\omega_r^2 MU_s(L)U_r(L) + \int_0^L EA(x)\frac{dU_s(x)}{dx}\frac{dU_r(x)}{dx}dx
\end{aligned} \tag{8.153}$$

Inserting Eq. (8.153) into Eq. (8.152) and rearranging, we obtain

$$\int_0^L EA(x)\frac{dU_r(x)}{dx}\frac{dU_s(x)}{dx}dx = \omega_r^2\left[\int_0^L m(x)U_r(x)U_s(x)dx + MU_r(L)U_s(L)\right] \tag{8.154}$$

In a similar fashion, multiplying Eq. (8.151) by $U_r(x)$, integrating over the length of the rod and following the same steps, we can write

$$\int_0^L EA(x)\frac{dU_r(x)}{dx}\frac{dU_s(x)}{dx}dx = \omega_s^2\left[\int_0^L m(x)U_r(x)U_s(x)dx + MU_r(L)U_s(L)\right] \tag{8.155}$$

Subtracting Eq. (8.155) from Eq. (8.154) and observing that the left side of both equations is identical, we have

$$(\omega_r^2 - \omega_s^2)\left[\int_0^L m(x)U_r(x)U_s(x)dx + MU_r(L)U_s(L)\right] = 0 \tag{8.156}$$

We conclude that, for distinct eigenvalues, Eq. (8.156) can be satisfied only if

$$\int_0^L m(x)U_r(x)U_s(x)dx + MU_r(L)U_s(L) = 0,\ r,s = 1,2,\ldots;\omega_r^2 \neq \omega_s^2 \tag{8.157}$$

Equations (8.157) represent the orthogonality relations for a rod in axial vibration with a lumped mass at $x = L$. For convenience, we normalize the natural modes so as to satisfy

$$\int_0^L m(x)U_r^2(x)dx + MU_r^2(L) = 1,\ r = 1,2,\ldots \tag{8.158}$$

so that, combining Eqs. (8.157) and (8.158), we obtain the orthonormality relations

$$\int_0^L m(x)U_r(x)U_s(x)dx + MU_r(L)U_s(L) = \delta_{rs},\ r,s = 1,2,\ldots \tag{8.159}$$

Next, we add $\omega_r^2 MU_r(L)U_s(L)$ to both sides of Eq. (8.152) consider boundary condition (8.149) and write

$$\begin{aligned} &-\int_0^L U_s(x)\frac{d}{dx}\left[EA(x)\frac{dU_r(x)}{dx}\right]dx + \omega_r^2 MU_r(L)U_s(L) \\ &= -\int_0^L U_s(x)\frac{d}{dx}\left[EA(x)\frac{dU_r(x)}{dx}\right]dx + \left[U_s(x)EA(x)\frac{dU_r(x)}{dx}\right]\Bigg|_{x=L} \\ &= \omega_r^2\left[\int_0^L m(x)U_s(x)U_r(x)dx + MU_s(L)U_r(L)\right],\ r,s = 1,2,\ldots \end{aligned} \tag{8.160}$$

so that, inserting Eqs. (8.159) into Eqs. (8.160), we obtain the companion orthonormality relations

$$\begin{aligned} &-\int_0^L U_s(x)\frac{d}{dx}\left[EA(x)\frac{dU_r(x)}{dx}\right]dx + \left[U_s(x)EA(x)\frac{dU_r(x)}{dx}\right]\Bigg|_{x=L} \\ &= \omega_r^2\delta_{rs},\ r,s = 1,2,\ldots \end{aligned} \tag{8.161}$$

Similarly, Eqs. (8.155) and (8.159) give the alternative companion orthonormality relations

$$\int_0^L EA(x)\frac{dU_r(x)}{dx}\frac{dU_s(x)}{dx}dx = \omega_r^2\delta_{rs},\ r,s = 1,2,\ldots \tag{8.162}$$

As in Sec. 8.5, we can state an expansion theorem for the response of the rod under consideration as follows: *Any function $U(x)$ representing a possible displacement of the rod, which implies that $U(x)$ satisfies boundary conditions* (8.148) *and* (8.149) *and is such that $(d/dx)[EA(x)dU(x)/dx]$ is a continuous function, can be expanded in the absolutely and uniformly convergent series of the eigenfunctions*

$$U(x) = \sum_{r=1}^{\infty} c_r U_r(x) \tag{8.163}$$

where the constant coefficients are defined by

$$c_r = \int_0^L m(x)U_r(x)U(x)dx + MU_r(L)U(L),\ r = 1,2,\ldots \tag{8.164}$$

and

$$\omega_r^2 c_r = -\int_0^L U_r(x)\frac{d}{dx}\left[EA(x)\frac{dU(x)}{dx}\right]dx + \left[U_r(x)EA(x)\frac{dU(x)}{dx}\right]\Bigg|_{x=L},$$
$$r = 1, 2, \ldots \tag{8.165}$$

Equations (8.164) and (8.165) can be verified by means of the orthonormality relations (8.159) and (8.161), respectively, in conjunction with Eq. (8.163) with r replaced by s.

Although the developments in this section were carried out for a rod in axial vibration, the developments are equally valid for a string in transverse vibration and a circular shaft in torsion, subject to the dependent variable and parameter changes listed in Table 8.1.

The orthogonality relations and expansion theorem for the cantilever beam with a lumped mass at the end depicted in Fig. 8.17a can be produced in the same manner as for the rod in axial vibration just discussed. Indeed, by analogy with Eq. (8.146), when $f(x,t) = 0$, the solution of Eqs. (8.140)-(8.142) can be assumed to have the form

$$y(x,t) = CY(x)\cos(\omega t - \phi) \tag{8.166}$$

where $Y(x)$ is the displacement profile. Then, dividing through by $C\cos(\omega t - \phi)$, we obtain the eigenvalue problem consisting of the differential equation

$$\frac{d^2}{dx^2}\left[EI(x)\frac{d^2Y(x)}{dx^2}\right] = \omega^2 m(x)Y(x),\ 0 < x < L \tag{8.167}$$

and the boundary conditions

$$Y(x) = 0,\quad \frac{dY(x)}{dx} = 0,\ x = 0 \tag{8.168}$$

and

$$EI(x)\frac{d^2Y(x)}{dx^2} = 0,\quad -\frac{d}{dx}\left[EI(x)\frac{d^2Y(x)}{dx^2}\right] = \omega^2 MY(x),\ x = L \tag{8.169}$$

At this point, we consider two distinct solutions of the eigenvalue problem, $Y_r(x)$ and $Y_s(x)$, and write

$$\frac{d^2}{dx^2}\left[EI(x)\frac{d^2Y_r(x)}{dx^2}\right] = \omega_r^2 m(x)Y_r(x),\ 0 < x < L \tag{8.170}$$

and

$$\frac{d^2}{dx^2}\left[EI(x)\frac{d^2Y_s(x)}{dx^2}\right] = \omega_s^2 m(x)Y_s(x),\ 0 < x < L \tag{8.171}$$

Multiplying Eq. (8.170) by $Y_s(x)$ and integrating over the length of the beam, we have

$$\int_0^L Y_s(x)\frac{d^2}{dx^2}\left[EI(x)\frac{d^2Y_r(x)}{dx^2}\right]dx = \omega_r^2\int_0^L m(x)Y_s(x)Y_r(x)dx \tag{8.172}$$

Two integrations by parts of the left side of Eq. (8.172), with due consideration of boundary conditions (8.168) and (8.169), yield

$$\int_0^L Y_s(x)\frac{d^2}{dx^2}\left[EI(x)\frac{d^2Y_r(x)}{dx^2}\right]dx$$

$$= \left\{Y_s(x)\frac{d}{dx}\left[EI(x)\frac{d^2Y_r(x)}{dx^2}\right]\right\}\Bigg|_0^L - \left[\frac{dY_s(x)}{dx}EI(x)\frac{d^2Y_r(x)}{dx^2}\right]\Bigg|_0^L$$

$$+ \int_0^L \frac{d^2Y_s(x)}{dx^2}EI(x)\frac{d^2Y_r(x)}{dx^2}dx$$

$$= -\omega_r^2 MY_s(L)Y_r(L) + \int_0^L EI(x)\frac{d^2Y_s(x)}{dx^2}\frac{d^2Y_r(x)}{dx^2}dx \tag{8.173}$$

Introducing Eq. (8.173) in Eq. (8.172) and rearranging, we obtain

$$\int_0^L EI(x)\frac{d^2Y_r(x)}{dx^2}\frac{d^2Y_s(x)}{dx^2}dx = \omega_r^2\left[\int_0^L m(x)Y_r(x)Y_s(x)dx + MY_r(L)Y_s(L)\right] \tag{8.174}$$

Similarly, multiplying Eq. (8.171) by $Y_r(x)$, integrating over the length of the beam and using the same process as that leading to Eq. (8.173), we can write

$$\int_0^L EI(x)\frac{d^2Y_r(x)}{dx^2}\frac{d^2Y_s(x)}{dx^2}dx = \omega_s^2\left[\int_0^L m(x)Y_r(x)Y_s(x)dx + MY_r(L)Y_s(L)\right] \tag{8.175}$$

But, the left side of Eqs. (8.174) and (8.175) is the same. Hence, subtracting Eq. (8.175) from Eq. (8.174), we have simply

$$(\omega_r^2 - \omega_s^2)\left[\int_0^L m(x)Y_r(x)Y_3(x)dx + MY_r(L)Y_s(L)\right] = 0 \tag{8.176}$$

so that, for $\omega_r^2 \neq \omega_s^2$, Eq. (8.176) can be satisfied only if

$$\int_0^L m(x)Y_r(x)Y_s(x)dx + MY_r(L)Y_s(L) = 0, \; r,s = 1,2,\ldots; \; \omega_r^2 \neq \omega_s^2 \tag{8.177}$$

Equations (8.177) represent the orthogonality conditions for a beam in bending with a lumped mass at $x = L$. They are essentially the same as the orthogonality relations for a rod in axial vibration with a lumped mass at $x = L$, Eqs. (8.157), except that the transverse displacement Y replaces the axial displacement U.

Taking the cue from Eq. (8.164), we normalize the natural modes so as to satisfy $\int_0^L m(x)Y_r^2(x)dx + MY_r^2(L) = 1$ $(r = 1,2,\ldots)$, so that the resulting modes satisfy the orthonormality relations

$$\int_0^L m(x)Y_r(x)Y_s(x)dx + MY_r(L)Y_s(L) = \delta_{rs}, \; r,s = 1,2,\ldots \tag{8.178}$$

Now, we add $\omega_r^2 MY_r(L)Y_s(L)$ to both sides of Eq. (8.172), use the second of boundary conditions (8.169) and write

$$\int_0^L Y_s(x)\frac{d^2}{dx^2}\left[EI(x)\frac{d^2Y_r(x)}{dx^2}\right]dx + \omega_r^2 MY_r(L)Y_s(L)$$
$$= \int_0^L Y_s(x)\frac{d^2}{dx^2}\left[EI(x)\frac{d^2Y_r(x)}{dx^2}\right]dx - \left\{Y_s(x)\frac{d}{dx}\left[EI(x)\frac{d^2Y_r(x)}{dx^2}\right]\right\}\Bigg|_{x=L}$$
$$= \omega_r^2\left[\int_0^L m(x)Y_r(x)Y_s(x)dx + MY_r(L)Y_s(L)\right] \tag{8.179}$$

It follows immediately from Eqs. (8.178) and (8.179) that the normal modes also satisfy the companion orthonormality relations

$$\int_0^L Y_s(x)\frac{d^2}{dx^2}\left[EI(x)\frac{d^2Y_r(x)}{dx^2}\right]dx - \left\{Y_s(x)\frac{d}{dx}\left[EI(x)\frac{d^2Y_r(x)}{dx^2}\right]\right\}\Bigg|_{x=L} = \omega_r^2\delta_{rs},$$
$$r, s = 1, 2, \ldots \tag{8.180}$$

and from Eq. (8.174) that they satisfy the alternative companion orthogonality relations

$$\int_0^L EI(x)\frac{d^2Y_r(x)}{dx^2}\frac{d^2Y_s(x)}{dx^2}dx = \omega_r^2\delta_{rs},\ r, s = 1, 2, \ldots \tag{8.181}$$

as well.

The preceding developments permit us to state the following expansion theorem concerning a cantilever beam with a mass at the end: *Any function* $Y(x)$ *representing a possible displacement of the beam, which implies that* $Y(x)$ *satisfies boundary conditions* (8.168) *and* (8.169) *and is such that* $(d^2/dx^2)[EI(x)d^2Y(x)/dx^2]$ *is a continuous function, can be expanded in the absolutely and uniformly convergent series of the eigenfunctions*

$$Y(x) = \sum_{r=1}^{\infty} c_r Y_r(x) \tag{8.182}$$

where the constant coefficients are defined by

$$c_r = \int_0^L m(x)Y_r(x)Y(x)dx + MY_r(L)Y(L),\ r = 1, 2, \ldots \tag{8.183}$$

and

$$\omega_r^2 c_r = \int_0^L Y_r(x)\frac{d^2}{dx^2}\left[EI(x)\frac{d^2Y(x)}{dx^2}\right]dx - \left\{Y_r(x)\frac{d}{dx}\left[EI(x)\frac{d^2Y(x)}{dx^2}\right]\right\}\Bigg|_{x=L},$$
$$r = 1, 2, \ldots \tag{8.184}$$

Equations (8.183) and (8.184) can be verified by means of Eqs. (8.178) and (8.180), respectively, in conjunction with Eq. (8.182) with r replaced by s.

The expansion theorem for rods in axial vibration with lumped masses at boundaries, and by implication for strings in transverse vibration and shafts in torsion, and that for beams in bending form the basis for a modal analysis for the response to initial excitations and applied forces. The corresponding theorems, in turn, are based on the orthogonality relations, which are more involved for systems with lumped masses at boundaries than for systems with all other types of boundaries. The orthogonality conditions involving the stiffness properties, Eqs. (8.161) for rods and Eqs. (8.184) for beams, appear particularly intimidating. In spite of this, modal analysis retains the same simplicity for systems with lumped masses at boundaries as for all other systems, as we shall see in Sec. 8.10.

Example 8.6. Solve the eigenvalue problem for a uniform circular shaft in torsion fixed at $x = 0$ and with a rigid disk at $x = L$ for the parameter ratio $IL/I_D = 1$, where I is the polar mass moment of inertia per unit length of shaft and I_D is the polar mass moment of inertia of the disk. Plot the three lowest modes.

By analogy with the eigenvalue problem for a rod in axial vibration, Eqs. (8.147)–(8.149), the eigenvalue problem for a shaft in torsion consists of the differential equation

$$-\frac{d}{dx}\left[GJ(x)\frac{d\Theta(x)}{dx}\right] = \omega^2 I(x)\Theta(x),\ 0 < x < L \tag{a}$$

where $\Theta(x)$ is the twist angle and $GJ(x)$ is the torsional stiffness, in which G is the shear modulus and $J(x)$ the area polar moment of inertia of the shaft, and the boundary conditions

$$\Theta(0) = 0 \tag{b}$$

and

$$GJ(x)\frac{d\theta(x)}{dx}\bigg|_{x=L} = \omega^2 I_D \Theta(L) \tag{c}$$

For a uniform shaft, $I(x) = I =$ constant, $GJ(x) = GJ =$ constant, the differential equation reduces to

$$\frac{d^2\Theta(x)}{dx^2} + \beta^2\Theta(x) = 0,\ 0 < x < L,\ \beta^2 = \frac{\omega^2 I}{GJ} \tag{d}$$

and boundary condition (c) can be written as

$$\frac{d\Theta(x)}{dx}\bigg|_{x=L} = \frac{\beta^2 I_D}{I}\Theta(L) \tag{e}$$

The solution of Eq. (d) is

$$\Theta(x) = A\sin\beta x + B\cos\beta x \tag{f}$$

Inserting Eq. (f) into boundary condition (b), we conclude that $B = 0$, so that the solution reduces to

$$\Theta(x) = A\sin\beta x \tag{g}$$

Then, introducing Eq. (g) in boundary condition (e), we obtain the characteristic equation

$$\tan\beta L = \frac{IL}{I_D}\frac{1}{\beta L} = \frac{1}{\beta L} \tag{h}$$

which represents a transcendental equation to be solved numerically for the eigenvalues $\beta_r L\ (r = 1, 2, \ldots)$. Inserting these eigenvalues into Eq. (g), the natural modes are simply

$$\Theta_r(x) = A_r \sin\beta_r x,\ r = 1, 2, \ldots \tag{i}$$

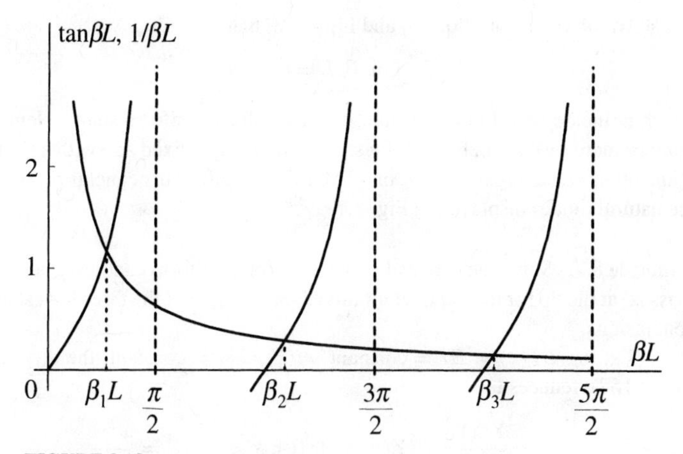

FIGURE 8.18
Graphical solution of the characteristic equation for a shaft fixed at $x = 0$ and with a rigid disk at $x = L$

Moreover, from the second of Eqs. (d), the natural frequencies are

$$\omega_r = \beta_r L \sqrt{\frac{GJ}{IL^2}}, \quad r = 1, 2, \ldots \tag{j}$$

A graphical solution of Eq. (h) is shown in Fig. 8.18. The first three eigenvalues are $\beta_1 L = 0.8903$, $\beta_2 L = 3.4256$, $\beta_3 L = 6.4373$. The first three modes, normalized so that $A_r = 1$ $(r = 1, 2, 3)$, are plotted in Fig. 8.19.

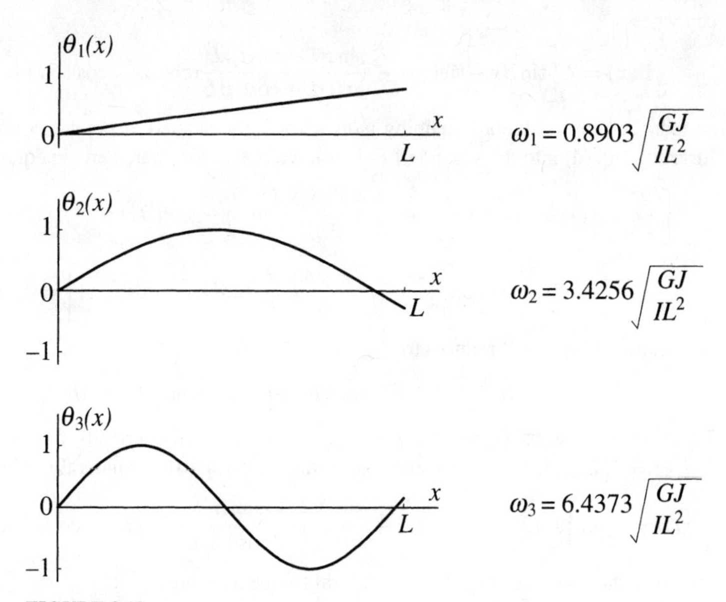

FIGURE 8.19
Natural modes for a shaft fixed at $x = 0$ and with a rigid disk at $x = L$

We observe from Eq. (h), and Fig. 8.18, that

$$\lim_{r\to\infty} \beta_r L = (r-1)\pi \tag{k}$$

which coincide with the eigenvalue of a fixed-fixed uniform shaft. Hence, as the mode number increases, the behavior of a shaft with one end fixed and with a rigid disk attached at the other end approaches that of a fixed-fixed shaft. This conclusion is corroborated by the natural modes displayed in Fig. 8.19.

Example 8.7. Solve the eigenvalue problem for a uniform cantilever beam with a lumped mass M at the tip for the parameter ratio $M/mL = 1$. Plot the three lowest modes and draw conclusions.

Letting $EI(x) = EI = \text{constant}$, $m(x) = m = \text{constant}$, the differential equation, Eq. (8.167), reduces to

$$\frac{d^4Y(x)}{dx^4} - \beta^4 Y(x) = 0,\ 0 < x < L,\ \beta^4 = \frac{\omega^2 m}{EI} \tag{a}$$

Moreover, boundary conditions (8.168) remain in the form

$$Y(x) = 0,\ \frac{dY(x)}{dx} = 0,\ x = 0 \tag{b}$$

whereas boundary conditions (8.169) become

$$\frac{d^2Y(x)}{dx^2} = 0,\ \frac{d^3Y(x)}{dx^3} + \frac{M}{m}\beta^4 Y(x) = 0,\ x = L \tag{c}$$

The eigenvalue problem for a uniform cantilever beam was discussed in Example 8.4. The only difference between the problem at hand and that in Example 8.4 lies in the second of boundary conditions (c). Hence, combining Eqs. (j) and (m) of Example 8.4, we can write

$$Y(x) = A\left[\sin\beta x - \sinh\beta x - \frac{\sin\beta L + \sinh\beta L}{\cos\beta L + \cosh\beta L}(\cos\beta x - \cosh\beta x)\right] \tag{d}$$

so that the only problem remaining is to enforce the second of boundary conditions (c). Inserting Eq. (d) into the second of Eqs. (c), we obtain the characteristic equation

$$\begin{aligned} &A\beta^3\left[-\cos\beta x - \cosh\beta x - \frac{\sin\beta L + \sinh\beta L}{\cos\beta L + \cosh\beta L}(\sin\beta x - \sinh\beta x)\right] \\ &+ \frac{M}{m}\beta^4 A\left[\sin\beta x - \sinh\beta x - \frac{\sin\beta L + \sinh\beta L}{\cos\beta L + \cosh\beta L}(\cos\beta x - \cosh\beta x)\right] = 0,\ x = L \end{aligned} \tag{e}$$

which, for $M/mL = 1$, reduces to

$$-(1 + \cos\beta L\cosh\beta L) + \beta L(\sin\beta L\cosh\beta L - \sinh\beta L\cos\beta L) = 0 \tag{f}$$

Equation (f) is a transcendental equation to be solved numerically for the eigenvalues $\beta_r L$ $(r = 1, 2, \ldots)$. Inserting these eigenvalues into Eq. (d), the natural modes are simply

$$Y_r(x) = A_r\left[\sin\beta_r x - \sinh\beta_r x - \frac{\sin\beta_r L + \sinh\beta_r L}{\cos\beta_r L + \cosh\beta_r L}(\cos\beta_r x - \cosh\beta_r x)\right] \tag{g}$$

and from the second of Eqs. (a), the natural frequencies are

$$\omega_r = (\beta_r L)^2\sqrt{\frac{EI}{mL^4}},\ r = 1, 2, \ldots \tag{h}$$

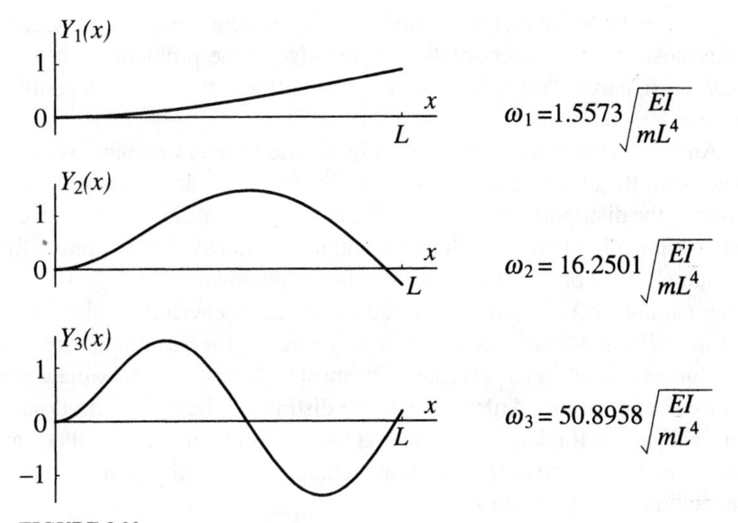

FIGURE 8.20
Natural modes for a cantilever beam with a lumped mass at the tip

The first three eigenvalues are $\beta_1 L = 1.2479$, $\beta_2 L = 4.0311$, $\beta_3 L = 7.1341$ and the corresponding three modes, normalized so as to satisfy Eqs. (8.178), are plotted in Fig. 8.20. Note that the coefficients have the values $A_1 = 0.7727$, $A_2 = 0.9946$ and $A_3 = 0.9956$. We observe that, as the mode number increases, the end $x = L$ acts more as a pinned end than a free end with a lumped mass.

8.8 RAYLEIGH'S QUOTIENT. THE VARIATIONAL APPROACH TO THE DIFFERENTIAL EIGENVALUE PROBLEM

Cases in which the differential eigenvalue problem admits a closed-form solution are very rare indeed, almost invariably involving uniformly distributed parameters and simple boundary conditions. Hence, for the most part, one must be content with approximate solutions. In this regard, Rayleigh's quotient plays a pivotal role.

We consider the eigenvalue problem for a rod in axial vibration fixed at $x = 0$ and with a spring of stiffness k at $x = L$. Using the analogy with the string in transverse vibration and referring to Eqs. (8.14)–(8.16), it is not difficult to verify that the eigenvalue problem consists of the differential equation

$$-\frac{d}{dx}\left[EA(x)\frac{dU(x)}{dx}\right] = \lambda m(x)U(x),\ 0 < x < L;\ \lambda = \omega^2 \tag{8.185}$$

and the boundary conditions

$$U(0) = 0,\ -EA(x)\frac{dU(x)}{dx}\bigg|_{x=L} = kU(L) \tag{8.186}$$

Equations (8.185) and (8.186) represent the *strong form* of the eigenvalue problem in question. The implication is that the solution $U(x)$ must satisfy the differential equation (8.185) at every point of the domain $0 < x < L$, as well as boundary conditions (8.186), and we observe that the first boundary condition is geometric and the second is natural. For the most part, an exact solution of the eigenvalue problem in the strong form is beyond reach, particularly when the mass and stiffness parameters depend on the spatial variable x.

An approximate solution of the eigenvalue problem requires some form of discretization of the differential equation, Eq. (8.185). The simplest discretization consists of lumping the distributed parameters. We discuss a number of such procedures in Ch. 9. Another approach is to replace the differential equation by a set of finite difference equations, namely, a set of algebraic equations in the unknowns $U_i = U(x_i)$, $(i = 1, 2, \ldots, n)$ and the parameter λ, thus replacing a differential eigenvalue problem by an algebraic one. Finite difference solutions are not very suitable for vibration problems, so that we do not pursue this subject any further. The most widely used approximate methods resort to series discretization. Among these, we distinguish between variational techniques, which include the Rayleigh-Ritz method and the finite element method, and weighted residuals methods, such as the Galerkin method and the collocation method. We discuss these methods in Chs. 9 and 10.

The variational approach to the series discretization of the differential eigenvalue problem makes use of Rayleigh's quotient to develop the algebraic eigenvalue problem. To derive the expression of Rayleigh's quotient for a given system, it is necessary to cast the differential eigenvalue problem in a *weak form*. To this end, we multiply Eq. (8.185) by a *test function* $V(x)$, integrate over the length of the rod and write

$$-\int_0^L V(x)\frac{d}{dx}\left[EA(x)\frac{dU(x)}{dx}\right]dx = \lambda\int_0^L m(x)V(x)U(x)dx \qquad (8.187)$$

Equation (8.187) implies that the solution of the differential eigenvalue problem is in a weighted average sense only, where the test function $V(x)$ plays the role of a weighting function. In the context of an approximate solution of the eigenvalue problem, it is common to refer to $U(x)$ as a *trial function*. In the same context, it is often advantageous to symmetrize the left side of Eq. (8.187) in $U(x)$ and $V(x)$. To this end, we insist that the test function $V(x)$ satisfy the geometric boundary condition at $x = 0$, i.e., that $V(0) = 0$. Assuming that this is the case, integrating the left side of Eq. (8.187) by parts and considering boundary conditions (8.186), we have

$$\begin{aligned}-\int_0^L V(x)\frac{d}{dx}\left[EA(x)\frac{dU(x)}{dx}\right]dx &= -V(x)EA(x)\frac{dU(x)}{dx}\bigg|_0^L \\ &\quad + \int_0^L EA(x)\frac{dV(x)}{dx}\frac{dU(x)}{dx}dx \\ &= kV(L)U(L) + \int_0^L EA(x)\frac{dV(x)}{dx}\frac{dU(x)}{dx}dx\end{aligned} \qquad (8.188)$$

so that, inserting Eq. (8.188) into Eq. (8.187), we obtain

$$\int_0^L EA(x)\frac{dV(x)}{dx}\frac{dU(x)}{dx}dx + kV(L)U(L) = \lambda\int_0^L m(x)V(x)U(x)dx \qquad (8.189)$$

Equation (8.189) represents the *weak form* of the eigenvalue problem. To examine the implications of Eq. (8.189), we first recall that the test function $V(x)$ must satisfy the geometric boundary condition at $x = 0$ and then observe that $V(x)$ must be differentiable once for the integral on the left side of Eq. (8.189) to be defined. We refer to the class of functions that satisfy the geometric boundary condition and are once differentiable as *admissible functions* for second-order systems of the type given by Eqs. (8.185) and (8.186). In view of this, Eq. (8.189) can be interpreted as follows: *Determine the values of the parameter λ and the associated functions $U(x)$ such that Eq. (8.189) is satisfied for all admissible functions $V(x)$.* In seeking approximate solutions to the eigenvalue problems, the weak form has many advantages over the strong form, as it places fewer demands on the trial function $U(x)$. In fact, we observe from Eq. (8.189) that the trial function need only be from the class of admissible functions, although this may not be always desirable. We return to this subject in Chs. 9 and 10, when we discuss the Rayleigh-Ritz method and the finite element method.

Next, we consider the case in which the test function is equal to the trial function. Letting $V(x) = U(x)$ in Eq. (8.189), we can write

$$R(U) = \lambda = \omega^2 = \frac{\displaystyle\int_0^L EA(x)\left[\frac{dU(x)}{dx}\right]^2 dx + kU^2(L)}{\displaystyle\int_0^L m(x)U^2(x)dx} \tag{8.190}$$

Equation (8.190) represents *Rayleigh's quotient* for the rod described by Eqs. (8.185) and (8.186). Clearly, the value of R depends on the trial function $U(x)$. One question of interest in vibrations is how the value of R behaves as $U(x)$ changes. To answer this question, we refer to the expansion theorem (Sec. 8.5), use the analogy with the string in transverse vibration and expand $U(x)$ in the series

$$U(x) = \sum_{i=1}^{\infty} c_i U_i(x) \tag{8.191}$$

where $U_i(x)$ $(i = 1, 2, \ldots)$ are the normal modes satisfying the orthonormality conditions

$$\int_0^L m(x)U_i(x)U_j(x)dx = \delta_{ij},\ i, j = 1, 2, \ldots \tag{8.192}$$

and

$$\int_0^L EA(x)\frac{dU_i(x)}{dx}\frac{dU_j(x)}{dx}dx + kU_i(L)U_j(L) = \lambda_i\delta_{ij},\ i, j = 1, 2, \ldots \tag{8.193}$$

in which λ_i are the system eigenvalues. Inserting Eq. (8.191) into Rayleigh's quotient,

Eq. (8.190), and considering Eqs. (8.192) and (8.193), we can write

$$R(c_1, c_2, \dots) = \lambda = \omega^2$$

$$= \frac{\int_0^L EA(x) \sum_{i=1}^{\infty} c_i \frac{dU_i(x)}{dx} \sum_{j=1}^{\infty} c_j \frac{dU_j(x)}{dx} dx + k \sum_{i=1}^{\infty} c_i U_i(L) \sum_{j=1}^{\infty} c_j U_j(L)}{\int_0^L m(x) \sum_{i=1}^{\infty} c_i U_i(x) \sum_{j=1}^{\infty} c_j U_j(x) dx}$$

$$= \frac{\sum_{i=1}^{\infty}\sum_{j=1}^{\infty} c_i c_j \left[\int_0^L EA(x) \frac{dU_i(x)}{dx}\frac{dU_j(x)}{dx} dx + kU_i(L)U_j(L)\right]}{\sum_{i=1}^{\infty}\sum_{j=1}^{\infty} c_i c_j \int_0^L m(x) U_i(x) U_j(x) dx}$$

$$= \frac{\sum_{i=1}^{\infty}\sum_{j=1}^{\infty} c_i c_j \lambda_i \delta_{ij}}{\sum_{i=1}^{\infty}\sum_{j=1}^{\infty} c_i c_j \delta_{ij}} = \frac{\sum_{i=1}^{\infty} c_i^2 \lambda_i}{\sum_{i=1}^{\infty} c_i^2} \tag{8.194}$$

In Sec. 7.13, it was demonstrated that the behavior of Rayleigh's quotient for discrete systems becomes interesting only when a trial vector enters the neighborhood of an eigenvector. But, an eigenfunction can be regarded as an eigenvector of infinite dimension. Hence, using the analogy with discrete systems, we consider the case in which the trial function $U(x)$ resembles closely a given eigenfunction $U_r(x)$. In terms of the series expansion for $U(x)$, Eq. (8.191), this implies that all coefficients c_i are small, with the exception of c_r; this can be expressed as

$$c_i = \epsilon_i c_r, \; i = 1, 2, \dots, r-1, r+1, \dots \tag{8.195}$$

where ϵ_i are small numbers. Inserting Eqs. (8.195) into Eq. (8.194), dividing top and bottom by c_r^2 and ignoring terms in ϵ_i of order larger than two, we obtain

$$R = \frac{c_r^2 \lambda_r + \sum_{\substack{i=1 \\ i\neq r}}^{\infty} c_i^2 \lambda_i}{c_r^2 + \sum_{\substack{i=1 \\ i\neq r}}^{\infty} c_i^2} = \frac{\lambda_r + \sum_{\substack{i=1 \\ i\neq r}}^{\infty} \epsilon_i^2 \lambda_i}{1 + \sum_{\substack{i=1 \\ 1\neq r}}^{\infty} \epsilon_i^2}$$

$$\cong \left(\lambda_r + \sum_{\substack{i=1 \\ i\neq r}}^{\infty} \epsilon_i^2 \lambda_i\right)\left(1 - \sum_{\substack{i=1 \\ i\neq r}}^{\infty} \epsilon_i^2\right) \cong \lambda_r + \sum_{i=1}^{\infty} (\lambda_i - \lambda_r)\epsilon_i^2 \tag{8.196}$$

But, our postulate was essentially that the trial function $U(x)$ differs from the rth eigenfunction $U_r(x)$ by a small quantity of first order in ϵ, or $U(x) = U_r(x) + O(\epsilon)$, and Eq. (8.196) states that in this case Rayleigh's quotient differs from the rth eigenvalue by a small quantity of second order in ϵ, or $R = \lambda_r + O(\epsilon^2)$. The implication is that *Rayleigh's quotient has a stationary value at an eigenfunction $U_r(x)$, where the stationary value is the associated eigenvalue $\lambda_r (r = 1, 2, \ldots)$*. This can be interpreted as stating that *the process of rendering Rayleigh's quotient stationary is equivalent to solving the weak form of the eigenvalue problem.* A more formal proof of this statement can be found in Ref. 13.

A case of particular interest is that in which $r = 1$, in which case Eq. (8.196) becomes

$$R \cong \lambda_1 + \sum_{i=2}^{\infty} (\lambda_i - \lambda_1)\epsilon_i^2 \tag{8.197}$$

In view of the fact that the eigenvalues satisfy the inequalities $\lambda_1 \leq \lambda_2 \leq \ldots$, we conclude that

$$R \geq \lambda_1 \tag{8.198}$$

Inequality (8.198) states that *Rayleigh's quotient provides an upper bound for the lowest eigenvalue* λ_1. Inequality (8.198) can be given a somewhat different interpretation by stating that

$$\lambda_1 = \omega_1^2 = \min R(U) = R(U_1) \tag{8.199}$$

or, *the lowest eigenvalue λ_1 is the minimum value that Rayleigh's quotient can take*, where the minimum value occurs at the lowest eigenfunction, $U(x) = U_1(x)$. Equation (8.199) is referred to at times as *Rayleigh's principle*.

For most structures, the lowest natural frequency is the most important one. Quite often, particularly in preliminary design, the interest lies in producing a quick estimate of this lowest natural frequency, a task for which Rayleigh's principle is ideally suited. To this end, all that is necessary is to introduce a trial function $U(x)$ closely resembling the lowest natural mode $U_1(x)$ in Rayleigh's quotient, Eq. (8.190), carry out the indicated integrations and calculate the value $R = \lambda = \omega^2$. Then, because of the fact that Rayleigh's quotient has a minimum at the lowest eigenfunction, the value $\omega = \sqrt{R}$ thus calculated will be one order of magnitude closer to the lowest natural frequency ω_1 than $U(x)$ is to $U_1(x)$. A trial function $U(x)$ resembling the lowest natural mode reasonably well consists of the static displacement curve due to the own weight of the structure. Another good choice for a trial function $U(x)$, although not likely to be as good as the static displacement curve, is the lowest eigenfunction of an intimately related but simpler structure. Examples of such simpler structures are structures with uniform parameter distributions, as opposed to nonuniform ones, and with simpler boundaries, such as a free end instead of spring-supported end or an end with a lumped mass.

Rayleigh's quotient expression given by Eq. (8.190) is for a rod in axial vibration with the end $x = 0$ fixed and the end $x = L$ attached to a spring of stiffness k. Yet, the stationarity of Rayleigh's quotient and Rayleigh's principle hold for a much larger class of systems, of which the rod in question is a mere example. Indeed, they hold for the very large class of conservative systems. In view of this, we propose to derive a generic form of Rayleigh's quotient, not restricted to any particular structural member. To this end, we first consider a rod with the end $x = 0$ fixed and with a lumped mass M at the end $x = L$. The corresponding eigenvalue problem was discussed in Sec. 8.7. Hence, using Eq. (8.154), we can write Rayleigh's quotient in the form

$$R(U) = \lambda = \omega^2 = \frac{\int_0^L EA(x)\left[\frac{dU(x)}{dx}\right]^2 dx}{\int_0^L m(x)U^2(x)dx + MU^2(L)} \tag{8.200}$$

Similarly, for a beam in bending supported by a spring at $x = 0$ and pinned at $x = L$, we conclude from Eq. (8.100) that Rayleigh's quotient can be written as

$$R(Y) = \lambda = \omega^2 = \frac{\int_0^L EI(x)\left[\frac{d^2Y(x)}{dx^2}\right]^2 dx + kY^2(0)}{\int_0^L m(x)Y^2(x)dx} \tag{8.201}$$

Moreover, using Eq. (8.174), Rayleigh's quotient for a cantilever beam with a lumped mass at $x = L$ is

$$R(Y) = \lambda = \omega^2 = \frac{\int_0^L EI(x)\left[\frac{d^2Y(x)}{dx^2}\right]^2 dx}{\int_0^L m(x)Y^2(x)dx + MY^2(L)} \tag{8.202}$$

Examining the Rayleigh's quotient for all the above systems, we conclude that they all have one thing in common, namely, the numerator is a measure of the potential energy and the denominator a measure of the kinetic energy. As an illustration, the potential energy for a rod in axial vibration fixed at $x = 0$ and restrained by a spring at $x = L$ has the expression

$$V(t) = \frac{1}{2}\int_0^L EA(x)\left[\frac{\partial u(x,t)}{\partial x}\right]^2 dx + \frac{1}{2}ku^2(L,t) \tag{8.203}$$

and the kinetic energy is simply

$$T(t) = \frac{1}{2}\int_0^L m(x)\left[\frac{\partial u(x,t)}{\partial t}\right]^2 dx \tag{8.204}$$

But, as established in Sec. 8.4, the free vibration of conservative systems is harmonic. Hence, by analogy with Eqs. (8.57) and (8.66), we can express the axial displacement $u(x,t)$ in the form

$$u(x,t) = U(x)\cos(\omega t - \phi) \tag{8.205}$$

Inserting Eq. (8.205) into Eq. (8.203), we can rewrite the potential energy as follows:

$$V(t) = \frac{1}{2}\left\{\int_0^L EA(x)\left[\frac{dU(x)}{dx}\right]^2 dx + kU^2(L)\right\}\cos^2(\omega t - \phi)$$
$$= V_{max}\cos^2(\omega t - \phi) \tag{8.206}$$

where

$$V_{max} = \frac{1}{2}\left\{\int_0^L EA(x)\left[\frac{dU(x)}{dx}\right]^2 dx + kU^2(L)\right\} \tag{8.207}$$

represents the maximum potential energy, obtained for $\cos(\omega t - \phi) = \pm 1$. Similarly, inserting Eq. (8.205) into Eq. (8.204), we have

$$T(t) = \frac{\omega^2}{2}\left[\int_0^L m(x)U^2(x)dx\right]\sin^2(\omega t - \phi) = \omega^2 T_{ref}\sin^2(\omega t - \phi) \tag{8.208}$$

in which

$$T_{ref} = \frac{1}{2}\int_0^L m(x)U^2(x)dx \tag{8.209}$$

represents a *reference kinetic energy*. It is easy to see that V_{max} is one half of the numerator of Rayleigh's quotient, Eq. (8.190), and T_{ref} is one half of the denominator. In view of this, we can express Eq. (8.190) in the form

$$R = \lambda = \omega^2 = \frac{V_{max}}{T_{ref}} \tag{8.210}$$

It is easy to see that the same form applies to the systems with the Rayleigh's quotient given by Eqs. (8.200)–(8.202), and to any conservative system in general. Clearly, Eq. (8.210) represents the generic form of Rayleigh's quotient sought. In fact, Eq. (8.210) applies not only to conservative distributed systems but to all conservative systems in general, including conservative discrete systems.

Example 8.8. Estimate the lowest eigenvalue of the string fixed at $x = 0$ and spring-supported at $x = L$ of Example 8.2 by means of Rayleigh's principle. Solve the problem in two ways: 1) using as a trial function the static displacement curve due to the string's own weight and 2) using the lowest eigenfunction of the fixed-free string. Compare results and draw conclusions.

Rayleigh's quotient for the string of Example 8.2 is

$$R = \omega^2 = \frac{T\int_0^L \left[\frac{dY(x)}{dx}\right]^2 dx + kY^2(L)}{\rho\int_0^L Y^2(x)dx} \tag{a}$$

where $Y(x)$ is the trial function, T the tension in the string, k the spring constant and ρ the mass density of the string. The estimates for the two cases are obtained as follows:

1) The static displacement curve as a trial function

Using Eqs. (a) and (b) of Example 8.2 as a guide, we conclude that the static displacement curve satisfies the boundary-value problem defined by the differential equation

$$T\frac{d^2Y(x)}{dx^2} = \rho g, \ 0 < x < L \tag{a}$$

where g is the gravitational constant, and the boundary conditions

$$Y(0) = 0, \ T\frac{dY(x)}{dx} + kY(x) = 0, \ x = L \tag{b}$$

The solution of Eq. (a) is simply

$$Y(x) = c_1 x + c_2 + \frac{1}{2}\frac{\rho g}{T}x^2 \tag{c}$$

in which c_1 and c_2 are constants of integration. Inserting Eq. (c) into Eqs. (b), the constants of integration can be shown to have the values

$$c_1 = -\frac{\rho g}{T}\frac{1+kL/2T}{1+kL/T}, \ c_2 = 0 \tag{d}$$

Hence, using the parameter ratio $kL/T = 0.5$ from Example 8.2, the static displacement curve has the expression

$$Y(x) = -\frac{5}{6}\frac{\rho g L}{T}x + \frac{1}{2}\frac{\rho g}{T}x^2 = \frac{\rho g L^2}{T}\left[-\frac{5}{6}\frac{x}{L} + \frac{1}{2}\left(\frac{x}{L}\right)^2\right] \tag{e}$$

Inserting Eq. (e) into Rayleigh's quotient, Eq. (a), and carrying out the indicated integrations, we obtain

$$\omega^2 = \frac{T\left(\frac{\rho g L}{T}\right)^2 \int_0^L -\left(\frac{5}{6} + \frac{x}{L}\right)^2 dx + k\left(\frac{\rho g L^2}{T}\right)^2 \left(-\frac{5}{6} + \frac{1}{2}\right)^2}{\rho\left(\frac{\rho g L^2}{T}\right)^2 \int_0^L \left[-\frac{5}{6}\frac{x}{L} + \frac{1}{2}\left(\frac{x}{L}\right)^2\right]^2 dx}$$

$$= \frac{\frac{T}{L^2}\frac{7}{36} + \frac{k}{L}\frac{1}{9}}{\rho\frac{79}{1080}} = 3.4177\frac{T}{\rho L^2} \tag{f}$$

so that, recalling that $\beta^2 = \omega^2\rho/T$, the estimate of the lowest eigenvalue is

$$\beta L = \sqrt{3.4177} = 1.8487 \tag{g}$$

which is slightly higher than the actual eigenvalue $\beta_1 L = 1.8366$, obtained in Example 8.2. In fact, the error is

$$e = \frac{\beta L - \beta_1 L}{\beta_1 L} = \frac{1.8487 - 1.8366}{1.8366} = 0.0066 = 0.66\% \tag{h}$$

which is extremely small.

2) The lowest eigenfunction of a fixed-free string as a trial function

The lowest eigenfunction of a fixed-free string is simply

$$Y(x) = \sin\frac{\pi x}{2L} \tag{i}$$

Inserting Eq. (i) into Eq. (a), we have

$$\omega^2 = \frac{T\left(\frac{\pi}{2L}\right)^2 \int_0^L \cos^2 \frac{\pi x}{2L} dx + k}{\rho \int_0^L \sin^2 \frac{\pi x}{2L} dx} = \frac{T\left(\frac{\pi}{2L}\right)^2 \frac{L}{2} + k}{\rho \frac{L}{2}}$$

$$= \left[\left(\frac{\pi}{2}\right)^2 + 1\right] \frac{T}{\rho L^2} = 3.4674 \frac{T}{\rho L^2} \tag{j}$$

so that the estimated lowest eigenvalue is

$$\beta L = \sqrt{3.4674} = 1.8621 \tag{k}$$

which is somewhat higher than the estimate obtained by using the static displacement curve, and hence not as good. The error is indeed higher,

$$e = \frac{\beta L - \beta_1 L}{\beta_1 L} = \frac{1.8621 - 1.8366}{1.8366} = 0.0139 = 1.39\% \tag{l}$$

The conclusion is that excellent results can be obtained by using the static displacement curve as a trial function. The disadvantage of this approach is that it may not be so easy at times to determine the static displacement curve. Somewhat poorer results, although still very good, can be expected with the lowest eigenfunction of a closely related but simpler system as a trial function. We observe that both estimated eigenvalues are higher than the actual eigenvalue, thus confirming the fact that Rayleigh's quotient provides an upper bound for the lowest eigenvalue.

8.9 RESPONSE TO INITIAL EXCITATIONS

As pointed out on several occasions, although various types of distributed-parameter systems exhibit similar vibrational characteristics, their mathematical description tends to differ in appearance. This difference is most obvious in the stiffness term. Hence, to discuss the system response, it is necessary to choose a certain elastic member.

From Sec. 8.4, the transverse displacement $y(x, t)$ of a string in free vibration is given by the partial differential equation

$$\frac{\partial}{\partial x}\left[T(x)\frac{\partial y(x,t)}{\partial x}\right] = \rho(x)\frac{\partial^2 y(x,t)}{\partial t^2}, \quad 0 < x < L \tag{8.211}$$

in which $T(x)$ is the tension and $\rho(x)$ the mass density. The solution of Eq. (8.211) must satisfy two boundary conditions, one at each end. In the case of discrete systems, the free vibration is caused by initial displacements and initial velocities of the individual masses. But, as shown in Sec. 8.1, distributed-parameter systems can be regarded as limiting cases of lumped-parameter systems whereby the lumped masses are spread over the entire domain of the system. Extending the analogy, we conclude that the free vibration of distributed systems is caused by initial excitations in the form of the initial displacement and initial velocity functions

$$y(x,0) = y_0(x), \quad \left.\frac{\partial y(x,t)}{\partial t}\right|_{t=0} = v_0(x) \tag{8.212}$$

We refer to Eqs. (8.212) as initial conditions.

In Sec. 8.4, it was indicated that the free vibration of a distributed-parameter system can be expressed as a linear combination of natural motions with amplitudes and phase angles depending on the initial conditions. The natural motions themselves consist of the natural modes multiplied by time-dependent harmonic functions with frequencies equal to the natural frequencies. This implies that, before we can solve for the response, we must solve the eigenvalue problem. The process can be formalized by considering the expansion theorem, which for a string in transverse vibration consists of Eqs. (8.93)–(8.95). Hence, consistent with this, we express the solution of Eq. (8.211) in the form

$$y(x,t) = \sum_{r=1}^{\infty} Y_r(x)\eta_r(t) \tag{8.213}$$

in which $Y_r(x)$ are the normal modes of the system and $\eta_r(t)$ are time-dependent functions. Introducing Eq. (8.213) in Eq. (8.211), we can write

$$\sum_{r=1}^{\infty} \frac{d}{dx}\left[T(x)\frac{dY_r(x)}{dx}\right]\eta_r(t) = \sum_{r=1}^{\infty} \rho(x)Y_r(x)\frac{d^2\eta_r(t)}{dt^2}, \quad 0 < x < L \tag{8.214}$$

Then, multiplying Eq. (8.214) by $Y_s(x)$ and integrating over the length of the string, we have

$$\sum_{r=1}^{\infty}\left\{\int_0^L Y_s(x)\frac{d}{dx}\left[T(x)\frac{dY_r(x)}{dx}\right]dx\right\}\eta_r(t) = \sum_{r=1}^{\infty}\left[\int_0^L \rho(x)Y_s(x)Y_r(x)dx\right]\frac{d^2\eta_r(t)}{dt^2} \tag{8.215}$$

Next, we consider the orthonormality relations, Eqs. (8.90) and (8.91), denote derivatives with respect to time by overdots and obtain the independent set of *modal equations*

$$\ddot{\eta}_r(t) + \omega_r^2\eta_r(t) = 0, \quad r = 1, 2, \ldots \tag{8.216}$$

where $\eta_r(t)$ $(r = 1, 2, \ldots)$ can now be identified as *modal coordinates*, and we observe that Eqs. (8.216) resemble entirely the equation of motion of an undamped single-degree-of-freedom system, Eq. (2.2). Hence, by analogy with Eq. (2.13), the solution of Eqs. (8.216) can be written as

$$\eta_r(t) = C_r\cos(\omega_r t - \phi_r) = \eta_r(0)\cos\omega_r t + \frac{\dot{\eta}_r(0)}{\omega_r}\sin\omega_r t, \quad r = 1, 2, \ldots \tag{8.217}$$

in which, from Eqs. (2.12), the amplitudes C_r and phase angles ϕ_r are related to the *initial modal displacements* $\eta_r(0)$ *and initial modal velocities* $\dot{\eta}_r(0)$ by

$$C_r = \sqrt{\eta_r^2(0) + [\dot{\eta}_r(0)/\omega_r]^2}, \quad \phi_r = \tan^{-1}\frac{\dot{\eta}_r(0)}{\omega_r\eta_r(0)}, \quad r = 1, 2, \ldots \tag{8.218}$$

where, in turn, $\eta_r(0)$ and $\dot{\eta}_r(0)$ are related to the actual initial conditions. To express the initial modal displacements in terms of the actual initial displacement function $y_0(x)$, we let $t = 0$ in Eq. (8.213), use the first of Eqs. (8.212) and write

$$y(x,0) = \sum_{r=1}^{\infty} Y_r(x)\eta_r(0) = y_0(x) \tag{8.219}$$

Then, multiplying Eq. (8.219) by $\rho(x)Y_s(x)$, integrating over the length of the string and using the orthonormality relations (8.90), we obtain

$$\eta_r(0) = \int_0^L \rho(x)Y_r(x)y_0(x)dx, \ r = 1, 2, \ldots \tag{8.220}$$

Using the same procedure, it is not difficult to see that

$$\dot{\eta}_r(0) = \int_0^L \rho(x)Y_r(x)v_0(x)dx, \ r = 1, 2, \ldots \tag{8.221}$$

The formal solution of the free vibration problem is completed by inserting Eqs. (8.217) into Eq. (8.213).

The process presented here is valid for arbitrary parameter distributions and boundary conditions, provided the eigenvalue problem admits an exact solution. Clearly, the same developments apply to all second-order systems, such as rods in axial vibration and shafts in torsion, and they are not restricted to strings in transverse vibration. All that is necessary is to replace $T(x)$ and $\rho(x)$ by the corresponding parameters listed in Table 8.1.

With some modifications, the same procedure can be used to obtain the response of beams in bending vibration. Indeed, in this case, from Eq. (8.69), the free vibration is described by the partial differential equation

$$-\frac{\partial^2}{\partial x^2}\left[EI(x)\frac{\partial^2 y(x,t)}{\partial x^2}\right] = m(x)\frac{\partial^2 y(x,t)}{\partial t^2}, \ 0 < x < L \tag{8.222}$$

where the solution $y(x, t)$ is subject to initial conditions analogous to those of Eqs. (8.212). Similarly, the solution can be expressed in a form resembling the series given by Eq. (8.213), which implies that it is necessary to solve the differential eigenvalue problem for the system. Following the procedure used earlier, we conclude that Eq. (8.214) must be replaced by

$$-\sum_{r=1}^{\infty}\frac{d^2}{dx^2}\left[EI(x)\frac{d^2Y_r(x)}{dx^2}\right]\eta_r(t) = \sum_{r=1}^{\infty} m(x)Y_r(x)\frac{d^2\eta_r(t)}{dt^2}, \ 0 < x < L \tag{8.223}$$

and Eq. (8.215) by

$$\begin{aligned} -\sum_{r=1}^{\infty}\left\{\int_0^L Y_s(x)\frac{d^2}{dx^2}\left[EI(x)\frac{d^2Y_r(x)}{dx^2}\right]dx\right\}\eta_r(t) \\ = \sum_{r=1}^{\infty}\left[\int_0^L m(x)Y_s(x)Y_r(x)dx\right]\frac{d^2\eta_r(t)}{dt^2} \end{aligned} \tag{8.224}$$

Then, Eqs. (8.216)–(8.221) remain the same, except that $m(x)$ replaces $\rho(x)$ in Eqs. (8.220) and (8.221).

At this point, we wish to demonstrate a statement made toward the end of Sec. 8.4 that every one of the natural modes can be excited independently of the other modes by proper initial conditions. To this end, we assume that the initial displacement of a string in transverse vibration is made to resemble the pth mode $Y_p(x)$ exactly. Hence, we have

$$y_0(x) = AY_p(x) \tag{8.225}$$

where A is a constant amplitude. Inserting Eq. (8.225) into Eqs. (8.220) and using the orthonormality relations (8.90), we can write

$$\eta_r(0) = A\int_0^L \rho(x)Y_r(x)Y_p(x)dx = \begin{cases} A \text{ for } r = p \\ 0 \text{ for } r = 1, 2, \ldots, p-1, p+1, \ldots \end{cases} \tag{8.226}$$

Moreover, the initial modal velocities $\dot{\eta}_r(0)$ are all zero $(r = 1, 2, \ldots)$. It follows from Eqs. (8.217) that

$$\eta_r(t) = \begin{cases} A\cos\omega_r t \text{ for } r = p \\ 0 \text{ for } r = 1, 2, \ldots, p-1, p+1, \ldots \end{cases} \tag{8.227}$$

Finally, inserting Eqs. (8.227) into Eq. (8.213), we obtain

$$y(x,t) = AY_p(x)\cos\omega_p t \tag{8.228}$$

which demonstrates that the ensuing motion is vibration of the string in the pth mode alone.

Next, we turn our attention to the response of systems with lumped masses at the boundaries to initial excitations. From Sec. 8.7, the boundary-value problem for a rod in free axial vibration fixed at $x = 0$ and with a lumped mass M at $x = L$ (Fig. 8.16a) is given by the partial differential equation

$$\frac{\partial}{\partial x}\left[EA(x)\frac{\partial u(x,t)}{\partial x}\right] = m(x)\frac{\partial^2 u(x,t)}{\partial t^2}, \quad 0 < x < L \tag{8.229}$$

and the boundary conditions

$$u(0,t) = 0 \tag{8.230}$$

and

$$-EA(x)\frac{\partial u(x,t)}{\partial x} = M\frac{\partial^2 u(x,t)}{\partial t^2}, \quad x = L \tag{8.231}$$

The interest lies in the response of the system to the initial displacement and initial velocity

$$u(x,0) = u_0(x), \quad \left.\frac{\partial u(x,t)}{\partial t}\right|_{t=0} = v_0(x) \tag{8.232}$$

respectively.

By analogy with Eq. (8.213), we assume a solution of Eq. (8.229) in the form

$$u(x,t) = \sum_{r=1}^{\infty} U_r(x)\eta_r(t) \tag{8.233}$$

where $U_r(x)$ are the system normal modes and $\eta_r(t)$ are the modal coordinates ($r = 1, 2, \dots$). Inserting Eq. (8.233) into Eq. (8.229), multiplying through by $U_s(x)$ and integrating over the length of the rod, we obtain

$$\sum_{r=1}^{\infty}\left\{\int_0^L U_s(x)\frac{d}{dx}\left[EA(x)\frac{dU_r(x)}{dx}\right]dx\right\}\eta_r(t) = \sum_{r=1}^{\infty}\left[\int_0^L m(x)U_s(x)U_r(x)dx\right]\ddot{\eta}_r(t),$$
$$s = 1, 2, \dots \qquad (8.234)$$

But, from Eqs. (8.160) and (8.161), we have

$$\int_0^L m(x)U_s(x)U_s(x)dx = \delta_{rs} - MU_r(L)U_s(L),\ r, s = 1, 2, \dots \qquad (8.235)$$

and

$$\int_0^L U_s(x)\frac{d}{dx}\left[EA(x)\frac{dU_r(x)}{dx}\right]dx = \left[U_s(x)EA(x)\frac{dU_r(x)}{dx}\right]\bigg|_{x=L} - \omega_r^2\delta_{rs},$$
$$r, s = 1, 2, \dots \qquad (8.236)$$

so that Eqs. (8.234) can be rewritten as

$$\ddot{\eta}_s(t) + \omega_s^2\eta_s(t) - \sum_{r=1}^{\infty}\left\{U_s(x)\left[MU_r(x)\ddot{\eta}_r(t) + EA(x)\frac{dU_r(x)}{dx}\eta_r(t)\right]\right\}\bigg|_{x=L} = 0,$$
$$s = 1, 2, \dots \qquad (8.237)$$

But, observing from boundary condition (8.231) and Eq. (8.233) that

$$\sum_{r=1}^{\infty}\left[MU_r(x)\ddot{\eta}_r(t) + EA(x)\frac{dU_r(x)}{dx}\eta_r(t)\right]\bigg|_{x=L}$$
$$= \left[M\frac{\partial^2 u(x,t)}{\partial t^2} + EA(x)\frac{\partial u(x,t)}{\partial x}\right]\bigg|_{x=L} = 0 \qquad (8.238)$$

we conclude that Eqs. (8.237) reduce to the standard independent modal equations

$$\ddot{\eta}_s(t) + \omega_s^2\eta_s(t) = 0,\ s = 1, 2, \dots \qquad (8.239)$$

Equations (8.239) have the familiar solution

$$\eta_s(t) = \eta_s(0)\cos\omega_s t + \frac{\dot{\eta}_s(0)}{\omega_s}\sin\omega_s t,\ s = 1, 2, \dots \qquad (8.240)$$

where $\eta_s(0)$ are initial modal displacements and $\dot{\eta}_s(0)$ are initial modal velocities ($s = 1, 2, \dots$). Their values can be obtained from the actual initial displacement $u_0(x)$ and initial velocity $v_0(x)$, respectively. To this end, we let $t = 0$ in Eq. (8.233) and write

$$u(x, 0) = \sum_{s=1}^{\infty} U_s(x)\eta_s(0) = u_0(x) \qquad (8.241)$$

Then, multiplying both sides of Eq. (8.241) by $m(x)U_r(x)$ and integrating over the length of the rod, multiplying both sides of Eq. (8.241) evaluated at $x = L$ by $MU_s(L)$, adding the two results and using the orthonormality relations (8.159), we obtain

$$\eta_s(0) = \int_0^L m(x)U_s(x)u_0(x)dx + MU_s(L)u_0(L), \; s = 1, 2, \ldots \tag{8.242}$$

Similarly, it is easy to see that

$$\dot{\eta}_s(0) = \int_0^L m(x)U_s(x)v_0(x)dx + MU_s(L)v_0(L), \; s = 1, 2, \ldots \tag{8.243}$$

The formal solution for the response to initial excitations is obtained by inserting Eqs. (8.240) in conjunction with Eqs. (8.242) and (8.243) into Eq. (8.233).

The response to initial excitations of a beam in bending cantilevered at $x = 0$ and with a lumped mass at $x = L$ has the same form as that of a rod in axial vibration fixed at $x = 0$ and with a lumped mass at $x = L$, Eqs. (8.233), (8.240), (8.242) and (8.243), the only difference being that the symbols $u(x,t)$, $u_0(x)$ and $U_s(x)$ must be replaced by $y(x,t)$, $y_0(x)$ and $Y_s(x)$, respectively.

Example 8.9. Determine the response of the uniform string of Example 8.1 to the initial displacement shown in Fig. 8.21. The initial velocity is zero. From Fig. 8.21, the initial displacement function has the analytical expression

$$y_0(x) = \begin{cases} \dfrac{Ax}{a}, & 0 < x < a \\ \dfrac{A}{L-a}(L-x), & a < x < L \end{cases} \tag{a}$$

From Eq. (8.213), the response of the system is given by

$$y(x,t) = \sum_{r=1}^{\infty} Y_r(x)\eta_r(t) \tag{b}$$

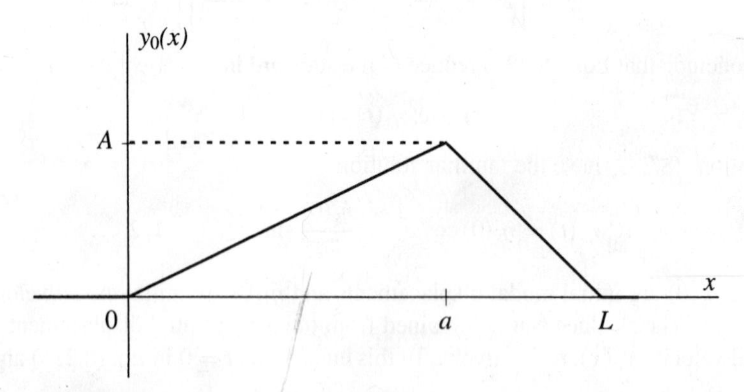

FIGURE 8.21
Initial displacement function for a string fixed at both ends

where $Y_r(x)$ are the normal modes satisfying $\int_0^L \rho(x)Y_r^2(x)dx = 1$ and $\eta_r(t)$ are the modal coordinates $(r = 1, 2, \ldots)$. The natural modes were obtained in Example 8.1. Hence, using Eqs. (k) of Example 8.1 and normalizing according to the formula just given, we obtain the normal modes

$$Y_r(x) = \sqrt{\frac{2}{\rho L}} \sin \frac{r\pi x}{L}, \ r = 1, 2, \ldots \tag{c}$$

where ρ is the constant mass density. Moreover, in view of the fact that $v_0(x) = 0$, we let $\dot{\eta}_r(0) = 0 \ (r = 1, 2, \ldots)$ in Eqs. (8.217), and obtain the modal coordinates

$$\eta_r(t) = \eta_r(0) \cos \omega_r t, \ r = 1, 2, \ldots \tag{d}$$

in which, from Eqs. (8.220),

$$\eta_r(0) = \int_0^L \rho Y_r(x) y_0(x) dx, \ r = 1, 2, \ldots \tag{e}$$

are the initial modal displacements and, from Eqs. (i) of Example 8.1,

$$\omega_r = r\pi \sqrt{\frac{T}{\rho L^2}}, \ r = 1, 2, \ldots \tag{f}$$

are the natural frequencies, where T is the string tension. Hence, inserting Eqs. (a) and (c) into Eqs. (e), we obtain the initial modal coordinates

$$\begin{aligned}
\eta_r(0) &= \frac{A}{a} \rho \sqrt{\frac{2}{\rho L}} \int_0^a x \sin \frac{r\pi x}{L} dx + \frac{A}{L-a} \rho \sqrt{\frac{2}{\rho L}} \int_a^L (L-x) \sin \frac{r\pi x}{L} dx \\
&= \frac{A\rho}{a} \sqrt{\frac{2}{\rho L}} \left(-\frac{L}{r\pi} x \cos \frac{r\pi x}{L} + \frac{L^2}{r^2\pi^2} \sin \frac{r\pi x}{L} \right) \Bigg|_0^a \\
&\quad + \frac{A\rho}{L-a} \sqrt{\frac{2}{\rho L}} \left[-\frac{L}{r\pi} (L-x) \cos \frac{r\pi x}{L} - \frac{L^2}{r^2\pi^2} \sin \frac{r\pi x}{L} \right] \Bigg|_a^L \\
&= A\sqrt{2\rho L} \frac{L^2}{r^2\pi^2 a(L-a)} \sin \frac{r\pi a}{L}, \ r = 1, 2, \ldots
\end{aligned} \tag{g}$$

Finally, inserting Eqs. (c)–(g) into Eq. (b), we obtain the response

$$y(x,t) = \frac{2AL^2}{\pi^2 a(L-a)} \sum_{r=1}^{\infty} \frac{1}{r^2} \sin \frac{r\pi a}{L} \sin \frac{r\pi x}{L} \cos r\pi \sqrt{\frac{T}{\rho L^2}} t \tag{h}$$

and we observe that the contribution of the higher mode is inversely proportional to the square of the mode number.

As a matter of interest, we consider the case in which $a = L/2$. To this end, we observe that

$$\sin \frac{r\pi a}{L} = \sin \frac{r\pi}{2} = \begin{cases} (-1)^{(r-1)/2} \text{ for } r \text{ odd} \\ 0 \text{ for } r \text{ even} \end{cases} \tag{i}$$

so that only the odd-numbered modes contribute to the response, or

$$y(x,t) = \frac{2AL^2}{\pi^2 a(L-a)} \sum_{r=1,3,\ldots}^{\infty} \frac{(-1)^{(r-1)/2}}{r^2} \sin \frac{r\pi x}{L} \cos r\pi \sqrt{\frac{T}{\rho L^2}} t \tag{j}$$

This can be easily explained by the fact that for $a = L/2$ the initial displacement function is symmetric with respect to $x = L/2$, so that the even-numbered modes, which are observed from Fig. 8.10 to be antisymmetric, cannot be excited.

Example 8.10. Determine the response of the cantilever beam with a lumped mass at the end of Example 8.7 to the initial velocity

$$v_0(x) = 13.72\left(\frac{x}{L}\right)^2 - 23.22\left(\frac{x}{L}\right)^3 + 9.26\left(\frac{x}{L}\right)^4 \tag{a}$$

The initial velocity function is plotted in Fig. 8.22. The initial displacement is zero.

From Sec. 8.6, the boundary-value problem for the free vibration of the system under consideration is given by the differential equation

$$-EI\frac{\partial^4 y(x,t)}{\partial x^4} = m\frac{\partial^2 y(x,t)}{\partial t^2}, \quad 0 < x < L \tag{b}$$

where the displacement $y(x,t)$ is subject to the boundary conditions

$$y(x,t) = 0, \quad \frac{\partial y(x,t)}{\partial x} = 0, \ x = 0 \tag{c}$$

and

$$EI\frac{\partial^2 y(x,t)}{\partial x^2} = 0, \ EI\frac{\partial^3 y(x,t)}{\partial x^3} = M\frac{\partial^2 y(x,t)}{\partial t^2}, \ x = L \tag{d}$$

The solution of Eq. (b) can be expressed in the form

$$y(x,t) = \sum_{r=1}^{\infty} Y_r(x)\eta_r(t) \tag{e}$$

where the modes $Y_r(x)$ are given by Eqs. (g) of Example 8.7. For convenience the coefficients A_r in Eqs. (g) are such that the modes satisfy the orthonormality relations

$$m\int_0^L Y_r(x)Y_s(x)dx + MY_r(L)Y_s(L) = \delta_{rs}, \ r,s = 1,2,\ldots \tag{f}$$

and

$$EI\left\{\int_0^L Y_s(x)\frac{d^4 Y_r(x)}{dx^4}dx - \left[Y_s(x)\frac{d^3 Y_r(x)}{dx^3}\right]\Bigg|_{x=L}\right\} = \omega_r^2\delta_{rs}, \ r,s = 1,2\ldots \tag{g}$$

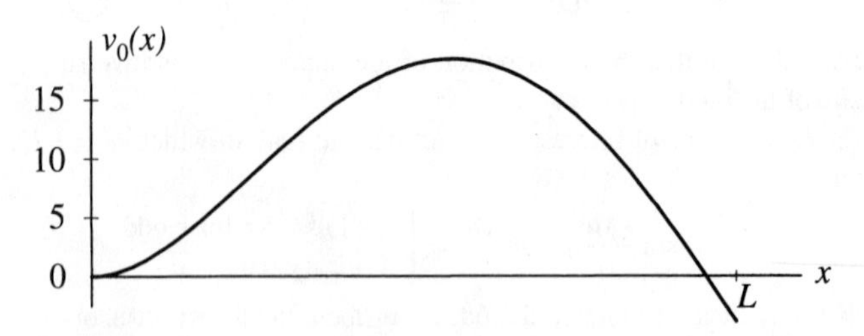

FIGURE 8.22
Initial velocity function for a cantilever beam with a lumped mass at the tip

as can be concluded from Eqs. (8.178) and (8.180), respectively. Inserting Eq. (e) into Eq. (b), multiplying through by $Y_s(x)$, integrating over the length of the beam and rearranging, we can write

$$\ddot{\eta}_s(t)+\omega_s^2\eta_s(t)-\sum_{r=1}^{\infty}\left\{Y_s(x)\left[MY_r(x)\ddot{\eta}_r(t)-EI\frac{d^3Y_r(x)}{dx^3}\eta_r(t)\right]\right\}\bigg|_{x=L}=0,$$

$$s=1,2,\ldots \tag{h}$$

which, in view of the second of boundary conditions (d), reduce to

$$\ddot{\eta}_c+\omega_s^2\eta_s(t)=0,\ s=1,2\ldots \tag{i}$$

Because the initial displacement is zero, the solution of Eqs. (i) is

$$\eta_s(t)=\frac{\dot{\eta}_s(0)}{\omega_s}\sin\omega_s t,\ s=1,2,\ldots \tag{j}$$

where, by analogy with Eqs. (8.243),

$$\dot{\eta}_s(0)=m\int_0^L Y_s(x)v_0(x)dx+MY_s(L)v_0(L)$$

$$=m\int_0^L Y_s(x)\left[13.72\left(\frac{x}{L}\right)^2-23.22\left(\frac{x}{L}\right)^3+9.26\left(\frac{x}{L}\right)^4\right]dx-0.24MY_s(L),$$

$$s=1,2,\ldots \tag{k}$$

Hence, combining Eqs. (e), (j) and (k), using Eq. (g) of Example 8.7 and recalling from that example that $M=mL$, we obtain the response

$$y(x,t)=\sum_{r=1}^{\infty}C_r\left[\sin\beta_r x-\sinh\beta_r x-\frac{\sin\beta_r L+\sinh\beta_r L}{\cos\beta_r L+\cosh\beta_r L}(\cos\beta_r x-\cosh\beta_r x)\right]\sin\omega_r t \tag{l}$$

The first three coefficients have the values $C_1=-0.0404$, $C_2=0.7761$, $C_3=-0.0003$, and we observe that C_2 has the largest value, which implies that the second mode has been excited the most. This should come as no surprise, as the initial velocity resembles the second mode, as can be concluded by comparing Figs. 8.20 and 8.22.

8.10 RESPONSE TO EXTERNAL EXCITATIONS

As pointed out in the beginning of Sec. 8.9, the various types of distributed-parameter systems differ more in appearance than in vibrational characteristics. Still, because of this difference in appearance, in discussing the response to external excitations, it is necessary to carry out the derivation of the response by means of a specific conservative vibrating system. With certain modifications, the same developments apply to all systems in this class.

We consider the response of a beam in bending supported by a spring of stiffness k at $x=0$ and pinned at $x=L$. From Sec. 8.3, we obtain the corresponding partial differential equation

$$-\frac{\partial^2}{\partial x^2}\left[EI(x)\frac{\partial^2 y(x,t)}{\partial x^2}\right]+f(x,t)=m(x)\frac{\partial^2 y(x,t)}{\partial t^2},\ 0<x<L \tag{8.244}$$

and, from Sec. 8.4, the boundary conditions

$$EI(x)\frac{\partial^2 y(x,t)}{\partial x^2}=0,\ \frac{\partial}{\partial x}\left[EI(x)\frac{\partial^2 y(x,t)}{\partial x^2}\right]+ky(x,t)=0,\ x=0 \tag{8.245}$$

and

$$y(x,t)=0,\ EI(x)\frac{\partial^2 y(x,t)}{\partial x^2}=0,\ x=L \tag{8.246}$$

To derive the response of the system under consideration, we must use modal analysis. To this end, we first solve the associated eigenvalue problem, Eqs. (8.72)–(8.74), which yields the natural modes $Y_r(x)$ and natural frequencies ω_r $(r=1,2,\ldots)$. The natural modes are orthogonal and we assume that they have been normalized so as to satisfy the orthonormality conditions

$$\int_0^L m(x)Y_r(x)Y_s(x)dx=\delta_{rs},\ r,s=1,2,\ldots \tag{8.247}$$

and

$$\int_0^L Y_s(x)\frac{d^2}{dx^2}\left[EI(x)\frac{d^2Y_r(x)}{dx^2}\right]dx=\omega_r^2\delta_{rs},\ r,s=1,2,\ldots \tag{8.248}$$

Then, we assume a solution of Eq. (8.244) in the form

$$y(x,t)=\sum_{r=1}^{\infty}Y_r(x)\eta_r(t) \tag{8.249}$$

where $\eta_r(t)$ $(r=1,2,\ldots)$ are modal coordinates. Next, we insert Eq. (8.249) into Eq. (8.244), multiply by $Y_s(x)$, integrate over the length of the beam, consider Eqs. (8.247) and (8.248) and obtain the independent modal equations

$$\ddot{\eta}_r(t)+\omega_r^2\eta_r(t)=N_r(t),\ r=1,2,\ldots \tag{8.250}$$

in which

$$N_r(t)=\int_0^L Y_r(x)f(x,t)dx,\ r=1,2,\ldots \tag{8.251}$$

are the modal forces $(r=1,2,\ldots)$.

Equations (8.250) resemble entirely the modal equations for discrete systems, Eqs. (7.206). Hence, to discuss their solution, we follow the pattern established in Sec. 7.15. First, we consider the case in which the excitation is harmonic and express it in the form

$$f(x,t)=F(x)\cos\Omega t \tag{8.252}$$

Introducing Eq. (8.252) in Eqs. (8.251), we obtain the modal forces

$$N_r(t) = \left[\int_0^L Y_r(x)F(x)dx\right]\cos\Omega t = F_r\cos\Omega t, \ r = 1, 2, \ldots \tag{8.253}$$

where

$$F_r = \int_0^L Y_r(x)F(x)dx, \ r = 1, 2, \ldots \tag{8.254}$$

are modal force amplitudes. Inserting Eqs. (8.253) into Eqs. (8.250), it is easy to verify that the steady-state solution is

$$\eta_r(t) = \frac{F_r}{\omega_r^2 - \Omega^2}\cos\Omega t, \ r = 1, 2, \ldots \tag{8.255}$$

so that, in view of Eq. (8.249), the steady-state harmonic response is

$$y(x, t) = \left[\sum_{r=1}^{\infty} \frac{F_r}{\omega_r^2 - \Omega^2} Y_r(x)\right]\cos\Omega t \tag{8.256}$$

From Eqs. (8.254) and (8.256), we conclude that, if the excitation amplitude function $F(x)$ resembles the product of the mass density $m(x)$ and one of the normal modes, say $F(x) = m(x)Y_k(x)$, then $F_r = 0, r \neq k$, so that the response reduces to

$$y(x, t) = \frac{F_k}{\omega_k^2 - \Omega^2} Y_k(x)\cos\Omega t \tag{8.257}$$

which implies that only the kth mode is excited. Moreover, if the excitation frequency is equal to one of the natural frequencies, then the system experiences resonance. These are basically the same conclusions as those reached in Sec. 7.15 in conjunction with discrete systems. This demonstrates that, whereas discrete and distributed systems constitute two different types of models, as long as they represent conservative systems, their behavior is entirely analogous.

In the case in which the external excitation $f(x, t)$ is arbitrary, the modal forces $N_r(t)$ $(r = 1, 2, \ldots, n)$ are arbitrary. Then, as in Sec. 7.15, the modal displacements, obtained by solving Eqs. (8.250), can be expressed in the form of the convolution integrals

$$\eta_r(t) = \frac{1}{\omega_r}\int_0^t N_r(t - \tau)\sin\omega_r\tau\, d\tau, \ r = 1, 2, \ldots \tag{8.258}$$

so that, using Eq. (8.249), the response of the distributed system to arbitrary excitations is

$$y(x, t) = \sum_{r=1}^{\infty} \frac{Y_r(x)}{\omega_r}\int_0^t N_r(t - \tau)\sin\omega_r\tau\, d\tau \tag{8.259}$$

If there are initial excitations, then, by virtue of the superposition principle, the response to initial excitations can be obtained separately by the approach of Sec. 8.9, and added to Eq. (8.259).

It should be stressed here that, although we derived the response using a beam supported by a spring at $x = 0$ and pinned at $x = L$, the developments remain essentially the same for all other boundary conditions, and the same can be said about other systems, such as strings in transverse vibration, rods in axial vibration and shafts in torsion.

Example 8.11. Derive the response of a uniform pinned-pinned beam to a concentrated force of amplitude F_0 acting at $x = L/2$ and having the form of a step function. The concentrated force can be treated as distributed by writing

$$f(x,t) = F_0\delta(x - L/2)\mathscr{u}(t) \tag{a}$$

where $\delta(x - L/2)$ is a spatial Dirac delta function having the property

$$\int_0^L g(x)\delta(x-a)dx = g(a) \tag{b}$$

The eigenvalue problem for a uniform pinned-pinned system was solved in Example 8.3, from which we obtain the natural modes

$$Y_r(x) = A_r \sin\frac{r\pi x}{L},\ r = 1,2,\ldots \tag{c}$$

and eigenvalues

$$\beta_r L = r\pi,\ r = 1,2,\ldots \tag{d}$$

If normalized so as to satisfy Eqs. (8.247), then the normal modes become

$$Y_r(x) = \sqrt{\frac{2}{mL}}\sin\frac{r\pi x}{L},\ r = 1,2,\ldots \tag{e}$$

Moreover, the natural frequencies are

$$\omega_r = \beta_r^2\sqrt{\frac{EI}{m}} = (r\pi)^2\sqrt{\frac{EI}{mL^4}} \tag{f}$$

The response is given by Eq. (8.259), which requires the modal forces $N_r(t)$ $(r = 1,2,\ldots)$. Inserting Eqs. (a) and (e) into Eq. (8.251) and considering Eq. (b), we obtain the modal forces

$$\begin{aligned} N_r(t) &= \int_0^L Y_r(x)f(x,t)dx = \sqrt{\frac{2}{mL}}F_0\mathscr{u}(t)\int_0^L \sin\frac{r\pi x}{L}\delta(x-L/2)dx \\ &= \sqrt{\frac{2}{mL}}F_0\mathscr{u}(t)\sin\frac{r\pi}{2} = (-1)^{(r-1)/2}\sqrt{\frac{2}{mL}}F_0\mathscr{u}(t),\ r = \text{odd} \end{aligned} \tag{g}$$

Then, introducing Eqs. (g) in Eqs. (8.258) and evaluating the convolution integrals, we obtain the modal coordinates

$$\begin{aligned} \eta_r(t) &= \frac{1}{\omega_r}\int_0^t N_r(t-\tau)\sin\omega_r\tau\,d\tau = \frac{(-1)^{(r-1)/2}F_0}{\omega_r}\sqrt{\frac{2}{mL}}\int_0^t \mathscr{u}(t-\tau)\sin\omega_r\tau\,d\tau \\ &= \frac{(-1)^{(r-1)/2}F_0}{\omega_r^2}\sqrt{\frac{2}{mL}}(1-\cos\omega_r t) \\ &= \frac{(-1)^{(r-1)/2}F_0}{(r\pi)^4}\frac{mL^4}{EI}\sqrt{\frac{2}{mL}}\left[1-\cos(r\pi)^2\sqrt{\frac{EI}{mL^4}}t\right],\ r = \text{odd} \end{aligned} \tag{h}$$

Finally, inserting Eqs. (e) and (h) into Eq. (8.249), we can write the response

$$y(x,t) = \sum_{r=1}^{\infty} Y_r(x)\eta_r(t) = \sum_{r=1,3,\ldots}^{\infty} \frac{(-1)^{(r-1)/2} F_0}{(r\pi)^4} \frac{mL^4}{EI} \frac{2}{mL} \sin\frac{r\pi x}{L}\left[1 - \cos(r\pi)^2\sqrt{\frac{EI}{mL^4}}t\right]$$

$$= \frac{2F_0L^3}{\pi^4 EI} \sum_{r=1,3,\ldots}^{\infty} \frac{(-1)^{(r-1)/2}}{r^4} \sin\frac{r\pi x}{L}\left[1 - \cos(r\pi)^2\sqrt{\frac{EI}{mL^4}}t\right] \tag{i}$$

We note that the even-numbered modes $\sin 2\pi x/L$, $\sin 4\pi x/L, \ldots$, which are antisymmetric, do not participate in the response. This can be easily explained by the fact that the load is concentrated at $x = L/2$, and the antisymmetric modes have a node at that point, so that they cannot be excited. Instead of skipping over the even integers in the summation, we replace r by $2j - 1$ and write the response in the form

$$y(x,t) = \frac{2F_0L^3}{\pi^4 EI} \sum_{j=1}^{\infty} \frac{(-1)^{(j-1)}}{(2j-1)^4} \sin\frac{(2j-1)\pi x}{L}\left[1 - \cos(2j-1)^2\pi^2\sqrt{\frac{EI}{mL^4}}t\right] \tag{j}$$

where now the summation is over all integers. We note that the mode contribution to the response is inversely proportional to $(2j-1)^4$, which indicates that the mode participation decreases rapidly as the mode number increases. Indeed, the contribution of the second participating mode, which is actually the third natural mode, is only a little over one percent compared to the contribution of the first mode.

8.11 SYSTEMS WITH EXTERNAL FORCES AT BOUNDARIES

In many cases, the system is subjected to external forces at the boundaries. Such systems differ from those encountered until now, because forces independent of displacements and/or velocities render the corresponding boundary conditions nonhomogeneous. Problems with nonhomogeneous boundary conditions do not lend themselves to the derivation of the response by modal analysis, so that difficulties can be expected. It turns out, however, that a reformulation of the problem can obviate these difficulties, thus permitting the use of modal analysis.

We consider a rod in axial vibration fixed at $x = 0$ and with an arbitrary force $F(t)$ applied at $x = L$, as shown in Fig. 8.23. For simplicity, we assume that the initial conditions are zero. The partial differential equation of motion for the system is

$$\frac{\partial}{\partial x}\left[EA(x)\frac{\partial u(x,t)}{\partial x}\right] = m(x)\frac{\partial^2 u(x,t)}{\partial t^2}, \quad 0 < x < L \tag{8.260}$$

where the axial displacement $u(x,t)$ must satisfy the boundary conditions

$$u(0,t) = 0, \quad EA(x)\frac{\partial u(x,t)}{\partial x}\bigg|_{x=L} = F(t) \tag{8.261}$$

Clearly, the second of boundary conditions (8.261) is nonhomogeneous, which precludes the use of modal analysis for the response.

The problem under consideration consists of a homogeneous differential equation, Eq. (8.260), and two boundary conditions, one homogeneous, the first of Eqs. (8.261),

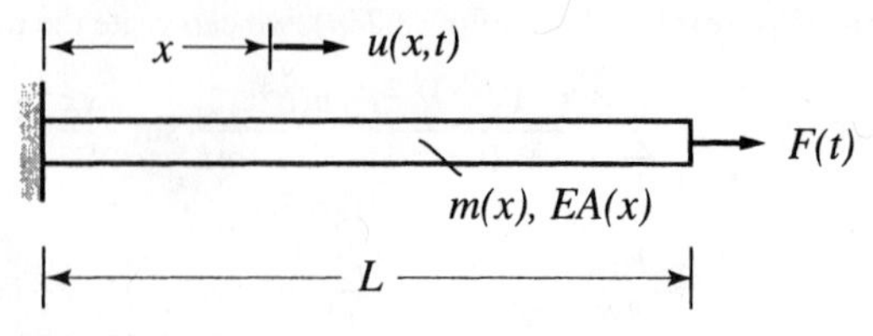

FIGURE 8.23
Rod fixed at $x = 0$ and with an axial force at $x = L$

and one nonhomogeneous, the second of Eqs. (8.261). A solution of the problem can be obtained by transforming it into a problem defined by a nonhomogeneous differential equation and two homogeneous boundary conditions, where the latter can be solved by modal analysis. Such an approach is used in Ref. 13, but the strictly analytical approach tends to be tedious, and so does the solution. A simpler and more intuitive approach achieving the same goal, namely, the transfer of the nonhomogeneity from the boundary condition at $x = L$ to the differential equation, is to treat the force $F(t)$ concentrated at the end $x = L$ as if it were distributed over a very small segment of the rod given by $L- < x < L$, where $L-$ denotes a point to the immediate left of $x = L$, as depicted in Fig. 8.24. For all practical purposes, Figs. 8.23 and 8.24 depict equivalent systems. Mathematically, the equivalent distributed force can be expressed as

$$f(x,t) = F(t)\delta(x - L) \tag{8.262}$$

where $\delta(x - L)$ is a spatial Diral delta function applied immediately to the left of $x = L$ and defined as

$$\begin{gathered} \delta(x - L) = 0, \; x \neq L \\ \int_0^L \delta(x - L)dx = 1 \end{gathered} \tag{8.263}$$

In view of this, we can reformulate the problem by rewriting the differential equation in the form

$$\frac{\partial}{\partial x}\left[EA(x)\frac{\partial u(x,t)}{\partial x}\right] + F(t)\delta(x - L) = m(x)\frac{\partial^2 u(x,t)}{\partial t^2}, \; 0 < x < L \tag{8.264}$$

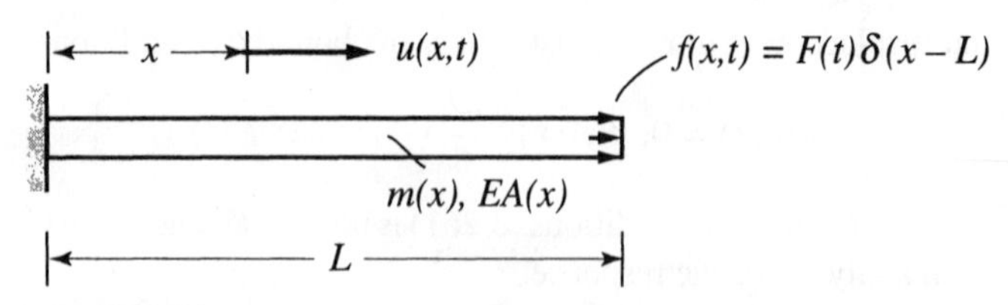

FIGURE 8.24
Concentrated axial force at $x = L$ treated as distributed

which is nonhomogeneous, and the boundary conditions as

$$u(0,t)=0, \quad \left[EA(x)\frac{\partial u(x,t)}{\partial x}\right]\Bigg|_{x=L}=0 \tag{8.265}$$

which are both homogeneous. The boundary-value problem defined by Eqs. (8.264) and (8.265) is of the standard type, and its solution can be obtained routinely by modal analysis, as described earlier in this chapter.

To obtain the response of the system described by Eqs. (8.264) and (8.265) by modal analysis, we must first solve the eigenvalue problem defined by the differential equation

$$-\frac{d}{dx}\left[EA(x)\frac{dU(x)}{dx}\right]=\omega^2 m(x)U(x), \; 0<x<L \tag{8.266}$$

and the boundary conditions

$$U(0)=0, \; EA(x)\frac{dU(x)}{dx}\bigg|_{x=L}=0 \tag{8.267}$$

The solution consists of the eigenfunctions $U_r(x)$ and the eigenvalues ω_r^2 $(r=1,2,\ldots)$. The eigenfunctions are orthogonal and are assumed to have been normalized so as to satisfy the orthonormality conditions

$$\int_0^L m(x)U_r(x)U_s(x)dx=\delta_{rs}, \; r,s=1,2,\ldots \tag{8.268}$$

and

$$-\int_0^L U_s(x)\frac{d}{dx}\left[EA(x)\frac{dU_r(xA)}{dx}\right]dx=\omega_r^2\delta_{rs}, \; r,s=1,2,\ldots \tag{8.269}$$

Next, we assume a solution in the form

$$u(x,t)=\sum_{r=1}^{\infty}U_r(x)\eta_r(t) \tag{8.270}$$

so that, inserting Eq. (8.270) into Eq. (8.264), multiplying by $U_s(x)$, integrating over the length of the rod and considering the orthonormality relations (8.268) and (8.269), we obtain the modal equations

$$\ddot{\eta}_r(t)+\omega_r^2\eta_r(t)=N_r(t), \; r=1,2,\ldots \tag{8.271}$$

where

$$N_r(t)=\int_0^L U_r(x)F(t)\delta(x-L)dx=U_r(L)F(t), \; r=1,2,\ldots \tag{8.272}$$

are the modal forces. Then, the solution of the modal equations can be written in the form of the convolution integrals

$$\eta_r(t)=\frac{1}{\omega_r}\int_0^t N_r(t-\tau)\sin\omega_r\tau d\tau=\frac{U_r(L)}{\omega_r}\int_0^t F(t-\tau)\sin\omega_r\tau d\tau, \; r=1,2,\ldots \tag{8.273}$$

The response of the rod to the boundary force $F(t)$ is completed by introducing Eqs. (8.273) in Eq. (8.270).

Example 8.12. Obtain the response of a uniform rod, $EA(x) = EA =$ constant, $m(x) = m =$ constant, fixed at $x = 0$ and subjected to a boundary force at $x = L$ in the form

$$F(t) = F_0 u(t) \tag{a}$$

where $u(t)$ is a unit step function.

From Eqs. (8.266) and (8.267), the eigenvalue problem for the uniform rod fixed at $x = 0$ and free at $x = L$ is given by the differential equation

$$\frac{d^2U(x)}{dx^2} + \beta^2 U(x) = 0,\ 0 < x < L,\ \beta^2 = \frac{\omega^2 m}{EA} \tag{b}$$

and the boundary conditions

$$U(0) = 0,\ \left.\frac{dU(x)}{dx}\right|_{x=L} = 0 \tag{c}$$

It can be verified that the solution of the eigenvalue problem consists of the orthonormal modes

$$U_r(x) = \sqrt{\frac{2}{mL}} \sin\frac{(2r-1)\pi x}{2L},\ r = 1, 2, \ldots \tag{d}$$

and the natural frequencies

$$\omega_r = \frac{(2r-1)\pi}{2}\sqrt{\frac{EA}{mL^2}},\ r = 1, 2, \ldots \tag{e}$$

Introducing Eq. (a) in Eqs. (8.273) and considering Eqs. (d) and (e), we can write the modal coordinates

$$\eta_r(t) = \frac{U_r(L)}{\omega_r}\int_0^t F_0 u(t-\tau)\sin\omega_r\tau\, d\tau = \frac{F_0 U_r(L)}{\omega_r^2}(1-\cos\omega_r t)$$

$$= \frac{4F_0\sqrt{2/mL}\sin\dfrac{(2r-1)\pi}{2}}{(2r-1)^2\pi^2}\frac{mL^2}{EA}\left[1-\cos\frac{(2r-1)\pi}{2}\sqrt{\frac{EA}{mL^2}}t\right]$$

$$= \frac{4F_0\sqrt{2/mL}(-1)^{r-1}}{(2r-1)^2\pi^2}\frac{mL^2}{EA}\left[1-\cos\frac{(2r-1)\pi}{2}\sqrt{\frac{EA}{mL^2}}t\right],\ r = 1, 2, \ldots \tag{f}$$

Finally, inserting Eqs. (d) and (f) into Eq. (8.270), we obtain the response of the rod

$$u(x,t) = \frac{8F_0L}{\pi^2 EA}\sum_{r=1}^{\infty}\frac{(-1)^{r-1}}{(2r-1)^2}\sin\frac{(2r-1)\pi x}{2L}\left[1-\cos\frac{(2r-1)\pi}{2}\sqrt{\frac{EA}{mL^2}}t\right] \tag{g}$$

The response at $x = 3L/4$ is shown in Fig. 8.25 as a function of time.

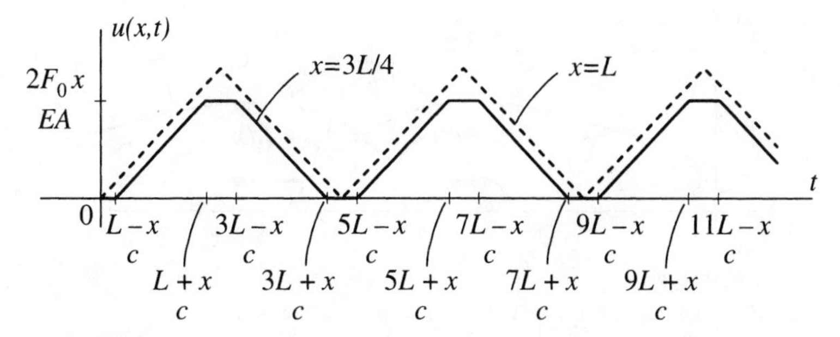

FIGURE 8.25
Axial displacement at $x = 3L/4$ due to a force in the form of a step function at $x = L$

8.12 THE WAVE EQUATION

A phenomenon intimately related to vibration is wave propagation. In fact, under certain circumstances vibration and wave propagation are two different representations of the same motion, namely, one in terms of standing waves and the other in terms of traveling waves.

The simplest form of wave motion is associated with second-order systems, such as strings in transverse vibration, rods in axial vibration and shafts in torsion, with uniformly distributed parameters. For easy visualization, we introduce the ideas by means of a string. For constant tension and uniform mass distribution, the free vibration of a string, Eq. (8.55), can be expressed in the form

$$\frac{\partial^2 y(x,t)}{\partial x^2} = \frac{1}{c^2}\frac{\partial^2 y(x,t)}{\partial t^2}, \quad c = \sqrt{\frac{T}{\rho}} \tag{8.274}$$

Equation (8.274) represents the one-dimensional *wave equation*, in which c is the *wave propagation velocity*. It is not difficult to verify by substitution that the general solution of Eq. (8.274) is

$$y(x,t) = f_1(x - ct) + f_2(x + ct) \tag{8.275}$$

where f_1 and f_2 are arbitrary functions of the arguments $x - ct$ and $x + ct$, respectively. We see that $f_1(x - ct)$ represents a displacement wave of arbitrary shape f_1 traveling in the positive x direction with the constant velocity c and *without altering the shape*. Similarly, $f_2(x + ct)$ is a displacement wave of shape f_2 traveling in the negative x direction. Hence, the most general type of motion of the string consists of a superposition of two waves of arbitrary shape traveling in opposite directions, as shown in Fig. 8.26.

A case of particular interest in vibrations is that of sinusoidal waves. One such wave having the amplitude A and traveling in the positive x direction can be expressed as

$$y(x,t) = A \sin \frac{2\pi(x - ct)}{\lambda} \tag{8.276}$$

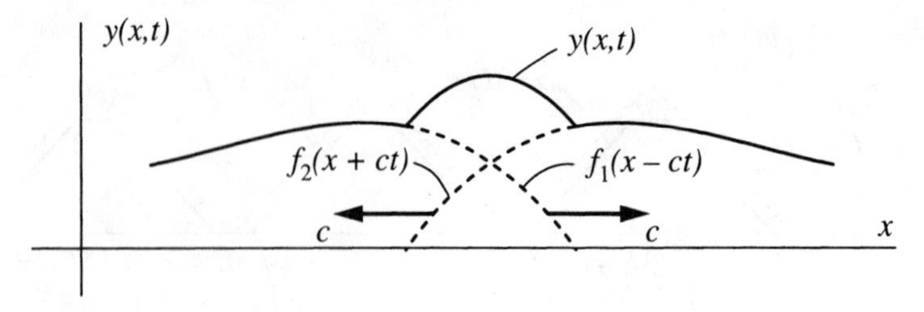

FIGURE 8.26
Superposition of two waves traveling in opposite directions

where λ is the *wavelength*, defined as the distance between two successive crests. Equation (8.276) can be rewritten in the form

$$y(x,t) = A\sin(2\pi kx - \omega t) \tag{8.277}$$

where

$$k = \frac{1}{\lambda} \tag{8.278}$$

is known as the *wave number*, defined as the number of waves per unit distance, and

$$\omega = c\frac{2\pi}{\lambda} \tag{8.279}$$

is the *frequency* of the wave. Moreover,

$$\tau = \frac{2\pi}{\omega} = \frac{\lambda}{c} \tag{8.280}$$

is the *period*, namely, the time necessary for a complete wave to pass through a given point.

Next, we consider a displacement consisting of two identical waves traveling in opposite directions. Recalling that $\sin(\alpha \pm \beta) = \sin\alpha\cos\beta \pm \cos\alpha\sin\beta$, we can write

$$\begin{aligned} y(x,t) &= A\sin(2\pi kx - \omega t) + A\sin(2\pi kx + \omega t) \\ &= 2A\sin 2\pi kx \cos\omega t \end{aligned} \tag{8.281}$$

Equation (8.281) states that the combination of two identical waves traveling in opposite directions represents a wave whose profile $2A\sin 2\pi kx$ no longer travels but oscillates harmonically about the zero position with the frequency ω. Such waves are known as *stationary*, or *standing waves*. At the points for which $2kx$ assumes integer values, $y(x,t)$ reduces to zero, with the implication that the two traveling waves cancel each other. Such points represent *nodes*. On the other hand, at points for which $2kx$ is equal to an odd integer multiple of $1/2$, $y(x,t)$ has the largest amplitude, with the implication that the two traveling waves reinforce each other. These latter points, which lie halfway between any two successive nodes, are called *loops*, or *antinodes*.

From Eq. (8.281), it is possible to conclude that there is some connection between vibration in a certain mode, Eq. (8.228), and standing waves. For a string of length L

fixed at both ends, Eq. (8.228) states that the frequency ω_p has a certain value, obtained by solving the differential eigenvalue problem. On the other hand, the frequency ω in Eq. (8.281) is arbitrary, which can be attributed to the fact that no boundary conditions have been imposed. However, if we assume that the string experiencing wave motions has nodes at $x = 0$ and $x = L$, as in the case of the vibrating string, then the wave number must satisfy the relation

$$2kL = r, \quad r = 1, 2, \ldots \tag{8.282}$$

Inserting the above values into Eq. (8.278) and considering the second of Eqs. (8.274) and Eq. (8.279), we obtain the natural frequencies

$$\omega_r = 2\pi kc = r\pi\frac{c}{L} = r\pi\sqrt{\frac{T}{\rho L^2}}, \quad r = 1, 2, \ldots \tag{8.283}$$

which are identical to the natural frequencies of a fixed-fixed string, as can be concluded from Eq. (i) of Example 8.1. Hence, the normal-mode vibration of a string of finite length can be regarded as consisting of standing waves, where the wave profile corresponding to the rth mode oscillates about the equilibrium position with the natural frequency ω_r.

8.13 TRAVELING WAVES IN RODS OF FINITE LENGTH

In Sec. 8.12, we discussed the subject of standing and traveling waves in the context of a string of unspecified length. The problem of wave propagation in infinite and semiinfinite strings is an interesting one, but not very pertinent to vibrations, as infinite and semiinfinite strings do not possess natural modes. On the other hand, the response of strings of finite length can be expressed in terms of either standing waves or traveling waves.

We consider a uniform rod fixed at $x = 0$ and with an arbitrary force $F(t)$ applied at $x = L$, as shown in Fig. 8.23; the initial excitations are assumed to be zero. In Sec. 8.11, we obtained the response of the system by modal analysis, which can be regarded as being in terms of standing waves. In this section, we wish to obtain the response of the same system in terms of traveling waves. To this end, we consider the formulation given by Eqs. (8.260) and (8.261), assume that $EA(x) = EA = \text{constant}$, $m(x) = m = \text{constant}$ and express the partial differential equation of motion in the form of the wave equation

$$\frac{\partial^2 u(x,t)}{\partial x^2} = \frac{1}{c^2}\frac{\partial^2 u(x,t)}{\partial t^2}, \quad 0 < x < L, \quad c = \sqrt{\frac{EA}{m}} \tag{8.284}$$

where the axial displacement $u(x,t)$ must satisfy the boundary conditions

$$u(0,t) = 0, \quad EA\left.\frac{\partial u(x,t)}{\partial x}\right|_{x=L} = F(t) \tag{8.285}$$

We propose to obtain a traveling wave solution of Eq. (8.284) in conjunction with Eqs. (8.285) by means of the Laplace transformation method. This represents a new use of the Laplace transformation method, which until now has been used to solve ordinary differential equations. In the first place, we define the Laplace transform of $u(x,t)$ as

$$U(x,s) = \mathscr{L}u(x,t) = \int_0^\infty e^{-st}u(x,t)dt \tag{8.286}$$

Moreover, assuming that $e^{-st}u(x,t)$ is such that differentiation with respect to x and integration with respect to t are interchangeable, we can write

$$\mathscr{L}\frac{\partial^2 u(x,t)}{\partial x^2} = \int_0^\infty e^{-st}\frac{\partial^2 u(x,t)}{\partial x^2}dt = \frac{d^2}{dx^2}\int_0^\infty e^{-st}u(x,t)dt = \frac{d^2 U(x,s)}{dx^2} \tag{8.287}$$

In addition, recalling that the initial conditions are zero, we have

$$\begin{aligned}\mathscr{L}\frac{\partial^2 u(x,t)}{\partial t^2} &= \int_0^\infty e^{-st}\frac{\partial^2 u(x,t)}{\partial t^2}dt \\ &= e^{-st}\frac{\partial u(x,t)}{\partial t}\bigg|_0^\infty + se^{-st}u(x,t)\bigg|_0^\infty + s^2\int_0^\infty e^{-st}u(x,t)dt \\ &= s^2 U(x,s) - \dot{u}(x,0) - su(x,0) = s^2 U(x,s)\end{aligned} \tag{8.288}$$

Hence, Laplace transforming Eq. (8.284) and using Eqs. (8.287) and (8.288), we obtain the ordinary differential equation

$$\frac{d^2 U(x,s)}{dx^2} - \left(\frac{s}{c}\right)^2 U(x,s) = 0,\ 0 < x < L \tag{8.289}$$

Similary, transforming Eqs. (8.285), we can write the transformed boundary conditions

$$U(0,s) = 0,\ EA\frac{dU(x,s)}{dx}\bigg|_{x=L} = F(s) \tag{8.290}$$

where

$$F(s) = \int_0^\infty e^{-st}F(t)dt \tag{8.291}$$

is the transformed boundary force.

The general solution of Eq. (8.289) is

$$U(x,s) = c_1 e^{(s/c)x} + c_2 e^{-(s/c)x} \tag{8.292}$$

Using the first of Eqs. (8.290), we have

$$U(0,s) = c_1 + c_2 = 0 \tag{8.293}$$

so that $c_2 = -c_1$ and

$$U(x,s) = c_1[e^{(s/c)x} - e^{-(s/c)x}] \tag{8.294}$$

Inserting Eq. (8.294) into the second of Eqs. (8.290), we can write

$$EAc_1\frac{s}{c}[e^{(s/c)L} + e^{-(s/c)L}] = F(s) \tag{8.295}$$

from which we conclude that

$$c_1 = \frac{c}{EA} \frac{F(s)}{s[e^{(s/c)L} + e^{-(s/c)L}]} \tag{8.296}$$

Hence, the transformed response is

$$U(x, s) = \frac{c}{EA} G(x, s) F(s) \tag{8.297}$$

where

$$G(x, s) = \frac{e^{(s/c)x} - e^{-(s/c)x}}{s[e^{(s/c)L} + e^{-(s/c)L}]} \tag{8.298}$$

The response is obtained by inverse transforming $U(x, s)$, Eq. (8.297). In view of the fact that $U(x, s)$ is the product of two functions of s, we use the convolution theorem (Sec. B.7) and write

$$u(x, t) = \frac{c}{EA} \mathscr{L}^{-1} G(x, s) F(s) = \frac{c}{EA} \int_0^t G(x, t - \tau) F(\tau) d\tau \tag{8.299}$$

Equation (8.299) implies that, before we can evaluate the integral, it is necessary to carry out the inverse Laplace transformation of $G(x, s)$ to obtain $G(x, t)$. It turns out that there are two distinct ways in which the inversion can be carried out. One leads to a response in terms of standing waves, thus duplicating the results of Sec. 8.11. The other one yields a response in terms of traveling waves, and is the one we pursue. To this end, we recall the binomial expansion

$$(1 + a)^{-1} = 1 - a + a^2 - a^3 + a^4 - a^5 + \dots, \quad |a| < 1 \tag{8.300}$$

and expand $G(x, s)$ as follows:

$$\begin{aligned} G(x, s) &= \frac{e^{(s/c)x} - e^{-(s/c)x}}{s[e^{(s/c)L} + e^{-(s/c)L}]} = \frac{e^{-(s/c)L}}{s} \frac{e^{(s/c)x} - e^{-(s/c)x}}{1 + e^{-2(s/c)L}} \\ &= \frac{e^{-(s/c)L}}{s} [e^{(s/c)x} - e^{-(s/c)x}][1 - e^{-2(s/c)L} + e^{-4(s/c)L} - e^{-6(s/c)L} + \dots] \\ &= \frac{1}{s} [e^{-(s/c)(L-x)} - e^{-(s/c)(L+x)} - e^{-(s/c)(3L-x)} \\ &\quad + e^{-(s/c)(3L+x)} + e^{-(s/c)(5L-x)} - \dots] \end{aligned} \tag{8.301}$$

Then, using the second shifting theorem (Sec. B.5), we have

$$\begin{aligned} G(x, t) = \mathscr{L}^{-1} G(x, s) &= \mathcal{u}\left(t - \frac{L - x}{c}\right) - \mathcal{u}\left(t - \frac{L + x}{c}\right) - \mathcal{u}\left(t - \frac{3L - x}{c}\right) \\ &\quad + \mathcal{u}\left(t - \frac{3L + x}{c}\right) + \mathcal{u}\left(t - \frac{5L - x}{c}\right) - \dots \end{aligned} \tag{8.302}$$

Next, we insert Eq. (8.302) into Eq. (8.299) and write

$$u(x,t) = \frac{c}{EA}\left[\int_0^t \mathscr{u}\left(t - \frac{L-x}{c} - \tau\right)F(\tau)d\tau - \int_0^t \mathscr{u}\left(t - \frac{L+x}{c} - \tau\right)F(\tau)d\tau\right.$$
$$- \int_0^t \mathscr{u}\left(t - \frac{3L-x}{c} - \tau\right)F(\tau)d\tau + \int_0^t \mathscr{u}\left(t - \frac{3L+x}{c} - \tau\right)F(\tau)d\tau$$
$$\left. + \int_0^t \mathscr{u}\left(t - \frac{5L-x}{c} - \tau\right)F(\tau)d\tau - \ldots\right] \tag{8.303}$$

Equation (8.303) can be expressed in a simpler form. To this end, we consider the integral $\int_0^t \mathscr{u}(t-\alpha-\tau)F(\tau)d\tau$ corresponding to a typical term $\mathscr{u}(t-\alpha)$ in Eq. (8.302) in which α is a positive quantity. Before we can evaluate the integral, some sketches revealing the effect of shifting in the argument of the unit step function should prove rewarding. Figure 8.27a shows the case in which $t > \alpha$, from which we conclude that

$$\mathscr{u}(t-\alpha-\tau) = \begin{cases} 1 & \text{for} \quad \tau < t-\alpha \\ 0 & \text{for} \quad \tau > t-\alpha \end{cases} \tag{8.304}$$

On the other hand, Fig. 8.27b shows the case in which $t < \alpha$, which yields simply

$$\mathscr{u}(t-\alpha-\tau) = 0, \ \tau > 0 \tag{8.305}$$

In view of this, we can write

$$\int_0^t \mathscr{u}(t-\alpha-\tau)F(\tau)d\tau = \begin{cases} \displaystyle\int_0^{t-\alpha} F(\tau)d\tau \text{ for } t > \alpha \\ 0 \text{ for } t < \alpha \end{cases} \tag{8.306}$$

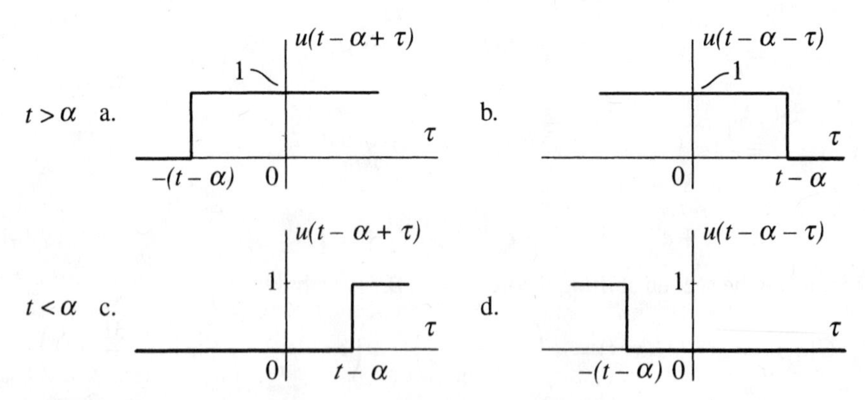

FIGURE 8.27
a. Unit step function shifted by $-(t-\alpha)$ for $t > \alpha$, b. Unit step function shifted by $-(t-\alpha)$ and folded for $t > \alpha$, c. Unit step function shifted by $-(t-\alpha)$ for $t < \alpha$, d. Unit step function shifted by $-(t-\alpha)$ and folded for $t < \alpha$

which can be expressed more compactly in the form

$$\int_0^t \mathcal{u}(t-\alpha-\tau)F(\tau)d\tau = \mathcal{u}(t-\alpha)\int_0^{t-\alpha} F(\tau)d\tau \tag{8.307}$$

Finally, we insert Eq. (8.307) into Eq. (8.303) and obtain the response

$$\begin{aligned} u(x,t) = \frac{c}{EA}\Bigg[& \mathcal{u}\left(t-\frac{L-x}{c}\right)\int_0^{t-(L-x)/c} F(\tau)d\tau - \mathcal{u}\left(t-\frac{L+x}{c}\right)\int_0^{t-(L+x)/c} F(\tau)d\tau \\ & - \mathcal{u}\left(t-\frac{3L-x}{c}\right)\int_0^{t-(3L-x)/c} F(\tau)d\tau + \mathcal{u}\left(t-\frac{3L+x}{c}\right)\int_0^{t-(3L+x)/c} F(\tau)d\tau \\ & + \mathcal{u}\left(t-\frac{5L-x}{c}\right)\int_0^{t-(5L-x)/c} F(\tau)d\tau - \ldots \Bigg] \end{aligned} \tag{8.308}$$

which represents a traveling wave solution.

The statement that Eq. (8.308) represents a traveling wave solution requires elaboration. In the first place, we observe that in contrast to a standing wave solution in which the wave profiles begin oscillating as soon as the excitation begins, a point at a distance x from the left end and $L-x$ from the right end remains at rest for $t < (L-x)/c$, and begins to move at $t = (L-x)/c$. Recognizing that $t = (L-x)/c$ represents the time necessary for a displacement wave to travel the distance $L-x$ from the right end to point x, we refer to $t = (L-x)/c$ as the *first arrival time*. Moreover, the first term on the right side of Eq. (8.308) is called the *incident wave*; its amplitude is given by the integral $\int_0^{t-(L-x)/c} F(\tau)d\tau$. The wave continues to travel past point x in the negative x direction until it reaches the fixed end $x = 0$, at which point it becomes a negative displacement wave traveling in the positive x direction. This wave is called a *reflected wave* and its amplitude is $\int_0^{t-(L+x)/c} F(\tau)d\tau$; it is represented by the second term on the right side of Eq. (8.308). The reflected wave arrives at point x at $t = (L+x)/c$. It reaches the end $x = L$ at $t = 2L/c$, at which point it is reflected as a negative displacement wave, defined by the third term in Eq. (8.308). Hence, the times $t = (L-x)/c,\ (L+x)/c,\ (3L-x)/c, \ldots$ can all be identified as arrival times, i.e., the times required for the incident wave and all the subsequent reflected waves to arrive at point x. In the time interval $0 \le t \le (L-x)/c$ nothing is sensed at point x, during the time interval $(L-x)/c \le t \le (L+x)/c$ only the incident wave is sensed, during the time interval $(L+x)/c \le t \le (3L-x)/c$ the incident wave and the first reflected wave are sensed, etc. At a fixed boundary a displacement wave is reflected as a displacement wave of opposite sign, thus canceling each other out. On the other hand, at a free boundary a displacement wave is reflected as a displacement wave of the same sign, thus doubling up. All waves travel without changing shape. Of course, the shape of the wave is determined by the force $F(t)$.

Another traveling wave of interest is the force wave, which can be obtained from Eq. (8.308) by writing

$$
\begin{aligned}
F(x,t) &= EA\frac{\partial u(x,t)}{\partial x} \\
&= c\left[\frac{1}{c}\delta\left(t-\frac{L-x}{c}\right)\int_0^{t-(L-x)/c}F(\tau)d\tau + \mathcal{u}\left(t-\frac{L-x}{c}\right)\frac{\partial}{\partial x}\int_0^{t-(L-x)/c}F(\tau)d\tau\right. \\
&\quad + \frac{1}{c}\delta\left(t-\frac{L+x}{c}\right)\int_0^{t-(L+x)/c}F(\tau)d\tau - \mathcal{u}\left(t-\frac{L+x}{c}\right)\frac{\partial}{\partial x}\int_0^{t-(L+x)/c}F(\tau)d\tau \\
&\quad - \frac{1}{c}\delta\left(t-\frac{3L-x}{c}\right)\int_0^{t-(3L-x)/c}F(\tau)d\tau - \mathcal{u}\left(t-\frac{3L-x}{c}\right)\frac{\partial}{\partial x}\int_0^{t-(3L-x)/c}F(\tau)d\tau \\
&\quad \left. - \dots \right]
\end{aligned}
\tag{8.309}
$$

From Leibnitz's rule for differentiation of definite integrals,[2] we can write

$$
\frac{\partial}{\partial x}\int_0^{t-(nL\mp x)/c}F(\tau)d\tau = \pm\frac{1}{c}F\left(t-\frac{nL\mp x}{c}\right), \quad n = 1, 3, \dots \tag{8.310}
$$

Moreover, at points for which $\delta[t-(nL\mp x)/c]$ is not zero, the upper limit of the integral, and hence the integral itself is zero. It follows that Eq. (8.309) reduces to

$$
\begin{aligned}
F(x,t) &= F\left(t-\frac{L-x}{c}\right)\mathcal{u}\left(t-\frac{L-x}{c}\right) + F\left(t-\frac{L+x}{c}\right)\mathcal{u}\left(t-\frac{L+x}{c}\right) \\
&\quad - F\left(t-\frac{3L-x}{c}\right)\mathcal{u}\left(t-\frac{3L-x}{c}\right) - F\left(t-\frac{3L+x}{c}\right)\mathcal{u}\left(t-\frac{3L+x}{c}\right) \\
&\quad + F\left(t-\frac{5L-x}{c}\right)\mathcal{u}\left(t-\frac{5L-x}{c}\right) + \dots
\end{aligned}
\tag{8.311}
$$

which expresses the force at any point x as a superposition of traveling waves. As with the displacement waves, the force waves travel without changing shape. In contrast with displacement waves, however, we conclude from Eq. (8.311) with $x = 0$ that a force wave is reflected at a fixed boundary as a force wave of the same sign, thus doubling up. Moreover, letting $x = L$ in Eq. (8.311), we see that a force wave is reflected at a free

[2] Pipes, L.A., *Applied Mathematics for Engineers and Physicists*, 2nd ed., McGraw-Hill Book Co., New York, 1958, Sec. 11.9.

boundary as a force wave of opposite sign, thus canceling out. This leaves the applied force $F(t)$ as the only force at $x = L$, as should be expected.

As an illustration, we consider the case in which the applied force has the form of the step function

$$F(t) = F_0 \mathscr{u}(t) \tag{8.312}$$

which is the same as the force in Example 8.12. Inserting Eq. (8.312) into Eq. (8.308), we obtain the displacement as a superposition of traveling waves in the form of ramp functions, as follows:

$$\begin{aligned} u(x,t) = \frac{cF_0}{EA}\bigg[&\left(t - \frac{L-x}{c}\right)\mathscr{u}\left(t - \frac{L-x}{c}\right) - \left(t - \frac{L+x}{c}\right)\mathscr{u}\left(t - \frac{L+x}{c}\right) \\ &- \left(t - \frac{3L-x}{c}\right)\mathscr{u}\left(t - \frac{3L-x}{c}\right) + \left(t - \frac{3L+x}{c}\right)\mathscr{u}\left(t - \frac{3L+x}{c}\right) \\ &+ \left(t - \frac{5L-x}{c}\right)\mathscr{u}\left(t - \frac{5L-x}{c}\right) - \dots\bigg] \end{aligned} \tag{8.313}$$

The response at $x = 3L/4$ is plotted in Fig. 8.28 as a function of time. We note that Fig. 8.28 is identical to Fig. 8.25, which demonstrates that the traveling waves solution obtained here is entirely equivalent to the solution obtained in Sec. 8.11 by modal analysis. Similarly, introducing Eq. (8.312) in Eq. (8.311), we can write simply

$$\begin{aligned} F(x,t) = F_0\bigg[&\mathscr{u}\left(t - \frac{L-x}{c}\right) + \mathscr{u}\left(t - \frac{L+x}{c}\right) - \mathscr{u}\left(t - \frac{3L-x}{c}\right) \\ &- \mathscr{u}\left(t - \frac{3L+x}{c}\right) + \mathscr{u}\left(t - \frac{5L-x}{c}\right) + \dots\bigg] \end{aligned} \tag{8.314}$$

which gives the force at point x as a superposition of traveling waves in the form of step functions. The force at $x = 3L/4$ is plotted in Fig. 8.29 as a function of time.

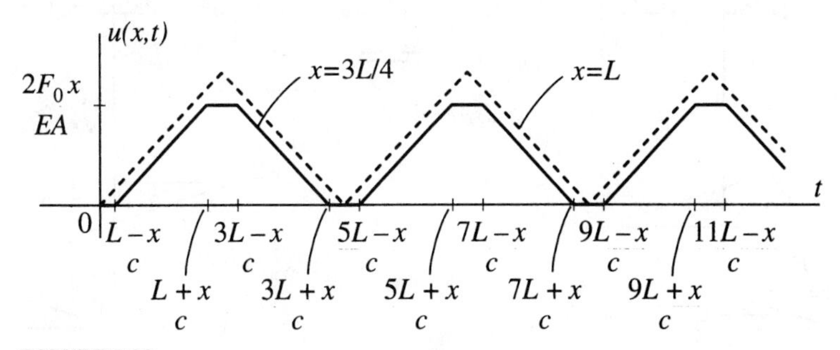

FIGURE 8.28
Axial displacement at $x = 3L/4$ due to a force in the form of a step function at $x = L$ as a superposition of traveling waves

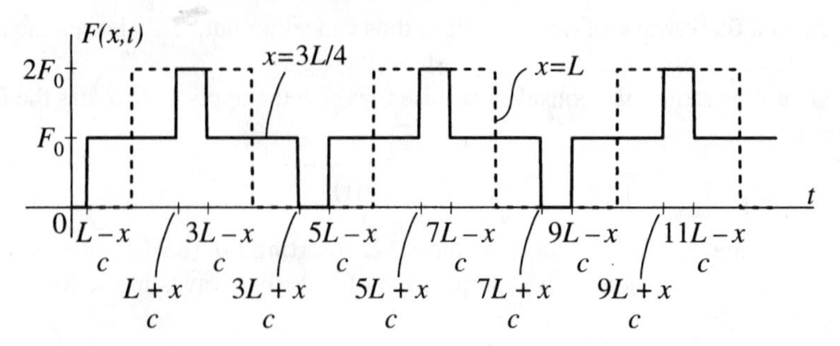

FIGURE 8.29
Axial force at $x = 3L/4$ due to α force in the form of a step function at $x = L$ as a superposition of traveling waves

Equation (8.313) can be given an interesting interpretation by considering the traveling waves shown in Fig. 8.30. The first wave on the right represents the incident wave I, the first wave on the left represents the first reflected wave R_1 at the fixed end $x = 0$, which is the reflection of the incident wave at $x = 0$, the second wave on the right represents the reflection R_2 at the end $x = L$ of the reflected wave R_1, etc. All waves start traveling simultaneously with the velocity c, waves $I,\ R_2,\ R_4,\dots$ to the left and waves $R_1,\ R_3,\ R_5,\ \dots$ to the right. The waves add up linearly as they arrive at any given point. We observe from Fig. 8.30 that the combination of the incident and reflected waves add up to zero displacement at $x = 0$. On the other hand, the waves add up to the maximum value of $2F_0L/EA$ at $t = 2L/c,\ 6L/c,\ \dots$ and to the minimum value of zero at $t = 0,\ 4L/c,\ 8L/c,\ \dots$. A similar scheme can be devised for the force waves (Problem 8.45).

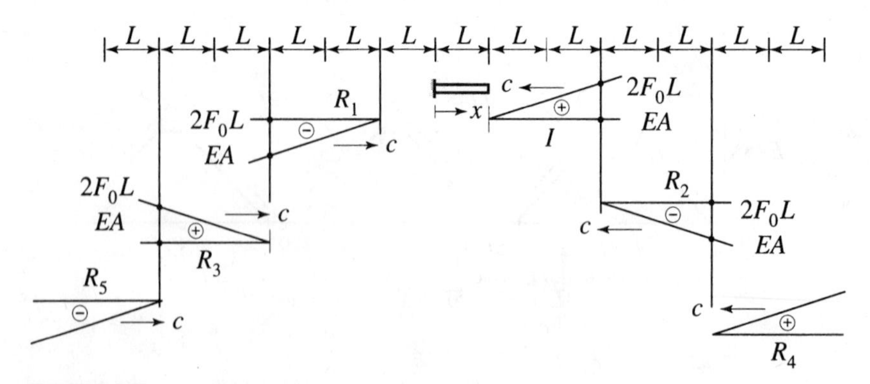

FIGURE 8.30
System response as a combination of the incident and reflected waves

8.14 SUMMARY

Systems with lumped, or discrete, parameters and systems with distributed parameters represent two distinct classes of models of vibrating systems from a mathematical point of view. Indeed, the vibration of discrete systems is described by sets of ordinary differential equations, whereas the vibration of distributed systems is described by boundary-value problems, which consists of a partial differential equation and given boundary conditions. Physically, this difference is more in appearance than in substance, as the two classes of systems exhibit analogous dynamic characteristics. This should come as no surprise, as a given physical system can be modeled either as discrete or as distributed. This fact was amply demonstrated in Sec. 8.1 in which we first derived the ordinary differential equations of motion for a system of lumped masses attached to a massless string and then derived the boundary-value problem for a string with distributed mass and stiffness by spreading the lumped masses over increments of length and letting the increments become infinitesimally small. A different type of analogy exists between strings in transverse vibration, rods in axial vibration and shafts in torsion, as the vibration of these three elastic members is described by a similar second-order (in the spatial variable) partial differential equation and boundary conditions, the only difference lying in the nature of the displacements and excitations, as well as in the system parameters. In the case of distributed-parameter systems, the extended Hamilton principle is clearly superior to the Newtonian approach, as it permits the derivation of boundary-value problems on the basis of three scalar quantities alone, the kinetic energy, potential energy and virtual work of the nonconservative forces. Because the first two represent quadratic expressions, and the steps leading to the boundary-value problem are well defined, the opportunity for errors is much smaller than in Newtonian mechanics, which relies a great deal on physical insight. This statement is even more true for beams in bending, which are defined by fourth-order partial differential equations and more complex and larger in number boundary conditions.

The parallels in dynamic characteristics between discrete and distributed systems become evident when the free vibration problem is considered. Indeed, for distributed systems as well the free vibration problem leads to an eigenvalue problem, albeit to a differential eigenvalue problem rather than an algebraic one. The solution of the eigenvalue problem consists once again of natural frequencies and natural modes, although for distributed systems their number is infinite rather than finite and the modes represent functions rather than vectors. Moreover, the modes are orthogonal with respect to the mass density and in some sense to the stiffness distribution, which forms the basis for a modal analysis. A Rayleigh quotient in the form of a ratio with the numerator being a measure of the potential energy and the denominator a measure of the kinetic energy can be defined as well, the only difference being that for distributed systems the quotient involves integrals rather than matrix products, which is consistent with the fact that summations represent discrete counterparts of integrations. Of course, the Rayleigh quotient possesses an analogous stationarity property, except that for distributed systems there is no highest eigenvalue. The response to initial excitations and external forces can also be obtained by modal analysis, thus completing the analogy with lumped systems.

The motion of strings, rods and shafts with uniformly distributed parameters lends itself to a representation in terms of waves traveling with constant velocity and without change of shape. When two identical sinusoidal waves travel in opposite directions in a uniform string, rod, or shaft, the two waves can be combined into a wave with a profile no longer traveling but oscillating harmonically. In view of this, for a string, rod or shaft, vibration in a given normal mode can be regarded as a standing wave.

PROBLEMS

8.1. Consider a lumped system in horizontal vibration under the action of applied forces, as shown in Fig. 8.31, and use the incremental approach of Sec. 8.1 to derive the boundary-value problem for a rod in axial vibration fixed at $x = 0$ and attached to a spring of stiffness k at $x = L$.

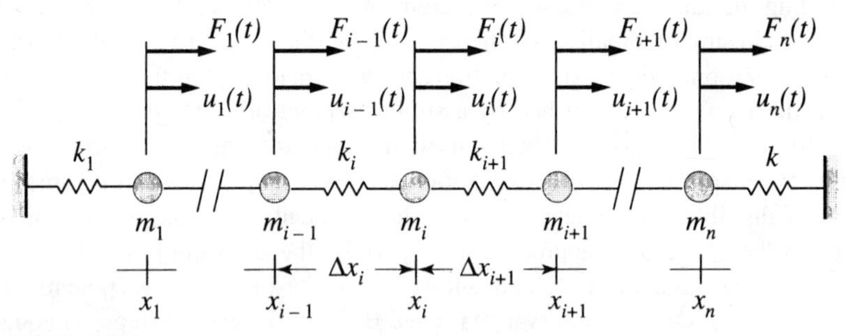

FIGURE 8.31
Lumped system in horizontal vibration

8.2. Consider a system consisting of n thin rigid disks connected by massless shafts in torsion, where the disks are subjected to external torques. The system is fixed at $x = 0$ and supported by a torsional spring at $x = L$. Draw a sketch of the system analogous to that shown in Fig. 8.31 and use the approach of Sec. 8.1 to derive the corresponding boundary-value problem.

8.3. Use the Newtonian approach to derive the boundary-value problem for a rod in axial vibration attached to a spring of stiffness k at $x = 0$ and free at $x = L$ by regarding the system as distributed from the onset (Fig. 8.32). The rod is subjected to the force per unit length $f(x, t)$, its mass per unit length is $m(x)$ and its axial stiffness is $EA(x)$, where E is the modulus of elasticity and $A(x)$ the cross-sectional area.

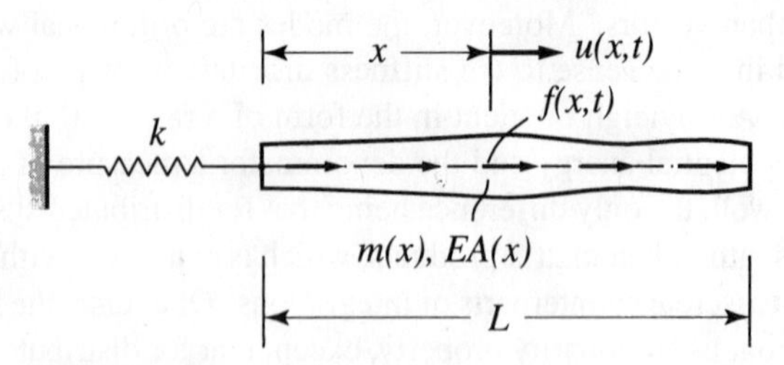

FIGURE 8.32
Rod in axial vibration attached to a spring at $x = 0$ and free at $x = L$

8.4. Use the Newtonian approach to derive the boundary-value problem for a shaft in torsional vibration restrained by torsional springs at both ends by regarding the system as distributed from the onset (Fig. 8.33). Explain the difference in signs in the two boundary conditions.

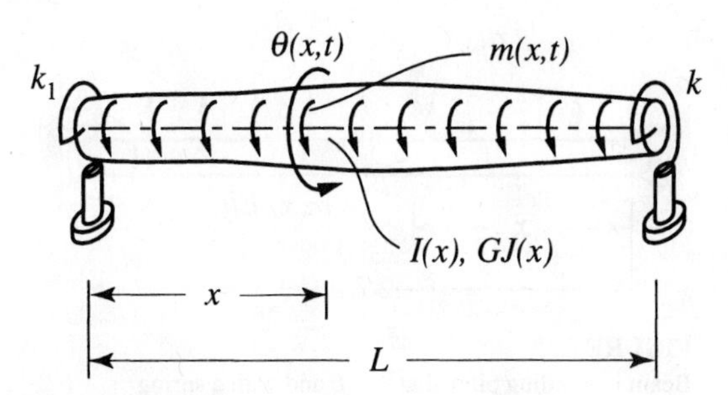

FIGURE 8.33
Shaft in torsional vibration restrained by torsional springs at both ends

8.5. Solve Problem 8.3 by the extended Hamilton principle. Compare results with those obtained in Problem 8.3 and draw conclusions as to the merits of each of the two approaches.

8.6. Solve Problem 8.4 by the extended Hamilton principle. Compare results with those obtained in Problem 8.4 and draw conclusions as to the merits of each of the two approaches.

8.7. A cable of uniform mass per unit length, $\rho(x) = \rho =$ constant, hangs freely from the ceiling, as shown in Fig. 8.34. Assume that the cable possesses no flexural stiffness and derive the boundary-value problem for the transverse vibration. **Hint:** The boundary condition at $x = 0$, ordinarily associated with a free end, is satisfied trivially in the case at hand, without involving the displacement. Hence, it must be replaced by a different boundary condition, based on physical considerations and the nature of the solution (see also Problem 8.13).

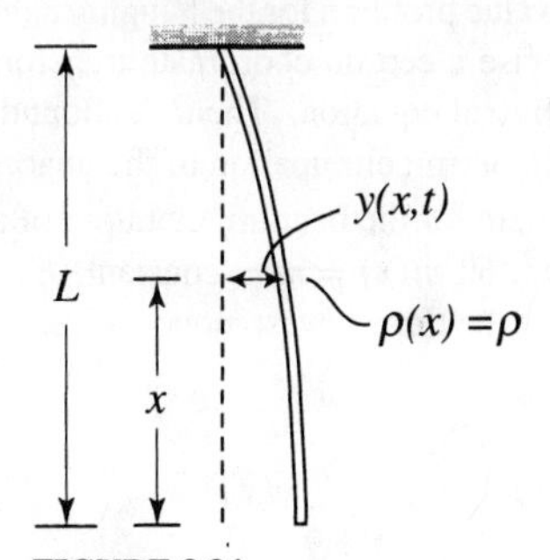

FIGURE 8.34
Cable in transverse vibration hanging freely from the ceiling

8.8. Use the Newtonian approach to derive the boundary-value problem for the bending vibration of a beam pinned at $x = 0$ and pinned but with the slope to the deflection curve restrained by a spring at $x = L$, as shown in Fig. 8.35.

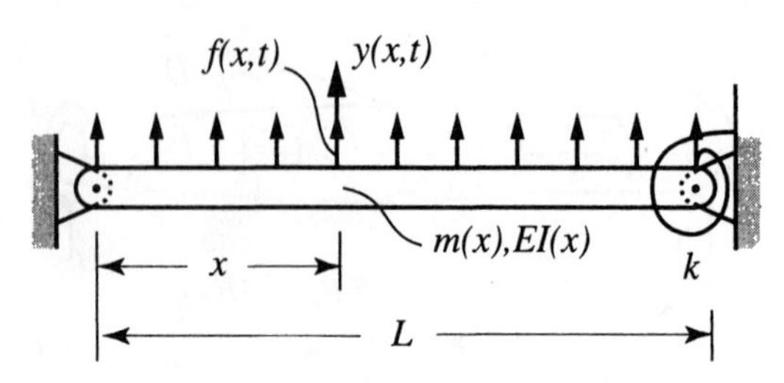

FIGURE 8.35
Beam in bending pinned at $x = 0$ and with a spring restraining rotation at $x = L$

8.9. Solve Problem 8.8 by the extended Hamilton principle.

8.10. Derive the eigenvalue problem for a shaft in torsional vibration free at both ends. Then, let $I(x) = I =$ constant, $GJ(x) = GJ =$ constant, solve the eigenvalue problem and plot the three lowest modes.

8.11. Derive the eigenvalue problem for the rod in axial vibration considered in Problem 8.3. Then, let $m(x) = m =$ constant, $EA(x) = EA =$ constant and solve the eigenvalue problem for the two cases: 1) $k = 0.5\,EA/L$ and 2) $k = 2\,EA/L$. Plot the three lowest modes for each of the two cases and draw conclusions as to the effect of the spring stiffness k on the system.

8.12. Derive the eigenvalue problem for the shaft in torsional vibration considered in Problem 8.4. Then, let $I(x) = I =$ constant, $GJ(x) = GJ =$ constant and solve the eigenvalue problem for the two cases: 1) $k_1 = k_2 = 0.5\,GJ/L$ and 2) $k_1 = 0.5\,GJ/L$, $k_2 = GJ/L$. Plot the three lowest modes for each of the two cases and explain the mode shapes in case 1.

8.13. Derive and solve the eigenvalue problem for the hanging cable of Problem 8.7. Plot the three lowest modes. **Hints:** Devise a certain coordinate transformation capable of reducing the differential equation to a Bessel equation. Then, the boundary condition at the free end of the cable must be such as to permit elimination of the unacceptable solution.

8.14. Derive the eigenvalue problem for the bending vibration of a beam pinned at $x = 0$ and free at $x = L$ (Fig. 8.36). Then, let $m(x) = m =$ constant, $EI(x) = EI =$ constant, solve the eigenvalue problem and plot the three lowest modes. **Hint:** The system is only semidefinite, as it admits a rigid-body rotation.

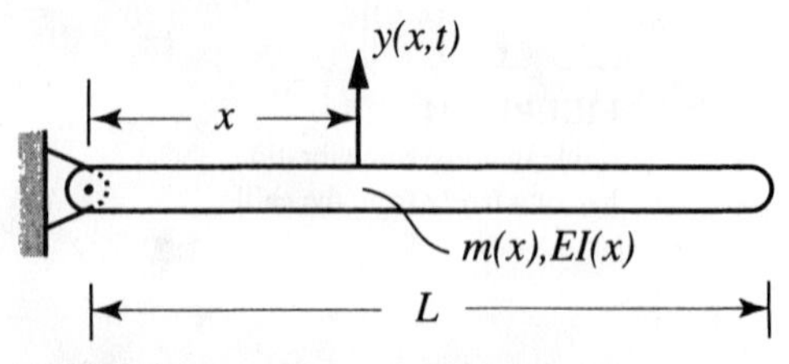

FIGURE 8.36
Beam in bending pinned at $x = 0$ and free at $x = L$

8.15. Derive the eigenvalue problem for the bending vibration of a beam free at both ends (Fig. 8.37). Then, let $m(x) = m =$ constant, $EI(x) = EI =$ constant, solve the eigenvalue problem and plot the four lowest modes. **Hint:** The system is only semidefinite and it admits two rigid-body motions, one representing transverse translation and the other rotation.

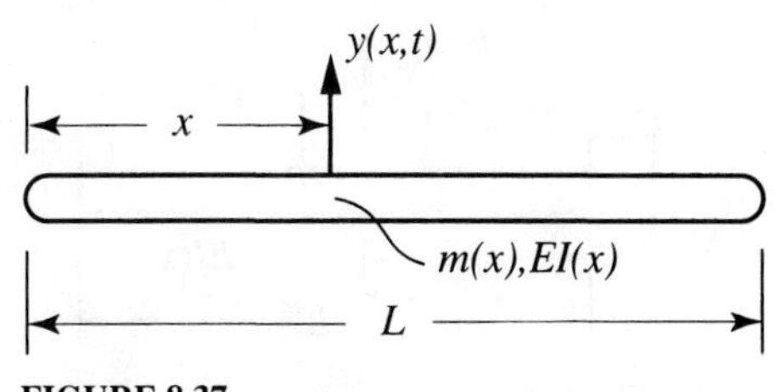

FIGURE 8.37
Free-free beam in bending

8.16. Derive the eigenvalue problem for the beam of Problem 8.8. Then, let $m(x) = m =$ constant, $EI(x) = EI =$ constant, $k = 0.5EI/L$, solve the eigenvalue problem and plot the three lowest modes.

8.17. Derive the orthogonality relations for the rod in axial vibration considered in Problem 8.11.

8.18. Derive the orthogonality relations for the shaft in torsional vibration considered in Problem 8.12.

8.19. Verify that the modes of the hanging cable obtained in Problem 8.13 are indeed orthogonal.

8.20. Derive the orthogonality relations for the pinned-free beam of Problem 8.14. Then, verify that the modes obtained in Problem 8.14 are indeed orthogonal and explain the meaning of the fact that the rigid-body mode is orthogonal to the remaining modes.

8.21. Derive the orthogonality relations for the free-free beam of Problem 8.15. Make sure that the modes obtained in Problem 8.15 are indeed orthogonal and explain the meaning of the fact that each of the two rigid-body modes is orthogonal to the remaining modes, including the other rigid-body mode.

8.22. Derive the orthogonality relations for the beam of Problem 8.16.

8.23. Derive the boundary-value problem for a rod in axial vibration with a lumped mass M at $x = 0$ and fixed at $x = L$ (Fig. 8.38). Compare the boundary condition involving the lumped mass obtained here with the one obtained in Sec. 8.6 and explain the sign difference.

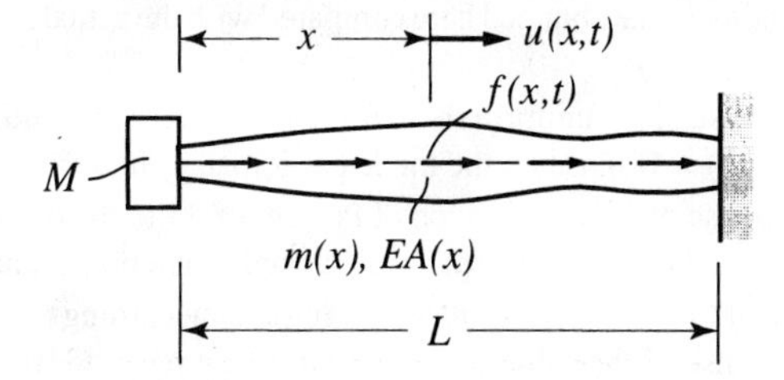

FIGURE 8.38
Rod in axial vibration with a lumped mass at $x = 0$ and fixed at $x = L$

8.24. Derive the boundary-value problem for a beam in bending vibration with a lumped mass M at $x = 0$ and fixed at $x = L$ (Fig. 8.39). Compare the boundary condition involving the lumped mass obtained here with that obtained in Sec. 8.6 and explain the sign difference.

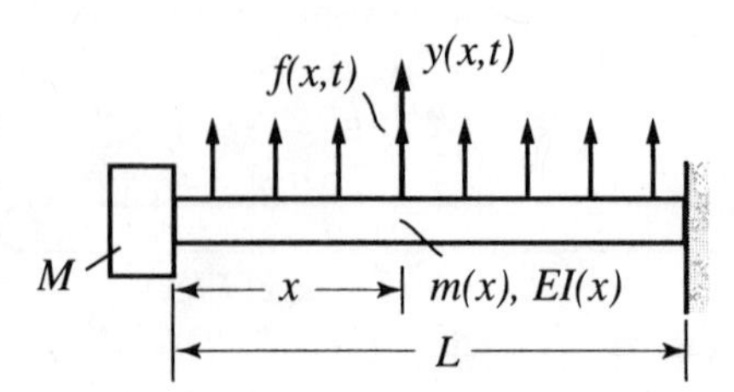

FIGURE 8.39
Beam in bending with a lumped mass at $x = 0$ and clamped at $x = L$

8.25. Derive the eigenvalue problem for the rod in axial vibration of Problem 8.23. Then, let $EA(x) = EA =$ constant, $m(x) = m =$ constant, $M = 0.5$ mL, solve the eigenvalue problem and plot the three lowest modes. Draw conclusions as to the effect of the lumped mass on the modes.

8.26. Derive the eigenvalue problem for the beam in bending vibration of Problem 8.24. Then, let $EI(x) = EI =$ constant, $m(x) = m =$ constant, $M = 0.5$ mL, solve the eigenvalue problem and plot the three lowest modes. Draw conclusions as to the effect of the lumped mass on the modes.

8.27. Estimate the lowest natural frequency of the hanging cable of Problem 8.13 by means of Rayleigh's principle in conjunction with the trial function $Y(x) = \cos \pi x/2L$. Calculate the error involved in the estimate obtained here compared to the actual lowest natural frequency from Problem 8.13.

8.28. Estimate the lowest natural frequency of the beam of Problem 8.16 by means of Rayleigh's principle in conjunction with the trial function $Y(x) = \sin \pi x/L$. Calculate the error involved in the estimate obtained here compared with the actual lowest natural frequency from Problem 8.16.

8.29. Estimate the lowest natural frequency of the beam of Problem 8.26 by means of Rayleigh's principle in conjunction with the trial function $Y(x) = 2 - 3(x/L) + (x/L)^3$. Calculate the error involved in the estimate obtained here compared with the actual lowest natural frequency from Problem 8.26.

8.30. Determine the response of the uniform shaft of Problem 8.10 to the initial excitation $\theta(x, 0) = \theta_0(1 - 2x/L)$, $\dot{\theta}(x, 0) = 0$. Discuss the mode participation in the response.

8.31. Determine the response of the uniform rod of Problem 8.11 to the initial excitation $u(x, 0) = 0$, $\dot{u}(x, 0) = v_0\delta(x)$, where $\delta(x)$ is a spatial Dirac delta function located at $x = 0$.

8.32. The hanging cable of Problem 8.13 is displaced initially according to $y(x, 0) = y_0(1 - x/L)$. Determine the response of the cable subsequent to being released from rest in the displaced position.

8.33. Determine the response of the uniform pinned-free beam of Problem 8.14 subsequent to being released from rest in the deformed configuration $y(x, 0) = y_0(x/L)^2$. Discuss the mode participation in the response.

8.34. Determine the response of the uniform beam of Problem 8.16 to the initial excitation $y(x, 0) = y_0[13(x/L) - 27(x/L)^3 + 14(x/L)^4]$, $\dot{y}(x, 0) = 0$. Discuss the mode participation in the response.

8.35. Determine the response of the rod of Problem 8.25 to the excitation $y(x, 0) = 0$, $\dot{y}(x, 0) = v_0(1 - x/L)$. Discuss the mode participation in the response.

8.36. Determine the response of the rod of Problem 8.11 to the uniformly distributed harmonic force $f(x, t) = f_0 \cos \Omega t$. Discuss the mode participation in the response.

8.37. Determine the response of the beam of Problem 8.14 to the concentrated harmonic force $F(t) = F_0 \cos \Omega t$ applied at $x = L$. Discuss the mode participation in the response. Note that the concentrated force can be represented as the distributed force $f(x, t) = F_0\delta(x - L) \cos \Omega t$, where $\delta(x - L)$ is a spatial Dirac delta function (see Eq. (8.263)).

8.38. Determine the response of a free-free uniform rod in axial vibration to a longitudinal impulsive force acting in the middle of the rod at $t = 0$. Discuss the mode participation in the response. Note that the concentrated force can be expressed as the distributed one $f(x, t) = \hat{F}_0\delta(x - L/2)\delta(t)$, where $\hat{F}_0$ is the impulse magnitude, $\delta(x - L/2)$ is a spatial Dirac delta function (see Eq. (8.263)) and $\delta(t)$ is the unit impulse.

8.39. Determine the response of the cantilever beam of Example 8.4 to the uniformly distributed force in the form of the rectangular pulse $f(x, t) = f_0[\mathcal{u}(t) - \mathcal{u}(t - T)]$, where $\mathcal{u}(t)$ is the unit step function. Discuss the mode participation in the response.

8.40. Determine the response of the beam of Problem 8.16 to a concentrated force expressed as distributed in the form $f(x, t) = F_0\delta(x - 3L/4)[r(t) - r(t - T)]$, where $\delta(x - 3L/4)$ is a spatial Dirac delta function located at $x = 3L/4$ and $r(t)$ is the unit ramp function. Discuss the mode participation in the response.

8.41. Determine the response of the rod with a lumped mass at the end considered in Problem 8.23 to the distributed sinusoidal pulse described by $f(x, t) = f_0(1 - x/L)[\sin \Omega t \mathcal{u}(t) + \sin \Omega (t - \pi/\Omega)\mathcal{u}(t - \pi/\Omega)]$. Discuss the mode participation in the response.

8.42. Determine the response of the rod of Problem 8.11 to the impulsive force $F(t) = \hat{F}_0\delta(t)$ applied at $x = L$. Discuss the mode participation in the response.

8.43. Determine the response of a uniform pinned-pinned beam to a moment in the form of the rectangular pulse $M(t) = M_0[\mathcal{u}(t) - \mathcal{u}(t - T)]$ applied at $x = 0$. Discuss the mode participation in the response. **Hint:** A concentrated moment $M(t)$ applied at $x = 0$ in the counterclockwise sense can be represented as the distributed force $f(x, t) = -M(t)\delta'(x)$, where $\delta'(x)$ is a *spatial unit doublet* (Ref. 13, p. 499), in which $\delta(x)$ is a spatial Dirac delta function and the prime denotes the derivative with respect to x. Then, the modal forces can be obtained through an integration by parts involving the spatial unit doublet with due consideration of the boundary conditions.

8.44. A free-free uniform rod in axial vibration is struck by the impulsive force $F(t) = \hat{F}_0\delta(t)$ at $x = 0$. Obtain a traveling wave solution, plot the displacement at $x = L$ as a function of time and interpret physically the motion of the rod.

8.45. Develop a traveling force solution for the free-free rod of Problem 8.44. Then, verify that the solution satisfies both boundary conditions.

CHAPTER
9

DISTRIBUTED-PARAMETER SYSTEMS: APPROXIMATE METHODS

Chapter 8 was devoted entirely to distributed-parameter systems admitting closed-form solutions. The implication is that the solutions could be expressed in terms of known functions. But, whereas many such solutions were obtained in Ch. 8, for the most part the solutions were for systems characterized by uniformly distributed parameters and simple boundaries. In real life, however, most systems do not possess these properties, so that closed-form solutions represent the exception rather than the rule. It follows that, more often than not, we must be content with approximate solutions. In this regard, the theory developed in Ch. 8 in conjunction with exact solutions is essential to the development of techniques for approximate solutions.

All approximate techniques have one thing in common, namely, they all model distributed-parameter systems as discrete systems, which amounts to spatial discretization and truncation. The approximate methods can be broadly divided into two classes, *lumped-parameter methods* and *series discretization methods*. The first is more physical in nature, but lacks rigor, and the second is more abstract, but has a solid mathematical foundation. The latter also tends to yield more predictable and accurate results.

In real life, parameters are distributed for the most part. In lumped-parameter methods, as the name suggests, the parameters are lumped at discrete points of the system. In particular, the length of the system is divided into small increments and the distributed mass within these increments is lumped at either the geometric center, or at the mass center if higher accuracy is desired. Then, any two lumped masses are assumed to be connected by massless springs of stiffness equivalent to the stiffness of the segment between the two masses. Consistent with this, the continuous displacement, say $y(x, t)$, is replaced by the discrete displacement $y_i(t)$, where i identifies the lumped mass m_i. The net result is a discrete model of a type encountered in Ch. 7. In a somewhat different approach, the stiffness is modeled by means of influence coefficients, whereas the mass is lumped as described above. Both approaches yield sets of ordinary differential equations generally associated with multi-degree-of-freedom systems.

Another lumped-parameter method is concerned primarily with the eigenvalue problem for shafts in torsion but can be equally applied to rods in axial vibration and strings in transverse vibration. It consists of dividing the mass of the shaft into rigid disks of appropriate mass moments of inertia connected by massless uniform shafts. Then, in a step-by-step approach, relations between the angular displacement and torque on both sides of a disk as well as on both sides of a shaft segment are established. Finally, imposing the boundary conditions at the two ends of the shaft, a characteristic equation is derived whose roots are the natural frequencies of the discrete model. This technique is associated with the name of Holzer. Another step-by-step approach, this time to the bending vibration of beams, was developed by Myklestad. The details of Myklestad's method are appreciably more involved than the details of Holzer's method, as now there are four variables of interest, the transverse displacement, angular displacement, bending moment and shearing force, instead of two.

An entirely different class of discretization techniques is based on Rayleigh's principle, according to which the lowest eigenvalue of a conservative system is the minimum value Rayleigh's quotient can take. In an attempt to minimize the estimate of the lowest eigenvalue, as well as to compute higher modes, Ritz conceived of the idea of representing an approximate solution of the differential eigenvalue problem as a finite series of trial functions multiplied by undetermined coefficients, and determining these coefficients by rendering Rayleigh's quotient stationary. This task reduces to the solution of an algebraic eigenvalue problem similar to that for multi-degree-of-freedom systems. This variational approach is commonly referred to as the Rayleigh-Ritz method, and amounts to series discretization of distributed-parameter systems. The Rayleigh-Ritz method is based on a rigorous mathematical theory and is capable of yielding very accurate results with only a limited number of degrees of freedom. Another series discretization method is Galerkin's method; it is based on the idea of reducing the weighted average error to zero. Galerkin's method is broader in scope than the Rayleigh-Ritz method, as it is applicable to both conservative and nonconservative systems. The collocation method is also a series discretization procedure, but instead of reducing the average error to zero, it reduces the error at discrete points to zero.

9.1 DISCRETIZATION OF DISTRIBUTED-PARAMETER SYSTEMS BY LUMPING

In Sec. 8.1, we derived the partial differential equation of motion for a string in transverse vibration by first writing the ordinary differential equations of motion for a system of lumped masses m_i on a massless string, expressing the equations in the form of difference equations and taking the limit by letting the distance between masses approach zero. In the process, the indexed nominal position x_i became the independent spatial variable x and the transverse displacement $y_i(t)$ became $y(x, t)$, where the latter depends on two independent variables, x and t. The net result was to replace a set of ordinary differential equations representing a discrete system by a partial differential equation representing a distributed system. This demonstrates clearly that lumped systems and distributed systems are intimately related and can be regarded merely as two different models of the same physical system.

In practice, it is simpler to solve ordinary differential equations than partial differential equations, so that it is more common to transform a distributed system into a discrete one. In this section, we consider the simplest approach, namely, lumping the distributed parameters. To illustrate the ideas, and to relate to some of the developments in Ch. 7, we consider a rod in axial vibration, such as that shown in Fig. 9.1, and divide it into n segments Δx_i $(i = 1, 2, \ldots, n)$. Then, we lump the mass and axial force in each of these segments by writing

$$m_i \cong m(x_i)\Delta x_i, \quad F_i(t) \cong f(x_i, t)\Delta x_i, \quad i = 1, 2, \ldots, n \tag{9.1}$$

In the process, the continuous independent variable x has been replaced by the index i and the continuous displacement $u(x, t)$ by the discrete ones $u_i(t)$ $(i = 1, 2, \ldots, n)$. Moreover, we represent the stiffness of the segment between m_{i-1} and m_i by the spring constant

$$k_i \cong \frac{EA(x_i)}{\Delta x_i}, \quad i = 1, 2, \ldots, n \tag{9.2}$$

The corresponding lumped system is displayed in Fig. 9.2. Note that there is some discrepancy in the discretization process, as the increment of rod used for lumping the mass is not the same as that used for lumping the stiffness. The stipulation is that the increments Δx_i are sufficiently small that no meaningful errors are incurred.

The equations of motion for the discretized system can be derived by means of Lagrange's equations using the model of Fig. 9.2 directly. Adapting the notation to that used in this section, Lagrange's equations, Eqs. (6.42), have the form

$$\frac{d}{dt}\left(\frac{\partial T}{\partial \dot{u}_i}\right) - \frac{\partial T}{\partial u_i} + \frac{\partial V}{\partial u_i} = F_i, \quad i = 1, 2, \ldots, n \tag{9.3}$$

where

$$T = \frac{1}{2}\sum_{j=1}^{n} m_j \dot{u}_j^2(t) \tag{9.4}$$

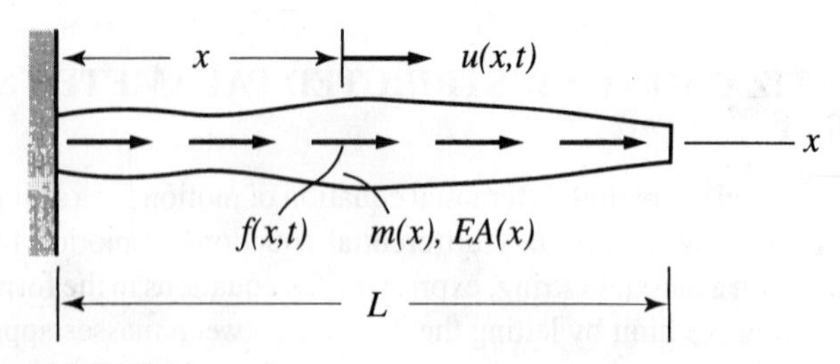

FIGURE 9.1
Rod in axial vibration

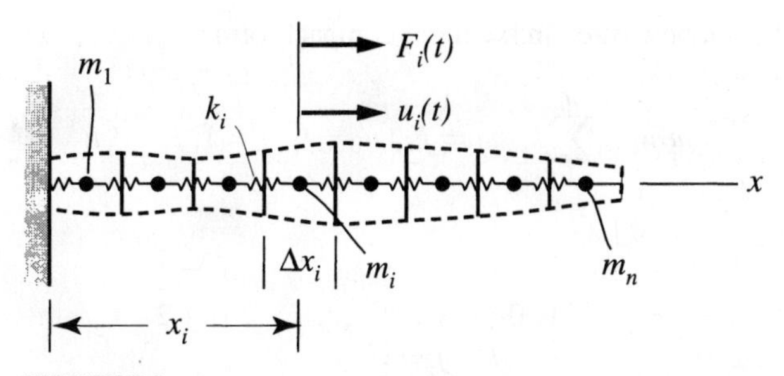

FIGURE 9.2
Lumped model for the rod of Fig. 9.1

is the kinetic energy and

$$V = \frac{1}{2} \sum_{j=1}^{n} k_j [u_j(t) - u_{j-1}(t)]^2, \ u_0 = 0 \tag{9.5}$$

is the potential energy. In the Lagrangian formulation the generalized nonconservative forces are obtained from the virtual work $\overline{\delta W}_{nc}$, Eq. (6.38). In the case at hand, the generalized forces coincide trivially with the discretized forces $F_i(t)$, a fact already taken into account in Eqs. (9.3). Next, we use Eq. (9.4) and form

$$\frac{\partial T}{\partial \dot{u}_i} = \sum_{j=1}^{n} m_j \dot{u}_j \frac{\partial \dot{u}_j}{\partial \dot{u}_i} = \sum_{j=1}^{n} m_j \dot{u}_j \delta_{ij} = m_i \dot{u}_i \tag{9.6}$$

so that

$$\frac{d}{dt}\left(\frac{\partial T}{\partial \dot{u}_i}\right) = m_i \ddot{u}_i \tag{9.7}$$

Similarly, we use Eq. (9.5) and write

$$\begin{aligned} \frac{\partial V}{\partial u_i} &= \sum_{j=1}^{n} k_j \left[(u_j - u_{j-1}) \frac{\partial u_j}{\partial u_i} - (u_j - u_{j-1}) \frac{\partial u_{j-1}}{\partial u_i} \right] \\ &= \sum_{j=1} k_j \left[(u_j - u_{j-1}) \delta_{ij} - (u_j - u_{j-1}) \delta_{i,j-1} \right] \\ &= k_i (u_i - u_{i-1}) - k_{i+1}(u_{i+1} - u_i) = -k_i u_{i-1} + (k_i + k_{i+1}) u_i - k_{i+1} u_{i+1}, \ u_0 = 0 \end{aligned} \tag{9.8}$$

Inserting Eqs. (9.7) and (9.8) into Eqs. (9.3) and observing that the kinetic energy does not depend on displacements, we obtain the equations of motion

$$m_i \ddot{u}_i - k_i u_{i-1} + (k_i + k_{i+1}) u_i - k_{i+1} u_{i+1} = F_i, \ u_0 = 0; \ i = 1, 2, \ldots, n \tag{9.9}$$

Equations (9.9) can be written in the more compact form

$$m_i \ddot{u}_i + \sum_{j=1}^{n} k_{ij} u_j = F_i, \; u_0 = 0; \; i = 1, 2, \ldots, n \tag{9.10}$$

where

$$k_{ij} = \begin{cases} 0, \; j = 1, 2, \ldots, \; i-2, \; i+2 \\ -k_i, \; j = i-1 \\ k_i + k_{i+1}, \; j = i \\ -k_{i+1}, \; j = i+1 \end{cases} \tag{9.11}$$

can be identified as stiffness influence coefficients, first encountered in Sec. 7.2.

In general, the lumping method just described tends to yield poor numerical results. To increase accuracy, it is necessary to increase the number of degrees of freedom, which implies a reduction in the size of the increments Δx_i $(i = 1, 2, \ldots, n)$. The method is not recommended, except when accuracy is not particularly important, such as in preliminary design.

9.2 LUMPED-PARAMETER METHOD USING INFLUENCE COEFFICIENTS

In Sec. 9.1, we introduced a very simple lumping method, whereby the elastic rod was divided into n increments Δx_i $(i = 1, 2, \ldots, n)$, the mass inside these increments was lumped at corresponding discrete points and the stiffness distributed parameter was lumped by considering equivalent springs corresponding to the rod segments between any two adjacent mass points. Then, in the context of the lumped model depicted in Fig. 9.2, it was stated that the accuracy of the method could be improved by increasing the number of increments.

Another kind of discrete model of distributed systems can be produced by lumping the distributed mass only, and continuing to regard the stiffness as distributed a little longer. In particular, the stiffness characteristics, or rather the flexibility characteristics, can be described in terms of a flexibility influence function $a(x, \xi)$, defined as the displacement at x due to a unit force at ξ. Then, the discretization of the flexibility characteristics can be carried out by calculating the flexibility influence coefficients $a_{ij} = a(x_i, x_j)$, which simply amounts to evaluating $a(x, \xi)$ at the locations x_i and x_j of the lumped masses m_i and m_j, respectively. In this regard, it should be recalled that the flexibility influence coefficients represent a static concept, and the masses m_i and m_j play no role in the definition of a_{ij}; they merely indicate the location of the points x_i and x_j, respectively. As an illustration, we consider the rod of Fig. 9.1 and assume that a concentrated unit force is applied at $x = \xi$, $F(\xi) = 1$. The force must be balanced by a unit reaction force at $x = 0$, as shown in Fig. 9.3. Then, using Eq. (1.93), we can write

$$EA(x)\frac{du(x)}{dx} = \begin{cases} 1, \; 0 < x < \xi \\ 0, \; \xi < x < L \end{cases} \tag{9.12}$$

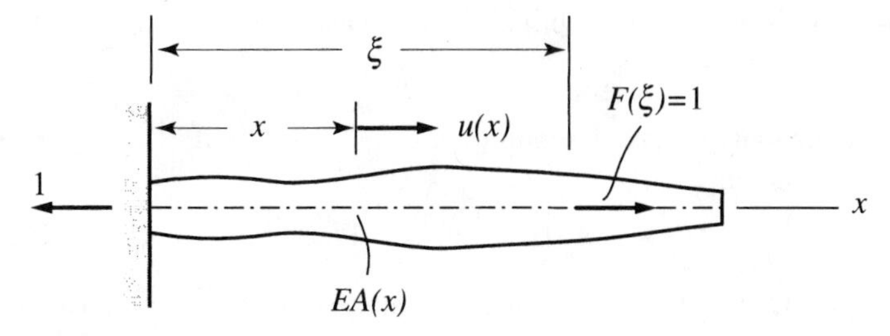

FIGURE 9.3
The rod of Fig. 9.1 subjected to a unit force at $x = \xi$

which upon integration yields

$$u(x) = \begin{cases} \displaystyle\int_0^x \frac{d\zeta}{EA(\zeta)}, & 0 < x < \xi \\ u(\xi) = \displaystyle\int_0^\xi \frac{d\zeta}{EA(\zeta)}, & \xi < x < L \end{cases} \tag{9.13}$$

For arbitrary axial stiffness $EA(x)$, the integrals in Eq. (9.13) may have to be evaluated numerically. Hence, the flexibility influence function can be written in the form

$$a(x, \xi) = \begin{cases} u(x), & 0 < x < \xi \\ u(\xi), & \xi < x < L \end{cases} \tag{9.14}$$

The flexibility influence coefficients $a_{ij} = a(x_i, x_j)$ can be derived by letting $\xi = x_j$ in Eq. (9.14) in conjunction with Eq. (9.13) and obtaining n displacement curves $a(x, x_j)$ $(j = 1, 2, \ldots, n)$. Then, the flexibility influence coefficients are obtained by sampling each of the functions $a(x, x_j)$ at $x = x_i$ $(i = 1, 2, \ldots, n)$. By Maxwell's reciprocity theorem, the influence function is symmetric, $a(x, \xi) = a(\xi, x)$. Moreover, the indices i and j correspond to the same set of points, so that the influence coefficients are symmetric $a_{ij} = a_{ji}$. The flexibility influence coefficients can be arranged in the $n \times n$ flexibility matrix $[a_{ij}] = A$. Then, the stiffness coefficients k_{ij} can be obtained from the stiffness matrix K by computing the inverse of A, or

$$[k_{ij}] = K = A^{-1} \tag{9.15}$$

Of course, the preceding discretization process applies equally well to strings in transverse vibration and shafts in torsion, subject to the replacement of the displacement, parameters and excitation indicated in Table 8.1.

The same approach can be used to develop a discretization process using influence coefficients for beams in bending. However, the process is more complex than for rods in axial vibration, which can be traced to the fact that the flexibility influence function is more difficult to obtain. The problem of determining the displacement curve for a beam in bending is discussed in Sec. 1.8.

Although Eqs. (9.10) imply the need for stiffness coefficients, this is not strictly necessary, and flexibility coefficients suffice. Indeed, to solve Eqs. (9.10), it is necessary to solve the associated eigenvalue problem, which can be written in the matrix form

$$K\mathbf{u} = \omega^2 M\mathbf{u} \tag{9.16}$$

where M is a diagonal matrix. Premultiplying both sides by $K^{-1} = A$, we can rewrite the eigenvalue problem as

$$AM\mathbf{u} = \lambda\mathbf{u}, \ \lambda = 1/\omega^2 \tag{9.17}$$

Example 9.1. A rod in axial vibration fixed at $x = 0$ and free at $x = L$ has the parameter distributions

$$m(x) = \frac{6m}{5}\left[1 - \frac{1}{2}\left(\frac{x}{L}\right)^2\right], \ EA(x) = \frac{6EA}{5}\left[1 - \frac{1}{2}\left(\frac{x}{L}\right)^2\right] \tag{a}$$

Construct a lumped model for the rod by dividing it into ten equal segments $\Delta x_i = \Delta x = L/10$ $(i = 1, 2, \ldots, 10)$ and assume that the mass within each of these segments is concentrated at $(2i-1)\Delta x/2 = (2i-1)L/20$. Moreover, assume that the lumped masses are connected with springs (Fig. 9.2) equivalent to uniform rods of length ℓ_i equal to the distance between masses and of stiffness EA_i equal to $EA(x)$ evaluated at one half that distance (Sec. 9.1). Then, construct a second model using influence coefficients based on the influence function given by Eq. (9.14) evaluated at the location of the lumped masses. Solve the eigenvalue problems corresponding to the two models and discuss the results.

The value of the lumped masses can be obtained using the first of Eqs. (a) and writing

$$\begin{aligned} m_i &= \int_{x_{i-1}}^{x_i} m(x)dx = \frac{6m}{5}\int_{x_{i-1}}^{x_i}\left[1 - \frac{1}{2}\left(\frac{x}{L}\right)^2\right]dx \\ &= \frac{6m}{5}\left(x - \frac{x^3}{6L^2}\right)\Bigg|_{x_{i-1}}^{x_i} = \frac{6m}{5}\left[x_i - x_{i-1} - \frac{1}{6L^2}\left(x_i^3 - x_{i-1}^3\right)\right] \\ &= \frac{6m}{5}\left[\Delta x_i - \frac{1}{6L^2}(x_i^3 - x_{i-1}^3)\right], \ i = 1, 2, \ldots, 10 \end{aligned} \tag{b}$$

which yields the explicit values

$$\begin{aligned} &m_1 = 0.1198mL, \ m_2 = 0.1186mL, \ m_3 = 0.1162mL, \ m_4 = 0.1126mL, \\ &m_5 = 0.1078mL, \ m_6 = 0.1018mL, \ m_7 = 0.0946mL, \ m_8 = 0.0862mL, \\ &m_9 = 0.0766mL, \ m_{10} = 0.0658mL \end{aligned} \tag{c}$$

Moreover, the equivalent spring constants are given by

$$k_1 = \frac{EA_1}{\Delta x/2} = \frac{EA(L/40)}{L/20}, \ k_i = \frac{EA_i}{\Delta x} = \frac{EA[(i-1)L/10]}{L/10}, \ i = 2, 3, \ldots, 10 \tag{d}$$

so that, using the second of Eqs. (a), we have

$$\begin{aligned} &k_1 = 23.9925\,EA/L, \ k_2 = 11.94\,EA/L, \ k_3 = 11.76\,EA/L, \ k_4 = 11.46\,EA/L, \\ &k_5 = 11.04\,EA/L, \ k_6 = 10.50\,EA/L, \ k_7 = 9.84\,EA/L, \ k_8 = 9.06\,EA/L, \\ &k_9 = 8.16\,EA/L, \ k_{10} = 7.14\,EA/L \end{aligned} \tag{e}$$

The eigenvalue problem is given by Eq. (9.16), in which the mass matrix is

$$\begin{aligned} M = \text{diag}[m_1 \ m_2 \ \ldots \ m_{10}] = mL\,\text{diag}[&0.1198 \ 0.1186 \ 0.1162 \ 0.1126 \ 0.1078 \\ &0.1018 \ 0.0964 \ 0.0862 \ 0.0766 \ 0.0658] \end{aligned} \tag{f}$$

and the stiffness matrix has the form

$$K = \begin{bmatrix} k_1+k_2 & -k_2 & 0 & \dots & 0 & 0 \\ -k_2 & k_2+k_3 & -k_3 & \dots & 0 & 0 \\ 0 & -k_3 & k_3+k_4 & \dots & 0 & 0 \\ \dots & \dots & \dots & \dots & \dots & \dots \\ 0 & 0 & 0 & \dots & k_9+k_{10} & -k_{10} \\ 0 & 0 & 0 & \dots & -k_{10} & k_{10} \end{bmatrix}$$

$$= \frac{12EA}{L} \begin{bmatrix} 2.9944 & -0.995 & & & & & & & & \\ -0.995 & 1.975 & -0.98 & & & & & & & \\ & -0.98 & 1.935 & -0.955 & & & & & & \\ & & -0.955 & 1.875 & -0.92 & & & & & \\ & & & -0.92 & 1.795 & -0.875 & & & & \\ & & & & -0.875 & 1.695 & -0.82 & & & \\ & & & & & -0.82 & 1.575 & -0.755 & & \\ & & & & & & -0.755 & 1.435 & -0.68 & \\ & & & & & & & -0.68 & 1.275 & -0.595 \\ & & & & & & & & -0.595 & 0.595 \end{bmatrix} \quad \text{(g)}$$

in which the entries not listed are zero.

The flexibility function $a(x, \xi)$ is given by Eq. (9.14) which requires the displacement $u(x)$. Inserting the second of Eqs. (a) into the top of Eq. (9.13) and integrating, we obtain

$$u(x) = \int_0^x \frac{d\zeta}{EA(\zeta)} = \int_0^x \frac{d\zeta}{\dfrac{6EA}{5}\left[1 - \dfrac{1}{2}\left(\dfrac{\zeta}{L}\right)^2\right]} = \frac{5L^2}{3EA}\int_0^x \frac{d\zeta}{(2L^2 - \zeta^2)}$$

$$= \frac{5L^2}{3EA}\frac{1}{2\sqrt{2}L}\ln\frac{\sqrt{2}L+\zeta}{\sqrt{2}L-\zeta}\bigg|_0^x = \frac{5L}{6\sqrt{2}EA}\ln\frac{\sqrt{2}L+x}{\sqrt{2}L-x} \quad \text{(h)}$$

Hence, inserting Eq. (h) into Eq. (9.14), the flexibility influence function is

$$a(x, \xi) = \begin{cases} \dfrac{5L}{6\sqrt{2}EA}\ln\dfrac{\sqrt{2}L+x}{\sqrt{2}L-x}, & 0 < x < \xi \\[2ex] \dfrac{5L}{6\sqrt{2}EA}\ln\dfrac{\sqrt{2}L+\xi}{\sqrt{2}L-\xi}, & \xi < x < L \end{cases} \quad \text{(i)}$$

which upon discretization yields

$$a_{ij} = a(x_i, x_j) = \begin{cases} \dfrac{5L}{6\sqrt{2}EA} \ln \dfrac{\sqrt{2}L + x_i}{\sqrt{2}L - x_i}, & 0 < x_i < x_j \\[2ex] \dfrac{5L}{6\sqrt{2}EA} \ln \dfrac{\sqrt{2}L + x_j}{\sqrt{2}L - x_j}, & x_j < x_i < L \end{cases} \tag{j}$$

We note that, to obtain all the flexibility coefficients, it is only necessary to calculate a_{ii} $(i = 1, 2, \ldots, 10)$, which amounts to evaluating the top line of Eq. (j) at $x_i = (2i - 1)L/20$ $(i = 1, 2, \ldots, 10)$. In view of this, the flexibility matrix can be verified to be

$$A = \frac{5L}{6\sqrt{2}EA} \left[\begin{array}{cccccccccc} 0.0707 & 0.0707 & 0.0707 & 0.0707 & 0.0707 & 0.0707 & 0.0707 & 0.0707 & 0.0707 & 0.0707 \\ & 0.2129 & 0.2129 & 0.2129 & 0.2129 & 0.2129 & 0.2129 & 0.2129 & 0.2129 & 0.2129 \\ & & 0.3573 & 0.3573 & 0.3573 & 0.3573 & 0.3753 & 0.3573 & 0.3573 & 0.3573 \\ & & & 0.5055 & 0.5055 & 0.5055 & 0.5055 & 0.5055 & 0.5055 & 0.5055 \\ & & & & 0.6593 & 0.6593 & 0.6593 & 0.6593 & 0.6593 & 0.6593 \\ & & & & & 0.8210 & 0.8210 & 0.8210 & 0.8210 & 0.8210 \\ & & & & & & 0.9937 & 0.9937 & 0.9937 & 0.9937 \\ & & \text{Symmetric} & & & & & 1.1812 & 1.1812 & 1.1812 \\ & & & & & & & & 1.3895 & 1.3895 \\ & & & & & & & & & 1.6279 \end{array} \right] \tag{k}$$

Of course, the mass matrix remains as that given by Eq. (f).

The eigenvalue problem has been solved twice, the first time using Eq. (9.16) in conjunction with Eqs. (f) and (g) and the second time using Eq. (9.17) in conjunction with Eqs. (f) and (k). For comparison purposes, we list the two sets of natural frequencies, normalized according to

$$\omega_i^* = \sqrt{\frac{mL^2}{EA}} \omega_i, \quad i = 1, 2, \ldots, 10 \tag{l}$$

The results are as follows:

1. Using the stiffness matrix, Eq. (g),

$$\omega_1^* = 1.7668, \ \omega_2^* = 4.7665, \ \omega_3^* = 7.7041, \ \omega_4^* = 10.4556, \ \omega_5^* = 13.0196,$$
$$\omega_6^* = 15.1636, \ \omega_7^* = 17.0534, \ \omega_8^* = 18.4433, \ \omega_9^* = 19.4050, \ \omega_{10}^* = 19.9062$$

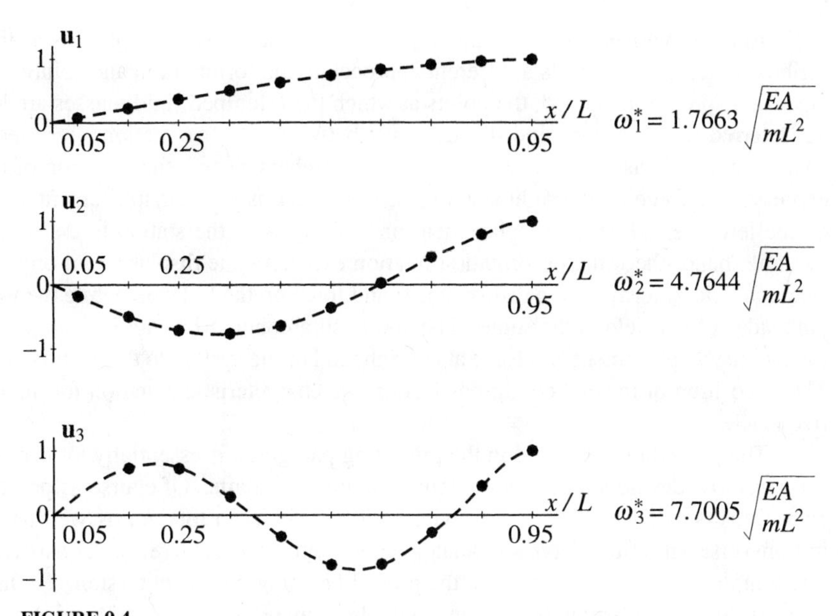

FIGURE 9.4
Modal vectors for a lumped model of a rod fixed at $x = 0$ and free at $x = L$

2. Using the flexibility matrix, Eq. (k),

$$\omega_1^* = 1.7663,\ \omega_2^* = 4.7644,\ \omega_3^* = 7.7005,\ \omega_4^* = 10.4495,\ \omega_5^* = 13.0124,$$
$$\omega_6^* = 15.1553,\ \omega_7^* = 17.0435,\ \omega_8^* = 18.4346,\ \omega_9^* = 19.3956,\ \omega_{10}^* = 19.9008$$

The frequencies were computed with 14 decimal places accuracy, but we listed only four decimal places for brevity. We observe that the two sets of natural frequencies are almost identical. This is not to be interpreted as both approaches being capable of giving very accurate results, but that the two models are virtually equivalent. In fact, the results are not accurate at all. Worse yet, there are no guidelines for predicting on what side of the exact natural frequencies the results are. As a matter of interest, the first three computed modal vectors are displayed in Fig. 9.4. To the resolution of the graphs, the results obtained by the two approaches are indistinguishable.

9.3 HOLZER'S METHOD FOR TORSIONAL VIBRATION

In Sec. 9.1, we introduced a technique for the discretization of a rod in axial vibration whereby the mass and stiffness distributed parameters are lumped at discrete points. The resulting discrete system resembles entirely an undamped multi-degree-of-freedom system of the type shown in Fig. 7.3. The equations of motion for such systems can be derived by Newton's second law in conjunction with one free-body diagram for each mass or by means of Lagrange's equations. As shown in Example 9.1, the dynamic characteristics of undamped multi-degree-of-freedom discretized systems can be obtained by solving a standard algebraic eigenvalue problem.

Another approximate technique, based on the same lumping process as that described in Sec. 9.1, suggests a different approach to the formulation and solution of the eigenvalue problem. Indeed, the points at which the n lumped rigid masses are located are referred to as *stations* and the segments between any two stations, referred to as *fields*, are treated as massless uniform elastic members. Then, the equation of motion expresses the force on the right side of a station in terms of the displacement and force on the left side, while the displacement on both sides of the station is the same. On the other hand, the force-deformation relation expresses the displacement on the right side of a field in terms of the displacement and force on the left side, while the force on both sides of the field is the same. The approach amounts to a *step-by-step* procedure relating the displacement and force at the right end of the system to those at the left end. The imposition of the end conditions results in a characteristic equation for the natural frequencies.

The procedure described in the preceding paragraph is essentially known as *Holzer's method*, developed for the torsional vibration of shafts. Of course as pointed out on various occasions, rods in axial vibration and shafts in torsion, as well as strings in transverse vibration, represent analogous systems (see Table 8.1). Consistent with tradition, in this section we present the procedure using a shaft in torsion. To this end, we consider the analogy with a rod in axial vibration, refer to Eq. (1.93) and express the relation between the angular displacement $\theta(x,t)$ and torque $M(x,t)$ as follows:

$$\frac{\partial\theta(x,t)}{\partial x} = \frac{M(x,t)}{GJ(x)} \tag{9.18}$$

where $GJ(x)$ is the torsional stiffness. Moreover, using the analogy with Eq. (8.55), and considering Eq. (9.18), the differential equation for the free vibration of a shaft in torsion can be written in the form

$$\frac{\partial M(x,t)}{\partial x} = I(x)\frac{\partial^2\theta(x,t)}{\partial t^2} \tag{9.19}$$

in which $I(x)$ is the mass polar moment of inertia per unit length. Because free vibration is harmonic, we can write

$$\theta(x,t) = \Theta(x)\cos(\omega t - \phi), \; M(x,t) = M(x)\cos(\omega t - \phi) \tag{9.20}$$

where ω is the frequency of oscillation, eliminate the time dependence and replace Eqs. (9.18) and (9.19) by

$$\frac{d\Theta(x)}{dx} = \frac{M(x)}{GJ(x)} \tag{9.21}$$

and

$$\frac{dM(x)}{dx} = -\omega^2 I(x)\Theta(x) \tag{9.22}$$

respectively. Equations (9.21) and (9.22) form the basis for an incremental approach to the problem.

Next, we consider the nonuniform shaft of Fig. 9.5a and represent it by $n+1$ rigid disks connected by n massless circular shafts of uniform stiffness, as shown in Fig. 9.5b.

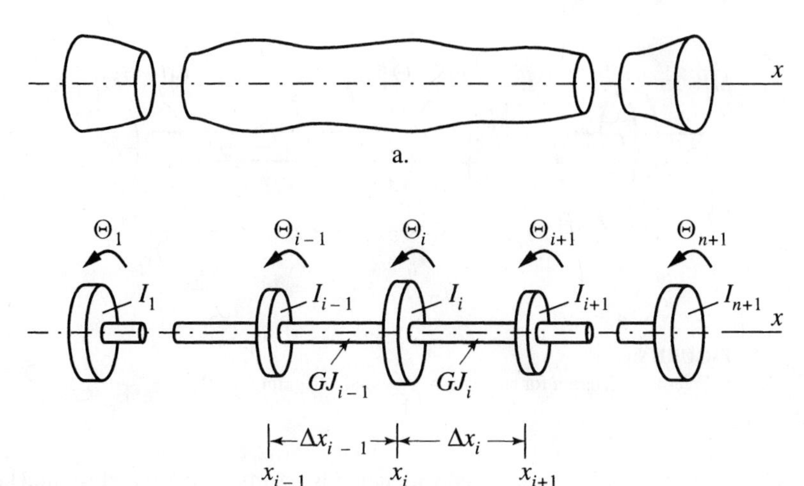

FIGURE 9.5
a. Nonuniform shaft in torsion, b. Lumped model of the shaft

The disks possess mass polar moments of inertia.

$$\begin{aligned} I_i &= \tfrac{1}{2} I(x_i)(\Delta x_{i-1} + \Delta x_i) \cong I(x_i)\Delta x_i, \; i = 2, 3, \ldots, n \\ I_1 &= \tfrac{1}{2} I(x_1)\Delta x_1, \; I_{n+1} = \tfrac{1}{2} I(x_{n+1})\Delta x_n \end{aligned} \tag{9.23}$$

where the increments Δx_i are sufficiently small that the above approximation can be justified. Moreover, we use the notation

$$GJ_i = GJ(x_i + \tfrac{1}{2}\Delta x_i), \; i = 1, 2, \ldots, n \tag{9.24}$$

Figures 9.6a and 9.6b show correspondingly free-body diagrams for station and field i. The superscripts R and L refer to the *right and left sides of a station*, respectively. In keeping with this notation, we observe that the left and right sides of field i use the notation corresponding to the right side of station i and left side of station $i+1$, respectively.

At this point, we invoke Eqs. (9.21) and (9.22) and write expressions relating the angular displacements and torques on both sides of station i and field i. Because the disks are rigid, the displacements on both sides of station i are the same,

$$\Theta_i^R = \Theta_i^L = \Theta_i \tag{9.25}$$

On the other hand, Eq. (9.22) can be written in the incremental form

$$\Delta M(x_i) = -\tfrac{1}{2}\omega^2 I(x_i)\Theta(x_i)(\Delta x_{i-1} + \Delta x_i) \cong -\omega^2 I(x_i)\Theta(x_i)\Delta x_i \tag{9.26}$$

so that, using Eqs. (9.23) and (9.25), Eq. (9.26) becomes

$$M_i^R = M_i^L - \omega^2 I_i \Theta_i^L \tag{9.27}$$

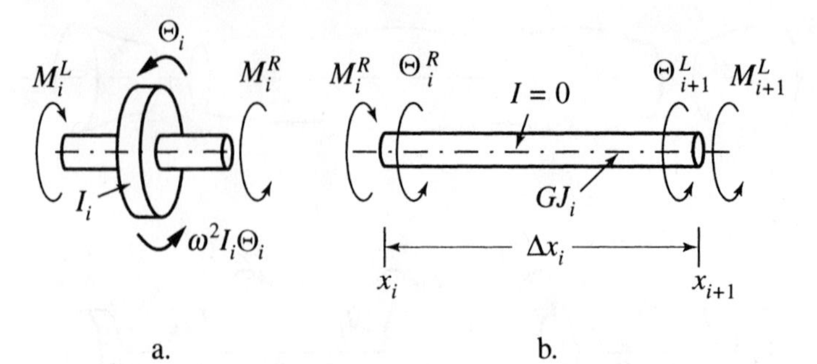

FIGURE 9.6
a. Free-body diagram for station i, b. Free-body diagram for field i

Because the segment of shaft associated with field i is assumed to be massless, and hence to possess no mass moment of inertia, Eq. (9.22) yields simply

$$M_{i+1}^L = M_i^R \tag{9.28}$$

Moreover, Eq. (9.21) can be expressed in the incremental form

$$\Delta\Theta(x_i + \tfrac{1}{2}\Delta x_i) = M(x_i + \tfrac{1}{2}\Delta x_i)\frac{\Delta x_i}{GJ(x_i + \frac{1}{2}\Delta x_i)} \cong \tfrac{1}{2}(M_{i+1}^L + M_i^R)\frac{\Delta x_i}{GJ_i} \tag{9.29}$$

which, upon using Eq. (9.28), reduces to

$$\Theta_{i+1}^L = \Theta_i^R + a_i M_i^R \tag{9.30}$$

where

$$a_i = \frac{\Delta x_i}{GJ_i} \tag{9.31}$$

represents a torsional flexibility influence coefficient, defined as the angular displacement of disk $i+1$ due to a unit moment $M_{i+1}^L = M_i^R = 1$ at station $i+1$ with disk i held fixed.

Equations (9.25) and (9.27) give the angular displacement and torque on the right side of station i in terms of the same quantities on the left side. The equations can be written in the matrix form

$$\begin{bmatrix} \Theta_i^R \\ M_i^R \end{bmatrix} = \begin{bmatrix} 1 & 0 \\ -\omega^2 I_i & 1 \end{bmatrix} \begin{bmatrix} \Theta_i^L \\ M_i^L \end{bmatrix} \tag{9.32}$$

Then, letting

$$\begin{bmatrix} \Theta_i^R \\ M_i^R \end{bmatrix} = \begin{bmatrix} \Theta \\ M \end{bmatrix}_i^R, \quad \begin{bmatrix} \Theta_i^L \\ M_i^L \end{bmatrix} = \begin{bmatrix} \Theta \\ M \end{bmatrix}_i^L \tag{9.33}$$

be the *station vectors* corresponding to the right side and left side of station i and introducing the *station transfer matrix*

$$T_{S,i} = \begin{bmatrix} 1 & 0 \\ -\omega^2 I_i & 1 \end{bmatrix} \tag{9.34}$$

relating these two station vectors, Eq. (9.32) can be rewritten as

$$\begin{bmatrix} \Theta \\ M \end{bmatrix}_i^R = T_{S,i} \begin{bmatrix} \Theta \\ M \end{bmatrix}_i^L \tag{9.35}$$

In a similar way, Eqs. (9.28) and (9.30) can be expressed in the matrix form

$$\begin{bmatrix} \Theta \\ M \end{bmatrix}_{i+1}^L = T_{F,i} \begin{bmatrix} \Theta \\ M \end{bmatrix}_i^R \tag{9.36}$$

where

$$T_{F,i} = \begin{bmatrix} 1 & a_i \\ 0 & 1 \end{bmatrix} \tag{9.37}$$

is referred to as a *field transfer matrix*. Inserting Eq. (9.35) into (9.36), we obtain

$$\begin{bmatrix} \Theta \\ M \end{bmatrix}_{i+1}^L = T_i \begin{bmatrix} \Theta \\ M \end{bmatrix}_i^L \tag{9.38}$$

in which

$$T_i = T_{F,i} T_{S,i} = \begin{bmatrix} 1 - \omega^2 a_i I_i & a_i \\ -\omega^2 I_i & 1 \end{bmatrix} \tag{9.39}$$

represents the *transfer matrix* relating the station vector on the left side of station $i+1$ to that on the left side of station i.

The transfer matrices T_i $(i = 1, 2, \ldots, n)$, $T_{F,0}$ and $T_{S,n}$ can be used as building blocks to construct an *overall transfer matrix* T relating the angular displacement and torque at one end of the system to the angular displacement and torque at the other end. Embedded in this overall transfer matrix is the frequency equation, which can be extracted by imposing the end conditions. Before this can be done, it is necessary to abandon generalities and consider specific cases, as follows:

1. *Clamped-free shaft*. In accordance with our convention, as implied in Fig. 9.5b, the notation is dictated by the stations, with field i lying to the right of station i. Hence, for a clamped-free shaft modeled as an n-degree-of-freedom system, we have n stations, $i = 1, 2, \ldots, n$, and n fields, $i = 0, 1, \ldots, n-1$, as shown in Fig. 9.7. Then, beginning

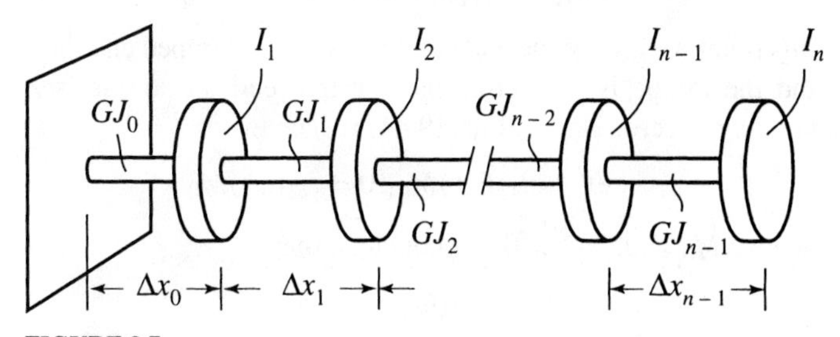

FIGURE 9.7
Lumped model of a clamped-free shaft in torsion

from the left end, we can write the recursive relations

$$\begin{bmatrix} \Theta \\ M \end{bmatrix}_1^L = T_{F,0} \begin{bmatrix} \Theta \\ M \end{bmatrix}_0$$

$$\begin{bmatrix} \Theta \\ M \end{bmatrix}_2^L = T_1 \begin{bmatrix} \Theta \\ M \end{bmatrix}_1^L = T_1 T_{F,0} \begin{bmatrix} \Theta \\ M \end{bmatrix}_0$$

$$\vdots$$

$$\begin{bmatrix} \Theta \\ M \end{bmatrix}_{i+1}^L = T_i \begin{bmatrix} \Theta \\ M \end{bmatrix}_i^L = T_i T_{i-1} \ldots T_2 T_1 T_{F,0} \begin{bmatrix} \Theta \\ M \end{bmatrix}_0 \tag{9.40}$$

$$\vdots$$

$$\begin{bmatrix} \Theta \\ M \end{bmatrix}_n^L = T_{n-1} \begin{bmatrix} \Theta \\ M \end{bmatrix}_{n-1}^L = T_{n-1} T_{n-2} \ldots T_2 T_1 T_{F,0} \begin{bmatrix} \Theta \\ M \end{bmatrix}_0$$

$$\begin{bmatrix} \Theta \\ M \end{bmatrix}_n^R = T_{S,n} \begin{bmatrix} \Theta \\ M \end{bmatrix}_n^L = T_{S,n} T_{n-1} T_{n-2} \ldots T_2 T_1 T_{F,0} \begin{bmatrix} \Theta \\ M \end{bmatrix}_0$$

The last of relations (9.40) can be expressed in the compact form

$$\begin{bmatrix} \Theta \\ M \end{bmatrix}_n^R = T \begin{bmatrix} \Theta \\ M \end{bmatrix}_0 \tag{9.41}$$

where

$$T = T_{S,n} T_{n-1} T_{n-2} \ldots T_2 T_1 T_{F,0} \tag{9.42}$$

is the overall transfer matrix for a clamped-free shaft. It is a 2×2 matrix of the form

$$T = \begin{bmatrix} t_{11}(\omega^2) & t_{12}(\omega^2) \\ t_{21}(\omega^2) & t_{22}(\omega^2) \end{bmatrix} \tag{9.43}$$

with every entry being a polynomial in ω^2. Inserting Eq. (9.43) into Eq. (9.41), we can write the matrix relation as the two scalar equations

$$\begin{aligned} \Theta_n^R &= t_{11}(\omega^2)\Theta_0 + t_{12}(\omega^2)M_0 \\ M_n^R &= t_{21}(\omega^2)\Theta_0 + t_{22}(\omega^2)M_0 \end{aligned} \tag{9.44}$$

At this point we invoke the end conditions. At a clamped end the displacement is zero, and the torque is not zero, and at a free end the torque is zero, and the displacement is not zero. Hence, Eqs. (9.44) reduce to

$$\Theta_n^R = t_{12}(\omega^2)M_0, \; 0 = t_{22}(\omega^2)M_0 \tag{9.45}$$

so that, because $M_0 \neq 0, \; \Theta_n^R \neq 0$, we conclude that

$$t_{22}(\omega^2) = 0 \tag{9.46}$$

Clearly, Eq. (9.46) represents the *frequency equation*, in which t_{22} is a polynomial of degree n in ω^2. It has n roots ω_r^2, where ω_r $(r = 1, 2, \ldots n)$ are recognized as the natural

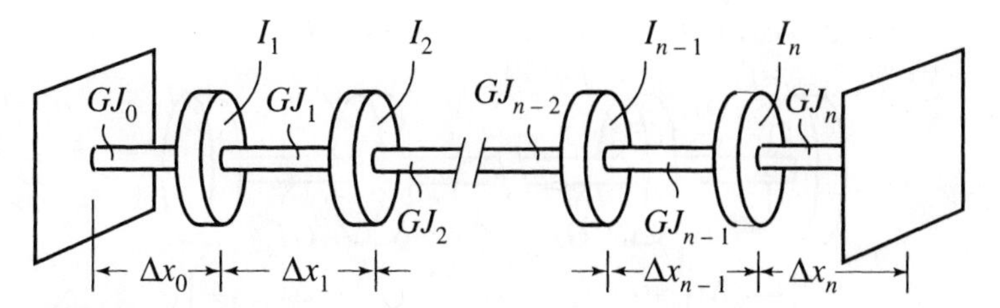

FIGURE 9.8
Lumped model of a clamped-clamped shaft in torsion

frequencies, which can be computed by a root-finding algorithm, such as the secant method, or the Newton-Raphson method. Upon solving the frequency equation, Eq. (9.46), the modal vectors $\mathbf{\Theta}_r$ $(r = 1, 2, \ldots, n)$ can be obtained by inserting ω_r^2 into the transfer matrices in the recursive relations (9.40), letting arbitrarily $M_0 = GJ/L = 1$, computing the station vectors on the left side of the equations, in sequence, and retaining the top component. The vector of torques in the same mode can be obtained by retaining the bottom component of the station vectors.

2. *Clamped-clamped shaft*. In this case, there are n stations and $n+1$ fields, as shown in Fig. 9.8. Following the same procedure as with the clamped-free shaft, we obtain a set of recursive relations, the last of which having the form

$$\begin{bmatrix} \Theta \\ M \end{bmatrix}_{n+1} = T \begin{bmatrix} \Theta \\ M \end{bmatrix}_0 \tag{9.47}$$

where this time the overall transfer matrix is given by

$$T = T_n T_{n-1} \ldots T_2 T_1 T_{F,0} \tag{9.48}$$

By analogy with Eqs. (9.44), Eq. (9.47) can be written in the scalar form

$$\begin{aligned} \Theta_{n+1} &= t_{11}(\omega^2)\Theta_0 + t_{12}(\omega^2)M_0 \\ M_{n+1} &= t_{21}(\omega^2)\Theta_0 + t_{22}(\omega^2)M_0 \end{aligned} \tag{9.49}$$

The end conditions for a clamped-clamped shaft are $\Theta_0 = \Theta_{n+1} = 0$, so that Eqs. (9.49) reduce to

$$\begin{aligned} 0 &= t_{12}(\omega^2)M_0 \\ M_{n+1} &= t_{22}(\omega^2)M_0 \end{aligned} \tag{9.50}$$

But, because $M_0 \neq 0,\ M_{n+1} \neq 0$, we conclude that the frequency equation is

$$t_{12}(\omega^2) = 0 \tag{9.51}$$

in which t_{12} is a polynomial of degree n in ω^2.

3. *Free-free shaft*. From Fig. 9.9, we observe that there are n stations and $n-1$ fields. Following the usual steps, we obtain a set of recursive relations, the last of which is

$$\begin{bmatrix} \Theta \\ M \end{bmatrix}_n^R = T \begin{bmatrix} \Theta \\ M \end{bmatrix}_0^L \tag{9.52}$$

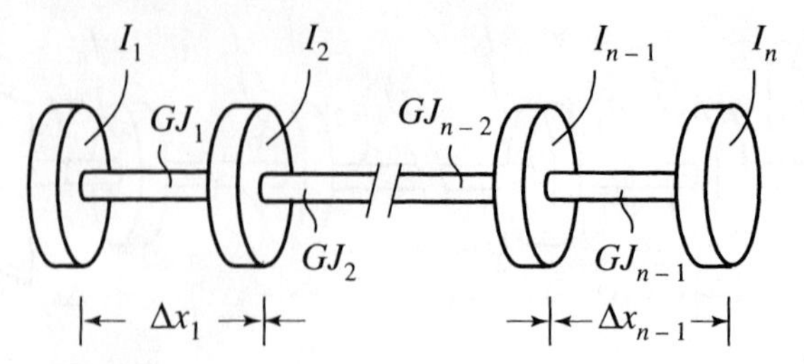

FIGURE 9.9
Lumped model of a free-free shaft in torsion

where the overall transfer matrix has the form

$$T = T_{S,n}\ T_{n-1}\ T_{n-2}\ \ldots\ T_2 T_1 \tag{9.53}$$

The end conditions are $M_0^L = M_n^R = 0$, which leads to the frequency equation

$$t_{21}(\omega^2) = 0 \tag{9.54}$$

Although t_{21} is a polynomial of degree n in ω^2, it can be verified that ω^2 can be factored out, so that Eq. (9.54) can be rewritten as

$$\omega^2 t_{21}^*(\omega^2) = 0 \tag{9.55}$$

where t_{21}^* is a polynomial of degree $n-1$ in ω^2. It follows that the frequency equation admits a solution $\omega_0^2 = 0$ and $n-1$ nonzero solutions ω_r^2 $(r = 1, 2, \ldots, n-1)$. This is consistent with the fact that the system, being unrestrained, admits a rigid-body mode with the natural frequency equal to zero and $n-1$ elastic modes.

4. *Clamped-spring restrained shaft.* This case is similar to the clamped-free case, the only difference being that the right end is restrained by a torsional spring k_T (Fig. 9.10). In fact, the recursive relations, the last of these relations and the overall transfer matrix have the same form as those given by Eqs. (9.40), (9.41) and (9.42), respectively. However, in the case at hand the end conditions are

$$\Theta_0 = 0,\ M_n^R = -k_T \Theta_n^R \tag{9.56}$$

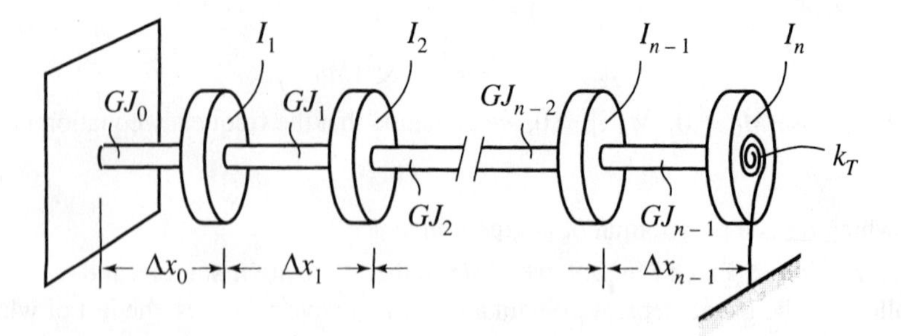

FIGURE 9.10
Lumped model of a clamped-spring supported shaft in torsion

so that Eq. (9.41) yields the two scalar equations

$$\begin{aligned} \Theta_n^R &= t_{12}(\omega^2)M_0 \\ M_n^R &= -k_T\Theta_n^R = t_{22}(\omega^2)M_0 \end{aligned} \tag{9.57}$$

Inserting the first of Eqs. (9.57) into the second and recognizing that $M_0 \neq 0$, $M_n^R \neq 0$, we conclude that the frequency equation is

$$k_T t_{12}(\omega^2) + t_{22}(\omega^2) = 0 \tag{9.58}$$

which is of degree n in ω^2.

Example 9.2. The polar mass moment of inertia per unit length and torsional stiffness of a tapered shaft clamped at $x = 0$ and free at $x = L$ are given by

$$I(x) = \frac{6I}{5}\left[1 - \frac{1}{2}\left(\frac{x}{L}\right)^2\right], \quad GJ(x) = \frac{6GJ}{5}\left[1 - \frac{1}{2}\left(\frac{x}{L}\right)^2\right] \tag{a}$$

Compute the three lowest natural frequencies and modes by means of Holzer's method. Plot the three modes, as well as the torque vector in the first mode.

The shaft under consideration is entirely analogous to the rod in axial vibration considered in Example 9.1. Hence, to permit a comparison of results, we use exactly the same lumping as in Example 9.1. The lumped model is exhibited in Fig. 9.11. The stations are located at $x_i = (2i-1)L/20$ $(i = 1, 2, \ldots 10)$, so that using Eqs. (c) of Example 9.1, the polar mass moments of inertia of the disks are

$$\begin{aligned} &I_1 = 0.1198\,IL,\ I_2 = 0.1186\,IL,\ I_3 = 0.1162\,IL,\ I_4 = 0.1126\,IL, \\ &I_5 = 0.1078\,IL,\ I_6 = 0.1018\,IL,\ I_7 = 0.0946\,IL,\ I_8 = 0.0862\,IL, \\ &I_9 = 0.0766\,IL,\ I_{10} = 0.0658\,IL \end{aligned} \tag{b}$$

Moreover, from Eqs. (9.31) and Eqs. (d) of Example 9.1, and recalling the fields notation from Fig. 9.11, we conclude that the torsional flexibility influence coefficients are the

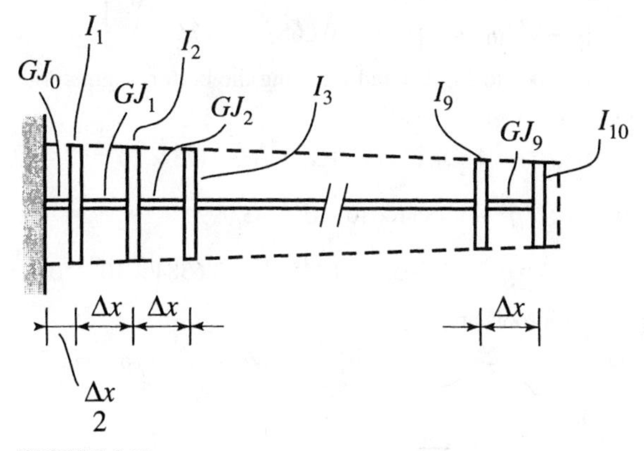

FIGURE 9.11
Lumped model of a clamped-free tapered shaft

reciprocals of the equivalent spring constants, or

$$a_0 = \frac{\Delta x/2}{GJ_0}, \quad a_i = \frac{\Delta x}{GJ_i}, \quad i = 1, 2, \ldots, 9 \tag{c}$$

so that, using Eqs. (e) of Example 9.1, we can write

$$\begin{aligned} a_0 &= 0.0418\frac{L}{GJ}, \quad a_1 = 0.0838\frac{L}{GJ}, \quad a_2 = 0.0850\frac{L}{GJ}, \quad a_3 = 0.0873\frac{L}{GJ}, \\ a_4 &= 0.0906\frac{L}{GJ}, \quad a_5 = 0.0952\frac{L}{GJ}, \quad a_6 = 0.1016\frac{L}{GJ}, \quad a_7 = 0.1110\frac{L}{GJ}, \\ a_8 &= 0.1225\frac{L}{GJ}, \quad a_9 = 0.1401\frac{L}{GJ} \end{aligned} \tag{d}$$

The overall transfer matrix is given by Eq. (9.42) in the form

$$T = T_{S,10}T_9T_8\ldots T_2T_1T_{F,0} \tag{e}$$

Introducing the notation

$$\frac{\omega^2 IL^2}{GJ} = \beta^2 \tag{f}$$

where β is recognized as being nondimensional, and using Eqs. (b) and (d), the constituent matrices in Eq. (e) can be shown to be

$$\begin{aligned} T_{F,0} &= \begin{bmatrix} 1 & a_0 \\ 0 & 1 \end{bmatrix} = \begin{bmatrix} 1 & 0.0418\,L/GJ \\ 0 & 1 \end{bmatrix} \\ T_1 &= \begin{bmatrix} 1-\omega^2 a_1 I_1 & a_1 \\ -\omega^2 I_1 & 1 \end{bmatrix} = \begin{bmatrix} 1-1.0039\times 10^{-2}\beta^2 & 0.0838\,L/GJ \\ -0.1198\beta^2\,GJ/L & 1 \end{bmatrix} \\ T_2 &= \begin{bmatrix} 1-\omega^2 a_2 I_2 & a_2 \\ -\omega^2 I_2 & 1 \end{bmatrix} = \begin{bmatrix} 1-1.0081\times 10^{-2}\beta^2 & 0.0850\,L/GJ \\ -0.1186\beta^2\,GJ/L & 1 \end{bmatrix} \\ &\vdots \\ T_9 &= \begin{bmatrix} 1-\omega^2 a_9 I_9 & a_9 \\ -\omega^2 I_9 & 1 \end{bmatrix} = \begin{bmatrix} 1-1.0732\times 10^{-2}\beta^2 & 0.1401\,L/GJ \\ -0.0766\beta^2 GJ/L & 1 \end{bmatrix} \\ T_{S,10} &= \begin{bmatrix} 1 & 0 \\ -\omega^2 I_{10} & 1 \end{bmatrix} = \begin{bmatrix} 1 & 0 \\ -0.0658\beta^2 GJ/L & 1 \end{bmatrix} \end{aligned} \tag{h}$$

Inserting Eqs. (h) into Eq. (e) and retaining the bottom right corner entry, we obtain the frequency equation

$$\begin{aligned} t_{22}(\omega^2) = t_{22}(\beta^2) &= 1 - 4.1025\times 10^{-1}\beta^2 + 3.2173\times 10^{-2}\beta^4 - 1.005\times 10^{-3}\beta^6 \\ &+ 1.6099\times 10^{-5}\beta^8 - 1.4884\times 10^{-7}\beta^{10} + 3.9994\times 10^{-11}\beta^{12} - 2.9389\times 10^{-12}\beta^{14} \\ &+ 6.2206\times 10^{-15}\beta^{16} - 7.2955\times 10^{-18}\beta^{18} + 3.6384\times 10^{-11}\beta^{20} = 0 \end{aligned} \tag{i}$$

The three lowest roots are

$$\beta_1^2 = 3.1377, \quad \beta_2^2 = 22.7697, \quad \beta_3^2 = 59.5335 \tag{j}$$

so that the three lowest natural frequencies are

$$\omega_1 = 1.7713\sqrt{\frac{GJ}{IL^2}}, \quad \omega_2 = 4.7718\sqrt{\frac{GJ}{IL^2}}, \quad \omega_3 = 7.7158\sqrt{\frac{GJ}{IL^2}} \tag{k}$$

To generate a given modal displacement vector Θ_r and associated torque vector $\mathbf{M}_r$, we must first compute the transition matrices $T_{F,0}, T_1, T_2, \ldots, T_9, T_{S,10}$. Then, recalling that $\Theta_0 = 0$, and choosing arbitrarily $M_0 = GJ/L = 1$, we obtain the station vectors from the recursive relations, Eqs. (9.40). As an illustration, to obtain the station vectors in the first mode, we use Eqs. (h) and compute the transition matrices

$$T_{F,0} = \begin{bmatrix} 1 & 0.0418 \\ 0 & 1 \end{bmatrix}$$

$$T_1(\beta_1^2) = \begin{bmatrix} 1 - 1.0039 \times 10^{-2}\beta_1^2 & 0.0838 \\ -0.1198\beta_1^2 & 1 \end{bmatrix} = \begin{bmatrix} 0.9685 & 0.0838 \\ -0.3759 & 1 \end{bmatrix}$$

$$T_2(\beta_1^2) = \begin{bmatrix} 1 - 1.0081 \times 10^{-2}\beta_1^2 & 0.0850 \\ -0.1186\beta_1^2 & 1 \end{bmatrix} = \begin{bmatrix} 0.9684 & 0.0850 \\ -0.3721 & 1 \end{bmatrix} \tag{l}$$

$$\vdots$$

$$T_9(\beta_1^2) = \begin{bmatrix} 1 - 1.0732 \times 10^{-2}\beta_1^2 & 0.1401 \\ -0.0766\beta_1^2 & 1 \end{bmatrix} = \begin{bmatrix} 0.9663 & 0.1401 \\ -0.2403 & 1 \end{bmatrix}$$

$$T_{S,10}(\beta_1^2) = \begin{bmatrix} 1 & 0 \\ -0.0658\beta_1^2 & 1 \end{bmatrix} = \begin{bmatrix} 1 & 0 \\ -0.2065 & 1 \end{bmatrix}$$

Hence, inserting Eqs. (l) into Eqs. (9.40) and considering the end conditions indicated above, we obtain

$$\begin{bmatrix} \Theta \\ M \end{bmatrix}_1^L = T_{F,0} \begin{bmatrix} \Theta \\ M \end{bmatrix}_0 = \begin{bmatrix} 1 & 0.0418 \\ 0 & 1 \end{bmatrix} \begin{bmatrix} 0 \\ 1 \end{bmatrix} = \begin{bmatrix} 0.0418 \\ 1 \end{bmatrix}$$

$$\begin{bmatrix} \Theta \\ M \end{bmatrix}_2^L = T_1(\beta_1^2) \begin{bmatrix} \Theta \\ M \end{bmatrix}_1^L = \begin{bmatrix} 0.9685 & 0.0838 \\ -0.3759 & 1 \end{bmatrix} \begin{bmatrix} 0.0418 \\ 1 \end{bmatrix} = \begin{bmatrix} 0.1243 \\ 0.9843 \end{bmatrix}$$

$$\begin{bmatrix} \Theta \\ M \end{bmatrix}_3^L = T_2(\beta_1^2) \begin{bmatrix} \Theta \\ M \end{bmatrix}_2^L = \begin{bmatrix} 0.9684 & 0.0850 \\ -0.3721 & 1 \end{bmatrix} \begin{bmatrix} 0.1243 \\ 0.9843 \end{bmatrix} = \begin{bmatrix} 0.2040 \\ 0.9380 \end{bmatrix}$$

$$\vdots$$

$$\begin{bmatrix} \Theta \\ M \end{bmatrix}_{10}^L = T_9(\beta_1^2) \begin{bmatrix} \Theta \\ M \end{bmatrix}_9^L = \begin{bmatrix} 0.9663 & 0.1401 \\ -0.2403 & 1 \end{bmatrix} \begin{bmatrix} 0.5344 \\ 0.2421 \end{bmatrix} = \begin{bmatrix} 0.5503 \\ 0.1136 \end{bmatrix}$$

$$\begin{bmatrix} \Theta \\ M \end{bmatrix}_{10}^R = T_{S,10}(\beta_1^2) \begin{bmatrix} \Theta \\ M \end{bmatrix}_{10}^L = \begin{bmatrix} 1 & 0 \\ -0.2065 & 1 \end{bmatrix} \begin{bmatrix} 0.5503 \\ 0.1136 \end{bmatrix} = \begin{bmatrix} 0.5503 \\ 0 \end{bmatrix} \tag{m}$$

As indicated earlier, the modal vectors consist of the top component of the station vectors corresponding to the system eigenvalues. In particular, taking the top component of the station vectors on the left side of Eqs. (m), which correspond to $\beta^2 = \beta_1^2$, we obtain the first modal vector

$$\mathbf{\Theta}_1 = [0.0418 \;\; 0.1243 \;\; 0.2040 \;\; 0.2794 \;\; 0.3487 \;\; 0.4103 \;\; 0.4627 \;\; 0.5048 \;\; 0.5344 \;\; 0.5503]^T \tag{n}$$

The three lowest modal vectors, $\mathbf{\Theta}_1$, $\mathbf{\Theta}_2$ and $\mathbf{\Theta}_3$, are displayed in Fig. 9.12. Note that the modal vectors have been normalized so that the last component is equal to 1. Similarly, taking the bottom component of the station vectors on the left side of Eqs. (m), we obtain

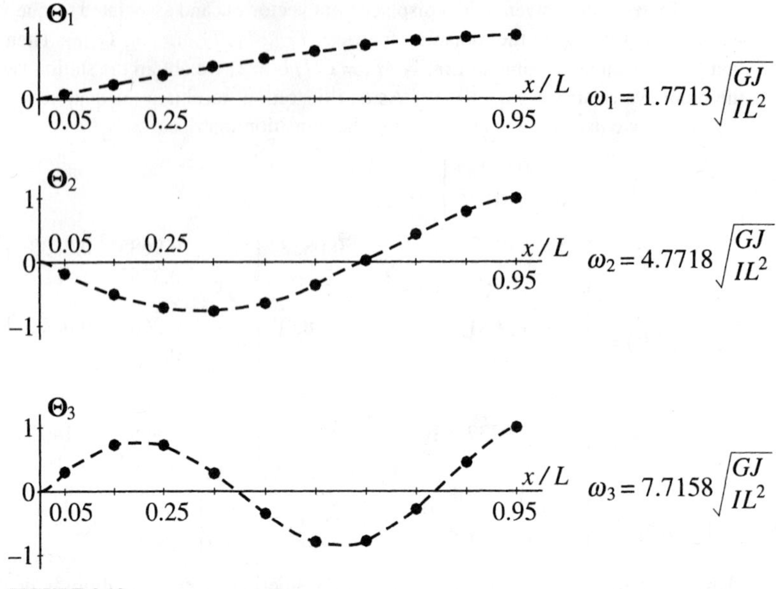

FIGURE 9.12
Modal vectors for a lumped model of the clamped-free shaft

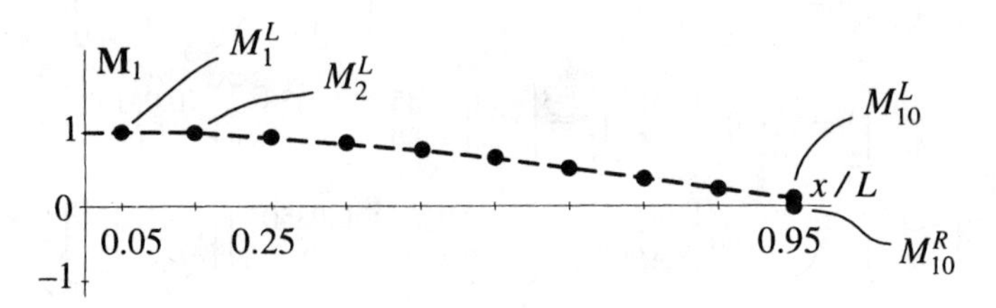

FIGURE 9.13
Torque vector for the first mode of the clamped-free shaft

the torque vector in the first mode, $\mathbf{M}_1$; it is shown in Fig. 9.13. We note that the plot consists of 11 points, rather than 10. The first 10 points represent the torques on the left side of the stations 1–10, $M_1^L, M_2^L, \ldots, M_{10}^L$, and the eleventh represents the torque on the right side of station 10, namely M_{10}^R; it is equal to zero, as it should be.

9.4 MYKLESTAD'S METHOD FOR BENDING VIBRATION

Myklestad's method for the formulation and solution of the eigenvalue problem for beams in bending vibration can be regarded as the counterpart of Holzer's method for shafts in torsional vibration. But, whereas the general ideas are similar, the extension from shafts in torsion to beams in bending is not as simple as one might expect.

As in Holzer's method, the first step is to generate a lumped model of the beam under consideration, generally a nonuniform beam. To this end, the distributed mass is

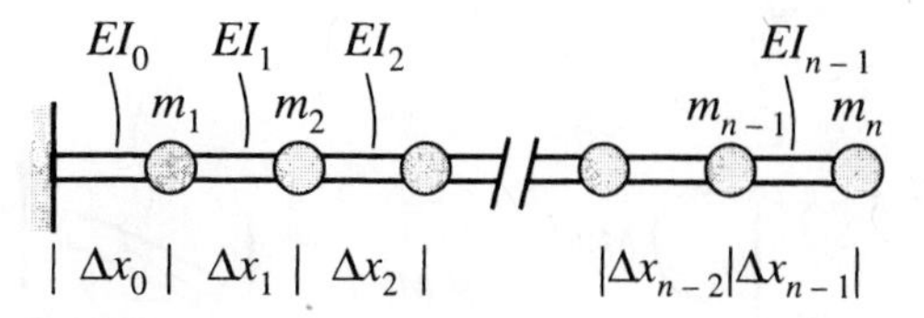

FIGURE 9.14
Lumped model for a beam in bending

lumped at individual points and the segments between the lumped masses are regarded as massless uniform beams. Here too, we refer to the locations of the lumped masses as stations and to the segments between the stations as fields. We do not dwell on the generation of a lumped model, which was covered adequately in Secs. 9.1 and 9.3, and concentrate instead on the eigenvalue problem formulation.

The general process for deriving and solving the eigenvalue problem for beams in bending remains essentially the same as for shafts in torsion, with one notable difference: for beams in bending the station vectors are two-dimensional and the transfer matrices are 4×4. We recall that for shafts in torsion the station vectors are two-dimensional, consisting of the angular displacements Θ and torque M. This is consistent with the fact that the differential eigenvalue problem for shafts in torsion (as well as for rods in axial vibration and strings in transverse vibration) are of order two. On the other hand, the differential eigenvalue problem for beams in bending is of order four, so that the station vectors must be four-dimensional. The components of the station vectors represent the translational displacement, angular displacement, bending moment and shearing force.

We assume that a nonuniform beam has been discretized through lumping, resulting in a system of lumped rigid masses connected by uniform massless beams (Fig. 9.14), and focus our attention on the free-body diagram for a typical station i (Fig. 9.15). The equation of motion for mass m_i is simply

$$Q_i^R(t) - Q_i^L(t) = m_i \ddot{y}_i(t) \tag{9.59}$$

Our interest lies in setting up recursive relations of the type given by Eqs. (9.40) for the purpose of solving eigenvalue problems. The implication is that Eq. (9.59) must represent free vibration, so that the displacement $y_i(t)$ is harmonic, as are the shearing

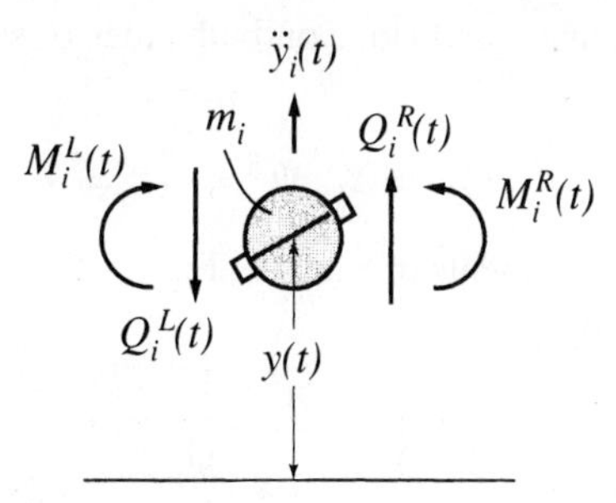

FIGURE 9.15
Free-body diagram for a typical station i

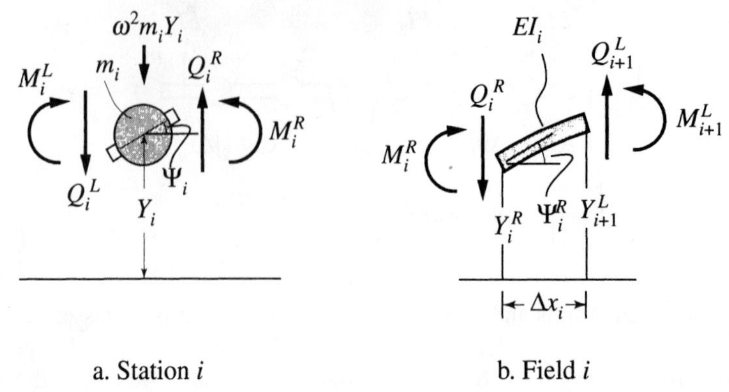

FIGURE 9.16
a. Free-body diagram for station i, b. Free-body diagram for field i

forces $Q_i^R(t)$ and $Q_i^L(t)$. Hence, letting

$$y_i(t) = Y_i \cos(\omega t - \phi),\ \ Q_i^R(t) = Q_i^R \cos(\omega t - \phi),\ \ Q_i^L(t) = Q_i^L \cos(\omega t - \phi) \quad (9.60)$$

in Eq. (9.59), where Y_i, Q_i^R and Q_i^L are amplitudes and ω is the free vibration frequency, and dividing through by $\cos(\omega t - \phi)$, we can rewrite Eq. (9.59) in the form

$$Q_i^R = Q_i^L - \omega^2 m_i Y_i \quad (9.61)$$

Equation (9.61) expresses the shearing force on the right side of station i in terms of the shearing force on the left side and the inertia force. In view of this, we can replace the free-body diagram of Fig. 9.15 by that shown in Fig. 9.16a, in which M_i^L and M_i^R represent the amplitudes of $M_i^L(t)$ and $M_i^R(t)$, respectively. Moreover, Ψ_i is the amplitude of the angular displacement $\psi_i(t)$, which is the same as the slope of the displacement curve. Because mass m_i is regarded as a point mass, so that its mass moment of inertia is zero, we must have

$$M_i^R = M_i^L \quad (9.62)$$

By continuity, the displacement and slope on both sides of station i must be the same, so that we can write

$$Y_i^R = Y_i^L = Y_i,\ \ \Psi_i^R = \Psi_i^L = \Psi_i \quad (9.63)$$

Taking a cue from the developments in Sec. 9.3, Eqs. (9.61)–(9.63) can be expressed in the matrix form

$$\begin{bmatrix} Y \\ \Psi \\ M \\ Q \end{bmatrix}_i^R = \begin{bmatrix} 1 & 0 & 0 & 0 \\ 0 & 1 & 0 & 0 \\ 0 & 0 & 1 & 0 \\ -\omega^2 m_i & 0 & 0 & 1 \end{bmatrix} \begin{bmatrix} Y \\ \Psi \\ M \\ Q \end{bmatrix}_i^L = T_{S,i} \begin{bmatrix} Y \\ \Psi \\ M \\ Q \end{bmatrix}_i^L \quad (9.64)$$

where the notation for the station vectors is obvious and

$$T_{S,i} = \begin{bmatrix} 1 & 0 & 0 & 0 \\ 0 & 1 & 0 & 0 \\ 0 & 0 & 1 & 0 \\ -\omega^2 m_i & 0 & 0 & 1 \end{bmatrix} \tag{9.65}$$

is the station transfer matrix for station i.

Next, we propose to derive the field transfer matrix, which is appreciably more involved than for shafts in torsion. To this end, we refer to the free-body diagram of Fig. 9.16b, in which we recall that the superscripts R and L refer to the right and left side of a station, and not of a field. Before we derive the internal loads-deformations relations, it is convenient to regard station i as clamped and introduce several definitions of flexibility influence coefficients, as follows:

$$a_i^{YM} = \text{displacement at } i+1 \text{ due to a unit moment at } i+1,\ M_{i+1}^L = 1$$

$$a_i^{YQ} = \text{displacement at } i+1 \text{ due to a unit force at } i+1,\ Q_{i+1}^L = 1$$

$$a_i^{\Psi M} = \text{slope at } i+1 \text{ due to a unit moment at } i+1,\ M_{i+1}^L = 1$$

$$a_i^{\Psi Q} = \text{slope at } i+1 \text{ due to a unit force at } i+1,\ Q_{i+1}^L = 1$$

Using developments from Sec. 1.8, it can be verified that the above flexibility coefficients represent the reciprocals of the corresponding stiffness coefficients for a cantilever beam. Hence, we have

$$a_i^{YM} = \frac{(\Delta x_i)^2}{2EI_i},\quad a_i^{YQ} = \frac{(\Delta x_i)^3}{3EI_i},\quad a_i^{\Psi M} = \frac{\Delta x_i}{EI_i},\quad a_i^{\Psi Q} = \frac{(\Delta x_i)^2}{2EI_i} \tag{9.66}$$

Then, from Fig. 9.16b, we can write

$$\begin{aligned} Y_{i+1}^L &= Y_i^R + \Delta x_i \Psi_i^R + a_i^{YM} M_{i+1}^L + a_i^{YQ} Q_{i+1}^L \\ &= Y_i^R + \Delta x_i \Psi_i^R + \frac{(\Delta x_i)^2}{2EI_i} M_{i+1}^L + \frac{(\Delta x_i)^3}{3EI_i} Q_{i+1}^L \\ \Psi_{i+1}^L &= \Psi_i^R + a_i^{\Psi M} M_{i+1}^L + a_i^{\Psi Q} Q_{i+1}^L \\ &= \Psi_i^R + \frac{\Delta x_i}{EI_i} M_{i+1}^L + \frac{(\Delta x_i)^2}{2EI_i} Q_{i+1}^L \end{aligned} \tag{9.67}$$

Moreover, because beam segments are assumed to be massless, we conclude from Fig. 9.16b that

$$\begin{aligned} M_{i+1}^L &= M_i^R - \Delta x_i Q_i^R \\ Q_{i+1}^L &= Q_i^R \end{aligned} \tag{9.68}$$

Inserting Eqs. (9.68) into Eqs. (9.67), we obtain

$$
\begin{aligned}
Y_{i+1}^L &= Y_i^R + \Delta x_i \Psi_i^R + \frac{(\Delta x_i)^2}{2EI_i} M_i^R - \frac{(\Delta x_i)^3}{6EI_i} Q_i^R \\
\Psi_{i+1}^L &= \Psi_i^R + \frac{\Delta x_i}{EI_i} M_i^R - \frac{(\Delta x_i)^2}{2EI_i} Q_i^R
\end{aligned}
\tag{9.69}
$$

Equations (9.68) and (9.69) relate the components of the station vector on the left side of station $i+1$ to those on the right side of station i. They can be cast in the matrix form

$$
\begin{bmatrix} Y \\ \Psi \\ M \\ Q \end{bmatrix}_{i+1}^L = \begin{bmatrix} 1 & \Delta x_i & (\Delta x_i)^2/2EI_i & -(\Delta x_i)^3/6EI_i \\ 0 & 1 & \Delta x_i/EI_i & -(\Delta x_i)^2/2EI_i \\ 0 & 0 & 1 & -\Delta x_i \\ 0 & 0 & 0 & 1 \end{bmatrix} \begin{bmatrix} Y \\ \Psi \\ M \\ Q \end{bmatrix}_i^R = T_{F,i} \begin{bmatrix} Y \\ \Psi \\ M \\ Q \end{bmatrix}_i^R
\tag{9.70}
$$

where

$$
T_{F,i} = \begin{bmatrix} 1 & \Delta x_i & (\Delta x_i)^2/2EI_i & -(\Delta x_i)^3/6EI_i \\ 0 & 1 & \Delta x_i/EI_i & -(\Delta x_i)^2/2EI_i \\ 0 & 0 & 1 & -\Delta x_i \\ 0 & 0 & 0 & 1 \end{bmatrix}
\tag{9.71}
$$

is recognized as the field transfer matrix for field i. Inserting Eq. (9.64) into Eq. (9.70), we can write

$$
\begin{bmatrix} Y \\ \Psi \\ M \\ Q \end{bmatrix}_{i+1}^L = T_i \begin{bmatrix} Y \\ \Psi \\ M \\ Q \end{bmatrix}_i^L
\tag{9.72}
$$

in which

$$
\begin{aligned}
T_i = T_{F,i} T_{S,i} &= \begin{bmatrix} 1 & \Delta x_i & (\Delta x_i)^2/2EI_i & -(\Delta x_i)^3/6EI_i \\ 0 & 1 & \Delta x_i/EI_i & -(\Delta x_i)^2/2EI_i \\ 0 & 0 & 1 & -\Delta x_i \\ 0 & 0 & 0 & 1 \end{bmatrix} \begin{bmatrix} 1 & 0 & 0 & 0 \\ 0 & 1 & 0 & 0 \\ 0 & 0 & 1 & 0 \\ -\omega^2 m_i & 0 & 0 & 1 \end{bmatrix} \\
&= \begin{bmatrix} 1+\omega^2 m_i (\Delta x_i)^3/6EI_i & \Delta x_i & (\Delta x_i)^2/2EI_i & -(\Delta x_i)^3/6EI_i \\ \omega^2 m_i (\Delta x_i)^2/2EI_i & 1 & \Delta x_i/EI_i & -(\Delta x_i)^2/2EI_i \\ \omega^2 m_i \Delta x_i & 0 & 1 & -\Delta x_i \\ -\omega^2 m_i & 0 & 0 & 1 \end{bmatrix}
\end{aligned}
\tag{9.73}
$$

is the transfer matrix relating the station vector on the left side of station $i+1$ to the station vector on the left side of station i.

Except for the fact that the station vectors are four-dimensional and the various transfer matrices are 4×4, the general process for deriving the frequency equation and the modal vectors remains essentially the same as for Holzer's method. Indeed, the recursive relations are as given by Eqs. (9.40) and the overall transfer matrix as given by Eq. (9.42), subject to changes in the first and last matrices, depending on the nature of the end conditions.

As an illustration, we consider the cantilever beam shown in Fig. 9.14. In this case, the overall transfer matrix is indeed as given by Eq. (9.42), namely

$$T = T_{S,n}\, T_{n-1}\, T_{n-2} \,\ldots\, T_2\, T_1\, T_{F,0} \tag{9.74}$$

To derive the frequency equation, we express the overall transfer matrix in the generic form

$$T = \begin{bmatrix} t_{11}(\omega^2) & t_{12}(\omega^2) & t_{13}(\omega^2) & t_{14}(\omega^2) \\ t_{21}(\omega^2) & t_{22}(\omega^2) & t_{23}(\omega^2) & t_{24}(\omega^2) \\ t_{31}(\omega^2) & t_{32}(\omega^2) & t_{33}(\omega^L) & t_{34}(\omega^2) \\ t_{41}(\omega^2) & t_{42}(\omega^2 & t_{43}(\omega^2) & t_{44}(\omega^2) \end{bmatrix} \tag{9.75}$$

The end conditions are

$$Y_0 = 0,\ \Psi_0 = 0,\ M_n^R = 0,\ Q_n^R = 0 \tag{9.76}$$

so that a relation analogous to the last of the recursive relations (9.40) leads to the four scalar equations

$$\begin{aligned} Y_n &= t_{13}(\omega^2)M_0 + t_{14}(\omega^2)Q_0 \\ \Psi_n &= t_{23}(\omega^2)M_0 + t_{24}(\omega^2)Q_0 \\ 0 &= t_{33}(\omega^2)M_0 + t_{34}(\omega^2)Q_0 \\ 0 &= t_{43}(\omega^2)M_0 + t_{44}(\omega^2)Q_0 \end{aligned} \tag{9.77}$$

The last two of Eqs. (9.77) have a nontrivial solution provided the determinant of the coefficients is equal to zero, which yields the frequency equation

$$\det\begin{bmatrix} t_{33}(\omega^2) & t_{34}(\omega^2) \\ t_{43}(\omega^2) & t_{44}(\omega^2) \end{bmatrix} = t_{33}(\omega^2)t_{44}(\omega^2) - t_{34}(\omega^2)t_{43}(\omega^2) = 0 \tag{9.78}$$

The solution of the frequency equation consists of the eigenvalues ω_r^2, whose square roots represent the natural frequencies ω_r $(r = 1, 2, \ldots, n)$. Then, as in Holzer's method, inserting the eigenvalues $\omega_1^2, \omega_2^2, \ldots, \omega_n^2$ into relations analogous to relations (9.40), in sequence, and retaining the top component of the station vectors, we obtain the modal vectors $\mathbf{Y}_1, \mathbf{Y}_2, \ldots, \mathbf{Y}_n$, respectively. Before we can initiate the process, however, we must choose a station vector for the left end. Of course, as stated by the first two of Eqs. (9.76), $Y_0 = 0$ and $\Psi_0 = 0$, but the question remains as to the other two components, M_0 and Q_0. In view of the fact that the magnitude of the station vector is arbitrary, we choose $M_0 = 1$. Then, from the third of Eqs. (9.77), we can write $Q_0 = -[t_{33}(\omega^2)/t_{34}(\omega^2)]M_0 = -t_{33}(\omega^2)/t_{34}(\omega^2)$. It should be noted that, had we used the fourth of Eqs. (9.77), the result would have been the same, because the third and fourth of Eqs. (9.77) are proportional to one another. Hence, to compute the station vectors corresponding to the rth mode, we choose as the left end station vector

$$\begin{bmatrix} Y \\ \Psi \\ M \\ Q \end{bmatrix}_0 = \begin{bmatrix} 0 \\ 0 \\ 1 \\ -t_{33}(\omega_r^2)/t_{34}(\omega_r^2) \end{bmatrix} \tag{9.79}$$

For models with a small to moderate number of degrees of freedom, say $n \leq 10$, the slopes, represented by the second component of the station vectors, can enhance the plots of the modal vectors. For large n, say $n \geq 100$, the graph resolution is likely to be such that displacements alone suffice for plotting the modes.

Another case of interest is the pinned-pinned beam, shown in lumped form in Fig. 9.17. In this case, the overall transfer matrix has the expression

$$T = T_n T_{n-1} \dots T_2 T_1 T_{F,0} \tag{9.80}$$

The end conditions are

$$Y_0 = 0, \ M_0 = 0, \ Y_{n+1} = 0, \ M_{n+1} = 0 \tag{9.81}$$

so that the four equations corresponding to the last of the recursive relations are

$$\begin{aligned} 0 &= t_{12}(\omega^2)\Psi_0 + t_{14}(\omega^2)Q_0 \\ \Psi_{n+1} &= t_{22}(\omega^2)\Psi_0 + t_{24}(\omega^2)Q_0 \\ 0 &= t_{32}(\omega^2)\Psi_0 + t_{34}(\omega^2)Q_0 \\ Q_{n+1} &= t_{42}(\omega^2)\Psi_0 + t_{44}(\omega^2)Q_0 \end{aligned} \tag{9.82}$$

It is not difficult to see that in this case the frequency equation is

$$\det \begin{bmatrix} t_{12}(\omega^2) & t_{14}(\omega^2) \\ t_{32}(\omega^2) & t_{34}(\omega^2) \end{bmatrix} = t_{12}(\omega^2)t_{34}(\omega^2) - t_{14}(\omega^2)t_{32}(\omega^2) = 0 \tag{9.83}$$

Moreover, from the first of Eqs. (9.82), we can write $Q_0 = -[t_{12}(\omega^2/t_{14}(\omega^2)]\Psi_0$, so that by choosing $\Psi_0 = 1$ and considering the first two of Eqs. (9.81), we can use as the left end station vector for the rth mode

$$\begin{bmatrix} Y \\ \Psi \\ M \\ Q \end{bmatrix}_0 = \begin{bmatrix} 0 \\ 1 \\ 0 \\ -t_{12}(\omega_r^2)/t_{14}(\omega_r^2) \end{bmatrix} \tag{9.84}$$

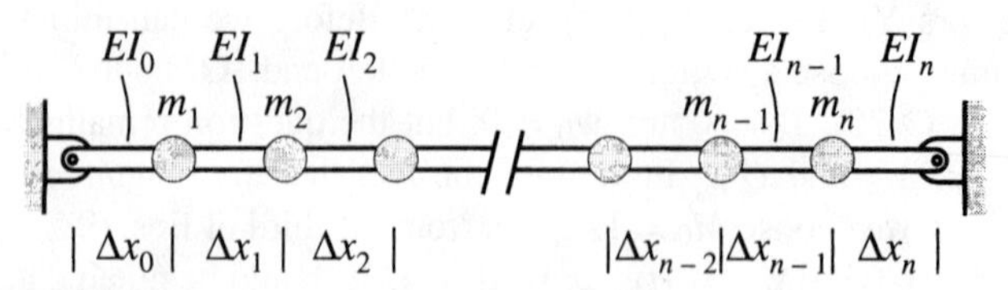

FIGURE 9.17
Lumped model for a pinned-pinned beam

Example 9.3. The pinned-pinned beam depicted in Fig. 9.18 has the mass distribution

$$m(x) = \begin{cases} m,\ 0 < x < 0.2L \text{ and } 0.8L < x < L \\ 1.2m,\ 0.2L < x < 0.4L \text{ and } 0.6L < x < 0.8L \\ 1.4m,\ 0.4L < x < 0.6L \end{cases} \tag{a}$$

and stiffness distribution

$$EI(x) = \begin{cases} EI,\ 0 < x < 0.2L \text{ and } 0.8L < x < L \\ 1.44EI,\ 0.2L < 0.4L \text{ and } 0.6L < x < 0.8L \\ 1.96EI,\ 0.4L < x < 0.6L \end{cases} \tag{b}$$

Construct a ten-degree-of-freedom lumped model as that shown in Fig. 9.17 and compute and plot the three lowest modes of vibration by means of Myklestad's method.

Dividing the beam into 10 equal increments and placing the lumped masses at the center of these increments, we conclude from Fig. 9.17 that

$$\Delta x_0 = \Delta x_{10} = L/20,\ \Delta x_1 = \Delta x_2 = \ldots = \Delta x_9 = L/10 \tag{c}$$

Then, from Eqs. (a) and Fig. 9.18, the lumped masses are given by

$$m_i = \begin{cases} mL/10,\ i = 1, 2, 9, 10 \\ 1.2mL/10,\ i = 3, 4, 7, 8 \\ 1.4mL/10,\ i = 5, 6 \end{cases} \tag{d}$$

Moreover, taking the average value in the increments in which the stiffness experiences a discontinuity, the lumped stiffnesses have the values

$$EI_i = \begin{cases} EI,\ i = 0, 1, 9, 10 \\ 1.22EI,\ i = 2, 8 \\ 1.44EI,\ i = 3, 7 \\ 1.70EI,\ i = 4, 6 \\ 1.96EI,\ i = 5 \end{cases} \tag{e}$$

Next, we assign some convenient values to the parameters. These values will affect the natural frequencies, but in a known manner. In particular, we let

$$m = 1\,\text{kg/m},\ L = 1\,\text{m},\ EI = 1\,\text{N}\cdot\text{m}^2 \tag{f}$$

In general, the natural frequencies can be expressed as

$$\omega_i = c_i\sqrt{\frac{EI}{mL^4}}, i = 1, 2, \ldots, n \tag{g}$$

where the coefficients c_i are computed by means of the frequency equation, Eq. (9.83). In the case at hand, the computed natural frequencies correspond to the parameter values given by Eqs. (f). In the case in which $L = 10\,\text{m}$, the natural frequencies are $\sqrt{1/L^4} = 10^{-2}$

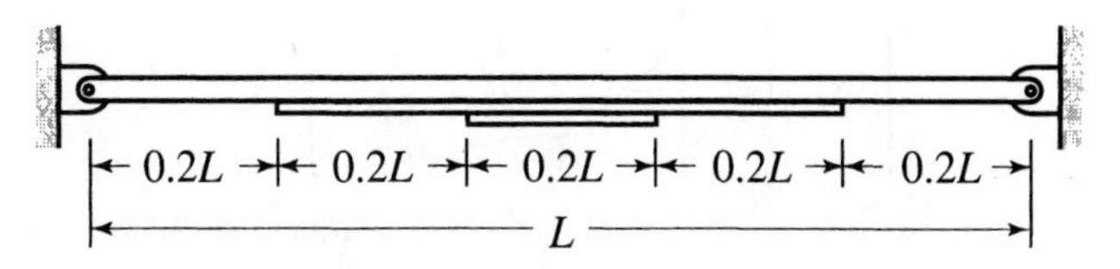

FIGURE 9.18
Nonuniform pinned-pinned beam

times the natural frequencies computed on the basis of $L = 1$ m, i.e., 100 times smaller. Of course, the postulated parameters do not affect the mode shapes.

Inserting $n = 10$ into Eq. (9.80), we can write the overall transfer matrix

$$T = T_{10}T_9 \dots T_2 T_1 T_{F,0} \tag{h}$$

where, using Eq. (9.71) with $i = 0$ in conjunction with $\Delta x_0 = 1/20$, we can write

$$T_{F,0} = \begin{bmatrix} 1 & 0.5\times10^{-1} & 0.125\times10^{-2} & -0.020833\times10^{-3} \\ 0 & 1 & 0.5\times10^{-1} & -0.125\times10^{-2} \\ 0 & 0 & 1 & -0.5\times10^{-1} \\ 0 & 0 & 0 & 1 \end{bmatrix} \tag{i}$$

Moreover, using Eq. (9.73), we have

$$T_1 = T_9$$

$$= \begin{bmatrix} 1+0.166667\times10^{-4}\omega^2 & 10^{-1} & 0.5\times10^{-2} & -0.166667\times10^{-3} \\ 0.5\times10^{-3}\omega^2 & 1 & 10^{-1} & -0.5\times10^{-2} \\ 10^{-2}\omega^2 & 0 & 1 & -10^{-1} \\ -10^{-1}\omega^2 & 0 & 0 & 1 \end{bmatrix}$$

$$T_2 = \begin{bmatrix} 1+0.136612\times10^{-4}\omega^2 & 10^{-1} & 0.409836\times10^{-2} & -0.136612\times10^{-3} \\ 0.409836\times10^{-3}\omega^2 & 1 & 10^{-1} & -0.409836\times10^{-2} \\ 10^{-2}\omega^2 & 0 & 1 & -10^{-1} \\ -10^{-1}\omega^2 & 0 & 0 & 1 \end{bmatrix}$$

$$\vdots$$

$$T_{10} = \begin{bmatrix} 1+0.020833\times10^{-4}\omega^2 & 0.5\times10^{-1} & 0.125\times10^{-2} & -0.020833\times10^{-3} \\ 0.125\times10^{-3}\omega^2 & 1 & 0.5\times10^{-1} & -0.125\times10^{-2} \\ 0.5\times10^{-2}\omega^2 & 0 & 1 & -0.5\times10^{-1} \\ -10^{-1}\omega^2 & 0 & 0 & 1 \end{bmatrix} \tag{j}$$

The frequency equation is obtained by introducing Eqs. (i) and (j) into Eq. (h) and using Eq. (9.83). The first three roots of the frequency equation are the three lowest natural frequencies

$$\omega_1 = 10.9230\sqrt{\frac{EI}{mL^4}}, \quad \omega_2 = 42.1541\sqrt{\frac{EI}{mL^4}}, \quad \omega_3 = 95.6730\sqrt{\frac{EI}{mL^4}} \tag{k}$$

To compute the modal vectors, we insert Eqs. (k) into Eq. (9.84), in sequence, and obtain the left end station vectors

$$\begin{bmatrix} Y \\ \Psi \\ M \\ Q \end{bmatrix}_0 = \begin{bmatrix} 0 \\ 1 \\ 0 \\ -t_{12}(\omega_r^2)/t_{14}(\omega_r^2) \end{bmatrix}, \quad r = 1, 2, 3 \tag{l}$$

Then, we use relations analogous to the recursive relations (9.40) to compute the station vectors for each of the three cases, $r = 1, 2$ and 3. The modal vectors $\mathbf{Y}_1, \mathbf{Y}_2$ and $\mathbf{Y}_3$ can be plotted by taking the top component of the corresponding station vectors, with some help from the second component; they are shown in Fig. 9.19.

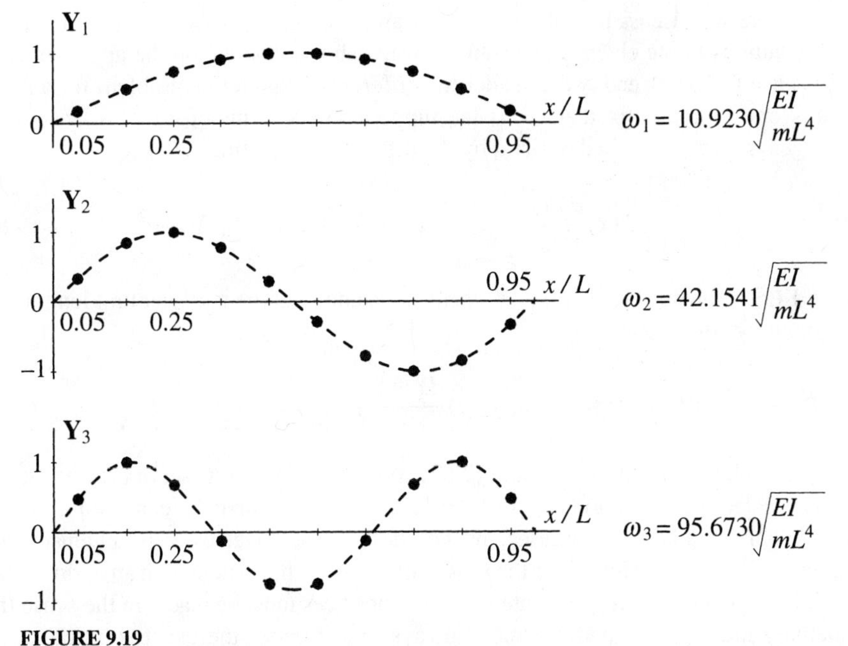

FIGURE 9.19
Modal vectors for the lumped model of the pinned-pinned beam

9.5 RAYLEIGH'S PRINCIPLE

The lumped-parameter methods discussed in Secs. 9.1–9.4 have the advantage that they are easy to understand physically. They have two main disadvantages, however; they lack mathematical rigor and the results tend to be inaccurate. What is worse is that, lumping being an arbitrary process, it is difficult to predict on what side of the actual natural frequencies the computed natural frequencies are. It is logical to assume that, as the number of degrees of freedom of the model is increased, the computed natural frequencies approach the actual ones, but there are no criteria for estimating how fast convergence is. Experience shows that good accuracy can be achieved only with a relatively large number of degrees of freedom.

An entirely different type of approximations of distributed-parameter systems by discrete models is based on the variational approach to the differential eigenvalue problem. As demonstrated in Sec. 8.8, Rayleigh's quotient has a stationary value in the neighborhood of an eigenfunction and this stationary value is actually a minimum at the lowest eigenfunction. The latter can be stated mathematically in the form

$$\lambda_1 = \omega_1^2 = \min R(Y) = R(Y_1) \tag{9.85}$$

or, in words, *the lowest eigenvalue is the minimum value that Rayleigh's quotient can take* by letting the trial function $Y(x)$ vary at will. Of course, *the minimum value is achieved when $Y(x)$ coincides with the lowest eigenfunction $Y_1(x)$*. Equation (9.85) represents *Rayleigh's principle*.

Before we discuss how the variational approach can be used to develop approximate solutions to the eigenvalue problem, we wish to elaborate on the approximation process itself. To this end, we consider the differential eigenvalue problem for a string in transverse vibration fixed at $x = 0$ and supported by a spring of stiffness k at $x = L$. The eigenvalue problem is described by the differential equation

$$-\frac{d}{dx}\left[T(x)\frac{dY(x)}{dx}\right] = \lambda\rho(x)Y(x),\ 0 < x < L,\ \lambda = \omega^2 \tag{9.86}$$

where $Y(x)$ is the displacement, $T(x)$ the tension and $\rho(x)$ the mass per unit length, and the boundary conditions

$$Y(x) = 0 \text{ at } x = 0,\ T(x)\frac{dY(x)}{dx} + kY(x) = 0 \text{ at } x = L \tag{9.87}$$

The solution of Eqs. (9.86) and (9.87) consists of an infinite set of eigenvalues λ_r and associated eigenfunctions $Y_r(x)$ $(r = 1, 2, \ldots)$. Unfortunately, exact solutions are possible only in relatively few cases, most of them characterized by constant tension and uniform mass density. Hence, for the most part, we must be content with an approximate solution. In seeking an approximate solution, sacrifices must be made, in the sense that something must be violated. Almost always, one forgoes the exact solution of the differential equation, Eq. (9.86), which will be satisfied only approximately, but insists on satisfying both boundary conditions (9.87) exactly.

Another differential eigenvalue problem likely to cause difficulties is that of a beam in bending fixed at $x = 0$ and with a lumped mass M at $x = L$. The eigenvalue problem for such a beam is defined by the differential equation

$$\frac{d^2}{dx^2}\left[EI(x)\frac{d^2Y(x)}{dx^2}\right] = \lambda m(x)Y(x),\ 0 < x < L,\ \lambda = \omega^2 \tag{9.88}$$

and the boundary conditions

$$Y(x) = 0,\quad \frac{dY(x)}{dx} = 0 \text{ at } x = 0 \tag{9.89}$$

$$EI(x)\frac{d^2Y(x)}{dx^2} = 0,\ \frac{d}{dx}\left[EI(x)\frac{d^2Y(x)}{dx^2}\right] + \omega^2 MY(x) = 0 \text{ at } x = L \tag{9.90}$$

Rayleigh's principle, Eq. (9.85), suggests a way of approximating the lowest eigenvalue, and hence the lowest natural frequency of a system without solving the differential eigenvalue problem directly. Indeed, if a trial function $Y(x)$ reasonably close to the lowest eigenfunction $Y_1(x)$ can be found, because of the minimum character of Rayleigh's quotient R, insertion of the trial function into Rayleigh's quotient will result in a value $R(Y)$ one order of magnitude closer to the lowest eigenvalue $R(Y_1) = \lambda_1 = \omega_1^2$ than $Y(x)$ is to the lowest eigenfunction $Y_1(x)$. The question remains as to the form of Rayleigh's quotient for a given system. To answer this question, we multiply Eq. (9.86) by $Y(x)$, integrate over the length of the string, rearrange and obtain Rayleigh's quotient for the string in the form

$$R(Y) = \lambda = \omega^2 = \frac{-\int_0^L Y(x)\frac{d}{dx}\left[T(x)\frac{dY(x)}{dx}\right]dx}{\int_0^L \rho(x)Y^2(x)dx} \tag{9.91}$$

A similar expression can be written for rods in axial vibration and shafts in torsion, subject to the changes in the dependent variable and parameters prescribed in Table 8.1. Following the same pattern as for the string in transverse vibration, Rayleigh's quotient for a beam in bending can be verified to be

$$R(Y) = \lambda = \omega^2 = \frac{\int_0^L Y(x)\frac{d^2}{dx^2}\left[EI(x)\frac{d^2Y(x)}{dx^2}\right]dx}{\int_0^L m(x)Y^2(x)dx} \tag{9.92}$$

Contrasting Eqs. (9.86) and (9.91), as well as Eqs. (9.88) and (9.92), we conclude that *minimizing Rayleigh's quotient is equivalent to solving the differential equation in a weighted average sense*, where the weighting function is $Y(x)$. Of course, the approximation $R(Y) = \lambda = \omega^2$ for $\lambda_1 = \omega_1^2$ arises from using $Y(x)$ instead of $Y_1(x)$ in Rayleigh's quotient.

Next, we wish to explore the nature of the trial functions. To this end, we observe that, in the form given by Eq. (9.91), or by Eq. (9.92), Rayleigh's quotient is defined for all trial functions $Y(x)$ that are differentiable twice, or four times, respectively, and satisfy all the boundary conditions. This differentiability requirement imposed on the trial functions coincides with the order of the differential equation, Eq. (9.86) or Eq. (9.88), which should come as no surprise as the numerator of Rayleigh's quotient involves the stiffness expression from the differential equation. In the absence of a more suitable term, it will prove convenient to refer to Rayleigh's quotient in the form given by Eqs. (9.91) and (9.92) as the *weighted average form of Rayleigh's quotient.* We observe that the boundary conditions do not appear explicitly in the weighted average form of Rayleigh's quotient. As a result, to ensure that the characteristics of the system are taken into account as much as possible, *the trial functions used in conjunction with the weighted average form of Rayleigh's quotient must satisfy all the boundary conditions of the problem.* We refer to trial functions that are as many times differentiable as the order of the system and satisfy all the boundary conditions as *comparison functions*. Hence, *when used in conjunction with the weighted average form of Rayleigh's quotient, the trial functions must be from the class of comparison functions.*

Generating comparison functions can cause problems at times. The differentiability of the trial functions is seldom an issue, but the satisfaction of all the boundary conditions, particularly the satisfaction of the natural boundary conditions (Sec. 8.1) can be. In view of this, we wish to examine the implications of violating the natural boundary conditions. We recall that the weighted average form of Rayleigh's quotient, e.g., Eqs. (9.91) and (9.92), does not account explicitly for the boundary conditions, so that a different form of Rayleigh's quotient is desirable. To this end, we integrate the numerator of Eq. (9.91) by parts, with due consideration to boundary conditions (9.87), and write

$$-\int_0^L Y(x)\frac{d}{dx}\left[T(x)\frac{dY(x)}{dx}\right]dx = -Y(x)T(x)\frac{dY(x)}{dx}\bigg|_0^L + \int_0^L T(x)\left[\frac{dY(x)}{dx}\right]^2 dx$$

$$= \int_0^L T(x)\left[\frac{dY(x)}{dx}\right]^2 dx + kY^2(L) \tag{9.93}$$

Inserting Eq. (9.93) into Eq. (9.91), we can express Rayleigh's quotient in the generic form (Sec. 8.8)

$$R(Y) = \lambda = \omega^2 = \frac{V_{\max}}{T_{\mathrm{ref}}} \tag{9.94}$$

where

$$V_{\max} = \frac{1}{2}\int_0^L T(x)\left[\frac{dY(x)}{dx}\right]^2 dx + \frac{1}{2}kY^2(L) \tag{9.95}$$

is the *maximum potential energy*, defined as the potential energy in which the time dependence has been eliminated, and

$$T_{\mathrm{ref}} = \frac{1}{2}\int_0^L \rho(x)Y^2(x)dx \tag{9.96}$$

is the *reference kinetic energy*, namely, the kinetic energy with $\dot{y}(x,t)$ replaced by $Y(x)$.

Equation (9.94) is valid for any distributed-parameter system and for any type of boundary conditions, provided they can be accounted for in $V_{\max}$ and T_{ref}. As an illustration of a case in which a boundary condition is accounted for in T_{ref}, we integrate the numerator in Eq. (9.92) by parts, consider boundary conditions (9.89) and (9.90) and write

$$\int_0^L Y(x)\frac{d^2}{dx^2}\left[EI(x)\frac{d^2Y(x)}{dx^2}\right]dx$$

$$= Y(x)\frac{d}{dx}\left[EI(x)\frac{d^2Y(x)}{dx^2}\right]\bigg|_0^L - \frac{dY(x)}{dx}EI(x)\frac{d^2Y(x)}{dx^2}\bigg|_0^L + \int_0^L EI(x)\left[\frac{d^2Y(x)}{dx^2}\right]^2 dx$$

$$= \int_0^L EI(x)\left[\frac{d^2Y(x)}{dx^2}\right]^2 dx - \omega^2 MY^2(L) \tag{9.97}$$

so that Eq. (9.92) can be rewritten as

$$\omega^2 = \frac{\displaystyle\int_0^L EI(x)\left[\frac{d^2Y(x)}{dx^2}\right]^2 dx - \omega^2 MY^2(L)}{\displaystyle\int_0^L m(x)Y^2(x)dx} \tag{9.98}$$

Upon rearranging Eq. (9.98), Rayleigh's quotient can be expressed once again in the generic form (9.94), in which

$$V_{\max} = \frac{1}{2}\int_0^L EI(x)\left[\frac{d^2Y(x)}{dx^2}\right]^2 dx \tag{9.99}$$

and

$$T_{\text{ref}} = \frac{1}{2}\int_0^L m(x)Y^2(x)dx + \frac{1}{2}MY^2(L) \tag{9.100}$$

For obvious reasons, we refer to the generic form of Rayleigh's quotient, Eq. (9.94), as the *energy form of Rayleigh's quotient.* We observe that Eq. (9.94) involves $V_{\max}$ and T_{ref}, which are defined for trial functions that are half as many times differentiable as the order of the system and need satisfy only the geometric boundary conditions, as the natural boundary conditions are accounted for in some fashion. We refer to trial functions that are half as many times differentiable as the order of the system and satisfy the geometric boundary conditions alone as *admissible functions.* We encountered admissible functions for the first time in Sec. 8.8. Hence, *when used in conjunction with the energy form of Rayleigh's quotient, the trial functions need be from the class of admissible functions only.* This does not preclude the use of comparison functions in conjunction with the energy form of Rayleigh's quotient, because when comparison functions are used, the weighted average form and the energy form of Rayleigh's quotient are equivalent. In fact, if comparison functions are available, then their use is preferable over the use of admissible functions, because the results are likely to be more accurate. Moreover, the use of comparison functions with the energy form of Rayleigh's quotient is advisable, because it requires simpler computations than the weighted average form. The conclusion is that *the energy form of Rayleigh's quotient is always the preferred choice, whether we use comparison functions or admissible functions.*

It should be pointed out that, in using admissible functions in conjunction with the energy form of Rayleigh's quotient, the natural boundary conditions are still violated. But, the deleterious effect of this violation is somewhat mitigated by the fact that the energy form of Rayleigh's quotient, Eq. (9.94), includes contributions to $V_{\max}$ from springs at boundaries and to T_{ref} from masses at boundaries.

Rayleigh's principle, Eq. (9.85), has practical as well as theoretical implications. In particular, it can be used to generate a quick estimate of the lowest natural frequency of a system, a fact already established in Sec. 8.8. To this end, all that is necessary is a reasonable guess of the lowest eigenfunction. Inserting this trial function into Rayleigh's quotient, in view of the stationarity property, we obtain a value of Rayleigh's quotient that is one order of magnitude closer to the lowest eigenvalue than the trial function is to the lowest eigenfunction. This procedure is known as *Rayleigh's energy method* and it implies the use of the energy form of Rayleigh's quotient in conjunction with either a comparison function or an admissible function; it can produce remarkably good estimates of the lowest natural frequency. A suitable trial function is the static deflection curve obtained by loading the system with a distributed force proportional to its own mass. Another suitable trial function may be the lowest eigenfunction of a closely related, but simpler system. For example, to estimate the lowest natural frequency of a nonuniform cantilever beam, a good trial function is likely to be the lowest eigenfunction of a uniform cantilever beam.

Our interest in Rayleigh's principle is not as much as a way of estimating the lowest natural frequency but in its role in the development of a mathematical theory for the discretization of distributed-parameter systems. Reference is made here to the theory behind the Rayleigh-Ritz method (Sec. 9.6). In turn, the Rayleigh-Ritz theory forms the mathematical foundation for the finite element method to be discussed in Ch. 10.

Example 9.4. Estimate the lowest natural frequency of the fixed-free tapered rod in axial vibration of Example 9.1 by means of Rayleigh's energy method. Use as a trial function the lowest eigenfunction of a uniform clamped-free rod.

From Example 9.1, the mass and stiffness distributions are given by

$$m(x) = \frac{6m}{5}\left[1 - \frac{1}{2}\left(\frac{x}{L}\right)^2\right], \quad EA(x) = \frac{6EA}{5}\left[1 - \frac{1}{2}\left(\frac{x}{L}\right)^2\right] \tag{a}$$

Moreover, from Example 8.12, the lowest eigenfunction of a uniform rod fixed at $x = 0$ and free at $x = L$ is

$$U(x) = \sin\frac{\pi x}{2L} \tag{b}$$

which represents a comparison function. Hence, inserting Eqs. (a) and (b) into Eqs. (9.94)–(9.96), letting $k = 0$ and carrying out the integrations, we obtain

$$\begin{aligned} R(U) = \omega^2 &= \frac{\displaystyle\int_0^L EA(x)\left[\frac{dU(x)}{dx}\right]^2 dx}{\displaystyle\int_0^L m(x)U^2(x)dx} \\ &= \frac{\displaystyle\frac{6EA}{5}\left(\frac{\pi}{2L}\right)^2 \int_0^L \left[1 - \frac{1}{2}\left(\frac{x}{L}\right)^2\right]\cos^2\frac{\pi x}{2L}dx}{\displaystyle\frac{6m}{5}\int_0^L \left[1 - \frac{1}{2}\left(\frac{x}{L}\right)^2\right]\sin^2\frac{\pi x}{2L}dx} \\ &= \frac{EA}{m}\left(\frac{\pi}{2L}\right)^2 \frac{(L/12\pi^2)(5\pi^2+6)}{(L/12\pi^2)(5\pi^2-6)} = 3.1504\frac{EA}{mL^2} \end{aligned} \tag{c}$$

so that the estimate of the lowest natural frequency is

$$\omega = 1.7749\sqrt{\frac{EA}{mL^2}} \tag{d}$$

For comparison purposes, we recall that in Examples 9.1 and 9.2 we solved essentially the same problem using three lumped-parameter methods, where we computed the values of $1.7668\sqrt{EA/mL^2}$, $1.7663\sqrt{EA/mL^2}$ and $1.7713\sqrt{EA/mL^2}$ for the lowest natural frequency. As we shall see in Example 9.5, the lowest natural frequency is $\omega_1 = 1.7742\sqrt{EA/mL^2}$, so that the estimate of the lowest natural frequency given by Eq. (d) is much closer to the actual value than those computed by the three lumped-parameter methods. Of course, the lumped-parameter models can be modified to bring the computed lowest natural frequency in closer agreement with the actual value. The difficulty with this argument is that the lumped-parameter methods give no clues as to where the true value lies. On the other hand, because the estimate obtained by Rayleigh's energy method is known to be larger than the actual value of the lowest natural frequency, it is always safe to lower the estimate, as this will cause the estimate to approach the actual value. Moreover, it is clear that the actual value can only be approached from above.

9.6 THE RAYLEIGH-RITZ METHOD

Rayleigh's principle states that Rayleigh's quotient has a minimum at the lowest eigenfunction of a conservative system, where the minimum value is the lowest eigenvalue. Rayleigh's energy method represents a technique for estimating the lowest eigenvalue, based on the idea behind Rayleigh's principle that, insertion of a trial function close to the lowest eigenfunction into Rayleigh's quotient results in a value one order of magnitude closer to the lowest eigenvalue than the trial function is to the lowest eigenfunction. Because the lowest eigenvalue is the minimum value of Rayleigh's quotient, any estimate is larger, or at least not smaller, than the lowest eigenvalue. Hence, it is only logical to look for ways of lowering the estimates, relying on the fact that no estimate can ever fall below the lowest natural frequency, where the latter acts as a safety net. The *Rayleigh-Ritz method* provides a rational approach toward this goal. However, matters do not stop there, as the Rayleigh-Ritz method yields estimates not only of the lowest natural frequency, but of a given number of lower natural frequencies. This is accomplished by assuming a solution in the form of a linear combination of trial function, rather than a single trial function, in a process known as *series discretization*. Essentially, the process represents a variational approach whereby a conservative distributed-parameter system is approximated by a discrete model.

The title Rayleigh-Ritz method implies some shared developments by two researchers, with Rayleigh being the main contributor, but this is not quite the case. The method was developed by Ritz as an extension of Rayleigh's energy method. Although Rayleigh claimed that the method originated with him, the form in which the method is generally used is due to Ritz. For this reason, the method is referred to at times as the Ritz method. Because the original developments on which the series discretization technique is based are due to Rayleigh, and because the method is almost universally referred to it as the Rayleigh-Ritz method, retaining the name can be justified.

The first step in the Rayleigh-Ritz method is to construct the *minimizing sequence*

$$
\begin{aligned}
Y^{(1)}(x) &= a_1\phi_1(x) \\
Y^{(2)}(x) &= a_1\phi_1(x) + a_2\phi_2(x) = \sum_{i=1}^{2} a_i\phi_i(x) \\
&\vdots \\
Y^{(n)}(x) &= a_1\phi_1(x) + a_2\phi_2(x) + \ldots + a_n\phi_n(x) = \sum_{i=1}^{n} a_i\phi_i(x)
\end{aligned}
\tag{9.101}
$$

where $\phi_1(x), \phi_2(x), \ldots, \phi_n(x)$ represent independent trial functions and $a_1, a_2, \ldots, a_n$ are undetermined coefficients. The next step is to introduce the minimizing sequence, Eqs. (9.101), in Rayleigh's quotient and carry out the indicated integrations, thus eliminating the spatial dependence. As a result, Rayleigh's quotient becomes a function of the undetermined coefficients $a_1, a_2, \ldots, a_n$ alone,

$$\lambda^{(n)} = R(Y^{(n)}) = R(a_1, a_2, \ldots, a_n) \tag{9.102}$$

We observe that, by approximating the solution $Y(x)$ of the eigenvalue problem by the function $Y^{(n)}(x)$, which represents a series consisting of n terms, we automatically reduce a distributed system, which can be regarded as having an infinite number of degrees of freedom, to a discrete system with n degrees of freedom.

Our objective is to produce an n-degree-of-freedom discrete model best approximating the distributed system under the assumption that the trial functions $\phi_1(x), \phi_2(x), \ldots, \phi_n(x)$ are given. This implies that the coefficients $a_1, a_2, \ldots, a_n$ must be so adjusted as to produce the desired approximate model. But, it is shown in Ref. 13 that rendering Rayleigh's quotient stationary is equivalent to solving the weak form of the differential eigenvalue problem. Hence, we insist that the coefficients $a_1, a_2, \ldots, a_n$ be determined so as to render Rayleigh's quotient stationary, which requires that the first variation of Rayleigh's quotient be zero. Because, according to Eq. (9.102), Rayleigh's quotient is a function of the coefficients $a_1, a_2, \ldots, a_n$ alone, the stationarity condition can be expressed as

$$\delta R = \frac{\partial R}{\partial a_1}\delta a_1 + \frac{\partial R}{\partial a_2}\delta a_2 + \ldots + \frac{\partial R}{\partial a_n}\delta a_n = 0 \tag{9.103}$$

where $\delta a_1, \delta a_2, \ldots, \delta a_n$ are variations in the undetermined coefficients. The independence of the trial functions $\phi_1(x), \phi_2(x), \ldots, \phi_n(x)$ implies the independence of the coefficients $a_1, a_2, \ldots, a_n$, which in turn implies the independence of the variations $\delta a_1, \delta a_2, \ldots, \delta a_n$. In view of this, Eq. (9.103) can only be satisfied if the quantities multiplying $\delta a_1, \delta a_2, \ldots, \delta a_n$ are all equal to zero, so that the necessary conditions for the stationarity of Rayleigh's quotient are

$$\frac{\partial R}{\partial a_i} = 0, \; i = 1, 2, \ldots, n \tag{9.104}$$

At this point, we recall that Rayleigh's quotient is really a ratio, so that it is convenient to write it in the form

$$\lambda^{(n)} = R(a_1, a_2, \ldots, a_n) = \frac{N(a_1, a_2, \ldots, a_n)}{D(a_1, a_2, \ldots, a_n)} \tag{9.105}$$

where N denotes the numerator and D the denominator of the quotient, both functions of the undetermined coefficients $a_1, a_2, \ldots, a_n$. Inserting Eq. (9.105) into Eqs. (9.104), we have

$$\begin{aligned} \frac{\partial R}{\partial a_i} &= \frac{(\partial N/\partial a_i)D - (\partial D/\partial a_i)N}{D^2} = \frac{1}{D}\left(\frac{\partial N}{\partial a_i} - \frac{N}{D}\frac{\partial D}{\partial a_i}\right) \\ &= \frac{1}{D}\left(\frac{\partial N}{\partial a_i} - \lambda^{(n)}\frac{\partial D}{\partial a_i}\right) = 0, \; i = 1, 2, \ldots, n \end{aligned} \tag{9.106}$$

Hence, the necessary conditions for the stationarity of Rayleigh's quotient are

$$\frac{\partial N}{\partial a_i} - \lambda^{(n)}\frac{\partial D}{\partial a_i} = 0, \; i = 1, 2, \ldots, n \tag{9.107}$$

Equations (9.107) represent n algebraic equations with the coefficients $a_1, a_2, \ldots, a_n$ as unknowns and with $\lambda^{(n)}$ as an unknown parameter, so that solving the equations amounts to determining the coefficients, which were undetermined until now, as well as to determining $\lambda^{(n)}$. Inserting the coefficients $a_1, a_2, \ldots, a_n$ thus determined into Eqs. (9.101), we obtain the approximate solution $Y^{(n)}(x)$ of the distributed-parameter problem. Before we discuss the details of the solution, we wish to explore the nature of the solution.

A question always arising in conjunction with approximate solutions is how good the approximation is. The answer to this question depends on the nature of the trial functions $\phi_1(x), \phi_2(x), \ldots, \phi_n(x)$, as the problem can be regarded as solved for all practical purposes as soon as the trial functions and their number have been selected. Of course, there are still the tasks of setting up Eqs. (9.107) and computing their solution, but these tasks follow an established pattern, which is shown later in this section to be a very familiar one. We recall from Sec. 9.5 that there are many advantages to the use of the energy form of Rayleigh's quotient. Hence, in the future, we will use the energy form of Rayleigh's quotient exclusively, and simply refer to it as Rayleigh's quotient. Then, we will use trial functions in the form of comparison functions if available, and in the form of admissible functions if comparison functions are not available.

To illustrate the Rayleigh-Ritz process, we consider the differential eigenvalue problem for the string in transverse vibration described by Eqs. (9.86) and (9.87), so that Rayleigh's quotient is given by Eqs. (9.94) to (9.96). Introducing the last of Eqs. (9.101) in Eq. (9.95), we can write the numerator of Rayleigh's quotient as

$$\begin{aligned} N = V_{\max} &= \frac{1}{2}\int_0^L T(x)\left[\frac{dY^{(n)}(x)}{dx}\right]^2 dx + \frac{1}{2}k[Y^{(n)}(L)]^2 \\ &= \frac{1}{2}\int_0^L T(x)\sum_{i=1}^{n} a_i \frac{d\phi_i(x)}{dx}\sum_{j=1}^{n} a_j \frac{d\phi_j(x)}{dx}dx + \frac{1}{2}k\sum_{i=1}^{n} a_i\phi_i(L)\sum_{j=1}^{n} a_j\phi_j(L) \\ &= \frac{1}{2}\sum_{i=1}^{n}\sum_{j=1}^{n} a_i a_j\left[\int_0^L T(x)\frac{d\phi_i(x)}{dx}\frac{d\phi_j(x)}{dx}dx + k\phi_i(L)\phi_j(L)\right] \\ &= \frac{1}{2}\sum_{i=1}^{n}\sum_{j=1}^{n} k_{ij}a_i a_j \end{aligned} \tag{9.108}$$

where

$$k_{ij} = k_{ji} = \int_0^L T(x)\frac{d\phi_i(x)}{dx}\frac{d\phi_j(x)}{dx}dx + k\phi_i(L)\phi_j(L), \; i, j = 1, 2, \ldots, n \tag{9.109}$$

are symmetric *stiffness coefficients*. On the other hand, the denominator of Rayleigh's quotient, Eq. (9.96), has the form

$$D = T_{\text{ref}} = \frac{1}{2}\int_0^L \rho(x)[Y^{(n)}(x)]^2 dx = \frac{1}{2}\sum_{i=1}^{n}\sum_{j=1}^{n} m_{ij}a_i a_j \tag{9.110}$$

in which

$$m_{ij} = m_{ji} = \int_0^L \rho(x)\phi_i(x)\phi_j(x)dx, \; i, j = 1, 2, \ldots, n \tag{9.111}$$

are symmetric *mass coefficients*. As pointed out earlier in this section, the trial functions can be either from the class of comparison functions or from the class of admissible functions.

Next, we return to Eqs. (9.107). From Eqs. (9.108), and (9.110), we see that the numerator N of Rayleigh's quotient is a quadratic form in terms of the stiffness coefficients k_{ij} and the denominator D is a quadratic form in terms of the mass coefficients m_{ij}. To permit proper differentiation with respect to a_i, we replace the indices in the quadratic forms and write

$$N = \frac{1}{2}\sum_{r=1}^{n}\sum_{s=1}^{n} k_{rs}a_r a_s \tag{9.112}$$

and

$$D = \frac{1}{2}\sum_{r=1}^{n}\sum_{s=1}^{n} m_{rs}a_r a_s \tag{9.113}$$

so that, recalling the symmetry of the stiffness coefficients, we have

$$\frac{\partial N}{\partial a_i} = \frac{1}{2}\sum_{r=1}^{n}\sum_{s=1}^{n} k_{rs}\left(\frac{\partial a_r}{\partial a_i}a_s + a_r\frac{\partial a_s}{\partial a_i}\right) = \frac{1}{2}\sum_{r=1}^{n}\sum_{s=1}^{n} k_{rs}(\delta_{ri}a_s + a_r\delta_{si})$$

$$= \frac{1}{2}\left(\sum_{s=1}^{n} k_{is}a_s + \sum_{r=1}^{n} k_{ri}a_r\right) = \sum_{s=1}^{n} k_{is}a_s \tag{9.114}$$

Similarly, invoking the symmetry of the mass coefficients, we can write

$$\frac{\partial D}{\partial a_i} = \sum_{s=1}^{n} m_{is}a_s \tag{9.115}$$

Hence, inserting Eqs. (9.114) and (9.115) into Eqs. (9.107), we obtain the set of algebraic equations

$$\sum_{s=1}^{n} k_{is}a_s = \lambda^{(n)}\sum_{s=1}^{n} m_{is}a_s, \; i = 1, 2, \ldots, n \tag{9.116}$$

Equations (9.116) are recognized as representing an *algebraic eigenvalue problem*, which is a very familiar problem (see Ch. 7).

The eigenvalue problem, Eqs. (9.116), can be cast in the matrix form

$$K^{(n)}\mathbf{a}^{(n)} = \lambda^{(n)}M^{(n)}\mathbf{a}^{(n)} \tag{9.117}$$

where $K^{(n)}$ is an $n \times n$ stiffness matrix, $M^{(n)}$ an $n \times n$ mass matrix, both symmetric, $\mathbf{a}^{(n)}$ an n-dimensional vector and $\lambda^{(n)}$ a scalar, in which the superscript (n) indicates that the eigenvalue problem corresponds to n terms in the approximating series, the last expression in the minimizing sequence given by Eqs. (9.101). The eigenvalue problem, Eqs. (9.116) or Eq. (9.117), resembles entirely the eigenvalue problem for a conservative n-degree-of-freedom discrete system, which justifies our statement in the beginning of this section that the Rayleigh-Ritz method is essentially a series discretization technique approximating a conservative distributed-parameter system by a discrete system.

The solution of the algebraic eigenvalue problem, Eq. (9.117), consists of the eigenvalues $\lambda_r^{(n)}$ and associated eigenvectors $\mathbf{a}_r^{(n)}$ $(r = 1, 2, \ldots, n)$, referred to as *Ritz eigenvalues* and *Ritz eigenvectors*, respectively. The Ritz eigenvalues $\lambda_r^{(n)}$ represent approximations to the actual eigenvalues λ_r $(r = 1, 2, \ldots, n)$ of the distributed-parameter system. To obtain the approximate eigenfunctions, referred to as *Ritz eigenfunctions*, we use the last of Eqs. (9.101) and write

$$Y_r^{(n)}(x) = \sum_{i=1}^{n} \phi_i(x) a_{ir}^{(n)} = \boldsymbol{\phi}^T(x)\mathbf{a}_r^{(n)} \tag{9.118}$$

in which $a_{ir}^{(n)}$ is the ith component of the Ritz eigenvector $\mathbf{a}_r^{(n)}$ and $\boldsymbol{\phi}(x) = [\phi_1(x)\ \phi_2(x)\ \ldots\ \phi_n(x)]^T$ is the vector of trial functions. From earlier studies in Ch. 7, we know that the eigenvectors $\mathbf{a}_r^{(n)}$ are orthogonal with respect to the mass matrix $M^{(n)}$, as well as with respect to the stiffness matrix $K^{(n)}$. Assuming that the eigenvectors have been normalized so that $(\mathbf{a}_r^{(n)})^T M^{(n)} \mathbf{a}_r^{(n)} = 1$ $(r = 1, 2, \ldots, n)$, the orthonormality conditions have the form

$$(\mathbf{a}_r^{(n)})^T M^{(n)} \mathbf{a}_s^{(n)} = \delta_{rs},\ (\mathbf{a}_r^{(n)})^T K^{(n)} \mathbf{a}_s^{(n)} = \lambda_r^{(n)} \delta_{rs},\ r, s = 1, 2, \ldots, n \tag{9.119}$$

Then, recalling the definition of the stiffness and mass coefficients, e.g., Eqs. (9.109) and (9.111), respectively, it is possible to derive orthonormality relations for the Ritz eigenfunctions $Y_r^{(n)}(x)$ $(r = 1, 2, \ldots, n)$ (see Problems 9.25–9.27).

The Rayleigh-Ritz method calls for a sequence of approximations obtained by letting $n = 2, 3, \ldots$ in the minimizing sequence, Eqs. (9.101). We note that the case $n = 1$ represents Rayleigh's energy method, which does not involve the solution of an eigenvalue problem at all (see Example 9.4). As the number n of terms in the series increases, there is steady improvement at the lower end of the eigenvalue spectrum, while new approximate eigenvalues are added at the higher end of the spectrum. The process is stopped when a desired number of eigenvalues reach sufficient accuracy, i.e., when the addition of terms to the series does not produce improvement in these eigenvalues within the specified accuracy level. This brings up an interesting peculiarity of the Rayleigh-Ritz method (or of any discretization method). In particular, only a fraction of the Ritz eigenvalues at the lower end of the spectrum tend to be accurate, with the newly added ones at the higher end being wildly in error. As a rough guideline, the number of terms in the series should be about twice as large as the number of accurate eigenvalues desired.

The question remains as to how the Ritz eigenvalues $\lambda_r^{(n)}$ relate to the actual eigenvalues λ_r $(r = 1, 2, \ldots n)$. For convenience, we order the Ritz and actual eigenvalues so as to satisfy $\lambda_1^{(n)} \leq \lambda_2^{(n)} \leq \ldots \leq \lambda_n^{(n)}$ and $\lambda_1 \leq \lambda_2 \leq \ldots$, respectively. Then, assuming that the trial functions $\phi_1(x),\ \phi_2(x),\ \ldots,\ \phi_n(x),\ \ldots$ are all from a *complete set*, which implies that the error incurred in using an approximate solution instead of the exact solution can be made as small as desired by simply increasing n in Eqs. (9.101), we conclude that the Ritz eigensolutions approach the actual eigensolutions as $n \to \infty$. Completeness is a mathematical concept having primarily negative implications, in the sense that it hurts convergence if the set of trial functions is not complete. But, the fact that completeness guarantees convergence as the number of terms in the series approaches infinity is not particularly meaningful, because in deriving approximate solutions the interest lies in convergence with as few terms as possible. It should be pointed out that the sets of trial functions to be considered in this text are complete almost by definition.

The approximation of a distributed-parameter system with an infinite number of degrees of freedom by a discrete system with n degrees of freedom implies *truncation*. In terms of the series given by the last of Eqs. (9.101), truncation is tantamount to the statement that the higher-order terms in the series have been ignored, so that the constraints

$$a_{n+1} = a_{n+2} = \ldots = 0 \tag{9.120}$$

have been imposed on the distributed system. Constraints tend to increase the stiffness of a system, without a commensurate increase in inertia. In terms of Rayleigh's quotient, this implies that the numerator tends to increase relative to the denominator, from which we conclude that

$$\lambda_r^{(n)} \geq \lambda_r, \quad r = 1, 2, \ldots, n \tag{9.121}$$

The nature of the Ritz eigenvalues requires further elaboration. To this end, we add one more term to the approximating series and write

$$Y^{(n+1)}(x) = \sum_{i=1}^{n+1} a_i \phi_i(x) \tag{9.122}$$

Following the usual steps, we obtain an $(n + 1)$-degree-of-freedom discrete system described by the eigenvalue problem

$$K^{(n+1)}\mathbf{a}^{(n+1)} = \lambda^{(n+1)} M^{(n+1)}\mathbf{a}^{(n+1)} \tag{9.123}$$

which is of order $n + 1$, as opposed to the nth-order eigenvalue problem given by Eq. (9.117). As a result, there are $n + 1$ eigensolutions $\lambda_r^{(n+1)}$, $\mathbf{a}_r^{(n+1)}$ $(r = 1, 2, \ldots, n + 1)$. A question of particular interest is how the eigenvalues $\lambda_r^{(n+1)}$ $(r = 1, 2, \ldots, n + 1)$ of the $(n + 1)$-degree-of-freedom approximation relate to the eigenvalues $\lambda_r^{(n)}$ $(r = 1, 2, \ldots, n)$ of the n-degree-of-freedom approximation. To answer this question, we observe that the extra term in series (9.122) does not affect the mass and stiffness coefficients computed on the basis of an n-term series. The implication is that the mass and stiffness matrices possess the *embedding property*, defined by

$$M^{(n+1)} = \begin{bmatrix} M^{(n)} & & x \\ & & x \\ x & x & x \end{bmatrix}, \quad K^{(n+1)} = \begin{bmatrix} K^{(n)} & & x \\ & & x \\ x & x & x \end{bmatrix} \tag{9.124}$$

where the x's imply one extra row and column, so that the symmetric matrices $M^{(n+1)}$ and $K^{(n+1)}$ are obtained by adding one row and one column to matrices $M^{(n)}$ and $K^{(n)}$, respectively. It is demonstrated in Ref. 13 that, for matrices satisfying Eqs. (9.124), the eigenvalues $\lambda_r^{(n+1)}$ $(r = 1, 2, \ldots, n+1)$ and $\lambda_r^{(n)}$ $(r = 1, 2, \ldots, n)$ satisfy the *separation theorem*, defined by the inequalities

$$\lambda_1^{(n+1)} \leq \lambda_1^{(n)} \leq \lambda_2^{(n+1)} \leq \lambda_2^{(n)} \leq \ldots \leq \lambda_n^{(n+1)} \leq \lambda_n^{(n)} \leq \lambda_{n+1}^{(n+1)} \tag{9.125}$$

which state that *the eigenvalues of the* $(n+1)$*-degree-of-freedom model bracket the eigenvalues of the n-degree-of-freedom model.* Inequalities (9.125) can be broken into inequalities for every eigenvalue as follows:

$$\lambda_r^{(n+1)} \leq \lambda_r^{(n)}, \; r = 1, 2, \ldots, n \tag{9.126}$$

which highlights the fact that the eigenvalues of the problem of order $n+1$ are generally lower, or at least never higher, than the eigenvalues of the problem of order n. Hence, as n increases, there is a steady decrease in the value of the Ritz eigenvalues, with the largest improvement taking place in the eigenvalues at the higher end of the spectrum. Coupled with the fact that the actual eigenvalues serve as lower bounds for the Ritz eigenvalues, we conclude that *as n increases, the Ritz eigenvalues approach the actual eigenvalues asymptotically and from above*, so that we can write

$$\lim_{n \to \infty} \lambda_r^{(n)} = \lambda_r, \; r = 1, 2, \ldots, n \tag{9.127}$$

The above results are illustrated in Fig. 9.20.

The separation theorem and the consequent statement on the asymptotic behavior of the Ritz eigenvalues make the Rayleigh-Ritz theory unique in vibration analysis. Unfortunately, there are no such precise statements concerning the Ritz eigenfunctions, although as n increases there is steady improvement in the eigenfunctions as well.

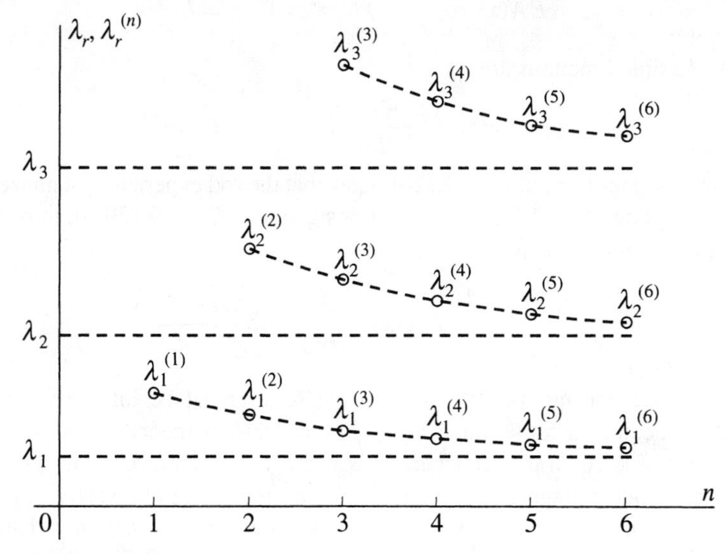

FIGURE 9.20
Convergence of Ritz eigenvalues to actual eigenvalues

Finally, we must address a frequently asked question in connection with the Rayleigh-Ritz method, namely, how to choose suitable comparison functions, or admissible functions, as the requirement that all boundary conditions, or merely the geometric boundary conditions be satisfied is too broad to serve as a guideline. This choice is more important than it may seem, because there may be several sets of functions that could be used and the rate of convergence tends to vary from set to set. Indeed, whereas all sets of comparison functions or admissible functions will lead to convergence, the rate of convergence can be unacceptably slow, particularly for admissible functions. This problem is aggravated by the fact that it is virtually impossible to predict the rate of convergence for a given set of functions. Still, some guidelines can be stated. In the first place, it is imperative that the functions be from a complete set, because otherwise convergence may not be possible. Among complete sets, we list power series, trigonometric functions, Bessel functions, Legendre polynomials, etc. Also complete for a given system are the eigenfunctions of a simpler but closely related system. For example, the eigenfunctions of a uniform cantilever beam are likely to be a suitable set of comparison functions for a nonuniform cantilever beam. Moreover, they are likely to match better the dynamic characteristics of the nonuniform cantilever beam than any other set. In this particular case, the natural boundary conditions at the free end are relatively easy to satisfy. Extreme care must be exercised when the end involves a discrete component, such as a spring or a lumped mass. As an illustration, we consider a rod in axial vibration fixed at $x = 0$ and restrained by a spring of stiffness k at $x = L$. In this case, if we choose as admissible functions the eigenfunctions of a uniform fixed-free rod, then the rate of convergence will be very poor. Indeed, the boundary condition at $x = L$ is

$$EA(x)\frac{dU(x)}{dx} + kU(x) = 0, \; x = L \tag{9.128}$$

and the admissible functions are

$$\phi_i(x) = \sin\frac{(2i-1)\pi x}{2L}, \; i = 1, 2, \ldots, n \tag{9.129}$$

But, whereas boundary condition (9.128) states that the rod experiences a nonzero slope at $x = L$, the slope of the admissible functions given by Eqs. (9.129) is zero at $x = L$. Assuming a solution in the form

$$U^{(n)}(x) = \sum_{i=1}^{n} a_i \phi_i(x) = \sum_{i=1}^{n} a_i \sin\frac{(2i-1)\pi x}{2L} \tag{9.130}$$

we conclude that the number of terms in the series must be infinite for the slope $dU^{(n)}/dx$ to acquire a finite value at $x = L$, at least in theory. In practice, for a given number of decimal places accuracy, convergence is achieved with a finite number of terms, albeit a large one. The rate of convergence can be vastly improved by using comparison functions, which can be generated for the problem at hand in the form

$$\phi_i(x) = \sin\beta_i x, \; i = 1, 2, \ldots, n \tag{9.131}$$

where the constants β_i are determined by requiring that ϕ_i satisfy boundary condition (9.128), or

$$EA(L)\beta_i \cos \beta_i L + k \sin \beta_i L = 0, \; i = 1, 2, \ldots, n \tag{9.132}$$

In the example at hand, it is relatively easy to generate comparison functions. Unfortunately, more often than not this is not the case. The task is considerably more difficult for two-dimensional members, such as membranes and plates, for which even admissible functions may be beyond reach, particularly if the boundaries are not simple, such as circular and rectangular.

For problems for which suitable trial functions can be found, the Rayleigh-Ritz method tends to produce excellent results with a relatively small number of degrees of freedom. The method is helpless, however, for problems with complicated boundary conditions, irregular boundaries (arising mostly in two-dimensional systems) and in general for very complex structures. A different version of the Rayleigh-Ritz method does not have these limitations, although it tends to require a large number of degrees of freedom for satisfactory accuracy. Reference is made here to the finite element method, discussed in detail in Ch. 10. As demonstrated there, the Rayleigh-Ritz theory is essential to a deep appreciation of the finite element method.

Example 9.5. Solve the eigenvalue problem for the fixed-free tapered rod in axial vibration of Example 9.4 by the Rayleigh-Ritz method using the comparison functions

$$\phi_i(x) = \sin \frac{(2i-1)\pi x}{2L}, \; i = 1, 2, \ldots, n \tag{a}$$

Give the natural frequencies and plot the modes for $n = 2$ and $n = 3$. Then, determine the number n of terms required for computing the lowest natural frequency with six decimal places accuracy.

In the case at hand, Rayleigh's quotient is given by Eq. (9.94), in which

$$V_{\max} = \frac{1}{2} \int_0^L EA(x) \left[\frac{dU(x)}{dx} \right]^2 dx \tag{b}$$

and

$$T_{\mathrm{ref}} = \frac{1}{2} \int_0^L m(x) U^2(x) dx \tag{c}$$

Inserting $EA(x)$ and $m(x)$, Eqs. (a) of Example 9.4, as well as the approximating series

$$U^{(n)}(x) = \sum_{i=1}^{n} a_i^{(n)} \phi_i(x) = \sum_{i=1}^{n} a_i^{(n)} \sin \frac{(2i-1)\pi x}{2L} \tag{d}$$

into Eqs. (b) and (c), we can write

$$V_{\max} \cong \frac{1}{2} \sum_{i=1}^{n} \sum_{j=1}^{n} k_{ij}^{(n)} a_i^{(n)} a_j^{(n)} \tag{e}$$

in which $k_{ij}^{(n)}$ are symmetric stiffness coefficients given by

$$
\begin{aligned}
k_{ij}^{(n)} &= \int_0^L EA(x)\frac{d\phi_i(x)}{dx}\frac{d\phi_j(x)}{dx}dx \\
&= \frac{6EA}{5}\frac{(2i-1)\pi}{2L}\frac{(2j-1)\pi}{2L}\int_0^L\left[1-\frac{1}{2}\left(\frac{x}{L}\right)^2\right]\cos\frac{(2i-1)\pi x}{2L}\cos\frac{(2j-1)\pi x}{2L}dx, \\
&\qquad i,j=1,2,\ldots,n
\end{aligned}
\tag{f}
$$

as well as

$$
T_{\text{ref}} \cong \frac{1}{2}\sum_{i=1}^{n}\sum_{j=1}^{n} m_{ij}^{(n)}a_i^{(n)}a_j^{(n)} \tag{g}
$$

where $m_{ij}^{(n)}$ are symmetric mass coefficients having the form

$$
\begin{aligned}
m_{ij}^{(n)} &= \int_0^L m(x)\phi_i(x)\phi_j(x)dx \\
&= \frac{6m}{5}\int_0^L\left[1-\frac{1}{2}\left(\frac{x}{L}\right)^2\right]\sin\frac{(2i-1)\pi x}{2L}\sin\frac{(2j-1)\pi x}{2L}dx,\quad i,j=1,2,\ldots,n
\end{aligned}
\tag{h}
$$

To obtain the natural frequencies and natural modes, we solve the algebraic eigenvalue problem given by Eq. (9.117) with the entries of the stiffness matrix $K^{(n)}$ and mass matrix $M^{(n)}$ given by Eqs. (f) and (h), respectively. For $n = 2$, the stiffness and mass matrices are

$$
K^{(2)} = \frac{EA}{L}\begin{bmatrix} 1.383701 & 0.337500 \\ 0.337500 & 11.253305 \end{bmatrix} \tag{i}
$$

and

$$
M^{(2)} = mL\begin{bmatrix} 0.439207 & 0.075991 \\ 0.075991 & 0.493245 \end{bmatrix} \tag{j}
$$

respectively. The solution of the eigenvalue problem consists of the Ritz natural frequencies and modal vectors

$$
\begin{aligned}
\omega_1^{(2)} &= 1.774312\sqrt{\frac{EA}{mL^2}},\quad \mathbf{a}_1^{(2)} = (mL)^{-1/2}\begin{bmatrix} 1.511481 \\ -0.015311 \end{bmatrix} \\
\omega_2^{(2)} &= 4.825444\sqrt{\frac{EA}{mL^2}},\quad \mathbf{a}_2^{(2)} = (mL)^{-1/2}\begin{bmatrix} -0.233683 \\ 1.443148 \end{bmatrix}
\end{aligned}
\tag{k}
$$

so that the Ritz natural modes are

$$
\begin{aligned}
U_1^{(2)}(x) &= 1.511481\sin\frac{\pi x}{2L} - 0.015311\sin\frac{3\pi x}{2L} \\
U_2^{(2)}(x) &= -0.233683\sin\frac{\pi x}{2L} + 1.443148\sin\frac{3\pi x}{2L}
\end{aligned}
\tag{l}
$$

in which $(mL)^{-1/2}$ was omitted. The modes, normalized so that $U_r^{(2)}(L) = 1$ $(r = 1, 2)$, are plotted in Fig. 9.21.

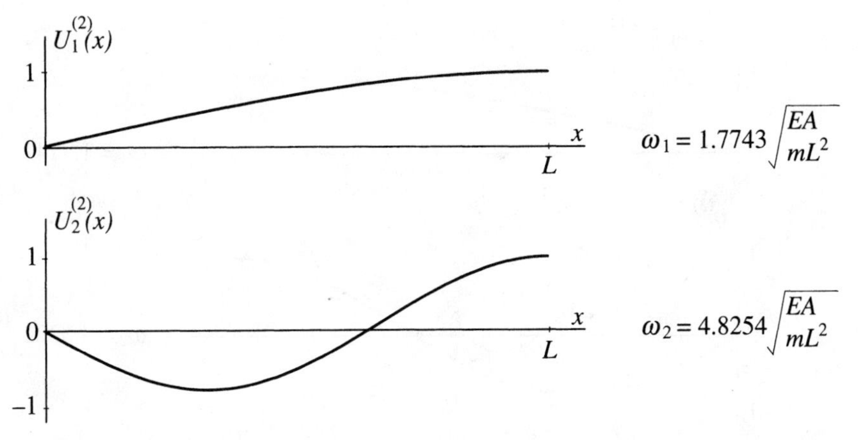

FIGURE 9.21
Two lowest modes of a fixed-free rod computed by the Rayleigh-Ritz method using two comparison functions

Similarly, for $n = 3$, the stiffness and mass matrices are

$$K^{(3)} = \frac{EA}{L}\begin{bmatrix} 1.383701 & 0.337500 & -0.104167 \\ 0.337500 & 11.253305 & 2.109375 \\ -0.104167 & 2.109375 & 30.992514 \end{bmatrix} \tag{m}$$

and

$$M^{(3)} = mL\begin{bmatrix} 0.439207 & 0.075991 & -0.021953 \\ 0.075991 & 0.493245 & 0.064592 \\ -0.021953 & 0.064592 & 0.497568 \end{bmatrix} \tag{n}$$

respectively. Solving once again the eigenvalue problem, we obtain the Ritz natural frequencies and modal vectors

$$\begin{aligned} \omega_1^{(3)} &= 1.774247\sqrt{\frac{EA}{mL^2}}, \quad \mathbf{a}_1^{(3)} = (mL)^{-1/2}\begin{bmatrix} 1.511715 \\ -0.015872 \\ 0.002829 \end{bmatrix} \\ \omega_2^{(3)} &= 4.822187\sqrt{\frac{EA}{mL^2}}, \quad \mathbf{a}_2^{(3)} = (mL)^{-1/2}\begin{bmatrix} -0.236352 \\ 1.448321 \\ -0.040348 \end{bmatrix} \\ \omega_3^{(3)} &= 7.931607\sqrt{\frac{EA}{mL^2}}, \quad \mathbf{a}_3^{(3)} = (mL)^{-1/2}\begin{bmatrix} 0.097373 \\ -0.163450 \\ 1.432793 \end{bmatrix} \end{aligned} \tag{o}$$

so that now, ignoring $(mL)^{-1/2}$, the Ritz natural modes are

$$\begin{aligned} U_1^{(3)}(x) &= 1.511715\sin\frac{\pi x}{2L} - 0.015872\sin\frac{3\pi x}{2L} + 0.002829\sin\frac{5\pi x}{2L} \\ U_2^{(2)}(x) &= -0.236352\sin\frac{\pi x}{2L} + 1.448321\sin\frac{3\pi x}{2L} - 0.040348\sin\frac{5\pi x}{2L} \\ U_3^{(3)}(x) &= 0.097373\sin\frac{\pi x}{2L} - 0.163450\sin\frac{3\pi x}{2L} + 1.432793\sin\frac{5\pi x}{2L} \end{aligned} \tag{p}$$

The natural modes, normalized so that $U_r^{(3)}(L) = 1$ $(r = 1, 2, 3)$, are plotted in Fig. 9.22.

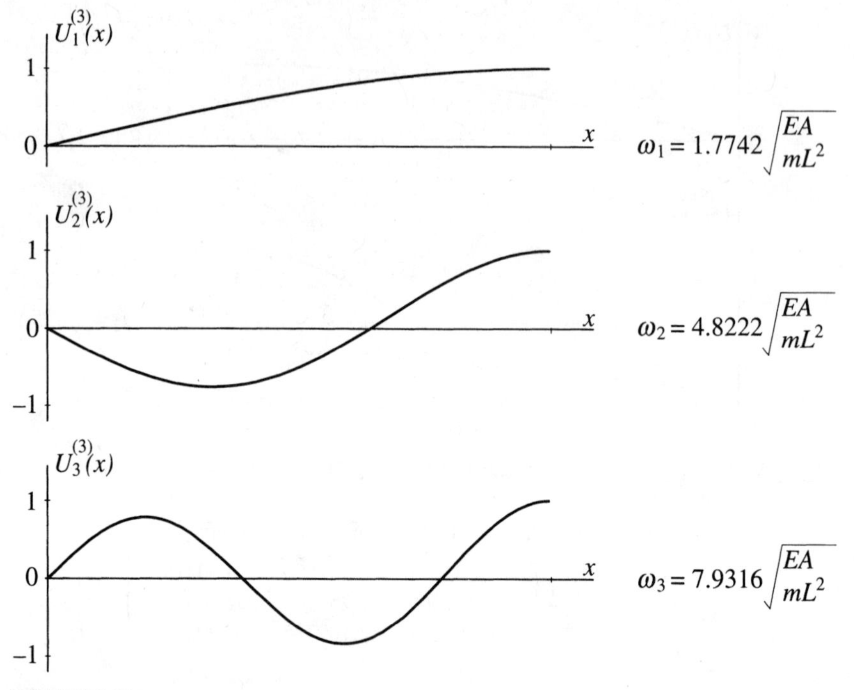

FIGURE 9.22
Three lowest modes of a fixed-free rod computed by the Rayleigh-Ritz method using three comparison functions

The Ritz eigenvalues for the two approximations are

$$\begin{aligned}&\lambda_1^{(2)} = 3.148183EA/mL^2,\ \lambda_2^{(2)} = 23.284913EA/mL^2\\&\lambda_1^{(3)} = 3.147951EA/mL^2,\ \lambda_2^{(3)} = 23.253490EA/mL^2,\ \lambda_3^{(3)} = 62.910394EA/mL^2\end{aligned} \tag{q}$$

and they can be verified to satisfy the separation theorem, inequalities (9.125). This comes as no surprise, as the mass and stiffness matrices possess the embedding property, Eqs. (9.124).

We observe from Eqs. (k) and (o) that improvement in the first two Ritz natural frequencies and natural modes from $n = 2$ to $n = 3$ is very small, which indicates that the chosen comparison functions, Eqs. (a), resemble very closely the actual natural modes. Another way of establishing this fact is by observing that the major contribution to the first, second and third mode is from the first, second and third comparison function, respectively, as can be concluded from the relative magnitude of the coefficients in the series given by Eqs. (p). The same pattern holds for the higher modes.

Convergence to the lowest eigenvalue is obtained with 11 terms in the approximating series. The value of the lowest eigenvalue is

$$\lambda_1^{(11)} = 3.147888EA/mL^2 \tag{r}$$

Example 9.6. Consider the case in which the end $x = L$ of the rod of Example 9.5 is restrained by a spring of stiffness $k = EA/L$ and obtain the solution of the eigenvalue problem derived by the Rayleigh-Ritz method in two ways: 1) using $\phi_i(x) = \sin(2i - 1)\pi x/2L$ $(i = 1, 2, \ldots, n)$ as admissible functions and 2) using the comparison functions defined by Eqs. (9.131) and (9.132). Give the Ritz natural frequencies and plot the modes for $n = 2$ and $n = 3$. Then, determine the number n of terms required for computing the three lowest natural frequencies with six decimal places accuracy.

In the first case, $\phi_i(x) = \sin(2i-1)\pi x/2L$ $(i = 1, 2, \dots, n)$, which are only admissible functions. Following the pattern of Example 9.5, we obtain the stiffness coefficients

$$\begin{aligned} k_{ij}^{(n)} &= \int_0^L EA(x)\frac{d\phi_i(x)}{dx}\frac{d\phi_j(x)}{dx}dx + k\phi_i(L)\phi_j(L) \\ &= \frac{6EA}{5}\frac{(2i-1)\pi}{2L}\frac{(2j-1)\pi}{2L}\int_0^L\left[1-\frac{1}{2}\left(\frac{x}{L}\right)^2\right]\cos\frac{(2i-1)\pi x}{2L}\cos\frac{(2j-1)\pi x}{2L}dx \\ &+ \frac{EA}{L}\sin\frac{(2i-1)\pi}{2}\sin\frac{(2j-1)\pi}{2}, \qquad i, j = 1, 2, \dots, n \end{aligned} \tag{a}$$

The mass coefficients remain as in Example 9.5, namely,

$$\begin{aligned} m_{ij}^{(n)} &= \int_0^L m(x)\phi_i(x)\phi_j(x)dx \\ &= \frac{6m}{5}\int_0^L\left[1-\frac{1}{2}\left(\frac{x}{L}\right)^2\right]\sin\frac{(2i-1)\pi x}{2L}\sin\frac{(2j-1)\pi x}{2L}dx, \qquad i, j = 1, 2, \dots, n \end{aligned} \tag{b}$$

For $n = 2$, the stiffness and mass matrices are

$$K^{(2)} = \frac{EA}{L}\begin{bmatrix} 2.383701 & -0.662500 \\ -0.662500 & 12.253305 \end{bmatrix} \tag{c}$$

and

$$M^{(2)} = mL\begin{bmatrix} 0.439207 & 0.075991 \\ 0.075991 & 0.493245 \end{bmatrix} \tag{d}$$

respectively. Solving the corresponding 2×2 eigenvalue problem, we obtain the Ritz natural frequencies and modal vectors

$$\begin{aligned} \omega_1^{(2)} &= 2.272911\sqrt{\frac{EA}{mL^2}}, \quad \mathbf{a}_1^{(2)} = (mL)^{-1/2}\begin{bmatrix} 1.471927 \\ 0.160018 \end{bmatrix} \\ \omega_2^{(2)} &= 5.139049\sqrt{\frac{EA}{mL^2}}, \quad \mathbf{a}_2^{(2)} = (mL)^{-1/2}\begin{bmatrix} -0.415467 \\ 1.434331 \end{bmatrix} \end{aligned} \tag{e}$$

so that, ignoring $(mL)^{-1/2}$, the Ritz natural modes are

$$\begin{aligned} U_1^{(2)}(x) &= 1.471927\sin\frac{\pi x}{2L} + 0.160018\sin\frac{3\pi x}{2L} \\ U_2^{(2)}(x) &= -0.415467\sin\frac{\pi x}{2L} + 1.434331\sin\frac{3\pi x}{2L} \end{aligned} \tag{f}$$

The modes, normalized so that $U_r^{(2)}(L) = 1 (r = 1, 2)$, are plotted in Fig. 9.23.

For $n = 3$, the stiffness and mass matrices are

$$K^{(3)} = \frac{EA}{L}\begin{bmatrix} 2.383701 & -0.662500 & 0.895833 \\ -0.662500 & 12.253305 & 1.109375 \\ 0.895833 & 1.109375 & 31.992514 \end{bmatrix} \tag{g}$$

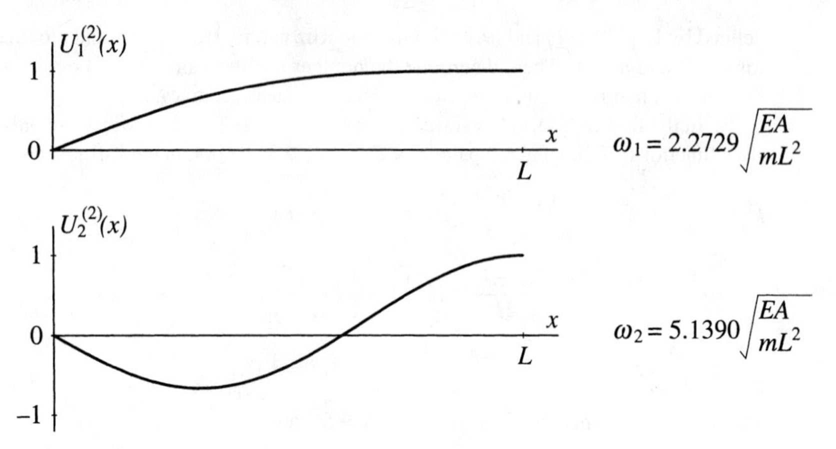

FIGURE 9.23
Two lowest modes of a fixed-spring supported rod computed by the Rayleigh-Ritz method using two admissible functions

and

$$M^{(3)} = mL\begin{bmatrix} 0.439207 & 0.075991 & -0.021953 \\ 0.075991 & 0.493245 & 0.064592 \\ -0.021953 & 0.064592 & 0.497568 \end{bmatrix} \tag{h}$$

yielding the Ritz natural frequencies and modal vectors

$$\begin{aligned} \omega_1^{(3)} &= 2.253516\sqrt{\frac{EA}{mL^2}}, \quad \mathbf{a}_1^{(3)} = (mL)^{-1/2}\begin{bmatrix} 1.468344 \\ 0.162283 \\ -0.054500 \end{bmatrix} \\ \omega_2^{(3)} &= 5.128225\sqrt{\frac{EA}{mL^2}}, \quad a_2^{(3)} = (mL)^{-1/2}\begin{bmatrix} -0.400771 \\ 1.422469 \\ 0.075563 \end{bmatrix} \\ \omega_3^{(3)} &= 8.131483\sqrt{\frac{EA}{mL^2}}, \quad \mathbf{a}_3^{(3)} = (mL)^{-1/2}\begin{bmatrix} 0.184319 \\ -0.273582 \\ 1.430333 \end{bmatrix} \end{aligned} \tag{i}$$

Hence, omitting $(mL)^{-1/2}$, the Ritz natural modes are

$$\begin{aligned} U_1^{(3)} &= 1.468344\sin\frac{\pi x}{2L} + 0.162283\sin\frac{3\pi x}{2L} - 0.054500\sin\frac{5\pi x}{2L} \\ U_2^{(3)} &= -0.400771\sin\frac{\pi x}{2L} + 1.422469\sin\frac{3\pi x}{2L} + 0.075563\sin\frac{5\pi x}{2L} \\ U_3^{(3)} &= 0.184319\sin\frac{\pi x}{2L} - 0.273582\sin\frac{3\pi x}{2L} + 1.430333\sin\frac{5\pi x}{2L} \end{aligned} \tag{j}$$

The modes, normalized so that $U_r^{(3)}(L) = 1 (r = 1, 2, 3)$, are plotted in Fig. 9.24.

The convergence using admissible functions is extremely slow. Using $n = 30$, none of the natural frequencies has reached convergence with six decimal places accuracy. For

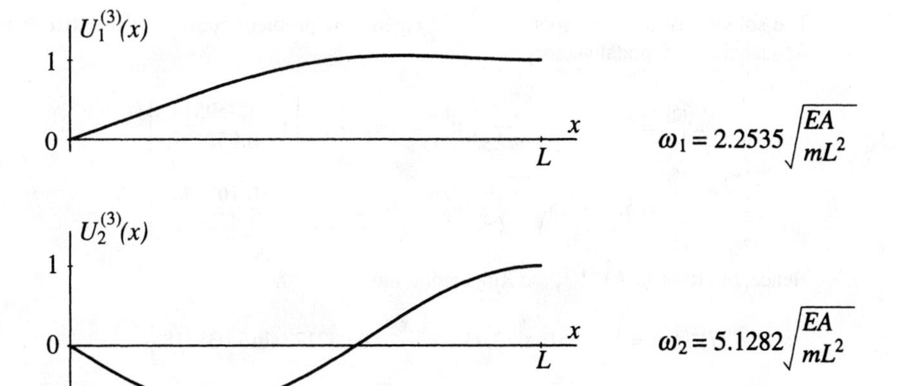

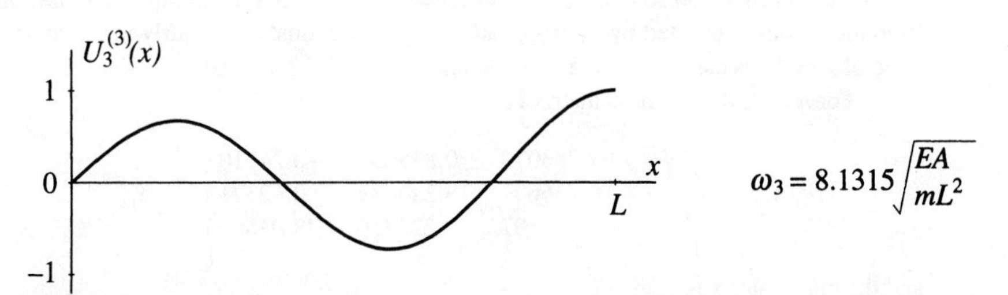

FIGURE 9.24
Three lowest modes of a fixed-spring supported rod computed by the Rayleigh-Ritz method using three admissible functions

later comparison, we list the values as follows:

$$\omega_1^{(30)} = 2.218950\sqrt{EA/mL^2},\ \omega_2^{(30)} = 5.102324\sqrt{EA/mL^2},\ \omega_3^{(30)} = 8.118398\sqrt{EA/mL^2} \tag{k}$$

In the second case, we use the comparison functions

$$\phi_i(x) = \sin\beta_i x,\ i = 1, 2, \ldots, n \tag{l}$$

where, from Eq. (9.132),

$$\beta_1 L = 2.215707,\ \beta_2 L = 5.032218,\ \beta_3 L = 8.057941, \ldots \tag{m}$$

The stiffness and mass matrices are obtained by inserting Eqs. (l) in conjunction with Eqs. (m) into Eqs. (a) and (b), respectively. For $n = 2$, the stiffness matrix is

$$K^{(2)} = \frac{EA}{L}\begin{bmatrix} 2.783074 & 0.836697 \\ 0.836697 & 13.223631 \end{bmatrix} \tag{n}$$

and the mass matrix is

$$M^{(2)} = mL\begin{bmatrix} 0.563196 & 0.085462 \\ 0.085462 & 0.513392 \end{bmatrix} \tag{o}$$

The solution of the corresponding 2×2 eigenvalue problem consists of the Ritz natural frequencies and modal vectors

$$\begin{aligned} \omega_1^{(2)} &= 2.216471\sqrt{\frac{EA}{mL^2}}, \quad \mathbf{a}_1^{(2)} = (mL)^{-1/2}\begin{bmatrix} 1.339519 \\ -0.052177 \end{bmatrix} \\ \omega_2^{(2)} &= 5.106305\sqrt{\frac{EA}{mL^2}}, \quad \mathbf{a}_2^{(2)} = (mL)^{-1/2}\begin{bmatrix} -0.165180 \\ 1.412652 \end{bmatrix} \end{aligned} \tag{p}$$

Hence, omitting $(mL)^{-1/2}$, the Ritz natural modes are

$$\begin{aligned} U_1^{(2)}(x) &= 1.339519 \sin 2.215707\frac{x}{L} - 0.052177 \sin 5.032218\frac{x}{L} \\ U_2^{(2)}(x) &= -0.165180 \sin 2.215707\frac{x}{L} + 1.412652 \sin 5.032218\frac{x}{L} \end{aligned} \tag{q}$$

The modes, normalized so that $U_r^{(2)}(L) = 1 (r = 1, 2,)$, are plotted in Fig. 9.25; they differ from the modes computed by means of admissible functions, particularly in the neighborhood of $x = L$, as can be observed by comparing Figs. 9.23 and 9.25.

For $n = 3$, the stiffness matrix is

$$K^{(3)} = \frac{EA}{L}\begin{bmatrix} 2.783074 & 0.836697 & -0.247107 \\ 0.836697 & 13.223631 & 2.623716 \\ -0.247107 & 2.623716 & 33.078693 \end{bmatrix} \tag{r}$$

and the mass matrix is

$$M^{(3)} = mL\begin{bmatrix} 0.563196 & 0.085462 & -0.020523 \\ 0.085462 & 0.513392 & 0.070501 \\ -0.020523 & 0.070501 & 0.505321 \end{bmatrix} \tag{s}$$

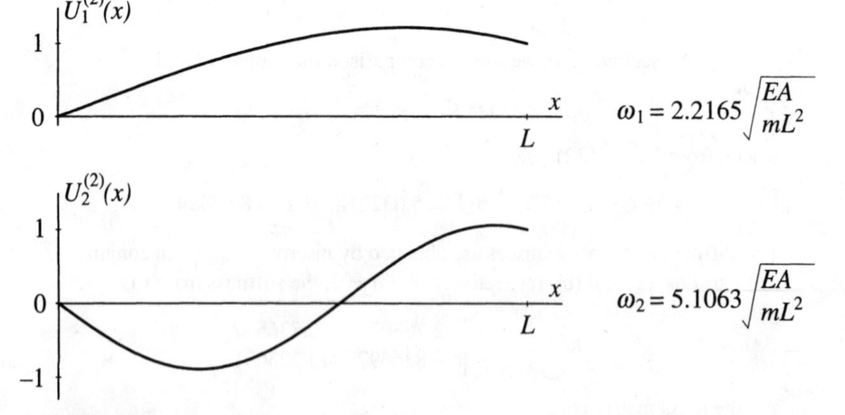

FIGURE 9.25
Two lowest modes of a fixed-spring supported rod computed by the Rayleigh-Ritz method using two comparison functions

yielding the Ritz natural frequencies and modal vectors

$$\omega_1^{(3)} = 2.215728\sqrt{\frac{EA}{mL^2}},\ \mathbf{a}_1^{(3)} = (mL)^{-1/2}\begin{bmatrix} 1.340184 \\ -0.054456 \\ 0.010464 \end{bmatrix}$$

$$\omega_2^{(3)} = 5.100701\sqrt{\frac{EA}{mL^2}},\ \mathbf{a}_2^{(3)} = (mL)^{-1/2}\begin{bmatrix} -0.1617149 \\ 1.419516 \\ -0.053821 \end{bmatrix} \tag{t}$$

$$\omega_3^{(3)} = 8.124264\sqrt{\frac{EA}{mL^2}},\ \mathbf{a}_3^{(3)} = (mL)^{-1/2}\begin{bmatrix} 0.067503 \\ -0.155385 \\ 1.422089 \end{bmatrix}$$

so that, ignoring $(mL)^{-1/2}$, the Ritz natural modes are

$$U_1^{(3)}(x) = 1.340184\sin 2.215707\frac{x}{L} - 0.054456\sin 5.032218\frac{x}{L} + 0.010464\sin 8.057941\frac{x}{L}$$

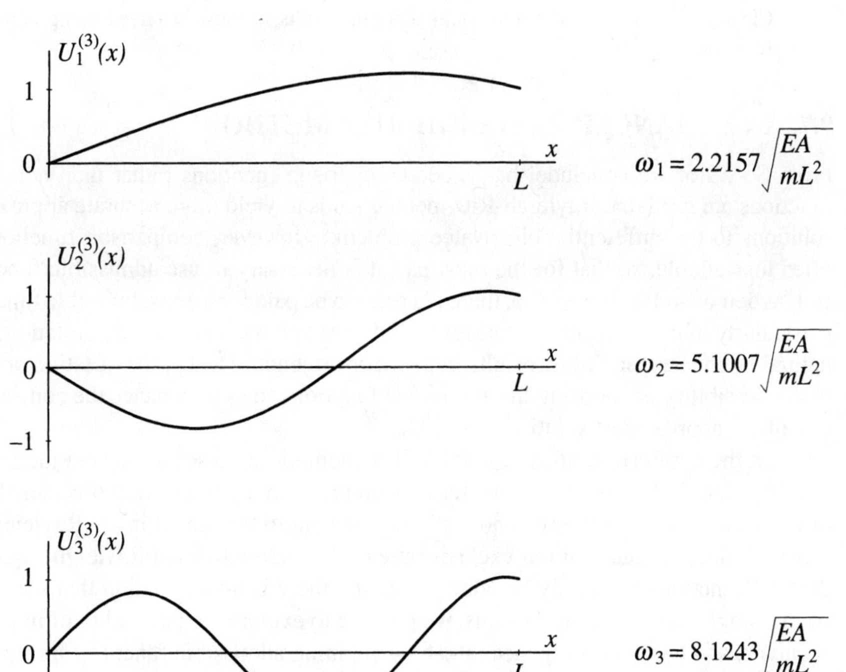

FIGURE 9.26
Three lowest modes of a fixed-spring supported rod computed by the Rayleigh-Ritz method using three comparison functions

$$
\begin{aligned}
U_2^{(3)}(x) = & -0.167149 \sin 2.215707 \frac{x}{L} + 1.419516 \sin 5.032218 \frac{x}{L} \\
& -0.053821 \sin 8.057941 \frac{x}{L} \\
U_3^{(3)}(x) = & \; 0.067503 \sin 2.215707 - 0.155385 \sin 5.032218 \frac{x}{L} \\
& + 1.422089 \sin 8.057941 \frac{x}{L}
\end{aligned} \tag{u}
$$

The modes, normalized so that $U_r^{(3)}(L) = 1$ $(r = 1, 2, 3)$, are plotted in Fig. 9.26, and once again we observe that they are more accurate than the modes obtained using admissible functions (Fig. 9.24), particularly in the neighborhood of $x = L$.

Convergence to six decimal places is reached by the three lowest natural frequencies as follows:

$$
\omega_1^{(14)} = 2.215524\sqrt{EA/mL^2}, \; \omega_2^{(14)} = 5.099525\sqrt{EA/mL^2}, \; \omega_3^{(20)} = 8.116318\sqrt{EA/mL^2} \tag{v}
$$

Clearly, comparison functions yield superior results to those obtained using admissible functions.

9.7 AN ENHANCED RAYLEIGH-RITZ METHOD

From Sec. 9.6, we conclude that, when comparison functions rather than admissible functions are used, the Rayleigh-Ritz method tends to yield more accurate approximate solutions to the differential eigenvalue problem. However, comparison functions are often unavailable, so that for the most part it is necessary to use admissible functions. But, as demonstrated in Sec. 9.6, there is a price to be paid for using admissible functions, particularly in cases involving springs and masses at boundaries, as the violation of the natural boundary conditions results in poor convergence. Hence, the question arises as to the possibility of choosing the admissible functions so as to enhance the convergence rate of the approximate solutions.

In the application of the Rayleigh-Ritz method, there seems to be a tacit understanding that all the trial functions for a given problem are to be from the same family of functions. However, there is no explicit statement to that effect in the Rayleigh-Ritz theory. Indeed, because of our exclusive use of the energy form of Rayleigh's quotient, the trial functions need only be admissible, and there is no stipulation that they all be from a single family. In view of this, we propose to explore the possibility of improving accuracy, and hence convergence rate, by combining admissible functions from several families, each family possessing different dynamic characteristics of the system under consideration. To illustrate the idea, we consider once again the nonuniform rod in axial vibration of Example 9.6; the rod is fixed at $x = 0$ and supported by a spring at $x = L$. We showed in Sec. 9.6 that a solution consisting of admissible functions representing the eigenfunctions of a uniform fixed-free rod is unable to satisfy the natural boundary condition at $x = L$ with a relatively small number of terms. In fact, the number of terms required to achieve satisfactory accuracy must approach infinity, at least in theory. The reason for this is that the spring force at the right end requires that the slope of the displacement curve be different from zero, whereas the slope of the admissible functions is zero at $x = L$, and only the product of infinity and zero can yield a finite number. Hence, the question is whether a linear combination of admissible functions can be found so as to satisfy boundary condition (9.128). To this end, we consider a linear combination of two functions as follows:

$$U(x) = a_1 \sin \frac{\pi x}{2L} + a_2 \sin \frac{\pi x}{L} \tag{9.133}$$

We observe that, although the two trigonometric functions may appear to be from a single family, they belong to two families with different dynamic characteristics as far as the system at hand is concerned. Indeed, $\sin \pi x/2L$ is equal to one and its slope is zero at $x = L$, which is typical of a free end, whereas $\sin \pi x/L$ is equal to zero and its slope is $-\pi/L$ at $x = L$, which is characteristic of a fixed end. Clearly, a boundary cannot be free and fixed at the same time. But, the linear combination (9.133) can be made to satisfy the boundary condition for a spring-supported end by merely adjusting the coefficients a_1 and a_2. Indeed, inserting Eq. (9.133) into Eq. (9.128), we have

$$\begin{aligned} EA(L) \left(a_1 \frac{\pi}{2L} \cos \frac{\pi x}{2L} + a_2 \frac{\pi}{L} \cos \frac{\pi x}{L} \right) \bigg|_{x=L} + k \left(a_1 \sin \frac{\pi x}{2L} + a_2 \sin \frac{\pi x}{L} \right) \bigg|_{x=L} \\ = -EA(L) a_2 \frac{\pi}{L} + k a_1 = 0 \end{aligned} \tag{9.134}$$

yielding

$$a_2 = \frac{kL}{\pi EA(L)} a_1 \tag{9.135}$$

so that Eq. (9.133) becomes

$$U(x) = a_1 \left[\sin \frac{\pi x}{2L} + \frac{kL}{\pi EA(L)} \sin \frac{\pi x}{L} \right] \tag{9.136}$$

Because $U(x)$, as given by Eq. (9.136), satisfies both the geometric boundary condition at $x = 0$ and the natural boundary condition at $x = L$, we conclude that two admissible functions from different families have been combined into a single comparison function. The comparison function $U(x)$ given by Eq. (9.136) will be used as a trial function in conjunction with Rayleigh's energy method in Example 9.7 at the end of this section to obtain an excellent approximation to the lowest natural frequency of the rod of Example 9.6.

Equation (9.134) represents a constraint equation defining a relation between the coefficients a_1 and a_2 ensuring the satisfaction of the natural boundary condition (9.128), thus determining the shape of the comparison function $U(x)$ uniquely. When used in conjunction with Rayleigh's quotient, $U(x)$ produces a unique estimate of the lowest eigenvalue, but there is no minimization process involved. Hence, the question arises whether it would not be better to regard Eq. (9.133) as part of a minimizing sequence, i.e., to regard a_1 and a_2 as independent undetermined coefficients, and let the Rayleigh-Ritz process determine these coefficients, or rather the ratio a_2/a_1. Of course, in this case the natural boundary condition (9.128) would not be satisfied exactly, but only approximately. However, by not imposing the constraint (9.134) on the two admissible functions $\sin \pi x/2L$ and $\sin \pi x/L$, the solution of the differential equation is likely to be approximated with better accuracy. Clearly, for independent a_1 and a_2, $U(x)$ is no longer a comparison function. This is not as important as the fact that the character of the admissible functions $\sin \pi x/2L$ and $\sin \pi x/L$ guarantees that the natural boundary condition can be satisfied exactly by merely adjusting the ratio a_2/a_1. This motivates us to create a new class of functions referred to as *quasi-comparison functions* (Ref. 13) and *defined as linear combinations of admissible functions capable of satisfying all the boundary conditions of the problem.* Note that, for $U(x)$ to qualify as a quasi-comparison function, it must possess a minimum number of terms, exceeding the number of constraint equations at least by one. As an example, in the case of the rod fixed at one end and supported by a spring at the other end, there is one constraint equation, so that the minimum number of terms is two. Clearly, the function $U(x)$ given by Eq. (9.133) in which a_1 and a_2 are independent represents a quasi-comparison function. It should be stressed here that no attempt should be made to satisfy the natural boundary conditions exactly, even if $U(x)$ has a sufficiently large number of terms, because this would reduce the number of degrees of freedom of the discrete model, thus reducing accuracy. Moreover, this would introduce complications in an otherwise uncomplicated process.

The preceding ideas can be generalized by observing that

$$U(x) = a_1 \sin \frac{\pi x}{2L} + a_2 \sin \frac{\pi x}{L} + a_3 \sin \frac{3\pi x}{2L} + \ldots + a_n \sin \frac{n\pi x}{2L}$$

$$= \sum_{i=1}^{n} a_i \sin \frac{i\pi x}{2L} \tag{9.137}$$

represents a quasi-comparison function for a rod fixed at $x = 0$ and supported by a spring at $x = L$. The admissible functions come from two families, the first one consisting of $\sin \pi x/2L$, $\sin 3\pi x/2L$, ..., and representing the eigenfunctions of a uniform fixed-free rod, and the second one consisting of $\sin \pi x/L$, $\sin 2\pi x/L$, ..., and representing the eigenfunctions of a uniform fixed-fixed rod. The quasi-comparison functions obtained by letting $n = 2, 3, \ldots$ are used in Example 9.8 to solve the problem of Example 9.6. The results show that quasi-comparison functions can yield faster convergence than comparison functions.

One word of caution is in order. Each of the two sets of admissible functions, $\sin \pi x/2L$, $\sin 3\pi/2L$, ... and $\sin \pi x/L$, $\sin 2\pi x/L$, ..., is complete, as each represents the eigenfunctions of a given system. As a result, a given function in one set can be expanded in terms of the functions in the other set. The implication is that, as the number of terms n increases, the two sets tend to become dependent, thus violating the requirement that the coefficients a_i $(i = 1, 2, \ldots n)$ be independent. When this happens, the mass and stiffness matrices tend to become singular and the eigensolutions meaningless. But, because convergence to the lower modes tends to be so fast, in general the singularity problem does not have the chance to materialize. The problem can arise, however, if the interest lies in a large number of modes.

Example 9.7. Use the comparison function given by Eq. (9.136) in conjunction with Rayleigh's energy method to estimate the lowest natural frequency of the rod of Example 9.6.

Ignoring the coefficient a_1, which is irrelevant when using Rayleigh's energy method, the comparison function has the expression

$$U(x) = \sin\frac{\pi x}{2L} + \frac{kL}{\pi EA(L)}\sin\frac{\pi x}{L} \tag{a}$$

But, from Example 9.6, $k = EA/L$. Moreover, $EA(L) = 0.6\,EA$. Hence,

$$U(x) = \sin\frac{\pi x}{2L} + \frac{1}{0.6\pi}\sin\frac{\pi x}{L} = \sin\frac{\pi x}{2L} + 0.530516\sin\frac{\pi x}{L} \tag{b}$$

For convenience, we write Rayleigh's quotient in the form

$$R(U(x)) = \omega^2 = \frac{V_{\text{max}}}{T_{\text{ref}}} \tag{c}$$

where, using Eq. (b),

$$\begin{aligned}
V_{\text{max}} &= \frac{1}{2}\int_0^L EA(x)\left[\frac{dU(x)}{dx}\right]^2 dx + \frac{1}{2}kU^2(L) \\
&= \frac{1}{2}\left\{\frac{6EA}{5}\int_0^L\left[1-\frac{1}{2}\left(\frac{x}{L}\right)^2\right]\left(\frac{\pi}{2L}\cos\frac{\pi x}{2L} + 0.530516\frac{\pi}{L}\cos\frac{\pi x}{L}\right)^2 dx + k\right\} \\
&= \frac{1}{2}\left\{\frac{6EA}{5}\int_0^L\left[1-\frac{1}{2}\left(\frac{x}{L}\right)^2\right]\left[\left(\frac{\pi}{2L}\right)^2\cos^2\frac{\pi x}{2L}\right.\right. \\
&\qquad \left.\left. + 2\times 0.530516\frac{\pi}{2L}\frac{\pi}{L}\cos\frac{\pi x}{2L}\cos\frac{\pi x}{L} + 0.530516^2\left(\frac{\pi}{L}\right)^2\cos^2\frac{\pi x}{L}\right]dx + \frac{EA}{L}\right\} \\
&= \frac{1}{2}(2.383701 + 2\times 0.530516\times 1.363968 + 0.530516^2\times 4.784802)\frac{EA}{L} \\
&= \frac{1}{2}\times 5.177584\frac{EA}{L}
\end{aligned} \tag{d}$$

and

$$\begin{aligned}
T_{\text{ref}} &= \frac{1}{2}\int_0^L m(x)U^2(x)dx = \frac{1}{2}\frac{6m}{5}\int_0^L\left[1-\frac{1}{2}\left(\frac{x}{L}\right)^2\right]\left(\sin\frac{\pi x}{2L} + 0.530516\sin\frac{\pi x}{L}\right)^2 dx \\
&= \frac{1}{2}\frac{6m}{5}\int_0^L\left[1-\frac{1}{2}\left(\frac{x}{L}\right)^2\right]\left(\sin^2\frac{\pi x}{2L} + 2\times 0.530516\sin\frac{\pi x}{2L}\sin\frac{\pi x}{L}\right. \\
&\qquad \left. + 0.530516^2\sin^2\frac{\pi x}{L}\right)dx \\
&= \frac{1}{2}(0.439207 + 2\times 0.530516\times 0.415189 + 0.530516^2\times 0.515198) \\
&= \frac{1}{2}\times 1.024737mL
\end{aligned} \tag{e}$$

Inserting Eqs. (d) and (e) into Eq. (c) and taking the square root, we obtain the estimate of the lowest frequency

$$\omega = \sqrt{\frac{5.177584}{1.024737}\frac{EA}{mL^2}} = 2.247798\sqrt{\frac{EA}{mL^2}} \tag{f}$$

Comparing $\omega = 2.247798\sqrt{EA/mL^2}$ with $\omega_1^{(14)} = 2.215524\sqrt{EA/mL^2}$ computed in Example 9.6, we conclude that the estimate is very good. The error is

$$\frac{\omega-\omega_1}{\omega} = \frac{2.247798 - 2.215524}{2.247798} = 0.014358 \cong 1.4\% \tag{g}$$

In fact, for a quick estimate, the result is excellent. Note that using the static displacement of the rod loaded with an axial force distribution proportional to the mass density as a trial function, which most likely would yield a more accurate estimate, would require in the case at hand considerably more effort than generating Eq. (a).

Example 9.8. Solve the problem of Example 9.6 using the quasi-comparison functions

$$U^{(n)}(x) = \sum_{i=1}^{n} a_i \phi_i(x) = \sum_{i=1}^{n} a_i \sin i\pi x/2L, \; n = 2, 3, \ldots \tag{a}$$

compare results with those obtained in Example 9.6 by means of comparison functions and draw conclusions.

Inserting Eqs. (a) into Eqs. (a) and (b) of Example 9.6, we can write correspondingly the stiffness coefficients

$$\begin{aligned} k_{ij}^{(n)} &= \int_0^L EA(x)\frac{d\phi_i(x)}{dx}\frac{d\phi_j(x)}{dx}dx + k\phi_i(L)\phi_j(L) \\ &= \frac{6EA}{5}\frac{i\pi}{2L}\frac{j\pi}{2L}\int_0^L \left[1 - \frac{1}{2}\left(\frac{x}{L}\right)^2\right]\cos\frac{i\pi x}{2L}\cos\frac{j\pi x}{2L}dx + \frac{EA}{L}\sin\frac{i\pi}{2}\sin\frac{j\pi}{2}, \\ &\qquad i, j = 1, 2, \ldots, n \end{aligned} \tag{b}$$

and the mass coefficients

$$m_{ij}^{(n)} = \int_0^L m(x)\phi_i(x)\phi_j(x)dx = \frac{6m}{5}\int_0^L \left[1 - \frac{1}{2}\left(\frac{x}{L}\right)^2\right]\sin\frac{i\pi x}{2L}\sin\frac{j\pi x}{2L}dx, \quad i, j = 1, 2, \ldots, n \tag{c}$$

For $n = 2$, the stiffness matrix is

$$K^{(2)} = \frac{EA}{L}\begin{bmatrix} 2.383701 & 1.363968 \\ 1.363968 & 4.784802 \end{bmatrix} \tag{d}$$

and the mass matrix is

$$M^{(2)} = mL\begin{bmatrix} 0.439207 & 0.415189 \\ 0.415189 & 0.515198 \end{bmatrix} \tag{e}$$

Solving the corresponding 2×2 eigenvalue problem, we obtain the Ritz natural frequencies and modal vectors

$$\begin{aligned} \omega_1^{(2)} &= 2.223595\sqrt{\frac{EA}{mL^2}}, \quad \mathbf{a}_1^{(2)} = (mL)^{-1/2}\begin{bmatrix} 1.159578 \\ 0.357015 \end{bmatrix} \\ \omega_2^{(2)} &= 5.984845\sqrt{\frac{EA}{mL^2}}, \quad \mathbf{a}_2^{(2)} = (mL)^{-1/2}\begin{bmatrix} 2.866064 \\ -2.832235 \end{bmatrix} \end{aligned} \tag{f}$$

The modal vectors can be used to obtain the Ritz natural modes

$$U_1^{(2)}(x) = 1.159578 \sin \frac{\pi x}{2L} + 0.357015 \sin \frac{\pi x}{L}$$
$$U_2^{(2)}(x) = 2.866064 \sin \frac{\pi x}{2L} - 2.832235 \sin \frac{\pi x}{L} \tag{g}$$

in which $(mL)^{-1/2}$ was ignored. The natural modes, normalized so that $U_r^{(2)}(L) = 1$ ($r = 1, 2$) are plotted in Fig. 9.27. It should be noted that $U_2^{(2)}(x)$ is grossly in error due to the small number of admissible functions.

For $n = 3$, the stiffness and mass matrices are

$$K^{(3)} = \frac{EA}{L}\begin{bmatrix} 2.383701 & 1.363968 & -0.662500 \\ 1.363968 & 4.784802 & 5.703086 \\ -0.662500 & 5.703086 & 12.253305 \end{bmatrix} \tag{h}$$

and

$$M^{(3)} = mL\begin{bmatrix} 0.439207 & 0.415189 & 0.075991 \\ 0.415189 & 0.515198 & 0.306358 \\ 0.075991 & 0.306358 & 0.493245 \end{bmatrix} \tag{i}$$

respectively. The corresponding 3×3 eigenvalue problem yields the Ritz natural frequencies and modal vectors

$$\omega_1^{(3)} = 2.216154\sqrt{\frac{EA}{mL^2}}, \quad \mathbf{a}_1^{(3)} = (mL)^{-1/2}\begin{bmatrix} 1.028923 \\ 0.519181 \\ -0.113326 \end{bmatrix}$$
$$\omega_2^{(3)} = 5.100072\sqrt{\frac{EA}{mL^2}}, \quad \mathbf{a}_2^{(3)} = (mL)^{-1/2}\begin{bmatrix} 0.217568 \\ -0.705970 \\ 1.778731 \end{bmatrix} \tag{j}$$
$$\omega_3^{(3)} = 11.092640\sqrt{\frac{EA}{mL^2}}, \quad \mathbf{a}_3^{(3)} = (mL)^{-1/2}\begin{bmatrix} -9.597960 \\ 11.040485 \\ -5.308067 \end{bmatrix}$$

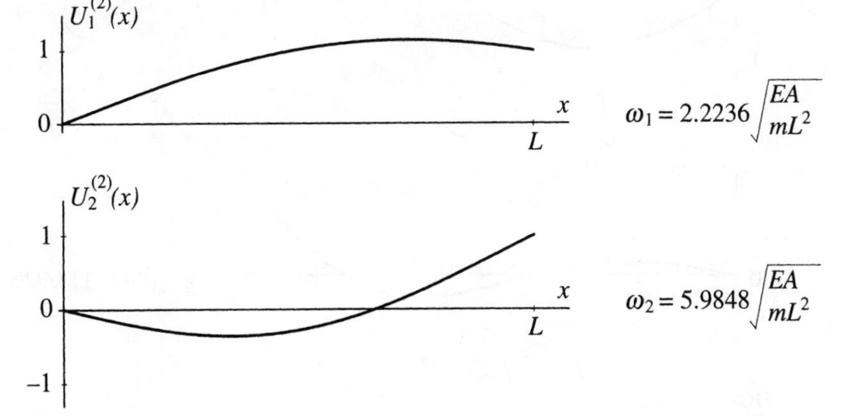

FIGURE 9.27
Two lowest modes of a fixed-spring supported rod computed by the enhanced Rayleigh-Ritz method using two quasi-comparison functions

Using the modal vectors, we obtain the Ritz natural modes

$$\begin{aligned} U_1^{(3)} &= 1.028923 \sin\frac{\pi x}{2L} + 0.519181 \sin\frac{\pi x}{L} - 0.113326 \sin\frac{3\pi x}{2L} \\ U_2^{(2)} &= 0.217568 \sin\frac{\pi x}{2L} - 0.705970 \sin\frac{\pi x}{L} + 1.778731 \sin\frac{3\pi x}{2L} \\ U_3^{(3)} &= -9.537960 \sin\frac{\pi x}{2L} + 11.040485 \sin\frac{\pi x}{L} - 5.308067 \sin\frac{3\pi x}{L} \end{aligned} \tag{k}$$

which are plotted in Fig. 9.28, following normalization so that $U_r^{(3)}(L) = 1$ $(r = 1, 2, 3)$. Note that $U_2^{(3)}(x)$ is vastly improved compared to $U_2^{(2)}(x)$, but $U_3^{(3)}(x)$ is grossly in error due to the inability of quasi-comparison functions to yield accurate highest eigenvalue and eigenfunction for discretized models of any order.

Convergence to six decimal places is very rapid, as can be concluded from Table 9.1 showing the three lowest Ritz natural frequencies.

A comparison with results obtained in Examples 9.6 and 9.7 is quite revealing. In the first place, we observe that the Rayleigh-Ritz method in conjunction with a quasi-comparison function consisting of two admissible functions yields a more accurate lowest natural frequency than by sacrificing one degree of freedom to generate one comparison

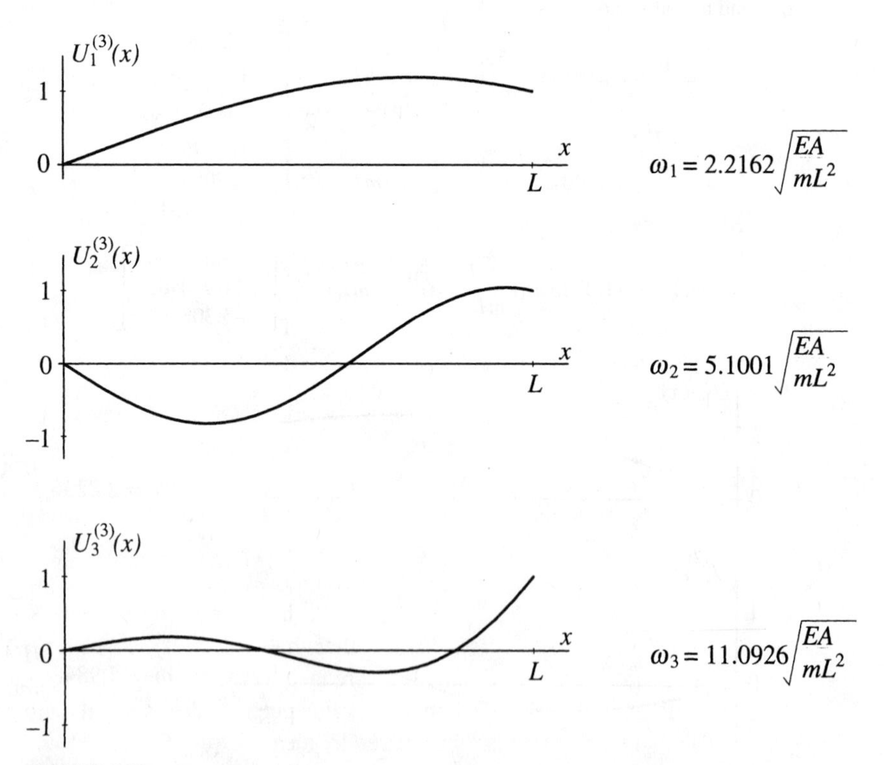

FIGURE 9.28
Three lowest modes of a fixed-spring supported rod computed by the enhanced Rayleigh-Ritz method using three quasi-comparison functions

Table 9.1 Normalized Ritz Natural Frequencies

n	$\omega_1^{(n)}\sqrt{mL^2/EA}$	$\omega_2^{(n)}\sqrt{mL^2/EA}$	$\omega_3^{(n)}\sqrt{mL^2/EA}$
1	2.329652	—	—
2	2.223595	5.984845	—
3	2.216154	5.100072	11.092640
4	2.215568	5.099571	8.153645
5	2.215527	5.099528	8.116320
6	2.215524	5.099525	8.116318

function, $\omega_1^{(2)} = 2.223595\sqrt{EA/mL^2}$ versus $\omega = 2.249798\sqrt{EAmL^2}$, the latter obtained in Example 9.7. For two, three and four degrees of freedom, the natural frequencies computed by means of comparison functions are slightly more accurate than those computed by means of quasi-comparison functions. Note that, although the highest natural frequency and mode computed by means of quasi-comparison functions are grossly in error, this is not important, as higher Ritz frequencies are typically unreliable, no matter what trial functions are used. What is remarkable about the quasi-comparison functions is the convergence rate. Indeed, from Table 9.1 we see that all three lowest natural frequencies achieve convergence with only six terms in the approximating series. By contrast, it took 14 comparison functions for the two lowest natural frequencies and 20 comparison functions for the third lowest to converge, as can be concluded from Example 9.6. To place matters in proper perspective, we note that the three lowest natural frequencies computed in Example 9.5 with 30 ordinary admissible functions were far from convergence, where "ordinary" is in the sense that linear combinations of these admissible functions did not represent quasi-comparison functions.

9.8 THE ASSUMED-MODES METHOD. SYSTEM RESPONSE

The assumed-modes method is a procedure for the discretization of distributed-parameter systems closely related to the Rayleigh-Ritz method. In fact, it is often referred to as the Rayleigh-Ritz method, which can be justified in view of the fact that the discrete model derived by the assumed-modes method is identical to that derived by the Rayleigh-Ritz method. The advantage of the assumed-modes method over the Rayleigh-Ritz method is that it is very easy to understand. Of course, the main reason why the assumed-modes method appears so easy to grasp is the earlier exposure to the Rayleigh-Ritz method. In fact, virtually all questions that could be raised in connection with the assumed-modes method find their answer in the Rayleigh-Ritz theory. Whereas the Rayleigh-Ritz method is essentially concerned with the discretization of the differential eigenvalue problem, the assumed-modes method begins with the discretization of the boundary-value problem, albeit in an implicit manner. This provides us with the opportunity to formulate the discretized system forced response problem, rather than merely the free response, a formulation that is equally valid for systems discretized by the Rayleigh-Ritz method.

As demonstrated in Sec. 8.2, the boundary-value problem describing the vibration of a distributed-parameter system can be derived by means of the extended Hamilton's principle, which requires the kinetic energy T, potential energy V and virtual work of the nonconservative forces $\overline{\delta W}_{nc}$. Because we are concerned with distributed systems for which no closed-form solutions are expected, we are not really interested in deriving boundary-value problems. Indeed, our interest lies in the equations of motion for discrete models of distributed systems. The assumed-modes method aims at deriving such equations of motion by first discretizing the kinetic energy, potential energy and virtual work and then making use of Lagrange's equations.

We consider a distributed system and approximate the displacement $y(x,t)$ by the finite series

$$y(x,t) = \sum_{i=1}^{n} \phi_i(x)q_i(t) \tag{9.138}$$

where $\phi_i(x)$ are known trial functions and $q_i(t)$ are unknown generalized coordinates $(i = 1, 2, \ldots, n)$. Assuming that there are no lumped masses at the boundaries, we can discretize the kinetic energy as follows:

$$\begin{aligned} T(t) &= \frac{1}{2}\int_0^L m(x)\dot{y}^2(x,t)dx = \frac{1}{2}\int_0^L m(x)\sum_{i=1}^{n}\phi_i(x)\dot{q}_i(t)\sum_{j=1}^{n}\phi_j(x)\dot{q}_j(t)dx \\ &= \frac{1}{2}\sum_{i=1}^{n}\sum_{j=1}^{n}\dot{q}_i(t)\dot{q}_j(t)\int_0^L m(x)\phi_i(x)\phi_j(x)dx = \frac{1}{2}\sum_{i=1}^{n}\sum_{j=1}^{n} m_{ij}\dot{q}_i(t)\dot{q}_j(t) \end{aligned} \tag{9.139}$$

where

$$m_{ij} = m_{ji} = \int_0^L m(x)\phi_i(x)\phi_j(x)dx,\ i,j = 1,2,\ldots,n \tag{9.140}$$

are symmetric mass coefficients. As can be concluded from Ch. 8 and this chapter, the potential energy expression varies from system to system. As an example, for a nonuniform beam in bending fixed at $x = 0$ and supported by a spring of stiffness k at $x = L$, the potential energy has the form

$$V(t) = \frac{1}{2}\int_0^L EI(x)\left[\frac{\partial^2 y(x,t)}{\partial x^2}\right]^2 dx + \frac{1}{2}ky^2(L,t) \tag{9.141}$$

so that inserting Eq. (9.138) into Eq. (9.141), we can write the discretized potential energy

$$\begin{aligned} V(t) &= \frac{1}{2}\int_0^L EI(x)\sum_{i=1}^{n}\frac{d^2\phi_i(x)}{dx^2}q_i(t)\sum_{j=1}^{n}\frac{d^2\phi_j(x)}{dx^2}q_j(t)dx \\ &\quad + \frac{1}{2}k\sum_{i=1}^{n}\phi_i(L)q_i(t)\sum_{j=1}^{n}\phi_j(L)q_j(t) \\ &= \frac{1}{2}\sum_{i=1}^{n}\sum_{j=1}^{n} q_i(t)q_j(t)\left[\int_0^L EI(x)\frac{d^2\phi_i(x)}{dx^2}\frac{d^2\phi_j(x)}{dx^2}dx + k\phi_i(L)\phi_j(L)\right] \\ &= \frac{1}{2}\sum_{i=1}^{n}\sum_{j=1}^{n} k_{ij}q_i(t)q_j(t) \end{aligned} \tag{9.142}$$

in which

$$k_{ij} = k_{ji} = \int_0^L EI(x)\frac{d^2\phi_i(x)}{dx^2}\frac{d^2\phi_j(x)}{dx^2}dx + k\phi_i(L)\phi_j(L),\ i,j = 1,2,\ldots,n \tag{9.143}$$

are symmetric stiffness coefficients. Finally, letting $f(x,t)$ be a distributed nonconservative force, we discretize the virtual work as follows:

$$\overline{\delta W}_{nc}(t) = \int_0^L f(x,t)\delta y(x,t)dx = \int_0^L f(x,t)\sum_{i=1}^{n}\phi_i(x)\delta q_i(t)dx$$
$$= \sum_{i=1}^{n} Q_{inc}(t)\delta q_i(t) \tag{9.144}$$

where

$$Q_{inc}(t) = \int_0^L f(x,t)\phi_i(x)dx,\ i = 1,2,\ldots,n \tag{9.145}$$

are generalized nonconservative forces, some abstract quantities depending on the trial functions $\phi_i(x)$.

From Sec. 6.5, Lagrange's equations have the form

$$\frac{d}{dt}\left(\frac{\partial T}{\partial \dot{q}_k}\right) - \frac{\partial T}{\partial q_k} + \frac{\partial V}{\partial q_k} = Q_k,\ k = 1,2,\ldots,n \tag{9.146}$$

in which the subscript "nc" was omitted from the generalized nonconservative forces. But, by analogy with Eqs. (9.115) and (9.114), we can write

$$\frac{\partial T}{\partial \dot{q}_k} = \sum_{j=1}^{n} m_{kj}\dot{q}_j,\quad \frac{\partial V}{\partial q_k} = \sum_{j=1}^{n} k_{kj}q_j,\ k = 1,2,\ldots,n \tag{9.147}$$

respectively. In addition, the kinetic energy does not depend on q_k, so that $\partial T/\partial q_k = 0\ (k = 1,2,\ldots,n)$. Hence, replacing the subscript k by i, we obtain the equations of motion

$$\sum_{j=1}^{n} m_{ij}\ddot{q}_j(t) + \sum_{j=1}^{n} k_{ij}q_j(t) = Q_i(t),\ i = 1,2,\ldots,n \tag{9.148}$$

which resemble entirely the equations of motion for an n-degree-of-freedom undamped system. It follows that the assumed-modes method is simply a series discretization technique yielding the discrete equations of motion directly from the kinetic energy, potential energy and virtual work of the nonconservative forces.

The question remains as to the nature of the trial functions. Of course, they are known functions, selected in advance by the analyst. From the structure of Eqs. (9.140) and (9.143), we observe that the mass and stiffness coefficients possess the same expressions as those obtained in Sec. 9.6 by means of the Rayleigh-Ritz method. Hence, the trial functions $\phi_i(x)$ can be chosen from the class of admissible functions, comparison functions, or quasi-comparison functions, as discussed in Secs. 9.6 and 9.7. It follows that the algebraic eigenvalue problem corresponding to Eqs. (9.148) coincides exactly with that obtained by the Rayleigh-Ritz method, so that the assumed-modes method can be regarded as merely another version of the Rayleigh-Ritz method. The significance of this is that the whole Rayleigh-Ritz theory, and in particular how the choice of trial functions affects accuracy and convergence, applies equally well to the assumed-modes method. It is perhaps appropriate to mention here that the term "assumed modes" represents a misnomer, as the trial functions $\phi_i(x)$ are no modes at all for the system under consideration. The term simply implies that the trial functions can be chosen at times as the modes of a related simpler system, which is done as a matter of course in the Rayleigh-Ritz method.

Finally, we must point out that the developments in this section represent precisely the steps to be followed in deriving the system response by the Rayleigh-Ritz method.

Example 9.9. Use the assumed-modes method in conjunction with a three-term series to obtain the response of the tapered rod of Example 9.6 to the uniformly distributed force

$$f(x,t) = f_0 \mathcal{u}(t) \tag{a}$$

where f_0 is a constant and $\mathcal{u}(t)$ is the unit step function. Use as trial functions the comparison functions

$$\phi_i(x) = \sin \beta_i x \tag{b}$$

in which, from Example 9.6,

$$\beta_1 L = 2.215707,\ \beta_2 L = 5.032218,\ \beta_3 L = 8.057941 \tag{c}$$

The equations of motion of the discretized model are given by Eqs. (9.148), which can be expressed in the matrix form

$$M^{(3)}\ddot{\mathbf{q}}(t) + K^{(3)}\mathbf{q}(t) = \mathbf{Q}(t) \tag{d}$$

where $\mathbf{q}(t)$ and $\mathbf{Q}(t)$ are the three-dimensional generalized displacement vector and nonconservative force vector, respectively. Moreover, from Eqs. (r) and (s) of Example 9.6,

$$K^{(3)} = \frac{EA}{L}\begin{bmatrix} 2.783074 & 0.836697 & -0.247107 \\ 0.836697 & 13.223631 & 2.623716 \\ -0.247107 & 2.623716 & 33.078693 \end{bmatrix} \tag{e}$$

and

$$M^{(3)} = mL\begin{bmatrix} 0.563196 & 0.085462 & -0.020523 \\ 0.085462 & 0.513392 & 0.070501 \\ -0.020523 & 0.070501 & 0.505321 \end{bmatrix} \tag{f}$$

are the stiffness matrix and mass matrix, respectively. Inserting Eqs. (a) into Eqs. (9.145), we can write the components of the three-dimensional generalized nonconservative force vector $\mathbf{Q}(t)$ in the general form

$$Q_i(t) = \int_0^L f(x,t)\phi_i(x)dx = f_0\mathcal{u}(t)\int_0^L \sin\beta_i x dx = \frac{f_0\mathcal{u}(t)(1-\cos\beta_i L)}{\beta_i},\quad i = 1,2,3 \tag{g}$$

which, upon using Eqs. (c), can be expressed in the explicit form

$$
\begin{aligned}
Q_1(t) &= \frac{1-\cos\beta_1 L}{\beta_1 L} f_0 \mathscr{u}(t) = 0.722626 f_0 \mathscr{u}(t) \\
Q_2(t) &= \frac{1-\cos\beta_2 L}{\beta_2 L} f_0 \mathscr{u}(t) = 0.136241 f_0 \mathscr{u}(t) \\
Q_3(t) &= \frac{1-\cos\beta_3 L}{\beta_3 L} f_0 \mathscr{u}(t) = 0.149238 f_0 \mathscr{u}(t)
\end{aligned}
\tag{h}
$$

To obtain the solution of Eq. (d), we use the linear transformation

$$\mathbf{q}(t) = U\boldsymbol{\eta}(t) \tag{i}$$

where, from Eqs. (t) of Example 9.6,

$$U = [\mathbf{a}_1^{(3)}\ \mathbf{a}_2^{(3)}\ \mathbf{a}_3^{(3)}] = (mL)^{-1/2}\begin{bmatrix} 1.340184 & -0.167149 & 0.067503 \\ -0.054456 & 1.419516 & -0.155385 \\ 0.010464 & -0.053821 & 1.422089 \end{bmatrix} \tag{j}$$

is the orthonormal modal matrix. Introducing Eq. (i) in Eq. (d) and premultiplying by U^T, we obtain the independent modal equations

$$\ddot{\boldsymbol{\eta}}(t) + \Lambda\boldsymbol{\eta}(t) = \mathbf{N}(t) \tag{k}$$

in which, using Eqs. (t) of Example 9.6,

$$\Lambda = \text{diag}[(\omega_1^{(3)})^2\ (\omega_2^{(3)})^2\ (\omega_3^{(2)})^2] = \text{diag}[4.909451\ 26.017151\ 66.003666]\frac{EA}{mL^2} \tag{l}$$

is a diagonal matrix of the Ritz natural frequencies $\omega_1^{(3)}$, $\omega_2^{(3)}$ and $\omega_3^{(3)}$ squared. Moreover,

$$
\begin{aligned}
\mathbf{N}(t) &= U^T\mathbf{Q}(t) \\
&= (mL)^{-1/2} f_0 L \mathscr{u}(t) \begin{bmatrix} 1.340184 & -0.167149 & 0.067503 \\ -0.054456 & 1.419516 & -0.155385 \\ 0.010464 & -0.053821 & 1.422089 \end{bmatrix}^T \begin{bmatrix} 0.722626 \\ 0.136241 \\ 0.149238 \end{bmatrix} \\
&= \frac{f_0 L^{1/2} \mathscr{u}(t)}{m^{1/2}} \begin{bmatrix} 0.955753 \\ 0.130856 \\ 0.212459 \end{bmatrix}
\end{aligned}
\tag{m}
$$

is a vector of modal forces.

The solution of Eq. (k) can be written by components in the form of convolution integrals, as follows:

$$
\begin{aligned}
\eta_1(t) &= \frac{1}{\omega_1}\int_0^t N_1(t-\tau)\sin\omega_1\tau\, d\tau = \frac{0.955753 f_0 L^{1/2}}{m^{1/2}\omega_1}\int_0^t \mathscr{u}(t-\tau)\sin\omega_1\tau\, d\tau \\
&= \frac{0.955753 f_0 L^{1/2}}{m^{1/2}\omega_1^2}(1-\cos\omega_1 t)
\end{aligned}
$$

$$= \frac{0.194676 f_0 m^{1/2} L^{5/2}}{EA}\left(1-\cos 2.215728\sqrt{\frac{EA}{mL^2}}t\right)$$

$$\eta_2(t) = \frac{1}{\omega_2}\int_0^t N_2(t-\tau)\sin\omega_2\tau d\tau = \frac{0.130856 f_0 L^{1/2}}{m^{1/2}\omega_2}\int_0^t u(t-\tau)\sin\omega_2\tau d\tau$$

$$= \frac{0.130856 f_0 L^{1/2}}{m^{1/2}\omega_2^2}(1-\cos\omega_2 t) \tag{n}$$

$$= \frac{0.005030 f_0 m^{1/2} L^{5/2}}{EA}\left(1-\cos 5.100701\sqrt{\frac{EA}{mL^2}}t\right)$$

$$\eta_3(t) = \frac{1}{\omega_3}\int_0^t N_3(t-\tau)\sin\omega_3\tau\, d\tau = \frac{0.212459 f_0 L^{1/2}}{m^{1/2}\omega_3}\int_0^t u(t-\tau)\sin\omega_3\tau\, d\tau$$

$$= \frac{0.212459 f_0 L^{1/2}}{m^{1/2}\omega_3^2}(1-\cos\omega_3 t)$$

$$= \frac{0.003219 f_0 m^{1/2} L^{5/2}}{EA}\left(1-\cos 8.124264\sqrt{\frac{EA}{mL^2}}t\right)$$

Finally, inserting Eqs. (j) and (n) into Eq. (i), and the result into Eq. (9.138) with $y(x,t)$ replaced by $u(x,t)$, we obtain the response of the rod in the form

$$u(x,t) = \sum_{i=1}^{3}\phi_i(x)q_i(t) = \sum_{i=1}^{3}\sin\beta_i x\sum_{j=1}^{3}U_{ij}\eta_j(t)$$

$$\begin{aligned}
= \frac{f_0L^2}{EA}\Bigg\{ & \sin 2.215707x\Big[0.260902\left(1-\cos 2.215728\sqrt{EA/mL^2}t\right) \\
& -0.000841\left(1-\cos 5.100701\sqrt{EA/mL^2}t\right) \\
& +0.000217\left(1-\cos 8.124264\sqrt{EA/mL^2}t\right)\Big] \\
& +\sin 5.032218x\Big[-0.010601\left(1-\cos 2.215728\sqrt{EA/mL^2}t\right) \\
& +0.007140\left(1-\cos 5.100701\sqrt{EA/mL^2}t\right) \\
& -0.000500\left(1-\cos 8.124264\sqrt{EA/mL^2}t\right)\Big] \\
& +\sin 8.057941x\Big[0.002037\left(1-\cos 2.215728\sqrt{EA/mL^2}t\right) \\
& -0.000271\left(1-\cos 5.100701\sqrt{EA/mL^2}t\right) \\
& +0.004578\left(1-\cos 8.124264\sqrt{EA/mL^2}t\right)\Big]\Bigg\}
\end{aligned} \tag{o}$$

We observe that the largest contribution to the response is from the first mode, with the contributions from the second and third modes being two orders of magnitude smaller than that from the first. This can be attributed to the fact that the external force is uniformly distributed, which tends to excite the lowest mode far more than the remaining modes.

9.9 THE GALERKIN METHOD

Galerkin's method belongs to a family of techniques for the approximate solution of differential eigenvalue problems known as weighted residual methods. As the Rayleigh-Ritz method, Galerkin's method also represents a series discretization technique whereby the approximate solution is assumed in the form

$$Y^{(n)}(x) = \sum_{j=1}^{n} a_j \phi_j(x) \tag{9.149}$$

where the trial functions $\phi_1(x), \phi_2(x), \ldots, \phi_n(x)$ are known independent comparison functions from a complete set and $a_1, a_2, \ldots a_n$ are undetermined coefficients. Solution (9.149) does not satisfy exactly the differential equation defining the eigenvalue problem, so that some error is incurred, where the error is denoted by $\mathcal{R}(Y^{(n)}(x), x)$ and referred to as *residual*. Because $Y^{(n)}(x)$ is a linear combination of comparison functions, the boundary conditions are satisfied exactly. To determine the coefficients, $a_1, a_2, \ldots, a_n$, we multiply the residual $\mathcal{R}(Y^{(n)}(x), x)$ by $\phi_1(x), \phi_2(x), \ldots, \phi_n(x)$, in sequence, integrate the result over the domain of the system and set the result equal to zero, or

$$\int_0^L \phi_i(x)\mathcal{R}(Y^{(n)}(x), x)dx = 0, \; i = 1, 2, \ldots, n \tag{9.150}$$

so that the comparison functions $\phi_1(x), \phi_2(x), \ldots, \phi_n(x)$ also play the role of test functions, or weighting functions (see Sec. 8.8). Following integration, Eqs. (9.150) become a set of algebraic equations in the unknowns $a_1, a_2, \ldots a_n$ with $\lambda = \lambda^{(n)}$ acting as a parameter; they are called *Galerkin's equations* and they represent an algebraic eigenvalue problem. The rest of the process is the same as for the Rayleigh-Ritz method.

There is a basic difference between the Raleigh-Ritz method and the Galerkin method. The Rayleigh-Ritz method represents a variational approach, whereby the eigenvalue problem is derived by rendering Raleigh's quotient stationary, and is restricted to conservative systems, whereas in the Galerkin method the eigenvalue problem is derived by setting the integrated weighted errors equal to zero. As a result, *Galerkin's method is more general in scope and can be used for both conservative and nonconservative systems.*

Equations (9.150) imply mathematically that the residual $\mathcal{R}$ is orthogonal to every trial function $\phi_i(x)$ $(i = 1, 2, \ldots, n)$. As n increases without bounds, $\mathcal{R}$ can remain orthogonal to an infinite set of independent functions only if it tends itself to zero, or

$$\lim_{n \to \infty} \mathcal{R}(Y^{(n)}(x), x) = 0, \; 0 < x < L \tag{9.151}$$

But, if the error tends to zero at every point, we must have

$$\lim_{n \to \infty} Y^{(n)}(x) = Y(x) \tag{9.152}$$

which demonstrates the convergence of Galerkin's method.

As an illustration, we first consider a conservative system in the form of a beam in transverse vibration. The eigenvalue problem is defined by the differential equation

$$\frac{d^2}{dx^2}\left[EI(x)\frac{d^2Y(x)}{dx^2}\right] = \lambda m(x)Y(x),\ 0 < x < L \tag{9.153}$$

and certain boundary conditions. For arbitrary mass and stiffness distributions and/or complex boundary conditions, no closed-form solution can be expected, so that we consider an approximate solution by the Galerkin method. To this end, we insert Eq. (9.149) into Eq. (9.153), and write the residual in the form

$$\begin{aligned}\mathcal{R}(Y^{(n)}(x), x) &= \frac{d^2}{dx^2}\left[EI(x)\frac{d^2Y^{(n)}(x)}{dx^2}\right] - \lambda^{(n)}m(x)Y^{(n)}(x)\\ &= \frac{d^2}{dx^2}\left[EI(x)\sum_{j=1}^{n}a_j\frac{d^2\phi_j(x)}{dx^2}\right] - \lambda^{(n)}m(x)\sum_{j=1}^{n}a_j\phi_j(x)\\ &= \sum_{j=1}^{n}a_j\left\{\frac{d^2}{dx^2}\left[EI(x)\frac{d^2\phi_j(x)}{dx^2}\right] - \lambda^{(n)}m(x)\phi_j(x)\right\}\end{aligned} \tag{9.154}$$

where we replaced the actual eigenvalue λ by the approximate eigenvalue $\lambda^{(n)}$. Introducing Eq. (9.154) in Eqs. (9.150), we have

$$\begin{aligned}&\int_0^L \phi_i(x)\sum_{j=1}^{n}a_j\left\{\frac{d^2}{dx^2}\left[EI(x)\frac{d^2\phi_j(x)}{dx^2}\right] - \lambda^{(n)}m(x)\phi_j(x)\right\}dx\\ &= \sum_{j=1}^{n}a_j\int_0^L\left\{\phi_i(x)\frac{d^2}{dx^2}\left[EI(x)\frac{d^2\phi_j(x)}{dx^2}\right] - \lambda^{(n)}m(x)\phi_i(x)\phi_j(x)\right\}dx\\ &= \sum_{j=1}^{n}k_{ij}a_j - \lambda^{(n)}\sum_{j=1}^{n}m_{ij}a_j = 0,\ i = 1, 2, \ldots, n\end{aligned} \tag{9.155}$$

in which

$$k_{ij} = \int_0^L \phi_i(x)\frac{d^2}{dx^2}\left[EI(x)\frac{d^2\phi_j(x)}{dx^2}\right]dx = \int_0^L \phi_j(x)\frac{d^2}{dx^2}\left[EI(x)\frac{d^2\phi_i(x)}{dx^2}\right]dx = k_{ji},\quad i, j = 1, 2, \ldots, n \tag{9.156}$$

are stiffness coefficients, which are symmetric because $\phi_i(x)$ $(i = 1, 2, \ldots, n)$ are comparison functions, and

$$m_{ij} = m_{ji} = \int_0^L m(x)\phi_i(x)\phi_j(x)dx,\ i, j = 1, 2, \ldots, n \tag{9.157}$$

are symmetric mass coefficients. It should be pointed out that these are the same coefficients as those obtained by the Rayleigh-Ritz method. Equations (9.155) can be written in the familiar matrix form

$$K^{(n)}\mathbf{a}^{(n)} = \lambda^{(n)}M^{(n)}\mathbf{a}^{(n)},\ \lambda^{(n)} = (\omega^{(n)})^2 \tag{9.158}$$

where the notation is obvious.

To demonstrate how the Galerkin method works for a nonconservative system, we consider a viscously damped beam in transverse vibration. The partial differential equation describing the free vibration can be written as

$$m(x)\frac{\partial^2 y(x,t)}{\partial t^2}+c(x)\frac{\partial y(x,t)}{\partial t}+\frac{\partial^2}{\partial x^2}\left[EI(x)\frac{\partial^2 y(x,t)}{\partial x^2}\right]=0,\ 0<x<L \quad (9.159)$$

in which $-c(x)\partial y(x,t)/\partial t$ is a viscous damping force density. The solution of Eq. (9.159) is subject to given boundary conditions. It has the exponential form

$$y(x,t)=e^{\lambda t}Y(x) \quad (9.160)$$

Inserting Eq. (9.160) into Eq. (9.159) and dividing through by $e^{\lambda t}$, we obtain the differential eigenvalue problem consisting of the differential equation

$$\lambda^2 m(x)Y(x)+\lambda c(x)Y(x)+\frac{d^2}{dx^2}\left[EI(x)\frac{d^2 Y(x)}{dx^2}\right]=0,\ 0<x<L \quad (9.161)$$

and two boundary conditions at each end.

Assuming that the eigenvalue problem does not admit an exact solution, we consider an approximate solution in the form of Eq. (9.149). Hence, inserting Eq. (9.149) into Eq. (9.161) and replacing λ by the approximate eigenvalue $\lambda^{(n)}$, we have simply

$$\mathcal{R}(Y^{(n)}(x),x)=(\lambda^{(n)})^2 m(x)\sum_{j=1}^{n}a_j\phi_j(x)+\lambda^{(n)}c(x)\sum_{j=1}^{n}a_j\phi_j(x)$$

$$+\sum_{j=1}^{n}a_j\frac{d^2}{dx^2}\left[EI(x)\frac{d^2\phi_j(x)}{dx^2}\right],\ 0<x<L \quad (9.162)$$

so that, multiplying Eq. (9.162) by $\phi_i(x)$ $(i=1,2,\ldots,n)$ and integrating over the length of the beam, we can write

$$(\lambda^{(n)})^2\sum_{j=1}^{n}a_j\int_0^L m(x)\phi_i(x)\phi_j(x)dx+\lambda^{(n)}\sum_{j=1}^{n}a_j\int_0^L c(x)\phi_i(x)\phi_j(x)dx$$

$$+\sum_{j=1}^{n}a_j\int_0^L \phi_i(x)\frac{d^2}{dx^2}\left[EI(x)\frac{d^2\phi_j(x)}{dx^2}\right]dx=0,\ i=1,2,\ldots,n \quad (9.163)$$

Then, introducing the damping coefficients

$$c_{ij}=c_{ji}=\int_0^L c(x)\phi_i(x)\phi_j(x)dx,\ i,j=1,2,\ldots,n \quad (9.164)$$

and using Eqs. (9.156) and (9.157), we obtain the algebraic eigenvalue problem

$$(\lambda^{(n)})^2\sum_{j=1}^{n}m_{ij}a_j+\lambda^{(n)}\sum_{j=1}^{n}c_{ij}a_j+\sum_{j=1}^{n}k_{ij}a_j=0,\ i=1,2,\ldots,n \quad (9.165)$$

which can be written in the matrix form

$$(\lambda^{(n)})^2 M^{(n)}\mathbf{a}^{(n)}+\lambda^{(n)}C^{(n)}\mathbf{a}^{(n)}+K^{(n)}\mathbf{a}^{(n)}=\mathbf{0} \quad (9.166)$$

We assume that $c(x)$ is such that the damping coefficients are symmetric, $c_{ij} = c_{ji}$ $(i, j = 1, 2, \dots, n)$, so that the damping matrix $C^{(n)}$ is symmetric.

The eigenvalue problem (9.166), although defined by symmetric matrices, differs materially from the eigenvalue problem given by Eq. (9.158). In fact, in general the problem cannot be solved in the form given by Eq. (9.166). An exception to this is the special case in which the damping term $c(x)Y(x)$ is a linear combination of the mass term $m(x)Y(x)$ and the stiffness term $(d^2/dx^2)[EI(x)d^2Y(x)/dx^2]$, or

$$c(x)Y(x) = \alpha m(x)Y(x) + \beta \frac{d^2}{dx^2}\left[EI(x)\frac{d^2Y(x)}{dx^2}\right], \quad 0 < x < L \tag{9.167}$$

In this case, it is not difficult to verify that the damping matrix is a linear combination of the mass matrix and stiffness matrix of the form

$$C^{(n)} = \alpha M^{(n)} + \beta K^{(n)} \tag{9.168}$$

Such damping is known as *proportional damping* and was discussed in Sec. 7.15.2, in which it was shown that the same matrix that diagonalizes the mass and stiffness matrices, namely, the modal matrix for the undamped system solving Eq. (9.158), also diagonalizes the damping matrix. We denote the modal matrix by $U = [\mathbf{a}_1^{(n)} \;\; \mathbf{a}_2^{(n)} \;\; \dots \;\; \mathbf{a}_n^{(n)}]$, where $\mathbf{a}_r^{(n)}$ $(r = 1, 2, \dots, n)$ are the modal vectors satisfying Eq. (9.158), and assume that the modal vectors have been normalized so that

$$U^T M^{(n)} U = I, \;\; U^T K^{(n)} U = \Lambda = \operatorname{diag}[(\omega_1^{(n)})^2 \;\; (\omega_2^{(n)})^2 \;\; \dots \;\; (\omega_n^{(n)})^2] \tag{9.169}$$

where $\omega_r^{(n)}$ $(r = 1, 2, \dots n)$ are the natural frequencies of the discretized undamped system. Then, introducing the linear transformation

$$\mathbf{a}^{(n)} = U\boldsymbol{\eta} \tag{9.170}$$

in Eq. (9.166), premultiplying the result by U^T and considering Eqs. (9.168) and (9.169), we obtain

$$(\lambda^{(n)})^2\boldsymbol{\eta} + \lambda^{(n)}(\alpha I + \beta\Lambda)\boldsymbol{\eta} + \Lambda\boldsymbol{\eta} = \mathbf{0} \tag{9.171}$$

which represents a set of independent quadratic equations in $\lambda^{(n)}$. Introducing the notation

$$\alpha + \beta(\omega_r^{(n)})^2 = 2\zeta_r\omega_r^{(n)}, \;\; r = 1, 2, \dots, n \tag{9.172}$$

the independent quadratic equations can be rewritten as

$$(\lambda^{(n)})^2 + 2\zeta_r\omega_r^{(n)}\lambda^{(n)} + (\omega_r^{(n)})^2 = 0, \;\; r = 1, 2, \dots, n \tag{9.173}$$

which can be solved for the approximate eigenvalues $\lambda_r^{(n)}$ of the proportionally damped system. Finally, inserting these eigenvalues into Eq. (9.166), we obtain sets of algebraic equations, which can be solved for the eigenvectors $\mathbf{a}_r^{(n)}$. Note that the eigenvalues and eigenvectors occur in general in pairs of complex conjugates.

In the general case of viscous damping, i.e., when the damping is nonproportional, the undamped modal matrix U does not diagonalize the damping matrix $C^{(n)}$, so that a different approach is necessary. This approach consists of transforming the eigenvalue problem, Eq. (9.166), to state form. To this end, we adjoin an obvious identity, rearrange Eq. (9.166) and obtain

$$\begin{aligned} \lambda^{(n)}\mathbf{a}^{(n)} &= \lambda^{(n)}\mathbf{a}^{(n)} \\ (\lambda^{(n)})^2\mathbf{a}^{(n)} &= -\lambda^{(n)}(M^{(n)})^{-1}C^{(n)}\mathbf{a}^{(n)} - (M^{(n)})^{-1}K^{(n)}\mathbf{a}^{(n)} \end{aligned} \tag{9.174}$$

Then, introducing the $2n$-dimensional state vector $\mathbf{x}^{(n)} = [(\mathbf{a}^{(n)})^T \;\; \lambda^{(n)}(\mathbf{a}^{(n)})^T]^T$, Eqs. (9.174) can be expressed in the form of the standard eigenvalue problem

$$A^{(n)}\mathbf{x}^{(n)} = \lambda^{(n)}\mathbf{x}^{(n)} \tag{9.175}$$

where the $2n \times 2n$ coefficient matrix has the familiar form (Sec. 7.16)

$$A^{(n)} = \left[\begin{array}{c|c} 0 & I \\ \hline -(M^{(n)})^{-1}K^{(n)} & -(M^{(n)})^{-1}C^{(n)} \end{array}\right] \tag{9.176}$$

Finally, it should be pointed out that, although the Galerkin method calls for the use of comparison functions, as with the Rayleigh-Ritz method, excellent results can be obtained through the use of quasi-comparison functions.[1]

9.10 THE COLLOCATION METHOD

The collocation method also belongs to the family of weighted residual methods, so that many of the concepts and developments in Sec. 9.9 remain the same. The main difference between the collocation method and Galerkin's method lies in the weighting functions, which in the case of the collocation method represent spatial Dirac delta functions. Hence, whereas Eq. (9.149) retains its form, Eqs. (9.150) must be replaced by

$$\int_0^L \delta(x - x_i)\mathcal{R}(Y^{(n)}(x), x)dx = 0, \; i = 1, 2, \ldots, n \tag{9.177}$$

Due to the sampling property of the Dirac delta function, Eqs. (9.177) require no integrations, and yield directly the set of n algebraic equations

$$\mathcal{R}(Y^{(n)}(x_i)) = 0, \; i = 1, 2, \ldots n \tag{9.178}$$

so that the process amounts to evaluating the residual at $x = x_i$.

Although it may not be evident at this point, Eqs. (9.178) represent the algebraic eigenvalue problem. To substantiate this statement, we consider the eigenvalue problem for a beam in transverse vibration with the differential equation given by Eq. (9.153). Moreover, the residual is as given by Eq. (9.154). Hence, letting $x = x_i$ in Eq. (9.154),

[1]Meirovitch, L. and Hagedorn, P., "A New Approach to the Modeling of Distributed Non-Self-Adjoint Systems," *Journal of Sound and Vibrations*, vol. 178, no. 2, 1994, pp. 227–241.

we can write simply

$$\mathcal{R}(Y^{(n)}(x_i)) = \sum_{j=1}^{n} a_j \left\{ \frac{d^2}{dx^2}\left[EI(x)\frac{d^2\phi_j(x)}{dx^2} \right] - \lambda^{(n)} m(x)\phi_j(x) \right\}\bigg|_{x=x_i} = 0, \tag{9.179}$$

in which $\phi_1(x),\ \phi_2(x),\ \ldots,\ \phi_n(x)$ are comparison functions. Equations (9.179) represent the algebraic eigenvalue problem

$$\sum_{j=1}^{n} k_{ij} a_j = \lambda^{(n)} \sum_{j=1}^{n} m_{ij} a_j, \ i = 1, 2, \ldots, n \tag{9.180}$$

where

$$k_{ij} = \frac{d^2}{dx^2}\left[EI(x)\frac{d^2\phi_j(x)}{dx^2} \right]\bigg|_{x=x_i}, \ i, j = 1, 2, \ldots, n \tag{9.181}$$

are the stiffness coefficients and

$$m_{ij} = m(x_i)\phi_j(x_i), \ i, j = 1, 2, \ldots, n \tag{9.182}$$

are the mass coefficients. The eigenvalue problem can be expressed in the matrix form

$$K^{(n)}\mathbf{a}^{(n)} = \lambda^{(n)} M^{(n)}\mathbf{a}^{(n)} \tag{9.183}$$

where the notation is obvious. We note that, in contrast with the Galerkin method, in the collocation method the stiffness matrix $K^{(n)}$ and mass matrix $M^{(n)}$ are not symmetric.

The main advantage of the collocation method is simplicity. Indeed, we observe from Eqs. (9.181) that the determination of the stiffness coefficients only requires differentiation and evaluation of the stiffness term at the chosen locations, $x = x_i$ $(i = 1, 2, \ldots, n)$. Moreover, the determination of the mass coefficients merely requires evaluation of the products of the mass density and the comparison functions at the chosen locations. However, the price to be paid for this simplicity is that the stiffness and mass matrices are not symmetric, in spite of the fact that this is a conservative system. As a result, the efficient computational algorithm typical of symmetric eigenvalue problems cannot be used. Although in general nonsymmetric eigenvalue problems possess complex solutions, for the most part the conservative nature of the system prevails, and the eigensolutions tend to be real. Still, if the eigensolutions are used to synthesize the system response, then it is necessary to obtain both right and left eigenvectors and to use a state space expansion theorem (Sec. 7.16), even though the eigenvalue problem, Eq. (9.183), is formulated in the configuration space and not in the state space.

Before the eigensolutions can be produced, it is necessary to cast the eigenvalue problem (9.183) into one in terms of a single matrix, of course a nonsymmetric one. To this end, we omit the superscript (n) for simplicity of notation, premultiply both sides of Eq. (9.183) by M^{-1} and write the eigenvalue problem in the standard form

$$A\mathbf{a} = \lambda\mathbf{a} \tag{9.184}$$

where

$$A = M^{-1}K \tag{9.185}$$

The solution of Eq. (9.184) consists of the eigenvalues λ_r and the right eigenvectors $\mathbf{a}_r$ $(r = 1, 2, \ldots, n)$. In addition, we must solve the adjoint eigenvalue problem

$$A^T \mathbf{b} = \lambda \mathbf{b} \tag{9.186}$$

which yields the same eigenvalues λ_s and the left eigenvectors $\mathbf{b}_s$ $(s = 1, 2, \ldots, n)$. We recall from Sec. 7.16 that the right and left eigenvectors are biorthogonal and can be normalized so as to satisfy the biorthonormality relations

$$\mathbf{b}_s^T \mathbf{a}_r = \delta_{rs}, \quad \mathbf{b}_s^T A \mathbf{a}_r = \lambda_r \delta_{rs}, \quad r, s = 1, 2, \ldots, n \tag{9.187}$$

The biorthonormality relations form the basis for the expansion theorem mentioned above.

Although both the Galerkin method and the collocation method are weighted residual methods, the fact that the weighting functions in the collocation method are spatial Dirac delta functions, as opposed to the weighting functions coinciding with the trial functions in the Galerkin method, makes the nature of the two methods entirely different. Indeed, in the Galerkin method the average weighted error is reduced to zero, whereas in the collocation method the error at a given number of individual points is annihilated. This makes the collocation method easier to visualize, and less abstract than the Galerkin method, but it also gives it a heuristic character. In addition, there is some arbitrariness in choosing the location of the points, which causes the results to differ from one choice to another. More importantly, the collocation method requires the solution of two nonsymmetric eigenvalue problems. Accuracy of the results can be improved by increasing the number of points at which the error is driven to zero, but this tends to be a slow process. Although convergence is assured as the number of points approaches infinity, it cannot be demonstrated rigorously as in the Rayleigh-Ritz method. Moreover, it cannot always be predicted from what side of the actual eigenvalues the approximate eigenvalues converge. Note that, for conservative systems, the Galerkin method has the same excellent convergence characteristics as the Rayleigh-Ritz method.

Example 9.10. Consider the eigenvalue problem of Example 9.6 for the tapered rod fixed at $x = 0$ and spring-supported at $x = L$. Solve the problem by the collocation method in two different ways: 1) using the locations $x_i = iL/n$ $(i = 1, 2, \ldots, n)$ and 2) using the locations $x_i = (2i - 1)L/2n$ $(i = 1, 2, \ldots, n)$; give results for $n = 2$ and $n = 3$. Then, list the three lowest natural frequencies for $n = 2, 3, \ldots, 30$ and discuss the nature of the convergence for both cases. Use the comparison functions from Example 9.6 throughout.

From Eqs. (1) of Example 9.6, the comparison functions are

$$\phi_i(x) = \sin \beta_i x, \quad i = 1, 2, \ldots, n \tag{a}$$

where the constants β_i represent solutions of Eq. (9.132). The first three are given by

$$\beta_1 L = 2.215707, \quad \beta_2 L = 5.032218, \quad \beta_3 L = 8.057941 \tag{b}$$

The stiffness coefficients for the problem at hand are

$$\begin{aligned} k_{ij} &= -\frac{d}{dx}\left[EA(x)\frac{d\phi_j(x)}{dx}\right]\Bigg|_{x=x_i} = -\frac{6EA}{5L}\frac{d}{dx}\left\{\left[1 - \frac{1}{2}\left(\frac{x}{L}\right)^2\right]\beta_j L \cos \beta_j x\right\}\Bigg|_{x=x_i} \\ &= \frac{6EA}{5L^2}\left\{\beta_j x_i \cos \beta_j x_i + \left[1 - \frac{1}{2}\left(\frac{x_i}{L}\right)^2\right](\beta_j L)^2 \sin \beta_j x_i\right\}, \quad i, j = 1, 2, \ldots, n \end{aligned} \tag{c}$$

and the mass coefficients are

$$m_{ij} = \frac{6m}{5}\left[1 - \frac{1}{2}\left(\frac{x_i}{L}\right)^2\right]\sin\beta_j x_i,\ i, j = 1, 2, \dots, n \tag{d}$$

The eigenvalue problem can be written in the general form

$$A\mathbf{a} = \lambda\mathbf{a},\ \lambda = \omega^2 mL^2/EA \tag{e}$$

in which

$$A = (mL^2/EA)M^{-1}K \tag{f}$$

where K and M are the stiffness and mass matrices with the entries given by Eqs. (c) and (d), respectively.

1. *Locations at* $x_i = iL/n$
   In this case, the stiffness coefficients, Eqs. (c), reduce to

$$k_{ij} = \frac{6EA}{5L^2}\left\{\frac{i\beta_j L}{n}\cos\frac{i\beta_j L}{n} + \left[1 - \frac{1}{2}\left(\frac{i}{n}\right)^2\right](\beta_j L)^2 \sin\frac{i\beta_j L}{n}\right\},\ i, j = 1, 2, \dots, n \tag{g}$$

and the mass coefficients, Eqs. (d), are simply

$$m_{ij} = \frac{6m}{5}\left[1 - \frac{1}{2}\left(\frac{i}{n}\right)^2\right]\sin\frac{i\beta_j L}{n},\ i, j = 1, 2, \dots, n \tag{h}$$

For $n = 2$, the stiffness matrix is

$$K = \frac{EA}{L^2}\begin{bmatrix} 5.205939 & 13.120091 \\ 0.755692 & -12.524855 \end{bmatrix} \tag{i}$$

and the mass matrix is

$$M = m\begin{bmatrix} 0.939479 & 0.614764 \\ 0.479492 & -0.569574 \end{bmatrix} \tag{j}$$

so that

$$A = \begin{bmatrix} 4.132826 & -0.273503 \\ 2.152427 & 21.759635 \end{bmatrix} \tag{k}$$

The eigenvalues, right eigenvectors and left eigenvectors of A, normalized so as to satisfy Eqs. (9.187), are

$$\begin{aligned} \lambda_1 &= 4.166287\frac{EA}{mL^2},\ \mathbf{a}_1 = \begin{bmatrix} 0.992599 \\ -0.121438 \end{bmatrix},\ \mathbf{b}_1 = \begin{bmatrix} 0.999879 \\ 0.015544 \end{bmatrix} \\ \lambda_2 &= 21.72617\frac{EA}{mL^2},\ \mathbf{a}_2 = \begin{bmatrix} 0.015544 \\ -0.999879 \end{bmatrix},\ \mathbf{b}_2 = \begin{bmatrix} 0.121438 \\ 0.992599 \end{bmatrix} \end{aligned} \tag{l}$$

Moreover, the natural frequencies are the square roots of the eigenvalues; they can be found in Table 9.2. Expanding in terms of the right eigenvectors, according to Eq. (9.149) with U replacing Y, the natural modes are

$$\begin{aligned} U_1(x) &= 0.992599\sin 2.215707x - 0.121438\sin 5.032218x \\ U_2(x) &= 0.015544\sin 2.215707x - 0.999879\sin 5.032218x \end{aligned} \tag{m}$$

They resemble those shown in Fig. 9.25.

Table 9.2 Normalized Natural Frequencies
$\omega_r^* = \omega_r\sqrt{mL^2/EA}$ **for** $x_i = iL/n$

r	ω_1^*	ω_2^*	ω_3^*
2	2.041149	4.661134	–
3	2.148223	4.950458	7.764421
4	2.180078	5.026274	7.974473
5	2.193677	5.055835	8.039231
6	2.200720	5.070436	8.067294
7	2.204835	5.078739	8.082175
8	2.207446	5.083920	8.091089
9	2.209205	5.087374	8.096877
10	2.210447	5.089793	8.100861
⋮	⋮	⋮	⋮
30	2.214987	5.098508	8.114744

For $n = 3$, the stiffness matrix is

$$K = \frac{EA}{L^2}\begin{bmatrix} 4.401161 & 28.322516 & 29.485150 \\ 4.727759 & -8.935946 & -43.940613 \\ 0.755692 & -12.524855 & 36.192188 \end{bmatrix} \tag{n}$$

and the mass matrix is

$$M = m\begin{bmatrix} 0.762995 & 1.126899 & 0.498681 \\ 0.929243 & -0.197500 & -0.737571 \\ 0.479492 & -0.569574 & 0.587563 \end{bmatrix} \tag{o}$$

yielding

$$A = \begin{bmatrix} 4.505092 & -1.492165 & 0.457863 \\ 1.338751 & 24.519460 & -0.866375 \\ -1.092557 & 3.669839 & 60.383582 \end{bmatrix} \tag{p}$$

The matrix A has the eigenvalues, right eigenvectors and left eigenvectors

$$\begin{aligned} \lambda_1 &= 4.614862\frac{EA}{mL^2}, \quad \mathbf{a}_1 = \begin{bmatrix} 0.997530 \\ -0.066052 \\ 0.023889 \end{bmatrix}, \quad \mathbf{b}_1 = \begin{bmatrix} 0.997080 \\ 0.076038 \\ -0.007005 \end{bmatrix} \\ \lambda_2 &= 24.507035\frac{EA}{mL^2}, \quad \mathbf{a}_2 = \begin{bmatrix} 0.076355 \\ -0.991667 \\ 0.103764 \end{bmatrix}, \quad \mathbf{b}_2 = \begin{bmatrix} 0.065499 \\ 0.997582 \\ 0.023254 \end{bmatrix} \\ \lambda_3 &= 60.286237\frac{EA}{mL^2}, \quad \mathbf{a}_3 = \begin{bmatrix} 0.008844 \\ -0.023884 \\ 0.999676 \end{bmatrix}, \quad \mathbf{b}_3 = \begin{bmatrix} -0.017014 \\ 0.102756 \\ 0.994561 \end{bmatrix} \end{aligned} \tag{q}$$

The normalized natural frequencies are displayed in Table 9.2 for $n = 2, 3, \ldots, 30$. The natural modes expanded in terms of the right eigenvectors are

$$U_1(x) = 0.997530 \sin 2.215707x - 0.066052 \sin 5.032218x + 0.023889 \sin 8.057941x$$

$$U_2(x) = 0.076355 \sin 2.215707x - 0.991667 \sin 5.032218x + 0.103764 \sin 8.057941x$$

$$U_3(x) = 0.008844 \sin 2.215707x - 0.023884 \sin 5.032218x + 0.999676 \sin 8.057941x \tag{r}$$

The modes resemble those displayed in Fig. 9.26.

2. *Locations at* $x_i = (2i-1)L/2n$
In this case, the stiffness coefficients, Eqs. (c), have the expressions

$$k_{ij} = \frac{6EA}{5L^2}\left\{\frac{(2i-1)\beta_j L}{2n}\cos\frac{(2i-1)\beta_j L}{2n} + \left[1 - \frac{1}{2}\left(\frac{2i-1}{2n}\right)^2\right](\beta_j L)^2 \sin\frac{(2i-1)\beta_j L}{2n}\right], \quad i, j = 1, 2, \ldots, n \tag{s}$$

and the mass coefficients, Eqs. (d), are given by

$$m_{ij} = \frac{6m}{5}\left[1 - \frac{1}{2}\left(\frac{2i-1}{2n}\right)^2\right]\sin\frac{(2i-1)\beta_j L}{2n}, i, j = 1, 2, \ldots, n \tag{t}$$

Following the same steps as for $x_i = iL/n$, it is possible to compute the eigenvalues, right eigenvectors and left eigenvectors for various values of n. For brevity, we omit these results and only list in Table 9.3 the three lowest normalized natural frequencies for $n = 2, 3, \ldots, 30$.

We observe with interest from Table 9.2 that for $x_i = iL/n$ $(i = 1, 2, \ldots, n)$ the natural frequencies increase as n increases. This can be explained by the fact that the specified locations tend to make the rod longer than it actually is. Because an increased length, while everything else remains the same, tends to reduce the stiffness, the approximate natural frequencies are lower than the actual natural frequencies and approach the latter from below. On the other hand, the locations $x_i = (2i-1)L/2n$ tend to make the rod shorter than it actually is, so that the stiffness of the model is larger

Table 9.3 Normalized Natural Frequencies
$\omega_r^* = \omega_r\sqrt{EA/mL^2}$ **for** $x_i = (2i-1)L/2n$

r	ω_1^*	ω_2^*	ω_3^*
2	2.245588	5.229317	–
3	2.231022	5.138904	8.255577
4	2.225251	5.121013	8.163432
5	2.222239	5.113512	8.142789
6	2.220445	5.109469	8.133890
7	2.219286	5.106992	8.129019
8	2.218494	5.105352	8.125996
9	2.217927	5.104204	8.123967
10	2.217509	5.103367	8.122529
⋮	⋮	⋮	⋮
30	2.215772	5.099994	8.117046

than the stiffness of the actual system. As a result, the approximate natural frequencies are larger than the actual natural frequencies and approach the latter from above, as can be concluded from Table 9.3. This points to the arbitrariness and lack of predictability inherent in the collocation method, with the nature of the results depending on the choice of locations. In this regard, the collocation method acts like a lumped-parameter method even though the parameters have not been lumped, only sampled.

Finally, it will prove instructive to compare some of the results obtained here with corresponding results obtained in Example 9.6 by means of the Rayleigh-Ritz method and in Example 9.8 by means of the enhanced Rayleigh-Ritz method. In particular, we observe that the three lowest natural frequencies computed in Example 9.6 using 30 ordinary admissible functions are not nearly as close to the actual values as the corresponding ones computed by means of the collocation method using the locations $x_i = iL/n$ and $x_i = (2i-1)L/2n$ $(i = 1, 2, \ldots, 30)$. These results should not be construed as an indication that the collocation method has good convergence characteristics, but as an indication that the use of admissible functions, which are unable to satisfy the natural boundary condition at $x = L$ even when taken in a linear combination, penalizes convergence greatly. Indeed, the fact is that neither the Rayleigh-Ritz method using ordinary admissible functions nor the collocation method achieves convergence with a discrete model possessing 30 degrees of freedom. By contrast, the two lowest natural frequencies computed by means of the Rayleigh-Ritz method using comparison functions converge with 14 terms and the third one converges with 20 terms. Convergence is even more dramatic using quasi-comparison functions, as demonstrated in Example 9.8. Indeed, all three lowest natural frequencies converge with only six terms.

9.11 MATLAB PROGRAM FOR THE SOLUTION OF THE EIGENVALUE PROBLEM BY THE RAYLEIGH-RITZ METHOD

The Rayleigh-Ritz method reduces differential eigenvalue problems for conservative systems to algebraic eigenvalue problems defined by the mass matrix M and stiffness matrix K. The latter resembles the eigenvalue problem for discrete systems, except that the entries of M and K for the Rayleigh-Ritz method consist of integrals in need of evaluation. As a result, the program 'evpc.m' given in Sec. 7.18, which assumes that the numerical values of M and K are given, cannot be used, and a new program must be written. The new program, entitled 'rayritz.m', carries out the integrations for the entries of M and K, using integrands provided by two separate programs, 'mass.m' and 'stiffness.m', respectively. The program 'rayritz.m' reads as follows:

```
% The program 'rayritz.m' sets up the eigenvalue problem for the fixed-free rod of
% Example 9.5 by the Rayleigh-Ritz method, solves it and plots the first three modes

clear
clf

N=3; % number of comparison functions in the approximating series

for m=1:N,
  for n=1:N,
    M(m,n)=quad8('mass', 0, 1, [ ], [ ], m, n); % mass matrix using program 'mass.m'
    % and MATLAB function
```

```
    K(m,n)=quad8('stiffness', 0, 1, [ ], [ ], m, n); % stiffness matrix using program
    % 'stiffness.m' and MATLAB function
  end
end
[v,W]=eig(K, M); % solution of the eigenvalue problem using MATLAB function
for i=1:N,
  w1(i)=sqrt(W(i, i)); % setting the approximate natural frequencies in an
  % N-dimensional vector
end
[w, I]=sort(w1); % arranging the approximate natural frequencies in ascending order
for j=1:N,
  U(:, j)=v(:, I(j)); % arranging the modal vectors in ascending order
end
  x=[0: 0.01: 1]; % origin of x-axis, nondimensional rod increment, nondimensional
  % rod length
for i=1:N,
  suma=0;
  for j=1:N,
    suma=suma+U(j, i)*sin((2*j-1)*x*pi/2); % approximate eigenfunctions
  end


  u(i, :)=suma/suma(size(suma, 2)); % normalization of the approximate eigenfunc-
  % tions so as to equal unity at the free end
end
for k=1:3,
  f=0.1; if k>=2; f=0.2; end
  axes('position',[0.3  0.8-0.35*(k-1)  0.4  f]); % positioning of the plots
  % in the workspace
  plot (x, u(k,:))
  st=num2str(w(k));
  st1=int2str(k);
  st2=int2str(N);
  st3=strcat('U_', st1, '^{(',st2,')} (x)');
  st4=strcat('Natural Mode ', {'  '}, st1, {' , '}, '\omega_', st1,
  '^{(',st2,')}' , '=', st, '[EA/mL ^2]{1/2}');


  title(st4)
  ylabel(st3)
  xlabel('x/L')
  end
```

The 'rayritz.m' program must be run in conjunction with

```
% The program 'mass.m' provides the entries to be integrated to obtain the mass
% matrix by the Rayleigh-Ritz method
function mij=mass(x,m,n)
mij=6*(1-(x.^2)/2).*sin((2*m-1)*pi*x/2). *sin((2*n-1)*pi*x/2)/5; % integrands for
% M(m,n) in program 'rayritz.m'
```

and

```
% The program 'stiffness.m' provides the entries to be integrated to obtain the
% stiffness matrix by the Rayleigh-Ritz method
function kij=stiffness(x,m,n)
kij=6*(2*m-1)*(2*n-1)*(pi^2)*(1-(x.^2)/2).*cos((2*m-1)*pi*x/2).*cos((2*n
    -1)*pi*x/2)/20; % integrands for K(m,n) in program 'rayritz.m'
```

9.12 SUMMARY

Systems with nonuniformly distributed parameters seldom admit closed-form solutions. Hence, one must be content with approximate solutions, which can only be obtained by means of discrete models acting as surrogates for distributed systems. Approximate techniques can be broadly divided into lumped-parameter methods and series discretization methods.

In the simplest of the lumped-parameter methods the system is divided into segments and the distributed mass and force over the segments are concentrated at certain discrete points inside these segments. Moreover, the stiffness is replaced by equivalent springs connecting these points. Another method differs from the one just described in that the stiffness distribution is represented by means of influence coefficients. In both cases, the equations of motion can be obtained by means of Lagrange's equations. Two different lumped-parameter procedures are concerned with computing the natural frequencies and natural modes only, Holzer's method for shafts in torsion and Myklestad's method for beams. They are step-by-step procedures, going from one lump to the next, and derive a frequency equation by invoking the boundary conditions; the natural frequencies squared are the roots of the corresponding characteristic polynomial. The lumped-parameter methods are easy to understand, but they lack rigor. Convergence improves as the number of lumped masses increases, but the process can be slow. Moreover, the lumping process is arbitrary, and it is difficult to predict whether the approximate natural frequencies are smaller or larger than the actual ones.

The origin of the series discretization methods can be traced to Rayleigh's energy method, a technique for approximating the lowest natural frequency of a conservative system. It is based on Rayleigh's principle, which states that Rayleigh's quotient has a minimum value equal to the lowest eigenvalue when the trial function used in conjunction with Rayleigh's quotient is the lowest eigenfunction. For convenience, the preferred form of Rayleigh's quotient is the energy form, namely, that in which the numerator is a measure of the potential energy and the denominator a measure of the kinetic energy. When the lowest eigenfunction is not known, a quick approximation to the lowest eigenvalue can be obtained by inserting into Rayleigh's quotient a trial function approximating the lowest eigenfunction closely. Then, because of the stationarity of Rayleigh's quotient, the resulting value of Rayleigh's quotient is an even better approximation to the lowest eigenvalue. Clearly, to improve the approximation, it is only necessary to find a trial function capable of lowering the approximation. This is the essence of the Rayleigh-Ritz method, which consists of constructing a sequence of improving approximations to the lowest eigenvalue by using a minimizing sequence of trial functions in the form of linear combinations of admissible functions multiplied by undetermined coefficients and determining the coefficients by rendering Rayleigh's quotient stationary. The process represents the variational approach to the differential eigenvalue problem, and yields not only improved approximations to the lowest eigenvalue, but also improved approximations to the higher eigenvalues. The approximate eigenvalues, known as Ritz eigenvalues, satisfy the separation theorem, which states that the eigenvalues obtained by using $n+1$ terms in the approximating series bracket the eigenvalues obtained with n terms in the series. The approximate eigenvalues converge to the actual eigenvalues from above. Convergence of the Rayleigh-Ritz method can be slow when there are natural boundary conditions, which are generally violated by admissible functions. In such cases, convergence can be improved significantly through the use of comparison functions and even more dramatically through the use of quasi-comparison functions. The Rayleigh-Ritz method is capable of producing accurate results with only a small number of terms in the approximating series, which translates into a discrete model with a small number of degrees of freedom. On the other hand, the Rayleigh-Ritz method can only handle systems with relatively simple geometry, and in particular one-dimensional structural members, or two-dimensional members of regular shape, such as rectangular and circular. A technique known as the assumed-modes method is regarded by some as the Rayleigh-Ritz method. It is more physically motivated and it obtains the same results as the Rayleigh-Ritz method. The assumed-modes method can be used to derive the equations of motion in conjunction with the Rayleigh-Ritz method.

Another class of series discretization techniques is known generically as weighted residuals methods and produce approximate solutions by satisfying the differential equation in some sense. The best known is the Galerkin method, and to a lesser extent the collocation method. As in the Rayleigh-Ritz method, an approximate solution is assumed in the form of a finite series of trial functions multiplied by undetermined coefficients. Then, the resulting error in the differential equation, known as a residual, is multiplied by a set of weighting functions, in sequence, and each of the weighted residuals is integrated over the domain of the system and set equal to zero. The result consists of a set of algebraic equations, which permits the determination of the coefficients. Because the procedure works with the differential equation alone, to ensure that all boundary conditions are satisfied, the trial functions must be from the class of comparison functions. In the Galerkin method, the weighting functions are the same as the trial functions. On the other hand, in the collocation method the trial functions are spatial Dirac delta functions located at judiciously chosen points of the domain. Hence, in the Galerkin method the differential equation is satisfied in an average sense and in the collocation method it is satisfied at individual points. The weighted residual methods are broader in scope than the Rayleigh-Ritz method, as they can be applied to both conservative and nonconservative systems. For conservative systems, the Galerkin method yields the same symmetric mass and stiffness matrices as the Rayleigh-Ritz method, but the collocation method yields nonsymmetric matrices. For nonconservative systems, such as systems with viscous damping, or with aerodynamic forces, both formulations must be cast in state form. In general, the usefulness of the weighted residuals methods is limited to the same kind of systems with simple geometry as the Rayleigh-Ritz method.

Although the Rayleigh-Ritz method reduces a differential eigenvalue problem to an algebraic eigenvalue problem in terms of two symmetric matrices resembling that for a multi-degree-of-freedom system, the eigenvalue problem cannot be solved by the MATLAB program 'evpc.m' developed in Sec. 7.18. The reason is that in the Rayleigh-Ritz method the entries of the mass and stiffness matrices represent integral expressions rather than simple combinations of parameters. A new MATLAB program, entitled 'rayritz.m', demonstrates how to address this problem.

PROBLEMS

9.1. The tubular shaft in torsional vibration shown in Fig. 9.29 has the radius $r(x) = r[1 + x(L - x)/L^2]$, thickness t, mass per unit volume ρ and shear modulus G, where the thickness t is small compared to the radius $r(x)$. Use the procedure described in Sec. 9.1 to construct a ten-degree-of-freedom lumped model approximating the distributed system and derive the corresponding ordinary differential equations of motion. Then, derive and solve the eigenvalue problem. Note that the shaft is symmetric with respect to $x = L/2$.

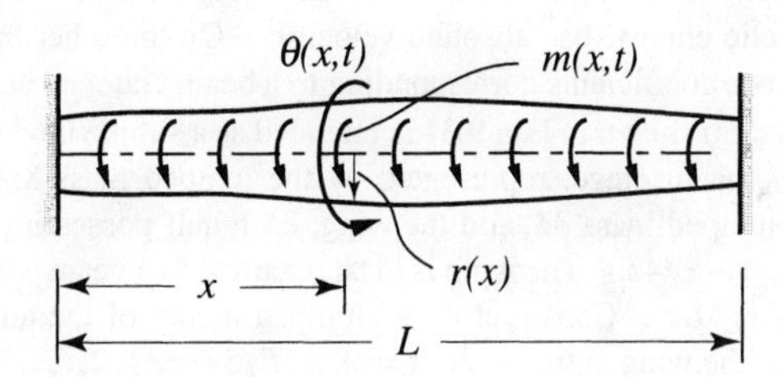

FIGURE 9.29
Tubular shaft in torsional vibration

9.2. The rod in axial vibration shown in Fig. 8.32 has the mass density $m(x) = m(1 - x/2L)$ and the axial rigidity $EA(x) = EA(1 - x/2L)$. Use the procedure described in Sec. 9.1 to construct a ten-degree-of-freedom lumped model approximating the distributed system and derive the corresponding ordinary differential equations of motion. Then, derive and solve the eigenvalue problem for $k = 0.5EA/L$.

9.3. Solve Problem 9.1 by the lumped-parameter method using influence coefficients.

9.4. Solve Problem 9.2 by the lumped-parameter method using influence coefficients. **Hint:** To formulate the problem, it is convenient to describe the absolute displacement of a given lumped mass as the sum of the displacement $u(0, t)$ of the end $x = 0$ and the displacement of the lumped mass relative to $u(0, t)$. Then, for the kinetic energy, use absolute velocities. On the other hand, regard the potential energy as consisting of two parts, one due to the deformation of the spring and one due to the elastic displacements of the lumped masses relative to $u(0, t)$. Define the latter potential energy in terms of influence coefficients obtained by holding the end $x = 0$ fixed.

9.5. Construct a ten-degree-of-freedom lumped model for the tapered cantilever beam shown in Fig. 9.30; the beam has unit width and its height is $h(x) = h(1 - x/2L)$. Derive the eigenvalue problem by the lumped-parameter method using influence coefficients.

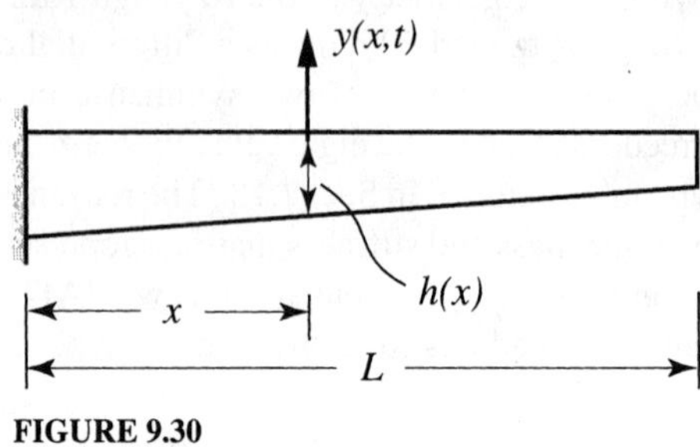

FIGURE 9.30
Tapered cantilever beam

9.6. Construct a ten-degree-of-freedom lumped model for the pinned-free uniformly distributed beam of Problem 8.14 based on the lumped-parameter method using influence coefficients. Then, derive and solve the eigenvalue problem, compare the three lowest natural frequencies and flexible modes with those obtained in Problem 8.14 and draw conclusions. **Hint:** To formulate the problem, it is convenient to describe the absolute displacement of any given mass as the sum of a displacement due to the rigid-body rotation of the beam about the pin and an elastic displacement of the mass obtained by regarding the end $x = 0$ as clamped. Then, for the kinetic energy, use absolute velocities. On the other hand, for the potential energy, use influence coefficients corresponding to a beam clamped at $x = 0$.

9.7. The symmetric system shown in Fig. 9.31 is intended as a simple model of a flexible aircraft and it consists of the fuselage, represented by the lumped mass $8M$, two engines, each represented by a lumped mass M, and the wing, each half possessing the distributed mass $m(x) = (8M/7L)(1 - x/4L)$. The wing is to be regarded as a beam with the flexural rigidity $EI(x) = EI(1 - x/4L)^3$. Construct an all lumped model of the aircraft by lumping the distributed mass of the wing at the points $x = 0, \pm iL/5$ $(i = 1, 2, \ldots, 5)$. Note that the wing mass lumped at $x = 0$ and $x = \pm 2L/5$ is to be added to the mass of the fuselage and engines, respectively. Derive the equations of motion for the aircraft based on the lumped-parameter method using influence coefficients. Then, derive and solve the eigenvalue problem and plot the two lowest flexible modes. **Hint:** Define the absolute displacement of a given lumped mass as the sum of three parts, one due to the vertical translation of the mass center C of the undeformed aircraft, a second due to the rotation of the aircraft axis x about the mass center C and a third due to the elastic displacement of the lumped mass relative to axis x. For the kinetic energy, use absolute velocities. For the potential energy, use influence coefficients obtained by regarding the flexible wing as clamped at $x = 0$.

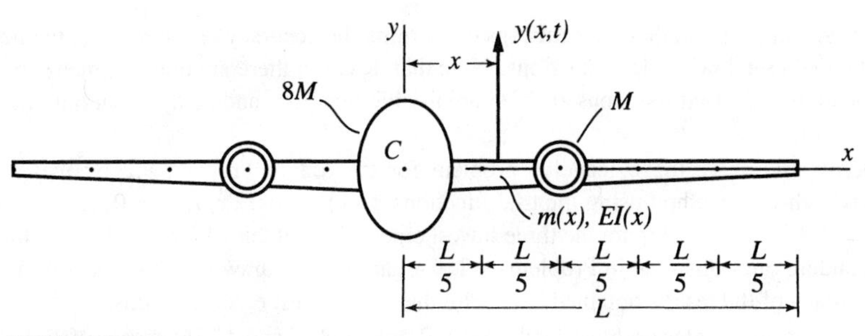

FIGURE 9.31
Model of a flexible aircraft

9.8. Solve the eigenvalue problem for the tubular shaft of Problem 9.1 by means of Holzer's method. Compare results with those obtained in Problem 9.1 and draw conclusions.

9.9. Formulate a lumped-parameter eigenvalue problem for rods in axial vibration analogous to Holzer's method for shafts in torsional vibration. Then, use the formulation to solve the eigenvalue problem for the rod of Problem 9.2. Compare results with those obtained in Problem 9.2 and draw conclusions.

9.10. Derive the eigenvalue problem for the tapered beam of Problem 9.5 by means of Myklestad's method. Solve the eigenvalue problem and plot the two lowest modes.

9.11. Derive the eigenvalue problem for the pinned-free beam of Problem 9.6 by means of Myklestad's method. Solve the eigenvalue problem for the three lowest flexible modes, compare results with those obtained in Problem 9.6 and draw conclusions concerning the accuracy of the two methods. Plot the modes.

9.12. Derive the eigenvalue problem for the aircraft of Problem 9.7 by means of Myklestad's method. Solve the eigenvalue problem for the two lowest flexible modes, compare results with those obtained in Problem 9.7 and draw conclusions concerning the accuracy of the two methods. Plot the modes.

9.13. Estimate the lowest natural frequency of the tubular shaft of Problem 9.1 by means of Rayleigh's energy method. Compare result with results obtained in Problems 9.1, 9.3 and 9.8 for the lowest natural frequency and draw conclusions concerning the accuracy of the estimate.

9.14. Estimate the lowest natural frequency of the rod of Problem 9.2 by means of Rayleigh's energy method. Compare result with results obtained in Problems 9.2, 9.4 and 9.9 for the lowest natural frequency and draw conclusions concerning the accuracy of the estimate.

9.15. Estimate the lowest natural frequency of the tapered beam of Problem 9.5 by means of Rayleigh's energy method. Compare result with results obtained in Problems 9.5 and 9.10 for the lowest natural frequency and draw conclusions concerning the accuracy of the estimate.

9.16. Estimate the lowest nonzero natural frequency of the pinned-free beam of Problem 9.6 by means of Rayleigh's energy method. Compare result with results obtained in Problems 9.6 and 9.11 for the same frequency and draw conclusions concerning the accuracy of the estimate.

9.17. Derive and solve the eigenvalue problem for the tubular shaft of Problem 9.1 by means of the Rayleigh-Ritz method using the trial functions $\phi_i(x) = \sin i\pi x/L$ $(i = 1, 2, \ldots, n)$. Let n increase until the lowest eigenvalue converges to six decimal places. Compare the three lowest eigenvalues obtained for $n = 10$ with the corresponding ones obtained in Problems

9.1, 9.3 and 9.8 and draw conclusions concerning the accuracy of the results obtained here. Plot the three lowest eigenfunctions. Note that, because there are only geometric boundary conditions, the trial functions are both admissible functions and comparison functions at the same time.

9.18. Derive and solve the eigenvalue problem for the rod of Problem 9.2 by means of the Rayleigh-Ritz method using the trial functions $\phi_i(x) = \cos i\pi x/L$ $(i = 0, 1, \dots, n-1)$ for $n = 2, 3, \dots, 10$. Compare the three lowest eigenvalues obtained for $n = 10$ with the corresponding ones obtained in Problems 9.2, 9.4 and 9.9 and draw conclusions concerning the accuracy of the results obtained here. Plot the three lowest eigenfunctions.

9.19. Derive the eigenvalue problem for the rod of Problems 8.23 and 8.25 by means of the Rayleigh-Ritz method using the trial functions $\phi_i(x) = \cos(2i-1)\pi x/2L$ $(i = 1, 2, \dots n)$. Solve the eigenvalue problem for $n = 2, 3, \dots, 6$, compare the two lowest eigenvalues obtained here with the exact ones obtained in Problem 8.25 and draw conclusions concerning convergence characteristics.

9.20. Derive the eigenvalue problem for the tapered beam of Problem 9.5 by means of the Rayleigh-Ritz method in the two ways: 1) using the trial functions $\phi_i(x) = x^{i+1}$ $(i = 1, 2, \dots n)$ and 2) using as trial functions the eigenfunctions of a *uniform* cantilever beam. Solve the eigenvalue problems through $n = 6$, compare results and draw conclusions concerning convergence characteristics.

9.21. Derive the eigenvalue problem for the beam of Problems 8.24 and 8.26 by means of the Rayleigh-Ritz method using as trial functions the eigenfunctions of a *uniform* free-fixed beam. Solve the eigenvalue problem for $n = 2, 3, \dots, 6$, compare the two lowest eigenvalues obtained here with the exact ones obtained in Problem 8.26 and draw conclusions concerning convergence characteristics.

9.22. Derive the eigenvalue problem for the aircraft of Problem 9.7 by means of the Rayleigh-Ritz method using the following trial functions: $\phi_i(x) = (x/L)^{i-1}, -L < x < L$ $(i = 1, 2, \dots, n)$. Solve the eigenvalue problems for $n = 4, 5, \dots, 10$ and discuss the convergence characteristics.

9.23. Derive the eigenvalue problem for the aircraft of Problem 9.7 by means of the Rayleigh-Ritz method using the following trial functions: $\phi_1(x) = 1$, $\phi_2(x) = x/L$, $-L < x < L$; $\phi_{2i-1}(x) = \psi_{i-1}(x)$ for $0 < x < L$ and 0 for $-L < x < 0$; $\phi_{2i}(x) = 0$ for $0 < x < L$ and $\psi_{i-1}(-x)$ for $-L < x < 0 (i = 2, 3, \dots, j)$, where $\psi_i(x)$ is the ith eigenfunction of a *uniform* cantilever beam. Solve the eigenvalue of a *uniform* cantilever beam. Solve the eigenvalue problems for $n = 2 + 2(j-1)$, $j = 1, 2, \dots, 5$, and discuss the convergence characteristics. Then, compare results obtained here with those obtained in Problem 9.22 and draw conclusions concerning the merits of the two sets of trial functions.

9.24. Derive the eigenvalue problem for the aircraft of Problem 9.7 by means of the Rayleigh-Ritz method using as trial functions the eigenfunctions of a *uniform* free-free beam. Note that the eigenfunctions of a uniform free-free beam were obtained in Problem 8.15 and normalized in Problem 8.21. Solve the eigenvalue problem for $n = 4, 5, \dots, 8$, where n includes the two rigid-body modes, compare results with those obtained in Problems 9.22 and 9.23 and draw conclusions concerning the accuracy of the results obtained through the use of the various trial functions.

9.25. Assume that the solution to the eigenvalue problem for the shaft of Problem 8.4, obtained by the Rayleigh-Ritz method using n terms in the approximating series, is given and it consists of the Ritz eigenvalues $\lambda_r^{(n)}$ and associated eigenvectors $\mathbf{a}_r^{(n)}$ $(r = 1, 2, \dots, n)$. Begin with the orthonormality relations for the Ritz eigenvectors, Eqs. (9.119), and derive the orthonormality relations for the Ritz eigenfunctions $\Theta_r^{(n)}(x)$ $(r = 1, 2, \dots, n)$.

9.26. Assume that the solution to the eigenvalue problem for the beam of Problem 8.8, obtained by the Rayleigh-Ritz method using n terms in the approximating series, is given and it consists of the Ritz eigenvalues $\lambda_r^{(n)}$ and associated eigenvectors $\mathbf{a}_r^{(n)}$ $(r = 1, 2, \dots, n)$. Begin with the orthonormality relations for the Ritz eigenvectors, Eqs. (9.119), and derive the orthonormality relations for the Ritz eigenfunctions $Y_r^{(n)}(x)$ $(r = 1, 2, \dots, n)$.

9.27. Assume that the solution to the eigenvalue problem for the aircraft of Problem 9.24, obtained by the Rayleigh-Ritz method using n terms in the approximating series, is given and it consists of the Ritz eigenvalues $\lambda_r^{(n)}$ and associated eigenvectors $\mathbf{a}_r^{(n)}$ $(r = 1, 2, \dots, n)$. Begin with the orthonormality relations for the Ritz eigenvectors, Eqs. (9.119), and derive the orthonormality relations for the Ritz eigenfunctions $Y_r^{(n)}(x)$ $(r = 1, 2, \dots n)$.

9.28. Derive the eigenvalue problem for the rod of Problem 9.2 by means of the Rayleigh-Ritz method using comparison functions in the form of the eigenfunctions of the associated uniform rod spring-supported at $x = 0$ and free at $x = L$. Then, derive the eigenvalue problem by the enhanced Rayleigh-Ritz method using the quasi-comparison functions $U(x) = \sum_{i=1}^{r} a_i \sin[(2i-1)\pi x/2L] + \sum_{j=1}^{s} b_j \cos j\pi x/L,\ r, s = 1, 2, \dots;\ r + s = n$. Solve the eigenvalue problems for $n = 2, 3, \dots, 6$ in each of the two cases, compare results and draw conclusions concerning convergence characteristics. Note that, for the enhanced Rayleigh-Ritz method, in the case $n = 3$ there are two quasi-comparison functions possible, one corresponding to $r = 1$ and $s = 2$ and the other corresponding to $r = 2$ and $s = 1$, and a similar statement can be made concerning the case $n = 5$.

9.29. Determine the response of the shaft of Problem 9.1 to the distributed torque $m(x,t) = mx(L - x)\delta(t)$ by means of the assumed-modes method using three terms in the series for the response. Note that the eigenvalue problem for the shaft was solved by means of the Rayleigh-Ritz method in Problem 9.17. Discuss the mode participation in the response.

9.30. Determine the response of the cantilever beam of Problem 9.5 to the uniformly distributed rectangular pulse $f(x,t) = f_0[\mathscr{u}(t) - \mathscr{u}(t - T)]$ by means of the assumed-modes method using three terms in the series for the response. Note that the associated eigenvalue problem was solved by means of the Rayleigh-Ritz method in Problem 9.20.

9.31. Derive the eigenvalue problem for the beam of Problems 8.16 by means of the Galerkin method using the comparison functions $\phi_i(x) = (L^2 - x^2)\sin\beta_i x$ $(i = 1, 2, \dots, n)$, where the values of β_i are to be obtained by satisfying the natural boundary condition at $x = L$. Solve the eigenvalue problem for $n = 2, 3, \dots, 8$, compare the computed eigenvalues with the exact ones obtained in Problem 8.16 and draw conclusions concerning accuracy.

9.32. Solve Problem 9.31 under the assumption that the beam is subjected to uniformly distributed viscous damping described by $cY(x) = 0.1\sqrt{EImL^4}d^4Y(x)/dx^4$.

9.33. Determine the response of the beam of Problem 9.32 to the excitation $f(x,t) = F_0\delta(x - L/2)\mathscr{u}(t)$ using three terms in the response series.

9.34. Derive the eigenvalue problem for the beam of Problem 9.31 by means of the collocation method using equally spaced locations x_i $(i = 1, 2, \dots, n)$. Solve the eigenvalue problem for $n = 2, 3, \dots$ and determine the number n of points x_i necessary to match the lowest eigenvalue computed in Problem 9.31 by means of the Galerkin method with $n = 8$. Draw conclusions concerning the accuracy of the collocation method.

9.35. Solve Problem 9.18 by MATLAB.

9.36. Solve Problem 9.19 by MATLAB.

9.37. Solve Problem 9.20 by MATLAB.

9.38. Solve Problem 9.22 by MATLAB.

9.39. Solve Problem 9.23 by MATLAB.

9.40. Solve Problem 9.24 by MATLAB.

9.41. Solve Problem 9.28 by MATLAB.

9.42. Write a MATLAB program for the system of Problem 9.30 and solve the problem.

9.43. Write a MATLAB program for the system of Problem 9.31 and solve the problem.

9.44. Write a MATLAB program for the system of Problem 9.32 and solve the problem.

9.45. Write a MATLAB program for the system of Problem 9.34 and solve the problem.

CHAPTER 10

THE FINITE ELEMENT METHOD

The finite element method is without a doubt the most important development in the static and dynamic analysis of structures in the second half of the twentieth century. The method was developed originally for the static stress analysis of complex distributed-parameter structures,[1] but has since broadened its scope significantly. It is basically a discretization technique that owes its enormous success to the development of the digital computer, which took place about the same time. Indeed, the method requires the solution of large sets of algebraic equations for static problems and large-order eigenvalue problems in the case of vibrations, which are numerical problems that would be impractical without a computer.

Although the finite element method was developed independently of any other method, and can be used without reference to any other method, it was soon recognized as the most important variant of the Rayleigh-Ritz method. To distinguish between the two, the latter is sometimes referred to as the classical Rayleigh-Ritz method. At this point, the popularity of the finite element method far exceeds that of the classical Rayleigh-Ritz method, to the extent that many regard the classical Rayleigh-Ritz method as mainly of academic interest. However, there are certain advantages to treating the finite element method as a Rayleigh-Ritz method. Indeed, the Rayleigh-Ritz theory lies on a solid mathematical foundation, providing a great deal of insight into the dynamic

[1] Turner, M. J., Clough, R. W., Martin, H. C. and Topp, L. J., "Stiffness and Deflection Analysis of Complex Structures," *Journal of Aeronautical Sciences*, vol. 23, 1956, pp. 805–823.

characteristics of a discretized model compared to the characteristics of the original distributed structure. Having the benefit of this theory, it is possible to develop a deeper understanding of the finite element method, and in particular of the implications of some of the current finite element practices.

As with the classical Rayleigh-Ritz method, the finite element method also envisions approximate solutions to problems of vibrating distributed systems in the form of linear combinations of known trial functions multiplied by undetermined coefficients, and determines the coefficients by solving corresponding eigenvalue problems. Moreover, the expressions for the stiffness and mass matrices defining the eigenvalue problem are the same as for the classical Rayleigh-Ritz method. The basic difference between the two approaches lies in the nature of the trial functions. Whereas in the classical Rayleigh-Ritz method the trial functions are global functions, in the sense that they extend over the entire domain of the system, in the finite element method they are local functions extending over small subdomains of the system, namely, over finite elements.

The use of local functions, commonly known as interpolation functions, gives the finite element method enormous versatility. Because the finite elements are in general very small, the differentiability tends to fade away as an issue, so that the interpolation functions can be low-degree polynomials, quite often satisfying the minimum differentiability requirements imposed on admissible functions. Moreover, the interpolation functions are the same for every element. As a result, the derivation of the stiffness and mass matrices can be carried out very efficiently by first deriving element stiffness and mass matrices and then assembling them into matrices for the whole system. In fact, the element stiffness and mass matrices for most structural members of interest can be readily found in any textbook on the subject, and the assembly task reduces to combining them so as to accommodate arbitrary parameter distributions and boundary conditions. Another advantage of the finite element method, of particular importance in two-dimensional domains such as membranes and plates, is that the finite elements can be made to fit any irregular boundary, thus permitting solutions where all other methods fail. Perhaps the most important feature of the finite element method is that the entire process lends itself to routine computer coding. In fact, many computer codes capable of accommodating a large variety of structures are available commercially.

In this chapter, the finite element method is presented in the context of the Rayleigh-Ritz theory, thus lending it the rigor lacking in the original developments of the method. When practical considerations dictate that the Rayleigh-Ritz requirements be violated, the effects of these violations on the nature of the solution are examined. The discussion includes the usual structural members, such as rods, shafts, strings and beams, as well as some that are not frequently discussed, such as trusses and frames.

10.1 THE FINITE ELEMENT METHOD AS A RAYLEIGH-RITZ METHOD

In Sec. 9.6, we demonstrated that the Rayleigh-Ritz method is essentially a discretization technique for deriving approximate solutions to differential eigenvalue problems whereby the displacement $Y(x)$ of a distributed elastic system is expressed as a linear combination of known trial functions $\phi_j(x)$ multiplied by undetermined coefficients

a_j $(j = 1, 2, \ldots, n)$, or

$$Y(x) = \sum_{j=1}^{n} a_j \phi_j(x) \tag{10.1}$$

The coefficients $a_1, a_2, \ldots, a_n$ can be determined by solving an algebraic eigenvalue problem obtained by inserting Eq. (10.1) into the energy form of Rayleigh's quotient

$$R(Y(x)) = \frac{V_{\max}}{T_{\text{ref}}} \tag{10.2}$$

where $V_{\max}$ is the elastic potential energy with $y(x, t)$ replaced by $Y(x)$ and T_{ref} is the kinetic energy with $\dot{y}(x, t)$ replaced by $Y(x)$, and rendering Rayleigh's quotient stationary. The algebraic eigenvalue problem represents an approximation to the differential eigenvalue problem.

The principal question is how well the solutions of the algebraic eigenvalue problem approximate the solutions of the differential eigenvalue problem. The answer to this question lies in the nature of the trial functions, as well as their number. In Secs. 9.6 and 9.7, we explored this question in great detail and, to this end, we introduced several classes of functions. The nature of the solution being different in the finite element method than in the classical Rayleigh-Ritz method, we focus our attention on only one class of functions, namely, one that is of particular interest in the finite element method. This is the class of admissible functions, defined as functions that are only half as many times differentiable as the order of the differential eigenvalue problem and satisfy the geometric boundary conditions alone. We note that admissible functions must satisfy only the conditions required for Rayleigh's quotient, Eq. (10.2), to be defined.

In the classical Rayleigh-Ritz method, the trial functions are often trigonometric functions, hyperbolic functions and products thereof, and they are defined over the entire domain of the system. The mass and stiffness coefficients are integrals involving squares and products of trial functions, and more often than not must be evaluated numerically. Moreover, in many cases the trial functions are not readily available and must be generated. These facts have tended to inhibit the wide use of the Rayleigh-Ritz method, even though in many cases the method can yield extremely accurate results with only a handful of degrees of freedom. Most objections to the classical Rayleigh-Ritz method have been overcome by the finite element method. Although the finite element method was originally developed independently of the Rayleigh-Ritz method, it can be regarded as a different version of the Rayleigh-Ritz method. It turns out that there are many advantages to treating the finite element method as a Rayleigh-Ritz method, the most important one being that it can claim a rigorous mathematical foundation. Hence, adopting the framework of the classical Rayleigh-Ritz method, in the finite element method as well we expand a solution as the linear combination given by Eq. (10.1) and determine the coefficients $a_1, a_2, \ldots, a_n$ by rendering Rayleigh's quotient, Eq. (10.2), stationary. Moreover, the mass and stiffness coefficients are given by the same expressions as in the Rayleigh-Ritz method. One significant difference between the two methods lies in the nature of the trial functions $\phi_j(x)$, and hence of the coefficients a_j $(j = 1, 2, \ldots, n)$. Indeed, in the finite element method the trial functions, known as *interpolation functions*, are defined over small subdomains of the system, called *finite elements*, and are zero everywhere else, where the set of finite elements is referred to as the *mesh*. Moreover, the interpolation functions represent low degree polynomials, sometimes the lowest degree admissible, and they are the same for each finite element. As a result, the computation of the mass and stiffness coefficients can be first carried out for each finite element separately, thus generating *element mass and stiffness matrices*, and then extended to the whole system to obtain *global mass and stiffness matrices*. In fact, for most problems of interest there is an inventory of element mass and stiffness matrices, which can be used to construct global mass and stiffness matrices for a variety of systems. In general, the finite element method is ideally suited for producing numerical solutions on a computer.

Some of the unique features of the finite element method can be conveniently illustrated by considering a string in transverse vibration fixed at $x = 0$ and supported by a spring at $x = L$, as shown in Fig. 10.1a. This being a second-order system, V_{max} is defined for functions that are only once differentiable with respect to x, so that a finite element approximation to the displacement curve can be generated by dividing the length L into n increments of length $h = L/n$ and connecting the corresponding displacements by straight lines, as depicted in Fig. 10.1b. The displacement curve can be described in the form of series (10.1), in which a typical trial function $\phi_j(x)$, known as a *roof function*, is as shown in Fig. 10.2. We note that the trial functions are linear in x, so that they represent admissible functions for the system. In fact, they represent the lowest-degree trial functions admissible, as lower-degree functions would be sectionally-constant. This would make the displacement profile look like a staircase,

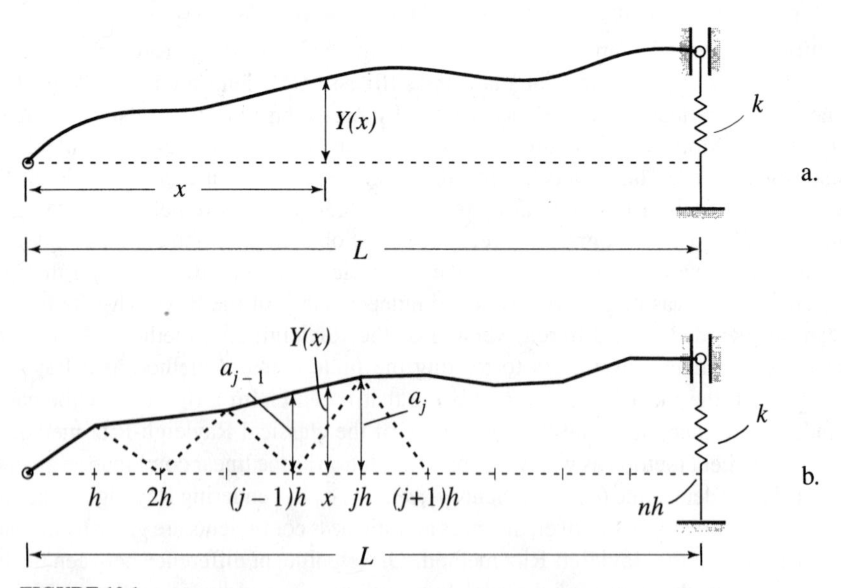

FIGURE 10.1
a. Displacement curve for a fixed-spring supported string in transverse vibration, b. Finite element approximation of the displacement curve

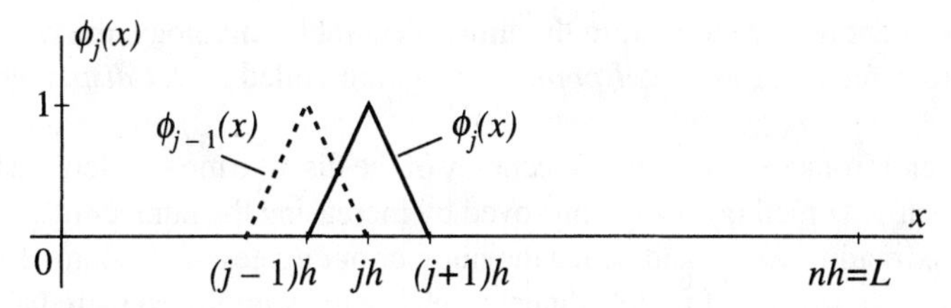

FIGURE 10.2
Roof functions

which is not admissible, because it implies infinite slope at the points of discontinuity, namely, at the boundary points between adjacent finite elements. The amplitude of all $\phi_j(x)$ was taken to be equal to unity, a very convenient choice not only because all trial functions are the same, except for locations, but also because this choice makes the coefficients a_j in Eq. (10.1) actual displacements at $x_j = jh$ $(j = 1, 2, \ldots, n)$, thus lending them important physical meaning. By contrast, in the Rayleigh-Ritz method, the coefficients a_j are abstract quantities, not unlike the coefficients in a Fourier series. Another matter of interest, and one with computational implications, is the fact that the trial functions $\phi_1(x), \phi_2(x), \ldots, \phi_n(x)$ are nearly orthogonal, as $\phi_j(x)$ overlaps only $\phi_{j-1}(x)$ and $\phi_{j+1}(x)$. As a result, the mass and stiffness matrices are *banded*, with the only nonzero entries lying in the principal diagonal and on the diagonals immediately above it and below it, independently of the system parameters.

We observe from Fig. 10.2 that the trial function $\phi_j(x)$ extends over two finite elements, $(j-1)h < x < jh$ and $jh < x < (j+1)h$. Actual computations are carried out over a single element at a time. For example, from Figs. 10.1b and 10.2, we can regard the displacement $Y(x)$ in the interval $(j-1)h < x < jh$ as consisting of contributions from the right half of $\phi_{j-1}(x)$ and the left half of $\phi_j(x)$, as shown in Fig. 10.3. Hence, we can express $Y(x)$ as the linear combination

$$Y(x) = a_{j-1}\phi_{j-1}(x) + a_j\phi_j(x), \quad (j-1)h < x < jh \tag{10.3}$$

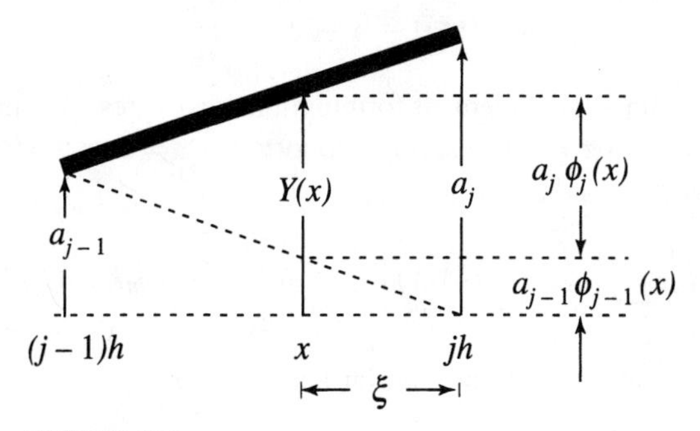

FIGURE 10.3
String displacement as a combination of two linear trial functions

It should be mentioned here that, in the finite element terminology, the nominal points $x_j = jh$ are referred to as *nodal points* and a_j are called *nodal displacements* ($j = 1, 2, \ldots, n$).

We recall from Sec. 9.6 that the accuracy of the discrete models derived by the classical Rayleigh-Ritz method can be improved by increasing the number of trial functions in the approximating series, and hence the number of degrees of freedom of the models. By contrast, the accuracy of finite element models can be improved by reducing the size h of the finite elements in a process known as refining the mesh, which is the same as increasing the number n of elements. Accuracy can also be improved by increasing the degree of the interpolation functions, but this approach is less common than refining the mesh. Both approaches result in an increase in the number of degrees of freedom of the finite element models.

In another departure from the classical Rayleigh-Ritz method, the *global coordinate* x is replaced by a *local coordinate* ξ. This simplifies the limits of the integrals for the element mass and stiffness matrices.

Beginning in Sec. 10.2, we present an efficient approach to the spatial discretization of vibrating distributed-parameter systems blending the mathematical rigor of the Rayleigh-Ritz theory with the systematic methodology of the traditional finite element method.

10.2 STRINGS, RODS AND SHAFTS

In Sec. 10.1, we introduced some of the basic ideas of the finite element method regarded as a Rayleigh-Ritz method. For easy visualization, we illustrated these ideas by means of a string in transverse vibration. As demonstrated in Ch. 8, strings represent second-order distributed systems, and so do rods in longitudinal vibration and shafts in torsion. Hence, subject to the analogies in displacement, excitation and parameters listed in Table 8.1, the developments in this section apply to all three systems. We choose to illustrate the development of the stiffness and mass matrices using a rod in axial vibration.

Using the variational approach to the eigenvalue problem, (Secs. 9.5 and 9.6) we consider the energy form of Rayleigh's quotient, Eq. (9.94), and write

$$R(U(x)) = \lambda = \omega^2 = \frac{V_{\max}}{T_{\text{ref}}} \tag{10.4}$$

where, consistent with a finite element formulation, we express the maximum potential energy for a rod fixed at $x = 0$ and spring-supported at $x = L$ in the form of a sum over the individual elements

$$V_{\max} = \frac{1}{2}\sum_{j=1}^{n}\left[\int_{(j-1)h}^{jh} EA(x)\left[\frac{dU(x)}{dx}\right]^2 dx + \delta_{jn}kU^2(L)\right] \tag{10.5}$$

and the reference kinetic energy in a similar fashion as

$$T_{\text{ref}} = \frac{1}{2}\sum_{j=1}^{n}\int_{(j-1)h}^{jh} m(x)U^2(x)dx \tag{10.6}$$

in which, by analogy with Eq. (10.3), the displacement $U(x)$ represents the vector product

$$U(x) = \boldsymbol{\phi}_j^T(x)\mathbf{a}_j,\ (j-1)h < x < jh \tag{10.7}$$

where $\boldsymbol{\phi}_j(x) = [\phi_{j-1}(x) \quad \phi_j(x)]^T$ is a vector of interpolation functions and $\mathbf{a}_j = [a_{j-1} \quad a_j]^T$ is a vector of nodal displacements for element j.

From the Rayleigh-Ritz theory (Sec. 9.6), admissible functions for the second-order problem at hand need be differentiable once only and they must satisfy the geometric boundary condition at $x = 0$. Hence, the minimum requirements are that they be linear and be zero at $x = 0$. Figure 10.3 with U replacing Y shows the typical finite element j with the displacement $U(x)$ varying linearly with x between the nodal points $x = (j-1)h$ and $x = jh$. Of course, the boundary condition at $x = 0$ is satisfied by setting $a_0 = 0$.

For convenience, we replace the *global coordinate* x by the *local coordinate*

$$\xi = (jh - x)/h \tag{10.8}$$

The objective is to simplify the integrals in Eqs. (10.5) and (10.6). To this end, we write

$$\frac{d}{dx} = \frac{d}{d\xi}\frac{d\xi}{dx} = -\frac{1}{h}\frac{d}{d\xi},\ dx = -h d\xi \tag{10.9}$$

Moreover, we must transform the limits of the integrals. From Eq. (10.8), we observe that $x = (j-1)h$ transforms into $\xi = 1$ and $x = jh$ transforms into $\xi = 0$. Next, we express the interpolation functions in terms of ξ. Referring to Fig. 10.3 and using the similarity of triangles, we have

$$\begin{aligned} \phi_{j-1}(x) &= \frac{a_{j-1}\phi_{j-1}(x)}{a_{j-1}} = \frac{jh - x}{h} \\ \phi_j(x) &= \frac{a_j\phi_j(x)}{a_j} = \frac{x - (j-1)h}{h} = 1 - \frac{jh - x}{h} \end{aligned} \tag{10.10}$$

Hence, denoting $\phi_{j-1}(x)$ and $\phi_j(x)$ by $\phi_1(\xi)$ and $\phi_2(\xi)$, respectively, and introducing Eq. (10.8) in Eqs. (10.10), we can write simply

$$\phi_1(\xi) = \xi,\ \phi_2(\xi) = 1 - \xi \tag{10.11}$$

The functions given by Eqs. (10.11) are commonly known as *linear interpolation functions*, and we observe that they apply to all finite elements, not merely to finite element j. Then, the displacement over the jth finite element, Eq. (10.7), can be rewritten as

$$U(\xi) = \boldsymbol{\phi}^T(\xi)\mathbf{a}_j = [\xi \quad 1-\xi]\mathbf{a}_j,\ 0 < \xi < 1 \tag{10.12}$$

so that, using the first of Eqs. (10.9), it follows that

$$\frac{dU(x)}{dx} = -\frac{1}{h}\frac{dU(\xi)}{d\xi} = -\frac{1}{h}\frac{d\boldsymbol{\phi}^T(\xi)}{d\xi}\mathbf{a}_j = -\frac{1}{h}[1 \quad -1]\mathbf{a}_j,\ 0 < \xi < 1 \tag{10.13}$$

Introducing Eqs. (10.9), (10.12) and (10.13) in the quantity inside brackets in Eq. (10.5), we obtain

$$\int_{(j-1)h}^{jh} EA(x)\left[\frac{dU(x)}{dx}\right]^2 dx + \delta_{jn}kU^2(L)$$

$$= \int_1^0 EA_j(\xi)\mathbf{a}_j^T\left(-\frac{1}{h}\right)^2 \frac{d\phi(\xi)}{d\xi}\frac{d\phi^T(\xi)}{d\xi}\mathbf{a}_j(-h)d\xi + \delta_{jn}k\mathbf{a}_j^T\phi(0)\phi^T(0)\mathbf{a}_j$$

$$= \mathbf{a}_j^T\left(\frac{1}{h}\int_0^1 EA_j(\xi)\begin{bmatrix} 1 \\ -1 \end{bmatrix}\begin{bmatrix} 1 \\ -1 \end{bmatrix}^T d\xi + \delta_{jn}k\begin{bmatrix} 0 \\ 1 \end{bmatrix}\begin{bmatrix} 0 \\ 1 \end{bmatrix}^T\right)\mathbf{a}_j$$

$$= \mathbf{a}_j^T\left(\frac{1}{h}\int_0^1 EA_j(\xi)\begin{bmatrix} 1 & -1 \\ -1 & 1 \end{bmatrix}^T d\xi + \delta_{jn}k\begin{bmatrix} 0 & 0 \\ 0 & 1 \end{bmatrix}\right)\mathbf{a}_j$$

$$= \mathbf{a}_j^T K_j \mathbf{a}_j, \ j = 1, 2, \ldots, n \tag{10.14}$$

where

$$K_j = \frac{1}{h}\int_0^1 EA_j(\xi)\begin{bmatrix} 1 & -1 \\ -1 & 1 \end{bmatrix} d\xi + \delta_{jn}k\begin{bmatrix} 0 & 0 \\ 0 & 1 \end{bmatrix}, \ j = 1, 2, \ldots, n \tag{10.15}$$

are the *element stiffness matrices*, in which, using Eq. (10.8),

$$EA(x) = EA[h(j-\xi)] = EA_j(\xi) \tag{10.16}$$

is the axial rigidity inside element j. Moreover, the integral in Eq. (10.6) can be evaluated as follows:

$$\int_{(j-1)h}^{jh} m(x)U^2(x)dx = \int_1^0 m_j(\xi)\mathbf{a}_j^T\phi(\xi)\phi^T(\xi)\mathbf{a}_j(-h)d\xi$$

$$= \mathbf{a}_j^T\left(h\int_0^1 m_j(\xi)\begin{bmatrix} \xi \\ 1-\xi \end{bmatrix}\begin{bmatrix} \xi \\ 1-\xi \end{bmatrix}^T d\xi\right)\mathbf{a}_j$$

$$= \mathbf{a}_j^T\left(h\int_0^1 m_j(\xi)\begin{bmatrix} \xi^2 & \xi(1-\xi) \\ \xi(1-\xi) & (1-\xi)^2 \end{bmatrix} d\xi\right)\mathbf{a}_j$$

$$= \mathbf{a}_j^T M_j \mathbf{a}_j, \ j = 1, 2, \ldots, n \tag{10.17}$$

where,

$$M_j = h\int_0^1 m_j(\xi)\begin{bmatrix} \xi^2 & \xi(1-\xi) \\ \xi(1-\xi) & (1-\xi)^2 \end{bmatrix} d\xi, \ j = 1, 2, \ldots, n \tag{10.18}$$

are the *element mass matrices*, in which

$$m(x) = m[h(j-\xi)] = m_j(\xi) \tag{10.19}$$

is the mass density inside element j.

It is common finite element practice to approximate the stiffness and mass distributions by assuming them to be sectionally constant, i.e., constant over each finite element. Strictly speaking, this approximation represents a violation of the Rayleigh-Ritz code tending to impart to the finite element method some of the characteristics of the lumped-parameter method. For practical purposes, this issue may not be as critical as it may appear, because accuracy considerations dictate the use of a large number n of elements, in which case the sectionally constant parameter distributions become nearly exact. Of course, when the parameters are in fact either uniformly distributed or sectionally constant, the finite element method can be safely regarded as a Rayleigh-Ritz method.

For constant axial rigidity over each element, $EA_j(\xi) = EA_j =$ constant, the element stiffness matrices reduce to

$$K_j = \frac{EA_j}{h}\begin{bmatrix} 1 & -1 \\ -1 & 1 \end{bmatrix} + \delta_{jn}k\begin{bmatrix} 0 & 0 \\ 0 & 1 \end{bmatrix}, \quad j = 1, 2, \ldots, n \tag{10.20}$$

Similarly for $m_j(\xi) = m_j =$ constant, the element mass matrices become

$$M_j = \frac{m_j h}{6}\begin{bmatrix} 2 & 1 \\ 1 & 2 \end{bmatrix}, \quad j = 1, 2, \ldots, n \tag{10.21}$$

Next, we use the element stiffness and mass matrices to construct the global stiffness and mass matrices in a process referred to at times as *assembly*. To this end, we insert Eqs. (10.5), (10.6), (10.14) and (10.17) into Eq. (10.4) and write Rayleigh's quotient in the form

$$R(\mathbf{a}) = \frac{\displaystyle\sum_{j=1}^{n} \mathbf{a}_j^T K_j \mathbf{a}_j}{\displaystyle\sum_{j=1}^{n} \mathbf{a}_j^T M_j \mathbf{a}_j} = \frac{\mathbf{a}^T K \mathbf{a}}{\mathbf{a}^T M \mathbf{a}} \tag{10.22}$$

where $\mathbf{a} = [a_1 \ a_2 \ \ldots \ a_n]^T$ is the *system nodal vector* and K and M are *global stiffness and mass matrices*. We observe that the nodal coordinate a_j appears twice, once as the bottom component in the element nodal vector $\mathbf{a}_j$ and once as the top component in the element nodal vector $\mathbf{a}_{j+1}$. Hence, in carrying out the summations in Eq. (10.22), the right bottom entry (2,2) of the element matrix K_j and the left top entry (1,1) of K_{j+1} add up, and the same can be said about the corresponding entries in matrices M_j and M_{j+1}. Moreover, because the rod is fixed at $x = 0$, $a_0 = 0$, a fact we already took into account in the nodal vector $\mathbf{a}$. For the same reason, we must cross out the first row and column in the element matrices K_1 and M_1. Hence, the global stiffness and mass matrices have the schematic form shown in Fig. 10.4, where the heavy symbols denote entries representing the sum of the (2,2) entry of K_j and M_j and the $(1, 1)$ entry of K_{j+1} and M_{j+1}, respectively. Clearly, K and M are *banded* due to the nature of the interpolation functions. Hence, using Eqs. (10.20) and (10.21) in conjunction with the scheme of Fig. 10.4, the global stiffness and mass matrices are

$$K = \frac{E}{h} \begin{bmatrix} A_1+A_2 & -A_2 & 0 & \dots & 0 & 0 \\ -A_2 & A_2+A_3 & -A_3 & \dots & 0 & 0 \\ 0 & -A_3 & A_3+A_4 & \dots & 0 & 0 \\ \cdots & \cdots & \cdots & \cdots & \cdots & \cdots \\ 0 & 0 & 0 & \dots & A_{n-1}+A_n & -A_n \\ 0 & 0 & 0 & \dots & -A_n & A_n+kh/E \end{bmatrix} \tag{10.23}$$

and

$$M = \frac{h}{6} \begin{bmatrix} 2(m_1+m_2) & m_2 & 0 & \dots & 0 & 0 \\ m_2 & 2(m_2+m_3) & m_3 & \dots & 0 & 0 \\ 0 & m_3 & 2(m_3+m_4) & \dots & 0 & 0 \\ \cdots & \cdots & \cdots & \cdots & \cdots & \cdots \\ 0 & 0 & 0 & \dots & 2(m_{n-1}+m_n) & m_n \\ 0 & 0 & 0 & \dots & m_n & 2m_n \end{bmatrix} \tag{10.24}$$

respectively. Matrices K and M are said to have *half-bandwidth* one, because there is one nonzero diagonal above and below the main diagonal.

In generating the global matrices, Eqs. (10.23) and (10.24), on a computer, it is convenient to implement the scheme of Fig. 10.4 sequentially, such as in a "do" loop or a "for" loop. As an illustration, to generate the global stiffness matrix K, we begin with an $n \times n$ null matrix K and add, in sequence, K_1 to the submatrix (actually a scalar in this case) K_{11} of K, K_2 to the 2×2 submatrix $[K_{11}\ K_{12};\ K_{21}\ K_{22}]$ of K, K_3 to the submatrix $[K_{22}\ K_{23};\ K_{32}\ K_{33}]$ of K, etc., where the semicolon indicates the end of a row.

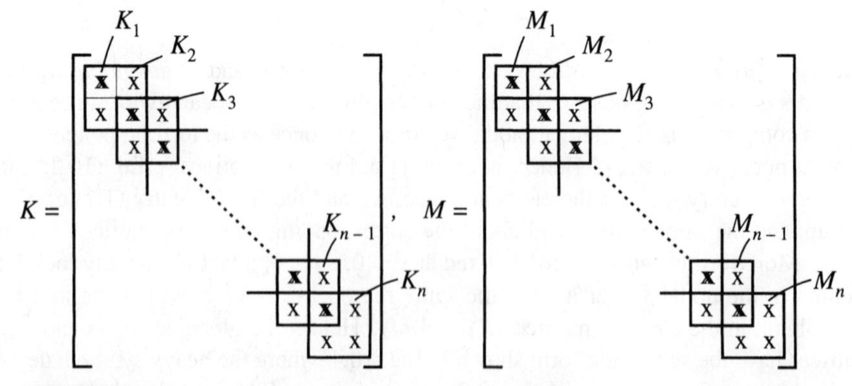

FIGURE 10.4
Scheme for the assembly of global matrices from element matrices for second-order systems using linear interpolation functions

Following the usual steps in the Rayleigh-Ritz method, in conjunction with Rayleigh's quotient in the form given by Eq. (10.22), we obtain the eigenvalue problem

$$K\mathbf{a} = \lambda M\mathbf{a}, \ \lambda = \omega^2 \tag{10.25}$$

which can be solved by one of the methods in Rev. 13.

The solution of the eigenvalue problem, Eq. (10.25), consists of the approximate eigenvalues $\lambda_r^{(n)}$ and associated eigenvectors $\mathbf{a}_r^{(n)}$ $(r = 1, 2, \ldots, n)$, where the superscript (n) was added to remove any ambiguity in notation. The approximate natural frequencies are simply $\omega_r^{(n)} = \sqrt{\lambda_r^{(n)}}$ $(r = 1, 2, \ldots, n)$. Moreover, the approximate natural modes $U_r^{(n)}(x)$ can be obtained by recognizing that $U_r^{(n)}(h), U_r^{(n)}(2h), \ldots, U_r^{(n)}(nh) = U_r^{(n)}(L)$ are simply the components $a_{1,r}^{(n)}, a_{2,r}^{(n)}, \ldots, a_{n,r}^{(n)}$ of the rth modal vector $\mathbf{a}_r^{(n)}$ $(r = 1, 2, \ldots, n)$.

Example 10.1. Solve the eigenvalue problem for the tapered rod in axial vibration of Example 9.6 by the finite element method in the two ways: (1) use the element stiffness and mass matrices given by Eqs. (10.15) and (10.18), respectively, and (2) approximate the stiffness and mass distributions over the finite elements as follows:

$$EA_j = \frac{6EA}{5}\left[1 - \frac{1}{2}\left(\frac{2j-1}{2n}\right)^2\right], \ m_j = \frac{6m}{5}\left[1 - \frac{1}{2}\left(\frac{2j-1}{2n}\right)^2\right], \ j = 1, 2, \ldots, n \tag{a}$$

Determine how the accuracy of the three lowest natural frequencies improves as the mesh is refined; begin the computations with $n = 10$. Compare results obtained by the finite element method in the two indicated ways, as well as with those obtained in Example 9.6 by the Rayleigh-Ritz method using both admissible and comparison functions and draw conclusions.

From Example 9.6, and using the transformation implied by Eqs. (10.16) and (10.19), the system parameters are

$$EA(\xi) = \frac{6EA}{5}\left[1 - \frac{(j-\xi)^2}{2n^2}\right], \ m(\xi) = \frac{6m}{5}\left[1 - \frac{(j-\xi)^2}{2n^2}\right], \ j = 1, 2, \ldots, n \tag{b}$$

Moreover, we recall that the spring at $x = L$ has the stiffness $k = EA/L$. Inserting the first of Eqs. (b) into Eqs. (10.15) and carrying out the integrations, we obtain the element stiffness matrices

$$K_1 = \frac{6EAn}{5L}\left(1 - \frac{1}{6n^2}\right)$$

$$K_j = \frac{6EAn}{5L}\left[1 - \frac{1 - 3j + 3j^2}{6n^2}\right]\begin{bmatrix} 1 & -1 \\ -1 & 1 \end{bmatrix} + \delta_{jn}\frac{EA}{L}\begin{bmatrix} 0 & 0 \\ 0 & 1 \end{bmatrix}, \ j = 2, 3, \ldots, n \tag{c}$$

where K_1 is really a scalar. Using Eqs. (c) in conjunction with the scheme for K shown in

Fig. 10.4, we obtain the global stiffness matrix

$$K = \frac{6EAn}{5L}\begin{bmatrix} 2 & -1 & 0 & \dots & 0 & 0 \\ -1 & 2 & -1 & \dots & 0 & 0 \\ 0 & -1 & 2 & \dots & 0 & 0 \\ \cdots & \cdots & \cdots & \cdots & \cdots & \cdots \\ 0 & 0 & 0 & \dots & 2 & -1 \\ 0 & 0 & 0 & \dots & -1 & 1+5/6n \end{bmatrix}$$

$$- \frac{EA}{5Ln}\begin{bmatrix} 8 & -7 & 0 & \dots & 0 & 0 \\ -7 & 26 & -19 & \dots & 0 & 0 \\ 0 & -19 & 56 & \dots & 0 & 0 \\ \cdots & \cdots & \cdots & \cdots & \cdots & \cdots \\ 0 & 0 & 0 & \dots & 2(4-6n+3n^2) & -(1-3n+3n^2) \\ 0 & 0 & 0 & \dots & -(1-3n+3n^2) & 1-3n+3n^2 \end{bmatrix} \quad \text{(d)}$$

Similarly, introducing the second of Eqs. (b) in Eqs. (10.18) and carrying out the integrations, the element mass matrices can be shown to be

$$M_1 = \frac{2mL}{5n}\left(1 - \frac{3}{10n^2}\right)$$

$$M_j = \frac{mL}{5n}\begin{bmatrix} 2 & 1 \\ 1 & 2 \end{bmatrix} - \frac{mL}{100n^3}\begin{bmatrix} 2(6-15j+10j^2) & 3-10j+10j^2 \\ 3-10j+10j^2 & 2(1-5j+10j^2) \end{bmatrix}, \quad j = 2, 3, \dots, n \quad \text{(e)}$$

Equations (e) can be combined as indicated by the scheme for M in Fig. 10.4 to yield the global mass matrix

$$M = \frac{mL}{5n}\begin{bmatrix} 4 & 1 & 0 & \dots & 0 & 0 \\ 1 & 4 & 1 & \dots & 0 & 0 \\ 0 & 1 & 4 & \dots & 0 & 0 \\ \cdots & \cdots & \cdots & \cdots & \cdots & \cdots \\ 0 & 0 & 0 & \dots & 4 & 1 \\ 0 & 0 & 0 & \dots & 1 & 2 \end{bmatrix}$$

$$- \frac{mL}{100n^3}\begin{bmatrix} 44 & 23 & 0 & \dots & 0 & 0 \\ 23 & 164 & 63 & \dots & 0 & 0 \\ 0 & 63 & 364 & \dots & 0 & 0 \\ \cdots & \cdots & \cdots & \cdots & \cdots & \cdots \\ 0 & 0 & 0 & \dots & 4(11-20n+10n^2) & 3-10n+10n^2 \\ 0 & 0 & 0 & \dots & 3-10n+10n^2 & 2(1-5n+10n^2) \end{bmatrix} \quad \text{(f)}$$

The eigenvalue problem, obtained by inserting Eqs. (d) and (f) into Eq. (10.25), was solved for $n = 10, 11, \dots, 75$. The three lowest normalized natural frequencies are displayed in Table 10.1. The three lowest natural modes corresponding to $n = 20$ are plotted in Fig. 10.5.

Next, we consider the case in which the stiffness and mass distributions are assumed to be constant over each finite element according to Eqs. (a). Inserting A_j from the first of

Table 10.1 Normalized Natural Frequencies for Linear Interpolation Functions—Exact Parameter Distributions

n	$\omega_1^{(n)}\sqrt{mL^2/EA}$	$\omega_2^{(n)}\sqrt{mL^2/EA}$	$\omega_3^{(n)}\sqrt{mL^2/EA}$
10	2.219979	5.152368	8.334965
11	2.219206	5.143180	8.296875
12	2.218619	5.136197	8.267934
13	2.218161	5.130764	8.245432
14	2.217798	5.126456	8.227593
⋮	⋮	⋮	⋮
20	2.216639	5.112713	8.170752
⋮	⋮	⋮	⋮
29	2.216054	5.105795	8.142184
30	2.216020	5.105384	8.140487
31	2.215988	5.105012	8.138952
⋮	⋮	⋮	⋮
73	2.215608	5.100514	8.120397
74	2.215606	5.100487	8.120288
75	2.215604	5.100462	8.120182

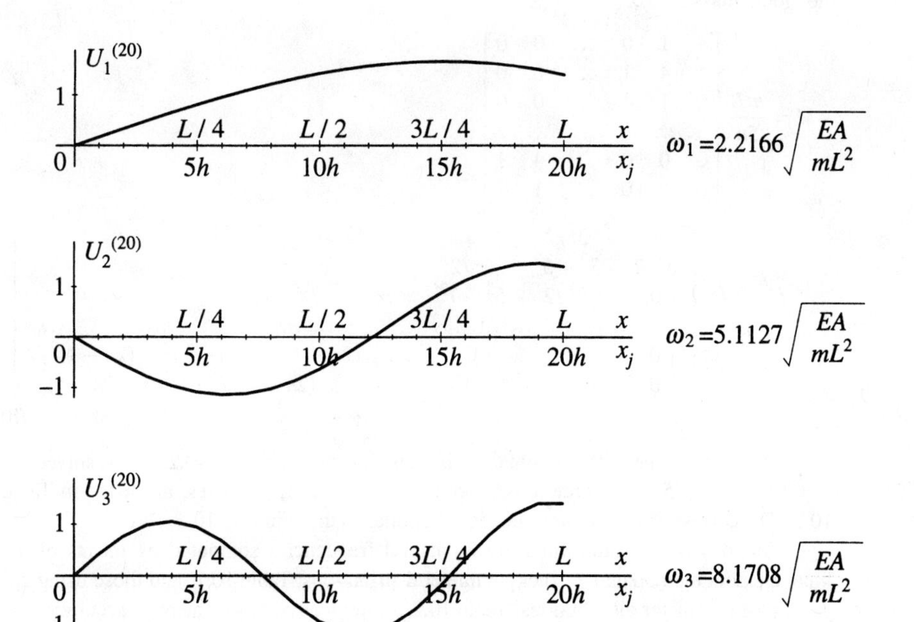

FIGURE 10.5
Three lowest modes computed using 20 finite elements in conjunction with linear interpolation functions

Eqs. (a), as well as $h = L/n$, into Eq. (10.23), we obtain the global stiffness matrix

$$K = \frac{6EAn}{5L}\begin{bmatrix} 2 & -1 & 0 & \dots & 0 & 0 \\ -1 & 2 & -1 & \dots & 0 & 0 \\ 0 & -1 & 2 & \dots & 0 & 0 \\ \dots & \dots & \dots & \dots & \dots & \dots \\ 0 & 0 & 0 & \dots & 2 & -1 \\ 0 & 0 & 0 & \dots & -1 & 1+5/6n \end{bmatrix}$$

$$-\frac{3EA}{20Ln}\begin{bmatrix} 1+3^2 & -3^2 & 0 & \dots & 0 & 0 \\ -3^2 & 3^2+5^2 & -5^2 & \dots & 0 & 0 \\ 0 & -5^2 & 5^2+7^2 & \dots & 0 & 0 \\ \dots & \dots & \dots & \dots & \dots & \dots \\ 0 & 0 & 0 & \dots & (2n-3)^2+(2n-1)^2 & -(2n-1)^2 \\ 0 & 0 & 0 & \dots & -(2n-1)^2 & (2n-1)^2 \end{bmatrix} \tag{g}$$

Similarly, introducing the second of Eqs. (a), as well as $h = L/n$, in Eq. (10.24), we obtain the global mass matrix

$$M = \frac{mL}{5n}\begin{bmatrix} 4 & 1 & 0 & \dots & 0 & 0 \\ 1 & 4 & 1 & \dots & 0 & 0 \\ 0 & 1 & 4 & \dots & 0 & 0 \\ \dots & \dots & \dots & \dots & \dots & \dots \\ 0 & 0 & 0 & \dots & 4 & 1 \\ 0 & 0 & 0 & \dots & 1 & 2 \end{bmatrix}$$

$$-\frac{mL}{20n^3}\begin{bmatrix} 1+3^2 & 3^2/2 & 0 & \dots & 0 & 0 \\ 3^2/2 & 3^2+5^2 & 5^2/2 & \dots & 0 & 0 \\ 0 & 5^2/2 & 5^2+7^2 & \dots & 0 & 0 \\ \dots & \dots & \dots & \dots & \dots & \dots \\ 0 & 0 & 0 & \dots & (2n-3)^2+(2n-1)^2 & (2n-1)^2/2 \\ 0 & 0 & 0 & \dots & (2n-1)^2/2 & (2n-1)^2 \end{bmatrix} \tag{h}$$

The eigenvalue problem, obtained by Eqs. (c) and (e) in Eq. (10.25), was solved for $n = 10, 11, \dots, 75$. The three lowest normalized natural frequencies, are listed in Table 10.2. The corresponding natural modes are similar to those in Fig. 10.5.

We observe that the three lowest natural frequencies computed by means of the finite element method using exact parameter distributions, Table 10.1, and those using approximate parameter distributions, Table 10.2, show steady improvement in accuracy as n increases. A comparison of the natural frequencies computed by means of the finite element method with those computed by the Rayleigh-Ritz method using admissible functions presents a mixed picture. Indeed, from Example 9.6, we have $\omega_1^{(30)} = 2.218950\sqrt{EA/mL^2}$, $\omega_2^{(30)} = 5.102324\sqrt{EA/mL^2}$ and $\omega_3^{(30)} = 8.118398\sqrt{EA/mL^2}$, so that the lowest natural

Table 10.2 Normalized Natural Frequencies for Linear Interpolation Functions—Approximate Parameter Distributions

n	$\omega_1^{(n)}\sqrt{mL^2/EA}$	$\omega_2^{(n)}\sqrt{mL^2/EA}$	$\omega_3^{(n)}\sqrt{mL^2/EA}$
10	2.219493	5.148365	8.325987
11	2.218807	5.139915	8.289660
12	2.218285	5.133480	8.262000
13	2.217877	5.128467	8.240460
14	2.217554	5.124487	8.223362
⋮	⋮	⋮	⋮
20	2.216520	5.111767	8.168765
⋮	⋮	⋮	⋮
29	2.215998	5.105350	8.141259
30	2.215967	5.104968	8.139624
31	2.215939	5.104623	8.138144
⋮	⋮	⋮	⋮
73	2.215599	5.100444	8.120254
74	2.215597	5.100420	8.120148
75	2.215595	5.100396	8.120046

frequency computed by means of the finite element method with a thirty-degree-of-freedom model is closer to the actual value than that computed by the Rayleigh-Ritz method using admissible functions, but the opposite is true for the second and third natural frequencies. This is not as comforting as it may seem, as the three lowest natural frequencies computed by the Rayleigh-Ritz method using comparison functions reach convergence as follows: $\omega_1^{(14)} = 2.215524\sqrt{EA/mL^2}$, $\omega_2^{(14)} = 5.099525\sqrt{EA/mL^2}$, $\omega_3^{(20)} = 8.116318\sqrt{EA/mL^2}$. On the other hand, as can be seen from Tables 10.1 and 10.2, the three lowest natural frequencies computed by means of the finite element method using 75 linear interpolation functions, i.e., 15 degrees of freedom, are quite far from convergence, as the rate of convergence is quite low.

10.3 HIGHER-DEGREE INTERPOLATION FUNCTIONS

In Sec. 10.2, we considered the derivation of approximate solutions to the eigenvalue problem for strings, rods and shafts by the finite element method using linear interpolation functions. In the process, we highlighted several attractive features of the finite element method, and in particular that it uses low-degree interpolation functions as admissible functions and that it renders the task of deriving the stiffness and mass matrices almost routine. The net result is a procedure ideally suited for computer coding. All this ease of implementation comes at a price, however, as the number of degrees of freedom of the discrete model tends to be very large compared to that required by the classical Rayleigh-Ritz method for the same level of accuracy. The advantages and disadvantages can be attributed to the use of linear interpolation functions, which are the lowest-degree admissible functions for second-order systems, such as strings, rods and shafts. Hence, the question arises as to whether the situation can be mitigated by using higher-degree interpolation functions.

The interpolation functions of next higher degree are *quadratic*. But, in attempting to use quadratic interpolation functions for strings, rods and shafts, we run immediately into some difficulty. For easy visualization of the problem, we go back to the string in transverse vibration considered in Sec. 10.1. From Fig. 10.3, we conclude that, by connecting two nodal displacements by a straight line, the string displacement at any point inside the finite element is defined uniquely. In fact, we can generate the two linear interpolation functions given by Eqs. (10.11) in this manner. To show this, and to help with the forthcoming generation of the quadratic interpolation functions, we express the linear displacement inside element j in terms of the local coordinate as follows:

$$Y(\xi) = c_1 + c_2\xi, \; 0 < \xi < 1 \tag{10.26}$$

where c_1 and c_2 are constants. They can be determined by observing from Fig. 10.3 that

$$Y(0) = a_j = c_1, \; Y(1) = a_{j-1} = c_1 + c_2 \tag{10.27}$$

Equations (10.27) have the solution

$$c_1 = a_j, \; c_2 = a_{j-1} - a_j \tag{10.28}$$

so that, inserting Eqs. (10.28) into Eq. (10.26), we obtain the equation of the straight line passing through the nodal displacements a_{j-1} and a_j in the form

$$Y(\xi) = a_{j-1}\xi + a_j(1-\xi), \; 0 < \xi < 1 \tag{10.29}$$

But, by analogy with Eq. (10.3), we can write

$$Y(\xi) = a_{j-1}\phi_1(\xi) + a_j\phi_2(\xi) \tag{10.30}$$

from which we conclude that

$$\phi_1(\xi) = \xi, \; \phi_2(\xi) = 1 - \xi, \; 0 < \xi < 1 \tag{10.31}$$

Equations (10.31) are identical to Eqs. (10.11) describing the linear interpolation functions for the problem at hand.

Next, we explore the possibility of approximating the displacement curve by means of quadratic interpolation functions. To this end, we express the displacement inside the finite element in the form of the generic quadratic function

$$Y(\xi) = c_1 + c_2\xi + c_3\xi^2, \; 0 < \xi < 1 \tag{10.32}$$

Following the same steps as earlier in this section, we write

$$Y(0) = a_j = c_1, \; Y(1) = a_{j-1} = c_1 + c_2 + c_3 \tag{10.33}$$

so that the difficulty is obvious: there are three unknowns, c_1, c_2 and c_3, and only two equations. It follows that two nodal displacements cannot be connected by a unique quadratic function. Indeed, there is an infinity of quadratic functions passing through two points. To render the quadratic function unique, we must have a third equation for c_1, c_2 and c_3, which necessitates that we specify a third point through which the curve must pass. Common sense dictates that we create a third nodal point located at $x = (j - 1/2)h$, which is halfway between the two original nodal points located at

$x = (j-1)h$ and $x = jh$; we denote the corresponding nodal displacement by $a_{j-1/2}$. We refer to $x = (j-1/2)h$ as an *internal node*, which makes the original two *external nodes*. Recognizing that the local coordinate of the internal node is $\xi = 1/2$, we can write the desired third equation as

$$Y\left(\frac{1}{2}\right) = a_{j-1/2} = c_1 + \frac{1}{2}c_2 + \frac{1}{4}c_3 \tag{10.34}$$

Solving Eqs. (10.33) and (10.34), we obtain

$$c_1 = a_j,\ c_2 = -a_{j-1} + 4a_{j-1/2} - 3a_j,\ c_3 = 2a_{j-1} - 4a_{j-1/2} + 2a_j \tag{10.35}$$

so that Eq. (10.32) becomes

$$Y(\xi) = a_{j-1}(-\xi + 2\xi^2) + a_{j-1/2}(4\xi - 4\xi^2) + a_j(1 - 3\xi + 2\xi^2),\ 0 < \xi < 1 \tag{10.36}$$

Then, using the analogy with Eq. (10.30) and writing

$$Y(\xi) = a_{j-1}\phi_1(\xi) + a_{j-1/2}\phi_2(\xi) + a_j\phi_3(\xi),\ 0 < \xi < 1 \tag{10.37}$$

we conclude that the quadratic interpolation functions, the same for every element, have the form

$$\phi_1(\xi) = \xi(2\xi - 1),\ \phi_2(\xi) = 4\xi(1-\xi),\ \phi_3(\xi) = (1-\xi)(1-2\xi) \tag{10.38}$$

They are displayed in Fig. 10.6.

Next, we return to the rod in axial vibration fixed at $x = 0$ and spring-supported at $x = L$ considered earlier in this section. To derive the element stiffness and mass matrices corresponding to quadratic interpolation functions, we use the analogy with Eq. (10.12) and write the displacement in the form

$$U(\xi) = \boldsymbol{\phi}^T(\xi)\mathbf{a}_j,\ 0 < \xi < 1 \tag{10.39}$$

where

$$\boldsymbol{\phi}(\xi) = [\xi(2\xi - 1)\ \ 4\xi(1-\xi)\ \ (1-\xi)(1-2\xi)]^T \tag{10.40}$$

is the vector of interpolation functions and

$$\mathbf{a}_j = [a_{j-1}\ \ a_{j-1/2}\ \ a_j]^T \tag{10.41}$$

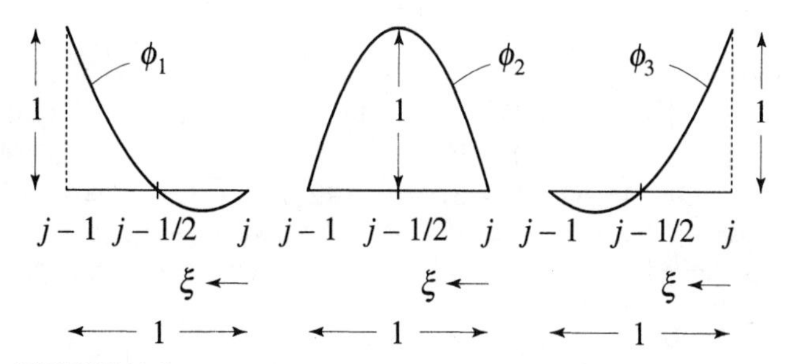

FIGURE 10.6
Quadratic interpolation functions

is the vector of nodal coordinates. Then, by analogy with Eq. (10.13), we write

$$\frac{dU(x)}{dx} = -\frac{1}{h}\frac{dU(\xi)}{d\xi} = -\frac{1}{h}\frac{d\boldsymbol{\phi}^T(\xi)}{d\xi}\mathbf{a}_j = -\frac{1}{h}[4\xi - 1 \;\; 4(1-2\xi) \;\; 4\xi - 3]^T \mathbf{a}_j \tag{10.42}$$

Inserting Eq. (10.42) into Eq. (10.14), recalling the second of Eqs. (10.9) and following the same steps as for linear interpolation functions, we obtain

$$\begin{aligned}
&\int_{(j-1)h}^{jh} EA(x)\left[\frac{dU(x)}{dx}\right]^2 dx + \delta_{jn}kU^2(L) \\
&= \int_1^0 EA_j(\xi)\mathbf{a}_j^T\left(-\frac{1}{h}\right)^2 \frac{d\boldsymbol{\phi}(\xi)}{d\xi}\frac{d\boldsymbol{\phi}^T(\xi)}{d\xi}\mathbf{a}_j(-h)d\xi + \delta_{jn}k\mathbf{a}_j^T\boldsymbol{\phi}(0)\boldsymbol{\phi}^T(0)\mathbf{a}_j \\
&= \mathbf{a}_j^T\left(\frac{1}{h}\int_0^1 EA_j(\xi)\begin{bmatrix} 4\xi - 1 \\ 4(1-2\xi) \\ 4\xi - 3 \end{bmatrix}\begin{bmatrix} 4\xi - 1 \\ 4(1-2\xi) \\ 4\xi - 3 \end{bmatrix}^T d\xi \right. \\
&\qquad \left. + \delta_{jn}k\begin{bmatrix} 0 \\ 0 \\ 1 \end{bmatrix}\begin{bmatrix} 0 \\ 0 \\ 1 \end{bmatrix}^T\right)\mathbf{a}_j \\
&= \mathbf{a}_j^T\left(\frac{1}{h}\int_0^1 EA_j(\xi)\begin{bmatrix} (4\xi-1)^2 & 4(4\xi-1)(1-2\xi) & (4\xi-1)(4\xi-3) \\ 4(4\xi-1)(1-2\xi) & 16(1-2\xi)^2 & 4(1-2\xi)(4\xi-3) \\ (4\xi-1)(4\xi-3) & 4(1-2\xi)(4\xi-3) & (4\xi-3)^2 \end{bmatrix} d\xi + \delta_{jn}k\begin{bmatrix} 0 & 0 & 0 \\ 0 & 0 & 0 \\ 0 & 0 & 1 \end{bmatrix}\right)\mathbf{a}_j \\
&= \mathbf{a}_j^T K_j \mathbf{a}_j, \quad j = 1, 2, \ldots, n
\end{aligned} \tag{10.43}$$

in which

$$K_j = \frac{1}{h}\int_0^1 EA_j(\xi)\begin{bmatrix} (4\xi-1)^2 & 4(4\xi-1)(1-2\xi) & (4\xi-1)(4\xi-3) \\ 4(4\xi-1)(1-2\xi) & 16(1-2\xi)^2 & 4(1-2\xi)(4\xi-3) \\ (4\xi-1)(4\xi-3) & 4(1-2\xi)(4\xi-3) & (4\xi-3)^2 \end{bmatrix} d\xi + \delta_{jn}k\begin{bmatrix} 0 & 0 & 0 \\ 0 & 0 & 0 \\ 0 & 0 & 1 \end{bmatrix}, \quad j = 1, 2, \ldots, n \tag{10.44}$$

are the element stiffness matrices. Similarly, by analogy with Eq. (10.17), we have

$$\int_{(j-1)h}^{jh} m(x)U^2(x)dx = \int_1^0 m_j(\xi)\mathbf{a}_j^T \phi(\xi)\phi^T(\xi)\mathbf{a}_j(-h)d\xi$$

$$= \mathbf{a}_j^T \left(h \int_0^1 m_j(\xi) \begin{bmatrix} \xi(2\xi-1) \\ 4\xi(1-\xi) \\ (1-\xi)(1-2\xi) \end{bmatrix} \begin{bmatrix} \xi(2\xi-1) \\ 4\xi(1-\xi) \\ (1-\xi)(1-2\xi) \end{bmatrix}^T d\xi \right) \mathbf{a}_j$$

$$= \mathbf{a}_j^T \left(h \int_0^1 m_j(\xi) \begin{bmatrix} \xi^2(2\xi-1)^2 & 4\xi^2(2\xi-1)(1-\xi) & \xi(2\xi-1)(1-\xi)(1-2\xi) \\ 4\xi^2(2\xi-1)(1-\xi) & 16\xi^2(1-\xi)^2 & 4\xi(1-\xi)^2(1-2\xi) \\ \xi(2\xi-1)(1-\xi)(1-2\xi) & 4\xi(1-\xi)^2(1-2\xi) & (1-\xi)^2(1-2\xi)^2 \end{bmatrix} d\xi \right) \mathbf{a}_j$$

$$= \mathbf{a}_j^T M_j \mathbf{a}_j, \quad j = 1, 2, \ldots, n \tag{10.45}$$

where

$$M_j = h \int_0^1 m_j(\xi) \begin{bmatrix} \xi^2(2\xi-1)^2 & 4\xi^2(2\xi-1)(1-\xi) & \xi(2\xi-1)(1-\xi)(1-2\xi) \\ 4\xi^2(2\xi-1)(1-\xi) & 16\xi^2(1-\xi)^2 & 4\xi(1-\xi)^2(1-2\xi) \\ \xi(2\xi-1)(1-\xi)(1-2\xi) & 4\xi(1-\xi)^2(1-2\xi) & (1-\xi)^2(1-2\xi)^2 \end{bmatrix} d\xi, \quad j = 1, 2, \ldots, n \tag{10.46}$$

are the element mass matrices.

For constant axial rigidity over the elements, $EA_j(\xi) = EA_j$ = constant, the element matrices reduce to

$$K_j = \frac{EA_j}{3h} \begin{bmatrix} 7 & -8 & 1 \\ -8 & 16 & -8 \\ 1 & -8 & 7 \end{bmatrix} + \delta_{jn}k \begin{bmatrix} 0 & 0 & 0 \\ 0 & 0 & 0 \\ 0 & 0 & 1 \end{bmatrix}, \quad j = 1, 2, \ldots, n \tag{10.47}$$

and, for constant mass density over the elements, $m_j(\xi) = m_j$ = constant, the mass matrices become

$$M_j = \frac{m_j h}{30} \begin{bmatrix} 4 & 2 & -1 \\ 2 & 16 & 2 \\ -1 & 2 & 4 \end{bmatrix}, \quad j = 1, 2, \ldots, n \tag{10.48}$$

The assembly process for the generation of the global stiffness and mass matrices is essentially the same as for linear interpolation functions, although there are some differences. In the first place, because of the end condition $a_0 = 0$, matrices K_1 and M_1 are only 2×2, obtained by removing the first row and column from Eqs. (10.44) and

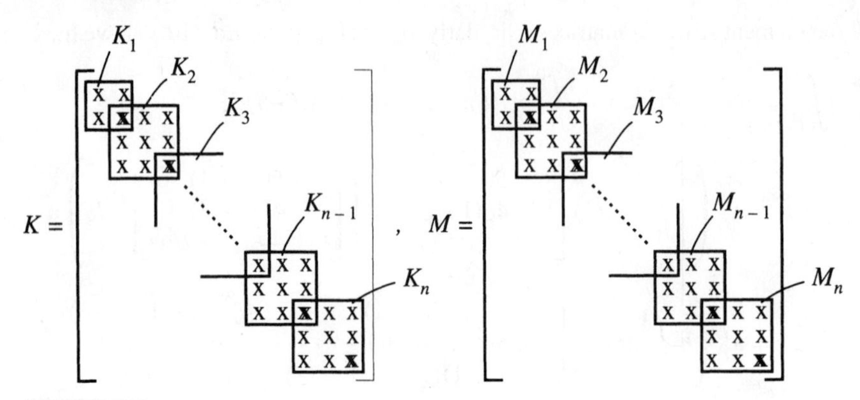

FIGURE 10.7
Scheme for the assembly of global matrices from element matrices for second-order systems using quadratic interpolation functions

(10.46), or Eqs. (10.47) and (10.48), respectively. Moreover, the summation of entries takes place only for external nodes. Hence, the global stiffness and mass matrices K and M, respectively, have the schematic form shown in Fig. 10.7, where the heavy symbols correspond to the sum of the entries (1,1) and (3,3) of the element matrices. Note that the global matrices K and M are $2n \times 2n$, which is consistent with the fact that a finite element model using quadratic interpolation functions in conjunction with n elements has $2n$ degrees of freedom; the corresponding $2n$-dimensional nodal vector is defined by $\mathbf{a} = [a_{1/2}\ a_1\ a_{3/2}\ \dots\ a_{n-1/2}\ a_n]^T$. The process is illustrated in the following example.

The generation of the global stiffness matrix K on a computer follows the same pattern as that described in Sec. 10.2, except that K_1 is added to the 2×2 submatrix $[K_{11}\ \ K_{12};\ K_{21}\ \ K_{22}]$ of K, K_2 is added to the 3×3 submatrix $[K_{22}\ \ K_{23}\ \ K_{24};\ K_{32}\ K_{33}\ K_{34};\ K_{42}\ K_{43}\ K_{44}]$ of K, etc.

Example 10.2. Solve the eigenvalue problem for the tapered rod of Example 10.1 by the finite element method using quadratic interpolation functions. Compare results with those obtained in Examples 9.6 and 10.1 and draw conclusions.

First, we consider the case in which the system parameters over the elements are accounted for exactly; they are given by Eqs. (b) of Example 10.1. Hence, inserting the first of these equations into Eqs. (10.44) and carrying out the indicated integrations, we obtain the element stiffness matrices

$$K_1 = \frac{2EAn}{5L}\begin{bmatrix} 16 & -8 \\ -8 & 7 \end{bmatrix} - \frac{EA}{25Ln}\begin{bmatrix} 32 & -26 \\ -26 & 23 \end{bmatrix}$$

$$K_j = \frac{2EAn}{5L}\begin{bmatrix} 7 & -8 & 1 \\ -8 & 16 & -8 \\ 1 & -8 & 7+\delta_{jn}5/2n \end{bmatrix} \tag{a}$$

$$-\frac{EA}{25Ln}\begin{bmatrix} 23-55j+35j^2 & -26+60j-40j^2 & 3-5j+5j^2 \\ -26+60j-40j^2 & 32-80j+80j^2 & -6+20j-40j^2 \\ 3-5j+5j^2 & -6+20j-40j^2 & 3-15j+35j^2 \end{bmatrix},$$

$$j = 2, 3, \dots, n$$

Then, using the scheme for K in Fig. 10.7, the global stiffness matrix can be shown to be

$$
K = \frac{2EAn}{5L}\begin{bmatrix}
16 & -8 & 0 & 0 & \dots & 0 & 0 & 0 \\
-8 & 14 & -8 & 1 & \dots & 0 & 0 & 0 \\
0 & -8 & 16 & -8 & \dots & 0 & 0 & 0 \\
0 & 1 & -8 & 14 & \dots & 0 & 0 & 0 \\
\cdots & \cdots & \cdots & \cdots & \cdots & \cdots & \cdots & \cdots \\
0 & 0 & 0 & 0 & \dots & 14 & -8 & 1 \\
0 & 0 & 0 & 0 & \dots & -8 & 16 & -8 \\
0 & 0 & 0 & 0 & \dots & 1 & -8 & 7+5/2n
\end{bmatrix}
$$

$$
-\frac{EA}{25Ln}\begin{bmatrix}
32 & -26 & 0 & 0 & \dots & 0 & 0 & 0 \\
-26 & 76 & -66 & 13 & \dots & 0 & 0 & 0 \\
0 & -66 & 192 & -126 & \dots & 0 & 0 & 0 \\
0 & 13 & -126 & 113 & \dots & 0 & 0 & 0 \\
\cdots & \cdots & \cdots & \cdots & \cdots & \cdots & \cdots & \cdots \\
0 & 0 & 0 & 0 & \dots & 76-140n+70n^2 & -26+60n-40n^2 & 3-5n+5n^2 \\
0 & 0 & 0 & 0 & \dots & -26+60n-40n^2 & 32-80n+80n^2 & -6+20n-40n^2 \\
0 & 0 & 0 & 0 & \dots & 3-5n+5n^2 & -6+20n-40n^2 & 3-15n+35n^2
\end{bmatrix} \tag{b}
$$

Similarly, introducing the second of Eqs. (b) of Example 10.1 in Eqs. (10.46), we obtain the element mass matrices

$$
M_1 = \frac{mL}{25n}\begin{bmatrix} 16 & 2 \\ 2 & 4 \end{bmatrix} - \frac{mL}{700n^2}\begin{bmatrix} 64 & 24 \\ 24 & 44 \end{bmatrix}
$$

$$
M_j = \frac{mL}{25n}\begin{bmatrix} 4 & 2 & -1 \\ 2 & 16 & 2 \\ -1 & 2 & 4 \end{bmatrix}
$$

$$
-\frac{mL}{700n^3}\begin{bmatrix}
44-98j+56j^2 & 24-56j+28j^2 & -5+14j-14j^2 \\
24-56j+28j^2 & 64-224j+224j^2 & -4+28j^2 \\
-5+14j-14j^2 & -4+28j^2 & 2-14j+56j^2
\end{bmatrix},
$$

$$
j = 2, 3, \dots, n \tag{c}
$$

so that, using the scheme for M in Fig. 10.7, we obtain the global mass matrix

$$M = \frac{mL}{25n}\begin{bmatrix} 16 & 2 & 0 & 0 & \dots & 0 & 0 & 0 \\ 2 & 8 & 2 & -1 & \dots & 0 & 0 & 0 \\ 0 & 2 & 16 & 2 & \dots & 0 & 0 & 0 \\ 0 & -1 & 2 & 8 & \dots & 0 & 0 & 0 \\ \cdots & \cdots & \cdots & \cdots & \cdots & \cdots & \cdots & \cdots \\ 0 & 0 & 0 & 0 & \dots & 8 & 2 & -1 \\ 0 & 0 & 0 & 0 & \dots & 2 & 16 & 2 \\ 0 & 0 & 0 & 0 & \dots & -1 & 2 & 4 \end{bmatrix}$$

$$-\frac{mL}{700n^3}\begin{bmatrix} 64 & 24 & 0 & 0 & \dots & 0 & 0 & 0 \\ 24 & 116 & 24 & -33 & \dots & 0 & 0 & 0 \\ 0 & 24 & 512 & 108 & \dots & 0 & 0 & 0 \\ 0 & -33 & 108 & 198 & \dots & 0 & 0 & 0 \\ \cdots & \cdots & \cdots & \cdots & \cdots & \cdots & \cdots & \cdots \\ 0 & 0 & 0 & 0 & \dots & 116-224n+112n^2 & 24-56n+28n^2 & -5+14n-14n^2 \\ 0 & 0 & 0 & 0 & \dots & 24-56n+28n^2 & 64-224n+224n^2 & -4+28n^2 \\ 0 & 0 & 0 & 0 & \dots & -5+14n-14n^2 & -4+28n^2 & 2-14n+56n^2 \end{bmatrix} \quad \text{(d)}$$

The eigenvalue problem, obtained by inserting Eqs. (b) and (d) into Eq. (10.25), was solved for $n = 5, 6, \dots, 75$. The three lowest normalized natural frequencies are displayed in Table 10.3. The corresponding natural modes for $n = 20$ are shown in Fig. 10.8.

Next, we solve the eigenvalue problem by approximating the stiffness and mass distributions. Using the sectionally constant axial stiffness from the first of Eqs. (a) of Example 10.1, the element stiffness matrices, Eqs. (10.47), become

$$K_1 = \frac{2EAn}{5L}\left(1 - \frac{1}{8n^2}\right)\begin{bmatrix} 16 & -8 \\ -8 & 7 \end{bmatrix}$$

$$K_j = \frac{2EAn}{5L}\left[1 - \frac{(2j-1)^2}{8n^2}\right]\begin{bmatrix} 7 & -8 & 1 \\ -8 & 16 & -8 \\ 1 & -8 & 7 \end{bmatrix} \quad \text{(e)}$$

$$+\delta_{jn}\frac{EA}{L}\begin{bmatrix} 0 & 0 & 0 \\ 0 & 0 & 0 \\ 0 & 0 & 1 \end{bmatrix}, \quad j = 2, 3, \dots, n$$

Table 10.3 Normalized Natural Frequencies for Quadratic Interpolation Functions—Exact Parameter Distributions

n	D.o.f.	$\omega_1^{(n)}\sqrt{mL^2/EA}$	$\omega_2^{(n)}\sqrt{mL^2/EA}$	$\omega_3^{(n)}\sqrt{mL^2/EA}$
5	10	2.215560	5.102570	8.148291
6	12	2.215543	5.101019	8.132307
7	14	2.215534	5.100341	8.125146
⋮	⋮	⋮	⋮	⋮
17	34	2.215525	5.099549	8.116587
18	36	2.215525	5.099544	8.116530
19	38	2.215524	5.099540	8.116491
⋮	⋮	⋮	⋮	⋮
39	78		5.099526	8.116328
40	80		5.099526	8.116327
41	82		5.099525	8.116326
⋮	⋮			⋮
73	146			8.116319
74	148			8.116319
75	150			8.116319

D.o.f. = degrees of freedom

Hence, using the first diagram in Fig. 10.7, the global stiffness matrix is

$$K = \frac{2EAn}{5L}\begin{bmatrix} 16 & -8 & 0 & 0 & \dots & 0 & 0 & 0 \\ -8 & 14 & -8 & 1 & \dots & 0 & 0 & 0 \\ 0 & -8 & 16 & -8 & \dots & 0 & 0 & 0 \\ 0 & 1 & -8 & 14 & \dots & 0 & 0 & 0 \\ \dots & \dots & \dots & \dots & \dots & \dots & \dots & \dots \\ 0 & 0 & 0 & 0 & \dots & 14 & -8 & 1 \\ 0 & 0 & 0 & 0 & \dots & -8 & 16 & -8 \\ 0 & 0 & 0 & 0 & \dots & 1 & -8 & 7+5/2n \end{bmatrix}$$

$$-\frac{EA}{20Ln}\begin{bmatrix} 16 & -8 & 0 & 0 & \dots & 0 & 0 & 0 \\ -8 & 7(1+3^2) & -8\times 3^2 & 3^2 & \dots & 0 & 0 & 0 \\ 0 & -8\times 3^2 & 16\times 3^2 & -8\times 3^2 & \dots & 0 & 0 & 0 \\ 0 & 3^2 & -8\times 3^2 & 7(3^2+5^2) & \dots & 0 & 0 & 0 \\ \dots & \dots & \dots & \dots & \dots & \dots & \dots & \dots \\ 0 & 0 & 0 & 0 & \dots & 7[(2n-3)^2+(2n-1)^2] & -8(2n-1)^2 & (2n-1)^2 \\ 0 & 0 & 0 & 0 & \dots & -8(2n-1)^2 & 16(2n-1)^2 & -8(2n-1)^2 \\ 0 & 0 & 0 & 0 & \dots & (2n-1)^2 & -8(2n-1)^2 & 7(2n-1)^2 \end{bmatrix} \quad \text{(f)}$$

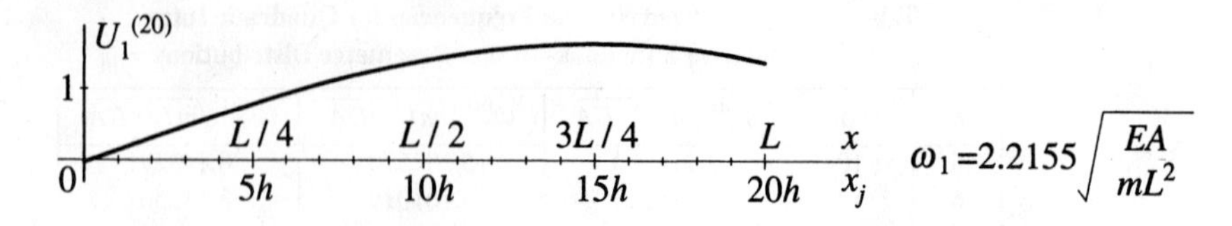

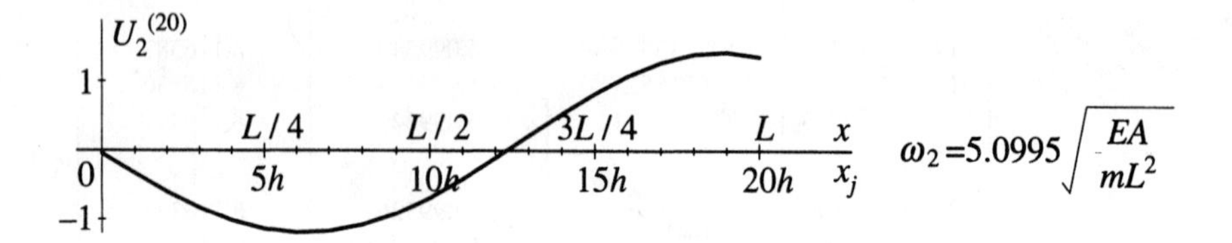

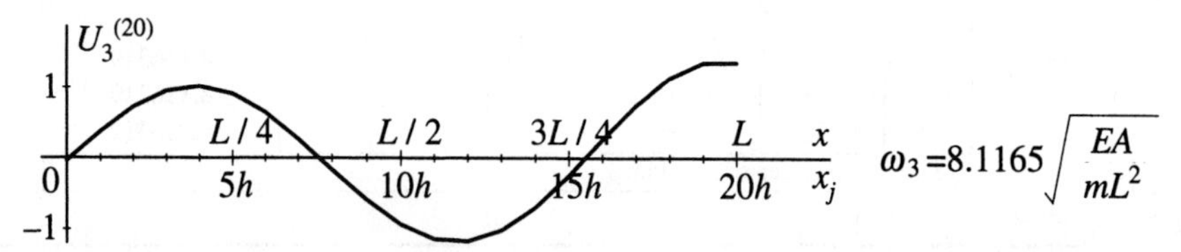

FIGURE 10.8
Three lowest modes computed using 20 finite elements in conjunction with quadratic interpolation functions

Similarly, from Eqs. (10.48), the element mass matrices are

$$
\begin{aligned}
M_1 &= \frac{mL}{25n}\left(1 - \frac{1}{8n^2}\right)\begin{bmatrix} 16 & 2 \\ 2 & 4 \end{bmatrix} \\
M_j &= \frac{mL}{25n}\left[1 - \frac{(2j-1)^2}{8n^2}\right]\begin{bmatrix} 4 & 2 & -1 \\ 2 & 16 & 2 \\ -1 & 2 & 4 \end{bmatrix}, \quad j = 2, 3, \ldots, n
\end{aligned} \tag{g}
$$

so that the global mass matrix is

$$
M = \frac{mL}{25n}\begin{bmatrix}
16 & 2 & 0 & 0 & \cdots & 0 & 0 & 0 \\
2 & 8 & 2 & -1 & \cdots & 0 & 0 & 0 \\
0 & 2 & 16 & 2 & \cdots & 0 & 0 & 0 \\
0 & -1 & 2 & 8 & \cdots & 0 & 0 & 0 \\
\cdots & \cdots & \cdots & \cdots & \cdots & \cdots & \cdots & \cdots \\
0 & 0 & 0 & 0 & \cdots & 8 & 2 & -1 \\
0 & 0 & 0 & 0 & \cdots & 2 & 16 & 2 \\
0 & 0 & 0 & 0 & \cdots & -1 & 2 & 4
\end{bmatrix}
$$

$$
- \frac{mL}{200n^3}\begin{bmatrix}
16 & 2 & 0 & 0 & \cdots \\
2 & 4(1+3^2) & 2\times 3^2 & -3^2 & \cdots \\
0 & 2\times 3^2 & 16\times 3^2 & 2\times 3^2 & \cdots \\
0 & -3^2 & 2\times 3^2 & 4(3^2+5^2) & \cdots \\
\cdots & \cdots & \cdots & \cdots & \cdots \\
0 & 0 & 0 & 0 & \cdots \\
0 & 0 & 0 & 0 & \cdots \\
0 & 0 & 0 & 0 & \cdots
\end{bmatrix}
$$

$$
\left.\begin{array}{ccc}
0 & 0 & 0 \\
0 & 0 & 0 \\
0 & 0 & 0 \\
0 & 0 & 0 \\
\cdots\cdots\cdots\cdots\cdots & \cdots\cdots\cdots\cdots & \cdots\cdots\cdots\cdots \\
4[(2n-3)^2+(2n-1)^2] & 2(2n-1)^2 & -(2n-1)^2 \\
2(2n-1)^2 & 16(2n-1)^2 & 2(2n-1)^2 \\
-(2n-1)^2 & 2(2n-1)^2 & 4(2n-1)^2
\end{array}\right] \tag{h}
$$

The eigenvalue problem is obtained by inserting Eqs. (f) and (h) into Eq. (10.25). The three lowest natural frequencies, computed for $n = 5, 6, \dots, 75$, are displayed in Table 10.4.

The results listed in Tables 10.3 and 10.4 reveal a number of important points. In the first place, we observe that quadratic interpolation functions tend to yield more accurate results than linear interpolation functions. More specifically, we observe from Table 10.3 that, using quadratic interpolation functions in conjunction with exact parameter distributions, the first natural frequency converges with six decimal places accuracy with 38 degrees of freedom, the second natural frequency converges with 82 degrees of freedom and the third natural frequency is very close to convergence with 150 degrees of freedom. We recall from Example 10.1 that the three lowest natural frequencies computed with linear interpolation functions and exact parameter distributions were far from convergence with 75 finite elements, which for linear interpolation functions is the same as 75 degrees of freedom. The results based on sectionally constant parameter approximations are equally interesting, but in a somewhat negative way. Indeed, from Table 10.4, we see in the first place that none of the natural frequencies converges with six decimal places accuracy with 150 degrees of freedom. More significantly, however, we see that *the computed natural frequencies approach the actual ones from below*. Hence, in this particular case, the violation

Table 10.4 Normalized Natural Frequencies for Quadratic Interpolation Functions—Approximate Parameter Distributions

n	D.o.f.	$\omega_1^{(n)}\sqrt{mL^2/EA}$	$\omega_2^{(n)}\sqrt{mL^2/EA}$	$\omega_3^{(n)}\sqrt{mL^2/EA}$
5	10	2.213034	5.074128	8.080075
6	12	2.213861	5.082038	8.089025
7	14	2.214332	5.086723	8.094932
⋮	⋮	⋮	⋮	⋮
17	34	2.215334	5.097362	8.111979
18	36	2.215355	5.097596	8.112431
19	38	2.215372	5.097794	8.112816
⋮	⋮	⋮	⋮	⋮
39	78	2.215489	5.099114	8.115466
40	80	2.215490	5.099134	8.115508
41	82	2.215492	5.099153	8.115547
⋮	⋮	⋮	⋮	⋮
73	146	2.215514	5.099407	8.116074
74	148	2.215514	5.099411	8.116080
75	150	2.215515	5.099414	8.116086

D.o.f. = degrees of freedom

of the Rayleigh-Ritz code causes the convergence to acquire characteristics more typical of a lumped-parameter process than a Rayleigh-Ritz process, where in the latter convergence is from above. Although these conclusions are based on a specific example, the results demonstrate what can be expected when the Rayleigh-Ritz code is violated.

10.4 BEAMS IN BENDING VIBRATION

From Chs. 8 and 9, we see that beams in bending vibration differ from strings in transverse vibration, rods in longitudinal vibration and shafts in torsion, although the differences are more in form than in substance. In particular, the differential equation of motion for beams is of order four, as opposed to that for strings, rods and shafts, which is of order two only. Hence, it should come as no surprise that the finite element formulation for beams is more involved than that for strings, rods and shafts. Because, the potential energy for beams involves second derivatives with respect to the spatial variable x, we conclude that the interpolation functions must be of minimum degree two. It turns out that other considerations make second-degree interpolation functions unsuitable, and the minimum degree is three. Still the discretization process for fourth-order systems is basically the same as for second-order systems.

Taking a cue from Myklestad's method for beams in bending vibration, we must formulate the problem in terms of two displacements at each nodal point $x_j = jh$, namely, the translation Y_j and the rotation Θ_j. Hence, the nodal displacement vector must have four components, which requires cubic interpolation functions. We note that, because there are two displacements at each nodal point, for a total of four, no internal nodes are necessary. Except for having two displacements at each nodal point, the process for deriving element and global mass and stiffness matrices remains as for strings, rods and shafts.

Figure 10.9 shows a typical finite element j for a beam in transverse vibration, where ξ represents the local coordinate defined by Eq. (10.8). Using the approach of Secs. 10.2 and 10.3, we define the displacement as follows:

$$Y(\xi) = \phi_1(\xi)Y_{j-1} + \phi_2(\xi)h\Theta_{j-1} + \phi_3(\xi)Y_j + \phi_4(\xi)h\Theta_j = \boldsymbol{\phi}^T(\xi)\mathbf{a}_j \tag{10.49}$$

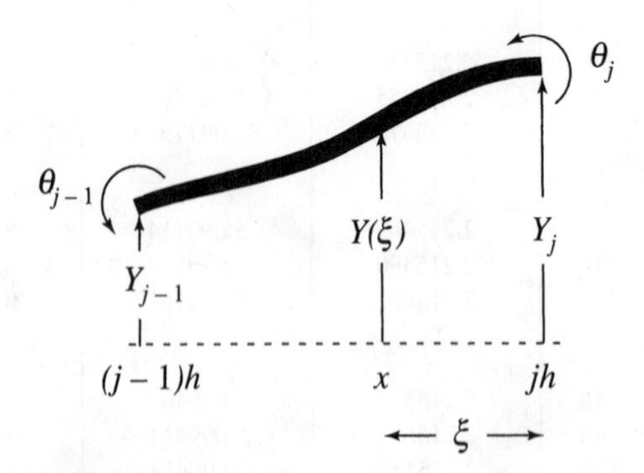

FIGURE 10.9
Finite element for a beam in bending showing the nodal displacements

where $\phi(\xi) = [\phi_1(\xi)\ \phi_2(\xi)\ \phi_3(\xi)\ \phi_4(\xi)]^T$ is a vector of interpolation functions and $\mathbf{a}_j = [Y_{j-1}\ h\Theta_{j-1}\ Y_j\ h\Theta_j]^T$ is a vector of nodal displacements. We note that, for convenience, we multiplied the angular displacements Θ_{j-1} and Θ_j by h so as to give all four components of $\mathbf{a}_j$ units of length. As a result, all four components of $\phi(\xi)$ are dimensionless. Following the pattern of Sec. 10.3, we represent the displacement by the cubic

$$Y(\xi) = c_1 + c_2\xi + c_3\xi^2 + c_4\xi^3 \tag{10.50}$$

where c_i $(i = 1, 2, 3, 4)$ are constants to be determined. In the first place, from Fig. 10.9, the transverse displacement at $\xi = 0$ is Y_j, so that

$$Y(0) = Y_j = c_1 \tag{10.51}$$

The rotation Θ_j is equal to the derivative of the displacement with respect to x at $x = jh$, which transforms as follows:

$$\Theta_j = \left.\frac{dY(x)}{dx}\right|_{x=jh} = -\frac{1}{h}\left.\frac{dY(\xi)}{d\xi}\right|_{\xi=0} \tag{10.52}$$

Hence, using Eqs. (10.52) and (10.50),

$$\left.\frac{dY(\xi)}{d\xi}\right|_{\xi=0} = -h\Theta_j = c_2 \tag{10.53}$$

At the other end, $\xi = 1$, we have

$$Y(1) = Y_{j-1} = c_1 + c_2 + c_3 + c_4 \tag{10.54}$$

and

$$\left.\frac{dY(\xi)}{d\xi}\right|_{\xi=1} = -h\Theta_{j-1} = c_2 + 2c_3 + 3c_4 \tag{10.55}$$

Solving Eqs. (10.51), (10.53), (10.54) and (10.55), we obtain

$$\begin{aligned} &c_1 = Y_j,\ c_2 = -h\Theta_j,\ c_3 = 3Y_{j-1} + h\Theta_{j-1} - 3Y_j + 2h\Theta_j, \\ &c_4 = -2Y_{j-1} - h\Theta_{j-1} + 2Y_j + 3h\Theta_j \end{aligned} \tag{10.56}$$

so that, inserting Eqs. (10.56) into Eq. (10.50) and rearranging, we can write

$$\begin{aligned} Y(\xi) &= Y_j - h\Theta_j\xi + (3Y_{j-1} + h\Theta_{j-1} - 3Y_j + 2h\Theta_j)\xi^2 \\ &\quad + (-2Y_{j-1} - h\Theta_{j-1} + 2Y_j + 3h\Theta_j)\xi^3 \\ &= (3\xi^2 - 2\xi^3)Y_{j-1} + (\xi^2 - \xi^3)h\Theta_{j-1} + (1 - 3\xi^2 + 2\xi^3)Y_j \\ &\quad + (-\xi + 2\xi^2 - \xi^3)h\Theta_j \end{aligned} \tag{10.57}$$

Finally, contrasting Eqs. (10.49) and (10.57), we conclude that the interpolation functions are

$$\phi_1(\xi) = 3\xi^2 - 2\xi^3,\ \phi_2(\xi) = \xi^2 - \xi^3,\ \phi_3(\xi) = 1 - 3\xi^2 + 2\xi^3,\ \phi_4(\xi) = -\xi + 2\xi^2 - \xi^3 \tag{10.58}$$

They are commonly known as *Hermite cubics*.

Next, we derive the element stiffness and mass matrices. Following the same steps as in Sec. 10.2, we first write the maximum potential energy as the sum over the elements

$$V_{\max} = \frac{1}{2}\sum_{j=1}^{n}\int_{(j-1)h}^{jh} EI(x)\left[\frac{d^2Y(x)}{dx^2}\right]^2 dx \tag{10.59}$$

But, using Eq. (10.8) and the first of Eqs. (10.9), we conclude that

$$\frac{d^2Y(x)}{dx^2} = \frac{1}{h^2}\frac{d^2Y(\xi)}{d\xi^2} \tag{10.60}$$

and, by analogy with Eq. (10.16), we have

$$EI(x) = EI[h(j-\xi)] = EI_j(\xi) \tag{10.61}$$

Then, using the second of Eqs. (10.9), we can write

$$\begin{aligned}\int_{(j-1)h}^{jh} EI(x)\left[\frac{d^2Y(x)}{dx^2}\right]^2 dx &= \int_1^0 EI_j(\xi)\left(\frac{1}{h^2}\right)^2\left[\frac{d^2Y(\xi)}{d\xi^2}\right]^2(-h)d\xi \\ &= \frac{1}{h^3}\int_0^1 EI_j(\xi)\mathbf{a}_j^T\frac{d^2\boldsymbol{\phi}(\xi)}{d\xi^2}\frac{d^2\boldsymbol{\phi}^T(\xi)}{d\xi^2}\mathbf{a}_j d\xi \\ &= \mathbf{a}_j^T K_j\mathbf{a}_j,\ j = 1, 2, \ldots, n\end{aligned} \tag{10.62}$$

where, using Eqs. (10.58),

$$\begin{aligned}K_j &= \frac{1}{h^3}\int_0^1 EI_j(\xi)\frac{d^2\boldsymbol{\phi}(\xi)}{d\xi^2}\frac{d^2\boldsymbol{\phi}^T(\xi)}{d\xi^2}d\xi \\ &= \frac{4}{h^3}\int_0^1 EI_j(\xi)\begin{bmatrix}3(1-2\xi)\\ 1-3\xi\\ -3(1-2\xi)\\ 2-3\xi\end{bmatrix}\begin{bmatrix}3(1-2\xi)\\ 1-3\xi\\ -3(1-2\xi)\\ 2-3\xi\end{bmatrix}^T d\xi \\ &= \frac{4}{h^3}\int_0^1 EI_j(\xi)\begin{bmatrix}9(1-2\xi)^2 & 3(1-2\xi)(1-3\xi) & -9(1-2\xi)^2 & 3(1-2\xi)(2-3\xi)\\ 3(1-2\xi)(1-3\xi) & (1-3\xi)^2 & -3(1-3\xi)(1-2\xi) & (1-3\xi)(2-3\xi)\\ -9(1-2\xi)^2 & -3(1-3\xi)(1-2\xi) & 9(1-2\xi)^2 & -3(1-2\xi)(2-3\xi)\\ 3(1-2\xi)(2-3\xi) & (1-3\xi)(2-3\xi) & -3(1-2\xi)(2-3\xi) & (2-3\xi)^2\end{bmatrix}d\xi, \\ &\qquad j = 1, 2, \ldots, n\end{aligned} \tag{10.63}$$

are the element stiffness matrices. To determine the element mass matrices, we write the reference kinetic energy in the form

$$T_{\text{ref}} = \frac{1}{2}\sum_{j=1}^{n}\int_{(j-1)h}^{jh} m(x)Y^2(x)dx = \frac{1}{2}\sum_{j=1}^{n}\int_{1}^{0} m_j(\xi)Y^2(\xi)(-h)d\xi$$

$$= \frac{1}{2}\sum_{j=1}^{n} h\int_{0}^{1} m_j(\xi)\mathbf{a}_j^T\phi(\xi)\phi^T(\xi)\mathbf{a}_j d\xi = \frac{1}{2}\sum_{j=1}^{n}\mathbf{a}_j^T M_j\mathbf{a}_j \tag{10.64}$$

in which, using Eqs. (10.58),

$$M_j = h\int_0^1 m_j(\xi)\begin{bmatrix} 3\xi^2-2\xi^3 \\ \xi^2-\xi^3 \\ 1-3\xi^2+2\xi^3 \\ -\xi+2\xi^2-\xi^3 \end{bmatrix}\begin{bmatrix} 3\xi^2-2\xi^3 \\ \xi^2-\xi^3 \\ 1-3\xi^2+2\xi^3 \\ -\xi+2\xi^2-\xi^3 \end{bmatrix}^T d\xi$$

$$= h\int_0^1 m_j(\xi)\begin{bmatrix} \xi^4(3-2\xi)^2 & \xi^4(3-2\xi)(1-\xi) & \xi^2(3-2\xi)(1-3\xi^2+2\xi^3) & -\xi^3(3-2\xi)(1-\xi)^2 \\ \xi^4(3-2\xi)(1-\xi) & \xi^4(1-\xi)^2 & \xi^2(1-\xi)(1-3\xi^2+2\xi^3) & -\xi^3(1-\xi)^3 \\ \xi^2(3-2\xi)(1-3\xi^2+2\xi^3) & \xi^2(1-\xi)(1-3\xi^2+2\xi^3) & (1-3\xi^2+2\xi^3)^2 & -\xi(1-3\xi^2+2\xi^3)(1-\xi)^2 \\ -\xi^3(3-2\xi)(1-\xi)^2 & -\xi^3(1-\xi)^3 & -\xi(1-3\xi^2+2\xi^3)(1-\xi)^2 & \xi^2(1-\xi)^4 \end{bmatrix} d\xi,$$

$$j = 1, 2, \ldots, n \tag{10.65}$$

are the element mass matrices.

For constant EI_j over the elements, following integration, Eqs. (10.63) yield the element stiffness matrices

$$K_j = \frac{EI_j n^3}{L^3}\begin{bmatrix} 12 & 6 & -12 & 6 \\ 6 & 4 & -6 & 2 \\ -12 & -6 & 12 & -6 \\ 6 & 2 & -6 & 4 \end{bmatrix}, \quad j = 1, 2, \ldots, n \tag{10.66}$$

and for constant m_j over the elements, upon integration of Eqs. (10.65), we obtain the element mass matrices

$$M_j = \frac{m_j L}{420n}\begin{bmatrix} 156 & 22 & 54 & -13 \\ 22 & 4 & 13 & -3 \\ 54 & 13 & 156 & -22 \\ -13 & -3 & -22 & 4 \end{bmatrix}, \quad j = 1, 2, \ldots, n \tag{10.67}$$

The assembly process for the derivation of the global stiffness and mass matrices, as well as their generation on a computer, follows the same pattern as for strings, rods and shafts, except that the element matrices are 4×4 and the sum involves 2×2 matrices, rather than scalars The global stiffness and mass matrices are shown schematically in Fig. 10.10. Of course, we must strike out rows and columns from the global matrices as the boundary conditions require.

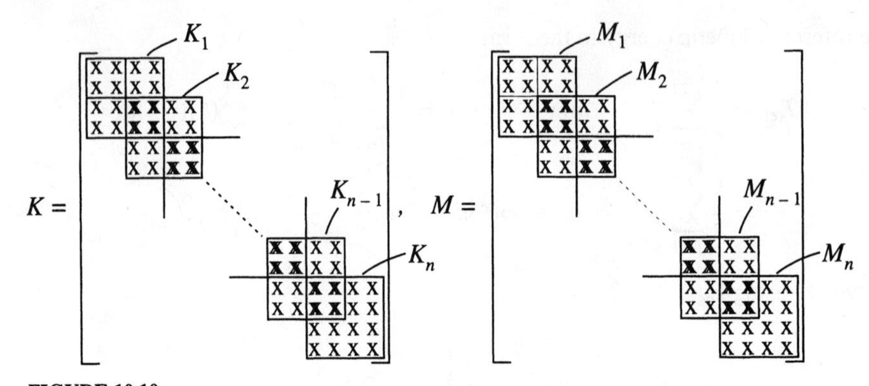

FIGURE 10.10
Scheme for the assembly of global matrices from element matrices for fourth-order systems

Finally, before we close this section, it should prove of interest to clear up a matter arising from time to time in connection with the mass matrix. To place things in perspective, it should be mentioned that the finite element method was first developed as a structural analysis method, and in particular as a procedure for static stress analysis of complex structures. For the most part, this amounts to deriving the stiffness matrix only. In extending the finite element method to problems of vibrations, it is expedient to use a lumped mass matrix. But, as can be concluded from Ch. 9 and this chapter, in the variational approach to the eigenvalue problem, as typified by the Rayleigh-Ritz method, the element and global stiffness and mass matrices are derived in a consistent manner. Because it is not consistent with the variational approach to lump parameters, a lumped mass matrix is referred to at times as an *inconsistent mass matrix*. In using an inconsistent mass matrix, the stationarity of Rayleigh's quotient no longer holds, so that errors can occur and convergence can suffer. Hence, the use of a lumped mass matrix in conjunction with the finite element method is not advised.

Example 10.3. Solve the eigenvalue problem for the pinned-pinned beam of Example 9.3 by means of the finite element method using 10 elements. Compare results with those obtained in Example 9.3 by means of Myklestad's method and draw conclusions.

The beam and the finite element mesh for $n = 10$ are shown in Fig. 10.11. Because the number of elements is the same as the number of increments used for the solution by Myklestad's method, we obtain from Example 9.3 the mass and stiffness distributions over the elements

$$\begin{aligned} m_j &= m, \quad EI_j = EI, \quad j = 1, 2, 9, 10 \\ m_j &= 1.2m, \quad EI_j = 1.44EI, \quad j = 3, 4, 7, 8 \\ m_j &= 1.4m, \quad EI_j = 1.96EI, \quad j = 5, 6 \end{aligned} \tag{a}$$

The element stiffness and mass matrices are given by Eqs. (10.66) and (10.67), respectively, in which provisions must be made for the boundary conditions. In particular, because the beam is pinned at both ends, we must set $Y_0 = Y_{10} = 0$, which amounts to striking out the first row and column from K_1 and M_1 and the third row and column from

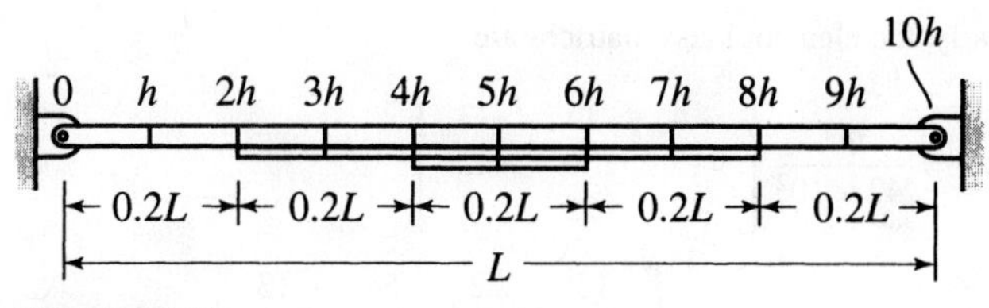

FIGURE 10.11
Finite elements for a nonuniform pinned-pinned beam

K_{10} and M_{10}. Hence, the element stiffness matrices are

$$K_1 = \frac{10^3 EI}{L^3}\begin{bmatrix} 4 & -6 & 2 \\ -6 & 12 & -6 \\ 2 & -6 & 4 \end{bmatrix}$$

$$K_2 = K_9 = \frac{10^3 EI}{L^3}\begin{bmatrix} 12 & 6 & -12 & 6 \\ 6 & 4 & -6 & 2 \\ -12 & -6 & 12 & -6 \\ 6 & 2 & -6 & 4 \end{bmatrix}$$

$$K_3 = K_4 = K_7 = K_8 = \frac{10^3 EI}{L^3}\begin{bmatrix} 17.28 & 8.64 & -17.28 & 8.64 \\ 8.64 & 5.76 & -8.64 & 2.88 \\ -17.28 & -8.64 & 17.28 & -8.64 \\ 8.64 & 2.88 & -8.64 & 5.76 \end{bmatrix} \tag{b}$$

$$K_5 = K_6 = \frac{10^3 EI}{L^3}\begin{bmatrix} 23.52 & 11.76 & -23.52 & 11.76 \\ 11.76 & 7.84 & -11.76 & 3.92 \\ -23.52 & -11.76 & 23.52 & -11.76 \\ 11.76 & 3.92 & -11.76 & 7.84 \end{bmatrix}$$

$$K_{10} = \frac{10^3 EI}{L^3}\begin{bmatrix} 12 & 6 & 6 \\ 6 & 4 & 2 \\ 6 & 2 & 4 \end{bmatrix}$$

so that, from Fig. 10.10, upon striking out the rows and columns corresponding to Y_0 and Y_{10}, the 20×20 global stiffness matrix is

$$K = \frac{10^3 EI}{L^3} \times$$

$$\begin{bmatrix} 4 & -6 & 2 & 0 & 0 & 0 & \cdots & 0 & 0 & 0 & 0 & 0 \\ -6 & 24 & 0 & -12 & 6 & 0 & \cdots & 0 & 0 & 0 & 0 & 0 \\ 2 & 0 & 8 & -6 & 2 & 0 & \cdots & 0 & 0 & 0 & 0 & 0 \\ 0 & -12 & -6 & 29.28 & 2.64 & -17.28 & \cdots & 0 & 0 & 0 & 0 & 0 \\ 0 & 6 & 2 & 2.64 & 9.76 & -8.64 & \cdots & 0 & 0 & 0 & 0 & 0 \\ 0 & 0 & 0 & -17.21 & -8.69 & 34.56 & \cdots & 0 & 0 & 0 & 0 & 0 \\ \cdots & \cdots & \cdots & \cdots & \cdots & \cdots & \cdots & \cdots & \cdots & \cdots & \cdots & \cdots \\ 0 & 0 & 0 & 0 & 0 & 0 & \cdots & 29.28 & -2.64 & -12 & 6 & 0 \\ 0 & 0 & 0 & 0 & 0 & 0 & \cdots & -2.64 & 9.76 & -6 & 2 & 0 \\ 0 & 0 & 0 & 0 & 0 & 0 & \cdots & -12 & -6 & 24 & 0 & 6 \\ 0 & 0 & 0 & 0 & 0 & 0 & \cdots & 6 & 2 & 0 & 8 & 2 \\ 0 & 0 & 0 & 0 & 0 & 0 & \cdots & 0 & 0 & 6 & 2 & 4 \end{bmatrix} \tag{c}$$

Similarly, the element mass matrices are

$$M_1 = \frac{mL}{42 \times 10^2}\begin{bmatrix} 4 & 13 & -3 \\ 13 & 156 & -22 \\ -3 & -22 & 4 \end{bmatrix}$$

$$M_2 = M_9 = \frac{mL}{42 \times 10^2}\begin{bmatrix} 156 & 22 & 54 & -13 \\ 22 & 4 & 13 & -3 \\ 54 & 13 & 156 & -22 \\ -13 & -3 & -22 & 4 \end{bmatrix}$$

$$M_3 = M_4 = M_7 = M_8 = \frac{mL}{42 \times 10^2}\begin{bmatrix} 187.2 & 26.4 & 64.8 & -15.6 \\ 26.4 & 4.8 & 15.6 & -3.6 \\ 64.8 & 15.6 & 187.2 & -26.4 \\ -15.6 & -3.6 & -26.4 & 4.8 \end{bmatrix} \tag{d}$$

$$M_5 = M_6 = \frac{mL}{42 \times 10^2}\begin{bmatrix} 218.4 & 30.8 & 75.6 & -18.2 \\ 30.8 & 5.6 & 18.2 & -4.2 \\ 75.6 & 18.2 & 218.4 & -30.8 \\ -18.2 & -4.2 & -30.8 & 5.6 \end{bmatrix}$$

$$M_{10} = \frac{mL}{42 \times 10^2}\begin{bmatrix} 156 & 22 & -13 \\ 22 & 4 & -3 \\ -13 & -3 & 4 \end{bmatrix}$$

so that, from Fig. 10.10 with the rows and columns corresponding to Y_0 and Y_{10} struck out, the 20×20 global mass matrix is

$$M = \frac{mL}{42 \times 10^2} \times$$

$$\begin{bmatrix} 4 & 13 & -3 & 0 & 0 & 0 & \dots & 0 & 0 & 0 & 0 & 0 \\ 13 & 312 & 0 & 54 & -13 & 0 & \dots & 0 & 0 & 0 & 0 & 0 \\ -3 & 0 & 8 & 13 & -3 & 0 & \dots & 0 & 0 & 0 & 0 & 0 \\ 0 & 54 & 13 & 343.2 & 4.4 & 64.8 & \dots & 0 & 0 & 0 & 0 & 0 \\ 0 & -13 & -3 & 4.4 & 8.8 & 15.6 & \dots & 0 & 0 & 0 & 0 & 0 \\ 0 & 0 & 0 & 64.8 & 15.6 & 374.4 & \dots & 0 & 0 & 0 & 0 & 0 \\ \cdots & \cdots & \cdots & \cdots & \cdots & \cdots & \cdots & \cdots & \cdots & \cdots & \cdots & \cdots \\ 0 & 0 & 0 & 0 & 0 & 0 & \dots & 343.2 & -8.8 & 54 & -13 & 0 \\ 0 & 0 & 0 & 0 & 0 & 0 & \dots & -8.8 & 9.6 & 13 & -3 & 0 \\ 0 & 0 & 0 & 0 & 0 & 0 & \dots & 54 & 13 & 312 & 0 & -13 \\ 0 & 0 & 0 & 0 & 0 & 0 & \dots & -13 & -3 & 0 & 8 & -3 \\ 0 & 0 & 0 & 0 & 0 & 0 & \dots & 0 & 0 & -13 & -3 & 4 \end{bmatrix} \tag{e}$$

Inserting Eqs. (c) and (e) into Eq. (10.25) and solving the eigenvalue problem, we obtain the approximate eigenvalues $\lambda_r^{(20)}$ and associated eigenvectors $\mathbf{a}_r^{(20)}$ $(r = 1, 2, \dots, 20)$. The eigenvalues can be used to compute the approximate natural frequencies by taking the square root. The three lowest approximate natural frequencies are as follows:

$$\omega_1^{(20)} = 10.909091\sqrt{\frac{EI}{mL^4}}, \quad \omega_2^{(20)} = 41.844423\sqrt{\frac{EI}{mL^4}}, \quad \omega_3^{(20)} = 95.321388\sqrt{\frac{EI}{mL^4}} \tag{f}$$

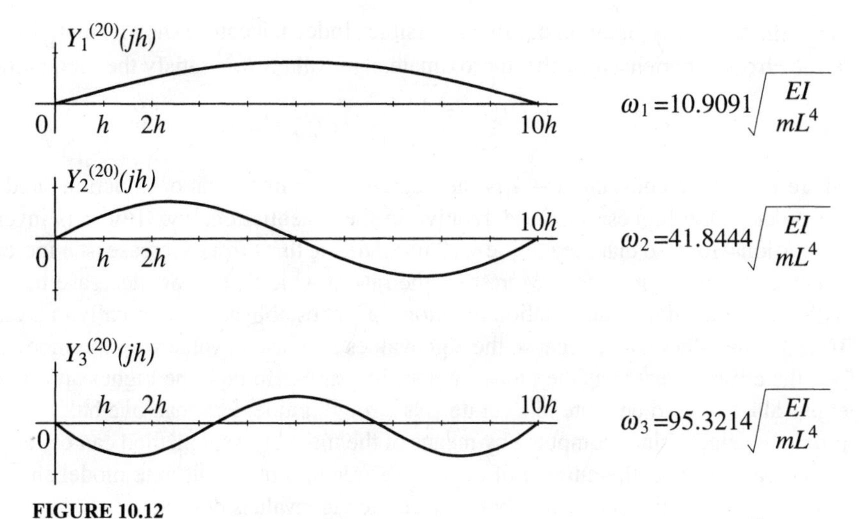

FIGURE 10.12
Three lowest modes for the pinned-pinned beam using 10 finite elements

We observe that all three are smaller than the corresponding approximate natural frequencies computed in Example 9.3 by Myklestad's method. Because in this case the mass and stiffness distributions were accounted for exactly, so that the Rayleigh-Ritz code was not violated, the actual natural frequencies are sure to be smaller than the approximate ones given by Eqs. (f). It follows that, at least in this example, the finite element method yielded more accurate approximations to the natural frequencies than Myklestad's method, where in both methods the same number of degrees of freedom was used.

To determine the approximate natural modes $Y_r^{(20)}(x)$, we recall that the eigenvectors $\mathbf{a}_r^{(20)}$ provide the displacements $Y_r^{(20)}(jh)$ and slopes $\Theta_r^{(20)}(jh)$ at the nodal points $x = jh$ $(j = 1, 2, \ldots, 20)$. The three lowest approximate natural modes are displayed in Fig. 10.12.

10.5 ERRORS IN THE EIGENVALUES

As for any discretization process, the eigenvalues and eigenfunctions computed by means of the finite element method experience errors. In general, these errors are difficult to estimate, particularly if the mass and stiffness parameter distributions are not accounted for properly within the framework of a variational process. The implication is that, strictly speaking, errors cannot be estimated if the Rayleigh-Ritz code is violated. Indeed, as demonstrated in Examples 10.1 and 10.2, when nonuniform mass and stiffness distributions are approximated by sectionally constant distributions, not only that errors in the approximate eigenvalues cannot be estimated but it is difficult to predict whether the approximate eigenvalues are larger or smaller than the actual eigenvalues.

We consider the case in which the finite element model does respect the Rayleigh-Ritz code, so that the finite element discretization can be regarded as a Rayleigh-Ritz method. This implies that the system is conservative and that the mass and stiffness parameters have been accounted for exactly. In this case, some estimates of the errors in the eigenvalues and eigenfunctions are possible. Indeed, it can be shown[2] that, for small h, the errors experienced by the approximate eigenvalues $\lambda_r^{(n)}$ satisfy the inequalities

$$\epsilon_r^{(n)} = \lambda_r^{(n)} - \lambda_r \le ch^{2(k-p)}(\lambda_r^{(n)})^{k/p}, \quad r = 1, 2, \dots, n \tag{10.68}$$

where c is some constant, $k - 1$ is the degree of the interpolation functions and p is the order of the highest spatial derivative in the potential energy. But h is inversely proportional to n, so that inequalities (10.68) indicate that errors decrease as n increases, which conforms to intuition. Moreover, the rate at which the errors decrease increases with the degree of the interpolation functions, a fact established numerically in Example 10.2. On the other hand, because the eigenvalues increase in value with the mode number, the error increases as the mode number increases. In fact, the higher approximate eigenvalues tend to be quite inaccurate. As a rough guide, less than one half of the approximate eigenvalues computed by means of the finite element method can be regarded as accurate. Hence, the number of degrees of freedom of the discrete model should be at least twice as large as the number of accurate eigenvalues desired.

Inequalities (10.68) tend to yield very conservative estimates, as the actual errors are in general much smaller than those suggested by the inequalities. As an illustration, we compare actual errors for the rod in axial vibration of Examples 10.1 and 10.2 with those prescribed by the inequalities. To this end, we first consider natural frequencies computed by means of linear interpolation functions, so that $k = 2$. From Table 10.1, for $n = 20$, we obtain the approximate lowest eigenvalue $\lambda_1^{(20)} = 4.913488EA/mL^2$, whereas the lowest eigenvalue is $\lambda_1 = 4.908547EA/mL^2$. Inserting these values into inequalities (10.68) with $r = 1$, observing that for rods in axial vibration $p = 1$, letting $c/L^2 = 1$ and ignoring EA/mL^2, we obtain

$$\begin{aligned} \epsilon_1^{(20)} = \lambda_1^{(20)} - \lambda_1 = 0.004941 &\le \left(\frac{1}{n}\right)^{2(k-p)} (\lambda_1^{(20)})^{k/p} \\ &= \left(\frac{1}{20}\right)^2 4.913488^2 = 0.060356 \end{aligned} \tag{10.69}$$

and we observe that the actual error is more than one order of magnitude smaller than the error implied by inequalities (10.68). Next, we consider the error in the lowest eigenvalue computed by means of quadratic interpolation functions, so that $k = 3$. From Table 10.3, for $n = 7$, we obtain the approximate lowest eigenvalue $\lambda_1^{(7)} = 4.908591$. Hence,

$$\epsilon_1^{(7)} = \lambda_1^{(7)} - \lambda_1 = 0.000044 \le \left(\frac{1}{7}\right)^4 4.908591^3 = 0.049258 \tag{10.70}$$

---

[2] Strang, G. and Fix, G. J., *An Analysis of the Finite Element Method*, Prentice-Hall, Englewood Cliffs, NJ, 1973, Sec. 6.3.

so that the actual error is over three orders of magnitude smaller than the error suggested by inequalities (10.68). We note here that $n = 7$ translates into a fourteen-degree-of-freedom discrete model. Comparing the actual errors in (10.69) and (10.70), we conclude that the error incurred using quadratic interpolation functions in conjunction with a fourteen-degree-of-freedom model is over two orders of magnitude smaller than the error experienced using linear interpolation functions in conjunction with a twenty-degree-of-freedom model. Of course, the use of inequalities (10.68) is intended for cases in which the actual eigenvalues are not known and some bounds on the error are desired.

Unfortunately, the approximate eigenfunctions do not lend themselves to the same type of meaningful error estimates as the approximate eigenvalues. In fact, error estimates are possible only in a weighted average sense, rather than error estimates at individual points of the elastic member.

The validity of inequalities (10.68) can be extended to cases in which the parameter distributions have been approximated in some fashion and the number of degrees of freedom of the finite element model is sufficiently large that the approximations are quite close to the actual distributions.

Errors of a different kind occur when the mass matrix is the result of some lumping process. In such cases the mass matrix represents an inconsistent mass matrix (Sec. 10.4) and, because the Rayleigh-Ritz code is violated, inequalities (10.68) can no longer be counted on to provide error estimates. In fact, errors can be negative, so that the approximate eigenvalues can drop below the corresponding actual ones. Another difficulty can arise in using lumped matrices for beams in bending. Indeed, if the lumped model makes no provisions for mass moments of inertia corresponding to the nodal angular displacements Θ_j, then the mass matrix is only positive semidefinite, which not only violates the physics of the problem but is also likely to cause computational problems. Moreover, whereas lumping the mass distribution into discrete masses is a relatively straightforward process, it is not entirely obvious how to generate discrete masses with equivalent mass moments of inertia. Hence, the use of lumped mass matrices in the context of the finite element method is not advised.

10.6 FINITE ELEMENT MODELING OF TRUSSES

Trusses represent two-dimensional structures consisting of assemblages of rods in axial vibration, where the ends of the rods are pinned at joints. Finite element modeling of trusses involves certain features not encountered before in this text. In particular, rods can have arbitrary orientations relative to a given reference frame and ordinarily several rods are pinned at one joint. As a result, the assembly process is appreciably more complicated than for single members.

We consider a typical truss member i making an angle β_i with respect to the x-direction and denote the two end joints by k and ℓ and the corresponding joint displacements by U_{kx} and U_{ky} and U_{lx} and U_{ly}, respectively, as shown in Fig. 10.13. We regard the truss member as a uniform rod undergoing elastic vibration in the axial direction and rigid-body motion in the transverse direction. The rigid-body motion does not affect the maximum potential energy, and hence the stiffness matrix. Hence, modeling

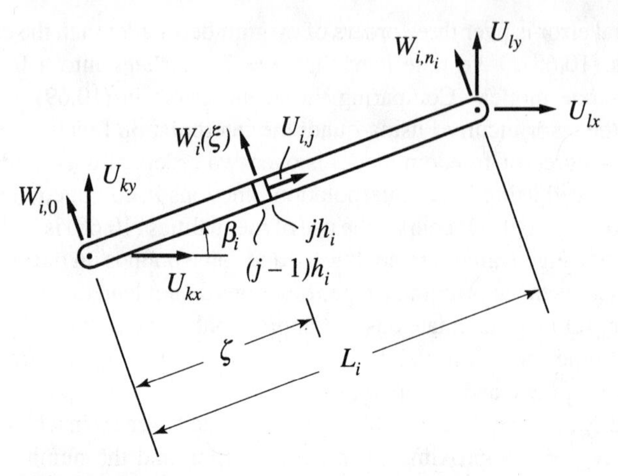

FIGURE 10.13
Typical truss member i making an angle β_i with respect to the x-direction

the rod by means of n_i finite elements, using the analogy with Eq. (10.23) and observing that the left end is not fixed, the disjoint global stiffness matrix for member i has the form

$$K_i = \frac{EA_i}{h_i}\begin{bmatrix} 1 & -1 & 0 & \dots & 0 & 0 \\ -1 & 2 & -1 & \dots & 0 & 0 \\ 0 & -1 & 2 & \dots & 0 & 0 \\ \dots & \dots & \dots & \dots & \dots & \dots \\ 0 & 0 & 0 & \dots & 2 & -1 \\ 0 & 0 & 0 & \dots & -1 & 1 \end{bmatrix} \tag{10.71}$$

where EA_i is the axial rigidity and h_i is the size of the finite element. We note that "disjoint" is in the sense that the global stiffness matrices are for individual members standing alone rather than for the members as parts of a truss. On the other hand, both the elastic vibration and the rigid-body motion affect the reference kinetic energy, so that we distinguish between a global mass matrix M_{ei} due to the axial elastic vibration and a global mass matrix M_{rbi} due to the transverse rigid-body motion. By analogy with Eq. (10.24), the disjoint global mass matrix for member i due to the elastic vibration is

$$M_{ei} = \frac{m_i h_i}{6}\begin{bmatrix} 2 & 1 & 0 & \dots & 0 & 0 \\ 1 & 4 & 1 & \dots & 0 & 0 \\ 0 & 1 & 4 & \dots & 0 & 0 \\ \dots & \dots & \dots & \dots & \dots & \dots \\ 0 & 0 & 0 & \dots & 4 & 1 \\ 0 & 0 & 0 & \dots & 1 & 2 \end{bmatrix} \tag{10.72}$$

in which m_i is the mass density. The corresponding nodal vector of axial displacements is $\mathbf{a}_i = [U_{i,0}\ U_{i,1}\ \dots\ U_{i,n_i}]^T$. But the nodal displacements at the joints must satisfy the boundary conditions

$$U_{i,0} = U_{kx}\cos\beta_i + U_{ky}\sin\beta_i,\ U_{i,n_i} = U_{lx}\cos\beta_i + U_{ly}\sin\beta_i \tag{10.73}$$

so that the nodal vector is

$$\mathbf{a}_i = [U_{kx}\cos\beta_i + U_{ky}\sin\beta_i\ \ U_{i,1}\ \ U_{i,2}\ \ \dots\ \ U_{i,n_i-1}\ \ U_{lx}\cos\beta_i + U_{ly}\sin\beta_i]^T \tag{10.74}$$

We observe from Eq. (10.74), however, that vector $\mathbf{a}_i$ has $n_i + 1$ components and it contains $n_i + 3$ unknowns, so that it is convenient to introduce an $(n_i + 3)$-dimensional vector of independent displacements as follows:

$$\bar{\mathbf{a}}_i = [U_{kx}\ \ U_{ky}\ \ U_{i,1}\ \ U_{i,2}\ \ \dots\ \ U_{i,n_i-1}\ \ U_{lx}\ \ U_{ly}]^T = [\mathbf{U}_{Jk}^T\ \ \mathbf{U}_i^T\ \ \mathbf{U}_{J\ell}^T]^T \tag{10.75}$$

in which $\mathbf{U}_i = [U_{i,1}\ \ U_{i,2}\ \ \dots\ \ U_{i,n_i-1}]^T$ is a vector of nodal displacements excluding the joint displacements and $\mathbf{U}_{Jk} = [\mathbf{U}_{kx}\ \ \mathbf{U}_{ky}]^T$ and $\mathbf{U}_{J\ell} = [U_{\ell x}\ \ U_{\ell y}]^T$ are vectors of joint displacements. From Eqs. (10.74) and (10.75), we conclude that the vector $\mathbf{a}_i$ is related to the vector $\bar{\mathbf{a}}_i$ by

$$\mathbf{a}_i = \begin{bmatrix} c\beta_i & s\beta_i & 0 & 0 & \dots & 0 & 0 & 0 \\ 0 & 0 & 1 & 0 & \dots & 0 & 0 & 0 \\ 0 & 0 & 0 & 1 & \dots & 0 & 0 & 0 \\ \dots & \dots & \dots & \dots & \dots & \dots & \dots & \dots \\ 0 & 0 & 0 & 0 & \dots & 1 & 0 & 0 \\ 0 & 0 & 0 & 0 & \dots & 0 & c\beta_i & s\beta_i \end{bmatrix} \bar{\mathbf{a}}_i = C_i \bar{\mathbf{a}}_i \tag{10.76}$$

where

$$C_i = \begin{bmatrix} c\beta_i & s\beta_i & 0 & 0 & \dots & 0 & 0 & 0 \\ 0 & 0 & 1 & 0 & \dots & 0 & 0 & 0 \\ 0 & 0 & 0 & 1 & \dots & 0 & 0 & 0 \\ \dots & \dots & \dots & \dots & \dots & \dots & \dots & \dots \\ 0 & 0 & 0 & 0 & \dots & 1 & 0 & 0 \\ 0 & 0 & 0 & 0 & \dots & 0 & c\beta_i & s\beta_i \end{bmatrix} \tag{10.77}$$

is an $(n_i + 1) \times (n_i + 3)$ transformation matrix playing the role of a constraint matrix, in which $c\beta_i = \cos\beta_i$ and $s\beta_i = \sin\beta_i$. The transformation from the constrained vector $\mathbf{a}_i$ to the independent vector $\bar{\mathbf{a}}_i$ requires a corresponding transformation in the disjoint global stiffness matrix K_i and mass matrix M_i. To this end, we write the maximum potential energy for member i in the form

$$V_{\max,i} = \frac{1}{2}\mathbf{a}_i^T K_i \mathbf{a}_i = \frac{1}{2}\bar{\mathbf{a}}_i^T C_i^T K_i C_i \bar{\mathbf{a}}_i = \frac{1}{2}\bar{\mathbf{a}}_i^T \bar{K}_i \bar{\mathbf{a}}_i \tag{10.78}$$

where, using Eqs. (10.71) and (10.77),

$$\bar{K}_i = C_i^T K_i C_i$$

$$= \frac{EA_i}{h_i}\begin{bmatrix} c^2\beta_i & s\beta_i c\beta_i & -c\beta_i & 0 & \dots & 0 & 0 & 0 \\ s\beta_i c\beta_i & s^2\beta_i & -s\beta_i & 0 & \dots & 0 & 0 & 0 \\ -c\beta_i & -s\beta_i & 2 & -1 & \dots & 0 & 0 & 0 \\ 0 & 0 & -1 & 2 & \dots & 0 & 0 & 0 \\ \cdots & \cdots & \cdots & \cdots & \cdots & \cdots & \cdots & \cdots \\ 0 & 0 & 0 & 0 & \dots & 2 & -c\beta_i & -s\beta_i \\ 0 & 0 & 0 & 0 & \dots & -c\beta_i & c^2\beta_i & s\beta_i c\beta_i \\ 0 & 0 & 0 & 0 & \dots & -s\beta_i & s\beta_i c\beta_i & s^2\beta_i \end{bmatrix} \tag{10.79}$$

is the desired $(n_i+3) \times (n_i+3)$ disjoint global stiffness matrix for member i. Similarly, we use Eq. (10.76) and write the reference kinetic energy due to elastic vibration

$$T_{\text{ref},ei} = \frac{1}{2}\mathbf{a}_i^T M_{ei}\mathbf{a}_i = \frac{1}{2}\bar{\mathbf{a}}_i^T C_i^T M_{ei} C_i \bar{\mathbf{a}}_i = \frac{1}{2}\bar{\mathbf{a}}_i^T \bar{M}_{ei}\bar{\mathbf{a}}_i \tag{10.80}$$

where, using Eqs. (10.72) and (10.77),

$$\bar{M}_{ei} = C_i^T M_{ei} C_i = \frac{m_i h_i}{6} \times$$

$$\begin{bmatrix} 2c^2\beta_i & 2s\beta_i c\beta_i & c\beta_i & 0 & \dots & 0 & 0 & 0 \\ 2s\beta_i c\beta_i & 2s^2\beta_i & s\beta_i & 0 & \dots & 0 & 0 & 0 \\ c\beta_i & s\beta_i & 4 & 1 & \dots & 0 & 0 & 0 \\ 0 & 0 & 1 & 4 & \dots & 0 & 0 & 0 \\ \cdots & \cdots & \cdots & \cdots & \cdots & \cdots & \cdots & \cdots \\ 0 & 0 & 0 & 0 & \dots & 4 & c\beta_i & s\beta_i \\ 0 & 0 & 0 & 0 & \dots & c\beta_i & 2c^2\beta_i & 2s\beta_i c\beta_i \\ 0 & 0 & 0 & 0 & \dots & s\beta_i & 2s\beta_i c\beta_i & 2s^2\beta_i \end{bmatrix} \tag{10.81}$$

is the $(n_i+3) \times (n_i+3)$ disjoint global mass matrix for member i due to the elastic vibrations.

Next, we derive the disjoint global mass matrix due to the transverse rigid-body motion. To this end, we write the corresponding reference kinetic energy. Denoting the transverse displacements of the end points of member i by $W_{i,0}$ and W_{i,n_i} (Fig. 10.13), the transverse displacement at a distance ζ from the left end is simply

$$W_i(\zeta) = W_{i,0} + \frac{W_{i,n_i} - W_{i,0}}{L_i}\zeta \tag{10.82}$$

in which L_i is the length of member i. Hence, the kinetic energy due to the rigid-body motion is

$$T_{\text{ref},rbi} = \frac{1}{2}\int_0^{L_i} W_i^2(\zeta)dm_i = \frac{1}{2}m_i \int_0^{L_i} \left(W_{i,0} + \frac{W_{i,n_i} - W_{i,0}}{L_i}\zeta\right)^2 d\zeta$$

$$= \frac{1}{2}m_i \left[W_{i,0}^2 \int_0^{L_i} \left(1 - \frac{\zeta}{L_i}\right)^2 d\zeta + 2W_{i,0}W_{i,n_i} \int_0^{L_i} \left(1 - \frac{\zeta}{L_i}\right)\frac{\zeta}{L_i} d\zeta \right.$$

$$\left. + W_{i,n_i}^2 \int_0^{L_i} \left(\frac{\zeta}{L_i}\right)^2 d\zeta \right]$$

$$= \frac{1}{2}\frac{m_i L_i}{3}(W_{i,0}^2 + W_{i,0}W_{i,n_i} + W_{i,n_i}^2) \tag{10.83}$$

But, from Fig. 10.13, the transverse displacements are related to the joint displacements by

$$W_{i,0} = U_{ky}\cos\beta_i - U_{kx}\sin\beta_i, \; W_{i,n_i} = U_{\ell y}\cos\beta_i - U_{\ell x}\sin\beta_i \tag{10.84}$$

so that the reference kinetic energy due to the rigid-body motion can be rewritten as

$$T_{\text{ref},rbi} = \frac{1}{2}\frac{m_i L_i}{3}(U_{kx}^2 s^2\beta_i - 2U_{kx}U_{ky}s\beta_i c\beta_i + U_{ky}^2 c^2\beta_i + U_{kx}U_{\ell x}s^2\beta_i - U_{kx}U_{\ell y}s\beta_i c\beta_i$$

$$+ U_{ky}U_{\ell y}c^2\beta_i - U_{ky}U_{\ell x}s\beta_i c\beta_i + U_{\ell x}^2 s^2\beta_i - 2U_{\ell x}U_{\ell y}s\beta_i c\beta_i + U_{\ell y}^2 c^2\beta_i)$$

$$= \frac{1}{2}\bar{\mathbf{a}}_i^T \bar{M}_{rbi}\bar{\mathbf{a}}_i \tag{10.85}$$

where once again the notation $c\beta_i = \cos\beta_i$, $s\beta_i = \sin\beta_i$ has been used and in which

$$\bar{M}_{rbi} = \frac{m_i L_i}{6}\begin{bmatrix} 2s^2\beta_i & -2s\beta_i c\beta_i & 0 & \dots & 0 & s^2\beta_i & -s\beta_i c\beta_i \\ -2s\beta_i c\beta_i & 2c^2\beta_i & 0 & \dots & 0 & -s\beta_i c\beta_i & c^2\beta_i \\ 0 & 0 & 0 & \dots & 0 & 0 & 0 \\ \hdotsfor{7} \\ 0 & 0 & 0 & \dots & 0 & 0 & 0 \\ s^2\beta_i & -s\beta_i c\beta_i & 0 & \dots & 0 & 2s^2\beta_i & -2s\beta_i\beta_i \\ -s\beta_i c\beta_i & c^2\beta_i & 0 & \dots & 0 & -2s\beta_i c\beta_i & 2c^2\beta_i \end{bmatrix} \tag{10.86}$$

is the disjoint global mass matrix for member i due to the rigid-body motion. But, the total reference kinetic energy is the sum of contributions from both the elastic vibration and the rigid-body motion, so that

$$T_{\text{ref},i} = T_{\text{ref},ei} + T_{\text{ref},rbi} = \frac{1}{2}\bar{\mathbf{a}}_i^T \bar{M}_{ei}\bar{\mathbf{a}}_i + \frac{1}{2}\bar{\mathbf{a}}_i^T \bar{M}_{rbi}\bar{\mathbf{a}}_i = \frac{1}{2}\bar{\mathbf{a}}_i^T \bar{M}_i\bar{\mathbf{a}}_i \tag{10.87}$$

in which

$$\bar{M}_i = \bar{M}_{ei} + \bar{M}_{rbi} = \frac{m_i h_i}{6} \left[\begin{array}{cccc|ccc} 2(c^2\beta_i + n_i s^2\beta_i) & 2(1-n_i)s\beta_i c\beta_i & c\beta_i & 0 & 0 & n_i s^2\beta_i & -n_i s\beta_i c\beta_i \\ 2(1-n_i)s\beta_i c\beta_i & 2(s^2\beta_i + n_i c^2\beta_i) & s\beta_i & 0 & 0 & -n_i s\beta_i c\beta_i & n_i c^2\beta_i \\ c\beta_i & s\beta_i & 4 & 1 & 0 & 0 & 0 \\ 0 & 0 & 1 & 4 & 0 & 0 & 0 \\ \hline 0 & 0 & 0 & 0 & 4 & c\beta_i & s\beta_i \\ n_i s^2\beta_i & -n_i s\beta_i c\beta_i & 0 & 0 & c\beta_i & 2(c^2\beta_i + n_i s^2\beta_i) & 2(1-n_i)s\beta_i c\beta_i \\ -n_i s\beta_i c\beta_i & n_i c^2\beta_i & 0 & 0 & s\beta_i & 2(1-n_i)s\beta_i c\beta_i & 2(s^2\beta_i + n_i c^2\beta_i) \end{array} \right] \tag{10.88}$$

is the disjoint global mass matrix for member i.

At this point, we turn our attention to the assembly process. To this end, we write the system maximum potential energy, i.e., the maximum potential energy for the whole truss, as follows:

$$V_{\max} = \sum_{i=1}^{N} V_{\max,i} = \frac{1}{2}\sum_{i=1}^{N} \bar{\mathbf{a}}_i^T \bar{K}_i \bar{\mathbf{a}}_i = \frac{1}{2}\mathbf{a}^T K \mathbf{a} \tag{10.89}$$

where N is the total number of members in the truss, $\mathbf{a}$ is the *system displacement vector* and K is the *system stiffness matrix.* Similarly, the system reference kinetic energy is

$$T_{\text{ref}} = \sum_{i=1}^{N} T_{\text{ref},i} = \frac{1}{2}\sum_{i=1}^{N} \bar{\mathbf{a}}_i^T \bar{M}_i \bar{\mathbf{a}}_i = \frac{1}{2}\mathbf{a}^T M \mathbf{a} \tag{10.90}$$

in which M is the *system mass matrix.* Hence, the assembly process reduces to the generation of the system stiffness and mass matrices corresponding to the system displacement vector.

In contrast to one-dimensional problems, in two-dimensional problems it is necessary to specify the system configuration before the assembly process can be carried out. This requires that we consider a specific truss configuration. We illustrate the process by means of the truss of Fig. 10.14, with the pertinent data summarized in Table 10.5.

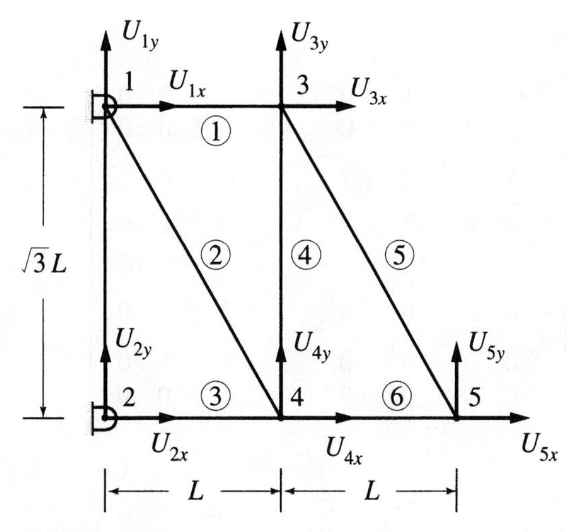

FIGURE 10.14
Specific truss configuration

Table 10.5 Truss Data

Member No.	Joint Nos.		No. of Elements	Element Size—h_i	Mass Density—m_i	Cross-Section Area—A_i	Angle w.r. to $x-\beta_i$
1	1	3	n_1	L/n_1	m_1	A_1	0
2	1	4	n_2	$2L/n_2$	m_2	A_2	$-60°$
3	2	4	n_3	L/n_3	m_3	A_3	0
4	3	4	n_4	$\sqrt{3}L/n_4$	m_4	A_4	$-90°$
5	3	5	n_5	$2L/n_5$	m_5	A_5	$-60°$
6	4	5	n_6	L/n_6	m_6	A_6	0

The disjoint global stiffness and mass matrices for member i are given by Eqs. (10.79) and (10.88), respectively. For convenience, we replace the size of the finite elements by the number of elements by simply writing

$$h_i = \begin{cases} L/n_i, \; i = 1,3,6 \\ 2L/n_i, \; i = 2,5 \\ \sqrt{3}L/n_i, \; i = 4 \end{cases} \tag{10.91}$$

Moreover, inserting the angles β_i from Table 10.5 into Eq. (10.79), the disjoint global stiffness matrices are

$$\bar{K}_i = \frac{EA_i n_i}{L}\begin{bmatrix} 1 & 0 & -1 & 0 & \dots & 0 & 0 & 0 \\ 0 & 0 & 0 & 0 & \dots & 0 & 0 & 0 \\ -1 & 0 & 2 & -1 & \dots & 0 & 0 & 0 \\ 0 & 0 & -1 & 2 & \dots & 0 & 0 & 0 \\ \dots & \dots & \dots & \dots & \dots & \dots & \dots & \dots \\ 0 & 0 & 0 & 0 & \dots & 2 & -1 & 0 \\ 0 & 0 & 0 & 0 & \dots & -1 & 1 & 0 \\ 0 & 0 & 0 & 0 & \dots & 0 & 0 & 0 \end{bmatrix}, \ i = 1, 3, 6$$

$$\bar{K}_i = \frac{EA_i n_1}{8L}\begin{bmatrix} 1 & -\sqrt{3} & -2 & 0 & \dots & 0 & 0 & 0 \\ -\sqrt{3} & 3 & 2\sqrt{3} & 0 & \dots & 0 & 0 & 0 \\ -2 & 2\sqrt{3} & 8 & -4 & \dots & 0 & 0 & 0 \\ 0 & 0 & -4 & 8 & \dots & 0 & 0 & 0 \\ \dots & \dots & \dots & \dots & \dots & \dots & \dots & \dots \\ 0 & 0 & 0 & 0 & \dots & 8 & -2 & 2\sqrt{3} \\ 0 & 0 & 0 & 0 & \dots & -2 & 1 & -\sqrt{3} \\ 0 & 0 & 0 & 0 & \dots & 2\sqrt{3} & -\sqrt{3} & 3 \end{bmatrix}, \ i = 2, 5 \tag{10.92}$$

$$\bar{K}_i = \frac{EA_i n_i}{\sqrt{3}L}\begin{bmatrix} 0 & 0 & 0 & 0 & \dots & 0 & 0 & 0 \\ 0 & 1 & 1 & 0 & \dots & 0 & 0 & 0 \\ 0 & 1 & 2 & -1 & \dots & 0 & 0 & 0 \\ 0 & 0 & -1 & 2 & \dots & 0 & 0 & 0 \\ \dots & \dots & \dots & \dots & \dots & \dots & \dots & \dots \\ 0 & 0 & 0 & 0 & \dots & 2 & 0 & 1 \\ 0 & 0 & 0 & 0 & \dots & 0 & 0 & 0 \\ 0 & 0 & 0 & 0 & \dots & 1 & 0 & 1 \end{bmatrix}, \ i = 4$$

and, using Eq. (10.88), the disjoint global mass matrices are

$$
\bar{M}_i = \frac{m_i L_i}{6n_i}\begin{bmatrix}
2 & 0 & 1 & 0 & \dots & 0 & 0 & 0 \\
0 & 2n_i & 0 & 0 & \dots & 0 & 0 & 0 \\
1 & 0 & 4 & 1 & \dots & 0 & 0 & 0 \\
0 & 0 & 1 & 4 & \dots & 0 & 0 & 0 \\
\cdots & \cdots & \cdots & \cdots & \cdots & \cdots & \cdots & \cdots \\
0 & 0 & 0 & 0 & \dots & 4 & 1 & 0 \\
0 & 0 & 0 & 0 & \dots & 1 & 2 & 0 \\
0 & 0 & 0 & 0 & \dots & 0 & 0 & 2n_i
\end{bmatrix}, \; i = 1, 3, 6
$$

$$
\bar{M}_i = \frac{m_i L_i}{6n_i}\begin{bmatrix}
1+3n_i & -(1-n_i)\sqrt{3} & 1 & 0 & \dots & 0 & 0 & 0 \\
-(1-n_i)\sqrt{3} & 3+n_i & -\sqrt{3} & 0 & \dots & 0 & 0 & 0 \\
1 & -\sqrt{3} & 8 & 2 & \dots & 0 & 0 & 0 \\
0 & 0 & 2 & 8 & \dots & 0 & 0 & 0 \\
\cdots & \cdots & \cdots & \cdots & \cdots & \cdots & \cdots & \cdots \\
0 & 0 & 0 & 0 & \dots & 8 & 1 & -\sqrt{3} \\
0 & 0 & 0 & 0 & \dots & 1 & 1+3n_i & -(1-n_i)\sqrt{3} \\
0 & 0 & 0 & 0 & \dots & -\sqrt{3} & -(1-n_i)\sqrt{3} & 3+n_i
\end{bmatrix}, \; i = 2, 5
$$

$$
\bar{M}_i = \frac{m_i L}{2\sqrt{3}n_i}\begin{bmatrix}
2n_i & 0 & 0 & 0 & \dots & 0 & 0 & 0 \\
0 & 2 & -1 & 0 & \dots & 0 & 0 & 0 \\
0 & -1 & 4 & 1 & \dots & 0 & 0 & 0 \\
0 & 0 & 1 & 4 & \dots & 0 & 0 & 0 \\
\cdots & \cdots & \cdots & \cdots & \cdots & \cdots & \cdots & \cdots \\
0 & 0 & 0 & 0 & \dots & 4 & 0 & -1 \\
0 & 0 & 0 & 0 & \dots & 0 & 2n_i & 0 \\
0 & 0 & 0 & 0 & \dots & -1 & 0 & 2
\end{bmatrix}, \; i = 4 \tag{10.93}
$$

The next step is the assembly process itself. This amounts to placing the entries of the member stiffness and mass matrices in the system stiffness and mass matrices, respectively. To this end, one must decide on the location of the joint and member displacements in the system displacement vector **a**. It is convenient to list first the joint displacement vectors U_{Jk} $(k = 1, 2, \ldots, 5)$ and then the member displacement vectors $\mathbf{U}_i$ $(i = 1, 2, \ldots 6)$, where the latter exclude the joint displacements. Hence, using the notation in Fig. 10.14, the system displacement vector can be written in the form

$$\mathbf{a} = [\mathbf{U}_{J1}^T \ \mathbf{U}_{J2}^T \ \cdots \ \mathbf{U}_{J5}^T \ \mathbf{U}_1^T \ \mathbf{U}_2^T \ \cdots \ \mathbf{U}_6^T]^T \tag{10.94}$$

Then, a scheme can be devised to generate the system stiffness and mass matrices from the disjoint member global stiffness and mass matrices. Before this can be done, it is necessary to choose the remaining truss data in Table 10.5, namely, the number n_i of elements per truss member, the cross-sectional area A_i and the mass density m_i $(i = 1, 2, \ldots, 6)$. The generation of the system stiffness and mass matrices is demonstrated in Example 10.4.

Example 10.4. Let $EA_i = EA$, $m_i = m$ $(i = 1, 2, \ldots, 6)$ and solve the eigenvalue problem for the truss of Fig. 10.14 by the finite element method for the two cases: one element per member and two elements per member. Compare results and draw conclusions.

The developments of this section apply to two or more elements per member. In the case of a single element per member, by analogy with Eq. (10.79), the disjoint global stiffness matrices can be shown to be

$$\bar{K}_i = C_i^T K_i C_i = \frac{EA}{h_i}\begin{bmatrix} c\beta_i & 0 \\ s\beta_i & 0 \\ 0 & c\beta_i \\ 0 & s\beta_i \end{bmatrix}\begin{bmatrix} 1 & -1 \\ -1 & 1 \end{bmatrix}\begin{bmatrix} c\beta_i & s\beta_i & 0 & 0 \\ 0 & 0 & c\beta_i & s\beta_i \end{bmatrix}$$

$$= \frac{EA}{h_i}\begin{bmatrix} c^2\beta_i & s\beta_i c\beta_i & -c^2\beta_i & -s\beta_i c\beta_i \\ s\beta_i c\beta_i & s^2\beta_i & -s\beta_i c\beta_i & -s^2\beta_i \\ -c^2\beta_i & -s\beta_i c\beta_i & c^2\beta_i & s\beta_i c\beta_i \\ -s\beta_i c\beta_i & -s^2\beta_i & s\beta_i c\beta_i & s^2\beta_i \end{bmatrix}, \quad i = 1, 2, \ldots, 6 \tag{a}$$

where h_i are given by Eqs. (10.91) and β_i are listed in Table 10.5. According to Eq. (10.88), the disjoint global mass matrices are the sum of the corresponding global mass matrices due to axial vibration and the global mass matrices due to transverse rigid-body motion. But, by analogy with Eq. (10.81), the disjoint global mass matrices due to axial vibration are

$$\bar{M}_{ei} = C_i^T M_{ei} C_i = \frac{mh_i}{6}\begin{bmatrix} c\beta_i & 0 \\ s\beta_i & 0 \\ 0 & c\beta_i \\ 0 & s\beta_i \end{bmatrix}\begin{bmatrix} 2 & 1 \\ 1 & 2 \end{bmatrix}\begin{bmatrix} c\beta_i & s\beta_i & 0 & 0 \\ 0 & 0 & c\beta_i & s\beta_i \end{bmatrix}$$

$$= \frac{mh_i}{6}\begin{bmatrix} 2c^2\beta_i & 2s\beta_i c\beta_i & c^2\beta_i & s\beta_i c\beta_i \\ 2s\beta_i c\beta_i & 2s^2\beta_i & s\beta_i c\beta_i & s^2\beta_i \\ c^2\beta_i & s\beta_i c\beta_i & 2c^2\beta_i & 2s\beta_i c\beta_i \\ s\beta_i c\beta_i & s^2\beta_i & 2s\beta_i c\beta_i & 2s^2\beta_i \end{bmatrix}, \quad i = 1, 2, \ldots, 6 \tag{b}$$

Moreover, using the analogy with Eq. (10.86), the disjoint global mass matrices due to the rigid-body motion are

$$\bar{M}_{rbi} = \frac{mL_i}{6}\begin{bmatrix} 2s^2\beta_i & -2s\beta_i c\beta_i & s^2\beta_i & -s\beta_i c\beta_i \\ -2s\beta_i c\beta_i & 2c^2\beta_i & -s\beta_i c\beta_i & c^2\beta_i \\ s^2\beta_i & -s\beta_i c\beta_i & 2s^2\beta_i & -2s\beta_i c\beta_i \\ -s\beta_i c\beta_i & c^2\beta_i & -2s\beta_i c\beta_i & 2c^2\beta_i \end{bmatrix}, \quad i = 1, 2, \ldots, 6 \tag{c}$$

Hence, letting $L_i = h_i$ and using Eq. (10.88), we obtain the total disjoint global mass matrices

$$\bar{M}_i = \bar{M}_{ei} + \bar{M}_{rbi} = \frac{mh_i}{6}\begin{bmatrix} 2 & 0 & 1 & 0 \\ 0 & 2 & 0 & 1 \\ 1 & 0 & 2 & 0 \\ 0 & 1 & 0 & 2 \end{bmatrix}, \ i = 1, 2, \ldots, 6 \tag{d}$$

Then, letting $n_i = 1$ in Eqs. (10.91) and using the values of β_i from Table 10.5 ($i = 1, 2, \ldots, 6$), Eqs. (a) yield the disjoint global stiffness matrices

$$\bar{K}_1 = \bar{K}_3 = \bar{K}_6 = \frac{EA}{L}\begin{bmatrix} 1 & 0 & -1 & 0 \\ 0 & 0 & 0 & 0 \\ -1 & 0 & 1 & 0 \\ 0 & 0 & 0 & 0 \end{bmatrix}$$

$$\bar{K}_2 = \bar{K}_5 = \frac{EA}{8L}\begin{bmatrix} 1 & -\sqrt{3} & -1 & \sqrt{3} \\ -\sqrt{3} & 3 & \sqrt{3} & -3 \\ -1 & \sqrt{3} & 1 & -\sqrt{3} \\ \sqrt{3} & -3 & -\sqrt{3} & 3 \end{bmatrix} \tag{e}$$

$$\bar{K}_4 = \frac{EA}{\sqrt{3}L}\begin{bmatrix} 0 & 0 & 0 & 0 \\ 0 & 1 & 0 & -1 \\ 0 & 0 & 0 & 0 \\ 0 & -1 & 0 & 1 \end{bmatrix}$$

Similarly, the disjoint global mass matrices are

$$\bar{M}_1 = \bar{M}_3 = \bar{M}_6 = \frac{mL}{6}\begin{bmatrix} 2 & 0 & 1 & 0 \\ 0 & 2 & 0 & 1 \\ 1 & 0 & 2 & 0 \\ 0 & 1 & 0 & 2 \end{bmatrix}$$

$$\bar{M}_2 = \bar{M}_5 = \frac{mL}{3}\begin{bmatrix} 2 & 0 & 1 & 0 \\ 0 & 2 & 0 & 1 \\ 1 & 0 & 2 & 0 \\ 0 & 1 & 0 & 2 \end{bmatrix} \tag{f}$$

$$\bar{M}_4 = \frac{mL}{2\sqrt{3}}\begin{bmatrix} 2 & 0 & 1 & 0 \\ 0 & 2 & 0 & 1 \\ 1 & 0 & 2 & 0 \\ 0 & 1 & 0 & 2 \end{bmatrix}$$

The transfer of the entries of the disjoint global stiffness matrices $\bar{K}_i$ $(i = 1, 2, \ldots, 6)$ to the system stiffness matrix K can be carried out by referring to the member and joint location in the truss of Fig. 10.14 and the matrix row and column in Table 10.6. As an illustration, the entry $\bar{K}_1(1,1)$ transfers to $K(1,1)$, $\bar{K}_1(1,3)$ to $K(1,5)$ and $\bar{K}_1(3,3)$ to $K(5,5)$. The system stiffness matrix K is 10×10, where the matrix was derived on the assumption that all joints are free. In reality, joints 1 and 2 are fixed, so that $U_{1x} = U_{1y} = U_{2x} = U_{2y} = 0$. In view of this, the first four rows and columns in K must be deleted. Hence, the resulting 6×6 system stiffness matrix is

$$K = \frac{EA}{8\sqrt{3}L}\begin{bmatrix} 9\sqrt{3} & -3 & 0 & 0 & -\sqrt{3} & 3 \\ -3 & 8+3\sqrt{3} & 0 & -8 & 3 & -3\sqrt{3} \\ 0 & 0 & 17\sqrt{3} & -3 & -8\sqrt{3} & 0 \\ 0 & -8 & -3 & 8+3\sqrt{3} & 0 & 0 \\ -\sqrt{3} & 3 & -8\sqrt{3} & 0 & 9\sqrt{3} & -3 \\ 3 & -3\sqrt{3} & 0 & 0 & -3 & 3\sqrt{3} \end{bmatrix} \tag{g}$$

Similarly, recognizing that Table 10.6 applies to mass matrices as well, the system mass matrix can be shown to be

$$M = \frac{mL}{6}\begin{bmatrix} 6+2\sqrt{3} & 0 & \sqrt{3} & 0 & 2 & 0 \\ 0 & 6+2\sqrt{3} & 0 & \sqrt{3} & 0 & 2 \\ \sqrt{3} & 0 & 8+2\sqrt{3} & 0 & 1 & 0 \\ 0 & \sqrt{3} & 0 & 8+2\sqrt{3} & 0 & 1 \\ 2 & 0 & 1 & 0 & 6 & 0 \\ 0 & 2 & 0 & 1 & 0 & 6 \end{bmatrix} \tag{h}$$

Next, we consider the case in which each member is divided into two elements, $n_i = 2(i = 1, 2, \ldots, 6)$. From Eqs. (10.92), we obtain the disjoint global stiffness matrices

$$\bar{K}_1 = \bar{K}_3 = \bar{K}_6 = \frac{2EA}{L}\begin{bmatrix} 1 & 0 & -1 & 0 & 0 \\ 0 & 0 & 0 & 0 & 0 \\ -1 & 0 & 2 & -1 & 0 \\ 0 & 0 & -1 & 1 & 0 \\ 0 & 0 & 0 & 0 & 0 \end{bmatrix}$$

$$\bar{K}_2 = \bar{K}_5 = \frac{EA}{4L}\begin{bmatrix} 1 & -\sqrt{3} & -2 & 0 & 0 \\ -\sqrt{3} & 3 & 2\sqrt{3} & 0 & 0 \\ -2 & 2\sqrt{3} & 8 & -2 & 2\sqrt{3} \\ 0 & 0 & -2 & 1 & -\sqrt{3} \\ 0 & 0 & 2\sqrt{3} & -\sqrt{3} & 3 \end{bmatrix} \tag{i}$$

$$\bar{K}_4 = \frac{2EA}{\sqrt{3}L}\begin{bmatrix} 0 & 0 & 0 & 0 & 0 \\ 0 & 1 & 1 & 0 & 0 \\ 0 & 1 & 2 & 0 & 1 \\ 0 & 0 & 0 & 0 & 0 \\ 0 & 0 & 1 & 0 & 1 \end{bmatrix}$$

Table 10.6 Transfer of Entries from $\bar{K}_i$ ($i = 1, 2, \ldots, 6$) to K. One Element per Member

	Of $\bar{K}_i$,	Of K for					
	$i = 1, 2, \ldots, 6$	$i = 1$	$i = 2$	$i = 3$	$i = 4$	$i = 5$	$i = 6$
Row and/or Column	1	1	1	3	5	5	7
	2	2	2	4	6	6	8
	3	5	7	7	7	9	9
	4	6	8	8	8	10	10

Table 10.7 Transfer of Entries from $\bar{K}_i$ ($i = 1, 2, \ldots, 6$) to K. Two Elements per Member

	Of $\bar{K}_i$,	Of K for					
	$i = 1, 2, \ldots, 6$	$i = 1$	$i = 2$	$i = 3$	$i = 4$	$i = 5$	$i = 6$
Row and/or Column	1	1	1	3	5	5	7
	2	2	2	4	6	6	8
	3	11	12	13	14	15	16
	4	5	7	7	7	9	9
	5	6	8	8	8	10	10

and, from Eqs. (10.88), the disjoint global mass matrices are

$$\bar{M}_1 = \bar{M}_3 = \bar{M}_6 = \frac{mL}{12}\begin{bmatrix} 2 & 0 & 1 & 0 & 0 \\ 0 & 4 & 0 & 0 & 2 \\ 1 & 0 & 4 & 1 & 0 \\ 0 & 0 & 1 & 2 & 0 \\ 0 & 2 & 0 & 0 & 4 \end{bmatrix}$$

$$\bar{M}_2 = \bar{M}_5 = \frac{mL}{12}\begin{bmatrix} 7 & \sqrt{3} & 1 & 3 & \sqrt{3} \\ \sqrt{3} & 5 & -\sqrt{3} & \sqrt{3} & 1 \\ 1 & -\sqrt{3} & 8 & 1 & -\sqrt{3} \\ 3 & \sqrt{3} & 1 & 7 & \sqrt{3} \\ \sqrt{3} & 1 & -\sqrt{3} & \sqrt{3} & 5 \end{bmatrix} \tag{j}$$

$$\bar{M}_4 = \frac{mL}{4\sqrt{3}}\begin{bmatrix} 4 & 0 & 0 & 2 & 0 \\ 0 & 2 & -1 & 0 & 0 \\ 0 & -1 & 4 & 0 & -1 \\ 2 & 0 & 0 & 4 & 0 \\ 0 & 0 & -1 & 0 & 2 \end{bmatrix}$$

The derivation of the system stiffness matrix from the disjoint global stiffness matrices follows the pattern established earlier in this example, as well as the scheme of Table 10.7. After striking out the first four rows and columns, the result is

$$K = \frac{EA}{4\sqrt{3}L} \begin{bmatrix} 9\sqrt{3} & -3 & 0 & 0 & 0 & 0 & -8\sqrt{3} & 0 & 0 & 0 & -2\sqrt{3} & 0 \\ -3 & 8+3\sqrt{3} & 0 & 0 & 0 & 0 & 0 & 0 & 0 & 8 & 6 & 0 \\ 0 & 0 & 17\sqrt{3} & -3 & 0 & 0 & 0 & -2\sqrt{3} & -8\sqrt{3} & 0 & 0 & -8\sqrt{3} \\ 0 & 0 & -3 & 8+3\sqrt{3} & 0 & 0 & 0 & 6 & 0 & 8 & 0 & 0 \\ 0 & 0 & 0 & 0 & 9\sqrt{3} & -3 & 0 & 0 & 0 & 0 & -2\sqrt{3} & -8\sqrt{3} \\ 0 & 0 & 0 & 0 & -3 & 3\sqrt{3} & 0 & 0 & 0 & 0 & 6 & 0 \\ -8\sqrt{3} & 0 & 0 & 0 & 0 & 0 & 16\sqrt{3} & 0 & 0 & 0 & 0 & 0 \\ 0 & 0 & -2\sqrt{3} & 6 & 0 & 0 & 0 & 8\sqrt{3} & 0 & 0 & 0 & 0 \\ 0 & 0 & -8\sqrt{3} & 0 & 0 & 0 & 0 & 0 & 16\sqrt{3} & 0 & 0 & 0 \\ 0 & 8 & 0 & 8 & 0 & 0 & 0 & 0 & 0 & 16 & 0 & 0 \\ -2\sqrt{3} & 6 & 0 & 0 & -2\sqrt{3} & 6 & 0 & 0 & 0 & 0 & 8\sqrt{3} & 0 \\ 0 & 0 & -8\sqrt{3} & 0 & -8\sqrt{3} & 0 & 0 & 0 & 0 & 0 & 0 & 16\sqrt{3} \end{bmatrix} \tag{k}$$

Similarly, the system mass matrix is

$$M = \frac{mL}{12} \begin{bmatrix} 9+4\sqrt{3} & \sqrt{3} & 2\sqrt{3} & 0 & 3 & \sqrt{3} & 1 & 0 & 0 & 0 & 1 & 0 \\ \sqrt{3} & 9+2\sqrt{3} & 0 & 0 & \sqrt{3} & 1 & 0 & 0 & 0 & -\sqrt{3} & -\sqrt{3} & 0 \\ 2\sqrt{3} & 0 & 11+4\sqrt{3} & \sqrt{3} & 0 & 0 & 0 & 1 & 1 & 0 & 0 & 1 \\ 0 & 0 & \sqrt{3} & 13+2\sqrt{3} & 0 & 2 & 0 & -\sqrt{3} & 0 & -\sqrt{3} & 0 & 0 \\ 3 & \sqrt{3} & 0 & 0 & 9 & \sqrt{3} & 0 & 0 & 0 & 0 & 1 & 1 \\ \sqrt{3} & 1 & 0 & 2 & \sqrt{3} & 9 & 0 & 0 & 0 & 0 & -\sqrt{3} & 0 \\ 1 & 0 & 0 & 0 & 0 & 0 & 4 & 0 & 0 & 0 & 0 & 0 \\ 0 & 0 & 1 & -\sqrt{3} & 0 & 0 & 0 & 8 & 0 & 0 & 0 & 0 \\ 0 & 0 & 1 & 0 & 0 & 0 & 0 & 0 & 4 & 0 & 0 & 0 \\ 0 & -\sqrt{3} & 0 & -\sqrt{3} & 0 & 0 & 0 & 0 & 0 & 4\sqrt{3} & 0 & 0 \\ 1 & -\sqrt{3} & 0 & 0 & 1 & -\sqrt{3} & 0 & 0 & 0 & 0 & 8 & 0 \\ 0 & 0 & 1 & 0 & 1 & 0 & 0 & 0 & 0 & 0 & 0 & 4 \end{bmatrix} \tag{l}$$

Table 10.8 The Six Lowest Natural Frequencies for the Truss of Fig. 10.14

r	$\omega_r\sqrt{mL^2/EA}$		
	$n_i = 1$	$n_i = 2$	$n_i = 3$
1	0.196632	0.196173	0.196087
2	0.529687	0.525177	0.524310
3	0.638737	0.625406	0.622611
4	0.884323	0.877491	0.875322
5	1.197316	1.148331	1.132328
6	1.494639	1.439026	1.421134

The eigenvalue problem was solved for one, two and three finite elements per member, $n_i = 1$, 2 and 3, yielding six, twelve and eighteen natural frequencies, respectively. The system stiffness and mass matrices for $n_i = 1$ are given by Eqs. (g) and (h) and those for $n_i = 2$ by Eqs. (k) and (l), whereas those for $n_i = 3$ are not listed. The six lowest natural frequencies for the three cases (the only ones actually for the first case) are displayed in Table 10.8. As expected, the accuracy of the approximate natural frequencies improves as the number of finite elements increases.

10.7 FINITE ELEMENT MODELING OF FRAMES

Frames represent two-dimensional structures consisting of assemblages of beams and columns, as shown in Fig. 10.15. Although the columns are acted upon by axial forces due to the structure's own weight, this effect is not central to the subject, so that it will be ignored. On the other hand, the effect of the horizontal inertia of the beam will be included. In view of the fact that the columns are regarded as structural members undergoing bending alone, we refer to all members as beams.

We assume that all three beams are uniform and with parameters as shown in Fig. 10.15. Then, modeling the beams by the finite element method with n_i elements

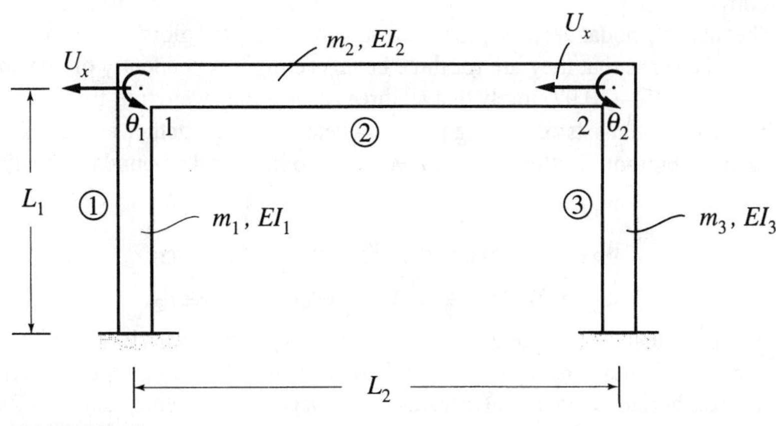

FIGURE 10.15
Frame structure

per member i $(i = 1, 2, 3)$, the nodal displacement vectors are simply

$$\mathbf{a}_i = [W_{i,0}\ h_i\Theta_{i,0}\ W_{i,1}\ h_i\Theta_{i,1}\ \dots\ W_{i,n_i}\ h_i\Theta_{i,n_i}]^T,\ i = 1, 2, 3 \tag{10.95}$$

Correspondingly, and using Eqs. (10.66) in conjunction with the scheme of Fig. 10.10, the member global stiffness matrices are

$$\bar{K}_i = \frac{EI_i}{h_i^3}\begin{bmatrix} 12 & 6 & -12 & 6 & 0 & 0 & \dots & 0 & 0 & 0 & 0 \\ & 4 & -6 & 2 & 0 & 0 & \dots & 0 & 0 & 0 & 0 \\ & & 24 & 0 & -12 & 6 & \dots & 0 & 0 & 0 & 0 \\ & & & 8 & -6 & 2 & \dots & 0 & 0 & 0 & 0 \\ & & & & \dots & \dots & \dots & \dots & \dots & \dots & \dots \\ & & & & & & & 24 & 0 & -12 & 6 \\ & \text{Symmetric} & & & & & & & 8 & -6 & 2 \\ & & & & & & & & & 12 & -6 \\ & & & & & & & & & & 4 \end{bmatrix},\ i = 1, 2, 3 \tag{10.96}$$

where EI_i is the flexural rigidity and h_i the size of the elements for member i $(i = 1, 2, 3)$. Similarly, the member global mass matrices are

$$\bar{M}_i = \frac{m_i h_i}{420}\begin{bmatrix} 156 & 22 & 54 & -13 & 0 & 0 & \dots & 0 & 0 & 0 & 0 \\ & 4 & 13 & -3 & 0 & 0 & \dots & 0 & 0 & 0 & 0 \\ & & 312 & 0 & 54 & -13 & \dots & 0 & 0 & 0 & 0 \\ & & & 8 & 13 & -3 & \dots & 0 & 0 & 0 & 0 \\ & & & & \dots & \dots & \dots & \dots & \dots & \dots & \dots \\ & & & & & & & 312 & 0 & 54 & -13 \\ & \text{Symmetric} & & & & & & & 8 & 13 & -3 \\ & & & & & & & & & 156 & -22 \\ & & & & & & & & & & 4 \end{bmatrix},$$

$$i = 1, 2, 3 \tag{10.97}$$

in which m_i is the mass density for member i $(i = 1, 2, 3)$. It should be pointed out here that provisions must yet be made for the horizontal inertia $m_2 L_2$ of beam 2; this will be done shortly.

The member nodal vectors, global stiffness matrices and global mass matrices are *disjoint*, in the sense that they are for three beams acting independently of one another. In fact, Eqs. (10.95)–(10.97) imply that all three act as if they were free-free beams. The assembly process consists of forcing the three beams to act together as a single frame clamped at the bottom. To this end, it is necessary to impose the boundary conditions

$$\begin{aligned} &W_{1,0} = 0,\ \Theta_{1,0} = 0,\ W_{1,n_1} = U_x,\ \Theta_{1,n_1} = \theta_1 \\ &W_{2,0} = 0,\ \Theta_{2,0} = \theta_1,\ W_{2,n_2} = 0,\ \Theta_{2,n_2} = \theta_2 \\ &W_{3,0} = 0,\ \Theta_{3,0} = 0,\ W_{3,n_3} = U_x,\ \Theta_{3,n_3} = \theta_2 \end{aligned} \tag{10.98}$$

We first observe that the third and eleventh of boundary conditions (10.98) state that the top of the two columns undergo the same horizontal translation as beam 2. Hence, to account for the horizontal inertia of beam 2, we add $m_2 L_2/2$ to the entry $(2n_i - 1,\ 2n_i - 1)$

of $\bar{M}_i$ $(i = 1, 3)$, which amounts to adding a lumped mass equal to one half of the mass of beam 2 to the top of each column. Then, in view of Eqs. (10.98), if we list the joint coordinates first, we can write the system nodal displacement vector as

$$\mathbf{a} = [U_x \;\; h_1\theta_1 \;\; h_3\theta_2 \;\; W_{1,1} \;\; h_1\Theta_{1,1} \;\; W_{1,2} \;\; h_1\Theta_{1,2} \; \dots \; W_{1,n_1-1} \;\; h_1\Theta_{1,n_1-1},$$
$$W_{2,1} \;\; h_2\Theta_{2,1} \;\; W_{2,2} \;\; h_2\Theta_{2,2} \; \dots \; W_{2,n_2-1} \;\; h_2\Theta_{2,n_2-1} \; \dots$$
$$W_{3,1} \;\; h_3\Theta_{3,1} \;\; W_{3,2} \;\; h_3\Theta_{3,2} \; \dots \; W_{3,n_3-1} \;\; h_3\Theta_{3,n_3-1}]^T \qquad (10.99)$$

The process of transferring the entries of the member stiffness matrices $\bar{K}_i$ and mass matrices $\bar{M}_i$ $(i = 1, 2, 3)$ to the system stiffness matrix K and mass matrix M, respectively, is similar to that for truss structures. In this regard, we must pay attention to the fact that the angular displacements in the member nodal displacement vectors are multiplied by the element size, so that if h_i differs from member to member some problem can arise. Indeed, the global stiffness matrices $\bar{K}_i$ and mass matrices $\bar{M}_i$ are computed on this basis, so that if there is a mismatch in the angular coordinates, the assembly will experience errors. To be more specific, in beam 1 the joint displacement is $h_1\theta_1$, whereas in beam 2 the joint displacement is $h_2\theta_1$. Hence, some alteration in the matrix entries affected must be made. To illustrate this problem, and in general to demonstrate the assembly process, it is necessary to consider a specific example.

Example 10.5. The parameters of the frame shown in Fig. 10.15 are as follows:

$$m_1 = m_3 = m, \; m_2 = 1.2m; \; I_1 = I_3 = I, \; I_2 = 2I; \; L_1 = L_3 = L, \; L_2 = 1.6L \qquad \text{(a)}$$

Derive and solve the eigenvalue problem by the finite element method for the two cases: 1) using one element per member and 2) using two elements per member. Compare results and draw conclusions.

As indicated earlier in this section, all three members are regarded as beams in bending. For a single element per member, $n_i = 1$ $(i = 1, 2, 3)$, Eqs. (10.96) and (10.97) do not apply, so that we must generate appropriate member stiffness and mass matrices. For a free-free member modeled by a single finite element, the stiffness matrix is given by Eq. (10.66). In the case of beam 1, in terms of the notation of Eqs. (10.95), the first row and column are associated with $W_{1,0}$, the second row and column with $h_1\Theta_{1,0}$, the third with $W_{1,1}$ and the fourth with $h_1\Theta_{1,1}$. But, according to the first two of Eqs. (10.98), $W_{1,0} = 0$ and $\Theta_{1,0} = 0$, so that we must strike out the first two rows and columns from Eqs. (10.66). Hence, using data from Eqs. (a), as well as $h_i = L_i$ $(i = 1, 2, 3)$, the stiffness matrix for beam 1 is

$$\bar{K}_1 = \frac{EI_1}{h_1^3}\begin{bmatrix} 12 & -6 \\ -6 & 4 \end{bmatrix} = \frac{EI}{L^3}\begin{bmatrix} 12 & -6 \\ -6 & 4 \end{bmatrix} \qquad \text{(b)}$$

In the case of beam 2, the first row and column are associated with $W_{2,0}$, the second with $h_2\Theta_{2,0}$, the third with $W_{2,1}$ and the fourth with $h_2\Theta_{2,1}$. But, from the fifth and seventh of Eqs. (10.98), $W_{2,0} = W_{2,1} = 0$, so that we must strike out the first and third rows and columns from Eq. (10.66). Moreover, to prevent a mismatch at the joints between beams 1 and 2 and beams 2 and 3, we must make the remaining second and fourth rows and columns correspond to $h_1\Theta_{2,0}$ and $h_1\Theta_{2,1}$, rather than to $h_2\Theta_{2,0}$ and $h_2\Theta_{2,1}$, respectively. This can be done through the maximum potential energy, as follows

$$2V_{\max,2} = \bar{K}_{2,11}(h_2\Theta_{2,0})^2 + 2\bar{K}_{2,12}(h_2\Theta_{2,0})(h_2\Theta_{2,1}) + \bar{K}_{2,22}(h_2\Theta_{2,1})^2$$

$$= \left(\frac{h_2}{h_1}\right)^2 [\bar{K}_{2,11}(h_1\Theta_{2,0})^2 + 2\bar{K}_{2,12}(h_1\Theta_{2,0})(h_1\Theta_{2,1}) + \bar{K}_{2,22}(h_1\Theta_{2,1})^2] \quad \text{(c)}$$

Hence, the stiffness matrix for beam 2 can be written as

$$\bar{K}_2 = \left(\frac{h_2}{h_1}\right)^2 \frac{EI_2}{h_2^3} \begin{bmatrix} 4 & 2 \\ 2 & 4 \end{bmatrix} = \frac{EI}{L^3} \begin{bmatrix} 5 & 2.5 \\ 2.5 & 5 \end{bmatrix} \quad \text{(d)}$$

It should be pointed out here that in this particular case all the entries of the stiffness matrix were multiplied by $(h_2/h_1)^2$, because both coordinates $\Theta_{2,0}$ and $\Theta_{2,1}$ represent angular displacements, and hence required adjustment. When $V_{\max,i}$ involves both ordinary displacements and angular displacements, and there is a mismatch between angular displacements in two beams or more, stiffness coefficients corresponding to displacements squared require no adjustment at all, those corresponding to products of a displacement and an angular displacement must be multiplied by h_2/h_1 and those corresponding to angular displacement squared must be multiplied by $(h_2/h_1)^2$, as in the case of Eq. (c). Beam 3 is equivalent to beam 1 in every respect, so that

$$\bar{K}_3 = \bar{K}_1 = \frac{EI}{L^3} \begin{bmatrix} 12 & -6 \\ -6 & 4 \end{bmatrix} \quad \text{(e)}$$

Using Eq. (10.67), the member mass matrices can be obtained in the same manner. Here, however, we recall that the effect of the mass m_2L_2 of the horizontal beam is to be taken into account through lumped masses $m_2L_2/2$ located at the top of beams 1 and 3, which amounts to adding $m_2L_2/2$ to the entry (1,1) of $\bar{M}_1$ and $\bar{M}_3$. Hence, striking out the first two rows and columns from Eq. (10.67) and considering Eqs. (a), the mass matrix for beam 1 is

$$\bar{M}_1 = \frac{m_1h_1}{420} \begin{bmatrix} 156 & -22 \\ -22 & 4 \end{bmatrix} + \frac{m_2L_2}{2} \begin{bmatrix} 1 & 0 \\ 0 & 0 \end{bmatrix} = \frac{mL}{420} \begin{bmatrix} 559.2 & -22 \\ -22 & 4 \end{bmatrix} \quad \text{(f)}$$

Using the same argument as for $\bar{K}_2$, the mass matrix for beam 2 can be shown to be

$$\bar{M}_2 = \left(\frac{h_2}{h_1}\right)^2 \frac{m_2h_2}{420} \begin{bmatrix} 4 & -3 \\ -3 & 4 \end{bmatrix} = \frac{mL}{420} \begin{bmatrix} 19.6608 & -14.7456 \\ -14.7456 & 19.6608 \end{bmatrix} \quad \text{(g)}$$

Moreover,

$$\bar{M}_3 = \bar{M}_1 = \frac{mL}{420} \begin{bmatrix} 559.2 & -22 \\ -22 & 4 \end{bmatrix} \quad \text{(h)}$$

Next, we turn our attention to the assembly process, which amounts to transferring the entries from the member stiffness and mass matrices to the system stiffness matrix and mass matrix, respectively. To this end, we note that the system nodal displacement vector is

$$\mathbf{a} = [U_x \;\; L\theta_1 \;\; L\theta_2]^T \quad \text{(i)}$$

and point out that the entries of the member matrices have been adjusted so that they all correspond to the same multiplying factor L for the angular displacements θ_1 and θ_2. To

Table 10.9 Transfer of Entries from $\bar{K}_i$ $(i = 1, 2, \ldots, 3)$ to K. One Element per Member

	Of $\bar{K}_i$,	Of K for		
	$i = 1, 2, \ldots, 3$	$i = 1$	$i = 2$	$i = 3$
Row and/or	1	1	2	1
Column	2	2	3	3

assemble the system stiffness matrix, we consider Table 10.9 and obtain

$$K = \frac{EI}{L^3}\begin{bmatrix} 24 & -6 & -6 \\ -6 & 9 & 2.5 \\ -6 & 2.5 & 9 \end{bmatrix} \tag{j}$$

Similarly, recognizing that Table 10.9 applies equally well to mass matrices, the system mass matrix can be shown to be

$$M = \frac{mL}{420}\begin{bmatrix} 1,118.4 & -22 & -22 \\ -22 & 23.6608 & -14.7456 \\ -22 & -14.7456 & 23.6608 \end{bmatrix} \tag{k}$$

Next, we consider the case of two finite elements per member, $n_i = 2,\ h_i = L_i/2$ $(i = 1, 2, 3)$. Recalling Eqs. (a), the nodal displacement vector for beam 1 is

$$\mathbf{a}_1 = [W_{1,1}\ \ h_1\Theta_{1,1}\ \ W_{1,2}\ \ h_1\Theta_{1,2}]^T = [W_{1,1}\ \ (L/2)\Theta_{1,1}\ \ U_x\ \ (L/2)\theta_1]^T \tag{l}$$

so that, striking out the first two rows and columns in Eq. (10.96), the stiffness matrix for beam 1 is

$$\bar{K}_1 = \frac{EI_1}{(L_1/2)^3}\begin{bmatrix} 24 & 0 & -12 & 6 \\ 0 & 8 & -6 & 2 \\ -12 & -6 & 12 & -6 \\ 6 & 2 & -6 & 4 \end{bmatrix} = \frac{8EI}{L^3}\begin{bmatrix} 24 & 0 & -12 & 6 \\ 0 & 8 & -6 & 2 \\ -12 & -6 & 12 & -6 \\ 6 & 2 & -6 & 4 \end{bmatrix} \tag{m}$$

To prevent mismatches, as in the case of one element per member, and recognizing that $W_{2,0} = W_{2,2} = 0$, we can write the maximum potential energy for beam 2 as

$$\begin{aligned}
2V_{\max,2} &= \bar{K}_{2,11}(h_2\Theta_{2,0})^2 + 2\bar{K}_{2,12}(h_2\Theta_{2,0})W_{2,1} + 2\bar{K}_{2,13}(h_2\Theta_{2,0})(h_2\Theta_{2,1}) \\
&\quad + 2\bar{K}_{2,14}(h_2\Theta_{2,0})(h_2\Theta_{2,2}) + \bar{K}_{2,22}W_{2,1}^2 + 2\bar{K}_{2,23}W_{2,1}(h_2\Theta_{2,1}) \\
&\quad + 2\bar{K}_{2,24}W_{2,1}(h_2\Theta_{2,2}) + \bar{K}_{2,33}(h_2\Theta_{2,1})^2 + 2\bar{K}_{2,34}(h_2\Theta_{2,1})(h_2\Theta_{2,2}) \\
&\quad + \bar{K}_{2,44}(h_2\Theta_{2,2})^2 \\
&= (h_2/h_1)^2\bar{K}_{2,11}(h_1\theta_1)^2 + 2(h_2/h_1)\bar{K}_{2,12}(h_1\theta_1)W_{2,1} \\
&\quad + 2(h_2/h_1)^2\bar{K}_{2,13}(h_1\theta_1)(h_1\Theta_{2,1}) + 2(h_2/h_1)^2\bar{K}_{2,14}(h_1\theta_1)(h_1\theta_2) \\
&\quad + \bar{K}_{2,22}W_{2,1}^2 + 2(h_2/h_1)\bar{K}_{2,23}W_{2,1}(h_1\Theta_{2,1}) + 2(h_2/h_1)\bar{K}_{2,24}W_{2,1}(h_1\theta_2) \\
&\quad + (h_2/h_1)^2\bar{K}_{2,33}(h_1\Theta_{2,1})^2 + 2(h_2/h_1)^2\bar{K}_{2,34}(h_1\Theta_{2,1})(h_1\theta_2) \\
&\quad + (h_2/h_1)^2\bar{K}_{2,44}(h_1\theta_2)^2
\end{aligned} \tag{n}$$

Hence, striking out the first and fifth rows and columns in Eq. (10.96) and letting $h_2/h_1 = 1.6,\ h_1 = L$, the stiffness matrix for beam 2 can be shown to be

$$\bar{K}_2 = \frac{EI_2}{(L_2/2)^3}\begin{bmatrix} (h_2/h_1)^2\times 4 & -(h_2/h_1)\times 6 & (h_2/h_1)^2\times 2 & 0 \\ -(h_2/h_1)\times 6 & 24 & 0 & (h_2/h_1)\times 6 \\ (h_2/h_1)^2\times 2 & 0 & (h_2/h_1)^2\times 8 & (h_2/h_1)^2\times 2 \\ 0 & (h_2/h_1)\times 6 & (h_2/h_1)^2\times 2 & (h_2/h_1)^2\times 4 \end{bmatrix}$$

$$= \frac{8EI}{L^3}\begin{bmatrix} 5 & -4.6875 & 2.5 & 0 \\ -4.6875 & 11.71875 & 0 & 4.6875 \\ 2.5 & 0 & 10 & 2.5 \\ 0 & 4.6875 & 2.5 & 5 \end{bmatrix} \tag{o}$$

and the corresponding nodal displacement vector is

$$\mathbf{a}_2 = [(L/2)\theta_1\ \ W_{2,1}\ \ (L/2)\Theta_{2,1}\ \ (L/2)\theta_2]^T \tag{p}$$

Similarly, the stiffness matrix for beam 3 is

$$\bar{K}_3 = \bar{K}_1 = \frac{8EI}{L^3}\begin{bmatrix} 24 & 0 & -12 & 6 \\ 0 & 8 & -6 & 2 \\ -12 & -6 & 12 & -6 \\ 6 & 2 & -6 & 4 \end{bmatrix} \tag{q}$$

and the associated nodal displacement vector is

$$\mathbf{a}_3 = [W_{3,1}\ \ (L/2)\Theta_{3,1}\ \ U_x\ \ (L/2)\theta_2]^T \tag{r}$$

Following the same procedure and using Eqs. (10.97), the member mass matrices can be written as

$$\bar{M}_1 = \bar{M}_3 = \frac{mL}{840}\begin{bmatrix} 312 & 0 & 54 & -13 \\ 0 & 8 & 13 & -3 \\ 54 & 13 & 962.4 & -22 \\ -13 & -3 & -22 & 4 \end{bmatrix}$$

$$\bar{M}_2 = \frac{mL}{840}\begin{bmatrix} 19.6608 & 39.936 & -14.7456 & 0 \\ 39.936 & 599.04 & 0 & -39.936 \\ -14.7546 & 0 & 39.3216 & -14.7456 \\ 0 & -39.936 & -14.7456 & 19.6608 \end{bmatrix} \tag{s}$$

where we added $m_2L_2/2 = 0.96\,mL$ to the (3,3) entry of $\bar{M}_1$ and $\bar{M}_3$, to account for the horizontal inertia of beam 2.

For the assembly process, we use the system nodal displacement vector

$$\mathbf{a} = [U_x\ \ (L/2)\theta_1\ \ (L/2)\theta_2\ \ W_{1,1}\ \ (L/2)\Theta_{1,1}\ \ W_{2,1}\ \ (L/2)\Theta_{2,1}\ \ W_{3,1}\ \ (L/2)\Theta_{3,1}]^T \tag{t}$$

Correspondingly, the transfer of entries from the member matrices to the system matrices is according to Table 10.10. Hence, using the scheme of Table 10.10, the system stiffness

Table 10.10 Transfer of Entries from $\bar{K}_i$ ($i = 1, 2, , 3$) to K. Two Elements per Member

	Of $\bar{K}_i$,	Of K for		
	$i = 1, 2, 3$	$i = 1$	$i = 2$	$i = 3$
Row and/or Column	1	4	2	8
	2	5	6	9
	3	1	7	1
	4	2	3	2

matrix can be shown to be

$$K = \frac{8EI}{L^3}\begin{bmatrix} 24 & -6 & -6 & -12 & -6 & 0 & 0 & -12 & -6 \\ -6 & 9 & 0 & 6 & 2 & -4.6875 & 2.5 & 0 & 0 \\ -6 & 0 & 9 & 0 & 0 & 4.6875 & 2.5 & 6 & 2 \\ -12 & 6 & 0 & 24 & 0 & 0 & 0 & 0 & 0 \\ -6 & 2 & 0 & 0 & 8 & 0 & 0 & 0 & 0 \\ 0 & -4.6875 & 4.6875 & 0 & 0 & 11.71875 & 0 & 0 & 0 \\ 0 & 2.5 & 2.5 & 0 & 0 & 0 & 10 & 0 & 0 \\ -12 & 0 & 6 & 0 & 0 & 0 & 0 & 24 & 0 \\ -6 & 0 & 2 & 0 & 0 & 0 & 0 & 0 & 8 \end{bmatrix} \tag{u}$$

Similarly, the system mass matrix is

$$M = \frac{mL}{840}\begin{bmatrix} 1{,}924.8 & -22 & -22 & 54 & 13 & 0 & 0 & 54 & 13 \\ -22 & 23.6608 & 0 & -13 & -3 & 39.936 & -14.7456 & 0 & 0 \\ -22 & 0 & 23.6608 & 0 & 0 & -39.936 & -14.7456 & -13 & -3 \\ 54 & -13 & 0 & 312 & 0 & 0 & 0 & 0 & 0 \\ 13 & -3 & 0 & 0 & 8 & 0 & 0 & 0 & 0 \\ 0 & 39.936 & -39.936 & 0 & 0 & 599.04 & 0 & 0 & 0 \\ 0 & -14.7456 & -14.7456 & 0 & 0 & 0 & 39.3216 & 0 & 0 \\ 54 & 0 & -13 & 0 & 0 & 0 & 0 & 312 & 0 \\ 13 & 0 & -3 & 0 & 0 & 0 & 0 & 0 & 8 \end{bmatrix} \tag{v}$$

The two eigenvalue problems, the first using one finite element per member and defined by the 3×3 stiffness and mass matrices given by Eqs. (j) and (k), respectively, and the second using two finite elements per member and defined by the 9×9 stiffness and mass matrices given by Eqs. (u) and (v), respectively, have been solved and the three lowest

Table 10.11 The Three Lowest Natural Frequencies for the Frame of Fig. 10.15

r	$\omega_r\sqrt{mL^4/EI}$	
	One Element per Member	Two Elements per Member
1	2.628685	2.626773
2	8.431008	6.903054
3	24.051269	18.750302

natural frequencies are listed in Table 10.11. Of course, in the first case the finite element model yields only three natural frequencies. We note that there is significant improvement in the approximate natural frequencies obtained by means of the model using two elements per member compared to those obtained by means of the model using only one element per member, particularly in the second and third lowest natural frequencies. Although the results show the correct trend, we conclude that only the lowest natural frequency exhibits any semblance of convergence. If frequencies higher than the first are of interest, then models using more than two finite elements per member must be used.

10.8 SYSTEM RESPONSE BY THE FINITE ELEMENT METHOD

The response of systems by the finite element method can be obtained in the same manner as the response by the Rayleigh-Ritz method, where the latter is the same as that by the assumed-modes method described in Sec. 9.8. We recall from Sec. 9.8 that the response by the assumed-modes method makes use of Lagrange's equations, which requires the kinetic energy, potential energy and virtual work of the nonconservative forces. Because the potential energy expression differs from system to system, we must discuss the response by means of a specific system.

We consider a string in transverse vibration, such as that shown in Fig. 8.3, refer to Sec. 8.2 and write the kinetic energy

$$T(t) = \frac{1}{2}\int_0^L \rho(x)\left[\frac{\partial y(x,t)}{\partial t}\right]^2 dx \tag{10.100}$$

the potential energy

$$V(t) = \frac{1}{2}\int_0^L T(x)\left[\frac{\partial y(x,t)}{\partial x}\right]^2 dx + \frac{1}{2}ky^2(L,t) \tag{10.101}$$

and the virtual work of the nonconservative forces

$$\overline{\delta W}_{nc}(t) = \int_0^L f(x,t)\delta y(x,t)dx \tag{10.102}$$

in which $\rho(x)$ is the mass per unit length of string, $y(x,t)$ the transverse displacement, $T(x)$ the string tension, k the spring constant and $f(x,t)$ the force per unit length.

To obtain the response by the finite element method, it is necessary to discretize Eqs. (10.100)–(10.102). To this end, we divide the domain $0 < x < L$ into n elements of width h. Then, by analogy with Eq. (10.3), we express the displacement over element j in terms of linear interpolation functions in the form

$$y(x,t) = a_{j-1}(t)\phi_{j-1}(x) + a_j(t)\phi_j(x), \ (j-1)h < x < jh \tag{10.103}$$

where $a_{j-1}(t)$ and $a_j(t)$ are string displacements at the nodal points $x = (j-1)h$ and $x = jh$, respectively, and, from Eqs. (10.10),

$$\phi_{j-1}(x) = j - x/h, \ \phi_j(x) = 1 - (j - x/h) \tag{10.104}$$

are the linear interpolation functions. The displacement $y(x,t)$ over element j, Eq. (10.103), can be expressed in terms of the nondimensional local coordinate $\xi = j - x/h$ as follows:

$$y(x,t) = y[h(j-\xi),t] = y_j(\xi,t) = \boldsymbol{\phi}^T(\xi)\mathbf{a}_j(t) \tag{10.105}$$

in which $\mathbf{a}_j(t) = [a_{j-1}(t) \ \ a_j(t)]^T$ is a vector of nodal displacements for element j and $\boldsymbol{\phi}(\xi) = [\phi_1(\xi) \ \ \phi_2(\xi)]^T$ is a vector of interpolation functions with components

$$\phi_1(\xi) = \xi, \ \phi_2(\xi) = 1 - \xi, \ \ 0 < \xi < 1 \tag{10.106}$$

which are the same for every element. Moreover, the derivatives required for the kinetic energy and potential energy are

$$\frac{\partial y(x,t)}{\partial t} = \boldsymbol{\phi}^T(\xi)\frac{d\mathbf{a}_j(t)}{dt} \tag{10.107}$$

and

$$\frac{\partial y(x,t)}{\partial x} = -\frac{1}{h}\frac{\partial y_j(\xi,t)}{\partial \xi} = -\frac{1}{h}\frac{d\boldsymbol{\phi}^T(\xi)}{d\xi}\mathbf{a}_j(t) \tag{10.108}$$

respectively. In addition, the distributed force can be expressed in terms of ξ over element j as follows:

$$f(x,t) = f[h(j-\xi),t] = f_j(\xi,t), \ 0 < \xi < 1 \tag{10.109}$$

The derivation of the equations of motion parallels the derivation of the eigenvalue problem in Sec. 10.2. Hence, by analogy with Eqs. (10.6) and (10.17), the kinetic energy can be written in the form

$$\begin{aligned} T(t) &= \frac{1}{2}\sum_{j=1}^{n}\int_{(j-1)h}^{jh} \rho(x)\left[\frac{\partial y(x,t)}{\partial t}\right]^2 dx \\ &= \frac{1}{2}\sum_{j=1}^{n}\dot{\mathbf{a}}_j^T(t)\left[h\int_0^1 \rho_j(\xi)\boldsymbol{\phi}(\xi)\boldsymbol{\phi}^T(\xi)d\xi\right]\dot{\mathbf{a}}_j(t) \\ &= \frac{1}{2}\sum_{j=1}^{n}\dot{\mathbf{a}}_j^T(t)M_j\dot{\mathbf{a}}_j(t) = \frac{1}{2}\dot{\mathbf{a}}^T(t)M\dot{\mathbf{a}}(t) \end{aligned} \tag{10.110}$$

where

$$M_j = h\int_0^1 \rho_j(\xi)\boldsymbol{\phi}(\xi)\boldsymbol{\phi}^T(\xi)d\xi = h\int_0^1 \rho_j(\xi)\begin{bmatrix} \xi^2 & \xi(1-\xi) \\ \xi(1-\xi) & (1-\xi)^2 \end{bmatrix} d\xi, \quad j = 1, 2, \ldots, n \tag{10.111}$$

are the element mass matrices, in which $\rho_j(\xi)$ is the mass density over element j and has an expression analogous to that given by Eq. (10.19). Moreover, $\mathbf{a}(t) = [a_1(t)\ a_2(t)\ \ldots\ a_n(t)]^T$ is the system nodal displacement vector, in which we considered the fact that $a_0(t) = 0$, because the string is fixed at $x = 0$. In addition, M is the global mass matrix, and is similar in structure to that given in Fig. 10.4. Similarly, by analogy with Eqs. (10.5) and (10.14), the potential energy can be written as

$$\begin{aligned} V(t) &= \frac{1}{2}\sum_{j=1}^{n}\left\{\int_{(j-1)h}^{jh} T(x)\left[\frac{\partial y(x,t)}{\partial x}\right]^2 dx + \delta_{jn}ky^2(L,t)\right\} \\ &= \frac{1}{2}\sum_{j=1}^{n}\mathbf{a}_j^T(t)\left[\frac{1}{h}\int_0^1 T_j(\xi)\frac{d\phi(\xi)}{d\xi}\frac{d\phi^T(\xi)}{d\xi}d\xi + \delta_{jn}k\phi(0)\phi^T(0)\right]\mathbf{a}_j(t) \\ &= \frac{1}{2}\sum_{j=1}^{n}\mathbf{a}_j^T(t)K_j\mathbf{a}_j(t) = \frac{1}{2}\mathbf{a}^T(t)K\mathbf{a}(t) \end{aligned} \tag{10.112}$$

where

$$\begin{aligned} K_j &= \frac{1}{h}\int_0^1 T_j(\xi)\frac{d\phi(\xi)}{d\xi}\frac{d\phi^T(\xi)}{d\xi}d\xi + \delta_{jn}k\phi(0)\phi^T(0) \\ &= \frac{1}{h}\int_0^1 T_j(\xi)\begin{bmatrix} 1 & -1 \\ -1 & 1 \end{bmatrix}d\xi + \delta_{jn}k\begin{bmatrix} 0 & 0 \\ 0 & 1 \end{bmatrix}, \quad j = 1, 2, \ldots, n \end{aligned} \tag{10.113}$$

are the element stiffness matrices, in which $T_j(\xi)$ is the string tension in the element j, and K is the global stiffness matrix having a structure similar to that in Fig. 10.4.

Finally, we must discretize the virtual work of the conservative forces. To this end, we insert Eq. (10.105) into Eq. (10.102), consider Eq. (10.109) and write

$$\begin{aligned} \overline{\delta W}_{nc}(t) &= \sum_{j=1}^{n}\int_{(j-1)h}^{jh} f(x,t)\delta y(x,t)dx = \sum_{j=1}^{n}\left[h\int_0^1 f_j(\xi,t)\phi^T(\xi)d\xi\right]\delta\mathbf{a}_j(t) \\ &= \sum_{j=1}^{n}\mathbf{F}_j^T(t)\delta\mathbf{a}_j(t) = \mathbf{F}^T(t)\delta\mathbf{a}(t) \end{aligned} \tag{10.114}$$

in which

$$\mathbf{F}_j(t) = h\int_0^1 f_j(\xi,t)\phi(\xi)d\xi = h\int_0^1 f_j(\xi,t)\begin{bmatrix} \xi \\ 1-\xi \end{bmatrix}d\xi, \quad j = 1, 2, \ldots, n \tag{10.115}$$

are the nodal forces and $\mathbf{F}(t) = [F_1(t)\ F_2(t)\ \ldots\ F_n(t)]^T$ is the global nodal force vector, and note that we omitted $F_0(t)$ from $\mathbf{F}(t)$ because $\delta a_0(t) = 0$. Hence, the effect of the discretization process given by Eq. (10.115) is to generate concentrated forces acting at the nodal points to replace the distributed force $f(x,t)$. We observe that, to obtain the global nodal force vector $\mathbf{F}(t)$, we must carry out an assembly process, which amounts to adding the bottom component of $\mathbf{F}_j(t)$ to the top component of $\mathbf{F}_{j+1}(t)$ to generate the nodal force $F_j(t)$.

The equations of motion for the discretized system can be obtained by inserting Eqs. (10.110) and (10.112) into Lagrange's equations, Eqs. (6.42), and carrying out the indicated differentiations. This is not really necessary, because from the quadratic forms of the kinetic energy and potential energy and the virtual work expression, we can write the equations of motion directly in the matrix form

$$M\ddot{\mathbf{a}}(t) + K\mathbf{a}(t) = \mathbf{F}(t) \tag{10.116}$$

Of course, the same developments are valid for rods in axial vibration and shafts in torsion, subject to the parameter substitution specified in Table 8.1.

With certain modifications, the preceding developments apply also to beams in bending (Sec. 10.4). In particular, the elements stiffness matrices K_j and mass matrices M_j are 4×4 and they are given by Eqs. (10.63) and (10.65), respectively, and the element nodal displacement vectors are four-dimensional and have the form $\mathbf{a}_j(t) = [y_{j-1}(t)\ \ h\theta_{j-1}(t)\ \ y_j(t)\ \ h\theta_j(t)]^T$ $(j = 1, 2, \ldots, n)$. Moreover, using the analogy with Eqs. (10.115) in conjunction with Eqs. (10.58), the element nodal force vectors are

$$\begin{aligned} \mathbf{F}_j(t) &= h\int_0^1 f_j(\xi, t)\phi(\xi)d\xi \\ &= h\int_0^1 f_j(\xi, t)[3\xi^2 - \xi^3\ \ \ \xi^2 - \xi^3\ \ \ 1 - 3\xi^2 + 2\xi^3\ \ \ -\xi + 2\xi^2 - \xi^3]^T d\xi, \\ & \qquad\qquad j = 1, 2, \ldots, n \end{aligned} \tag{10.117}$$

We observe that the first and third components in $\mathbf{F}_j(t)$ represent forces, whereas the second and fourth components represent moments, although they have units of force, because the rotations $\theta_{j-1}(t)$ and $\theta_j(t)$ in the element modal displacement vector $\mathbf{a}_j(t)$ have been multiplied by h so that all the components of $\mathbf{a}_j(t)$ have the same units.

Consistent with the above, the global stiffness and mass matrices have structures similar to those shown in Fig. 10.10, the exact form depending on the boundary conditions. Finally, the third component of $\mathbf{F}_j(t)$ is to be added to the first component of $\mathbf{F}_{j+1}(t)$ to obtain the nodal force $F_j(t)$ and the fourth component of $\mathbf{F}_j(t)$ is to be added to the second component of $\mathbf{F}_{j+1}(t)$ to obtain the nodal moment $M_j(t)$, where we note that $M_j(t)$ actually has units of force.

Example 10.6. Derive the equations of motion for the pinned-pinned beam of Example 10.3. The beam is acted upon by the triangularly distributed force $f(x, t) = (x/L)f(t)$.

The global mass and stiffness matrices entering into Eq. (10.116) were derived in Example 10.3 using 10 finite elements. Hence, the only task remaining is to derive the nodal force vector $\mathbf{F}(t)$ corresponding to the system nodal displacement vector

$$\mathbf{a}(t) = [h\theta_0(t)\ \ y_1(t)\ \ h\theta_1(t)\ \ y_2(t)\ \ h\theta_2(t)\ \ldots\ y_9(t)\ \ h\theta_9(t)\ \ h\theta_{10}(t)]^T \tag{a}$$

To this end, we first use Eqs. (10.117) and derive the element nodal force vectors $\mathbf{F}_j(t)$ $(j = 1, 2, \ldots, 10)$. Hence, recalling Eq. (10.109), we can write

$$f(x, t) = f_j[h(j - \xi), t] = \frac{h(j - \xi)f(t)}{L} = \frac{f(t)}{10}(j - \xi),\ 0 < \xi < 1 \tag{b}$$

so that, inserting Eq. (b) into Eqs. (10.117) and carrying out the indicated integrations, we obtain

$$\mathbf{F}_j(t) = \frac{hf(t)}{10}\int_0^1 (j-\xi)[3\xi^2-2\xi^3 \;\; \xi^2-\xi^3 \;\; 1-3\xi^2+2\xi^3 \;\; -\xi+2\xi^2-\xi^3]^T d\xi$$

$$= \frac{Lf(t)}{1.2\times 10^4}[6(10j-7) \;\; 2(5j-3) \;\; 6(10j-3) \;\; -2(5j-2)]^T, \; j=1,2,\ldots,10 \tag{c}$$

Introducing the notation

$$\mathbf{F}_j(t) = \frac{Lf(t)}{1.2\times 10^4}\mathbf{F}_{0j}, \; j=1,2,\ldots,10 \tag{d}$$

we can write the vectors $\mathbf{F}_{0j}$ in the more explicit form

$$\mathbf{F}_{01} = \begin{bmatrix} 18 \\ 4 \\ 42 \\ -6 \end{bmatrix}, \; \mathbf{F}_{02} = \begin{bmatrix} 78 \\ 14 \\ 102 \\ -16 \end{bmatrix}, \; \mathbf{F}_{03} = \begin{bmatrix} 138 \\ 24 \\ 162 \\ -26 \end{bmatrix}, \; \mathbf{F}_{04} = \begin{bmatrix} 198 \\ 34 \\ 222 \\ -36 \end{bmatrix},$$

$$\mathbf{F}_{05} = \begin{bmatrix} 258 \\ 44 \\ 282 \\ -46 \end{bmatrix}, \; \mathbf{F}_{06} = \begin{bmatrix} 318 \\ 54 \\ 342 \\ -56 \end{bmatrix}, \; \mathbf{F}_{07} = \begin{bmatrix} 378 \\ 64 \\ 402 \\ -56 \end{bmatrix}, \; \mathbf{F}_{08} = \begin{bmatrix} 438 \\ 74 \\ 462 \\ -76 \end{bmatrix}, \tag{e}$$

$$\mathbf{F}_{09} = \begin{bmatrix} 498 \\ 84 \\ 522 \\ -86 \end{bmatrix}, \; \mathbf{F}_{10} = \begin{bmatrix} 558 \\ 94 \\ 582 \\ -96 \end{bmatrix}$$

Next, we carry out the assembly process. To this end, we recall that the translation of nodal point 1 is obtained by adding the third component of $\mathbf{F}_1(t)$ and the first component of $\mathbf{F}_2(t)$ and the rotation at the nodal point 1 is the sum of the fourth component of $\mathbf{F}_1(t)$ and the second component of $\mathbf{F}_2(t)$, etc. Moreover, the first component of $\mathbf{F}_1(t)$ and the third component of $\mathbf{F}_{10}(t)$ must be deleted, because they correspond to $y_0(t) = y_{10}(t) = 0$. The resulting system nodal force vector is

$$\mathbf{F}(t) = \frac{Lf(t)}{1.2\times 10^4}[4 \;\; 120 \;\; 8 \;\; 240 \;\; 8 \;\; 360 \;\; 8 \;\; 480 \;\; 8 \;\; 600 \;\; 8 \;\; 720$$

$$8 \;\; 840 \;\; 8 \;\; 960 \;\; 8 \;\; 1080 \;\; 8 \;\; -96]^T \tag{f}$$

The desired equations of motion are obtained by inserting Eqs. (a) and (f), together with the global stiffness and mass matrices from Example 10.3, into Eq. (10.116).

10.9 MATLAB PROGRAM FOR THE SOLUTION OF THE EIGENVALUE PROBLEM BY THE FINITE ELEMENT METHOD

Although the formalism for the derivation of the eigenvalue problem by the finite element method is the same as for the Rayleigh-Ritz method, the details for generating the mass and stiffness matrices are different. In fact, as far as MATLAB programming is concerned, it turns out that the generation of the mass and stiffness matrices by the finite element method requires a more elaborate program than that by the Rayleigh-Ritz method. Following is a MATLAB program for solving the eigenvalue problem for the pinned-pinned beam of Example 10.3.

```
% The program 'femppb.m' computes the natural frequencies and modes of the pinned-
% pinned beam of Example 10.3 and plots the modes

clear
clf

n=10; % number of finite elements
Ke=[12 6 -12 6 ; 6 4 -6 2 ; -12 -6 12 -6 ; 6 2 -6 4]*n^3; % element stiffness
% matrix
Me=[156 22 54 -13 ; 22 4 13 -3 ; 54 13 156 -22 ; -13 -3 -22 4]/(420*n);
% element mass matrix
ke=[1 1 1.44 1.44 1.96 1.96 1.44 1.44 1 1]; % stiffness distribution over the
% elements arranged in an ten-dimensional vector
me=[1 1 1.2 1.2 1.4 1.4 1.2 1.2 1 1]; % mass distribution over the elements
% arranged in an n-dimensional vector

for k=1:n,
  IC(k,1)=k; % generation of a connectivity array placing the element stiffness and
  IC(k,2)=k+1; % mass matrices in the global stiffness and mass matrices
end

K1=zeros (2*n+2,2*n+2); % nulling the global stiffness and mass matrices before
M1=zeros (2*n+2,2*n+2); % enforcement of the boundary conditions
K=zeros (2*n,2*n); % nulling the stiffness matrix
M=zeros (2*n,2*n); % nulling the mass matrix

for k=1:n,
  for p=1:2,
    for q=1:2,
      K1(2*IC(k,p)-1,2*IC(k,q) -1)=K1(2*IC(k,p)-1,2*IC(k,q)-1) +ke(k)*Ke(2*p-1,2
*q-1);
      K1(2*IC(k,p) -1,2*IC(k,q))=K1(2*IC(k,p) -1,2*IC(k,q))+ke(k)*Ke(2*p -1,2*q);
      K1(2*IC(k,p),2*IC(k,q)-1)=K1(2*IC(k,p),2*IC(k,q)-1) +ke(k)*Ke(2*p,2*q-1);
      K1(2*IC(k,p),2*IC(k,q))=K1(2*IC(k,p),2*IC(k,q)) +ke(k)*Ke(2*p,2*q);
      M1(2*IC(k,p)-1,2*IC(k,q) -1)=M1(2*IC(k,p) -1,2*IC(k,q)-1)+me(k)*Me(2*p-
1,2*q-1);
      M1(2*IC(k,p) -1,2*IC(k,q))=M1(2*IC(k,p)-1,2*IC(k,q))+me(k)*Me(2*p-1,2*q);
      M1(2*IC(k,p), 2*IC(k,q)-1)=M1(2*IC(k,p),2*IC(k,q)-1)+me(k)*Me(2*p,2*q-1);
      M1(2*IC(k,p),2*IC(k,q))=M1(2*IC(k,p),2*IC(k,q))+me(k)*Me(2*p,2*q) ;
      % assembly of the global matrices before enforcement of the boundary
      % conditions
    end
  end
end

for i=2:(2*n),
  for j=2: (2*n),
  K(i-1,j-1)=K1(i,j);
  M(i-1,j-1)=M1(i,j);
```

```
end
end

for i=2: (2*n),
  K(2*n,i-1)=K1(2*n+2,i);
  K(i-1,2*n)=K1(i,2*n+2);
  M(2*n,i-1)=M1(2*n+2,i);
  M(i-1,2*n)=M1(i,2*n+2);
end
K(2*n,2*n)=K1(2*n+2,2*n+2);
M(2*n,2*n)=M1(2*n+2,2*n+2);
% enforcement of the boundary conditions by retaining and relabeling the
% nonzero entries

[v,W]=eig(K,M); % solution of the algebraic eigenvalue problem using MATLAB
% function

for i=1:2*n,
  w1(i)=sqrt(W(i,i)); % natural frequencies arranged in a 2n-dimensional vector
  a=1/sqrt(v(:,j)'*M*v(:,j)); % normalization factors
  V(:,i)=a*v(:,i); normalization of the eigenvectors
end

[w,I]=sort(w1); % arranging the natural frequencies in ascending order
for j=1:2*n,
  U1(:,j)=V(:,I(j)); % arranging the modal vectors in ascending order
  end

  for i=1:2*n-1,
    for j=1:2*n-1,
      U(i+1,j+1)=U1(i,j);
    end
  end

  for j=1:2*n-1,
  U(2*n+2,j+1)=U1(2*n,j);
  U(j+1,2*n+2)=U1(j,2*n);
end
U(2*n+2,2*n+2)=U1(2*n,2*n);

s=[1:-0.05:0]; % local coordinate increments
phi1=3*s.^2-2*s.^3; phi2=s.^2-s.^3; phi3=1-3*s.^2+2*s.^3; phi4=-s+2*s.^2-s.^3;

for m=2:4,
  y=[ ];
  x=[ ];
  f=0.1;
  if m>=3; f=0.2; end
  axes('position',[0.3 0.8-0.35*(m-2) 0.4 f])
```

```
for k=1:n,
  y1=phi1*U(2*IC(k,1)-1,m)+phi2*U(2*IC(k,1),m)+phi3*U(2*IC(k,2)-1,m )+phi4*
U(2*IC(k,2),m);
  x1=(k-s)/n;
  y=[y  y1];
  x=[x  x1];
end
plot(x,-y/max(abs(y)))
set(gca,'XTick',[0:0.1:1], 'XTickLabel' , '0|h|2h||||||9h|10h')
st=num2str(w(m-1));
st1=int2str(m-1);
st2=int2str(2*n);
st3=strcat('Y_',st1,'^{(',st2,')}(x)');
st4=strcat('Natural Mode',{' '},st1,{' , '} , '\omega_',st1, '^{(',st2,')}' , '=',st, '[EI/mL
^4]^{1/2}');
title(st4)
ylabel(st3)
end
```

10.10 SUMMARY

In this textbook, we subscribe to the point of view that the finite element method is a Rayleigh-Ritz method, so that the finite element method can be assumed to share the formalism and theory developed for the Rayleigh-Ritz method. The most important difference between the two spatial discretization techniques lies in the nature of the trial functions. In the finite element method, the trial functions, known as interpolation functions, are low-degree polynomials defined over small finite elements, whereas in the Rayleigh-Ritz method they tend to be involved functions defined over the entire domain. It is the use of small finite elements that gives the finite element method enormous versatility in solving problems with complex geometry.

Strings in transverse vibration, rods in axial vibration and shafts in torsion represent second-order systems. The interpolation functions most commonly used in such cases are linear, which are the lowest-degree polynomials admissible. The derivation of the mass and stiffness matrices follows a well-established pattern, which consists of first deriving element matrices and then assembling them into global matrices. At some point, it is necessary to enforce the boundary conditions. Improper enforcement of the boundary conditions can result in singular global matrices. The components of the displacement vector for systems discretized by the finite element method represent actual displacements at the nodes, where nodes are defined as boundary points shared by adjacent finite elements (not points of zero displacement, as commonly defined in vibrations). Moreover, the computation of the element matrices can be greatly simplified by the use of nondimensional local coordinates. Although quadratic and cubic interpolation functions yield better accuracy than linear interpolation functions, this advantage is negated by the simplicity of the latter. Note that, if a sufficient number of elements is used in conjunction with linear interpolation functions, computer plots appear as continuous. For beams in bending, which represent fourth-order systems, the lowest-degree polynomials that qualify as interpolation functions are cubics. In particular, the Hermite cubics are almost universally used. The displacement vector includes, in addition to actual nodal translational displacements, also nodal rotations. The derivation of the global mass and stiffness matrices follows the same pattern as for second-order systems, but the details are more involved. Moreover, the number of degrees of freedom for the same number of elements is twice as large, as there are two nodal displacements per element. Trusses represent assemblages of rods undergoing axial deformations, as well as transverse rigid-body translations. On the other hand, frames represent assemblages of structural members in bending, some subjected to axial forces, although the effect of the latter can be neglected at times. The assembly of element matrices into global matrices is significantly more involved for trusses and frames than for one-dimensional systems.

Determination of the system response by the finite element method requires the derivation of the force vector, in addition to the mass and stiffness matrices. To this end, it is convenient to begin with the virtual work expression and follow the same steps as for the generation of global matrices, namely, derivation of the element nodal force vectors, global nodal force vector and enforcement of the boundary conditions, where the latter ordinarily amounts to deleting appropriate global nodal force vector components.

A MATLAB program for the finite element solution of the eigenvalue problem for the pinned-pinned beam of Example 10.3 is included. Whereas the computational details involved in the finite element method are relatively simple, the computer program is more involved than one would expect.

PROBLEMS

10.1. Derive the eigenvalue problem for the tubular shaft of Problem 9.1 by means of the finite element method using linear interpolation functions. Solve the eigenvalue problem using 10 finite elements and compare the three lowest eigenvalues with those obtained in Problem 9.17 by means of the Rayleigh-Ritz method and draw conclusions concerning the accuracy of the two sets of results. Plot the three lowest modes obtained here.

10.2. Derive the eigenvalue problem for the rod of Problem 9.2 by means of the finite element method using linear interpolation functions. Solve the eigenvalue problem using 10 finite elements and compare the three lowest eigenvalues with those obtained in Problem 9.18 by means of the Rayleigh-Ritz method and in Problem 9.28 by means of the enhanced Rayleigh-Ritz method and draw conclusions concerning the accuracy of the three sets of eigenvalues. Plot the three lowest modes obtained here.

10.3. Solve Problem 10.1 with quadratic interpolation functions instead of linear.

10.4. Solve Problem 10.2 with quadratic interpolation functions instead of linear.

10.5. Solve Problem 10.1 with cubic interpolation functions instead of linear.

10.6. Solve Problem 10.2 with cubic interpolation functions instead of linear.

10.7. Derive the eigenvalue problem for the rod of Problem 9.19 by means of the finite element method using linear interpolation functions. Solve the eigenvalue problem corresponding to 10 finite elements and compare the three lowest eigenvalues obtained here with those obtained in Problem 9.19 by means of the Rayleigh-Ritz method with $n = 6$ and draw conclusions. Plot the three lowest modes.

10.8. Derive the eigenvalue problem for the tapered beam of Problem 9.5 by means of the finite element method using Hermite cubics as interpolation functions. Determine the number of finite elements required to compute the lowest eigenvalue with the same accuracy as that obtained in Problem 9.20 by means of the Rayleigh-Ritz method using six trial functions. Begin computations with six finite elements. Plot the three lowest modes.

10.9. Derive the eigenvalue problem for the beam of Problem 8.16 by means of the finite element method using Hermite cubics as interpolation functions. Then, solve the eigenvalue problem using an increasing number of elements and determine the number of elements required to compute the lowest eigenvalue with the same accuracy as that obtained in Problem 9.31 by means of the Galerkin method using eight comparison functions. Begin computations with eight finite elements. Plot the three lowest modes.

10.10. Derive the eigenvalue problem for the beam of Problem 9.21 by means of the finite element method using Hermite cubics as interpolation functions. Solve the eigenvalue problem corresponding to 10 finite elements and compare the three lowest eigenvalues obtained here with those obtained in Problem 9.21 by means of the Rayleigh-Ritz method with $n = 6$ and draw conclusions. Plot the three lowest modes.

10.11. Derive the eigenvalue problem for the aircraft of Problem 9.24 by means of the finite element method using Hermite cubics as interpolation functions. Solve the eigenvalue problem corresponding to 10 finite elements and compare the two lowest nonzero eigenvalues obtained here with those obtained in Problem 9.24 by means of the Rayleigh-Ritz method with $n = 8$ and draw conclusions. Plot the two lowest elastic modes.

10.12. The truss shown in Fig. 10.16 is pinned at all joints and has the following parameters: $m_i = m\ (i = 1, 2, \ldots 5)$, $EA_1 = EA_4 = EA$, $EA_2 = EA_3 = EA_5 = 1.5EA$. Derive and solve the eigenvalue problem modeling the truss by means of one finite element per structural member and using linear interpolation functions.

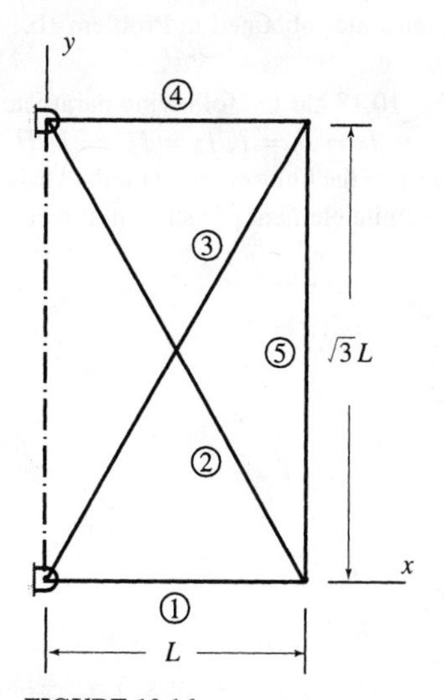

FIGURE 10.16
Truss for Problem 10.12

10.13. The truss shown in Fig. 10.17 is pinned at all joints and has the following parameters: $m_i = m\ (i = 1, 2, \ldots 7)$, $EA_1 = EA_2 = EA_7 = EA$, $EA_3 = EA_4 = EA_5 = EA_6 = 1.5EA$.

Derive and solve the eigenvalue problem modeling the truss by means of one finite element per structural member and using linear interpolation functions.

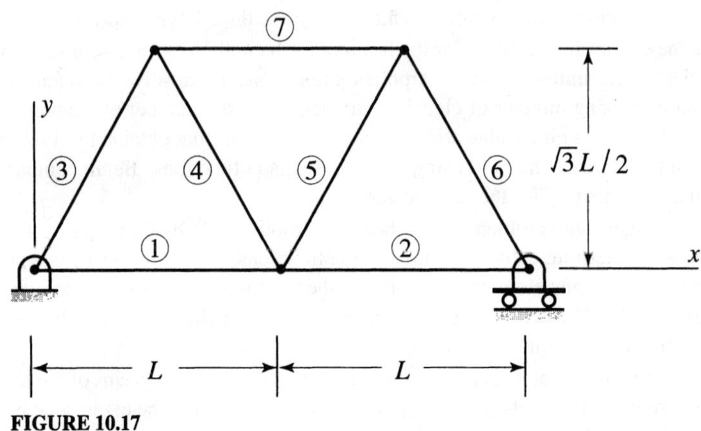

FIGURE 10.17
Truss for Problem 10.13

10.14. Solve Problem 10.12 modeling the truss by means of two finite elements per structural member. Compare the eigenvalues obtained in Problem 10.12 with those obtained here and draw conclusions.

10.15. Solve Problem 10.13 modeling the truss by means of two finite elements per structural member. Compare the eigenvalues obtained in Problem 10.13 with those obtained here and draw conclusions.

10.16. The two-story frame of Fig. 10.18 has the following parameters: $m_1 = m_3 = m_4 = m_6 = m$, $m_2 = m_5 = 1.2m$, $I_1 = I_3 = I_4 = I_6 = I$, $I_2 = I_5 = 2I$, $H = 0.8L$. Derive and solve the eigenvalue problem by means of the finite element method using Hermite cubics and modeling the frame by means of one finite element per structural member.

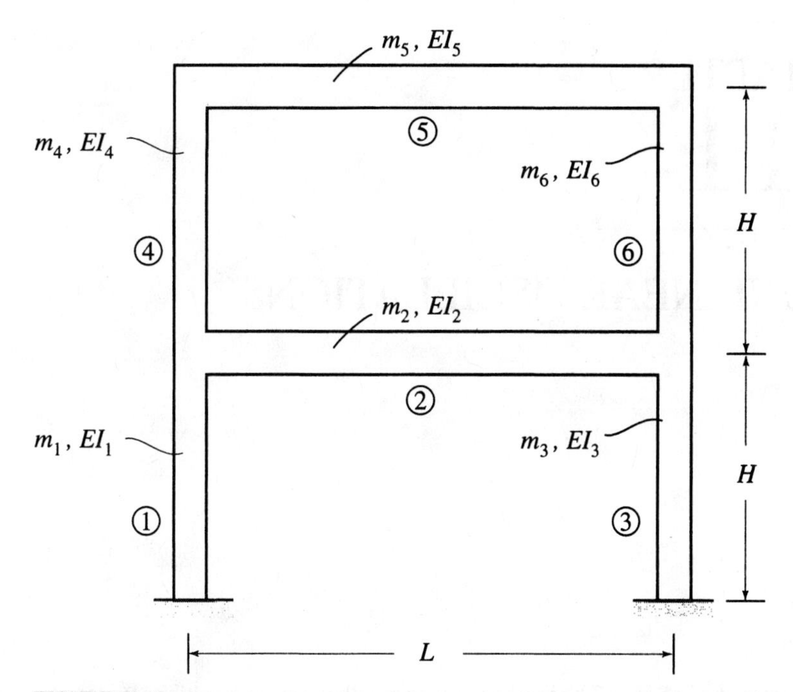

FIGURE 10.18
Two-story frame for Problem 10.16

10.17. Solve Problem 10.16 modeling the frame by means of two finite elements per member. Compare the eigenvalues obtained in Problem 10.16 with those obtained here and draw conclusions.

10.18. Obtain the response of the shaft of Problem 10.1 to the distributed torque $m(x,t) = m_0\{(2x/L)\mathcal{u}(x) - 2[(2x-L)/L]\mathcal{u}(x-L/2)\}\delta(t)$, where $\mathcal{u}(x-a)$ is a spatial unit step functions beginning at $x = a$ and $\delta(t)$ is the unit impulse.

10.19. Obtain the response of the beam of Problem 10.8 to the distributed vertical force $f(x,t) = f_0[1-(x/L)^2/2]\mathcal{u}(t)$ using six finite elements, where $\mathcal{u}(t)$ is the unit step function.

10.20. Solve Problem 10.1 by MATLAB.

10.21. Solve Problem 10.2 by MATLAB.

10.22. Solve Problem 10.7 by MATLAB.

10.23. Solve Problem 10.8 by MATLAB.

10.24. Solve Problem 10.9 by MATLAB.

10.25. Solve Problem 10.10 by MATLAB.

10.26. Solve Problem 10.11 by MATLAB.

10.27. Solve Problem 10.12 by MATLAB.

10.28. Solve Problem 10.13 by MATLAB.

10.29. Solve Problem 10.16 by MATLAB.

10.30. Solve Problem 10.18 by MATLAB.

10.31. Solve Problem 10.19 by MATLAB.

CHAPTER

11

NONLINEAR OSCILLATIONS

As pointed out in Sec. 1.7, the question as to whether a system exhibits linear or nonlinear behavior cannot be answered unequivocally, unless the range over which the system operates is specified. Indeed, the answer to this question depends on the relation between the excitation and response, which in turn depends on that range. When a spring is subjected to a tensile force it stretches. Over a given range, generally corresponding to small deformations, the deformation is proportional to the restoring force in the spring, so that the relation between the two is linear, in which case the range is said to be linear. Beyond a certain point, however, the deformation ceases to be proportional to the force, so that the relation is nonlinear, and so is the range. The restoring force increases at a higher rate than the deformation for a "hardening spring" and at a lower rate for a "softening spring." Hence, the same simple mass-spring oscillator must be regarded as a linear system if its motion is confined to the linear range of the spring, and as a nonlinear system if its motion exceeds the linear range. Quite often, the point separating the linear from the nonlinear range is somewhat arbitrary, as it depends on the accuracy with which the response is measured. Perhaps the concept can be illustrated better by means of a simple pendulum. From Example 1.2, we see that the restoring moment about the point of support O is $M_o = -mgL\ \sin\theta$, where mg is the weight of the bob, L the length of the string and θ the angle from the equilibrium position, where the latter coincides with the vertical through point O. Hence, the linear range can be identified as that in which $\sin\theta$ can be approximated by θ, which raises immediately the question of accuracy of the approximation. In this regard, we recall that $\sin\theta$ represents an infinite series, which must be truncated for practical purposes, where the truncation is dictated by the number of accurate decimal places desired. Indeed, $\sin\theta$ can be regarded as being equal to θ when it is computed with four decimal places accuracy and different from θ with six decimal places. The preceding discussion was mainly to convey the idea that under certain circumstances a system can be regarded as linear and under other circumstances the same system must be treated as nonlinear. As pointed out in Sec. 2.1, a simple pendulum can be regarded as linear for surprisingly large angular displacements, as the error in approximating $\sin\theta$ by θ for $\theta = 20°$ is about 2% and for $\theta = 30°$ is less than 5%.

The fact that a system is linear has profound implications on the analysis of vibrations problems, because the principle of superposition holds for linear systems alone, and does not hold for nonlinear systems. We recall from Sec. 1.12 that the principle of superposition states that the response of a linear system to a linear combination of excitations can be obtained by first obtaining the response to the individual excitations separately and then combining the individual responses linearly. There is no counterpart to the principle of superposition for nonlinear systems. This explains why the linear system theory is so well developed, and why the treatment of nonlinear systems often requires ad hoc methods of attack. In fact, because of the relative ease with which linear systems can be treated, one approach to nonlinear systems consists of assuming that the motion is confined to the neighborhood of known solutions, a process referred to as linearization. To be sure, caution must be exercised in using linearization, as demonstrated in this chapter.

There are many approaches to the study of nonlinear systems, but they can all be divided into two broad classes, qualitative and quantitative. In the qualitative approach, the interest lies not so much in the explicit time history of the motion of the system as in a statement whether the motion in the neighborhood of a known solution is stable or unstable. For the most part, the known solution represents an equilibrium position. On the other hand, the quantitative approach is concerned with just these time histories. When the system nonlinearity is relatively small, a solution may be possible by a perturbation technique whereby the response is expanded in a series in a small parameter representing a measure of the magnitude of the nonlinearity. Of special interest here is the case in which the actual solution is known to be periodic. When the nonlinearity is not small, time histories can be obtained by numerical integration, which requires that the equations of motion be cast in state form. In this chapter, we discuss both qualitative and quantitative techniques.

11.1 FUNDAMENTAL CONCEPTS IN STABILITY. EQUILIBRIUM POINTS

In Sec. 1.13, we introduced definitions of equilibrium points and stability of motion about equilibrium points by means of a vibrating single-degree-of-freedom system. In this section, we extend these concepts to general multi-degree-of-freedom dynamical system.

We are concerned with the motion of an arbitrary n-degree-of-freedom system described by the differential equations

$$\ddot{q}_i(t) = f_i[q_1(t), q_2(t), \ldots, q_n(t), \dot{q}_1(t), \dot{q}_2(t), \ldots, \dot{q}_n(t)],\ i = 1, 2, \ldots, n \tag{11.1}$$

where f_i are in general nonlinear functions of the generalized coordinates $q_i(t)$ and generalized velocities $\dot{q}_i(t)$ $(i = 1, 2, \ldots, n)$; they represent generalized forces per unit mass and include both conservative and nonconservative forces. Assuming that the functions f_i and the initial displacements $q_i(0)$ and initial velocities $\dot{q}_i(0)$ are given, Eqs. (11.1) can be integrated, at least in theory, to obtain the coordinates $q_i(t)$ $(i = 1, 2, \ldots, n)$ as explicit functions of time.

The solution of Eqs. (11.1) can be given a geometric interpretation by conceiving of an n-dimensional space defined by the coordinates q_i and known as the *configuration space*. For a given value of time, the solution $q_i(t)$ $(i = 1, 2, \ldots, n)$ can be represented by an n-dimensional vector $\mathbf{q}(t) = [q_1(t)\ q_2(t)\ \ldots\ q_n(t)]^T$ in the configuration space, with the tip of the vector defining a point P called the *representative point*. As time unwinds, point P traces a *path* in the configuration space (Fig. 11.1) showing how the solution varies with time, although the time appears only implicitly. Whereas this geometric description is intuitively appealing, it is not very useful, because it does not describe the motion uniquely. As a simple illustration, we can envision the motion in the x, y-plane of a projectile leaving the gun muzzle with the velocity $\mathbf{v} = [\dot{x}\ \dot{y}]^T$ in a direction making an angle β relative to the horizontal axis. Neglecting air resistance, so that the projectile moves under the gravitational force alone, the path traced by the projectile in the x, y-plane is a parabola, with the time t playing the role of a parameter. The shape of the parabola depends on the initial conditions $x(0)$, $y(0)$, $\dot{x}(0)$ and $\dot{y}(0)$; Fig. 11.2 shows two parabolas, both corresponding to the case in which $x(0) = y(0) = 0$. By changing the initial velocities, the parabolas change, and the possibility exists that the two parabolas will intersect. At the point of intersection between two parabolas, unless more information than the position of the intersection point is available, it is not possible to say which path the projectile will follow. In particular, the same position corresponds to different velocities, and hence to different slopes (Fig. 11.2). It follows that, unless both the position and velocity vectors are specified, the motion is not described uniquely. In the following, we consider a geometric description of the motion that does not suffer from this drawback.

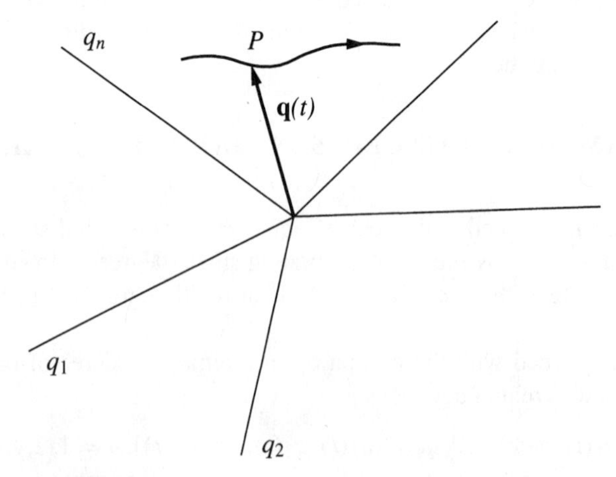

FIGURE 11.1
Point P tracing a path in the configuration space

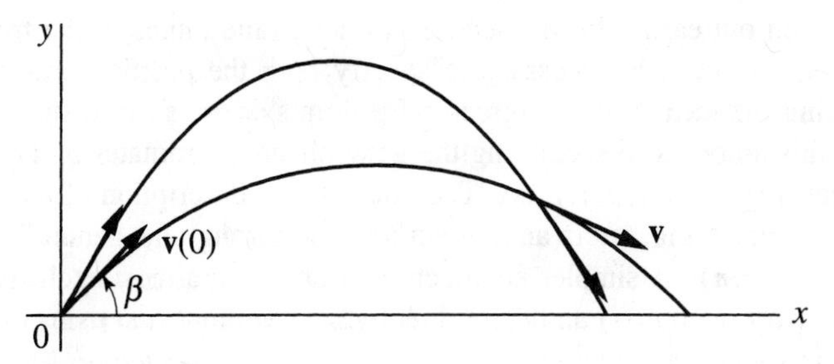

FIGURE 11.2
Intersecting paths in the configuration plane

Equations (11.1) represent a set of second-order Lagrangian equations of motion in terms of the generalized coordinates $q_i(t)$ $(i = 1, 2, \ldots, n)$. Quite often, the equations are nonlinear, and no closed-form solution to nonlinear equations can be expected in general. Hence, for the most part, it is necessary to integrate the equations numerically. It is another drawback of second-order equations, and one intimately related to the difficulties in describing the motion geometrically in the configuration space, that second-order equations are not very suitable for numerical integration. Indeed, first-order differential equations are not only quite suitable for numerical integration but also permit a satisfactory geometric description of the solution. Hence, it is only natural to seek a transformation from the set of n second-order Lagrangian equations, Eqs. (11.1), to a set of first-order equations. To this end, we introduce the *generalized momenta*

$$p_i = \partial L / \partial \dot{q}_i, \; i = 1, 2, \ldots, n \tag{11.2}$$

as a set of auxiliary variables, where L is the Lagrangian. But, from Sec. 6.4, and Example 6.1, we conclude that the Lagrangian is quadratic in the generalized velocities $\dot{q}_i$, so that p_i are linear combinations of $\dot{q}_i$ with the coefficients of $\dot{q}_i$ being in general nonlinear functions of q_i $(i = 1, 2, \ldots, n)$. By the same token, $\dot{q}_i$ are linear combinations of p_i. Hence, eliminating $\dot{q}_i$ in favor of p_i in Eqs. (11.1) and using Eqs. (11.2), it is possible to obtain a set of $2n$ first-order equations in q_i and p_i $(i = 1, 2, \ldots, n)$ replacing the n second-order equations, Eqs. (11.1).

The transformation from n second-order equations to $2n$ first-order equations can be carried out more efficiently by beginning with a set of generic Lagrange's equations, Eqs. (6.42), in which $T - V$ was replaced by L, and ending with a set of generic $2n$ first-order equations known as *Hamilton's equations* (Ref. 13). Consistent with this formulation, it is possible to conceive of a $2n$-dimensional space defined by the generalized coordinates q_i and generalized momenta p_i $(i = 1, 2, \ldots, n)$ and known as the *phase space*. Then, the solution of Hamilton's equations for a given set of initial conditions $q_i(0)$, $p_i(0)$ $(i = 1, 2, \ldots, n)$ can be described geometrically as a *trajectory* in the phase space. The advantage of the motion representation in the $2n$-dimensional phase space over the representation in the n-dimensional configuration space is that trajectories in the phase space corresponding to different initial conditions do not intersect and have the appearance of a steady fluid flow. The implication is that each trajectory is unique to a given set of initial conditions. Exceptions to this rule are certain points representing special solutions, as discussed later in this section.

As pointed out earlier in this section, to determine uniquely the trajectory of a projectile at any time, it is necessary to specify both the position and velocity vectors. Extending the idea to an n-degree-of-freedom system, such as that described by Eqs. (11.1), this amounts to specifying the generalized coordinates $q_i(t)$ and generalized velocities $\dot{q}_i(t)$ $(i = 1, 2, \ldots, n)$. The phase space description of the motion does essentially the same thing, but in an indirect manner, through the generalized momenta $p_i(t)$ $(i = 1, 2, \ldots, n)$. A simpler approach, and one used almost exclusively in engineering, is to work with $q_i(t)$ and $\dot{q}_i(t)$ directly, i.e., without first using $p_i(t)$. To this end, we first introduce the notation

$$\begin{aligned} q_i = x_i, \ \dot{q}_i = x_{n+i} \\ x_{n+i} = X_i, \ f_i = X_{n+i} \end{aligned} \qquad i = 1, 2, \ldots, n \tag{11.3}$$

Then, we combine the first two groups in Eqs. (11.3) into a set of identities playing the role of n auxiliary equations, transform the dynamical equations, Eqs. (11.1), into a set n first-order equations using the notation of Eqs. (11.3) and obtain the set of $2n$ first-order equations

$$\dot{x}_i(t) = X_i[x_1(t), \ x_2(t), \ldots, x_{2n}(t)], \ i = 1, 2, \ldots, 2n \tag{11.4}$$

which are known as *state equations*; correspondingly, $x_i(t)$ $(i = 1, 2, \ldots, 2n)$ represent *state variables*. The variables $x_i(t)$ can be used to define a $2n$-dimensional vector $\mathbf{x}(t) = [x_1(t) \ x_2(t) \ \ldots \ x_{2n}(t)]^T$ called the *state vector*. It is the same state vector first introduced in a narrower context in Sec. 7.16. Similarly, the quantities $X_i(t)$ can be regarded as the components of the $2n$-dimensional vector $\mathbf{X}(t) = [X_1(t) \ \ X_2(t) \ \ldots \ X_{2n}(t)]^T$, so that Eqs. (11.4) can be written in the compact vector form

$$\dot{\mathbf{x}}(t) = \mathbf{X}[\mathbf{x}(t)] \tag{11.5}$$

It is convenient to think of $\mathbf{X}(t)$ as an excitation vector, although only the bottom half of the vector can be regarded as such. The components x_i of the state vector define a $2n$-dimensional space called the *state space*. At any time t, the solution $\mathbf{x}(t)$ of Eq. (11.5) represents a vector in the state space. As time unwinds, the tips of the state vectors $\mathbf{x}(t)$ corresponding to different inertial conditions trace unique, nonintersecting trajectories in the state space resembling stream lines in fluid flow, as shown in Fig. 11.3. Exceptions are certain points, as discussed in the following paragraph. The trajectories in the state space are topologically equivalent to those traced by the solutions of Hamilton's equations in the phase space.

A point for which $\mathbf{X}^T\mathbf{X} = \sum_{i=1}^{2n} X_i^2 > 0$ is referred to as an *ordinary point*, or a *regular point*. On the other hand, a point for which $\mathbf{X} = \mathbf{0}$ is called a *singular point*, or an *equilibrium point*. Recognizing from Eq. (11.5) that $\dot{\mathbf{x}}$ vanishes at points for which $\mathbf{X}$ is zero, and recalling that the upper half of the state vector $\mathbf{x}$ consists of displacements and the lower half consists of velocities, we conclude that the vanishing of $\dot{\mathbf{x}}$ implies that all velocity and acceleration components are zero, which explains why a singular point is called an equilibrium point. It should be pointed out that, although it may not be immediately obvious, the concept of equilibrium points just defined is the same as that discussed in Secs. 1.10 and 1.13. In view of the fact that $\dot{\mathbf{x}}$ is zero at a singular point, which implies that both the magnitude and direction of $\dot{\mathbf{x}}$ are zero, there can be more

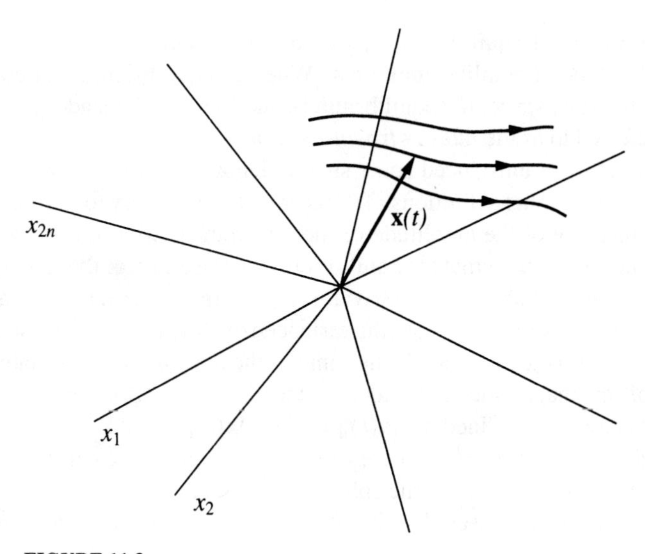

FIGURE 11.3
Nonintersecting trajectories in the state space

than one trajectory beginning from or terminating at a singular point. This identifies the singular points as the special points mentioned earlier at which trajectories are not unique. However, not too much importance should be attached to this fact. Indeed, we observe that, because velocities reduce to zero at singular points, trajectories can only reach such points as $t \to \pm\infty$. It follows that the lack of uniqueness of trajectories at singular points does not compromise in any way the geometrical description of the motion in the state space.

Because $\dot{\mathbf{x}}$ is zero at an equilibrium point, the state vector at such a point must be constant; we denote that constant vector by $\mathbf{x}_e$, where $\mathbf{x}_e$ must satisfy the *equilibrium equation*

$$\mathbf{X}(\mathbf{x}_e) = \mathbf{0} \tag{11.6}$$

Recalling that the upper half of the state vector $\mathbf{x}$ consists of displacements and the lower half consists of velocities, it follows that equilibrium points are such that

$$\mathbf{q}_e = \text{constant}, \quad \dot{\mathbf{q}}_e = \mathbf{0} \tag{11.7}$$

which implies that *all equilibrium points lie in the configuration space*. But, from Eqs. (11.3), we conclude that the upper half of Eq. (11.6) is identically zero and the lower half represents the set of n algebraic equilibrium equations

$$f_i(q_{1e}, q_{2e}, \ldots, q_{ne}) = 0, \; i = 1, 2, \ldots, n \tag{11.8}$$

with the components $q_{1e}, q_{2e}, \ldots, q_{ne}$ of the equilibrium displacement vector $\mathbf{q}_e$ as the unknowns. If f_i are linear functions of $q_{1e}, q_{2e}, \ldots, q_{ne}$, there is only one equilibrium point, and if f_i are nonlinear functions of $q_{1e}, q_{2e}, \ldots, q_{ne}$, there is in general more than one equilibrium point. An important question in dynamics is how the solution behaves in the neighborhood of equilibrium points, and in particular how stable is the motion in the neighborhood of equilibrium points. When an equilibrium point coincides with the origin of the state space, the equilibrium is said to be *trivial*, and the corresponding solution is referred to as the *null*, or *trivial solution*.

In Sec. 1.13, we introduced some simple definitions of stability. In this section, we provide some rigorous definitions. To this end, it is necessary to introduce a quantity serving as a measure of the amplitude of motion from equilibrium in a general sense. For convenience we assume that the equilibrium is at the origin of the state space, so that our interest lies in the stability of the trivial solution. In this case, a reasonable measure of the amplitude of motion is simply the distance from the origin of the state space to a point on the trajectory $\mathbf{x}(t)$, which is the same as the magnitude of the state vector $\mathbf{x}(t)$. A measure of this magnitude is provided by the *Euclidean norm*, or *Euclidean length*, of the state vector $\mathbf{x}(t)$, defined as $\|\mathbf{x}(t)\| = [\mathbf{x}^T(t)\mathbf{x}(t)]^{1/2} = [\Sigma_{i=1}^{2n} x_i^2(t)]^{1/2}$. Then, a sphere of radius r with the center at the origin of the state space can be written simply as $\|\mathbf{x}\| = r$ and the domain inside the sphere as $\|\mathbf{x}\| < r$.

The most frequently used definitions of stability are due to Liapunov and can be stated as follows:

1. The trivial solution is *stable in the sense of Liapunov* if for any arbitrary positive quantity ε there exists a positive quantity δ such that the satisfaction of the inequality

$$\|\mathbf{x}_0\| < \delta \tag{11.9}$$

implies the satisfaction of the inequality

$$\|\mathbf{x}(t)\| < \varepsilon, \; 0 \leq t < \infty \tag{11.10}$$

where $\mathbf{x}_0 = \mathbf{x}(0)$ is the initial state vector.

2. The trivial solution is *asymptotically stable* if it is Liapunov stable and in addition

$$\lim_{t \to \infty} \|\mathbf{x}(t)\| = 0 \tag{11.11}$$

3. The trivial solution is *unstable* if it is not stable.

Geometrically, the trivial solution is stable if any motion initiated inside the sphere $\|\mathbf{x}\| = \delta$ remains inside the sphere $\|\mathbf{x}\| = \varepsilon$ for all times. If in addition the motion approaches the origin as $t \to \infty$, the trivial solution is asymptotically stable. The trivial solution is unstable if $\mathbf{x}(t)$ reaches the boundary of the sphere $\|\mathbf{x}\| = \varepsilon$ in finite time. The three possibilities are illustrated in Fig. 11.4. The solution labeled as I is merely stable, solution II is asymptotically stable and solution III is unstable.

The Liapunov stability definitions are very precise, but they are mere definitions; they do not provide ways of ascertaining stability or instability. Perhaps the most effective way of investigating stability is to linearize the state equations about equilibrium, solve the associated eigenvalue problem and draw stability conclusions based on the nature of the eigenvalues. To this end, we assume that the state variables can be written in the form

$$x_i(t) = x_{ie} + \tilde{x}_i(t), \; i = 1, 2, \ldots, 2n \tag{11.12}$$

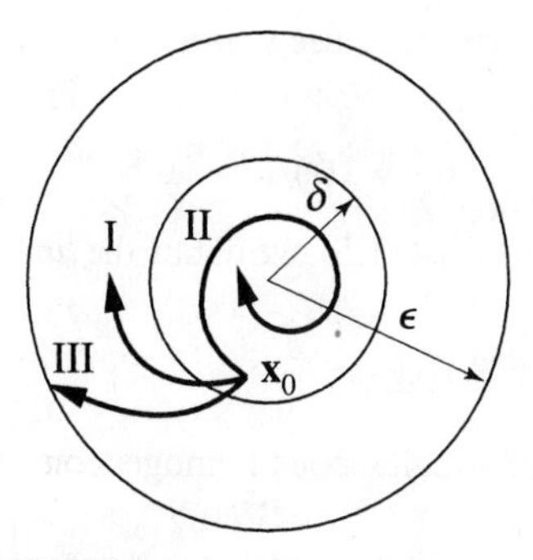

FIGURE 11.4
Trajectories in the state space illustrating mere stability (I), asymptotic stability (II) and instability (III)

where x_{ie} are constants representing the values of the state variables at a given equilibrium point and $\tilde{x}_i(t)$ are small perturbations from the equilibrium point, and expand the right side of Eqs. (11.4) in the Taylor's series

$$X_i[x_1(t),\ x_2(t),\ \ldots,\ x_{2n}(t)] = X_i(x_{1e}, x_{2e}, \ldots, x_{2ne}) + \sum_{j=1}^{2n} \left.\frac{\partial X_i}{\partial x_j}\right|_{\mathbf{x}=\mathbf{x}_e} \tilde{x}_j(t) + O(\tilde{\mathbf{x}}^2),$$

$$i = 1, 2, \ldots, 2n \tag{11.13}$$

where $O(\tilde{\mathbf{x}}^2)$ denotes terms of degree two and higher in the state variables perturbations $\tilde{x}_1(t), \tilde{x}_2(t), \ldots, \tilde{x}_{2n}(t)$, i.e., nonlinear terms. But, by virtue of the equilibrium equations, Eq. (11.6), the first term on the right side of Eqs. (11.13) is zero. Hence, introducing the notation

$$\left.\frac{\partial X_i}{\partial x_j}\right|_{\mathbf{x}=\mathbf{x}_e} = a_{ij},\ i, j = 1, 2, \ldots, 2n \tag{11.14}$$

in which a_{ij} are constant coefficients, and ignoring the nonlinear terms, Eqs. (11.13) reduce to

$$X_i[x_1(t), x_2(t), \ldots, x_{2n}(t)] = \sum_{j=1}^{2n} a_{ij}\tilde{x}_j(t),\ i = 1, 2, \ldots, 2n \tag{11.15}$$

But, Eqs. (11.12) can be combined into

$$\mathbf{x}(t) = \mathbf{x}_e + \tilde{\mathbf{x}}(t) \tag{11.16}$$

where $\tilde{\mathbf{x}}(t) = [\tilde{x}_1(t)\ \tilde{x}_2(t)\ \ldots\ \tilde{x}_{2n}(t)]^T$ is the perturbation in the state vector. In view of this, Eqs. (11.15) can be written in the compact matrix form

$$\mathbf{X}[\mathbf{x}(t)] = A\tilde{\mathbf{x}}(t) \tag{11.17}$$

where $A = [a_{ij}]$ is the coefficient matrix. Because $\mathbf{x}_e$ is a constant vector, it follows from Eq. (11.16) that

$$\dot{\mathbf{x}}(t) = \dot{\tilde{\mathbf{x}}}(t) \tag{11.18}$$

Inserting Eqs. (11.17) and (11.18) into Eq. (11.5), we obtain the *linearized state equations* in the matrix form

$$\dot{\tilde{\mathbf{x}}}(t) = A\tilde{\mathbf{x}}(t) \tag{11.19}$$

Equation (11.19) represents a set of simultaneous homogeneous differential equations having the exponential solution

$$\tilde{\mathbf{x}}(t) = \tilde{\mathbf{x}}e^{\lambda t} \tag{11.20}$$

in which $\tilde{\mathbf{x}}$ is a constant $2n$-dimensional vector and λ is a constant scalar. Introducing Eq. (11.20) in Eq. (11.19) and dividing through by $e^{\lambda t}$, we obtain the algebraic eigenvalue problem

$$A\tilde{\mathbf{x}} = \lambda\tilde{\mathbf{x}} \tag{11.21}$$

The solution of the eigenvalue problem, Eq. (11.21), consists of $2n$ eigenvalues λ_r and eigenvectors $\tilde{\mathbf{x}}_r$ $(r = 1, 2, \dots, 2n)$. As can be concluded from Eq. (11.20), the behavior of the solution depends on the eigenvalues. Because the coefficient matrix A is real, if some or all of the eigenvalues are complex, then they must occur in pairs of complex conjugates. We consider the following cases:

1. **All the eigenvalues are pure imaginary**. In this case, it is convenient to introduce the notation

$$\lambda_r = i\omega_r, \quad \bar{\lambda}_r = -i\omega_r, \quad r = 1, 2, \dots, n \tag{11.22}$$

so that, from Eq. (11.20), the solution of Eq. (11.19) has the general form

$$\begin{aligned}\tilde{\mathbf{x}}(t) &= \sum_{r=1}^{n}(c_r\tilde{\mathbf{x}}_r e^{i\omega_r t} + \bar{c}_r\bar{\tilde{\mathbf{x}}}_r e^{-i\omega_r t}) \\ &= 2\sum_{r=1}^{n}\mathrm{Re}(c_r\tilde{\mathbf{x}}_r e^{i\omega_r t}) = 2\sum_{r=1}^{n}[\mathrm{Re}(c_r\tilde{\mathbf{x}}_r)\cos\omega_r t - \mathrm{Im}(c_r\tilde{\mathbf{x}}_r)\sin\omega_r t]\end{aligned} \tag{11.23}$$

Hence, the response consists of a superposition of pure oscillatory terms, in which ω_r $(r = 1, 2, \dots, n)$ are recognized as frequencies of oscillation. Because the response neither goes to zero with time nor does it increase without bounds, the motion in the neighborhood of equilibrium is merely *stable*, such as typified by curve I in Fig. 11.4.

2. **Some eigenvalues are complex and the rest are real, and all the complex eigenvalues possess negative real part and all the real eigenvalues are negative.** In this case, from Eq. (11.20), the response tends to zero as $t \to \infty$. If all the eigenvalues are complex the motion represents oscillatory decay, and if all the eigenvalues are real the motion represents aperiodic decay. In all these cases, the equilibrium is *asymptotically stable*, and the trajectories resemble curve II in Fig. 11.4.

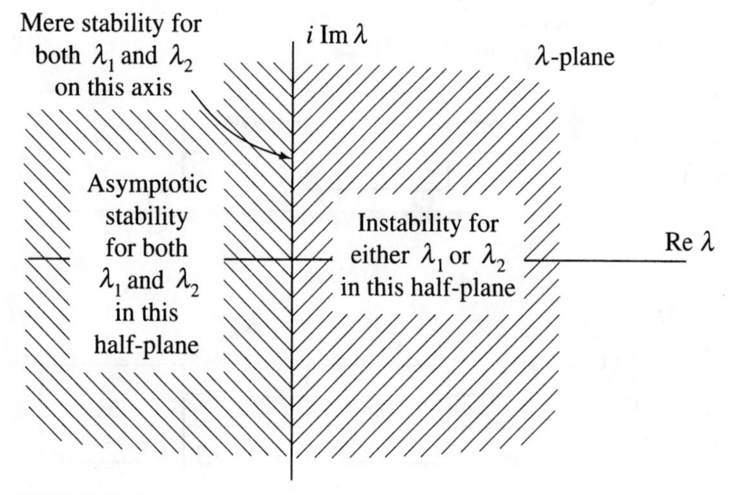

FIGURE 11.5
λ-plane divided into regions of mere stability, asymptotic stabilitiy and instability

3. **Some eigenvalues are complex and the rest are real, and at least one complex eigenvalue pair possesses positive real part, or at least one real eigenvalue is positive.** In this case, the response increases with time, so that the trivial solution is *unstable*; the corresponding trajectories resemble curve III in Fig. 11.4.

It is convenient to represent the eigenvalues by points in a complex plane referred to as the λ-plane (Fig. 11.5). For mere stability, all the eigenvalues must lie on the imaginary axis. On the other hand, for asymptotic stability, all the eigenvalues must lie in the left half of the λ-plane (excluding the imaginary axis). Finally, for instability, at least one eigenvalue must lie in the right half of the λ-plane (excluding the imaginary axis).

Stability statements concerning nonlinear systems, but based on linearized equations are qualified by referring to them as *infinitesimal stability* statements. If the linearized system is judged as asymptotically stable, or unstable, it is said to exhibit *significant behavior*. On the other hand, if the linearized system is found to be merely *stable*, it is said to exhibit *critical behavior. If the linearized system possesses significant behavior, then the stability characteristics of the nonlinear system are the same as those of the linearized system.* On the other hand, *if the linearized system possesses critical behavior, the stability conclusions do not necessarily extend to the full nonlinear system*, so that the nonlinear terms must be taken into account.

Example 11.1. Consider the two-degree-of-freedom system with a nonlinear spring shown in Fig. 11.6a, derive the state equations of motion, determine the equilibrium points and investigate the stability in the neighborhood of the equilibrium points. The force in the nonlinear spring has the expression

$$f(q_1) = -kq_1\left[1 - \left(\frac{q_1}{a}\right)^2\right] \tag{a}$$

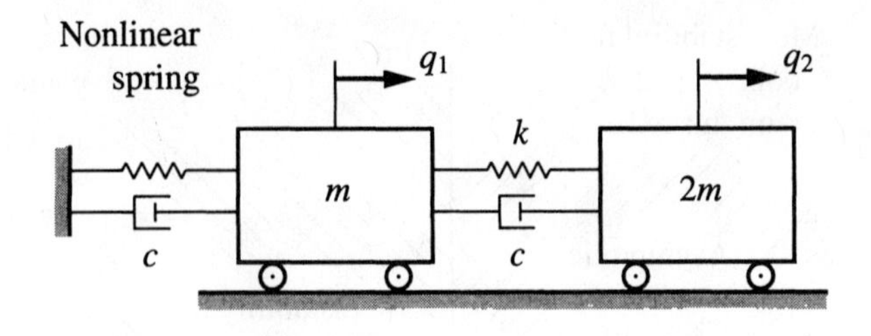

a.

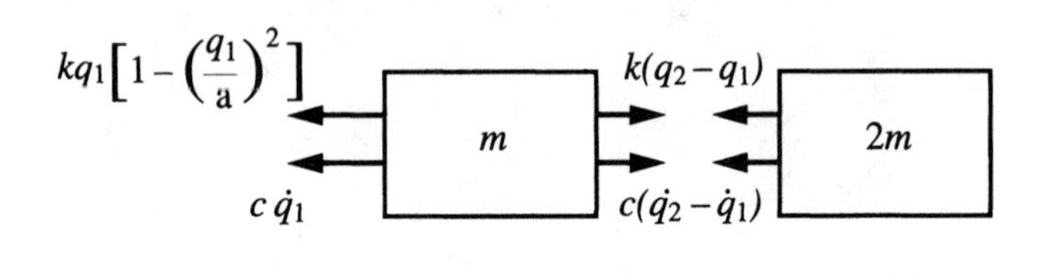

b.

FIGURE 11.6
a. Damped two-degree of freedom system, b. Free-body diagrams

Using Newton's second law in conjunction with the free-body diagrams of Fig. 11.6b, we can write

$$\begin{aligned} m\ddot{q}_1 &= -2c\dot{q}_1 + c\dot{q}_2 - kq_1\left[2-\left(\frac{q_1}{a}\right)^2\right] + kq_2 \\ 2m\ddot{q}_2 &= c\dot{q}_1 - c\dot{q}_2 + kq_1 - kq_2 \end{aligned} \tag{b}$$

so that the equations of motion are as given by Eqs. (11.1), in which

$$\begin{aligned} f_1(q_1,q_2,\dot{q}_1,\dot{q}_2) &= \frac{1}{m}\left\{-2c\dot{q}_1 + c\dot{q}_2 - kq_1\left[2-\left(\frac{q_1}{a}\right)^2\right] + kq_2\right\} \\ f_2(q_1,q_2,\dot{q}_1,\dot{q}_2) &= \frac{1}{2m}(c\dot{q}_1 - c\dot{q}_2 + kq_1 - kq_2) \end{aligned} \tag{c}$$

Then, using the notation $q_1 = x_1$, $q_2 = x_2$, $\dot{q}_1 = x_3$, $\dot{q}_2 = x_4$ in conjunction with Eqs. (11.3), the state equations are

$$\begin{aligned} \dot{x}_1 &= x_3 \\ \dot{x}_2 &= x_4 \\ \dot{x}_3 &= \frac{1}{m}\left\{-2cx_3 + cx_4 - kx_1\left[2-\left(\frac{x_1}{a}\right)^2\right] + kx_2\right\} \\ \dot{x}_4 &= \frac{1}{2m}(cx_3 - cx_4 + kx_1 - kx_2) \end{aligned} \tag{d}$$

The state equations have the matrix form given by Eq. (11.5) in which the state vector is

$\mathbf{x} = [x_1 \ x_2 \ x_3 \ x_4]^T$ and the "excitation vector" is

$$\mathbf{X} = \left[x_3 \ \ x_4 \ \ \frac{1}{m}\left\{-2cx_3 + cx_4 - kx_1\left[2 - \left(\frac{x_1}{a}\right)^2\right] + kx_2\right\} \right.$$
$$\left. \frac{1}{2m}(cx_3 - cx_4 + kx_1 - kx_2)\right]^T \tag{e}$$

According to Eq. (11.6), the equilibrium points are obtained by setting **X** equal to zero, which yields the four equilibrium equations

$$\begin{aligned} &X_1 = x_3 = 0, \ X_2 = x_4 = 0 \\ &X_3 = \frac{1}{m}\left\{-2cx_3 + cx_4 - kx_1\left[2 - \left(\frac{x_1}{a}\right)^2\right] + kx_2\right\} = 0 \\ &X_4 = \frac{1}{2m}(cx_3 - cx_4 + kx_1 - kx_2) = 0 \end{aligned} \tag{f}$$

Solving Eqs. (f), we obtain three equilibrium points, the trivial one and two nontrivial ones, as follows:

$$\begin{aligned} &e_1 : x_1 = x_2 = x_3 = x_4 = 0 \\ &e_2 : x_1 = a, \ x_2 = x_3 = x_4 = 0 \\ &e_3 : x_1 = -a, \ x_2 = x_3 = x_4 = 0 \end{aligned} \tag{g}$$

In terms of physical coordinates and velocities, they are given by

$$\begin{aligned} &e_1 : q_1 = q_2 = 0, \ \dot{q}_1 = \dot{q}_2 = 0 \\ &e_2 : q_1 = a, \ q_2 = 0, \ \dot{q}_1 = \dot{q}_2 = 0 \\ &e_3 : q_1 = -a, \ q_2 = 0, \ \dot{q}_1 = \dot{q}_2 = 0 \end{aligned} \tag{h}$$

confirming a statement made earlier that all equilibrium points lie in the configuration space.

To investigate the stability of the equilibrium points, we first write

$$\begin{aligned} &\frac{\partial X_1}{\partial x_1} = 0, \ \frac{\partial X_1}{\partial x_2} = 0, \ \frac{\partial X_1}{\partial x_3} = 1, \ \frac{\partial X_1}{\partial x_4} = 0 \\ &\frac{\partial X_2}{\partial x_1} = 0, \ \frac{\partial X_2}{\partial x_2} = 0, \ \frac{\partial X_2}{\partial x_3} = 0, \ \frac{\partial X_2}{\partial x_4} = 1 \\ &\frac{\partial X_3}{\partial x_1} = -\frac{k}{m}\left[2 - 3\left(\frac{x_1}{a}\right)^2\right], \ \frac{\partial X_3}{\partial x_2} = \frac{k}{m}, \ \frac{\partial X_3}{\partial x_3} = -\frac{2c}{m}, \ \frac{\partial X_3}{\partial x_4} = \frac{c}{m} \\ &\frac{\partial X_4}{\partial x_1} = \frac{k}{2m}, \ \frac{\partial X_4}{\partial x_2} = -\frac{k}{2m}, \ \frac{\partial X_4}{\partial x_3} = \frac{c}{2m}, \ \frac{\partial X_4}{\partial x_4} = -\frac{c}{2m} \end{aligned} \tag{i}$$

Equations (i) can be used to derive the entries a_{ij} of the coefficient matrix A corresponding to the individual equilibrium points. Introducing the first of Eqs. (g) in Eqs. (i), we obtain the coefficient matrix for the *equilibrium point* e_1

$$A = \begin{bmatrix} 0 & 0 & 1 & 0 \\ 0 & 0 & 0 & 1 \\ -2k/m & k/m & -2c/m & c/m \\ k/2m & -k/2m & c/2m & -c/2m \end{bmatrix} \tag{j}$$

Similarly, using the second of Eqs. (g), the coefficient matrix for the *equilibrium point* e_2 is simply

$$A = \begin{bmatrix} 0 & 0 & 1 & 0 \\ 0 & 0 & 0 & 1 \\ k/m & k/m & -2c/m & c/m \\ k/2m & -k/2m & c/2m & -c/2m \end{bmatrix} \tag{k}$$

The coefficient matrix for the equilibrium point e_3 is the same as for e_2, which indicates that the two equilibrium points are dynamically equivalent.

The eigenvalue problems for both equilibrium points have been solved for the parameter ratios $k/m = 1$ and $c/m = 0.1$. The eigenvalues for the *equilibrium point* e_1 are

$$\begin{matrix}\lambda_1 \\ \lambda_2\end{matrix} = -0.0110 \pm 0.4681i, \quad \begin{matrix}\lambda_3 \\ \lambda_4\end{matrix} = -0.1140 \pm 1.5059i \tag{l}$$

All four eigenvalues have negative real part, so that the *equilibrium point* e_1 *is asymptotically stable*. On the other hand, the eigenvalues for the equilibrium points e_2 and e_3 are

$$\begin{matrix}\lambda_1 \\ \lambda_2\end{matrix} = -0.0592 \pm 0.8794i, \quad \lambda_3 = 1.0707, \quad \lambda_4 = -1.2024 \tag{m}$$

Because λ_3 is positive, *the equilibrium points* e_2 *and* e_3 *are unstable*. We observe that all three equilibrium points exhibit significant behavior, so that the stability conclusions based on the linearized equations remain valid for the nonlinear equations.

11.2 SMALL MOTIONS OF SINGLE-DEGREE-OF-FREEDOM SYSTEMS FROM EQUILIBRIUM

In Sec. 11.1, we discussed ways of representing the motion of dynamical systems geometrically in the state space. Whereas the idea is appealing, dimensionality prevents the approach from providing more than a general qualitative picture of the system behavior, as even a two-degree-of-freedom system implies plots in a four-dimensional state space. More quantitative statements can be made by linearizing the state equations about equilibrium points and checking the stability of these points by examining the eigenvalues of the linearized system. A combination of the two approaches can provide more information concerning the system behavior, but dimensionality remains a problem. The problem does not exist for single-degree-of-freedom systems, for which state plane plots can be used in combination with eigenvalue information to shed a great deal of light into the nature of the motion in the neighborhood of equilibrium points.

We consider a single-degree-of-freedom system, assume without loss of generality that the trivial solution is an equilibrium point and write the linearized equation for small motions from the equilibrium in the generic form

$$\ddot{q}(t) + a\dot{q}(t) + bq(t) = 0 \tag{11.24}$$

where a and b are constants. Letting $q = x_1$, $\dot{q} = x_2$, we obtain the corresponding state equations

$$\begin{aligned} \dot{x}_1(t) &= x_2(t) \\ \dot{x}_2(t) &= -bx_1(t) - ax_2(t) \end{aligned} \tag{11.25}$$

which can be expressed in the matrix form

$$\dot{\mathbf{x}}(t) = A\mathbf{x}(t) \tag{11.26}$$

where $\mathbf{x}(t) = [x_1(t) \;\; x_2(t)]^T$ is the state vector and

$$A = \begin{bmatrix} 0 & 1 \\ -b & -a \end{bmatrix} \tag{11.27}$$

is the coefficient matrix. Clearly, the behavior of the system depends on the coefficient matrix A, which in turn depends on the constants a and b.

To examine the behavior of the system, it is convenient to transform the state equations to modal form. To this end, we recall from Sec. 11.1 that the solution of a homogeneous equation such as Eq. (11.26) has the exponential form

$$\mathbf{x}(t) = \mathbf{x}e^{\lambda t} \tag{11.28}$$

in which $\mathbf{x}$ is a constant vector and λ is a constant scalar; their values are obtained by solving the eigenvalue problem

$$A\mathbf{x} = \lambda\mathbf{x} \tag{11.29}$$

The solution of Eq. (11.29) consists of the eigenvalues λ_r and right eigenvectors $\mathbf{x}_r$ $(r = 1, 2)$. Because A is nonsymmetric, it is necessary to solve the adjoint eigenvalue problem

$$A^T\mathbf{y} = \lambda\mathbf{y} \tag{11.30}$$

as well; its solution consists of the same eigenvalues λ_r and left eigenvectors $\mathbf{y}_r$ $(r = 1, 2)$. The right and left eigenvectors are biorthogonal and can be normalized so as to satisfy the biorthonormality relations

$$\mathbf{y}_s^T\mathbf{x}_r = \delta_{rs}, \;\; \mathbf{y}_s^T A\mathbf{x}_r = \lambda_r\delta_{rs}, \;\; r, s = 1, 2 \tag{11.31}$$

Next, we consider the linear transformation

$$\mathbf{x}(t) = \mathbf{x}_1 z_1(t) + \mathbf{x}_2 z_2(t) \tag{11.32}$$

where $z_1(t)$ and $z_2(t)$ play the role of modal coordinates. Inserting Eq. (11.32) into Eq. (11.26), premultiplying by $\mathbf{y}_1^T$ and $\mathbf{y}_2^T$, in sequence, and using the biorthonormality relations, Eqs. (11.31), we obtain the independent modal equations

$$\dot{z}_1(t) = \lambda_1 z_1(t), \;\; \dot{z}_2(t) = \lambda_2 z_2(t) \tag{11.33}$$

which have the solution

$$z_1(t) = z_{10}e^{\lambda_1 t}, \;\; z_2(t) = z_{20}e^{\lambda_2 t} \tag{11.34}$$

where z_{10} and z_{20} are the initial values of $z_1(t)$ and $z_2(t)$, respectively.

The behavior of the system in the neighborhood of the equilibrium depends on the nature of the eigenvalues λ_1 and λ_2, which in turn depend on the constants a and b. To examine this dependence, we insert Eq. (11.27) into Eq. (11.29) and write the characteristic equation

$$\det[A - \lambda I] = \det\begin{bmatrix} -\lambda & 1 \\ -b & -a-\lambda \end{bmatrix} = \lambda^2 + a\lambda + b = 0 \tag{11.35}$$

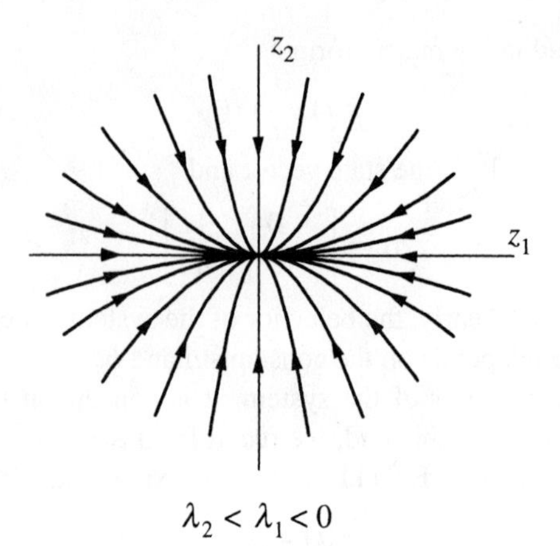

FIGURE 11.7
Trajectories in the case of a stable node

which has the roots

$$\begin{matrix}\lambda_1\\ \lambda_2\end{matrix} = -\frac{a}{2} \pm \sqrt{\left(\frac{a}{2}\right)^2 - b} \tag{11.36}$$

We distinguish the following cases:

1. **λ_1 and λ_2 are real and of the same sign.**
   Solutions (11.34) can be rewritten as

$$\ln\frac{z_1}{z_{10}} = \lambda_1 t, \ \ln\frac{z_2}{z_{20}} = \lambda_2 t \tag{11.37}$$

   so that, eliminating the time t, we have

$$\ln\frac{z_2}{z_{20}} = \frac{\lambda_2}{\lambda_1}\ln\frac{z_1}{z_{10}} \tag{11.38}$$

   Equation (11.38) can be used to generate state plane trajectories, which are simply plots z_2 versus z_1 for various values of z_{10} and z_{20}, with the time t playing the role of an implicit parameter. The trajectories are shown in Fig. 11.7 for the case in which λ_1 and λ_2 are both negative and satisfy the inequality $\lambda_2 < \lambda_1 < 0$. An equilibrium point of the type shown in Fig. 11.7 is called a *node*. Because the trajectories approach the equilibrium as $t \to \infty$, this is a *stable node*. Note that stable nodes are by definition *asymptotically stable*. When both λ_1 and λ_2 are positive and satisfy $\lambda_1 > \lambda_2 > 0$, the arrowheads in Fig. 11.7 simply reverse directions, and the equilibrium point becomes an *unstable node*.

   Stable nodes occur when $a > 0$, $b > 0$ and $a^2 > 4b$. On the other hand, unstable nodes occur when $a < 0$, $b > 0$ and $a^2 > 4b$.

2. **λ_1 and λ_2 are real and of opposite signs.**
   We consider the case in which λ_1 is positive and λ_2 is negative, or $\lambda_2 < 0 < \lambda_1$. In

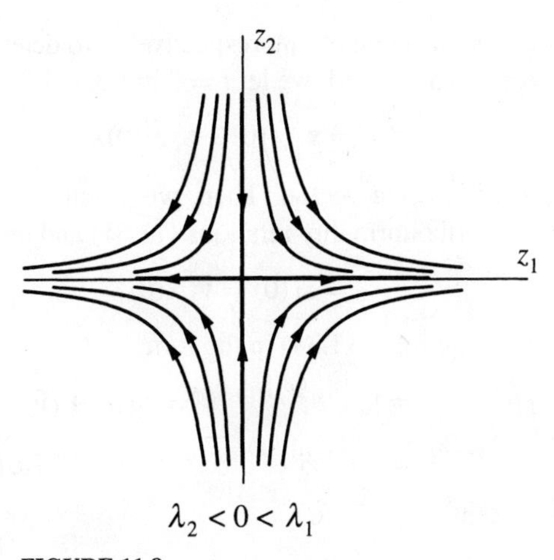

FIGURE 11.8
Trajectories in the case of a saddle point

this case, by analogy with Eq. (11.38), the state plane trajectories are given by

$$\ln \frac{z_2}{z_{20}} = -\frac{|\lambda_2|}{\lambda_1} \ln \frac{z_1}{z_{10}} \tag{11.39}$$

They are plotted in Fig. 11.8 for various values of the initial conditions z_{10} and z_{20}. Note that, to decide on the direction of the arrowheads, we consider Eqs. (11.34) and conclude that

$$\lim_{t \to \infty} z_1(t) = \begin{cases} \infty \text{ for } z_{10} > 0 \\ -\infty \text{ for } z_{10} < 0 \end{cases}$$
$$\lim_{t \to \infty} z_2(t) = 0 \tag{11.40}$$

In this case, the equilibrium point is called a *saddle point*, which is *unstable*.

Saddle points occur when $b < 0$, irrespective of the value of a.

3. **λ_1 and λ_2 are complex conjugates, $\lambda_2 = \bar{\lambda}_1$.**
In this case, the eigenvectors are also complex conjugates, $\mathbf{x}_2 = \bar{\mathbf{x}}_1$. Then, because the state vector $\mathbf{x}(t)$ must be real, we conclude from Eq. (11.32) that the modal coordinates are complex conjugates as well, $z_2(t) = \bar{z}_1(t)$, and so are their initial values, $z_{20} = \bar{z}_{10}$.

Next, we introduce the notation

$$\lambda_1 = \alpha + i\beta, \ \lambda_2 = \bar{\lambda}_1 = \alpha - i\beta \tag{11.41}$$

as well as

$$z_{10} = |z_{10}| e^{-i\phi} \tag{11.42}$$

where

$$|z_{10}| = \sqrt{(\text{Re}\, z_{10})^2 + (\text{Im}\, z_{10})^2}, \ \phi = \tan^{-1} \frac{-\text{Im}\, z_{10}}{\text{Re}\, z_{10}} \tag{11.43}$$

are the magnitude and phase angle of z_{10}, respectively. To determine $|z_{10}|$ and ϕ, we must first determine z_{10}. To this end, we let $t = 0$ in Eq. (11.32) and write

$$\mathbf{x}(0) = \mathbf{x}_1 z_1(0) + \mathbf{x}_2 z_2(0) \tag{11.44}$$

in which $\mathbf{x}(0)$ is the initial state vector. Then, we premultiply Eq. (11.44) by $\mathbf{y}_1^T$, consider the first of the orthonormality relations (11.31) and obtain

$$z_{10} = z_1(0) = \mathbf{y}_1^T \mathbf{x}(0) \tag{11.45}$$

Next, we rewrite solutions (11.34) in the form

$$\begin{aligned} z_1(t) &= z_{10} e^{(\alpha+i\beta)t} = |z_{10}| e^{\alpha t} e^{i(\beta t - \phi)} = v_1(t) + i v_2(t) \\ z_2(t) &= \bar{z}_{10} e^{(\alpha-i\beta t)} = |z_{10}| e^{\alpha t} e^{-i(\beta t-\phi)} = v_1(t) - i v_2(t) \end{aligned} \tag{11.46}$$

from which it follows that

$$v_1(t) = |z_{10}| e^{\alpha t} \cos(\beta t - \phi), \quad v_2(t) = |z_{10}| e^{\alpha t} \sin(\beta t - \phi) \tag{11.47}$$

For any initial values $v_1(0), v_2(0)$, the curve v_2 versus v_1 represents a trajectory having the form of a logarithmic spiral. Note that the initial values $v_1(0), v_2(0)$ correspond to a given initial modal coordinate z_{10}, which in turn corresponds to the initial state vector $\mathbf{x}(0)$, as can be concluded from Eq. (11.45).

To plot the trajectory in the v_1, v_2-plane, it is convenient to introduce the definitions

$$r(t) = \sqrt{v_1^2(t) + v_2^2(t)} = |z_{10}| e^{\alpha t}, \quad \tan(\beta t - \phi) = \frac{v_2(t)}{v_1(t)} \tag{11.48}$$

where $r(t)$ can be identified as the radius vector from the origin of the v_1, v_2-plane to a point on the trajectory and $\beta t - \phi$ is the angle from axis v_1 to $r(t)$. The spiral is shown in Fig. 11.9 for $\alpha < 0$ and $\beta > 0$. For $\alpha > 0$ and $\beta > 0$, the spiral unwinds,

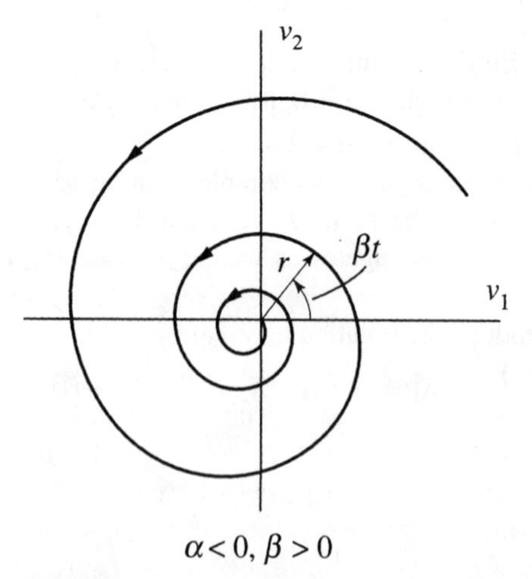

FIGURE 11.9
Trajectory in the case of a stable focus

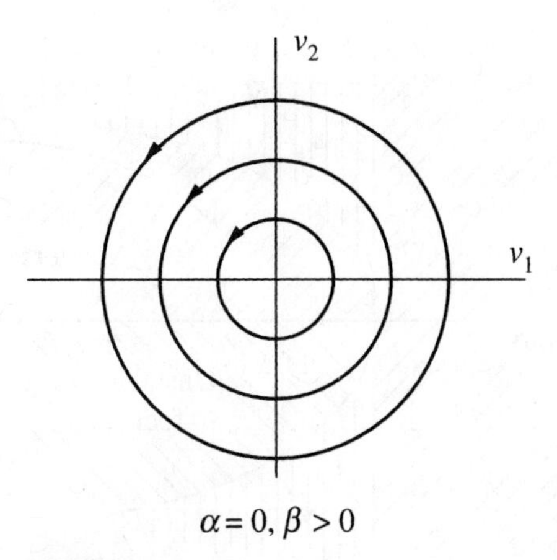

FIGURE 11.10
Trajectories in the case of a center

moving away from the origin. In both cases, the equilibrium point is known as a *spiral point*, or a *focus*. In the first case, $\alpha < 0$, the focus is *asymptotically stable*, and in the second case, $\alpha > 0$, the focus is *unstable*.

Stable foci occur when $a > 0$, $b > 0$ and $a^2 < 4b$, and unstable ones when $a < 0$, $b > 0$ and $a^2 < 4b$.

4. **λ_1 and λ_2 are pure imaginary complex conjugates, $\lambda_1 = i\beta,\ \lambda_2 = \bar{\lambda}_1 = -i\beta$.** This is a special case of the preceding case, obtained by letting $\alpha = 0$. The trajectories corresponding to different initial values $v_1(0)$, $v_2(0)$ represent circles centered at the origin, as shown in Fig. 11.10. In this case the equilibrium point is known as a *center*, which is merely *stable*.

   Centers occur when $a = 0$ and $b > 0$.

A great deal of insight into the system behavior can be gained by conceiving of a plane defined by the parameters a and b and divided into regions according to the equilibrium type. Such a parameter plane is shown in Fig. 11.11, from which we observe that stability is obtained only in the first quadrant. Indeed, in the region between the positive a-axis and the parabola $a^2 = 4b$ we obtain asympotically stable nodes, which imply aperiodically decaying response, between the parabola and the positive b-axis we obtain asymptotically stable foci, characterized by decaying oscillation, and on the positive b-axis itself, $a = 0$, we obtain merely stable centers, which imply pure harmonic oscillation at the frequency $\omega = \sqrt{b}$. The whole second and third quadrants, $b < 0$, represent a region of saddle points, which imply unstable motion by definition. In the fourth quadrant, we have an exact mirror image of the first quadrant, except that the nodes and foci are unstable. As a matter of interest, we note that the parameter plane of Fig. 11.11, in addition to the stability information contained in Fig. 1.46, ties together the stability information and the various types of state plane trajectories shown in Figs. 11.7-11.10.

The trajectories of Figs. 11.7 and 11.8 are plotted in the z_1, z_2-plane and those of Figs. 11.9 and 11.10 are plotted in the v_1, v_2-plane. Hence, the question can be raised

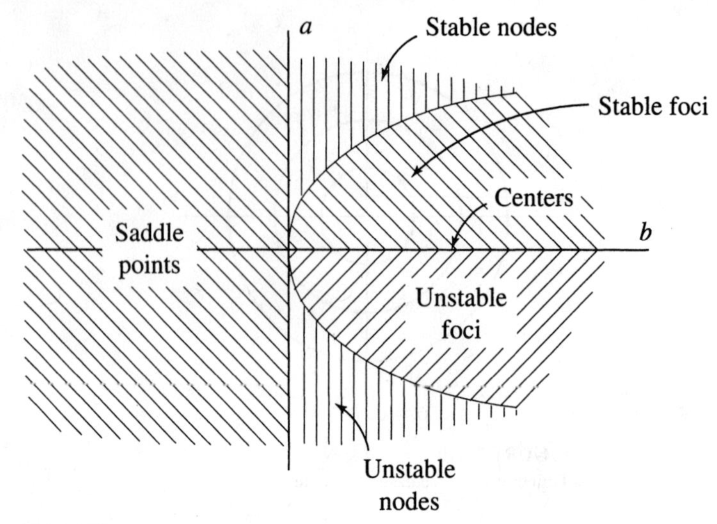

FIGURE 11.11
Equilibrium points in the parameter plane

as to whether the same characteristics of the equilibrium points would be present if the trajectories were plotted in the x_1, x_2-state plane. To answer this question, we observe that x_1 and x_2 can be obtained from z_1 and z_2, or from v_1 and v_2, through a linear transformation. Such a transformation tends to change the shape of the trajectories, but does not change the nature of the equilibrium points. As an example, to plot the spiral trajectory depicted in Fig. 11.9 in the x_1, x_2-state plane, we insert the first of Eqs. (11.46)

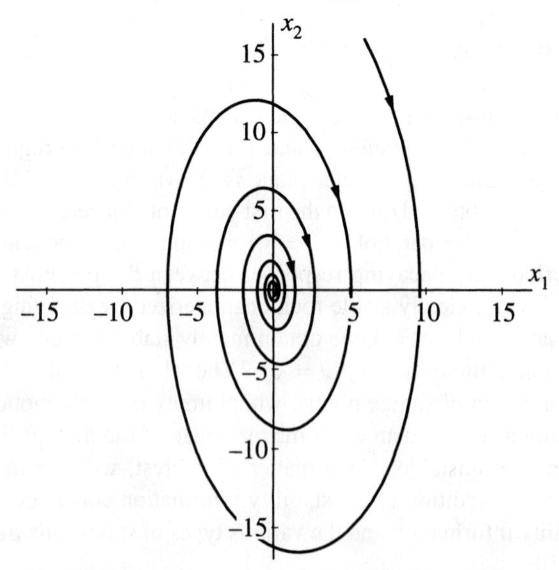

FIGURE 11.12
State-plane trajectory corresponding to a stable focus

into Eq. (11.32) and write

$$\mathbf{x}(t) = \mathbf{x}_1 z_1(t) + \bar{\mathbf{x}}_1 \bar{\mathbf{z}}_1(t) = 2\text{Re}\,[\mathbf{x}_1 z_1(t)] = 2|z_{10}|e^{\alpha t}\text{Re}\,[\mathbf{x}_1 e^{i(\beta t - \phi)}] \tag{11.49}$$

The state-plane trajectory is shown in Fig. 11.12. Clearly, a stable focus remains a stable focus.

Example 11.2. The response of a viscously damped single-degree-of-freedom system is described by the differential equation (Sec. 2.2)

$$\ddot{q}(t) + 2\zeta\omega_n\dot{q}(t) + \omega_n^2 q(t) = 0 \tag{a}$$

where $\zeta = c/2m\omega_n$ is the viscous damping factor and $\omega_n = \sqrt{k/m}$ is the frequency of undamped oscillation. Determine the nature of the equilibrium points and plot the state-plane trajectory for each of the following cases: i) $\zeta = 0,\ \omega_n = 2\,\text{rad/s}$, ii) $\zeta = 0.1,\ \omega_n = 2\,\text{rad/s}$, iii) $\zeta = 1,\ \omega_n = 2\,\text{rad/s}$ and iv) $\zeta = 1.5,\ \omega_n = 2\,\text{rad/s}$. The initial conditions are $q(0) = 6$ cm, $\dot{q}(0) = 16$ cm/s.

Comparing Eq. (a) to Eq. (11.24), we conclude that the parameters are given by

$$a = 2\zeta\omega_n,\ b = \omega_n^2 \tag{b}$$

so that the state equations are

$$\begin{aligned} \dot{x}_1(t) &= x_2(t) \\ \dot{x}_2(t) &= -\omega_n^2 x_1(t) - 2\zeta\omega_n x_2(t) \end{aligned} \tag{c}$$

in which $x_1(t) = q(t),\ x_2(t) = \dot{q}(t)$. The state equations are subject to the initial conditions $x_1(0) = 6\text{cm},\ x_2(0) = 16\text{cm/s}$. The coefficient matrix in Eq. (11.26) is

$$A = \begin{bmatrix} 0 & 1 \\ -b & -a \end{bmatrix} = \begin{bmatrix} 0 & 1 \\ -\omega_n^2 & -2\zeta\omega_n \end{bmatrix} \tag{d}$$

which has the eigenvalues

$$\begin{matrix}\lambda_1 \\ \lambda_2\end{matrix} = -\frac{a}{2} \pm \sqrt{\left(\frac{a}{2}\right)^2 - b} = -\zeta\omega_n \pm \sqrt{\zeta^2 - 1}\omega_n \tag{e}$$

Equating the right side of the state equations to zero, we conclude that there is only one equilibrium point, namely, the trivial one.

In case i, $\zeta = 0,\ \omega_n = 2\,\text{rad/s}$, the coefficients are $a = 0,\ b = 4(\text{rad/s})^2$, so that the eigenvalues are pure imaginary complex conjugates

$$\begin{matrix}\lambda_1 \\ \lambda_2\end{matrix} = \pm\sqrt{-1}\omega_n = \pm 2i\,\text{rad/s} \tag{f}$$

Hence, the equilibrium point is a center, which is merely stable. The trajectory in the v_1, v_2-plane is a circle and in the x_1, x_2-plane is an ellipse. The latter can be obtained through a coordinate transformation from axes v_1, v_2 to axes x_1, x_2. Because in the case at hand $a = 0$, it is perhaps simpler to integrate the state equations directly. Dividing the second of Eqs. (c) by the first and letting $\zeta = 0$, we obtain

$$\frac{dx_2}{dx_1} = -\frac{\omega_n^2 x_1}{x_2} \tag{g}$$

which, upon integration, yields the equation of the ellipse

$$\omega_n^2 x_1^2(t) + x_2^2(t) = \omega_n^2 x_1^2(0) + x_2^2(0) \tag{h}$$

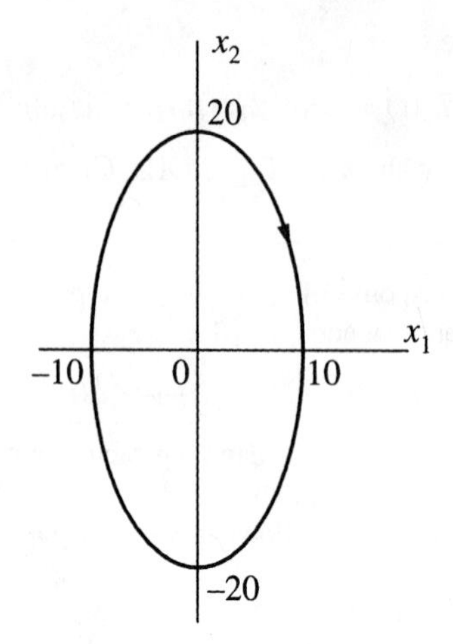

FIGURE 11.13
State-plane trajectory for a harmonic oscillator

or more explicitly

$$4x_1^2 + x_2^2 = 400 \tag{i}$$

The ellipse is plotted in Fig. 11.13. Of course, the motion is pure harmonic oscillation.

In case ii, $\zeta = 0.1,\ \omega_n = 2\,\text{rad/s}$, the coefficients are $a = 0.4\,\text{rad/s},\ b = 4(\text{rad/s})^2$, so that the eigenvalues are the complex conjugates

$$\begin{matrix}\lambda_1\\ \lambda_2\end{matrix} = -\frac{a}{2} \pm \sqrt{\left(\frac{a}{2}\right)^2 - b} = \alpha \pm i\beta = -0.2 \pm 1.99i\,\text{rad/s} \tag{j}$$

and the first right and left eigenvectors are

$$\mathbf{x}_1 = \begin{bmatrix} 0.5590 - 0.0562i \\ 1.1237i \end{bmatrix}, \quad \mathbf{y}_1 = \begin{bmatrix} 0.8944 \\ 0.0447 - 0.4450i \end{bmatrix} \tag{k}$$

and we note that the eigenvectors have been normalized so as to satisfy $\mathbf{y}_1^T\mathbf{x}_1 = 1$. The other eigenvectors are the complex conjugates $\mathbf{x}_2 = \bar{\mathbf{x}}_1,\ \mathbf{y}_2 = \bar{\mathbf{y}}_1$, but they are not really needed. From Eqs. (11.47), in conjunction with Eqs. (j), it follows that the equilibrium point is a focus with the corresponding trajectory being defined in the v_1, v_2-plane by

$$\begin{aligned} v_1(t) &= |z_{10}|e^{\alpha t}\cos(\beta t - \phi) = |z_{10}|e^{-0.2t}\cos(1.99t - \phi) \\ v_2(t) &= |z_{10}|e^{\alpha t}\sin(\beta t - \phi) = |z_{10}|e^{-0.2t}\sin(1.99t - \phi) \end{aligned} \tag{l}$$

To complete the determination of $v_1(t)$ and $v_2(t)$, we must first obtain $|z_{10}|$ and ϕ. To this end, we use Eq. (11.45) and write

$$z_{10} = \mathbf{y}_1^T\mathbf{x}(0) = \mathbf{y}_1^T \begin{bmatrix} q(0) \\ \dot{q}(0) \end{bmatrix} = \begin{bmatrix} 0.8944 \\ 0.0447 - 0.4450i \end{bmatrix}^T \begin{bmatrix} 6 \\ 16 \end{bmatrix} = 6.0821 - 7.1196i \tag{m}$$

Then, from Eqs. (11.43), we obtain

$$\begin{aligned} |z_{10}| &= \sqrt{(\text{Re}\,z_{10})^2 + (\text{Im}\,z_{10})^2} = \sqrt{6.0821^2 + (-7.1196)^2} = 9.3638 \\ \phi &= \tan^{-1}\frac{-\text{Im}\,z_{10}}{\text{Re}\,z_{10}} = \tan^{-1}\frac{-(-7.1196)}{6.0821} = 0.8638\,\text{rad} \end{aligned} \tag{n}$$

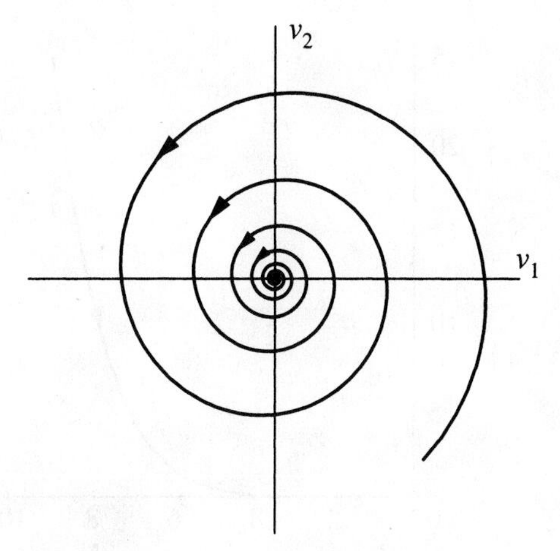

FIGURE 11.14
State-plane trajectory for an underdamped single-degree-of-freedom system

so that the spiral is defined in the v_1, v_2-plane by

$$\begin{aligned} v_1(t) &= 9.3638e^{-0.2t}\cos(1.99t - 0.8638) \\ v_2(t) &= 9.3638e^{-0.2t}\sin(1.99t - 0.8638) \end{aligned} \tag{o}$$

It is plotted in Fig. 11.14.

In case iii, $\zeta = 1,\ \omega_n = 2\,\text{rad/s}$, the coefficients are $a = 4\,\text{rad/s},\ b = 4(\text{rad/s})^2$, so that the eigenvalues are real, negative and equal in magnitude

$$\lambda_1 = \lambda_2 = -2\,\text{rad/s} \tag{p}$$

From Eqs. (11.34) and (11.38), we conclude that the trajectory is a straight line through the origin

$$z_2(t) = \frac{z_{20}}{z_{10}} z_1(t) \tag{q}$$

which can be interpreted as the limiting case of a trajectory corresponding to an asymptotically stable node. We recall from Sec. 3.2 that $\zeta = 1$ represents the so-called "critical damping" corresponding to a borderline case between decaying oscillations and decaying aperiodic motions.

In case iv, $\zeta = 1.5,\ \omega_n = 2\,\text{rad/s}$, the coefficients are $a = 6\,\text{rad/s},\ b = 4(\text{rad/s})^2$, so that both eigenvalues are real and negative

$$\lambda_1 = -0,7639\,\text{rad/s},\ \lambda_2 = -5.2361\,\text{rad/s} \tag{r}$$

Moreover, the right and left eigenvectors are

$$\begin{aligned} \mathbf{x}_1 &= \begin{bmatrix} 1.1921 \\ -0.9107 \end{bmatrix},\ \mathbf{x}_2 = \begin{bmatrix} -0.1876 \\ 0.9822 \end{bmatrix} \\ \mathbf{y}_1 &= \begin{bmatrix} 0.9822 \\ 0.1876 \end{bmatrix},\ \mathbf{y}_2 = \begin{bmatrix} 0.9107 \\ 1.1921 \end{bmatrix} \end{aligned} \tag{s}$$

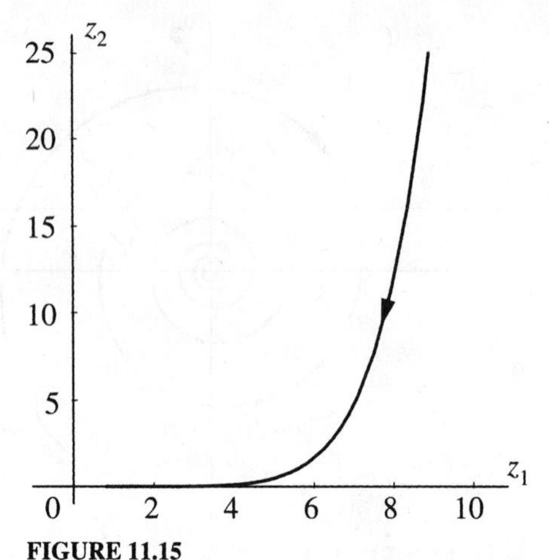

FIGURE 11.15
Trajectory for an overdamped single-degree-of-freedom system in the z_1, z_2-plane

From Eq. (11.38), the trajectory in the z_1, z_2-plane is given by

$$\ln\frac{z_2(t)}{z_{20}} = \frac{\lambda_2}{\lambda_1}\ln\frac{z_1(t)}{z_{10}} = \frac{-5.2361}{-0.7639}\ln\frac{z_1(t)}{z_{10}} = 6.8541\ln\frac{z_1(t)}{z_{10}} \tag{t}$$

where, multiplying Eq. (11.44) by $\mathbf{y}_1^T$ and $\mathbf{y}_2^T$, in sequence, and using the first of the orthonormality relations (11.31),

$$\begin{aligned} z_{10} = z_1(0) = \mathbf{y}_1^T\mathbf{x}(0) &= \begin{bmatrix} 0.9822 \\ 0.1876 \end{bmatrix}^T \begin{bmatrix} 6 \\ 16 \end{bmatrix} = 8.8948 \\ z_{20} = z_2(0) = \mathbf{y}_2^T\mathbf{x}(0) &= \begin{bmatrix} 0.9107 \\ 1.1921 \end{bmatrix}^T \begin{bmatrix} 6 \\ 16 \end{bmatrix} = 24.5367 \end{aligned} \tag{u}$$

Hence, Eq. (t) becomes

$$\ln\frac{z_2(t)}{24.5367} = 6.8541\ln\frac{z_1(t)}{8.8948} \tag{v}$$

from which we conclude that the trajectory equation is

$$z_2(t) = 24.5367\left(\frac{z_1(t)}{8.8948}\right)^{6.8541} \tag{w}$$

The trajectory in the z_1, z_2-plane is plotted in Fig. 11.15, where we obtained guidance concerning the direction of the arrowhead from Eq. (11.34). Clearly, the equilibrium point is an asymptotically stable node.

As a matter of interest, we propose to plot the trajectory of Fig. 11.15 in the x_1, x_2-state plane. To this end, we use Eqs. (11.32) and (11.34) and write

$$\begin{aligned} \mathbf{x}(t) &= \mathbf{x}_1 z_1(t) + \mathbf{x}_2 z_2(t) = \mathbf{x}_1 z_{10} e^{\lambda_1 t} + \mathbf{x}_2 z_{20} e^{\lambda_2 t} \\ &= \begin{bmatrix} 1.1921 \\ -0.9107 \end{bmatrix} 8.8948 e^{-0.7639t} + \begin{bmatrix} -0.1876 \\ 0.9822 \end{bmatrix} 24.5367 e^{-5.2361t} \\ &= \begin{bmatrix} 10.6030 \\ -8.1000 \end{bmatrix} e^{-0.7639t} + \begin{bmatrix} -4.6030 \\ 24.1000 \end{bmatrix} e^{-5.2361t} \end{aligned} \tag{x}$$

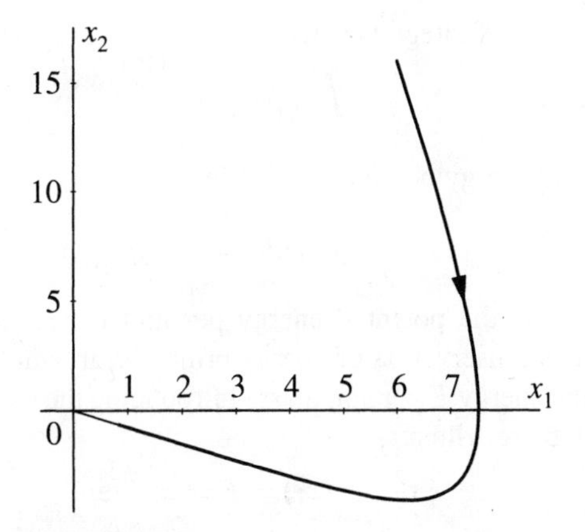

FIGURE 11.16
Trajectory for an overdamped single-degree-of-freedom system in the x_1, x_2-plane

and we observe that the state matches the initial condition $\mathbf{x}(0) = [6 \;\; 16]^T$. The trajectory is plotted in Fig. 11.16 using the components of $\mathbf{x}(t)$.

11.3 CONSERVATIVE SYSTEMS. MOTIONS IN THE LARGE

In Sec. 11.2, we concerned ourselves with motions in the neighborhood of equilibrium points. Such motions are sometimes referred to as *motions in the small.* By contrast, motions at some distance away from equilibrium points are called *motions in the large.* Large motions generally imply the use of the full nonlinear equations of motion. Nonlinear differential equations do not admit closed-form solutions for the most part, so that one must be content with numerical solutions. But, numerical solutions have the disadvantage that they do not reveal readily the dynamic characteristics of the system, particularly for multi-degree-of-freedom systems and/or nonconservative systems. The situation is considerably better in the case of single-degree-of-freedom conservative systems, in which case some general discussion of system behavior for large motions is possible.

We consider a conservative single-degree-of-freedom system and express the equation of motion in the form

$$\ddot{q} = f(q) \tag{11.50}$$

where $f(q)$ is a generally nonlinear conservative force per unit mass. From kinematics, we can write the relation

$$\ddot{q} = \frac{d\dot{q}}{dt} = \frac{d\dot{q}}{dq}\frac{dq}{dt} = \dot{q}\frac{d\dot{q}}{dq} \tag{11.51}$$

Multiplying Eq. (11.50) by dq and considering Eq. (11.51), we have

$$\int_0^q \ddot{q}dq = \int_0^{\dot{q}} \dot{q}d\dot{q} = \int_0^q f(q)dq + c \tag{11.52}$$

in which c is a constant of integration. But,

$$\int_0^{\dot{q}} \dot{q}\,d\dot{q} = \tfrac{1}{2}\dot{q}^2 \tag{11.53}$$

represents the kinetic energy per unit mass. Moreover,

$$\int_0^{q} f(q)dq = -V(q) \tag{11.54}$$

where $V(q)$ represents the potential energy per unit mass. Hence, Eq. (11.52) can be recognized as the conservation of energy principle, in which the constant c can be identified as the total energy E per unit mass. Introducing the usual notation $q = x_1,\ \dot{q} = x_2$, Eq. (11.52) can be rewritten as

$$\tfrac{1}{2}x_2^2 + V(x_1) = E = \text{constant} \tag{11.55}$$

If we introduce an axis E normal to x_1 and x_2, then Eq. (11.55) can be interpreted as a three-dimensional surface symmetric with respect to the x_1, E-plane. For any given value of E, the equation $E =$ constant represents a plane parallel to the x_1, x_2-state plane and intersecting the three-dimensional surface along a curve whose projection on the state plane is defined by

$$x_2 = \pm\sqrt{2[E - V(x_1)]} \tag{11.56}$$

as can be concluded from Eq. (11.55). To interpret Eq. (11.56) geometrically, we observe that any potential energy function $V(x_1)$ represents a curve in the x_1, E-plane, as shown in Fig. 11.17a. Then, according to Eq. (11.56), for any point x_1, the difference $E_i - V(x_1)$ between the level line $E = E_i$ and the potential energy at x_1 defines two values of x_2 in the state plane. By varying x_1, we obtain a curve in the state plane corresponding to $E = E_i$, where the curve can be identified as a trajectory for the single-degree-of-freedom system described by Eq. (11.50). Figure 11.17b shows four trajectories, two corresponding to E_1, one to E_2 and one to E_3, where the values were chosen so as to illustrate various types of motion. To determine the sense of the arrowheads, we differentiate Eq. (11.55) and obtain the slope of the trajectories in the form

$$\frac{dx_2}{dx_1} = -\frac{1}{x_2}\frac{dV(x_1)}{dx_1} \tag{11.57}$$

Observing that the slope dx_2/dx_1 is positive for $x_2 > 0$ and $dV/dx_1 < 0$, we conclude from Figs. 11.17a and 11.17b that the sense indicated in Fig. 11.17b is the correct choice.

To determine the location and nature of the equilibrium points, we consider Eq. (11.54) and write the state equations corresponding to Eq. (11.50) as follows:

$$\begin{aligned} \dot{x}_1 &= x_2 \\ \dot{x}_2 &= f(x_1) = -\frac{dV(x_1)}{dx_1} \end{aligned} \tag{11.58}$$

Hence, the equilibrium points, obtained by setting the right side of Eqs. (11.58) to zero, are given by

$$x_2 = 0,\quad \left.\frac{dV}{dx_1}\right|_{x_1 = x_{1e}} = 0 \tag{11.59}$$

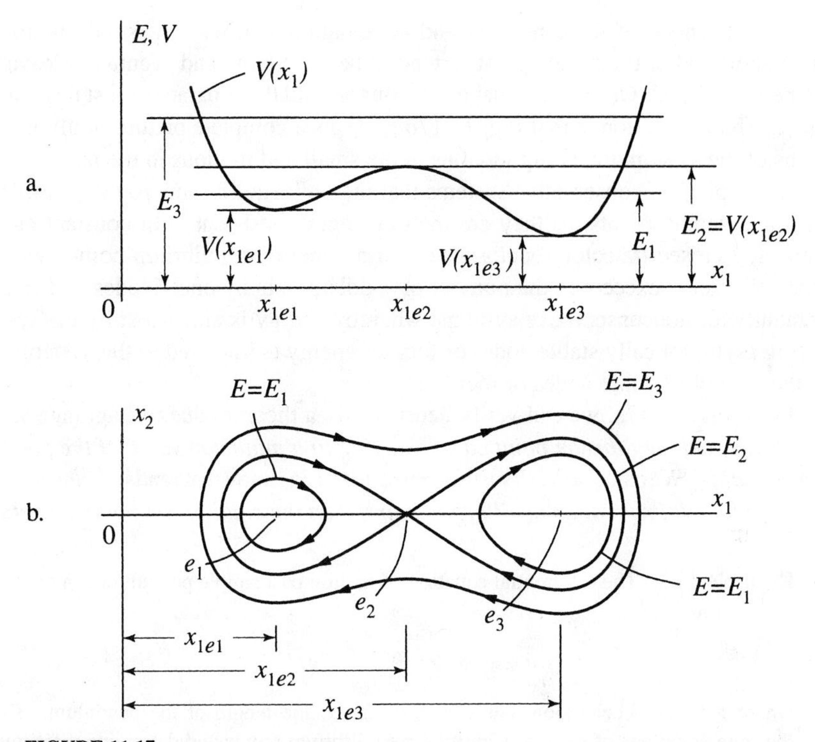

FIGURE 11.17
a. Plot of $V(x_1)$ versus x_1 with lines of constant energy, b. State-plane trajectories corresponding to the lines of constant energy

As expected, all the equilibrium points lie on the x_1-axis. Moreover, they occur at points for which the slope of the curve $V(x_1)$ versus x_1 is zero. In the case at hand, we conclude from Fig. 11.17a that there are three equilibrium points; they are denoted by e_i and occur at $x_{1i} = x_{1ei}$ $(i = 1, 2, 3)$. The nature of the equilibrium points depends on the nature of the trajectories, which in turn depends on the energy level E. For $E < V(x_{1e3})$, the plane $E =$ constant does not intersect the three-dimensional surface, so that no motion is possible. For $V(x_{1e3}) < E < V(x_{1e1})$, stable motion along a closed trajectory takes place in the clockwise sense around the equilibrium point e_3, making e_3 a center. For $V(x_{1e1}) < E < E_2 = V(x_{1e2})$, stable motion in a clockwise sense can take place along one of two closed trajectories, one around center e_1 and one around center e_3, depending on how the motion was initiated. For $E = E_2 = V(x_{1e2})$, there is an equilibrium point e_2 in the form of a saddle point, which is unstable. For $E > E_2$, once again stable motion takes place along a closed trajectory in a clockwise sense, this time surrounding all three equilibrium points, two centers and one saddle point. Hence, the trajectory corresponding to $E = E_2$ separates two different types of periodic motions, one around a single center, whether e_1 or e_3, and one around two centers and a saddle point. For this reason, this trajectory is referred to as a *separatrix*. We observe that small motions in the neighborhood of the centers e_1 and e_3 remain small, whereas small motions in the neighborhood of the saddle point e_2 tend to become large and eventually leave that small neighborhood. On the other hand, motions around the separatrix must be regarded as large. The conclusion is that Fig. 11.17b presents a complete picture of all possible motions of the system, including motions in the small and motions in the large.

It is typical of conservative systems that the only equilibrium points possible are centers and saddle points, as they are the only ones consistent with constant energy. Moreover, a closed trajectory encloses an odd number of equilibrium points, with the number of centers exceeding the number of saddle points by one. Nodes and foci are consistent with nonconservative systems, whereby energy is either lost by the system, implying asymptotically stable nodes or foci, or energy is imparted to the system, such as in the case of unstable nodes or foci.

From Fig. 11.17a, we can verify heuristically a theorem due to Lagrange stating that: *An isolated equilibrium point corresponding to a minimum value of the potential energy is stable*. We can also verify a theorem due to Liapunov that reads: *If the potential energy has no minimum at an equilibrium point, then the equilibrium point is unstable*.

Example 11.3. The differential equation of motion of a simple pendulum can be written in the form

$$\ddot{\theta}+\omega^2\sin\theta=0,\ \omega^2=g/L \tag{a}$$

where g is the acceleration due to gravity and L the length of the pendulum. Derive the state equations of motion, identify the equilibrium points and determine the nature of the equilibrium points. Then, plot the trajectories corresponding to $E=\omega^2, 2\omega^2, 3\omega^2$ and discuss the type of motion associated with the regions defined by $0\leq E\leq 2\omega^2$ and $E>2\omega^2$, where E is the total energy per unit moment of inertia of the bob about the point of support.

Introducing the notation $\theta=x_1,\ \dot{\theta}=x_2$, the state equations are simply

$$\begin{aligned}\dot{x}_1&=x_2\\ \dot{x}_2&=-\omega^2\sin x_1\end{aligned} \tag{b}$$

Equating the right side of Eqs. (b) to zero, we obtain the equilibrium equations

$$x_2=0,\ \sin x_1=0 \tag{c}$$

so that the equilibrium points are given by

$$x_1=\pm j\pi,\ j=0,\ 1,2,\ldots;\ x_2=0 \tag{d}$$

Although mathematically there is an infinite number of equilibrium points, physically there are only two

$$\begin{aligned}e_1&:\ x_1=0,\ x_2=0\\ e_2&:\ x_1=\pi,\ x_2=0\end{aligned} \tag{e}$$

The equilibrium point e_1 is the common one, in which the pendulum is at rest hanging down. On the other hand, the equilibrium point e_2 is that in which the pendulum is at rest in the upright position.

In the neighborhood of $x_1 = 0$, we have the approximation $\sin x_1 \cong x_1$, so that the coefficient matrix is

$$A = \begin{bmatrix} 0 & 1 \\ -\omega^2 & 0 \end{bmatrix} \tag{f}$$

yielding the eigenvalues

$$\begin{matrix} \lambda_1 \\ \lambda_2 \end{matrix} = \pm i\omega \tag{g}$$

Because both eigenvalues are pure imaginary, the equilibrium point e_1 is a center, and hence merely stable. On the other hand, in the neighborhood of $x_1 = \pi$, the approximation is $\sin x_1 \cong -x_1$, so that the coefficient matrix is

$$A = \begin{bmatrix} 0 & 1 \\ \omega^2 & 0 \end{bmatrix} \tag{h}$$

from which we conclude that the eigenvalues are

$$\begin{matrix} \lambda_1 \\ \lambda_2 \end{matrix} = \pm \omega \tag{i}$$

Because both eigenvalues are real, one being positive and the other one negative, the equilibrium point e_2 is a saddle point, and hence unstable.

The trajectories for $E = \omega^2, 2\omega^2, 3\omega^2$ are shown in Fig. 11.18. We note that the range $-\pi \le x_1 \le \pi$ covers all possible positions of the bob, as points such that $x_1 > \pi$ and $x_1 < -\pi$ simply cover the same physical positions. For $E = \omega^2$, the trajectory is closed, indicating periodic motion. Indeed, for $0 < E < \omega^2$, all trajectories are closed. For small E, i.e., in the neighborhood of the equilibrium point e_1, the trajectory is an ellipse and the motion is pure harmonic oscillation. As E increases, the motion ceases to be harmonic, but it remains periodic. For $E = 3\omega^2$, the trajectory is open and the motion is nonuniformly rotary with, the bob going over the top. Of course, for $E = 2\omega^2$ the trajectory represents a separatrix, separating different types of periodic motions, oscillatory for $0 < E < 2\omega^2$ and rotary for $E > 2\omega^2$.

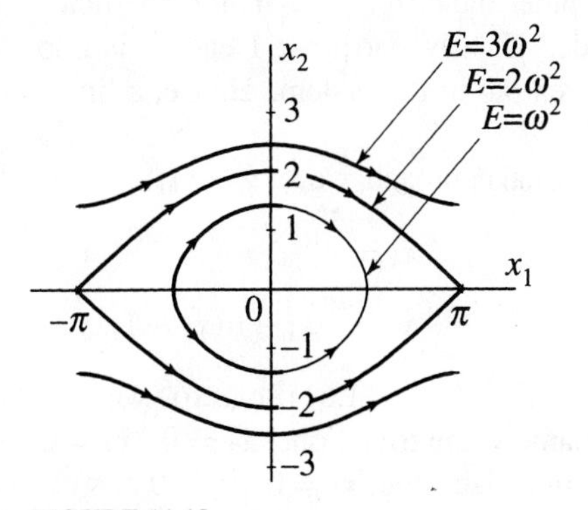

FIGURE 11.18
Trajectories for a simple pendulum corresponding to different energy levels

11.4 LIMIT CYCLES. THE VAN DER POL OSCILLATOR

A question of particular interest in nonlinear systems is whether they exhibit closed trajectories, as such trajectories imply periodic motion. In Sec. 11.3, we encountered closed trajectories in conjunction with conservative systems, with the closed trajectories enclosing an odd number of equilibrium points. The equilibrium points in the case of conservative systems are centers and saddle points and the number of centers enclosed by a given trajectory exceeds the number of saddle points by one.

Closed trajectories can occur also in nonconservative nonlinear systems, provided the net energy change at the completion of a full cycle is zero. This implies that the system dissipates energy over some parts of the cycle and acquires energy over the balance of the cycle. Closed trajectories exhibiting this type of characteristics are referred to as *limit cycles of Poincaré*, or simply *limit cycles*. Limit cycles can be regarded as "equilibrium motions" in which the system performs periodic motions, as opposed to equilibrium points at which the system is at rest. One basic difference between limit cycles experienced by nonconservative systems and closed trajectories occurring in the case of conservative systems is that the amplitude of a given limit cycle depends on the system parameters alone and that of a mere closed trajectory depends on the energy imparted to the system initially. In the case of limit cycles we speak of *orbital stability*, as opposed to stability in the sense of Liapunov (Sec. 11.1) in the case of equilibrium points. It is very difficult in general to establish the existence of a limit cycle for a given system. The situation is considerably better when it is known in advance that the system does possess a limit cycle.

A classical example of a system known to possess a limit cycle is the *van der Pol oscillator*, described by the differential equation

$$\ddot{q} + \mu(q^2 - 1)\dot{q} + q = 0, \ \mu > 0 \tag{11.60}$$

It can be regarded as an oscillator with variable damping, as the term $\mu(q^2 - 1)$ represents an amplitude-dependent damping coefficient; such a system is both nonconservative and nonlinear. For $|q| < 1$, the coefficient is negative, which tends to increase the motion amplitude. On the other hand, for $|q| > 1$, the coefficient is positive, which tends to reduce the amplitude. Clearly, for $|q| < 1$ energy is imparted to the system, and for $|q| > 1$ energy is taken out of the system. Hence, a limit cycle can be expected and is indeed obtained.

Following the usual approach, we let $q = x_1$, $\dot{q} = x_2$ and obtain the state equations

$$\begin{aligned} \dot{x}_1 &= x_2 \\ \dot{x}_2 &= -x_1 - \mu(x_1^2 - 1)x_2 \end{aligned} \tag{11.61}$$

Equating the right side of Eqs. (11.61) to zero, we conclude that there is only one equilibrium point, namely, the trivial one, $x_1 = 0, \ x_2 = 0$. To determine the nature of the equilibrium, we linearize about $x_1 = 0, \ x_2 = 0$ and obtain the coefficient matrix

$$A = \begin{bmatrix} 0 & 1 \\ -1 & \mu \end{bmatrix} \tag{11.62}$$

so that the eigenvalues are

$$\frac{\lambda_1}{\lambda_2} = \frac{\mu}{2} \pm \sqrt{\left(\frac{\mu}{2}\right)^2 - 1} \tag{11.63}$$

For $\mu > 2$ both roots λ_1 and λ_2 are real and positive, so that the origin is an unstable node. On the other hand, for $\mu < 2$ roots λ_1 and λ_2 are complex conjugates with positive real part, so that the origin is an unstable focus. Hence, regardless of the value of μ, the origin is an unstable equilibrium point, so that any motion initiated in a small neighborhood of the origin will eventually leave this neighborhood and reach the limit cycle.

To obtain the equation of the trajectories, we divide the second of Eqs. (11.61) by the first and write

$$\frac{dx_2}{dx_1} = \mu(1 - x_1^2) - \frac{x_1}{x_2} \tag{11.64}$$

Equation (11.64) does not lend itself to closed-form solution. The equation was integrated numerically for $\mu = 0.2$ and $\mu = 3$, and trajectories corresponding to several sets of initial conditions are plotted in Figs. 11.19a and 11.19b, respectively. Clearly, the shape of the limit cycle depends on the parameter μ. In fact, for $\mu \to 0$ the limit cycle tends to resemble a circle. Because all trajectories approach the limit cycle, either from the inside or from the outside, *the limit cycle is stable*. We observe that a stable limit cycle encloses an unstable equilibrium point.

Finally, we must point out that the van der Pol oscillator is a very good example of a nonlinear system for which linearization about the trivial equilibrium is totally inadequate. Indeed, a linearized analysis would have predicted instability, with the

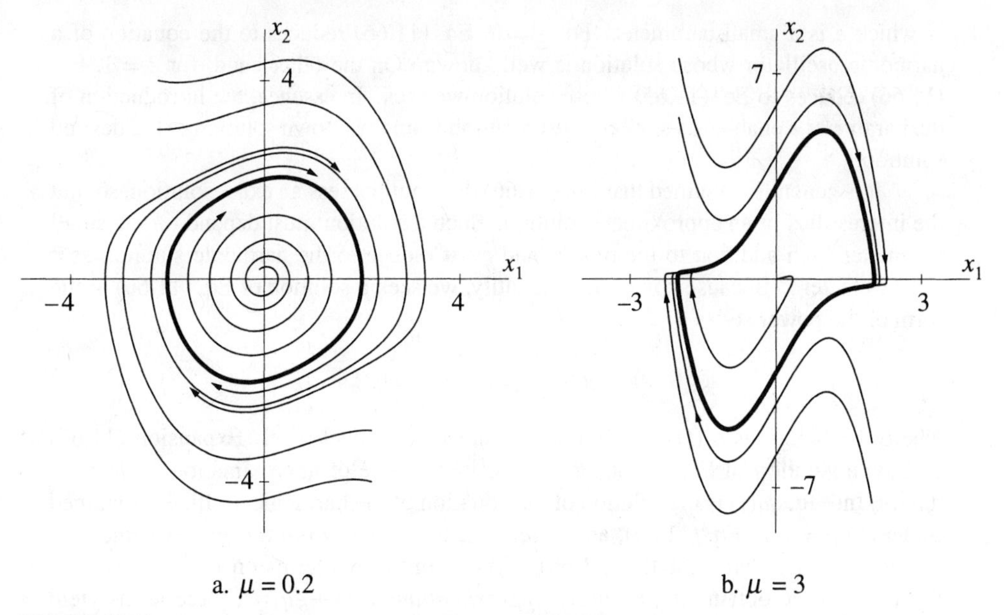

FIGURE 11.19
Trajectories for the van der Pol oscillator for the cases: a. $\mu = 0.2$ and b. $\mu = 3$

motion increasing without bounds, thus ignoring entirely the existence of a limit cycle. We must also point out that the shape of the limit cycle depends on the parameter μ alone, so that the initial conditions have no effect on the amplitude of motion after the system has reached the limit cycle. A MATLAB program for plotting trajectories in the state plane for the van der Pol oscillator is given in Sec. 11.12.

11.5 THE FUNDAMENTAL PERTURBATION TECHNIQUE

Many physical systems are described by differential equations consisting of two parts, one part containing linear terms with constant coefficients and a second part containing nonlinear terms and/or terms with time-dependent coefficients, where the second part is relatively small compared to the first. We refer to the small terms rendering the system nonlinear and/or possessing time-dependent coefficients as *perturbations*. We are interested in the important class of problems in which the linear part represents a harmonic oscillator, and we refer to such systems as *quasi-harmonic*.

We consider the quasi-harmonic system described by the differential equation

$$\ddot{q} + \omega_0^2 q = f(q, \dot{q}) \tag{11.65}$$

where $f(q, \dot{q})$ is a sufficiently small nonlinear function of the displacement q and velocity $\dot{q}$ that it can be regarded as a perturbation. To emphasize the fact that $f(q, \dot{q})$ represents a perturbation, it is convenient to rewrite Eq. (11.65) in the form

$$\ddot{q} + \omega_0^2 q = \varepsilon f(q, \dot{q}) \tag{11.66}$$

in which ε is a small parameter. For $\varepsilon = 0$, Eq. (11.66) reduces to the equation of a harmonic oscillator whose solution is well known. On the other hand, for $\varepsilon = 1$, Eq. (11.66) reduces to Eq. (11.65) whose solution we seek. In essence, the introduction of the parameter ε enables us to effect the transition from the known solution to the desired solution.

It is generally assumed that Eq. (11.66) does not possess an exact solution, so that the interest lies in an approximate solution. Such a solution must depend on the small parameter ε, in addition to the time t, and must reduce to the harmonic solution as ε reduces to zero. Because ε is a small quantity, we seek a solution of Eq. (11.66) in the form of the power series

$$q(t, \varepsilon) = q_0(t) + \varepsilon q_1(t) + \varepsilon^2 q_2(t) + \dots \tag{11.67}$$

where the functions $q_i(t)$ $(i = 0, 1, 2, \dots)$ are independent of ε. Expansion (11.67) permits a solution of Eq. (11.66) to any desired degree of approximation, at least in theory. Indeed, $q_0(t)$ is the solution of the equation of the harmonic oscillator, obtained by letting $\varepsilon = 0$ in Eq. (11.66), and is referred to as the *zero-order approximation*, or the *generating solution* of Eq. (11.66). By retaining two terms on the right side of Eq. (11.67), we obtain the *first-order-approximation* $q_0(t) + \varepsilon q_1(t)$, three terms yield the *second-order approximation* $q_0(t) + \varepsilon q_1(t) + \varepsilon^2 q_2(t)$, etc.

Next, we derive the differential equation for every level of approximation. To this end, we insert Eq. (11.67) into the left side of Eq. (11.66) and obtain simply

$$\begin{aligned}\ddot{q}+\omega_0^2 q &= \ddot{q}_0+\varepsilon\ddot{q}_1+\varepsilon^2\ddot{q}_2+\ldots+\omega_0^2(q_0+\varepsilon q_1+\varepsilon^2 q_2+\ldots)\\ &= \ddot{q}_0+\omega_0^2 q_0+\varepsilon(\ddot{q}_1+\omega_0^2 q_1)+\varepsilon^2(\ddot{q}_2+\omega_0^2 q_2)+\ldots \qquad (11.68)\end{aligned}$$

Moreover, we expand $f(q,\dot{q})$ in a power series in ε about the generating solution $(q_0,\dot{q}_0)$ as follows:

$$\begin{aligned}f(q,\dot{q}) &= f(q_0+\varepsilon q_1+\varepsilon^2 q_2+\ldots,\dot{q}_0+\varepsilon\dot{q}_1+\varepsilon^2\dot{q}_2+\ldots)\\
&= f(q_0,\dot{q}_0)+\left.\frac{\partial f(q,\dot{q})}{\partial q}\right|_{q=q_0,\dot{q}=\dot{q}_0}\varepsilon q_1+\left.\frac{\partial f(q,\dot{q})}{\partial \dot{q}}\right|_{q=q_0,\dot{q}=\dot{q}_0}\varepsilon\dot{q}_1\\
&\quad+\left.\frac{\partial f(q,\dot{q})}{\partial q}\right|_{q=q_0,\dot{q}=\dot{q}_0}\varepsilon^2 q_2+\left.\frac{\partial f(q,\dot{q})}{\partial \dot{q}}\right|_{q=q_0,\dot{q}=\dot{q}_0}\varepsilon^2\dot{q}_2\\
&\quad+\frac{1}{2!}\left.\frac{\partial^2 f(q,\dot{q})}{\partial q^2}\right|_{q=q_0,\dot{q}=\dot{q}_0}(\varepsilon q_1)^2+\frac{1}{2!}\left.\frac{\partial^2 f(q,\dot{q})}{\partial \dot{q}^2}\right|_{q=q_0,\dot{q}=\dot{q}_0}(\varepsilon\dot{q}_1)^2\\
&\quad+\frac{2}{2!}\left.\frac{\partial^2 f(q,\dot{q})}{\partial q\partial\dot{q}}\right|_{q=q_0,\dot{q}=\dot{q}_0}(\varepsilon q_1)(\varepsilon\dot{q}_1)+\ldots\\
&= f(q_0,\dot{q}_0)+\varepsilon\left[\left.\frac{\partial f(q,\dot{q})}{\partial q}\right|_{q=q_0,\dot{q}=\dot{q}_0}q_1+\left.\frac{\partial f(q,\dot{q})}{\partial \dot{q}}\right|_{q=q_0,\dot{q}=\dot{q}_0}\dot{q}_1\right]\\
&\quad+\varepsilon^2\left[\left.\frac{\partial f(q,\dot{q})}{\partial q}\right|_{q=q_0,\dot{q}=\dot{q}_0}q_2+\left.\frac{\partial f(q,\dot{q})}{\partial \dot{q}}\right|_{q=q_0,\dot{q}=\dot{q}_0}\dot{q}_2\right.\\
&\quad\quad+\frac{1}{2!}\left.\frac{\partial f(q,\dot{q})}{\partial q^2}\right|_{q=q_0,\dot{q}=\dot{q}_0}q_1^2+\frac{1}{2!}\left.\frac{\partial f(q,\dot{q})}{\partial \dot{q}^2}\right|_{q=q_0,\dot{q}=\dot{q}_0}\dot{q}_1^2\\
&\quad\quad\left.+\frac{2}{2!}\left.\frac{\partial f(q,\dot{q})}{\partial q\partial\dot{q}}\right|_{q=q_0,\dot{q}=\dot{q}_0}q_1\dot{q}_1\right]+\ldots \qquad (11.69)\end{aligned}$$

Inserting Eqs. (11.68) and (11.69) into Eq. (11.66), recognizing that the resulting equation must be satisfied for all values of ε and recalling that the functions q_i $(i=0,1,2,\ldots)$ are independent of ε, it follows that the coefficients of like powers of ε on both sides must be equal to one another. This amounts to separating terms according to the order of magnitude as follows:

$$\begin{aligned}&O(\varepsilon^0): \ \ddot{q}_0+\omega_0^2 q_0=0\\
&O(\varepsilon^1): \ \ddot{q}_1+\omega_0^2 q_1=f(q_0,\dot{q}_0)\\
&O(\varepsilon^2): \ \ddot{q}_2+\omega_0^2 q_2=\left.\frac{\partial f(q,\dot{q})}{\partial q}\right|_{q=q_0,\dot{q}=\dot{q}_0}q_1+\left.\frac{\partial f(q_1\dot{q})}{\partial \dot{q}}\right|_{q=q_0,\dot{q}=\dot{q}_0}\dot{q}_1 \qquad (11.70)\\
&\cdots\cdots \quad \cdots\cdots\cdots\cdots\cdots\cdots\cdots\cdots\end{aligned}$$

The process of using expansion (11.69) to derive Eqs. (11.70) can be simplified a great deal when the function $f(q, \dot{q})$ is given explicitly. Indeed, in such cases, it is significantly simpler to insert Eq. (11.67) directly into the explicit form of $f(q, \dot{q})$ and separate orders of magnitude, as shown in Example 11.4.

We observe that Eqs. (11.70) are all linear. Moreover, of equal importance is the fact that Eqs. (11.70) can be solved recursively. Indeed, the zero-order approximation equation, namely the first of Eqs. (11.70), is simply the equation of a harmonic oscillator whose solution is (Sec. 2.1)

$$q_0(t) = A_0 \cos(\omega_0 t - \phi_0) \tag{11.71}$$

where A_0 and ϕ_0 are the amplitude and phase angle in the zero-order approximation, constants depending on the initial conditions. Then, inserting Eq. (11.71) into the second of Eqs. (11.70), the first-order perturbation equation becomes

$$\ddot{q}_1 + \omega_0^2 q_1 = f[A_0 \cos(\omega_0 t - \phi_0),\ -\omega_0 A_0 \sin(\omega_0 t - \phi_0)] \tag{11.72}$$

which represents the equation of a harmonic oscillator subjected to a known time-dependent excitation. The form of the excitation depends on the function f, but for the most part the excitation is likely to be a linear combination of trigonometric functions of frequencies equal to integer multiples of ω_0. Hence, its solution can be obtained with relative ease by the methods of Secs. 3.1 and 3.2, thus completing the first-order approximation solution $q_0 + \varepsilon q_1$. The next step is to solve the third of Eqs. (11.70) for the second-order perturbation q_2. To this end, we observe that the right side of this third equation depends on $q_0, \dot{q}_0, q_1$ and $\dot{q}_1$, by now all known functions of time. The solution for the second-order perturbation q_2 essentially completes the second-order approximation solution $q_0 + \varepsilon q_1 + \varepsilon^2 q_2$. The process continues in the same fashion, with the equation for the nth-order perturbation q_n representing a harmonic oscillator with the right side depending on the preceding solutions $q_0, \dot{q}_0, q_1, \dot{q}_1, \dots, q_{n-1}, \dot{q}_{n-1}$. Hence, at least in theory, there is no obstacle to obtaining its solution. In practice, however, the process becomes increasingly laborious as the approximation order increases. Fortunately, for sufficiently small ε, the significance of the higher-order terms decreases rapidly, so that it is seldom necessary to go beyond the second-order approximation.

Equation (11.67) expresses the solution of Eq. (11.66) as a power series in the small parameter ε and is referred to as a *formal solution*. As discussed in the preceding paragraph, the recursive solution of Eqs. (11.70) gives rise to increasingly higher-order approximation solutions of Eq. (11.66). Then, the formal solution of Eq. (11.65) is obtained by setting $\varepsilon = 1$, which stipulates, of course, that the function $f(q, \dot{q})$ on the right side of Eq. (11.65) is itself small.

Formal solutions need not converge, and in fact there is a real possibility that they diverge. Nevertheless, such solutions tend to be quite useful for numerical calculations. Indeed, such power series in ε may give a good approximation with a limited number of terms.

Example 11.4. Consider the case in which the parameter μ in the van der Pol oscillator of Sec. 11.4 is small, $\mu = \varepsilon$, and use Eqs. (11.70) to derive the perturbation equations through the third order.

The van der Pol oscillator is described by the differential equation (11.60). Consistent with the formulation of this section, we rewrite Eq. (11.60) in the form

$$\ddot{q} + q = \varepsilon f(q, \dot{q}) = \varepsilon(1 - q^2)\dot{q} \tag{a}$$

Because, $f(q, \dot{q})$ is multiplied by ε, to derive the perturbation equations through the third order, it is only necessary to include in the expansion for $f(q, \dot{q})$ small terms in ε through the second order. Hence, using Eq. (11.67) and retaining small terms in ε through second order, we write

$$\begin{aligned} f(q, \dot{q}) = (1 - q^2)\dot{q} &\cong [1 - (q_0 + \varepsilon q_1 + \varepsilon^2 q_2)^2](\dot{q}_0 + \varepsilon \dot{q}_1 + \varepsilon^2 \dot{q}_2) \\ &\cong (1 - q_0^2)\dot{q}_0 + \varepsilon[-2q_0 q_1 \dot{q}_0 + (1 - q_0^2)\dot{q}_1] \\ &\quad + \varepsilon^2[-q_1^2 \dot{q}_0 - 2q_0 q_2 \dot{q}_0 - 2q_0 q_1 \dot{q}_1 + (1 - q_0^2)\dot{q}_2] \end{aligned} \tag{b}$$

Then, inserting Eqs. (11.67) and (b) into Eq. (a) and separating terms of different orders of magnitude, we obtain the desired perturbation equations

$$\begin{aligned} &O(\varepsilon^0): \ \ddot{q}_0 + q_0 = 0 \\ &O(\varepsilon^1): \ \ddot{q}_1 + q_1 = (1 - q_0^2)\dot{q}_0 \\ &O(\varepsilon^2): \ \ddot{q}_2 + q_2 = -2q_0 q_1 \dot{q}_0 + (1 - q_0^2)\dot{q}_1 \\ &O(\varepsilon^3): \ \ddot{q}_3 + q_3 = -q_1^2 \dot{q}_0 - 2q_0 q_2 \dot{q}_0 - 2q_0 q_1 \dot{q}_1 + (1 - q_0^2)\dot{q}_2 \end{aligned} \tag{c}$$

11.6 SECULAR TERMS

In seeking a perturbation solution to nonlinear differential equations, practical considerations dictate that we limit expansion (11.67) to the first several terms, as the equations for the higher-order perturbations become progressively more complicated and their solution more difficult to obtain. This practice can be easily justified when the parameter ε is small compared 1, say of the order of 10^{-1}. Indeed, in this case the equation for the first-order perturbation produces a correction to the zero-order solution of order 10^{-1}, the equation for the second-order perturbation yields a correction of order 10^{-2}, etc. But, retention of a limited number of terms is capable of creating a problem of a different kind in that it can produce terms increasing indefinitely with time, and hence a divergent solution. Such diverging terms are commonly referred to as *secular terms*, and they can appear in systems known to possess bounded solutions, such as stable conservative systems characterized by periodic solutions. Hence, a modification of the formal perturbation solution designed to prevent the formation of secular terms for systems known to possess bounded solutions demands itself. Before discussing ways of generating bounded perturbation solutions, a closer examination of the nature of secular terms, and how they enter into solutions, is in order.

A classical example of a nonlinear conservative system known to possess periodic solutions consists of a mass m attached to a stiffening, or hardening spring (Sec. 1.7). We consider a spring with a restoring force in the form of the sum of two terms, one proportional to the elongation and the other varying as the third power of the elongation. We are concerned with the case in which the cubic term is appreciably smaller than the linear term, so that the spring is nearly linear; the spring force can be expressed in the form

$$F(q) = -k(q + \varepsilon q^3) \tag{11.73}$$

where k is the slope of the force-elongation curve at $q = 0$ and ε is a small parameter. It is not difficult to see that the system is quasi-harmonic. Indeed, dividing through by m, the system differential equation can be shown to be

$$\ddot{q} + \omega_0^2(q + \varepsilon q^3) = 0, \ \varepsilon \ll 1 \tag{11.74}$$

where $\omega_0 = \sqrt{k/m}$ can be identified as the natural frequency of the harmonic oscillator obtained by letting $\varepsilon = 0$. Equation (11.74) is the well-known *Duffing's equation.*

Following the pattern of Sec. 11.5, we assume a solution of Eq. (11.74) in the form of Eq. (11.67), separate orders of magnitude and obtain the following set of differential equations

$$\begin{aligned} &\ddot{q}_0 + \omega_0^2 q_0 = 0 \\ &\ddot{q}_1 + \omega_0^2 q_1 = -\omega_0^2 q_0^3 \\ &\ddot{q}_2 + \omega_0^2 q_2 = -3\omega_0^2 q_0^2 q_1 \\ &\dots\dots\dots\dots\dots\dots \end{aligned} \tag{11.75}$$

which can be solved recursively. Indeed, solving the first of Eqs. (11.75), we obtain the zero-order solution

$$q_0 = A_0 \cos(\omega_0 t - \phi_0) \tag{11.76}$$

where A_0 and ϕ_0 are constants representing the zero-order amplitude and phase angle, respectively; they are related to the initial conditions, as shown in Sec. 11.7. Substituting Eq. (11.76) into the second of Eqs. (11.75) and recalling from trigonometry that $\cos^3\alpha = \frac{1}{4}(3\cos\alpha + \cos 3\alpha)$, we can write

$$\begin{aligned} \ddot{q}_1 + \omega_0^2 q_1 &= -\omega_0^2 A_0^3 \cos^3(\omega_0 t - \phi_0) \\ &= -\tfrac{3}{4}\omega_0^2 A_0^3 \cos(\omega_0 t - \phi_0) - \tfrac{1}{4}\omega_0^2 A_0^3 \cos 3(\omega_0 t - \phi_0) \end{aligned} \tag{11.77}$$

We observe that the first term on the right side of Eq. (11.77) represents a harmonic excitation with the same frequency ω_0 as the natural frequency of the harmonic oscillator on the left side, so that a resonance condition (Sec. 3.2) has been created. Hence, using results from Secs. 3.1 and 3.2, we obtain the solution

$$q_1 = -\tfrac{3}{8}\omega_0 t A_0^3 \sin(\omega_0 t - \phi_0) + \tfrac{1}{32} A_0^3 \cos 3(\omega_0 t - \phi_0) \tag{11.78}$$

Examining solution (11.78), we conclude that the first term increases indefinitely with time, so that the term is secular.

The system described by Eq. (11.74), however, is known to admit only bounded solutions. In fact, the system is of the type studied in Sec. 11.2. Hence, letting $q = x_1$, $\dot{q} = x_2$ and following the pattern of Sec. 11.2, we can write the conservation of energy statement in the form

$$\tfrac{1}{2}mx_2^2 + V(x_1) = E = \text{constant} \tag{11.79}$$

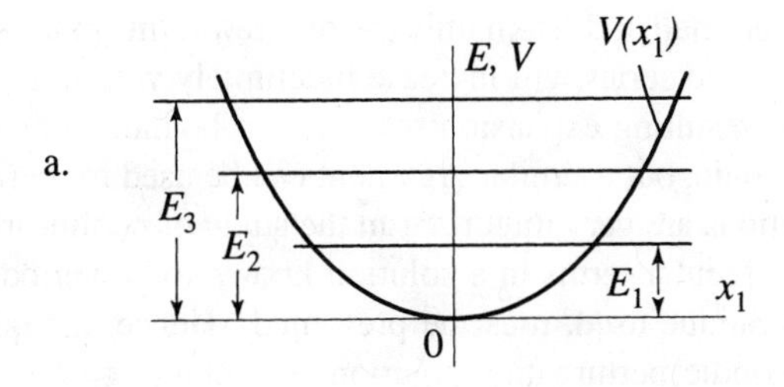

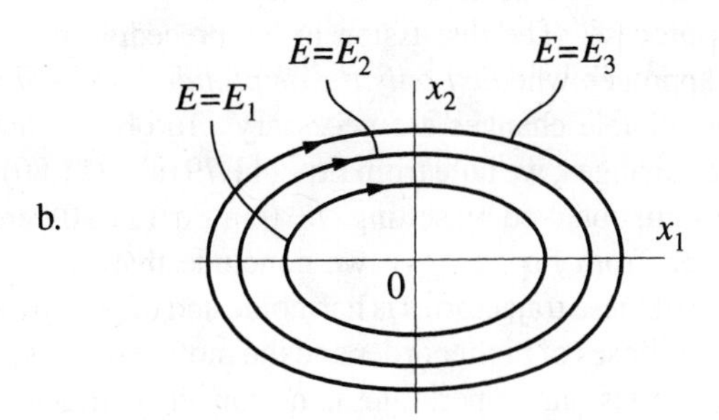

FIGURE 11.20
a. Plot of $V(x_1)$ versus x_1 for Duffing's equation,
b. Trajectories for Duffing's equation

in which the first term represents the kinetic energy,

$$V(x_1) = \int_{x_1}^{0} F(\zeta)d\zeta = -k\int_{x_1}^{0}(\zeta + \varepsilon\zeta^3)d\zeta = \tfrac{1}{2}k(x_1^2 + \tfrac{1}{2}\varepsilon x_1^4) \tag{11.80}$$

is the potential energy and E is the total energy. It is not difficult to show that the only equilibrium point of the system is at the origin of the state plane, $x_1 = 0$, $x_2 = 0$, and it is a center. For any given value of E, a value depending on corresponding initial conditions, the equation $E =$ constant represents a closed trajectory enclosing the center. The potential energy function is shown in Fig. 11.20a and several trajectories are displayed in Fig. 11.20b. Because the trajectories are closed, the motion of the system must be periodic, and hence bounded, so that the presence of secular terms in the solution demands an explanation.

Although secular terms increase indefinitely with time, their presence in the context of a perturbation solution does not necessarily imply unbounded behavior. To explain this apparent paradox, we consider the expansion

$$\begin{aligned}\sin(\omega_0 + \varepsilon)t &= \sin\omega_0 t\cos\varepsilon t + \cos\omega_0 t\sin\varepsilon t \\ &= \left(1 - \frac{1}{2!}\varepsilon^2 t^2 + \frac{1}{4!}\varepsilon^4 t^4 - \ldots\right)\sin\omega_0 t \\ &\quad + \left(\varepsilon t - \frac{1}{3!}\varepsilon^3 t^3 + \frac{1}{5!}\varepsilon^5 t^5 - \ldots\right)\cos\omega_0 t\end{aligned} \tag{11.81}$$

If we assume that ε is small and retain only the first few terms in the series for $\sin \varepsilon t$ and $\cos \varepsilon t$, then the truncated series will increase indefinitely with time, making it difficult to conclude that the resulting expansion represents a bounded function. The function $\sin(\omega_0 + \varepsilon)t$ is harmonic, but a similar argument can be used for periodic functions.

Periodic solutions are very important in the study of nonlinear vibrating systems, and the presence of secular terms in a solution known to be periodic, a peculiarity of the perturbation technique used, must be prevented. Hence, the question arises as to how to produce periodic perturbation solutions by retaining only the first few terms in the expansion, which is the same as asking how to prevent the resonance induced by the perturbation process. The discussion in the preceding paragraph demonstrates that a perturbation approach whereby *only the amplitude is altered* is unlikely to prove satisfactory, so that suitable changes are necessary. To obtain some insights into the nature of the required changes, we note from Eqs. (11.79) and (11.80) that the trajectories of the linearized system, obtained by setting $\varepsilon = 0$ in Eq. (11.80), are ellipses enclosing the origin. Moreover, from Eq. (11.74), we conclude that the motion of the system corresponding to any of these trajectories is harmonic and of period $T_0 = 2\pi/\omega_0$. For $\varepsilon \neq 0$, the trajectories are ellipses of higher order and the motion of the system corresponding to any of these trajectories, albeit periodic, is no longer harmonic and the period T is not equal to T_0. In fact, the period differs from trajectory to trajectory and the period for any given trajectory depends on the small parameter ε, as well as on the associated total energy E; the latter in turn depends on the initial conditions. It follows that a perturbation method designed to produce periodic solutions to nonlinear problems, and in the process to prevent the formation of secular terms, *must include an alteration of both the amplitude and period of oscillation.* To this end, both the amplitude and the period of the solution must be regarded as unknowns, and the period must be determined by insisting that the response be periodic. We present such a perturbation technique in Sec. 11.7.

11.7 LINDSTEDT'S METHOD

As demonstrated in Sec. 11.6, in the case of quasi-harmonic nonlinear systems, the fundamental perturbation technique, whereby the system response is expanded in a power series in a small parameter ε, tends to produce solutions containing secular terms even for systems known to possess periodic, and hence bounded, solutions. The reason for this erroneous result lies in the fact that, unlike linear harmonic systems characterized by a constant period of oscillation, regardless of the initial conditions, the period of nonlinear quasi-harmonic systems does depend on the initial conditions. Hence, to remedy the situation, we must treat the period as an unknown quantity and seek a perturbation solution for both the amplitude and the period at the same time, which amounts to expanding the period in a power series in ε as well. Then, the perturbations in the period are determined by imposing a certain periodicity condition on the perturbation in the response corresponding to each order of approximation, which is the same as suppressing secular terms for every order. This is the essence of *Lindstedt's method*.

We consider a generic quasi-harmonic system described by the differential equation

$$\ddot{q} + \omega_0^2 q = \varepsilon f(q, \dot{q}) \tag{11.82}$$

in which $f(q,\dot{q})$ is a nonlinear function of q and $\dot{q}$. The response of the associated linear system, obtained by letting $\varepsilon = 0$ in Eq. (11.82), is harmonic and of period $T_0 = 2\pi/\omega_0$. In the presence of the nonlinear term $\varepsilon f(q,\dot{q})$, the response is periodic and of period $T = 2\pi/\omega$, where ω is an unknown fundamental frequency depending on ε, $\omega = \omega(\varepsilon)$. The implication is that the solution is a superposition of harmonic terms of frequencies equal to integer multiples of the fundamental frequency ω, reminding of a Fourier series expansion of periodic functions. According to Lindstedt's method, we assume a solution of Eq. (11.82) in the form

$$q(t) = q_0(t) + \varepsilon q_1(t) + \varepsilon^2 q_2(t) + \ldots \tag{11.83}$$

where $q(t)$ is periodic with the period $T = 2\pi/\omega$, in which ω is the fundamental frequency. The period T, and hence the fundamental frequency ω, is not known at this time. We assume that the fundamental frequency is given by

$$\omega = \omega_0 + \varepsilon\omega_1 + \varepsilon^2\omega_2 + \ldots \tag{11.84}$$

in which the parameters $\omega_1, \omega_2, \ldots$ are yet to be determined; they are determined by insisting that the perturbations $q_1(t), q_2(t), \ldots$ be periodic.

Instead of working with an unknown period T, it is more convenient to change the time scale by replacing the independent variable t by τ, where *the period of oscillation in terms of the new variable τ is equal to 2π*. This amounts to the change of variables

$$\tau = \omega t, \; d/dt = \omega d/d\tau \tag{11.85}$$

and we observe that τ plays the role of a dimensionless time variable. Introducing Eqs. (11.85) in Eq. (11.82) and denoting differentiations with respect to τ by primes, we can write

$$\omega^2 q'' + \omega_0^2 q = \varepsilon f(q, \omega q') \tag{11.86}$$

Next, we consider Eqs. (11.83) and (11.84), use the analogy with Eq. (11.69) and expand $f(q,\omega q')$ as follows:

$$\begin{aligned} f(q,\omega q') = & f(q_0,\omega_0 q_0') + \varepsilon \left[q_1 \left. \frac{\partial f(q,\omega q')}{\partial q} \right|_{\substack{q=q_0,\, q'=q_0', \\ \omega=\omega_0}} \right. \\ & + q_1' \left. \frac{\partial f(q,\omega q')}{\partial q'} \right|_{\substack{q=q_0,\, q'=q_0', \\ \omega=\omega_0}} \\ & \left. + \omega_1 \left. \frac{\partial f(q,\omega q')}{\partial \omega} \right|_{\substack{q=q_0,\, q'=q_0', \\ \omega=\omega_0}} \right] + \varepsilon^2[\ldots] + \ldots \end{aligned} \tag{11.87}$$

Hence, inserting Eqs. (11.83), (11.84) and (11.87) into Eq. (11.86) and separating orders of magnitude, we obtain the perturbation equations

$$
\begin{aligned}
&q_0'' + q_0 = 0 \\
&q_1'' + q_1 = \frac{1}{\omega_0^2} f(q_0, \omega_0 q_0') - 2\frac{\omega_1}{\omega_0} q_0'' \\
&q_2'' + q_2 = \frac{1}{\omega_0^2}\left[q_1 \left.\frac{\partial f(q,\omega q')}{\partial q}\right|_{\substack{q=q_0,\, q'=q_0',\\ \omega=\omega_0}} + q_1' \left.\frac{\partial f(q,\omega q')}{\partial q'}\right|_{\substack{q=q_0,\, q'=q_0',\\ \omega=\omega_0}} \right. \\
&\qquad \left. + \omega_1 \left.\frac{\partial f(q,\omega q')}{\partial \omega}\right|_{\substack{q=q_0,\, q'=q_0',\\ \omega=\omega_0}} \right] - \left[2\frac{\omega_2}{\omega_0} + \left(\frac{\omega_1}{\omega_0}\right)^2 \right] q_0'' - 2\frac{\omega_1}{\omega_0} q_1'' \\
&\cdots\cdots\cdots\cdots\cdots\cdots\cdots\cdots\cdots\cdots\cdots\cdots
\end{aligned} \tag{11.88}
$$

Equations (11.88) are to be solved recursively, as in Sec. 11.6. But, in contrast with the process used in Sec. 11.6, here we have the additional task of determining the corrections $\omega_1, \omega_2, \ldots$ required for the calculation of the fundamental frequency ω. To this end, we require that every perturbation $q_1(t), q_2(t), \ldots$ in the response be periodic and of period 2π, so that the periodicity conditions are

$$q_i(\tau + 2\pi) = q_i(\tau), \; i = 1, 2, \ldots \tag{11.89}$$

The functions $q_i(\tau)$ can be periodic only if there are no secular terms. But, as pointed out in Sec. 11.6, $q_i(\tau)$ are free of secular terms only in the absence of resonance, which amounts to requiring that the right side of every one of Eqs. (11.88), beginning with the second, does not contain harmonic terms in τ of unit frequency. This is guaranteed if the corrections $\omega_1, \omega_2, \ldots$ are such that the coefficients of the trigonometric terms with unit frequency in $q_1(t), q_2(t), \ldots$ are all equal to zero.

Example 11.5. Obtain a periodic solution of Duffing's equation (Sec. 11.6) to a second-order approximation.

Using the notation of Eq. (11.86), the small part of the restoring force is

$$f(q, \omega q') = f(q) = -\omega_0^2 q^3 \tag{a}$$

so that from Eqs. (11.88), the perturbation equations through second order are

$$
\begin{aligned}
&q_0'' + q_0 = 0 \\
&q_1'' + q_1 = -q_0^3 - 2\frac{\omega_1}{\omega_0} q_0'' \\
&q_2'' + q_2 = -3q_0^2 q_1 - \left[2\frac{\omega_2}{\omega_0} + \left(\frac{\omega_1}{\omega_0}\right)^2 \right] q_0'' - 2\frac{\omega_1}{\omega_0} q_1''
\end{aligned} \tag{b}
$$

where we recall that q_0, q_1 and q_2 are functions of the dimensionless time τ; the solutions q_1 and q_2 are subject to the periodicity conditions (11.89). For simplicity, we consider the case in which the initial velocity is zero, which results in the conditions

$$q_0'(0) = q_1'(0) = q_2'(0) = 0 \tag{c}$$

If the initial velocity is not zero, then the solution must be modified by adding a phase angle to τ, a process that does not affect the nature of the solution.

In view of the first of the initial conditions (c), the solution of the first of Eqs. (b) is simply

$$q_0 = A_0 \cos\tau \tag{d}$$

Inserting solution (d) into the second of Eqs. (b) and recalling the trigonometric relation $\cos^3\tau = \frac{1}{4}(3\cos\tau + \cos 3\tau)$, we obtain

$$q_1'' + q_1 = \frac{A_0}{4\omega_0}(8\omega_1 - 3\omega_0 A_0^2)\cos\tau - \tfrac{1}{4}A_0^3\cos 3\tau \tag{e}$$

It is easy to see that the first term on the right side of Eq. (e) is likely to cause resonance, and hence to produce a secular term. To suppress this term, we invoke the periodicity condition, Eq. (11.89) corresponding to $i = 1$, which simply amounts to setting the coefficient of $\cos\tau$ in Eq. (e) to zero. This permits us to solve for ω_1 with the result

$$\omega_1 = \tfrac{3}{8}\omega_0 A_0^2 \tag{f}$$

Then, if we consider the second of initial conditions (c), the particular solution of Eq. (e) is simply

$$q_1 = \tfrac{1}{32}A_0^3\cos 3\tau \tag{g}$$

Next, we insert Eqs. (f) and (g) into the third of Eqs. (b) and write

$$q_2'' + q_2 = -\tfrac{3}{32}A_0^5\cos^2\tau\cos 3\tau + \left(2\frac{\omega_2}{\omega_0} + \frac{9}{64}A_0^4\right)A_0\cos\tau + \tfrac{27}{128}A_0^5\cos 3\tau \tag{h}$$

But, from trigonometry, $\cos^2\tau\cos 3\tau = \frac{1}{4}(\cos\tau + 2\cos 3\tau + \cos 5\tau)$, so that Eq. (h) reduces to

$$q_2'' + q_2 = \frac{A_0}{128\omega_0}(256\omega_2 + 15\omega_0 A_0^4)\cos\tau + \tfrac{21}{128}A_0^5\cos 3\tau - \tfrac{3}{128}A_0^5\cos 5\tau \tag{i}$$

Once again, to prevent the formation of secular terms, we must enforce the periodicity condition, Eq. (11.89) corresponding to $i = 2$. This amounts to equating the coefficient of $\cos\tau$ in Eq. (i) to zero, which yields

$$\omega_2 = -\tfrac{15}{256}\omega_0 A_0^4 \tag{j}$$

Then, the solution of Eq. (i) is simply

$$q_2 = -\tfrac{21}{1024}A_0^5\cos 3\tau + \tfrac{1}{1024}A_0^5\cos 5\tau \tag{k}$$

Hence, combining Eqs. (d), (g) and (k), the response to a second-order approximation is

$$\begin{aligned} q(t) &= A_0\cos\tau + \varepsilon\tfrac{1}{32}A_0^3\cos 3\tau - \varepsilon^2\tfrac{1}{1024}A_0^5(21\cos 3\tau - \cos 5\tau) \\ &= A_0\cos\omega t + \varepsilon\tfrac{1}{32}A_0^3\left(1 - \varepsilon\tfrac{21}{32}A_0^2\right)\cos 3\omega t + \varepsilon^2\tfrac{1}{1024}A_0^5\cos 5\omega t \end{aligned} \tag{l}$$

Moreover, inserting Eqs. (f) and (j) into Eq. (11.84), the fundamental frequency to a second-order approximation is

$$\omega = \omega_0\left(1 + \varepsilon\tfrac{3}{8}A_0^2 - \varepsilon^2\tfrac{15}{256}A_0^4\right) \tag{m}$$

Both the periodic response $q(t)$, Eq. (l), and the fundamental frequency ω, Eq. (m), depend on the initial displacement A_0 in the zero-order approximation, which differs from the actual initial displacement A. Hence, it appears desirable to express $q(t)$ and ω in terms of A rather than in terms of A_0. To this end, we let $t = 0$ in Eq. (l) and write

$$q(0) = A = A_0 + \varepsilon \tfrac{1}{32} A_0^3 \left(1 - \varepsilon \tfrac{21}{32} A_0^2\right) + \varepsilon^2 \tfrac{1}{1024} A_0^5$$
$$= A_0 + \varepsilon \tfrac{1}{32} A_0^3 - \varepsilon^2 \tfrac{5}{256} A_0^5 \tag{n}$$

Then, if we let

$$A_0 = A + \varepsilon A_1 + \varepsilon^2 A_2 \tag{o}$$

Eq. (n) can be rewritten as

$$A = A + \varepsilon A_1 + \varepsilon^2 A_2 + \varepsilon \tfrac{1}{32}(A + \varepsilon A_1 + \varepsilon^2 A_2)^3 - \varepsilon^2 \tfrac{5}{256}(A + \varepsilon A_1 + \varepsilon^2 A_2)^5$$
$$\cong A + \varepsilon(A_1 + \tfrac{1}{32} A^3) + \varepsilon^2 \left(A_2 + \tfrac{3}{32} A^2 A_1 - \tfrac{5}{256} A^5\right) \tag{p}$$

from which we conclude that

$$A_1 = -\tfrac{1}{32} A^3, \quad A_2 = \tfrac{23}{1024} A^5 \tag{q}$$

so that

$$A_0 = A - \varepsilon \tfrac{1}{32} A^3 + \varepsilon^2 \tfrac{23}{1024} A^5 \tag{r}$$

Hence, inserting Eq. (r) into Eq. (l), ignoring third-order terms and recalling the first of Eqs. (11.85), the response to the second-order approximation can be shown to be

$$q(t) = A \cos \omega t - \varepsilon \tfrac{1}{32} A^3 (\cos \omega t - \cos 3\omega t)$$
$$+ \varepsilon^2 \tfrac{1}{1024} A^5 (23 \cos \omega t - 24 \cos 3\omega t + \cos 5\omega t) \tag{s}$$

Similarly, inserting Eq. (r) into Eq. (m) and ignoring third-order terms, the fundamental frequency to the second-order approximation is

$$\omega = \omega_0 \left(1 + \varepsilon \tfrac{3}{8} A^2 - \varepsilon^2 \tfrac{21}{256} A^4\right) \tag{t}$$

Equation (t) justifies an earlier statement that the period of oscillation of nonlinear conservative systems depends on the initial conditions, as well as on the system parameters, in contrast with the period of linear conservative systems, which is not affected by the initial conditions.

11.8 FORCED OSCILLATION OF QUASI-HARMONIC SYSTEMS. JUMP PHENOMENON

We consider a quasi-harmonic system consisting of a mass and a nonlinear spring subjected to an external harmonic force. We assume that the restoring force in the spring is the sum of a linear term and a cubic term and write the differential equation of motion in the form

$$\ddot{q} + \omega^2 q = \varepsilon[-\omega^2(\alpha q + \beta q^3) + F \cos \Omega t], \; \varepsilon \ll 1 \tag{11.90}$$

in which ω is the natural frequency for $\varepsilon = 0$, α and β are given parameters, εF is the amplitude of the harmonic force and Ω is the driving frequency. Equation (11.90) is recognized as Duffing's equation (Secs. 11.6 and 11.7) with a small harmonic excitation.

Our interest lies in exploring the circumstances under which Eq. (11.90) admits a periodic solution of period $T = 2\pi/\Omega$. As in Sec. 11.7, it is convenient to change the time scale so that the period of oscillation becomes 2π. To this end, we introduce the change of variables

$$\Omega t = \tau + \phi, \ d/dt = \Omega d/d\tau \tag{11.91}$$

where τ is the new time variable and ϕ is a phase angle yet to be determined. In terms of the new time variable, Eq. (11.90) becomes

$$\Omega^2 q'' + \omega^2 q = \varepsilon[-\omega^2(\alpha q + \beta q^3) + F\cos(\tau + \phi)], \ \varepsilon \ll 1 \tag{11.92}$$

in which primes denote differentiations with respect to τ. To prevent the formation of secular terms, the solution of Eq. (11.92) must satisfy the periodicity condition

$$q(\tau + 2\pi) = q(\tau) \tag{11.93}$$

Moreover, for convenience, we choose the initial condition

$$q'(0) = 0 \tag{11.94}$$

In view of the fact that ε is a small parameter, a perturbation solution of Eq. (11.92) is advised. To this end, we expand $q(\tau)$ and ϕ in the power series

$$q(\tau) = q_0(\tau) + \varepsilon q_1(\tau) + \varepsilon^2 q_2(\tau) + \ldots \tag{11.95}$$

and

$$\phi = \phi_0 + \varepsilon\phi_1 + \varepsilon^2\phi_2 + \ldots \tag{11.96}$$

respectively, where $q_i(\tau)$ $(i = 0, 1, 2, \ldots)$ are subject to the periodicity conditions

$$q_i(\tau + 2\pi) = q_i(\tau), \ i = 0, 1, 2, \ldots \tag{11.97}$$

as well as the initial conditions

$$q_i'(0) = 0, \ i = 0, 1, 2, \ldots \tag{11.98}$$

Inserting Eqs. (11.95) and (11.96) into Eq. (11.92) and equating the coefficients of like powers of ε on both sides, we obtain the set of perturbation equations

$$\begin{aligned}
&\Omega^2 q_0'' + \omega^2 q_0 = 0 \\
&\Omega^2 q_1'' + \omega^2 q_1 = -\omega^2(\alpha q_0 + \beta q_0^3) + F\cos(\tau + \phi_0) \\
&\Omega^2 q_2'' + \omega^2 q_2 = -\omega^2(\alpha q_1 + 3\beta q_0^2 q_1) - F\phi_1 \sin(\tau + \phi_0) \\
&\cdots\cdots\cdots\cdots\cdots\cdots\cdots\cdots\cdots\cdots\cdots\cdots
\end{aligned} \tag{11.99}$$

Equations (11.99) are to be solved recursively for $q_i(\tau)$ $(i = 0, 1, 2, \ldots)$, subject to the periodicity conditions (11.97) and initial conditions (11.98).

In view of the initial condition, Eq. (11.98) corresponding to $i = 0$, the solution of the first of Eqs. (11.99) is simply

$$q_0(\tau) = A_0 \cos \frac{\omega \tau}{\Omega} \tag{11.100}$$

where A_0 is a constant amplitude. Solution (11.100) must satisfy the periodicity condition, Eq. (11.97) corresponding to $i = 0$, which is possible only if

$$\omega = \Omega \tag{11.101}$$

We assume that this is the case and replace Ω by ω wherever it appears. Introducing Eq. (11.100) in the second of Eqs. (11.99), dividing through by ω^2 and using the trigonometric relation $\cos^3 \tau = \frac{1}{4}(3\cos\tau + \cos 3\tau)$, we obtain

$$q_1'' + q_1 = -\frac{F}{\omega^2} \sin\phi_0 \sin\tau - (\alpha A_0 + \tfrac{3}{4}\beta A_0^3 - \frac{F}{\omega^2}\cos\phi_0)\cos\tau - \tfrac{1}{4}\beta A_0^3 \cos 3\tau \tag{11.102}$$

To satisfy the periodicity condition, Eq. (11.97) corresponding to $i = 1$, which amounts to preventing resonance (Sec. 3.2), we must set the coefficients of $\sin\tau$ and $\cos\tau$ equal to zero. This can be accomplished in two ways, namely,

$$\phi_0 = 0, \quad \alpha A_0 + \tfrac{3}{4}\beta A_0^3 - \frac{F}{\omega^2} = 0 \tag{11.103}$$

or

$$\phi_0 = \pi, \quad \alpha A_0 + \tfrac{3}{4}\beta A_0^3 + \frac{F}{\omega^2} = 0 \tag{11.104}$$

From Eqs. (11.103), we conclude that for $\phi_0 = 0$ the amplitude A_0 of the zero-order response q_0 and the amplitude F of the external force have the same sign, so that the zero-order response and the external force are in phase. Moreover, from Eqs. (11.104), we deduce that for $\phi_0 = \pi$ the amplitude of the zero-order response and the amplitude of the external force have opposite signs, so that the zero-order response and the external force are 180° out of phase. But, because a 180° out-of-phase response is equivalent to an in-phase response of negative amplitude, we conclude that Eqs. (11.104) do not yield any information that cannot be obtained from Eqs. (11.103). Hence, we will base further discussions on Eqs. (11.103), and we note that A_0 can be regarded as being fully determined by the second of these equations for any given value of F.

Considering Eqs. (11.103), as well as the initial condition, Eq. (11.98) corresponding to $i = 1$, the solution of Eq. (11.102) becomes

$$q_1(\tau) = A_1 \cos\tau + \tfrac{1}{32}\beta A_0^3 \cos 3\tau \tag{11.105}$$

where the constant A_1 is determined by requiring that the second-order perturbation $q_2(\tau)$ be periodic. Introducing Eqs. (11.100) and (11.105) in the third of Eqs. (11.99) and recalling from above that $\cos^3\tau = \frac{1}{4}(3\cos\tau + \cos 3\tau)$ and from Sec. 11.7 that $\cos^2\tau \cos 3\tau = \frac{1}{4}(\cos\tau + 2\cos 3\tau + \cos 5\tau)$, we obtain

$$q_2'' + q_2 = -\frac{F\phi_1}{\omega^2}\sin\tau - \left(\alpha A_1 + \tfrac{9}{4}\beta A_0^2 A_1 + \tfrac{3}{128}\beta^2 A_0^5\right)\cos\tau$$
$$- \tfrac{1}{4}\beta A_0^2 \left(3A_1 + \tfrac{1}{8}\alpha A_0 + \tfrac{3}{16}\beta A_0^3\right)\cos 3\tau - \tfrac{3}{128}\beta^2 A_0^5 \cos 5\tau \tag{11.106}$$

For $q_2(\tau)$ to be periodic, the coefficients of $\sin\tau$ and $\cos\tau$ must be zero, from which we conclude that

$$\phi_1 = 0, \quad A_1 = -\frac{3\beta^2 A_0^5}{32(4\alpha + 9\beta A_0^2)} \tag{11.107}$$

Hence, inserting Eqs. (11.107) into Eq. (11.106) and considering the initial condition, Eq. (11.98) corresponding to $i = 2$, the solution of Eq. (11.106) is simply

$$q_2(\tau) = A_2 \cos\tau + \tfrac{1}{256}\beta A_0^2(48A_1 + 2\alpha A_0 + 3\beta A_0^3)\cos 3\tau + \tfrac{1}{1024}\beta^2 A_0^5 \cos 5\tau \tag{11.108}$$

where A_2 is obtained by requiring that $q_3(\tau)$ be periodic.

Higher-order perturbations can be obtained by following the same pattern, but this is seldom necessary. Hence, using Eqs. (11.100), (11.105) and (11.108) and recalling the first of Eqs. (11.91), we can express the second-order approximation solution of Eq. (11.90) in the form

$$\begin{aligned} q(t) &\cong q_0(t) + \varepsilon q_1(t) + \varepsilon^2 q_2(t) \\ &= A_0 \cos\omega t + \varepsilon\left(A_1 \cos\omega t + \tfrac{1}{32}\beta A_0^3 \cos 3\omega t\right) \\ &\quad + \varepsilon^2\left[A_2 \cos\omega t + \tfrac{1}{256}\beta A_0^2(48A_1 + 2\alpha A_0 + 3\beta A_0^3)\cos 3\omega t + \tfrac{1}{1024}\beta^2 A_0^5 \cos 5\omega t\right] \\ &= (A_0 + \varepsilon A_1 + \varepsilon^2 A_2)\cos\omega t + \tfrac{1}{32}\varepsilon\beta A_0^2\left[A_0 + \tfrac{1}{16}\varepsilon(48A_1 + 2\alpha A_0 + 3\beta A_0^3\right]\cos 3\omega t \\ &\quad + \tfrac{1}{1024}\varepsilon^2\beta^2 A_0^5 \cos 5\omega t \end{aligned} \tag{11.109}$$

We observe that the phase angle ϕ is missing from Eq. (11.109). From the first of Eqs. (11.103) and (11.107), however, we conclude that to first-order approximation the phase angle is zero, $\phi = \phi_0 + \varepsilon\phi_1 = 0$. It turns out that the phase angle is zero to every order of approximation, a result that can be attributed to the fact that the system is undamped. When the system is viscously damped the phase angle is not zero, as we shall see later in this section.

Next, we return to Eqs. (11.103) and observe that the second of them represents a relation between the response amplitude and the excitation amplitude, with the driving frequency ω playing the role of a parameter. From Sec. 3.1, however, we recall that for linear systems the frequency response $G(i\omega)$ represents a relation between the response amplitude and the harmonic excitation amplitude. Hence, we can expect the second of Eqs. (11.103) to represent an analogous relation for nonlinear systems. This is indeed the case, and such an interpretation helps reveal a phenomenon typical of Duffing's equation, Eq. (11.90). To demonstrate this phenomenon, we introduce the notation

$$\omega_0^2 = (1 + \varepsilon\alpha)\omega^2 \tag{11.110}$$

where ω_0 can be identified as the natural frequency of the corresponding linearized system, obtained by letting $\beta = 0$ in Eq. (11.90). Using Eq. (11.110) to eliminate α from the second of Eqs. (11.103) and recalling that ε is small, so that second-order terms in ε

can be ignored, we obtain

$$\omega^2 = \omega_0^2 \left(1 + \tfrac{3}{4}\varepsilon\beta A_0^2\right) - \frac{\varepsilon F}{A_0} \tag{11.111}$$

Then, regarding $\varepsilon\beta$ as a given quantity, we can use Eq. (11.111) to plot A_0 versus ω with εF as a parameter and with ω measured in units of ω_0. We note that for $\beta = 0$ the plot A_0 versus ω has two branches, one above and one below the ω-axis, where both branches approach the vertical line $\omega = \omega_0$ asymptotically (see Fig. 11.21); the plot is analogous to the frequency response plot for an undamped linear system. The vertical line $\omega = \omega_0$ in Fig. 11.21 corresponds to the free-vibration case of the linearized system, $F = 0$, $\beta = 0$. When $\varepsilon\beta \neq 0$, but still a small quantity, the case $F = 0$ no longer represents the vertical line $\omega = \omega_0$ but a parabola intersecting the ω-axis at $\omega = \omega_0$. The plots A_0 versus ω corresponding to different values of εF consist of two branches, one above the parabola and one between the ω-axis and the lower half of the parabola, where both branches approach the parabola asymptotically, as shown in Fig. 11.21; the plots are for $\varepsilon\beta = 0.1$. Hence, the nonlinearity of the spring causes the asymptote $\omega = \omega_0$ to bend into a parabola. Moreover, the plots A_0 versus ω corresponding to different values of εF also bend and approach the parabola asymptotically. In the case of a hardening spring the parabola bends to the right, as shown in Fig. 11.21. The analogy with the linear oscillator becomes more evident when we plot $|A_0|$ versus ω, which amounts to replacing A_0 in Fig. 11.21 by its magnitude $|A_0|$. In fact, the plot $|A_0|$ versus ω can be obtained from the plot A_0 versus ω of Fig. 11.21 by folding the part below the ω-axis about this axis. The resulting $|A_0|$ versus ω plot is displayed in the Fig. 11.22a. The case

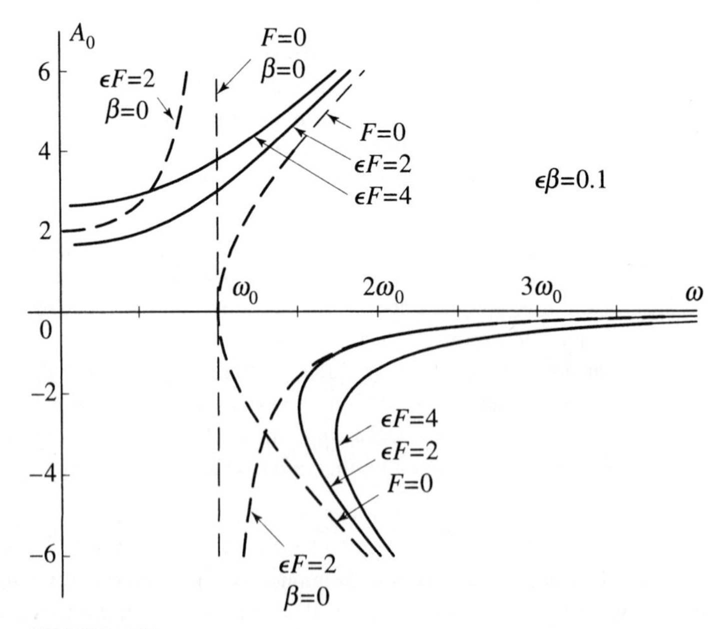

FIGURE 11.21
Frequency response plots for Duffing's equation with $\varepsilon\beta = 0.1$

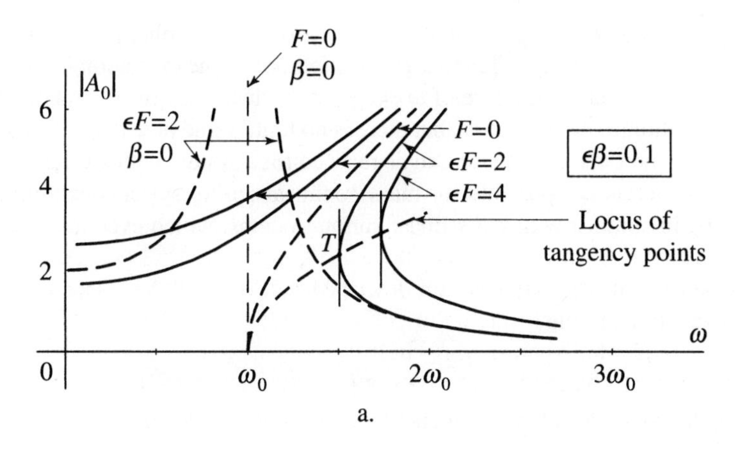

a.

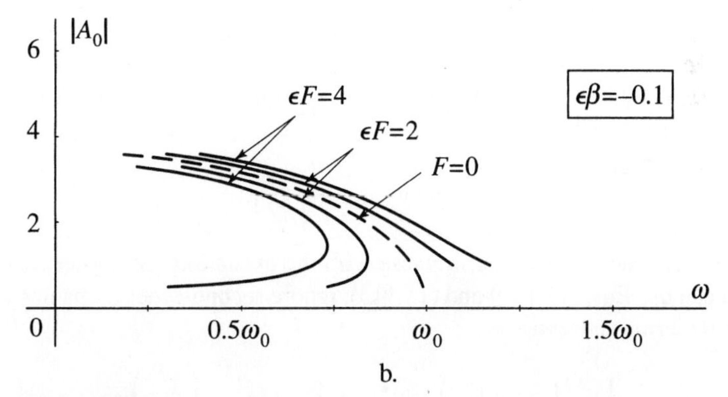

b.

FIGURE 11.22
Frequency response magnitude plots for Duffing's equation for the cases:
a. $\varepsilon\beta = 0.1$ and b. $\varepsilon\beta = -0.1$

of a softening spring simply corresponds to negative values of $\varepsilon\beta$. Plots $|A_0|$ versus ω for different values of εF and for $\varepsilon\beta = -0.1$ are shown in Fig. 11.22b. We observe that the curves bend to the left for a softening spring, $\varepsilon\beta < 0$, as opposed to bending to the right for a hardening spring, $\varepsilon\beta > 0$.

In contrast with linear systems, the mass-nonlinear spring system exhibits no resonance. To elaborate on this point, we consider again Fig. 11.22a and denote by T the point at which a vertical line is tangent to a given $|A_0|$ versus ω curve and by ω_T the corresponding frequency. From Fig. 11.22a we conclude that for a hardening spring such a tangency point can only exist on the right branch of the plot. A vertical line corresponding to any ω such that $\omega < \omega_T$ intersects only the left branch of the plot and only at one point. Hence, for $\omega < \omega_T$, Eq. (11.111) has only one real root A_0 and two complex roots. On the other hand, for $\omega > \omega_T$, Eq. (11.111) has three distinct real roots, one on the left branch and two on the right branch. It follows that, in a certain frequency range, the nonlinear theory predicts the existence of three distinct possibilities for the response amplitude corresponding to a given amplitude of the excitation force. The two roots on the right branch coalesce of $\omega = \omega_T$. As ω increases from a relatively small value, the amplitude $|A_0|$ increases, but there is no finite value of ω that renders $|A_0|$ infinitely large. The same conclusion can be reached for a system with a softening spring. Hence, resonance is not possible for mass-nonlinear spring systems exhibiting cubic nonlinearity, in contrast with mass-linear spring systems, which experience resonance at $\omega = \omega_0$.

In the case in which the system described by Eq. (11.90) is viscously damped, Duffing's equation has the form

$$\ddot{q} + \omega^2 q = \varepsilon[-2\zeta\omega\dot{q} - \omega^2(\alpha q + \beta q^3) + F\cos\Omega t], \ \varepsilon \ll 1 \tag{11.112}$$

Following the same procedure as for undamped systems, we conclude that q_1 is periodic provided the relations

$$\left(\alpha + \tfrac{3}{4}\beta A_0^2\right) A_0 - \frac{F}{\omega^2}\cos\phi_0 = 0, \ 2\zeta A_0 - \frac{F}{\omega^2}\sin\phi_0 = 0 \tag{11.113}$$

are satisfied. It follows from Eqs. (11.113) that, to the zero-order approximation, the phase angle has the value

$$\phi_0 = \tan^{-1}\frac{2\zeta}{\alpha + \frac{3}{4}\beta A_0^2} \tag{11.114}$$

so that the response is no longer in phase with the excitation. Moreover, because ε is small, we can use Eqs. (11.110) and (11.113), ignore second-order terms in ε compared to first-order terms and write

$$\left[\omega_0^2\left(1 + \tfrac{3}{4}\varepsilon\beta A_0^2\right) - \omega^2\right]^2 + (2\varepsilon\zeta\omega_0^2)^2 = \left(\frac{\varepsilon F}{A_0}\right)^2 \tag{11.115}$$

Equation (11.115) can be used to plot $|A_0|$ versus ω, which is shown in Fig. 11.23 for a damped system with a hardening spring with the values $\varepsilon F = 6$, $\varepsilon\beta = 0.2$. It is easy to see from Fig. 11.23 that, in the presence of damping, the amplitude magnitude $|A_0|$ does not increase indefinitely with the driving frequency ω. Although the plot $|A_0|$ versus ω is now continuous, in the sense that it no longer consists of two branches, the possibility of discontinuities in the response remains. Indeed, as the driving frequency ω is increased from a relatively small value, the amplitude $|A_0|$ increases until it reaches point 1, at which point the tangent to the curve $|A_0|$ versus ω is vertical and the amplitude experiences a sudden "jump" down to point 2 on the lower limb of the response curve. From that point on, it decreases with an increase in the driving frequency, approaching zero asymptotically. On the other hand, if the driving frequency ω is decreased from a relatively large value, the amplitude $|A_0|$ increases until it reaches point 3, at which point the tangent to the curve $|A_0|$ versus ω is again vertical and the amplitude jumps up to point 4 on the upper limb, from which point it decreases with a decrease in the frequency. The portion of the response curve between 1 and 3 is never traversed and must be regarded as unstable. Whether the system traverses the arc between 4 and 1 or that between 2 and 3 depends on the limb on which the system moves just prior to entering

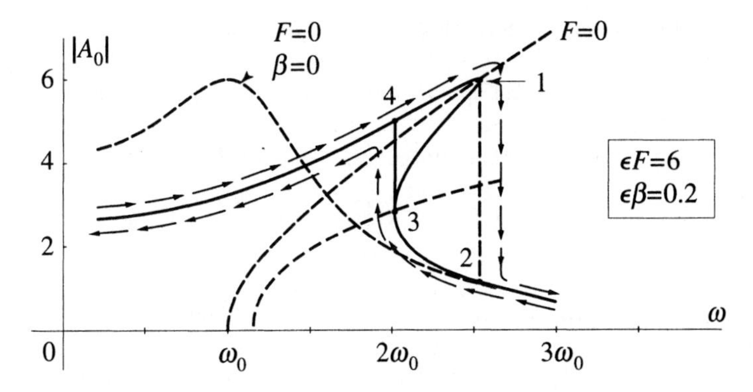

FIGURE 11.23
Frequency response magnitude plot demonstrating the jump phenomenon for Duffing's equation

one of the two arcs, as the jump takes place after one of these two arcs is traversed. Whereas the jump from 3 up to 4 can take place also for undamped systems, the jump from 1 down to 2 has no counterpart in undamped systems. The jump phenomenon can also occur for damped systems with a softening spring, in which cases the jump in amplitude takes place in reverse directions.

11.9 SUBHARMONICS AND COMBINATION HARMONICS

The response of a linear oscillator to a harmonic force is harmonic and has the same frequency as the excitation. On the other hand, in Sec. 11.8, we demonstrated that a mass-nonlinear spring system of the type described by Duffing's equation and subjected to a harmonic force is capable of periodic response if the fundamental frequency of the response is equal to the driving frequency.

The undamped Duffing equation possesses another periodic solution with the fundamental frequency equal to one third of the driving frequency. To substantiate this statement, we consider the equation

$$\ddot{q}+\omega^2 q=-\varepsilon\omega^2(\alpha q+\beta q^3)+F\cos\Omega t,\ \varepsilon\ll 1 \tag{11.116}$$

Equation (11.116) has the same form as Eq. (11.90), with the exception that the force amplitude is not necessarily small. Because the nonlinearity is due to the cubic term in q, we wish to explore the possibility of a periodic solution of Eq. (11.116) with the fundamental frequency $\omega=\Omega/3$. Hence, letting $\omega=\Omega/3$ in Eq. (11.116), assuming a solution of the form

$$q(t)=q_0(t)+\varepsilon q_1(t)+\varepsilon^2 q_2(t)+\ldots \tag{11.117}$$

and equating coefficients of like powers of ε on both sides of the resulting equation, we obtain the set of equations of increasing order of approximation

$$\begin{aligned}
&\ddot{q}_0 + \left(\tfrac{1}{3}\Omega\right)^2 q_0 = F\cos\Omega t \\
&\ddot{q}_1 + \left(\tfrac{1}{3}\Omega\right)^2 q_1 = -\left(\tfrac{1}{3}\Omega\right)^2 \left(\alpha q_0 + \beta q_0^3\right) \\
&\ddot{q}_2 + \left(\tfrac{1}{3}\Omega\right)^2 q_2 = -\left(\tfrac{1}{3}\Omega\right)^2 \left(\alpha q_1 + 3\beta q_0^2 q_1\right) \\
&\cdots\cdots\cdots\cdots\cdots\cdots\cdots\cdots\cdots\cdots
\end{aligned} \tag{11.118}$$

Equations (11.118) are to be solved recursively, where the solutions $q_i(t)$ $(i = 0, 1, 2, \ldots)$ are subject to the periodicity conditions

$$q_i\left(\tfrac{1}{3}\Omega t + 2\pi\right) = q_i\left(\tfrac{1}{3}\Omega t\right), \; i = 0, 1, 2, \ldots \tag{11.119}$$

and the initial conditions

$$\dot{q}_i(0) = 0, \; i = 0, 1, 2, \ldots \tag{11.120}$$

Taking into account the appropriate initial condition, the solution of the first of Eqs. (11.118) is simply

$$q_0(t) = A_0 \cos\tfrac{1}{3}\Omega t - \frac{9F}{8\Omega^2}\cos\Omega t \tag{11.121}$$

where A_0 is a constant amplitude to be determined so as to ensure a periodic $q_1(t)$. Introducing solution (11.121) in the second of Eqs. (11.118) and using the trigonometric relation $\cos a \cos b = \frac{1}{2}[\cos(a+b) + \cos(a-b)]$, we obtain

$$\begin{aligned}
\ddot{q}_1 + \left(\tfrac{1}{3}\Omega\right)^2 q_1 = -\left(\tfrac{1}{3}\Omega\right)^2 \Bigg\{ & A_0\left[\alpha + \tfrac{3}{4}\beta A_0^2 - \tfrac{3}{4}\beta A_0 \frac{9F}{8\Omega^2} + \tfrac{3}{2}\left(\frac{9F}{8\Omega^2}\right)^2\right]\cos\tfrac{1}{3}\Omega t \\
& - \left[\alpha\frac{9F}{8\Omega^2} - \tfrac{1}{4}\beta A_0^3 + \beta A_0^2 \frac{9F}{8\Omega^2} + \tfrac{3}{4}\beta\left(\frac{9F}{8\Omega^2}\right)^2\right]\cos\Omega t \\
& - \tfrac{3}{4}\beta A_0 \frac{9F}{8\Omega^2}\left(A_0 - \frac{9F}{8\Omega^2}\right)\cos\tfrac{5}{3}\Omega t \\
& + \tfrac{3}{4}\beta A_0 \left(\frac{9F}{8\Omega^2}\right)^2 \cos\tfrac{7}{3}\Omega t - \tfrac{1}{4}\beta\left(\frac{9F}{8\Omega^2}\right)^3 \cos 3\Omega t \Bigg\}
\end{aligned} \tag{11.122}$$

To ensure a periodic solution of Eq. (11.122), we must prevent the formation of secular terms. To this end, we set the coefficient of $\cos\frac{1}{3}\Omega t$ equal to zero, which yields the quadratic equation in A_0

$$A_0^2 - \frac{9F}{8\Omega^2}A_0 + 2\left(\frac{9F}{8\Omega^2}\right)^2 + \frac{4\alpha}{3\beta} = 0 \tag{11.123}$$

having the roots

$$A_0 = \frac{1}{2}\frac{9F}{8\Omega^2} \pm \frac{1}{2}\sqrt{\left(\frac{9F}{8\Omega^2}\right)^2 - 8\left(\frac{9F}{8\Omega^2}\right)^2 - \frac{16\alpha}{3\beta}}$$
$$= \frac{1}{2}\frac{9F}{8\Omega^2} \pm \frac{1}{2}\sqrt{-7\left(\frac{9F}{8\Omega^2}\right)^2 - \frac{16\alpha}{3\beta}} \quad (11.124)$$

Because A_0 is a real quantity by definition, a periodic solution of Eq. (11.122) with the fundamental frequency $\Omega/3$ is possible only if the expression under the radical is nonnegative, i.e., either positive or zero. To explore this possibility, we let $\omega = \Omega/3$ in Eq. (11.110) and obtain the relation

$$\Omega^2 = \frac{9}{\varepsilon\alpha}\left(\omega_0^2 - \tfrac{1}{9}\Omega^2\right) \quad (11.125)$$

where, as in Sec. 11.8, ω_0 is the natural frequency of the associated linearized system, obtained by setting $\beta = 0$ in Eq. (11.116). In view of Eq. (11.125), the expression under the radical can be shown to be nonnegative provided

$$\Omega^2 \geq 9\left[\omega_0^2 + \tfrac{21}{16}\varepsilon\beta\left(\frac{3F}{8\Omega}\right)^2\right] \quad (11.126)$$

Periodic oscillations with fundamental frequency equal to a fraction of the driving frequency are known as *subharmonic oscillations*. Hence, if inequality (11.126) is satisfied, the undamped Duffing equation, Eq. (11.116), admits a subharmonic solution with the fundamental frequency equal to $\Omega/3$. The subharmonic is said to be of order 3, and we note that the order of the subharmonic coincides with the power of the nonlinear term in the restoring force in the spring.

When a linear harmonic oscillator is subjected to two harmonic forces with distinct frequencies, say Ω_1 and Ω_2, the response is a superposition of two harmonic components with frequencies Ω_1 and Ω_2. On the other hand, when a mass-nonlinear spring system is excited by a combination of two harmonic forces with distinct frequencies Ω_1 and Ω_2, the response is a superposition of harmonic components with frequencies in the form of integer multiples of Ω_1 and Ω_2, as well as of harmonic components with frequencies equal to linear combinations of Ω_1 and Ω_2, where the type of harmonics obtained depends on the power of the nonlinear term. To substantiate this statement, we consider the slightly different form of Duffing's equation

$$\ddot{q} + \omega^2 q = -\varepsilon\beta_0 q^3 + F_1\cos\Omega_1 t + F_2\cos\Omega_2 t, \ \varepsilon \ll 1 \quad (11.127)$$

which differs from Eq. (11.116) only to the extent that $\alpha = 0$ and $\beta_0 = \beta\omega^2 = \beta\omega_0^2$. Of course, now the excitation consists of the sum of two harmonic forces instead of one harmonic force. Assuming a solution in the form of Eq. (11.117) and separating orders of magnitude, we obtain the set of equations

$$\begin{aligned}
\ddot{q}_0 + \omega_0^2 q_0 &= F_1 \cos \Omega_1 t + F_2 \cos \Omega_2 t \\
\ddot{q}_1 + \omega_0^2 q_1 &= -\beta_0 q_0^3 \\
\ddot{q}_2 + \omega_0^2 q_2 &= -3\beta_0 q_0^2 q_1 \\
&\cdots\cdots\cdots\cdots\cdots\cdots
\end{aligned} \tag{11.128}$$

which is to be solved recursively. For convenience, we require that the solutions $q_i(t)$ $(i = 0, 1, 2, \ldots)$ satisfy the initial conditions (11.120). Because our interest is in demonstrating the existence of harmonic solutions with frequencies equal to integer multiples of Ω_1 and Ω_2, as well as linear combinations of Ω_1 and Ω_2, it is possible to ignore the homogeneous solutions.

The solution of the first of Eqs. (11.128) is simply

$$q_0(t) = G_1 \cos \Omega_1 t + G_2 \cos \Omega_2 t \tag{11.129}$$

in which

$$G_1 = \frac{F_1}{\omega_0^2 - \Omega_1^2}, \quad G_2 = \frac{F_2}{\omega_0^2 - \Omega_2^2} \tag{11.130}$$

so that Eq. (11.129) represents the steady-state response of a harmonic oscillator to a superposition of two harmonic forces. Inserting Eq. (11.129) into the second of Eqs. (11.128) and recalling the trigonometric relation $\cos a \cos b = \frac{1}{2}[\cos(a+b) + \cos(a-b)]$, we can write

$$\begin{aligned}
\ddot{q}_1 + \omega_0^2 q_1 = {} & H_1 \cos \Omega_1 t + H_2 \cos \Omega_2 t + H_3[\cos(2\Omega_1 + \Omega_2)t + \cos(2\Omega_1 - \Omega_2)t] \\
& + H_4[\cos(\Omega_1 + 2\Omega_2)t + \cos(\Omega_1 - 2\Omega_2)t] + H_5 \cos 3\Omega_1 t + H_6 \cos 3\Omega_2 t
\end{aligned} \tag{11.131}$$

in which

$$\begin{aligned}
& H_1 = -\tfrac{3}{4}\beta_0 G_1(G_1^2 + 2G_2^2), \; H_2 = -\tfrac{3}{4}\beta_0 G_2(2G_1^2 + G_2^2), \; H_3 = -\tfrac{3}{4}\beta_0 G_1^2 G_2 \\
& H_4 = -\tfrac{3}{4}\beta_0 G_1 G_2^2, \; H_5 = -\tfrac{1}{4}\beta_0 G_1^3, \; H_6 = -\tfrac{1}{4}\beta_0 G_2^3
\end{aligned} \tag{11.132}$$

It is evident from the nature of the excitation in Eq. (11.131) that the response $q_1(t)$, i.e., the solution of Eq. (11.131), consists of a linear combination of harmonic components with frequencies equal to $\Omega_1, \Omega_2, 2\Omega_1 \pm \Omega_2, \Omega_1 \pm 2\Omega_2, 3\Omega_1$ and $3\Omega_2$. Hence, in contrast with linear systems, the response of the mass-nonlinear spring system described by Eq. (11.127) consists not only of harmonic components with frequencies Ω_1 and Ω_2 but also of harmonic components with the higher frequencies $3\Omega_1$ and $3\Omega_2$, as well as the frequencies $2\Omega_1 \pm \Omega_2$ and $\Omega_1 \pm 2\Omega_2$, where the latter are known as *combination harmonics*. Because the terms involving higher harmonics and combination harmonics appear only in the first-order perturbation $q_1(t)$ and not in the zero-order solution $q_0(t)$, they tend to be one order of magnitude smaller than the terms with frequencies equal to the driving frequencies Ω_1 and Ω_2. However, when one of the frequencies

$2\Omega_1 \pm \Omega_2, \Omega_1 \pm 2\Omega_2$, $3\Omega_1$ and $3\Omega_2$ is close in value to ω_0 higher amplitudes can be expected.

It should be pointed out that the frequencies $2\Omega_1 \pm \Omega_2, \Omega_1 \pm 2\Omega_2$, $3\Omega_1$ and $3\Omega_2$ are peculiar to Eq. (11.127), because the nonlinear term is cubic in q. For systems with other than cubic nonlinearity, harmonic components in the response with different higher frequencies and combination frequencies are obtained.

11.10 SYSTEMS WITH TIME-DEPENDENT COEFFICIENTS. MATHIEU'S EQUATION

Under certain circumstances, the differential equation describing the vibration of a system contains time-dependent coefficients. Even when the system is linear, time-dependent coefficients tend to cause great difficulties, as the methods of solution commonly used for systems with constant coefficient no longer work. The situation is considerably better when the terms involving the time-dependent coefficients are relatively small, because this opens the possibility for a perturbation solution.

We consider a pendulum whose support is acted upon by a vertical force, and denote by θ the angular displacement of the pendulum, by u the vertical displacement of the support and by F the vertical force, as shown in Fig. 11.24. The interest lies in the case in which the support executes harmonic motion of the form

$$u = A\cos\omega t \tag{11.133}$$

We derive the equations of motion by first regarding θ and u as unknowns and then use

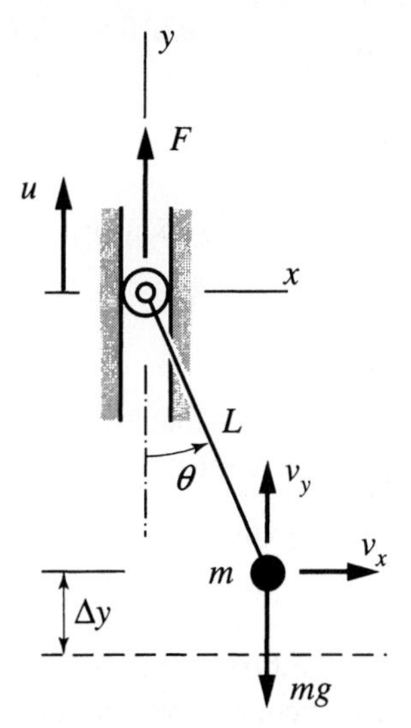

FIGURE 11.24
Pendulum with vertically moving support

Eq. (11.133) to eliminate u. It is convenient to derive the equations of motion by means of Lagrange's equations, which have the general form

$$\frac{d}{dt}\left(\frac{\partial T}{\partial \dot{\theta}}\right) - \frac{\partial T}{\partial \theta} + \frac{\partial V}{\partial \theta} = \Theta$$
$$\frac{d}{dt}\left(\frac{\partial T}{\partial \dot{u}}\right) - \frac{\partial T}{\partial u} + \frac{\partial V}{\partial u} = U \tag{11.134}$$

where T is the kinetic energy, V is the potential energy and Θ and U are generalized nonconservative forces. With reference to Fig. 11.24, the kinetic energy can be shown to be

$$T = \frac{1}{2}m(v_x^2 + v_y^2) = \frac{1}{2}m[(L\dot{\theta}\cos\theta)^2 + (\dot{u} + L\dot{\theta}\sin\theta)^2] = \frac{1}{2}m(L^2\dot{\theta}^2 + 2L\dot{u}\dot{\theta}\sin\theta + \dot{u}^2) \tag{11.135}$$

and the potential energy is

$$V = mg\Delta y = mg[L(1-\cos\theta) + u] \tag{11.136}$$

Moreover, the virtual work of the nonconservative forces is

$$\overline{\delta W} = \Theta\delta\theta + U\delta u = F\delta u \tag{11.137}$$

so that the generalized nonconservative forces are

$$\Theta = 0,\ U = F \tag{11.138}$$

Next, we write

$$\frac{\partial T}{\partial \dot{\theta}} = mL^2\dot{\theta} + mL\dot{u}\sin\theta,\quad \frac{d}{dt}\left(\frac{\partial T}{\partial \dot{\theta}}\right) = mL^2\ddot{\theta} + mL(\ddot{u}\sin\theta + \dot{u}\dot{\theta}\cos\theta)$$
$$\frac{\partial T}{\partial \theta} = mL\dot{u}\dot{\theta}\cos\theta,\quad \frac{\partial V}{\partial \theta} = mgL\sin\theta$$
$$\frac{\partial T}{\partial \dot{u}} = m(L\dot{\theta}\sin\theta + \dot{u}),\quad \frac{d}{dt}\left(\frac{\partial T}{\partial \dot{u}}\right) = mL(\ddot{\theta}\sin\theta + \dot{\theta}^2\cos\theta) + m\ddot{u} \tag{11.139}$$
$$\frac{\partial T}{\partial u} = 0,\quad \frac{\partial V}{\partial u} = mg$$

Hence, inserting Eqs. (11.138) and (11.139) into Eqs. (11.134) and canceling appropriate terms, we obtain the explicit Lagrange's equations

$$mL^2\ddot{\theta} + mL\ddot{u}\sin\theta + mgL\sin\theta = 0$$
$$mL(\ddot{\theta}\sin\theta + \dot{\theta}^2\cos\theta) + m\ddot{u} + mg = F \tag{11.140}$$

Inserting Eq. (11.133) into the first of Eqs. (11.140), we obtain a nonlinear differential equation with one time-dependent coefficient, which can be solved for $\theta(t)$, at least in theory. Then, introducing $\theta(t)$ thus obtained and Eq. (11.133) in the second of Eqs. (11.140), we obtain the force F necessary for generating the harmonic motion u of the support. In practice, the problem is intractable for arbitrarily large $\theta(t)$. The situation is considerably better when the angular motion is confined to a small neighborhood of $\theta = 0$. In this case, Eqs. (11.140) reduce to the linearized form

$$\begin{aligned} mL^2\ddot{\theta} + mL(g+\ddot{u})\theta &= 0 \\ m\ddot{u} + mg &= F \end{aligned} \tag{11.141}$$

The first of Eqs. (11.141) may appear nonlinear due to the product $\ddot{u}\theta$, but it is not, because u is a known quantity. Indeed, introducing Eq. (11.133) in the first of Eqs. (11.141), we obtain

$$\ddot{\theta} + \left(\frac{g}{L} - \frac{A\omega^2}{L}\cos\omega t\right)\theta = 0 \tag{11.142}$$

Although Eq. (11.142) is linear, it is by no means a simple equation, because it has a coefficient varying harmonically with time. The equation is known in mathematical physics as *Mathieu's equation*, with the pendulum with a harmonically moving support representing merely one example of many systems described by the equation. Moreover, from the second of Eqs. (11.141), we obtain the force producing the harmonic motion of the support by writing simply

$$F = m(g - A\omega^2\cos\omega t) \tag{11.143}$$

The interest lies not so much in the response of the harmonically excited pendulum described by Mathieu's equation, Eq. (11.142), as in the stability of the system. Of course, if the system is unstable, the angle θ will not remain for long in the small neighborhood of $\theta = 0$, so that Eq. (11.142) will soon cease to be valid. When the support is fixed, $A = 0$, Eq. (11.142) reduces to that of a simple harmonic oscillator, whose motion is known to be stable in the neighborhood of $\theta = 0$. On the other hand, for $A \neq 0$ it is possible to render the position $\theta = 0$ unstable by exciting the support. By contrast, for $A = 0$, the upright position, $\theta = \pi$, is known to be unstable. Under certain circumstances, the same upright position can be stabilized by harmonic excitation of the support. To induce the behavior just described, the excitation force F need not be very large. In view of this, there is some advantage in assuming that the excitation is relatively small, as this permits a perturbation solution. To this end, it is convenient to introduce the notation

$$\frac{g}{L} = \delta, \quad -\frac{A\omega^2}{L} = 2\varepsilon \tag{11.144}$$

Moreover, it is customary to let $\omega = 2\,\mathrm{rad/s}$, so that Eq. (11.142) reduces to the standard form of Mathieu's equation

$$\ddot{\theta} + (\delta + 2\varepsilon\cos 2t)\theta = 0 \tag{11.145}$$

where in the case at hand, $\varepsilon \ll 1$. Equation (11.145) represents a quasi-harmonic system. The stability characteristics of Eq. (11.145) can be studied conveniently by means of the parameter plane δ, ε. The plane is divided into regions of stability and instability by the so-called *boundary curves*, or *transition curves*. These transition curves, separating the stability regions from the instability regions, are such that a point belonging to any of these curves is characterized by a periodic solution of Eq. (11.145). But, from Sec. 11.7, we can obtain periodic solutions of a quasi-harmonic system by means of Lindstedt's method. To this end, we assume a solution of Eq. (11.145) in the form

$$\theta(t) = \theta_0(t) + \varepsilon\theta_1(t) + \varepsilon^2\theta_2(t) + \dots \tag{11.146}$$

Moreover, we assume that

$$\delta = n^2 + \varepsilon\delta_1 + \varepsilon^2\delta_2 + \dots, \; n = 0, 1, 2, \dots \tag{11.147}$$

with the implication that δ differs from an integer squared by a small quantity. Inserting Eqs. (11.146) and (11.147) into Eq. (11.145) and equating coefficients of like powers of ε, we obtain the sets of equations

$$\begin{aligned} &\ddot{\theta}_0 + n^2\theta_0 = 0 \\ &\ddot{\theta}_1 + n^2\theta_1 = -(\delta_1 + 2\cos 2t)\theta_0 \\ &\ddot{\theta}_2 + n^2\theta_2 = -(\delta_1 + 2\cos 2t)\theta_1 - \delta_2\theta_0 \\ &\dots\dots\dots\dots\dots\dots\dots\dots\dots \end{aligned} \qquad n = 0, 1, 2, \dots \tag{11.148}$$

one set for every n. Equations (11.148) must be solved recursively for the various values of n $(n = 0, 1, 2, \dots)$. From the first of Eqs. (11.148), the zero-order approximation is given by

$$\theta_0 = \begin{cases} \cos nt, \\ \sin nt, \end{cases} \qquad n = 0, 1, 2, \dots \tag{11.149}$$

The transition curves are obtained by introducing solutions $\theta_0 = \cos nt$ and $\theta_0 = \sin nt$ $(n = 0, 1, 2, \dots)$ in Eqs. (11.148) and insisting that the solutions $\theta_i(t)$ $(i = 1, 2, \dots)$ be periodic. Equations (11.148) yield an infinite number of solution pairs, one pair for every value of n, with the exception of the case $n = 0$ for which there is only one solution.

Considering first the case $n = 0$, in which case $\theta_0 = 1$, the second of Eqs. (11.148) reduces to

$$\ddot{\theta}_1 = -\delta_1 - 2\cos 2t \tag{11.150}$$

For θ_1 to be periodic, δ_1 must be equal to zero, in which case the solution of Eq. (11.150) is simply

$$\theta_1 = \tfrac{1}{2}\cos 2t \tag{11.151}$$

In view of Eq. (11.151), the third of Eqs. (11.148) becomes

$$\ddot{\theta}_2 = (-2\cos 2t)\left(\tfrac{1}{2}\cos 2t\right) - \delta_2 = -\left(\tfrac{1}{2} + \delta_2\right) - \tfrac{1}{2}\cos 4t \tag{11.152}$$

in which we used the trigonometric relation $\cos^2 2t = \frac{1}{2}(1 + \cos 4t)$. For θ_2 to be periodic, the constant term on the right side of Eq. (11.152) must be equal to zero, which yields $\delta_2 = -1/2$. Hence, corresponding to $n = 0$ there is only one transition curve, namely,

$$\delta = -\tfrac{1}{2}\varepsilon^2 + \dots \tag{11.153}$$

which, to a second-order approximation, is a parabola passing through the origin of the parameter plane δ, ε.

Next, we consider the case $n = 1$, in which case, from Eqs. (11.149), there are two zero-order solutions, $\theta_0 = \cos t$ and $\theta_0 = \sin t$. Corresponding to $\theta_0 = \cos t$, the second of Eqs. (11.148) becomes

$$\ddot{\theta}_1 + \theta_1 = -(\delta_1 + 2\cos 2t)\cos t = -(\delta_1 + 1)\cos t - \cos 3t \tag{11.154}$$

where we used the relation $2\cos 2t \cos t = \cos 3t + \cos t$. To prevent resonance, and hence the formation of secular terms in θ_1, we must set $\delta_1 = -1$, from which it follows that the solution of Eq. (11.154) is

$$\theta_1 = \tfrac{1}{8}\cos 3t \tag{11.155}$$

Inserting θ_0, θ_1 and δ_1 into the third of Eqs. (11.148) corresponding to $n = 1$ and considering the relation $2\cos 2t \cos 3t = \cos 5t + \cos t$, we obtain

$$\begin{aligned}\ddot{\theta}_2 + \theta_2 &= -\tfrac{1}{8}(-1 + 2\cos 2t)\cos 3t - \delta_2 \cos t \\ &= -\left(\tfrac{1}{8} + \delta_2\right)\cos t + \tfrac{1}{8}\cos 3t - \tfrac{1}{8}\cos 5t\end{aligned} \tag{11.156}$$

Using the same argument as with θ_1, for θ_2 to be periodic, the coefficient of $\cos t$ must be zero, which yields $\delta_2 = -1/8$. Hence, using Eq. (11.147), the transition curve corresponding to $\theta_0 = \cos t$ is

$$\delta = 1 - \varepsilon - \tfrac{1}{8}\varepsilon^2 + \ldots \tag{11.157}$$

Corresponding to $\theta_0 = \sin t$, the second of Eqs. (11.148) becomes

$$\ddot{\theta}_1 + \theta_1 = -(\delta_1 + 2\cos 2t)\sin t = -(\delta_1 - 1)\sin t - \sin 3t \tag{11.158}$$

where we used the relation $2\cos 2t \sin t = \sin 3t - \sin t$. The solution of the Eq. (11.158) is periodic provided $\delta_1 = 1$, and has the form

$$\theta_1 = \tfrac{1}{8}\sin 3t \tag{11.159}$$

Inserting Eq. (11.159) into the third of Eqs. (11.148) corresponding to $n = 1$ and using the relation $2\cos 2t \sin 3t = \sin 5t + \sin t$, we obtain

$$\begin{aligned}\ddot{\theta}_2 + \theta_2 &= -\tfrac{1}{8}(1 + 2\cos 2t)\sin 3t - \delta_2 \sin t \\ &= -\left(\tfrac{1}{8} + \delta_2\right)\sin t - \tfrac{1}{8}\sin 3t + \tfrac{1}{8}\sin 5t\end{aligned} \tag{11.160}$$

so that, for θ_2 to be periodic, we must have $\delta_2 = -1/8$. Hence, the transition curve corresponding to $\theta_0 = \sin t$ is

$$\delta = 1 + \varepsilon - \tfrac{1}{8}\varepsilon^2 + \ldots \tag{11.161}$$

Following the same pattern, it can be shown that the transition curve corresponding to $n = 2$ and $\theta_0 = \cos 2t$ is

$$\delta = 4 + \tfrac{5}{12}\varepsilon^2 + \ldots \tag{11.162}$$

and that corresponding to $n = 2$ and $\theta_0 = \sin 2t$ is

$$\delta = 4 - \tfrac{1}{12}\varepsilon^2 + \ldots \tag{11.163}$$

Transition curves for $n = 3, 4\ldots$ can be obtained in a similar fashion.

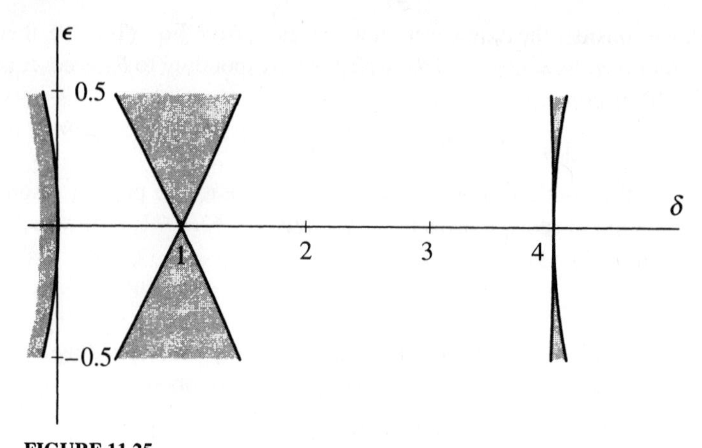

FIGURE 11.25
Stability and instability regions for a pendulum with harmonically moving support

The transition curves defined by Eqs. (11.153), (11.157), (11.161), (11.162) and (11.163) can be used to define regions of stability and instability in the parameter plane δ, ε. They are shown in Fig. 11.25, with the shaded areas defining the regions of instability. The region terminating at $\delta = 1,\ \varepsilon = 0$ is known as the *principal instability region*, and is appreciably wider than the regions terminating at $\delta = n^2,\ \varepsilon = 0\ (n = 2, 3, \ldots)$, as the latter become progressively narrower as n increases. Although we treated ε as a positive quantity, no such restriction was actually placed on ε, and the results are equally valid for negative values of ε. In fact, the transition curves of Fig. 11.25 have been plotted for both positive and negative values of ε. As a matter of interest, it should be mentioned that Fig. 11.25 is commonly known as a *Strutt diagram*.

We observe from Fig. 11.25 that stability is possible also for negative values of δ, which corresponds to the upright equilibrium position, $\theta = 180°$. Although this region is small, for the right choice of parameters, the pendulum can be stabilized in the upright position by moving the support harmonically. Perhaps more important is the fact that the harmonic motion of the support can destabilize the equilibrium position $\theta = 0$.

11.11 NUMERICAL INTEGRATION OF THE EQUATIONS OF MOTION. THE RUNGE-KUTTA METHODS

In Chs. 5–7, we discussed techniques for deriving the response of linear systems with constant coefficients to initial excitations and external forces. Of course, the fact that the principle of superposition holds for such systems makes the derivation of the response almost routine. The situation is entirely different in the case of nonlinear systems, for which the principle of superposition does not hold. In fact, for the most part, analytical solutions for the response of nonlinear systems do not exist. For this reason, in the earlier part of this chapter, we confined ourselves to a qualitative determination of the stability characteristics in the neighborhood of equilibrium points. Then, later in this chapter, we considered a perturbation approach to the response of weakly nonlinear systems, i.e., systems for which the nonlinearity is sufficiently small that it can be regarded as a higher-order effect. In this section, we finally address the response of nonlinear systems for which the nonlinearity is not necessarily small. Evaluation of the response of arbitrary nonlinear systems almost invariably involves some kind of numerical integration, which requires that we convert the time into a discrete variable, not unlike the discrete-time process used in Chs. 4, 5 and 7 to derive the response of linear systems. Moreover, we recall that for the most part we obtained the response using first-order state equations. Hence, it should come as no surprise that numerical integration for nonlinear systems as well is carried out most conveniently in terms of first-order state equations.

Numerical integration provides only an approximate solution whose accuracy depends on the order of approximation, among other things. We consider here solutions by the *Runge-Kutta methods*, a family of algorithms with members characterized by different orders of approximation, where the order is related to the number of terms retained in a truncated Taylor series expansion of the nonlinear force. Derivation of higher-order Runge-Kutta algorithms is very tedious, and serves no useful purpose. To develop a feel for the approach, we derive the second-order Runge-Kutta method and merely list the equations for the widely used fourth-order Runge-Kutta method, as well as for an efficient modified version known as the Runge-Kutta-Fehlberg method. To introduce the ideas, we consider a first-order nonlinear system described by the differential equation

$$\dot{x}(t) = f[x(t), t] \tag{11.164}$$

where f is a nonlinear function of $x(t)$ and t. Actually, in most application f does not involve the time t explicitly, but only implicitly through $x(t)$. The solution of Eq. (11.164) can be expanded in the Taylor series

$$x(t+T) = x(t) + Tx^{(1)}(t) + \frac{T^2}{2!}x^{(2)}(t) + \frac{T^3}{3!}x^{(3)}(t) + \ldots \tag{11.165}$$

in which T is a small time increment and the superscript (i) denotes the ith derivative with respect to time $(i = 1, 2, \ldots)$. For numerical computation, we must limit the number of terms in series (11.165), which amounts to using a finite series to approximate $x(t+T)$. Assuming that the solution $x(t)$ has $N+1$ continuous derivatives, we can rewrite Eq. (11.165) in terms of an Nth-degree Taylor polynomial about t as follows:

$$x(t+T) = x(t) + Tx^{(1)}(t) + \frac{T^2}{2!}x^{(2)}(t) + \ldots + \frac{T^N}{N!}x^{(N)}(t) + \frac{T^{N+1}}{(N+1)!}x^{(N+1)}(\xi) \tag{11.166}$$

for some ξ such that $t < \xi < t+T$, where the last term is known as the remainder. But, in view of Eq. (11.164), the various derivatives of $x(t)$ with respect to time can be expressed as

$$x^{(i)}(t) = \frac{d^i x(t)}{dt^i} = \frac{d^{(i-1)} f[x(t), t]}{dt^{i-1}} = f^{(i-1)}[x(t), t], \; i = 1, 2, \ldots, N \tag{11.167}$$

Moreover, the interest lies in a discrete-time version of Eq. (11.166). To this end, we consider the discrete times $t = t_k$, $t + T = t_{k+1} = t_k + T$ $(k = 0, 1, 2, \ldots)$, where T is known as the *step size*, and introduce the notation

$$x(t) = x(t_k),\ x(t+T) = x(t_{k+1}),\ k = 0, 1, 2, \ldots$$
$$f^{(i)}[x(t), t] = f^{(i)}[x(t_k), t_k],\ i = 0, 1, \ldots, N-1;\ k = 0, 1, 2 \ldots \tag{11.168}$$

Then, inserting Eqs. (11.167) and (11.168) into Eq. (11.166), we have

$$x(t_{k+1}) = x(t_k) + \sum_{j=1}^{N} \frac{T^j}{j!} f^{(j-1)}[x(t_k), t_k] + \frac{T^{N+1}}{(N+1)!} f^{(N)}[x(\xi_k), \xi_k] \tag{11.169}$$

where $t_k < \xi_k < t_{k+1}$.

An approximate solution of Eq. (11.164) is obtained by ignoring the remainder in Eq. (11.169), i.e., the term involving ξ_k. Hence, introducing the notation

$$x(t_k) = x(k);\ f^{(i)}[x(t_k), t_k] = f^{(i)}(k),\ i = 0, 1, 2, \ldots \tag{11.170}$$

we can write the Nth-order Taylor series in the discrete-time form

$$x(k+1) = x(k) + Tf(k) + \frac{T^2}{2!} f^{(1)}(k) + \ldots + \frac{T^N}{N!} f^{(N-1)}(k),\ k = 0, 1, \ldots \tag{11.171}$$

in which $x(0)$ is the initial value of $x(t)$. The method for computing the numerical solution of Eq. (11.164) by means of Eqs. (11.171) is called the *Taylor method of order* N. By making the order N sufficiently large, or the step size T for a given N sufficiently small, the truncation error can be made as small as desirable. The Taylor methods form the basis for the Runge-Kutta methods.

The Taylor methods have a serious drawback in that they require derivatives of $f(x, t)$, which tends to make the process very tedious, thus limiting their appeal. The Runge-Kutta methods remove the need for derivatives of f while retaining the desirable error characteristics of the Taylor methods. To derive the Runge-Kutta methods, it is necessary to rewrite Eqs. (11.171) in a more suitable form. To this end, we consider Eq. (11.164) and write

$$f^{(1)}(x, t) = \frac{df(x,t)}{dt} = \frac{\partial f(x,t)}{\partial x}\frac{dx(t)}{dt} + \frac{\partial f(x,t)}{\partial t} = f(x,t)\frac{\partial f(x,t)}{\partial x} + \frac{\partial f(x,t)}{\partial t}$$

$$f^{(2)}(x, t) = \frac{df^{(1)}(x,t)}{dt} = \frac{\partial}{\partial x}\left[f(x,t)\frac{\partial f(x,t)}{\partial x} + \frac{\partial f(x,t)}{\partial t}\right]\frac{dx(t)}{dt} + \frac{\partial}{\partial t}\left[f(x,t)\frac{\partial f(x,t)}{\partial x} + \frac{\partial f(x,t)}{\partial t}\right]$$

$$= \left\{ \left[\frac{\partial f(x,t)}{\partial x} \right]^2 + f(x,t)\frac{\partial^2 f(x,t)}{\partial x^2} + \frac{\partial^2 f(x,t)}{\partial x \partial t} \right\} f(x,t)$$

$$+ \frac{\partial f(x,t)}{\partial t}\frac{\partial f(x,t)}{\partial x} + f(x,t)\frac{\partial^2 f(x,t)}{\partial t \partial x} + \frac{\partial^2 f(x,t)}{\partial t^2}$$

$$= f^2(x,t)\frac{\partial^2 f(x,t)}{\partial x^2} + f(x,t)\left[\frac{\partial f(x,t)}{\partial x} \right]^2 + 2f(x,t)\frac{\partial^2 f(x,t)}{\partial x \partial t}$$

$$+ \frac{\partial f(x,t)}{\partial x}\frac{\partial f(x,t)}{\partial t} + \frac{\partial^2 f(x,t)}{\partial t^2}$$

. . . .

(11.172)

so that, letting $x(t) = x(t_k) = x(k),\ f(x,t) = f[x(t_k), t_k] = f(k)$ in Eqs. (11.172) and inserting the results in Eqs. (11.171), we obtain

$$x(k+1) = x(k) + Tf(k) + \frac{T^2}{2!}\left[f(k)\frac{\partial f(k)}{\partial x} + \frac{\partial f(k)}{\partial t} \right] + \frac{T^3}{3!}\left\{ f^2(k)\frac{\partial^2 f(k)}{\partial x^2} \right.$$

$$\left. + f(k)\left[\frac{\partial f(k)}{\partial x} \right]^2 + 2f(k)\frac{\partial^2 f(k)}{\partial x \partial t} + \frac{\partial f(k)}{\partial x}\frac{\partial f(k)}{\partial t} + \frac{\partial^2 f(k)}{\partial t^2} \right\} + \ldots$$

$$k = 0, 1, 2, \ldots \qquad (11.173)$$

More often than not f does not depend explicitly on time, in which case Eqs. (11.173) reduce to

$$x(k+1) = x(k) + Tf(k) + \frac{T^2}{2!}f(k)\frac{\partial f(k)}{\partial x} + \frac{T^3}{3!}\left\{ f(k)\frac{\partial^2 f(k)}{\partial x^2} + f(k)\left[\frac{\partial f(k)}{\partial x} \right]^2 \right\} + \ldots,$$

$$k = 0, 1, 2, \ldots \qquad (11.174)$$

We illustrate the derivation of the Runge-Kutta methods from the Taylor methods by means of the second-order Runge-Kutta method. To this end, we assume an approximation of the form

$$x(k+1) = x(k) + c_1 g_1(k) + c_2 g_2(k),\ k = 0, 1, 2, \ldots \qquad (11.175)$$

where c_1 and c_2 are constants and

$$g_1(k) = Tf(k),\ g_2(k) = Tf[x(k) + \alpha g_1(k)] \qquad (11.176)$$

It should be pointed out that $x(k) + \alpha g_1(k)$ merely represents the argument of the function f in the expression for $g_2(k)$, in which α is a constant. The constants c_1, c_2 and α are determined by insisting that Eqs. (11.174) and (11.175) agree through terms of second order in T. Using Eqs. (11.176), we can write the Taylor series expansion

$$g_2(k) = Tf[x(k) + \alpha g_1(k)] = Tf[x(k) + \alpha Tf(k)]$$

$$= T[f(k) + \alpha Tf(k)\frac{\partial f(k)}{\partial x} + \ldots] \qquad (11.177)$$

so that, using the first of Eqs. (11.176) and Eq. (11.177), Eqs. (11.175) become

$$x(k+1) = x(k) + c_1 T f(k) + c_2 T \left[f(k) + \alpha T f(k) \frac{\partial f(k)}{\partial x} + \dots \right]$$
$$= x(k) + (c_1 + c_2) T f(k) + c_2 \alpha T^2 f(k) \frac{\partial f(k)}{\partial x} + \dots,$$
$$k = 0, 1, 2, \dots \tag{11.178}$$

Equating terms through second order in T in Eqs. (11.174) and (11.178), we conclude that the constants c_1, c_2 and α must satisfy

$$c_1 + c_2 = 1, \quad c_2 \alpha = \tfrac{1}{2} \tag{11.179}$$

Because there are two equations and three unknowns, Eqs. (11.179) do not have a unique solution. This implies that one of the unknowns can be chosen arbitrarily, provided the choice $c_2 = 0$ is excluded, for obvious reasons. One satisfactory choice is

$$c_1 = c_2 = \tfrac{1}{2}, \quad \alpha = 1 \tag{11.180}$$

Hence, inserting Eqs. (11.180) into Eqs. (11.175) and (11.176), we obtain a computational algorithm defining the *second-order Runge-Kutta method*, often referred to as the RK2 *method*, in the form

$$x(k+1) = x(k) + \tfrac{1}{2}[g_1(k) + g_2(k)], \quad k = 0, 1, 2, \dots \tag{11.181}$$

where

$$g_1(k) = T f(k), \quad g_2(k) = T f[x(k) + g_1(k)], \quad k = 0, 1, 2, \dots \tag{11.182}$$

Other choices are possible, but the choice given by Eqs. (11.180) has the advantage that it yields a symmetric form for the algorithm. We observe that $g_1(k)$ and $g_2(k)$ are to be evaluated in sequence, as the computation of $g_2(k)$ depends on $g_1(k)$.

Following the same pattern, we can derive higher-order Runge-Kutta approximations. The derivations become increasingly complex, however, without providing additional insights. In view of this, we omit the derivations and merely list the results. The most widely used is the *fourth-order Runge-Kutta* method, commonly known as the RK4 method, defined by the algorithm

$$x(k+1) = x(k) + \tfrac{1}{6}[g_1(k) + 2g_2(k) + 2g_3(k) + g_4(k)], \quad k = 0, 1, 2, \dots \tag{11.183}$$

where

$$\begin{aligned} &g_1(k) = T f(k), \quad g_2(k) = T f[x(k) + \tfrac{1}{2} g_1(k)], \\ &g_3(k) = T f[x(k) + \tfrac{1}{2} g_2(k)], \quad g_4(k) = T f[x(k) + g_3(k)], \end{aligned} \quad k = 0.1, 2, \dots \tag{11.184}$$

The method is easy to implement and has good accuracy, provided the step size T is sufficiently small. One way of ensuring that is to solve the problem twice, once using the step size T and the other using the step size $T/2$. If the results do not agree within the desired accuracy level, the computations must be repeated with $T/2$ and $T/4$. This process is not particularly efficient, as it requires a large amount of computation.

The *Runge-Kutta-Fehlberg method*, denoted by RKF45, resolves the above problem in an efficient manner by determining an optimal step size for a given accuracy. It requires two different approximations at each step, an RK4 approximation and an RK5 approximation. The RKF45 algorithm forms the basis for the MATLAB function 'ode45'.

Under certain circumstances, solutions obtained by the MATLAB function 'ode45' are not sufficiently accurate, which can be attributed to a relatively large step size. In this regard, it should be pointed out that the step size is determined optimally by the routine, and cannot be changed. Inaccurate solutions can occur when the 'ode45' routine is used for *stiff systems*, in which the solution can change on a time scale that is very short relative to the interval of integration, but the solution of interest changes on a much longer time scale. In such cases, a MATLAB function capable of handling stiff problems, such as 'ode23s', may produce more accurate solutions. Stiffness is a qualitative property that more often than not cannot be ascertained before choosing an algorithm. In view of the interactive nature of MATLAB, the question of stiffness need not be addressed directly, as it is relatively easy to try different integration routines and make a choice as to the one to use based on the smoothness of the solution.

The Runge-Kutta methods are *one-step methods*, in the sense that information from step k alone is used to compute $x(k+1)$. Such methods are said to be *self-starting*.

The above developments are based on a single first-order scalar equation, Eq. (11.164). In vibrations, however, even a single-degree-of-freedom system requires two first-order equations. We recall from Sec. 4.8 that a single second-order differential equation can be replaced by two first-order state equations. Similarly, the vibration of an n-degree-of-freedom system can be described by $2n$ first-order state equations. Hence, for vibration problems, we must replace the single first-order scalar equation used to introduce the Runge-Kutta methods by a set of $2n$ first-order equations, which can be done with relative ease using vector notation. Indeed, using the analogy with Eq. (11.164), the vibration of an n-degree-of-freedom nonlinear system can be described by the vector equation

$$\dot{\mathbf{x}}(t) = \mathbf{f}[\mathbf{x}(t)] \tag{11.185}$$

where $\mathbf{x}(t)$ is the $2n$-dimensional state vector and $\mathbf{f}$ is a $2n$-dimensional excitation vector depending on the state vector $\mathbf{x}(t)$. Then, by analogy with Eqs. (11.183) and (11.184), the fourth-order Runge-Kutta (RK4) method is defined by the algorithm

$$\mathbf{x}(k+1) = \mathbf{x}(k) + \tfrac{1}{6}[\mathbf{g}_1(k) + 2\mathbf{g}_2(k) + 2\mathbf{g}_3(k) + \mathbf{g}_4(k)], \; k = 0, 1, 2, \ldots \tag{11.186}$$

where

$$\begin{aligned} &\mathbf{g}_1(k) = T\mathbf{f}(k), \; \mathbf{g}_2(k) = Tf[\mathbf{x}(k) + \tfrac{1}{2}\mathbf{g}_1(k)], \\ &\mathbf{g}_3(k) = T\mathbf{f}[\mathbf{x}(k) + \tfrac{1}{2}\mathbf{g}_2(k)], \; \mathbf{g}_4(k) = T\mathbf{f}[\mathbf{x}(k) + \mathbf{g}_3(k)], \end{aligned} \quad k = 0, 1, 2, \ldots \tag{11.187}$$

are $2n$-dimensional vectors. The RK4 method involves four computations of the vector $\mathbf{f}$ for each integration step, which represents a large amount of computation. The method is extremely accurate, however, and it requires fewer steps for a desired accuracy level than other methods. Moreover, the method is very stable. These advantages make RK4 a favorite for the numerical integration of nonlinear differential equations.

As can be concluded from Eqs. (11.173), when the excitation depends explicitly on time, $f = f[x(t), t]$ the computation of the response by the Runge-Kutta methods is likely to be very complex. Discussions of the second-order and fourth-order Runge-Kutta methods, as well as of the Runge-Kutta-Fehlberg method, are presented in Ref. 13.

Example 11.6. The motion of a mass-nonlinear spring system is described by the differential equation

$$\ddot{q}(t) + 4[q(t) + q^3(t)] = 0 \tag{a}$$

Obtain the response to initial conditions by the fourth-order Runge-Kutta method using the sampling period $T = 0.1$ s for the two cases: 1) $q(0) = 0.8$ cm, $\dot{q}(0) = 0$ and 2) $q(0) = 1.2$ cm, $\dot{q}(0) = 0$. Plot $q(t)$ versus t for the two cases over the time interval $0 < t < 5$ s and draw conclusions concerning the period.

In the first place, we transform Eq. (a) to state form. To this end, we introduce the notation

$$q(t) = x_1(t),\ \dot{q}(t) = x_2(t) \tag{b}$$

and replace the scalar equation (a) by the vector state equations

$$\dot{\mathbf{x}}(t) = \mathbf{f}[\mathbf{x}(t)] \tag{c}$$

in which the state vector and excitation vector are

$$\mathbf{x}(t) = \begin{bmatrix} x_1(t) \\ x_2(t) \end{bmatrix},\ \mathbf{f}[\mathbf{x}(t)] = \begin{bmatrix} x_2(t) \\ -4[x_1(t) + x_1^3(t)] \end{bmatrix} \tag{d}$$

The vector form of the fourth-order Runge-Kutta method is given by Eqs. (11.186) and (11.187), which can be written by components as follows:

$$\begin{aligned} x_1(k+1) &= x_1(k) + \tfrac{1}{6}[g_{11}(k) + 2g_{21}(k) + 2g_{31}(k) + g_{41}(k)], \\ &\qquad\qquad\qquad k = 0, 1, 2\ldots \\ x_2(k+1) &= x_2(k) + \tfrac{1}{6}[g_{12}(k) + 2g_{22}(k) + 2g_{32}(k) + g_{42}(k)], \end{aligned} \tag{e}$$

and

$$\begin{aligned} g_{11}(k) &= Tf_1[x_1(k), x_2(k)] = Tx_2(k), \\ g_{12}(k) &= Tf_2[x_1(k), x_2(k)] = -4T[x_1(k) + x_1^3(k)], \\ g_{21}(k) &= Tf_1[x_1(k) + \tfrac{1}{2}g_{11}(k),\ x_2(k) + \tfrac{1}{2}g_{12}(k)] = T[x_2(k) + \tfrac{1}{2}g_{12}(k)], \\ g_{22}(k) &= Tf_2[x_1(k) + \tfrac{1}{2}g_{11}(k),\ x_2(k) + \tfrac{1}{2}g_{12}(k)] \\ &= -4T\{x_1(k) + \tfrac{1}{2}g_{11}(k) + [x_1(k) + \tfrac{1}{2}g_{11}(k)]^3\}, \\ g_{31}(k) &= Tf_1[x_1(k) + \tfrac{1}{2}g_{21}(k),\ x_2(k) + \tfrac{1}{2}g_{22}(k)] = T[x_2(k) + \tfrac{1}{2}g_{22}(k)], \\ g_{32}(k) &= Tf_2[x_1(k) + \tfrac{1}{2}g_{21}(k),\ x_2(k) + \tfrac{1}{2}g_{22}(k)] \\ &= -4T\{x_1(k) + \tfrac{1}{2}g_{21}(k) + [x_1(k) + \tfrac{1}{2}g_{21}(k)]^3\}, \\ g_{41}(k) &= Tf_1[x_1(k) + g_{31}(k),\ x_2(k) + g_{32}(k)] = T[x_2(k) + g_{32}(k)], \end{aligned}$$

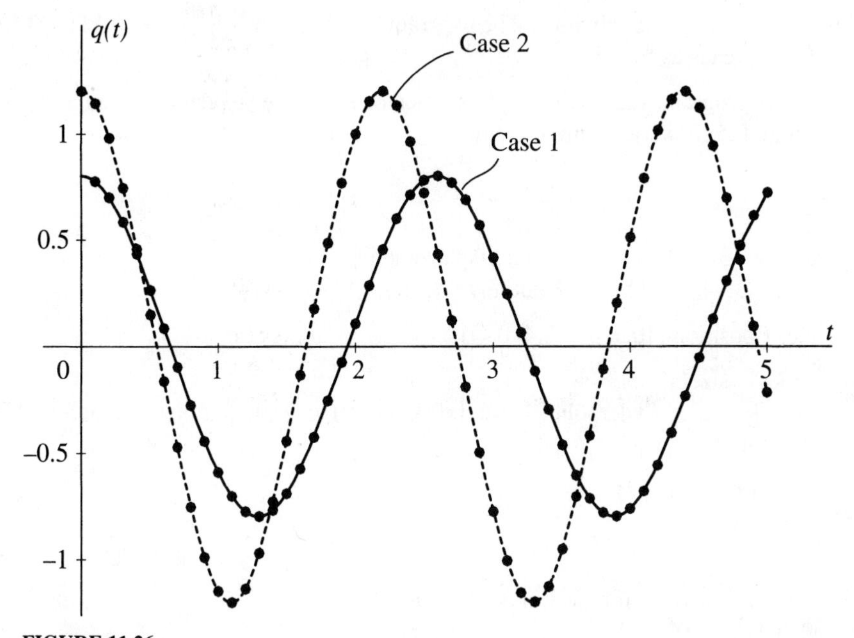

FIGURE 11.26
Response of a mass-nonlinear spring system computed by the fourth-order Runge-Kutta method

$$
\begin{aligned}
g_{42}(k) &= Tf_2[x_1(k)+g_{31}(k),\ x_2(k)+g_{32}(k)] \\
&= -4T\{x_1(k)+g_{31}(k)+[x_1(k)+g_{31}(k)]^3\}, \\
&\qquad\qquad k = 0.1, 2, \ldots \qquad \text{(f)}
\end{aligned}
$$

respectively. Equations (f) are subject to the initial conditions: (1) $x_1(0) = 0.8$ cm, $x_2(0) = 0$ and 2) $x_1(0) = 1.2$ cm, $x_2(0) = 0$. The response is shown in Fig. 11.26, and it is clear that the period depends on the motion amplitude, which in turn depends on the initial conditions.

11.12 TRAJECTORIES FOR THE VAN DER POL OSCILLATOR BY MATLAB

For the most part, the solution of nonlinear ordinary differential equations must be obtained numerically, as discussed in Sec. 11.11. To this end, the Runge-Kutta-Fehlberg algorithm, and in particular the MATLAB routine 'ode45', appears as a suitable candidate. Based on this routine, a MATLAB program for plotting trajectories in the state plane for the van der Pol oscillator has been written. As it turns out, the trajectories obtained using 'ode45' are not particularly smooth, an indication that the system may be relatively stiff (Sec. 11.11), with the implication that the step size may be too large. Because the step size is determined automatically by the routine, and cannot be changed, a different routine is advised. Indeed, the MATLAB routine 'ode23s', designed to handle stiff systems, and hence using a smaller step size, yields smoother trajectories, making the choice of routines obvious. The program 'vanderpol.m', using the MATLAB routine 'ode23s', reads as follows:

```
% The program 'vanderpol.m' plots trajectories for the van der Pol oscillator with
% mu=0.5 for various initial conditions

clear
clf

x0=[-3 -3 3 3 0 0]; % initial displacements
v0=[4 3 -4 -3 0.5 1]; % initial velocities

axes ('position', [0.3 0.3 0.4 0.4])

for k=1: size(x0, 2),
  [t, x]=ode23s ('vderpol', [0, 20], [x0(k) v0(k)]); %integration using MATLAB
  % function

  plot (x(:, 1),x(:, 2))
  hold on
end

title ('Trajectories for the van der Pol Oscillator')
ylabel ('x_2(t)')
xlabel ('x_1(t)')
axis ([-5 5 -4 4])
```

For the program to work, it is necessary to provide the equation of the system under consideration. This is done through the following program entitled 'vderpol', which contains the van der Pol equation with $\mu = 0.5$:

```
% The program 'vderpol' provides the equation to be integrated in the program
% 'vanderpol.m'

function xp=vderpol (t,x)
xp=[x(2); -x(1)-0.5*(x(1).^2-1).*x(2)];
```

11.13 SUMMARY

Nonlinear systems differ from linear systems in a very important respect, namely, they do not satisfy the conditions imposed by the superposition principle. As a result, except for certain cases, the techniques for solving the differential equations of motion developed for linear systems do not work for nonlinear systems. It is possible to conclude that a system is nonlinear when at least one of the dependent variables and their derivatives appears in the differential equations of motion to a power different from one. Because for the most part linear analysis does not apply to nonlinear systems, new concepts and methods of approach are required.

An important concept in dynamics is that of equilibrium points, defined as points in the state space having constant value. The implication is that the displacements are constant and the velocities are zero at equilibrium points, which further implies that the accelerations and the excitations are all zero there. Whereas linear systems have only one equilibrium point, nonlinear systems have one or more, depending on the nature of the nonlinearity. On occasions, particularly in preliminary design, the interest lies more in the stability of motion in the neighborhood of equilibrium points than in the complete system response. Under the small-motions assumption, such stability statements can be made by linearizing the equations of motion about a certain equilibrium point and reaching conclusions concerning the stability of motion in the neighborhood of that point by solving the eigenvalue problem for the linearized system and examining the nature of the associated eigenvalues.

For conservative single-degree-of-freedom systems, there exists a motion integral in the form of the total energy, which can be used to plot trajectories in the state plane away from equilibrium points. These are closed trajectories corresponding to given energy levels, and representing periodic motions. For nonconservative systems, trajectories can be obtained through numerical integration, but in general the trajectories are not closed. Closed trajectories can occur in nonconservative systems also with nonlinear damping such that energy is dissipated over part of the trajectory and gained over the remaining part in a way that the net energy change over the full period is equal to zero. Such closed trajectories represent equilibrium motions, in the sense that trajectories on both sides of the closed trajectory tend to it (or away from it), and are called limit cycles; they enclose an equilibrium point. A typical example of a system exhibiting a limit cycle is the van der Pol oscillator.

For certain systems, the differential equation of motion consists of a linear part and a nonlinear part, where the latter is so much smaller than the former that it can be regarded as a perturbation on the linear system. In such cases, an analytical solution can be obtained by a perturbation approach whereby the nonlinear part is identified by a small parameter ε and the solution is expanded in a power series in ε. Then, a separation of the terms multiplying like powers of ε gives rise to a set of linear equations that can be solved in sequence. The perturbation solution just described can contain secular terms, i.e., terms increasing indefinitely with time, even for systems known to possess periodic solutions. To prevent the formation of secular terms, the solution for every order of approximation must be forced to be periodic. This can be done by assuming that the period, or rather the fundamental frequency, which is not known a priori, can also be expanded in a power series in ε. Then, the perturbations of increasing order in the fundamental frequency can be determined by insisting that the perturbation solutions be periodic. This is the essence of Lindstedt's method.

The equation of motion of a harmonic oscillator with a small cubic nonlinearity is known as Duffing's equation. Unlike a harmonic oscillator, a system described by Duffing's equation subjected to a small harmonic excitation does not experience resonance. Indeed, the vertical asymptote at resonance in the frequency response magnitude plot for a harmonic oscillator bends into a parabola pointing to the right for a hardening spring and to the left for a softening spring. Moreover, conforming to the bent asymptote, the two branches of the frequency response plot also bend, approaching the parabola from both sides, with the response following one branch or the other, depending on the starting value of the driving frequency. If in addition the system possesses small damping, then the frequency response plots bend in a similar fashion relative to the plots for the linear system. Now, however, the plots are continuous and consist of three parts, an upper limb, a lower limb and a section connecting the two. As the driving frequency varies, the response can jump from one limb to another, but the in-between section is never traversed. When the harmonic excitation is not small, Duffing's equation admits subharmonic solutions. Moreover, when the excitation is a linear combination of two harmonic forces, the response is the sum of harmonic components with frequencies equal to the two driving frequencies, as well as of harmonic components with frequencies equal to certain combinations of the two driving frequencies.

Linear systems with time-dependent coefficients are considerably more difficult to handle than those with constant coefficients. Although such systems do not really belong in this chapter, when the time-dependent terms are relatively small, solutions can be obtained by the same perturbation techniques ordinarily used for systems with small nonlinearities. A typical example is that of a pendulum whose support is acted upon by a small harmonic force. Under certain circumstances, the equation of motion reduces to Mathieu's equation, a well-known equation in mathematical physics. A perturbation solution demonstrates that the harmonic motion of the support can render a stable equilibrium position of the pendulum unstable, and vice versa.

The situation is dramatically different when the interest lies in large motions, and the nonlinearities are not small. In such cases, solutions must be obtained numerically, which involves some degree of approximation. In this text, we consider solutions by the Runge-Kutta methods, a family of algorithms characterized by different orders of approximation. In particular, we consider efficient modifications of these methods known as Runge-Kutta-Fehlberg methods. In general, to establish convergence with a given step size, it is necessary to compare the solution obtained by means of a method of a given order with that obtained using a method one order higher. In the Runge-Kutta-Fehlberg methods, the comparison requires only a fraction of the computational effort required by a full solution by the higher-order method. Note that numerical integration of nonlinear equations is generally carried out in the state space. To this end, computer programs based on the MATLAB functions 'ode45' and 'ode23s', which in turn are based on Runge-Kutta-Fehlberg methods, are quite effective. The function 'ode45' uses a larger step size and is suitable for nonstiff problems, and the function 'ode23s' uses a smaller step size and is to be used for stiff problems. The computer program 'vanderpol.m', (Sec. 11.12), which plots trajectories for the van der Pol equation, is based on 'ode23s'.

PROBLEMS

11.1. The differential equation of a viscously damped pendulum undergoing large angular displacements can be written in the form

$$\ddot{\theta}(t) + 2\zeta\omega\dot{\theta}(t) + \omega^2 \sin\theta(t) = 0$$

where ζ is the viscous damping factor and ω the natural frequency of small undamped oscillation. Derive the corresponding state equations and determine the equilibrium points. Then, derive the linearized equations about each of the equilibrium points, solve the associated eigenvalue problem for $\omega = 1$ rad/s in the two cases $\zeta = 0.1$ and $\zeta = 2$, use the eigenvalues to determine the nature of the stability in the neighborhood of the equilibrium points and state in each case whether the linearized system possesses significant or critical behavior.

11.2. The motion of an undamped single-degree-of-freedom system with a nonlinear spring is described by the differential equation

$$\ddot{q}(t)+q(t)-\frac{\pi}{2}\sin q(t)=0$$

Derive the corresponding state equations and determine the equilibrium points. Then, derive the linearized equations about each of the equilibrium points, solve the associated eigenvalue problem, use the eigenvalues to determine the nature of the stability in the neighborhood of the equilibrium points and state in each case whether the linearized system possesses significant or critical behavior.

11.3. The differential equation of motion for a bead of mass m sliding freely along a smooth circular hoop of radius R rotating about a vertical axis with the constant angular velocity Ω was derived in Problem 1.2 by means of Newton's second law and in Problem 6.12 by means of Lagrange's equation. Derive the corresponding state equations and determine the equilibrium points for the two cases: 1) $R\Omega^2 > g$ and 2) $R\Omega^2 < g$, where g is the acceleration due to gravity. Then, derive the linearized equations about each of the equilibrium points, solve the associated eigenvalue problem, use the eigenvalues to determine the nature of the stability in the neighborhood of the equilibrium points and state in each case whether the linearized system possesses significant or critical behavior.

11.4. Consider the damped pendulum of Problem 11.1, let $\omega = 1$ rad/s and plot trajectories in the neighborhood of each of the equilibrium points for both cases, $\zeta = 0.1$ and $\zeta = 2$. Do the trajectories confirm the conclusions concerning the nature of the equilibrium points reached in Problem 11.1?

11.5. Consider the undamped system of Problem 11.2 and plot trajectories in the neighborhood of each of the equilibrium points. Do the trajectories confirm the conclusions concerning the nature of the equilibrium points reached in Problem 11.2?

11.6. Consider the bead on a rotating hoop of Problem 11.3 and plot trajectories in the neighborhood of the equilibrium points for both cases, $R\Omega^2 < g$ and $R\Omega^2 > g$. Do the trajectories confirm the conclusions concerning the nature of the equilibrium points reached in Problem 11.3?

11.7. Use the approach of Sec. 11.3 to generate an integral of the equation of motion for the system of Problem 11.2. Then, use the integral to plot a sufficient number of trajectories so as to illustrate the nature of the motion in the neighborhood of each of the equilibrium points, as well as to illustrate the motion in the large.

11.8. Use the approach of Sec. 11.3 to generate an integral of the equation of motion for the system of Problem 11.3. Then, use the integral to plot a sufficient number of trajectories so as to illustrate the nature of the motion in the neighborhood of each of the equilibrium points, as well as to illustrate the motion in the large. Consider the two cases $R\Omega^2 = 0.25\text{g}$ and $R\Omega^2 = 2\text{g}$.

11.9. Consider the system of equations

$$\dot{x}_1 = x_2 + x_1(1-x_1^2-x_2^2),\ \dot{x}_2 = -x_1 + x_2(1-x_1^2-x_2^2)$$

use the coordinate transformation $x_1 = r\cos\theta,\ x_2 = r\sin\theta$ and derive the equation of the trajectories in terms of the polar coordinates r and θ. Integrate the equation, express r as a function of θ and verify that $r = 1$ is a limit cycle of the system. Determine whether $r = 1$ is a stable or an unstable limit cycle and establish the stability of the equilibrium point at the origin. Plot a sufficient number of trajectories to verify your conclusions.

11.10. Consider the damped linear oscillator

$$\ddot{q}(t)+2\varepsilon\omega_0\dot{q}(t)+\omega_0^2 q(t)=0,\ \varepsilon \ll 1$$

and obtain a perturbation solution of the form (11.67). Include in the solution terms through second order in ε, compare the result with Eq. (2.32) and draw conclusions. Note that, before a comparison can be made, Eq. (2.32) must be expanded in a power series in ζ under the assumption that ζ is small.

11.11. Consider the quasi-harmonic system described by the differential equation

$$\ddot{q}(t)+q(t)=\varepsilon q^2(t),\ \varepsilon \ll 1$$

and use Lindstedt's method to obtain a periodic solution approximate to the second order. Let the initial conditions by $q(0)=A_0,\ \dot{q}(0)=0$.

11.12. Consider the van der Pol equation

$$\ddot{q}(t)+\omega^2 q(t)=\varepsilon\dot{q}(t)[1-q^2(t)],\ \varepsilon \ll 1$$

and obtain a periodic solution approximate to the first order by means of Lindstedt's method. Note that the amplitude is not arbitrary but determined by the periodicity condition. Let $\varepsilon=0.2$, plot the solution in the state plane and draw conclusions as to the meaning of the plot.

11.13. The differential equation describing the behavior of a van der Pol oscillator subjected to a harmonic excitation can be written in the form

$$\ddot{q}(t)+\omega^2 q(t)=\varepsilon\{-\omega^2\alpha q(t)+\dot{q}(t)[1-q^2(t)]+F\cos\Omega t\},\ \varepsilon \ll 1$$

Use the method of Sec. 11.8 to obtain a periodic solution with period $2\pi/\Omega$ approximate to the first order.

11.14. Equation (11.112) is known as Duffing's equation with small damping. Use the method of Sec. 11.8 to obtain a periodic solution with period $2\pi/\Omega$ approximate to the first order, and in the process verify relations (11.113), (11.114) and (11.115). Use (11.115) to plot the response for the parameters $\varepsilon\zeta=0.1,\ \varepsilon\beta=-0.2$ and $\varepsilon F=4$.

11.15. Obtain a subharmonic solution of the differential equation

$$\ddot{q}(t)+\omega^2 q(t)=-\varepsilon\omega^2[\alpha q(t)-\beta q^2(t)]+F\cos\Omega t,\ \varepsilon \ll 1$$

11.16. Use the method of Sec. 11.10 to verify Eqs. (11.162) and (11.163).

11.17. The behavior of a pendulum undergoing large motions is described by the nonlinear differential equation

$$\ddot{\theta}(t)+4\sin\theta(t)=0$$

Obtain a fourth-order Runge-Kutta solution for the initial conditions $\theta(0)=\pi/3,\ \dot{\theta}(0)=0$. Plot $\theta(t)$ versus t over the time interval $0<t<3\pi$.

11.18. Write a MATLAB program for solving Problem 11.9. Use the state equations in terms of rectangular coordinates.

11.19. Write a MATLAB program for plotting trajectories for the system of Problem 11.11 for $\varepsilon=0.1$.

11.20. Write a MATLAB program for plotting trajectories for the system of Problem 11.12.

11.21. Write a MATLAB program for plotting trajectories for the system of Problem 11.17. Plot trajectories for two cases, first using the MATLAB routine 'ode45' and then using 'ode23s', compare results and draw conclusions concerning the suitability of the two routines.

CHAPTER 12

RANDOM VIBRATIONS

In our preceding study of vibrations, it was possible to distinguish between three types of excitation functions, namely, harmonic, periodic, and nonperiodic, where the latter is also known as transient. The common characteristic of these functions is that their values can be determined for any future time t. Such functions are said to be *deterministic*, and typical examples are shown in Fig. 12.1a, b, and c. The response of systems to deterministic excitations is also deterministic. For linear systems, there is no difficulty in expressing the response to any arbitrary determinisitc excitation in some closed form, such as the convolution integral, although the integral may not always be easy to evaluate. The theory of nonlinear systems is not nearly as well developed, and the response to arbitrary excitations cannot be obtained even in the form of a convolution integral. Nevertheless, even for nonlinear systems, the response can be obtained in terms of time by means of numerical integration.

There are many physical phenomena, however, that do not lend themselves to explicit time description. Examples of such phenomena are jet engine noise, the height of waves in a rough sea, the intensity of an earthquake, etc. The implication is that the value at some future time of the variables describing these phenomena cannot be predicted. If the intensity of earth tremors is measured as a function of time, then the record of one tremor will be different from that of another one. The reasons for the difference are many and varied, and they may have little or nothing to do with the measuring instrument. The main reason may be that there are simply too many factors affecting the outcome. Phenomena whose outcome at a future instant of time cannot be predicted are classified as *nondeterministic*, and referred to as *random*. A typical random function is shown in Fig. 12.1d.

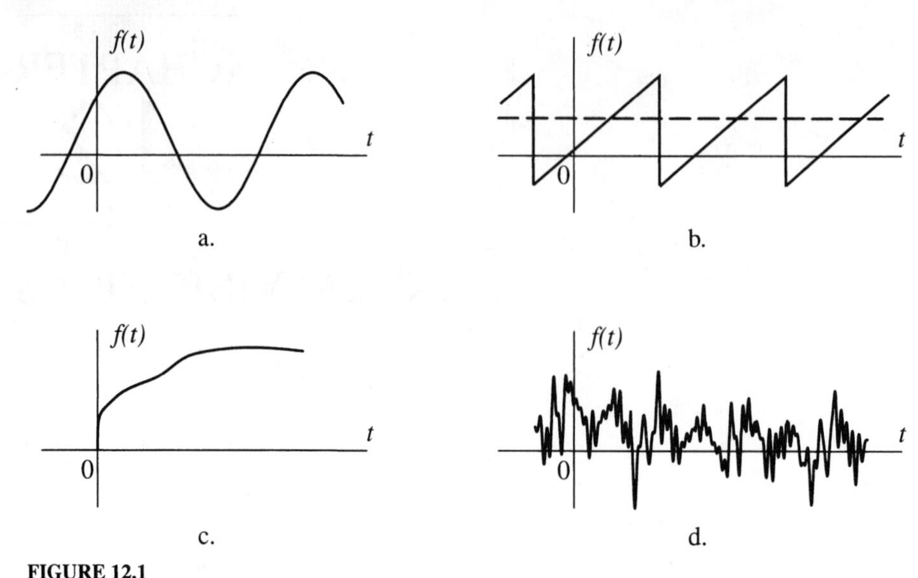

FIGURE 12.1
a. Harmonic excitation, b. Periodic excitation, c. Nonperiodic excitation, d. Random excitation

The response of a system to a random excitation is also a random phenomenon. Because of the complexity involved, the description of random phenomena as functions of time does not appear particularly meaningful, and new methods of analysis must be adopted. Many random phenomena exhibit a certain pattern, in the sense that the data can be described in terms of certain averages. This characteristic of random phenomena is called *statistical regularity*. If the excitation exhibits statistical regularity, so does the response. In such cases it is more feasible to describe the excitation and response in terms of *probabilities* of occurrence than to seek a deterministic description. In this chapter we develop the tools for the statistical approach to vibration analysis, and then use these tools to derive the response of linear systems to random excitations.

12.1 ENSEMBLE AVERAGES. STATIONARY RANDOM PROCESSES

The interest lies in determining the force exerted by the landing gear on an aircraft taxiing on a rough runway subsequent to landing. To this end, we measure the displacement of the landing gear axle of a certain aircraft taxiing on the same runway a large number of times and denote the corresponding displacement time histories by $x_1(t), x_2(t), \ldots, x_n(t)$. Because the aircraft does not follow the same path on the runway every time it lands, the payload varies from flight to flight and because of many other factors too difficult to identify, the time histories differ from one another, as shown in Fig. 12.2. It is clear from Fig. 12.2 that such time histories cannot be expressed explicitly in terms of known functions of time. As a result, it is impossible to use any of these functions to predict the displacement of the axle at some future time. Such displacements are said to represent a *nondeterministic phenomenon*, or a *random phenomenon*. An individual time history

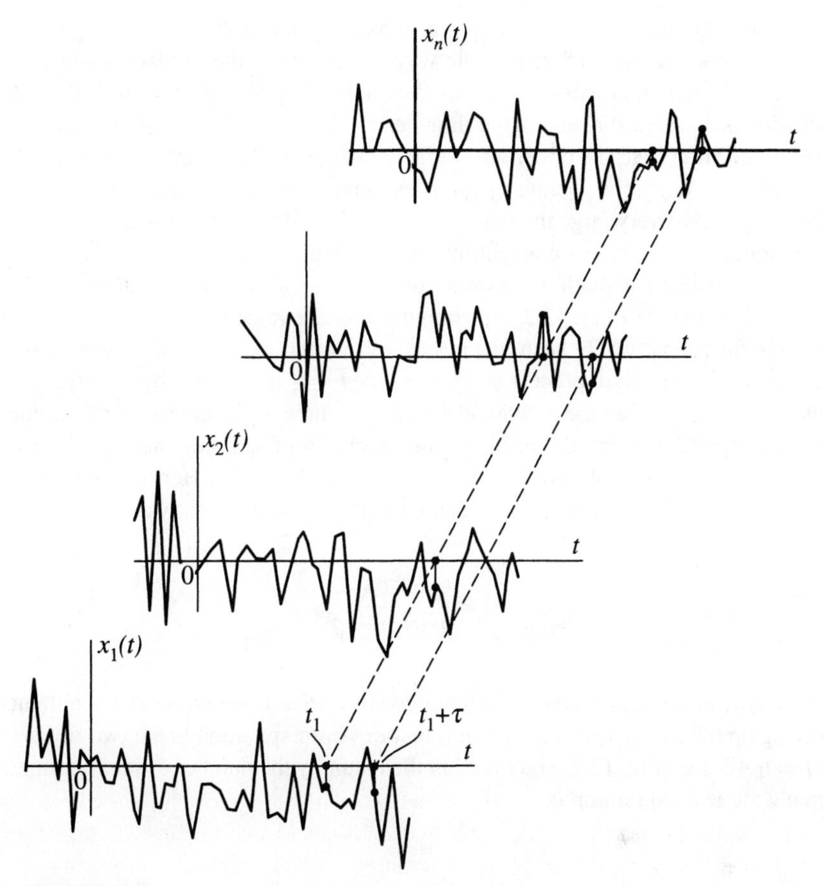

FIGURE 12.2
Displacement time histories of a landing gear axle

describing a random phenomenon, say $x_k(t)$, is called a *sample function* and the variable $x_k(t)$ itself is referred to as a *random variable*. The entire collection of all possible time histories, known as an *ensemble*, represents a *random process*, or a *stochastic process*. Random phenomena are quite common in the physical world, and the question arises as to whether it is possible to treat them mathematically. The answer to this question depends on the nature of the random phenomenon under consideration, and in particular on whether the phenomenon exhibits certain regularity, as discussed shortly.

The displacement of the landing gear axle discussed in the preceding paragraph plays the role of one of the excitations on the aircraft. Because the excitation represents a random process, rather than a deterministic function of time, the question arises as to how to calculate the response. The most direct approach would be to calculate the response to every sample function in the ensemble numerically, as if they were ordinary functions of time. Such an approach, however, would not be very efficient, and very likely not meaningful, as there may be numerous sample functions in the excitation random process, and there is some question as to how to interpret the response to all these functions. Hence, a more suitable way of describing the excitation and response in the case of random processes is highly desirable. This line of thought leads us to the conclusion that we must abandon the time description of random processes and replace it by a description in terms of certain averages, where the latter are sometimes referred to as *statistics*. When the averages tend to recognizable limits as the number of sample functions becomes very large the random process is said to exhibit *statistical regularity*. Throughout this text, we assume that this is indeed the case.

We consider the random process consisting of n sample functions $x_k(t)$ $(k = 1, 2, \ldots, n)$ depicted in Fig. 12.2 and compute average values over the entire collection of sample functions, where such quantities are referred to as *ensemble averages*. The *mean value* of the random process at a given time $t = t_1$ is obtained by simply summing up the values $x_k(t_1)$ of all the individual sample functions in the ensemble corresponding to the time t_1 and dividing the result by the number n of sample functions. The implication is that every sample function is assigned equal weight. Hence, the mean value corresponding to the arbitrary time t_1 can be written mathematically as

$$\mu_x(t_1) = \lim_{n \to \infty} \frac{1}{n} \sum_{k=1}^{n} x_k(t_1) \tag{12.1}$$

Another type of ensemble average is the *autocorrelation function*, which is obtained by summing up the products of the sample functions corresponding to the two times $t = t_1$ and $t = t_1 + \tau$ (see Fig. 12.2) and dividing the result by the number of sample functions; its mathematical expression is

$$R_x(t_1, t_1 + \tau) = \lim_{n \to \infty} \frac{1}{n} \sum_{k=1}^{n} x_k(t_1) x_k(t_1 + \tau) \tag{12.2}$$

By considering three or more times, such as t_1, $t_1 + \tau$, $t_1 + \sigma$, etc., it is possible to calculate higher-order averages than the autocorrelation function. Such averages are seldom needed, however.

In general, the mean value $\mu_x(t_1)$ and the autocorrelation function $R_x(t_1, t_1 + \tau)$ depend on the time t_1, in which case the random process is said to be *nonstationary*. In the special case in which $\mu_x(t_1)$ and $R_x(t_1, t_1 + \tau)$ do not depend on t_1 the random process is said to be *weakly stationary*. Hence, for a weakly stationary random process the mean value is constant, $\mu_x(t_1) = \mu_x =$ constant, and the autocorrelation function depends on the time shift τ alone, $R_x(t_1, t_1 + \tau) = R_x(\tau)$. If all possible ensemble averages are independent of t_1, the random process is said to be *strongly stationary*. In many practical applications, strong stationarity can be assumed if weak stationarity is established. This will be shown to be the case for a large class of random processes, namely, Gaussian random processes (Sec. 12.4). In view of this, we will not insist on distinguishing between weak and strong stationarity, and refer to random processes as simply *stationary*.

12.2 TIME AVERAGES. ERGODIC RANDOM PROCESSES

In general, ensemble averages, such as the mean value, Eq. (12.1) and autocorrelation function, Eq. (12.2), require a large number of sample functions, so that the computation of ensemble averages can create difficulties. Indeed, in the first place, it is necessary to collect a great deal of data to generate the sample functions. Then, it is necessary to process this data. However, under certain circumstances, it is possible to avoid these difficulties by calculating the mean value and autocorrelation function by means of a single representative sample function and averaging over the time t, instead of averaging over the ensemble. Such averages are called *time averages*, or *temporal averages*, as opposed to ensemble averages. Concentrating on a given sample function $x_k(t)$ from a random process, we define the *temporal mean value* as

$$\mu_x(k) = \lim_{T\to\infty} \frac{1}{T} \int_{-T/2}^{T/2} x_k(t)dt \tag{12.3}$$

and the *temporal autocorrelation function* as

$$R_x(k,\tau) = \lim_{T\to\infty} \frac{1}{T} \int_{-T/2}^{T/2} x_k(t)x_k(t+\tau)dt \tag{12.4}$$

If the random process under consideration is stationary and if the temporal mean value $\mu_x(k)$ and the temporal autocorrelation function $R_x(k,\tau)$ are the same, irrespective of which time history from the entire process is used to calculate these averages, then the process is said to be *ergodic*. It follows that, for ergodic processes the temporal mean value and autocorrelation function calculated over a representative sample function must by necessity be equal to the ensemble mean value and autocorrelation function, respectively, so that

$$\mu_x(k) = \mu_x = \text{constant}, \quad R_x(k,\tau) = R_x(\tau) \tag{12.5}$$

As with the stationarity property, we can distinguish between *weakly ergodic* processes, for which the mean value and autocorrelation function are the same regardless of the sample function used, and *strongly ergodic* processes, for which all possible statistics are the same. Here too, there is no distinction between weakly and strongly ergodic processes for Gaussian random processes (see Sec. 12.4), so that weak ergodicity can be assumed to imply strong ergodicity, or simply ergodicity. It should be pointed that *an ergodic process is by necessity stationary, but a stationary process is not necessarily ergodic.*

The ergodicity property permits the use of a single sample function to calculate averages for a given random process, thus obviating the need to use the entire ensemble. Because the chosen sample function can be regarded as being representative of the whole random process, in subsequent discussions we will drop the subscript k identifying a particular time history. Many stationary random processes describing the behavior of physical systems are ergodic, so that our study will concentrate on ergodic processes. If a given process is not ergodic but merely stationary, then we use ensemble averages instead of temporal averages.

It should perhaps be pointed out that, although we defined the temporal mean value and autocorrelation function in conjunction with a sample function describing

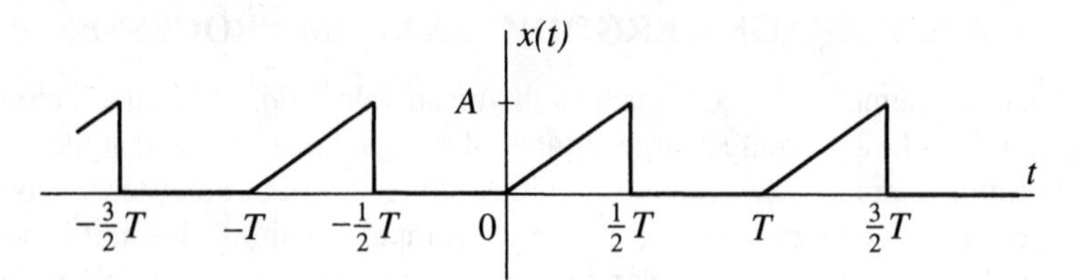

FIGURE 12.3
Periodic function

an ergodic random process, such averages apply to all functions of time, including deterministic functions. In fact, to illustrate the calculation of averages, we will use primarily deterministic functions.

Example 12.1. Calculate the temporal mean value and autocorrelation function for the function depicted in Fig. 12.3 and plot the autocorrelation function.

Because the function is periodic, averages calculated over a long time interval approach those calculated by considering one period alone. Concentrating on the period $-T/2 < t < T/2$, the function can be described by

$$x(t) = \begin{cases} 0, & -\dfrac{T}{2} < t < 0 \\ \dfrac{2A}{T}t, & 0 < t < \dfrac{T}{2} \end{cases} \tag{a}$$

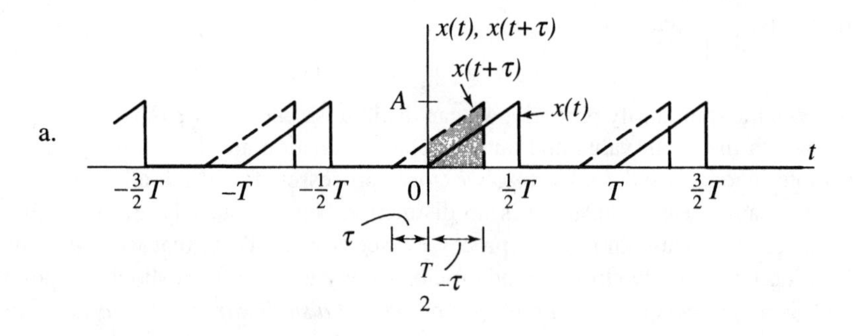

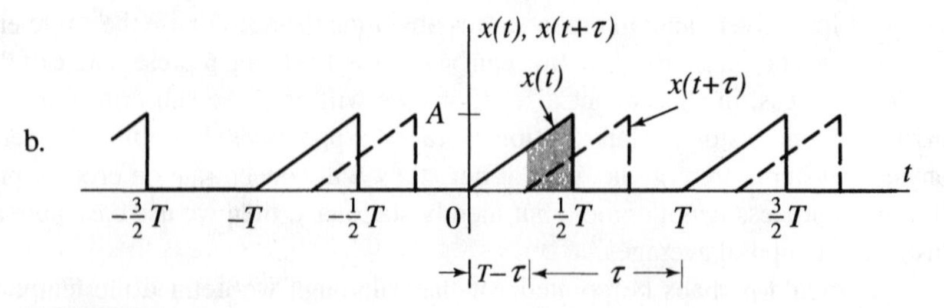

FIGURE 12.4
a. Functions $f(t)$ and $f(t+\tau)$ for the cases: a. $0 < \tau < T/2$ and b. $T/2 < \tau < T$

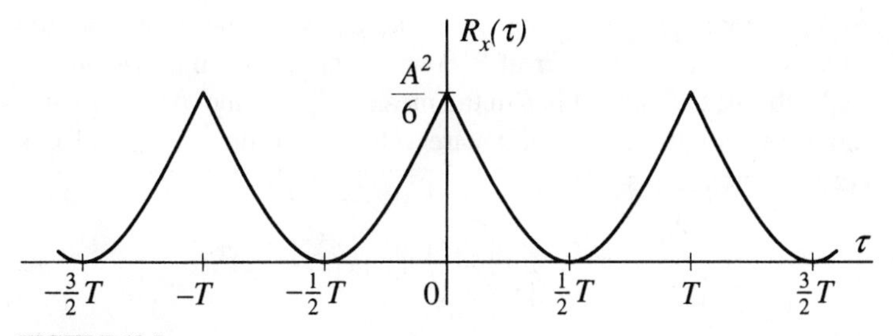

FIGURE 12.5
Autocorrelation function for the periodic function of Fig. 12.3

Hence, using Eq. (12.3), the mean values is simply

$$\mu_x = \frac{1}{T}\int_{-T/2}^{T/2} x(t)dt = \frac{1}{T}\int_0^{T/2} \frac{2A}{T}t\,dt = \frac{A}{4} \tag{b}$$

To calculate the autocorrelation function, we distinguish between the time shifts $0 < \tau < T/2$ and $T/2 < \tau < T$, as shown in Figs. 12.4a and b, respectively. Using Eq. (12.4), and considering Fig. 12.4a, we obtain for $0 < \tau < T/2$

$$\begin{aligned} R_x(\tau) &= \frac{1}{T}\int_{-T/2}^{T/2} x(t)x(t+\tau)dt = \frac{1}{T}\int_0^{(T/2)-\tau} \frac{2A}{T}t\frac{2A}{T}(t+\tau)dt \\ &= \frac{A^2}{6}\left[1 - 3\frac{\tau}{T} + 4\left(\frac{\tau}{T}\right)^3\right], \quad 0 < \tau < \frac{T}{2} \end{aligned} \tag{c}$$

where the limits of integration are defined by the overlapping portions of $x(t)$ and $x(t+\tau)$ (note shaded area in Fig. 12.4a). Similarly, from Fig. 12.4b, we obtain for $T/2 < \tau < T$

$$\begin{aligned} R_x(\tau) &= \frac{1}{T}\int_{T-\tau}^{T/2} \frac{2A}{T}t\frac{2A}{T}[t-(T-\tau)]dt \\ &= \frac{A^2}{6}\left[1 - \frac{3}{T}(T-\tau) + \frac{4}{T^3}(T-\tau)^3\right], \quad \frac{T}{2} < \tau < T \end{aligned} \tag{d}$$

The expressions for any other time shifts can be deduced from those above. Indeed, from Figs. 12.4a and b, it is not difficult to conclude that the autocorrelation function $R_x(\tau)$ must be periodic in τ with period T. Hence, from Eqs. (c) and (d), and the fact that $R_x(\tau)$ is periodic, we obtained the autocorrelation function plotted in Fig. 12.5.

12.3 MEAN SQUARE VALUES AND STANDARD DEVIATION

The *mean square value* of a random variable $x(t)$ is defined as

$$\psi_x^2 = \lim_{T\to\infty} \frac{1}{T}\int_{-T/2}^{T/2} x^2(t)dt \tag{12.6}$$

The positive square root of the mean square value is known as the *root mean square value*, or the rms *value*. Note that definition (12.6) applies to any arbitrary function $x(t)$, although our interest lies in sample functions from an ergodic random process.

For an ergodic process, the mean value μ_x is constant. In vibrations, μ_x can be regarded as the *static component* of $x(t)$ and $x(t) - \mu_x$ as the *dynamic component*. In many applications, the interest lies in the mean square value of the dynamic component. This quantity is simply the mean square value about the mean, and is known as the *variance*; its expression is

$$\sigma_x^2 = \lim_{T\to\infty} \frac{1}{T} \int_{-T/2}^{T/2} [x(t) - \mu_x]^2 dt \tag{12.7}$$

The positive square root of the variance is known as the *standard deviation*. Expanding Eq. (12.7), we obtain

$$\sigma_x^2 = \lim_{T\to\infty} \frac{1}{T} \int_{-T/2}^{T/2} x^2(t)dt - 2\mu_x \lim_{T\to\infty} \frac{1}{T} \int_{-T/2}^{T/2} x(t)dt + \mu_x^2 \tag{12.8}$$

and, in view of definitions (12.3) and (12.6), Eq. (12.8) reduces to

$$\sigma_x^2 = \psi_x^2 - \mu_x^2 \tag{12.9}$$

or *the variance is equal to the mean square value minus the square of the mean value.*

Example 12.2. Calculate the mean square value, the variance and the standard deviation for the function of Example 12.1.

Comparing Eqs. (12.4) and (12.6), we conclude that $\psi_x^2 = R_x(0)$, or the mean square value is equal to the autocorrelation function evaluated at $\tau = 0$. Hence, from Eq. (c) of Example 12.1, we simply obtain the mean square value

$$\psi_x^2 = R_x(0) = \frac{A^2}{6} \tag{a}$$

Introducing the above Eq. (a) and Eq. (b) of Example 12.1 into Eq. (12.9), we obtain the variance

$$\sigma_x^2 = \psi_x^2 - \mu_x^2 = \frac{A^2}{6} - \left(\frac{A}{4}\right)^2 = \tfrac{5}{48}A^2 \tag{b}$$

so that the standard deviation is

$$\sigma_x = \sqrt{\tfrac{5}{48}}A \tag{c}$$

12.4 PROBABILITY DENSITY FUNCTIONS

We have indicated in Sec. 12.2 that averages describing a given ergodic random process can be calculated by using a single representative sample function from the ensemble. Information concerning the properties of the random variable in the amplitude domain can be gained by means of *probability density functions*. To introduce the concept, we consider the time history $x(t)$ depicted in Fig. 12.6a and denote by $\Delta t_1, \Delta t_2, \ldots$ the time intervals during which the amplitude $x(t)$ is smaller than a given value x. Denoting by $\text{Prob}[x(t) < x]$ the probability that $x(t)$ is smaller than x, we observe that $\text{Prob}[x(t) < x]$ is equal to the probability that t lies in one of the time intervals $\Delta t_1, \Delta t_2, \ldots$. Considering a given large time interval T such that $0 < t < T$ and assuming that t has an equal chance

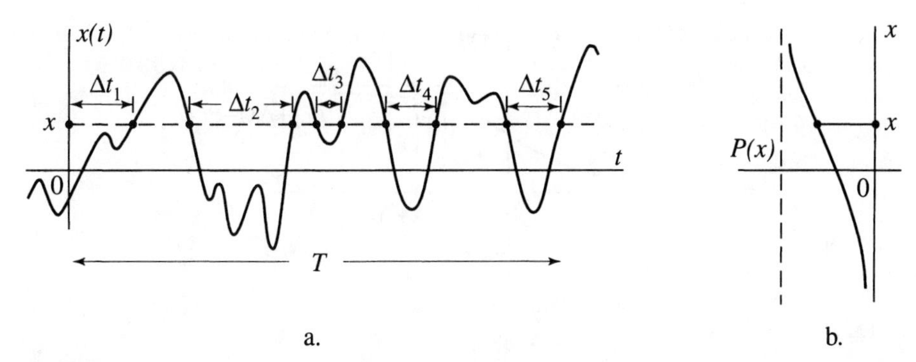

FIGURE 12.6
a. Time intervals for which $x(t) < x$, b. Probability distribution function

of taking any value from 0 to T, we obtain an estimate of the desired probability in the form

$$\text{Prob}[x(t) < x] = \lim_{T\to\infty} \frac{1}{T} \sum_i \Delta t_i \tag{12.10}$$

Letting x vary, we obtain the function

$$P(x) = \text{Prob}[x(t) < x] \tag{12.11}$$

which is known as the *probability distribution function* associated with the random variable $x(t)$. The function $P(x)$ is plotted in Fig. 12.6b as a function of x. The probability distribution function is a monotonically increasing function possessing the properties

$$P(-\infty) = 0,\ 0 \le P(x) \le 1,\ P(\infty) = 1 \tag{12.12}$$

Next, we consider the probability that the amplitude of the random variable is smaller than the value $x + \Delta x$ and denote that probability by $P(x + \Delta x)$. Clearly, the probability that $x(t)$ takes values between x and $x + \Delta x$ is $P(x + \Delta x) - P(x)$. This enables us to introduce the *probability density function*, defined as

$$p(x) = \lim_{\Delta x\to 0} \frac{P(x+\Delta x) - P(x)}{\Delta x} = \frac{dP(x)}{dx} \tag{12.13}$$

Geometrically, $p(x)$ represents the slope of the probability distribution function $P(x)$. Typical functions $P(x)$ and $p(x)$ are shown in Figs. 12.7a and b, respectively. From Eq. (12.13) and Figs. 12.7a and b, we conclude that the area under the curve $p(x)$ versus x corresponding to the amplitude increment Δx is equal to the change in $P(x)$ corresponding to the same increment. From Eq. (12.13), it is clear that the probability that $x(t)$ lies between the values x_1 and x_2 is

$$\text{Prob}(x_1 < x < x_2) = \int_{x_1}^{x_2} p(x)dx \tag{12.14}$$

which is equivalent to saying that the probability in question is equal to the area under the curve $p(x)$ versus x bounded by the vertical lines through $x = x_1$ and $x = x_2$. The

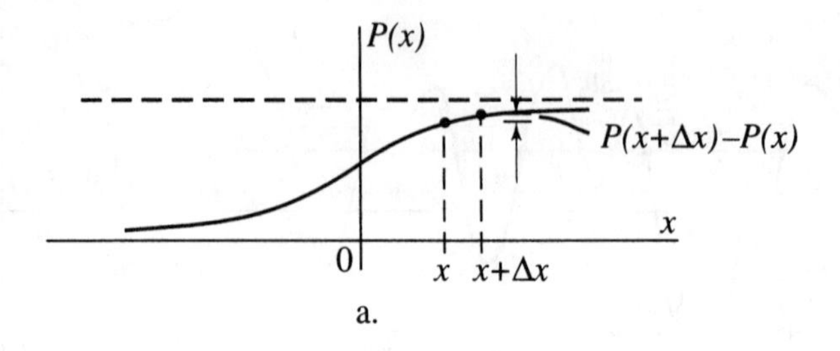

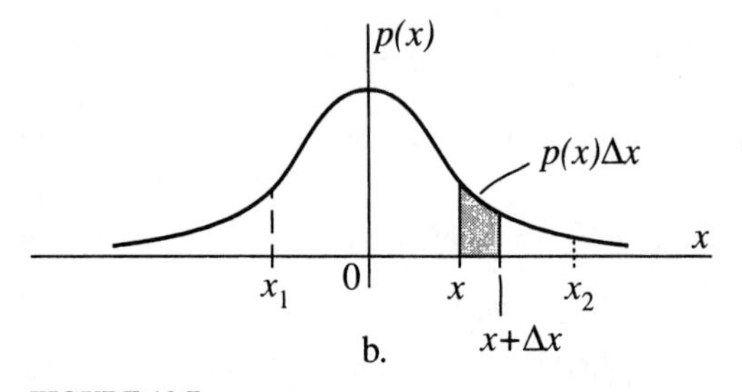

FIGURE 12.7
a. Probability distribution function, b. Probability density function

function $p(x)$ has the properties

$$p(x) \geq 0, \ p(-\infty) = 0, \ p(\infty) = 0,$$
$$P(x) = \int_{-\infty}^{x} p(\xi)d\xi, \ P(\infty) = \int_{-\infty}^{\infty} p(x)dx = 1 \tag{12.15}$$

where ξ is a mere dummy variable of integration.

As an illustration, we consider first the function $x(t)$ depicted in Fig. 12.8a. The fact that the function is deterministic does not detract from the usefulness of the example. From Fig. 12.8a, we conclude that the probability that $x(t)$ takes values smaller than $-A$ is zero. Similarly, the probability that $x(t)$ takes values smaller than A is equal to unity, because the event is a certainty. Due to the nature of the function $x(t)$, the probability increases linearly from zero at $x = -A$ to unity at $x = A$. The plot $P(x)$ versus x is shown in Fig. 12.8b. Using Eq. (12.13), it is possible to plot $p(x)$ versus x, as shown in Fig. 12.8c. The probability density function $p(x)$ is known as the *rectangular distribution*, or *uniform distribution*, for obvious reasons.

The probability distribution associated with a random variable such as that shown in Fig. 12.9a is of particular interest in our study. According to the *central limit theorem*,[1] if the random variable is the sum of a large number of independent random variables,

[1]See, for example, W. Feller, *Probability Theory and Its Applications*, vol. 1, John Wiley & Sons, Inc., New York, 1950, p. 202.

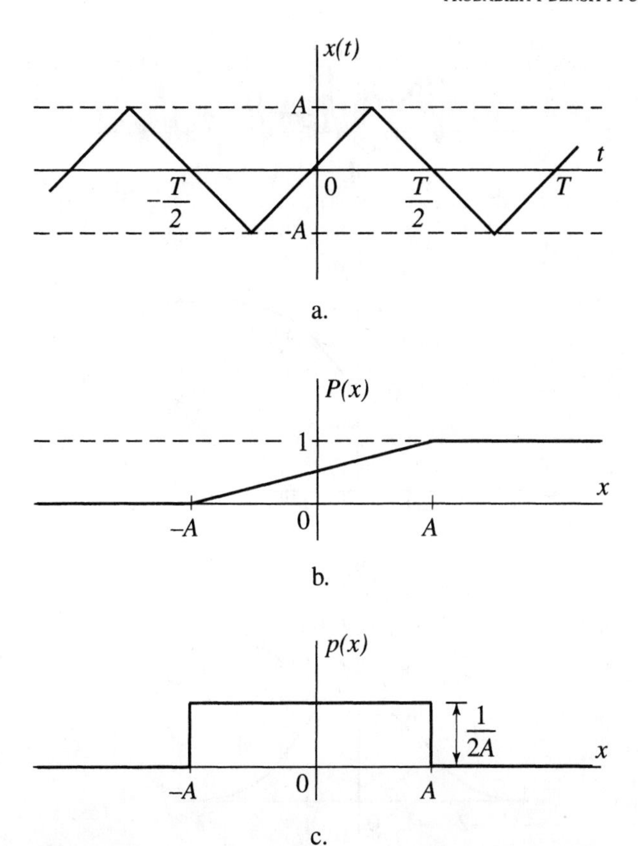

FIGURE 12.8
a. Time history, b. Probability distribution function, c. Probability density function

none of which contributes significantly to the sum, then under very general conditions the distribution approaches the *normal*, or *Gaussian distribution.* This is true even when the individual distributions of the independent random variables may not be specified, may all be different and may not be Gaussian. The normal distribution is described by the expressions

$$P(x) = \frac{1}{\sqrt{2\pi}} \int_{-\infty}^{x} e^{-\xi^2/2} d\xi, \; p(x) = \frac{1}{\sqrt{2\pi}} e^{-x^2/2} \tag{12.16}$$

The functions $P(x)$ versus x and $p(x)$ versus x given by Eqs. (12.16) are plotted in Figs. 12.9b and c, respectively. Figure 12.9c represents the so-called "standardized" normal distribution, in the sense that its mean value is zero and its standard deviation is unity. Normal distributions that are not standardized will be discussed later in this chapter. The probability distribution function $P(x)$ is also known as the *error function*, and is given in tabulated form in many mathematical handbooks, although the definition may vary slightly from table to table.

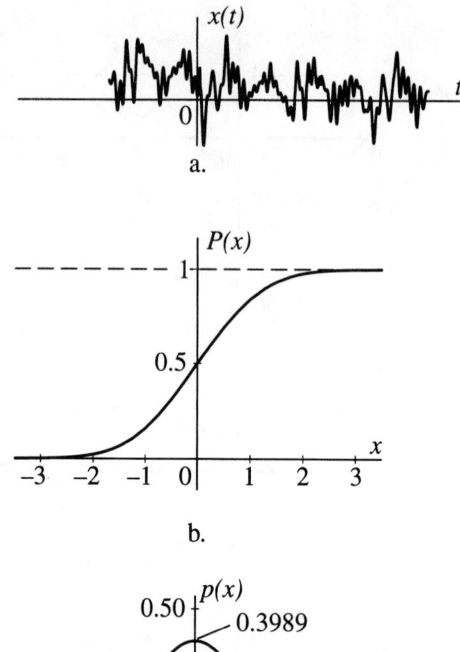

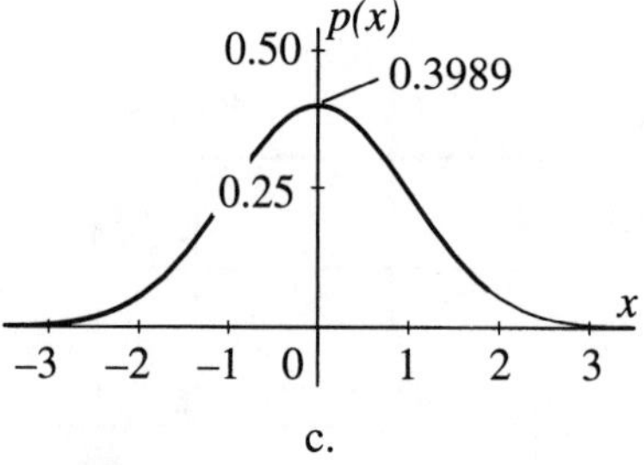

FIGURE 12.9
a. Random function, b. Normal, or Gaussian probability distribution function, c. Normal, or Gaussian probability density function

Another probability distribution of interest is the *Rayleigh distribution*, obtained when the random variable is restricted to positive values only. The Rayleigh distribution is defined by

$$P(x) = \begin{cases} 1 - e^{-x^2/2}, & x > 0 \\ 0, & x < 0 \end{cases}$$
$$p(x) = \begin{cases} xe^{-x^2/2}, & x > 0 \\ 0, & x < 0 \end{cases} \tag{12.17}$$

The functions $P(x)$ versus x and $p(x)$ versus x are plotted in Figs. 12.10a and b, respectively. The Rayleigh distribution discussed here can also be regarded as standardized.

On occasions, the random variable x is given as a function of another random variable y, $x = x(y)$, where y has the known probability density function $p(y)$, and the

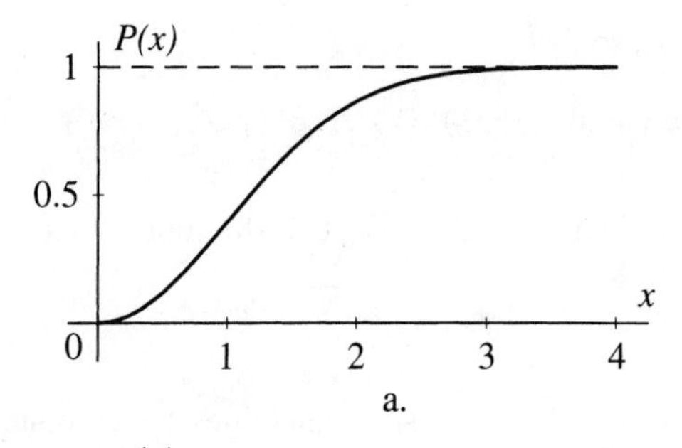

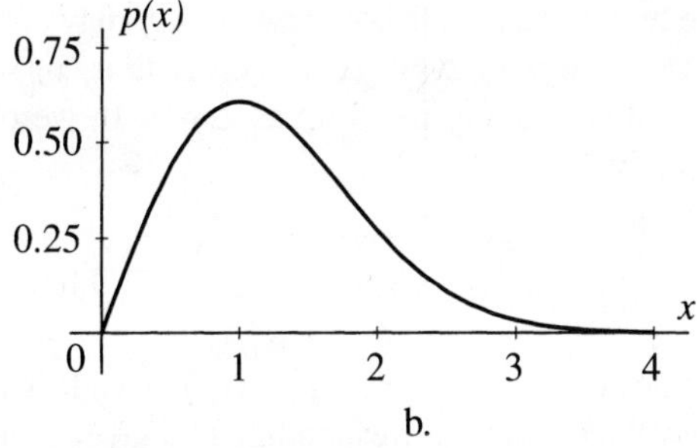

FIGURE 12.10
a. Rayleigh probability distribution function,
b. Rayleigh probability density function

interest lies in determining the probability density function $p(x)$, which necessitates a certain transformation of variables. To derive the required transformation, we consider the random variable $x(y)$ depicted in Fig. 12.11 and draw horizontal lines corresponding to $x = x_0$ and $x = x_0 + \Delta x_0$. The intersections of these lines with the curve $x(y)$ versus y define the increments of y bounded by y_1 and $y_1 + \Delta y_1$, y_2 and $y_2 + \Delta y_2$, etc. But the probability that $x(y)$ lies in the interval bounded by x_0 and $x_0 + \Delta x_0$ must be equal to the probability that y lies in any one of the increments bounded by y_i and

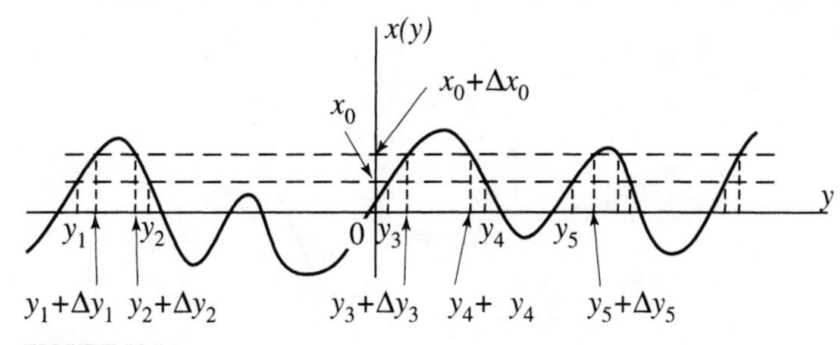

FIGURE 12.11
Random variable $x(y)$ versus y

$y_i + \Delta y_i$ $(i = 1, 2, \ldots)$, so that

$$\text{Prob}(x_0 < x < x_0 + \Delta x_0) = \sum_i \text{Prob}(y_i < y < y_i + \Delta y_i) \tag{12.18}$$

For a sufficiently small increment Δx_0, Eq. (12.18) implies that

$$p(x_0)\Delta x_0 = \sum_i p(y_i)|\Delta y_i| \tag{12.19}$$

where, because $p(x_0)$ and $p(y_i)$ are positive quantities, the absolute values $|\Delta y_i|$ must be used to account for the fact that a negative increment Δy_i may correspond to a positive increment Δx_0, as is the case with Δy_2, Δy_4, etc. Letting x_0 vary, dropping the no longer needed subscript 0 and taking the limit as $\Delta x \to 0$, we obtain the probability density function $p(x)$ in the form

$$p(x) = \lim_{\Delta x \to 0} \sum_i \frac{p(y_i)}{\Delta x / |\Delta y_i|} = \sum_i \left. \frac{p(y)}{|dx/dy|} \right|_{y=y_i} \tag{12.20}$$

where y_i are all the values of y corresponding to $x(y) = x$. It is clear from Fig. 12.11 that there can be many values $y = y_i$ corresponding to a given value $x(y) = x$.

As an illustration, we consider a sine wave of given amplitude A and frequency ω but random phase angle ϕ. For a fixed value t_0 of the time t, the sine wave can be regarded as a random function of ϕ and represented as follows:

$$x(\phi) = A \sin(\omega t_0 + \phi) \tag{12.21}$$

The function $x(\phi)$ is plotted in Fig. 12.12. Assuming that ϕ has a uniform probability density function, as defined by Fig. 12.8c, and considering only the interval $0 < \phi < 2\pi$, we can write

$$p(\phi) = \begin{cases} \dfrac{1}{2\pi}, & 0 < \phi < 2\pi \\ 0, & \phi < 0 \text{ and } \phi > 2\pi \end{cases} \tag{12.22}$$

But from Fig. 12.12 we see that there are two values of ϕ in the interval $0 < \phi < 2\pi$ for each value of x. Moreover, because the magnitude of the slope at one point is equal to

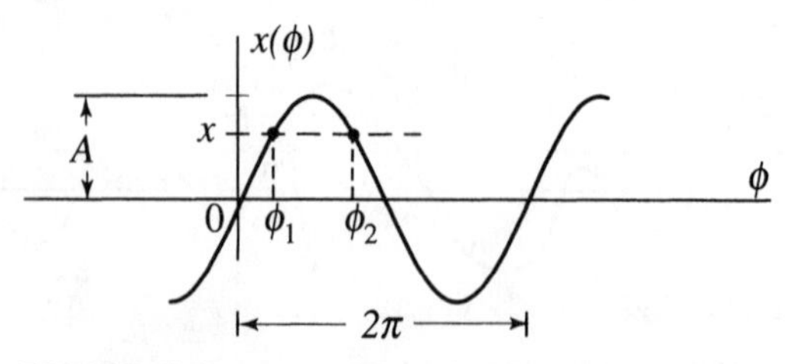

FIGURE 12.12
Sine wave with random phase angle ϕ

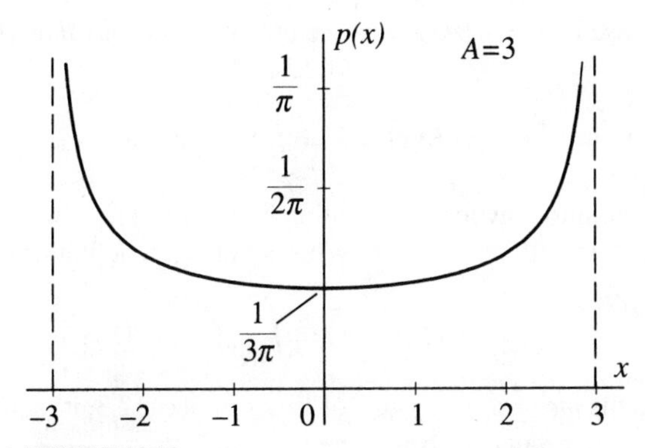

FIGURE 12.13
Probability density function for a sine wave with a random phase angle

that at the other point, we have

$$p(x) = 2\frac{1}{2\pi}\frac{1}{|dx/d\phi|} = \frac{1}{\pi}\frac{1}{A\cos(\omega t_0 + \phi)} = \frac{1}{\pi}\frac{1}{A[1 - \sin^2(\omega t_0 + \phi)]^{1/2}} \tag{12.23}$$

Inserting Eq. (12.21) into (11.23), and considering the fact that x cannot exceed A in magnitude, we obtain

$$p(x) = \begin{cases} \dfrac{1}{\pi}\dfrac{1}{(A^2 - x^2)^{1/2}}, & |x| < A \\ 0, \ |x| > A \end{cases} \tag{12.24}$$

The probability density function $p(x)$ is plotted in Fig. 12.13 for $A = 3$.

12.5 DESCRIPTION OF RANDOM DATA IN TERMS OF PROBABILITY DENSITY FUNCTIONS

If a sample time history $x(t)$ from a stationary random process is given, it is often convenient to reduce it to a probability density function $p(x)$. This is done by converting the function $x(t)$ into a voltage signal and feeding it into an analog amplitude probability density analyzer.[2] Then, having the probability density function $p(x)$, various averages can be calculated.

Next, we consider a real single-valued continuous function $g(x)$ of the random variable $x(t)$. Then, by definition, the *mathematical expectation of* $g(x)$, or the *expected value of* $g(x)$, is given by

$$E[g(x)] = \overline{g(x)} = \int_{-\infty}^{\infty} g(x)p(x)dx \tag{12.25}$$

[2]See J. S. Bendat and A. G. Piersol, *Random Data: Analysis and Measurement Procedures*, sec. 8.2, Interscience-Wiley, New York, 1971.

In the special case in which $g(x) = x$, we obtain the *mean value*, or *expected value*, of x in the form

$$E[x] = \bar{x} = \int_{-\infty}^{\infty} xp(x)dx \tag{12.26}$$

Note that this definition involves integration with respect to x, whereas definition (12.3) involves integration with respect to t. When $g(x) = x^2$, definition (12.25) yields

$$E[x^2] = \overline{x^2} = \int_{-\infty}^{\infty} x^2 p(x)dx \tag{12.27}$$

which is called the *mean square value* of x. As in Sec. 12.3, its square root is known as the *root mean square* value, or rms value.

Following the same pattern, the *variance* of x is

$$\begin{aligned}\sigma_x^2 &= E[(x-\bar{x})^2] = \int_{-\infty}^{\infty} (x-\bar{x})^2 p(x)dx \\ &= \int_{-\infty}^{\infty} x^2 p(x)dx - 2\bar{x}\int_{-\infty}^{\infty} xp(x)dx + (\bar{x})^2 \int_{-\infty}^{\infty} p(x)dx \end{aligned} \tag{12.28}$$

Considering Eqs. (12.26) and (12.27), as well as the fact that $\int_{-\infty}^{\infty} p(x)dx = 1$, Eq. (12.28) yields

$$\sigma_x^2 = \overline{x^2} - (\bar{x})^2 \tag{12.29}$$

As in Sec. 12.3, the square root of the variance is known as the *standard deviation*.

The above results can be given a geometric interpretation. To this end, we consider Fig. 12.14 showing the plot $p(x)$ versus x and recall that the area under the curve is equal to unity. Then, if $p(x)dx = dA$ is a differential element of area, as indicated in Fig. 12.14, $\bar{x}$ is simply the centroidal distance of the total area under the curve. It also follows that the variance σ_x^2 is equal to the centroidal moment of the area and the standard deviation σ_x plays the role of the radius of gyration. Moreover, Eq. (12.29) represents the "parallel axis theorem," according to which the centroidal moment of the area is

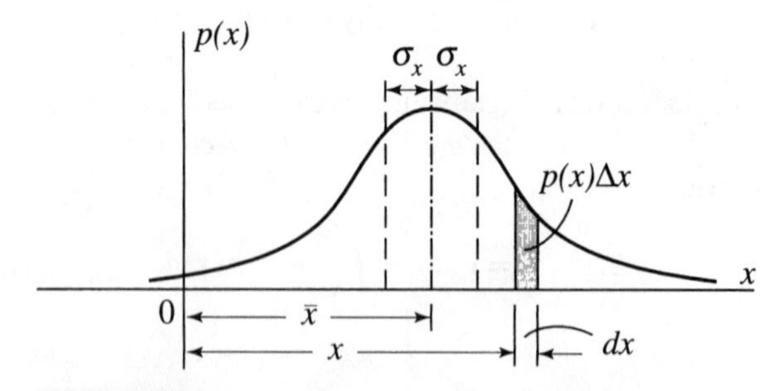

FIGURE 12.14
Probability density function showing the centroidal distance $\bar{x}$ and the standard deviation σ_x

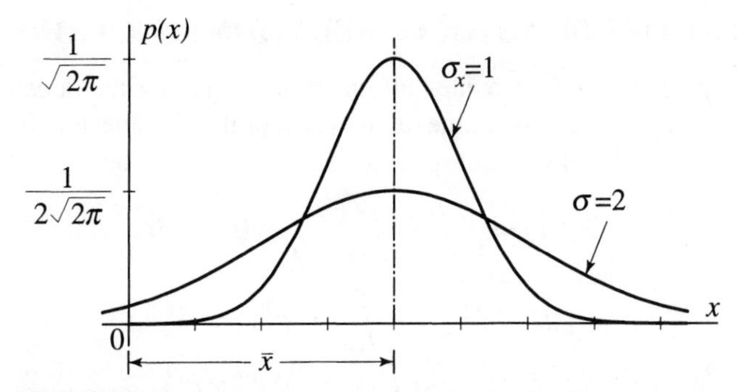

FIGURE 12.15
Normal probability density functions for $\sigma_x = 1$ and $\sigma_x = 2$

equal to the moment of the area about point 0 minus the total area times the centroidal distance squared.

The normal probability density function can be expressed in terms of the mean value $\bar{x}$ and standard deviation σ_x in the form

$$p(x) = \frac{1}{\sigma_x\sqrt{2\pi}}\exp\left[-\frac{(x-\bar{x})^2}{2\sigma_x^2}\right] \tag{12.30}$$

From Eq. (12.30), we conclude that for small σ_x the curve $p(x)$ versus x has a sharp peak at $x = \bar{x}$, whereas for large σ_x the curve tends to flatten and spread out. Plots of $p(x)$ versus x are shown in Fig. 12.15 for $\sigma_x = 1$ and $\sigma_x = 2$.

Example 12.3. Calculate the mean value and mean square value of the function $x(t)$ of Example 12.1 by using the probability density function of $x(t)$.

Using the analogy with the function of Fig. 12.8a, it can be shown that the function of Example 12.1 has the probability density function (see Prob. 12.10)

$$p(x) = \begin{cases} \dfrac{1}{2A}[A\delta(x)+1], & 0 \le x < A \\ 0 & \text{everywhere else} \end{cases} \tag{a}$$

where $\delta(x)$ is the Dirac delta function.

Inserting Eq. (a) into Eq. (12.26), we obtain the mean value

$$E[x] = \int_{-\infty}^{\infty} xp(x)dx = \int_0^A x\frac{1}{2A}[A\delta(x)+1]dx = \frac{A}{4} \tag{b}$$

Moreover, introducing Eq. (a) into Eq. (12.27), we arrive at the mean square value

$$E[x^2] = \int_{-\infty}^{\infty} x^2p(x)dx = \int_0^A x^2\frac{1}{2A}[A\delta(x)+1]dx = \frac{A^2}{6} \tag{c}$$

Note that the mean value and mean square value obtained here agree with those obtained in Example 12.2 by means of time averages, which is to be expected.

12.6 PROPERTIES OF AUTOCORRELATION FUNCTIONS

The autocorrelation function provides information concerning the dependence of the value of a random variable at one time on the value of the variable at another time. We recall from Sec. 12.2 that the definition of the autocorrelation function is

$$R_x(\tau) = \lim_{T\to\infty} \frac{1}{T} \int_{-T/2}^{T/2} x(t)x(t+\tau)dt \tag{12.31}$$

Next, we consider

$$\begin{aligned} R_x(-\tau) &= \lim_{T\to\infty} \frac{1}{T} \int_{-T/2}^{T/2} x(t)x(t-\tau)dt \\ &= \lim_{T\to\infty} \frac{1}{T} \int_{(-T/2)-\tau}^{(T/2)-\tau} x(\lambda)x(\tau+\lambda)d\lambda \end{aligned} \tag{12.32}$$

where we made the substitution $t-\tau=\lambda$, $dt=d\lambda$. Because both limits of integration in the last integral are shifted in the same direction and by the same amount τ, the interval of integration remains equal to T. It is easy to see that, as $T\to\infty$, the shift in the location of the interval of integration becomes inconsequential, so that

$$R_x(-\tau) = \lim_{T\to\infty} \frac{1}{T} \int_{-T/2}^{T/2} x(t)x(t+\tau)dt \tag{12.33}$$

Comparing Eqs. (12.31) and (12.33), we conclude that

$$R_x(\tau) = R_x(-\tau) \tag{12.34}$$

or *the autocorrelation is an even function of* τ.

Another property of the autocorrelation function can be revealed by considering

$$\begin{aligned} &\lim_{T\to\infty} \frac{1}{T} \int_{-T/2}^{T/2} [x(t) \pm x(t+\tau)]^2 dt \\ &= \lim_{T\to\infty} \frac{1}{T} \int_{-T/2}^{T/2} [x^2(t) \pm 2x(t)x(t+\tau) + x^2(t+\tau)]dt \\ &= \lim_{T\to\infty} \frac{2}{T} \int_{-T/2}^{T/2} x^2(t)dt \pm \lim_{T\to\infty} \frac{2}{T} \int_{-T/2}^{T/2} x(t)x(t+\tau)dt \\ &= 2R_x(0) \pm 2R_x(\tau) \geq 0 \end{aligned} \tag{12.35}$$

The above inequality is true because the first integral cannot be negative. From inequality (12.35), it follows that

$$R_x(0) \geq |R_x(\tau)| \tag{12.36}$$

which implies that the *the maximum value of the autocorrelation function is obtained for* $\tau = 0$. From definition (12.6) we conclude that $R_x(0)$ is equal to the mean square value of the random variable $x(t)$, namely,

$$R_x(0) = \psi_x^2 \tag{12.37}$$

Hence, *the maximum value of the autocorrelation function is equal to the mean square value*. Note that if $x(t)$ is periodic, then $R_x(\tau)$ is also periodic, and the maximum value of $R_x(\tau)$ is obtained not only at $\tau = 0$ but also for values of τ equal to integer multiples of the period. An illustration of this fact can be seen in Fig. 12.5.

12.7 RESPONSE TO ARBITRARY EXCITATIONS BY FOURIER TRANSFORMS

In Sec. 4.4, we derived the response to arbitrary excitations in the time domain by means of the convolution integral. Another approach to the derivation of the response to arbitrary excitations is by mean of the Fourier transformation, which is a frequency domain technique. In the case of deterministic problems, time domain solutions tend to have an edge over frequency domain solutions. In the case of stochastic problems, however, it is advisable to work in the frequency domain. In this section, we introduce the Fourier transforms in the context of the response to arbitrary excitations, but our real interest in the approach is to apply it to stochastic problems. Indeed, Fourier transforms are indispensable to the characterization of the response to random excitations, as can be concluded from the remainder of this chapter.

In Sec. 3.9, we demonstrated that a periodic function of period T, such as that shown in Fig. 3.25, can be represented by a Fourier series, namely, an infinite series of harmonic functions of frequencies $p\omega_0$ $(p = 0, \pm 1, \pm 2, \ldots)$, where $\omega_0 = 2\pi/T$ is the fundamental frequency. Letting the period T approach infinity, the function becomes nonperiodic. In the process, the discrete frequencies $p\omega_0$ draw closer and closer together until they become continuous, at which time the Fourier series becomes a Fourier integral. To substantiate the preceding statement, we represent a periodic function, such as that illustrated in Fig. 3.25,by the Fourier series in its complex form

$$f(t) = \sum_{p=-\infty}^{\infty} C_p e^{ip\omega_0 t}, \quad \omega_0 = \frac{2\pi}{T} \tag{12.38}$$

where the coefficients C_p are given by

$$C_p = \frac{1}{T}\int_{-T/2}^{T/2} f(t)e^{-ip\omega_0 t}dt, \quad p = 0, \pm 1, \pm 2, \ldots \tag{12.39}$$

provided the integrals exist. The Fourier expansion, Eqs. (12.38) and (12.39), provides the information concerning the frequency composition of the periodic function $f(t)$. Introducing the notation $p\omega_0 = \omega_p$, $(p+1)\omega_0 - p\omega_0 = \omega_0 = 2\pi/T = \Delta\omega_p$, Eqs. (12.38) and (12.39) can be rewritten as

$$f(t) = \sum_{p=-\infty}^{\infty} \frac{1}{T}(TC_p)e^{i\omega_p t} = \frac{1}{2\pi}\sum_{p=-\infty}^{\infty}(TC_p)e^{i\omega_p t}\Delta\omega_p \tag{12.40}$$

$$TC_p = \int_{-T/2}^{T/2} f(t)e^{-i\omega_p t}dt \tag{12.41}$$

Letting the period increase without bounds, $T \to \infty$, dropping the subscript p, so that the discrete variable ω_p simply becomes the continuous variable ω, and taking the limit, we can replace the summation in Eq. (12.40) by integration and obtain

$$f(t) = \lim_{\substack{T\to\infty \\ \Delta\omega_p \to 0}} \frac{1}{2\pi} \sum_{p=-\infty}^{\infty} (TC_p) e^{i\omega_p t} \Delta\omega_p = \frac{1}{2\pi} \int_{-\infty}^{\infty} F(\omega) e^{i\omega t} d\omega \tag{12.42}$$

$$F(\omega) = \lim_{\substack{T\to\infty \\ \Delta\omega_p \to 0}} (TC_p) = \int_{-\infty}^{\infty} f(t) e^{-i\omega t} dt \tag{12.43}$$

Equation (12.42) implies that any arbitrary function $f(t)$ can be described by an integral representing contributions of harmonic components having a *continuous frequency spectrum* ranging from $-\infty$ to $+\infty$. The quantity $F(\omega)d\omega$ can be regarded as the contribution to the function $f(t)$ of the harmonics in the frequency interval from ω to $\omega + d\omega$.

Equation (12.42) is the Fourier integral representation of an arbitrary function $f(t)$, such as that shown in Fig. 12.16. Moreover, the function $F(\omega)$ in Eq. (12.43) is known as the *Fourier transform of* $f(t)$, so that the integrals

$$F(\omega) = \int_{-\infty}^{\infty} f(t) e^{-i\omega t} dt \tag{12.44}$$

$$f(t) = \frac{1}{2\pi} \int_{-\infty}^{\infty} F(\omega) e^{i\omega t} d\omega \tag{12.45}$$

represent simply a Fourier transform pair, where $f(t)$ is known as the *inverse Fourier transform of* $F(\omega)$. By analogy with the Fourier series expansion of a periodic function, Eqs. (12.38) and (12.39), the Fourier transform pair, Eqs. (12.44) and (12.45), also provides the information concerning the frequency composition of $f(t)$, where this time $f(t)$ is nonperiodic.

The representation of $f(t)$ by an integral is possible provided the integral (12.44) exists. The existence is ensured if $f(t)$ satisfies Dirichlet's conditions[3] in the domain $-\infty < t < \infty$ and if the integral $\int_{-\infty}^{\infty} |f(t)| dt$ is convergent. If the integral $\int_{-\infty}^{\infty} |f(t)| dt$ is not convergent, then the Fourier transform $F(\omega)$ need not exist. This is indeed the case for $f(t) = \sin \alpha t$, for which the integral $\int_{-\infty}^{\infty} |f(t)| dt$ does not converge.

From Sec. 3.9, we conclude that if Eq. (12.38) represents an excitation function of the form $f(t) = F(t)/k$, where $F(t)$ is the applied force and k the spring constant, according to Eq. (3.2), then the response of the system can be written in the form

$$x(t) = \sum_{p=-\infty}^{\infty} C_p G_p e^{ip\omega_0 t} \tag{12.46}$$

[3]The function $f(t)$ is said to satisfy Dirichlet's conditions in the interval (a, b) if 1) $f(t)$ has only a finite number of maxima and minima in (a,b) and 2) $f(t)$ has only a finite number of finite discontinuities in (a, b), and no infinite discontinuities.

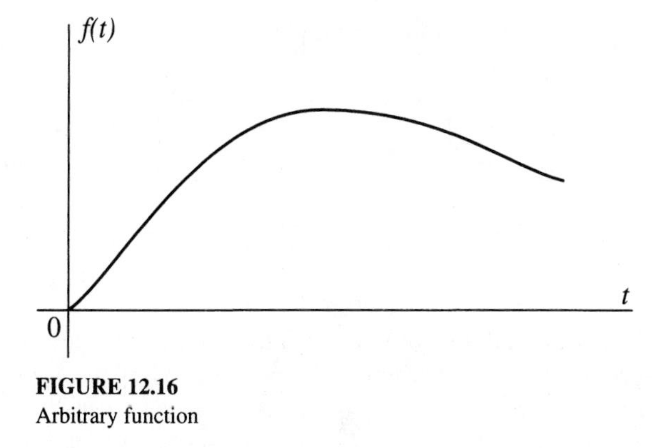

FIGURE 12.16
Arbitrary function

where G_p is the frequency response associated with the frequency $p\omega_0$. Following a procedure similar to that used for $f(t)$, we conclude that the response of the system to an arbitrary excitation of the type shown in Fig. 12.16 can also be written in the form of a Fourier transform pair, as follows:

$$X(\omega) = \int_{-\infty}^{\infty} x(t) e^{-i\omega t} dt \tag{12.47}$$

$$x(t) = \frac{1}{2\pi} \int_{-\infty}^{\infty} X(\omega) e^{i\omega t} d\omega \tag{12.48}$$

where the Fourier transform of the response is

$$X(\omega) = G(\omega) F(\omega) \tag{12.49}$$

which is simply the product of the frequency response and the Fourier transform of the excitation. Note that, for consistency of notation, we dropped i from the argument of G.

To obtain the system response as a function of time, it is necessary to evaluate the definite integral in Eq. (12.48), which can lead to contour integrations in the complex plane, a delicate task at best. However, in a manner reminiscent of the discrete-time approach introduced in Sec. 4.9, Fourier transforms can be evaluated numerically by means of discrete Fourier transforms. Then, the computational effort can be reduced significantly by means of the fast Fourier transform, an efficient algorithm for the computer evaluation of discrete Fourier transforms.[4]

Example 12.4. Calculate the response $x(t)$ of an undamped single-degree-of-freedom system to the excitation $f(t)$ in the form of the rectangular pulse shown in Fig. 12.17a using an approach based on the Fourier transform. Plot the frequency spectra associated with $f(t)$ and $x(t)$.

---

[4]E. D. Brigham, *The Fast Fourier Transform*, Prentice-Hall, Inc., Englewood Cliffs, NJ, 1974.

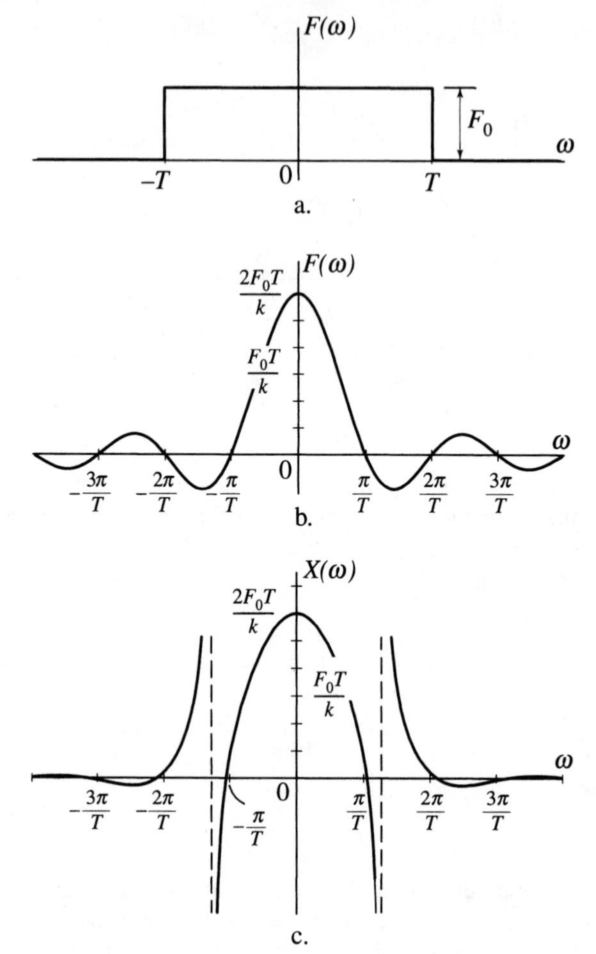

FIGURE 12.17
a. Excitation in the form of a rectangular pulse, b. Fourier transform of the rectangular pulse, c. Response frequency spectrum

Recalling that $f(t) = F(t)/k$, the function $f(t)$ depicted in Fig. 12.17a can be defined by

$$f(t) = \begin{cases} \dfrac{F_0}{k}, & -T < t < T \\ 0 \text{ everywhere else} \end{cases} \tag{a}$$

and we note that $f(t)$ has only two finite discontinuities and no infinite discontinuities, so that $f(t)$ satisfies Dirichlet's conditions. Hence, it is possible to write a Fourier transform for $f(t)$ as follows:

$$F(\omega) = \int_{-\infty}^{\infty} f(t)e^{-i\omega t}dt = \frac{F_0}{k}\int_{-T}^{T} e^{-i\omega t}dt = \frac{F_0}{k}\frac{1}{i\omega}(e^{i\omega T} - e^{-i\omega T}) \tag{b}$$

For an undamped system, $\zeta = 0$, the frequency response, Eq. (3.20), reduces to

$$G(\omega) = \frac{1}{1-(\omega/\omega_n)^2} \tag{c}$$

so that, inserting Eqs. (b) and (c) into Eq. (12.49), we obtain

$$X(\omega) = G(\omega)F(\omega) = \frac{F_0}{k}\frac{e^{i\omega T} - e^{-i\omega T}}{i\omega[1-(\omega/\omega_n)^2]} \tag{d}$$

Hence, the response $x(t)$ can be written in the form of the inverse Fourier transform

$$x(t) = \frac{1}{2\pi}\int_{-\infty}^{\infty} X(\omega)e^{i\omega t}d\omega = \frac{F_0}{k}\frac{1}{2\pi i}\int_{-\infty}^{\infty}\frac{e^{i\omega T} - e^{-i\omega T}}{\omega[1-(\omega/\omega_n)^2]}e^{i\omega t}d\omega \tag{e}$$

Before attempting the evaluation of the above integral, it is convenient to consider the following partial fractions expansion:

$$\frac{1}{\omega[1-(\omega/\omega_n)^2]} = \frac{1}{\omega} - \frac{1}{2(\omega-\omega_n)} - \frac{1}{2(\omega+\omega_n)} \tag{f}$$

so that Eq. (e) becomes

$$x(t) = \frac{F_0}{k}\frac{1}{2\pi i}\int_{-\infty}^{\infty}\left[\frac{1}{\omega} - \frac{1}{2(\omega-\omega_n)} - \frac{1}{2(\omega+\omega_n)}\right][e^{i\omega(t+T)} - e^{i\omega(t-T)}]d\omega \tag{g}$$

To evaluate the integral involved in (g), it is necessary to perform contour integrations in the complex plane. As this exceeds the scope of this text, we present here pertinent results only, namely,

$$\begin{aligned}
&\int_{-\infty}^{\infty}\frac{e^{i\omega\lambda}}{\omega}d\omega = \begin{cases} 0, & \lambda < 0 \\ 2\pi i, & \lambda > 0 \end{cases} \\
&\int_{-\infty}^{\infty}\frac{e^{i\omega\lambda}}{\omega-\omega_n}d\omega = \begin{cases} 0, & \lambda < 0 \\ 2\pi i e^{i\omega_n\lambda}, & \lambda > 0 \end{cases} \\
&\int_{-\infty}^{\infty}\frac{e^{i\omega\lambda}}{\omega+\omega_n}d\omega = \begin{cases} 0, & \lambda < 0 \\ 2\pi i e^{-i\omega_n\lambda}, & \lambda > 0 \end{cases}
\end{aligned} \tag{h}$$

From Eq. (g) we note that λ takes the values $t+T$ and $t-T$. Hence, we must distinguish between the time domains defined by $t+T<0$ and $t-T<0$, $t+T>0$ and $t-T<0$ and $t+T>0$ and $t-T>0$, which are the same as the domains $t<-T$, $-T<t<T$ and $t>T$, respectively. Inserting the integrals (h) with proper λ into (g), we obtain

$$\begin{aligned}
x(t) &= 0, \quad t < -T \\
x(t) &= \frac{F_0}{k}\frac{1}{2\pi i}\left(2\pi i - \tfrac{1}{2}2\pi i e^{i\omega_n(t+T)} - \tfrac{1}{2}2\pi i e^{-i\omega_n(t+T)}\right) \\
&= \frac{F_0}{k}[1-\cos\omega_n(t+T)], \quad -T<t<T \\
x(t) &= \frac{F_0}{k}\frac{1}{2\pi i}[(2\pi i - \tfrac{1}{2}2\pi i e^{i\omega_n(t+T)} - \tfrac{1}{2}2\pi i e^{-i\omega_n(t+T)}) \\
&\qquad - (2\pi i - \tfrac{1}{2}2\pi i e^{i\omega_n(t-T)} - \tfrac{1}{2}2\pi i e^{-i\omega_n(t-T)})] \\
&= \frac{F_0}{k}[\cos\omega_n(t-T) - \cos\omega_n(t+T)], \quad t > T
\end{aligned} \tag{i}$$

Note that essentially the same problem was solved in Example 4.4 by means of the convolution integral in the time domain, except that in Example 4.4 the system was damped and the rectangular pulse started at $t = 0$.

The frequency spectrum associated with $f(t)$ is given by Eq. (b). Recalling that $(e^{i\omega T} - e^{-i\omega T})/2i = \sin\omega T$, Eq. (b) becomes

$$F(\omega) = \frac{2F_0}{k}\frac{\sin\omega T}{\omega} \tag{j}$$

Figure 12.17b shows the plot $F(\omega)$ versus ω. Moreover, the frequency spectrum associated with $x(t)$ is given by Eq. (d). In a similar manner, the equation can be reduced to

$$X(\omega) = \frac{2F_0}{k}\frac{\sin\omega T}{\omega[1-(\omega/\omega_n)^2]} \tag{k}$$

Figure 12.17c shows the plot $X(\omega)$ versus ω. Note that Figs. 12.17b and c represent continuous frequency spectra, as opposed to Figs. 3.26a and b, which represent discrete frequency spectra.

Comparing the method of solution of this example to that of Example 4.4, it is easy to see that the use of the convolution integral provides a simpler approach to the problem of obtaining the response $x(t)$ than the Fourier transform approach. This is particularly true in view of the fact that the question of contour integrations in the complex plane has not really been addressed in this example. In random vibrations, however, the time-domain response plays no particular role and the interest lies primarily in frequency-domain analyses for which Fourier transforms are indispensable. The preceding statement refers to spectral analysis, a basic tool in the treatment of random vibrations.

12.8 POWER SPECTRAL DENSITY FUNCTIONS

The autocorrelation function provides information concerning properties of a random variable in the time domain. On the other hand, the *power spectral density function* provides similar information in the frequency domain. Although for ergodic random processes the power spectral density function furnishes essentially no information not provided by the autocorrelation function, in certain applications the first form is more convenient than the second.

We consider the representative sample function $f(t)$ from an ergodic random process and write the autocorrelation function of the process in the form

$$R_f(\tau) = \lim_{T\to\infty}\frac{1}{T}\int_{-T/2}^{T/2} f(t)f(t+\tau)dt \tag{12.50}$$

Then, we define the power spectral density function $S_f(\omega)$ as the Fourier transform of $R_f(\tau)$, namely,

$$S_f(\omega) = \int_{-\infty}^{\infty} R_f(\tau)e^{-i\omega\tau}d\tau \tag{12.51}$$

which implies that the autocorrelation function can be obtained in the form of the inverse Fourier transform

$$R_f(\tau) = \frac{1}{2\pi}\int_{-\infty}^{\infty} S_f(\omega)e^{i\omega\tau}d\omega \tag{12.52}$$

The conditions for the existence of the power spectral density function $S_f(\omega)$ are that the function $R_f(\tau)$ satisfy Dirichlet's conditions and that the integral $\int_{-\infty}^{\infty}|R_f(\tau)|d\tau$ be convergent (see Sec. 12.7). Various analysts define $S_f(\omega)$ as the quantity given by Eq. (12.51) divided by 2π. As will be seen shortly, this latter definition has certain advantages. However, in this case $S_f(\omega)$ would no longer be the Fourier transform of $R_f(\tau)$.

Next, we explore the physical significance of the function $S_f(\omega)$. To this end, we let $\tau = 0$ in Eqs. (12.50) and (12.52), and write the mean square value of $f(t)$ in the two forms

$$R_f(0) = \lim_{T\to\infty} \frac{1}{T}\int_{-T/2}^{T/2} f^2(t)dt = \frac{1}{2\pi}\int_{-\infty}^{\infty} S_f(\omega)d\omega \tag{12.53}$$

Assuming that $f(t)$ describes a voltage, the mean square value of $f(t)$ represents the mean power dissipated in a 1-ohm resistor. In view of this, Eq. (12.53) can be interpreted as stating that the integral of $S_f(\omega)/2\pi$ with respect to ω over the entire range of frequencies, $-\infty < \omega < \infty$, gives the total mean power of $f(t)$. Hence, it follows that $S_f(\omega)$ (divided by 2π) is the *power spectral density function*, or the *power density spectrum of* $f(t)$. The function $S_f(\omega)$ is also known as the *mean square spectral density*. As can be inferred from the name, the power spectral density function represents a continuous spectrum, so that in terms of electrical terminology the average power dissipated in a 1-ohm resistor by the frequency components of a voltage lying in an infinitesimal band between ω and $\omega + d\omega$ is proportional to $S_f(\omega)d\omega$ (again divided by the factor 2π). If for a given random process the mean square spectral density $S_f(\omega)$ is known, perhaps obtained through measurement, then Eq. (12.53) can be used to evaluate the mean square value of an ergodic random process. The function $S_f(\omega)$ has certain properties that can be used to render the evaluation of averages easier. These properties will now be discussed.

In view of its physical interpretation, we must conclude that $S_f(\omega)$ *is always nonnegative*, i.e., it is either positive or zero, $S_f(\omega) \geq 0$. We have shown in Sec. 12.6 that $R_f(\tau)$ is an even function of τ, $R_f(\tau) = R_f(-\tau)$. From Eq. (12.51), it follows that

$$\begin{aligned} S_f(\omega) &= \int_{-\infty}^{\infty} R_f(\tau)e^{-i\omega\tau}d\tau = \int_{-\infty}^{\infty} R_f(-\tau)e^{-i\omega\tau}d\tau \\ &= -\int_{\infty}^{-\infty} R_f(\sigma)e^{i\omega\sigma}d\sigma = S_f(-\omega) \end{aligned} \tag{12.54}$$

where σ is a dummy variable of integration, so that *the power spectral density* $S_f(\omega)$ *is an even function of* ω. Because $R_f(\tau)$ is an even function of τ, Eq. (12.51) leads to

$$\begin{aligned} S_f(\omega) &= \int_{-\infty}^{\infty} R_f(\tau)e^{-i\omega\tau}d\tau = \int_{-\infty}^{\infty} R_f(\tau)(\cos\omega\tau - i\sin\omega\tau)d\tau \\ &= \int_{-\infty}^{\infty} R_f(\tau)\cos\omega\tau d\tau = 2\int_{0}^{\infty} R_f(\tau)\cos\omega\tau d\tau \end{aligned} \tag{12.55}$$

But the autocorrelation $R_f(\tau)$ is a real function, so that from the last integral in Eq. (12.55) it follows that $S_f(\omega)$ *is a real function*. As a result of $S_f(\omega)$ being an even, real function of ω, Eq. (12.52) can be reduced to

$$R_f(\tau) = \frac{1}{\pi}\int_0^{\infty} S_f(\omega)\cos\omega\tau d\omega \tag{12.56}$$

Equations (12.55) and (12.56) are called the *Wiener-Khintchine equations*, and except for a factor of 2 they represent what is known as a Fourier cosine transform pair. It follows from Eq. (12.56) that

$$R_f(0) = \frac{1}{\pi}\int_0^{\infty} S_f(\omega) d\omega \tag{12.57}$$

which provides a convenient formula for the calculation of the mean square value of a stationary random process if the power spectral density is given. The advantage of Eqs. (12.56) and (12.57) over Eqs. (12.52) and (12.53), respectively, is that Eqs. (12.56) and (12.57) contain no negative frequencies.

12.9 NARROWBAND AND WIDEBAND RANDOM PROCESSES

The mean square spectral density provides a measure of the representation of given frequencies in a random process. For convenience, we present our discussion in terms of ergodic random processes. Random processes are often identified by the shape of the power density spectra. In particular, we distinguish between narrowband and wideband random processes. The terminology used is not precise, and it provides only a qualitative description of a given process. A *narrowband process* is characterized by a sharply peaked power density spectrum $S_f(\omega)$, in the senses that $S_f(\omega)$ has significant values only in a short band of frequencies centered around the frequency corresponding to the peak. A sample time history representative of a narrowband process contains only a narrow range of frequencies. In the case of a *wideband process*, on the other hand, the power density spectrum $S_f(\omega)$ has significant values over a wide band of frequencies whose width is of the same order of magnitude as the center frequency of the band. A sample time history representative of a wideband process contains a wide range of frequencies. At the two extremes we find a power density spectrum consisting of two symmetrically placed delta functions, corresponding to a sinusoidal sample function, and a uniform power density spectrum, corresponding to a sample function in which all the frequencies are equally represented. The first, of course, is a deterministic function, but it can be regarded as random if the phase angle is randomly distributed (see Sec. 12.4). The second is known as *white noise* by analogy with white light, which has a flat spectrum over the visible range. If the frequency band is infinite, then we speak of *ideal white noise*. This concept represents a physical impossibility because it implies an infinite mean square value, and hence infinite power. A judicious use of the concept, however, can lead to meaningful results. For comparison purposes, it may prove of interest to plot some sample functions and the autocorrelation functions, probability density functions and power density spectra corresponding to these sample functions.

Figure 12.18 shows plots of possible time histories. Figure 12.18a shows the simple sinusoidal function $f(t) = A\sin(\omega_0 t + \phi)$, whereas Figs. 12.18b, c and d show time histories corresponding to a narrowband random process, a wideband random process

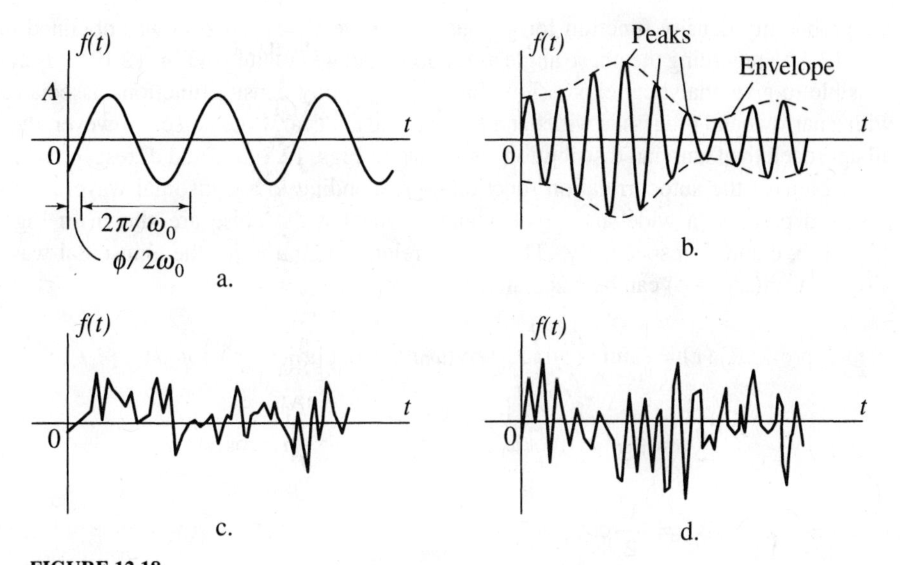

FIGURE 12.18
a. Sinusoidal function, b. Narrowband random process, c. Wideband random process, d. Ideal white noise

and an ideal white noise, respectively. Note that Fig. 12.18b has the appearance of a sinusoidal function with randomly varying amplitude. Figures 12.18c and d look somewhat similar because both time histories contain a wide range of frequencies. Figure 12.19 shows plots of corresponding probability density functions. Figure 12.19a depicts

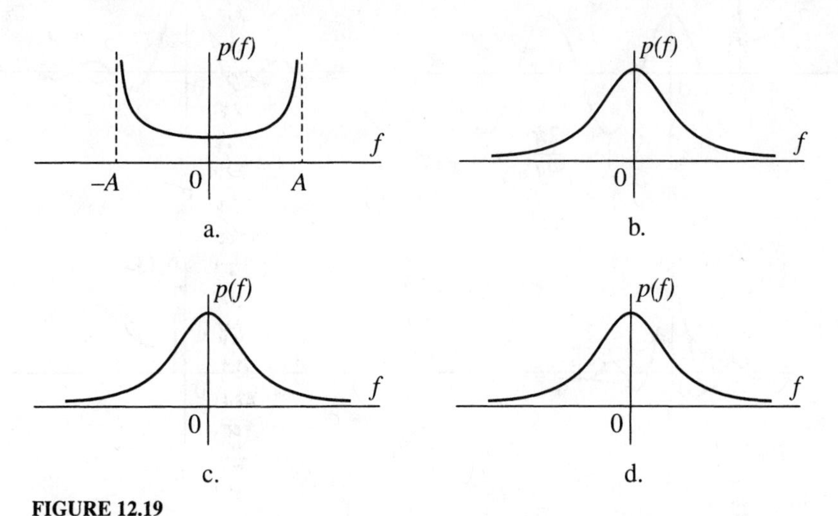

FIGURE 12.19
Probability density function for: a. Sinusoidal function, b. Narrowband process, c. Wideband process and d. Ideal white noise

the probability density function for a sinusoidal wave. This function was obtained in Sec. 12.4 by regarding the phase angle as random, and was plotted in Fig. 12.13. It is not possible to give analytical expressions for the probability density functions associated with a narrowband process, a wideband process and an ideal white noise. However, they all approach the Gaussian distribution, as shown in Figs. 12.19b, c and d, respectively.

Plots of the autocorrelation function corresponding to a sinusoidal wave, a narrowband process, a wideband process and an ideal white noise are shown in Figs. 12.20a, b, c and d, respectively. The autocorrelation function for the sinusoidal wave $f(t) = A\sin(\omega_0 t + \phi)$ can be calculated as follows:

$$\begin{aligned} R_f(\tau) &= \lim_{T\to\infty} \frac{A^2}{T}\int_{-T/2}^{T/2} \sin(\omega_0 t + \phi)\sin[\omega_0(t+\tau)+\phi]dt \\ &= \frac{A^2}{2\pi}\int_{-\pi}^{\pi} (\cos\omega_0\tau\sin^2\alpha + \sin\omega_0\tau\sin\alpha\cos\alpha)d\alpha \\ &= \frac{A^2}{2}\cos\omega_0\tau \qquad (12.58) \end{aligned}$$

which is a cosine function with the same frequency as the sine wave but with zero phase angle. The autocorrelation function for the narrowband process appears as a cosine function of decaying amplitude, and that of a wideband process appears sharply peaked and decaying rapidly to zero. In the limit, as the width of the frequency band increases indefinitely, the autocorrelation function reduces to that for the ideal white noise, having

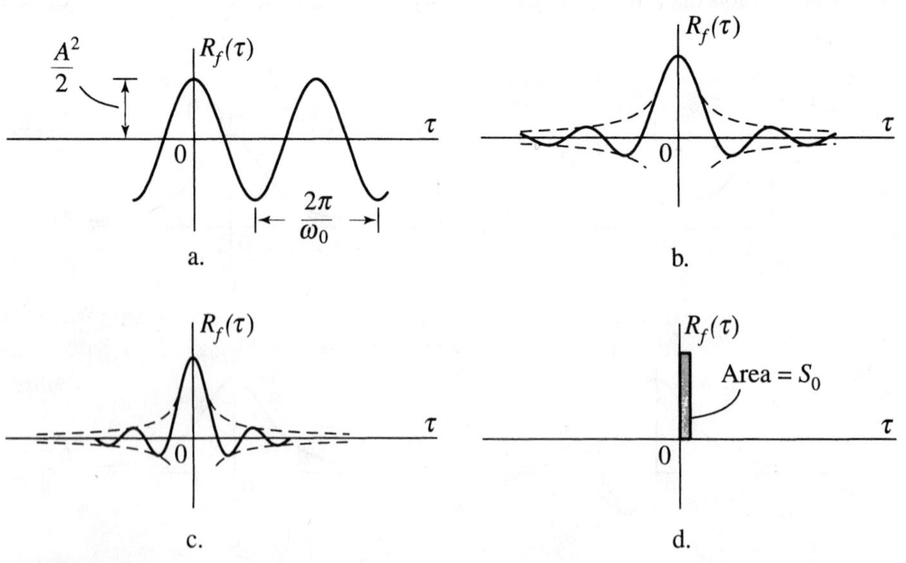

FIGURE 12.20
Autocorrelation function for: a. Sinosoidal function, b. Narrowband process, c. Wideband process and d. Ideal white noise

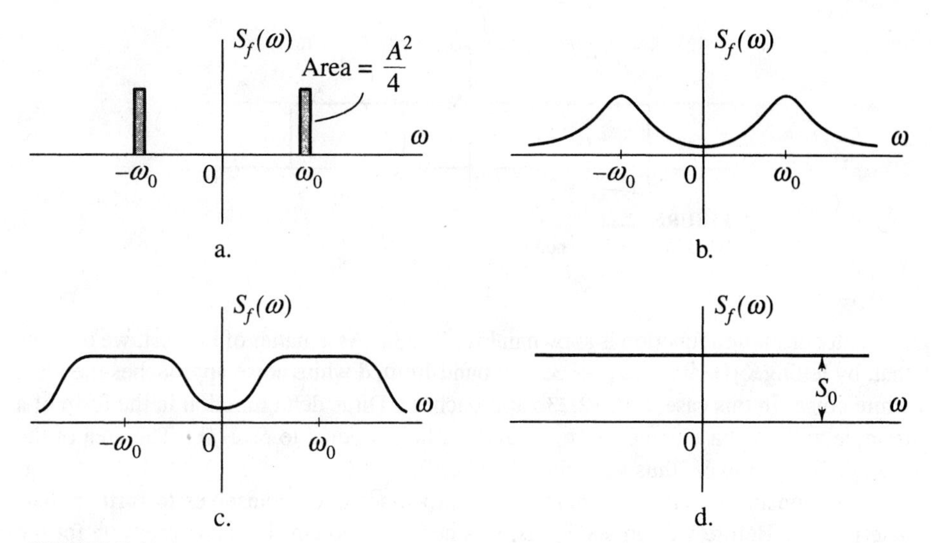

FIGURE 12.21
Power spectral density for: a. Sinusoidal function, b. Narrowband process, c. Wideband process and d. Ideal white noise

the form

$$R_f(\tau) = S_0\delta(\tau) \tag{12.59}$$

where $\delta(\tau)$ is the Dirac delta function. This can be verified by substituting Eq. (12.59) into Eq. (12.51).

Figure 12.21a shows a plot of the power density spectrum for the sine wave. It can be verified that its mathematical expression is

$$S_f(\omega) = \frac{\pi A^2}{2}[\delta(\omega+\omega_0)+\delta(\omega-\omega_0)] \tag{12.60}$$

The power spectral densities for the narrowband and wideband process are shown in Figs. 12.21b and c, respectively, which justifies the terminology used to describe these processes. Figure 12.21d depicts the power density spectrum for the ideal white noise, indicating that all frequencies are equally represented.

A more realistic random process than the ideal white noise is the *band-limited white noise*. The corresponding power density spectrum, shown in Fig. 12.22, is flat over the band of frequencies $\omega_1 < \omega < \omega_2$ (and $-\omega_2 < \omega < -\omega_1$), where ω_1 and ω_2 are known as the *lower cutoff* and *upper cutoff* frequencies, respectively. The band-limited white noise can serve at times as a reasonable approximation for the power density spectrum of a wideband process. The associated autocorrelation function can be obtained from Eq. (12.56) in the form

$$R_f(\tau) = \frac{1}{\pi}\int_0^{\infty} S_f(\omega)\cos\omega\tau d\omega = \frac{S_0}{\pi}\int_{\omega_1}^{\omega_2}\cos\omega\tau d\omega = \frac{S_0}{\pi}\frac{\sin\omega_2\tau - \sin\omega_1\tau}{\tau} \tag{12.61}$$

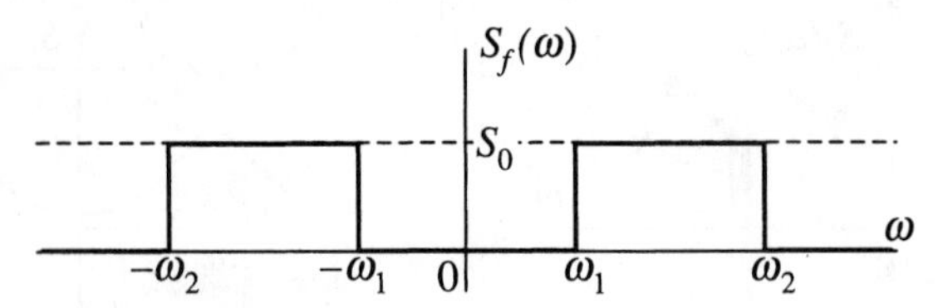

FIGURE 12.22
Band-limited white noise

The autocorrelation function is shown in Fig. 12.23a. As a matter of interest, we observe that, by letting $\omega_1 = 0$ and $\omega_2 \to \infty$, the band-limited white noise approaches the ideal white noise. In this case, Fig. 12.23b approaches a Dirac delta function in the form of a triangle with the base equal to $2\pi/\omega_2$ and the height equal to $S_0\omega_2/\pi$. The area of the triangle is equal to S_0, thus verifying Eq. (12.59).

Stationary and Gaussian narrowband processes lend themselves to further characterization. Before we can show this, it is necessary to develop an expression for the power spectral density of a derived process. In particular, the interest lies in an expression for the power spectral density $S_{\dot{f}}(\omega)$ of a stationary process $\dot{f}(t)$ under the assumption that the power spectral density $S_f(\omega)$ of the stationary process $f(t)$ is known. To this end, we recall Eq. (12.2) and recognize that the autocorrelation function for a stationary process does not depend on the time t_1, so that replacing t_1 by the arbitrary time t the ensemble autocorrelation function can be written in the form

$$R_f(\tau) = \lim_{n\to\infty} \frac{1}{n} \sum_{k=1}^{n} f_k(t) f_k(t+\tau) \tag{12.62}$$

Differentiating Eq. (12.62) with respect to τ, we obtain

$$\frac{dR_f(\tau)}{d\tau} = \lim_{n\to\infty} \frac{1}{n} \sum_{k=1}^{n} \frac{d}{d\tau}[f_k(t) f_k(t+\tau)] \tag{12.63}$$

But,

$$\begin{aligned}\frac{d}{d\tau}[f_k(t) f_k(t+\tau)] &= f_k(t)\frac{d}{d\tau}[f_k(t+\tau)] \\ &= f_k(t)\frac{d}{d(t+\tau)}[f_k(t+\tau)]\frac{d(t+\tau)}{d\tau} = f_k(t)\dot{f}_k(t+\tau)\end{aligned} \tag{12.64}$$

so that

$$\frac{dR_f(\tau)}{d\tau} = \lim_{n\to\infty} \frac{1}{n} \sum_{k=1}^{n} f_k(t)\dot{f}_k(t+\tau) \tag{12.65}$$

For stationary processes, however, the value of the sum is independent of time, so that we can also write

$$\frac{dR_f(\tau)}{d\tau} = \lim_{n\to\infty} \frac{1}{n} \sum_{k=1}^{n} f_k(t-\tau)\dot{f}_k(t) \tag{12.66}$$

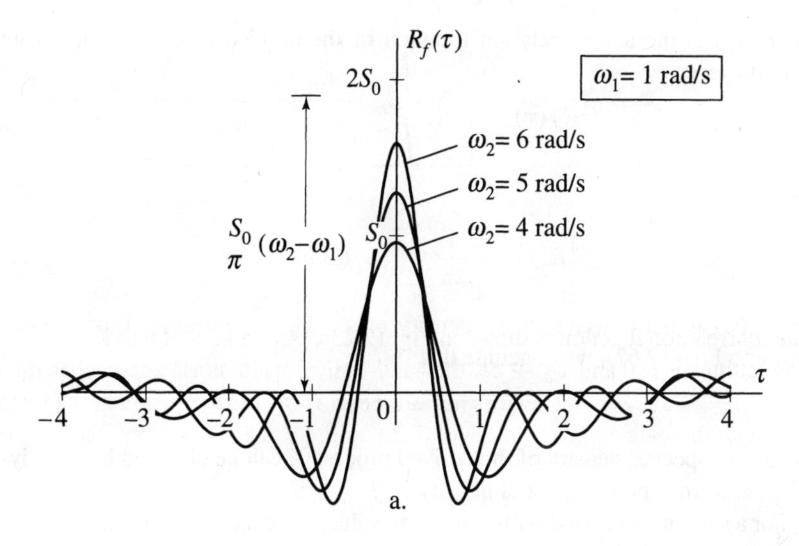

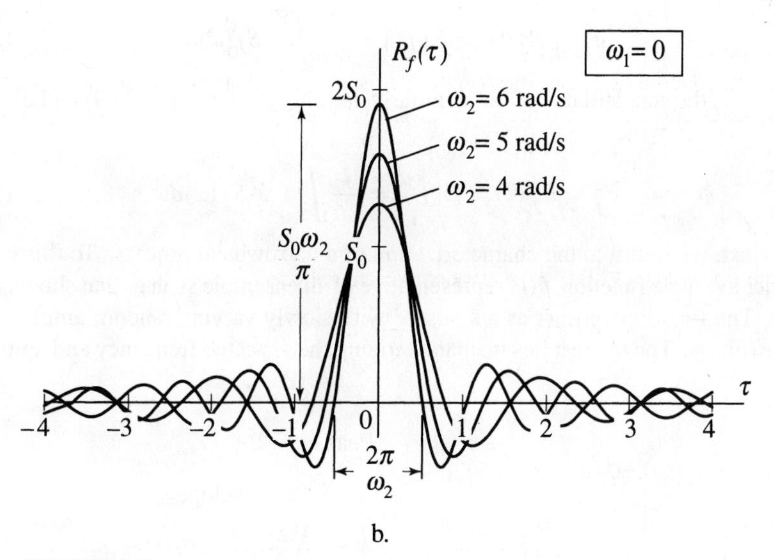

FIGURE 12.23
Autocorrelation function for: a. Band-limited white noise and b. Ideal white noise

Using the above procedure once more, it is not difficult to show that

$$\frac{d^2R_f(\tau)}{d\tau^2} = -\lim_{n\to\infty}\frac{1}{n}\sum_{k=1}^{n}\dot{f}_k(t-\tau)\dot{f}_k(t)$$

$$= -\lim_{n\to\infty}\frac{1}{n}\sum_{k=1}^{n}\dot{f}_k(t)\dot{f}_k(t+\tau) = -R_{\dot{f}}(\tau) \qquad (12.67)$$

where $R_{\dot{f}}(\tau)$ is the autocorrelation function of the derived process $\dot{f}(t)$. From Eq. (12.52), however, we can write

$$\frac{d^2R_f(\tau)}{d\tau^2} = -\frac{1}{2\pi}\int_{-\infty}^{\infty}\omega^2 S_f(\omega)e^{i\omega\tau}d\omega \tag{12.68}$$

Moreover,

$$R_{\dot{f}}(\tau) = \frac{1}{2\pi}\int_{-\infty}^{\infty} S_{\dot{f}}(\omega)e^{i\omega\tau}d\omega \tag{12.69}$$

where $S_{\dot{f}}(\omega)$ is the power spectral density of $\dot{f}$. Hence, inserting Eqs. (12.68) and (12.69) into Eq. (12.67), we conclude that

$$S_{\dot{f}}(\omega) = \omega^2 S_f(\omega) \tag{12.70}$$

or, the power spectral density of the derived process $\dot{f}$ can be obtained by merely multiplying the known power spectral density of f by ω^2.

For a stationary process with zero mean value, if we let $\tau = 0$ and use Eqs. (12.27), (12.29), (12.53) and (12.62), we obtain

$$\sigma_f^2 = R_f(0) = E[f^2] = \frac{1}{2\pi}\int_{-\infty}^{\infty} S_f(\omega)d\omega \tag{12.71}$$

where σ_f is the standard deviation. Similarly, letting $\tau = 0$ in Eq. (12.69) and using Eq. (12.70), we can write

$$\sigma_{\dot{f}}^2 = R_{\dot{f}}(0) = E[\dot{f}^2] = \frac{1}{2\pi}\int_{-\infty}^{\infty}\omega^2 S_f(\omega)d\omega \tag{12.72}$$

Next, we return to the characterization of a narrowband process. To this end, we consider a sample function $f(t)$ representative of an ensemble such as that shown in Fig. 12.24. The function appears as a sinusoid with slowly varying random amplitude and random phase. The interest lies in characterizing the expected frequency and amplitude.

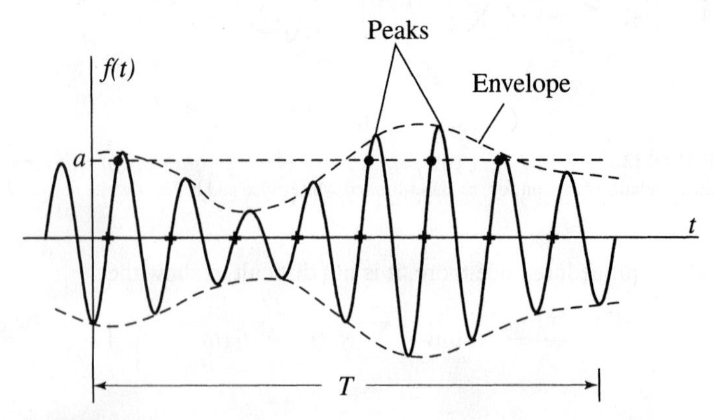

FIGURE 12.24
Sample function from a narrowband process

To characterize the expected frequency, we define the expected number of *crossings with positive slope* per unit time at the level $f = a$ as follows:

$$\nu_a^+ = \lim_{n\to\infty} \frac{1}{n} \sum_{k=1}^{n} \frac{1}{T} N_{ka}^+(T) \tag{12.73}$$

where $N_{ka}^+(T)$ represents the number of crossing with positive slope in the time interval T. Note that crossings with positive slope at $f = 0$ are marked by crosses in Fig. 12.24. It can be shown[5] that for a stationary process

$$\nu_a^+ = \int_0^\infty \dot{f} p(a, \dot{f}) d\dot{f} \tag{12.74}$$

where $p(a, \dot{f})$ is the intersection of the joint probability density function $p(f, \dot{f})$ and the plane $f = a$ (see Sec. 12.12). Equation (12.74) is valid for any arbitrary stationary process. If the process is Gaussian, then the joint probability density function has the form

$$p(f, \dot{f}) = \frac{1}{2\pi\sigma_f\sigma_{\dot{f}}} \exp\left[-\frac{1}{2}\left(\frac{f^2}{\sigma_f^2} + \frac{\dot{f}^2}{\sigma_{\dot{f}}^2}\right)\right] \tag{12.75}$$

where the standard deviations σ_f and $\sigma_{\dot{f}}$ can be obtained from the power spectral density $S_f(\omega)$ by means of Eqs. (12.71) and (12.72), and we note that Eq. (12.75) reflects the fact that f and $\dot{f}$ are uncorrelated. Inserting Eq. (12.75) with $f = a$ into Eq. (12.74) and carrying out the integration, we obtain

$$\nu_a^+ = \frac{1}{2\pi} \frac{\sigma_{\dot{f}}}{\sigma_f} e^{-a^2/2\sigma_f^2} \tag{12.76}$$

Then, the *average frequency*, or *expected frequency*, ω_0 is defined as the expected number of *zero crossings with positive slope* per unit time multiplied by 2π, so that letting $a = 0$ in Eq. (12.76) and considering Eqs. (12.71) and (12.72), we can write the expected frequency

$$\omega_0 = 2\pi\nu_0^+ = \frac{\sigma_{\dot{f}}}{\sigma_f} = \left[\frac{\int_{-\infty}^{\infty} \omega^2 S_f(\omega) d\omega}{\int_{-\infty}^{\infty} S_f(\omega) d\omega}\right]^{1/2} \tag{12.77}$$

It can also be shown[6] that the *probability density function of the envelope* for a narrowband stationary Gaussian random process is

$$p(a) = \frac{a}{\sigma_f^2} e^{-a^2/2\sigma_f^2} \tag{12.78}$$

---

[5] See S. H. Crandall and W. D. Mark, *Random Vibration in Mechanical Systems*, Academic Press, Inc., New York, 1963, p. 47.

[6] See S. H. Crandall and W. D. Mark, op. cit. pp. 48–53.

which can be identified as the Rayleigh distribution. The probability density function of the peaks is also given by the Rayleigh distribution of Eq. (12.78).

12.10 RESPONSE OF LINEAR SYSTEMS TO STATIONARY RANDOM EXCITATIONS

In Chap. 4, we showed that the response $x(t)$ of a linear system to the arbitrary excitation $f(t)$ can be written in the form of the convolution integral

$$x(t) = \int_0^t f(\lambda)g(t-\lambda)d\lambda \tag{12.79}$$

where $g(t)$ is the impulse response and λ merely a dummy variable. The function $f(t)$ is defined only for $t > 0$ and is zero for $t < 0$. Likewise, Eq. (12.79) defines the response $x(t)$ only for $t > 0$. Random variables, however, are not restricted to positive times, so that we wish to modify Eq. (12.79) to accommodate functions $f(t)$ of negative argument. To this end, using the same argument as that used in Sec. 4.4 to derive the convolution integral, Eq. (12.79), it is easy to see that the lower limit of the convolution integral can be merely extended to $-\infty$, so that

$$x(t) = \int_{-\infty}^t f(\lambda)g(t-\lambda)d\lambda \tag{12.80}$$

However, from the definition of the impulse response (see Sec. 4.1), $g(t-\lambda)$ is zero for $t < \lambda$. Because the variable of integration in (12.80) is λ and not t, a slight change in perspective permits us to restate the above by saying that $g(t-\lambda)$ is zero for $\lambda > t$. It follows that the upper limit of the integral in (12.80) can be changed to any value larger than t without affecting the value of the integral. Choosing the upper limit as infinity, for convenience, we can write the convolution integral in the form

$$x(t) = \int_{-\infty}^{\infty} f(\lambda)g(t-\lambda)d\lambda \tag{12.81}$$

Using the change of variable $t-\lambda = \tau$, $d\lambda = -d\tau$, with an appropriate change in the integration limits, it is easy to demonstrate that the convolution integral remains symmetric in $f(t)$ and $g(t)$, or

$$x(t) = \int_{-\infty}^{\infty} f(\lambda)g(t-\lambda)d\lambda = \int_{-\infty}^{\infty} g(\lambda)f(t-\lambda)d\lambda \tag{12.82}$$

Next, we denote the Fourier transform of $x(t)$ by $X(\omega)$, so that using Eq. (12.81), as well as the substitution $t-\lambda = \sigma$, $dt = d\sigma$, we can write

$$\begin{aligned} X(\omega) &= \int_{-\infty}^{\infty} x(t)e^{-i\omega t}dt = \int_{-\infty}^{\infty} f(\lambda)\left[\int_{-\infty}^{\infty} g(t-\lambda)e^{-i\omega t}dt\right]d\lambda \\ &= \int_{-\infty}^{\infty} f(\lambda)e^{-i\omega\lambda}d\lambda \int_{-\infty}^{\infty} g(\sigma)e^{-i\omega\sigma}d\sigma \end{aligned} \tag{12.83}$$

But,

$$\int_{-\infty}^{\infty} f(\lambda)e^{-i\omega\lambda}d\lambda = F(\omega) \tag{12.84}$$

is the Fourier transform of the excitation and

$$\int_{-\infty}^{\infty} g(\sigma)e^{-i\omega\sigma}d\sigma = G(\omega) \tag{12.85}$$

is the Fourier transform of the impulse response, so that Eq. (12.83) yields

$$X(\omega) = G(\omega)F(\omega) \tag{12.86}$$

Comparing Eq. (12.86) with Eq. (12.49), we conclude that *the frequency response $G(\omega)$ can be identified as the Fourier transform of the impulse response*. Equations (12.82) and (12.86) state that *the convolution of $f(t)$ and $g(t)$ and the product $G(\omega)F(\omega)$ represent a Fourier transform pair*. This statement is known as the *time-domain convolution theorem*.

The above relations are valid for any arbitrary excitation $f(t)$. Our interest lies in the case in which the excitation is a stationary random process. Then, the response random process also is stationary. We are interested in calculating first- and second-order statistics for the response random process, given the corresponding statistics for the excitation random process. Averaging the second form of the convolution integral, Eq. (12.82), over the ensemble, we can write the mean value of the response random process as

$$E[x(t)] = E\left[\int_{-\infty}^{\infty} g(\lambda)f(t-\lambda)d\lambda\right] \tag{12.87}$$

Assuming that the order of the ensemble averaging and integration operations are interchangeable, Eq. (12.87) can be written as

$$E[x(t)] = \int_{-\infty}^{\infty} g(\lambda)E[f(t-\lambda)]d\lambda \tag{12.88}$$

But for stationary random processes, the mean value of the process is constant, $E[f(t-\lambda)] = E[f(t)] = \text{constant}$, so that

$$E[x(t)] = E[f(t)]\int_{-\infty}^{\infty} g(\lambda)d\lambda \tag{12.89}$$

Letting $\omega = 0$ in Eq. (12.85) and changing the dummy variable from σ to λ, we obtain

$$\int_{-\infty}^{\infty} g(\lambda)d\lambda = G(0) \tag{12.90}$$

so that Eq. (12.89) reduces to

$$E[x(t)] = G(0)E[f(t)] = \text{constant} \tag{12.91}$$

which implies that the mean value of the response to an excitation in the form of a stationary random process is constant and proportional to the mean value of the excitation process. It follows that *if the excitation mean value is zero, then the response mean value is also zero*.

Next, we evaluate the autocorrelation function of the response random process. To this end, it is convenient to introduce two new dummy variables, λ_1 and λ_2, and write the convolution integrals

$$x(t) = \int_{-\infty}^{\infty} g(\lambda_1) f(t-\lambda_1) d\lambda_1, \quad x(t+\tau) = \int_{-\infty}^{\infty} g(\lambda_2) f(t+\tau-\lambda_2) d\lambda_2 \tag{12.92}$$

Using Eqs. (12.92) to form the response autocorrelation function $R_x(\tau)$ and assuming once again that the order of ensemble averaging and integration is interchangeable, we can write

$$\begin{aligned} R_x(\tau) &= E[x(t)x(t+\tau)] \\ &= E\left[\int_{-\infty}^{\infty} g(\lambda_1) f(t-\lambda_1) d\lambda_1 \int_{-\infty}^{\infty} g(\lambda_2) f(t+\tau-\lambda_2) d\lambda_2\right] \\ &= E\left[\int_{-\infty}^{\infty}\int_{-\infty}^{\infty} g(\lambda_1) g(\lambda_2) f(t-\lambda_1) f(t+\tau-\lambda_2) d\lambda_1 d\lambda_2\right] \\ &= \int_{-\infty}^{\infty}\int_{-\infty}^{\infty} g(\lambda_1) g(\lambda_2) E[f(t-\lambda_1) f(t+\tau-\lambda_2)] d\lambda_1 d\lambda_2 \end{aligned} \tag{12.93}$$

Because the excitation random process is stationary, we have

$$\begin{aligned} E[f(t-\lambda_1) f(t+\tau-\lambda_2)] &= E[f(t) f(t+\tau+\lambda_1-\lambda_2)] \\ &= R_f(\tau+\lambda_1-\lambda_2) \end{aligned} \tag{12.94}$$

where $R_f(\tau+\lambda_1-\lambda_2)$ is the autocorrelation function of the excitation process. Hence, the response autocorrelation function, Eq. (12.93), reduces to

$$R_x(\tau) = \int_{-\infty}^{\infty}\int_{-\infty}^{\infty} g(\lambda_1) g(\lambda_2) R_f(\tau+\lambda_1-\lambda_2) d\lambda_1 d\lambda_2 \tag{12.95}$$

We note that Eq. (12.95) does not depend on t, which implies that the value of the response autocorrelation function is also insensitive to a translation in time, thus corroborating the statement made earlier that *if for a linear system the excitation is a stationary random process, then the response is also a stationary random process.*

Quite often information concerning the response random process can be obtained more readily by calculating first the response power spectral density instead of the response autocorrelation function, particularly if the excitation random process is given in terms of the power spectral density. To demonstrate this, we use Eq. (12.95) and express the response mean square spectral density as the Fourier transform of the response autocorrelation function in the form

$$\begin{aligned} S_x(\omega) &= \int_{-\infty}^{\infty} R_x(\tau) e^{-i\omega\tau} d\tau \\ &= \int_{-\infty}^{\infty} e^{-i\omega\tau}\left[\int_{-\infty}^{\infty}\int_{-\infty}^{\infty} g(\lambda_1) g(\lambda_2) R_f(\tau+\lambda_1-\lambda_2) d\lambda_1 d\lambda_2\right] d\tau \end{aligned} \tag{12.96}$$

But $R_f(\tau+\lambda_1-\lambda_2)$ can be expressed as the inverse Fourier transform

$$R_f(\tau+\lambda_1-\lambda_2) = \frac{1}{2\pi}\int_{-\infty}^{\infty} S_f(\omega) e^{i\omega(\tau+\lambda_1-\lambda_2)} d\omega \tag{12.97}$$

so that, inserting Eq. (12.97) into Eq. (12.96), considering Eq. (12.85), interchanging the order of integration and rearranging, we obtain

$$\begin{aligned} S_x(\omega) &= \int_{-\infty}^{\infty} e^{-i\omega\tau} \left\{ \int_{-\infty}^{\infty}\int_{-\infty}^{\infty} g(\lambda_1)g(\lambda_2) \right. \\ &\quad \left. \times \left[\frac{1}{2\pi} \int_{-\infty}^{\infty} S_f(\omega)e^{i\omega(\tau+\lambda_1-\lambda_2)}d\omega \right] d\lambda_1 d\lambda_2 \right\} d\tau \\ &= \int_{-\infty}^{\infty} e^{-i\omega\tau} \left\{ \frac{1}{2\pi} \int_{-\infty}^{\infty} S_f(\omega) \left[\int_{-\infty}^{\infty} g(\lambda_1)e^{i\omega\lambda_1}d\lambda_1 \right. \right. \\ &\quad \left. \left. \times \int_{-\infty}^{\infty} g(\lambda_2)e^{-i\omega\lambda_2}d\lambda_2 \right] e^{i\omega\tau}d\omega \right\} d\tau \\ &= \int_{-\infty}^{\infty} e^{-i\omega\tau} \left[\frac{1}{2\pi} \int_{-\infty}^{\infty} S_f(\omega)G(-\omega)G(\omega)e^{i\omega\tau}d\omega \right] d\tau \\ &= \int_{-\infty}^{\infty} e^{-i\omega\tau} \left[\frac{1}{2\pi} \int_{-\infty}^{\infty} S_f(\omega)|G(\omega)|^2 e^{i\omega\tau}d\omega \right] d\tau \end{aligned} \tag{12.98}$$

where use has been made of the fact that $G(-\omega) = \bar{G}(\omega)$ is the complex conjugate of the frequency response $G(\omega)$. Comparing the first integral in Eq. (12.96) with the last in Eq. (12.98), and recognizing that the response autocorrelation function $R_x(\tau)$ must be equal to the inverse Fourier transform of the response mean square spectral density $S_x(\omega)$, we conclude that

$$S_x(\omega) = |G(\omega)|^2 S_f(\omega) \tag{12.99}$$

and

$$R_x(\tau) = \frac{1}{2\pi}\int_{-\infty}^{\infty} S_x(\omega)e^{i\omega\tau}d\omega = \frac{1}{2\pi}\int_{-\infty}^{\infty} |G(\omega)|^2 S_f(\omega)e^{i\omega\tau}d\omega \tag{12.100}$$

constitute a Fourier transform pair. Equation (12.99) represents a simple algebraic expression relating the power spectral densities of the excitation and response random processes, whereas Eq. (12.100) gives the response autocorrelation function in the form of an inverse Fourier transform involving the excitation power spectral density. From Eq. (12.99), we conclude that in the case of a lightly damped single-degree-of-freedom system, for which the frequency response has a sharp peak at $\omega = \omega_n(1-2\zeta^2)^{1/2}$, where ζ is the damping factor and ω_n the frequency of undamped oscillation, if the excitation power spectral density function represents a wideband random process, then the response power spectral density function is a narrowband random process.

The mean square value of the response random process can be obtained by letting $\tau = 0$ in Eq. (12.100). The result is simply

$$R_x(0) = E[x^2(t)] = \frac{1}{2\pi}\int_{-\infty}^{\infty} |G(\omega)|^2 S_f(\omega)d\omega \tag{12.101}$$

Examining Eqs. (12.99), (12.100) and (12.101), it appears that if the system is linear and the excitation random process is stationary, then the response mean square spectral density, autocorrelation function and mean square value can all be calculated from the mean square spectral density $S_f(\omega)$ of the excitation random process and the magnitude $|G(\omega)|$ of the frequency response.

It should be pointed out that *if the excitation random process is Gaussian and the system is linear, then the response random process is also Gaussian. But, from Eq. (12.30), the Gaussian probability density function depends on the mean value and standard deviation alone, where, from Eq. (12.29), the standard deviation depends on the mean value and mean square value. It follows that, for Gaussian random processes, the response probability distribution is completely defined by the response mean value and mean square value.*

It is not difficult to show that the above relations and conclusions concerning response random processes remain valid if the excitation random process is not merely stationary but ergodic. The only difference is that for ergodic random processes the averages are time averages, calculated by using a single representative sample function from the entire process, instead of ensemble averages over the collection of sample functions.

12.11 RESPONSE OF SINGLE-DEGREE-OF-FREEDOM SYSTEMS TO RANDOM EXCITATIONS

We consider a mass-damper-spring system traveling with the uniform velocity v on a rough road, so that its support is imparted a vertical motion, as shown in Fig. 12.25. If the road roughness is described by the random variable $y(s)$, in which $s = vt$, we conclude that the differential equation of motion for the mass m is (see Sec. 3.5)

$$\ddot{x}(t) + 2\zeta\omega_n\dot{x}(t) + \omega_n^2 x(t) = \omega_n^2 f(t) \tag{12.102}$$

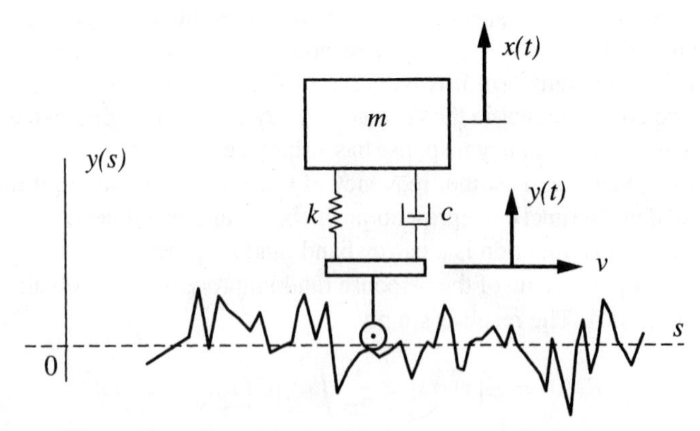

FIGURE 12.25
Mass-damper-spring system traveling on a rough road

where

$$f(t) = \frac{2\zeta}{\omega_n}\dot{y}(t) + y(t) \tag{12.103}$$

is an equivalent displacement excitation, in which ζ is the damping factor and ω_n the undamped frequency of oscillation. We assume that the random process associated with $f(t)$ is ergodic and Gaussian, so that the response $x(t)$ is also an ergodic and Gaussian process. Hence, both the excitation and response random processes are fully described by the mean value and mean square value.

For a stationary process the mean value is constant. Because a constant component of the excitation merely leads to a constant component of the response, a problem that can be treated separately, we can assume without loss of generality that this constant is zero,

$$E[f(t)] = 0 \tag{12.104}$$

It follows immediately that the response mean value is also zero.

$$E[x(t)] = 0 \tag{12.105}$$

Next, we calculate various basic statistics describing the response random process, such as the autocorrelation function, the power spectral density function and the mean square value. This requires certain statistics describing the excitation random process. We consider two related cases, namely, the ideal white noise and the band-limited white noise. In Sec. 12.9, we indicated that the autocorrelation function corresponding to the ideal white noise power spectral density $S_f(\omega) = S_0$ is

$$R_f(\tau) = S_0\delta(\tau) \tag{12.106}$$

where $\delta(\tau)$ is the Dirac delta function. Moreover, from Eq. (4.13), we conclude that the impulse response of the single-degree-of-freedom system described by Eq. (12.102) has the form

$$g(t) = \frac{\omega_n^2}{\omega_d} e^{-\zeta\omega_n t} \sin\omega_d t \, u(t) \tag{12.107}$$

where the unit step function $u(t)$ ensures that $g(t) = 0$ for $t < 0$. Hence, introducing Eqs. (12.106) and (12.107) into Eq. (12.95), we obtain the response autocorrelation function

$$\begin{aligned} R_x(\tau) &= \frac{S_0\omega_n^4}{\omega_d^2}\int_{-\infty}^{\infty}\int_{-\infty}^{\infty}\delta(\tau+\lambda_1-\lambda_2)e^{-\zeta\omega_n(\lambda_1+\lambda_2)} \\ &\qquad \times \sin\omega_d\lambda_1 \sin\omega_d\lambda_2 u(\lambda_1)u(\lambda_2)d\lambda_1 d\lambda_2 \\ &= \frac{S_0\omega_n^4}{\omega_d^2}\int_0^{\infty}\int_0^{\infty}\delta(\tau+\lambda_1-\lambda_2)e^{-\zeta\omega_n(\lambda_1+\lambda_2)}\sin\omega_d\lambda_1 \sin\omega_d\lambda_2 d\lambda_1 d\lambda_2 \end{aligned} \tag{12.108}$$

In our evaluation of $R_x(\tau)$, we assume that $\tau > 0$. The value of $R_x(-\tau)$ can be obtained by using the fact that the autocorrelation is an even function of τ. Due to the nature of the delta function, if we integrate with respect to λ_2, we obtain

$$R_x(\tau) = \frac{S_0\omega_n^4}{\omega_d^2}\int_0^\infty e^{-\zeta\omega_n(\tau+2\lambda_1)}\sin\omega_d\lambda_1 \sin\omega_d(\tau+\lambda_1)d\lambda_1$$

$$= \frac{S_0\omega_n^4}{\omega_d^2}e^{-\zeta\omega_n\tau}\left(\sin\omega_d\tau\int_0^\infty e^{-2\zeta\omega_n\lambda_1}\sin\omega_d\lambda_1\cos\omega_d\lambda_1 d\lambda_1\right.$$

$$\left.+\cos\omega_d\tau\int_0^\infty e^{-2\zeta\omega_n\lambda_1}\sin^2\omega_d\lambda_1 d\lambda_1\right), \quad \tau > 0 \qquad (12.109)$$

But the value of the integrals in Eq. (12.109) can be found in standard integral tables,[7] so that Eq. (12.109) reduces to

$$R_x(\tau) = \frac{S_0\omega_n}{4\zeta}e^{-\zeta\omega_n\tau}\left[\cos\omega_d\tau + \frac{\zeta}{(1-\zeta^2)^{1/2}}\sin\omega_d\tau\right], \quad \tau > 0 \qquad (12.110)$$

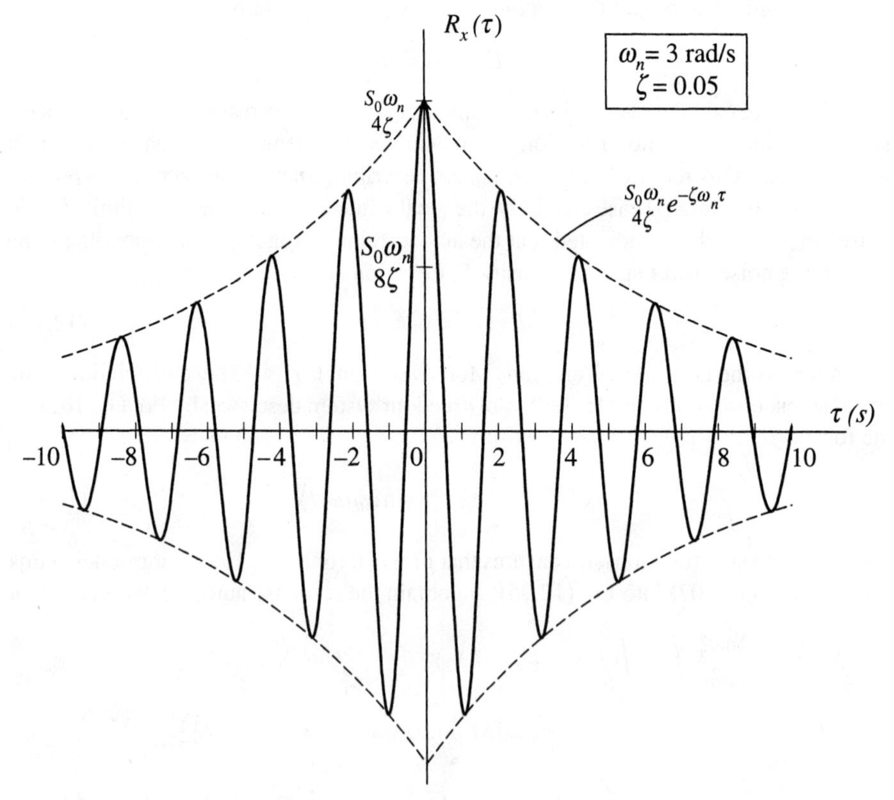

FIGURE 12.26
Response autocorrelation function for a lightly damped single-degree-of-freedom system

[7] See, for example, B. O. Pierce and R. M. Foster, *A Short Table of Integrals*, nos. 430 and 435, Ginn and Company, Boston, 1956.

Using the fact that $R_x(-\tau) = R_x(\tau)$, we can write directly

$$R_x(\tau) = \frac{S_0\omega_n}{4\zeta} e^{\zeta\omega_n\tau} \left[\cos\omega_d\tau - \frac{\zeta}{(1-\zeta^2)^{1/2}} \sin\omega_d\tau \right], \quad \tau < 0 \tag{12.111}$$

The autocorrelation function is plotted in Fig. 12.26 for the case of light damping. It is easy to see that the response autocorrelation function is that of a narrowband process (see Sec. 12.9).

The response power density spectrum is relatively easy to obtain. We recall that the frequency response for the system in question was obtained in Sec. 3.1. Hence, inserting $S_f(\omega) = S_0$ and Eq. (3.20) into Eq. (12.99), we obtain simply

$$S_x(\omega) = |G(\omega)|^2 S_f(\omega) = \frac{S_0}{[1-(\omega/\omega_n)^2]^2 + (2\zeta\omega/\omega_n)^2} \tag{12.112}$$

The response power spectral density $S_x(\omega)$ is plotted in Fig. 12.27. Once again we conclude that the plot $S_x(\omega)$ versus ω is typical of a narrowband process. Because, according to Eqs. (12.99) and (12.100), $R_x(\tau)$ and $S_x(\omega)$ represent a Fourier transform pair, no essentially new information can be derived from $S_x(\omega)$ that cannot be derived from $R_x(\tau)$, or from $R_x(\tau)$ that cannot be derived from $S_x(\omega)$.

The mean square value can be obtained by letting $\tau = 0$ in Eq. (12.110) and writing simply

$$R_x(0) = E[x^2(t)] = \frac{S_0\omega_n}{4\zeta} \tag{12.113}$$

It can also be obtained by inserting Eq. (12.112) into (12.101) and writing

$$R_x(0) = E[x^2(t)] = \frac{S_0}{2\pi} \int_{-\infty}^{\infty} \frac{d\omega}{[1-(\omega/\omega_n)^2]^2 + (2\zeta\omega/\omega_n)^2} \tag{12.114}$$

The integration in Eq. (12.114) can be performed by converting the real variable ω into a complex variable and the real line integral into a contour integral in the complex plane, where the latter can be evaluated by the residue theorem. Following this procedure, it

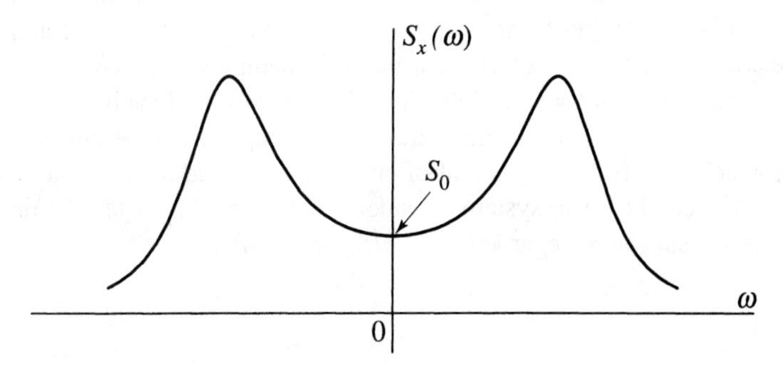

FIGURE 12.27
Response power spectral density for a damped single-degree-of-freedom system

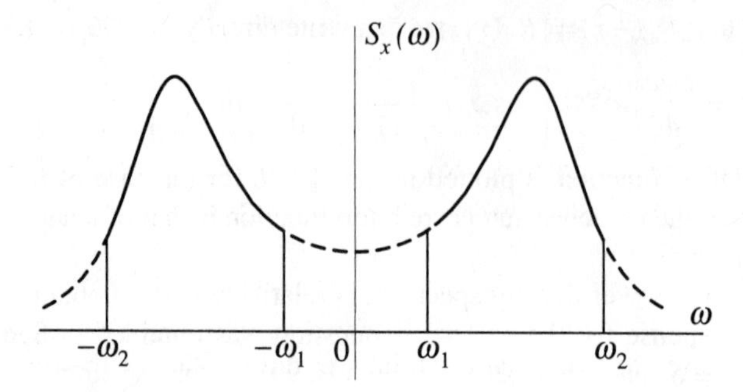

FIGURE 12.28
Response power spectral density corresponding to a band-limited white noise excitation power spectral density

can be shown[8] that Eq. (12.114) yields the same mean square value as that given by Eq. (12.113).

Because the random process is Gaussian with zero mean value, the mean square value, Eq. (12.113), is sufficient to determine the shape of the response probability density function, thus making it possible to evaluate the probability that the response $x(t)$ might exceed a given displacement. The mean square value also determines the probability density function of the Rayleigh distribution for the envelope and peaks of the response (see Fig. 12.24).

When the excitation power spectral density has the form of a band-limited white noise with lower and upper cutoff frequencies ω_1 and ω_2, respectively, the response power spectral density has the form depicted in Fig. 12.28. Then, if the system is lightly damped and the excitation frequency band $\omega_1 < \omega < \omega_2$ includes the system natural frequency ω_n as well as its bandwidth $\Delta\omega = 2\zeta\omega_n$ (see Sec. 3.2) and if the excitation bandwidth is large compared to the system bandwidth, the response mean square value, which is equal to the area under the curve $S_x(\omega)$ versus ω divided by 2π, can be approximated by $S_0\omega_n/4\zeta$. Hence, under these circumstances, the ideal white noise assumption leads to meaningful results.

Returning to Eq. (12.112), we observe that, whereas the excitation power spectral density $S_f(\omega)$ is flat, the response power spectral density $S_x(\omega)$ is not, and in fact is sharply peaked in the vicinity of $\omega = \omega_n$ for light damping. Moreover, the response spectrum has the value S_0 for relatively small frequencies, and vanishes for very large frequencies, as can be seen from Fig. 12.28. This behavior can be attributed entirely to $|G(\omega)|$, which prescribes the amount of energy transmitted by the system at various frequencies. Hence, the linear system considered acts like a *linear filter*. For very light damping, the system can be regarded as a *narrowband filter*.

[8]See L. Meirovitch, *Analytical Methods in Vibrations*, The Macmillan Co., New York, 1967, pp. 503–505.

12.12 JOINT PROBABILITY DISTRIBUTION OF TWO RANDOM VARIABLES

The preceding discussion was confined to properties of a single random process. Yet in many instances it is necessary to describe certain joint properties of two or more random processes. For example, these random processes may consist of the vibration of two or more distinct points in a structure. The statistics discussed in Secs. 12.1–12.9 can be calculated independently for the various random processes involved, but in addition there may be important information contained in certain joint statistics. In this section we confine ourselves to two random variables, and in Sec. 12.13 we discuss random processes.

There are three basic types of statistical functions describing joint properties of sample time histories representative of two random processes, namely, joint probability density functions, cross-correlation functions and cross-spectral density functions. These functions provide information concerning joint properties of two processes in the amplitude domain, time domain and frequency domain, respectively.

We consider the two random variables $x(t)$ and $y(t)$, and define the *joint*, or *second-order, probability distribution function* $P(x, y)$ associated with the probability that $x(t) \leq x$ and $y(t) \leq y$ as follows:

$$P(x, y) = \text{Prob}[x(t) \leq x;\ y(t) \leq y] \tag{12.115}$$

The above joint probability distribution function can be described in terms of a *joint probability density function* $p(x, y)$ according to

$$P(x, y) = \int_{-\infty}^{x} \int_{-\infty}^{y} p(\xi, \eta) d\xi d\eta \tag{12.116}$$

where the function $p(x, y)$ is given by the surface shown in Fig. 12.29. Note that ξ and η in Eq. (12.116) are mere dummy variables. The probability that $x_1 < x \leq x_2$ and $y_1 < y \leq y_2$ is given by

$$\text{Prob}(x_1 < x \leq x_2;\ y_1 < y \leq y_2) = \int_{x_1}^{x_2} \int_{y_1}^{y_2} p(x, y) dx\, dy \tag{12.117}$$

and is represented by the shaded volume in Fig. 12.29.

The joint probability density function $p(x, y)$ possesses the property

$$p(x, y) \geq 0 \tag{12.118}$$

which implies that the joint probability is a *nonnegative* number. Moreover, the probability that both x and y are any real numbers is unity because the event is a certainty. This is expressed by

$$\int_{-\infty}^{\infty} \int_{-\infty}^{\infty} p(x, y) dx\, dy = 1 \tag{12.119}$$

First-order probabilities can be obtained from second-order joint probabilities. Indeed, the probability that x lies within the open interval $x_1 < x < x_2$ regardless of the

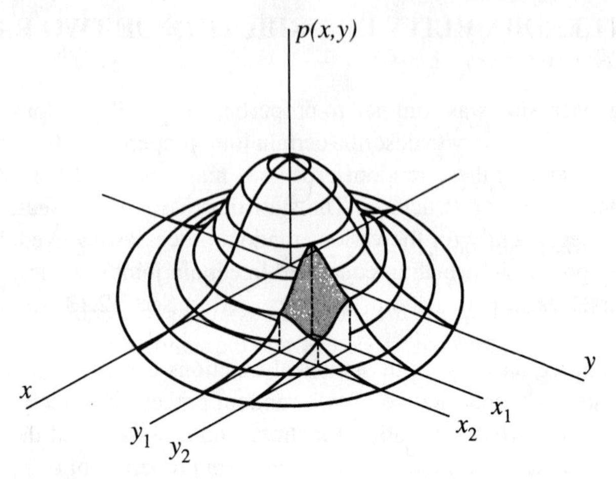

FIGURE 12.29
Joint probability density function

value of y is

$$\text{Prob}(x_1 < x < x_2;\ -\infty < y < \infty) = \int_{x_1}^{x_2} \left[\int_{-\infty}^{\infty} p(x, y) dy \right] dx = \int_{x_1}^{x_2} p(x) dx \tag{12.120}$$

where

$$p(x) = \int_{-\infty}^{\infty} p(x, y) dy \tag{12.121}$$

is the first-order probability density of x alone. Similarly,

$$p(y) = \int_{-\infty}^{\infty} p(x, y) dx \tag{12.122}$$

is the first-order probability density of y alone. The two random variables x and y are said to be *statistically independent* if

$$p(x, y) = p(x)p(y) \tag{12.123}$$

Next, we define the *mathematical expectation* of a real continuous function $g(x, y)$ of the random variables $x(t)$ and $y(t)$ in the form

$$E[g(x, y)] = \int_{-\infty}^{\infty} \int_{-\infty}^{\infty} g(x, y) p(x, y) dx\, dy \tag{12.124}$$

The *mean values* of $x(t)$ and $y(t)$ alone are simply

$$\begin{aligned} \bar{x} = E[x] &= \int_{-\infty}^{\infty} \int_{-\infty}^{\infty} x p(x, y) dx\, dy = \int_{-\infty}^{\infty} x p(x) dx \\ \bar{y} = E[y] &= \int_{-\infty}^{\infty} \int_{-\infty}^{\infty} y p(x, y) dx\, dy = \int_{-\infty}^{\infty} y p(y) dy \end{aligned} \tag{12.125}$$

In the case in which $g(x, y) = (x - \bar{x})(y - \bar{y})$, Eq. (12.124) defines the *covariance* between x and y in the form

$$C_{xy} = E[(x - \bar{x})(y - \bar{y})] = \int_{-\infty}^{\infty}\int_{-\infty}^{\infty}(x - \bar{x})(y - \bar{y})p(x, y)dx\,dy$$
$$= E[xy] - E[x]E[y] \tag{12.126}$$

Recalling Eq. (12.28), we conclude that $C_x = E[(x - \bar{x})^2] = \sigma_x^2$ represents the variance of x, whereas $C_y = E[(y - \bar{y})^2] = \sigma_y^2$ is the variance of y. The square roots of the variances, namely, σ_x and σ_y, are the standard deviations of x and y, respectively.

A relation between the covariance C_{xy} and the standard deviations σ_x and σ_y can be revealed by considering the integral

$$\int_{-\infty}^{\infty}\int_{-\infty}^{\infty}\left(\frac{x - \bar{x}}{\sigma_x} \pm \frac{y - \bar{y}}{\sigma_y}\right)^2 p(x, y)dx\,dy$$
$$= \frac{1}{\sigma_x^2}\int_{-\infty}^{\infty}\int_{-\infty}^{\infty}(x - \bar{x})^2 p(x, y)dx\,dy$$
$$\pm \frac{2}{\sigma_x\sigma_y}\int_{-\infty}^{\infty}\int_{-\infty}^{\infty}(x - \bar{x})(y - \bar{y})p(x, y)dx\,dy$$
$$+ \frac{1}{\sigma_y^2}\int_{-\infty}^{\infty}\int_{-\infty}^{\infty}(y - \bar{y})^2 p(x, y)dx\,dy = 2 \pm 2\frac{C_{xy}}{\sigma_x\sigma_y} \geq 0 \tag{12.127}$$

where the inequality is valid because the first integral cannot be negative. It follows that

$$\sigma_x\sigma_y \geq |C_{xy}| \tag{12.128}$$

or the product of the standard deviations of x and y is larger than or equal to the magnitude of the covariance between x and y. The normalized quantity

$$\rho_{xy} = \frac{C_{xy}}{\sigma_x\sigma_y} \tag{12.129}$$

is known as the *correlation coefficient*. Its value lies between -1 and $+1$, as can be concluded from inequality (12.127).

When the covariance C_{xy} is equal to zero the random variables x and y are said to be *uncorrelated. Statistically independent random variables are also uncorrelated, but uncorrelated random variables are not necessarily statistically independent*, although they can be. To show this, we introduce $p(x, y) = p(x)p(y)$ into Eq. (12.126) and obtain

$$C_{xy} = \int_{-\infty}^{\infty}\int_{-\infty}^{\infty}(x - \bar{x})(y - \bar{y})p(x, y)dx\,dy$$
$$= \int_{-\infty}^{\infty}xp(x)dx\int_{-\infty}^{\infty}yp(y)dy - E[x]E[y] = 0 \tag{12.130}$$

On the other hand, in the general case in which $p(x, y) \neq p(x)p(y)$ the fact that the covariance is zero merely implies that

$$E[xy] = E[x]E[y] \tag{12.131}$$

However, *in the very important case in which $p(x, y)$ represents the joint normal, or Gaussian, probability density function, uncorrelated random variables are also statistically independent.* Indeed, the joint normal probability density function has the expression

$$p(x, y) = \frac{1}{2\pi\sigma_x\sigma_y\sqrt{1-\rho_{xy}^2}}\exp\left\{-\frac{1}{2\sqrt{1-\rho_{xy}^2}}\left[\left(\frac{x-\bar{x}}{\sigma_x}\right)^2 - 2\rho_{xy}\frac{x-\bar{x}}{\sigma_x}\frac{y-\bar{y}}{\sigma_y} + \left(\frac{y-\bar{y}}{\sigma_y}\right)^2\right]\right\} \quad (12.132)$$

so that when the correlation coefficient ρ_{xy} is zero, Eq. (12.132) reduces to the product of the individual normal probability density functions

$$p(x) = \frac{1}{\sqrt{2\pi}\sigma_x}\exp\left[-\frac{(x-\bar{x})^2}{2\sigma_x^2}\right], \quad p(y) = \frac{1}{\sqrt{2\pi}\sigma_y}\exp\left[-\frac{(y-\bar{y})^2}{2\sigma_y^2}\right] \quad (12.133)$$

thus satisfying Eq. (12.123), with the implication that the random variables x and y are statistically independent. Note that this result is valid for joint normal probability density functions alone, and is not valid for arbitrary joint probability density functions.

12.13 JOINT PROPERTIES OF STATIONARY RANDOM PROCESSES

We consider two arbitrary random processes $x_k(t)$ and $y_k(t)$ $(k = 1, 2, \dots)$ of the type discussed in Sec. 12.1. The time histories $x_k(t)$ and $y_k(t)$ $(k = 1, 2, \dots)$ resemble those depicted in Fig. 12.2. The object is to calculate certain *ensemble averages*. In particular, we calculate the *mean values* at the arbitrary fixed time t_1 as follows:

$$\mu_x(t_1) = \lim_{n\to\infty}\frac{1}{n}\sum_{k=1}^{n} x_k(t_1), \quad \mu_y(t_1) = \lim_{n\to\infty}\frac{1}{n}\sum_{k=1}^{n} y_k(t_1) \quad (12.134)$$

For arbitrary random processes, the mean values at different times, say $t_1 \neq t_2$, are different, so that

$$\mu_x(t_1) \neq \mu_x(t_2), \quad \mu_y(t_1) \neq \mu_y(t_2) \quad (12.135)$$

Next, we calculate the *covariance functions* at the arbitrary fixed times t_1 and $t_1 + \tau$ as follows:

$$C_x(t_1, t_1+\tau) = \lim_{n\to\infty}\frac{1}{n}\sum_{k=1}^{n}[x_k(t_1) - \mu_x(t_1)][x_k(t_1+\tau) - \mu_x(t_1+\tau)]$$

$$C_y(t_1, t_1+\tau) = \lim_{n\to\infty}\frac{1}{n}\sum_{k=1}^{n}[y_k(t_1) - \mu_y(t_1)][y_k(t_1+\tau) - \mu_y(t_1+\tau)] \quad (12.136)$$

$$C_{xy}(t_1, t_1+\tau) = \lim_{n\to\infty}\frac{1}{n}\sum_{k=1}^{n}[x_k(t_1) - \mu_x(t_1)][y_k(t_1+\tau) - \mu_y(t_1+\tau)]$$

The values of the covariance functions depend in general on the times t_1 and $t_1 + \tau$.

To provide a more detailed description of the random processes, higher-order statistics should be calculated, which involves the values of the time histories evaluated at three or more times, such as $t_1,\ t_1+\tau,\ t_1+\sigma$, etc. For reasons to be explained shortly, this is not always necessary.

In the special case in which the mean values $\mu_x(t_1)$ and $\mu_y(t_1)$ and the covariance functions $C_x(t_1, t_1+\tau),\ C_y(t_1, t_1+\tau)$ and $C_{xy}(t_1, t_1+\tau)$ do not depend on t_1, the random processes $x_k(t)$ and $y_k(t)$ $(k = 1, 2, \dots)$ are said to be *weakly stationary*. Otherwise they are nonstationary. Hence, for weakly stationary random processes the mean values are constant, $\mu_x(t_1) = \mu_x =$ constant and $\mu_y(t_1) = \mu_y =$ constant, and the covariance functions depend on the time shift τ alone, $C_x(t_1, t_1+\tau) = C_x(\tau)$, $C_y(t_1, t_1+\tau) = C_y(\tau)$ and $C_{xy}(t_1, t_1+\tau) = C_{xy}(\tau)$. If all possible statistics are independent of t_1, then the random processes $x_k(t)$ and $y_k(t)$ $(k = 1, 2, \dots)$ are said to be *strongly stationary*. For normal, or Gaussian, random processes, however, higher-order averages can be derived from the mean values and covariance functions alone. It follows that *for Gaussian random processes weak stationarity implies also strong stationarity*. Because our interest lies primarily in normal random processes, there is no need to calculate higher-order statistics, and random processes will be referred to as merely *stationary* if the mean values and covariance functions are insensitive to a translation in the time t_1. The remainder of this section is devoted exclusively to stationary random processes.

Ensemble averages can be calculated conveniently in terms of probability density functions. To this end, we introduce the notation $x_1 = x_k(t),\ x_2 = x_k(t+\tau),\ y_1 = y_k(t),\ y_2 = y_k(t+\tau)$, where x_1 and x_2 represent random variables from the stationary random process $x_k(t)$ and y_1 and y_2 represent random variables from the stationary random process $y_k(t)$ $(k = 1, 2, \dots)$. Then, the joint probability density functions $p(x_1, x_2),\ p(y_1, y_2)$ and $p(x_1, y_2)$ are independent of t. In view of this, the *mean values* can be written as

$$\begin{aligned}
\mu_x = E[x] &= \int_{-\infty}^{\infty}\int_{-\infty}^{\infty} x_1 p(x_1, x_2)\,dx_1 dx_2 = \int_{-\infty}^{\infty} x_1 p(x_1)\,dx_1 = \text{constant} \\
\mu_y = E[y] &= \int_{-\infty}^{\infty}\int_{-\infty}^{\infty} y_1 p(y_1, y_2)\,dy_1 dy_2 = \int_{-\infty}^{\infty} y_1 p(y_1)\,dy_1 = \text{constant}
\end{aligned} \tag{12.137}$$

and the *correlation functions* have the expressions

$$\begin{aligned}
R_x(\tau) = E[x_1 x_2] &= \int_{-\infty}^{\infty}\int_{-\infty}^{\infty} x_1 x_2 p(x_1, x_2)\,dx_1 dx_2 \\
R_y(\tau) = E[y_1 y_2] &= \int_{-\infty}^{\infty}\int_{-\infty}^{\infty} y_1 y_2 p(y_1, y_2)\,dy_1 dy_2 \\
R_{xy}(\tau) = E[x_1 y_2] &= \int_{-\infty}^{\infty}\int_{-\infty}^{\infty} x_1 y_2 p(x_1, y_2)\,dx_1 dy_2
\end{aligned} \tag{12.138}$$

where $R_x(\tau)$ and $R_y(\tau)$ represent *autocorrelation functions* and $R_{xy}(\tau)$ is a *cross-*

correlation function. Moreover, the *covariance functions* can be written as

$$\begin{aligned} C_x(\tau) &= E[(x_1-\mu_x)(x_2-\mu_x)] \\ &= \int_{-\infty}^{\infty}\int_{-\infty}^{\infty}(x_1-\mu_x)(x_2-\mu_x)p(x_1,x_2)dx_1dx_2 = R_x(\tau)-\mu_x^2 \\ C_y(\tau) &= E[(y_1-\mu_y)(y_2-\mu_y)] \\ &= \int_{-\infty}^{\infty}\int_{-\infty}^{\infty}(y_1-\mu_y)(y_2-\mu_y)p(y_1,y_2)dy_1dy_2 = R_y(\tau)-\mu_y^2 \\ C_{xy}(\tau) &= E[(x_1-\mu_x)(y_2-\mu_y)] \\ &= \int_{-\infty}^{\infty}\int_{-\infty}^{\infty}(x_1-\mu_x)(y_2-\mu_y)p(x_1,y_2)dx_1dy_2 = R_{xy}(\tau)-\mu_x\mu_y \end{aligned} \tag{12.139}$$

From Eqs. (12.139), we conclude that the covariance functions are identical to the correlation functions only when the mean values are zero. When the covariance function $C_{xy}(\tau)$ is equal to zero for all τ, the stationary random processes $x_k(t)$ and $y_k(t)$ $(k=1,2,\ldots)$ are said to be uncorrelated. From the last of Eqs. (12.139), we conclude that this can happen only if the cross-correlation function $R_{xy}(\tau)$ is equal to zero for all τ and, in addition, either μ_x or μ_y is equal to zero.

Next, we denote $x_1 = x_k(t-\tau)$, $x_2 = x_k(t)$, $y_1 = y_k(t-\tau)$ and $y_2 = y_k(t)$. Then, because for stationary random processes $p(x_1,x_2)$, $p(y_1,y_2)$ and $p(x_1,y_2)$ are independent of a translation in the time t, it follows that the autocorrelation functions are even functions of τ, that is,

$$R_x(-\tau) = R_x(\tau), \quad R_y(-\tau) = R_y(\tau) \tag{12.140}$$

whereas the cross-correlation function merely satisfies

$$R_{xy}(-\tau) = R_{yx}(\tau) \tag{12.141}$$

Using the same approach as that used in Sec. 12.7, it can be shown that

$$R_x(0) \ge |R_x(\tau)|, \quad R_y(0) \ge |R_y(\tau)| \tag{12.142}$$

In contrast, however, $R_{xy}(\tau)$ does not necessarily have a maximum at $\tau=0$. Bounds on the cross-correlation function can be established by considering

$$\begin{aligned} &\int_{-\infty}^{\infty}\int_{-\infty}^{\infty}(x_1\pm y_2)^2p(x_1,y_2)dx_1dy_2 \\ &\quad = \int_{-\infty}^{\infty}\int_{-\infty}^{\infty}x_1^2p(x_1,y_2)dx_1dy_2 \pm 2\int_{-\infty}^{\infty}\int_{-\infty}^{\infty}x_1y_2p(x_1,y_2)dx_1dy_2 \\ &\qquad + \int_{-\infty}^{\infty}\int_{-\infty}^{\infty}y_2^2p(x_1,y_2)dx_1dy_2 \\ &\quad = R_x(0) \pm 2R_{xy}(\tau) + R_y(0) \ge 0 \end{aligned} \tag{12.143}$$

where the inequality is valid because the first integral in Eq. (12.143) cannot be negative. Note that the dependence on the time shift τ appears only when the variables with different subscripts are involved. It follows from Eq. (12.143) that

$$R_x(0) + R_y(0) \geq 2|R_{xy}(\tau)| \tag{12.144}$$

Moreover, considering the integral

$$\int_{-\infty}^{\infty}\int_{-\infty}^{\infty}\left[\frac{x_1}{\sqrt{R_x(0)}} \pm \frac{y_2}{\sqrt{R_y(0)}}\right]^2 p(x_1, y_2)dx_1dy_2 \tag{12.145}$$

which is also nonnegative, it can be shown that

$$R_x(0)R_y(0) \geq |R_{xy}(\tau)|^2 \tag{12.146}$$

From the above, we conclude that the correlation properties of the two stationary random processes $x_k(t)$ and $y_k(t)$ $(k = 1, 2, \ldots)$ can be described by the correlation functions $R_x(\tau)$, $R_y(\tau)$, $R_{xy}(\tau)$ and $R_{yx}(\tau)$. Moreover, in view of Eqs. (12.140) and (12.141), these functions need be calculated only for values of τ larger than or equal to zero.

At this point, it is possible to introduce power spectral densities and cross-spectral densities associated with the two random processes $x_k(t)$ and $y_k(t)$ $(k = 1, 2, \ldots)$. We defer the discussion to the next section, however, when these concepts are discussed in the context of ergodic random processes.

12.14 JOINT PROPERTIES OF ERGODIC RANDOM PROCESSES

We consider the two stationary random processes $x_k(t)$ and $y_k(t)$ $(k = 1, 2, \ldots)$ of Sec. 12.13, but instead of calculating ensemble averages, we select two arbitrary time histories $x_k(t)$ and $y_k(t)$ from these processes and calculate time averages. In general, the averages calculated by using these sample functions will be different for different $x_k(t)$ and $y_k(t)$, so that we identify these averages by the index k.

The *temporal mean values* can be written in the form

$$\mu_x(k) = \lim_{T\to\infty}\frac{1}{T}\int_{-T/2}^{T/2} x_k(t)dt, \quad \mu_y(k) = \lim_{T\to\infty}\frac{1}{T}\int_{-T/2}^{T/2} y_k(t)dt \tag{12.147}$$

whereas the *temporal covariance functions* have the expressions

$$\begin{aligned}
C_x(\tau, k) &= \lim_{T\to\infty}\frac{1}{T}\int_{-T/2}^{T/2}[x_k(t) - \mu_x(k)][x_k(t+\tau) - \mu_x(k)]dt \\
C_y(\tau, k) &= \lim_{T\to\infty}\frac{1}{T}\int_{-T/2}^{T/2}[y_k(t) - \mu_y(k)][y_k(t+\tau) - \mu_y(k)]dt \\
C_{xy}(\tau, k) &= \lim_{t\to\infty}\frac{1}{T}\int_{-T/2}^{T/2}[x_k(t) - \mu_x(k)][y_k(t+\tau) - \mu_y(k)]dt
\end{aligned} \tag{12.148}$$

If the temporal mean values and covariance functions calculated by using the sample functions $x_k(t)$ and $y_k(t)$ are equal to the ensemble mean values and covariance functions, regardless of the pair of sample functions used, then the stationary random processes $x_k(t)$ and $y_k(t)$ $(k = 1, 2, \ldots)$ are said to be *weakly ergodic*. If all ensemble averages can be deduced from temporal averages, then the stationary random processes are said to be *strongly ergodic*. Because Gaussian processes are fully described by first- and second-order statistics alone, no distinction need be made for such processes, and we refer to them as merely *ergodic*. Again, *ergodicity implies stationarity, but stationarity does not imply ergodicity*. Hence, the processes $x_k(t)$ and $y_k(t)$ $(k = 1, 2, \ldots)$ are ergodic if

$$\mu_x(k) = \mu_x = \text{constant}, \quad \mu_y(k) = \mu_y = \text{constant} \tag{12.149}$$

and

$$C_x(\tau, k) = C_x(\tau), \quad C_y(\tau, k) = C_y(\tau), \quad C_{xy}(\tau, k) = C_{xy}(\tau) \tag{12.150}$$

The covariance functions are related to the correlation functions $R_x(\tau)$, $R_y(\tau)$ and $R_{xy}(\tau)$ by

$$C_x(\tau) = R_x(\tau) - \mu_x^2, \quad C_y(\tau) = R_y(\tau) - \mu_y^2, \quad C_{xy}(\tau) = R_{xy}(\tau) - \mu_x \mu_y \tag{12.151}$$

in which the correlation functions have the expressions

$$\begin{aligned} R_x(\tau) &= \lim_{T \to \infty} \frac{1}{T} \int_{-T/2}^{T/2} x(t)x(t+\tau)dt, \; R_y(\tau) = \lim_{T \to \infty} \frac{1}{T} \int_{-T/2}^{T/2} y(t)y(t+\tau)dt, \\ R_{xy}(\tau) &= \lim_{T \to \infty} \frac{1}{T} \int_{-T/2}^{T/2} x(t)y(t+\tau)dt \end{aligned} \tag{12.152}$$

where the index identifying the sample functions $x_k(t)$ and $y_k(t)$ has been omitted because the correlation functions are the same for any pair of sample functions. In view of the fact that ergodicity implies stationarity, Eqs. (12.140) and (12.141) and inequalities (12.142), (12.144) and (12.146) continue to be valid.

Next, we assume that the autocorrelation functions $R_x(\tau)$ and $R_y(\tau)$ and the cross-correlation function $R_{xy}(\tau)$ exist and define the power spectral density functions as the Fourier transforms

$$S_x(\omega) = \int_{-\infty}^{\infty} R_x(\tau)e^{-i\omega\tau}d\tau, \quad S_y(\omega) = \int_{-\infty}^{\infty} R_y(\tau)e^{-i\omega\tau}d\tau \tag{12.153}$$

and the cross-spectral density function as the Fourier transform

$$S_{xy}(\omega) = \int_{-\infty}^{\infty} R_{xy}(\tau)e^{-i\omega\tau}d\tau \tag{12.154}$$

Then, if the power spectral and cross-spectral density functions are given for the two processes, the autocorrelation and cross-correlation functions can be obtained from the inverse Fourier transforms

$$\begin{aligned} R_x(\tau) &= \frac{1}{2\pi} \int_{-\infty}^{\infty} S_x(\omega)e^{i\omega\tau}d\omega, \quad R_y(\tau) = \frac{1}{2\pi} \int_{-\infty}^{\infty} S_y(\omega)e^{i\omega\tau}d\omega, \\ R_{xy}(\tau) &= \frac{1}{2\pi} \int_{-\infty}^{\infty} S_{xy}(\omega)e^{i\omega\tau}d\omega \end{aligned} \tag{12.155}$$

Using Eqs. (12.140), it can be shown that the power spectral density functions are even functions of ω,

$$S_x(-\omega) = S_x(\omega), \quad S_y(-\omega) = S_y(\omega) \tag{12.156}$$

whereas using Eq. (12.141) it follows that

$$S_{xy}(-\omega) = S_{yx}(\omega) \tag{12.157}$$

from which we conclude that if $S_{xy}(\omega)$ and $S_{yx}(\omega)$ are given for $\omega > 0$, then Eq. (12.157) can be used to obtain $S_{yx}(\omega)$ and $S_{xy}(\omega)$ for $\omega < 0$, respectively. In view of Eqs. (12.156), Eqs. (12.153) reduce to

$$S_x(\omega) = 2\int_0^\infty R_x(\tau)\cos\omega\tau\, d\tau, \quad S_y(\omega) = 2\int_0^\infty R_y(\tau)\cos\omega\tau\, d\tau \tag{12.158}$$

and the first two of Eqs. (12.155) become

$$R_x(\tau) = \frac{1}{\pi}\int_0^\infty S_x(\omega)\cos\omega\tau\, d\omega, \quad R_y(\tau) = \frac{1}{\pi}\int_0^\infty S_y(\omega)\cos\omega\tau\, d\omega \tag{12.159}$$

Equations (12.158) and (12.159) are known as the *Wiener-Khintchine equations*. Note that $S_x(\omega)$ and $S_y(\omega)$ are nonnegative on physical grounds, and they are real because $R_x(\tau)$ and $R_y(\tau)$ are real.

12.15 RESPONSE CROSS-CORRELATION FUNCTIONS FOR LINEAR SYSTEMS

We consider two linear systems defined in the time domain by the impulse responses $g_r(t)$ and $g_s(t)$ and in the frequency domain by the frequency responses $G_r(\omega)$ and $G_s(\omega)$, where the latter are the Fourier transforms of the former, namely,

$$G_r(\omega) = \int_{-\infty}^\infty g_r(t)e^{-i\omega t}dt, \quad G_s(\omega) = \int_{-\infty}^\infty g_s(t)e^{-i\omega t}dt \tag{12.160}$$

The relations between the excitations $N_r(t)$ and $N_s(t)$ and the corresponding responses $\eta_r(t)$ and $\eta_s(t)$ are given in the form of the block diagrams of Fig. 12.30a, whereas those between the transformed excitations $N_r(\omega)$ and $N_s(\omega)$ and the corresponding transformed response $\eta_r(\omega)$ and $\eta_s(\omega)$ are given in the form of the block diagrams of Fig. 12.30b, where $N_r(\omega)$ is the Fourier transform of $N_r(t)$, etc.

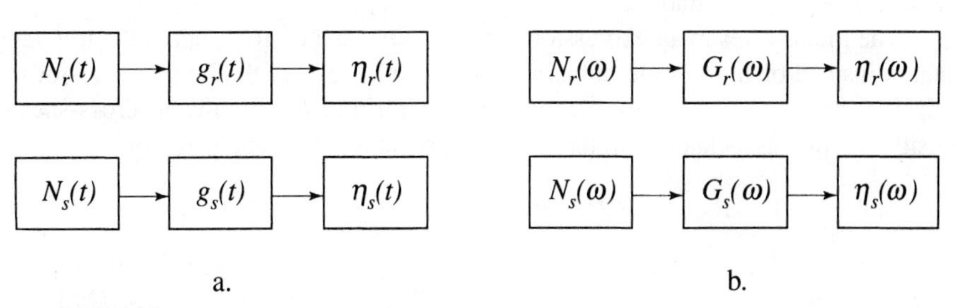

FIGURE 12.30
Block diagrams relating responses to excitations a. in the time domain and b. in the frequency domain

Assuming that the excitation and response processes are ergodic, the cross-correlation function between the response processes $\eta_r(t)$ and $\eta_s(t)$ can be written in the form

$$R_{\eta_r\eta_s}(\tau) = \lim_{T\to\infty}\int_{-T/2}^{T/2} \eta_r(t)\eta_s(t+\tau)dt \tag{12.161}$$

But, for linear systems the relation between the excitation and response can be expressed in terms of the convolution integral, Eq. (12.82). Hence, we can write

$$\eta_r(t) = \int_{-\infty}^{\infty} g_r(\lambda_r)N_r(t-\lambda_r)d\lambda_r,\ \eta_s(t) = \int_{-\infty}^{\infty} g_s(\lambda_s)N_s(t-\lambda_s)d\lambda_s \tag{12.162}$$

where λ_r and λ_s are corresponding dummy variables. Inserting Eqs. (12.162) into Eq. (12.161) and changing the order of integration, we obtain

$$\begin{aligned} R_{\eta_r\eta_s}(\tau) &= \lim_{T\to\infty}\frac{1}{T}\int_{-T/2}^{T/2}\left[\int_{-\infty}^{\infty} g_r(\lambda_r)N_r(t-\lambda_r)d\lambda_r \int_{-\infty}^{\infty} g_s(\lambda_s)N_s(t+\tau-\lambda_s)d\lambda_s\right]dt \\ &= \int_{-\infty}^{\infty}\int_{-\infty}^{\infty} g_r(\lambda_r)g_s(\lambda_s)\left[\lim_{T\to\infty}\frac{1}{T}\int_{-T/2}^{T/2} N_r(t-\lambda_r)N_s(t+\tau-\lambda_s)dt\right]d\lambda_r d\lambda_s \end{aligned} \tag{12.163}$$

Because the excitation processes are ergodic, and hence stationary, we recognize that

$$\begin{aligned} &\lim_{T\to\infty}\frac{1}{T}\int_{-T/2}^{T/2} N_r(t-\lambda_r)N_s(t+\tau-\lambda_s)dt \\ &\quad = \lim_{T\to\infty}\frac{1}{T}\int_{-T/2}^{T/2} N_r(t)N_s(t+\tau+\lambda_r-\lambda_s)dt = R_{N_rN_s}(\tau+\lambda_r-\lambda_s) \end{aligned} \tag{12.164}$$

is the cross-correlation function between the excitation processes. Hence, Eq. (12.163) can be written in the form

$$R_{\eta_r\eta_s}(\tau) = \int_{-\infty}^{\infty}\int_{-\infty}^{\infty} g_r(\lambda_r)g_s(\lambda_s)R_{N_rN_s}(\tau+\lambda_r-\lambda_s)d\lambda_r d\lambda_s \tag{12.165}$$

which represents a time-domain expression relating the cross-correlation function between the response processes to the cross-correlation function between the excitation processes. Note the analogy between Eq. (12.165) and Eq. (12.95), where the latter is an expression relating the autocorrelation function of a single response to the autocorrelation function of a single excitation.

The interest lies in an expression analogous to Eq. (12.165), but in the frequency domain instead of in the time domain. To this end, we take the Fourier transform of both sides of Eq. (12.165). But, the Fourier transform of $R_{\eta_r\eta_s}(\tau)$ is the cross-spectral density function associated with the response processes $\eta_r(t)$ and $\eta_s(t)$, or

$$\begin{aligned} S_{\eta_r\eta_s}(\omega) &= \int_{-\infty}^{\infty} R_{\eta_r\eta_s}(\tau)e^{-i\omega\tau}d\tau \\ &= \int_{-\infty}^{\infty} e^{-i\omega\tau}\left[\int_{-\infty}^{\infty}\int_{-\infty}^{\infty} g_r(\lambda_r)g_s(\lambda_s)R_{N_rN_s}(\tau+\lambda_r-\lambda_s)d\lambda_r d\lambda_s\right]d\tau \end{aligned} \tag{12.166}$$

Moreover, $R_{N_rN_s}(\tau+\lambda_r-\lambda_s)$ can be expressed as the inverse Fourier transform

$$R_{N_rN_s}(\tau+\lambda_r-\lambda_s)=\frac{1}{2\pi}\int_{-\infty}^{\infty}S_{N_rN_s}(\omega)e^{i\omega(\tau+\lambda_r-\lambda_s)}d\omega \tag{12.167}$$

where $S_{N_rN_s}(\omega)$ is the cross-spectral density function associated with the excitation processes $N_r(t)$ and $N_s(t)$. Inserting Eq. (12.167) into Eq. (12.166), considering Eqs. (12.160), changing the integration order and rearranging, we obtain

$$\begin{aligned}
S_{\eta_r\eta_s}(\omega) &= \int_{-\infty}^{\infty}e^{-i\omega\tau}\left\{\int_{-\infty}^{\infty}\int_{-\infty}^{\infty}g_r(\lambda_r)g_s(\lambda_s)\right.\\
&\qquad\left.\times\left[\frac{1}{2\pi}\int_{-\infty}^{\infty}S_{N_rN_s}(\omega)e^{i\omega(\tau+\lambda_r-\lambda_s)}d\omega\right]d\lambda_r d\lambda_s\right\}d\tau\\
&= \int_{-\infty}^{\infty}e^{-i\omega\tau}\left\{\frac{1}{2\pi}\int_{-\infty}^{\infty}S_{N_rN_s}(\omega)\left[\int_{-\infty}^{\infty}g_r(\lambda_r)e^{i\omega\lambda_r}d\lambda_r\right.\right.\\
&\qquad\left.\left.\times\int_{-\infty}^{\infty}g_s(\lambda_s)e^{-i\omega\lambda_s}d\lambda_s\right]e^{i\omega\tau}d\omega\right\}d\tau\\
&= \int_{-\infty}^{\infty}e^{-i\omega\tau}\left[\frac{1}{2\pi}\int_{-\infty}^{\infty}S_{N_rN_s}(\omega)\bar{G}_r(\omega)G_s(\omega)e^{i\omega\tau}d\omega\right]d\tau
\end{aligned} \tag{12.168}$$

where $\bar{G}_r(\omega)=G_r(-\omega)$ is the complex conjugate of $G_r(\omega)$. Comparing the first integral in Eq. (12.166) with the last one in Eq. (12.168), and recognizing that the cross-correlation function $R_{\eta_r\eta_s}(\tau)$ between the response processes $\eta_r(t)$ and $\eta_s(t)$ must be equal to the inverse Fourier transform of the cross-spectral density function $S_{\eta_r\eta_s}(\omega)$ associated with these response processes, we must conclude that

$$S_{\eta_r\eta_s}(\omega)=\bar{G}_r(\omega)G_s(\omega)S_{N_rN_s}(\omega) \tag{12.169}$$

and

$$R_{\eta_r\eta_s}(\tau)=\frac{1}{2\pi}\int_{-\infty}^{\infty}S_{\eta_r\eta_s}(\omega)e^{i\omega\tau}d\omega=\frac{1}{2\pi}\int_{-\infty}^{\infty}\bar{G}_r(\omega)G_s(\omega)S_{N_rN_s}(\omega)e^{i\omega\tau}d\omega \tag{12.170}$$

represent a Fourier transform pair. The algebraic expression (12.169) relates the cross-spectral density functions associated with the excitation and response processes in the frequency domain. Note the analogy between Eq. (12.169) and Eq. (12.99).

For any two time histories $N_r(t)$ and $N_s(t)$ corresponding to two stationary random signals, the cross-spectral density function $S_{N_rN_s}(\omega)$ can be obtained by means of an analog cross-spectral density analyzer.[9]

---

[9]See Bendat and Piersol, op. cit., sec. 8.5.

12.16 RESPONSE OF MULTI-DEGREE-OF-FREEDOM SYSTEMS TO RANDOM EXCITATIONS

We showed in Sec. 7.1 that the equations of motion of a damped n-degree-of-freedom system can be written in the matrix form

$$M\ddot{\mathbf{q}}(t)+C\dot{\mathbf{q}}(t)+K\mathbf{q}(t)=\mathbf{Q}(t) \tag{12.171}$$

where M, C and K are $n\times n$ symmetric inertia, damping and stiffness matrices, respectively. The n-dimensional vector $\mathbf{q}(t)$ contains the generalized coordinates $q_i(t)$, whereas the n-dimensional vector $\mathbf{Q}(t)$ contains the associated generalized forces $Q_i(t)$ $(i=1,2,\dots,n)$. The interest lies in the case in which the excitations $Q_i(t)$ represent ergodic random processes, from which it follows that the responses $q_i(t)$ are also ergodic random processes.

The general response of a damped multi-degree-of-freedom system to external excitations cannot be obtained readily, even when the excitation is deterministic. The difficulty lies in the fact that classical modal analysis cannot generally be used to uncouple the system of equations (12.171), and one must formulate the problem in the state space (Sec. 7.16). However, as shown in Sec. 7.15, in the special case in which the damping matrix is a linear combination of the inertia and stiffness matrices, the modal matrix associated with the undamped linear system can be used as a linear transformation uncoupling the system of equations. For simplicity, we confine ourselves to the case in which the classical modal matrix $U=[u_1\ u_2\dots u_n]$ associated with the undamped system can be used as a transformation matrix uncoupling the set (12.171). Following the procedure of Sec. 7.15, we write the solution of Eq. (12.171) in the form

$$\mathbf{q}(t)=U\boldsymbol{\eta}(t) \tag{12.172}$$

where the components $\eta_r(t)$ $(r=1,2,\dots,n)$ of the vector $\boldsymbol{\eta}(t)$ are generalized coordinates consisting of linear combinations of the random variables $q_i(t)$ $(i=1,2,\dots,n)$. Inserting Eq. (12.172) into Eq. (12.171), premultiplying the result by U^T, using the orthonormality relations (see Eqs. (7.202))

$$U^TMU=I,\quad U^TKU=\Omega \tag{12.173}$$

in which I is the identity matrix and $\Omega=\text{diag}[\omega_1^2\ \omega_2^2\ \dots\ \omega_n^2]$ is the diagonal matrix of the eigenvalues, as well as assuming that

$$U^TCU=2\Xi\Omega^{1/2} \tag{12.174}$$

where $\Xi=\text{diag}[\zeta_1\ \zeta_2\ \dots\ \zeta_n]$, we obtain the set of independent equations for the natural coordinates

$$\ddot{\eta}_r(t)+2\zeta_r\omega_r\dot{\eta}_r(t)+\omega_r^2\eta_r(t)=\omega_r^2N_r(t),\ r=1,2,\dots,n \tag{12.175}$$

where ζ_r is a damping factor associated with the rth mode, ω_r is the rth frequency of the undamped system and

$$N_r(t)=\frac{1}{\omega_r^2}\mathbf{u}_r^T\mathbf{Q}(t),\ r=1,2,\dots,n \tag{12.176}$$

is a generalized random force, in which $\mathbf{u}_r$ represents the rth modal vector of the undamped system. Note that $N_r(t)$ actually has units $LM^{1/2}$, where L denotes length and M denotes mass.

Our first objective is to calculate the cross-correlation function between two response processes. To this end, we introduce the Fourier transforms of $\eta_r(t)$ and $N_r(t)$, respectively, in the form

$$\begin{aligned} \eta_r(\omega) &= \int_{-\infty}^{\infty} \eta_r(t)e^{-i\omega t}dt \\ N_r(\omega) &= \int_{-\infty}^{\infty} N_r(t)e^{-i\omega t}dt = \frac{1}{\omega_r^2}\mathbf{u}_r^T \int_{-\infty}^{\infty} \mathbf{Q}(t)e^{-i\omega t}dt \end{aligned} \tag{12.177}$$

Then, transforming both sides of Eqs. (12.175), we obtain

$$(-\omega^2 + i2\zeta_r\omega\omega_r + \omega_r^2)\eta_r(\omega) = \omega_r^2 N_r(\omega), \ r = 1, 2, \ldots, n \tag{12.178}$$

Equations (12.178) can be solved for $\eta_r(\omega)$ with the result

$$\eta_r(\omega) = G_r(\omega)N_r(\omega), \ r = 1, 2, \ldots, n \tag{12.179}$$

where

$$G_r(\omega) = \frac{1}{1 - (\omega/\omega_r)^2 + i2\zeta_r\omega/\omega_r}, \ r = 1, 2, \ldots, n \tag{12.180}$$

is the frequency response associated with the rth natural mode. Note the analogy between Eqs. (12.179) and Eq. (12.86).

Next, we calculate the cross-correlation function $R_{q_iq_j}(\tau)$ between the two components $q_i(t)$ and $q_j(t)$ $(i, j = 1, 2, \ldots, n)$ of the response process. There are n^2 such cross-correlation functions, which can be readily arranged in an $n \times n$ matrix. Hence, using Eq. (12.172), we introduce the *response correlation matrix* in the form

$$\begin{aligned} R_q(\tau) &= [R_{q_iq_j}(\tau)] \\ &= \lim_{T\to\infty} \frac{1}{T}\int_{-T/2}^{T/2} \begin{bmatrix} q_1(t)q_1(t+\tau) & q_1(t)q_2(t+\tau) & \ldots & q_1(t)q_n(t+\tau) \\ q_2(t)q_1(t+\tau) & q_2(t)q_2(t+\tau) & \ldots & q_2(t)q_n(t+\tau) \\ \ldots & \ldots & \ldots & \ldots \\ q_n(t)q_1(t+\tau) & q_n(t)q_2(t+\tau) & \ldots & q_n(t)q_n(t+\tau) \end{bmatrix} dt \\ &= \lim_{T\to\infty} \frac{1}{T}\int_{-T/2}^{T/2} \mathbf{q}(t)\mathbf{q}^T(t+\tau)dt \\ &= \lim_{T\to\infty} \frac{1}{T}\int_{-T/2}^{T/2} U\boldsymbol{\eta}(t)\boldsymbol{\eta}^T(t+\tau)U^T dt = UR_\eta(\tau)U^T \end{aligned} \tag{12.181}$$

where

$$R_\eta(\tau) = \lim_{T\to\infty} \frac{1}{T}\int_{-T/2}^{T/2} \boldsymbol{\eta}(t)\boldsymbol{\eta}^T(t+\tau)dt \tag{12.182}$$

is the *modal response correlation matrix*. Our objective is to express the response correlation matrix in terms of quantities defining the excitation random process. To this end, we use Eq. (12.170) and rewrite the modal response correlation matrix as

$$R_\eta(\tau) = \frac{1}{2\pi}\int_{-\infty}^{\infty} \bar{G}(\omega)S_N(\omega)G(\omega)e^{i\omega\tau}d\omega \tag{12.183}$$

in which $G(\omega) = \text{diag}[G_1(\omega) \;\; G_2(\omega) \;\; \ldots \;\; G_n(\omega)]$ is the diagonal matrix of the modal frequency responses, Eqs. (12.180), $\bar{G}(\omega)$ is the complex conjugate of $G(\omega)$ and $S_N(\omega) = [S_{N_iN_j}(\omega)]$ is the $n \times n$ modal excitation spectral density matrix. It follows immediately from Eq. (12.181) that the response correlation matrix is

$$R_q(\tau) = \frac{1}{2\pi}U\left[\int_{-\infty}^{\infty} \bar{G}(\omega)S_N(\omega)G(\omega)e^{i\omega\tau}d\omega\right]U^T \tag{12.184}$$

where the modal excitation spectral density matrix represents the Fourier transform of the modal excitations correlation matrix $R_N(\tau)$, or

$$S_N(\omega) = \int_{-\infty}^{\infty} R_N(\omega)e^{-i\omega\tau}d\tau \tag{12.185}$$

At this point, we turn our attention to the characterization of the spectral density matrix in terms of actual forces, instead of modal forces. To this end, we begin by writing the modal excitation correlation matrix in the form

$$R_N(\tau) = \lim_{T\to\infty}\frac{1}{T}\int_{-T/2}^{T/2} \mathbf{N}(t)\mathbf{N}^T(t+\tau)dt \tag{12.186}$$

in which, from Eqs. (12.176), the modal force vector is

$$\mathbf{N}(t) = \Omega^{-1}U^T\mathbf{Q}(t) \tag{12.187}$$

Inserting Eq. (12.187) into Eq. (12.186), we have

$$\begin{aligned} R_N(\tau) &= \lim_{T\to\infty}\frac{1}{T}\int_{-T/2}^{T/2} \Omega^{-1}U^T\mathbf{Q}(t)\mathbf{Q}^T(t+\tau)U\Omega^{-1}dt \\ &= \Omega^{-1}U^TR_Q(\tau)U\Omega^{-1} \end{aligned} \tag{12.188}$$

where

$$R_Q(\tau) = \lim_{T\to\infty}\frac{1}{T}\int_{-T/2}^{T/2} \mathbf{Q}(t)\mathbf{Q}^T(t+\tau)dt \tag{12.189}$$

is the actual force correlation matrix. Hence, introducing Eq. (12.188) in Eq. (12.185), we can write

$$S_N(\omega) = \Omega^{-1}U^T\left[\int_{-\infty}^{\infty} R_Q(\tau)e^{-i\omega\tau}d\tau\right]U\Omega^{-1} = \Omega^{-1}U^TS_Q(\omega)U\Omega^{-1} \tag{12.190}$$

in which

$$S_Q(\omega) = \int_{-\infty}^{\infty} R_Q(\tau)e^{-i\omega\tau}d\tau \tag{12.191}$$

is the desired excitation spectral density matrix expressed in terms of actual forces. For stationary random processes, the entries of $S_Q(\omega)$ can be obtained by means of an analog cross-spectral density analyzer.[10] Then, inserting Eq. (12.190) into Eq. (12.184), we obtain the response correlation matrix

$$R_q(\tau) = \frac{1}{2\pi} U \left[\int_{-\infty}^{\infty} \bar{G}(\omega)\Omega^{-1}U^T S_Q(\omega) U \Omega^{-1} G(\omega) e^{i\omega\tau} d\omega \right] U^T \tag{12.192}$$

where $S_Q(\omega)$ is given by Eq. (12.191).

Finally, we derive the autocorrelation function associated with the response random process $q_i(t)$. To this end, we denote the ith row of the modal matrix U as follows:

$$\mathbf{u}_i' = [u_{i1}\ u_{i2}\ \dots\ u_{in}],\ i = 1, 2, \dots, n \tag{12.193}$$

Then, the response autocorrelation function associated with $q_i(t)$ is simply

$$R_{qi}(\tau) = \frac{1}{2\pi} \mathbf{u}_i' \left[\int_{-\infty}^{\infty} \bar{G}(\omega)\Omega^{-1}U^T S_Q(\omega) U \Omega^{-1} G(\omega) e^{i\omega\tau} d\omega \right] (\mathbf{u}_i')^T,\ i = 1, 2, \dots, n \tag{12.194}$$

which for $\tau = 0$ reduces to the mean square value

$$R_{qi}(0) = \frac{1}{2\pi} \mathbf{u}_i' \left[\int_{-\infty}^{\infty} \bar{G}(\omega)\Omega^{-1}U^T S_Q(\omega) U \Omega^{-1} G(\omega) d\omega \right] (\mathbf{u}_i')^T,\ i = 1, 2, \dots, n \tag{12.195}$$

Example 12.5. Consider the system shown in Fig. 12.31, where the force $Q_1(t)$ can be regarded as an ergodic random process with zero mean and with ideal white noise power spectral density, $S_{Q_1}(\omega) = S_0$, and obtain the mean square values associated with the responses $q_1(t)$ and $q_2(t)$.

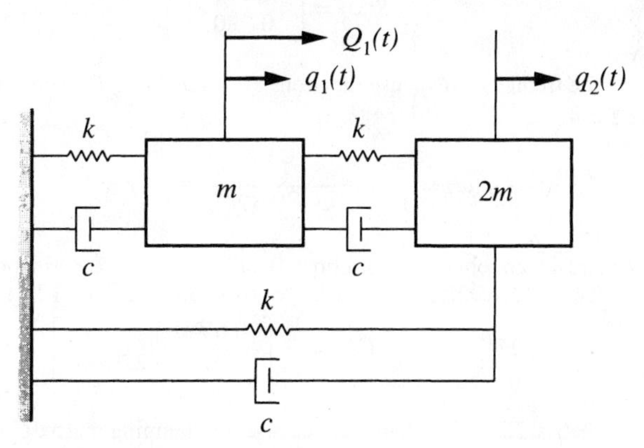

FIGURE 12.31
Proportionally damped two-degree-of-freedom system

[10]See Bendat and Piersol, op. cit., sec. 8.5.

The mean square values associated with $q_1(t)$ and $q_2(t)$ will be obtained by the modal analysis outlined in this section. The differential equations of motion associated with the system can be shown to be

$$\begin{aligned} m\ddot{q}_1 + 2c\dot{q}_1 - c\dot{q}_2 + 2kq_1 - kq_2 &= Q_1(t) \\ 2m\ddot{q}_2 - c\dot{q}_1 + 2c\dot{q}_2 - kq_1 + 2kq_2 &= 0 \end{aligned} \tag{a}$$

from which we conclude that damping is of the proportional type. Hence, the classical modal matrix does uncouple the equations of motion. To obtain the modal matrix, we must solve the eigenvalue problem associated with the undamped free vibration of the system, which has the form

$$\omega^2 m \begin{bmatrix} 1 & 0 \\ 0 & 2 \end{bmatrix} \begin{bmatrix} u_1 \\ u_2 \end{bmatrix} = k \begin{bmatrix} 2 & -1 \\ -1 & 2 \end{bmatrix} \begin{bmatrix} u_1 \\ u_2 \end{bmatrix} \tag{b}$$

The solution of the eigenvalue problem (b) consists of the modal matrix

$$U = \frac{1}{\sqrt{m}} \begin{bmatrix} 0.4597 & 0.8881 \\ 0.6280 & -0.3251 \end{bmatrix} \tag{c}$$

and the matrix of the natural frequencies squared

$$\Omega = \frac{k}{m} \begin{bmatrix} 0.6340 & 0 \\ 0 & 2.3660 \end{bmatrix} \tag{d}$$

both normalized so as to satisfy Eqs. (12.173).

The excitation spectral matrix associated with the actual coordinates $q_1(t)$ and $q_2(t)$ is

$$S(\omega) = \begin{bmatrix} S_0 & 0 \\ 0 & 0 \end{bmatrix} \tag{e}$$

Moreover, the frequency response functions associated with the coordinates $\eta_1(t)$ and $\eta_2(t)$ have the form

$$G_r(\omega) = \frac{1}{1 - (\omega/\omega_r)^2 + i2\zeta_r\omega/\omega_r}, \quad r = 1, 2 \tag{f}$$

where ω_1^2 and ω_2^2 are obtained from Eq. (d) and $2\zeta_1\omega_1$ and $2\zeta_2\omega_2$ from

$$2\Xi\Omega^{-1} = U^T C U = \frac{c}{m} \begin{bmatrix} 0.6340 & 0 \\ 0 & 2.3660 \end{bmatrix} \tag{g}$$

where the matrix $2\Xi\Omega^{-1}$ is diagonal because the damping matrix C is proportional to the stiffness matrix K.

The response mean square values are given by Eqs. (12.195), in which, from Eqs. (12.193) and (c),

$$\mathbf{u}_1' = \frac{1}{\sqrt{m}}[0.4597 \;\; 0.8881], \quad \mathbf{u}_2' = \frac{1}{\sqrt{m}}[0.6280 \;\; -0.3251] \tag{h}$$

Equations (12.195) involve the terms

$$\begin{aligned} &\mathbf{u}_1'\bar{G}(\omega)\Omega^{-1}S_Q(\omega)U\Omega^{-1}G(\omega)(\mathbf{u}_1')^T \\ &\quad = \frac{0.1111S_0}{k^2}(|G_1|^2+\bar{G}_1G_2+G_1\bar{G}_2+|G_2|^2) \\ &\mathbf{u}_2'\bar{G}(\omega)\Omega^{-1}S_Q(\omega)U\Omega^{-1}G(\omega)(\mathbf{u}_2')^T \\ &\quad = \frac{S_0}{k^2}[0.2074|G_1|^2-0.0556(\bar{G}_1G_2+G_1\bar{G}_2)+0.0149|G_2|^2] \end{aligned} \tag{i}$$

Hence, using Eqs. (12.195), we can write the mean square values

$$\begin{aligned} R_{q_1}(0) &= \frac{0.1111S_0}{2\pi k^2}\int_{-\infty}^{\infty}(|G_1|^2+\bar{G}_1G_2+G_1\bar{G}_2+|G_2|^2)d\omega \\ R_{q_2}(0) &= \frac{S_0}{2\pi k^2}\left[0.2074\int_{-\infty}^{\infty}|G_1|^2d\omega-0.0556\int_{-\infty}^{\infty}(\bar{G}_1G_2+G_1\bar{G}_2)d\omega \right. \\ &\qquad \left. +0.0149\int_{-\infty}^{\infty}|G_2|^2d\omega\right] \end{aligned} \tag{j}$$

Equations (j) give the mean square values $R_{q_i}(0)$ $(i=1,2)$ in terms of integrals involving the frequency response functions $G_1(\omega)$ and $G_2(\omega)$ and their complex conjugates. The integrals are as follows:

$$\begin{aligned} &\int_{-\infty}^{\infty}|G_r|^2d\omega = \int_{-\infty}^{\infty}\frac{d\omega}{[1-[\omega/\omega_r)^2]^2+[2\zeta_r\omega/\omega_r]^2},\quad r=1,2 \\ &\int_{-\infty}^{\infty}(\bar{G}_1G_2+G_1\bar{G}_2)d\omega \\ &\quad = \int_{-\infty}^{\infty}\frac{\{[1-(\omega/\omega_1)^2][1-(\omega/\omega_2)^2]+(2\zeta_1\omega/\omega_1)(2\zeta_2\omega/\omega_2)\}d\omega}{\{[1-(\omega/\omega_1)^2]^2+(2\zeta_1\omega/\omega_1)^2\}\{[1-(\omega/\omega_2)^2]^2+(2\zeta_2\omega/\omega_2)^2\}} \end{aligned} \tag{k}$$

The first integral can be evaluated using results from Sec. 12.11, but the second integral is likely to cause a great deal of difficulties. Because no new knowledge is gained from the evaluation of the integrals, we do not pursue the subject any further.

12.17 RESPONSE OF DISTRIBUTED-PARAMETER SYSTEMS TO RANDOM EXCITATIONS

The response of distributed-parameter systems to random excitations can also be conveniently obtained by means of modal analysis. In fact, the procedure is entirely analogous to that for discrete systems, and it can be best illustrated by considering a specific example. In particular, we choose a beam in bending (Sec. 8.10). For convenience, we assume that the beam is uniform. If in addition the beam is subject to viscous damping, the boundary-value problem is described by the differential equation

$$m\frac{\partial^2 y(x,t)}{\partial t^2}+c\frac{\partial y(x,t)}{\partial t}+EI\frac{\partial^4 y(x,t)}{\partial x^4}=f(x,t),\ 0<x<L \tag{12.196}$$

where $f(x,t)$ is an ergodic distributed random excitation and $y(x,t)$ is the ergodic random response. Note that the second term on the left side of Eq. (12.196) represents a uniformly distributed damping force. Moreover, the vibration $y(x,t)$ is subject to four boundary conditions, two at each end. We assume that the solution of the eigenvalue problem associated with the undamped system consists of the natural frequencies ω_r and natural modes $Y_r(x)$ $(r = 1, 2\ldots)$, and that the solution is known; the modes are orthogonal and we assume that they have been normalized so as to satisfy the orthonormality relations

$$\int_0^L mY_r(x)Y_s(x)dx = \delta_{rs}, \quad \int_0^L Y_r(x)EI\frac{d^4Y_s(x)}{dx^4}dx = \omega_r^2\delta_{rs}, \quad r,s = 1,2,\ldots \tag{12.197}$$

where δ_{rs} is the Kronecker delta. In addition, the damping is such that

$$\int_0^L cY_r(x)Y_s(x)dx = 2\zeta_r\omega_r\delta_{rs}, \quad r,s = 1,2,\ldots \tag{12.198}$$

Then, letting the solution of Eq. (12.196) have the form

$$y(x,t) = \sum_{r=1}^{\infty} Y_r(x)\eta_r(t) \tag{12.199}$$

and using the standard modal analysis, we obtain the independent set of ordinary differential equations

$$\ddot{\eta}_r(t) + 2\zeta_r\omega_r\dot{\eta}_r(t) + \omega_r^2\eta_r(t) = \omega_r^2 N_r(t), \quad r = 1,2,\ldots \tag{12.200}$$

where

$$N_r(t) = \frac{1}{\omega_r^2}\int_0^L Y_r(x)f(x,t)dx, \quad r = 1,2,\ldots \tag{12.201}$$

are generalized random forces. As for the discrete systems of Sec. 12.16, the forces $N_r(t)$ actually have units $LM^{1/2}$.

Equations (12.200) for the distributed system possess precisely the same structure as Eqs. (12.175) for the discrete system. Hence, the remaining part of the analysis resembles entirely that of Sec. 12.16. Indeed, using Eq. (12.201) and a similar equation for $N_s(t+\tau)$, we can write

$$\begin{aligned}
R_{N_rN_s}(\tau) &= \lim_{T\to\infty}\frac{1}{T}\int_{-T/2}^{T/2} N_r(t)N_s(t+\tau)dt \\
&= \lim_{T\to\infty}\frac{1}{T}\int_{-T/2}^{T/2}\left[\frac{1}{\omega_r^2}\int_0^L Y_r(x)f(x,,t)dx\frac{1}{\omega_s^2}\int_0^L Y_s(x')f(x',t+\tau)dx'\right]dt \\
&= \frac{1}{\omega_r^2}\frac{1}{\omega_s^2}\int_0^L\int_0^L Y_r(x)Y_s(x')\left[\lim_{T\to\infty}\frac{1}{T}\int_{-T/2}^{T/2} f(x,t)f(x',t+\tau)dt\right]dx\,dx' \\
&= \frac{1}{\omega_r^2}\frac{1}{\omega_s^2}\int_0^L\int_0^L Y_r(x)Y_s(x')R_{f_xf_{x'}}(x,x',\tau)dx\,dx'
\end{aligned} \tag{12.202}$$

where x and x' are dummy variables denoting different points of the domain $0 < x < L$, and

$$R_{f_x f_{x'}}(x, x', \tau) = \lim_{T\to\infty} \frac{1}{T} \int_{-T/2}^{T/2} f(x,t) f(x', t+\tau) dt \tag{12.203}$$

is the *distributed cross-correlation function* between the distributed forces $f(x,t)$ and $f(x',t)$. Note that $R_{f_x f_{x'}}(x, x', \tau)$ has units of distributed force squared. Fourier transforming Eq. (12.202), we obtain the cross-spectral density function

$$\begin{aligned} S_{N_r N_s}(\omega) &= \int_{-\infty}^{\infty} \left[\frac{1}{\omega_r^2} \frac{1}{\omega_s^2} \int_0^L \int_0^L Y_r(x) Y_s(x') R_{f_x f_{x'}}(x, x', \tau) dx\, dx' \right] e^{-i\omega\tau} d\tau \\ &= \frac{1}{\omega_r^2} \frac{1}{\omega_s^2} \int_0^L \int_0^L Y_r(x) Y_s(x') \left[\int_{-\infty}^{\infty} R_{f_x f_{x'}}(x, x', \tau) e^{-i\omega\tau} d\tau \right] dx\, dx' \\ &= \frac{1}{\omega_r^2} \frac{1}{\omega_s^2} \int_0^L \int_0^L Y_r(x) Y_s(x') S_{f_x f_{x'}}(x, x', \omega) dx\, dx' \end{aligned} \tag{12.204}$$

where

$$S_{f_x f_{x'}}(x, x', \omega) = \int_{-\infty}^{\infty} R_{f_x f_{x'}}(x, x', \tau) e^{-i\omega\tau} d\tau \tag{12.205}$$

is the *distributed cross-spectral density function* between the excitation processes $f(x,t)$ and $f(x',t)$.

The cross-correlation function between the response at x and x' can be written in the form

$$\begin{aligned} R_{y_x y_{x'}}(x, x', \tau) &= \lim_{T\to\infty} \frac{1}{T} \int_{-T/2}^{T/2} y(x,t) y(x', t+\tau) dt \\ &= \lim_{T\to\infty} \frac{1}{T} \int_{-T/2}^{T/2} \left[\sum_{r=1}^{\infty} Y_r(x) \eta_r(t) \right] \left[\sum_{s=1}^{\infty} Y_s(x') \eta_s(t+\tau) \right] dt \\ &= \sum_{r=1}^{\infty} \sum_{s=1}^{\infty} Y_r(x) Y_s(x') R_{\eta_r \eta_s}(\tau) \end{aligned} \tag{12.206}$$

where $R_{\eta_r \eta_s}(\tau)$ is the cross-correlation function between the generalized responses $\eta_r(t)$ and $\eta_s(t)$ and resembles the (r,s) entry of $R_\eta(\tau)$, Eq. (12.182). However, $R_{\eta_r \eta_s}(\tau)$ is related to the cross-spectral density function $S_{N_r N_s}(\omega)$ between the generalized excitations $N_r(t)$ and $N_s(t)$ by the (r,s) entry of $R_N(\tau)$, Eq. (12.188), so that, inserting that entry into Eq. (12.206), we obtain

$$R_{y_x y_{x'}}(x, x', \tau) = \frac{1}{2\pi} \sum_{r=1}^{\infty} \sum_{s=1}^{\infty} Y_r(x) Y_s(x') \int_{-\infty}^{\infty} \bar{G}_r(\omega) G_s(\omega) S_{N_r N_s}(\omega) e^{i\omega\tau} d\omega \tag{12.207}$$

where $S_{N_r N_s}(\omega)$ is given by Eq. (12.204). Note that in Eq. (12.204) x and x' play the role of dummy variables of integration, whereas in Eq. (12.207) x and x' identify the points between which the cross-correlation function is evaluated.

For $x = x'$, the response cross-correlation function reduces to the autocorrelation function

$$R_y(x,\tau) = \frac{1}{2\pi}\sum_{r=1}^{\infty}\sum_{s=1}^{\infty} Y_r(x)Y_s(x)\int_{-\infty}^{\infty} \bar{G}_r(\omega)G_s(\omega)S_{N_rN_s}(\omega)e^{i\omega\tau}d\omega \tag{12.208}$$

Then letting $\tau = 0$ in Eq. (12.208), we obtain the mean square value of the response at point x in the form

$$R_y(x,0) = \frac{1}{2\pi}\sum_{r=1}^{\infty}\sum_{s=1}^{\infty} Y_r(x)Y_s(x)\int_{-\infty}^{\infty} \bar{G}_r(\omega)G_s(\omega)S_{N_rN_s}(\omega)d\omega \tag{12.209}$$

The square root of $R_y(x,0)$ is the standard deviation associated with the probability density function of $y(x,t)$. Hence, assuming that $S_{f_x f_{x'}}(x,x',\omega)$ is given, Eq. (12.209) can be used in conjunction with Eq. (12.204) to calculate the standard deviation.

If the excitation process is Gaussian with zero mean, then so is the response process. In this case, the standard deviation $\sqrt{R_y(x,0)}$ determines fully the probability density function associated with the vibration $y(x,t)$.

The above formulation calls for an infinite number of natural modes $Y_r(x)$ $(r = 1,2,\dots)$. Of course, in practice only a finite number of modes need and should be included, as Eq. (12.196) ceases to be valid for higher modes (see Sec. 8.3). It was implicit in the above discussion that a closed-form solution of the eigenvalue problem of the system is available. A similar approach can be used also when only an approximate solution of the eigenvalue problem can be obtained. In such a case, the classical Rayleigh-Ritz method, or the finite element method, leads to a formulation resembling in structure that of a multi-degree-of-freedom system (see Prob. 12.24).

12.18 SUMMARY

Complex phenomena described by variables whose value at some future time cannot be predicted are known as nondeterministic, or random. Examples of these are rocket engine noise, earthquake intensity, etc. Yet many of these phenomena exhibit such a large degree of statistical regularity that their behavior can be described in terms of certain averages. In vibrations, there is a great deal of interest in the manner in which systems respond to random excitations.

If the ground displacement at a given location is measured during a number of earthquakes, the collection of records, or time histories, is referred to as an ensemble. An individual time history from the ensemble is called a sample function, the dependent variable itself, the ground displacement in the case at hand, is known as a random variable and the ensemble represents a random process. A random process is characterized by means of certain averages over the ensemble. If the averages tend to be recognizable limits as the number of sample functions increases, the process is said to exhibit statistical regularity. The most frequently used averages are the mean value and the autocorrelation function. In general, the mean value depends on the time t_1 for which it was computed, and the autocorrelation function depends on t_1 and a time shift τ. In the special case in which the mean value does not depend on t_1, i.e., the mean value is constant, and the autocorrelation function depends on τ alone the process is said to be stationary. More often than not, averages computed over the time variable using a single typical sample function, and known as the temporal averages, are more convenient than ensemble averages. If the temporal averages are equal to the corresponding ensemble averages, the random process is said to be ergodic. In this text, ergodicity is assumed for the most part. When the shift τ is taken to be equal to zero the autocorrelation function reduces to the mean square value; its positive square root is the root mean square value. For ergodic processes, as for stationary processes, the mean value is constant, so that it can be regarded as the static part of the random variable. Consistent with this, the difference between the random variable and its mean value represents the dynamic part. The quantity obtained by computing the mean square value using the dynamic part alone, instead of all of the random variable, is known as the variance, and the square root of the variance is the standard deviation.

In view of the fact that the response to random excitations cannot be given explicitly in terms of time, one must be content with other ways of describing it. One way is through the probability that the response will remain below a certain value. For a given random variable, this probability can be expressed by means of the probability distribution function. Of wider use is the probability density function, which represents the derivative of the probability distribution function. Indeed, the probability density function can be used to compute such statistics as the mean value, mean square value and the standard deviation. A probability density function widely used in random vibrations is the normal, or Gaussian one; it represents a bell-shaped curve with the area under the curve equal to unity, as is the case with all probability density functions. The Gaussian probability density function has the advantage that it is defined uniquely by two statistics alone, the mean value and the standard deviation. The mean value represents the distance between the origin of the reference axes and the curve symmetry axis and the standard deviation determines the peak value of the curve.

In random vibrations, it is more convenient to derive the response in the frequency domain, rather than in the time domain, which can be done by means of Fourier transform techniques. However, unlike the case of deterministic processes, in the case of Gaussian random processes the interest lies in the response mean value and mean square value, rather than in the response itself. Of course, it is a simple matter to compute the response standard deviation from the mean value and mean square value, thus defining uniquely the response probability density. The response mean value is simply the product of the excitation mean value and the frequency response evaluated at $\omega = 0$. To obtain the response mean square value, we first note that the autocorrelation function and the mean square spectral density represent a Fourier transform pair. Hence, the first step is to obtain the excitation mean square spectral density by Fourier transforming the excitation autocorrelation function. But, the response mean square spectral density is equal to the product of the magnitude of the frequency response squared and the excitation mean square spectral density. Then, the response autocorrelation function can be obtained by inverse Fourier transforming the response mean square spectral density. However, in general the mean square value is equal to the autocorrelation function evaluated at $\tau = 0$. Hence, to obtain the response man square value, it is not really necessary to evaluate the inverse Fourier transform of the response mean square spectral density. Indeed, it is only necessary to integrate it. Of course, having the response mean square value, as well as the response mean value, it is a simple matter to compute the standard deviation, thus defining uniquely the response probability density function.

The preceding discussion was concerned with single-degree-of-freedom systems, but applies equally well to multi-degree-of-freedom systems and to distributed-parameter systems. In this regard, it must be recognized that the intrinsic characteristics of a system are not affected by the nature of the excitations. Hence, the same modal analysis presented in Chs. 7 and 8 can be used to decouple and ultimately to compute the mean square values corresponding to the discrete displacements for a multi-degree-of-freedom system, or the mean square value corresponding to the displacement of a typical point of a distributed-parameter system.

PROBLEMS

12.1. Calculate and plot the temporal autocorrelation function for the sinusoid $x(t) = A\sin 2\pi t/T$.

12.2. Calculate the temporal mean value and autocorrelation function for the periodic function shown in Fig. 12.32. Plot the autocorrelation function.

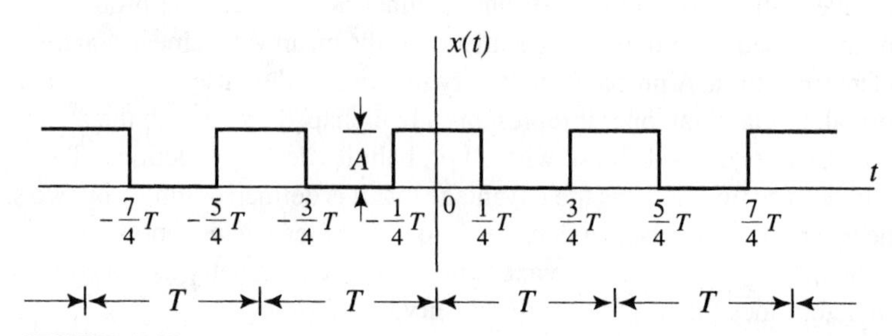

FIGURE 12.32
Periodic function

12.3. Calculate and plot the temporal autocorrelation function for the periodic function shown in Fig. 12.8a.

12.4. The function $x(t) = A|\sin 2\pi t/T|$ is known as a *rectified sinusoid*. Its period is $T/2$, as opposed to T for the ordinary sinusoid, as can be seen from Fig. 12.33. Calculate the mean value and the autocorrelation function for the rectified sinusoid.

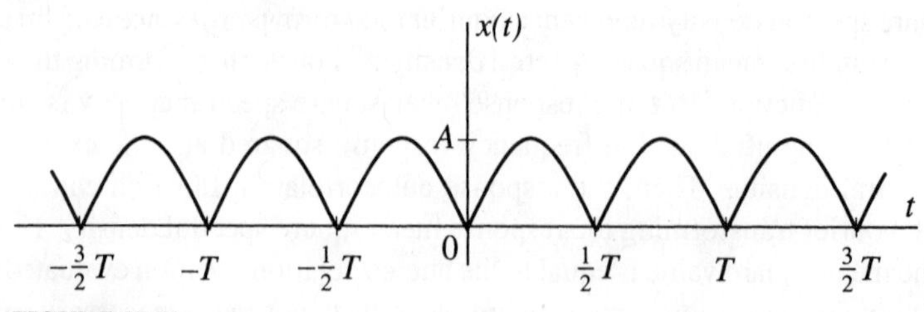

FIGURE 12.33
Rectified sinusoid

12.5. Calculate and plot the autocorrelation function for the pulse-width-modulated wave shown in Fig. 12.34.

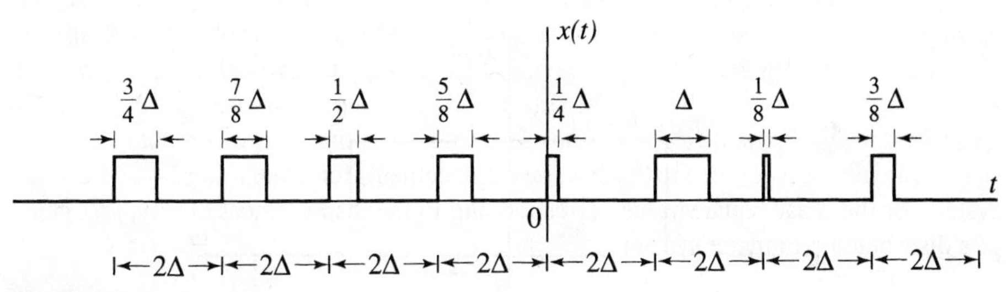

FIGURE 12.34
Pulse-width-modulated wave

12.6. Calculate the mean square value for the function of Prob. 12.1.

12.7. Calculate the mean square value, the variance and the standard deviation for the function of Prob. 12.2.

12.8. Calculate the mean square value for the function of Prob. 12.3.

12.9. Calculate the mean value, the mean square value, the variance and the standard deviation for the rectified sinusoid of Prob. 12.4.

12.10. Use definition (12.10) and obtain the probability distribution $P(x)$ for the function of Example 12.1. Then use Eq. (12.13) and derive the probability density function $p(x)$. Plot $P(x)$ versus x and $p(x)$ versus x.

12.11. Assume that the time t is uniformly distributed, and use Eq. (12.20) to verify the probability density function shown in Fig. 12.8c.

12.12. Consider a rectified sinusoid with the constant amplitude A and constant frequency ω but random phase angle ϕ. For a fixed value t_0 of time, the rectified sinusoid can be regarded as a function of the random variable ϕ given by $x(\phi) = A|\sin(\omega t_0 + \phi)|$. Let ϕ have a uniform probability density function $p(\phi)$, and calculate the probability density function $p(x)$ by the method of Sec. 12.4.

12.13. Calculate the mean square value for the function shown in Fig. 12.8a by using Eq. (12.27).

12.14. Calculate the mean value and the mean square value for the rectified sinusoid by using the probability density function $p(x)$ derived in Prob. 12.12.

12.15. Calculate the power spectral density for the function of Example 12.2.

12.16. Consider an ergodic random process with zero power spectral density at $\omega = 0$, and show that the autocorrelation function $R_f(\tau)$ must satisfy $\int_{-\infty}^{\infty} R_f(\tau)d\tau = 0$.

12.17. Verify that the mathematical expression for the power density spectrum of the sine wave $f(t) = A\sin 2\pi t/T$ is $S_f(\omega) = (\pi A^2/2)[\delta(\omega + 2\pi/T) + \delta(\omega - 2\pi/T)]$, where $\delta(\omega + 2\pi/T)$ and $(\delta(\omega - 2\pi/T)$ are Dirac delta functions acting at $\omega = -2\pi/T$ and $\omega = 2\pi/T$, respectively.

12.18. A damped single-degree-of-freedom system is excited by a random process whose power density spectrum is as shown in Fig. 12.35. Let $\zeta = 0.05$ and $\omega_n = \omega_0/2$, and plot the response power density spectrum.

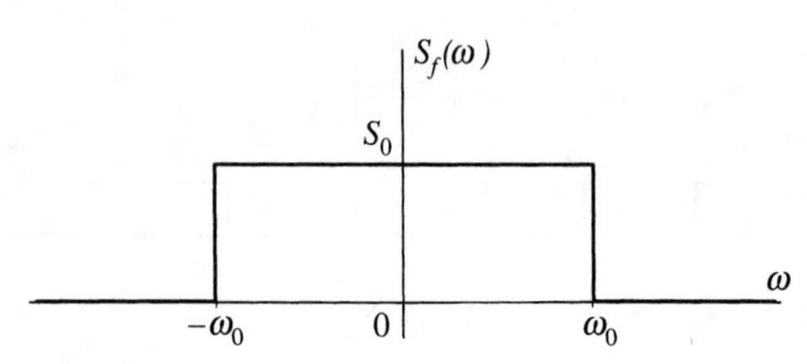

FIGURE 12.35
Excitation power density spectrum

12.19. Prove inequality (12.146).

12.20. Calculate the cross-correlation function between the functions of Prob. 12.1 and 12.2.

12.21. Let $x(t)$ be the response of a linear system to the excitation $f(t)$ and show that $S_{fx}(\omega) = G(\omega)S_f(\omega)$, where $S_{fx}(\omega)$ is the cross-spectral density function between the excitation and response, $G(\omega)$ is the frequency response and $S_f(\omega)$ is the excitation power spectral density function. [*Hint:* Begin by writing the cross-correlation function between the excitation and response in the form

$$R_{fx}(\tau) = \lim_{T\to\infty} \frac{1}{T} \int_{-T/2}^{T/2} f(t)x(t+\tau)dt$$

and recall that the response is related to the excitation by the convolution integral, Eq. (12.82)].

12.22. Consider the system shown in Fig. 12.36, and derive the equations of motion. Let $c = 0.02\sqrt{km}$, and derive general equations for the cross-correlation function between $q_1(t)$ and $q_2(t)$ by observing that the modal matrix uncouples the equations of motion. Obtain the response mean square values for $q_1(t)$ and $q_2(t)$. The excitation $Q_1(t)$ can be assumed to be an ergodic random process possessing an ideal white noise power density spectrum, whereas $Q_2(t) = 0$.

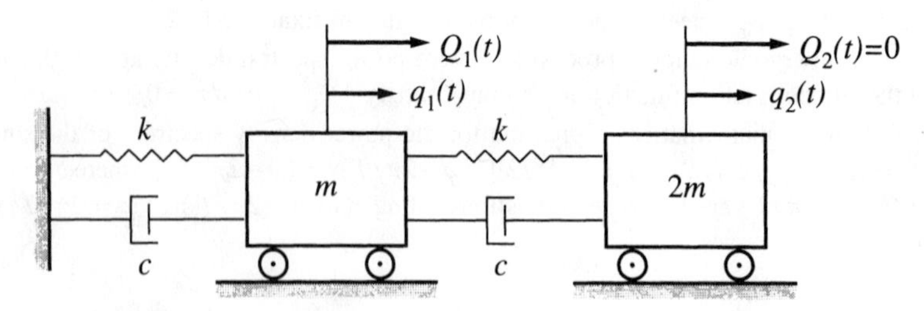

FIGURE 12.36
Damped two-degree-of-freedom system

12.23. Consider a uniform bar in bending simply supported at both ends and subjected to the excitation $f(x,t) = F(t)\delta(x - L/2)$ where $F(t)$ is an ergodic random process with ideal white noise power spectral density, and $\delta(x - L/2)$ a spatial Dirac delta function. Use the method of Sec. 12.17, and derive expressions for the cross-correlation function between the responses at the points $x = L/4$ and $x' = 3L/4$ and for the mean square value of the response at $x = L/4$.

12.24. Formulate the response problem of a continuous system to random excitation by means of an approximate method whereby the formulation is reduced to that of a multi-degree-of-freedom discrete system.

APPENDIX A

FOURIER SERIES

In many problems of engineering analysis it is necessary to work with periodic functions, i.e., with functions that repeat themselves every given interval, where the interval is known as the *period*. Periodic functions satisfy a relation of the type

$$f(t) = f(t + T) \tag{A.1}$$

where T represents the period. Some of the simplest and most commonly encountered periodic functions are the trigonometric functions. Indeed, the functions $\sin nt$ and $\cos nt$ $(n = 1, 2, \ldots)$, being harmonic, are periodic by definition; their period is $2\pi/n$. Clearly, trigonometric functions are special cases of periodic functions. Because trigonometric functions are relatively easy to work with, and possess the orthogonality property, a most important and useful property, they are more desirable than arbitrary periodic functions. Owing to these facts, it is advantageous at times to expand arbitrary periodic functions in series of trigonometric functions, where the expansions are known as *Fourier series*.

A.1 ORTHOGONAL SETS OF FUNCTIONS

We consider a set of functions $\psi_r(t)$ $(r = 1, 2, \ldots)$ defined over the interval $0 \leq t \leq T$. Then, if for any two distinct functions $\psi_r(t)$ and $\psi_s(t)$,

$$\int_0^T \psi_r(t)\psi_s(t)dt = 0, \; r, s = 1, 2, \ldots; \; r \neq s \tag{A.2}$$

the set $\psi_r(t)$ is said to be *orthogonal* in the interval $0 \leq t \leq T$, or more generally in any interval of length T. If the functions $\psi_r(t)$ are such that, in addition to satisfying Eq.

(A.2), they satisfy

$$\int_0^T \psi_r^2(t)dt = 1, \; r = 1, 2, \ldots \tag{A.3}$$

then the set is referred to as *orthonormal*. Hence, for an orthonormal set of functions we have

$$\int_0^T \psi_r(t)\psi_s(t)dt = \delta_{rs}, \; r, s = 1, 2, \ldots \tag{A.4}$$

where δ_{rs} is the Kronecker delta, defined as being equal to unity for $r = s$ and equal to zero for $r \neq s$. It is easy to verify that the set of functions

$$\frac{1}{\sqrt{2\pi}}, \frac{\sin t}{\sqrt{\pi}}, \frac{\cos t}{\sqrt{\pi}}, \frac{\sin 2t}{\sqrt{\pi}}, \frac{\cos 2t}{\sqrt{\pi}}, \frac{\sin 3t}{\sqrt{\pi}}, \ldots \tag{A.5}$$

constitutes an orthonormal set. Indeed, we can write

$$\int_0^{2\pi} \frac{1}{\sqrt{2\pi}} \frac{\sin rt}{\sqrt{\pi}} dt = -\frac{1}{\sqrt{2\pi}} \frac{\cos rt}{r}\Bigg|_0^{2\pi} = 0,$$

$$r = 1, 2, \ldots \tag{A.6}$$

$$\int_0^{2\pi} \frac{1}{\sqrt{2\pi}} \frac{\cos rt}{\sqrt{\pi}} dt = \frac{1}{\sqrt{2\pi}} \frac{\sin rt}{r}\Bigg|_0^{2\pi} = 0,$$

Moreover, for $r \neq s$, we have

$$\int_0^{2\pi} \frac{\sin rt}{\sqrt{\pi}} \frac{\cos st}{\sqrt{\pi}} dt = \frac{1}{2\pi}\int_0^{2\pi} [\sin(r+s)t + \sin(r-s)t]dt$$

$$= -\frac{1}{2\pi}\left[\frac{\cos(r+s)t}{r+s} + \frac{\cos(r-s)t}{r-s}\right]_0^{2\pi} = 0,$$

$$r, s = 1, 2, \ldots \tag{A.7}$$

and for $r = s$, we obtain

$$\int_0^{2\pi} \frac{\sin rt}{\sqrt{\pi}} \frac{\cos rt}{\sqrt{\pi}} dt = \frac{1}{2\pi}\int_0^{2\pi} \sin 2rt\, dt = -\frac{1}{4r\pi}\cos 2rt\Bigg|_0^{2\pi} = 0,$$

$$r = 1, 2, \ldots \tag{A.8}$$

so that the set (A.5) satisfies Eqs. (A.2); hence, it is orthogonal. On the other hand, because

$$\int_0^{2\pi} \left(\frac{1}{\sqrt{2\pi}}\right)^2 dt = 1$$

$$\int_0^{2\pi} \left(\frac{\sin rt}{\sqrt{\pi}}\right)^2 dt = \frac{1}{r\pi}\left[\frac{rt}{2} - \frac{\sin 2rt}{4}\right]_0^{2\pi} = 1, \; r = 1, 2, \ldots \tag{A.9}$$

$$\int_0^{2\pi} \left(\frac{\cos rt}{\sqrt{\pi}}\right)^2 dt = \frac{1}{r\pi}\left[\frac{rt}{2} + \frac{\sin 2rt}{4}\right]_0^{2\pi} = 1, \; r = 1, 2, \ldots$$

the set (A.5) is not only orthogonal but *orthonormal*.

If for a set of constants c_r $(r = 1, 2, \ldots)$, not all equal to zero, there exists a homogeneous linear relation

$$\sum_{r=1}^{n} c_r \psi_r(t) = 0 \tag{A.10}$$

for all t, then the set of functions $\psi_r(t)$ $(r = 1, 2, \ldots)$ is said to be *linearly dependent*. If no relation of the type (A.10) exists, then the set is said to be *linearly independent*. The set (A.5) can be shown to be linearly independent. Indeed, if we write the series

$$c_0 \frac{1}{\sqrt{2\pi}} + c_1 \frac{\sin t}{\sqrt{\pi}} + c_2 \frac{\cos t}{\sqrt{\pi}} + c_3 \frac{\sin 2t}{\sqrt{\pi}} + c_4 \frac{\cos 2t}{\sqrt{\pi}} + \cdots + c_{2p} \frac{\cos pt}{\sqrt{\pi}} = 0 \tag{A.11}$$

multiply the series by any of the functions in (A.5), say $\cos 2t/\sqrt{\pi}$ and integrate with respect to t over the interval $0 \le t \le 2\pi$, we obtain $c_4 = 0$. The procedure can be repeated for all constants, with the conclusion that Eq. (A.11) can hold only if all the coefficients are zero, $c_0 = c_1 = c_2 = \cdots = c_{2p} = 0$. Because this contradicts the stipulation that not all constants be zero, we must conclude that the set is linearly independent. Note that an orthogonal set is by definition linearly independent.

A.2 TRIGONOMETRIC SERIES

An orthonormal set of functions $\psi_r(t)$ $(r = 1, 2, \ldots)$ is said to be *complete* if any piecewise continuous function $f(t)$ can be approximated in the mean to any desired degree of accuracy by the series $\sum_{r=1}^{n} c_r \psi_r(t)$ by choosing the integer n large enough. In view of this, because the set (A.5) is complete in the interval $0 \le t \le 2\pi$ every function $f(t)$ which is continuous in that interval can be represented by the *Fourier series*

$$f(t) = \tfrac{1}{2} a_0 + \sum_{r=1}^{\infty} (a_r \cos rt + b_r \sin rt) \tag{A.12}$$

where the constants a_r $(r = 0, 1, 2, \ldots)$ and b_r $(r = 1, 2, \ldots)$ are known as *Fourier coefficients*.

To establish the exact composition of the trigonometric representation of a given periodic function, it is necessary to calculate the Fourier coefficients. To this end, we list the following results derived in Sec. A.1

$$\int_0^{2\pi} \cos rt \cos st \, dt = 0, \quad \int_0^{2\pi} \sin rt \sin st \, dt = 0, \quad r, s = 1, 2, \ldots; \; r \ne s \tag{A.13}$$

and

$$\int_0^{2\pi} \cos rt \sin st \, dt = \int_0^{2\pi} \sin rt \cos st \, dt = 0, \quad r, s = 1, 2, \ldots \tag{A.14}$$

where Eqs. (A.14) are valid whether r and s are distinct or not. On the other hand, when $r = s$, the integrals in Eqs. (A.13) are not zero but have the values

$$\int_0^{2\pi} \cos^2 rt \, dt = \begin{cases} 2\pi, & r = 0 \\ \pi, & r = 1, 2, \ldots \end{cases} \tag{A.15}$$

and

$$\int_0^{2\pi} \sin^2 rt\,dt = \pi, \quad r = 1, 2, \ldots \tag{A.16}$$

Moreover, we can write

$$\int_0^{2\pi} \cos rt\,dt = \begin{cases} 2\pi, & r = 0 \\ 0, & r = 1, 2, \ldots \end{cases} \tag{A.17}$$

and

$$\int_0^{2\pi} \sin rt\,dt = 0, \quad r = 1, 2, \ldots \tag{A.18}$$

Next, we multiply Eq. (A.12) by $\cos st$, integrate over the interval $0 \le t \le 2\pi$, interchange the order of integration and summation and obtain

$$\int_0^{2\pi} f(t)\cos st\,dt = \tfrac{1}{2}a_0 \int_0^{2\pi} \cos st dt + \sum_{r=1}^{\infty} a_r \int_0^{2\pi} \cos rt \cos st\,dt$$
$$+ \sum_{r=1}^{\infty} b_r \int_0^{2\pi} \sin rt \cos st\,dt \tag{A.19}$$

For $s = 0$, Eq. (A.19) in conjunction with Eqs. (A.17) and (A.18) yields

$$a_0 = \frac{1}{\pi}\int_0^{2\pi} f(t)dt \tag{A.20}$$

so that $\frac{1}{2}a_0$ can be identified as the *average value* of $f(t)$. If $s \neq 0$, we conclude that only one term survives from the series in Eq. (A.19), namely, that corresponding to the integral $\int_0^{2\pi} \cos rt \cos st\,dt$ with $r = s$. Indeed, considering Eqs. (A.13)–(A.15), we conclude that Eq. (A.19) reduces to

$$a_r = \frac{1}{\pi}\int_0^{2\pi} f(t)\cos rt\,dt, \; r = 1, 2, \ldots \tag{A.21}$$

Similarly, multiplying series (A.12) by $\sin st$, integrating over the interval $0 \le t \le 2\pi$, and considering Eqs. (A.13), (A.14), (A.16) and (A.18), we obtain

$$b_r = \frac{1}{\pi}\int_0^{2\pi} f(t)\sin rt\,dt, \quad r = 1, 2, \ldots \tag{A.22}$$

thus determining the series (A.12) uniquely.

When $f(t)$ is an *even function*, i.e., when $f(t) = f(-t)$, the coefficients b_r $(r = 1, 2, \ldots)$ vanish and the series is known as a *Fourier cosine series*. On the other hand, when $f(t)$ is an *odd function*, i.e., when $f(t) = -f(-t)$, the coefficients a_r $(r = 0, 1, 2, \ldots)$ vanish and the series is called a *Fourier sine series*. This can be more conveniently demonstrated by considering the interval $-\pi \le t \le \pi$ instead of $0 \le t \le 2\pi$.

If the function $f(t)$ is only piecewise continuous in a given interval, then a Fourier series representation using a finite number of terms approaches $f(t)$ in every interval that does not contain discontinuities. In the immediate neighborhood of a jump discontinuity,

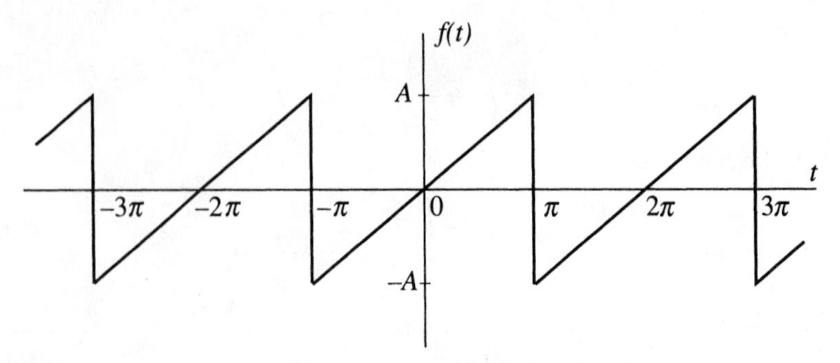

FIGURE A.1
Periodic function

convergence is not uniform and, as the number of terms increases, the finite series approximation contains increasingly high-frequency oscillations which move closer and closer to the discontinuity point. However, the total oscillation of the approximating curve does not approach the jump of $f(t)$, a fact known as the *Gibbs phenomenon*.

As an illustration, we consider the periodic function $f(t)$ shown in Fig. A.1, where the function repeats itself every 2π. The function is recognized as being an odd function of t, which by definition implies that its average value is zero, so that $f(t)$ can be represented by a Fourier sine series of the form

$$f(t) = \sum_{r=1}^{\infty} b_r \sin rt \tag{A.23}$$

The proof that $a_r = 0$ $(r = 0, 1, 2, \ldots)$ is left as an exercise to the reader. The function $f(t)$ can be described mathematically by

$$f(t) = \frac{A}{\pi} t, \quad -\pi \le t \le \pi \tag{A.24}$$

so that the coefficients become

$$\begin{aligned} b_r &= \frac{1}{\pi} \int_{-\pi}^{\pi} f(t) \sin rt \, dt = \frac{A}{\pi^2} \int_{-\pi}^{\pi} t \sin rt \, dt \\ &= \frac{A}{\pi^2 r^2} (\sin rt - rt \cos rt) \Big|_{-\pi}^{\pi} = \frac{2A}{\pi r} (-1)^{r+1}, \quad r = 1, 2, \ldots \end{aligned} \tag{A.25}$$

Hence, the series becomes

$$f(t) = \frac{2A}{\pi} \sum_{r=1}^{\infty} \frac{(-1)^{r+1}}{r} \sin rt \tag{A.26}$$

Fourier series are infinite series and on occasions they must be approximated by finite ones, as intimated earlier. This is done by replacing the upper limit in the series by a finite integer n, a process known as *truncation*. Figure A.2 shows the series representation for $n = 1, 2, \ldots, 6$. It is clear that the approximation improves with increasing n. Of

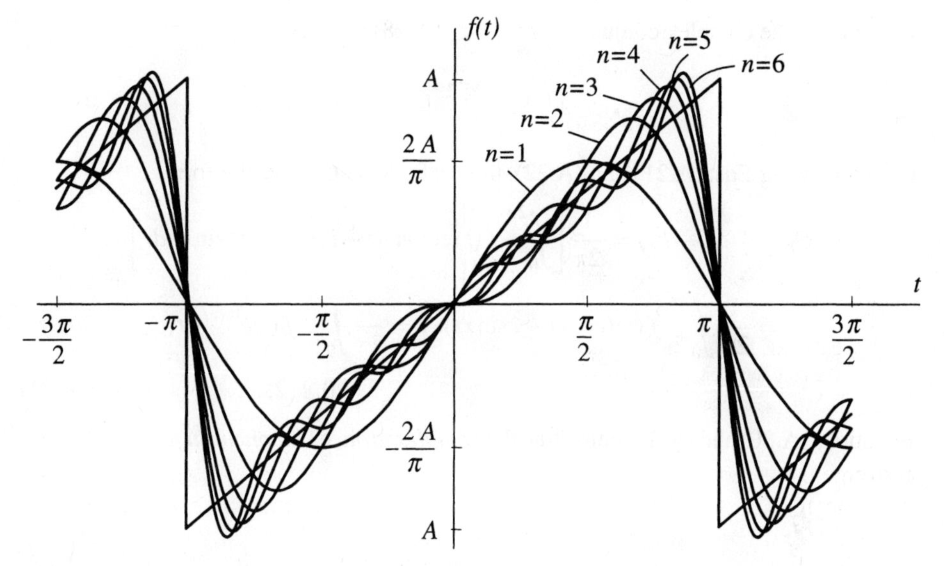

FIGURE A.2
Fourier series representation of the periodic function of Fig. A.1 showing the Gibbs pheneomenon

course, if the accuracy of Fig. A.2 is not satisfactory, then additional terms must be included to bring the series representation to the desired level of accuracy. As $n \to \infty$, the series approaches $f(t)$, except at the discontinuity points $t = \pm\pi,\ \pm 3\pi,\ \ldots$, where we encounter the Gibbs phenomenon, as can be concluded from Fig. A.2.

A.3 COMPLEX FORM OF FOURIER SERIES

The Fourier series can also be expressed in terms of exponential functions. Indeed, the trigonometric functions are related to exponential functions as follows:

$$\cos rt = \frac{e^{irt} + e^{-irt}}{2}, \quad \sin rt = \frac{e^{irt} - e^{-irt}}{2i} \tag{A.27}$$

Inserting Eqs. (A.27) into Eq. (A.12), we obtain

$$\begin{aligned} f(t) &= \tfrac{1}{2}a_0 + \frac{1}{2}\sum_{r=1}^{\infty}[a_r(e^{irt} + e^{-irt}) - ib_r(e^{irt} - e^{-irt})] \\ &= \tfrac{1}{2}a_0 + \frac{1}{2}\sum_{r=1}^{\infty}[(a_r - ib_r)e^{irt} + (a_r + ib_r)e^{-irt}] \end{aligned} \tag{A.28}$$

Introducing the notation

$$\begin{aligned} &C_0 = \tfrac{1}{2}a_0 \\ &C_r = \tfrac{1}{2}(a_r - ib_r), \quad C_{-r} = \bar{C}_r = \tfrac{1}{2}(a_r + ib_r), \quad r = 1, 2, \ldots \end{aligned} \tag{A.29}$$

where $\bar{C}_r$ is the complex conjugate of C_r, Eq. (A.28) reduces to

$$f(t) = \sum_{r=-\infty}^{\infty} C_r e^{irt} \tag{A.30}$$

in which, using Eqs. (A.21) and (A.22), the coefficients C_r have the form

$$\begin{aligned} C_r &= \tfrac{1}{2}(a_r - ib_r) = \frac{1}{2\pi}\left[\int_0^{2\pi} f(t)\cos rt\,dt - i\int_0^{2\pi} f(t)\sin rt\,dt\right] \\ &= \frac{1}{2\pi}\int_0^{2\pi} f(t)(\cos rt - i\sin rt)dt = \frac{1}{2\pi}\int_0^{2\pi} f(t)e^{-irt}dt, \\ & \qquad r = 0, 1, 2, \ldots \end{aligned} \tag{A.31}$$

Equations (A.30) and (A.31) constitute the *complex form*, or *exponential form, of Fourier series*.

APPENDIX B

LAPLACE TRANSFORMATION

The Laplace transformation is an important tool in the study of the response of linear systems with constant coefficients, particularly when the excitation is in the form of discontinuous functions, for which other techniques tend to experience difficulties. This introduction to the Laplace transformation method is modest in scope, and its main purpose is to provide a rudimentary knowledge of the method and a certain degree of familiarity with the terminology.

The idea behind the Laplace transformation method is to transform a relatively complicated problem into a simpler one, solve the simpler problem and then perform an inverse transformation to obtain the solution to the original problem. The most common use of the method is to solve initial-value problems, namely, problems in which the system behavior is defined by ordinary differential equations, to be satisfied for all positive times, and by a given set of initial conditions. In such cases, the transformed problem involves algebraic expressions alone, with the initial conditions being taken into account automatically.

B.1 DEFINITION OF THE LAPLACE TRANSFORMATION

We consider a function $f(t)$ defined for all values of time larger than zero, $t > 0$, and define the (one-sided) Laplace transformation of $f(t)$ by the integral

$$\mathcal{L}f(t) = F(s) = \int_0^\infty e^{-st} f(t)dt \tag{B.1}$$

where e^{-st} is known as the kernel of the transformation and s represents a subsidiary variable. The variable s is in general a complex quantity, and the associated complex plane is called the s-plane, and at times the Laplace plane. Because transformation (B.1) is defined in terms of an integral, it is said to be an integral transformation, commonly referred to as an integral transform.

The function $f(t)$ must be such that the integral in Eq. (B.1) exists, which places on $f(t)$ the restriction

$$[e^{-st} f(t)] < Ce^{-(s-a)t}, \quad \text{Re}\, s > a \tag{B.2}$$

where C is a constant, Condition (B.2) implies that $f(t)$ must not increase with time more rapidly than the exponential function Ce^{at}. Another restriction on $f(t)$ is that it must be piecewise continuous. Most functions describing physical phenomena satisfy these conditions.

B.2 TRANSFORMATION OF DERIVATIVES

Because our interest lies in using the Laplace transformation method to solve differential equations, it necessary to transform derivatives of functions. Considering the transform of $df(t)/dt$, and integrating by parts, we obtain

$$\begin{aligned}\mathcal{L}\frac{df(t)}{dt} &= \int_0^\infty e^{-st}\frac{df(t)}{dt}dt = e^{-st} f(t)\Big|_0^\infty - \int_0^\infty (-se^{-st}) f(t)dt \\ &= -f(0) + sF(s)\end{aligned} \tag{B.3}$$

where $f(0)$ is the initial value of $f(t)$, namely, the value of $f(t)$ at $t = 0$.

Following the same pattern, the transform of $d^2 f(t)/dt^2$ can be shown to be

$$\mathcal{L}\frac{d^2 f(t)}{dt^2} = \int_0^\infty e^{-st}\frac{d^2 f(t)}{dt^2}dt = -\dot{f}(0) - sf(0) + s^2 F(s) \tag{B.4}$$

where $\dot{f}(0)$ is the value of $df(t)/dt$ at $t = 0$.

B.3 TRANSFORMATION OF ORDINARY DIFFERENTIAL EQUATIONS

The differential equation of motion of a viscously damped single-degree-of-freedom system was shown in Sec. 3.1 to be

$$m\frac{d^2x(t)}{dt^2} + c\frac{dx(t)}{dt} + kx(t) = f(t) \tag{B.5}$$

Introducing the notation $\mathcal{L}x(t) = X(s)$, $\mathcal{L}f(t) = F(s)$, transforming both sides of Eq. (B.5) and considering Eqs. (B.3) and (B.4), we obtain the algebraic equation

$$m[s^2 X(s) - sx(0) - \dot{x}(0)] + c[sX(s) - x(0)] + kX(s) = F(s) \tag{B.6}$$

where $x(0)$ and $\dot{x}(0)$ are the initial displacement and velocity, respectively. Recalling from Ch. 2 that $c/m = 2\zeta\omega_n$, $k/m = \omega_n^2$ and solving Eq. (B.6) for the transformed response $X(s)$, we can write

$$X(s) = \frac{1}{m(s^2 + 2\zeta\omega_n s + \omega_n^2)} F(s) + \frac{s + 2\zeta\omega_n}{s^2 + 2\zeta\omega_n s + \omega_n^2} x(0) + \frac{1}{s^2 + 2\zeta\omega_n s + \omega_n^2} \dot{x}(0) \tag{B.7}$$

To obtain the actual response $x(t)$, we must evaluate the inverse Laplace transformation of $X(s)$. It is clear from Eq. (B.7) that the Laplace transformation method yields both the particular and the homogeneous solution simultaneously, with the implication that the method takes the initial conditions into account automatically.

B.4 THE INVERSE LAPLACE TRANSFORMATION

As can be concluded from Eq. (B.7), the transformed response $X(s)$ is a function of the subsidiary variable s. To obtain the time-dependent response $x(t)$, we must evaluate the inverse Laplace transform of $X(s)$. The operation is denoted symbolically by

$$\mathcal{L}^{-1} X(s) = x(t) \tag{B.8}$$

The rigorous definition (not given here) of the inverse transform (B.8) involves the evaluation of a line integral in the s-plane. In many cases, the integral can be replaced by a closed contour integral, which in turn can be evaluated by the residue theorem of complex algebra. By far the simplest way to evaluate inverse transformations is to decompose the function $X(s)$ into a sum of simple functions with known inverse transformations. This is the essence of the method of partial fractions, to be described in Sec. B.6. To expedite the inversion process, a table of commonly encountered Laplace transform pairs can be found in Sec. B.8.

B.5 SHIFTING THEOREMS

A frequently encountered function has the form

$$f_1(t) = f(t)e^{at} \tag{B.9}$$

where a is a real or complex number. The Laplace transform of $f_1(t)$ is given by

$$F_1(s) = \int_0^\infty [f(t)e^{at}]e^{-st}\,dt = \int_0^\infty f(t)e^{-(s-a)t}\,dt = F(s-a) \tag{B.10}$$

from which it follows that

$$\mathcal{L}[f(t)e^{at}] = F(s-a) \tag{B.11}$$

Equation (B.11) states that the effect of multiplying a function $f(t)$ by e^{at} in the time domain is to shift the Laplace transform $F(s)$ of $f(t)$ by the amount a in the s-domain. Because the s-domain is the complex plane, the statement embodied by Eq. (B.11) is called the *complex shifting theorem*.

Next, we consider the Laplace transform

$$F(s) = \int_0^\infty e^{-s\tau} f(\tau) d\tau \tag{B.12}$$

and let $\tau = t - a,\ d\tau = dt$, so that

$$F(s) = \int_0^\infty e^{-s(t-a)} f(t-a) dt = e^{as} \int_0^\infty e^{-st} f(t-a) u(t-a) dt \tag{B.13}$$

where we multiplied the integrand by the unit step function initiated at $t = a$, $u(t-a)$, in recognition of the fact that $f(t-a)$ must be set equal to zero for $t < a$, because $f(\tau)$ is zero for $\tau < 0$. Multiplying Eq. (B.13) through by e^{-as}, we obtain

$$F(s)e^{-as} = \int_0^\infty e^{-st} [f(t-a) u(t-a)] dt \tag{B.14}$$

from which it follows that

$$\mathcal{L}^{-1} F(s) e^{-as} = f(t-a) u(t-a) \tag{B.15}$$

Equation (B.15) represents the *shifting theorem in the real domain.*

B.6 METHOD OF PARTIAL FRACTIONS

We consider the case in which $X(s)$ can be written as the ratio

$$X(s) = \frac{A(s)}{B(s)} \tag{B.16}$$

where both $A(s)$ and $B(s)$ are polynomials in s. Generally $B(s)$ is a polynomial of higher degree than $A(s)$. Denoting by $s = a_k$ $(k = 1, 2, \ldots, n)$ the roots of $B(s)$, the polynomial can be written as the product

$$B(s) = (s - a_1)(s - a_2) \cdots (s - a_k) \cdots (s - a_n) = \prod_{k=1}^{n} (s - a_k) \tag{B.17}$$

where $\prod$ is the product symbol. The roots $s = a_k$ are known as *simple poles* of $X(s)$. The partial fractions expansion of Eq. (B.16) has the form

$$X(s) = \frac{c_1}{s - a_1} + \frac{c_2}{s - a_2} + \cdots + \frac{c_k}{s - a_k} + \cdots + \frac{c_n}{s - a_n} = \sum_{k=1}^{n} \frac{c_k}{s - a_k} \tag{B.18}$$

where the coefficients c_k are given by the formula

$$c_k = [(s - a_k) X(s)] \bigg|_{s=a_k} = \frac{A(s)}{B'(s)} \bigg|_{s=a_k} \tag{B.19}$$

With a view to the inversion of $X(s)$, we write

$$\mathcal{L} e^{a_k t} = \int_0^\infty e^{a_k t} e^{-st} dt = \int_0^\infty e^{(a_k - s)t} dt = \frac{e^{(a_k - s)t}}{a_k - s} \bigg|_0^\infty = \frac{1}{s - a_k} \tag{B.20}$$

from which it follows that

$$\mathcal{L}^{-1}\frac{1}{s-a_k} = e^{a_k t} \tag{B.21}$$

Equations (B.20) and (B.21) constitute a Laplace transform pair. In view of Eqs. (B.19) and (B.21), the inverse transform of $X(s)$, Eq. (B.18), becomes

$$x(t) = \sum_{k=1}^{n} [(s-a_k)X(s)]\Big|_{s=a_k} e^{a_k t} = \sum_{k=1}^{n} [(s-a_k)X(s)e^{st}]\Big|_{s=a_k} \tag{B.22}$$

Quite often, however, it is simpler to consider Eq. (B.18) and write $A(s)$ in the form

$$A(s) = c_1 \prod_{i=2}^{n}(s-a_i) + c_2 \prod_{\substack{i=1 \\ i\neq 2}}^{n} (s-a_i) + \ldots + c_n \prod_{i=1}^{n-1}(s-a_i)$$

$$= \sum_{k=1}^{n} c_k \prod_{\substack{i=1 \\ i\neq k}}^{n} (s-a_i) \tag{B.23}$$

Comparing the coefficients of s^{j-1} $(j = 1, 2, \ldots, n)$ on both sides of Eq. (B.23), we derive a set of algebraic equations that can be solved for the coefficients c_k $(k = 1, 2, \ldots, n)$, as demonstrated in Example B.1.

Next, we consider the case in which $B(s)$ has a multiple root of order k, which implies that $X(s)$ has a pole of order k, in addition to the simple poles considered above. Hence, $B(s)$ can be written as

$$B(s) = (s-a_1)^k(s-a_2)(s-a_3)\ldots(s-a_n) \tag{B.24}$$

In this case, the partial fractions expansion has the form

$$X(s) = \frac{A(s)}{B(s)} = \frac{c_{11}}{(s-a_1)^k} + \frac{c_{12}}{(s-a_1)^{k-1}} + \ldots + \frac{c_{1k}}{s-a_1}$$
$$+ \frac{c_2}{s-a_2} + \frac{c_3}{s-a_3} + \ldots + \frac{c_n}{s-a_n} \tag{B.25}$$

It is not difficult to verify that the coefficients $c_{11}, c_{12}, \ldots, c_{1k}$ are given by the formula

$$c_{1r} = \frac{1}{(r-1)!}\frac{d^{r-1}}{ds^{r-1}}[(s-a_1)^k X(s)]\Big|_{s=a_1}, \quad r = 1, 2, \ldots, k \tag{B.26}$$

To carry out the inverse Laplace transformation of the terms due to the higher-order pole, we observe that

$$\mathcal{L}t^{r-1} = \frac{(r-1)!}{s^r} \tag{B.27}$$

so that, using the complex shifting theorem, Eq. (B.11), we can write

$$\mathcal{L}^{-1}\frac{1}{(s-a_1)^r} = \frac{t^{r-1}}{(r-1)!}e^{a_1 t} \tag{B.28}$$

Hence, the inverse Laplace transform of $X(s)$, Eq. (B.25), is

$$x(t) = \left[c_{11}\frac{t^{k-1}}{(k-1)!} + c_{12}\frac{t^{k-2}}{(k-2)!} + \ldots + c_{1k} \right] e^{a_1 t} + c_2 e^{a_2 t} + c_3 e^{a_3 t} + \ldots + c_n e^{a_n t} \tag{B.29}$$

Finally, using Eqs. (B.22) and (B.26), it can be shown that Eq. (B.29) can be written in the compact form

$$x(t) = \frac{1}{(k-1)!}\frac{d^{k-1}}{ds^{k-1}}[(s-a_1)^k X(s)e^{st}]\Big|_{s=a_1} + \sum_{i=2}^{n}[(s-a_i)X(s)e^{st}]\Big|_{s=a_i} \tag{B.30}$$

Example B.1. Evaluate the inverse Laplace transform of

$$X(s) = \frac{s + 2\zeta\omega_n}{s^2 + 2\zeta\omega_n s + \omega_n^2} = \frac{A(s)}{B(s)} \tag{a}$$

Assuming that $\zeta < 1$, the roots of $B(s)$ are

$$\begin{matrix} a_1 \\ a_2 \end{matrix} = -\zeta\omega_n \pm i(1-\zeta^2)^{1/2}\omega_n \tag{b}$$

so that the poles of $X(s)$ are simple. Using the approach implied by Eq. (B.23), we write

$$\begin{aligned} X(s) &= \frac{c_1}{s-a_1} + \frac{c_2}{s-a_2} \\ &= \frac{c_1[s + \zeta\omega_n + i(1-\zeta^2)^{1/2}\omega_n] + c_2[s + \zeta\omega_n - i(1-\zeta^2)^{1/2}\omega_n]}{s^2 + 2\zeta\omega_n s + \omega_n^2} \end{aligned} \tag{c}$$

Then, equating the numerator in Eqs. (a) and (c), we conclude that

$$\begin{aligned} A(s) &= c_1[s + \zeta\omega_n + i(1-\zeta^2)^{1/2}\omega_n] + c_2[s + \zeta\omega_n - i(1-\zeta^2)^{1/2}\omega_n] \\ &= (c_1 + c_2)s + c_1[\zeta\omega_n + i(1-\zeta^2)^{1/2}\omega_n] + c_2[\zeta\omega_n - i(1-\zeta^2)^{1/2}\omega_n] \\ &= s + 2\zeta\omega_n \end{aligned} \tag{d}$$

Equating the coefficients of s^{j-1} $(j = 1, 2)$ in the second and third lines of Eq. (d), we obtain two equations in c_1 and c_2, as follows:

$$\begin{aligned} c_1 + c_2 &= 1 \\ c_1[\zeta\omega_n + i(1-\zeta^2)^{1/2}\omega_n] + c_2[\zeta\omega_n - i(1-\zeta^2)^{1/2}\omega_n] &= 2\zeta\omega_n \end{aligned} \tag{e}$$

The solution of Eqs. (e) is simply

$$c_1 = \frac{1}{2}\left[1 + \frac{\zeta}{i(1-\zeta^2)^{1/2}}\right] \quad c_2 = \frac{1}{2}\left[1 - \frac{\zeta}{i(1-\zeta^2)^{1/2}}\right] \tag{f}$$

Hence,

$$\begin{aligned} X(s) = \frac{1}{2}\Bigg[1 + \frac{\zeta}{i(1-\zeta^2)^{1/2}}\Bigg]\frac{1}{s + \zeta\omega_n - i(1-\zeta^2)^{1/2}\omega_n} \\ + \Bigg[1 - \frac{\zeta}{i(1-\zeta^2)^{1/2}}\Bigg]\frac{1}{s + \zeta\omega_n + i(1-\zeta^2)^{1/2}\omega_n} \end{aligned} \tag{g}$$

so that, considering Eq. (B.21), we obtain the inverse Laplace transformation

$$\begin{aligned} x(t) &= \frac{1}{2}\left[1+\frac{\zeta}{i(1-\zeta^2)^{1/2}}\right]e^{-[\zeta-i(1-\zeta^2)^{1/2}]\omega_n t} \\ &\quad +\frac{1}{2}\left[1-\frac{\zeta}{i(1-\zeta^2)^{1/2}}\right]e^{-[\zeta+i(1-\zeta^2)^{1/2}]\omega_n t} \\ &= e^{-\zeta\omega_n t}\left[\cos(1-\zeta^2)^{1/2}\omega_n t+\frac{\zeta}{(1-\zeta^2)^{1/2}}\sin(1-\zeta^2)^{1/2}\omega_n t\right] \end{aligned} \tag{h}$$

From Eqs. (B.7) and (a), we conclude that Eq. (h) represents the response of a damped single-degree-of-freedom system to an initial unit displacement, $x(0) = 1$.

Example B.2. Evaluate the inverse Laplace transform of

$$X(s) = \frac{A(s)}{B(s)} = \frac{\omega^3}{s^2(s^2+\omega^2)} \tag{a}$$

We observe that $X(s)$ has a pole of order 2 at $s = 0,\ k = 2$, and two simple poles at $s = \pm i\omega$. Hence, using formula (B.30), we obtain

$$\begin{aligned} x(t) &= \frac{d}{ds}[s^2X(s)e^{st}]\Big|_{s=0} + [(s-i\omega)X(s)e^{st}]\Big|_{s=i\omega} + [(s+i\omega)X(s)e^{st}]\Big|_{s=-i\omega} \\ &= \frac{d}{ds}\left(\frac{\omega^3}{s^2+\omega^2}e^{st}\right)\Big|_{s=0} + \frac{\omega^3}{s^2(s+i\omega)}e^{st}\Big|_{s=i\omega} + \frac{\omega^3}{s^2(s-i\omega)}e^{st}\Big|_{s=-i\omega} \\ &= \left[\frac{-2s\omega^3}{(s^2+\omega^2)^2}+\frac{t\omega^3}{s^2+\omega^2}\right]e^{st}\Big|_{s=0} + \frac{\omega^3}{-\omega^2(2i\omega)}e^{i\omega t} + \frac{\omega^3}{-\omega^2(-2i\omega)}e^{-i\omega t} \\ &= \omega t - \frac{e^{i\omega t}-e^{-i\omega t}}{2i} = \omega t - \sin\omega t \end{aligned} \tag{b}$$

B.7 THE CONVOLUTION INTEGRAL. BOREL'S THEOREM

We consider two functions $f_1(t)$ and $f_2(t)$, both defined for $t > 0$. Moreover, we assume that $f_1(t)$ and $f_2(t)$ possess Laplace transforms $F_1(s)$ and $F_2(s)$, respectively, and consider the integral

$$x(t) = \int_0^t f_1(\tau)f_2(t-\tau)d\tau = \int_0^\infty f_1(\tau)f_2(t-\tau)d\tau \tag{B.31}$$

The function $x(t)$ defined by Eq. (B.31), sometimes denoted by $x(t) = f_1(t) * f_2(t)$, is called the *convolution* of the functions f_1 and f_2 over the interval $0 < t < \infty$. The upper limits of the integrals in Eq. (B.31) are interchangeable because $f_2(t-\tau) = 0$ for $\tau > t$, which is the same as $t - \tau < 0$. Transforming both sides of Eq. (B.31), we obtain

$$\begin{aligned} X(s) &= \int_0^\infty e^{-st}\left[\int_0^\infty f_1(\tau)f_2(t-\tau)d\tau\right]dt \\ &= \int_0^\infty f_1(\tau)\left[\int_0^\infty e^{-st}f_2(t-\tau)dt\right]d\tau \end{aligned}$$

$$= \int_0^\infty f_1(\tau) \left[\int_\tau^\infty e^{-st} f_2(t-\tau) dt \right] d\tau \tag{B.32}$$

where the lower limit of the second integral was changed without affecting the result because $f_2(t-\tau) = 0$ for $t < \tau$. Next, we introduce the transformation $t - \tau = \lambda$ in the last integral, observe that to $t = \tau$ corresponds $\lambda = 0$ and write

$$X(s) = \int_0^\infty f_1(\tau) \left[\int_0^\infty e^{-s(\tau+\lambda)} f_2(\lambda) d\lambda \right] d\tau$$

$$= \int_0^\infty e^{-s\tau} f_1(\tau) d\tau \int_0^\infty e^{-s\lambda} f_2(\lambda) d\lambda = F_1(s) F_2(s) \tag{B.33}$$

From Eqs. (B.31) and (B.33), it follows that

$$x(t) = \mathcal{L}^{-1} X(s) = \mathcal{L}^{-1} F_1(s) F_2(s)$$

$$= \int_0^t f_1(\tau) f_2(t-\tau) d\tau = \int_0^t f_1(t-\tau) f_2(\tau) d\tau \tag{B.34}$$

The second integral in Eq. (B.34) is valid because it does not matter in which function the time is shifted. The integrals are known as *convolution integrals*. This enables us to state *Borel's theorem*, or *convolution theorem: The inverse Laplace transformation of the product of two transforms is equal to the convolution of their inverse transforms.*

We recall that in Sec. 4.4 we derived a special case of the convolution integral without reference to Laplace transforms, in which one of the functions in the convolution was the impulse response and the other was the applied force.

B.8 TABLE OF LAPLACE TRANSFORM PAIRS

$f(t)$	$F(s)$
$\delta(t)$ (Dirac delta function)	1
$u(t)$ (unit step function)	$\frac{1}{s}$
$t^n \quad n = 1, 2, \ldots$	$\frac{n!}{s^{n+1}}$
$e^{-\omega t}$	$\frac{1}{s+\omega}$
$te^{-\omega t}$	$\frac{1}{(s+\omega)^2}$
$\cos \omega t$	$\frac{s}{s^2+\omega^2}$
$\sin \omega t$	$\frac{\omega}{s^2+\omega^2}$
$\cosh \omega t$	$\frac{s}{s^2-\omega^2}$
$\sinh \omega t$	$\frac{\omega}{s^2-\omega^2}$
$1 - e^{-\omega t}$	$\frac{\omega}{s(s+\omega)}$
$1 - \cos \omega t$	$\frac{\omega^2}{s(s^2+\omega^2)}$
$\omega t - \sin \omega t$	$\frac{\omega^3}{s^2(s^2+\omega^2)}$
$\omega t \cos \omega t$	$\frac{\omega(s^2-\omega^2)}{(s^2+\omega^2)^2}$
$\omega t \sin \omega t$	$\frac{2\omega^2 s}{(s^2+\omega^2)^2}$
$\frac{1}{(1-\zeta^2)^{1/2}\omega} e^{-\zeta\omega t} \sin(1-\zeta^2)^{1/2}\omega t$	$\frac{1}{s^2+2\zeta\omega s+\omega^2}$
$e^{-\zeta\omega t}\left[\cos(1-\zeta^2)^{1/2}\omega t + \frac{\zeta}{(1-\zeta^2)^{1/2}} \sin(1-\zeta^2)^{1/2}\omega t\right]$	$\frac{s+2\zeta\omega}{s^2+2\zeta\omega s+\omega^2}$

APPENDIX C

LINEAR ALGEBRA

Linear algebra is concerned with three types of mathematical concepts, namely, matrices, vector spaces and algebraic forms. Problems in mechanics, and particularly vibration problems, involve all three concepts. Vibration problems involve algebraic forms, which can be conveniently formulated in terms of matrices. The concept of vector spaces is quite helpful in providing a deeper understanding of linear transformations and their properties.

Our particular interest in linear algebra lies in the fact that it permits us to formulate problems associated with the vibration of multi-degree-of-freedom systems in a compact form, it enables us to draw general conclusions concerning the dynamical characteristics of such systems and is indispensable to the solution of the differential equations describing the vibration of these systems. The discussion of linear algebra presented here is relatively modest in nature, and its main purpose is to introduce fundamental concepts of particular interest in vibrations.

C.1 MATRICES

C.1.1 Definitions

Many problems in vibrations can be formulated in terms of rectangular arrays of scalars of the form

$$A = \begin{bmatrix} a_{11} & a_{12} & \dots & a_{1n} \\ a_{21} & a_{22} & \dots & a_{2n} \\ \dots & \dots & \dots & \dots \\ a_{m1} & a_{m2} & \dots & a_{mn} \end{bmatrix} \tag{C.1}$$

where A is called an $m \times n$ *matrix* because it contains m rows and n columns. It is also customary to say that the *dimensions* of A are $m \times n$. Each element[1] a_{ij} ($i = 1, 2, \ldots, m$; $j = 1, 2, \ldots, n$) of the matrix A represents a scalar. For the most part, the scalars represent real numbers, although in general they can be complex. The position of element a_{ij} in matrix A is in the ith row and jth column, so that i is referred to as the row index and j as the column index.

In the special case in which $m = n$, matrix A reduces to a *square matrix of order* n. The elements a_{ii} in a square matrix A are called the *main diagonal elements of* A. The remaining elements are referred to as the *off-diagonal elements of* A. If all the off-diagonal elements of A are zero, then A is said to be a *diagonal matrix.* If A is a diagonal matrix and all its diagonal elements are equal to unity, $a_{ii} = 1$, then the matrix is called a *unit matrix*, or *identity matrix*, and denoted by I. Introducing the *Kronecker delta* symbol δ_{ij}, defined as being equal to unity if $i = j$ and equal to zero if $i \neq j$, a diagonal matrix can be written in the form $[a_{ij}\delta_{ij}]$. Similarly, the identity matrix can be written in terms of the Kronecker delta as $[\delta_{ij}]$.

A matrix with all its rows and columns interchanged is known as the *transpose* of A and denoted by A^T, so that

$$A^T = \begin{bmatrix} a_{11} & a_{21} & \cdots & a_{m1} \\ a_{12} & a_{22} & \cdots & a_{m2} \\ \cdots & \cdots & \cdots & \cdots \\ a_{1n} & a_{2n} & \cdots & a_{mn} \end{bmatrix} \tag{C.2}$$

Clearly, if A is an $m \times n$ matrix, then A^T is an $n \times m$ matrix.

When all the elements of a matrix A are such that $a_{ij} = a_{ji}$, with the implication that the matrix is equal to its transpose, $A = A^T$, the matrix A is said to be *symmetric.* When the elements of A are such that $a_{ij} = -a_{ji}$ for $i \neq j$ and $a_{ii} = 0$, the matrix is said to be *skew symmetric*. Hence, A is a skew symmetric if $A = -A^T$. Clearly, symmetric and skew symmetric matrices must be square.

A matrix consisting of one column and n rows is called a *column matrix* and denoted by

$$\mathbf{x} = \begin{bmatrix} x_1 \\ x_2 \\ \vdots \\ x_n \end{bmatrix} \tag{C.3}$$

The transpose of the column matrix $\mathbf{x}$ is the *row matrix* $\mathbf{x}^T$. They are also known as a *column vector* and a *row vector*, respectively.

A matrix with all its elements equal to zero is called a *null matrix* and denoted by 0, $\mathbf{0}$, or $\mathbf{0}^T$, depending on whether it is a rectangular, a column, or a row matrix, respectively.

---

[1] In discussing computational algorithms, matrix elements are often referred to as entries.

C.1.2 Matrix algebra

Having defined various types of matrices, we are now in a position to present some basic matrix operations. Two matrices A and B are said to be equal if and only if they have the same number of rows and columns, and $a_{ij} = b_{ij}$ for all pairs of subscripts i and j. Hence, considering two $m \times n$ matrices, the statement

$$A = B \tag{C.4}$$

implies that

$$a_{ij} = b_{ij}, \; i = 1, 2, \ldots, m; \; j = 1, 2, \ldots, n \tag{C.5}$$

Addition and subtraction of matrices can be performed if and only if the matrices have the same number of rows and columns. If A, B and C are three $m \times n$ matrices, then the statement

$$C = A \pm B \tag{C.6}$$

implies that, for every pair of subscripts i and j,

$$c_{ij} = a_{ij} \pm b_{ij}, \; i = 1, 2, \ldots, m; \; j = 1, 2, \ldots, n \tag{C.7}$$

Matrix addition, or subtraction, is *commutative* and *associative*, namely,

$$A + B = B + A \tag{C.8}$$

and

$$(A + B) + C = A + (B + C) \tag{C.9}$$

respectively.

The *product of a matrix and a scalar* implies that every element of the matrix in question is multiplied by the same scalar. Hence, if A is any arbitrary $m \times n$ matrix and s an arbitrary scalar, then the statement

$$C = sA \tag{C.10}$$

implies that, for every pair of subscripts i and j,

$$c_{ij} = s a_{ij}, \; i = 1, 2, \ldots, m; \; j = 1, 2, \ldots, n \tag{C.11}$$

The *product of two matrices* is generally *not a commutative process*. Hence, the relative position of the matrices is important, and indeed it must be specified. For example, the product AB can be described by the statement that A is *postmultiplied* by B, or that B is *premultiplied* by A. It is also customary to describe the product by the statement that A is *multiplied on the right* by B, or that B is *multiplied on the left* by A. For a product of two matrices to be possible the number of columns of the first matrix must be equal to the number of rows of the second matrix. If A is an $m \times n$ matrix and B an $n \times p$ matrix, then the product of the two matrices is defined as

$$C = AB \tag{C.12}$$

where C is an $m \times p$ matrix whose elements are given by

$$c_{ij} = a_{i1}b_{1j} + a_{i2}b_{2j} + \ldots + a_{in}b_{nj} = \sum_{k=1}^{n} a_{ik}b_{kj} \tag{C.13}$$

in which k is a dummy index. We note that the element c_{ij} is obtained by multiplying the elements in the ith row of A by the corresponding elements in the jth column of B and summing the products. It must be pointed out that the product BA is not defined, except in the special case in which the number of columns of B is equal to the number of rows of A, $p = m$. Still, even if $p = m$, and the product exists, BA is not equal to AB. In this case AB is an $m \times m$ matrix and BA is an $n \times n$ matrix. In fact, the matrix product in general is not commutative, even when both matrices are square.

As an illustration, we evaluate the following matrix product:

$$\begin{bmatrix} 5 & 2 & 4 \\ 4 & -1 & 1 \\ 1 & 3 & -2 \end{bmatrix} \begin{bmatrix} 3 & 2 \\ 1 & 7 \\ -5 & 4 \end{bmatrix}$$

$$= \begin{bmatrix} 5 \times 3 + 2 \times 1 + 4(-5) & 5 \times 2 + 2 \times 7 + 4 \times 4 \\ 4 \times 3 + (-1) \times 1 + 1 \times (-5) & 4 \times 2 + (-1) \times 7 + 1 \times 4 \\ 1 \times 3 + 3 \times 1 + (-2) \times (-5) & 1 \times 2 + 3 \times 7 + (-2) \times 4 \end{bmatrix}$$

$$= \begin{bmatrix} -3 & 40 \\ 6 & 5 \\ 16 & 15 \end{bmatrix}$$

In the above example, it is clear that the product is not commutative because the number of columns of the second matrix is 2, whereas the number of rows of the matrix is 3. Hence, when the position of the matrices is reversed the matrix product cannot be defined. As an illustration of the case in which both matrix products can be defined and the process is still not commutative, we consider the simple example

$$\begin{bmatrix} 3 & 2 \\ 1 & -5 \end{bmatrix} \begin{bmatrix} 5 & 7 \\ 9 & 3 \end{bmatrix} = \begin{bmatrix} 3 \times 5 + 2 \times 9 & 3 \times 7 + 2 \times 3 \\ 1 \times 5 + (-5) \times 9 & 1 \times 7 + (-5) \times 3 \end{bmatrix}$$

$$= \begin{bmatrix} 33 & 27 \\ -40 & -8 \end{bmatrix}$$

$$\begin{bmatrix} 5 & 7 \\ 9 & 3 \end{bmatrix} \begin{bmatrix} 3 & 2 \\ 1 & -5 \end{bmatrix} = \begin{bmatrix} 5 \times 3 + 7 \times 1 & 5 \times 2 + 7 \times (-5) \\ 9 \times 3 + 3 \times 1 & 9 \times 2 + 3 \times (-5) \end{bmatrix}$$

$$= \begin{bmatrix} 22 & -25 \\ 30 & 3 \end{bmatrix}$$

Although there may be cases when a particular matrix product is commutative, these are exceptions and not the rule. One notable exception is the case in which one of the matrices in the product is the unit matrix, as in this case

$$AI = IA = A \tag{C.14}$$

where A must clearly be a square matrix of the same order as I.

The matrix product satisfies *associative laws*. Indeed, considering the $m \times n$ matrix A, the $n \times p$ matrix B and the $p \times q$ matrix C, it can be shown that

$$D = (AB)C = A(BC) \tag{C.15}$$

were D is an $m \times q$ matrix whose elements are given by

$$d_{ij} = \sum_{l=1}^{p} \sum_{k=1}^{n} a_{ik} b_{kl} c_{lj} = \sum_{k=1}^{n} \sum_{l=1}^{p} a_{ik} b_{kl} c_{lj} \tag{C.16}$$

The matrix product satisfies *distributive laws*. If A and B are $m \times n$ matrices, C is a $p \times m$ matrix and D is an $n \times q$ matrix, then it is easy to show that

$$C(A+B) = CA + CB \tag{C.17}$$

$$(A+B)D = AD + BD \tag{C.18}$$

The matrix product

$$AB = 0 \tag{C.19}$$

does not imply that either A or B, or both A and B, are null matrices. The above statement can be easily verified by considering the example

$$\begin{bmatrix} 1 & 1 \\ 1 & 1 \end{bmatrix} \begin{bmatrix} 1 & -1 \\ -1 & 1 \end{bmatrix} = \begin{bmatrix} 0 & 0 \\ 0 & 0 \end{bmatrix}$$

From the above discussion, we conclude that matrix algebra differs from ordinary algebra on two major counts: (1) matrix products are not commutative and (2) the fact that the product of two matrices is equal to a null matrix cannot be construed to mean that either multiplicand (or both) is a null matrix. Both these rules hold in ordinary algebra.

C.1.3 Determinant of a square matrix

The determinant of the square matrix A, denoted by $\det A$ or by $|A|$ is defined as

$$\det A = |A| = \begin{vmatrix} a_{11} & a_{12} & \cdots & a_{1n} \\ a_{21} & a_{22} & \cdots & a_{2n} \\ \cdots & \cdots & \cdots & \cdots \\ a_{n1} & a_{n2} & \cdots & a_{nn} \end{vmatrix} \tag{C.20}$$

where $|A|$ is said to be of *order n*. Unlike the matrix A, representing a given array of numbers, the determinant $|A|$ represents a number with a unique value that can be evaluated by following certain rules for the expansion of a determinant. Although determinants have very interesting properties, we do not study them in detail but confine ourselves to certain pertinent aspects only.

We denote by $|M_{rs}|$ the *minor determinant* corresponding to the element a_{rs} where $|M_{rs}|$ is obtained by taking the determinant of A with the rth row and sth column struck out. Hence, $|M_{rs}|$ is of order $n-1$. The signed minor determinant corresponding to the element a_{rs} is called the *cofactor* of a_{rs} and is given by

$$|A_{rs}| = (-1)^{r+s} |M_{rs}| \tag{C.21}$$

With this definition in mind, the value of the determinant can be obtained by expanding the determinant in terms of cofactors by the rth row as follows:

$$|A| = \sum_{s=1}^{n} a_{rs}|A_{rs}| \tag{C.22}$$

or by the sth column in the form

$$|A| = \sum_{r=1}^{n} a_{rs}|A_{rs}| \tag{C.23}$$

where the value of $|A|$ is the same regardless of whether the determinant is expanded by a row or a column, any row or column. The expansions by cofactors are called *Laplace expansions*. The cofactors $|A_{rs}|$ are determinants of order $n-1$, and if $n \geq 2$ they can be further expanded in terms of their own cofactors. The procedure can be continued until the minor determinants are of order 2, in which case their cofactors are simply scalars. As an illustration, we calculate the value of a determinant of order 3 by expanding by the first row, as follows:

$$\begin{aligned}
\begin{vmatrix} a_{11} & a_{12} & a_{13} \\ a_{21} & a_{22} & a_{23} \\ a_{31} & a_{32} & a_{33} \end{vmatrix} &= a_{11}|A_{11}| + a_{12}|A_{12}| + a_{13}|A_{13}| \\
&= a_{11}\begin{vmatrix} a_{22} & a_{23} \\ a_{32} & a_{33} \end{vmatrix} - a_{12}\begin{vmatrix} a_{21} & a_{23} \\ a_{31} & a_{33} \end{vmatrix} + a_{13}\begin{vmatrix} a_{21} & a_{22} \\ a_{31} & a_{32} \end{vmatrix} \\
&= a_{11}(a_{22}a_{33} - a_{23}a_{32}) - a_{12}(a_{21}a_{33} - a_{23}a_{31}) \\
&\quad + a_{13}(a_{21}a_{32} - a_{22}a_{31})
\end{aligned} \tag{C.24}$$

From Eqs. (C.22) and (C.23) we conclude that

$$|A| = \det A = \det A^T \tag{C.25}$$

or the determinant of a matrix is equal to the determinant of the transposed matrix. It is easy to verify that the determinant of a diagonal matrix is equal to the product of the diagonal elements. In particular, the determinant of the identity matrix is equal to 1.

If the value of $\det A$ is equal to zero, then matrix A is said to be *singular*, otherwise it is said to be *nonsingular*. Clearly, $\det A = 0$ if all the elements in one row or column are zero. It is easy to verify that the value of a determinant does not change if one row, or one column, is added to or subtracted from another. Hence, if a determinant possesses two identical rows, or two identical columns, its value is zero. Moreover, if a main diagonal element of a diagonal matrix is zero, then the determinant of the matrix is zero.

By definition, the *adjoint* $[A_{ji}]$ of the matrix A is the transposed matrix of the cofactors of A, namely,

$$\text{adj}\, A = [A_{ji}] = [(-1)^{i+j}|M_{ij}|]^T \tag{C.26}$$

C.1.4 Inverse of a matrix

If A and B are $n \times n$ matrices such that

$$AB = BA = I \tag{C.27}$$

then B is said to be the *inverse* of A and denoted by

$$B = A^{-1} \tag{C.28}$$

To obtain the inverse A^{-1}, provided the matrix A is given, we consider the product

$$A \operatorname{adj} A = \begin{bmatrix} a_{11} & a_{12} & \dots & a_{1n} \\ a_{21} & a_{22} & \dots & a_{2n} \\ \dots & \dots & \dots & \dots \\ a_{n1} & a_{n2} & \dots & a_{nn} \end{bmatrix}$$
$$\times \begin{bmatrix} |M_{11}| & -|M_{21}| & \dots & (-1)^{1+n}|M_{n1}| \\ -|M_{12}| & |M_{22}| & \dots & (-1)^{2+n}|M_{n2}| \\ \dots & \dots & \dots & \dots \\ (-1)^{1+n}|M_{1n}| & (-1)^{2+n}|M_{2n}| & \dots & |M_{nn}| \end{bmatrix}$$
$$= \left[\sum_{j=1}^{n} (-1)^{i+j} a_{kj} |M_{ij}| \right] \tag{C.29}$$

But a typical element of the matrix on the right side of Eq. (C.29) has the value

$$\sum_{j=1}^{n} (-1)^{i+j} a_{kj} |M_{ij}| = \begin{vmatrix} a_{11} & a_{12} & \dots & a_{1n} \\ a_{21} & a_{22} & \dots & a_{2n} \\ \dots & \dots & \dots & \dots \\ a_{n1} & a_{n2} & \dots & a_{nn} \end{vmatrix} = |A|, \; i = k \tag{C.30}$$

On the other hand, if $i \neq k$ the determinant possesses two identical rows. This is because the determinant corresponding to $i \neq k$ is obtained from the matrix A by replacing the ith row by the kth row and keeping the kth row intact. Hence, if $i \neq k$ the value of the element is zero.

Considering the above, Eq. (C.29) can be written in the form

$$A \operatorname{adj} A = |A| I \tag{C.31}$$

Premultiplying both sides of Eq. (C.31) by A^{-1} and dividing the result by $|A|$, we obtain

$$A^{-1} = \frac{\operatorname{adj} A}{\det A} \tag{C.32}$$

so that the inverse of a matrix A is obtained by dividing its adjoint matrix by its determinant.

If $\det A$ is equal to zero, then the elements of A^{-1} approach infinity (or are indeterminate at best), in which case the inverse A^{-1} is said *not to exist*, and the matrix A is said to be *singular*. Hence, for the inverse of a matrix to exist its determinant must be different from zero, which is equivalent to the statement that the matrix must be *nonsingular*.

As the order of the matrix A increases, formula (C.32) for the calculation of A^{-1} ceases to be practical, because the computation of $|A|$ requires a rapidly increasing number of multiplications, so that other methods must be used. We present later a more efficient method of obtaining the inverse of a matrix, namely, the method based on Gaussian elimination in conjunction with back substitution.

C.1.5 Transpose, inverse and determinant of a product of matrices

If A is an $m \times n$ matrix and B an $n \times p$ matrix, according to Eq. (C.13), $C = AB$ is an $m \times p$ matrix with its elements given by

$$c_{ij} = \sum_{k=1}^{n} a_{ik} b_{kj} \tag{C.33}$$

Next, we consider the product $B^T A^T$. Because to any element a_{ik} in A corresponds the element a_{ki} in A^T, and to any element b_{kj} in B corresponds the element b_{jk} in B^T, we have

$$\sum_{k=1}^{n} b_{jk} a_{ki} = c_{ji} \tag{C.34}$$

from which we conclude that

$$C^T = B^T A^T \tag{C.35}$$

or the *transpose of a product of matrices is equal to the product of the transposed matrices in reversed order*. This statement can be generalized to a product of several matrices. Hence, if

$$C = A_1 A_2 \ldots A_{s-1} A_s \tag{C.36}$$

then

$$C^T = A_s^T A_{s-1}^T \ldots A_2^T A_1^T \tag{C.37}$$

We consider again the product

$$C = AB \tag{C.38}$$

but this time A and B are square matrices of order n. Then, premultiplying Eq. (C.38) by $B^{-1}A^{-1}$ and postmultiplying the result by C^{-1}, we obtain simply

$$C^{-1} = B^{-1} A^{-1} \tag{C.39}$$

or the *inverse of a product of matrices is equal to the product of the inverse matrices in reversed order*. Equation (C.39) can be generalized by considering the product (C.36) in which all matrices A_i $(i = 1, 2, \ldots, s)$ are square matrices of order n. Following the same procedure as that used to obtain Eq. (C.39), it is easy to show that

$$C^{-1} = A_s^{-1} A_{s-1}^{-1} \ldots A_2^{-1} A_1^{-1} \tag{C.40}$$

We state here without proof[2] that *the determinant of a product of two matrices is equal to the product of the determinants of the matrices in question*. The statement can be extended to the determinant of the product of any number of matrices. Hence, considering the product of matrices (C.36) in which A_i $(i = 1, 2, \ldots, s)$ are all square matrices, we have

$$\det C = \det A_1 \det A_2 \ldots \det A_s \tag{C.41}$$

In view of Eqs. (C.27), (C.28) and (C.41), we conclude that *the value of* $\det(A^{-1})$ *is equal to the reciprocal of the value of* $\det A$.

C.1.6 Partitioned matrices

At times it proves convenient to partition a matrix into submatrices and regard the submatrices as the elements of the matrix. As an example, a 3×4 matrix A can be partitioned as follows:

$$A = \left[\begin{array}{cc|cc} a_{11} & a_{12} & a_{13} & a_{14} \\ a_{21} & a_{22} & a_{23} & a_{24} \\ \hline a_{31} & a_{32} & a_{33} & a_{34} \end{array}\right] = \left[\begin{array}{c|c} A_{11} & A_{12} \\ \hline A_{21} & A_{22} \end{array}\right] \tag{C.42}$$

where

$$A_{11} = \begin{bmatrix} a_{11} & a_{12} \\ a_{21} & a_{22} \end{bmatrix}, \quad A_{12} = \begin{bmatrix} a_{13} & a_{14} \\ a_{23} & a_{24} \end{bmatrix}$$
$$A_{21} = \begin{bmatrix} a_{31} & a_{32} \end{bmatrix}, \quad A_{22} = \begin{bmatrix} a_{33} & a_{34} \end{bmatrix} \tag{C.43}$$

are submatrices of A. Then if a second 4×4 matrix B is partitioned in the form

$$B = \left[\begin{array}{cc|cc} b_{11} & b_{12} & b_{13} & b_{14} \\ b_{21} & b_{22} & b_{23} & b_{24} \\ \hline b_{31} & b_{32} & b_{33} & b_{34} \\ b_{41} & b_{42} & b_{43} & b_{44} \end{array}\right] = \left[\begin{array}{c|c} B_{11} & B_{12} \\ \hline B_{21} & B_{22} \end{array}\right] \tag{C.44}$$

where

$$B_{11} = \begin{bmatrix} b_{11} & b_{12} \\ b_{21} & b_{22} \end{bmatrix}, \quad B_{12} = \begin{bmatrix} b_{13} & b_{14} \\ b_{23} & b_{24} \end{bmatrix}$$
$$B_{21} = \begin{bmatrix} b_{31} & b_{32} \\ b_{41} & b_{42} \end{bmatrix}, \quad B_{22} = \begin{bmatrix} b_{33} & b_{34} \\ b_{43} & b_{44} \end{bmatrix} \tag{C.45}$$

[2]For the proof, see B. Noble and J. W. Daniel, *Applied Linear Algebra*, 2nd ed., Prentice-Hall, Inc., Englewood Cliffs, NJ, 1977, p. 203.

the matrix produce AB can be treated as if the submatrices were ordinary elements, namely,

$$AB = \left[\begin{array}{c|c} A_{11} & A_{12} \\ \hline A_{21} & A_{22} \end{array}\right]\left[\begin{array}{c|c} B_{11} & B_{12} \\ \hline B_{21} & B_{22} \end{array}\right]$$

$$= \left[\begin{array}{c|c} A_{11}B_{11}+A_{12}B_{21} & A_{11}B_{12}+A_{12}B_{22} \\ \hline A_{21}B_{11}+A_{22}B_{21} & A_{21}B_{12}+A_{22}B_{22} \end{array}\right] \tag{C.46}$$

Note that $A_{11}B_{11}+A_{12}B_{21}$ and $A_{11}B_{12}+A_{12}B_{22}$ are 2×2 matrices, whereas $A_{21}B_{11}+A_{22}B_{21}$ and $A_{21}B_{12}+A_{22}B_{22}$ are 1×2 matrices, so that the product AB is a 3×4 matrix, as is to be expected.

If the off-diagonal submatrices of a square matrix are null matrices, then the matrix is said to be *block-diagonal*. In this case the determinant of the matrix is equal to the product of the determinants of the submatrices on the main diagonal. Considering the matrix (C.44), with B_{12} and B_{21} being identically equal to zero, we have

$$\det B = \det B_{11} \det B_{22} \tag{C.47}$$

C.2 VECTOR SPACES

C.2.1 Definitions

Let $\mathbf{V}$ be a set of objects called *vectors* and R any *field* with its elements consisting of a set of scalars possessing certain algebraic properties. Then, if $\mathbf{V}$ and R are such that two operations, namely, *vector addition* and *scalar multiplication*, are defined for $\mathbf{V}$ and R, the set of vectors together with the two operations are called a *vector space* $\mathbf{V}$ *over a field* R. A vector space is also referred to as a *linear space*.

We have considerable interest in *vector spaces of n-tuples*, i.e., the vectors in the space possess n components from a field R. For two such vectors

$$\mathbf{u} = \begin{bmatrix} u_1 \\ u_2 \\ \vdots \\ u_n \end{bmatrix}, \quad \mathbf{v} = \begin{bmatrix} v_1 \\ v_2 \\ \vdots \\ v_n \end{bmatrix} \tag{C.48}$$

and a scalar c in R, the addition and multiplication are defined as follows:

$$\mathbf{u}+\mathbf{v} = \begin{bmatrix} u_1+v_1 \\ u_2+v_2 \\ \vdots \\ u_n+v_n \end{bmatrix}, \quad c\mathbf{u} = \begin{bmatrix} cu_1 \\ cu_2 \\ \vdots \\ cu_n \end{bmatrix} \tag{C.49}$$

The vector space of n-tuples over R is denoted by $\mathbf{V}^{(n)}(R)$; it consists of all column vectors with n components. The first three vector spaces lend themselves to geometric interpretation. Indeed, the one-dimensional space $\mathbf{V}^{(1)}$ is a line, the two-dimensional space $\mathbf{V}^{(2)}$ is a plane and $\mathbf{V}^{(3)}$ is the usual three-dimensional space. There is no difficulty in conceiving of vector spaces with the number of components larger than three, although such spaces are more abstract and defy physical interpretation. Still the physical interpretation of the vector space is not really necessary, so that no distinction need be made between the cases $1 \leq n \leq 3$ and $n > 3$. Consistent with this, for the most part we will omit the superscript (n) from the vector space notation.

C.2.2 Linear dependence

Vector spaces are very useful in vibrations, as the response of multi-degree-of-freedom systems can be conveniently represented in a vector space. We consider a vector space $\mathbf{V}$ over R and let $\mathbf{u}_1, \mathbf{u}_2, \ldots, \mathbf{u}_n$ and $c_1, c_2, \ldots, c_n$ be n vectors in $\mathbf{V}$ and n scalars in R, respectively. Then, the vector $\mathbf{u}$ given by

$$\mathbf{u} = c_1\mathbf{u}_1 + c_2\mathbf{u}_2 + \cdots + c_n\mathbf{u}_n \tag{C.50}$$

is called a *linear combination* of $\mathbf{u}_1, \mathbf{u}_2, \ldots, \mathbf{u}_n$ with *coefficients* $c_1, c_2, \ldots, c_n$. The totality of linear combinations of $\mathbf{u}_1, \mathbf{u}_2, \ldots, \mathbf{u}_n$ obtained by letting $c_1, c_2, \ldots, c_n$ vary over R is a vector space. The space of all linear combinations of $\mathbf{u}_1, \mathbf{u}_2, \ldots, \mathbf{u}_n$ is said to be *spanned* by $\mathbf{u}_1, \mathbf{u}_2, \ldots, \mathbf{u}_n$. If the relation

$$c_1\mathbf{u}_1 + c_2\mathbf{u}_2 + \cdots + c_n\mathbf{u}_n = \mathbf{0} \tag{C.51}$$

can be satisfied only for the *trivial case*, namely, when all the coefficients $c_1, c_2, \ldots, c_n$ are identically zero, then the vectors $\mathbf{u}_1, \mathbf{u}_2, \ldots, \mathbf{u}_n$ are said to be *linearly independent*. If at least one of the coefficients $c_1, c_2, \ldots, c_n$ is different from zero, the vectors $\mathbf{u}_1, \mathbf{u}_2, \ldots, \mathbf{u}_n$ are said to be *linearly dependent*, implying that one vector is a linear combination of the remaining vectors.

C.2.3 Bases and dimension of vector spaces

A vector space $\mathbf{V}$ over R is said to be *finite dimensional* if there exists a finite set of vectors $\mathbf{u}_1, \mathbf{u}_2, \ldots, \mathbf{u}_n$ which span $\mathbf{V}$, with the implication that every vector in $\mathbf{V}$ is a linear combination of $\mathbf{u}_1, \mathbf{u}_2, \ldots, \mathbf{u}_n$. For example, the space $\mathbf{V}^{(n)}(R)$ is finite dimensional because it can be spanned by a set of n vectors, where n is a finite integer.

Let $\mathbf{V}$ be a vector space over R. A set of vectors $\mathbf{u}_1, \mathbf{u}_2, \ldots, \mathbf{u}_n$ which span $\mathbf{V}$ is called a *generating system* for $\mathbf{V}$. If $\mathbf{u}_1, \mathbf{u}_2, \ldots, \mathbf{u}_n$ are linearly independent and span $\mathbf{V}$, then the generating system is called a *basis* for $\mathbf{V}$. If $\mathbf{V}$ is a finite-dimensional vector space, any two bases for $\mathbf{V}$ contain the same number of vectors.

If $\mathbf{V}$ is a finite-dimensional vector space over R, then the *dimension* of $\mathbf{V}$ is defined as the number of vectors in any basis for $\mathbf{V}$. This integer is denoted by dim $\mathbf{V}$. In particular, the vector space $\mathbf{V}^{(n)}(R)$ has dimension n, because a basis for $\mathbf{V}^{(n)}(R)$ contains n linearly independent vectors.

Let $\mathbf{u}$ be an arbitrary n-dimensional vector with components $u_1, u_2, \ldots, u_n$, where $\mathbf{u}$ is in $\mathbf{V}^{(n)}(R)$, and introduce a set of n-dimensional vectors given by

$$\mathbf{e}_1 = \begin{bmatrix} 1 \\ 0 \\ \vdots \\ 0 \end{bmatrix}, \quad \mathbf{e}_2 = \begin{bmatrix} 0 \\ 1 \\ \vdots \\ 0 \end{bmatrix}, \quad \ldots, \quad \mathbf{e}_n = \begin{bmatrix} 0 \\ 0 \\ \vdots \\ 1 \end{bmatrix} \tag{C.52}$$

Then, the vector $\mathbf{u}$ can be written in terms of the vectors $\mathbf{e}_i$ $(i = 1, 2, \ldots, n)$ as follows:

$$\mathbf{u} = u_1\mathbf{e}_1 + u_2\mathbf{e}_2 + \cdots + u_n\mathbf{e}_n = \sum_{i=1}^{n} u_i\mathbf{e}_i \tag{C.53}$$

Hence, $\mathbf{V}^{(n)}(R)$ is spanned by the set of vectors $\mathbf{e}_i$ $(i = 1, 2, \ldots, n)$. Clearly, the set $\mathbf{e}_i$ is a generating system for $\mathbf{V}^{(n)}(R)$ and is generally referred to as *the standard basis for* $\mathbf{V}^{(n)}(R)$.

C.3 LINEAR TRANSFORMATIONS

C.3.1 The concept of linear transformations

We consider a vector $\mathbf{x}$ in $\mathbf{V}^{(n)}(R)$ and write it in the form

$$\mathbf{x} = x_1\mathbf{e}_1 + x_2\mathbf{e}_2 + \cdots + x_n\mathbf{e}_n = \sum_{i=1}^{n} x_i\mathbf{e}_i \tag{C.54}$$

where x_i are scalars belonging to R and $\mathbf{e}_i$ are the standard unit vectors $(i = 1, 2, \ldots, n)$. The scalars x_i are called the *coordinates* of the vector $\mathbf{x}$ with respect to the basis $\mathbf{e}_1, \mathbf{e}_2, \ldots, \mathbf{e}_n$. Equation (C.54) is entirely analogous to the equation

$$\mathbf{x} = x_1\mathbf{i} + x_2\mathbf{j} + x_3\mathbf{k} \tag{C.55}$$

expressing a three-dimensional vector $\mathbf{x}$ in terms of the components x_1, x_2, x_3, where $\mathbf{i}, \mathbf{j}, \mathbf{k}$ are unit vectors along rectangular axes. Next, we consider an $n \times n$ matrix A and write

$$\mathbf{x}' = A\mathbf{x} \tag{C.56}$$

The resulting vector $\mathbf{x}'$ is another vector in $\mathbf{V}^{(n)}(R)$, so that Eq. (C.56) can be regarded as representing a *linear transformation* on the vector space $\mathbf{V}^{(n)}(R)$ which maps the vector $\mathbf{x}$ into a vector $\mathbf{x}'$.

Equation (C.54) expresses the vector $\mathbf{x}$ in terms of the standard basis. In many applications, the interest lies in expressing $\mathbf{x}$ in terms of any arbitrary basis $\mathbf{p}_1, \mathbf{p}_2, \ldots, \mathbf{p}_n$ for $\mathbf{V}^{(n)}(R)$ as follows:

$$\mathbf{x} = y_1\mathbf{p}_1 + y_2\mathbf{p}_2 + \cdots + y_n\mathbf{p}_n = \sum_{i=1}^{n} y_i\mathbf{p}_i = P\mathbf{y} \tag{C.57}$$

where

$$P = [\mathbf{p}_1 \ \mathbf{p}_2 \ \cdots \ \mathbf{p}_n] \tag{C.58}$$

is an $n \times n$ matrix of the basis vectors and

$$\mathbf{y} = \begin{bmatrix} y_1 \\ y_2 \\ \vdots \\ y_n \end{bmatrix} \tag{C.59}$$

is an n-dimensional vector whose components y_i are the coordinates of $\mathbf{x}$ with respect to the basis $\mathbf{p}_1, \mathbf{p}_2, \ldots, \mathbf{p}_n$. By the definition of a basis, the vectors $\mathbf{p}_1, \mathbf{p}_2, \ldots, \mathbf{p}_n$ are linearly independent, so that the matrix P is nonsingular. Similarly, denoting by $y_1', y_2', \ldots, y_n'$ the coordinates of $\mathbf{x}'$ with respect to the basis $\mathbf{p}_1, \mathbf{p}_2, \ldots, \mathbf{p}_n$, we can write

$$\mathbf{x}' = P\mathbf{y}' \tag{C.60}$$

where

$$\mathbf{y}' = \begin{bmatrix} y_1' \\ y_2' \\ \vdots \\ y_n' \end{bmatrix} \tag{C.61}$$

Inserting Eqs. (C.57) and (C.60) into Eq. (C.56), we can write

$$P\mathbf{y}' = AP\mathbf{y} \tag{C.62}$$

so that, premultiplying both sides of Eq. (C.62) by P^{-1}, we obtain

$$\mathbf{y}' = B\mathbf{y} \tag{C.63}$$

where

$$B = P^{-1}AP \tag{C.64}$$

Note that P^{-1} exists by virtue of the fact that P is nonsingular. The matrix B represents the same linear transformation as A, but in a different coordinate system. Two matrices A and B related by an equation of the type (C.64) are said to be *similar* and the relationship (C.64) itself is known as a *similarity transformation*

Next, we define the *eigenvalue problem* for the $n \times n$ matrix A as follows:

$$A\mathbf{x} = \lambda\mathbf{x} \tag{C.65}$$

which can be stated in words as *the problem of determining the values of the parameter* λ *such that Eq.* (C.65) *admits nontrivial solutions* $\mathbf{x}$. The interest lies in expressing Eq. (C.65) in terms of a different set of coordinates. To this end, we introduce transformation (C.57) in Eq. (C.65), premultiply by P^{-1} and rewrite the eigenvalue problem in the form

$$B\mathbf{y} = \lambda\mathbf{y} \tag{C.66}$$

in which B is given by Eq. (C.64). At this point, we consider the *characteristic determinant* associated with B, recall Eq. (C.64) and write

$$\det(B - \lambda I) = \det(P^{-1}AP - \lambda P^{-1}P)$$

$$= \det(P^{-1}(A - \lambda I)P)$$

$$= \det P^{-1} \det(A - \lambda I) \det P \tag{C.67}$$

But

$$\det P^{-1} \det P = \det(P^{-1}P) = \det I = 1 \tag{C.68}$$

so that

$$\det(B - \lambda I) = \det(A - \lambda I) \tag{C.69}$$

Equation (C.69) states that matrices A and B possess the same characteristic determinant, and hence the same characteristic equation. It follows that *similar matrices possess the same eigenvalues.*

One similarity transformation of particular interest is the *orthogonal transformation.* A matrix P is said to be *orthonormal* if it satisfies

$$P^T P = I \tag{C.70}$$

from which it follows that an orthonormal matrix also satisfies

$$P^{-1} = P^T \tag{C.71}$$

Introducing Eq. (C.71) into Eq. (C.64), we obtain

$$B = P^T A P \tag{C.72}$$

Equation (C.72) represents an orthogonal transformation. One orthogonal transformation of particular interest in vibrations is the one in which A is symmetric and the resulting matrix B is diagonal. Many computational algorithms for the eigenvalue problem of a real symmetric matrix A are based on the use of orthogonal transformations to diagonalize the matrix A.

C.3.2 Solution of algebraic equations. Matrix inversion

A basic problem in linear algebra is the solution of sets of nonhomogeneous algebraic equations. Of particular interest here is the case in which the number of equations is equal to the number of unknowns. Hence, we consider the system of equations

$$\begin{aligned} a_{11}x_1 + a_{12}x_2 + \cdots + a_{1n}x_n &= b_1 \\ a_{21}x_1 + a_{22}x_2 + \cdots + a_{2n}x_n &= b_2 \\ \cdots\cdots\cdots\cdots\cdots\cdots\cdots\cdots \\ a_{n1}x_1 + a_{n2}x_2 + \cdots + a_{nn}x_n &= b_n \end{aligned} \tag{C.73}$$

Equations (C.73) can be written in the compact matrix form

$$A\mathbf{x} = \mathbf{b} \tag{C.74}$$

where $A = [a_{ij}]$ is the $n \times n$ coefficient matrix, $\mathbf{x} = [x_1 \ x_2 \ \ldots \ x_n]^T$ the n-vector of unknowns and $\mathbf{b} = [b_1 \ b_2 \ \ldots \ b_n]^T$ the n-vector of nonhomogeneous terms. We assume that Eq. (C.74) has a unique solution, which implies that the matrix A is nonsingular, and write the solution simply as

$$\mathbf{x} = A^{-1}\mathbf{b} \tag{C.75}$$

where the inverse A^{-1} is given by Eq. (C.32).

Equation (C.75) gives the impression that the matter of solving linear algebraic equations of the type (C.73) is closed, and for small n this is indeed the case. But, from Eq. (C.26), we observe that the computation of adj A involves the computation of n^2 determinants of order $n-1$, a task requiring a rapidly increasing effort. As an illustration, the computation of a relatively moderate 10×10 determinant requires 3,628,800 multiplications. Hence, as n increases, the use of Eq. (C.32) becomes impractical, so that a different approach demands itself. A very efficient approach to the solution of algebraic equations of the type (C.73) does indeed exist and it consists of *Gaussian elimination* in conjunction with *back substitution.* Of course, the approach amounts to an efficient inversion of a matrix.

The Gaussian elimination is basically a procedure for solving sets of linear algebraic equations through elementary operations. The net effect of these elementary operations is to carry out a linear transformation on Eq. (C.74), which amounts to premultiplying Eq. (C.74) by the $n \times n$ matrix P, so that

$$PA\mathbf{x} = P\mathbf{b} \tag{C.76}$$

The transformation matrix P is such that PA is an upper triangular matrix U, i.e., a matrix with all the entries below the main diagonal equal to zero. Hence, introducing the notation

$$PA = U, \quad P\mathbf{b} = \mathbf{c} \tag{C.77}$$

where U is an upper triangular matrix, Eq. (C.76) can be rewritten as

$$U\mathbf{x} = \mathbf{c} \tag{C.78}$$

The question remains as to how to generate the transformation matrix P required for the computation of U and $\mathbf{c}$. The process involves $n-1$ steps, which implies that P is the product of $n-1$ matrices. To demonstrate the process, we introduce the notation

$$A = A_0, \quad \mathbf{b} = \mathbf{a}_{n+1}^{(0)} \tag{C.79}$$

premultiply Eq. (C.74) by an $n \times n$ transformation matrix P_1 and obtain

$$A_1\mathbf{x} = \mathbf{a}_{n+1}^{(1)} \tag{C.80}$$

where the coefficient matrix A_1 and the vector $\mathbf{a}_{n+1}^{(1)}$ are obtained from A_0 and $\mathbf{a}_{n+1}^{(0)}$, respectively, by writing

$$A_1 = P_1 A_0, \quad \mathbf{a}_{n+1}^{(1)} = P_1\mathbf{a}_{n+1}^{(0)} \tag{C.81}$$

in which the transformation matrix P_1 has the form

$$P_1 = \begin{bmatrix} 1 & 0 & 0 & \dots & 0 \\ -p_{21} & 1 & 0 & \dots & 0 \\ -p_{31} & 0 & 1 & \dots & 0 \\ \dots & \dots & \dots & \dots & \dots \\ -p_{n1} & 0 & 0 & \dots & 1 \end{bmatrix} \tag{C.82}$$

where

$$p_{i1} = \frac{a_{i1}^{(0)}}{a_{11}^{(0)}}, \quad i = 2, 3, \dots, n \tag{C.83}$$

Inserting Eqs. (C.82) and (C.83) into Eqs. (C.81), we can write the matrix A_1 and vector $\mathbf{a}_{n+1}^{(1)}$ in the general form

$$A_1 = \begin{bmatrix} a_{11}^{(0)} & a_{12}^{(0)} & a_{13}^{(0)} & \dots & a_{1n}^{(0)} \\ 0 & a_{22}^{(1)} & a_{23}^{(1)} & \dots & a_{2n}^{(1)} \\ 0 & a_{32}^{(1)} & a_{33}^{(1)} & \dots & a_{3n}^{(1)} \\ \dots & \dots & \dots & \dots & \dots \\ 0 & a_{n2}^{(1)} & a_{n3}^{(1)} & \dots & a_{nn}^{(1)} \end{bmatrix}, \quad \mathbf{a}_{n+1}^{(1)} = \begin{bmatrix} a_{1,n+1}^{(0)} \\ a_{2,n+1}^{(1)} \\ a_{3,n+1}^{(1)} \\ \dots \\ a_{n,n+1}^{(1)} \end{bmatrix} \tag{C.84}$$

In a similar fashion, premultiplication of Eq. (C.80) by P_2 yields

$$A_2\mathbf{x} = \mathbf{a}_{n+1}^{(2)} \tag{C.85}$$

in which

$$A_2 = P_2 A_1, \quad \mathbf{a}_{n+1}^{(2)} = P_2 \mathbf{a}_{n+1}^{(1)} \tag{C.86}$$

where

$$P_2 = \begin{bmatrix} 1 & 0 & 0 & \dots & 0 \\ 0 & 1 & 0 & \dots & 0 \\ 0 & -p_{32} & 1 & \dots & 0 \\ \dots & \dots & \dots & \dots & \dots \\ 0 & -p_{n2} & 0 & \dots & 1 \end{bmatrix} \tag{C.87}$$

in which

$$p_{i2} = \frac{a_{i2}^{(1)}}{a_{22}^{(1)}}, \quad i = 3, 4, \dots, n \tag{C.88}$$

Moreover, the matrix A_2 and vector $\mathbf{a}_{n+1}^{(2)}$ have the general form

$$A_2 = \begin{bmatrix} a_{11}^{(0)} & a_{12}^{(0)} & a_{13}^{(0)} & \dots & a_{1n}^{(0)} \\ 0 & a_{22}^{(1)} & a_{23}^{(1)} & \dots & a_{2n}^{(1)} \\ 0 & 0 & a_{33}^{(2)} & \dots & a_{3n}^{(2)} \\ \dots & \dots & \dots & \dots & \dots \\ 0 & 0 & a_{n3}^{(2)} & \dots & a_{nn}^{(2)} \end{bmatrix}, \quad \mathbf{a}_{n+1}^{(2)} \begin{bmatrix} a_{1,n+1}^{(0)} \\ a_{2,n+1}^{(1)} \\ a_{3,n+1}^{(2)} \\ \dots \\ a_{n,n+1}^{(2)} \end{bmatrix} \tag{C.89}$$

The process continues in the same fashion and ends after $n-1$ steps with the result

$$U = A_{n-1} = P_{n-1}A_{n-2}, \quad \mathbf{c} = \mathbf{a}_{n+1}^{(n-1)} = P_{n-1}\mathbf{a}_{n+1}^{(n-2)} \tag{C.90}$$

By induction, Eqs. (C.90) yield

$$\begin{aligned} U &= P_{n-1}A_{n-2} = P_{n-1}P_{n-2}A_{n-3} = \dots = P_{n-1}P_{n-2}\dots P_2P_1A \\ \mathbf{c} &= P_{n-1}\mathbf{a}_{n+1}^{(n-2)} = P_{n-1}P_{n-2}\mathbf{a}_{n+1}^{(n-3)} = \dots = P_{n-1}P_{n-2}\dots P_2P_1\mathbf{b} \end{aligned} \tag{C.91}$$

in which we replaced A_0 by A and $\mathbf{a}_{n+1}^{(0)}$ by $\mathbf{b}$, according to Eqs. (C.77). Then, comparing Eqs. (C.77) and (C.91), we conclude that the transformation matrix has the form of the continuous matrix product

$$P = P_{n-1}P_{n-2}\dots P_2P_1 \tag{C.92}$$

which indicates that the transformation matrix P can be generated in $n-1$ steps. This is merely of academic interest, however, as P is never computed explicitly, because U and $\mathbf{c}$ are determined by means of Eqs. (C.81), (C.86) ... (C.90) and not through Eqs. (C.77).

With U and $\mathbf{c}$ obtained from Eqs. (C.90), Eq. (C.78) can be solved with ease by *back substitution*. Indeed, the bottom equation involves x_n alone, and can be solved with the result

$$x_n = c_n/u_{nn} \tag{C.93}$$

Then, having x_n, the $(n-1)$th equation can be solved to obtain

$$x_{n-1} = \frac{1}{u_{n-1,n-1}}(c_{n-1} - u_{n-1,n}x_n) \tag{C.94}$$

Next, upon substitution of x_{n-1} and x_n into the $(n-2)$th equation, we are able to solve for x_{n-2}. The procedure continues by solving in sequence for $x_{n-3}, \dots, x_3$, x_2 and terminates with

$$x_1 = \frac{c_1 - u_{12}x_2 - u_{13}x_3 - \dots - u_{in}x_n}{u_{11}} \tag{C.95}$$

From the second of Eqs. (C.75) and Eq. (C.78), the solution of Eq. (C.74) is simply

$$\mathbf{x} = U^{-1}P\mathbf{b} \tag{C.96}$$

so that implicit in back substitution is the calculation of the inverse of the triangular matrix U. Comparing Eq. (C.96) with Eq. (C.75), we conclude that the inverse of A is

$$A^{-1} = U^{-1}P \tag{C.97}$$

But, unlike the process given by Eq. (C.32), Eq. (C.97) represents a very attractive algorithm for computing the inverse of an $n \times n$ matrix for large n, because the inverse of a triangular matrix is a relatively simple task. In this regard, we observe that, even when the process given by Eq. (C.32) is used to compute U^{-1}, the determinant of the triangular matrix U is merely equal to the product of the diagonal entries of U.

Example C.1. Solve the algebraic equations

$$\begin{aligned} 2x_1 - x_2 \qquad &= -1 \\ -x_1 + 3x_2 - 2x_3 &= 6.5 \\ -2x_2 + 2x_3 &= -5 \end{aligned} \tag{a}$$

by Gaussian elimination with back substitution.

Equations (a) can be arranged in the matrix form (C.74), in which

$$A = A_0 = \begin{bmatrix} 2 & -1 & 0 \\ -1 & 3 & -2 \\ 0 & -2 & 2 \end{bmatrix}, \quad \mathbf{b} = \mathbf{a}_4^{(0)} = \begin{bmatrix} -1 \\ 6.5 \\ -5 \end{bmatrix} \tag{b}$$

Using Eqs. (C.82) and (C.83), the transformation matrix P_1 is

$$P_1 = \begin{bmatrix} 1 & 0 & 0 \\ 0.5 & 1 & 0 \\ 0 & 0 & 1 \end{bmatrix} \tag{c}$$

Hence, using Eqs. (C.81), we obtain

$$\begin{aligned} A_1 = P_1 A_0 &= \begin{bmatrix} 1 & 0 & 0 \\ 0.5 & 1 & 0 \\ 0 & 0 & 1 \end{bmatrix} \begin{bmatrix} 2 & -1 & 0 \\ -1 & 3 & -2 \\ 0 & -2 & 2 \end{bmatrix} = \begin{bmatrix} 2 & -1 & 0 \\ 0 & 2.5 & -2 \\ 0 & -2 & 2 \end{bmatrix} \\ \mathbf{a}_4^{(1)} = P_1 \mathbf{a}_4^{(0)} &= \begin{bmatrix} 1 & 0 & 0 \\ 0.5 & 1 & 0 \\ 0 & 0 & 1 \end{bmatrix} \begin{bmatrix} -1 \\ 6.5 \\ -5 \end{bmatrix} = \begin{bmatrix} -1 \\ 6 \\ -5 \end{bmatrix} \end{aligned} \tag{d}$$

Next, we use Eqs. (C.87) and (C.88) and write

$$P_2 = \begin{bmatrix} 1 & 0 & 0 \\ 0 & 1 & 0 \\ 0 & 0.8 & 1 \end{bmatrix} \tag{e}$$

so that, using Eqs. (C.90), we obtain

$$\begin{aligned} U = A_2 = P_2 A_1 &= \begin{bmatrix} 1 & 0 & 0 \\ 0 & 1 & 0 \\ 0 & 0.8 & 1 \end{bmatrix} \begin{bmatrix} 2 & -1 & 0 \\ 0 & 2.5 & -2 \\ 0 & -2 & 2 \end{bmatrix} = \begin{bmatrix} 2 & -1 & 0 \\ 0 & 2.5 & -2 \\ 0 & 0 & 0.4 \end{bmatrix} \\ \mathbf{c} = \mathbf{a}_4^{(2)} = P_2 \mathbf{a}_4^{(1)} &= \begin{bmatrix} 1 & 0 & 0 \\ 0 & 1 & 0 \\ 0 & 0.8 & 1 \end{bmatrix} \begin{bmatrix} -1 \\ 6 \\ -5 \end{bmatrix} = \begin{bmatrix} -1 \\ 6 \\ -0.2 \end{bmatrix} \end{aligned} \tag{f}$$

At this point, we begin the back substitution with Eq. (C.93) and write

$$x_3 = \frac{c_3}{u_{33}} = \frac{-0.2}{0.4} = -0.5 \tag{g}$$

Then, Eq. (C.94) yields

$$x_2 = \frac{c_2 - u_{23}x_3}{u_{22}} = \frac{6-(-2)\times(-0.5)}{2.5} = 2 \tag{h}$$

Finally, from Eq. (C.95), we have

$$x_1 = \frac{c_1 - u_{12}x_2 - u_{13}x_2}{u_{11}} = \frac{-1-(-1)\times 2 - 0\times(-0.5)}{2} = 0.5 \tag{i}$$

which completes the solution.

As a matter of interest, we carry out the solution by first obtaining A^{-1} using Eq. (C.97). To this end, we first compute

$$U^{-1} = \frac{\operatorname{adj} U}{\det U} = \frac{\begin{bmatrix} 1 & 0.4 & 2 \\ 0 & 0.8 & 4 \\ 0 & 0 & 5 \end{bmatrix}}{2} = \begin{bmatrix} 0.5 & 0.2 & 1 \\ 0 & 0.4 & 2 \\ 0 & 0 & 2.5 \end{bmatrix} \tag{j}$$

and

$$P = P_2 P_1 = \begin{bmatrix} 1 & 0 & 0 \\ 0 & 1 & 0 \\ 0 & 0.8 & 1 \end{bmatrix} \begin{bmatrix} 1 & 0 & 0 \\ 0.5 & 1 & 0 \\ 0 & 0 & 1 \end{bmatrix} = \begin{bmatrix} 1 & 0 & 0 \\ 0.5 & 1 & 0 \\ 0.4 & 0.8 & 1 \end{bmatrix} \tag{k}$$

so that

$$A^{-1} = U^{-1}P = \begin{bmatrix} 0.5 & 0.2 & 1 \\ 0 & 0.4 & 2 \\ 0 & 0 & 2.5 \end{bmatrix} \begin{bmatrix} 1 & 0 & 0 \\ 0.5 & 1 & 0 \\ 0.4 & 0.8 & 1 \end{bmatrix} = \begin{bmatrix} 1 & 1 & 1 \\ 1 & 2 & 2 \\ 1 & 2 & 2.5 \end{bmatrix} \tag{l}$$

Hence, the solution is

$$\mathbf{x} = A^{-1}\mathbf{b} = \begin{bmatrix} 1 & 1 & 1 \\ 1 & 2 & 2 \\ 1 & 2 & 2.5 \end{bmatrix} \begin{bmatrix} -1 \\ 6.5 \\ -5 \end{bmatrix} = \begin{bmatrix} 0.5 \\ 2 \\ -0.5 \end{bmatrix} \tag{m}$$

which is the same as that given by Eqs. (g)–(i).

BIBLIOGRAPHY

1. Beer, F. P. and Johnston, E. R. Jr., *Mechanics of Materials*, 2nd ed., McGraw-Hill, New York, 1992.
2. Bendat, J. S. and Piersol, A. G., *Random Data: Analysis and Measurement Procedures*, Wiley-Interscience, New York, 1971.
3. Burden, R. L. and Faires, J. D., *Numerical Analysis*, 5th ed., Prindle, Weber and Schmidt, Boston, 1993.
4. Courant, R., "Variational Methods for the Solution of Problems of Equilibrium and Vibrations," *Bulletin of the American Mathematical Society*, Vol. 49, January 1943, pp. 1–23.
5. Crandall, S. H. and Mark, W. D., *Random Vibration in Mechanical Systems*, Academic Press, Inc., New York, 1963.
6. Forsythe, G. E., Malcolm, M. A. and Moller, C. B., *Computer Methods for Mathematical Computations*, Prentice-Hall, Englewood Cliffs, NJ, 1977.
7. Huebner, K. H. and Thornton, E. A., *The Finite Element Method for Engineers*, 2nd ed., Wiley, New York, 1983.
8. Hurty, W. C., "Vibration of Structural Systems by Component Modes Synthesis," *Journal of Engineering Mechanics Division, ASCE*, Vol. 86, August 1960, pp. 51–69.
9. Mathews, J. H., *Numerical Methods for Computer Science, Engineering, and Mathematics*, 2nd ed., Prentice Hall, Englewood Cliffs, NJ, 1992.
10. Meirovitch, L., *Methods of Analytical Dynamics*, McGraw-Hill, New York, 1970.
11. Meirovitch, L., *Introduction to Dynamics and Control*, Wiley, New York, 1985.
12. Meirovitch, L., *Elements of Vibration Analysis*, 2nd ed., McGraw-Hill, New York, 1986.
13. Meirovitch, L., *Principles and Techniques of Vibrations*, Prentice-Hall, Upper Saddle River, NJ, 1997.
14. Murdoch, D. C., *Linear Algebra*, Wiley, New York, 1970.
15. Noble, B. and Daniel, J. W., *Applied Linear Algebra*, 2nd ed., Prentice-Hall, Englewood Cliffs, NJ, 1977.
16. Rayleigh (Lord), *Theory of Sound*, Vol. 1, Dover, New York, 1945 (first American edition of the 1894 edition).
17. Strang, G. and Fix, G. I., *An Analysis of the Finite Element Method*, Prentice-Hall, Englewood Cliffs, NJ, 1973.

18. Strang, G., *Linear Algebra and Its Applications*, 3rd ed., Harcourt, Brace, Jovanovich, San Diego, 1988.
19. Turner, M. I., Clough, R. W., Martin, H. C. and Topp, L. I., "Stiffness and Deflection Analysis of Complex Structures," *Journal of Aeronautical Sciences*, Vol. 23, 1956, pp. 805–823.
20. Zienkiewicz, O. C. and Taylor, R. I., *The Finite Element Method*, 4th ed., McGraw-Hill, London, 1991.

INDEX